PRINCIPLES OF NEURAL SCIENCE

Sixth Edition

神经科学原理

（英文版·原书第6版）

下册

埃里克·R. 坎德尔 　 约翰·D. 凯斯特
(Eric R. Kandel) 　 (John D. Koester)

[美] 　 　 编著

萨拉·H. 麦克 　 史蒂文·A. 西格尔鲍姆
(Sarah H. Mack) 　 (Steven A. Siegelbaum)

机械工业出版社
CHINA MACHINE PRESS

图书在版编目（CIP）数据

神经科学原理：原书第 6 版 . 下册：英文 /（美）埃里克・R. 坎德尔（Eric R. Kandel）等编著 . —北京：机械工业出版社，2024.3
书名原文：Principles of Neural Science, Sixth Edition
ISBN 978-7-111-75299-8

I. ①神… II. ①埃… III. ①神经科学 – 英文 IV. ① Q189

中国国家版本馆 CIP 数据核字（2024）第 051539 号

机械工业出版社（北京市百万庄大街22 号　邮政编码100037）
策划编辑：向睿洋　　责任编辑：向睿洋　　欧阳智
责任校对：彭　箫　　责任印制：单爱军
保定市中画美凯印刷有限公司印刷
2025年1月第1版第1次印刷
214mm×275mm・60印张・1插页・2528千字
标准书号：ISBN 978-7-111-75299-8
定价：299.00 元

电话服务　　　　　　　　　网络服务
客服电话：010-88361066　　机　工　官　网：www.cmpbook.com
　　　　　010-88379833　　机　工　官　博：weibo.com/cmp1952
　　　　　010-68326294　　金　书　网：www.golden-book.com
封底无防伪标均为盗版　　　机工教育服务网：www.cmpedu.com

Contents in Brief

Part IX
Diseases of the Nervous System

Contents

34   Voluntary Movement: Motor Cortices 815

Stephen H. Scott, John F. Kalaska

35   The Control of Gaze 860

Michael E. Goldberg, Mark F. Walker

41   The Hypothalamus: Autonomic, Hormonal, and Behavioral Control of Survival 1010

Bradford B. Lowell, Larry W. Swanson, John P. Horn

42   Emotion 1045

C. Daniel Salzman, Ralph Adolphs

Part VIII
Learning, Memory, Language and Cognition

Part V

Preceding Page

Fresco of dancing Peucetian women from the Tomb of the Dancers in the Corso Cotugno necropolis of Ruvo di Puglia, 4th–5th century BC. The tomb has a semichamber design. Its six painted panels depict 30 dancing women, moving from left to right with arms interlocked as though they were dancing in a circle around the interior of the tomb. The skeletal remains of the deceased in the tomb clearly belonged to a distinguished male warrior. The tomb is named after the dancing women that appear on the frescoes in the tomb. The panels with the frescoes are now exhibited in the Naples National Archaeological Museum, inv. 9353. (Source: https://en.wikipedia.org/wiki/Tomb_of_the_Dancers.)

V Movement

THE CAPACITY FOR MOVEMENT, as many dictionaries remind us, is a defining feature of animal life. As Sherrington, who pioneered the study of the motor system pointed out, "to move things is all that mankind can do, for such the sole executant is muscle, whether in whispering a syllable or in felling a forest."[*]

The immense repertoire of motions that humans are capable of stems from the activity of some 640 skeletal muscles—all under the control of the central nervous system. After processing sensory information about the body and its surroundings, the motor centers of the brain and spinal cord issue neural commands that effect coordinated, purposeful movements.

The task of the motor systems is the reverse of the task of the sensory systems. Sensory processing generates an internal representation in the brain of the outside world or of the state of the body. Motor processing begins with an internal representation: the desired purpose of movement. Critically, however, this internal representation needs to be continuously updated by internally generated information (efference copy) and external sensory information to maintain accuracy as the movement unfolds.

Just as psychophysical analysis of sensory processing tells us about the capabilities and limitations of the sensory systems, psychophysical analyses of motor performance reveal the control rules used by the motor system.

Because many of the motor acts of daily life are unconscious, we are often unaware of their complexity. Simply standing upright, for example, requires continual adjustments of numerous postural muscles in response to the vestibular signals evoked by miniscule swaying. Walking, running, and other forms of locomotion involve the combined action of central pattern generators, gated sensory information, and descending commands, which together generate the complex patterns of alternating excitation and inhibition to the appropriate sets of muscles. Many actions, such as serving a tennis

---

[*]Sherrington CS. 1979. 1924 Linacre lecture. In: JC Eccles, WC Gibson (eds). *Sherrington: His Life and Thought*, p. 59. New York: Springer-Verlag.

ball or executing an arpeggio on a piano, occur far too quickly to be shaped by sensory feedback. Instead, centers, such as the cerebellum, make use of predictive models that simulate the consequences of the outgoing commands and allow very short latency corrections. Motor learning provides one of the most fruitful subjects for studies of neural plasticity.

Motor systems are organized in a functional hierarchy, with each level concerned with a different decision. The highest and most abstract level, likely requiring the prefrontal cortex, deals with the purpose of a movement or series of motor actions. The next level, which is concerned with the formation of a motor plan, involves interactions between the posterior parietal and premotor areas of the cerebral cortex. The premotor cortex specifies the spatiotemporal characteristics of a movement based on sensory information from the posterior parietal cortex about the environment and about the position of the body in space. The lowest level of the hierarchy coordinates the spatiotemporal details of the muscle contractions needed to execute the planned movement. This coordination is executed by the primary motor cortex, brain stem, and spinal cord. This serial view has heuristic value, but evidence suggests that many of these processes can occur in parallel.

Some functions of the motor systems and their disturbance by disease have now been described at the level of the biochemistry of specific transmitter systems. In fact, the discovery that neurons in the basal ganglia of parkinsonian patients are deficient in dopamine was the first important clue that neurological disorders in the central nervous system can result from altered chemical transmission. Neurophysiological studies have provided information as to how such transmitters play a critical role in action selection and the reinforcement of successful movements.

Understanding the functional properties of the motor system is not only fundamental in its own right, but it is of further importance in helping us to understand disorders of this system and explore the possibilities for treatment and recovery. As would be expected for such a complex apparatus, the motor system is subject to various malfunctions. Disruptions at different levels in the motor hierarchy produce distinctive symptoms, including the movement-slowing characteristic of disorders of the basal ganglia, such as Parkinson disease, the incoordination seen with cerebellar disease, and the spasticity and weakness typical of spinal cord damage. For this reason, the neurological examination of a patient inevitably includes tests of reflexes, gait, and dexterity, all of which provide information about the status of the nervous system. In addition to pharmacological therapies, the treatment of motor system disorders has been augmented by two new approaches. First, focal stimulation of the basal ganglia has been shown to restore motility to certain patients with Parkinson disease; such deep-brain stimulation is also being tested in the context of other neurological and psychiatric conditions. And second, the motor systems have become a target for the application of neural prosthetics; neural signals are decoded and used to drive devices that aid patients with paralysis caused by spinal cord injury and stroke.

Part Editors: Daniel M. Wolpert and Thomas M. Jessell

Part V

30

Principles of Sensorimotor Control

THE PRECEDING CHAPTERS IN THIS BOOK consider how the brain constructs internal representations of the world around us. These representations are behaviorally meaningful when used to guide movement. Thus, an important function of the sensory representations is to shape the actions of the motor systems. This chapter describes the principles that govern the neural control of movement using concepts derived from behavioral studies and computational models of the brain and musculoskeletal system.

We start by considering the challenges motor systems face in generating skillful actions. We then examine some of the neural mechanisms that have evolved to meet these challenges and produce smooth, accurate, and efficient movements. Finally, we see how motor learning improves our performance and allows us to adapt to new mechanical conditions, such as when using a tool, or to learn novel correspondences between sensory and motor events, such as when using a computer mouse to control a cursor. This chapter focuses on voluntary movement; reflexes and rhythmic movements are discussed in further detail in Chapters 32 and 33.

Voluntary movements are generated by neural circuits that span different levels of the sensory and motor hierarchies, including regions of the cerebral cortex, subcortical areas such as the basal ganglia and cerebellum, and the brain stem and spinal networks. These different structures have unique patterns of neural activity. Moreover, focal damage to different structures can cause distinct motor deficits. Although it is tempting to suggest that these individual structures have distinct functions, these brain and spinal areas normally work together as a network, such that damage to one component likely affects the function of all others. Many of the principles discussed in this chapter cannot be easily attributed to a single brain or spinal area. Instead, distributed neural processing is likely to underlie the computational mechanisms that subserve sensorimotor control.

The Control of Movement Poses Challenges for the Nervous System

Motor systems produce neural commands that act on the muscles, causing them to contract and generate movement. The ease with which we move, from tying our shoelaces to returning a tennis serve, masks the complexity of the control processes involved. Many factors inherent in sensorimotor control are responsible for this complexity, which becomes clearly evident when we try to build machines to perform human-like movement (Chapter 39). Although computers can now beat the world's best players at chess and Go, no robot can manipulate a chess piece with the dexterity of a 6-year-old child.

The act of returning a tennis serve illustrates why the control of movement is challenging for the brain (Figure 30–1). First, motor systems have to contend with different forms of uncertainty, such as our incomplete knowledge with regard to the state of the world and the rewards we might gain. On the sensory side, although the player may see the serve, she cannot be certain where her opponent will aim or where the ball might strike the racket. On the motor side, there is

uncertainty as to the likely success of different possible returns. Skilled performance requires reducing uncertainty by anticipating events we may encounter (the trajectory of an opponent's tennis serve) and by motor planning (adopting an appropriate stance to return the expected serve).

Second, even if the player can reliably estimate the ball's trajectory, she must determine from sensory signals which of the 600 muscles she will use in order to move her body and racket to intercept the ball. Controlling such a system can be challenging as it is hard to explore all possible actions effectively in a system with many degrees of freedom (eg, a large number of individual muscles), making learning difficult. We will see how the motor system reduces the degrees of freedom of the musculoskeletal system by controlling groups of muscles, termed synergies, to simplify control.

Third, unwanted disturbances, termed noise, corrupt many signals and are present at all stages of sensorimotor control, from sensory processing, through planning, to the outputs of the motor system. For example, in a tennis serve, such noise will cause the ball to land in different places even when the server is trying to hit the same location on the court. Both sensory feedback, reflecting the ball's location, and motor outputs are contaminated with noise. The variability inherent in such noise limits our ability to perceive accurately and act precisely. The amount of noise in our motor commands tends to increase with stronger commands (ie, more force). This limits our ability to move rapidly and accurately at the same time and thus

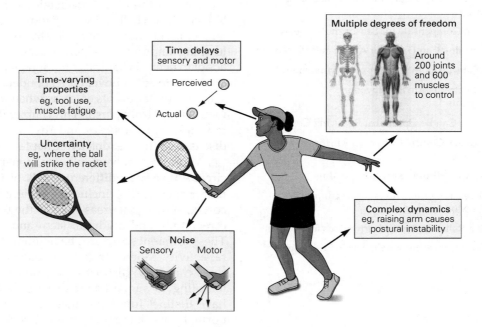

Figure 30–1 The challenges of sensorimotor control.

leads to a trade-off between speed and accuracy. We will see how efficient planning of movement can minimize the deleterious effects of noise on task success.

Fourth, time delays are present at all stages of the sensorimotor system, including the delays arising from receptor dynamics, conduction delays along nerve fibers and synapses, and delays in the contraction of muscles in response to motor commands. Together, these delays, which can be on the order of 100 ms, depend on the particular sensory modality (eg, longer for vision than proprioception) and complexity of processing (eg, longer for face recognition than motion perception). Therefore, we effectively live in the past, with the control system only having access to out-of-date information about the world and our own body. Such delays can result in instability when trying to make fast movements, as we try to correct for errors we perceive but that no longer exist. We will see how the brain makes predictions of the future states of the body and environment to reduce the negative consequences of such delays.

Fifth, the body and environment change both on a short and a long timescale. For example, within the relatively short period of a game, a player must correct for weakening muscles as she fatigues and changes in the court surface when it rains. On a longer timescale, the properties of our motor system change dramatically during growth as our limbs lengthen and increase in weight. As we will see, the ever-changing properties of the motor system place a premium on our ability to use motor learning to adapt control appropriately.

Finally, the relation between motor command and the ensuing action is highly complex. The motion of each body segment produces torques, and potentially motions, at all other body segments through mechanical interactions. For example, when a player raises the racket to hit the ball, she must anticipate destabilizing forces and counteract them to maintain balance. Indeed, when we raise our arms forward while standing, the first muscle to activate is an ankle flexor ensuring you remain upright. We will see how the sensorimotor system controls movement of different segments to maintain fine coordination of actions.

Actions Can Be Controlled Voluntarily, Rhythmically, or Reflexively

Although movements are often classified according to function—as eye movement, prehension (reach and grasp), posture, locomotion, breathing, and speech—many of these functions are subserved by overlapping groups of muscles. Moreover, the same groups of muscles can be controlled voluntarily, rhythmically, or reflexively. For example, the muscles that control respiration can be used to take a deep breath voluntarily before diving under water, to breathe automatically and rhythmically in a regular cycle of inspiration and expiration, or to act reflexively in response to a noxious stimulus in the throat, producing a cough.

Voluntary movements are those that are under conscious control. Rhythmic movements can also be controlled voluntarily but differ from voluntary movements in that their timing and spatial organization are to a larger extent controlled autonomously by spinal or brain stem circuitry. Reflexes are stereotyped responses to specific stimuli that are generated by neural circuits in the spinal cord or brain stem (although some reflexes involve pathways through cortex). These responses occur on a shorter timescale than voluntary responses.

Although we may consciously intend to perform a task or plan a certain sequence of actions, and at times are aware of deciding to move at a particular moment, movements generally seem to occur automatically. Conscious processes are not necessary for moment-to-moment control of movement. We carry out the most complicated movements without a thought to the actual joint motions or muscle contractions required. The tennis player does not consciously decide which muscles to use to return a serve with a backhand or which body parts must be moved to intercept the ball. In fact, thinking about each body movement before it takes place can disrupt the player's performance.

Motor Commands Arise Through a Hierarchy of Sensorimotor Processes

Although the final output to the musculoskeletal system is via motor neurons in the spinal cord, the motor control of muscles for a specific action occurs through a hierarchy of control centers. This arrangement can simplify control: Higher levels can plan more global goals, whereas lower levels are concerned with how these goals are implemented.

At the lowest level, muscles themselves have properties that can contribute to control even without any change in the motor command. Unlike the electric motors of a robot, muscles have substantial passive properties that depend on both the motor command acting on the muscle as well as the muscle's length and rate of change of length (Chapter 31). As a simple approximation, a muscle can be seen as acting like a spring (increasing tension as it is stretched and reducing tension as it is shortened) and damper (increasing tension as the rate of stretch increases). For small perturbations, these properties tend to act to stabilize the length of a muscle and hence stabilize the joint on which the muscle acts. For example, if an external perturbation extends a joint, the flexor muscles will be stretched, increasing their tension, while the extensor muscles will be shortened, reducing their tension, and the imbalance in tension will tend to bring the joint back toward its original position. A particular advantage of such control is that, unlike higher levels in the motor hierarchy, such changes in force act with minimal delay as they are simply an effect of passive physical properties of the muscles.

In addition to the passive properties of muscles, sensory inputs can cause motor output directly without the intervention of higher brain centers. Sensorimotor responses, such as spinal reflexes, control for local disturbance or noxious stimuli. Reflexes are stereotyped responses to specific stimuli that are generated by simple neural circuits in the spinal cord or brain stem. For example, a spinal flexor withdrawal reflex can remove your hand from a hot stove without any descending input from the brain. The advantage of such reflexes is that they are fast; the disadvantage is they are less flexible than voluntary control systems (Chapter 32). Again, there is a hierarchy of reflex circuits. The fastest is the monosynaptic stretch reflex, which drives contraction of a stretched muscle. In this reflex circuit, sensory neurons that are activated by stretch receptors in the muscle (the muscle spindle) directly synapse onto motor neurons that cause the same muscle to contract. The time from the stimulus to the response is around 25 ms. This reflex can be tested clinically by striking the quadriceps muscle tendon just below the patella.

While this monosynaptic stretch reflex is not adaptable on short timescales, multisynaptic reflexes, which involve higher level structures such as motor cortex, can produce responses at around 70 ms. Unlike the monosynaptic reflex, multisynaptic reflexes are adaptable to changes in behavioral goals because the circuit connecting sensory and motor neurons can be modified by task-dependent properties. The strength of a reflex tends to increase with the tension in a muscle (called gain-scaling), and therefore, reflexes can be amplified by co-contracting the set of muscles around a joint so as to respond to perturbations with a greater force. In fact, we use such co-contraction when holding the hand of a rebellious child when crossing a road. Such a strategy can amplify the reflexes, thereby reducing deviations of the arm caused by random external forces.

Finally, voluntary movements are those that are under conscious control by the brain. Voluntary movements can be generated in the absence of a stimulus or used to compensate for a perturbation. The time to generate a voluntary movement in response to a physical perturbation depends both on the nature of the perturbation (modality and size) as well as whether the response can be specified before the perturbation occurs. For example, a voluntary correction to a small physical perturbation can occur with a latency of about 110 ms.

Although we have described clear distinctions between the different levels of the motor hierarchy, from reflexes through to voluntary control, in reality, such distinctions are blurred in a continuum of responses spanning different latencies. Increasing the response time permits additional neural circuitry to be involved in the sensorimotor loop and tends to increase the sophistication and adaptability of the response, leading to a trade-off between the speed of the response and the sophistication of processing as one ascends the motor hierarchy.

Motor Signals Are Subject to Feedforward and Feedback Control

In this section, we will first illustrate some principles of control that are important for dealing with the problem of sensory delays, sensory noise, and motor noise. For simplicity, we confine our discussion to relatively simple movements, such as moving the eyes in response to head movements or moving the hand from one location to another. We consider two broad classes of control, feedforward and feedback, which differ in their reliance on sensory feedback during the movement.

Feedforward Control Is Required for Rapid Movements

Some movements are executed without monitoring the sensory feedback that arises from the action. In such feedforward control situations, the motor command is generated without regard to the sensory consequences. Such commands are therefore also termed *open-loop*, reflecting the fact that the sensorimotor loop is not completed by sensory feedback (Figure 30–2A).

Open-loop control requires some information about the body so that the appropriate command can be generated. For example, it should include information about the dynamics of the motor system. Here, "dynamics" refers to the relation between the motor command (or the torques or forces) applied and the ensuing motion of the body, for example, joint rotations. For perfect open-loop control, one needs to

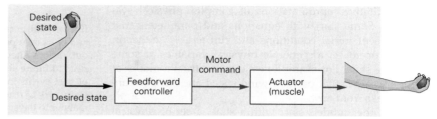

A Feedforward control

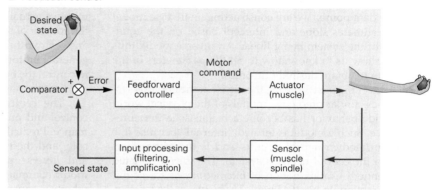

B Feedback control

Figure 30–2 Feedforward and feedback control.

A. A feedforward control motor command is based on a desired state. Any errors that arise during the movement are not monitored. Although we illustrate the elements of feedforward control for the arm, only the initial portion of any arm movement is driven by feedforward signals and the movement typically involves feedback control.

B. With feedback control, the desired and sensed states are compared (at the comparator) to generate an error signal, which helps shape the motor command. There can be considerable delay in the feedback of sensory information to the comparator.

invert the dynamics so as to calculate the motor command that will generate the desired motion. The neural mechanism that performs this inversion is called an inverse model, a type of internal model (Box 30–1). An inverse model coupled to open-loop control can determine what motor commands are needed to produce the particular movements necessary to achieve a goal.

Although not monitoring the consequences of an action may seem counterproductive, there are good reasons for not doing so. The main reason, as discussed earlier, is that there are delays in both sensing and acting. That is, the conversion of a stimulus into neural signals by sensory receptors and conveyance of these signals to central neurons take time. For example, visual inputs can take around 60 ms to be processed in the retina and transmitted to the visual cortex. In addition to delays in afferent sensory systems, there are also delays in central processing, in the transmission of efferent signals to motor neurons, and in the response of muscles. In all, the combined sensorimotor loop delay is appreciable, approximately 120 to 150 ms for a motor response to a visual stimulus. This delay means that movements like saccades, which redirect gaze within 30 ms, cannot use sensory feedback to guide movement. Even for slower movements like reaching, which takes on the order of 500 ms, sensory information cannot be used to guide the initial part of a movement, so open-loop control must be used.

Open-loop control also has disadvantages. Any movement errors caused by inaccuracies in planning or execution will not be corrected, and therefore will compound themselves over time or successive movements. The more complex the system under control, the more difficult it is to arrive at an accurate inverse model through learning.

An example of a purely open-loop control system is the control of the eye in response to head rotation. The vestibulo-ocular reflex (Chapter 27) uses open-loop control to fix gaze on an object during head rotation. The vestibular labyrinth senses the head rotation and drives appropriate movements of the eyes through a three-synapse circuit. The reflex does not require (or use) vision during the movement (the eyes maintain a stable gaze when the head is rotated in the dark). Sensory information from the vestibular system does drive the eye movement, but the control is feedforward (any errors that arise are not corrected during the movement). Such precise open-loop control is possible because the dynamic properties of the eye are relatively simple, the rotation of the head can be directly sensed by the vestibular labyrinth, and the eye tends not to be substantially perturbed by external events. In contrast, it is very difficult to optimize an inverse model for a complex musculoskeletal system such as the arm, and thus, the control of arm movement requires some form of error correction.

Box 30–1 Internal Models

The utility of numerical models in the physical sciences has a long history. Numerical models are abstract quantitative representations of complex physical systems. Some start with equations and parameters that represent initial conditions and run *forward*, either in time or space, to generate physical variables at some future state. For example, we can construct a model of the weather that predicts wind speed and temperature 2 weeks from now.

Other models start with a state, a set of physical variables with specific values, and run in the *inverse* direction to determine what parameters in the system account for that state. When we fit a straight line to a set of data points, we are constructing an inverse model that estimates slope and intercept based on the equations of the system being linear. An inverse model thus may allow us to know how to set the parameters of the system to obtain desired outcomes.

Over the past 50 years, the idea that the nervous system has similar predictive models of the physical world to guide behavior has become a major issue in neuroscience. Such a model is termed "internal" because it is instantiated in neural circuits and is therefore internal to the central nervous system. The idea originated in Kenneth Craik's notion of *internal models* for cognitive function. In his 1943 book *The Nature of Explanation*, Craik was perhaps the first to suggest that organisms make use of internal representations of the external world:

If the organism carries a "small-scale model" of external reality and of its own possible actions within its head, it is able to try out various alternatives, conclude which is the best of them, react to future situations before they arise, use the knowledge of past events in dealing with the present and future, and in every way to react in a much fuller, safer, and more competent manner to the emergencies that face it.

In this view, an internal model allows an organism to contemplate the consequences of potential actions without actually committing itself to those actions. In the context of sensorimotor control, internal models can answer two fundamental questions. First, how can we generate motor commands that act on our muscles so as to control the behavior of our body? Second, how can we predict the consequences of our own motor commands?

The central nervous system must exercise both control and prediction to achieve skilled motor performance. Prediction and control are two sides of the same coin, and the two processes map exactly onto forward and inverse models (Figure 30–3). Prediction turns motor commands into expected sensory consequences, whereas control turns desired sensory consequences into motor commands.

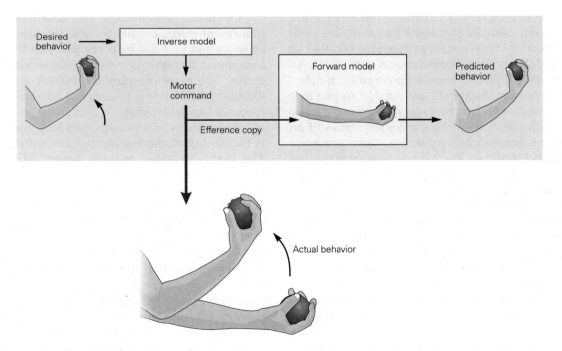

Figure 30–3 Internal sensorimotor models represent relationships of the body and external world. The inverse model determines the motor commands that will produce a behavioral goal, such as raising the arm while holding a ball. A descending motor command acts on the musculoskeletal system to produce the movement. A copy of the motor command is passed to a forward model that simulates the interaction of the motor system and the world and thus can predict behaviors. If both forward and inverse models are accurate, the output of the forward model (the predicted behavior) will be the same as the input to the inverse model (the desired behavior).

Feedback Control Uses Sensory Signals to Correct Movements

To correct movement errors as they arise, movement must be monitored. Systems that perform error correction are known as feedback or closed-loop control because the sensorimotor loop is complete (Figure 30–2B).

The simplest form of feedback control is one in which the control system generates a fixed response when the error exceeds some threshold. Such a system is seen in most central heating systems in which a thermostat is set to a desired temperature. When the house temperature falls below the specified level, the heating is turned on until the temperature reaches that level. Although such a system is simple and can be effective, it has the drawback that the amount of heat being put into the house does not relate to the discrepancy between the actual and desired temperature (the error). A better system is one in which the control signal is proportional to the error.

Such proportional control of movement involves sensing the error between the actual and desired position of, for example, the hand. The size of the corrective motor command is in proportion to the size of the error and in a direction to reduce the error. The amount by which the corrective motor command is increased or decreased per unit of positional error is called the gain. By continuously correcting a movement, feedback control can be robust both to noise in the sensorimotor system and to environmental perturbations.

While feedback control can update commands in response to deviations that arise during the movement, it is sensitive to feedback delays. Without any delay, as the gain of the feedback controller increases, the system will track a desired position with increasing fidelity (Figure 30–4). However, as feedback delay increases, the control system may start to oscillate and eventually become unstable. This is because with a delay the system may respond to errors that no longer exist and may therefore even correct in the wrong direction.

Smooth pursuit eye movement, used to track a moving object, is an example of a movement driven primarily by feedback. Smooth pursuit uses feedback to minimize the velocity error on the retina (the difference between the gaze and target velocity). We can compare the efficiency of feedforward and feedback control in minimizing error. Compare how easy it is to fixate on your outstretched stationary finger when quickly rotating your head back and forth versus trying to track your finger when it is moving it rapidly sinusoidally left and right while your head remains stationary. Although the relative motion of finger to head is the same in both conditions, the former is precise because it uses the vestibulo-ocular reflex, whereas the latter uses feedback (requiring an error in velocity to drive the eye movement) and thus is less precise, particularly as the frequency of motion increases.

In most motor systems, movement control is achieved through a combination of feedforward and feedback processes. We will see later that these two components arise naturally in a unified model of movement production.

Estimation of the Body's Current State Relies on Sensory and Motor Signals

Accurate control of movement requires information about our body's current state, for example, the positions and velocities of the different segments of the body. To grasp an object, we need to know not only the location, shape, and surface properties of the object but also the current configuration of our arm and fingers so as to appropriately shape and position the hand.

Estimating the state of the body is not a trivial problem. First, as we have seen, sensory signals are delayed due to sensory transduction and conduction time. Therefore, signals from our muscles, joints, and vision are all out of date by the time they reach the central nervous system. Second, the sensory signals we receive are often imprecise and corrupted by neural noise. For example, if you touch the underside of a table with the finger of one hand and try to estimate its location on the top of the table with your other hand, you can be off by a considerable distance. Third, we often do not have sensors that directly communicate relevant information. For example, although we have sensors that report muscle length and joint angle, we have no sensors within the limb that directly determine the location of the hand in space. Therefore, sophisticated computation is required to estimate current body states as accurately as possible. Several principles have emerged as to how the brain estimates state.

First, state estimation relies on internal models of sensorimotor transformations. Given the fixed lengths of our limb segments, there is a mathematical relation between the muscle lengths or joint angles of the arm and the location of the hand in space. A neural representation of this relation allows the central nervous system to estimate hand position if it knows the joint angles and segment lengths. Neural circuits that compute such sensorimotor transformations are examples of internal models (Box 30–1).

Second, state estimation can be improved by combining multiple sensory modalities. For example, information about the state of our limbs arrives from

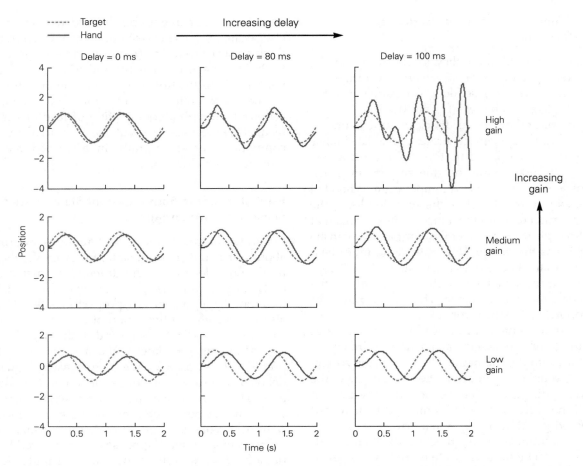

Figure 30–4 The interplay of gain and delay in feedback control. Performance of a feedback controller trying to track a target moving sinusoidally in one dimension. The sensory feedback signal that conveys error in the position arrives after some period of time (the delay), and the motor system tries to correct for the error by increasing or decreasing the size of its command relative to the error (the gain).

The plots show the performance in which there is either instantaneous feedback (no delay) of error (**left column**) or feedback with delay of 80 or 100 ms (**middle and right columns**). When the gain is high and the delay is low, tracking is very good. However, when the delay increases, because the controller is compensating for errors that existed 80 or 100 ms earlier, the correction may be inappropriate for the current error. The gain can be lowered to maintain stability, but as the feedback controller corrects errors only slowly, tracking becomes inaccurate.

At low gain (**bottom row**), the feedback controller corrects errors only slowly and tracking is inaccurate. As the gain increases (**middle row**), the feedback controller corrects errors more rapidly and tracking performance improves. At high gain (**top row**), the system corrects rapidly but is prone to overcorrect, leading to instability when the time delay in feedback is on the order of physiological time delays (**top right**). Because the controller is compensating for errors that existed 100 ms earlier, the correction may therefore be inappropriate for the current error. This overcorrection leads to oscillations and is one mechanism proposed to account for some forms of oscillatory tremor seen in neurological disease.

proprioceptive information from muscle spindles, the stretch of the skin, and the sight of the arm. These modalities have different amounts of variability (or noise) associated with them, and just as we average a set of experimental data to reduce measurement error, these sensory modalities can be combined to reduce the overall uncertainty in the state estimate.

The optimal way to combine these sources is for higher brain centers to take the uncertainty of each modality into account and rely on the more certain

modalities. For example, the location of the hand can be sensed both by proprioception and vision. The sight of your hand in front of you tends to be more reliable than proprioception for estimating location along the azimuth (right–left) but less reliable for depth (forward–back). Therefore, visual input should be given greater weight than proprioceptive input when estimating the azimuth location of your hand, and vice versa for depth. By measuring the precision of each modality when used alone, it is possible to predict the increased

Box 30–2 Bayesian Inference

Bayesian inference is a mathematical framework for making estimates about the world based on uncertain information. The fundamental idea is that probabilities (between 0 and 1) can be used to represent the degree of belief in different alternatives, such as the belief that the chance of your rolling a six with fair dice is 1 in 6.

The beauty of Bayesian inference is that by using the rules of probability we can specify how beliefs should be formed and updated based on our experience and new information from sensory input. For example, when playing tennis, we want to estimate where the ball will land. Because vision does not provide perfect information about the ball's position and velocity in flight, there is uncertainty as to the landing location. However, if we know the level of noise in our sensory system, then the current sensory input can be used to compute the *likelihood* (ie, probability) of the particular sensory input for different potential landing locations.

We can learn additional information from repeated experience of the game: The position where the ball lands is not equally probable over the court. For example, bounce locations are likely to be concentrated near the boundary lines where it is most difficult to return the ball. This distribution is termed the *prior*.

The Bayes rule defines how to combine the prior and likelihood to make an optimal estimate of the bounce location. While the Bayesian approach was originally developed in statistics, it now provides a unifying framework to understand how the brain deals with uncertainty in the perceptual, motor, and cognitive domains.

---

precision when both are used at the same time. Experiments have shown that this process is often close to optimal. Precision can also be improved by combining prior knowledge with sensory inputs using the mathematics of Bayesian inference (Box 30–2).

Third, the motor command can also provide valuable information. If both the current state of the body and the descending motor command are known, the next state of the body can be estimated. This estimate can be derived from an internal model that represents the causal relation between actions and their consequences. This is called a forward model because it estimates future sensory inputs based on motor outputs (Box 30–1). Thus, a forward model can be used to anticipate how the motor system's state will change as the result of a motor command. A copy of a descending motor command is passed into a forward model that acts as a neural simulator of the musculoskeletal system moving in the environment. This copy of the motor command is known as an efference copy (or corollary discharge). Forward and inverse models can be better understood if we place the two in series. If the structure and parameter values of each model are correct, the output of the forward model (the predicted behavior) will be the same as the input to the inverse model (the desired behavior) (Figure 30–3).

Using the motor command to estimate the state of the body is advantageous as, unlike sensory information that is delayed, the motor command is available before it acts on the musculoskeletal system and therefore can be used to anticipate changes in the state.

However, this estimate will tend to drift over time if the forward model is not perfectly accurate, and therefore, sensory feedback is used to correct the state estimate, albeit with a delay.

It may seem surprising that the motor command is used in state estimation. In fact, the first demonstration of a forward model used a motor system that relies on only the motor command to estimate state, that is, the position of the eye within the orbit. The concept of motor prediction was first considered by Helmholtz when trying to understand how we localize visual objects. To calculate the location of an object relative to the head, the central nervous system must know both the retinal location of the object and the gaze direction of the eye. Helmholtz's ingenious suggestion was that the brain, rather than sensing the gaze direction, predicted it based on a copy of the motor command to the eye muscles.

Helmholtz used a simple experiment on himself to demonstrate this. If you move your own eye without using the eye muscles (cover one eye and gently press with your finger on your open eye through the eyelid), the retinal locations of visual objects change. Because the motor command to the eye muscles is required to update the estimate of the eye's state, the predicted eye position is not updated. However, because the retinal image has changed, this leads to the false percept that the world must have moved. A more dramatic example is that if the eye muscles are temporarily paralyzed with curare, then trying to move the eyes leads to a percept that the world is moving. This is because the command leads to a state estimate that the eye has moved, but with a fixed retinal input (due to the paralysis), the only consistent interpretation is that the world has moved.

Finally, the best estimate of state is achieved by combining sensory modalities with motor commands. The drawbacks of using only sensory feedback or only motor prediction can be ameliorated by monitoring both and using a forward model to estimate the current state. A neural apparatus that does this is known as an *observer model*. The major objectives of the observer model are to compensate for sensorimotor delays and

to reduce uncertainty in the estimate of current state arising from noise in both the sensory and motor signals (Figure 30–5). Such a model has been supported by empirical studies of how the nervous system estimates hand position, posture, and head orientation. We will see how such models are used to decode neural signals in brain–machine interfaces (Chapter 39).

State estimation is not a passive process. Skilled performance requires the effective and efficient gathering and processing of sensory information relevant to an action. The quality of sensory information depends on our actions because what we see, hear, and touch is

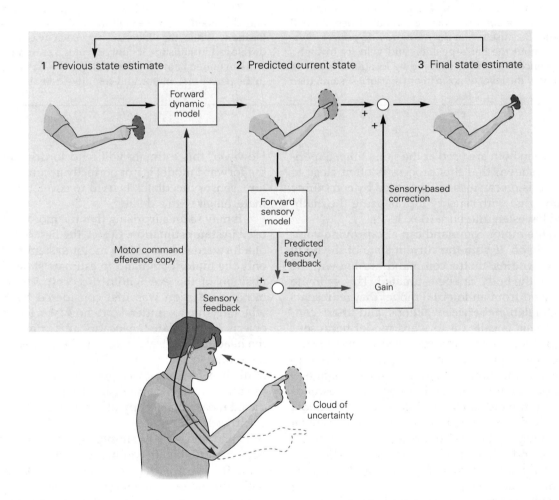

Figure 30–5 An observer model. The model is being used to estimate the finger's location during movement of the arm. A previous estimate of the distribution of possible finger positions (**1, blue cloud**) is updated (**2, yellow cloud**) using an efference copy of the motor command and a forward model of the dynamics. The updated distribution of finger positions is larger than that of the previous estimate. The model then uses a forward sensory model to predict the sensory feedback that would occur for these new finger positions, and the error between the predicted and actual sensory feedback is used to

correct the estimate of current finger position. This correction changes the sensory error into state errors and also determines the relative reliance on the efference copy and sensory feedback.

The final estimate of current finger position (**3, purple cloud**) has less uncertainty. This estimate will become the new previous estimate for subsequent movement as this sequence is repeated many times. Delays in sensory feedback that must be compensated have been omitted from the diagram for clarity.

influenced by our movements. For example, the ocular motor system controls the eyes' sensory input by orienting the fovea to points of interest within the visual scene. Thus, movement can be used to efficiently gather information, a process termed *active sensing*. Active sensing involves two main processes: perception, by which we process sensory information and make inferences about the world, and action, by which we choose how to sample the world to obtain useful sensory information. Eye movements can betray the difference between skilled and amateur performers. For example, a batsman in the game of cricket will make a predictive saccade to the place where he expects a bowled ball to hit the ground, wait for it to bounce, and use a pursuit eye movement to follow the ball's trajectory after the bounce. A shorter latency for this first saccade distinguishes expert from amateur batsmen. Therefore, the motor system can also be used to improve our sensing of the world so as to gather information that, in turn, helps us achieve our motor goals.

Prediction Can Compensate for Sensorimotor Delays

As we have seen, delays in feedback can lead to problems during a movement, as the delayed information does not reflect the present state of the body and world. Two strategies, intermittency and prediction, can compensate for such delays and thus increase accuracy of information during movement. With intermittency, movement is momentarily interrupted by rest, as in eye saccades and manual tracking. Provided the interval of rest is greater than the time delay of the sensorimotor loop, intermittency fosters more accurate sensory feedback. Prediction is a better strategy and, as we have seen, can form a major component of a state estimator.

The nervous system uses different modes of control that depend on prediction and sensory feedback to different extents. These modes are nicely illustrated by differences in object manipulation under different conditions. When an object's behavior is unpredictable, sensory feedback provides the most useful signal for estimating load. For example, when flying a kite, we need to adjust our grip almost continuously in response to unpredictable wind currents. When dealing with such unpredictability, grip force needs to be high to prevent slippage because adjustments to grip tend to lag behind changes in load force (Figure 30–6A).

However, when handling objects with stable properties, predictive control mechanisms can be effective. For example, when the load is increased by a self-generated action, such as moving the arm, the grip force increases instantaneously with load force (Figure 30–6B). Sensory detection of the load would be too slow to account for this rapid increase in grip force.

Such predictive control is essential for the rapid movements commonly observed in dexterous behavior. Indeed, this predictive ability can be demonstrated easily with the "waiter task." Hold a weighty book on the palm of your hand with an outstretched arm. If you then use your other hand to remove the book (like a waiter removing objects from a tray), the supporting hand remains stationary. This shows our ability to anticipate a change in load caused by our own action and thus generate an appropriate and exquisitely timed change in muscle activity. In contrast, if someone else removes the book from your hand, even though you are watching the removal, it is close to impossible to maintain the hand stationary. We will see how cerebellar lesions affect this ability to predict, leading to a lack of such a coordinated response (Chapter 37).

Detecting any discrepancies between predicted and actual sensory feedback is also essential in motor control. This discrepancy, termed sensory prediction error, can drive learning of internal models and also be used for control. For example, when we pick up an object, we anticipate when the object will lift off the surface. The brain is particularly sensitive to the occurrence of unexpected events or the nonoccurrence of expected events (ie, to sensory prediction errors). Thus, if an object is lighter or heavier than expected and therefore lifts off too early or cannot be lifted, reactive responses are initiated.

In addition to its use in compensating for delays, prediction is a key element in sensory processing. Sensory feedback can arise as a consequence of both external events and our own movements. In the sensory receptors, these two sources are not distinguishable, as sensory signals do not carry a label of "external stimulus" or "internal stimulus." Sensitivity to external events can be amplified by reducing the feedback from our own movement. Thus, subtracting predictions of sensory signals that arise from our own movements from the total sensory feedback enhances the signals that carry information about external events. Such a mechanism is responsible for the fact that self-tickling is a less intense experience than tickling by another. When subjects were asked to tickle themselves using a robotic interface, but a time delay was introduced between the motor command and the resulting tactile input, the ticklishness increased. With such delayed tactile input, the predictions become inaccurate and thus fails to cancel the sensory feedback, resulting in the increased tickle sensation. Such predictive modulation of sensory

A Robot controls movement

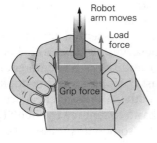

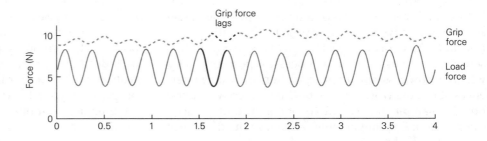

B Hand controls movement

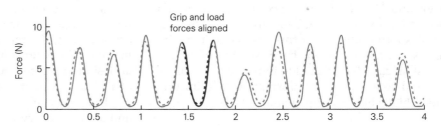

Figure 30–6 Anticipatory control of self-generated actions. (Adapted, with permission, from Blakemore, Goodbody, and Wolpert 1998. Copyright © 1998 Society for Neuroscience.

A. When a subject is instructed to hold an object to which a sinusoidal load force is mechanically applied, the grip force of the fingers is high to prevent slippage, and the grip force

modulation lags behind the changes in load force. This is high-lighted for a portion of the load force modulation (**dark red solid line**) that leads to a corresponding grip force (**dark red dashed line**), which is delayed. (Trial duration 4 s).

B. When a subject generates a similar load profile by pulling down on the fixed object, the load force can be anticipated, and thus, the grip force is lower and also tracks the load force without delay.

signals by motor actions is a fundamental property of many sensory systems.

Sensory Processing Can Differ for Action and Perception

A growing body of research supports the idea that the sensory information used to control actions is processed in neural pathways that are distinct from the afferent pathways that contribute to perception. It has been proposed that visual information flows in two streams in the brain (Chapter 25). A dorsal stream that projects to the posterior parietal cortex is particularly involved in the use of vision for action (Chapter 34), while a ventral stream that projects to the inferotemporal cortex is involved in conscious visual perception.

This distinction between the uses of vision for action and perception is based on a double dissociation seen in patient studies. For example, the patient D.F. developed visual agnosia after damage to her ventral stream. She is unable, for example, to indicate the orientation of a slot either verbally or with her hand. However, when asked to perform a simple action, such as putting a card

through the slot, she has no difficulty orienting her hand appropriately to put the card through the slot (Chapter 59). Conversely, patients with damage to the dorsal stream can develop optic ataxia in which perception is intact but control is affected.

Although the distinction between perception and action arose from clinical observations, it can also be seen in normal people, as in the size–weight illusion. When lifting two objects of different size but equal weight, people report that the smaller object feels heavier. This illusion, first documented more than 100 years ago, is both powerful and robust. It does not lessen when a person is informed that the objects are of equal weight and does not weaken with repeated lifting.

When subjects begin to lift large and small objects that weigh the same, they generate larger grip and load forces for the larger object because they assume that larger objects are heavier. After alternating between the two objects, they rapidly learn to scale their finger-tip forces precisely for the true object weight (Figure 30–7). This shows that the sensorimotor system recognizes that the two weights are equal. Nevertheless, the size–weight illusion persists, suggesting not only that

Figure 30–7 The size–weight illusion.

A. In each trial, subjects alternately lifted a large object and a small object that weighed the same. Subjects thought the smaller object felt heavier than it actually was.

B. In the first trial, subjects generated greater grip and load forces for the bigger object (**orange traces**) as it was expected to be heavier than the small object. In the eighth trial, the grip and load forces are the same for the two objects, showing that the sensorimotor system for this action generates grip and load forces appropriate to the weights of the two objects despite the persistent conscious perception of a difference in weight. (Adapted with permission, from Flanagan and Beltzner 2000. Copyright © 2000 Springer Nature.)

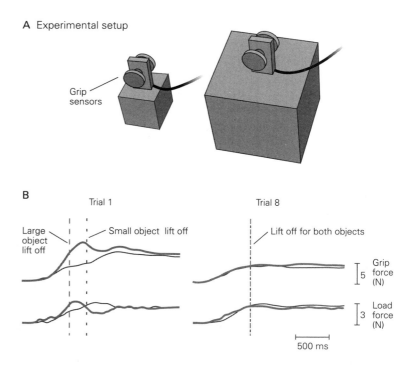

the illusion is a result of high-level cognitive centers in the brain but also that the sensorimotor system can operate independently of these centers.

Motor Plans Translate Tasks Into Purposeful Movement

Real-world tasks are expressed as goals: I want to pick up a glass, dance, or have lunch. However, action requires a detailed specification of the temporal sequence of movements powered by the 600 or so muscles in the human body. There is clearly a gap between the statement of a goal and a motor plan that recruits specific muscles in pursuit of that goal.

Stereotypical Patterns Are Employed in Many Movements

The ability of the motor systems to achieve the same task in many different ways is called redundancy. If one way of achieving a task is not practical, there is usually an alternative. For example, the simplest of all tasks, reaching for an object, can be achieved in infinitely many ways.

The duration of the movement can be freely selected from a wide range and, given a particular choice of duration, the path and speed profile of the

hand along the path (ie, trajectory) can take on many different patterns. Even selecting one trajectory still allows for infinitely many joint configurations to hold the hand on any given point of the path. Finally, holding the arm in a fixed posture can be achieved with a wide range of muscular co-contraction levels. Therefore, for any movement, a choice must be made from a large number of alternatives.

Do we all choose to move in our own way? The answer is clearly no. Repetitions of the same behavior by one individual as well as comparisons between individuals have shown that the patterns of movement are very stereotypical.

Invariance in stereotypical patterns of movement tells us something about the principles the brain uses when planning and controlling our actions. For example, when reaching, our hand tends to follow roughly a straight path and the hand speed over time is typically smooth, unimodal, and roughly symmetric (bell-shaped, Figure 30–8). The tendency to make straight-line movements characterizes a large class of movements and is surprising given that the muscles act to rotate joints.

To achieve such a straight-line movement of the hand requires complex joint rotations. The motions of the joints in series (the shoulder, elbow, and wrist) are complicated and vary greatly with different initial and final positions. Because rotation at a single joint would

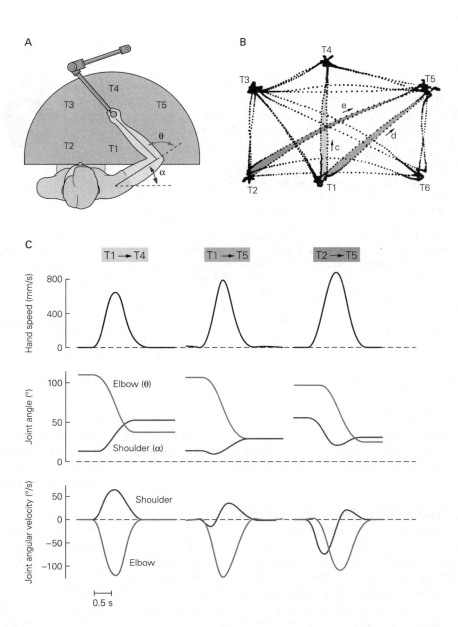

Figure 30–8 Hand path and velocity have stereotypical features. (Adapted, with permission, from Morasso 1981. Copyright © 1981 Springer Nature.)

A. The subject sits in front of a semicircular plate and grasps the handle of a two-jointed apparatus that moves in the horizontal plane and records hand position. The subject is instructed to move the hand between various targets (**T1–T6**).

B. The paths traced by one subject while moving his hand between targets.

C. Kinematic data for three hand paths (**c**, **d**, and **e**) shown in panel **B**. All paths are roughly straight, and all hand speed profiles have the same shape and scale in proportion to the distance covered. In contrast, the profiles for the angular velocity of the elbow and shoulder for the three hand paths differ. The straight hand paths and common profiles for speed suggest that planning is done with reference to the hand because these parameters can be linearly scaled. Planning with reference to joints would require computing nonlinear combinations of joint angles.

produce an arc at the hand, both elbow and shoulder joints must be rotated concurrently to produce a straight path. In some directions, the elbow moves more than the shoulder; in others, the reverse occurs. When the hand is moved from one side of the body to the other (Figure 30–8, movement from T2 to T5), one or both joints may have to reverse direction in midcourse. The fact that hand trajectories are more invariant than joint trajectories suggests that the motor system is more concerned with controlling the hand, even at the cost of generating complex patterns of joint rotations.

Such task-centered motor plans can account for our ability to perform a specific action, such as writing, in different ways with more or less the same result.

Handwriting is structurally similar regardless of the size of the letters or the limb or body segment used to produce it (Figure 30–9). This phenomenon, termed motor equivalence, suggests that purposeful movements are represented in the brain abstractly rather than as sets of specific joint motions or muscle contractions. Such abstract representations of movement, which are able to drive different effectors, provide a degree of flexibility of action not practical with preset motor programs.

Motor Planning Can Be Optimal at Reducing Costs

Why do humans choose one particular manner of performing a task out of the infinite number of possibilities?

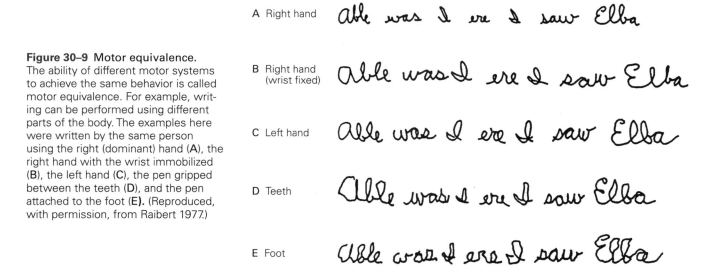

A Right hand

B Right hand
   (wrist fixed)

C Left hand

D Teeth

E Foot

Figure 30–9 Motor equivalence. The ability of different motor systems to achieve the same behavior is called motor equivalence. For example, writing can be performed using different parts of the body. The examples here were written by the same person using the right (dominant) hand (**A**), the right hand with the wrist immobilized (**B**), the left hand (**C**), the pen gripped between the teeth (**D**), and the pen attached to the foot (**E**). (Reproduced, with permission, from Raibert 1977.)

Extensive research has attempted to answer this question, and the fundamental idea that has emerged is that planning can be equated with choosing the best way to achieve a task. Mathematically, this is equivalent to the process of optimizing (ie, minimizing) a cost associated with the movement. The cost is a way of quantifying what is good or bad about a movement (eg, energy, accuracy, stability) with a single number.

Different ways of achieving a task will lead to different costs. This allows all possible solutions to be ranked, thus identifying the one with the lowest cost. Invariances in our movements will reflect the particular cost we care about for that type of movement. Many costs have been proposed, but currently, most successful theories propose that there are two main components to movement cost: task success and effort. The effort component means that we want to achieve success but with minimal energetic cost.

To understand how task success is a component of the cost, it is useful to understand what leads to lack of success. Having inaccurate internal models or processing clearly limits our ability to complete tasks, and motor learning is designed to keep these processes accurate. However, low-level components in the motor system, such as motor noise, limit success. Movements tend to be variable, and the variability tends to increase with the speed or force of the movement. Part of this increase is caused by random variation in both the excitability of motor neurons and the recruitment of additional motor units needed to increase force. Incremental increases in force are produced by progressively smaller sets of motor neurons, each of which produces disproportionately greater increments

of force (Chapter 31). Therefore, as force increases, fluctuations in the number of motor neurons lead to greater fluctuations in force.

The consequences of this can be observed experimentally by asking subjects to generate a constant force. The variability of such force production increases with the level of the force. Over a large range, this increase in variability is captured by a constant coefficient of variation (the standard deviation divided by the mean force). This dependence of variability on force also increases the variability of pointing movements as the speed of movement increases (as greater speed requires greater muscle force). The decrease in movement accuracy with increasing speed is known as the speed–accuracy trade-off (Figure 30–10). This relationship is not fixed, and part of skill learning, such as learning to play the piano, involves being able to increase speed without sacrificing accuracy.

In general, effort and accuracy are in conflict. Accuracy requires energy because corrections require muscular activity and thus comes at some cost. The trade-off between accuracy and energy varies for different movements. When walking, we could choose to step gingerly to ensure we never trip, but this would require substantial energy use. Therefore, we are willing to save energy by allowing ourselves the risk of occasionally tripping. In contrast, while eating with a knife and fork, we prioritize accuracy over energy to ensure the fork does not end up in our cheek.

The optimal movement is thus the one that minimizes the bad consequences of noise while saving energy. One way to do this is to specify a desired

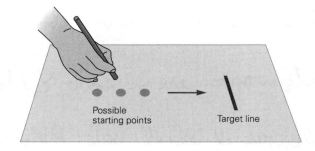

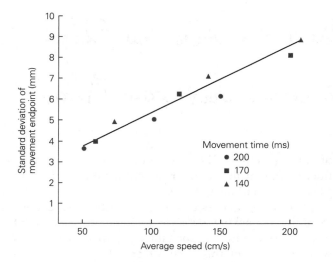

Figure 30–10 Accuracy of movement varies in direct proportion to its speed. Subjects held a stylus and had to hit a straight line lying perpendicular to the direction in which they moved the stylus. Subjects started from one of three different initial positions and were required to complete the movement within three different times (140, 170, or 200 ms). A trial was successful if the subject completed the movement within 10% of the required time. Only successful trials were used for analysis. Subjects were informed when a trial was not successful. The variability in the motion of the subjects' arm movements is shown in the plot as the standard deviation of the movement endpoint plotted against average speed (for each of three movement starting points and three movement times, giving nine data points). The variability in movement increases in proportion to the speed and therefore to the force producing the movement. (Adapted, with permission, from Schmidt et al. 1979.)

movement trajectory or sequence of states that can be considered optimal. Although noise and environmental disturbances can cause the motor system to deviate from the desired behavior, the role of feedback is simply to return the movement back to the desired trajectory. However, this approach is not necessarily computationally efficient. Rather than specifying the desired state of the body, we can specify an optimal feedback controller to generate the movement.

Optimal Feedback Control Corrects for Errors in a Task-Dependent Manner

Optimal feedback control aims to minimize a cost such as a combination of energy and task inaccuracy (Chapter 34). This type of feedback control is based on the idea that people do not plan a trajectory given a particular cost. Instead, the cost is used to create a feedback controller that specifies, for example, how the feedback gain for positional errors (and other errors such as velocity and force) changes over time. Therefore, given the goal of the task, the controller specifies the motor command suitable for different possible states of the body. The trajectory is then simply a consequence of applying the feedback control law to the current estimate of the state of the body (Figure 30–11). The feedback controller is optimal in that it can minimize the cost even in the presence of potential disturbances.

Optimal feedback control, therefore, does not make a hard distinction between feedforward and feedback control. Rather, during a task, the balance between feedforward and feedback control varies along a continuum that depends on the extent to which the estimate of current body state is influenced by predictions (feedforward) or by sensory input (feedback).

An important feature of optimal feedback control is that it will correct only for deviations that are task relevant and allow variation in task-irrelevant deviations. For example, when reaching to open an exit door that has a long horizontal handle, it is of little importance where along the handle one makes contact, so deviations in the horizontal direction can be ignored. Such considerations lead naturally to the minimal intervention principle that one should only intervene in an ongoing task if deviations will affect task success.

Intervening will generally add noise into the system (and require an increased effort), so intervening unnecessarily will lead to a decrement in performance. The aim of optimal feedback control is not to eliminate all variability, but to allow it to accumulate in dimensions that do not interfere with the task while minimizing it in the dimensions relevant for the task completion. The minimal intervention principle is supported by studies that show that feedback does not always return the system to the unperturbed trajectory but often acts in a manner to reduce the effect of the disturbance on the achievement of the task goal and to ensure that corrections are task-dependent.

Optimal feedback control emphasizes the setting of feedback gains, which can be partially instantiated by reflexes that generate rapid motor responses. Optimal feedback control proposes that these rapid responses should be highly tuned to the task at hand.

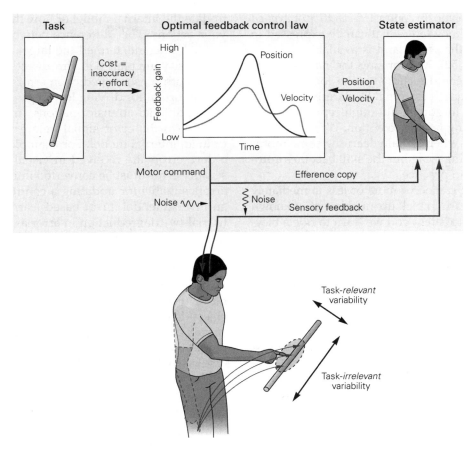

Figure 30–11 Optimal feedback control. In order to generate a movement for a given task, such as touching a horizontal bar, the sensorimotor system specifies a cost that is a combination of accuracy (eg, distance of the finger to the bar) and effort. To generate a movement that minimizes this cost, the sensorimotor system sets up an optimal feedback control rule that specifies the time-varying gains. These gains specify how the motor command should depend on states such as positional error and hand velocity. The form of this feedback control law assures that the movement is the best it can be in the presence of internal noise and external perturbations. The optimal behavior tends to let variability (**blue ellipsoid**, showing the possible final locations of the hand) accumulate in dimensions that do not affect task success (task-irrelevant variability), such as along the axis of the bar, while controlling variability that would lead to the hand missing the bar (task-relevant variability). Three paths for reaching from the same starting point are shown; corrections are made only in the task-relevant dimension.

Although the short-latency (monosynaptic) stretch reflex responds only to muscle stretch, the long-latency response has long been known to respond to task-dependent factors (Chapter 32). Optimal feedback control is important because it combines trajectory generation, noise, and motor cost and provides a clear comparison for the results of experimental work.

Multiple Processes Contribute to Motor Learning

Animals have a remarkable capacity for learning new motor skills simply through everyday interaction with their environment. Although evolution can hard wire some innate behaviors, such as the ability of a foal to stand or a spider to spin a web, motor learning is required to adapt to new and varying environments.

New motor skills cannot be acquired by fixed neural systems. Sensorimotor systems must constantly adapt over a lifetime as body size and proportions change, thereby maintaining an appropriate relationship between motor commands and body mechanics. In addition, learning is the only way to acquire motor skills that are defined by social convention, such as writing or dancing.

Most forms of motor learning involve *procedural* or *implicit* learning, so-called because subjects are generally unable to express what it is they have learned. Implicit learning often takes place without consciously thinking about it and can be retained for extended periods of time without practice (Chapter 52). Typical examples of procedural learning are learning to ride a bicycle or play the piano. In contrast, *declarative* or *explicit* learning refers to knowledge that can be expressed in statements about the world and is available to introspection (Chapter 52). Memorizing the names of the cranial nerves or the directions to the local hospital are examples of explicit learning. Declarative memory tends to be easily forgotten, although repeated exposure can lead to long-lasting retention. We use explicit learning strategies when initially learning some motor tasks, such as driving a car, but the skill becomes automatic with time and practice.

Motor learning can occur more or less immediately or over time. We learn to pick up an object of unknown weight almost immediately, and we learn to ride a bicycle after a few weeks of practice, but mastering the piano requires years. These different timescales may reflect the intrinsic difficulty of the task as well as evolutionary constraints that have to be unlearned to perform the task. For example, piano playing requires learning precise control of individual fingers, whereas in normal movements, such as reaching and grasping, individuated finger movements are rare. Sensorimotor learning can be divided into two broad, but overlapping, classes: adaptations to alterations in the properties of sensorimotor systems and learning new skills. We focus on each in turn.

Error-Based Learning Involves Adapting Internal Sensorimotor Models

Error-based learning is the driving force behind many well-studied sensorimotor adaptation paradigms. For example, the relation between the visual and proprioceptive location of a limb can be altered by wearing prismatic glasses (or even spectacles). This shifts the visual input so that a person's reach for an object is misdirected. Over repeated attempts, the reach trajectories are adjusted to account for the discrepancy between vision and proprioception, a process termed visuomotor learning. Similarly, to control a computer mouse, we must learn the kinematic relation between the movement of the mouse and the cursor on the screen. In addition, the properties of the limbs change with both growth and tool use. The brain must adapt to such changes by reorganizing or adjusting motor commands.

In error-based learning, the sensorimotor system senses the outcome of each movement and compares this to both the desired outcome and the predicted outcome. For example, when shooting a basketball the desired outcome is for the ball to go through the hoop. However, once you let go of the ball you may predict that the ball will miss to the right of the hoop. The difference between the prediction and actual outcome, termed the sensory prediction error, can be used to

update the internal model of how the ball responds to your actions. The difference between the actual and desired outcome, termed the target error, can be used to adjust your plan (i.e. aim direction) to reduce the error. Both sensory prediction errors and target errors are important for driving learning.

Additional transformations may have to be applied to the error signal before it can be used to train an internal model. For example, when we throw a dart, errors are received in visual coordinates. This sensory error must be converted into motor command errors suitable for updating a control process such as an inverse model. Error-based learning tends to lead to trial-by-trial reduction in error as the motor system learns the novel sensorimotor properties.

An example of such error-based learning in reaching occurs when the dynamics of the arm are unexpectedly changed. As we saw earlier, we normally move the hand with a straight-line path to reach an object. Unexpected dynamic interactions can produce curved paths, but subjects learn to anticipate and compensate for these effects. This learning is conveniently studied by having subjects make reaching movements while holding the end of a robotic apparatus that can introduce novel forces on the arm (Figure 30–12A–C). Applying a force that is proportional to the speed of the hand but that acts at right angles to the direction of movement will produce a curved movement before finally reaching the target. Over time, the subject adapts to this perturbation and is able to maintain a straight-line movement (Figure 30–12D).

Subjects might adapt to such a situation in either of two ways. Subjects could co-contract the muscles in their arm, thereby stiffening the arm and reducing the impact of the perturbation, or they could learn an internal model that compensates for the anticipated force. By examining the aftereffects (movements after the robot is turned off), we can distinguish between these two forms of learning. If the arm simply stiffens, it should continue to move in a straight path. If a new internal model is learned, the new model should compensate for a force that no longer exists, thereby producing a path in the direction opposite from the earlier perturbation. Early in learning, co-contraction is used to reduce the errors before an internal model can be learned, but the co-contraction then decreases as the internal model is able to compensate for the perturbation. Therefore, when the force is turned off after learning, subjects normally show a large aftereffect in the opposite direction, demonstrating that they have compensated for the perturbation (Figure 30–12D).

Such error-based processes appear to underlie adaptation across a number of different movement

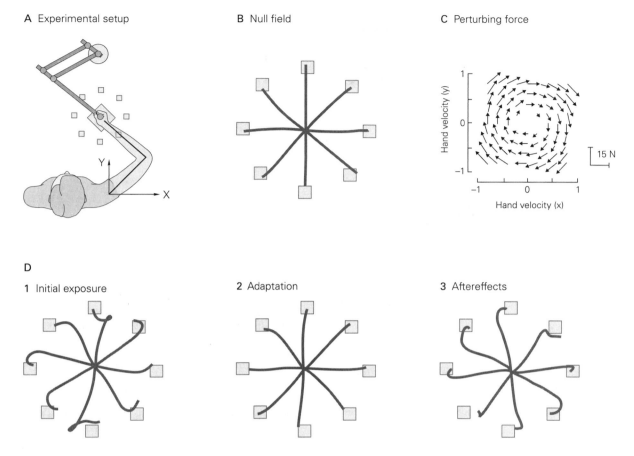

A Experimental setup

B Null field

C Perturbing force

D

1 Initial exposure

2 Adaptation

3 Aftereffects

Figure 30–12 Learning improves the accuracy of reaching in a novel dynamic environment. (Adapted, with permission, from Brashers-Krug, Shadmehr, and Bizzi 1996. Copyright © 1996 Springer Nature.)

A. A subject holds a robotic apparatus that measures the position and velocity of the hand and applies forces to the hand.

B. When the motors are off (null field), the subject makes approximately straight movements from the center of the workspace to targets arrayed in a circle.

C. A clockwise force is then applied to the hand, shown as a function of hand velocity. This field produces a force proportional to the speed of the hand that always acts at right angles to the current direction of motion.

D. Initially, the hand paths are severely perturbed in response to the perturbing force (**1**). After some time, the subject adapts and can again follow a straight path during the entire movement (**2**). When the motors are then turned off, movement is again perturbed, but in a direction opposite to the earlier perturbation (**3**).

types and effectors, from the eye to whole-body movements. For example, our normal symmetric pattern of gait seems to rely on error-based learning. When the gait pattern of subjects is perturbed by walking on a split-belt treadmill in which one belt moves faster than the other, they initially limp. However, step by step the gait pattern naturally regains its symmetry (Figure 30–13), thus showing that error-based learning can drive complex whole-body coordinated movements. There is extensive evidence that fast trial-by-trial error-based learning relies on the cerebellum (Chapter 37).

Motor adaptation may not be a single unitary process. Recent evidence suggests that adaptation is driven by interacting processes whose outputs are combined. These interacting processes could have different temporal properties: one process quickly adapting to perturbations but also rapidly forgetting what was learned and the other learning more slowly but retaining learning for a longer period (Figure 30–13B). The advantage of such a mechanism is that the learning processes can be matched to the temporal properties of the perturbations, which can range from short-lived (fatigue) to long-lasting (growth).

Although motor learning often takes much practice, once a task is no longer performed, deadaptation is typically faster. However, the sensory inputs associated with the particular action can be enough to switch behavior. When subjects wear prismatic glasses

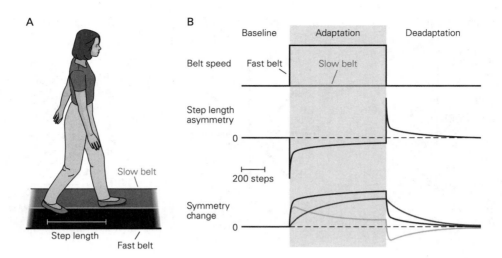

Figure 30–13 Learning new coordination patterns in walking.

A. A subject walks on a split-belt treadmill. When the two belts move at the same speed, subjects have a symmetric gait pattern with equal step lengths.

B. In an adaptation study, the speeds of the belts are initially the same, then become split so that the right belt moves faster than the left, and then finally return to the same speed (**top**). Step length symmetry is initially lost when the belts move at different speeds, causing the subject to limp. Over time, the symmetry is restored and the limping is abolished. When the belts are once again moving at the same speed, an aftereffect is seen (**middle**).

Maurice Smith and colleagues have shown that this type of adaptation is composed of multiple underlying processes that adapt on different timescales (**bottom**). The change in the step length symmetry is composed of two processes: a fast process (**light green line**) that adapts quickly but also rapidly forgets what has been learned, and a slow process (**dark green line**) that learns more slowly but has better retention. These processes both adapt to learn from the error, and the sum of these processes is the final adaptation (**blue line**). This dual-rate learning system gives rise to the typical double exponential learning curves seen in many forms of adaptation in which adaptation is initially fast but tends to slow down as learning proceeds. (Adapted, with permission, from Roemmich, Long, and Bastian 2016.)

that shift visual space, for example, they initially miss when reaching to targets but soon learn to reach correctly. After repeated trials, the mere feel of the glasses, without the prisms in place, is sufficient to evoke the adaptive behavior appropriate for the prisms.

In general, we can quantify performance with two measurements, accuracy and precision. Accuracy is a measure of systematic errors or biases, for example, on average how far a series of thrown darts are away from the target. In contrast, precision is a measure of random errors, or statistical variability, in our actions. Both accuracy and precision contribute to performance. In general, accuracy can be improved by adapting or calibrating motor commands so as to reduce systematic errors. Although there is always some variability in movement arising from irreducible sensory and motor noise, the variability, as we have seen, can be reduced through planning so as to have minimal impact on task success. Most motor learning tends to become automatic (ie, implicit) with time, but early learning of some tasks can be aided by explicit learning (ie, strategy), such as a verbal instruction on how best to approach the task.

Not all sensory modalities are equally important in learning all motor tasks. In learning dynamic tasks, proprioception and tactile input are more important than vision. We normally learn dynamic tasks equally well with or without vision. However, individuals who have lost proprioception and tactile input have particular difficulty controlling the dynamic properties of their limbs or learning new dynamic tasks without vision (Box 30–3).

Skill Learning Relies on Multiple Processes for Success

In contrast to error-based learning in which the sensorimotor system adapts to a perturbation to return to pre-perturbation performance, learning skills such as tying one's shoelaces, juggling, typing, or playing the piano instead involves improving performance in the absence of a perturbation. Such learning tends to improve the speed–accuracy trade-off. Initially, we may be able to hit the correct keys on a keyboard when paced 1 second apart, but with practice, the same accuracy can be achieved at an increasingly quickening pace.

Box 30–3 Proprioception and Tactile Sense Are Critical for Sensorimotor Control

While visual impairment certainly has limiting effects on sensorimotor control, blind people are able to walk normally and reach and grasp known objects with ease. This is in stark contrast to the rare loss of proprioceptive and tactile sense.

Some sensory neuropathies selectively damage the large-diameter sensory fibers in peripheral nerves and dorsal roots that carry most proprioceptive information. Impairments in motor control resulting from loss of proprioception have fascinated neurologists and physiologists for well over a century. Studies of patients with sensory neuropathies provide invaluable insight into the interactions between sensation and movement planning.

As expected, such patients lose joint position sense, vibration sense, and fine tactile discrimination (as well as tendon reflexes), but both pain and temperature senses are fully preserved. Patients with peripheral neuropathies are unable to maintain a steady posture, for example, while holding a cup or standing, with the eyes closed. Movements also become clumsy, uncoordinated, and inaccurate.

Some recovery of function may occur over many months as the patient learns to use vision as a substitute for proprioception, but this compensation still leaves patients completely incapacitated in the dark. Some of this difficulty reflects an inability to detect errors that develop during unseen movements, as occurs if the weight of an object differs from expectation.

Peripheral neuropathies are particularly incapacitating when patients try to make movements with rapid direction reversals. Analyses of the joint torques during these movements show that subjects with intact proprioception anticipate intersegmental torques, whereas those without proprioception fail to do so (Figure 30–14).

However, the same patients easily adapt to drastic kinematic changes, such as tracing a drawing while viewing their hand in a mirror. In fact, they perform better than normal subjects, perhaps because they have learned to guide their movements visually and, because they lack proprioception, do not experience any conflict between vision and proprioception.

Even in normal subjects, the relative importance of tactile input in manipulation tasks can be easily demonstrated. It is relatively easy to light a match with one's eyes closed. However, if the tips of the digits are made numb with local anesthetic, then even under full vision the task is remarkably hard because the match tends to slip from the fingers.

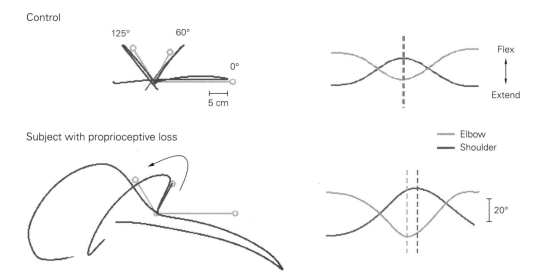

Figure 30–14 Patients lacking proprioception cannot make an accurate movement that requires a rapid reversal in path. (*Left*) The subject tries to trace a template (**gray line**) while her hand is hidden from view. The joint angles for the elbow and shoulder of a normal subject show good alignment (*Right top*), leading to an accurate reversal (*Left top*). In contrast, the timing of the joint reversal is poor in subjects who lack proprioceptive input (*Right bottom*), leading to large errors in the path (*Left bottom*). These patients cannot anticipate and correct for the intersegmental dynamics that occur around the path reversal. (Adapted, with permission, from Sainburg et al. 1995.)

For some skills, there can be a complex relation between the actions performed and success or failure at the task. For example, when children first sit on a swing, they have to learn the complex sequence of leg and body movements required to make the swing go higher. In contrast to error-based learning, there is no readily available error signal that can be used to adjust the current action because the swing's height is not directly determined by the current action but by a long history of body and leg motion. Learning in such complex scenarios can be achieved using reinforcement learning in which the sensorimotor system adjusts its commands in an effort to maximize reward, that is, task success. In the most general form, the performance measure that the reinforcement learning tries to maximize is the sum of all future rewards. However, as we tend to favor an immediate versus time-delayed reward, the sum is typically weighted to reflect this by progressively discounting future rewards.

Reinforcement learning is more general than error-based learning in that the training signal is success or failure, rather than an error at each point in time. Another distinguishing property of reinforcement learning is that the success or failure that the learning system receives can depend in nontrivial ways on the history of the actions taken. For tasks that require a complex sequence of actions to take place to achieve a goal, such as tying one's shoelaces, and the outcome or reward is removed in time from the action, error-based learning cannot easily be applied. A key problem that reinforcement solves is that of credit assignment: Which action within a sequence should we credit or blame when we eventually succeed or fail? This is just the sort of problem reinforcement learning algorithms are good at solving.

There are two main classes of reinforcement learning, those that depend on an internal model and those that do not. Model-based reinforcement builds a model of the task (eg, the structure of a maze). With such a model, the learner can efficiently plan in a goal-directed manner. In contrast, with model-free reinforcement learning, the learner simply associates movements with success or failure; those that lead to success are more likely to be performed again. Such learning can lead to motor habits. While model-free learning avoids the computational burden of building a model, it is also less able to generalize to novel situations. These two types of reinforcement learning can even act together, and different tasks can rely on them to different extents. Dopaminergic systems in the basal ganglia have been tied to signals that one would expect in reinforcement learning, such as expected reward. Moreover, dysfunction in these systems is related to movement disorders, addiction, and other problems that could be related to reinforcement signals (Chapter 38).

Finally, the development of efficient strategies plays a key part in motor skill acquisition. Skill learning for real-world tasks typically involves a sequence of decision-making processes at different spatiotemporal scales. The skill of a tennis player, for example, is not only determined by the precision with which she can strike the ball but also by the speed with which she can make the correct decision on where to aim it and how well she uses her senses to extract task-relevant information.

Sensorimotor Representations Constrain Learning

The information obtained during a single movement is often too sparse or noisy to unambiguously determine the source of error. For example, if a tennis player hits a shot into the net on the serve, the problem could be that the ball was not thrown high enough, the ball was hit too early, the racquet strings are loose, there was a gust of wind, or the player is fatigued. If the racquet dynamics have changed, the player would do well to adapt to these for the next shot. If a temporary gust of wind was the problem, then no adjustment is needed. To resolve this issue, the sensorimotor learning system constrains the way in which the system is updated in response to errors. These constraints reflect the internal assumptions about the task structure and the source of errors and determine how the system represents the task. Indeed, on a slower timescale, learning itself can alter the representation.

While the final output of the motor system is the contraction of its 600 or so muscles, it is not the case that the brain controls each independently. In current models of sensorimotor control, motor commands are generated by multiple modules that can be selectively engaged depending on the requirements of the task. Examples of modular architectures include multiple internal models, motor primitives, and motor synergies (Chapter 36).

Motor primitives can be thought of as neural control modules that can be flexibly combined to generate a large repertory of behaviors. A primitive might represent the temporal profile of a particular muscle activity or a set of muscles that are activated together, termed a synergy. The overall motor output will be the sum of all primitives, weighted by the level of the activation of each module. The makeup of the population of such primitives then determines which structural constraints are imposed on learning. For example, a behavior for which the motor system has many primitives will be easy to learn, whereas a behavior that cannot be approximated by any existing primitives would be impossible to learn.

Highlights

1. The primary purpose of the elaborate information processing and storage that occurs in the brain is to enable us to interact with our environment through our motor system.

2. Our infinitely varied and purposeful motor behaviors are governed by the integrated actions of the motor systems, including the motor cortex, spinal cord, cerebellum, and basal ganglia.

3. To control action, the central nervous system uses a hierarchy of sensorimotor transformations that convert incoming sensory information into motor outputs.

4. There is a trade-off in the speed versus sophistication of the different levels of sensorimotor response from rapid reflexes to slower voluntary control.

5. The motor systems generate commands using feedforward circuits or error-correcting feedback circuits; most movement involves both types of control.

6. The brain uses internal models of the sensorimotor system to facilitate control.

7. The state of the body is estimated using both sensory and motor signals together with a forward predictive mode to reduce the adverse effects of delays in feedback.

8. Variability in the sensory inputs and motor outputs together with inaccuracies in sensorimotor transformations underlie the errors and variability in movement, leading to the trade-off between speed and accuracy.

9. Motor planning can use the redundancy of the motor system to move in such a way as to reduce the negative consequences of motor noise while reducing effort.

10. Motor control circuits are not static but undergo continual modification and recalibration throughout life.

11. Motor learning improves motor control in novel situations, and different forms of sensory information are vital for learning. Error-based learning is particularly important for adapting to simple sensorimotor perturbations. Reinforcement learning is particularly important for more complex skill learning and can rely on a model (model-based) or on simply reinforcing motor actions directly (model-free).

12. The motor representations used by the brain constrain the way the sensorimotor system updates during learning.

13. Studies of sensorimotor control have focused on developing a detailed understanding of relatively simple tasks, such as reaching and walking. Although these tasks are amenable to analysis and modeling, they do not capture the full complexity of real-world motor control. The challenge will be to determine if these principles can be generalized to tasks such as tying shoelaces and learning to skateboard.

Daniel M. Wolpert
Amy J. Bastian

Selected Reading

Diedrichsen J, Shadmehr R, Ivry R B. 2010. The coordination of movement: optimal feedback control and beyond. Trends Cogn Sci 14:31–39.

Roemmich RT, Bastian AJ. 2018. Closing the loop: From motor neuroscience to neurorehabilitation. Annu Rev Neurosci 41:415–429.

Scott SH. 2016. A functional taxonomy of bottom-up sensory feedback processing for motor actions. Trends Neurosci 39:512–526.

Shadmehr R, Smith MA, Krakauer JW. 2010. Error correction, sensory prediction, and adaptation in motor control. Annu Rev Neurosci 33:89–108.

Wolpert DM, Diedrichsen J, Flanagan JR. 2011. Principles of sensorimotor learning. Nat Rev Neurosci 12:739–751.

Wolpert DM, Flanagan JR. 2016. Computations underlying sensorimotor learning. Curr Opin Neurobiol 37:7–11.

References

Blakemore SJ, Frith CD, Wolpert DM. 1999. Spatio-temporal prediction modulates the perception of self-produced stimuli. J Cogn Neurosci 11:551–559.

Blakemore SJ, Goodbody S, Wolpert DM. 1998. Predicting the consequences of our own actions: the role of sensorimotor context estimation. J Neurosci 18:7511–7518.

Brashers-Krug T, Shadmehr R, Bizzi E. 1996. Consolidation in human motor memory. Nature 382:252–255.

Burdet E, Osu R, Franklin DW, Milner TE, Kawato M. 2001. The central nervous system stabilizes unstable dynamics by learning optimal impedance. Nature 414:446–449.

Craik KJW. 1943. *The Nature of Explanation.* Cambridge: Cambridge Univ. Press.

Crapse TB, Sommer MA. 2008. Corollary discharge across the animal kingdom. Nature Rev Neurosci 9:587.

Crevecoeur F, Scott SH. 2013. Priors engaged in long-latency responses to mechanical perturbations suggest a rapid update in state estimation. PLoS Comput Biol 9:e1003177.

Crevecoeur F, Scott SH. 2014. Beyond muscles stiffness: importance of state-estimation to account for very fast motor corrections. PLoS Comput Biol 10:e1003869.

Diedrichsen J, Kornysheva K. 2015. Motor skill learning between selection and execution. Trends Cogn Sci 19:227–233.

Ernst MO, Bulthoff HH. 2004. Merging the senses into a robust percept. Trends Cogn Sci 8:162–169.

Faisal AA, Selen LP, Wolpert DM. 2008. Noise in the nervous system. Nat Rev Neurosci 9:292–303.

Flanagan JR, Beltzner MA. 2000. Independence of perceptual and sensorimotor predictions in the size-weight illusion. Nat Neurosci 3:737–741.

Goodale MA, Milner AD. 1992. Separate visual pathways for perception and action. Trends Neurosci 15:20–25.

Harris CM, Wolpert DM. 1998. Signal-dependent noise determines motor planning. Nature 394:780–784.

Huberdeau DM, Krakauer JW, Haith AM. 2015. Dual-process decomposition in human sensorimotor adaptation. Curr Opin Neurobiol 33:71–77.

Krakauer JW, Mazzoni P. 2011. Human sensorimotor learning: adaptation, skill, and beyond. Curr Opin Neurobiol 21:636–644.

Land MF, McLeod P. 2000. From eye movements to actions: how batsmen hit the ball. Nat Neurosci 3:1340–1345.

McDougle SD, Ivry RB, Taylor JA. 2016. Taking aim at the cognitive side of learning in sensorimotor adaptation tasks. Trends Cogn Sci 20:535–544.

Morasso P. 1981. Spatial control of arm movements. Exp Brain Res 42:223–227.

Muller H, Sternad D. 2004. Decomposition of variability in the execution of goal-oriented tasks: three components of skill improvement. J Exp Psychol Hum Percept Perform 30:212–233.

O'Doherty JP, Lee SW, McNamee D. 2015. The structure of reinforcement-learning mechanisms in the human brain. Curr Opin Behav Sci 1:94–100.

Pruszynski JA, Scott SH. 2012. Optimal feedback control and the long-latency stretch response. Exp Brain Res 218:341–359.

Raibert MH. 1977. Motor control and learning by the state space model. Ph.D. Dissertation. Cambridge, MA: Artificial Intelligence Laboratory, MIT.

Reisman DS, Block HJ, Bastian AJ. 2005. Interlimb coordination during locomotion: what can be adapted and stored? J Neurophysiol 94:2403–2415.

Roemmich RT, Long AW, Bastian AJ. 2016. Seeing the errors you feel enhances locomotor performance but not learning. Curr Biol 26:1–10.

Rothwell JC, Traub MM, Day BL, Obeso JA, Thomas PK, Marsden CD. 1982. Manual motor performance in a deafferented man. Brain 105:515–542.

Sainburg RL, Ghilardi MF, Poizner H, Ghez C. 1995. Control of limb dynamics in normal subjects and patients without proprioception. J Neurophysiol 73:820–835.

Schmidt RA, Zelaznik H, Hawkins B, Frank JS, Quinn JT. 1979. Motor-output variability: a theory for the accuracy of rapid motor acts. Psychol Rev 47:415–451.

Scott SH, Cluff T, Lowrey CR, Takei T. 2015. Feedback control during voluntary motor actions. Curr Opin Neurobiol 33:85–94.

Sing GC, Joiner WM, Nanayakkara T, Brayanov JB, Smith MA. 2009. Primitives for motor adaptation reflect correlated neural tuning to position and velocity. Neuron 64:575–589.

Smith MA, Ghazizadeh A, Shadmehr R. 2006. Interacting adaptive processes with different timescales underlie short-term motor learning. PLoS Biol 4:e179.

Todorov E, Jordan MI. 2002. Optimal feedback control as a theory of motor coordination. Nat Neurosci 5:1226–1235.

Torres-Oviedo G, Macpherson JM, Ting LH. 2006. Muscle synergy organization is robust across a variety of postural perturbations. J Neurophysiol 96:1530–1546.

Valero-Cuevas FJ, Venkadesan M, Todorov E. 2009. Structured variability of muscle activations supports the minimal intervention principle of motor control. J Neurophysiol 102:59–68.

van Beers RJ, Sittig AC, Gon JJ. 1999. Integration of proprioceptive and visual position-information: an experimentally supported model. J Neurophysiol 81:1355–1364.

Wolpert DM, Flanagan JR. 2001. Motor prediction. Curr Biol 11:R729–732.

Yang SC-H, Wolpert DM, Lengyel M. 2016. Theoretical perspectives on active sensing. Curr Opin Behav Sci 11:100–108.

31

The Motor Unit and Muscle Action

ANY ACTION—ASCENDING A FLIGHT of stairs, typing on a keyboard, even holding a pose—requires coordinating the movement of body parts. This is accomplished by the interaction of the nervous system with muscle. The role of the nervous system is to activate the muscles that provide the forces needed to move in a particular way. This is not a simple task. Not only must the nervous system decide which muscles to activate, how much to activate them, and the sequence in which they must be activated in order to move one part of the body, but it must also control the influence of the resultant muscle forces on other body parts and maintain the required posture.

This chapter examines how the nervous system controls muscle force and how the force exerted by a limb depends on muscle structure. We also describe how muscle activation changes to perform different types of movement.

The Motor Unit Is the Elementary Unit of Motor Control

A Motor Unit Consists of a Motor Neuron and Multiple Muscle Fibers

The nervous system controls muscle force with signals sent from motor neurons in the spinal cord or brain stem to the muscle fibers. A motor neuron and the muscle fibers it innervates are known as a motor unit, the basic functional unit by which the nervous system controls movement, a concept proposed by Charles Sherrington in 1925.

A typical muscle is controlled by a few hundred motor neurons whose cell bodies are clustered in a motor nucleus in the spinal cord or brain stem. The axon of each motor neuron exits the spinal cord through the ventral root or through a cranial nerve in the brain stem and runs in a peripheral nerve to the muscle. When the axon reaches the muscle, it branches and innervates from a few to several thousand muscle fibers (Figure 31–1).

Once synaptic input depolarizes the membrane potential of a motor neuron above threshold, the neuron generates an action potential that is propagated along the axon to its terminals in the muscle. The action potential releases acetylcholine at the neuromuscular synapse, triggering an action potential at the sarcolemma of the muscle fiber (Chapter 12). A muscle fiber has electrical properties similar to those of a large-diameter, unmyelinated axon, and thus, action potentials propagate along the sarcolemma, although more slowly due to the higher capacitance of the fiber resulting from the transverse tubules (see Figure 31–9). Because the action potentials in all the muscle fibers of a motor unit occur at approximately the same time, they contribute to extracellular currents that sum to generate a field potential near the active muscle fibers.

Most muscle contractions involve the activation of many motor units, whose currents sum to produce signals (*compound action potentials*) that can be detected by electromyography. The electromyogram (EMG) is typically large and can be easily recorded with electrodes

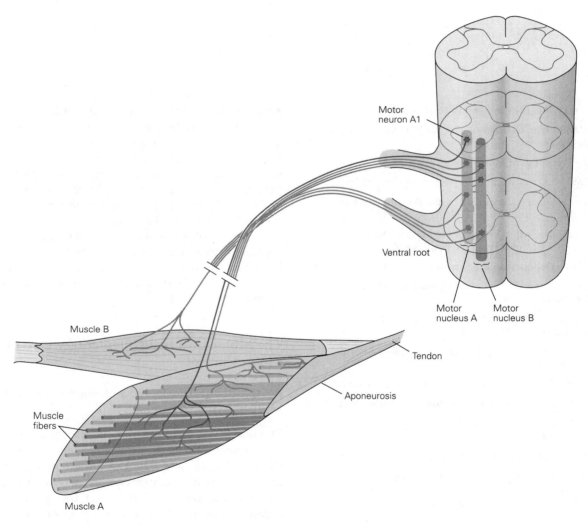

Figure 31–1 A typical muscle consists of many thousands of muscle fibers working in parallel and organized into a smaller number of motor units. A motor unit comprises a motor neuron and the muscle fibers it innervates, illustrated here by motor neuron A1. The motor neurons innervating one muscle are usually clustered into an elongated motor nucleus that may extend over one to four segments within the ventral spinal cord. The axons from a motor nucleus exit the spinal cord in several ventral roots and peripheral nerves but are collected into one nerve bundle near the target muscle. In the figure, motor nucleus A includes all those motor neurons innervating muscle A; likewise, motor nucleus B includes all the motor neurons that innervate muscle B. The extensively branched dendrites of each motor neuron (not shown in the figure) tend to intermingle with those of motor neurons from other nuclei.

placed on the skin over the muscle. The timing and amplitude of EMG activity, therefore, reflect the activation of muscle fibers by the motor neurons. EMG signals are useful for studying the neural control of movement and for diagnosing pathology (Chapter 57).

Each fiber in most mature vertebrate muscles is innervated by a single motor neuron. The number of muscle fibers innervated by one motor neuron, the *innervation number*, varies across muscles. In human skeletal muscles, the innervation number ranges from average values of 5 for an eye muscle to 1,800 for a leg muscle (Table 31–1). Because innervation number denotes the number of muscle fibers within a motor unit, differences in innervation number determine the differences in increments in force produced by activation of different motor units in the same muscle. Thus, the innervation number also indicates the fineness of control of the muscle at low forces; the smaller the innervation number, the finer the control achieved by varying the number of activated motor units.

The differences in innervation numbers between motor units in the same muscle can be substantial. For example, motor units of the first dorsal interosseous muscle of the hand have innervation numbers ranging

Table 31–1 Innervation Numbers in Human Skeletal Muscles

Muscle	Alpha motor axons	Muscle fibers	Average innervation number
Biceps brachii	774	580,000	750
Brachioradialis	333	129,200	410
Cricothyroid	112	18,550	155
Gastrocnemius (medial)	579	1,042,000	1,800
Interossei dorsales (1)	119	40,500	340
Lumbricales (1)	96	10,269	107
Masseter	1,452	929,000	640
Opponens pollicis	133	79,000	595
Platysma	1,096	27,100	25
Posterior cricoarytenoid	140	16,200	116
Rectus lateralis	4,150	22,000	5
Temporalis	1,331	1,247,000	936
Tensor tympani	146	1,100	8
Tibialis anterior	445	272,850	613
Transverse arytenoid	139	34,470	247

Source: Adapted, with permission, from Enoka 2015. © Human Kinetics, Inc.

from approximately 21 to 1,770. The strongest motor unit in the hand's first dorsal interosseous muscle can exert approximately the same force as the average motor unit in the leg's medial gastrocnemius muscle due to different ranges of innervation numbers in the two muscles.

The muscle fibers of a single motor unit are distributed throughout the muscle and intermingle with fibers innervated by other motor neurons. The muscle fibers innervated by a single motor unit can be distributed across 8% to 75% of the volume in a limb muscle, with 2 to 5 muscle fibers belonging to the same motor unit among 100 muscle fibers. Therefore, the muscle fibers in a cross-section through the middle of an entire muscle are associated with 20 to 50 different motor units. This distribution and even the number of motor units change with age and with some neuromuscular disorders (Chapter 57). For example, muscle fibers that lose their innervation after the death of a motor neuron can be reinnervated by collateral sprouts from neighboring axons.

Some muscles comprise discrete compartments that are each innervated by a different primary branch of the muscle nerve. Branches of the median and ulnar nerves in the forearm, for example, innervate distinct compartments in three multitendon extrinsic hand muscles that enable the fingers to be moved relatively independently. The muscle fibers belonging to each motor unit in such muscles tend to be confined to one compartment. A muscle can therefore consist of several functionally distinct regions.

The Properties of Motor Units Vary

The force exerted by a muscle depends not only on the number of motor units that are activated during a contraction but also on three properties of motor units: contraction speed, maximal force, and fatigability. These properties are assessed by examining the force exerted by individual motor units in response to variations in the number and rate of evoked action potentials.

The mechanical response to a single action potential is known as a *twitch contraction*. The time it takes the twitch to reach its peak force, the *contraction time*, is one measure of the contraction speed of the muscle fibers that compose a motor unit. The motor units in a muscle typically exhibit a range of contraction times from slow to fast contracting. The mechanical response to a series of action potentials that produce overlapping twitches is known as a *tetanic contraction* or *tetanus*.

The force exerted during a tetanic contraction depends on the extent to which the twitches overlap and summate (ie, the force varies with the contraction time of the motor unit and the rate at which the action potentials are evoked). At lower rates of stimulation, the ripples in the tetanus denote the peaks of individual twitches (Figure 31–2A). The peak force achieved during a tetanic contraction varies as a sigmoidal function of action potential rate, with the shape of the curve depending on the contraction time of the motor unit (Figure 31–2B). Maximal force is reached at lower

action potential rates for slow-contracting motor units than the rates needed to achieve maximal force in fast-contracting units.

The functional properties of motor units vary across the population and between muscles. At one end of the distribution, motor units have long twitch contraction times and produce small forces, but are less fatigable. At the other end of the distribution, motor units have short contraction times, produce large forces, and are

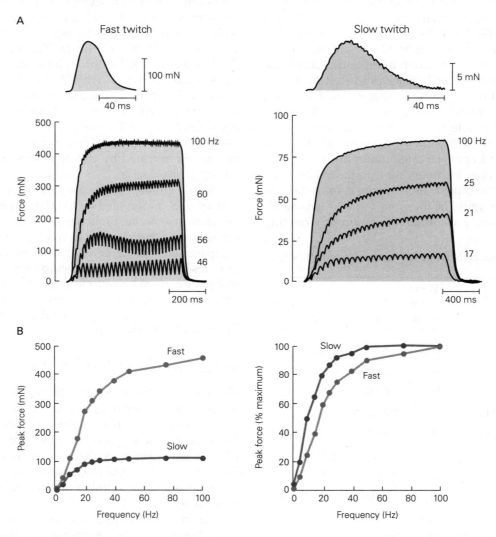

Figure 31–2 The force exerted by a motor unit varies with the rate at which its neuron generates action potentials.

A. Traces show the forces exerted by fast- and slow-contracting motor units in response to a single action potential (**top trace**) and a series of action potentials (set of **four traces below**). The time to the peak twitch force, or contraction time, is briefer in the faster unit. The rates of the action potentials used to evoke the tetanic contractions range from 17 to 100 Hz in the slow-contracting unit to 46 to 100 Hz in the fast-contracting unit. The peak tetanic force evoked by 100-Hz stimulation is greater for the fast-contracting unit. Note the different force scales for the

two sets of traces. (Adapted, with permission, from Botterman, Iwamoto, and Gonyea 1986; adapted from Fuglevand, Macefield, and Bigland-Ritchie 1999; and Macefield, Fuglevand, and Bigland-Ritchie 1996.)

B. Relation between peak force and the rate of action potentials for fast- and slow-contracting motor units. The absolute force (*left plot*) is greater for the fast-contracting motor unit at all frequencies. At lower stimulus rates (*right plot*), the force evoked in the slow-contracting motor unit (longer contraction time) sums to a greater relative force (percent of peak force) than in the fast-contracting motor unit (shorter contraction time).

A Twitch torques

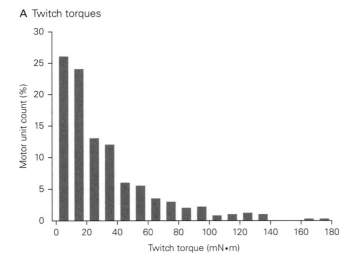

B Twitch contraction times

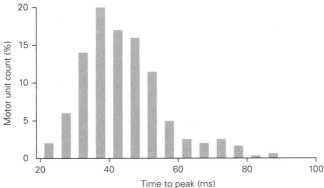

Figure 31–3 Most human motor units produce low forces and have intermediate contraction times. (Reproduced, with permission, from Van Cutsem et al. 1997. © Canadian Science Publishing.)

A. Distribution of twitch torques for 528 motor units in the tibialis anterior muscle obtained from 10 subjects.

B. Distribution of twitch contraction times for 528 motor units in the tibialis anterior muscle.

more fatigable. The order in which motor units are recruited during a voluntary contraction begins with the slow-contracting, low-force units and proceeds up to the fast-contracting, high-force units. As observed by Jacques Duchateau and colleagues, most motor units in humans produce low forces and have intermediate contraction times (Figure 31–3).

The range of contractile properties exhibited by motor units is partly attributable to differences in the structural specializations and metabolic properties of muscle fibers. One commonly used scheme to characterize muscle fibers is based on their reactivity to histochemical assays for the enzyme myosin adenosine triphosphatase (ATPase), which is used as an index

of contractile speed. Histochemical stains for myosin ATPase can identify two types of muscle fibers: type I (low levels of myosin ATPase) and type II (high levels of myosin ATPase). Slow-contracting motor units contain type I muscle fibers, and fast-contracting units include type II fibers. The type II fibers can be further classified as being less fatigable (type IIa) or more fatigable (type IIb, IIx, or IId), due to the association between myosin ATPase content and the relative abundance of oxidative enzymes. Another commonly used scheme distinguishes muscle fibers on the basis of genetically defined isoforms of the myosin heavy chain (MHC). Muscle fibers in slow-contracting motor units express MHC-I, those in the less fatigable fast-contracting units express MHC-IIA, and those in the more fatigable fast-contracting units express MHC-IIX.

In actuality, the contractile properties of single muscle fibers are less distinct than the two classification schemes suggest (Figure 31–4). In addition to the variability in the contractile properties of each type of muscle fiber (MHC-I, -IIA, or -IIX), some muscle fibers co-express more than one MHC isoform. Such hybrid muscle fibers exhibit contractile properties that are intermediate between the muscle fibers that compose a

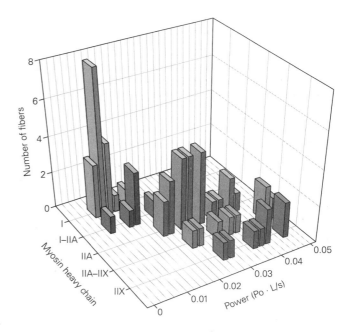

Figure 31–4 The contractile properties of muscle fiber types are distributed continuously. Peak power produced by segments of single muscle fibers from the vastus lateralis muscle with different types of myosin heavy chain (MHC) isoforms. Two types of hybrid fibers (I-IIA and IIA-IIX) contain isoforms of both types of MHCs. Power is calculated as the product of peak tetanic force (P_O) and maximal shortening velocity (segment length per second [L/s]). (Adapted, with permission, from Bottinelli et al. 1996. Copyright © 1996 The Physiological Society.)

single isoform. The relative proportion of hybrid fibers in a muscle increases with age. As with the distribution of contractile properties across motor units (Figure 31–3), the distribution across individual muscle fibers is also continuous, from slow to fast contracting and from least to most powerful (Figure 31–4).

Physical Activity Can Alter Motor Unit Properties

Alterations in habitual levels of physical activity can influence the three contractile properties of motor units (contraction speed, maximal force, and fatigability). A decrease in muscle activity, such as occurs with aging, bed rest, limb immobilization, or space flight, reduces the maximal capabilities of all three properties. The effects of increased physical activity vary with the intensity and duration of the activity. Brief sets of strong contractions performed a few times each week can increase motor unit force (strength training); brief sets of rapid contractions performed a few times each week can increase motor unit discharge rate (power training); and prolonged periods of weaker contractions can reduce motor unit fatigability (endurance training).

Changes in the contractile properties of motor units involve adaptations in the structural specializations and biochemical properties of muscle fibers. The improvement in contraction speed caused by power training, for example, is associated with an increase in the maximal shortening velocity of a muscle fiber caused by an increase in the quantity of myosin ATPase in the fiber. Similarly, the increase in maximal force is associated with the enlarged size and increased intrinsic force capacity of the muscle fibers produced by an increase in the number and density of the contractile proteins.

In contrast, decreases in the fatigability of a muscle fiber can be caused by many different adaptations, such as increases in capillary density, number of mitochondria, efficiency of the processes involved in activating the contractile proteins (excitation-contraction coupling), and oxidative capacity of the muscle fibers. Although the adaptive capabilities of muscle fibers decline with age, the muscles remain responsive to exercise even at 90 years of age.

Despite the efficacy of strength, power, and endurance training in altering the contractile properties of muscle fibers, these training regimens have little effect on the composition of a muscle's fibers. Although several weeks of exercise can change the relative proportion of type IIA and IIX fibers, it produces no change in the proportion of type I fibers. All fiber types adapt in response to exercise, although to varying extents depending on the type of exercise. For example, strength training of leg muscles for 2 to 3 months can increase the cross-sectional area of type I fibers by 0% to 20% and of type II fibers by 20% to 60%, increase the proportion of type IIa fibers by approximately 10%, and decrease the proportion of type IIx fibers by a similar amount. Furthermore, endurance training may increase the enzyme activities of oxidative metabolic pathways without noticeable changes in the proportions of type I and type II fibers, but the relative proportions of type IIa and IIx fibers do change as a function of the duration of each exercise session. Conversely, although several weeks of bed rest or limb immobilization do not change the proportions of fiber types in a muscle, they do decrease the size and intrinsic force capacity of muscle fibers. Adaptations in fiber type properties and proportions in turn alter the distribution of contractile properties in muscle fibers (Figure 31–4) and motor units (Figure 31–3).

Although physical activity has little influence on the proportion of type I fibers in a muscle, more substantial interventions can have an effect. Space flight, for example, exposes muscles to a sustained decrease in gravity that reduces the proportion of type I fibers in some leg muscles and decreases contractile properties. Similarly, surgically changing the nerve that innervates a muscle alters the pattern of activation and eventually causes the muscle to exhibit properties similar to those of the muscle that was originally innervated by the transplanted nerve. Connecting a nerve that originally innervated a rapidly contracting leg muscle to a slowly contracting leg muscle, for example, will cause the slower muscle to become more like a faster muscle. In contrast, a history of performing powerful contractions with leg muscles is associated with a modest reduction in the proportion of type I fibers, a marked increase in the proportion of type IIx fibers, and a huge increase in the power that can be produced by the type IIa and IIx fibers.

Muscle Force Is Controlled by the Recruitment and Discharge Rate of Motor Units

The force exerted by a muscle during a contraction depends on the number of motor units that are activated and the rate at which each of the active motor neurons discharges action potentials. Force is increased during a muscle contraction by the activation of additional motor units, which are recruited progressively from the weakest to the strongest (Figure 31–5). A motor unit's recruitment threshold is the force during the contraction at which the motor unit is activated. Muscle force decreases gradually by terminating the activity of motor units in the reverse order from strongest to weakest.

The order in which motor units are recruited is highly correlated with several indices of motor unit size, including the size of the motor neuron cell bodies, the diameter and conduction velocity of the axons, and the amount of force that the muscle fibers can exert. Because individual sources of synaptic input are broadly distributed across most neurons in a motor nucleus, the orderly recruitment of motor neurons is not accomplished by the sequential activation of different sets of synaptic inputs that target specific motor neurons. Rather, recruitment order is determined by intrinsic differences in the responsiveness of individual motor neurons to relatively uniform synaptic input.

One of these factors is the anatomical size of a neuron's soma and dendrites. Smaller neurons have a higher input resistance (R_{in}) to current and, due to Ohm's law ($\Delta V_m = I_{syn} \times R_{in}$), experience a greater change in membrane potential (ΔV_m) in response to a given synaptic current (I_{syn}). Consequently, increases in the net excitatory input to a motor nucleus cause the levels of depolarization to reach threshold in an ascending order of motor neuron size: Contraction force is increased by recruiting the smallest motor neuron first and the largest motor neuron last (Figure 31–6). This effect is known as the size principle of motor neuron recruitment, a concept enunciated by Elwood Henneman in 1957.

The size principle has two important consequences for the control of movement by the nervous system. First, the sequence of motor neuron recruitment is determined by the properties of the spinal neurons and not by supraspinal regions of the nervous system. This means that the brain cannot selectively activate specific motor units. Second, the axons arising from small motor neurons are thinner than those associated with

A Action potentials in two motor units

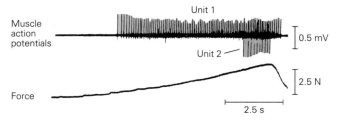

B Force produced by the two units

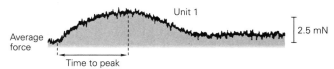

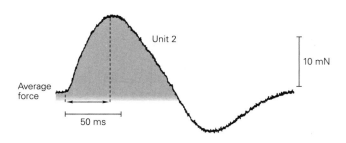

C Recruitment of 64 motor units in one muscle

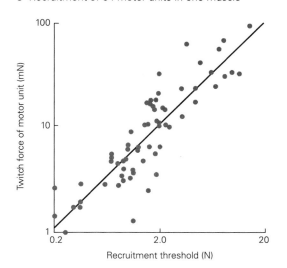

Figure 31–5 (Left) Motor units that exert low forces are recruited before those that exert greater forces. (Adapted, with permission, from Desmedt and Godaux 1977; Milner-Brown, Stein, and Yemm 1973. Copyright © 1973 The Physiological Society.)

A. Action potentials in two motor units were recorded concurrently with a single intramuscular electrode while the subject gradually increased muscle force. Motor unit 1 began discharging action potentials near the beginning of the voluntary contraction, and its discharge rate increased during the contraction. Motor unit 2 began discharging action potentials near the end of the contraction.

B. Average twitch forces for motor units 1 and 2 as extracted with an averaging procedure during the voluntary contraction.

C. The plot shows the net muscle forces at which 64 motor units in a hand muscle of one person were recruited (recruitment threshold) during a voluntary contraction relative to the twitch forces of the individual motor units.

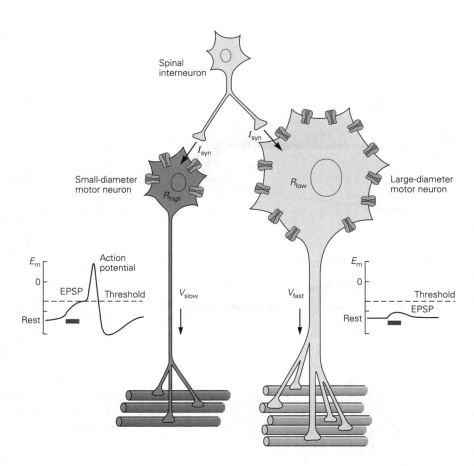

Figure 31–6 The size principle of motor neuron recruitment. Two motor neurons of different sizes have the same resting membrane potential (V_r) and receive the same excitatory synaptic current (I_{syn}) from a spinal interneuron. Because the small motor neuron has a smaller surface area, it has fewer parallel ion channels and therefore a higher input resistance (R_{high}). According to Ohm's law ($V = IR$), I_{syn} in the small neuron produces a large excitatory postsynaptic potential (**EPSP**) that reaches threshold, resulting in the discharge of an action potential. However, the axon of the small motor neuron has a small diameter and thus conducts the action potential at a relatively low velocity (V_{slow}) and to fewer muscle fibers. In contrast, the large motor neuron has a larger surface area, which results in a lower transmembrane resistance (R_{low}) and a smaller EPSP that does not reach threshold in response to I_{syn}; however, when synaptic input does reach threshold, the action potential is conducted relatively rapidly (V_{fast}) (Chapter 9).

large motor neurons and innervate fewer muscle fibers. Because the number of muscle fibers innervated by a motor neuron is a key determinant of motor unit force, motor units are activated in order of increasing strength, so the earliest recruited motor units are the weakest ones.

As suggested by Edgar Adrian in the 1920s, the muscle force at which the last motor unit in a motor nucleus is recruited varies between muscles. In some hand muscles, all the motor units have been recruited when the force reaches approximately 60% of maximum during a slow muscle contraction. In the biceps brachii, deltoid, and tibialis anterior muscles, recruitment continues up to approximately 85% of the maximal force. Beyond the upper limit of motor unit recruitment, changes in muscle force depend solely on variations in the rate at which motor neurons generate action potentials. Over most of the operating range of a muscle, the force it exerts depends on concurrent changes in discharge rate and the number of active motor units (Figure 31–7). Except at low forces, however, variation in

Figure 31–7 Muscle force can be adjusted by varying the number of active motor units and their discharge rate. Each line shows the discharge rate (pulses per second [**pps**]) for a single motor unit in a hand muscle over a range of finger forces (maximal voluntary contraction [**MVC**]). The finger force was produced by the action of a single hand muscle. The leftmost point of each line indicates the threshold force at which the motor unit is recruited, whereas the rightmost point corresponds to the peak force at which the motor unit could be identified. The range of discharge rates was often less for motor units with lower recruitment thresholds. Increases in finger force were produced by concurrent increases in discharge rate and the number of activated motor units. (Adapted, with permission, from Moritz et al. 2005.)

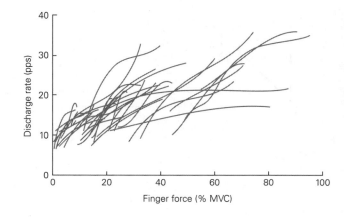

discharge rate has a greater influence on muscle force than does changes in the number of active motor units.

The order in which motor units are recruited does not change with contraction speed. Due to the time involved in excitation-contraction coupling, faster contractions require the action potential for each motor unit to be generated earlier than during a slow contraction. As a result of this adjustment, the upper limit of motor unit recruitment during the fastest muscle contractions is approximately 40% of maximum. Consequently, it is possible to manipulate the rate at which motor units are recruited by varying contraction speed.

The Input–Output Properties of Motor Neurons Are Modified by Input From the Brain Stem

The discharge rate of motor neurons depends on the magnitude of the depolarization generated by excitatory inputs and the intrinsic membrane properties of the motor neurons in the spinal cord. These properties can be profoundly modified by input from monoaminergic neurons in the brain stem (Chapter 40). In the absence of this input, the dendrites of motor neurons passively transmit synaptic current to the cell body, resulting in a modest depolarization that immediately ceases when the input stops. Under these conditions, the relation between input current and discharge rate is linear over a wide range.

The input–output relation becomes nonlinear, however, when the monoamines serotonin and norepinephrine induce a huge increase in conductance by activating L-type Ca^{2+} channels that are located on the dendrites of the motor neurons. The resulting inward Ca^{2+} currents can enhance synaptic currents by three- to five-fold (Figure 31–8). In an active motor neuron, this augmented current can sustain an elevated discharge rate after a brief depolarizing input has ended, a behavior known as *self-sustained firing*. A subsequent brief inhibitory input, such as from a spinal reflex pathway, can terminate such self-sustained firing.

Because the properties of motor neurons are strongly influenced by monoamines, the excitability of the pool of motor neurons innervating a single muscle is partly under control of the brain stem. In the awake state, moderate levels of monoaminergic input to the motor neurons of slowly contracting motor units promote self-sustained firing. This is probably the source of the sustained force exerted by slower motor units to maintain posture (Chapter 36). Conversely, the withdrawal of monoaminergic drive during sleep decreases excitability and helps ensure a relaxed motor state. Thus, monoaminergic input from the brain stem can adjust the gain of the motor unit pool to meet the demands of different tasks. This flexibility does not compromise the size principle of orderly recruitment because the threshold for activation of the persistent inward currents is lowest in the motor neurons of slower contracting motor units, which are the first recruited even in the absence of monoamines.

Muscle Force Depends on the Structure of Muscle

Muscle force depends not only on the amount of motor neuron activity but also on the arrangement of the fibers in the muscle. Because movement involves the controlled variation of muscle force, the nervous system must take into account the structure of muscle to achieve specific movements.

The Sarcomere Is the Basic Organizational Unit of Contractile Proteins

Individual muscles contain thousands of fibers that vary from 1 to 50 mm in length and from 10 to 60 µm in diameter. The variation in fiber dimensions reflects differences in the quantity of contractile protein. Despite this quantitative variation, the organization of contractile proteins is similar in all muscle fibers. The proteins are arranged in repeating sets of thick and thin filaments, each set known as a *sarcomere* (Figure 31–9). The in vivo length of a sarcomere, which is bounded by Z disks, ranges from 1.5 to 3.5 µm within and across muscles. Sarcomeres are arranged in series to form a *myofibril*, and the myofibrils are aligned in parallel to form a muscle fiber (myocyte).

The force that each sarcomere can generate arises from the interaction of the contractile thick and thin filaments. The thick filament consists of several hundred myosin molecules arranged in a structured sequence. Each myosin molecule comprises paired coiled-coil domains that terminate in a pair of globular heads. The myosin molecules in the two halves of a thick filament point in opposite directions and are progressively displaced so that the heads, which extend away from the filament, protrude around the thick filament (Figure 31–9C). The thick filament is anchored in the middle of the sarcomere by the protein titin, which connects each end of the thick filament with neighboring strands of actin in the thin filament and with the Z-disc. To maximize the interaction between the globular heads of myosin and the thin filaments, six thin filaments surround each thick filament.

The primary components of the thin filament are two helical strands of fibrous F-actin, each of which

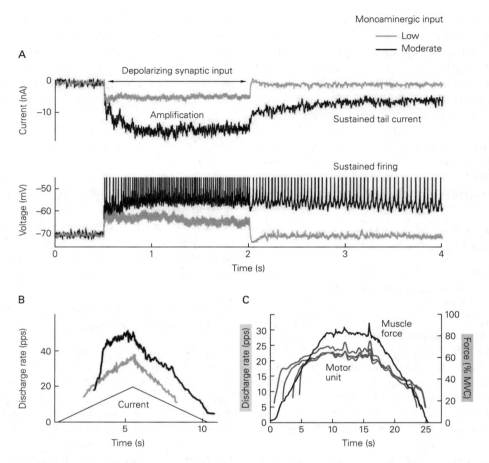

Figure 31–8 Monoaminergic input enhances the excitability of motor neurons. (Part **A**, adapted, with permission, of Heckman et al. 2009. Copyright © 2009 International Federation of Clinical Neurophysiology; Part **B**, data from CJ Heckman; Part **C**, adapted, with permission, from Erim et al. 1996. Copyright © 1996 John Wiley & Sons, Inc.)

A. Membrane currents and potentials in spinal motor neurons of adult cats that were either deeply anesthetized (low monoaminergic drive) or decerebrate (moderate monoaminergic drive). When monoaminergic input is absent or low, a brief excitatory input produces an equally brief synaptic current during voltage clamp (**upper record**). This current is not sufficient to bring the membrane potential of the neuron to threshold for generating action potentials in the unclamped condition (**lower record**). The same brief excitatory input during moderate levels of monoaminergic input activates a persistent inward current in the dendrites, which amplifies the excitatory synaptic current and decays slowly following cessation of synaptic input (**upper record**). This persistent inward current causes a high discharge

rate during the input and sustains a lesser discharge rate after the input ceases (**lower record**). A brief inhibitory input will return the neuron to its resting state.

B. High levels of monoaminergic input to a motor neuron give rise to a persistent inward current in response to injected current, resulting in a much greater discharge rate for a given amount of current.

C. The **blue** trace represents the force exerted by the dorsiflexor muscle during a contraction that gradually increased to 80% of maximal voluntary isometric contraction (**MVC**) force in a human subject. Each of the four **pink** traces indicates the change in the rate at which a single motor unit discharged action potentials during the contraction. The leftmost point (start) of each of these four traces shows the time when the motor unit was recruited, and the rightmost point (end) denotes the time at which the motor neuron stopped discharging action potentials. The rapid increase in discharge rate during the increase in muscle force is similar to the change in rate observed in the presence of moderate levels of monoaminergic input (see part **B**).

contains approximately 200 actin monomers. Superimposed on F-actin are tropomyosin and troponin, proteins that control the interaction between actin and myosin. Tropomyosin consists of two coiled strands that lie in the groove of the F-actin helix; troponin is a small molecular complex that is attached to tropomyosin at regular intervals (Figure 31–9C).

The thin filaments are anchored to the Z disk at each end of the sarcomere, whereas the thick filaments occupy the middle of the sarcomere (Figure 31–9B). This organization accounts for the alternating light and dark bands of striated muscle. The light band contains only thin filaments, whereas the dark band contains both thick and thin filaments. When a muscle is activated, the width of the light band decreases but the width of the dark band does not change, suggesting that the thick and thin filaments slide relative to one another during a contraction. This led to the *sliding filament hypothesis* of muscle contraction proposed by A. F. Huxley and H. E. Huxley in the 1950s.

The sliding of the thick and thin filaments is triggered by the release of Ca^{2+} from within the sarcoplasm of a muscle fiber in response to an action potential that travels along the fiber's membrane, the sarcolemma. Varying the amount of Ca^{2+} in the sarcoplasm controls the interaction between the thick and thin filaments. The Ca^{2+} concentration in the sarcoplasm is kept low under resting conditions by active pumping of Ca^{2+} into the sarcoplasmic reticulum, a network of longitudinal tubules and chambers of smooth endoplasmic reticulum. Calcium is stored in the terminal cisternae, which are located next to intracellular extensions of the sarcolemma known as transverse tubules (T-tubules). The transverse tubules, terminal cisternae, and sarcoplasmic reticulum constitute an activation system that transforms an action potential into the sliding of the thick and thin filaments (Figure 31–9A).

As an action potential propagates along the sarcolemma, it invades the transverse tubules and causes the rapid release of Ca^{2+} from the terminal cisternae into the sarcoplasm. Once in the sarcoplasm, Ca^{2+} diffuses among the filaments and binds reversibly to troponin, which results in the displacement of the troponin–tropomyosin complex and enables the sliding of the thick and thin filaments. Because a single action potential does not release enough Ca^{2+} to bind all available troponin sites in skeletal muscle, the strength of a contraction increases with the action potential rate.

The sliding of the filaments depends on mechanical work performed by the globular heads of myosin, work that uses chemical energy contained in adenosine triphosphate (ATP). The actions of the myosin heads are regulated by the *cross-bridge cycle*, a sequence of detachment, activation, and attachment (Figure 31–10). In each cycle, a globular head undergoes a displacement of 5 to 10 nm. Contractile activity continues as long as Ca^{2+} and ATP are present in the cytoplasm in sufficient amounts.

Once the contractile proteins have been activated by the release of Ca^{2+}, sarcomere length may increase, remain the same, or decrease depending on the magnitude of the load against which the muscle is acting. The force generated by an activated sarcomere when its length does not change or decreases can be explained by the cross-bridge cycle involving the thick and thin filaments. When the length of the activated sarcomere increases, however, the force developed by the

extension of titin adds significantly to the sarcomere force. The force produced by titin during the stretch of an activated sarcomere is augmented by its ability to increase stiffness, which is accomplished when titin binds Ca^{2+} and then attaches at specific locations on actin to reduce the length that it can be stretched. The force produced by activated sarcomeres therefore depends on the interactions of three filaments (actin, myosin, and titin).

Noncontractile Elements Provide Essential Structural Support

Structural elements of the muscle fiber maintain the alignment of the contractile proteins within the fiber and facilitate the transmission of force from the sarcomeres to the skeleton. A network of proteins (nebulin, titin) maintains the orientation of the thick and thin filaments within the sarcomere, whereas other proteins (desmin, skelemins) constrain the lateral alignment of the myofibrils (Figure 31–9B). These proteins contribute to the elasticity of muscle and maintain the appropriate alignment of cellular structures when the muscle acts against an external load.

Although some of the force generated by the cross bridges is transmitted along the sarcomeres in series, most of it travels laterally from the thin filaments to an extracellular matrix that surrounds each muscle fiber, through a group of transmembrane and membrane-associated proteins called a *costamere* (see inset for Figure 31–9B). The lateral transmission of force follows two pathways through the costamere, one through a dystrophin–glycoprotein complex and the other through vinculin and members of the integrin family. Mutations of genes that encode components of the dystrophin–glycoprotein complex cause muscular dystrophies in humans, which are associated with substantial decreases in muscle force.

Contractile Force Depends on Muscle Fiber Activation, Length, and Velocity

The force that a muscle fiber can exert depends on the number of cross bridges formed and the force produced by each cross bridge. These two factors are influenced by the Ca^{2+} concentration in the sarcoplasm, the amount of overlap between the thick and thin filaments, and the velocity with which the thick and thin filaments slide past one another.

The influx of Ca^{2+} that activates formation of the cross bridges is transitory because continuous pump activity quickly returns Ca^{2+} to the sarcoplasmic reticulum. The release and reuptake of Ca^{2+} in response to a single action potential occurs so quickly that only some

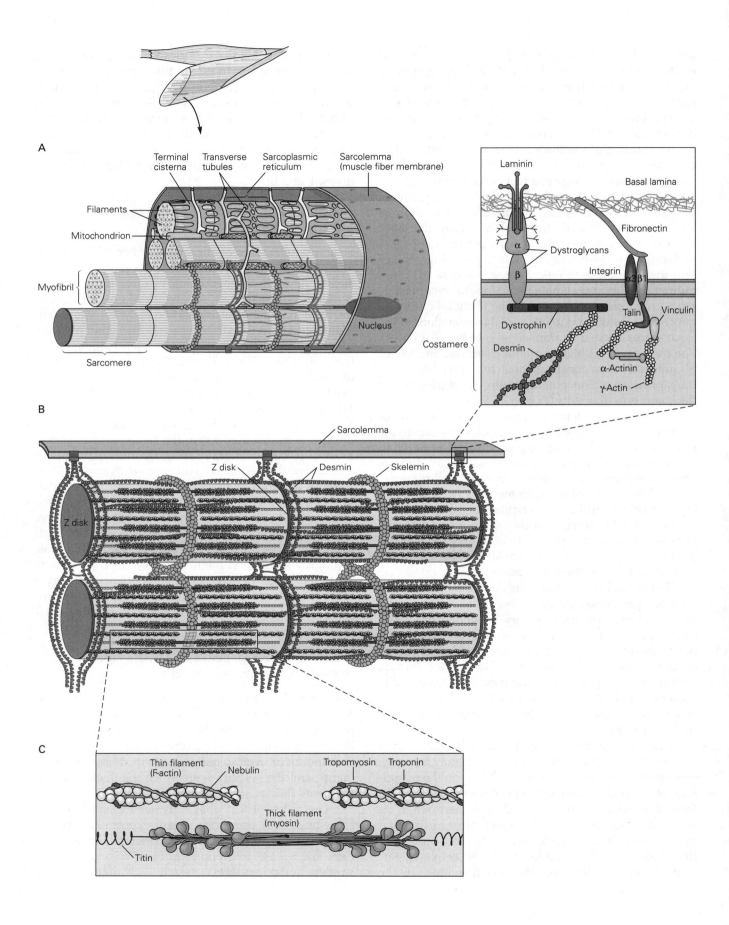

A

Terminal cisterna
Transverse tubules
Sarcoplasmic reticulum
Sarcolemma (muscle fiber membrane)

Filaments
Mitochondrion
Myofibril
Sarcomere
Nucleus
Costamere

Laminin
Basal lamina
Fibronectin
α
β
Dystroglycans
Integrin
α3β1
Dystrophin
Talin
Vinculin
Desmin
α-Actinin
γ-Actin

B

Sarcolemma

Z disk
Desmin
Skelemin

Z disk

C

Thin filament (F-actin)
Nebulin
Tropomyosin
Troponin
Thick filament (myosin)
Titin

of the potential cross bridges are formed. This explains why the peak force of a twitch is less than the maximal force of the muscle fiber (see Figure 31–2A). Maximal force can be achieved only with a series of action potentials that sustains the Ca^{2+} concentration in the sarcoplasm, thus maximizing cross bridge formation.

Although Ca^{2+} activates formation of the cross bridges, cross bridges can be formed only when the thick and thin filaments overlap. This overlap varies as the filaments slide relative to one another (Figure 31–11A). The amount of overlap between actin and myosin is optimal at an intermediate sarcomere length (L_o), and the relative force is maximal. Increases in sarcomere length reduce the overlap between actin and myosin and the force that can be developed. Decreases in sarcomere length cause the thin filaments to overlap, reducing the number of binding sites available to the myosin heads. Although many muscles operate over a narrow range of sarcomere lengths (approximately $94 \pm 13\%$ L_o, mean $\pm$ standard deviation), among muscles, there is considerable diversity in sarcomere lengths during movement.

Because structures that connect the contractile proteins to the skeleton also influence the force that a muscle can exert, muscle force increases with length over its operating range. This property enables muscle to function like a spring and to resist changes in length. Muscle stiffness, which corresponds to the slope of the relation between muscle force and muscle length (N/m), depends on the structure of the muscle. A stiffer muscle, similar to a stronger spring, is more resistant to changes in length.

Once activated, cross bridges perform work and cause the thick and thin filaments to slide relative to one another. Due to the elasticity of intracellular cytoskeletal proteins and the extracellular matrix, sarcomeres will shorten when the cross bridges are activated and the length of the muscle fiber is held fixed (*isometric contraction*). When the length of the muscle fiber is not kept constant, the direction and rate of change in sarcomere length depend on the amount of muscle fiber force relative to the magnitude of the load against which the fiber acts. Sarcomere length decreases when the muscle fiber force exceeds the load (*shortening contraction*) but increases when the force is less than the load (*lengthening contraction*). The maximal force that a muscle fiber can exert decreases as shortening velocity increases but increases as lengthening velocity increases (Figure 31–11B).

The maximal rate at which a muscle fiber can shorten is limited by the peak rate at which cross bridges can form. The variation in fiber force as contraction velocity changes is largely caused by differences in the average force exerted by each cross bridge. For example, the decrease in force during a shortening contraction is attributable to a reduction in cross-bridge displacement during each power stroke and the failure of some myosin heads to find attachment sites. Conversely, the increase in force during a lengthening contraction reflects the stretching of incompletely activated sarcomeres, the more rapid reattachment of cross bridges after they have been pulled apart, and the attachment of Ca^{2+} to titin.

The rate of cross-bridge cycling depends not only on contraction velocity but also on the preceding activity of the muscle. For example, the rate of cross-bridge cycling increases after a brief isometric contraction. When a muscle is stretched while in this state, such as would occur during a postural disturbance, muscle stiffness is enhanced, and the muscle is more effective at resisting the change in length. This property is known as *short-range stiffness*. Conversely, the cross-bridge cycling rate decreases after shortening contractions, and the muscle does not exhibit short-range stiffness.

Figure 31–9 (Opposite) The sarcomere is the basic functional unit of muscle. (Adapted from Bloom and Fawcett 1975.)

A. This section of a muscle fiber shows its anatomical organization. Several myofibrils lie side by side in a fiber, and each myofibril is made up of sarcomeres arranged end to end and separated by Z disks (see part B). The myofibrils are surrounded by an activation system (the transverse tubules, terminal cisternae, and sarcoplasmic reticulum) that initiates muscle contraction.

B. Sarcomeres are connected to one another and to the muscle fiber membrane by the cytoskeletal lattice. The cytoskeleton influences the length of the contractile elements, the thick and thin filaments (see part C). It maintains the alignment of these filaments within a sarcomere, connects adjacent myofibrils, and transmits force to the extracellular matrix of connective tissue through costameres. One consequence of this organization is that the force generated by the contractile elements in a sarcomere can be transmitted along and across sarcomeres (through desmin and skelemin), within and between sarcomeres (through nebulin and titin), and to the sarcolemma through the costameres. The Z disk is a focal point for many of these connections.

C. The thick and thin filaments comprise different contractile proteins. The thin filament includes polymerized actin along with the regulatory proteins tropomyosin and troponin. The thick filament is an array of myosin molecules; each molecule includes a stem that terminates in a pair of globular heads. The protein titin maintains the position of each thick filament in the middle of the sarcomere.

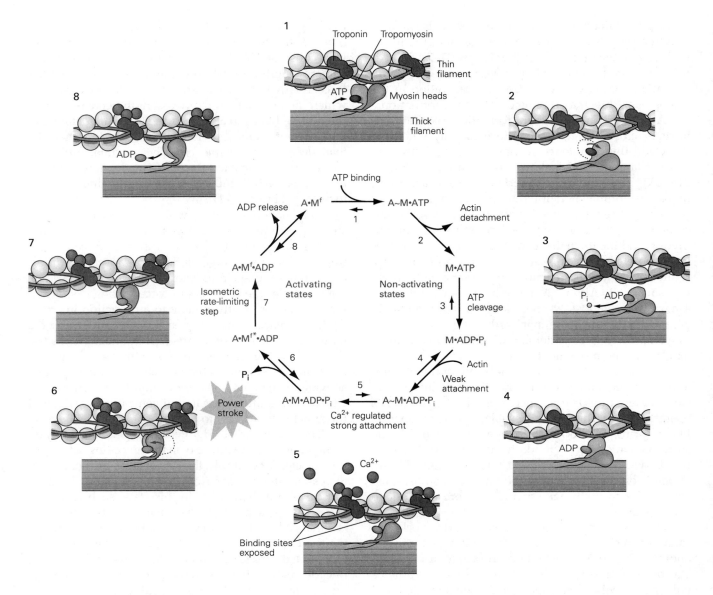

Figure 31–10 The cross-bridge cycle. Several nonactivating states are followed by several activating states triggered by Ca²⁺. The cycle begins at the top (**step 1**) with the binding of adenosine triphosphate (**ATP**) to the myosin head. The myosin head detaches from actin (**step 2**), ATP is cleaved to phosphate (**P$_i$**) and adenosine diphosphate (**ADP**) (**step 3**), and the myosin becomes weakly bound to actin (**step 4**). The binding of Ca²⁺ to troponin causes tropomyosin to slide over actin and enables the two myosin heads to close (**step 5**). This results in the release of P$_i$ and the extension of the myosin neck, the power stroke of the cross-bridge cycle (**step 6**). Each cross-bridge exerts a force of approximately 2 pN during a structural change (**step 7**) and the release of ADP (**step 8**). (•, strong binding; ~, weak binding; **M^f**, cross-bridge force of myosin; and **M^{f*}**, force-bearing state of myosin.) (Adapted, with permission, from Gordon, Regnier, and Homsher 2001.)

Muscle Torque Depends on Musculoskeletal Geometry

The anatomy of a muscle has a pronounced effect on its force capacity, range of motion, and shortening velocity. The anatomical features that influence function include the arrangement of the sarcomeres in each muscle fiber, the organization of the muscle fibers within the muscle, and the location of the muscle's attachments on the skeleton. These features vary widely among muscles.

At the level of a single muscle fiber, the number of sarcomeres in series and in parallel can vary. The number of sarcomeres in series determines the length

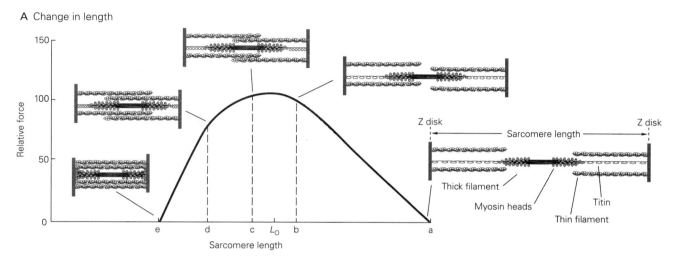

A Change in length

B Rate of change

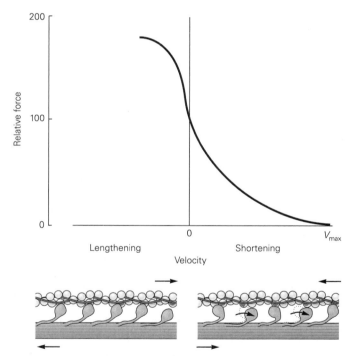

Figure 31–11 Contractile force varies with the change in sarcomere length and velocity.

A. At an intermediate sarcomere length, L_o, the amount of overlap between actin and myosin is optimal and the relative force is maximal. When the sarcomere is stretched beyond the length at which the thick and thin filaments overlap (length **a**), cross bridges cannot form and no force is exerted. As sarcomere length decreases and the overlap of the thick and thin filaments increases (between lengths **a** and **b**), the force increases because the number of cross bridges increases. With further reductions in length (between lengths **c** and **e**), the extreme overlap of the thin filaments with each other occludes potential attachment sites and the force decreases.

B. Contractile force varies with the rate of change in sarcomere length. Relative to the force that a sarcomere can exert during an isometric contraction (zero velocity), the peak force declines as the rate of shortening increases. Muscle force reaches a minimum at the maximal shortening velocity (V_{max}). In contrast, when the sarcomere is lengthened while being activated, the peak force increases to values greater than those during an isometric contraction. Shortening causes the myosin heads to spend more time near the end of their power stroke, where they produce less contractile force, and more time detaching, recocking, and reattaching, during which they produce no force. When the muscle is actively lengthened, the myosin heads spend more time stretched beyond their angle of attachment and little time unattached because they do not need to be recocked after being pulled away from the actin in this manner. Titin also contributes significantly to sarcomere force during lengthening contractions.

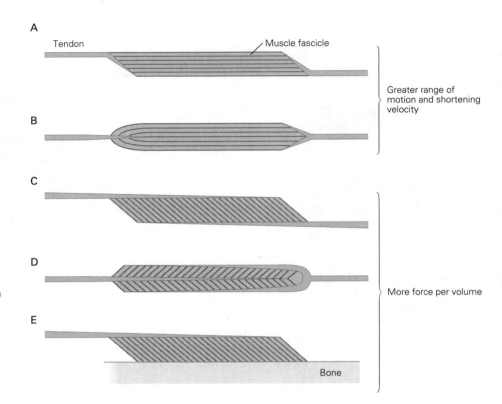

Figure 31–12 Five common arrangements of tendon and muscle. The fundamental distinction between these arrangements is whether or not the muscle fascicles are aligned with the line of pull of the muscles. The fascicles in muscles A and B are parallel to the line of pull (longitudinal axis of the muscle), whereas the fascicles in muscles C, D, and E are rotated away from the line of pull. The magnitude of this rotation is expressed as the pennation angle. (Reproduced, with permission, from Alexander and Ker 1990.)

of the myofibril and thus the length of the muscle fiber. Because one sarcomere can shorten by a certain length with a given maximal velocity, both the range of motion and the maximal shortening velocity of a muscle fiber are proportional to the number of sarcomeres in series. The force that a myofibril can exert is equal to the average sarcomere force and is not influenced by the number of sarcomeres in series. Rather, the force capacity of a fiber depends on the number of sarcomeres in parallel and hence on the diameter or cross-sectional area of the fiber. At the level of the muscle, the functional attributes of the fibers are modified by the orientation of the fascicles (bundles of muscle fibers) to the line of pull of the muscle and the length of the fiber relative to the muscle length. In most muscles, the fascicles are not parallel to the line of pull but fan out in feather-like (pennate) arrangements (Figure 31–12).

The relative orientation, or pennation angle of the fascicles, ranges from close to 0° (biceps brachii, sartorius) to approximately 30° (soleus). Because more fibers can fit into a given volume as the pennation angle increases, muscles with large pennation angles typically have more fascicles in parallel and hence large cross-sectional areas when measured perpendicular to the long axis of individual muscle fibers. Given the linear relation between cross-sectional area (quantity

of contractile proteins in parallel) and maximal force (~22.5 N $\cdot$ cm^{-2}), these muscles are capable of a greater maximal force. However, the fibers in pennate muscles are generally short and have a lesser maximal shortening velocity than those in nonpennate muscles.

The functional consequences of this anatomical arrangement can be seen by comparing the contractile properties of two muscles with different numbers of fibers and fiber lengths. If the two muscles have identical fiber lengths but one has twice as many fibers, the range of motion of the two muscles will be similar because it is a function of fiber length, but the maximal force capacity will vary in proportion to the number of muscle fibers. If the two muscles have identical numbers of fibers but the fibers in one muscle are twice as long, the muscle with the longer fibers will have a greater range of motion and a greater maximal shortening velocity, even though the two muscles have a similar force capacity. Because of this effect, the muscle with longer fibers is able to exert more force and produce more power (the product of force and velocity) at a given absolute shortening velocity (Figure 31–13).

Muscle fiber lengths and cross-sectional areas vary substantially throughout the human body, which suggests that the contractile properties of individual muscles also differ markedly (Table 31–2). In the leg,

A Different number of fibers

B Different fiber lengths

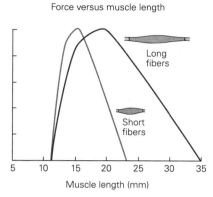

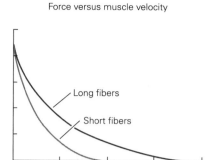

Figure 31–13 Muscle dimensions influence the peak force and maximal shortening velocity. (Reproduced, with permission, from Lieber and Fridén 2000. Copyright © 2000 John Wiley & Sons, Inc.)

A. Muscle force at various muscle lengths for two muscles with similar fiber lengths but different numbers of muscle fibers (different cross-sectional area). The muscle with twice as many fibers exerts greater force.

B. Muscle force at various muscle lengths for two muscles with the same cross-sectional area but different fiber lengths. The muscle with longer fibers (approximately twice as long as those of the other muscle) has an increased range of motion (left plot). It also has a greater maximal shortening velocity and exerts greater force at a given absolute velocity (right plot).

Table 31–2 Average Architectural Properties for Some Human Skeletal Leg Muscles

Muscle	Mass (g)	Muscle length (cm)	Fiber length (cm)	Pennation angle (°)	Cross-sectional area (cm²)
Thigh					
Sartorius	78	45	40	1	2
Rectus femoris	111	36	8	14	14
Vastus lateralis	376	27	10	18	35
Vastus intermedius	172	41	10	5	17
Vastus medialis	239	44	10	30	21
Gracilis	53	29	23	8	2
Adductor longus	75	22	11	7	7
Adductor brevis	55	15	10	6	5
Adductor magnus	325	38	14	16	21
Biceps femoris (long)	113	35	10	12	11
Biceps femoris (short)	60	22	11	12	5
Semitendinosus	100	30	19	13	5
Semimembranosus	134	29	7	15	18
Lower leg					
Tibialis anterior	80	26	7	10	11
Extensor hallucis longus	21	24	7	9	3
Extensor digitorum longus	41	29	7	11	6
Peroneus longus	58	27	5	14	10
Peroneus brevis	24	24	5	11	5
Gastrocnemius (medial)	113	27	5	10	21
Gastrocnemius (lateral)	62	22	6	12	10
Soleus	276	41	4	28	52
Flexor hallucis longus	39	27	5	17	7
Flexor digitorum longus	20	27	4	14	4
Tibialis posterior	58	31	4	14	14

Source: Adapted, with permission, from Ward et al. 2009.

for example, pennation angle ranges from 1° (sartorius) to 30° (vastus medialis), fiber length ranges from 4 mm (soleus) to 40 mm (sartorius), and cross-sectional area ranges from 2 cm² (sartorius) to 52 cm² (soleus). In addition, the fact that muscle fiber length is usually less than muscle length indicates that muscle fibers are connected serially within a muscle. Functionally coupled muscles tend to have complementary combinations of these properties. For example, the three vasti muscles have similar muscle fiber lengths (10 cm), but they differ in pennation angle (intermedius is the smallest) and cross-sectional area (lateralis is the largest). A similar relation exists for soleus and the two heads (medial and lateral) of gastrocnemius.

Movement involves the muscle-controlled rotation of adjacent body segments, which means that the capacity of a muscle to contribute to a movement also depends on its location relative to the joint that it spans. The rotary force exerted by a muscle about a joint is referred to as *muscle torque* and is calculated as the product of the muscle force and the *moment arm*, the shortest perpendicular distance from the line of pull of the muscle to the joint's center (Figure 31–14).

The moment arm usually changes as a joint rotates through its range of motion; the amount of change depends on where the muscle is attached to the skeleton

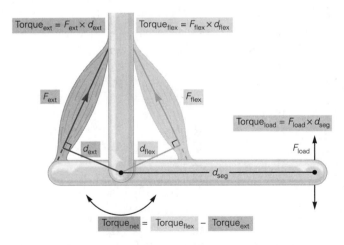

Figure 31–14 Muscle torque varies over the range of motion about a joint. A muscle's torque about a joint is the product of its contractile force (F) and its moment arm relative to the joint (d). The moment arm is the shortest perpendicular distance from the line of pull of the muscle to the center of rotation of the joint. Because the moment arm changes when the joint rotates, muscle torque varies with angular displacement about the joint. The net torque about a joint, which determines the mechanical action, is the difference in the torques exerted by opposing muscles, such as extensors (ext) and flexors (flex). Similarly, a force applied to the limb (F_{load}) will exert a torque about the joint that depends on F_{load} and its distance from the joint (d_{seg}).

relative to the joint. If the force exerted by a muscle remains relatively constant throughout the joint's range of motion, muscle torque varies in direct proportion to the change in the moment arm. For many muscles, the moment arm is maximal in the middle of the range of motion, which usually corresponds to the position of maximal muscle force and hence greatest muscle torque.

Different Movements Require Different Activation Strategies

The human body has approximately 600 muscles, each with a distinct torque profile about one or more joints. To perform a desired movement, the nervous system must activate an appropriate combination of muscles with adequate intensity and timing of activity. The activation must be appropriate for the contractile properties and musculoskeletal geometry of many muscles, as well as the mechanical interactions between body segments. As a result of these demands, activation strategies differ with the details of the movement.

Contraction Velocity Can Vary in Magnitude and Direction

Movement speed depends on the contraction velocity of a muscle. The only ways to vary contraction velocity are to alter either the number of motor units recruited or the rates at which they discharge action potentials. The velocity of a contraction can vary in both magnitude and direction (see Figure 31–11B). To control the velocity of a contraction, the nervous system must scale the magnitude of the net muscle torque relative to the load torque (Figure 31–14), which includes both the weight of the body part and any external load acting on the body.

When muscle torque exceeds load torque, the muscle shortens as it performs a shortening contraction. When muscle torque is less than load torque, the muscle lengthens as it performs a lengthening contraction. For the example shown in Figure 31–14, the load is lifted with a shortening contraction of the flexor and lowered with a lengthening contraction of the flexor. Both types of contractions are common in daily activities.

Shortening and lengthening contractions are not simply the result of adjusting motor unit activity so that the net muscle torque is greater or less than the load torque. When the task involves lifting a load with a prescribed trajectory, activation of the motor units must be aligned so that the sum of the rise times produce the appropriate torque so as to match the desired trajectory while lifting (shortening contractions), whereas while lowering the load (lengthening contractions), the sum of the decay times must be similarly controlled. The nervous system accomplishes this with different descending input and sensory feedback during the two types of contractions. Because of these differences in required motor unit activity, the control of the two types of contraction respond differently to stresses imposed on the system. Declines in the capacity to control motor unit activity, such as observed in older adults and persons performing rehabilitation exercises after an orthopedic procedure, are associated with greater difficulty in performing lengthening contractions.

The amount of motor unit activity relative to the load also influences the contraction velocity. This effect

depends on both the number of motor units recruited and the maximal rates at which the motor units can discharge action potentials. As described previously, physical training with rapid contractions, such as power training, increases the rate at which motor units can discharge trains of action potentials, which can be mimicked by step injections of current into a motor neuron. Changes in the maximal shortening velocity of a muscle after a change in the habitual level of physical activity are the result, at least partly, of factors that influence the ability of motor neurons to discharge action potentials at high rates.

Movements Involve the Coordination of Many Muscles

In the simplest case, muscles span a single joint and cause the attached body segments to accelerate about a single axis of rotation. Because muscles can exert only a pulling force, motion about a single axis of rotation requires at least two muscles or groups of muscles when the action involves shortening contractions (Figure 31–15A).

Because most muscles attach to the skeleton slightly off center from the axis of rotation, they can cause movement about more than one axis of rotation. If one of the actions is not required, the nervous system must activate other muscles to control the unwanted movement. For example, activation of the radial flexor muscle of the wrist can cause the wrist to flex and abduct. If the intended action is only wrist flexion, then the abduction action must be opposed by another muscle, such as the ulnar flexor muscle, which causes wrist flexion and adduction. Depending on the geometry of the articulating surfaces and the attachment sites of the muscles, the multiple muscles that span a joint are capable of producing movements about one to three axes of rotation. Furthermore, some structures

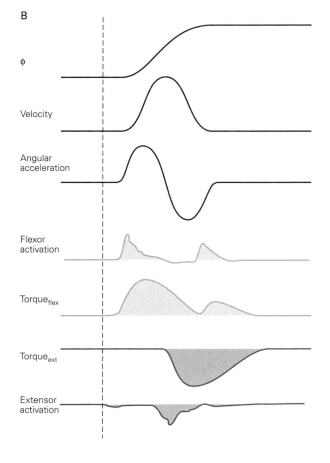

Figure 31–15 (Left) Antagonist muscles spanning a single joint control movement of a limb about a single axis of rotation.

A. According to Newton's law of acceleration (force = mass × acceleration), force is required to change the velocity of a mass. Muscles exert a torque to accelerate the inertial mass of the skeletal segment around a joint. For angular motion, Newton's law is written as torque = rotational inertia × angular acceleration.

B. The angular velocity for movement of a limb from one position to another has a bell-shaped profile. Acceleration in one direction is followed by acceleration in the opposite direction—the flexor and extensor muscles are activated in succession. The records here show the activation profiles and associated muscle torques for a fast elbow flexion movement. Because contractile force decays relatively slowly, the flexor muscle is usually activated a second time to counter the prolonged acceleration generated by the extensor muscle and to stop the limb at the intended joint angle.

can be displaced linearly (eg, the scapula on the trunk), adding to the degrees of freedom about a joint.

The off-axis attachment of muscles enhances the flexibility of the skeletal motor system; the same movement can be achieved by activating different combinations of muscles. However, this additional flexibility requires the nervous system to control the unwanted actions. A solution used by the nervous system is to organize relations among selected muscles to produce specific actions. A particular sequence of muscle activations is known as a *muscle synergy*, and movement is produced through the coordinated activation of these synergies. For example, EMG recordings of human subjects suggest that variations of movements with the same general purpose, such as grasping various objects with the hand, reaching and pointing in different directions, or walking and running at several speeds, are controlled by approximately five muscle synergies.

The number of muscles that participate in a movement also varies with the speed of the movement. For example, slow lifting of a load requires only that the

muscle torque slightly exceed the load torque (see Figure 31–14), and thus, only the flexor muscle is activated. This strategy is used when lifting a handheld weight with the elbow flexor muscles. In contrast, to perform this movement rapidly with an abrupt termination at an intended joint angle, both the flexor and extensor muscles must be activated. First, the flexor muscle is activated to accelerate the limb in the direction of flexion, followed by activation of the extensor muscle to accelerate the limb in the direction of extension, and finally a burst of activity by the flexor muscle to increase the angular momentum of the limb and the handheld weight in the direction of flexion so that it arrives at the desired joint angle (Figure 31–15B). The amount of extensor muscle activity increases with the speed of the movement.

Increases in movement speed introduce another factor that the nervous system must control: unwanted accelerations in other body segments. Because body parts are connected to one another, motion in one part can induce motion in another. The induced motion is often controlled with lengthening contractions, such as

Figure 31–16 A single muscle can influence the motion about many joints.

A. Muscles that cross one joint can accelerate an adjacent body segment. For example, at the beginning of the swing phase while running, the hip flexor muscles are activated to accelerate the thigh forward (**red arrow**). This action causes the lower leg to rotate backward (**blue arrow**) and the knee joint to flex. To control the knee joint flexion during the first part of the swing phase, the knee extensor muscles are activated and undergo a lengthening contraction to accelerate the lower leg forward (**red arrow**) while it continues to rotate backward (**blue arrow**).

B. Many muscles cross more than one joint to exert an effect on more than one body segment. For example, the hamstring muscles of the leg accelerate the hip in the direction of extension and the knee in the direction of flexion (**red arrows**). During running, at the end of the swing phase, the hamstring muscles are activated and undergo lengthening contractions to control the forward rotation of the leg (hip flexion and knee extension). This strategy is more economical than activating individual muscles at the hip and knee joints to control the forward rotation of the leg.

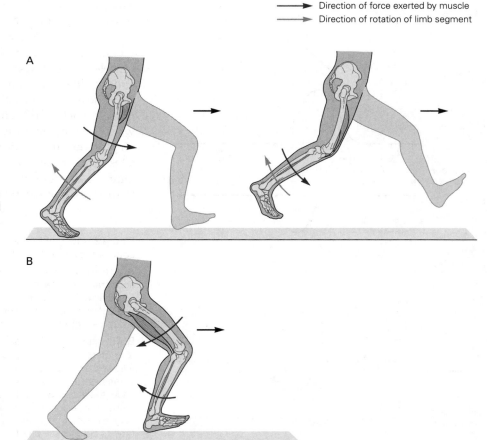

→ Direction of force exerted by muscle
→ Direction of rotation of limb segment

those experienced by thigh muscles during the swing phase of running (Figure 31–16A).

Muscles that span more than one joint can be used to control these motion-dependent interactions between body parts. At the end of the swing phase in running, activation of the hamstring muscles causes both the thigh and lower leg to accelerate backward (Figure 31–16B). If a hip extensor muscle is used to accelerate the thigh backward instead of the hamstring muscles, the lower leg would accelerate forward, requiring activation of a knee flexor muscle to control the unwanted lower leg motion so that the foot could be placed on the ground. Use of the two-joint hamstring muscles is a more economical strategy, but one that can subject the hamstrings to high stresses during fast movements, such as sprinting. The control of such

motion-dependent interactions often involves lengthening contractions, which maximize muscle stiffness and the ability of muscle to resist changes in length.

For most movements, the nervous system must establish rigid connections between some body parts for two reasons. First, as expressed in Newton's law of action and reaction, a reaction force must provide a foundation for the acceleration of a body part. For example, in a reaching movement performed by a person standing upright, the ground must provide a reaction force against the feet. The muscle actions that produce the arm movement exert forces that are transmitted through the body to the feet and are opposed by the ground. Different substrates provide different amounts of reaction force, which is why ice or sand can influence movement capabilities.

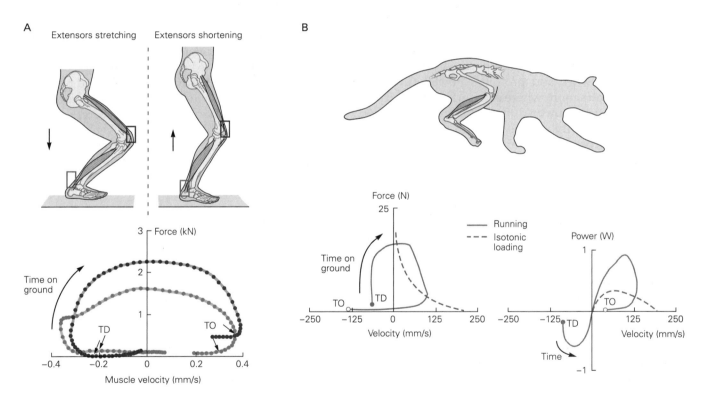

Figure 31–17 An initial phase of negative work augments subsequent positive work performed by the muscle. (Reproduced, with permission, from Finni, Komi, and Lepola 2000. Copyright © 2000, Springer-Verlag; Gregor et al. 1988.)

A. The force in the Achilles tendon (**orange**) and patellar tendon (**purple**) vary during the ground-contact phase of two-legged hopping. The feet contact the ground at touchdown (**TD**) and leave the ground at toe-off (**TO**). For approximately the first half of the movement, the quadriceps and triceps surae muscles lengthen, performing negative work (negative velocity). The muscles perform positive work when they shorten (positive velocity). The sites of force transducer measurements are indicated by rectangles.

B. The force exerted by the soleus muscle of a cat running at moderate speed varies from the instant the paw touches the ground (**TD**) until it leaves the ground (**TO**). The force exerted by the muscle during the shortening contraction (positive velocity) is greater than the peak forces measured when the muscle contracts maximally against various constant loads (isotonic loading). Negative velocity reflects a lengthening contraction in the soleus muscle. The power produced by the soleus muscle of the cat during running is greater than that produced in an isolated-muscle experiment (**dashed line**). The phase of negative power corresponds to the lengthening contraction just after the paw is placed on the ground (**TD**), when the muscle performs negative work.

Second, uncertain conditions are usually accommodated by stiffening the joints through concurrent activation of the muscles that produce force in opposite directions. Coactivation of opposing muscles occurs often when a support surface is unsteady, when the body might experience an unexpected perturbation, or when lifting a heavy load. Because coactivation increases the energetic cost of performing a task, one characteristic of skilled performance is the ability to accomplish a task with minimal activation of muscles that produce opposing actions.

Muscle Work Depends on the Pattern of Activation

Limb muscles in healthy young adults are active 10% to 20% of the time during waking hours. For much of this time, the muscles perform constant-length (*isometric*) contractions to maintain a variety of static body postures. In contrast, muscle length has to change during a movement so that the muscle can perform work to displace body parts. A muscle performs positive work and produces power during a shortening contraction, whereas it performs negative work and absorbs power during a lengthening contraction. The capacity of muscle to do positive work establishes performance capabilities, such as the maximal height that can be jumped.

The nervous system can augment the positive work capacity of muscle by commanding a brief period of negative work before performing positive work. This activation sequence, the *stretch–shorten cycle*, occurs in many movements. When a person jumps in place on two feet, for example, the support phase involves an initial stretch (lengthening) and subsequent shortening of the ankle extensor and knee extensor muscles (Figure 31–17A). The forces in the Achilles and patellar tendons increase during the stretch of the lengthening contraction and reach a maximum at the onset of the shortening phase. As a result, the muscles can perform more positive work and produce more power during the shortening contraction (Figure 31–17B).

Although negative work involves an increase in the length of the muscle, the length of the fascicles in the muscle often remains relatively constant, which indicates that the connective tissue structures are stretched prior to the shortening contraction. Thus, the capacity of the muscle to perform more positive work comes from strain energy that can be stored in the elastic elements of muscle and tendon during the stretch phase and released during the subsequent shortening phase. More strain energy can be stored in long tendons, but short tendons are more advantageous when the movement requires the rapid release of strain energy.

Highlights

1. The basic functional unit for the control of movement by the nervous system is the motor unit, which comprises a motor neuron and the muscle fibers it innervates.

2. The force exerted by a muscle depends in part on the number and properties of the motor units that are activated and the rates at which they discharge action potentials. The key motor unit properties include contraction speed, maximal force, and fatigability, all of which can be altered by physical activity. Motor unit properties vary continuously across the population that innervates each muscle; that is, there are not distinct types of motor units. Due to technological advances, it is becoming possible to characterize the adaptations exhibited by populations of motor units in response to different types of changes in physical activity.

3. Motor units tend to be activated in a stereotypical order that is highly correlated with motor neuron size. The rate at which motor units are recruited during a voluntary contraction increases with contraction speed.

4. The rate at which a motor unit discharges action potentials in response to a given synaptic input can be modulated by descending inputs from the brain stem. The modulatory input is likely critical for establishing the level of excitation in spinal pathways, but this has been difficult to demonstrate in humans.

5. Except at low muscle forces, variation in discharge rate has a greater influence on muscle force than does the number of activated motor units. Moreover, the variability in discharge rate of the motor unit population influences the level of fine motor control.

6. The sarcomere is the smallest element of muscle to include a complete set of contractile proteins. A transient connection between the contractile proteins myosin and actin, known as the cross-bridge cycle, enables muscle to exert a force. The organization of the sarcomeres within a muscle varies substantially and, in addition to motor unit activity, has a major effect on the contractile properties of the muscle.

7. For a given arrangement of sarcomeres, the force a muscle can exert depends on the activation of the cross bridges by Ca^{2+}, the amount of overlap between the thick and thin filaments, and the velocity of the moving filaments. Sarcomere force during lengthening contractions is augmented by a Ca^{2+}-mediated increase in titin stiffness. The force

produced by activated sarcomeres depends on the interactions of three filaments: actin, myosin, and titin.

8. Most of the force generated by activated sarcomeres is transmitted laterally through a network of non-contractile proteins that maintains the alignment of the thick and thin filaments.

9. The functional capability of a muscle depends on the torque that it can exert, which is influenced both by its contractile properties and by the location of its attachments on the skeleton relative to the joint that it spans.

10. To perform a movement, the nervous system activates multiple muscles and controls the torque exerted about the involved joints. The nervous system can vary the magnitude and direction of a movement by altering the amount of motor unit activity, and hence muscle torque, relative to the load acting on the body.

11. Although muscle exerts only a pulling force on the skeleton, it can do so whether the activated muscle shortens or is lengthened by a load torque that exceeds the muscle torque. The force capacity of muscle is greater during lengthening contractions. Motor unit activity differs during shortening and lengthening contractions, but little is known about how the synaptic inputs to motor neurons differ during these two types of contractions.

12. Faster movements elicit motion-dependent interactions between body parts that produce unwanted accelerations. These actions must be controlled by the nervous system to produce an intended movement.

13. The nervous system must coordinate the activity of multiple muscles to provide a mechanical link between moving body parts and the required support from the surroundings. The muscles engaged for each action, such as grasping, reaching, running, and walking, are organized into a few sets that exhibit a stereotypical pattern of activation, but it is not known why particular patterns are preferred.

14. The patterns of muscle activity vary substantially between movements and often include strategies that augment the work capacity of muscles. The patterns can be modified by experience, but little is known about the locus of the adaptations other than that both spinal and supraspinal pathways are involved.

Roger M. Enoka

Selected Reading

Booth FW. 2015. Muscle adaptation to exercise: new Saltin's paradigms. Scand J Med Sci Sports Suppl 4:49–52.

Duchateau J, Enoka RM. 2011. Human motor unit recordings: origins and insights into the integrated motor system. Brain Res 1409:42–62.

Enoka RM. 2015. *Neuromechanics of Human Movement*. 5th ed. Champaign, IL: Human Kinetics.

Farina D, Negro F, Muceli S, Enoka RM. 2016. Principles of motor unit physiology evolve with advances in technology. Physiology 31:83–94.

Gordon AM, Regnier M, Homsher E. 2001. Skeletal and cardiac muscle contractile activation: tropomyosin "rocks and rolls." News Physiol Sci 16:49–55.

Heckman CJ, Enoka RM. 2012. Motor unit. Compr Physiol 2:2629–2682.

Herzog W, Schappacher G, DuVall M, Leonard TR, Herzog JA. 2016. Residual force enhancement following eccentric contractions: a new mechanism involving titin. Physiology 31:300–312.

Hunter SK, Pereira HM, Keenan KG. 2016. The aging neuromuscular system and motor performance. J Appl Physiol 121:982–995.

Huxley AF. 2000. Mechanics and models of the myosin motor. Philos Trans R Soc Lond B Biol Sci 355:433–440.

Lieber RL, Ward SR. 2011. Skeletal muscle design to meet functional demands. Philos Trans R Soc Lond B Biol Sci 366:1466–1476.

Merletti R, Muceli S. 2020. Tutorial. Surface EMG detection in space and time: best practices. J Electromyogr Kinesiol 49:102363.

Narici M, Franchi M, Maganaris C. 2016. Muscle structural assembly and functional consequences. J Exp Biol 219:276–284.

References

Alexander RM, Ker RF. 1990. The architecture of leg muscles. In: JM Winters, SL-Y Woo (eds). *Multiple Muscle Systems: Biomechanics and Movement Organization*, pp. 568–577. New York: Springer-Verlag.

Azizi E, Roberts TJ. 2012. Geared up to stretch: pennate muscle behavior during active lengthening. J Exp Biol 217:376–381.

Bloom W, Fawcett DW. 1975. *A Textbook of Histology*. 10th ed. Philadelphia, PA: Saunders.

Botterman BR, Iwamoto GA, Gonyea WJ. 1986. Gradation of isometric tension by different activation rates in motor units of cat flexor carpi radialis muscle. J Neurophysiol 56:494–506.

Bottinelli R, Canepari M, Pellegrino MA, Reggiani C. 1996. Force-velocity properties of human skeletal muscle fibres: myosin heavy chain isoform and temperature dependence. J Physiol 495:573–586.

Del Vecchio A, Negro F, Felici F, Farina D. 2017. Associations between motor unit action potential parameters and surface EMG features. J Appl Physiol 123:835–843.

Desmedt JE, Godaux E. 1977. Ballistic contractions in man: characteristic recruitment pattern of single motor units of the tibialis anterior muscle. J Physiol 264:673–693.

Duchateau J, Enoka RM. 2016. Neural control of lengthening contractions. J Exp Biol 219:197–204.

Enoka RM, Duchateau J. 2016. Translating muscle fatigue to human performance. Med Sci Sport Exerc 48:2228–2238.

Enoka RM, Duchateau J. 2017. Rate coding and the control of muscle force. Cold Spring Harb Perspect Med 27:1–12.

Enoka RM, Fuglevand AJ. 2001. Motor unit physiology: some unresolved issues. Muscle Nerve 24:4–17.

Erim Z, De Luca CJ, Mineo K, Aoki T. 1996. Rank-ordered regulation of motor units. Muscle Nerve 19:563–573.

Farina D, Merletti R, Enoka RM. 2014. The extraction of neural strategies from the surface EMG: an update. J Appl Physiol 117:1215–1230.

Finni T, Komi PV, Lepola V. 2000. In vivo human triceps surae and quadriceps muscle function in a squat jump and counter movement jump. Eur J Appl Physiol 83:416–426.

Fuglevand AJ, Macefield VG, Bigland-Ritchie B. 1999. Force-frequency and fatigue properties of motor units in muscles that control digits of the human hand. J Neurophysiol 81:1718–1729.

Gordon AM, Regnier M, Homsher R. 2001. Skeletal and cardiac muscle contractile activation: tropomyosin "rocks and rolls." News Physiol Sci 16:49–55.

Gregor RJ, Roy RR, Whiting WC, Lovely RG, Hodgson JA, Edgerton VR. 1988. Mechanical output of the cat soleus during treadmill locomotion: in vivo vs in situ characteristics. J Biomech 21:721–732.

Heckman CJ, Mottram C, Quinlan K, Theiss R, Schuster J. 2009. Motoneuron excitability: the importance of neuromodulatory inputs. Clin Neurophysiol. 120:2040–2054.

Henneman E, Somjen G, Carpenter DO. 1965. Functional significance of cell size in spinal motoneurons. J Neurophysiol 28:560–580.

Hepple RT, Rice CL. 2016. Innervation and neuromuscular control in ageing skeletal muscle. J Physiol 594:1965–1978.

Huxley AF, Simmons RM. 1971. Proposed mechanism of force generation in striated muscle. Nature 233:533–538.

Kubo K, Miyazaki D, Shimoju S, Tsunoda N. 2015. Relationship between elastic properties of tendon structures and performance in long distance runners. Eur J Appl Physiol 115:1725–1733.

Lai A, Schache AG, Lin YC, Pandy MG. 2014. Tendon elastic strain energy in the human ankle plantar-flexors and its role with increased running speed. J Exp Biol 217:3159–3168.

Lieber RL, Fridén J. 2000. Functional and clinical significance of skeletal muscle architecture. Muscle Nerve 23:1647–1666.

Lieber RL, Ward SR. 2011. Skeletal muscle design to meet functional demands. Philos Trans R Soc Lond B Biol Sci 366:1466–1476.

Liddell EGT, Sherrington CS. 1925. Recruitment and some other factors of reflex inhibition. Proc R Soc Lond B Biol Sci 97:488–518.

Maas H, Finni T. 2018. Mechanical coupling between muscle-tendon units reduces peak stresses. Exerc Sport Sci Rev 46:26–33.

Macefield VG, Fuglevand AJ, Bigland-Ritchie B. 1996. Contractile properties of single motor units in human toe extensors assessed by intraneural motor axon stimulation. J Neurophysiol 75:2509–2519.

Milner-Brown HS, Stein RB, Yemm R. 1973. The orderly recruitment of human motor units during voluntary isometric contraction. J Physiol 230:359–370.

Moritz CT, Barry BK, Pascoe MA, Enoka RM. 2005. Discharge rate variability influences the variation in force fluctuations across the working range of a hand muscle. J Neurophysiol 93:2449–2459.

O'Connor SM, Cheng EJ, Young KW, Ward SR, Lieber RL. 2016. Quantification of sarcomere length distribution in whole muscle frozen sections. J Exp Biol 219:1432–1436.

Overduin SA, d'Avella A, Roh J, Carmena JM, Bizzi E. 2015. Representation of muscle synergies in the primate brain. J Neurosci 35:12615–12624.

Palmisano MG, Bremner SN, Homberger TA, et al. 2015. Skeletal muscle intermediate filaments form a stress-transmitting and stress-signalling network. J Cell Sci 128: 219–224.

Person RS, Kudina LP. 1972. Discharge frequency and discharge pattern of human motor units during voluntary contraction of muscle. Electroencephalogr Clin Neurophysiol 32:471–483.

Sawicki GS, Robertson BD, Azizi E, Roberts TJ. 2015. Timing matters: tuning the mechanics of a muscle-tendon unit by adjusting stimulation phase during cyclic contractions. J Exp Biol 218:3150–3159.

Schache AG, Dorn TW, Williams GP, Brown NA, Pandy MG. 2014. Lower-limb muscular strategies for increasing running speed. J Orthop Sports Phys Ther 44:813–824.

Sherrington CS. 1925. Remarks on some aspects of reflex inhibition. Proc R Soc Lond B Biol Sci 97:519–545.

Trappe S, Costill D, Gallagher P, et al. 2009. Exercise in space: human skeletal muscle after 6 months aboard the International Space Station. J Appl Physiol 106:1159–1168.

Trappe S, Luden N, Minchev K, Raue U, Jemiolo B, Trappe TA. 2015. Skeletal muscle signature of a champion sprint runner. J Appl Physiol 118:1460–1466.

Van Cutsem M, Duchateau J, Hainaut K. 1998. Changes in single motor unit behaviour contribute to the increase in contraction speed after dynamic training in humans. J Physiol 513:295–305.

Van Cutsem M, Feiereisen P, Duchateau J, Hainaut K. 1997. Mechanical properties and behaviour of motor units in the tibialis anterior during voluntary contractions. Can J Appl Physiol 22:585–597.

Ward SR, Eng CM Smallwood LH, Lieber RL. 2009. Are current measurements of lower extremity muscle architecture accurate? Clin Orthop Rel Res 467:1074–1082.

32

Sensory-Motor Integration in the Spinal Cord

D URING PURPOSEFUL MOVEMENTS the central nervous system uses information from a vast array of sensory receptors to ensure that the pattern of muscle activity suits the purpose. Without this sensory information, movements tend to be imprecise, and tasks requiring fine coordination in the hands, such as buttoning one's shirt, are difficult. The sensory-motor integration that makes the ongoing regulation of movement possible takes place at many levels of the nervous system, but the spinal cord has a special role because of the close coupling in the cord between sensory input and the motor output to the muscles.

Charles Sherrington was among the first to recognize the importance of sensory information in regulating movements. In 1906, he proposed that simple reflexes—stereotyped movements elicited by activation of receptors in skin or muscle—are the basic units for movement. He also emphasized that all parts of the nervous system are connected and that no part is ever capable of activation without affecting or being affected by other parts. In his words, the simple reflex is a convenient if not a probable fiction.

Laboratory studies of reflexes in animals from the 1950s and onward demonstrated that descending motor pathways and afferent sensory pathways converge on common interneurons in the spinal cord. Later research in intact animals and in humans engaged in normal behavior confirmed that the neural circuitries in the spinal cord take part in conveying and shaping the motor command to the muscles by integrating descending motor commands and sensory feedback signals. Nevertheless, the idea of simple reflexes is convenient for understanding the principles of organization of sensory-motor integration in the spinal cord and of how sensory input to different spinal circuits contributes to movement control.

In this chapter, we explain the principles underlying sensory-motor integration in the spinal cord and describe how this integration regulates movement. For this purpose, we must first have a thorough knowledge of how reflex pathways in the spinal cord are organized.

Reflex Pathways in the Spinal Cord Produce Coordinated Patterns of Muscle Contraction

The sensory stimuli that activate spinal reflex pathways act outside the spinal cord, on receptors in muscles, joints, and skin. By contrast, the neural circuitry responsible for the motor response is entirely contained within the spinal cord. The interneurons in the reflex pathways and the resulting reflexes have traditionally been classified based on the sensory modality and type of sensory fiber that activates the interneurons. As we shall see, this classification is inconsistent with the significant convergence of multiple modalities on common interneurons, but as a starting point, it is still useful to distinguish reflex pathways based on whether the principal sensory input originates from muscle or skin.

The Stretch Reflex Acts to Resist the Lengthening of a Muscle

The simplest and certainly the most studied spinal reflex is the *stretch reflex*, a reflex muscle contraction elicited by lengthening of the muscle. Stretch reflexes

were originally thought to be an intrinsic property of muscles. Early in the 20th century, however, Liddell and Sherrington showed that the stretch reflex could be abolished by cutting either the dorsal or ventral root, thus establishing that these reflexes require sensory input from muscle to spinal cord and a return path to muscle (Figure 32–1A).

We now know that the receptor that senses the change of length is the muscle spindle (Box 32–1) and that the type Ia sensory axon from this receptor makes direct excitatory connections with motor neurons. (The classification of sensory fibers from muscle is discussed in Box 32–2.) The afferent axon also connects to interneurons that inhibit the motor neurons innervating antagonist muscles, an arrangement called reciprocal innervation. This inhibition prevents muscle contractions that might otherwise resist the movements produced by the stretch reflexes.

Sherrington developed an experimental model for investigating spinal circuitry that is especially valuable in the study of stretch reflexes. He conducted his experiments on cats whose brain stems had been surgically transected at the level of the midbrain, between the superior and inferior colliculi. This is referred to as a *decerebrate preparation*. The effect of this procedure is to disconnect the rest of the brain from the spinal cord, thus blocking sensations of pain as well as interrupting normal modulation of reflexes by higher brain centers. A decerebrate animal has stereotyped and usually heightened stretch reflexes, making it easier to examine the factors controlling their expression.

Without control by higher brain centers, descending pathways from the brain stem powerfully facilitate the neuronal circuits involved in the stretch reflexes of extensor muscles. This results in a dramatic increase in tone of the extensor muscle that sometimes can suffice to support the animal in a standing position. In normal animals and humans, owing to the balance between facilitation and inhibition, stretch reflexes are weaker and considerably more variable in strength than those in decerebrate animals.

Neuronal Networks in the Spinal Cord Contribute to the Coordination of Reflex Responses

The Stretch Reflex Involves a Monosynaptic Pathway

The neural circuit responsible for the stretch reflex was one of the first reflex pathways to be examined in detail. The physiological basis of this reflex was examined by measuring the latency of the response in

A Monosynaptic pathways (stretch reflex)

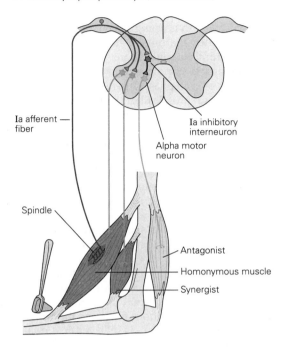

B Polysynaptic pathways (flexion reflex)

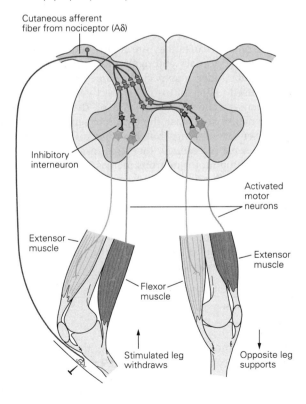

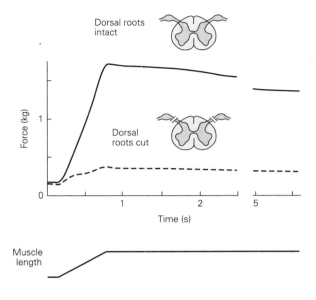

Figure 32–1 Spinal reflexes involve coordinated contractions of numerous muscles in the limbs.

A. In monosynaptic pathways, Ia sensory axons from muscle spindles make excitatory connections on two sets of motor neurons: alpha motor neurons that innervate the same (homonymous) muscle from which they arise and motor neurons that innervate synergist muscles. They also act through interneurons to inhibit the motor neurons that innervate antagonist muscles. When a muscle is stretched by a tendon tap with a reflex hammer, the firing rate in the sensory fiber from the spindle increases. This leads to contraction of the same muscle and its synergists and relaxation of the antagonist. The reflex therefore tends to counteract the stretch, enhancing the spring-like properties of the muscles.

The records on the right demonstrate the reflex nature of contractions produced by muscle stretch in a decerebrate cat. When an extensor muscle is stretched, it normally produces a large force, but it produces a very small force (**dashed line**) after the sensory afferents in the dorsal roots have been severed. (Adapted, with permission, from Liddell and Sherrington 1924.)

B. In polysynaptic pathways, one excitatory pathway activates motor neurons that innervate ipsilateral flexor muscles, which withdraw the limb from noxious stimuli, while another pathway simultaneously excites motor neurons that innervate contralateral extensor muscles, providing support during withdrawal of the limb. Inhibitory interneurons ensure that the motor neurons supplying antagonist muscles are inactive during the reflex response. (Adapted, with permission, from Schmidt 1983.)

Box 32–1 Muscle Spindles

Muscle spindles are small encapsulated sensory receptors that have a spindle-like or fusiform shape and are located within the fleshy part of a muscle. Their main function is to signal changes in the length of the muscle within which they reside. Changes in length of muscles are closely associated with changes in the angles of the joints that the muscles cross. Thus, muscle spindles are used by the central nervous system to sense relative positions of the body segments.

Each spindle has three main components: (1) a group of specialized *intrafusal* muscle fibers with non-contractile central regions; (2) sensory fibers that terminate on the central regions of the intrafusal fibers; and (3) motor axons that terminate on the contractile polar regions of the intrafusal fibers (Figure 32–2A,B).

When the intrafusal fibers are stretched, often referred to as "loading the spindle," the sensory axon endings are also stretched and increase their firing rate. Because muscle spindles are arranged in parallel with the *extrafusal* muscle fibers that make up the main body of the muscle, the intrafusal fibers change in length as the whole muscle changes. Thus, when a muscle is stretched, activity in the sensory axons of muscle spindles increases. When a muscle shortens, the spindle is unloaded and the activity decreases.

The intrafusal muscle fibers are innervated by *gamma* motor neurons, which have small-diameter myelinated axons, whereas the extrafusal muscle fibers are innervated by *alpha* motor neurons, with large-diameter myelinated axons. Activation of gamma motor neurons causes shortening of the polar regions of the intrafusal fibers. This in turn stretches the central region from both ends, leading to an increase in firing rate of the sensory axons or to a greater likelihood that the axons will fire in response to stretch of the muscle. Thus, the gamma motor neurons adjust the sensitivity of the muscle spindles. Contraction of the intrafusal muscle fibers does not contribute significantly to the force of muscle contraction.

The structure and functional behavior of muscle spindles is considerably more complex than this simple description depicts. As a muscle is stretched, the change in length has two phases: a dynamic phase, the period during which length is changing, and a static or steady-state phase, when the muscle has stabilized at a new length. Structural specializations within each component of the muscle spindle enable the sensory axons to signal aspects of each phase separately.

The intrafusal muscle fibers include nuclear bag fibers and nuclear chain fibers. The bag fibers can be classified as dynamic or static. A typical spindle has two or three bag fibers and a variable number of chain fibers, usually about five. Furthermore, the intrafusal fibers receive two types of sensory endings. A single Ia (large diameter) axon spirals around the central region of all intrafusal muscle fibers and serves as the *primary sensory ending* (Figure 32–2B). A variable number of type II (medium diameter) axons spiral around the static bag and chain fibers near their central regions and serve as *secondary sensory endings*.

The gamma motor neurons can also be divided into two classes: Dynamic gamma motor neurons innervate the dynamic bag fibers, whereas the static gamma motor neurons innervate the static bag fibers and the chain fibers.

This duality of structure is reflected in a duality of function. The tonic discharge of both primary and secondary sensory endings signals the steady-state length of the muscle. The primary sensory endings are, in addition, highly sensitive to the velocity of stretch, allowing them to provide information about the speed of movements. Because they are highly sensitive to small changes, the primary endings rapidly provide information about sudden unexpected changes in length, which can be used to generate quick corrective reactions.

Increases in the firing rate of dynamic gamma motor neurons increase the dynamic sensitivity of primary sensory endings but have no influence on secondary sensory endings. Increases in the firing rate of static gamma motor neurons increase the tonic level of activity in both primary and secondary sensory endings, decrease the dynamic sensitivity of primary endings (Figure 32–2C), and can prevent the silencing of primary endings when a muscle is released from stretch. Thus, the central nervous system can independently adjust the dynamic and static sensitivity of the different sensory endings in muscle spindles.

---

ventral roots to electrical stimulation of dorsal roots. When the Ia sensory axons innervating the muscle spindles were selectively activated, the reflex latency through the spinal cord was less than 1 ms. This demonstrated that the Ia fibers make direct connections on the alpha motor neurons because the delay at a single synapse is typically 0.5 ms to 0.9 ms (Figure 32–3B). In humans, an analog of the monosynaptic stretch reflex, the Hoffmann reflex, may be elicited by electrical stimulation of peripheral nerves (Box 32–3).

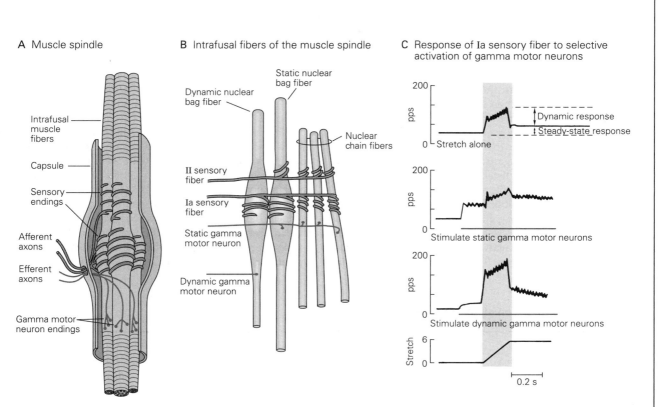

A Muscle spindle

B Intrafusal fibers of the muscle spindle

C Response of Ia sensory fiber to selective activation of gamma motor neurons

Figure 32–2 The muscle spindle detects changes in muscle length.

A. The main components of the muscle spindle are intrafusal muscle fibers, sensory axon endings, and motor axon endings. The intrafusal fibers are specialized muscle fibers with central regions that are not contractile. Gamma motor neurons innervate the contractile polar regions of the intrafusal fibers. Contraction of the polar regions pulls on the central regions of the intrafusal fiber from both ends. The sensory endings spiral around the central regions of the intrafusal fibers and are responsive to stretch of these fibers. (Adapted, with permission, from Hulliger 1984. Copyright © Springer-Verlag 1984.)

B. The muscle spindle contains three types of intrafusal fibers: dynamic nuclear bag, static nuclear bag, and nuclear chain fibers. A single Ia sensory axon innervates all three types of fibers, forming a primary sensory ending. Type II sensory axons innervate the nuclear chain fibers

and static bag fibers, forming a secondary sensory ending. Two types of motor neurons innervate different intrafusal fibers. Dynamic gamma motor neurons innervate only dynamic bag fibers; static gamma motor neurons innervate various combinations of chain and static bag fibers. (Adapted, with permission, from Boyd 1980. Copyright © 1980. Published by Elsevier Ltd.)

C. Selective stimulation of the two types of gamma motor neurons has different effects on the firing of the Ia sensory fibers from the spindle. Without gamma stimulation, the Ia fiber shows a small dynamic response to muscle stretch and a modest increase in steady-state firing. When a static gamma motor neuron is stimulated, the steady-state response of the Ia fiber increases but the dynamic response decreases. When a dynamic gamma motor neuron is stimulated, the dynamic response of the Ia fiber is markedly enhanced, but the steady-state response gradually returns to its original level. (Adapted, with permission, from Brown and Matthews 1966.)

The pattern of connections of Ia fibers to motor neurons can be shown directly by intracellular recording. Ia fibers from a given muscle excite not only the motor neurons innervating that same (*homonymous*) muscle but also the motor neurons innervating other (*heteronymous*) muscles with a similar mechanical action.

Lorne Mendell and Elwood Henneman used a computer enhancement technique called *spike-triggered averaging* to determine the extent to which the action

---

Box 32–2 Classification of Sensory Fibers From Muscle

Sensory fibers are classified according to their diameter. Axons with larger diameters conduct action potentials more rapidly than those with smaller diameters (Chapters 9 and 18). Because each class of sensory receptors is innervated by fibers with diameters within a restricted range, this method of classification distinguishes to some extent the fibers that arise from different types of receptor organs. The main groups of sensory fibers from muscle are listed in Table 32–1.

The organization of reflex pathways in the spinal cord has been established primarily by electrically stimulating the sensory fibers and recording evoked responses in different classes of neurons in the spinal cord. This method of activation has three advantages over natural stimulation. The timing of afferent input

can be precisely established; the responses evoked in motor neurons and other neurons by different classes of sensory fibers can be assessed by grading the strength of the electrical stimulus; and certain classes of receptors can be selectively activated.

The strength of the electrical stimulus required to activate a sensory fiber is measured relative to the strength required to activate the fibers with the largest diameter because these fibers have the lowest threshold for electrical activation. The thresholds of most type I fibers usually range from one to two times that of the largest fibers (with Ia fibers having, on average, a slightly lower threshold than Ib fibers). For most type II fibers, the threshold is 2 to 5 times higher, whereas types III and IV have thresholds in the range of 10 to 50 times that of the largest sensory fibers.

Table 32–1 Classification of Sensory Fibers From Muscle

Type	Axon	Receptor	Sensitivity to
Ia	12–20 µm myelinated	Primary spindle ending	Muscle length and rate of change of length
Ib	12–20 µm myelinated	Golgi tendon organ	Muscle tension
II	6–12 µm myelinated	Secondary spindle ending	Muscle length (little rate sensitivity)
II	6–12 µm myelinated	Nonspindle endings	Deep pressure
III	2–6 µm myelinated	Free nerve endings	Pain, chemical stimuli, and temperature (important for physiological responses to exercise)
IV	0.5–2 µm nonmyelinated	Free nerve endings	Pain, chemical stimuli, and temperature

---

potentials in single Ia fibers are transmitted to a population of spinal motor neurons. They found that individual Ia axons make excitatory synapses with all homonymous motor neurons innervating the medial gastrocnemius of the cat. This widespread divergence effectively amplifies the signals of individual Ia fibers, leading to a strong excitatory drive to the muscle from which they originate (*autogenic excitation*).

The Ia axons in reflex pathways also provide excitatory inputs to many of the motor neurons innervating synergist muscles (up to 60% of the motor neurons of some synergists) (Figure 32–1A). Although widespread, these connections are not as strong as the connections to homonymous motor neurons.

The Ia fibers also send inhibitory signals via the *Ia inhibitory interneurons* to the alpha motor neurons innervating antagonistic muscles. This disynaptic inhibitory pathway is the basis for reciprocal innervation: When a muscle is stretched, its antagonists relax.

Gamma Motor Neurons Adjust the Sensitivity of Muscle Spindles

Activity of muscle spindles may be modulated by changing the level of activity in the gamma motor neurons, which innervate the intrafusal muscle fibers of muscle spindles (Box 32–1). This function of gamma motor neurons, often referred to as the fusimotor

A Experimental setup

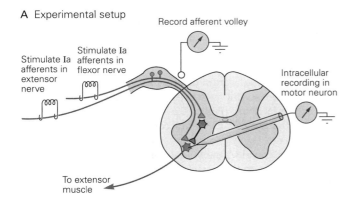

B Inferring the number of synapses in a pathway

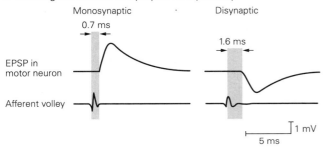

Figure 32-3 The number of synapses in a reflex pathway can be inferred from intracellular recordings.

A. An intracellular recording electrode is inserted into the cell body of a spinal motor neuron that innervates an extensor muscle. Stimulation of Ia sensory fibers from flexor or extensor muscles produces a volley of action potentials at the dorsal root.

B. *Left:* When Ia fibers from an extensor muscle are stimulated, the latency between the recording of the afferent volley and the excitatory postsynaptic potential (**EPSP**) in the motor neuron is only 0.7 ms, approximately equal to the duration of signal transmission across a single synapse. Thus, it can be inferred that the excitatory action of the stretch reflex pathway is monosynaptic. *Right:* When Ia fibers from an antagonist flexor muscle are stimulated, the latency between the recording of the afferent volley and the inhibitory postsynaptic potential in the motor neuron is 1.6 ms, approximately twice the duration of signal transmission across a single synapse. Thus, it can be inferred that the inhibitory action of the stretch reflex pathway is disynaptic.

system, can be demonstrated by selectively stimulating the alpha and gamma motor neurons under experimental conditions.

When only alpha motor neurons are stimulated, the firing of the Ia fiber from the muscle spindle pauses during contraction of the muscle because the muscle is shortening and therefore unloading (slackening) the spindle. However, if gamma motor neurons are activated at the same time as alpha motor neurons, the pause is eliminated. The contraction of the intrafusal fibers by the gamma motor neurons keeps the spindle

under tension, thus maintaining the firing rate of the Ia fibers within an optimal range for signaling changes in length, whatever the actual length of the muscle (Figure 32–5). This *alpha-gamma co-activation* is recruited for many voluntary movements because it stabilizes the sensitivity of the muscle spindles.

In addition to the axons of gamma motor neurons, collaterals of alpha motor neuron axons sometimes innervate the intrafusal fibers. Axons that innervate both intrafusal and extrafusal muscle fibers are referred to as *beta* axons. Beta axon collaterals provide the equivalent of alpha-gamma coactivation. Beta innervation in spindles exists in both cats and humans, although it is unquantified for most muscles.

The forced linkage of extrafusal and intrafusal contraction by the beta fusimotor system highlights the importance of the independent fusimotor system (the gamma motor neurons). Indeed, in lower vertebrates, such as amphibians, beta efferents are the only source of intrafusal innervation. Mammals have evolved a mechanism that frees muscle spindles from complete dependence on the behavior of their parent muscles. In principle, this uncoupling allows greater flexibility in controlling spindle sensitivity for different types of motor tasks.

This conclusion is supported by recordings in spindle sensory axons during a variety of natural movements in cats. The amount and type of activity in gamma motor neurons are set at steady levels, which vary according to the specific task or context. In general, activity levels in both static and dynamic gamma motor neurons (Figure 32–2B) are set at progressively higher levels as the speed and difficulty of the movement increase. Unpredictable conditions, such as when the cat is picked up or handled, lead to marked increases in activity in dynamic gamma motor neurons and thus increased spindle responsiveness when muscles are stretched. When an animal is performing a difficult task, such as walking across a narrow beam, both static and dynamic gamma activation are at high levels (Figure 32–6).

Thus, the nervous system uses the fusimotor system to fine-tune muscle spindles so that the ensemble output of the spindles provides information most appropriate for a task. The task conditions under which independent control of alpha and gamma motor neurons occurs in humans have not yet been clearly established.

The Stretch Reflex Also Involves Polysynaptic Pathways

The monosynaptic Ia pathway is not the only spinal reflex pathway activated when a muscle is stretched. Type II sensory fibers from muscle spindles are also activated. These discharge tonically depending on

Box 32–3 The Hoffmann Reflex

The characteristics of the monosynaptic connections from Ia sensory fibers to spinal motor neurons in humans can be studied using an important technique introduced in the 1950s and based on early work by Paul Hoffmann. This technique involves electrically stimulating the Ia sensory fibers in a peripheral nerve and recording the reflex electromyogram (EMG) response in the homonymous muscle. The response is known as the *Hoffmann reflex*, or H-reflex.

The H-reflex is readily measured in the soleus muscle, an ankle extensor. The Ia fibers from the soleus and its synergists are excited by an electrode placed above the tibial nerve behind the knee (Figure 32–4A). The response recorded from the soleus muscle depends on stimulus strength. At low stimulus strengths, a pure H-reflex is evoked, for the threshold for activation of the Ia fibers is lower than the threshold for motor axons. Increasing the stimulus strength excites the motor axons innervating the soleus, producing two successive responses.

The first results from direct activation of the motor axons, and the second is the H-reflex evoked by stimulation of the Ia fibers (Figure 32–4B). These two components of the evoked EMG are called the M-wave and H-wave. The H-wave occurs later because it results from signals that travel to the spinal cord, across a synapse, and back again to the muscle. The M-wave, in contrast, results from direct stimulation of the motor axon innervating the muscle.

As the stimulus strength is increased still further, the M-wave continues to become larger and the H-wave progressively declines (Figure 32–4C). The decline in the H-wave amplitude occurs because action potentials in the motor axons propagate toward the cell body (antidromic conduction) and cancel reflexively evoked action potentials in the same motor axons. At very high stimulus strengths, only the M-wave persists.

Figure 32–4 The Hoffmann reflex.

A. The Hoffmann reflex (H-reflex) is evoked by electrically stimulating Ia sensory fibers from muscle spindles. The sensory fibers excite alpha motor neurons, which in turn activate the muscle. When a mixed nerve is used, the motor neurons axons may also be activated directly.

B. At intermediate stimulus strengths, an M-wave precedes the H-wave (H-reflex) in the electromyogram (**EMG**).

C. As the stimulus strength increases, the orthodromic motor neuron spikes generated reflexively by the spindle sensory fibers are obliterated by antidromic spikes initiated by the electrical stimulus in the same motor axons. (Adapted, with permission, from Schieppati 1987. Copyright © 1987. Published by Elsevier Ltd.)

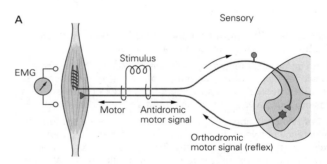

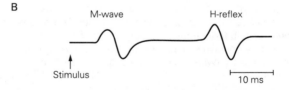

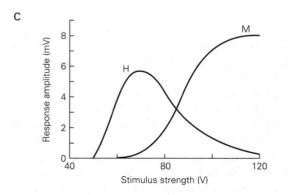

A Sustained stretch of muscle

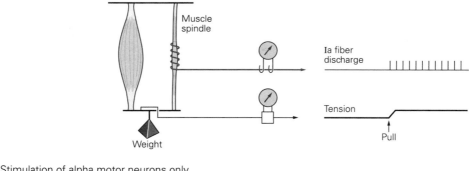

B Stimulation of alpha motor neurons only

C Stimulation of alpha and gamma motor neurons

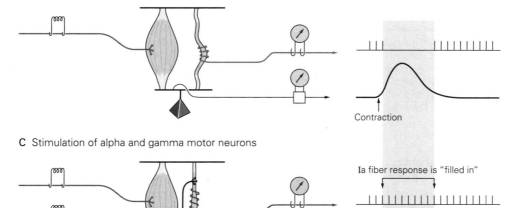

Figure 32–5 Activation of gamma motor neurons during active muscle contraction maintains muscle spindle sensitivity to muscle length. (Adapted, with permission, from Hunt and Kuffler 1951.)

A. Sustained tension elicits steady firing in the Ia sensory fiber from the muscle spindle (the two muscle fibers are shown separately for illustration only).

B. A characteristic pause occurs in the discharge of the Ia fiber when the alpha motor neuron is stimulated, causing a brief

contraction of the muscle. The Ia fiber stops firing because the spindle is unloaded by the contraction.

C. Gamma motor neurons innervate the contractile polar regions of the intrafusal fibers of muscle spindles (see Figure 32–2A). If a gamma motor neuron is stimulated at the same time as the alpha motor neuron, the spindle is not unloaded during the contraction. As a result, the pause in discharge of the Ia sensory fiber that occurs when only the alpha motor neuron is stimulated is "filled in" by the response of the fiber to stimulation of the gamma motor neuron.

muscle length and gamma motor neuron activity (Box 32–1) and connect to different populations of excitatory and inhibitory interneurons in the spinal cord.

Some of the interneurons project directly to the spinal motor neurons, whereas others have more indirect connections. Because of the slower conduction velocity of type II sensory fibers and the signal relay through interneurons, the muscular responses elicited by group II fibers are smaller, more variable, and delayed compared to the monosynaptic stretch reflex. Some of the interneurons activated by group II fibers send

axons across the midline of the spinal cord and give rise to crossed reflexes. Such connections that cross the midline are important for coordination of bilateral muscle activity in functional motor tasks.

Golgi Tendon Organs Provide Force-Sensitive Feedback to the Spinal Cord

Stimulation of Golgi tendon organs or their Ib sensory fibers in passive animals produces disynaptic inhibition of homonymous motor neurons (*autogenic*

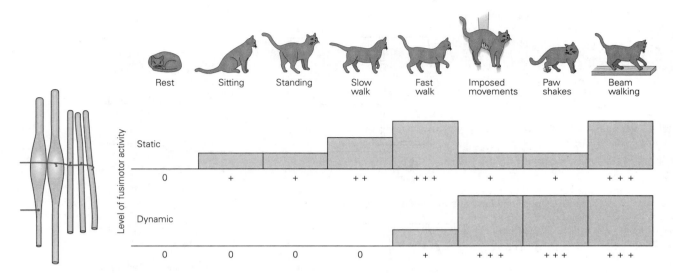

Figure 32–6 The level of activity in the fusimotor system varies with the type of behavior. Only static gamma motor neurons are active during activities in which muscle length changes slowly and predictably. Dynamic gamma motor neurons are activated during behaviors in which muscle length may change rapidly and unpredictably. (Adapted, with permission, from Prochazka et al. 1988.)

inhibition) and excitation of antagonist motor neurons (reciprocal excitation). Thus, these effects are the exact opposite of the responses evoked by muscle stretch or stimulation of Ia sensory axons.

This autogenic inhibition is mediated by *Ib inhibitory interneurons.* These inhibiting interneurons receive their principal input from Golgi tendon organs, sensory receptors that signal the tension in a muscle (Box 32–4), and they make inhibitory connections with homonymous motor neurons. However, stimulation of the Ib sensory fibers from tendon organs in active animals does not always inhibit homonymous motor neurons. Indeed, as we shall see later, stimulation of tendon organs may in certain conditions excite homonymous motor neurons.

One reason that the reflex actions of the sensory axons from tendon organs are complex in natural situations is that the Ib inhibitory interneurons also receive input from the muscle spindles, cutaneous receptors, and joint receptors (Figure 32–8A). In addition, they receive both excitatory and inhibitory input from various descending pathways.

Golgi tendon organs were first thought to have a protective function, preventing damage to muscle. It was assumed that they always inhibited homonymous motor neurons and that they fired only when tension in the muscle was high. We now know that these receptors signal minute changes in muscle tension, thus providing the nervous system with precise information about the state of a muscle's contraction.

The convergent sensory input from tendon organs, cutaneous receptors, and joint receptors to the Ib inhibitory interneurons (Figure 32–8A) may allow for precise spinal control of muscle force in activities such as grasping a delicate object. Additional input from cutaneous receptors may facilitate activity in the Ib inhibitory interneurons when the hand reaches an object, thus reducing the level of muscle contraction and permitting a soft grasp.

As is the case with the Ia fibers from muscle spindles, the Ib fibers from tendon organs form widespread connections with motor neurons that innervate muscles acting at different joints. Therefore, the connections of the sensory fibers from tendon organs with the Ib inhibitory interneurons are part of spinal networks that regulate movements of whole limbs.

Cutaneous Reflexes Produce Complex Movements That Serve Protective and Postural Functions

Most reflex pathways involve interneurons. One such reflex pathway is that of the flexion-withdrawal reflex, in which a limb is quickly withdrawn from a painful stimulus. Flexion-withdrawal is a protective reflex in which a discrete stimulus causes all the flexor muscles in that limb to contract coordinately. We know that this is a spinal reflex because it persists after complete transection of the spinal cord.

The sensory signal of the flexion-withdrawal reflex activates divergent polysynaptic reflex pathways. One

Box 32–4 Golgi Tendon Organs

Golgi tendon organs are slender encapsulated structures approximately 1 mm long and 0.1 mm in diameter located at the junction between skeletal muscle fibers and tendon. Each capsule encloses several braided collagen fibers connected in series to a group of muscle fibers.

Each tendon organ is innervated by a single Ib axon that branches into many fine endings inside the capsule; these endings become intertwined with the collagen fascicles (Figure 32–7A).

Stretching of the tendon organ straightens the collagen fibers, thus compressing the Ib nerve endings and causing them to fire. Because the nerve endings are so closely associated with the collagen fibers, even very small stretches of the tendons can compress the nerve endings.

Whereas muscle spindles are most sensitive to changes in length of a muscle, tendon organs are most sensitive to changes in muscle tension. Contraction of the muscle fibers connected to the collagen fiber bundle containing the receptor is a particularly potent stimulus to a tendon organ. The tendon organs are thus readily activated during normal movements. This has been demonstrated by recordings from single Ib axons in humans making voluntary finger movements and in cats walking normally.

Studies in anesthetized animal preparations have shown that the average level of activity in the population of tendon organs in a muscle is a good index of the total force in a contracting muscle (Figure 32–7B). This close agreement between firing frequency, and force is consistent with the view that the tendon organs continuously measure the force in a contracting muscle.

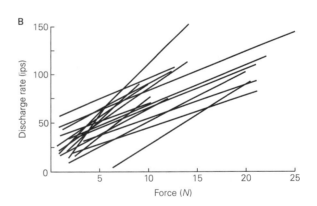

A

Ib afferent axon

Muscle fibers

Capsule

Ib axon

Collagen fiber

Tendon

250 µm

B

Discharge rate (ips)

Force (N)

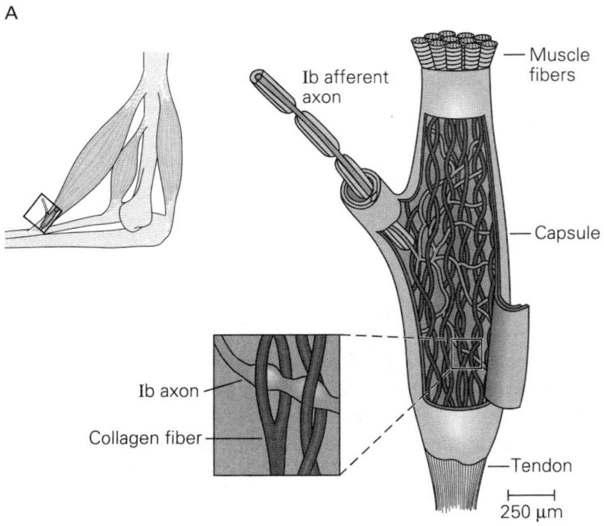

Figure 32–7A When the Golgi tendon organ is stretched (usually because of contraction of the muscle), the Ib afferent axon is compressed by collagen fibers (see enlargement) and its rate of firing increases. (Adapted, with permission, from Schmidt 1983; inset adapted, with permission, from Swett and Schoultz 1975.)

Figure 32–7B The discharge rate of a population of Golgi tendon organs signals the force in a muscle. Linear regression lines show the relationship between discharge rate and force for Golgi tendon organs of the soleus muscle of the cat. (Adapted, with permission, from Crago, Houk, and Rymer 1982.)

excites motor neurons that innervate flexor muscles of the stimulated limb, whereas another inhibits motor neurons that innervate the limb's extensor muscles (Figure 32–1B). This reflex can produce an opposite effect in the contralateral limb, that is, excitation of extensor motor neurons and inhibition of flexor motor neurons. This *crossed-extension reflex* serves to enhance postural support during withdrawal of a foot from a painful stimulus. Activation of the extensor muscles in the opposite leg counteracts the increased load caused by lifting the stimulated limb. Thus, flexion-withdrawal is a complete, albeit simple, motor act.

Although flexion reflexes are relatively stereotyped, both the spatial extent and the force of muscle contraction depend on stimulus intensity. Touching a stove that is slightly hot may produce moderately

A Convergence onto Ib interneurons

B Reversal of action of Ib afferents

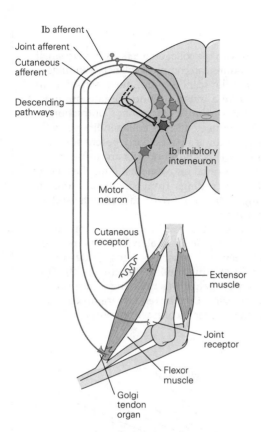

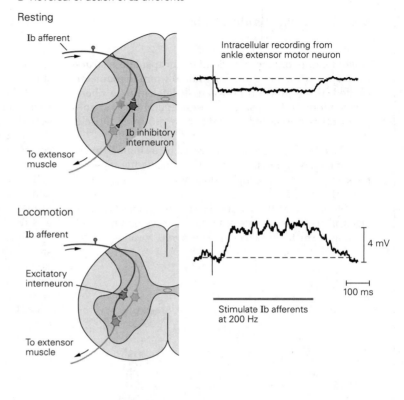

Figure 32–8 The reflex actions of Ib sensory fibers from Golgi tendon organs are modulated during locomotion.

A. The Ib inhibitory interneuron receives input from tendon organs, muscle spindles (not shown), joint and cutaneous receptors, and descending pathways.

B. The action of Ib sensory fibers on extensor motor neurons is reversed from inhibition to excitation when walking is initiated. When the animal is resting, stimulation of Ib fibers

from the ankle extensor muscle inhibits ankle extensor motor neurons through Ib inhibitory interneurons, as shown by the hyperpolarization in the record. During walking, the Ib inhibitory interneurons are inhibited while excitatory interneurons that receive input from Ib sensory fibers are facilitated by the command system for walking, thus opening a Ib excitatory pathway from the Golgi tendon organs to motor neurons.

fast withdrawal only at the wrist and elbow, whereas touching a very hot stove invariably leads to a forceful contraction at all joints, leading to rapid withdrawal of the entire limb. The duration of the reflex usually increases with stimulus intensity, and the contractions produced in a flexion reflex always outlast the stimulus.

Because of the similarity of the flexion-withdrawal reflex to stepping, it was once thought that the flexion reflex is important in producing contractions of flexor muscles during walking. We now know, however, that a major component of the neural control system for walking is a set of intrinsic spinal circuits that do not require sensory stimuli (Chapter 33). Nevertheless, in mammals, the intrinsic spinal circuits that control

walking share many of the interneurons involved in flexion reflexes.

Convergence of Sensory Inputs on Interneurons Increases the Flexibility of Reflex Contributions to Movement

The Ib inhibitory interneuron is not the only interneuron that receives convergent input from many different sensory modalities. An enormous diversity of sensory information converges on interneurons in the spinal cord, enabling them to integrate information from muscle, joints, and skin.

Interneurons activated by groups I and II sensory fibers have received special attention. It was thought for some time that excitatory and inhibitory interneurons activated by group II fibers could be distinguished from those activated by group Ib afferents, but it is now believed that this distinction has to be abandoned and that groups I and II fibers converge on common populations of interneurons that integrate force and length information from the active muscle and thereby help coordinate muscle activity according to the length of the muscle, its activity level, and the external load.

Sensory Feedback and Descending Motor Commands Interact at Common Spinal Neurons to Produce Voluntary Movements

As pointed out by Michael Foster in his 1879 physiology textbook, it must be an "economy to the body" that the will should make use of the networks in the spinal cord to generate coordinated movements "rather than it should have recourse to an apparatus of its own of a similar kind." Research in the subsequent 140 years has confirmed this conjecture.

The first evidence came from intracellular recordings of synaptic potentials elicited in cat spinal motor neurons by combined and separate stimulation of sensory fibers and descending pathways. When separate stimuli are reduced in intensity to just below threshold for evoking a synaptic potential, combining the stimulations at appropriate intervals makes the synaptic potential reappear. This provides evidence of convergence of the sensory fibers and the descending pathways onto common interneurons in the reflex pathway (see Figure 13–14). Direct recordings from spinal interneurons have confirmed this, as have noninvasive Hoffmann reflex tests in human subjects (Figure 32–9).

Direct evidence that sensory feedback helps to shape voluntary motor commands through spinal reflex networks in humans comes from experiments in which sensory activity in length- and force-sensitive afferents has suddenly been reduced or abolished. This can be done by suddenly unloading or shortening a muscle during a voluntary contraction. The short latency of the consequent reduction in muscle activity can only be explained by sensory activity through a reflex pathway that directly contributes to the muscle activity.

Muscle Spindle Sensory Afferent Activity Reinforces Central Commands for Movements Through the Ia Monosynaptic Reflex Pathway

Stretch reflex pathways can contribute to the regulation of motor neurons during voluntary movements and during maintenance of posture because they form closed feedback loops. For example, stretching a muscle increases activity in spindle sensory afferents, leading to muscle contraction and consequent shortening of the muscle. Muscle shortening in turn leads to decreased activity in spindle afferents, reduction of muscle contraction, and lengthening of the muscle.

The stretch reflex loop thus acts continuously—the output of the system, a change in muscle length, becomes the input—tending to keep the muscle close to a desired or reference length. The stretch reflex pathway is a negative feedback system, or *servomechanism*, because it tends to counteract or reduce deviations from the reference value of the regulated variable.

In 1963, Ragnar Granit proposed that the reference value in voluntary movements is set by descending signals that act on both alpha and gamma motor neurons. The rate of firing of alpha motor neurons is set to produce the desired shortening of the muscle, and the rate of firing of gamma motor neurons is set to produce an equivalent shortening of the intrafusal fibers of the muscle spindle. If the shortening of the whole muscle is less than what is required by a task, as when the load is greater than anticipated, the sensory fibers increase their firing rate because the contracting intrafusal fibers are stretched (loaded) by the relatively greater length of the whole muscle. If shortening is greater than necessary, the sensory fibers decrease their firing rate because the intrafusal fibers are relatively slackened (unloaded) (Figure 32–10A).

In theory, this mechanism could permit the nervous system to produce movements of a given distance without having to know in advance the actual load or weight being moved. In practice, however, the stretch reflex pathways do not have sufficient control over motor neurons to overcome large unexpected loads. This is immediately obvious if we consider what happens when we attempt to lift a heavy suitcase that we believe to be empty. Automatic compensation for the greater-than-anticipated load does not occur. Instead, we have to pause briefly to plan a new movement with much greater muscle activation.

Strong evidence that alpha and gamma motor neurons are co-activated during voluntary human movement comes from direct measurements of the activity of the sensory fibers from muscle spindles. In the late 1960s, Åke Vallbo and Karl-Erik Hagbarth developed microneurography, a technique for recording from the largest afferent fibers in peripheral nerves. Vallbo later found that during slow movements of the fingers the large-diameter Ia fibers from spindles in the contracting muscles increase their rate of firing even when the muscle shortens as it contracts (Figure 32–10B). This occurs because the gamma motor neurons, which have

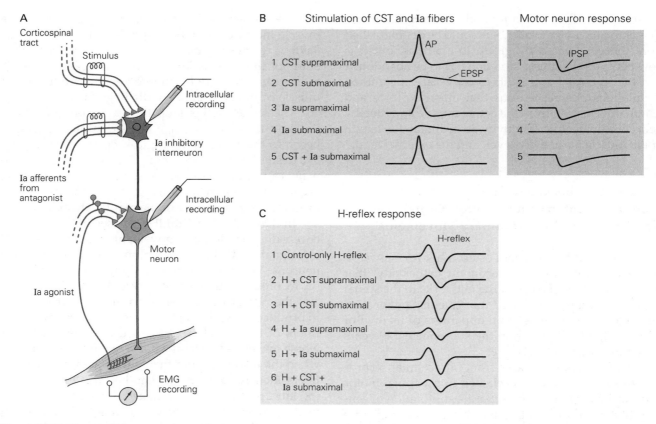

Figure 32–9 The spatial summation technique demonstrates how signals from descending inputs and spinal networks are integrated. This technique was introduced originally for investigation of spinal circuits in the cat in the 1950s, but it is also the basis of later investigations of the human spinal mechanisms of motor control. It relies on the spatial summation of synaptic inputs (see Figure 13–14), as illustrated here using the reciprocal Ia inhibitory pathway and corticospinal tract (CST).

A. The diagram shows the experimental setups for testing for convergence of excitatory reciprocal Ia and corticospinal pathways onto Ia inhibitory interneurons in the spinal cord.

B. In acute experiments on the cat spinal cord, supramaximal stimuli were applied separately to corticospinal fiber tracts (1) and Ia axons (3); each stimulus elicited an inhibitory postsynaptic potential (IPSP) in the motor neuron. Next, the intensities of the two stimuli were reduced to just submaximal levels, at which point each pathway failed to elicit an IPSP in the motor neuron (2, 4). Then, when the two sets of submaximal stimuli were paired, they elicited an IPSP in the motor neuron (5), leading to the conclusion that the two input pathways converge on

the same interneurons. This was confirmed by direct recording from a Ia inhibitory interneuron. (**AP,** action potential).

C. In humans, direct intracellular recording from interneurons and motor neurons is not possible, but recording of H-reflexes (Box 32–4, Figure 32–4) and transcutaneous stimulation of the corticospinal tract have provided indirect evidence for convergence similar to that demonstrated in cats (see part B). The electromyogram (**EMG**) record of the H-reflex provides a measure of the excitability of the spinal motor neurons (1). When the CST and antagonist Ia fibers were stimulated separately at supramaximal levels, the H-reflex amplitude was diminished due to the compound IPSPs elicited in the motor neurons (2, 4). Next, the stimuli to these two excitatory pathways to the inhibitory interneurons were reduced until neither stimulus alone elicited a reduction in amplitude of the H-reflex (3, 5). Then, the two submaximal stimuli were timed to produce synchronous subthreshold excitatory postsynaptic potentials (**EPSPs**) in the inhibitory interneurons (6). Because this protocol caused suppression of the H-reflex, one may conclude that the CST and Ia afferents converge on the same Ia inhibitory interneurons.

direct excitatory connections with spindles, are co-activated with alpha motor neurons.

Furthermore, when subjects attempt to make slow movements at a constant velocity, the firing of the Ia fibers mirrors the small deviations in velocity in the trajectory of the movements (sometimes the muscle

shortens quickly and at other times more slowly). When the velocity of flexion increases transiently, the rate of firing in the fibers decreases because the muscle is shortening more rapidly and therefore exerts less tension on the intrafusal fibers. When the velocity decreases, firing increases because the muscle is

A Alpha-gamma co-activation reinforces alpha motor activity

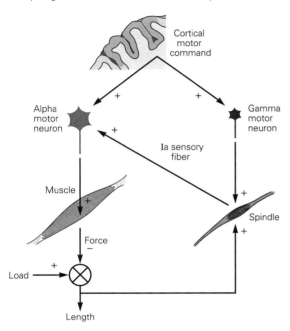

B Spindle activity increases during muscle shortening

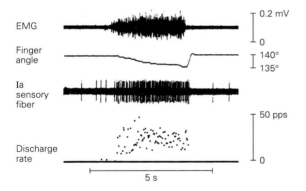

Figure 32–10 Co-activation of alpha and gamma motor neurons.

A. Co-activation of alpha and gamma motor neurons by a cortical motor command allows feedback from muscle spindles to reinforce activation in the alpha motor neurons. Any disturbance during a movement alters the length of the muscle and thus changes the activity in the sensory fibers from the spindles. The changed spindle input to the alpha motor neuron compensates for the disturbance.

B. The discharge rate in the Ia sensory fiber from a spindle increases during slow flexion of a finger. This increase depends on alpha-gamma co-activation. If the gamma motor neurons were not active, the spindle would slacken, and its discharge rate would decrease as the muscle shortened. (**EMG**, electromyogram; **PPS**, pulses/s) (Adapted, with permission, from Vallbo 1981.)

shortening more slowly, and therefore, the relative tension on the intrafusal fibers increases. This information can be used by the nervous system to compensate for irregularities in the movement trajectory by exciting the alpha motor neurons.

Modulation of Ia inhibitory Interneurons and Renshaw Cells by Descending Inputs Coordinate Muscle Activity at Joints

Reciprocal innervation is useful not only in stretch reflexes but also in voluntary movements. Relaxation of the antagonist muscle during a movement enhances speed and efficiency because the muscles that act as prime movers are not working against the contraction of opposing muscles.

The Ia inhibitory interneurons receive inputs from collaterals of the axons of neurons in the motor cortex that make direct excitatory connections with spinal motor neurons. This organizational feature simplifies the control of voluntary movements, because higher centers do not have to send separate commands to the opposing muscles.

It is sometimes advantageous to contract both the prime mover and the antagonist at the same time. Such *co-contraction* has the effect of stiffening the joint and is most useful when precision and joint stabilization are critical. An example of this phenomenon is the co-contraction of flexor and extensor muscles of the elbow immediately before catching a ball. The Ia inhibitory interneurons receive both excitatory and inhibitory signals from all of the major descending pathways (Figure 32–11A). By changing the balance of excitatory and inhibitory inputs onto these interneurons, supraspinal centers can modulate reciprocal inhibition of muscles and enable co-contraction, thus controlling the relative amount of joint stiffness to meet the requirements of the motor act.

The activity of spinal motor neurons is also regulated by another important class of inhibitory interneurons, the *Renshaw cells*. Excited by collaterals of the axons of motor neurons and receiving significant synaptic input from descending pathways, Renshaw cells make inhibitory synaptic connections with several populations of motor neurons, including the motor neurons that excite them, as well as Ia inhibitory interneurons (Figure 32–11B). The connections with motor neurons form a negative feedback system that regulates the firing rate of the motor neurons, whereas the connections with the Ia inhibitory interneurons regulate the strength of inhibition of antagonistic motor neurons, for instance in relation to co-contraction of antagonists. The distribution of projections from Renshaw

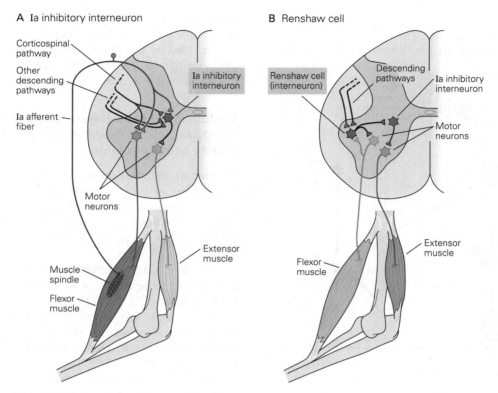

A Ia inhibitory interneuron

B Renshaw cell

Figure 32–11 Inhibitory spinal interneurons coordinate reflex actions.

A. The Ia inhibitory interneuron regulates contraction in antagonist muscles in stretch reflex circuits through its divergent contacts with motor neurons. In addition, the interneuron receives excitatory and inhibitory inputs from corticospinal and other descending pathways. A change in the balance of these supraspinal signals allows the interneuron to coordinate co-contractions in antagonist muscles at a joint.

B. The Renshaw cell produces recurrent inhibition of motor neurons. These interneurons are excited by collaterals from motor neurons and inhibit those same motor neurons. This negative feedback system regulates motor neuron excitability and stabilizes firing rates. Renshaw cells also send collaterals to synergist motor neurons (not shown) and Ia inhibitory interneurons that synapse on antagonist motor neurons. Thus, descending inputs that modulate the excitability of the Renshaw cells adjust the excitability of all the motor neurons that control movement around a joint.

cells to different motor nuclei also facilitate that muscle activity is coordinated in functional synergies during movement.

Transmission in Reflex Pathways May Be Facilitated or Inhibited by Descending Motor Commands

As we have seen, in an animal at rest, the Ib sensory fibers from extensor muscles have an inhibitory effect on homonymous motor neurons. During locomotion, they produce an excitatory effect on those same motor neurons because transmission in the disynaptic inhibitory pathway is depressed (Figure 32–8B), while at the same time transmission through excitatory interneurons is facilitated.

This phenomenon, called *state-dependent reflex reversal*, illustrates how transmission in spinal circuit is regulated by descending motor commands to meet the changing requirements during movement. By favoring transmission through excitatory pathways from the load-sensitive Golgi tendon organs, the descending motor commands ensure that feedback from the active muscles automatically facilitates the activation of the muscles, thereby greatly simplifying the task for supraspinal centers.

State-dependent reflex reversal has also been demonstrated in humans. Stimulation of skin and muscle afferents from the foot produces facilitation of muscles that lift the foot early in the swing phase, but suppresses activity of the same muscles late in the swing phase. Both effects make good functional sense. Early in the swing phase, positive feedback from the foot will help to lift the foot over an obstacle, whereas suppression of the same muscles in late swing will help to lower the foot quickly to the ground so that the obstacle may be passed using the opposite leg first.

Descending Inputs Modulate Sensory Input to the Spinal Cord by Changing the Synaptic Efficiency of Primary Sensory Fibers

In the 1950s and early 1960s, John C. Eccles and his collaborators demonstrated that monosynaptic excitatory postsynaptic potentials (EPSPs) elicited in cat spinal motor neurons by stimulation of Ia sensory fibers become smaller when other Ia fibers are stimulated. This led to the discovery in the spinal cord of several groups of GABAergic inhibitory interneurons that exert presynaptic inhibition of primary sensory neurons (Figure 32–12). Some interneurons inhibit mainly Ia sensory axons, whereas others inhibit mainly Ib axons or sensory fibers from skin.

The principal mechanism responsible for sensory inhibition is a depolarization of the primary terminal caused by an inward Cl⁻ current when GABAergic receptors on the terminal are activated. This depolarization inactivates some of the Na⁺ channels in the terminal, so the action potentials that reach the synapse are reduced in size. The effect of this is that release of neurotransmitter from the sensory afferent is diminished.

When tested by stimulation of peripheral afferents, presynaptic inhibition is widespread in the spinal cord and affects primary afferents from all muscles in a limb. However, similar to other interneurons, the neurons responsible for presynaptic inhibition are also controlled by descending pathways, making possible a much more focused modulation of presynaptic inhibition in relation to movement. Presynaptic inhibition at the synapse of Ia axons with motor neurons of the muscles that are activated as part of a movement is reduced at the onset of movement. In contrast, presynaptic inhibition of Ia axons on motor neurons connected to inactive muscles is increased. One example of this selective modulation is increased presynaptic inhibition of Ia axons at their synapse with antagonist motor neurons, which explains part of the reduction of stretch reflexes in antagonist muscles at the onset of agonist contraction. In this way, the nervous system takes advantage of the widespread connectivity of Ia axons, using presynaptic inhibition to shape activity in the Ia afferent network to facilitate activation of specific muscles.

Presynaptic inhibition provides a mechanism by which the nervous system can reduce sensory feedback predicted by the motor command, while allowing unexpected feedback access to the spinal motor circuit and the rest of the nervous system. In line with this function, presynaptic inhibition of Ia sensory axons from muscle spindles generally increases during

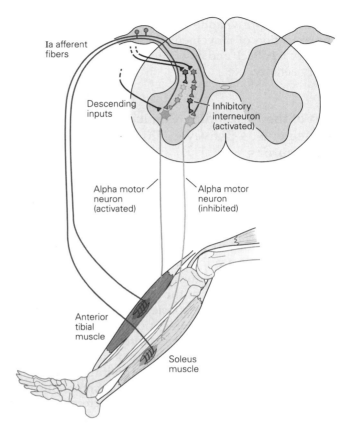

Figure 32–12 Selective modulation of primary sensory axon terminals by descending presynaptic inhibitory inputs contributes to generation of coordinated limb movements. Inhibitory interneurons (**blue**) activated by descending inputs can have either pre- or postsynaptic effects. Some interneurons releasing the inhibitory neurotransmitter γ-aminobutyric acid (GABA) form axo-axonic synapses with the primary sensory fibers. The principal inhibitory mechanism involves activation of GABAergic receptors on the terminals of the presynaptic Ia sensory axons, resulting in depolarization of the terminals and reduced transmitter release. Such presynaptic inhibition is widely distributed in the spinal cord. Stimulation of Ia sensory fibers from one flexor muscle may elicit presynaptic inhibition of both flexor and extensor Ia axon terminals on motor neurons innervating muscles throughout the limb. However, several different populations of interneurons mediating presynaptic inhibition exist, which allows a very specific regulation of presynaptic inhibition in relation to voluntary movements. Interaction of sensory inputs with descending motor commands in the corticospinal tract may thus *decrease* presynaptic inhibition of Ia axon terminals on agonist motor neurons (eg, anterior tibial motor neurons) and at the same time *increase* presynaptic inhibition of Ia terminals on antagonist motor neurons (eg, soleus motor neurons). Regulation of presynaptic inhibition may thus simultaneously facilitate the sensory feedback to the activation of agonist motor neurons and at the same time diminish the risk that stretch of the antagonist muscles will elicit a stretch reflex that would counteract the movement.

movements that are highly predictable, such as walking and running.

Finally, presynaptic inhibition may help stabilize the execution of movements by preventing excessive sensory feedback and associated self-reinforcing oscillatory activity.

Part of the Descending Command for Voluntary Movements Is Conveyed Through Spinal Interneurons

In cats as well as most other vertebrates, the corticospinal tract has no direct connections to spinal motor neurons; all the descending commands have to be channeled through spinal interneurons that are also part of reflex pathways. Humans and Old World monkeys are the only species in which corticospinal neurons make direct connections with the spinal motor neurons in the ventral horn of the spinal cord. Even in these species, a considerable fraction of the corticospinal tract fibers terminate in the intermediate nucleus on spinal interneurons, and the corticospinal fibers that terminate on motor neurons also have collaterals that synapse on interneurons. A considerable part of each descending command for movement in the corticospinal tract therefore has to be conveyed through spinal interneurons—and integrated with sensory activity—before reaching the motor neurons.

Propriospinal Neurons in the C3–C4 Segments Mediate Part of the Corticospinal Command for Movement of the Upper Limb

In the 1970s, Anders Lundberg and his collaborators demonstrated that a group of neurons in the C3–C4 spinal segments of the cat spinal cord send their axons to motor neurons located in more caudal cervical segments (Figure 32–13). Since the neurons in the C3–C4 segments project to motor neurons that innervate a range of forelimb muscles controlling different joints, and receive input from both skin and muscles throughout the forelimb, they are named *propriospinal neurons*. In addition to sensory input from skin and muscle afferents, the C3–C4 propriospinal neurons are activated by collaterals from the corticospinal tract and thereby relay disynaptic excitation from the motor cortex to the spinal motor neurons.

Subsequent experiments by Bror Alstermark in Sweden and Tadashi Isa in Japan have confirmed that similar propriospinal neurons also exist in the C3–C4

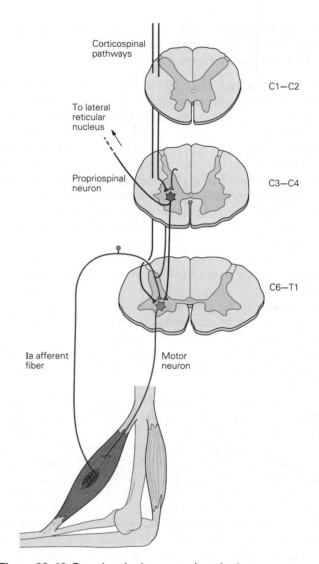

Figure 32–13 Propriospinal neurons in spinal segments C3–C4 mediate part of the descending motor command to cervical motor neurons. Some corticospinal fibers (**green**) send collaterals to propriospinal neurons in the C3–C4 segments (**blue**). These C3–C4 propriospinal neurons project to motor neurons located in more caudal cervical segments. They also receive excitatory input from muscle afferents and send collaterals to the lateral reticular nucleus.

segments of the monkey spinal cord and are involved in mediating at least part of the motor command for reaching. Noninvasive experiments have also provided indirect evidence of the existence of C3–C4 propriospinal neurons in the human spinal cord. With the evolution of direct monosynaptic corticomotor connections in monkeys and humans, the corticospinal transmission through this disynaptic pathway may have become less important.

Lumbar interneurons that receive input from groups I and II sensory axons from muscle also receive significant input from descending motor tracts and provide excitatory projections to spinal motor neurons. These interneurons thus convey part of the indirect motor command for voluntary movements to the spinal motor neurons that control leg muscles and may be a lumbar equivalent of the C3–C4 propriospinal neurons in the cervical spinal cord.

Neurons in Spinal Reflex Pathways Are Activated Prior to Movement

Synaptic transmission in spinal reflex pathways may change in response to the intention to move, independent of movement. Intracellular recordings from active monkeys have demonstrated that the intention to make a movement modifies activity in interneurons in the spinal cord and alters transmission in spinal reflex pathways. Similarly, in human subjects who have been prevented from contracting a muscle (by injection of lidocaine into the peripheral nerve supplying the muscle), the voluntary effort to contract the muscle still changes transmission in reflex pathways as if the movement had actually taken place.

In both humans and monkeys, spinal interneurons also change their activity well in advance of the actual movement. For example, in human subjects, Hoffmann reflexes elicited in a muscle that is about to be activated are facilitated fully 50 ms prior to the onset of contraction and remain facilitated throughout the movement. Conversely, reflexes in the antagonist muscles are suppressed. The suppression of stretch reflexes in the antagonist muscle prior to the onset of movement is an efficient way of preventing the antagonist from being reflexively activated when it is stretched at the onset of the agonist contraction.

Transmission in spinal reflex pathways can also be modified in connection with higher cognitive functions. Two examples are (1) an increase in the tendon jerk reflex in the soleus muscle of a human subject imagining pressing a foot pedal and (2) modulation of the Hoffmann reflex in arm and leg muscles while a subject observes grasping and walking movements, respectively.

Proprioceptive Reflexes Play an Important Role in Regulating Both Voluntary and Automatic Movements

All movements activate receptors in muscles, joints, and skin. Sensory signals generated by the body's own movements were termed *proprioceptive* by Sherrington, who proposed that they control important aspects of normal movements. A good example is the Hering-Breuer reflex, which regulates the amplitude of inspiration. Stretch receptors in the lungs are activated during inspiration, and the Hering-Breuer reflex eventually triggers the transition from inspiration to expiration when the lungs are expanded.

A similar situation exists in the walking systems of many animals; sensory signals generated near the end of the stance phase initiate the onset of the swing phase (Chapter 33). Proprioceptive signals can also contribute to the regulation of motor activity during voluntary movements, as shown in studies of individuals with sensory neuropathy of the arms. These patients display abnormal reaching movements and have difficulty in positioning the limb accurately because the lack of proprioception results in a failure to compensate for the complex inertial properties of the human arm.

Therefore, a primary function of proprioceptive reflexes in regulating voluntary movements is to adjust the motor output according to the changing biomechanical state of the body and limbs. This adjustment ensures a coordinated pattern of motor activity during an evolving movement and compensates for the intrinsic variability of motor output.

Spinal Reflex Pathways Undergo Long-Term Changes

Transmission in spinal reflex pathways is modulated not only to suit the immediate requirements of the movement but also to adapt the motor command to the motor experience of the individual. For example, transmission in the reciprocal Ia inhibitory pathway shows a gradual change when subjects improve their ability in coordinating agonist and antagonist contraction. Inactivity following long periods of bedrest or immobilization also results in changes in stretch reflexes and H-reflexes. Conversely, the soleus stretch reflex is low in highly trained ballet dancers and varies among different kinds of athletes.

Extensive studies of humans, monkeys, and rats by Jonathan Wolpaw and his colleagues have found that stretch reflexes can be operantly conditioned to either increase or decrease. The mechanisms underlying these changes are complex and involve alterations at multiple sites including changes in the properties of motor neurons. A general prerequisite for these changes is that corticospinal control of the spinal motor circuits must be intact.

Damage to the Central Nervous System Produces Characteristic Alterations in Reflex Responses

Stretch reflexes are routinely used in clinical examinations of patients with neurological disorders. They are typically elicited by sharply tapping the tendon of a muscle with a reflex hammer. Although the responses are often called tendon reflexes or tendon jerks, the receptor that is stimulated, the muscle spindle, actually lies in the muscle rather than the tendon. Only the primary sensory fibers in the spindle participate in the tendon reflex, for these are selectively activated by a rapid stretch of the muscle produced by the tendon tap.

Measuring alterations in the strength of the stretch reflex can assist in the diagnosis of certain conditions and in localizing injury or disease in the central nervous system. Absent or hypoactive stretch reflexes often indicate a disorder of one or more of the components of the peripheral reflex pathway: sensory or motor axons, the cell bodies of motor neurons, or the muscle itself (Chapter 57). Nevertheless, because the excitability of motor neurons is dependent on descending excitatory and inhibitory signals, absent or hypoactive stretch reflexes can also result from lesions of the central nervous system. Hyperactive stretch reflexes, conversely, always indicate that the lesion is in the central nervous system.

Interruption of Descending Pathways to the Spinal Cord Frequently Produces Spasticity

The force with which a muscle resists being lengthened depends on the muscle's intrinsic elasticity, or stiffness. Because a muscle has elastic elements in series and parallel that resist lengthening, it behaves like a spring (Chapter 31). In addition, connective tissue in and around the muscle may also contribute to its stiffness. These elastic elements may be pathologically altered following brain and spinal cord injury and thereby cause contractures and abnormal joint positions. However, there is also a neural contribution to the resistance of a muscle to stretch; the feedback loop inherent in the stretch reflex pathway acts to resist lengthening of the muscle.

Spasticity is characterized by hyperactive tendon jerks and an increase in resistance to rapid stretching of the muscle. Slow movement of a joint elicits only passive resistance, which is caused by the elastic properties of the joint, tendon, muscle, and connective tissues. As the speed of the stretch is increased, resistance to the stretch rises progressively. This phasic relation

is what characterizes spasticity; an active reflex contraction occurs only during a rapid stretch, and when the muscle is held in a lengthened position, the reflex contraction subsides.

Spasticity is seen following lesion of descending motor pathways caused by stroke, injuries of the brain or spinal cord, and degenerative diseases such as multiple sclerosis. It is also seen in individuals with brain damage that occurs before, during, or shortly after birth, resulting in *cerebral palsy*.

Spasticity is not seen immediately following lesions of descending pathways, but develops over days, weeks, and even months. This parallels plastic changes at multiple sites in the stretch reflex circuitry. Sensory group Ia axons release more transmitter substance when active, and the alpha motor neurons change their intrinsic properties and their morphology (dendritic sprouting and denervation hypersensitivity) so that they become more excitable. Changes in excitatory and inhibitory interneurons that project to the motor neurons also take place and probably contribute to the increased excitability.

Whatever the precise mechanisms that produce spasticity, the effect is a strong facilitation of transmission in the monosynaptic reflex pathway. It is not the only reflex pathway affected by lesions of descending motor pathways. Pathways involving group I/II interneurons and sensory fibers from skin are also affected and exhibit the symptomatology observed in patients with central motor lesions. In the clinic, spasticity is therefore used in a broader sense and does not only relate to stretch reflex hyperexcitability. It is still debated whether reflex hyperexcitability contributes to the movement disorder following lesion of descending pathways or whether it may be a pertinent adaptation that helps to activate the muscles when descending input is diminished.

Lesion of the Spinal Cord in Humans Leads to a Period of Spinal Shock Followed by Hyperreflexia

Damage to the spinal cord can cause large changes in the strength of spinal reflexes. Each year, approximately 11,000 Americans sustain spinal cord injuries, and many more suffer from strokes. More than half of these injuries produce permanent disability, including impairment of motor and sensory functions and loss of voluntary bowel and bladder control. Approximately 250,000 people in the United States today have some permanent disability from spinal cord injury.

When the spinal cord is completely transected, there is usually a period immediately after the injury when all spinal reflexes below the level of the transection are reduced or completely suppressed, a condition known as *spinal shock*. During the course of weeks and months, spinal reflexes gradually return, often greatly exaggerated. For example, a light touch to the skin of the foot may elicit strong flexion withdrawal of the leg.

Highlights

1. Reflexes are coordinated, involuntary motor responses initiated by a stimulus applied to peripheral receptors.
2. Many groups of interneurons in spinal reflex pathways are also involved in producing complex movements such as walking and transmitting voluntary commands from the brain.
3. Some components of reflex responses, particularly those involving the limbs, are mediated by supraspinal centers, such as brain stem nuclei, the cerebellum, and the motor cortex.
4. Reflexes are smoothly integrated into centrally generated motor commands because of the convergence of sensory signals onto spinal and supraspinal interneuronal systems involved in initiating movements. Establishing the details of these integrative events is one of the major challenges of contemporary research on sensory-motor integration in the spinal cord.
5. Because of the role of supraspinal centers in spinal reflex pathways, injury to or disease of the central nervous system often results in significant alterations in the strength of spinal reflexes. The pattern of changes provides an important aid to diagnosis of patients with neurological disorders.

Jens Bo Nielsen
Thomas M. Jessell

Selected Reading

Alstermark B, Isa T. 2012. Circuits for skilled reaching and grasping. Annu Rev Neurosci 35:559–578.

Baldissera F, Hultborn H, Illert M. 1981. Integration in spinal neuronal systems. In: JM Brookhart, VB Mountcastle, VB Brooks, SR Geiger (eds). *Handbook of Physiology: The Nervous System*, pp. 509–595. Bethesda, MD: American Physiological Society.

Boyd IA. 1980. The isolated mammalian muscle spindle. Trends Neurosci 3:258–265.

Fetz EE, Perlmutter SI, Orut Y. 2000. Functions of spinal interneurons during movement. Curr Opin Neurobiol 10:699–707.

Jankowska E. 1992. Interneuronal relay in spinal pathways from proprioceptors. Prog Neurobiol 38:335–378.

Nielsen JB. 2016. Human spinal motor control. Annu Rev Neurosci 39:81–101.

Pierrot-Deseilligny E, Burke D. 2005. *The Circuitry of the Human Spinal Cord. Its Role in Motor Control and Movement Disorders*. Cambridge: Cambridge Univ. Press.

Prochazka A. 1996. Proprioceptive feedback and movement regulation. In: L Rowell, JT Sheperd (eds). *Handbook of Physiology: Regulation and Integration of Multiple Systems*, pp. 89–127. New York: American Physiological Society.

Windhorst U. 2007. Muscle proprioceptive feedback and spinal networks. Brain Res Bull 73:155–202.

Wolpaw JR. 2007. Spinal cord plasticity in acquisition and maintenance of motor skills. Acta Physiol (Oxf) 189:155–169.

References

Appenteng K, Prochazka A. 1984. Tendon organ firing during active muscle lengthening in normal cats. J Physiol (Lond) 353:81–92.

Brown MC, Matthews PBC. 1966. On the sub-division of the efferent fibres to muscle spindles into static and dynamic fusimotor fibres. In: BL Andrew (ed). *Control and Innervation of Skeletal Muscle*, pp. 18–31. Dundee, Scotland: University of St. Andrews.

Crago A, Houk JC, Rymer WZ. 1982. Sampling of total muscle force by tendon organs. J Neurophysiol 47:1069–1083.

Gossard JP, Brownstone RM, Barajon I, Hultborn H. 1994. Transmission in a locomotor-related group Ib pathway from hind limb extensor muscles in the cat. Exp Brain Res 98:213–228.

Granit R. 1970. *Basis of Motor Control*. London: Academic.

Hagbarth KE, Kunesch EJ, Nordin M, Schmidt R, Wallin EU. 1986. Gamma loop contributing to maximal voluntary contractions in man. J Physiol (Lond) 380:575–591.

Hoffmann P. 1922. *Untersuchungen über die Eigenreflexe (Sehnenreflexe) menschlicher Muskeln*. Berlin: Springer.

Hulliger M. 1984. The mammalian muscle spindle and its central control. Rev Physiol Biochem Pharmacol 101:1–110.

Hunt CC, Kuffler SW. 1951. Stretch receptor discharges during muscle contraction. J Physiol (Lond) 113:298–315.

Liddell EGT, Sherrington C. 1924. Reflexes in response to stretch (myotatic reflexes). Proc R Soc Lond B Biol Sci 96:212–242.

Marsden CD, Merton PA, Morton HB. 1981. Human postural responses. Brain 104:513–534.

Matthews PBC. 1972. *Muscle Receptors*. London: Edward Arnold.

Mendell LM, Henneman E. 1971. Terminals of single Ia fibers: location, density, and distribution within a pool of 300 homonymous motoneurons. J Neurophysiol 34:171–187.

Pearson KG, Collins DF. 1993. Reversal of the influence of group Ib afferents from plantaris on activity in model gastrocnemius activity during locomotor activity. J Neurophysiol 70:1009–1017.

Prochazka A, Hulliger M, Trend P, Dürmüller N. 1988. Dynamic and static fusimotor set in various behavioural contexts. In: P Hnik, T Soukup, R Vejsada, J Zelena (eds). *Mechanoreceptors: Development, Structure and Function*, pp. 417–430. New York: Plenum.

Schieppati M. 1987. The Hoffmann reflex: a means of assessing spinal reflex excitability and its descending control in man. Prog Neurobiol 28:345–376.

Schmidt RF. 1983. Motor systems. In: RF Schmidt, G Thews (eds), MA Biederman-Thorson (transl). *Human Physiology*, pp. 81–110. Berlin: Springer.

Sherrington CS. 1906. *Integrative Actions of the Nervous System*. New Haven, CT: Yale Univ. Press.

Swett JE, Schoultz TW. 1975. Mechanical transduction in the Golgi tendon organ: a hypothesis. Arch Ital Biol 113:374–382.

Vallbo ÅB. 1981. Basic patterns of muscle spindle discharge in man. In: A Taylor, A Prochazka (eds). *Muscle Receptors and Movement*, pp. 263–275. London: Macmillan.

Vallbo ÅB, Hagbarth KE, Torebjörk HE, Wallin BG. 1979. Somatosensory, proprioceptive, and sympathetic activity in human peripheral nerves. Physiol Rev 59:919–957.

Wickens DD. 1938. The transference of conditioned excitation and conditioned inhibition form one muscle group to the antagonist muscle group. J Exp Psychol 22:101–123.

33

Locomotion

LOCOMOTION IS ONE OF THE MOST FUNDAMENTAL of animal behaviors and is common to all members of the animal kingdom. As one might expect of such an essential behavior, the neural mechanisms responsible for the basic alternating rhythmicity that underlies locomotion are highly conserved throughout the animal kingdom, from invertebrates to vertebrates, and from the early vertebrates to primates. However, while the basic locomotor-generating circuits have been conserved, the evolution of limbs, and then of ever more complex patterns of behavior, has resulted in the development of progressively more complex spinal and supraspinal circuits (Figure 33–1).

Scientists have been intrigued with the neural mechanisms of locomotion since the beginning of the 20th century, when pioneering work by Charles Sherrington and Thomas Graham Brown showed that the isolated spinal cord of the cat is able to generate the basic aspects of locomotor activity and subsequently that this capacity was intrinsic to the spinal cord. Throughout the 20th century, major advances were made in detailing both the rhythm- and pattern-producing capacities of the spinal cord, leading ultimately to the groundbreaking concept of a central pattern generator for locomotion in the spinal cord. This single concept, more than any other, has driven research into the mechanisms underlying locomotor control since the 1970s, allowing a detailed electrophysiological examination of the neuronal mechanisms involved in the control of locomotion that is not possible for most other motor acts.

Most research throughout the 20th century on the spinal mechanisms mediating locomotion was performed on the cat, which remains an important model

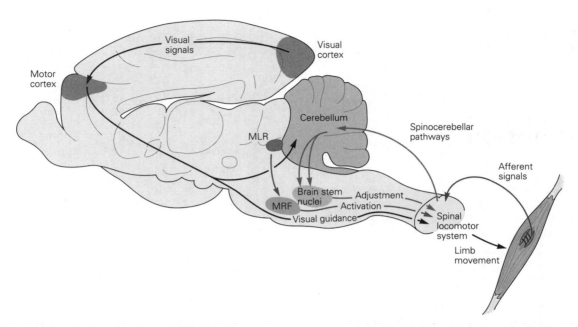

Figure 33–1 The locomotor system. Multiple regions of the central nervous system interact to initiate and regulate locomotion. Locomotor networks in the spinal cord—the central pattern generators (CPGs)—generate the precise timing and patterning of locomotion. Proprioceptive sensory feedback modulates the activity of the locomotor CPG. The initiation of locomotion is mediated by neurons in the mesencephalic locomotor region (**MLR**) that project to neurons in the medial reticular formation (**MRF**) in the lower brain stem, which in turn project to the spinal cord. Descending fibers from the vestibular nuclei, pontomedullary reticular formation, and the red nucleus (**brain stem nuclei**) maintain equilibrium and modulate the ongoing locomotor activity. Cortical activity from the posterior parietal cortex (not illustrated) and the motor cortex is involved in the planning and execution of visually guided locomotion, while the basal ganglia (not illustrated) and cerebellum are important for the selection and coordination of locomotor activity.

for studying many aspects of locomotor control. However, the complexity of the spinal circuits in mammals led to the search for simpler preparations that would allow a better understanding of the synaptic connectivity and neuronal properties responsible for the generation of locomotion. This search led to the development of the lamprey and the tadpole models (Box 33–1; Figures 33–2 and 33–3). Experiments using these species have led to a detailed understanding of the neuronal circuits responsible for generating swimming. Influential work on understanding the processes underlying locomotion has also come from other experimental models, including mouse, rat, turtle, salamander, and zebrafish.

More recently, the development of molecular-genetic techniques has provided a powerful tool to probe the spinal circuits responsible for locomotion in preparations as diverse as zebrafish and mouse. These techniques have allowed researchers to explore more thoroughly both the neuronal circuits in the mammalian spinal cord responsible for rhythmic, alternating patterns of activity that define over-ground locomotion and those responsible for swimming.

The rhythmic pattern of activity is only one element of the complex locomotor behavior observed in most vertebrates, especially mammals, which have evolved to allow them to move quickly and elegantly. This flexibility is provided via feedback and feedforward modification of the locomotor patterns generated by spinal networks.

Feedback information from the body and limbs in the form of cutaneous and proprioceptive inputs is important for regulating aspects of the locomotor cycle, including bending of the body, stride length, and the force produced during propulsion. This information is equally critical in assuring that animals can rapidly and efficiently react to unexpected perturbations in the environment, such as when hitting a branch during walking or stepping on an unstable surface.

Feedforward information from supraspinal systems modifies activity according to the goals of the animal and the environment in which it moves. Information from defined structures in the brain stem is important for both the initiation of locomotion and for regulating general aspects of locomotor activity, including the speed of locomotion, level of muscle activity, and interlimb coupling in animals with limbs. Information from cortical structures

Box 33–1 Preparations Used to Study the Neuronal Control of Locomotion

The neuronal control of locomotion is studied experimentally in diverse vertebrate species that produce swimming or over-ground locomotion, or both. The prevailing experimental models used for studying swimming are the lamprey, the tadpole, and the zebrafish; for over-ground locomotion, the cat, rat, or mouse; and for both swimming and locomotion, the turtle, salamander, and frog.

Semi-intact preparations—in which influences from parts of the brain, all supraspinal inputs, and/or afferent inputs to the spinal cord have been removed—are also commonly used in studies of the neuronal control of locomotion in vertebrates (Figure 33–2A). Finally, in vitro preparations of the spinal cord or of the brain stem and spinal cord from young animals or adult and anoxia-resistant animals are extensively used for circuit analysis (Figure 33–2C).

Intact Preparations Are Used to Study the Behavioral Output

In intact preparations, locomotion is studied either during walking over ground or on a motorized treadmill. Chronic electromyographic (EMG) recordings of limb muscles, coupled with video recordings of the movement, reveal details of the rhythm of locomotion, the pattern of muscle or joint activation, and interlimb coordination (Figure 33–2B). Such studies allow researchers to understand how normal locomotion behavior is expressed.

These behavioral studies are often combined with experimental manipulations that modify the supraspinal or afferent control of locomotion. Such experiments may use electrical stimulation or surgical ablation of circumscribed areas in the central nervous system, genetic inactivation or activation of defined populations of nerve cells, or perturbation of the afferent input to the spinal cord using genetic techniques or electrical stimulation. Finally, single-cell activity in the brain can be recorded from identified populations of neurons and correlated with specific aspects of the locomotor behavior (eg, speed, postural adjustments, gait modifications, flexor-extensor muscle activity). Cells are identified by their anatomical location, their projection pattern, transmitter content, and molecular markers.

Semi-intact Preparations Are Commonly Used to Study the Central Control of Locomotion in the Absence of Cortical Influence or Sensory Feedback

Decerebrate Preparations

In the decerebrate preparation, the brain stem is completely transected at the level of the midbrain (Figure 33–2A), disconnecting rostral brain centers, including the cortex, basal ganglia, and thalamus, from locomotor-initiating centers in the brain stem and spinal cord. These preparations allow investigation of the role of cerebellum and brain stem structures in controlling locomotion in the absence of influence from higher brain centers.

Locomotion is generally evoked by electrical stimulation of locomotor regions in the brain stem, as described in the text. To increase recording stability, the animals are often paralyzed by blocking transmission at the neuromuscular junction. When locomotion is initiated in such an immobilized preparation, often referred to as *fictive locomotion*, the motor nerves to flexors and extensor muscles discharge (recorded as an electroneurogram), but no movement takes place.

Spinal Preparations

In spinal preparations, the spinal cord is completely transected, generally at the lower thoracic level, thus isolating the spinal segments that control the hindlimb musculature from the rest of the central nervous system (Figure 33–2A). This procedure allows investigations of the spinal locomotor circuits without any influence from supraspinal structures.

Two types of spinal preparation are used: acute spinal preparations, in which studies are performed immediately after the spinalization, and chronic spinal preparations, in which the animals are allowed to recover from the surgery and are then studied over a period of time.

In acute spinal preparations, locomotion is frequently induced chemically, either by intravenous administration of drugs that stimulate monoaminergic and/or serotonergic receptors or by local application of glutamatergic receptor agonists. These drugs increase the excitability in the spinal locomotor circuits,

(continued)

contributes primarily to the planning and execution of locomotion in situations in which vision is used to make anticipatory modifications of gait. Finally, two structures with no direct spinal connections, the basal ganglia and the cerebellum, contribute to the

selection of locomotor activity and to its coordination (Figure 33–1).

The way in which all of these structures interact and permit diverse modes of locomotion is the subject of this chapter.

Box 33–1 Preparations Used to Study the Neuronal Control of Locomotion (continued)

mimicking the locomotor-initiating drive from the brain stem. Alternatively, locomotion is induced electrically, by stimulation of the dorsal roots or dorsal columns. Acute spinal preparations are often paralyzed in order to increase recording stability from motor neurons and interneurons in the spinal cord, as well as to discriminate between central and peripheral effects.

In chronic spinal preparations, animals are studied for weeks or months after transection, often with the aim of finding better ways to improve the locomotor capability after spinal cord injury. In both young and adult cats and in young rodents, the hindlimb locomotor capability can often return following training but with no further treatment. In all animals, the locomotor capability is improved dramatically by drug treatments that activate the spinal central pattern generator. Electromyographic activity, together with behavioral measures, can be recorded before and after transection (Figure 33–2B).

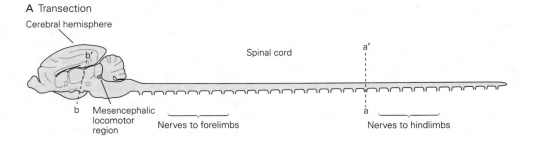

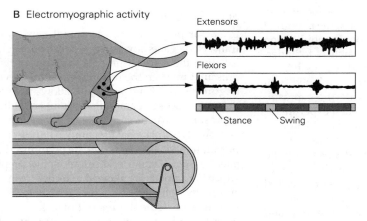

Figure 33–2 Selected animal models used to study locomotor control systems.

A. Schematic of the cat cerebral hemispheres, brain stem, and spinal cord showing the level of transection for spinalization (a'-a) and decerebration (b'-b). Decerebration isolates the brain stem and spinal cord from the cerebral

hemispheres. Transection at a'-a isolates the lumbar spinal cord from all descending inputs.

B. The electromyogram can be used to record locomotor activity during actual movement in intact, decerebrate, or spinal animals.

Locomotion Requires the Production of a Precise and Coordinated Pattern of Muscle Activation

Locomotion requires the production of activity in many muscles that need to be coordinated in a precise rhythm and pattern. The rhythm defines the frequency of the cyclic activity, whereas the pattern defines the spatiotemporal activation of muscle groups within a cycle. In swimming animals, such as the lamprey or the tadpole, locomotion is expressed as a traveling

In Vitro Preparations Are Used to Study Central Organization of Networks

With in vitro preparations, the spinal cord or brain stem is removed from the animal and placed in a bath that is perfused with artificial cerebrospinal fluid (rodent, lamprey, and turtle) (Figure 33–2C). Alternatively, the brain stem and spinal cord are left in situ in the animal that is paralyzed or immobilized and kept in vitro (tadpole and zebrafish) (Figure 33–2D).

In all cases, no rhythmic afferent inputs occur in the cord, and motor activity is recorded in peripheral nerves or, more often, in the ventral roots where the motor neurons have their axons leaving the spinal cord.

Locomotion is induced chemically, either by application of glutamatergic or serotonergic receptor agonists or a combination of both, or electrically by stimulating the brain stem or peripheral afferents. Rhythm and pattern generation, circuit connectivity, cellular properties of interneurons and motor neurons, and circuit neuromodulation are studied with conventional electrophysiological methods, imaging, and anatomical tracing, or with molecular genetic methods that allow manipulation and recording of identified populations of neurons.

C Isolated spinal cord

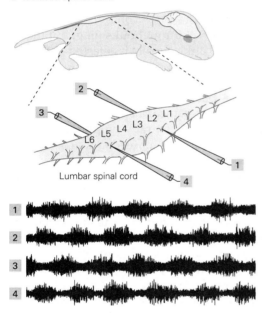

D In situ spinal cord

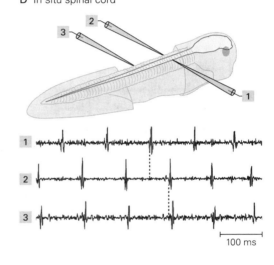

100 ms

C. The isolated lumbar (L1–L6) spinal cord from a newborn rat or mouse. Motor activity is recorded in flexor-related L2 ventral roots and extensor-related L5 ventral roots on either side of the cord. Locomotor-like activity is induced by application of N-methyl-D-aspartate (NMDA) and serotonin (5-hydroxytryptamine, 5-HT) to the bathing solution. Flexor-extensor alternation is seen as out-of-phase activity between L2 and L5 ventral roots on the same side of the cord (1 and 4; 2 and 3), and left–right alternations are seen as out-of-phase activity between L2–L2 and L5–L5 ventral roots on either side of the cord (1 and 2; 3 and 4). (Adapted, with permission, from Kiehn et al. 1999; data from O Kiehn.)

D. In vitro tadpole preparation, in which the spinal cord remains in situ, showing ventral root recordings on the right side (1) and on the left side (2 and 3). side of the spinal cord. The swimming rhythm in the nervous system of the paralyzed animal was induced by a brief stimulation of the skin on the head. (Data from L Picton and KT Silar.)

wave of activity (Figure 33–3A) that propagates from rostral to caudal body segments during forward progression. This pattern can be recorded as an electromyogram (EMG) during locomotion in the intact animal (Figure 33–3B) and as an electroneurogram in the isolated spinal cord (Figure 33–3C). Activity in more caudal roots occurs later than that in more rostral roots, and the activity on each side of the body is reciprocal.

In limbed animals, the pattern of muscle activity is more complex and serves to support the body as well

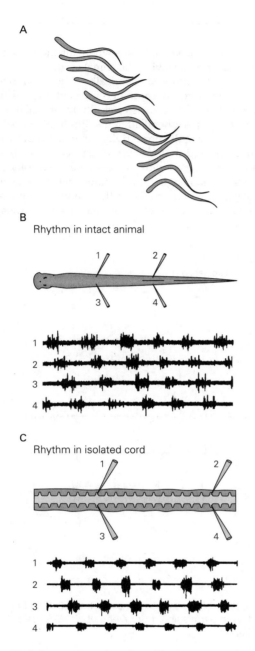

Figure 33–3 Lamprey swimming. The lamprey swims by means of a wave of muscle contractions traveling down one side of the body 180° out of phase with a similar traveling wave on the opposite side (**A**). This pattern is evident in electromyogram recordings from four locations along the animal during normal swimming (**B**). A similar pattern is recorded from four ventral roots in an isolated cord (**C**). (Data from S Grillner.)

as to transport it forward. The general unit of measure of locomotion in limbed vertebrates is the *step cycle*, which is defined as the time between any two successive events (eg, foot or paw contact of a given limb). The step cycle is divided into a *swing* phase, when the foot is off the ground and being transferred forward, and a *stance* phase, when the foot is in contact with the ground and propelling the body forward. Based on measures of changes in joint angle, each of these phases can be further divided into a period of flexion (F) followed by an initial period of extension (E_1) during swing and two additional periods of extension (E_2 and E_3) during stance (Figure 33–4A; see below).

Muscles within a single limb must be activated and coordinated in a precise spatiotemporal pattern (Figure 33–4B) so that the relative time of activation of different muscles, the duration of their activity, and the magnitude of that activity are coordinated to meet the demands of the environment (*intralimb coordination*).

In the hindlimb, swing is initiated by flexion of the knee produced by activation of muscles such as the semitendinosus, followed shortly by activation of hip and ankle flexors (the F phase). The hip flexors continue to contract throughout swing, but the activity in the knee and ankle flexors is arrested as the leg extends in preparation for contact with the support surface (the E_1 phase). Activity in most extensor muscles begins at this stage, before the foot contacts the ground. This preparatory prestance phase signifies that the extensor muscle activity is centrally programmed and not simply the result of afferent feedback arising from contact of the foot with the ground.

Stance begins with contact of the foot or paw with the ground. During early stance (the E_2 phase), the knee and ankle joints flex due to the acceptance of the weight of the body, causing extensor muscles to lengthen at the same time they are contracting strongly (eccentric contraction). The spring-like yielding of these muscles as weight is accepted allows the body to move smoothly over the foot and is essential for establishing an efficient gait. During late stance (the E_3 phase), the hip, knee, and ankle all extend as the extensor muscles provide a propulsive force to move the body forward.

There is also a requirement for *interlimb coordination*, the precise coupling between different limbs. The coupling between the four legs in quadrupeds, for example, can vary quite substantially, dependent on both the speed of locomotion and the adopted gait (a walk, pace, trot, gallop, or bound). This is particularly true of the pattern of coupling between muscles of limbs of the same side (homolateral limbs) and for the diagonal limbs. The relation between limbs can be characterized by the phase difference, with 0 reflecting limbs that move together in phase and 0.5 limbs that move fully out of phase (ie, in opposite directions). During walking, activity between the homolateral limbs varies by a phase value of 0.25, and three legs are always in contact

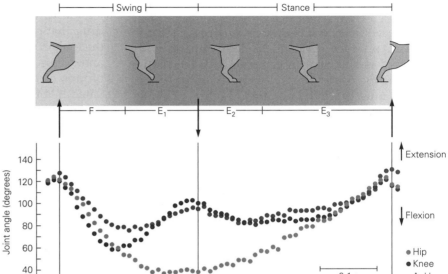

A Four phases of the step cycle

Figure 33–4 Stepping is produced by complex patterns of contractions in leg muscles.

A. The step cycle is divided into four phases. The flexion (F) and first extension (E_1) phases occur during the swing phase, when the foot is off the ground, whereas second extension (E_2) and third extension (E_3) occur during the stance phase, when the foot contacts the ground. E_2 is characterized by flexion at the knee and ankle as the leg begins to bear the animal's weight. The contracting knee and ankle extensor muscles lengthen during this phase. (Adapted, with permission, from Engberg and Lundberg 1969.)

B. Profiles of electrical activity in some of the hind leg flexor (**yellow**) and extensor (**blue**) muscles in the cat during stepping. Although flexor and extensor muscles are generally active during the swing and stance phases, respectively, the overall pattern of activity is complex in both timing and amplitude. (Muscles: **IP**, iliopsoas; **LG** and **MG**, lateral and medial gastrocnemius; **PB**, posterior biceps; **RF**, rectus femoris; **Sart_m** and **Sart_a**, medial and anterior sartorius; **SOL**, soleus; **ST**, semitendinosus; **TA**, tibialis anterior; **VL**, **VM**, and **VI**, vastus lateralis, medialis, and intermedius.)

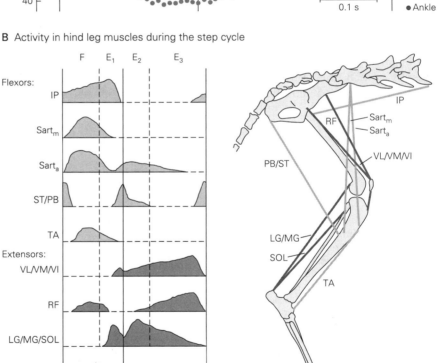

B Activity in hind leg muscles during the step cycle

with the ground. During a trot, the diagonal limbs (eg, the left hindlimb and the right forelimb) are in phase, and the phase difference between homolateral limbs is 0.5. Phase relationships between limbs of the same girdle (ie, the forelimbs or hindlimbs) are more stable during gaits produced by activation of alternating limbs, such as a walk or trot (generally out of phase by 0.5 cycle), compared to synchronous locomotion like a gallop or bound (generally in-phase).

The appropriate generation of the intra- and inter-limb coordination of activity and the adaptation of these patterns of activity according to circumstance is one of the major functions of the central nervous system during locomotion.

The Motor Pattern of Stepping Is Organized at the Spinal Level

While the entire nervous system is necessary for an animal to produce a rich behavioral repertory, the spinal cord is sufficient to generate both the rhythm underlying locomotion as well as much of the specific pattern of muscle activity required for intra- and inter-limb coordination.

At the beginning of the 20th century, Graham Brown showed that the isolated spinal cord had the intrinsic capacity to generate a rudimentary alternating locomotor pattern around the ankle joint in the absence of sensory inputs to the spinal cord (Figure 33–5). He proposed that locomotor networks controlling flexor and extensor activity in the spinal cord were organized as half-centers such that when half of the circuit was active it would inhibit the other half. The center would be released from inhibition through some sort of synaptic or neuronal fatigue.

This ground-breaking observation was mostly ignored until the mid-1960s and early 1970s, when there began a period of intense study of the mechanisms by which the spinal cord could generate a rhythmical pattern of activity. Initial studies showed that stimulation of sensory fibers in spinal cats treated with L-DOPA (a precursor of the monoamine transmitters dopamine and norepinephrine) and nialamide (a drug that prolongs the action of L-DOPA) could produce short sequences of rhythmic activity in flexor and extensor motor neurons. It was further found that groups of interneurons in the spinal cord were activated in a reciprocal flexor and extensor pattern. This organizational feature was consistent with Graham Brown's theory that mutually inhibiting half-centers produced the alternating burst activity in flexor and extensor motor neurons.

In the half-center model, the spinal cord produces only the locomotor rhythm, while the pattern is sculpted by afferent feedback caused by the movement. However, this view was changed by experiments that demonstrated that a well-organized locomotor pattern could be observed in decerebrate and spinal cats walking on a treadmill after section of the dorsal roots, thus removing the afferent feedback (Figure 33–6A,B). Later experiments in chronic spinal cats in which rhythmic afferent feedback was abolished by preventing movement (Figure 33–6C) showed that spinal circuits were not only able to intrinsically produce a locomotor rhythm but could also produce some of

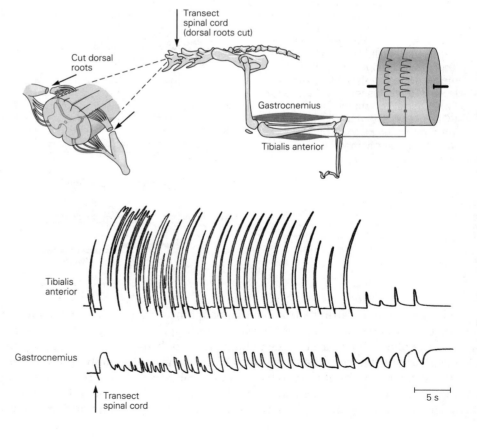

Figure 33–5 Rhythmic stepping is generated by spinal networks. The existence of intrinsic spinal networks was first demonstrated in 1911 by Thomas Graham Brown who developed an experimental preparation in which the dorsal roots were cut so that sensory information from the limb could not reach the spinal cord. The lower figure shows an original record from Graham Brown's study. Rhythmic alternating contractions of an ankle flexor (tibialis anterior) and an ankle extensor (gastrocnemius) are generated by the isolated spinal cord and persist for some time after the transection.

A Decerebrate, deafferented, walking

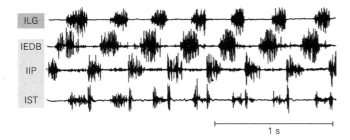

C Chronic spinal, paralyzed

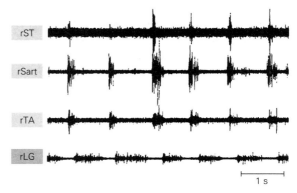

B Spinal, deafferented, walking

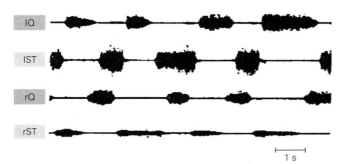

D Locomotor pattern generator

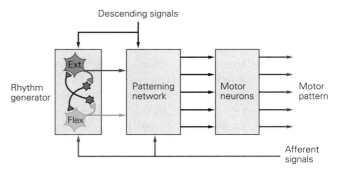

Figure 33–6 Spinal circuits generate both a rhythm and a pattern.

A. Even after removal of all sensory input to the spinal cord by cutting the dorsal roots, a decerebrate cat walking on a treadmill exhibits a complex motor pattern that is not just a simple alternation of flexor and extensor activity. (Abbreviations: **I**, left; **EDB**, extensor digitorum brevis; **LG**, lateral gastrocnemius; **IP**, iliopsoas; **ST**, semitendinosus.) (Adapted, with permission, from Grillner and Zangger 1984.)

B. Intravenous injection of L-DOPA and nialamide produces a well-organized locomotor pattern in an acute spinal cat with the dorsal roots cut. (Abbreviation: **I**, left; **Q**, quadriceps;

r, right.) (Adapted with permission from Grillner and Zangger 1979. Copyright © 1979 Springer Nature.)

C. Fictive locomotion in a chronic spinal paralyzed cat, demonstrating the typical pattern of activity in the semitendinosus, tibialis anterior (**TA**), lateral gastrocnemius (**LG**), and sartorius (**Sart**) muscles in intact cats. (**I**, left; **r**, right.) (Adapted from Pearson and Rossignol 1991.)

D. Conceptual model of a spinal locomotor central pattern generator (CPG) based on studies in decerebrate cats. The CPG model is formed of separate rhythm- and pattern-generating layers. Each of these layers can be modified by descending inputs and peripheral afferent information. (Adapted from Rybak et al. 2006.)

the spatiotemporal details of the pattern of activity observed in the intact cat (Figure 33–6C).

These observations led to the important concept of a *central pattern generator* (CPG) that can generate both the rhythm and the pattern, independent of sensory inputs. Subsequent experiments led to the idea that separate components of the CPG are responsible for generating the underlying rhythm of locomotion within a limb and the spatiotemporal pattern of muscle action in the limb (Figure 33–6D). This notion was based on the observation that changes in rhythm and pattern can be influenced independently. Other studies have led to the concept that the CPG is modular,

allowing independent control of activity around different joints.

Experiments in a variety of species have suggested that there are probably separate CPGs for each limb. For example, experiments using split belts, in which either the fore- and hindlimbs or the left and right limbs walk on separate treadmill belts, show that animals can independently modify step cycle duration in each pair of limbs. This organization would allow relatively simple descending commands to modify the coupling between each CPG and so to alter the pattern of the gait.

CPGs have now been identified and analyzed in many rhythmic motor systems, including those controlling over-ground locomotion, swimming, flying, respiration, and swallowing, in both invertebrates and vertebrates. In all vertebrates except higher primates and humans, a prominent locomotor pattern can be observed immediately after spinal transection when the spinal cord below the transection is activated with neuroactive drugs that function as a substitute for the descending drive that normally activates the spinal locomotor networks (Box 33–1).

The Spinal Circuits Responsible for Locomotion Can Be Modified by Experience

Lesion of the spinal cord in otherwise intact adult mammals leads to paralysis. In the absence of any further intervention, such animals will regain only minimal locomotion. However, when quadrupedal animals with complete lesions of the thoracic spinal cord are trained daily, they regain a remarkable ability to use their hindlimbs to walk on a treadmill.

A similar improvement in locomotion can also be obtained from the application of noradrenergic agonists. Indeed, recordings of hindlimb joint angles and EMG activity from these animals show that the spinal cord isolated from all descending systems can generate most of the coordinating features in the hindlimb that are observed in intact animals. This training effect is believed to occur because of an activity-dependent reorganization of both internal spinal circuits and the modification of synaptic inputs from peripheral afferents that is specific to the training regimen. Indeed, cats can be trained specifically to either support their weight or to walk, without a transfer of motor skills between the two behaviors.

Spinal Locomotor Networks Are Organized Into Rhythm- and Pattern-Generation Circuits

The question of how the spinal cord generates the complex activity underlying locomotion has been one of intense study that has followed three complementary paths. The earliest experiments directed at this issue were performed in the cat and provided important information on the functional characteristics of different interneuronal populations. However, the complex nature of the mammalian spinal cord led researchers to identify models with fewer neurons in the spinal cord, such as the turtle and two aquatic preparations, the tadpole and the lamprey (Box 33–1). These latter two models have provided an excellent window into the organization of the spinal circuits involved in swimming and a foundation for studying

rhythm and pattern generation in limbed animals. Last, the development of important molecular-genetic models in the mouse and the zebrafish has provided additional insights not available by more traditional methods.

The Swimming Central Pattern Generator

The lamprey—a jawless fish—swims like an eel with a wave of left–right bending traveling from front to back (Figure 33–3A). The spinal cord is made up of about 100 spinal segments, each containing neurons that can generate the rhythm and produce alternation between the two sides of the body. The rhythm is generated by interconnected glutamatergic excitatory neurons endowed with active membrane properties supporting rhythm generation. These glutamatergic neurons, which are the kernel in the swimming network, excite commissural inhibitory neurons, local inhibitory neurons, and motor neurons on the same side of the cord (Figure 33–7A).

The commissural interneurons, whose axons cross the midline, inhibit the contralateral interneurons involved in generating the alternating rhythm as well as contralateral motor neurons (Figure 33–7A). Cellular mechanisms contribute to phase switching in the network (Box 33–2). For example, Ca^{2+} entry triggered by bursting in glutamatergic neurons activates their calcium-activated K^+ channels. The opening of these channels hyperpolarizes the cells and enables termination of the burst. The termination of bursting on one side activates the other side by the commissural interneurons, thus allowing the contralateral rhythm-generating interneurons and motor neurons to become active. To enable coordination along the body, the segmental networks are connected through long-distance descending projections of excitatory and inhibitory neurons. This basic organization of interconnected excitatory neurons, inhibitory commissural neurons, and a rostrocaudal connectivity gradient for intersegmental coordination is also found in the tadpole and is possibly common to other swimming species.

Molecular and genetic approaches have expanded our understanding of the functional organization of CPGs in fish and identified two groups of glutamatergic interneurons—a group of commissural neurons and a group of ipsilaterally projecting neurons—that are involved in rhythm generation but at different speeds of locomotion. In adult zebrafish, the rhythm-generating circuit is composed of three functional classes of excitatory neurons that drive slow, intermediate, and fast pools of motor neurons that are selectively recruited as the speed of swimming increases.

The Quadrupedal Central Pattern Generator

The CPG controlling quadrupedal locomotion has added organizational complexity compared to the swimming CPG since it must generate both the rhythm and the pattern that involves the sequential flexor-extensor alternation of muscles around different joints within a limb (Figure 33–4B), as well as left–right coordination and coordination between the forelimbs and hindlimbs. Circuits controlling the forelimb are located in the cervical enlargement, whereas circuits controlling the hindlimb are located in the lower thoracic and lumbar spinal cord.

As in the CPG that generates rhythmical swimming activity, glutamatergic excitatory interneurons are involved in quadrupedal rhythm generation. Using advanced mouse genetics together with a molecular code that builds on expression of gene-regulating transcription factors that differentiate spinal neurons into classes with specific projection and transmitter phenotypes (Box 33–3), it has now been shown that the core of the rhythm-generating circuits in rodents includes two nonoverlapping groups of molecularly distinct glutamatergic neurons (Shox2^{ON} and Hb9; Figure 33–7B1).

The flexor (f) and extensor (e) rhythm-generating (R) circuits, which are connected by reciprocal inhibition (Figure 33–7B), drive other neurons in the locomotor network into rhythmicity and provide the rhythmic excitation for motor neurons (Figure 33–7B). As has been observed in the swimming CPG, ionic channels are also likely to contribute to rhythm generation and phase switching in the quadrupedal CPG.

The Flexor and Extensor Coordination Circuit

Flexor and extensor activity must be coordinated around joints (eg, hip-knee-ankle-toe in the hindlimb) to control the limb movement in a precise manner. Accordingly, the flexor-extensor alternation around the different joints is not simultaneous but has a sequential pattern, which suggests that multiple flexor-extensor alternating circuits are needed to time muscle actions in a limb. The basic flexor-extensor alternation circuits are organized in flexor and extensor modules composed of inhibitory and excitatory interneurons that are one synapse away from the flexor and extensor motor neurons they control (Figure 33–7B,B1).

Inhibitory and excitatory neurons in the module provide alternating inhibition and excitation of motor neurons. The reciprocally connected inhibitory Ia interneurons (Chapter 32) are part of the flexor and extensor modules providing the direct motor neuron inhibition in a reciprocal fashion (rIa in Figure 33–7B1). The rIas belong to the molecularly defined inhibitory

V1 and V2b neurons (Figure 33–7B1). The excitatory neurons that directly excite motor neurons during locomotion are likely to belong to multiple classes of neurons in the spinal cord, including V2a-Shox2^{ON} and the dI3 neurons (Figure 33–7B1).

In this basic scheme, the flexor-extensor modules are driven by flexor (fR in Figure 33–7B1) and extensor rhythm-generating circuits (eR in Figure 33–7B1), which themselves are reciprocally connected via inhibitory neurons (Figure 33–7B), resulting in their out-of-phase activity.

Left–Right Coordination

Left–right alternation, for both swimming and overground locomotion, depends on crossed inhibition produced in two ways: directly by inhibitory commissural neurons or indirectly by excitatory commissural neurons, each of which acts on premotor inhibitory neurons (Figure 33–7B2). This dual inhibitory system has a counterpart in one specific neuronal population, the V0 commissural neurons (Figure 33–7B2). Ablation of V0 neurons results in loss of left–right alternation at all speeds of locomotion. The inhibitory dorsal class of V0 neurons (V0$_D$), which makes up about half of the V0 population, controls alternating locomotion during walking, whereas the excitatory ventral class of V0 neurons (V0$_V$), which makes up the remaining half of V0 neurons, controls alternating locomotion during trot. The dual system thus serves a speed-dependent role in coordinating alternating gaits (walk and trot). Separate excitatory non-V0 commissural neurons—possibly the ventral V3 neurons (Box 33–3)—are responsible for synchrony in gaits such as bound and gallop (Figure 33–7B2).

The dual-mode left–right alternating pathways are driven directly by the rhythm-generating neurons or indirectly by other non–rhythm-generating excitatory neurons, including the V2a-Shox2^{Off} neurons that are recruited at high speeds of locomotion and synaptically connect to the V0$_V$ neurons. The left–right synchronous pathways are active at higher speeds of locomotion when the alternating system is suppressed or less active.

The speed-dependent changes in the left–right alternation circuits in the rodent are an example of functional reorganization of the vertebrate locomotor network needed to produce diverse motor outputs. Similar dynamic circuit reorganization has also been demonstrated in zebrafish and in studies of rhythmic networks in invertebrates, such as the stomatogastric ganglion controlling gut movements in crustaceans, where different functional networks emerge from a common CPG network.

A Swimming CPG: Rhythm and left–right coordination circuits

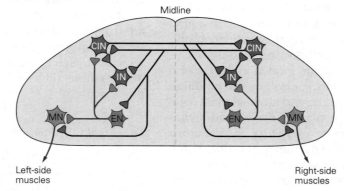

B Quadrupedal CPG

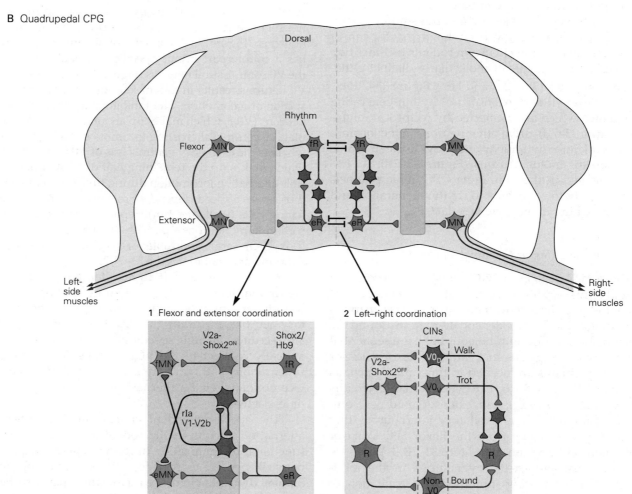

Interlimb Coordination

The organization of the networks that couple fore- and hindlimbs is not known in detail, but experiments using both lesion and genetic ablation suggest that these pathways involve both inhibitory and excitatory intersegmental connections.

Somatosensory Inputs From Moving Limbs Modulate Locomotion

Even though the CPG can produce the precise timing and phasing of the muscle activity needed to walk, this central pattern is normally modulated by sensory signals from the moving limbs. Two types of sensory input modulate the CPG activity: proprioceptive information generated by the active movement of the limb and tactile information generated when the moving limb meets an obstacle in the surrounding environment.

Proprioception Regulates the Timing and Amplitude of Stepping

One of the clearest indications that somatosensory signals from moving limbs regulate the locomotor cycle is that the rate of locomotion in spinal and decerebrate cats matches the speed of the motorized treadmill belt on which they walk. As the stepping rate increases, the stance phase becomes shorter while the swing phase remains relatively constant.

This observation suggests that some form of sensory input from the moving limb signals the end of the stance phase and thus leads to the initiation of the swing phase. The sensory information from the moving limb is generated by proprioceptors in the muscles and joints. These proprioceptors include stretch-sensitive muscle spindles in the hip and force-sensitive Golgi tendon organs in the ankle that are particularly important for facilitating locomotor phase transition.

The influence from the hip was noticed already by Sherrington, who showed that rapid extension at the hip joint leads to contractions in the hip flexor muscles of chronic spinal cats and dogs. More recent studies have found that preventing hip extension in a limb suppresses stepping in that limb, whereas rhythmically moving the hip in an immobilized cat can entrain the locomotor rhythm; that is, the stretching of the hip muscles causes the timing of the motor output to match the rhythm of the externally imposed movements (Figure 33–8A). The stretching also activates flexor muscle spindles and mimics the lengthening that occurs at the end of the stance phase, thus inhibiting extensor activity and facilitating activation of the flexor rhythm-generating circuits in the spinal cord (Figure 33–8B).

Activation of sensory fibers from Golgi tendon organs and muscle spindles in ankle extensor muscles prolongs the stance phase, often delaying the onset of

Figure 33–7 (Opposite) Spinal locomotor networks are organized into rhythm- and pattern-generation circuits with distinct cellular identities.

A. Circuit diagram of swimming central pattern generator (**CPG**) in the lamprey. Rhythm-generating circuits include excitatory interneurons (**EN**) that drive motor neurons (**MN**), inhibitory commissural interneurons (**CIN**) whose axons project to the other half of the cord, and local inhibitory interneurons (**IN**) with axons projecting on the same side of the cord. A single neuron in the diagram represents multiple neurons in the animal. **Gray neurons**, inhibitory; **red neurons**, excitatory. The vertical **dashed line** indicates the midline. (Data from Grillner 2006.)

B. General circuit diagram for limbed locomotion. Rhythm-generating circuits (**fR** and **eR**) composed of excitatory neurons on either side of the spinal cord drive flexor and extensor muscles on the same side through a pattern-generating layer (empty box). Rhythm-generating flexor (**fR**) and extensor (**eR**) neurons are reciprocally connected via inhibitory neurons and are connected across the midline via commissural interneurons (not shown) that mediate left–right coordination. The diagram shows one spinal segment. (Abbreviation: **MN**, motor neurons.) (Data from Kiehn 2016.)

B1. Flexor and extensor alternation is controlled at multiple levels in the locomotor network. One synapse away from flexor (**f**) and extensor (**e**) motor neurons (**MN**) are Ia-inhibitory interneurons, which reciprocally innervate antagonist motor neurons and each other (Chapter 32). The rIa neurons belong to two major groups of molecularly defined inhibitory neurons, V1 and V2b, in the ventral spinal cord. Excitatory neurons with different molecular markers (including V2a-Shox2^{ON}) provide premotor rhythmic excitation of motor neurons. Rhythm-generating Shox2^{ON} or Hb9 neurons (**fR** and **eR**) drive both inhibitory and excitatory premotor neurons. (Data from Kiehn 2016.)

B2. Rhythm-generating circuits drive left–right coordinating circuits composed of a dual inhibitory pathway involved in alternation and a single excitatory pathway involved in synchrony. The dual inhibitory pathway is composed of inhibitory V0$_D$ commissural neurons that directly inhibit rhythm generation on the other side and excitatory V0$_V$ commissural neurons that indirectly inhibit locomotor networks on the other side. The inhibitory V0$_D$ commissural neuron pathway controls the alternating gait walk. A population of V2a excitatory neurons is part of the left–right alternating circuit and connects to V0$_V$ commissural neurons. This pathway controls the alternating gait trot. Rhythm-generation circuits also drive a left–right synchronizing circuit possibly involved in bound, composed of non-V0 neurons. Only the projections from the left to the right side are shown. (Data from Kiehn 2016.)

Box 33–2 Neuronal Ion Channels Contribute to Central Pattern Generator Function

Neuronal membrane properties make an important contribution to the function of the central pattern generator (CPG). Neurons have a variety of K^+, Na^+, and Ca^{2+} channels that determine their activity and response to synaptic inputs. Studies of CPGs in diverse experimental models have shown that ion channels may be important for promoting rhythmicity, through bursting properties, or patterning, through ion channels that affect phase transitions or the rate of neuronal discharge.

Bursting and Plateau Properties Amplify Cellular Responses

Membrane properties that produce bursting allow cells to produce sustained oscillations in the absence of synaptic inputs. These properties are either intrinsic, as in cells in the sinusoidal node in the heart, or conditional, dependent on the presence of certain neurotransmitters. In some small motor CPGs (such as the pyloric network in the stomatogastric ganglion, which controls rhythmic movements in the gut of crustaceans), intrinsic bursting properties are essential for generating the rhythm.

Conditional bursting triggered by glutaminergic activation of N-methyl-D-aspartate (NMDA) receptors has been described in spinal cord interneurons and motor neurons in lamprey, rodents, and amphibians. In the lamprey, bursting due to NMDA receptor activation plays a role in generating swimming. In mammals, it is as yet uncertain whether NMDA receptor–induced bursting is essential for rhythm generation, although it may facilitate excitatory synaptic inputs in the circuit.

Plateau potential is another membrane property that may cause a neuron's membrane potential to jump to a depolarized state that will support action potential firing without further increase in the excitatory drive. Plateau properties amplify and prolong the effect of synaptic excitatory inputs and may promote rhythm generation and motor output. Plateau properties are generated by activation of slowly inactivating L-type Ca^{2+} channels or slowly inactivating Na^+ channels. These channels have been found in vertebrate interneurons and motor neurons. The expression of plateau properties mediated by L-type Ca^{2+} channels in motor neurons is controlled by neuromodulatory neurotransmitters, such as serotonin and norepinephrine. The slowly inactivating Na^+ channels are generally not regulated by neurotransmitters. Blockage of these channels decreases rhythm generation.

Phase Transitions May Be Regulated by Voltage-Gated Ion Channel Activation

Reciprocal inhibition between neurons is a common design in locomotor circuits; ion channels activated in the subthreshold spike range may enhance or delay phase transitions by such inhibition. Three types of voltage-gated channels are involved: a transient low threshold Ca^{2+} channel, cation-nonselective permeable hyperpolarization-activated cyclic nucleotide-gated (HCN) channels, and transient K^+ channels.

The transient low-threshold Ca^{2+} channels are inactivated at membrane potentials around rest. Transient inhibitory synaptic inputs remove the inactivation. When released from synaptic inhibition, activation of low-threshold Ca^{2+} channels will cause a short-lasting rebound excitation before the channels inactivate again. In the lamprey, spinal cord activation of metabotropic $GABA_B$ receptors depresses low-threshold Ca^{2+} channels involved in producing the swimming motor pattern. The suppression leads to a longer hyperpolarized phase and therefore to a slower alternation between antagonistic muscles, a possible mechanism for the slowing of swimming seen following $GABA_B$ receptor activation.

HCN channels are found in many CPG neurons and motor neurons and may help neurons escape from inhibition. They are activated by hyperpolarization, such as occurs during synaptic inhibition. Their activation depolarizes the cell, counteracting the hyperpolarization. Finally, the kinetics of their activation and deactivation are slow, so they stay open for some time after the hyperpolarization is released. The channel kinetics affect the integrative properties of the cell in two important ways. First, the depolarization caused by the channel opening limits the effect of sustained inhibitory inputs and helps the cell escape from inhibition. Second, the slow closing following synaptic inhibition leads to a rebound excitation promoting the next burst.

Voltage-gated A-type transient K^+ channels are usually inactivated at resting membrane potential. Hyperpolarization removes the resting inactivation, and subsequent depolarization will cause a transient activation of the channel. Their activation will therefore delay the onset of the next burst.

Regulation of Spiking Controls How Much Cells Are Activated

A number of different ion channels play a role in regulating the firing rate of a cell. Activation and inactivation kinetics of Na^+ channels are factors. Other important channels are sodium- and calcium-activated K^+ channels. The effect of activation of these K^+ channels is often seen as a slow after-hyperpolarization following an action potential or a train of action potentials. Activation of these channels therefore causes spike train adaptation and postactivation inhibition, which contribute to burst termination.

Box 33–3 Molecular-Genetics Combined With Anatomical, Electrophysiological, and Behavioral Analyses Are Used to Unravel the Locomotor Network Organization

To unravel the functional organization of the large neuronal networks in the spinal cord, researchers have used molecular-genetic–driven network analysis to take advantage of a molecular code that determines the spatial layout of the spinal locomotor networks.

It has been well documented that motor neurons develop and differentiate according to a genetic code expressed in the embryonic spinal cord (Chapter 45). This feature extends also to the development of spinal interneurons, which can be identified by different transcription factors (Table 33–1). The cardinal classes of interneuronal types belong to dorsally located interneurons (dI1–dI6) and ventrally located interneurons (V0–V3), with further subdivision within these categories (eg, $V0_D$ and $V0_V$, V2a-Shox2^{Off}, V2a-Shox^{On}) where a combination

of transcription factors defines these subtypes (Table 33–1). Each group of interneurons has specific transmitter content and characteristic axonal projection patterns.

The ability to manipulate these specific interneuron types gives an unparalleled opportunity to examine the functional contribution of specific subsets of interneurons in the mouse or zebrafish that is not possible in species such as the cat. The molecular code of the spinal cord neurons is used to mark cells with a marker protein such as green fluorescent protein or for the expression of proteins that allow for cell type-specific ablation or activation/inactivation of cells types. Such studies have ascribed specific locomotor functions to the dI3, V0–V3, and Hb9 cells, all molecularly differentiated classes of neurons (Table 33–1).

Table 33–1 Developmental Molecular Codes Specify the Identity of Spinal Neurons in the Spinal Cord

Postmitotic transcription factors	Neuron type	Transmitters
Islt1/Tlx3	dI3	Glutamate
Pax2/7	$V0_D$	GABA/glycine
Evx1	$V0_V$	Glutamate
Evx1/Pitx2	$V0_C$	Acetylcholine
Evx1/Pitx2	$V0_D$	Glutamate
En1	V1	GABA/glycine
Chx10	V2a-Shox2^{Off}	Glutamate
Chx10/Shox2	V2a-Shox2^{ON}	Glutamate
GATA2/3	V2b	GABA/glycine
Sox1	V2c	GABA/glycine
Shox2	V2d	Glutamate
Hb9/Islt1-2	MN	Acetylcholine
Hb9	Hb9	Glutamate
Sim1	$V3_D$	Glutamate
Sim1	$V3_V$	Glutamate

Chx10, Ceh-10 homeodomain-containing homolog; Evx1, even skipped homeobox 1; En1, engrailed 1; GABA, γ-aminobutyric acid; GATA2/3, gata protein; Hb9, homeobox 9; Islt1-2, ISL1-2 transcription factor; Pax, paired box gene; Pitx2, paired-like homeodomain transcription factor 2; Sim1, single-minded homolog 1; MN, motor neuron; Shox2, Short stature homeobox 2; Sox1, SRY box-containing gene 1; Tlx1/3, T cell leukemia, homeobox 1/3.
Source: Adapted from Jessell 2000, Goulding 2009, Dougherty et al. 2013.

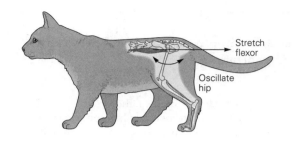

A Oscillate hip

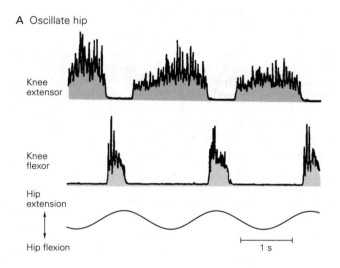

Knee extensor

Knee flexor

Hip extension

Hip flexion

1 s

B Stretch hip flexor

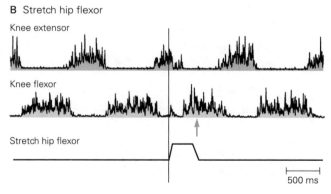

Knee extensor

Knee flexor

Stretch hip flexor

500 ms

Figure 33–8 Hip extension initiates the transition from stance to swing phase of walking.

A. In an immobilized decerebrate cat, passive oscillating movement around the hip joint initiates and entrains the fictive locomotor pattern in knee extensor and flexor motor neurons. The flexor electromyogram (EMG) bursts correspond to the swing phase and are generated when the hip is extended. (Adapted, with permission, from Kriellaars et al. 1994.)

B. In a walking decerebrate cat, stretching of the hip flexor muscle (iliopsoas) inhibits knee extensor EMG activity, allowing knee flexor activity to begin earlier. The **arrow** in the knee flexor record indicates when activity in the muscle would have begun had the hip flexor muscle not been stretched. Activation of sensory fibers from muscle spindles in the hip flexor muscle is responsible for this effect. (Adapted, with permission, from Hiebert et al. 1996.)

the swing phase until the stimulus has ended (Figure 33–9A). Sensory fibers from both types of receptors are active during stance, with the intensity of the signal from the Golgi tendon organs being strongly related to the load carried by the leg. Golgi tendon organs have inhibitory actions on ankle extensor motor neurons when the body is at rest (Chapter 32) but an excitatory action during walking. This reversal of the sign of the reflex is caused by inhibition of inhibitory interneuron pathways together with a release of excitatory pathways during locomotion. The functional consequence of this reflex reversal during locomotion is that the swing phase is not initiated until the extensor muscles are unloaded and the forces exerted by these muscles are low, as signaled by a decrease in activity from the Golgi tendon organs near the end of stance.

In sum, proprioceptive signals from the ankle extensor muscles and hip flexor muscles work synergistically to facilitate the stance-to-swing phase transition. In the late stance phase, when the limb is unloaded, as inhibitory signals from Golgi tendon organs wane, their effects on extensor rhythm generation declines, while at the same time the activity in muscle afferents around the hip joint is increased, facilitating activity in flexor rhythm generation.

At least three excitatory pathways transmit sensory information from extensor muscles to extensor motor neurons during walking: a monosynaptic pathway from primary muscle spindles (group Ia afferents), a disynaptic pathway from primary muscle spindles and Golgi tendon organs (group Ia and Ib afferents), and a polysynaptic pathway from primary muscle spindles and Golgi tendon organs that includes interneurons in the extensor rhythm generator (Figure 33–9B). These pathways all contribute to phase transition from stance to swing when the ankle is unloaded and maintain extensors in stance phase when the ankle is loaded.

In addition to regulating the transition from stance to swing, proprioceptive information from muscle spindles and Golgi tendon organs contributes significantly to the generation of burst activity in extensor motor neurons. Reducing this sensory input in cats diminishes the level of extensor activity by more than half; in humans, it has been estimated that up to 30% of the activity of ankle extensor motor neurons is caused by feedback from the extensor muscles.

Mechanoreceptors in the Skin Allow Stepping to Adjust to Unexpected Obstacles

Mechanoreceptors in the skin, including some nociceptors, have a powerful influence on the CPG for

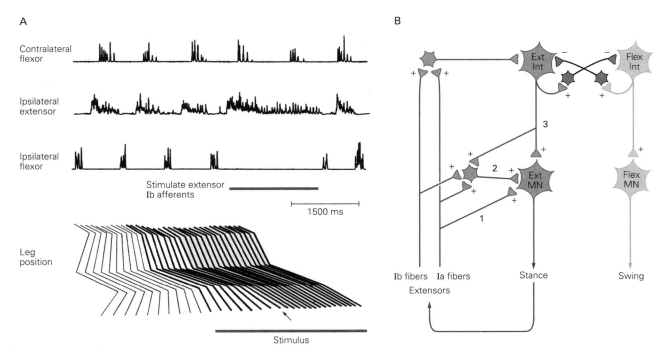

Figure 33–9 The swing phase of walking is initiated by sensory feedback from extensor muscles.

A. In a decerebrate cat, electrical stimulation of group I sensory fibers from ankle extensor muscles inhibits the electromyogram burst in ipsilateral flexors and prolongs the burst in the ipsilateral extensors during walking. The timing of contralateral flexor activity is not altered. Stimulating group I fibers from ankle extensors prevents initiation of the swing phase, as can be seen in the position of the leg during the time the fibers were stimulated. The **arrow** shows the point at which the swing phase would normally have occurred had the ankle extensor afferents not been stimulated. (Adapted, with permission, from Whelan, Hiebert, and Pearson 1995. Copyright © Springer-Verlag 1995.)

B. Mutually inhibiting groups of extensor (**Ext**) and flexor (**Flex**) interneurons (**Int**) constitute a rhythm generator in the afferent pathway regulating the stance phase. Feedback from extensor muscles increases the level of activity in extensor motor neurons (**MN**) during the stance phase and maintains extensor activity when the extensor muscles are loaded. The feedback is relayed through three excitatory (**+**) pathways: (**1**) monosynaptic connections from Ia fibers to extensor motor neurons; (**2**) disynaptic connections from Ia and Ib fibers to extensor motor neurons; and (**3**) polysynaptic excitatory pathways that act through the extensor rhythm generator to maintain the extensor motor neurons active in stance phase.

walking. One important function of these receptors is to detect obstacles and adjust stepping movements to avoid them. A well-studied example is the corrective reaction to stumbling in cats.

A mild mechanical stimulus applied to the dorsal part of the paw during the swing phase produces excitation of flexor motor neurons and inhibition of extensor motor neurons, leading to rapid flexion of the paw away from the stimulus and elevation of the leg in an attempt to step over the object. Because this corrective response is readily observed in spinal cats, it must be produced to a large extent by circuits entirely contained within the spinal cord.

One of the interesting features of the corrective reaction is that corrective flexion movements are produced only if the paw is stimulated during the swing phase. An identical stimulus applied during the stance phase produces the opposite response—excitation of extensor muscles that reinforces the ongoing extensor activity. This extensor action is appropriate; if a flexion reflex were produced during the stance phase, the animal might collapse because it is being supported by the limb. This is an example of a phase-dependent reflex reversal. The same stimulus can excite one group of motor neurons during one phase of locomotion while activating the antagonist motor neurons during another phase.

Supraspinal Structures Are Responsible for Initiation and Adaptive Control of Stepping

Although the basic motor patterns for locomotion are generated in the spinal cord, the initiation, selection, and planning of locomotion require activation of supraspinal structures, including the brain stem, the basal ganglia, cerebellum, and cerebral cortex. Supraspinal regulation of stepping provides a number of behavioral modifications that cannot be mediated by spinal circuits alone. These include the voluntary initiation of locomotion and the regulation of speed; postural regulation, including weight support, balance, and interlimb coordination; and the planning and execution of anticipatory modifications of gait, particularly visually guided modifications.

Midbrain Nuclei Initiate and Maintain Locomotion and Control Speed

The locomotor networks in the spinal cord require a command or start signal from supraspinal regions to initiate and maintain their activity. The major neuronal structure involved in the initiation in vertebrates is a region in the midbrain called the mesencephalic locomotor region (MLR). The MLR was first identified in cats as a unitary region localized in or around the cuneiform nucleus, just below the inferior colliculus. Tonic electrical stimulation in this area in the resting animal increased postural tonus so that the animal stood up and then started to walk. As the intensity of stimulation rose, the speed of locomotion increased and alternating gaits switched to synchronous gaits such as gallop or bound (Figure 33–10).

Later studies with electrical stimulation confirmed the presence of the MLR in all vertebrates, suggesting that the MLR is evolutionarily conserved from the oldest vertebrates to humans. These studies have pointed to two midbrain structures as part of the MLR (Figure 33–11A): the cuneiform nucleus (CNF) and the more ventrally located pedunculopontine nucleus (PPN) (Figure 33–11A). These two nuclei differ in the types of neurons they contain.

Long-range projection neurons in the CNF are excitatory and use glutamate as their neurotransmitter, whereas those in the PPN are both glutamatergic and cholinergic. In both nuclei, the excitatory neurons are intermingled with local GABAergic interneurons. Electrical stimulation has, however, been unable to determine which nucleus or which types of neurons are involved in the initiation of locomotion and speed control. However, the use of selective activation and inactivation of neurotransmitter-specific CNF and PPN neurons suggests that the two nuclei play specific roles in speed control and gait selection of locomotion (Figure 33–11B). Glutamatergic neurons in both PPN and CNF are sufficient for supporting alternating locomotion at slower speeds, such as walking and trot, while glutamatergic neurons in the CNF are necessary

for high-speed locomotion, such as gallop and bound, characteristic of escape locomotion. Expression of these gaits is dependent on the stimulation frequency, possibly reflecting the effect of firing frequency in the intact animal.

The role of cholinergic PPN neurons for locomotion is less well understood. In mammals, they do not seem to have a strong role in maintaining locomotion.

These roles of glutamatergic CNF and PPN neurons in locomotor control may also be reflected in the different inputs. PPN neurons receive strong input from the basal ganglia, specifically the substantia nigra pars reticulata, globus pallidus pars interna, and subthalamic nucleus, as well as from sensorimotor and frontal cortex. Additionally, the PPN receives sensorimotor information from many nuclei in the midbrain and brain stem. The nucleus may therefore serve as a hub for integrating information from many brain structures, possibly leading to the release of slower exploratory locomotion. In contrast, the input to neurons in CNF is much more restricted and arises principally from structures that may be involved in escape responses. The MLR is therefore composed of two regions that act together to select context-dependent locomotor behavior.

Another brain area that evokes locomotion when stimulated is the subthalamic (or diencephalic) locomotor region (to be distinguished from the subthalamic nucleus). This region includes nuclei in the dorsal and lateral hypothalamus involved in various homeostatic features such as regulating feeding. Neurons in these areas project to neurons in the reticular formation and bypass the PPN and CNF, suggesting a parallel pathway for initiating locomotion, possibly driven by the need to find food.

Midbrain Nuclei That Initiate Locomotion Project to Brain Stem Neurons

The excitatory signals from CNF and PPN are relayed indirectly to the spinal cord by way of neurons in the brain stem reticular formation, which provide the final command signal to the locomotor networks in the spinal cord. The identity of these neurons is only partly known. In general terms, two transmitter-defined pathways are involved: glutamatergic and serotonergic.

The glutamatergic locomotor pathways probably have multiple origins in the brain stem reticular formation, forming parallel descending pathways. They project directly or indirectly via a chain of intersegmental (propriospinal) glutamatergic interneurons to locomotor neurons in the spinal cord (Figure 33–10A). Reticulospinal neurons also participate in regulating

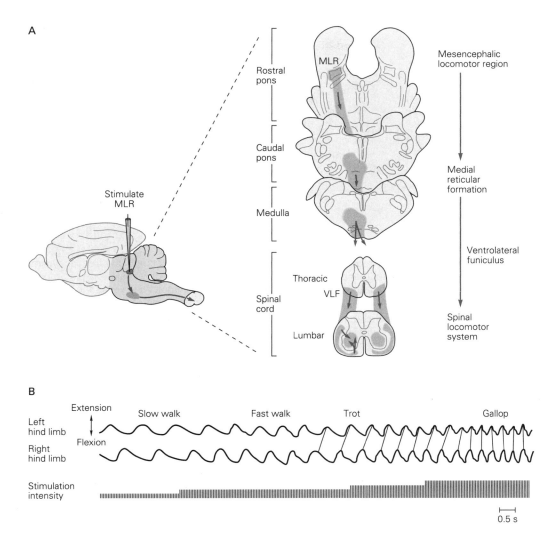

Figure 33–10 The mesencephalic locomotor region initiates locomotion.

A. Electrical stimulation of the mesencephalic locomotor region (**MLR**) in the cat initiates locomotion by activating neurons in the medial reticular formation whose axons descend in the ventrolateral funiculus (**VLF**) to the spinal locomotor system.

B. When the strength of electrical stimulation of the MLR in a decerebrate cat walking on a treadmill is gradually increased, the gait and rate of stepping change from slow walking to trotting and finally to galloping. As the cat progresses from trotting to galloping, the hind limbs shift from alternating to in-phase activity. (Adapted from Shik et al. 1966.)

the postural activity that is needed for the animal to locomote (see later discussion).

Evidence for the existence of a serotonergic locomotor pathway in mammals is restricted to experiments in rats that have shown the involvement of serotonergic neurons in the caudal brain stem. The mechanisms by which the final command signals from the brain stem to the spinal cord activate the spinal locomotor networks, maintain their activity, and allow the expression of different gaits are unknown.

The episodic nature of locomotion indicates that the initiating signals may be complemented by stop commands to allow for sudden locomotor arrest. Such signals have been found in *Xenopus* tadpole, in which head contact with obstacles activates GABAergic descending pathways that immediately terminate swimming. Likewise, in decerebrate cats, tonic electrical stimulation of the medullary and caudal pontine reticular formation leads to a general motor inhibition. Studies in mice have identified a restricted contingent of V2a neurons in the reticular formation that mediate an immediate arrest of ongoing locomotor activity. Such "V2a stop neurons" send a behaviorally relevant stop signal via descending projections to inhibitory

A

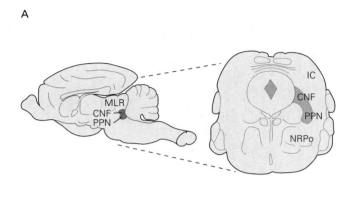

B

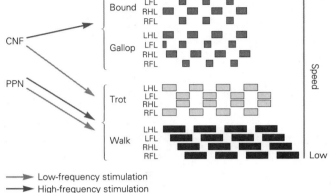

Low-frequency stimulation
High-frequency stimulation

Figure 33–11 The mesencephalic locomotor region is composed of dual midbrain glutamatergic nuclei that control initiation of locomotion, speed and gait regulation, and context-dependent selection of locomotion.

A. *Left:* The site of the localization of mesencephalic locomotor region (**MLR**) in the midbrain of the mouse. *Right:* Transverse section shows that the MLR is composed of the cuneiform nucleus (**CNF**) and the pedunculopontine nucleus (**PPN**) in the midbrain, lateral to the cerebral aqueduct, and dorsal to the nucleus reticularis pontis oralis (**NRPo**). Glutamatergic, GABAergic, and cholinergic neurons are intermingled in the CNF and PPN. (Abbreviation: **IC**, inferior colliculus).

B. Effect in mice of optical stimulation of glutamatergic cells in the CNF or PPN that have been transfected with the light-sensitive channel, channelrhodopsin 2. Stimulation at low and high frequencies in the PPN leads only to alternating

gaits—walking and trotting. Low-frequency simulation in the CNF likewise results only in slow, exploratory locomotion, while high-frequency stimulation evokes the synchronous gaits gallop and bound corresponding to escape locomotion.

The different types of gaits are shown as idealized diagrams from low to high speeds of locomotion. Filled boxes represent the stance phase; open spaces the swing phase. Walk is characterized by periods of support by three or four feet simultaneously. Trot is characterized by simultaneous activity in the diagonal fore and hindlimbs. Gallop is characterized by the forelimbs moving slightly out of phase and hind limbs being almost in phase. Bound is characterized by hind limbs and forelimbs moving simultaneously and forelimb and hindlimb out of phase. (Abbreviations: **LFL**, left forelimb; **LHL**, left hindlimb; **RFL**, right forelimb; **RHL**, right hindlimb.) (Adapted from data in Caggiano et al. 2018.)

interneurons in the ventral lumbar spinal cord that inhibits rhythm generation. A similar stop signal arrests swimming in the lamprey.

The Brain Stem Nuclei Regulate Posture During Locomotion

An important aspect of locomotor control is the regulation of posture. This general term encompasses several types of behavior, including the production of the postural support on which locomotion is superimposed, the control of balance, the regulation of interlimb coordination in quadrupeds, and the modification of muscle tonus required to adapt to locomotion on slopes or during turning. In addition, anticipatory changes in posture precede changes in voluntary gait modifications, and compensatory changes in posture follow unexpected perturbations. These functions are largely subserved by two descending systems originating from the brain stem: the vestibulospinal tract (VST), originating in the lateral vestibular nucleus (LVN), and the reticulospinal

tract (RST), originating in the pontomedullary reticular formation (PMRF). Both pathways are phylogenetically old and found in all vertebrates.

Lesions of the LVN, the PMRF, or their descending axons in the spinal cord lead to a loss of weight support and the control of equilibrium, expressed as a crouched gait and swaying of the hindquarters to one side or the other. Lesions of these nuclei are also followed by large changes in the interlimb coordination between the fore- and hindlimbs. Likewise, tonic electrical or chemical stimulation of the pons and the medulla modulates the level of muscle tonus in the limbs and can either facilitate or suppress locomotion depending on the exact site stimulated (Figure 33–12).

Activity in the VST and RST, together with activity in the rubrospinal tract, which originates from the red nucleus, also phasically modifies the level of muscle tonus during each step. Weak electrical stimulation of any of these three structures produces phase-dependent modulation of locomotor activity. Brief activation of these pathways with short trains of stimuli produces

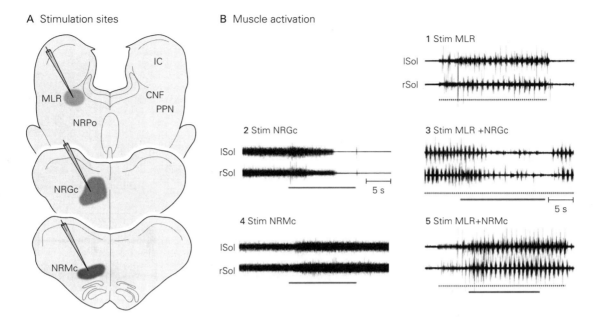

A Stimulation sites

B Muscle activation

Figure 33–12 Locomotor activity is modified by the level of postural tone. (Adapted, with permission, from Takakusaki et al. 2016.)

A. Transverse sections of the brain stem of the cat at three different rostrocaudal levels. Colored areas indicate the regions stimulated during the trials shown in part **B.** (Abbreviations: **CNF,** cuneiform nucleus; **IC,** inferior colliculus; **MLR,** mesencephalic locomotor region; **NRPo,** nucleus reticularis pontis oralis; **NRGc,** nucleus reticularis gigantocellularis; **NRMc,** nucleus reticularis magnocellularis; **PPN,** pedunculo-pontine nucleus.)

B. Effects of stimulating the different regions of the brain stem indicated in part **A** in the decerebrate cat.

1. Stimulation of the MLR (CNF/PPN) (**green bar**) produces rhythmic activation in the left and right hindlimb extensor soleus muscles (**Sol**).

2. Tonic stimulation of the NRGc (**red bar**) in the medulla results in a loss of muscle tone in the extensor muscles.

3. Stimulation of the NRGc during CNF-induced locomotion reduces muscle tone and thereby inhibits locomotion.

4. Tonic stimulation of the NRMc (**blue bar**) in the ventral medulla produces an increase in muscle tone.

5. Stimulation of the NRMc during MLR stimulation results in increased vigor of locomotion.

transient changes in the amplitude of the muscle bursts but rarely produces any changes in the timing of the step cycle. Activation of the LVN primarily enhances responses in ipsilateral extensor muscles during their natural period of activity in the stance phase. In contrast, stimulation of the red nucleus generally produces transient increases in activity in contralateral flexor muscles, again during their natural period of activity in the swing phase.

Stimulation of the PMRF produces more complex and widespread responses that may modify activity in flexor muscles during the swing phase and in extensor muscles during the stance phase across all four limbs in a coordinated pattern (Figure 33–13). In flexor muscles, activity is generally facilitated by PMRF stimulation, but in extensor muscles, it may be facilitated or suppressed depending on the exact site stimulated. This phase-dependent nature of the responses is thought to

be mediated by activation of interneurons in the spinal CPG. Stimulation of these three structures at higher strengths, or with longer trains, may produce changes in the timing of the step cycle as well as in the magnitude of EMG activity.

During locomotion, neurons within the LVN, PMRF, and red nucleus are phasically modulated at the frequency of the step cycle. Neurons in the LVN are generally activated in phase with ipsilateral extensor muscles, whereas neurons in the red nucleus are generally active during the contralateral swing phase. Neurons in the PMRF have more complicated periods of activity and may discharge in relation to ipsilateral or contralateral flexor or extensor muscles.

Brain stem structures also contribute to more complex activities during locomotion. For example, the red nucleus contributes to the complex modifications in muscle activity required for precise modifications of

Figure 33–13 Microstimulation of the pontomedullary reticular formation (PMRF) produces phase-dependent responses in flexor and extensor muscles. (Data from T. Drew.)

A. Stimulation of the left PMRF during the swing phase of the left limb produces a transient increase in the electromyogram activity of the left flexor muscles (**IF**) and a simultaneous decrease in activity in the right extensor muscles (**rE**) (**red arrows**). There is little stimulus-evoked activity in the left extensor (**IE**) or right flexor (**rF**) muscles, which are inactive at this phase of the step cycle.

B. Stimulation at the same location in the PMRF during the swing phase of the right limb produces the inverse responses.

C. The phase-dependent nature of the responses is likely determined by the cyclical nature of the level of excitability in interneurons that are part of the locomotor central pattern generator (**CPG**). Responses are gated by activity in the flexor (**F**) and extensor (**E**) parts of the locomotor CPG. When the first stimulation arrives, flexor interneurons in the left CPG (**IF**) are active, whereas those in the right CPG (**rF**) are inactive. The stimulation therefore produces a response only in the left flexor motor neurons (**IFmn**). When the second stimulation arrives, flexor interneurons in the right CPG (**rF**) are active, whereas those on the left side are inactive, and therefore, the stimulation elicits a response only in the flexor motor neurons on the right (**rFmn**).

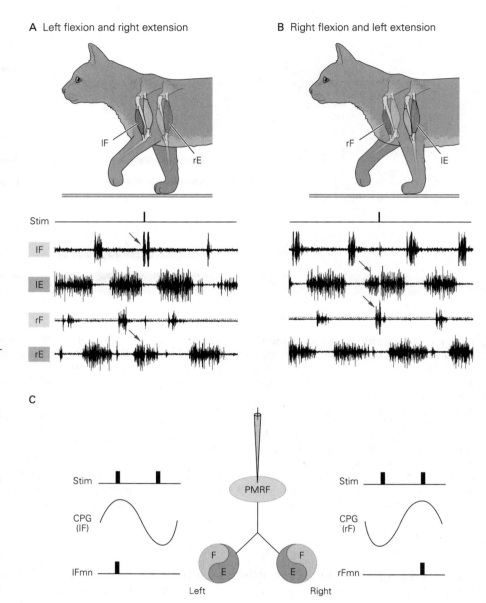

A Left flexion and right extension

B Right flexion and left extension

gait (see below). In a complementary manner, the widespread effects of the PMRF on multiple limbs allow it to produce the coordinated changes in postural activity that accompany gait modifications. The coordination between gait modifications and postural activity is assured by the strong connections from the motor cortex to the PMRF in the same manner as for discrete voluntary movements (Chapter 34). The PMRF also contributes to the compensatory changes in posture that occur as a consequence of perturbations. In this situation, it forms part of a spino-bulbo-spinal reflex that contributes to the widespread postural responses that follow the immediate spinal reflexes activated by a sudden perturbation.

Visually Guided Locomotion Involves the Motor Cortex

Walking is most often guided by vision, and the motor cortex is largely essential for visually guided movement, especially when gait must be modified to ensure precise control over limb trajectory and foot placement. In mammals, lesions of the motor cortex do not prevent animals from walking on a smooth floor, but they severely impair "precision locomotion," which requires a high degree of visuomotor coordination, such as walking on the rungs of a horizontal ladder, stepping over a series of barriers, and stepping over single objects placed on a treadmill belt.

Experiments in intact cats trained to step over obstacles attached to a moving treadmill belt show that precision locomotion is associated with considerable modulation of the activity of numerous neurons in the motor cortex (Figure 33–14). Other neurons in the motor cortex show a more discrete pattern of activity and are activated sequentially during different parts of the swing phase. The activity of these cortical neurons correlates with the periods of modified muscle activity required to produce the gait modifications

in a similar manner to what occurs during reaching (see Figure 34–21). Such subpopulations of neurons may serve to modify the activity of the groups of synergistic muscles required to produce flexible changes in limb trajectory.

Many of these cortical neurons project directly to the spinal cord (corticospinal neurons) and thus may regulate the activity of spinal interneurons, including those within the CPG, thereby adapting the timing and magnitude of motor activity to a specific locomotor task. Brief trains of electrical stimulation applied to either the motor cortex or the corticospinal tract in normal walking cats produce transient responses in the contralateral limb in a phase-dependent manner, similar to that produced by activity in various brain stem structures. However, in contrast to the situation

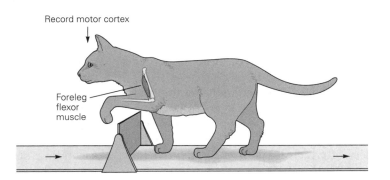

Figure 33–14 Stepping movements are adapted by the motor cortex in response to visual inputs. When a cat steps over a visible object fixed to a treadmill, neurons in the motor cortex increase in activity. This increase in cortical activity is associated with enhanced activity in foreleg muscles, as seen in the electromyograms (**EMG**). (Adapted, with permission, from Drew 1988.)

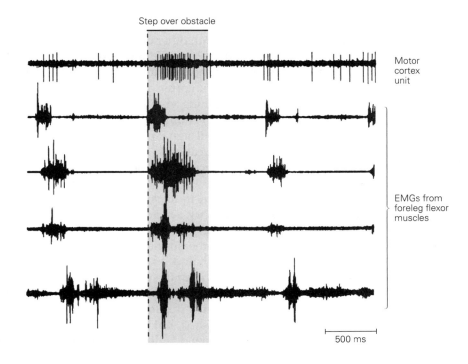

observed with brain stem structures, increasing the duration of the stimulation train applied to the motor cortex frequently results in a reset of the locomotor rhythm, characterized as an interruption of the ongoing step cycle and the initiation of a new step cycle. This suggests that in mammals the corticospinal tract has privileged access to the rhythm generator of the CPG.

Planning of Locomotion Involves the Posterior Parietal Cortex

When humans and animals approach an obstacle in their pathway, they must adjust their walking pattern to move around the obstacle or step over it. Planning of these adjustments begins two or three steps before the obstacle is reached. Recent experiments suggest that the posterior parietal cortex (PPC) is particularly involved in planning gait modifications. Lesions in this region cause walking cats to misplace the positioning of their paws as they approach an obstacle and increase the probability that one or more legs contact the obstacle as they step over it.

In contrast to what is observed in the motor cortex, recordings in PPC show that many neurons increase their activity in advance of the step over the obstacle. Moreover, many cells in the PPC discharge similarly regardless of which leg is first to step over the obstacle (Figure 33–15A,B). Such cells may provide an estimation of the position of the body with respect to objects in the environment (Figure 33–15B), allowing animals to modify gait as they approach the obstacle. The manner in which the PPC interacts with other cortical and subcortical structures generally considered to be involved in motor planning is unknown. However, recent work shows that the premotor cortex also makes an important contribution to planning visually guided gait modifications (Figure 33–15C) and may be implicated in the transformation of a global signal providing information concerning obstacle location to the muscle-based signal necessary for the execution of the step over the obstacle.

Visual information about the size and location of an obstacle is also stored in working memory, a form of short-term memory (Chapter 52). This information is used to ensure that gait modifications in the hindlimb are coordinated with those of the forelimb and is necessary because the obstacle is no longer within the visual field by the time the hindlimbs are stepping over it. The neurobiological mechanisms underlying this form of working memory remain to be established, but the persistence of the memory appears to depend, at least

in part, on neuronal systems in the PPC. With bilateral lesions or cooling of the medial PPC, the memory is completely abolished (Figure 33–16A). Complementing this observation is the finding that the activity of some neurons in the PPC is elevated during a step over an obstacle, as well as throughout the time the animal straddles the obstacle (Figure 33–16B). This activity could represent the working memory of key features of the obstacle such as height.

The Cerebellum Regulates the Timing and Intensity of Descending Signals

Damage to the cerebellum results in marked abnormalities in locomotor movements, including the need for a widened base of support, impaired coordination of joints, and abnormal coupling between limbs during stepping. These symptoms, which are characteristic of *ataxia* (Chapter 37), indicate that the cerebellum contributes importantly to the regulation of locomotion.

A major function of the cerebellum is to correct movement based on a comparison of the motor signals sent to the spinal cord and the movement produced by that motor command (Chapter 37). In the context of locomotion, the motor signal is generated by neurons in the motor cortex and brain stem nuclei. Information about the movement comes from the ascending spinocerebellar pathways. For the hind legs of the cat, these are the dorsal and ventral spinocerebellar tracts. Neurons in the dorsal spinocerebellar tract (DSCT neurons) are strongly activated by numerous leg proprioceptors and thus provide the cerebellum with detailed information about the mechanical state of the hind legs. In contrast, neurons in the ventral tract (VSCT neurons) are activated primarily by interneurons in the CPG, thus providing the cerebellum with information about the state of the spinal locomotor network.

During locomotion, the motor command (the central efference copy), the movement (the afference copy, via the DSCT), and the state of the spinal networks (the spinal efference copy, via the VSCT) are integrated within the cerebellum and expressed as changes in the pattern of rhythmical discharge of Purkinje cells in the cerebellar cortex and neurons in the deep cerebellar nuclei. These signals from the deep cerebellar nuclei are then sent to the motor cortex and the various brain stem nuclei where they modulate descending signals to the spinal cord to correct any motor errors.

Behavioral experiments show that the cerebellum also plays an important role in the adaptation of gait. For example, when subjects walk on a split treadmill,

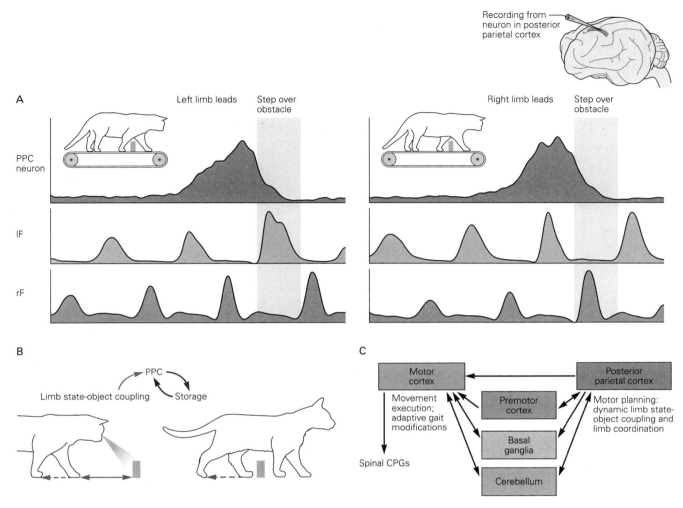

Figure 33–15 Neurons in the posterior parietal cortex (PPC) are involved in planning voluntary gait modifications.

A. Activity of a PPC neuron in the right cortex during a step over the obstacle when the left or right forelimb is the first to step over the obstacle. In each situation, the cell in the PPC discharges two to three steps in advance of the step over the obstacle.

B. The observation that PPC neurons discharge independent of which limb is the first to step over the obstacle suggests a global function of PPC in the planning of locomotion. In a general scheme, the PPC neurons are involved in the estimation of

the relative location of an object with respect to the body (limb state–object coupling [**double arrow**]) and storage information in the PPC for later retrieval.

C. The PPC does not act alone in planning gait modifications. It is part of a cortical and subcortical network that includes, among other structures, the premotor cortex, the basal ganglia, and the cerebellum. Connections exist between each of these structures as well as between each of them and the motor cortex, which is responsible for the execution of the gait modification. (Abbreviation: **CPG**, central pattern generator.) (Adapted, with permission, from Drew and Marigold 2015.)

so that each leg walks at a different speed, they initially show a very asymmetric gait before adapting over time to a more asymmetric one. When the two treadmill belts are reset to the same speed, they again show an asymmetric gait, demonstrating that the experimental condition had produced adaptation (see Figure 30–13). Patients with damage to the cerebellum are not able to adapt to this condition.

The Basal Ganglia Modify Cortical and Brain Stem Circuits

The basal ganglia are found in all vertebrates from the oldest vertebrates to primates and probably contribute to the selection of different motor patterns. The importance of the basal ganglia to the control of locomotion is clearly demonstrated by the deficits in

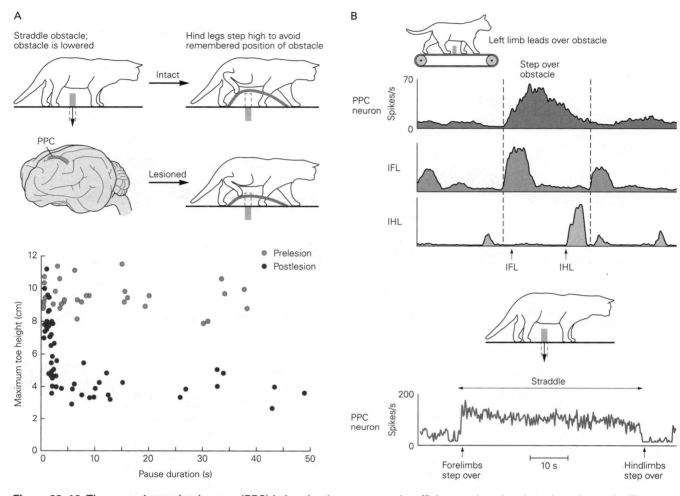

Figure 33–16 The posterior parietal cortex (PPC) is involved in maintaining an estimate of an obstacle in working memory during locomotion.

A. Upper figure: Normal animals were trained to walk forward, step over an obstacle, and then pause. While the animal paused, the obstacle was removed. When walking resumed, the hind legs stepped high to avoid the remembered obstacle. This memory lasted for more than 30 seconds. The trajectory of the hindlimbs was scaled appropriately for the height of the obstacle and for the relative position of the hind paws. Bilateral lesions of the PPC led to an impairment in the memory, making it impossible for the animal to pass the obstacle without hitting it. **Lower figure:** Following the lesion, animals stored the memory for only 1 to 2 seconds, and the maximum height of the

toe was insufficient to clear the obstacle and was significantly lower than in the prelesion condition. (Adapted from McVea and Pearson 2009.)

B. Upper figure: Neurons recorded in an intact animal in the PPC on the right side discharged in the period between the passage of the left forelimb (**IFL**) and hindlimb (**IHL**) over an obstacle (represented by electromyogram activity from representative flexor muscles in each limb). This discharge may be used to coordinate the movement of the hindlimb with that of the forelimb during the visually guided gait modification. **Lower figure:** When the cat steps over an obstacle and pauses, as in part **A**, cells in the PPC show a maintained discharge that could provide the neural representation of the working memory. (Adapted, with permission, from Lajoie et al. 2010.)

locomotion observed in patients with Parkinson disease, which disrupts the normal functioning of the basal ganglia due to degradation of their dopaminergic inputs from the substantia nigra (Chapter 38).

Such patients show a characteristic slow, shuffling gait and, in later stages of the disease, can also show "freezing" of gait. Patients with Parkinson disease also show problems with balance during locomotion and

with the anticipatory postural adjustments that occur at the initiation of a gait pattern. These deficits suggest that the basal ganglia contribute to the initiation, regulation, and modification of gait patterns. This regulation is mediated by the two major projections of the basal ganglia to the brain stem pathways and cortical structures.

The basal ganglia influence brain stem activity through their projections to the PPN. The PPN receives inhibitory inputs from GABAergic inhibitory neurons in the substantia nigra pars reticulata (SNr) as well as from the globus pallidus pars interna (GPi); it also receives glutamatergic input from neurons in the subthalamic nucleus (STN). Decreased inhibitory input and increased glutamatergic input to PPN from the basal ganglia are thought to promote activity in PPN and favor exploratory locomotion. The STN and GPi are major targets of deep brain stimulation for improvement of motor symptoms such as rigidity and reduced mobility in patients with Parkinson disease.

The basal ganglia influence cortical activity by means of its connections via the thalamus to different parts of the frontal cortex, including the supplementary motor regions. These connections allow the basal ganglia to exert a modulatory effect on visually guided locomotion, possibly by selecting the appropriate motor patterns required in different behavioral situations.

Computational Neuroscience Provides Insights Into Locomotor Circuits

While functional studies have revealed much about the organization of the locomotor networks, their overall complexity makes it difficult to capture the integrative function of synaptic and cellular properties of the circuit. Computational network modeling, however, allows one to simulate the circuit activity and to investigate the dynamic interactions between the circuit elements. Computational models can be developed at many levels: to study the ionic basis of neural activity within a given circuit, to study the connectivity between different groups of neurons in a particular circuit, or to better understand the interactions between different structures in the locomotor network. Computational models at each of these levels have been developed to study rhythm and pattern generation in both invertebrates and vertebrates and in the latter, ranging from the lamprey to mammals. As in other domains, approaches combining experimental manipulation and computational modeling are likely to increase in the coming years and have the potential to advance our understanding of the complex systems and interconnections between structures that are required to produce the full locomotor repertoire.

Neuronal Control of Human Locomotion Is Similar to That of Quadrupeds

By necessity, most of our understanding of the neural mechanisms underlying the control of locomotion comes from experiments on quadrupedal animals.

Nonetheless, the available evidence suggests that all the major principles concerning the origin and regulation of walking in quadrupeds also pertain to locomotion in humans. Although the issue of whether CPGs exist in humans remains contentious, several observations are compatible with the view that CPGs are important for human locomotion.

For example, observations of some patients with spinal cord injury parallel the findings from studies of spinal cats. Striking cases of patients with nearly complete transection of the spinal cord have shown uncontrollable, spontaneous, rhythmic movements of the legs when the hips were extended. This behavior closely parallels the rhythmic stepping movements in chronic spinal cats. Moreover, tonic electrical stimulation of the spinal cord below the injury can evoke locomotor-like activity, as in other mammals.

Parallels between human and quadrupedal walking have also been found in patients trained after spinal cord injury. Daily training combined with drug treatments restores stepping in spinal cats and improves stepping in patients with chronic spinal injuries. People with severe spinal cord injury who have been exposed to both treadmill-induced stepping and drug treatments similar to those that have been shown to activate the CPG in cats have demonstrated dramatic improvements in the ability to produce locomotion (Box 33–4). These results suggest that CPGs are present in humans and share functional similarities with CPGs found in other vertebrates.

Compelling evidence for the existence of spinal CPGs in humans also comes from studies in human infants who make rhythmic stepping movements immediately after birth if held upright and moved over a horizontal surface. This strongly suggests that some of the basic neuronal circuits for locomotion are innate and present at birth when descending control systems are not well developed. Because stepping can also occur in infants who lack cerebral hemispheres (*anencephaly*), these circuits must be located at or below the brain stem, perhaps entirely within the spinal cord.

During the first year of life, as automatic stepping is transformed into functional walking, these basic circuits are thought to be brought under supraspinal control. In particular, the stepping pattern gradually develops from a more primitive flexion-extension pattern that generates little effective forward movement to the mature pattern of complex movements. It is plausible, based on studies of cats, that this adaptation reflects maturation of descending systems that originate in the motor cortex and brain stem nuclei and are modulated by the cerebellum.

Box 33–4 Rehabilitative Training Improves Walking After Spinal Cord Injury in Humans

According to the World Health Organization, between 250,000 and 500,000 people worldwide incur spinal cords injuries annually. For many, this results in permanent loss of sensation, movement, and autonomic function. The devastating loss of functional abilities, together with the enormous cost of treatment and care, creates an urgent need for effective methods to repair the injured spinal cord and to facilitate functional recovery.

Over the past decades, progress has been made in animal research aimed at preventing secondary damage after injury, repairing the axons of lesioned neurons in the spinal cord, and promoting the regeneration of severed axons through and beyond the site of injury. In many instances, the regeneration of axons has been associated with modest recovery of locomotor function. However, none of the regeneration strategies has reached the point where they can be confidently used in humans with spinal cord injury.

Thus, rehabilitative training is the preferred treatment for people with spinal cord injury. One especially successful technique for enhancing walking in patients with partial damage to the spinal cord is repetitive, weight-supported stepping on a treadmill (Figure 33–17). This technique is based on the observation that spinal cats and rodents can be trained to step with their hind legs on a moving treadmill.

For humans, partial support of the body weight through a harness system is critical to the success of training; presumably, it facilitates the training of spinal cord circuits by reducing the requirements for supraspinal control of posture and balance.

Although the neural basis for the improvement in locomotor function with treadmill training has not been established, it is thought to depend on synaptic plasticity in local spinal circuits as well as successful transmission of at least some command signals from the brain through preserved descending pathways if the spinal cord injury is only partial.

Locomotor training is sometimes combined with other treatments. These include different types of medication designed to reduce spasticity, seen as involuntary muscle contractions, and facilitation of activity in spinal circuits by electrical transcutaneous activation of spinal circuits and/or activation of corticospinal pathways by transcranial magnetic stimulation.

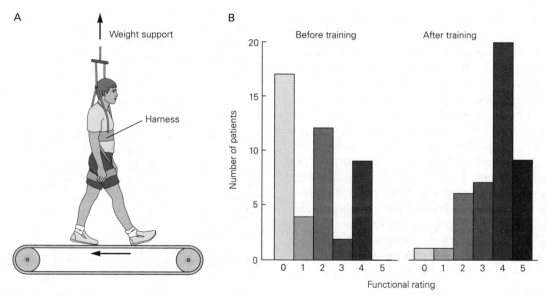

Figure 33–17 Treadmill training improves locomotor function in patients with partial spinal cord injury.

A. The patient is partially supported on a moving treadmill by a harness, and stepping movements are assisted by therapists.

B. Locomotor function improvement in 44 patients with chronic spinal cord injury after daily training lasting from 3 to 20 weeks. The functional rating ranges from 0 (unable to stand or walk) to 5 (walking without devices for more than five steps). (Adapted, with permission, from Wernig et al. 1995. Copyright © 2006, John Wiley and Sons.)

At the cortical level, stroke involving the motor cortex or damage to the corticospinal tracts leads to deficits in locomotion, as in cats. However, the deficits in humans are much stronger than in cats or even nonhuman primates, suggesting that the motor cortex in humans plays a more important role in locomotion than in other mammals. Studies using transcranial magnetic stimulation (TMS) to modulate motor cortical activity also show that the motor cortex contributes importantly to the control of human locomotion. TMS parameters that result in cortical inactivation, for example, produce a decrease in the level of muscle activity during locomotion. In contrast, TMS parameters that activate the motor cortex improve the recovery of locomotion following incomplete spinal cord injury.

Imaging studies, together with high-resolution electroencephalogram recordings, show changes in the activity of several cortical regions, including the motor cortex, premotor cortex, and PPC, during locomotion and particularly during imagined locomotion over obstacles. Imaging studies have also shown increased activity during locomotion in those parts of the midbrain shown to be important for the initiation and speed control of locomotion in animals. Similarly, neurons in the pedunculopontine nucleus can be affected in Parkinson disease, contributing to the severe gait disturbances seen in the late phase of the disease.

Highlights

1. Locomotion is a highly conserved behavior that is essential for the survival of the species. Our understanding of the neuronal mechanisms involved in the generation and control of locomotion came initially from the study of phylogenetically older animals, such as the lamprey and the tadpole. More recently, in mammals, with their more complex nervous systems, the organization of the different neural pathways involved in the generation and regulation of locomotion has also been determined in significant detail.

2. The spinal cord, in isolation from descending and rhythmical peripheral afferent inputs, can generate a complex locomotor pattern that contains elements of the rhythms and patterns observed in intact animals. The circuits responsible for producing this activity are referred to as central pattern generators (CPGs). Activity in spinal circuits can be modified by experience.

3. The basic components of CPGs controlling swimming are excitatory rhythm-generating neurons together with commissural inhibitory neurons responsible for left–right alternation. This organizational principle is also found in CPGs controlling limbed movements with the addition of flexor-extensor pattern-generating circuits and additional commissural neuronal networks. The circuits in the locomotor networks have a modular organization with distinct transmitter and molecular codes for the constituent neurons. Descending command signals act on these circuit elements to produce the diverse aspects of locomotor behavior.

4. Ionic membrane properties in interneurons and motor neurons contribute to rhythm and pattern generation. Cell-specific manipulation of these properties will enable a precise understanding of their relative contributions to locomotor production.

5. Peripheral afferent inputs modulate the function of spinal locomotor circuits. Proprioceptive sensors are used to stabilize phase transitions between stance and swing (and vice versa), whereas input from exteroceptors is used to modify limb activity in response to unexpected perturbations.

6. Circuits that are involved in initiating locomotion, controlling speed of locomotion, and selecting gaits are localized in the midbrain and encompass excitatory neurons in the pedunculopontine and cuneiform nuclei. These excitatory nuclei serve diverse roles in controlling either slow explorative locomotion or the full range of speeds and gaits including fast escape locomotion. Molecular-genetically driven cell-specific approaches allow unparalleled access to the organization of these pathways in the brain stem and how they integrate with spinal locomotor networks.

7. Activity in the three main structures in the brain stem with axons that descend to the spinal cord (the pontomedullary reticular formation, the lateral vestibular nucleus, and the red nucleus) contributes to the control of posture and interlimb coordination. Signals from these structures modify the level of muscle activity in a structure-specific manner.

8. The motor cortex provides precise control of muscle activity patterns to allow animals to make visually guided anticipatory adjustments of their gait. The signal from the motor cortex is integrated into the ongoing rhythm.

9. The posterior parietal cortex (PPC) is part of a network that contributes to the advanced planning

of gait based on visual information. PPC neurons estimate the relative location of objects with respect to the body and retain information in working memory to facilitate coordination of the limbs. The contribution of other cortical and subcortical areas to locomotor planning remains little studied.

10. Inputs from the cerebellum and the basal ganglia are used to correct motor errors and select the appropriate patterns of motor activity. The contribution of the basal ganglia to the control of locomotion is complex and is only now being determined.

11. The available evidence suggests that the neural control mechanisms determined from experiments in animals are also used to control locomotion in humans, including the existence of a CPG. Major advances remain to be made in understanding the mechanisms of spinal and supraspinal influences on human locomotor control.

12. Recent technological advances now give us an unparalleled opportunity to investigate the control mechanisms underlying locomotion. Molecular and genetic advances provide the ability to manipulate behavior at both the cellular and systems level and allow detailed study of the contributions of brain stem and spinal circuits to the initiation and regulation of locomotion. Advances in multineuronal recording techniques in animals, as well as the development of high-resolution recordings of human brain activity, will facilitate our understanding of the contribution of cortical structures to the control of locomotion.

Trevor Drew

Ole Kiehn

Suggested Reading

Armstrong DM. 1988. The supraspinal control of mammalian locomotion. J Physiol 405:1–37.

Brown T. 1911. The intrinsic factors in the act of progression in mammals. Proc R Soc B 84:308–319.

Drew T, Andujar JE, Lajoie K, Yakovenko S. 2008. Cortical mechanisms involved in visuomotor coordination during precision walking. Brain Res Rev 57:199–211.

Grillner S. 2006. Biological pattern generation: the cellular and computational logic of networks in motion. Neuron 52:751–766.

Jankowska E. 2008. Spinal interneuronal networks in the cat: elementary components. Brain Res Rev 57:46–55.

Kiehn O. 2016. Decoding the organization of spinal circuits that control locomotion. Nat Rev Neurosci 17:224–238.

Orlovsky G, Deliagina TG, Grillner S. 1999. *Neuronal Control of Locomotion: From Mollusc to Man.* Oxford: Oxford Univ. Press.

Pearson KG. 2008. Role of sensory feedback in the control of stance duration in walking cats. Brain Res Rev 57:222–227.

Rossignol S. 1996. Neural control of stereotypic limb movements. Supplement 29. In: *Handbook of Physiology, Exercise: Regulation and Integration of Multiple Systems.* New York: Wiley-Blackwell.

Sherrington CS. 1913. Further observations on the production of reflex stepping by combination of reflex excitation with reflex inhibition. J Physiol 47:196–214.

Takakusaki K, Chiba R, Nozu T, Okumura T. 2016. Brainstem control of locomotion and muscle tone with special reference to the role of the mesopontine tegmentum and medullary reticulospinal systems. J Neural Transm (Vienna) 123:695–729.

References

Ampatzis K, Song J, Ausborn J, El Manira A. 2014. Separate microcircuit modules of distinct v2a interneurons and motoneurons control the speed of locomotion. Neuron 83:934–943.

Bouvier J, Caggiano V, Leiras R, et al. 2015. Descending command neurons in the brain stem that halt locomotion. Cell 163:1191–1203.

Brocard F, Tazerart S, Vinay L. 2010. Do pacemakers drive the central pattern generator for locomotion in mammals? Neuroscientist 16:139–155.

Buchanan JT, Grillner S. 1987. Newly identified "glutamate interneurons" and their role in locomotion in the lamprey spinal cord. Science 236:312–314.

Butt SJ, Kiehn O. 2003. Functional identification of interneurons responsible for left-right coordination of hindlimbs in mammals. Neuron 38:953–963.

Caggiano V, Leiras R, Goni-Erro H, et al. 2018. Midbrain circuits that set locomotor speed and gait selection. Nature 553:455–460.

Capelli P, Pivetta C, Soledad Esposito M, Arber S. 2017. Locomotor speed control circuits in the caudal brainstem. Nature 551:373–377.

Choi JT, Bouyer LJ, Nielsen JB. 2015. Disruption of locomotor adaptation with repetitive transcranial magnetic stimulation over the motor cortex. Cereb Cortex 25:1981–1986.

Conway BA, Hultborn H, Kiehn O. 1987. Proprioceptive input resets central locomotor rhythm in the spinal cat. Exp Brain Res 68:643–656.

Crone SA, Quinlan KA, Zagoraiou L, et al. 2008. Genetic ablation of V2a ipsilateral interneurons disrupts left-right locomotor coordination in mammalian spinal cord. Neuron 60:70–83.

Dougherty KJ, Zagoraiou L, Satoh D, et al. 2013. Locomotor rhythm generation linked to the output of spinal shox2 excitatory interneurons. Neuron 80:920–933.

Drew T. 1988. Motor cortical cell discharge during voluntary gait modification. Brain Res 457:181–187.

Drew T. 1991. Functional organization within the medullary reticular formation of the intact unanesthetized cat. III. Microstimulation during locomotion. J Neurophysiol 66:919–938.

Drew T, Dubuc R, Rossignol S. 1986. Discharge patterns of reticulospinal and other reticular neurons in chronic, unrestrained cats walking on a treadmill. J Neurophysiol 55:375–401.

Drew T, Marigold DS. 2015. Taking the next step: cortical contributions to the control of locomotion. Curr Opin Neurobiol 33C:25–33.

Dubuc R, Brocard F, Antri M, et al. 2008. Initiation of locomotion in lampreys. Brain Res Rev 57:172–182.

Edgerton VR, Leon RD, Harkema SJ, et al. 2001. Retraining the injured spinal cord. J Physiol 533:15–22.

Engberg I, Lundberg A. 1969. An electromyographic analysis of muscular activity in the hindlimb of the cat during unrestrained locomotion. Acta Physiol Scand 75:614–630.

Forssberg H. 1985. Ontogeny of human locomotor control. I. Infant stepping, supported locomotion and transition to independent locomotion. Exp Brain Res 57:480–493.

Goulding M. 2009. Circuits controlling vertebrate locomotion: moving in a new direction. Nat Rev Neurosci 10:507–518.

Grillner S. 2006. Biological pattern generation: the cellular and computational logic of networks in motion. Neuron 52:751–766.

Grillner S. 1981. Control of locomotion in bipeds, tetrapods, and fish. In: V Brooks (ed). *Handbook of Physiology.* Rockville, MD: American Physiological Society.

Grillner S, Jessell TM. 2009. Measured motion: searching for simplicity in spinal locomotor networks. Curr Opin Neurobiol 19:572–586.

Grillner S, Rossignol S. 1978. On the initiation of the swing phase of locomotion in chronic spinal cats. Brain Res 146:269–277.

Grillner S, Zangger P. 1979. On the central generation of locomotion in the low spinal cat. Exp Brain Res 34:241–261.

Grillner S, Zangger P. 1984. The effect of dorsal root transection on the efferent motor pattern in the cat's hindlimb during locomotion. Acta Physiol Scand 120:393–405.

Hagglund M, Borgius L, Dougherty KJ, Kiehn O. 2010. Activation of groups of excitatory neurons in the mammalian spinal cord or hindbrain evokes locomotion. Nat Neurosci 13:246–252.

Harris-Warrick RM. 2011. Neuromodulation and flexibility in central pattern generator networks. Curr Opin Neurobiol 21:685–692.

Hiebert GW, Whelan PJ, Prochazka A, Pearson KG. 1996. Contribution of hind limb flexor muscle afferents to the timing of phase transitions in the cat step cycle. J Neurophysiol 75:1126–1137.

Hounsgaard J, Hultborn H, Kiehn O. 1986. Transmitter-controlled properties of alpha-motoneurones causing long-lasting motor discharge to brief excitatory inputs. Prog Brain Res 64:39–49.

Hultborn H, Nielsen JB. 2007. Spinal control of locomotion—from cat to man. Acta Physiol (Oxf) 189:111–121.

Jessell TM. 2000. Neuronal specification in the spinal cord: inductive signals and transcriptional codes. Nat Rev Genet 1:20–29.

Jordan LM, Liu J, Hedlund PB, Akay T, Pearson KG. 2008. Descending command systems for the initiation of locomotion in mammals. Brain Res Rev 57:183–191.

Juvin L, Gratsch S, Trillaud-Doppia E, Gariepy JF, Buschges A, Dubuc R. 2016. A specific population of reticulospinal neurons controls the termination of locomotion. Cell Rep 15:2377–2386.

Kiehn O. 2006. Locomotor circuits in the mammalian spinal cord. Annu Rev Neurosci 29:279–306.

Kiehn O, Sillar KT, Kjaerulff O, McDearmid JR. 1999. Effects of noradrenaline on locomotor rhythm-generating networks in the isolated neonatal rat spinal cord. J Neurophysiol 82:741–746.

Kriellaars DJ, Brownstone RM, Noga BR, Jordan LM. 1994. Mechanical entrainment of fictive locomotion in the decerebrate cat. J Neurophysiol 71:2074–2086.

Lajoie K, Andujar JE, Pearson K, Drew T. 2010. Neurons in area 5 of the posterior parietal cortex in the cat contribute to interlimb coordination during visually guided locomotion: a role in working memory. J Neurophysiol 103:2234–2254.

Lanuza GM, Gosgnach S, Pierani A, Jessell TM, Goulding M. 2004. Genetic identification of spinal interneurons that coordinate left-right locomotor activity necessary for walking movements. Neuron 42:375–386.

McCrea DA, Rybak IA. 2007. Modeling the mammalian locomotor CPG: insights from mistakes and perturbations. Prog Brain Res 165:235–253.

McLean DL, Masino MA, Koh IY, Lindquist WB, Fetcho JR. 2008. Continuous shifts in the active set of spinal interneurons during changes in locomotor speed. Nat Neurosci 11:1419–1429.

McVea DA, Pearson KG. 2009. Object avoidance during locomotion. Adv Exp Med Biol 629:293–315.

Pearson KG, Rossignol S. 1991. Fictive motor patterns in chronic spinal cats. J Neurophysiol 66:1874–1887.

Picton LD, Sillar KT. 2016. Mechanisms underlying the endogenous dopaminergic inhibition of spinal locomotor circuit function in Xenopus tadpoles. Sci Rep 6:35749.

Rossignol S, Frigon A. 2011. Recovery of locomotion after spinal cord injury: some facts and mechanisms. Annu Rev Neurosci 34:413–440.

Rybak IA, Stecina K, Shevtsova NA, McCrea DA. 2006. Modelling spinal circuitry involved in locomotor pattern generation: insights from the effects of afferent stimulation. J Physiol 577:641–658.

Ryczko D, Dubuc R. 2013. The multifunctional mesencephalic locomotor region. Curr Pharm Des 19:4448–4470.

Shik M, Severin F, Orlovskii G. 1966. [Control of walking and running by means of electric stimulation of the midbrain]. Biofizika 11:659–666. [article in Russian]

Soffe SR, Roberts A, Li WC. 2009. Defining the excitatory neurons that drive the locomotor rhythm in a simple vertebrate: insights into the origin of reticulospinal control. J Physiol 587:4829–4844.

Talpalar AE, Bouvier J, Borgius L, Fortin G, Pierani A, Kiehn O. 2013. Dual-mode operation of neuronal networks involved in left-right alternation. Nature 500:85–88.

Talpalar AE, Endo T, Low P, et al. 2011. Identification of minimal neuronal networks involved in flexor-extensor alternation in the mammalian spinal cord. Neuron 71: 1071–1084.

Wernig A, Muller S, Nanassy A, Cagol E. 1995. Laufband therapy based on 'rules of spinal locomotion' is effective in spinal cord injured persons. Eur J Neurosci 7:823–829.

Whelan PJ, Hiebert GW, Pearson KG. 1995. Stimulation of the group I extensor afferents prolongs the stance phase in walking cats. Exp Brain Res 103:20–30.

Yang JF, Mitton M, Musselman KE, Patrick SK, Tajino J. 2015. Characteristics of the developing human locomotor system: similarities to other mammals. Dev Psychobiol 57:397–408.

Zagoraiou L, Akay T, Martin JF, Brownstone RM, Jessell TM, Miles GB. 2009. A cluster of cholinergic premotor interneurons modulates mouse locomotor activity. Neuron 64:645–662.

Zhang J, Lanuza GM, Britz O, et al. 2014. V1 and v2b interneurons secure the alternating flexor-extensor motor activity mice require for limbed locomotion. Neuron 82:138–150.

Zhang Y, Narayan S, Geiman E, et al. 2008. V3 spinal neurons establish a robust and balanced locomotor rhythm during walking. Neuron 60:84–96.

34

Voluntary Movement: Motor Cortices

I N THIS CHAPTER, WE DESCRIBE HOW the cerebral cortex uses sensory information from the external world to guide motor actions that allow the individual to interact with the surrounding environment. We begin with a general description of what we mean by the term voluntary movement and some theoretical frameworks for understanding its control, followed by the basic anatomy of the cortical circuits involved in voluntary motor behavior. We then consider how information related to the body, external space, and behavioral goals is combined and processed in parietal cortical regions. This is followed by a discussion of the role of premotor cortical regions in selecting and planning motor actions. Finally, we examine the role played by the primary motor cortex in motor execution.

Voluntary Movement Is the Physical Manifestation of an Intention to Act

Animals, including humans, have a nervous system not just so that they can sense their world or think about it, but primarily to interact with it to survive and reproduce. Understanding how purposeful actions are achieved is one of the great challenges in neuroscience. We focus here on the cerebral cortical control of voluntary motor behavior, in particular voluntary arm and hand movements in primates.

In contrast to stereotypical fixed-latency reflexive responses that are automatically triggered by incoming sensory stimuli (Chapter 32), voluntary movements are purposeful, intentional, and context-dependent, and are typically accompanied, at least in humans, by a sense of "ownership" of the actions, the sense that the actions have been willfully caused by the individual. Decisions to act are often made without an external trigger stimulus. Moreover, the continuous flux of events and conditions in the world presents changing opportunities for action, and thus voluntary action involves choices between alternatives, including the choice not to act. Finally, the same object or event can evoke different actions at different times, depending on the current context.

Throughout evolution, these features of voluntary behavior have become increasingly prominent in higher primates, especially in humans, indicating that the neural circuits controlling voluntary behavior in primates are adaptive. In particular, evolution has resulted in an increasing degree of dissociation of the physical properties of sensory inputs from their behavioral salience to the individual. Adaptation of the control circuits also enhances the repertoire of voluntary motor actions available to a species by allowing individuals to remember and learn from prior experience, to predict the future outcomes of different action choices, and to adopt new strategies and find new solutions to attain their desired goals. Volitional self-control over how, when, and even whether to act endows primate voluntary behavior with much of its richness and flexibility and prevents behavior from becoming impulsive, compulsive, or even harmful.

Voluntary behavior is the physical manifestation of an individual's intention to act on the environment, usually to achieve a goal immediately or at some point in the future. This may require single nonstereotypical movements or sequences of actions tailored to current conditions and to the longer-term objectives of the individual. The ability to use fingers, hands, and arms independent of locomotion further helps primates, and especially humans, exploit their environment. Most animals must search their environment for food when hungry. In contrast, humans can "forage" by using their hands to cook a meal or simply enter a few numbers on a cellphone to order food for delivery. Because large areas of the cerebral cortex are implicated in various aspects of voluntary motor control, the study of the cortical control of voluntary movement provides important insights into the purposive functional organization of the cerebral cortex as a whole.

Theoretical Frameworks Help Interpret Behavior and the Neural Basis of Voluntary Control

The neural processes by which individuals acquire information about their environment and the relationship of their body to it, decide how to interact with the environment to achieve short- or long-term goals, and organize and execute the voluntary movement(s) that will fulfill their goals are traditionally partitioned into three analytic components: Perceptual mechanisms generate an internal representation of the external world and the individual within it, cognitive processes use this internal model of the world to select a course of action to interact with its environment, and the chosen plan of action is then relayed to the motor systems for implementation. This serial view of the brain's overall functional organization has long dominated neuroscience; this textbook, for example, has separate sections dedicated to perception, cognition, and movement.

The brain must transform a goal into motor commands that realize the goal. For example, taking a sip of coffee requires the brain to convert visual information about the coffee cup and somatic information about the current posture and motion of your arm and hand into a pattern of muscle contractions that moves your hand to the cup, grasps it, and then lifts it to your mouth. Many behavioral and modeling studies suggest that this could be accomplished by a series of transformations of sensorimotor coordinates that convert the retinal image of the cup into motor commands (Figure 34–1A).

Variants of this sensorimotor transformation model have guided the design and interpretation of many studies on the control of voluntary arm movements. Neural recording studies, including many that will be described here, have found possible neural correlates of the motor parameters and sensorimotor transformations presumed to underlie movement

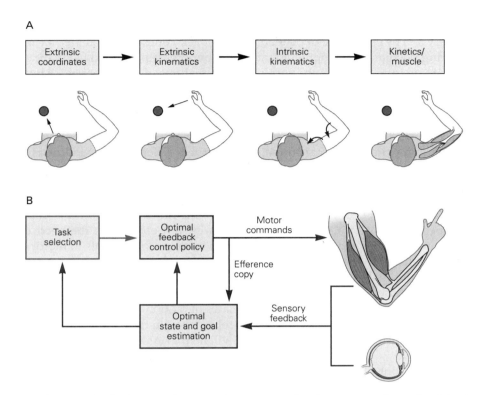

Figure 34–1 Theoretical frameworks for interpreting neural processing during voluntary motor actions.

A. The concept of sensorimotor transformations addresses the basic problem that tasks such as reaching to a visual target require the brain and spinal cord to convert sensory information about the spatial location of the target, initially represented in retinal coordinates, into patterns of muscle activity to move the limb to the target object. It is assumed that this sensorimotor transformation involves the use of intermediary representations—representation of the location of the target object relative to the body, the spatiotemporal trajectory of the hand (extrinsic kinematics), and motion of the joints (intrinsic kinematics) necessary to reach and grasp the object—before generating the patterns of neural activity that specify the causal forces (kinetics) or muscle activity.

B. Optimal feedback control recognizes three key processes for control. Optimal state and goal estimation (**red box**) integrates sensory feedback from various modalities along with an efference copy of motor commands to estimate the present position and motion of the body and objects in the world. Task selection (**blue box**) involves processes that identify behavioral goals based on internal desires and information about the state of the body and the world. Control policy (**green box**) determines the feedback gains, operations, and processes necessary to generate motor commands to control movement.

planning and execution. This conceptual framework is an example of a *representational model* of brain function. Just as the activity of neurons in primary sensory areas appears to encode specific physical properties of stimuli, the sensorimotor transformation model assumes that the activity of neurons in the motor system explicitly encodes or represents specific properties and parameters of the intended movement.

However, the sensorimotor transformation model has important limitations. Among them, the parameters and coordinate systems typically used in such models were imported from physics and engineering, rather than derived from the physiological properties of biological sensors and effectors. Furthermore, the model places all emphasis on strictly serial feedforward computations and relegates feedback circuits primarily to the detection and correction of performance errors after they are committed. The model also requires that every temporal detail of a movement be explicitly calculated before the motor system can generate any motor commands. Another limitation is its rigidity; it assumes that the same sequence of computations controls every movement in every context. Finally, this approach has not addressed how the proposed sensorimotor transformations could be implemented by neurons.

In recent years, theoretical studies of the motor system have been moving away from strictly representational models to more dynamical causal models. This approach begins with the premise that the functional architecture of motor control circuits evolved to generate movements, not to represent their parameters.

These circuit properties were acquired by evolutionary changes in neural circuitry and by experience-dependent adaptive processes during postnatal development that produce the patterns of synaptic connectivity within the neural circuits that are necessary to generate the desired movements. Spinal and supraspinal motor circuits ensure that spinal motor neurons generate the appropriate muscle contraction signals across task conditions without relying on computational formalisms such as coordinate transformations.

One such theoretical framework is optimal feedback control (Figure 34–1B; and see Chapter 30). There are many different forms of optimal control, and each captures important aspects of control. Optimal feedback control, as the name implies, emphasizes the importance of feedback signals for the planning and control of movements. It is optimal in the sense that it emphasizes the importance of the behavioral goal and the current context in determining how best to plan and control movements. This flexibility can explain how human motor performance can be both highly variable and yet successful.

The optimal feedback control framework also divides the control of voluntary movements into three key processes: state estimation, task selection, and control policy (Figure 34–1B). State estimation involves forward internal models that use efference copies of motor commands and external sensory feedback to provide the best estimate of the present state of the body and the environment (Chapter 30). Task selection involves the neural processes by which the brain chooses a behavioral goal in the current context and what motor action(s) might best attain that goal. This selection can be based on the sensory evidence supporting alternative actions and alternate options to attain the goal, and on other factors that influence the optimal response such as motivational state, task urgency, preferences, relative benefits versus risks, the mechanical properties of the body and environment, and even the biomechanical costs of different action choices. Finally, the control policy provides the set of rules and computations that establish how to generate the motor commands to attain the behavioral goal given the present state of the body and the environment. Importantly, the control policy process in optimal feedback control is not a series of pure feedforward computations to calculate every instantaneous detail of a desired movement trajectory and associated muscle activity patterns before movement onset. Instead, it involves context- and time-dependent adjustments to feedback circuit gains that allow the spatiotemporal form of muscle activity to emerge dynamically in real time as part of the control process underlying movement generation.

The sensorimotor transformation and optimal feedback control models are not mutually incompatible hypotheses. Optimal feedback control explains certain features of motor behavior but is largely agnostic as to the neural implementation for control. It assumes that motor circuits are dynamical systems that attain desired goals under varying task constraints. As a result, a given neuron may contribute to sensorimotor control in different task conditions, but its activity may not correspond to a specific movement parameter in a definable coordinate framework. In contrast, sensorimotor transformation models do not fully explain how real-time movement control is implemented by motor circuits, but emphasize the need to convert information from sensory signals to motor commands.

Even if the neural control system is dynamical, the system it controls—the musculoskeletal plant—is a physical object that must obey the universal physical laws of motion. Thus, neural activity should show correlations with those physical parameters and laws that will help to infer how those neurons are contributing to voluntary motor control, even if they are not attempting to encode those terms. Indeed, experimental tasks that dissociate different types of movement-related information have revealed important differences in how neural activity in different cortical motor regions correlates with different movement properties and different aspects of movement planning and execution. Finally, we can impose arbitrary volitional control on how we move. For example, we can choose to make an unobstructed reaching movement efficiently along a straight path to the target or whimsically along a complex curved path even though there is no obstacle to avoid and the movement is energetically costly. The experimental challenge is to reveal how the brain can implement this willful control with neurons and neural circuits.

Many Frontal and Parietal Cortical Regions Are Involved in Voluntary Control

Here we describe the regions of frontal and parietal cortex that convert sensory inputs into motor commands to produce voluntary movement. We then examine the neural circuits involved in the voluntary control of arm and hand movements that are prominent components of the motor repertory of primates. We focus on studies in the rhesus monkey (*Macaca mulatta*), as much of our knowledge of the cortical control of the arm and hand comes from this species and the neural circuitry underlying human voluntary control appears to have a similar organization. Many other neural structures, including the prefrontal cortex, the basal ganglia, and cerebellum, also play critical roles in the global organization of goal-directed voluntary behavior (Chapters 37 and 38).

Several different nomenclatures have been used in partitioning the precentral, postcentral, and parietal cortex, based on regional differences in cytoarchitectonic and myeloarchitectonic details, cortico-cortical connectivity, the distribution of different marker molecules, and regional differences in neural response properties. Here we will use some of the more widely accepted terminology without describing approximate homologies among the various nomenclatures.

Based on the pioneering cytoarchitectonic studies of humans by Brodmann, the different lobes of the monkey's cerebral cortex were divided into smaller regions, including two in precentral cortex (areas 4 and 6), four in the postcentral cortex (areas 1, 2, 3a, and 3b), and at least two in the superior and inferior parietal cortex (areas 5 and 7). While these cytoarchitectonic divisions persist in the literature, subsequent anatomical and functional studies have radically changed the view of how the precentral and parietal cortices are organized (Figure 34–2).

Current maps usually place the *primary motor cortex* (M1), the cortical region most directly involved in motor execution in primates, in Brodmann's area 4. Brodmann's area 6 is now typically divided into five or six functional areas that are principally involved in different aspects of the planning and control of motor actions of different parts of the body. Arm-control regions include the *dorsal premotor cortex* (PMd) and *predorsal premotor cortex* (pre-PMd), in the caudal and rostral parts of the dorsal convexity of lateral area 6, respectively. Hand-control regions include the *ventral premotor cortex* (PMv), found on the ventral convexity of area 6, which has been further divided into two or three smaller subregions. A variety of functions related to motor selection, sequencing, and initiation have been found in medial premotor cortical regions. These include a region on the medial surface of the cortical hemisphere that was originally called the secondary motor cortex by Woolsey and colleagues, who discovered it, but is now called the *supplementary motor area*. This region is in turn split into two regions, a *supplementary motor area proper* (SMA) in the caudal part and a *presupplementary motor area* (pre-SMA) in the rostral part. Outside of Brodmann's area 6, three additional motor areas, the dorsal, ventral, and rostral cingulate motor areas (CMAd, CMAv, and CMAr, respectively), are also involved in motor selection but have not been as well studied as more lateral premotor areas.

The *primary somatosensory cortex* (S-I; including areas 1, 2, 3a, and 3b) is located in the anterior postcentral gyrus. It processes cutaneous and muscle mechanoreceptor signals from the periphery and transmits that information to other parietal and precentral cortical regions (Chapter 19). Like area 6, Brodmann's parietal areas 5 and 7 are now divided into several regions within and adjacent to the intraparietal sulcus (IPS), each of which integrates various types of sensory information about the body or spatial goals for voluntary motor control. These include parietal lobe areas PE and PEc on the rostral or superior bank, and PF, PFG, PG, and OPT on the caudal, inferior bank. Areas inside the IPS include the anterior, lateral, medial, and ventral intraparietal areas (AIP, LIP, MIP, and VIP, respectively) as well as intraparietal area PEip and higher visual area V6A.

These precentral, postcentral, and parietal cortical regions are interconnected by complex patterns of reciprocal, convergent, and divergent projections. The SMA, PMd, and PMv have somatotopically organized reciprocal connections not only with M1 but also with each other. Both the SMA and M1 receive somatotopically organized input from S-I and the dorsorostral parietal cortex, whereas PMd and PMv are reciprocally connected with progressively more caudal, medial, and lateral parts of the parietal cortex. These somatosensory and parietal inputs provide the primary motor and caudal premotor regions with sensory information related to behavioral goals, target objects, and the position and motion of the body that is used to plan and guide motor acts.

In contrast, pre-SMA and pre-PMd project to SMA and PMd but do not project to M1 and are only weakly connected with the parietal lobe. They instead have reciprocal connections with prefrontal cortex and so may impose more arbitrary context-dependent control over voluntary behavior. Prefrontal cortex is also connected with other premotor cortical regions.

The control of hand and arm motor actions is implemented by partially segregated parallel circuits distributed across several parietal and precentral motor areas. Hand motor function is generally supported by frontoparietal circuits that are located more laterally, notably AIP and PMv. In contrast, proximal arm motor function is supported by circuits that are more medial, notably parietal areas PE and MIP and precentral areas PMd, SMA, and pre-SMA.

Descending Motor Commands Are Principally Transmitted by the Corticospinal Tract

Older textbooks often referred to the primary motor cortex (M1) as the "final common path." Other

A Human

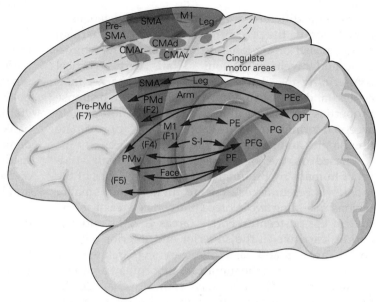

B Macaque monkey

Areas on the cortical convexity

Areas inside the parietal sulcus

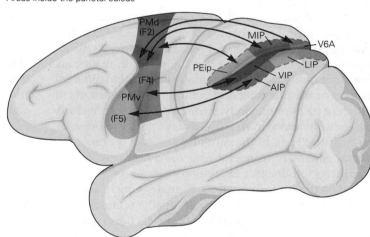

Figure 34–2 Parietal and frontal motor areas that support voluntary control. For illustration purposes, the intraparietal sulcus is opened in the bottom panel. The parietal areas are designated in Constantin von Economo's terminology by the letter **P** (parietal), followed by letters instead of numbers to indicate the cytoarchitectonically different areas. Areas PF and PFG roughly correspond to Brodmann's area 7b, and areas PG and OPT to Brodmann's area 7a. Areas inside the intraparietal sulcus include the anterior, lateral, medial, and ventral intraparietal areas (**AIP, LIP, MIP, VIP,** respectively), as well as the PE intraparietal area

(**PEip**) and visual area 6A (**V6A**). **Arrows** show the patterns of the principal reciprocal connections between functionally related parietal and frontal motor areas. (Abbreviations: **CMAr**, rostral cingulate motor area; **CMAv**, ventral cingulate motor area; **CMd**, dorsal cingulate motor area; **F**, frontal; **M1**, primary motor cortex; **OPT**, occipito-parieto-temporal; **P**, parietal; **PE, PF,** and **PFG** are parietal areas according to the nomenclature of von Economo; **PMd**, dorsal premotor cortex; **PMv**, ventral premotor cortex; **Pre-PMd**, predorsal premotor cortex; **S-I**, primary somatosensory cortex; **SMA**, supplemental motor area.)

cortical motor areas were thought to influence voluntary movements via their projections to M1, which then formulated the descending motor command that was transmitted to the spinal cord. This is not correct.

Several cortical motor regions outside of M1 project to subcortical areas of the brain as well as to the spinal cord in parallel with the descending projections from M1. The key descending pathway for voluntary control is the *pyramidal tract*, which originates in cortical layer V in a number of precentral and parietal areas. The pyramidal tract contains axons that terminate in brain stem motor structures (the *corticobulbar tract*) and axons that project down to the spinal cord (*corticospinal tract*). Precentral areas include not only M1 but also SMA, PMd, PMv, and the cingulate motor areas (Figure 34–3). Descending fibers from S-I and parietal areas, including PE and PFG, also travel in the pyramidal tract. The pre-SMA and pre-PMd do not send axons directly to the spinal cord; instead, their descending outputs reach the spinal cord indirectly through projections to other subcortical structures.

Most corticospinal tract axons originating in one hemisphere cross to the other side of the midline (decussate) at the pyramid in the caudal medulla, and from there project to the spinal cord itself, forming the lateral corticospinal tract. A small portion does not decussate and forms the ventral corticospinal tract. Many corticospinal axons in primates, and virtually all corticospinal axons in other mammals, terminate only on spinal interneurons and exert their influence on voluntary movement indirectly through spinal interneuronal and reflex pathways. In monkeys, all corticospinal axons from premotor cortical areas and many from M1 terminate on interneurons in the spinal intermediate zone, whereas postcentral and parietal areas target interneurons in the dorsal horn. The terminal endings of a sizeable portion of the corticospinal axons arising from M1 in primates, but not other mammals, arborize at their targets and synapse directly on spinal alpha motor neurons that in turn innervate muscles; these M1 neurons with direct monosynaptic projections to spinal motor neurons are called *corticomotoneuronal cells*.

Any voluntary arm movement can have destabilizing effects on the rest of the body due to mechanical interactions between body segments. Thus, control of voluntary arm movements requires coordination with neural circuits responsible for the control of posture and balance. This is mediated by descending projections from cortical motor areas to the reticular formation, which in turn project to the spinal cord via the reticulospinal tract (Chapters 33 and 36).

Imposing a Delay Period Before the Onset of Movement Isolates the Neural Activity Associated With Planning From That Associated With Executing the Action

Voluntary movement requires the intervention of a number of neural processes between the arrival of salient sensory inputs and the initiation of an appropriate motor response. With the development in the 1960s of single-cell recording in the cerebral cortex of awake animals, tasks that experimentally manipulate different attributes of movements have been used to study every cortical area involved in the control of arm and hand movements to try to identify neural correlates of the presumed control processes in each area.

In "reaction-time" tasks, animals make a prespecified response when they detect a particular stimulus, such as reaching to a target when it appears (Figure 34–4A). The stimulus informs the animal both what movement to make and when to make it. However, reaction times in such tasks are typically short, often less than 300 ms, and most or all putative planning stages leading up to the initiation of the movement are accomplished within that brief time. This makes it very difficult to discern what kinds of information are represented in the activity of the neurons at each given moment and thus to what processes they are contributing (Figure 34–4B).

However, a critical feature of voluntary behavior is that movement initiation is not obligatory the instant an intention to act is formed. This volitional control over the timing of movement has been exploited by so-called "instructed-delay" motor tasks (Figure 34–4A), in which an instructional cue informs the animal about specific aspects of an impending movement such as the location of a target, but the animal must withhold the response until a delayed stimulus signals when to make the movement. This protocol allows researchers to dissociate in time the neural processes associated with the early stages of planning the intended act from those that are directly coupled in real time to the initiation and control of the movement.

As expected, neurons in all the movement-related cortical areas discharge prior to and during movement execution in reaction-time tasks (Figure 34–4B), and their activity correlates systematically with different properties of movements, such as their direction, velocity, spatial trajectory, and causal forces and muscle activity. Critically, however, many neurons in the same areas also signal information about an intended motor act during an instructed-delay period long before its initiation (Figure 34–4B). Thus, even though planning and execution are distinct serial stages in voluntary motor control, they are not implemented by distinct neural populations in different cortical areas. Moreover, even a well-trained monkey will occasionally make the wrong movement in response to an instructional cue. In those trials, the activity during the delay period generally predicts the erroneous motor response that the monkey will eventually make. This is compelling evidence that the activity is a neural correlate of the monkey's motor intentions, not a passive sensory response to the instructional cues.

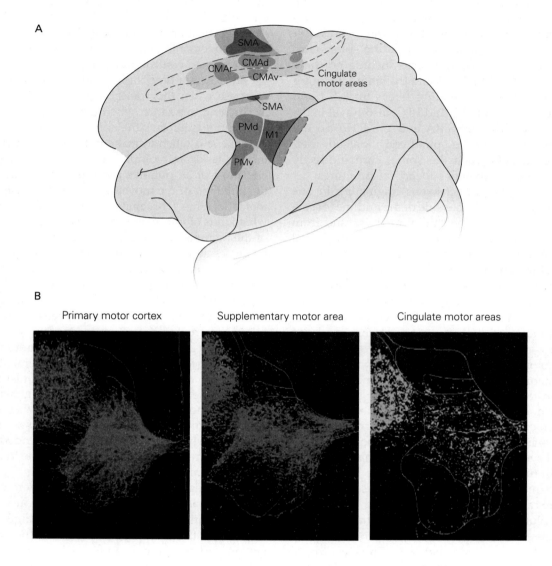

Figure 34–3 Cortical origins of the corticospinal tract. (Reproduced, with permission, from Dum and Strick 2002. Copyright © 2002 Elsevier Science Inc.)

A. Corticospinal neurons that modulate muscle activity in the contralateral arm and hand originate in the parts of the primary motor cortex (**M1**) motor map and many subdivisions of the premotor cortex (**PMd, PMv, SMA**) that are related to arm and hand movements (indicated by the darker zones). The axons from these areas project into the spinal cord cervical enlargement (see part **B**). Corticospinal fibers projecting to the leg, trunk, and other somatotopic parts of the brain stem and spinal motor system originate in the other parts of the motor and premotor cortex, indicated by the lighter zones. (Abbreviations: **CMAd**, dorsal cingulate motor area; **CMAr**, rostral cingulate motor area; **CMAv**, ventral cingulate motor area; **M1**, primary motor cortex; **PMd**, dorsal premotor cortex; **PMv**, ventral premotor cortex; **SMA**, supplementary motor area.)

B. Transverse sections of the spinal cord at the level of the cervical enlargement in monkeys after injection of the anterograde tracer horseradish peroxidase into different arm-related cortical motor regions to label the distribution of corticospinal axons arising from each cortical region. The corticospinal axons from the primary motor cortex (*left*), supplementary motor area (*middle*), and cingulate motor areas (*right*) all terminate on interneuronal networks in the intermediate laminae (V–VIII) of the spinal cord. Only the primary motor cortex contains corticospinal neurons (corticomotoneuronal cells) whose axons terminate directly on spinal motor neurons in the most ventral and lateral part of the spinal ventral horn (Rexed's lamina IX). Rexed's laminae I to IX of the dorsal and ventral horns are shown in faint outline in each section. The dense cluster of labeled axons adjacent to the dorsal horn (*upper left*) in each section are corticospinal axons descending in the dorsolateral funiculus, before entering the spinal intermediate and ventral laminae.

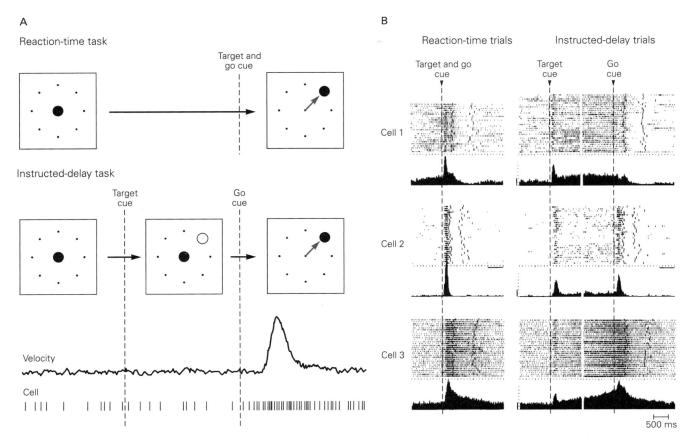

Figure 34–4 Neural processes related to movement planning and movement execution can be dissociated in time. (Reproduced, with permission, from Crammond and Kalaska 2000.)

A. In a *reaction-time task*, a sensory cue instructs the subject both where to move (target cue) and when to move (go cue). All neural operations required to plan and initiate the execution of the movement are performed in the brief time between the appearance of the cue and the onset of movement. In an *instructed-delay task*, an initial cue tells the subject where to move, and only later is the go cue given. The knowledge provided by the first cue permits the subject to plan the upcoming movement. Any changes in activity that occur after the first cue but before the second are presumed to be neural correlates of the planning stage.

B. Movement planning and execution are not completely segregated at the level of single neurons or neural populations in a given cortical area. Raster plots and cumulative histograms show the responses of three premotor cortex neurons to movements in each cell's preferred direction during reaction-time trials and instructed-delay trials. In the raster plots, each row represents activity in a single trial. The thin tics in each raster row represent action potentials, and the two thicker tics show the onset and end of movement. In reaction-time trials, the monkey does not know in which direction to move until the target appears. In contrast, in instructed-delay trials, an initial cue informs the monkey where the target lies well in advance of the appearance of a second signal to initiate the movement. During the delay period, activity in many premotor cells shows directionally tuned changes that signal the direction of the impending delayed movement. The activity in cell 1 appears to be strictly related to the planning phase of the task, for there is no execution-related activity after the go signal in the instructed-delay task. The other two cells show different degrees of activity related to both planning and execution.

Parietal Cortex Provides Information About the World and the Body for State Estimation to Plan and Execute Motor Actions

Sensory information is essential for selecting appropriate and effective actions. Before drinking from a cup, the brain uses visual input to identify which object is the cup, where it is located relative to the body, and its physical properties such as size, shape, and handle orientation.

In addition, information about the current posture and motion of the arm and hand is provided by integrating proprioceptive signals from the limb with efference copies of motor commands (Chapter 30). Finally, cutaneous signals are critical when interacting manually with objects, such as grasping and lifting the cup.

Several lines of evidence implicate the parietal cortex as a key brain region in sensory processing for motor action. The parietal lobe, especially PE, PEip, and MIP, receives strong somatic sensory inputs about body posture and movement from S-I. Several parietal regions along and within the IPS are major components of the dorsal visual pathway, which processes visuospatial information about objects that guides arm and hand movements while reaching to, grasping, and manipulating them. The parietal lobe is also reciprocally interconnected with precentral cortical motor areas to provide the precentral cortex with signals for the sensory guidance of movement and to receive efference copies of motor commands from those same precentral areas. Finally, human subjects with lesions of the posterior parietal cortex often demonstrate specific impairments in using sensory information to guide motor action (Box 34–1).

The Parietal Cortex Links Sensory Information to Motor Actions

We experience the space that surrounds us as a single unified environment within which objects have specific locations relative to each other and to ourselves. Classical neurology suggested that the parietal lobe constructed a unified multimodal neural representation of the world by integration of inputs from different sensory modalities. This single map of space was assumed to provide all the information necessary both for spatial perception and for the sensory guidance of movement, and so was shared by the different motor circuits that controlled different parts of the body, such as the eyes, arm, and hand.

However, the idea that the parietal cortex contains a single topographically organized representation of space is incorrect. Instead, the posterior parietal cortex contains several distinct functional areas that work in parallel and receive different combinations of sensory and motor inputs related to the guidance of movement of different effectors, such as the eyes, arm, and hand. Neurons in these areas are often multimodal, with both visual and somatic sensory receptive fields, and also discharge preferentially prior to and during movements of a specific effector. Each functional area is connected to frontal motor regions involved in control of the same effectors. Finally, each region is not topographically organized in the familiar sense of a faithful point-to-point representation of surrounding space, but rather comprises a complex mixture of neurons with different sensory inputs that may contribute to the multisensory integration required to guide motor actions with the environment.

Body Position and Motion Are Represented in Several Areas of Posterior Parietal Cortex

The S-I and adjacent superior parietal cortex regions PE, MIP, and PEip are a major source of proprioceptive and tactile sensory information about the position and motion of body parts. Neurons in S-I areas 1 and 2

Box 34–1 Lesion Studies of Posterior Parietal Cortex Lead to Deficits in the Use of Sensory Information to Guide Action

Naturally occurring or experimentally induced lesions have long been used to infer the roles of different neural structures. However, the effects of lesions must always be interpreted with caution. It is often incorrect to conclude that the function perturbed by an insult to a part of the motor system resides uniquely in the damaged structure or that the injured neurons explicitly perform that function. Furthermore, the adverse effects of lesions can be masked or altered by compensatory mechanisms in remaining, intact structures. Nevertheless, lesion experiments have been fundamental in differentiating the functional roles of brain regions.

Behavioral studies by Goodale, Milner, Rossetti, and others on patients with parietal cortical damage have led to the conclusion that a primary function of the parietal lobe is to extract sensory information about the external world and one's own body for the planning and guidance of movements. Such studies have shown that patients with

lesions of certain parts of the parietal lobe suffer specific deficits in the ability to direct their arm and hand accurately to the spatial location of objects and to shape the orientation and grip aperture of the hand prior to grasping it.

They have also shown a particularly severe deficit in the ability to make rapid adjustments to their ongoing reach and grasping actions in response to unexpected changes in the location or orientation of the target object. This visual guidance of action is provided by visual signals that are routed through the dorsal visual stream and may operate in parallel with and independently of perceptual processes evoked by the visual inputs that are routed simultaneously through the ventral visual stream in the temporal lobe. For instance, whereas our visual perception of the size and orientation of objects can be deceived by certain perceptual illusions, the motor system often behaves as if it is not fooled and makes accurate movements.

typically respond to tactile input from a limited part of the contralateral body or to movements of one or a few adjacent joints in specific directions.

In contrast, many PE and MIP neurons discharge during passive and active movements of multiple joints. Some cells also respond during combined movements of multiple body parts, including bilateral movements of both arms. Many PE and MIP neurons also have large tactile receptive fields whose responses are modulated by context during limb movement or posture. For instance, a neuron with a tactile receptive field that covers the entire glabrous (palmar) surface of the hand may only respond to physical contact with an object when the hand is close to the body and not when it touches the object with the arm fully extended.

These findings indicate that while area 1 and 2 neurons encode the positions and movements of specific body parts, superior parietal neurons integrate information on the positions of individual joints as well as the positions of limb segments with respect to the body. This integration creates a neural "body schema" that provides information on where the arm is located with respect to the body and how different arm segments are positioned and moving with respect to one another. This body schema is critical for selecting how to attain behavioral goals and for ongoing control of movement.

For instance, a key requirement for efficient reaching is knowledge of where the arm is before and during the reach. Monkeys with experimental lesions in Brodmann's area 2 and the adjacent superior parietal lobule (area 5 or PE) show deficits in reaching to and manipulating objects under proprioceptive and tactile guidance without vision. Human patients with similar lesions show the same deficit, without the spatial neglect that is a common consequence of more lateral lesions in the inferior parietal lobe.

Spatial Goals Are Represented in Several Areas of Posterior Parietal Cortex

Functional areas within the IPS are strongly implicated in the processing of spatial, especially visual, information relevant to action. Each of these areas has unique ways of representing objects and spatial goals relative to the body and contributes to controlling motor actions of different parts of the body. For instance, many neurons in the lateral intraparietal area (LIP) receive visual input from extrastriate cortical areas. Their receptive fields are fixed in retinal coordinates and shift to new spatial locations whenever the monkey changes its direction of gaze. Neural responses also often increase when the animal attends to a stimulus within the receptive field even without looking

at it, and they often discharge prior to a saccade that is directed toward a visual stimulus in their receptive field (Figure 34–5A; and see Chapter 35).

Several parietal regions are preferentially implicated in the control of arm and hand movements. For instance, the most medial regions of the superior parietal cortex, areas V6A and PEc, receive input from extrastriate visual areas V2 and V3. Many V6A and PEc neurons have visual receptive fields in retinal coordinates, but their activity is also frequently modulated by the direction of gaze, the current arm posture, and the direction of reaching movements.

The ventral intraparietal area (VIP) in the fundus of the IPS receives inputs from two components of the dorsal visual stream, the medial temporal cortex and medial superior temporal cortex, which are involved in the analysis of optic flow and visual motion. Many VIP neurons respond to visual stimuli and somatosensory stimuli with receptive fields on the face or head and, in some cases, on the arm or trunk. Neural activity is in head-centered coordinates, as somatosensory and visual information remains in register even if the eyes move to fixate different spatial locations (Figure 34–5B). Some VIP neurons respond to both visual and tactile stimuli moving in the same direction, whereas others are strongly activated by visual stimuli that move toward their tactile receptive field but only if the path of motion will eventually intersect the tactile receptive field. These neurons may allow the monkeys to link the location and motion of objects in their immediate peripersonal space with different parts of their body.

Another area of parietal cortex related to reaching is the parietal reach region (PRR). The PRR likely corresponds to the medial intraparietal cortex (MIP) and adjacent arm-control parts of the superior and inferior parietal cortex. The activity of many PRR neurons varies with the location of reach targets relative to the hand. However, this signal is not fixed to the current location of the hand or target but rather on the current direction of gaze (Figure 34–5C). Each time the monkey looks in a different direction, the reach-related activity of PRR neurons changes, even if the location of the target and hand and the required reach trajectory do not change. In contrast, the reach-related activity of many neurons in areas PE and PEip is less related to gaze and more strongly related to the current hand position and arm posture. PE and PEip neurons thus provide a more stable signal about the location of the reach target relative to the current position of the hand compared to PRR.

Finally, neurons in the anterior intraparietal area (AIP) are primarily implicated in object grasping and manipulation by movements of the hand. Many AIP neurons are preferentially active while reaching for

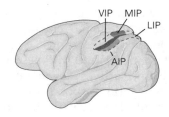

A Lateral intraparietal area (LIP)

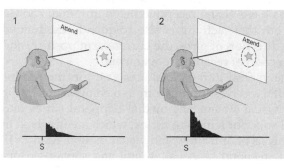
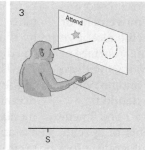
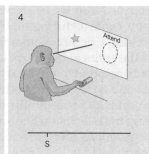

Receptive field
characteristics

Retina-centered,
attention sensitive

B Ventral intraparietal area (VIP)

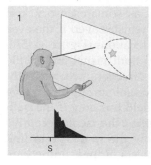
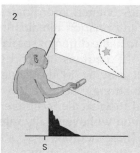
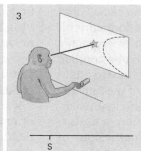
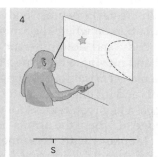

Head-centered

C Medial intraparietal area (MIP)

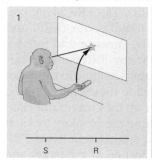
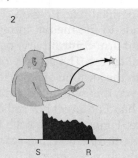
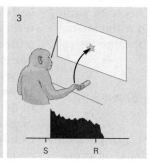
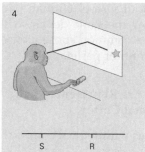

Retina-centered
direction of reach;
preparation to
reach

D Anterior intraparietal area (AIP)

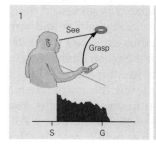
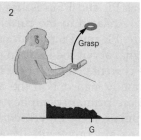
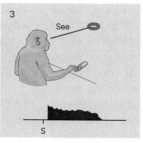
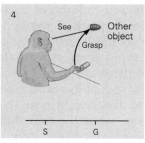

Retina-centered,
object-specific
viewing, grasping

and grasping objects of particular shapes, sizes, and spatial orientations, and often even while viewing those objects before grasping them (Figure 34–5D). There is a broad range of neural response properties, from neurons that respond almost exclusively to visual input about the objects but not to the grasping actions to neurons that discharge only during the hand movements themselves even in the dark. This suggests that the AIP contains neural circuits that begin to transform visual information about the physical properties of an object that are relevant to how it could be handled—what James Gibson has called the object's *affordances*—into appropriate hand actions (Chapter 56).

A fascinating discovery about the parietal cortex is that the receptive fields of neurons can be altered by individual experience, such as tool use. Monkeys were trained to retrieve food pellets that were out of normal reach of the arm and hand by using a rake-shaped tool. Many VIP neurons normally respond to visual objects when they are either located near the current position of the hand or anywhere within reach with the arm. After training, their visual receptive fields transiently expand to incorporate the tool when the monkey grasps it, as if the distal end of the tool had become a functional extension of the monkey's own hand and arm (Figure 34–6).

Internally Generated Feedback May Influence Parietal Cortex Activity

The delays involved in the transmission of visual and somatic feedback about arm movements from the periphery to cortical circuits can lead to oscillations or even instabilities in real-time sensorimotor control. One theoretical solution to this problem is to use a forward internal model to make predictive estimates of body motion based on internal efference copies of outgoing motor commands as well as from slower peripheral feedback signals (Chapter 30).

Several lines of evidence suggest that parietal cortex circuits, along with the cerebellum (Chapter 37), may implement a similar solution. Many reach-related neurons in PE, MIP, and PRR are active not only in response to passive sensory inputs but also before the onset of movement and during the instructed-delay period of delayed-reaching tasks. These responses suggest that these neurons process centrally generated signals about motor intentions prior to movement onset. This premovement activity is often interpreted as evidence that the parietal cortex generates feedforward signals that contribute to the early planning of movements. However, an alternate interpretation is that the premovement activity is driven by an efference copy of the motor command for the intended movement that is transmitted into the parietal cortex via its reciprocal connections with precentral motor areas. This combination of peripheral sensory inputs and central efference copies could permit some parietal reach-related circuits to compute a continuously updated estimate of the current state of the arm and its position relative to the behavioral goal. This estimate could be used to make rapid corrections for errors in ongoing arm movements.

Whether the parietal circuits are primarily involved in the formation of a subject's motor intentions or in

Figure 34–5 (Opposite) Neurons in the parietal cortex of the monkey are selective for the location of objects in the visual field relative to particular parts of the body. Each histogram represents the firing rate of a representative neuron as a function of time following presentation of a stimulus. In each diagram, the line emanating from the eyes indicates where the monkey is looking.

A. Neurons in the lateral intraparietal area have *retina-centered* receptive fields. The strength of the visual response depends on whether the monkey is paying attention to the stimulus (**S**). The neuron fires when a light is flashed inside its receptive field (**dotted circle**) (1). The response is more robust if the monkey is instructed to attend to the location of the stimulus (2). The neuron does not fire if the stimulus is presented outside the receptive field, regardless of where attention is directed (3, 4).

B. In the ventral intraparietal area, some neurons have *head-centered* receptive fields. This is determined by keeping the head in a fixed position while the monkey is instructed to shift its gaze to various locations. This neuron fires when a light appears to the right of the midline of the head (1, 2). It does not fire when the light appears at another location relative to the head, such as the midline or to the left. (3, 4). The critical

contrast is between situations 1 and 4. The retinal location of the light is the same in both (slightly to the right of the fixation point), yet the neuron fires in 1, when the stimulus is to the right of the head, but not in 4, when the stimulus is to the left of the head.

C. In the medial intraparietal area, neurons are selective for the retina-centered direction of the reach (**R**) and fire when the monkey is preparing to reach for a visual target. This neuron fires when the monkey reaches for a target to the right of where he is looking (2, 3). It does not fire when he reaches for a target at which he is looking (1) or when he moves only his eyes to the target at the right (4). The physical direction of the reach is not a factor in the neuron's firing: It is the same in 1 and 3, and yet, the neuron fires only in 3.

D. In the anterior intraparietal area, neurons are selective for objects of particular shapes and fire when the monkey is looking at or preparing to grasp (**G**) an object. This neuron fires when the monkey is viewing a ring (3) or making a memory-guided reach to it in the dark (2). It fires especially strongly when the monkey is grasping the ring under visual guidance (1). It does not fire during viewing or grasping of other objects (4).

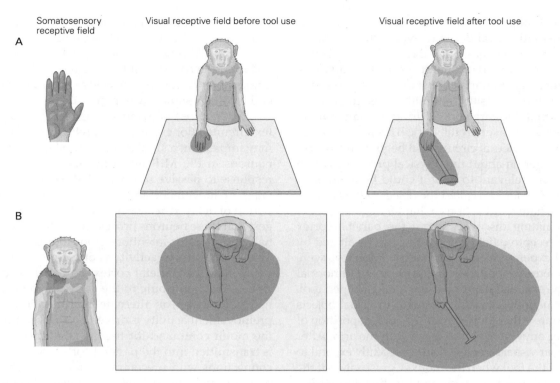

Somatosensory receptive field

Visual receptive field before tool use

Visual receptive field after tool use

A

B

Figure 34–6 Some neurons in the parietal cortex of the monkey have receptive fields that dynamically expand once a tool is grasped. (Adapted from Maravita and Iriki, 2004. Copyright © 2003 Published by Elsevier Ltd.)

A. The **orange** area on the hand (*left*) indicates the somatosensory receptive field for a neuron. The **purple** area (*middle*) indicates the neuron's visual receptive field (vRF) region around the hand. The vRF is anchored to the hand and changes spatial location whenever the monkey moves its arm. The vRF expands

when the monkey grasps a rake after it has learned how to use the rake to reach for objects in the workspace (*right*).

B. A single neuron that has a shoulder-centered bimodal somatosensory (**orange**) and visual (**purple**) receptive field is illustrated. The vRF for this neuron (*middle*) is larger than the one shown in part **A**, possibly reflecting the potential workspace related to whole-arm function. The vRF also expands to incorporate the extended workspace permitted by use of a rake (*right*).

state estimation will depend on the origin of its premovement activity. If it is mainly generated within the parietal cortex itself, this will strongly implicate the parietal cortex in the planning of intended movements. In contrast, if it is primarily driven by an efference copy relayed from precentral motor areas, this would strongly implicate the parietal circuits in state estimation, including predicting how the arm should move in response to the motor command.

Premotor Cortex Supports Motor Selection and Planning

As outlined at the beginning of this chapter, a decision to act in a particular way in a given situation is shaped by many factors, including sensory information about objects, events, and opportunities for action from the

environment, body position and motion, internal motivational states, prior experiences, reward preferences, and learned arbitrary rules and strategies linking sensory inputs to motor actions. There can be many reasons why you want to drink some coffee, and that desire can be fulfilled by actions ranging from simply reaching out to your full coffee cup to making coffee at home or going to a café.

Frontal premotor cortical regions just rostral to M1 play an important role in early movement planning or task-selection processes. Many neurons in those areas, such as the PMd neurons shown in Figure 34–4, generate activity during instructed-delay tasks that reflect the motor intentions of the monkey and even the factors that influenced those action choices. The different premotor cortical regions are presumed to make different but overlapping contributions to motor selection and planning. For instance, the lateral premotor cortex, including PMd and PMv, have traditionally been implicated in actions initiated and guided by external sensory inputs. In contrast, medial premotor areas, including SMA, pre-SMA, and CMA, have been implicated in the control of self-initiated movements as well as the suppression of actions. However, the distinction between their respective contributions is not absolute.

Medial Premotor Cortex Is Involved in the Contextual Control of Voluntary Actions

Clinton Woolsey's pioneering electrical stimulation studies showed that, in addition to the motor map in M1, the medial wall of the frontal cortex contains an array of neurons that also regulate body movements. This medial motor map, now called the supplementary motor area (SMA), includes the entire contralateral body but is coarser than the detailed map in M1, as described later. Strong stimulus currents are required to evoke movements, which are often complex actions such as postural adjustments or stepping

and climbing and can involve both sides of the body. Today, there is agreement that this region contains two areas that have distinct cytoarchitectonic characteristics, axonal connections, and functional properties: a more caudal SMA proper and a more rostral presupplementary motor area (pre-SMA), which we will collectively call the supplementary motor cortex (SMC).

The SMC has been implicated in many aspects of voluntary behavior, although its contribution remains controversial. Several lines of evidence support a role in self-initiated behavior. In humans, electrical stimulation of SMC below the threshold for movement initiation can evoke an introspective sense of an urge to move that does not arise during M1 stimulation. Lesions of SMC produce problems initiating desired movements or suppressing undesirable movements (Box 34–2). Moreover, recordings of slow cortical potentials at the surface of the skull during the execution of self-paced movements show that the initial potential arises in the frontal cortex as much as 0.8 to 1.0 second before the onset of movement. This signal, named the *readiness*

Box 34–2 Lesions of Premotor Cortex Lead to Impairments in the Selection, Initiation, and Suppression of Voluntary Behavior

Lesions of the supplementary motor area (SMA) and presupplementary motor area (pre-SMA) and the prefrontal areas connected with them produce deficits in the initiation and suppression of movements. Initiation deficits manifest themselves as loss of self-initiated arm movements, even though the patient can move when adequately prompted. This deficit can involve movement of parts of the body (*akinesia*) contralateral to the region and speech (*mutism*).

Deficits in movement suppression, in contrast, include the inability to suppress behaviors that are socially inappropriate. These include compulsive grasping of an object when the hand touches it (*forced grasping*), irrepressible reaching and searching movements aimed at an object that has been presented visually (*groping movements*), and impulsive arm and hand movements to grab nearby objects and even people without conscious awareness of the intention to do so (*alien-hand* or *anarchic-hand syndrome*).

Another striking syndrome is *utilization behavior*, in which a patient compulsively grabs and uses objects without consideration of need or the social context. Examples are picking up and putting on multiple pairs of glasses or reaching for and eating food when the

patient is not hungry or when the food is clearly part of someone else's meal.

These deficits in the initiation and suppression of actions may represent opposite facets of the same functional role for SMA and especially pre-SMA in the conditional or context-dependent control of voluntary behavior.

Lesions affecting premotor cortex also lead to impairments in the selection of motor actions. For example, when a normal monkey sees a tasty food treat behind a small transparent barrier, it readily reaches around the barrier to grasp it. However, after a large premotor cortex lesion, the monkey may persistently try to reach directly toward the treat and so repeatedly strikes the barrier with its hand, rather than making a detour around the barrier.

More focal lesions or inactivation of the ventral premotor cortex perturbs the ability to use visual information about an object to shape the hand appropriately for the object's size, shape, and orientation before grasping it. Focal lesions of the dorsal premotor cortex affect the ability to learn and recall arbitrary sensorimotor mappings or conditional stimulus–response associations, whereas supplementary motor cortex lesions impede the ability to learn and recall temporal sequences of movement.

potential, has its peak in the cortex centered in SMC. Because it occurs well before movement, the readiness potential has been widely interpreted as evidence that neural activity in this region is involved in forming the intention to move, not just in executing movement.

Neurons in both SMA and pre-SMA discharge before and during voluntary movements. Unlike M1 neurons, the activity of most SMA neurons is less tightly coupled to particular actions of a body part and appears instead to be associated with more complex, coordinated motor acts of the hand, arm, head, or trunk. Compared to SMA neurons, pre-SMA neurons often begin to discharge much earlier in advance of movement onset and are less tightly coupled to the execution of movements.

The SMC has been implicated in the so-called *executive control* of behavior, such as operations required to switch between different actions, plans, and strategies. For example, in monkeys, some SMC neurons discharge strongly when a subject is presented with a cue instructing it to change movement targets or to suppress a previously intended movement. The

SMC may therefore contain a system that can override motor plans when they are no longer appropriate.

The SMC has also been implicated in the organization and execution of movement sequences. Some SMC neurons discharge before the start of a particular sequence of three movements but not before a different sequence of the same three movements (Figure 34–7). Other neurons discharge only when a particular movement occurs in a specific position in a sequence or when a particular pair of consecutive movements occurs regardless of their position in the sequence. In contrast, some other SMC neurons discharge only when the monkey makes the movement that occurs in a particular ordinal position of a sequence (eg, only the third) irrespective of its nature or how many movements remain to be executed in the sequence.

These seemingly disparate functions may reflect a more general role of the SMC in *contextual control* of voluntary behavior. Contextual control involves selecting and executing those actions deemed appropriate on the basis of different combinations of internal and external cues as well as withholding inappropriate actions in

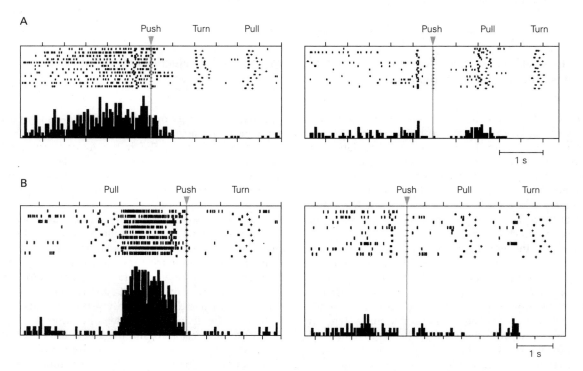

Figure 34–7 Some neurons in the supplementary motor complex of monkeys encode a specific sequence of motor acts. (Adapted, with permission, from Tanji 2001. Copyright © 2001 by Annual Reviews.)

A. A neuron discharges selectively during the waiting period before the first movement of the memorized sequence push-turn-pull (*left*). When the sequence is push-pull-turn (*right*), the cell remains relatively silent, even though the first movement in

both sequences is the same (push). Triangles at the top of each raster plot indicate the start of the push movement.

B. Records of a neuron whose activity increases selectively during the interval between completion of one motor act, a pull, and the initiation of another act, a push. The cell is not active when a push is the first movement in the sequence or when pull is followed by turn.

a specific environmental or social context. It also can involve organizing the sequence of actions required to achieve a particular goal. Contextual control likely also involves contributions from other neural circuits such as regions of the prefrontal cortex and the basal ganglia.

The cingulate motor areas (CMA) may also contribute to the contextual control of behavior. CMA appears to be involved in selecting alternate actions following motor errors or in response to changing reward contingencies. For example, monkeys were trained to push or turn a handle in response to a noninstructive trigger signal. Initially, the monkeys received a large reward if they made the same movement (pushing or turning the handle) in sequential trials. After several trials, the reward size began to decrease. If the monkeys then switched to the other movement, the reward size returned to maximum once that movement was repeated for several trials. The best strategy for the monkeys, therefore, was to switch between repetitions of either pushing or turning the handle as soon as they detected a reduction in reward size.

In this task, some neurons in the rostral CMA responded during the interval between the reception of reward and the start of the next trial. On trials with a reduced reward, task-related activity in these neurons did not change when the monkeys made the same movement in the next trial; their activity only changed when the monkeys switched to the other movement in the next trial. Importantly, those same neurons did not show the same response change when a visual cue instructed the monkeys to change movements in the next trial. This suggests that these rostral CMA neurons were preferentially involved in the voluntary decision to switch and move to the alternate goal based on action outcomes (reward size), but not by visual instructions to switch.

Dorsal Premotor Cortex Is Involved in Planning Sensory-Guided Movement of the Arm

Some of the first neural evidence that the lateral premotor cortex, including PMd and PMv, plays a crucial role in the selection and planning of sensory-guided motor actions came from recording studies by Ed Evarts, Steven Wise, and colleagues in the 1980s. These studies showed that many premotor neurons emitted brief short-latency discharge bursts in response to instructional cues that signaled specific movements, or sustained activity during the instructed-delay period between the appearance of the instructional cue and a second cue that permitted the instructed movement (Figure 34–4).

This activity reflects information about the intended act, including the spatial location of the target, the direction of arm movement, and other movement attributes. Importantly, PMd delay-period activity can reflect the intention to reach to a particular location with either the contralateral or ipsilateral arm, even though the biomechanical details of the two arm movements are very different. This suggests that PMd activity can signal the intention to generate a motor act independent of the effector used to generate the action, in an extrinsic spatial coordinate framework consistent with a prediction of the sensorimotor coordinate transformation model of motor planning. Imaging studies have likewise found evidence for an extrinsic spatial representation of finger-tapping sequences made with either hand in human premotor cortex.

Selection of an appropriate action from among multiple alternatives is a critical aspect of voluntary control. Delay-period activity in PMd can reflect that process. For example, in one experiment, recordings were made from PMd neurons in monkeys during a task in which the animals first received two colored spatial cues that identified two potential targets for reaching in opposite directions. After a memorized-delay period, a new centrally-located color cue informed the monkeys which of the spatial cues was the correct target. Following the first instruction, neural activity in PMd signaled both potential-reaching movements, but immediately after the second instruction, activity in PMd signaled only the monkeys' reaching choice (Figure 34–8A). This showed that PMd can prepare multiple potential motor actions prior to the final decision about which action to take. Subsequent studies suggest that this might be limited to no more than three to four simultaneous potential actions. Reach-related neurons in parietal area PRR also contribute to the preparation for two potential motor actions before the final action decision is made (Figure 34–8B), revealing how this process is distributed across multiple arm movement–related cortical neural populations.

PMd neurons can also signal a deliberate decision not to move. Many PMd neurons generate directionally tuned activity during an instructed-delay period when a colored visual cue at a target location instructs a monkey to reach to the target, but decrease their activity when a different colored cue at the same location instructs the monkey to refrain from reaching to it. This differential activity is an unequivocal signal, seconds before the action is executed, about the monkey's intention to reach in a particular direction or not to move in response to an instructional cue (Figure 34–9). Interestingly, many neurons in the parietal area PE/MIP studied in the same task continue to generate directionally tuned activity during the delay period even after the instructional cue to withhold

Figure 34–8 Activity of reach-related cortical neurons in monkeys during a target selection task reflects potential movements to different targets as well as the chosen direction of reach.

A. The three-dimensional colored surface depicts the mean level of activity of a population of dorsal premotor cortex (**PMd**) neurons with respect to baseline in a task in which a monkey must choose one of two color-coded reach targets in each trial. Cells are sorted along one axis (labeled "cells") based on their preferred movement direction (neurons located at the **red** and **blue circles** prefer movements at 45° and 215°, respectively). Diagrams beside the neural response profile display the stimuli presented to the monkey at different times during the trial. **Red** and **blue** cues provide information about potential actions; **green** cues guide the monkeys through different stages of each trial but provide no information about what reach to make. Shortly after the start of each trial, two potential reach targets (**blue** and **red** spatial cues) appear in opposite locations relative to the starting position of the arm (**green circle**) for 500 ms and then disappear. After a memorized delay period, the color of the starting circle changes to either **red** or **blue** (color cue), indicating to the monkey which is the correct target, in this case at 45°. After a further delay period, the go signal (**green circles** at all eight possible target locations) instructs the monkey to begin reaching to its chosen target. During the period of target uncertainty between the appearance of the two spatial cues and the central color cue, PMd neurons that prefer the two potential reach movements (**red** and **blue circles**) are simultaneously activated, whereas neurons that prefer other movements are inactive or suppressed, so that the entire PMd population encodes the two potential reach actions. As soon as the color cue appears to identify the correct target, the PMd neural activity changes rapidly to signal the reach movement chosen by the monkey. Had the color cue designated the target at 215°, the neurons preferring that target (**blue circle**) would increase their activity, and the neurons preferring the target at 45° (**red circle**) would decrease their activity (not shown). (Reproduced, with permission, from Cisek and Kalaska 2010. Copyright © 2010 by Annual Reviews.)

B. In a second study of neural activity in the parietal reach region (**PRR**), the format of data is the same as in part **A**. In this study, the monkey is presented with a single spatial cue that instructs it to prepare to reach either to the cue's location (**PD**) or in the opposite direction (**OD**). After a random memorized delay period, a color cue specifies whether the reach should be to the remembered location of the spatial cue (**green; PD**) or in the OD (**blue**). PRR neural activity is sorted according to the preferred movement direction of each neuron, as in part **A**. Population activity initially specifies the spatial cue location but then reflects both potential movement directions during the remainder of the memorized delay period. Shortly after the color cue appears, the activity quickly shifts to reflect the chosen reach direction, either the PD or OD. (Reproduced, with permission, from Klaes et al. 2011. Copyright © 2011 Elsevier Inc.)

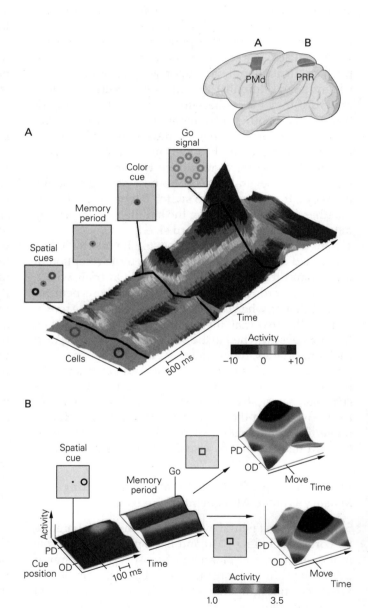

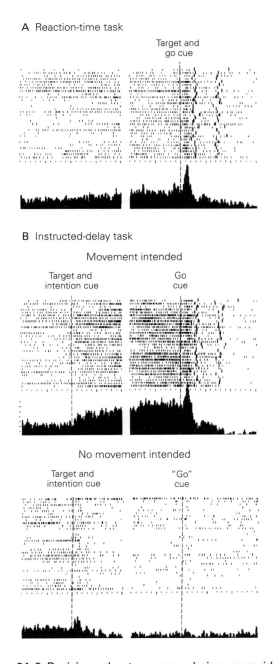

A Reaction-time task

Target and
go cue

B Instructed-delay task

Movement intended

Target and   Go
intention cue   cue

No movement intended

Target and   "Go"
intention cue   cue

Figure 34–9 Decisions about response choices are evident in the activity of premotor cortex neurons in the monkey. (Reproduced, with permission, from Crammond and Kalaska 2000.)

A. In a reaction-time task (reaching), a cell exhibits gradually increasing tonic firing while waiting for the appearance of a target. When the target appears (go cue), the cell generates a directionally tuned response.

B. In an instructed-delay task, when a monkey is shown the target and instructed to move once the go cue appears, the cell generates a strong directionally tuned signal for the duration of the delay period before the go cue (**top**). When the monkey is shown the target and instructed not to move when the go cue appears, the cell's activity decreases (**bottom**).

reaching, suggesting that the parietal cortex retains a representation of potential actions that ultimately are not executed.

Many neurons in premotor cortex also discharge during movement execution. Given this close proximity of planning- and execution-related activity, even at the level of individual neurons, a major question is why planning-related neural activity does not immediately initiate a movement. What prevents the movement from being executed prematurely? It does not appear that planning-related activity simply fails to exceed a minimum threshold required to initiate the movement or that there is a separate overt braking mechanism that must be released to allow the movement to begin.

A different way to interpret neural processing during the planning and execution of reaching that might provide answers to such questions comes from a dynamical-systems perspective. The idea is that cortical motor circuits form a dynamical system whose distributed activity patterns evolve in time as a function of their initial state, input signals, and stochastic neural response variability ("noise"). Activity patterns during different stages of planning and execution thus reflect different states of the network, including a specific state during the delay period that can prepare the movement but not activate muscles (Figure 34–10). The overall similarity of the population-level activity patterns during repetitions of the same movement shows that the entire population undergoes a coordinated pattern of co-modulation of activity during the planning and execution of the movement, determined by the synaptic connectivity within the neural circuit.

Dorsal Premotor Cortex Is Involved in Applying Rules (Associations) That Govern Behavior

Behavior is often guided by arbitrary rules that link specific symbolic cues to particular actions. When driving your car, you must perform different actions depending on whether a traffic light is green, amber, or red. In monkeys that have learned to associate arbitrary cues with specific movements, many cells in premotor areas respond selectively to specific cues. For instance, in order to select the correct target in the two-target study in Figure 34–8, the monkeys had to apply a rule that mapped color to target location provided by the two sequential instructional cues.

The PMd is implicated in the acquisition of new movement-related associations or rules. In one experiment, recordings from PMd neurons were made while the monkeys learned the association between four unfamiliar visual cues and four different movement

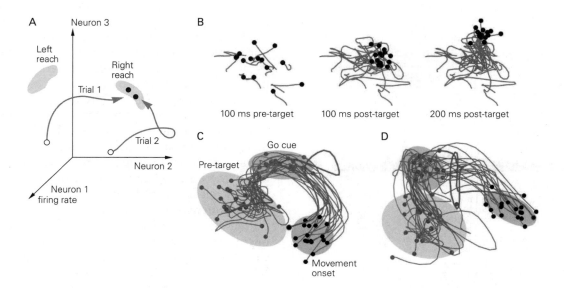

Figure 34–10 The time-varying neural activity in the dorsal premotor cortex of monkeys during different stages of the planning and execution of a movement can be viewed as transitions between different activation states. (Adapted, with permission, from Churchland MM et al. 2010. Stimulus onset quenches neural variability: a widespread cortical phenomenon. Nat Neurosci 13:369-378. Copyright © Springer Nature.)

A. A schematic illustration of how the simultaneous activity of neurons can be viewed as a trajectory through a multi-neuron activity "state space." The time-varying activity level of three simultaneously recorded neurons is represented along three axes, which defines a three-neuron state space. A specific plan (reach left or reach right) requires different combinations of preparatory firing rates for the three neurons (**gray zones**). Prior to the formation of the intention to move left or right, the baseline activity of the three neurons occupies a region in state space that is associated with holding the arm in its current position (**open circles**, for two different trials). When an instruction appears to make a reach to the right, the combined activity of the three neurons changes in a coordinated fashion, creating time-varying "neural trajectories" (**gray arrows**) that converge on the region of state space that is associated with generating a rightward movement (**filled circles** within the "right reach" gray zone).

B. Projection of the simultaneous activity of a large population of dorsal premotor cortex (PMd) neurons onto a two-dimensional state space shortly before (pre-target) and after (post-target) the appearance of a reach target cue in a task in which the reach movement must be delayed until a subsequent go cue is presented. **Gray lines** show the temporal evolution of the neural trajectories during the earliest part of movement preparation from 200 ms before target cue until the specified pre- or post-target time (**black dots**) in 15 different trials to the same target location. Neural activity initially meanders randomly within the region of state space associated with the starting posture of the arm (*left*). It then begins to converge onto a smaller region of the state space shortly after the reach target instruction appears (*center*) and begins to evolve along

the neural trajectory associated with entering the preparatory state for the reach (*right*).

C. A more complete illustration of the neural trajectories recorded during 18 different repeated trials to the same target in this delayed reaching task from the initial pre-target postural state to the onset of movement. **Blue dots** indicate activity while holding the arm in the starting posture 100 ms before appearance of the target instruction onset. Once the target instruction appears, the neural trajectories evolve toward a region of state space associated with the preparatory activity state during the delay period (**green zone**), where it dwells until a go cue appears that allows the monkey to initiate the withheld movement (**green dots**). While in this reach-preparatory part of the state space during the delay period, the arm stays at the start position because PMd activity in that part of state space is not capable of activating muscles (ie, it is "output-null"). When the go cue appears, the neural trajectories unfold toward a different region of state space associated with the initiation of the intended reach movement (**gray zone** and **black dots**). The neural activity can only cause the muscle activity for the intended movement when it enters this "output-potent" zone of state space. The trial-to-trial variability of the neural trajectories can account for intertrial variability in movement kinematics and reaction times. One outlier trial (**red**) had a long reaction time and followed a more complex and time-consuming neural trajectory from the **green** to the **gray** zone. The output-null preparatory (**green**) and output-potent movement-initiation (**gray**) zones for reaches to different target locations occupy different regions of the total population state space distinct from those associated with this reach target.

D. Data are for the same target location as in part **C** but were recorded on a different day. The neural trajectory structure is fundamentally similar for the same movements between recording sessions. Differences in the overall pattern of activity can be explained by interday differences in the activity of individual neurons and differences in the composition of the recorded neural population between sessions.

directions. Although the monkeys' choices were initially random, they learned the rules within a few dozen trials. The monkeys made an arm movement in response to each cue; during the early "guessing" phase of learning, the activity of many PMd neurons was weak but gradually increased in strength and directional tuning as the monkeys learned which cue signaled which movement. Other neurons showed a reciprocal decline in activity as the rules were acquired. These changes in activity during learning reflected both the movement choices and the rising level of knowledge of the rules linking cues with actions.

The nature of the rule can also have a strong effect on neural responses. In monkeys that have been trained to choose between several possible movements based on a spatial rule (a visual cue's location) or a semantic rule (a cue's arbitrarily designated meaning independent of its location), many prefrontal and PMd neurons are preferentially active when the animal chooses a movement using one rule but not the other. This shows that the neural activity is related not just to a particular cue or action but also to the association between them.

Premotor areas are involved in the implementation of even abstract rules. For example, monkeys were trained in a task that required two decisions, one perceptual and the other behavioral, that had no prior association. In each trial, the monkeys first had to decide whether two sequentially presented visual images were the same or different (a *match/nonmatch perceptual decision*). In some trials, a *rule cue* presented at the same time as the sample visual image instructed the monkeys to move their hand if the two images were identical and to refrain from moving if they differed (a *go/no-go motor decision*); in other trials, the rule was reversed—move if the images differ and do not move if they match. Neural activity in PMd after the test visual images were presented was correlated more strongly to the motor decision than the perceptual decision in each trial, but both decisions were expressed in PMd. More strikingly, PMd activity was also correlated with the match/non-match *behavioral rule* during the delay period between the two visual images that guided the motor decision after the test image appeared (Figure 34–11). These results suggest that PMd has a major role in applying rules that govern the appropriateness of a behavior and in making behavioral decisions according to the prevailing rules. Neural recordings in prefrontal cortex during the same task (not shown) found a strong representation of the physical identity of the visual images, but weaker and later correlates of the behavioral rule and the motor decision than in PMd.

Ventral Premotor Cortex Is Involved in Planning Motor Actions of the Hand

The most lateral part of the premotor cortex, area PMv, is reciprocally connected with parietal cortex areas AIP, PF, and PFG and the secondary somatosensory area. Electrical stimulation shows that PMv contains extensively overlapping circuits that control hand and mouth movements.

Like AIP neurons, many PMv neurons appear to contribute to the control of hand actions based on the physical affordances offered by target objects. These neurons tend to fire preferentially during certain stereotypical hand actions, such as grasping, holding, tearing, or manipulating objects. Many neurons discharge only if the monkey uses a specific type of grip, such as a precision grip, whole-hand prehension, or finger prehension (Figure 34–12). Precision grip is the type most often represented. Some PMv neurons discharge throughout the entire action, while others discharge selectively at particular stages of one type of prehension, such as during the opening or closing of the fingers.

Another striking property of PMv neurons is that their discharge often correlates with the goal of a motor act and not with the individual movements forming it. Thus, many PMv neurons discharge when grasping an object is executed with effectors as different as the right hand, the left hand, and even the mouth. Conversely, a PMv neuron may be active when an index finger is flexed to grasp an object but not when the animal flexes the same finger to scratch itself.

Premotor Cortex May Contribute to Perceptual Decisions That Guide Motor Actions

A series of studies provide evidence that cortical motor areas not only represent the sensory information that guides voluntary movements but also express the neural operations necessary to make and act on perceptual decisions. Monkeys were trained to discriminate the difference in frequency between two brief vibratory stimuli applied to one finger and separated in time by a few seconds. The animals had to decide whether the frequency of the second stimulus was higher or lower than the first and to report their perceptual decision by reaching out to push one of two buttons with the other hand.

The decision-making process in this task can be conceived as a chain of neural operations: (1) encode the first stimulus frequency (f1) when it is presented; (2) maintain a representation of f1 in working memory during the interval between the two stimuli; (3) encode the second stimulus frequency (f2) when it is

A Delayed match-to-sample task

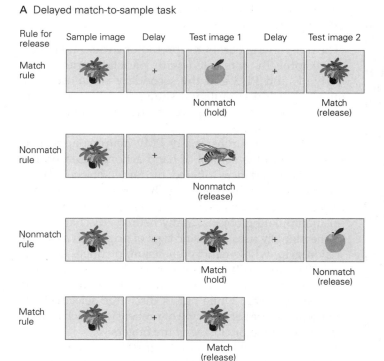

B Premotor neurons show rule-dependent activity

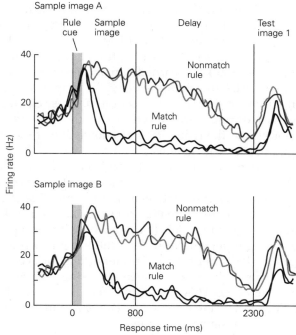

Figure 34–11 Premotor cortex neurons in the monkey choose particular voluntary behaviors based on decisional rules. (Reproduced, with permission, from Wallis and Miller 2003.)

A. A monkey must make a decision about whether to release a lever or keep holding it based on two prior decisions: a perceptual choice, whether a test image is the same as or different from a sample image presented earlier, and a behavioral choice, whether the current rule is to release the lever when the test image is the same as the sample (match rule) or when it is different (nonmatch rule). The monkey is informed of the behavioral rule that applies in each trial by a rule cue, such as an auditory tone or juice drops, which is presented for 100 ms at the same time as the onset of the sample image at the start of the trial.

B. A neuron in the dorsal premotor cortex has a higher discharge rate whenever the nonmatch rule is in effect during the delay between the presentation of the first and second images. The responses to two different sample images (upper and lower plots) were recorded from the same cell, indicating that the rule-dependent activity is not altered by changing the images. Nor, as shown by the pairs of curves associated with each rule, does activity depend on the type of rule cue (auditory tone or juice drops). (Tone cue trials: **orange** and **blue curves**; juice cue trials: **red** and **black curves**). Other dorsal premotor cortex cells (not shown) respond preferentially to the match rule over the nonmatch rule. The differential activity of the neuron up to presentation of the test image reflects the rule that will guide the animal's motor response to the test image, not the physical properties of the visual stimuli or the motor response.

presented; (4) compare f2 to the memory trace of f1; (5) decide whether the frequency of f2 is higher or lower than that of f1; and finally, (6) use that decision to choose the appropriate movement of the other hand. Everything prior to the last step would appear to fall entirely within the domain of sensory discriminative processing.

While the monkeys performed the task, neurons in the primary (S-I) and secondary (S-II) somatosensory cortices encoded the frequencies of the stimuli while they were presented. During the interval between f1 and f2, there was no sustained activity in S-I representing the memorized f1 and only a transient representation in S-II, which vanished before f2 was presented.

Strikingly, however, the activity of many neurons in the prefrontal cortex, SMC, and PMv scaled with the frequencies of f1 and f2 while they were being delivered. Furthermore, some prefrontal and premotor neurons showed sustained activity proportional to the frequency of f1 during the delay period between f1 and f2. Most remarkably, many neurons in those areas, especially in PMv, encoded the *difference* in frequency between f2 and f1 independently of their actual frequencies when f2 was delivered (Figure 34–13). This centrally generated signal is appropriate to mediate the perceptual discrimination that determines which button to push. Neurons that encoded the f2–f1 difference were absent in S-I and were far more common in SMC and PMv than in S-II.

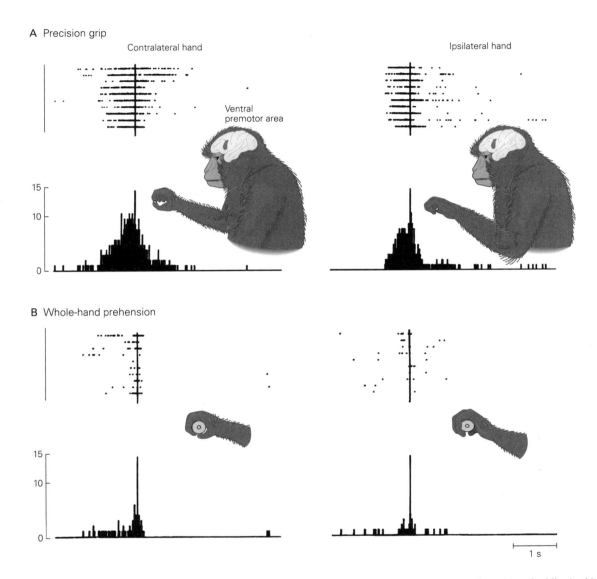

Figure 34–12 Some neurons in the ventral premotor cortex of a monkey discharge selectively during one type of grasping. This neuron discharges vigorously during a precision grip with the thumb and index finger of either the right or the left hand but very weakly during whole-hand prehension with either hand.

Raster plots and histograms are aligned (**vertical line**) with the moment the monkey touches the food (**A**) or grasps the handle (**B**). (Reproduced, with permission, from Rizzolatti et al. 1988. Copyright © Springer-Verlag 1988.)

Several Cortical Motor Areas Are Active When the Motor Actions of Others Are Being Observed

Some premotor and parietal areas can be activated when no overt action is intended, such as when an individual is asked to imagine performing a certain motor act. This phenomenon, termed *motor imagery,* has been demonstrated in humans using functional brain imaging. The neural activity evoked by motor imagery presumably reflects brain mechanisms associated with motor planning and preparation that have been disassociated from its overt execution.

A second condition in which cortical motor circuits are activated without intending overt action is when an individual observes another individual performing motor acts that are part of her own motor repertoire. The control of behavior and social interaction depends greatly on the ability to recognize and understand what others are doing and why they are doing it. Such understanding could result from a high-order visual perceptual analysis of the nature of the observed behavior and by drawing inferences about the motivation and purpose of the behavior based on one's own experience. An alternative explanation is the *direct-matching hypothesis*, the idea that observation of the actions of others activates motor circuits in the observer that control similar motor actions. According to this hypothesis, empathetic activation of motor circuits could provide a link between the observed actions and the observer's stored knowledge of the nature, motives, and consequences of similar actions that they had performed in the past.

Striking evidence in support of the direct-matching hypothesis was provided by the discovery of a remarkable population of neurons called mirror neurons, first in PMv and later in the parietal AIP of monkeys. Mirror neurons discharge both when the monkey actively grasps and manipulates objects and when it observes similar actions performed by another monkey or the experimenter (Figure 34–14). Mirror neurons typically do not respond when a monkey simply observes a potential target object or when it observes mimed arm and hand actions without a target object. Some parietal mirror neurons can even differentiate the ultimate goal of similar observed actions, such as grasping and picking up food to eat it versus putting it into a cup.

Figure 34–13 (Right) Neural activity in ventral premotor cortex in monkeys expresses the operations required to choose a motor response based on sensory information. (Adapted, with permission, from Romo, Hernández, and Zainos 2004. Copyright © 2004 Cell Press.)

A. These records of three neurons in the ventral premotor cortex of a monkey were made while the animal performed a task in which it had to decide whether the second of two vibration stimuli (**f1** and **f2**, applied to the index finger of one hand) was of higher or lower frequency than the first. The choice was signaled by pushing one of two buttons with the nonstimulated hand. The frequencies of f1 and f2 are indicated by the numbers on the left of each set of raster plots. Cell 1 encoded the frequencies of both f1 and f2 while the stimuli were being presented but was not active at any other time. This response profile resembles that of many neurons in the primary somatosensory cortex. Cell 2 encoded the frequency of f1 and sustained its response during the delay period. During the presentation of f2, the neuron's response was enhanced when f1 was higher than f2 and suppressed when it was lower. Cell 3 responded to f1 during stimulation and was weakly active during the delay period. However, during exposure to f2, the cell's activity robustly signaled the difference f2–f1 independently of the specific frequencies f1 and f2.

B. Histograms show the percentage of neurons in different cortical areas whose activity correlated at each instant with different parameters during the tactile discrimination task. **Green** shows the correlation with f1, **red** the correlation with f2, **black** the interaction between f1 and f2, and **blue** the correlation with the difference between f2–f1. (Abbreviations: **M1**, primary motor cortex; **PMv**, ventral premotor cortex; **S-I**, primary somatosensory cortex; **S-II**, secondary somatosensory cortex; **SMA**, supplementary motor area.)

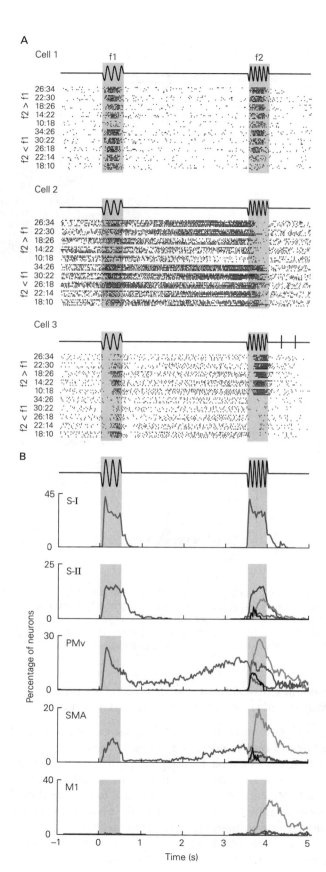

Neural-recording and brain-imaging studies show that humans are also endowed with a mirror-like mechanism to match observed actions with actions encoded in their motor system. This activity arises in various areas of cortex, including the rostral inferior parietal lobule, IPS, PMv, and posterior sector of the inferior frontal gyrus.

Cortical motor circuits appear to be involved in understanding and predicting the outcomes of observed events. In one experiment, PMd neurons implicated in the selection of reaching targets using visual cues (Figure 34–8) also discharged when monkeys simply watched the same cues and cursor motions on the monitor while an unseen party performed the task. The monkeys received a free juice reward when the cursor moved to the correct target but not if it moved to the wrong target. The monkeys began to lick the juice tube shortly after the cursor started to move to the correct target well before the juice was actually delivered, but quickly removed their mouth from the tube when the cursor moved toward the wrong target. This behavior showed that the monkeys correctly interpreted what they saw and accurately predicted its consequences.

Remarkably, the activity of most of the task-related PMd neurons was strikingly similar whether the monkeys used visual cues to plan and make arm movements or simply observed the visual events and predicted their outcome. Those neurons stopped responding during observation if no reward was delivered after correct trials or if the animal was sated and not interested in drinking juice. This showed that the neurons were not simply responding to the sensory inputs, but instead were processing the observed sensory events to predict their ultimate outcome for the monkey, namely the likelihood of a free juice reward.

This activation in connection with passive observation supports the idea that activation of premotor

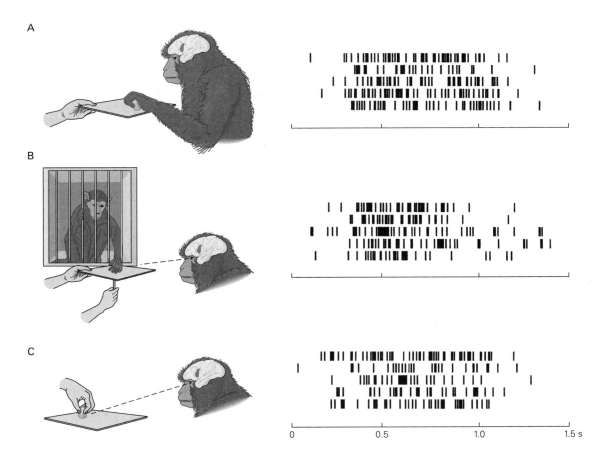

Figure 34–14 A mirror neuron in the ventral premotor cortex (area F5) of a monkey. (Reproduced, with permission, from Rizzolatti et al. 1996. Copyright © 1996 Elsevier Science B.V.)

A. The neuron is active when the monkey grasps an object.

B. The same neuron is also excited when the monkey observes another monkey grasping the object.

C. The neuron is similarly activated when the monkey observes the human experimenter grasping the object.

Time zero in the cell activity rasters corresponds approximately to the time of presentation of the object to grasp (panel A) or the onset of the observed grasping actions (panels B and C).

circuits in nonmotor contexts may contribute to understanding the nature and consequences of observed events in the environment. It has also been implicated in the ability of human subjects to learn new motor skills simply by observing a skilled person perform the same actions. Moreover, dysfunction of the mirror-neuron system in young children may contribute to some of the symptoms of autism.

Many Aspects of Voluntary Control Are Distributed Across Parietal and Premotor Cortex

While we have described the roles of premotor areas in parietal and precentral cortex separately, it must be emphasized that major sensorimotor control processes are shared across multiple cortical regions via their reciprocal interconnections.

For instance, the neural processes that link the physical affordances of target objects to appropriate hand actions are distributed across parietal area AIP, premotor area PMv, and M1, with visuospatial aspects of the process more prominent in AIP and motor components more prevalent in precentral cortex (Figure 34–15). Likewise, as already noted, neural correlates of reach target selection in PRR (Figure 34–8B) strikingly resemble those reported in PMd (Figure 34–8A).

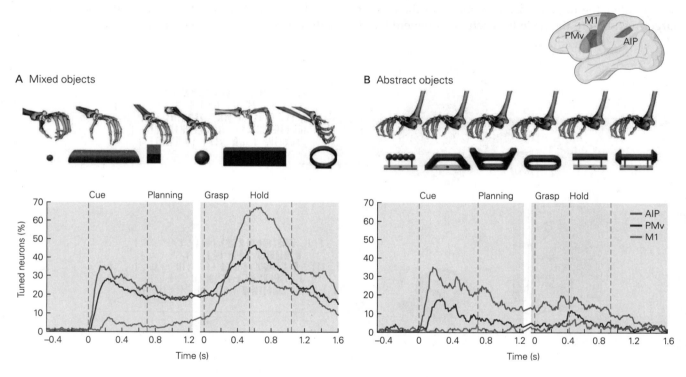

Figure 34–15 Visuomotor processing of object shape is distributed across several cortical areas in the monkey. (Reproduced, with permission, from Schaffelhofer and Scherberger 2016.)

A. A set of "mixed" objects elicit different visual responses and require different motor responses to grasp them. The plots show the percentages of neurons in the anterior intraparietal areas (AIP; **orange**), ventral premotor cortex (PMv; F5; **dark green**), and primary motor cortex (M1; **light green**) that significantly modulated their response as a function of object identity across time. Monkeys were first shown the object to grasp (cue and planning periods) and then allowed to reach to, grasp, and hold the object (grasp and hold periods). The proportion of neurons that varied their activity across object types (tuned neurons) during the cue and planning periods was greatest in

AIP and least in M1, indicating that sensitivity to object visual shape was most prominent in AIP. During motor action (grasp and hold periods), the reverse pattern was observed, with many neurons in PMv and especially M1 displaying a strong dependence on the different grasping actions required to hold onto the different objects.

B. A set of "abstract" objects elicit different visual responses but require similar motor responses to grasp them. As with the "mixed" object set, many AIP neurons varied their activity as a function of object shape during the cue and planning periods, but fewer PMv and almost no M1 neurons showed sensitivity to observed object shape. During motor action (grasp and hold periods), very few PMv and M1 neurons showed any difference in activity as a function of the shape of the different objects, all of which required the same grasping action.

The Primary Motor Cortex Plays an Important Role in Motor Execution

Once an individual has decided on a behavioral goal, motor commands must then be communicated to muscles to move the body. The complexity of this problem cannot be underestimated as it requires precise control of the spatiotemporal patterns of activity of large numbers of muscles acting across many joints to achieve the behavioral goal, while also accounting for the complex, nonlinear mechanical properties of the musculoskeletal system and forces and loads imposed by the environment. These detailed patterns of muscle activity are coordinated by spinal motor neurons and interneuronal circuits (Chapter 32). However, the primary motor cortex (M1) plays an important role in generating the motor commands that control that spinal activity, including essential information necessary to select and control the timing and magnitude of muscle activity.

The Primary Motor Cortex Includes a Detailed Map of the Motor Periphery

The idea that a local region of the cerebral cortex contains a motor map of the body dedicated to voluntary motor control dates back to the work of the English neurologist John Hughlings Jackson in the middle of the 19th century. He reached this conclusion while treating patients with epileptic seizures that were characterized by recurring spasmodic involuntary movements that sometimes resembled fragments of purposive voluntary actions and that progressed systematically to include different parts of the body during each seizure episode (Chapter 58). Later in the 19th century, improved anesthesia and aseptic surgical techniques allowed direct experimental study of the cerebral cortex in experimental animals. Using those new methods, Gustav Fritsch and Eduard Hitzig in Berlin and David Ferrier in England showed that electrical stimulation of the surface of a limited area of cortex in different anesthetized mammalian species evoked movements of parts of the contralateral body. In monkeys, the electric currents needed to evoke movements were lowest in a narrow strip along the rostral bank of the central sulcus, the same region now called primary motor cortex.

Their experiments demonstrated that within this strip of tissue stimulation of adjacent sites evoked movements in adjacent body parts, starting with the foot, leg, and tail medially, and proceeding to the trunk, arm, hand, face, mouth, and tongue more laterally. When they lesioned a cortical site at which stimulation had evoked movements of a part of the body, movement of that body part was perturbed or lost after the animal recovered from surgery. These early experiments showed that the motor cortex contains an orderly motor map of major parts of the contralateral body and that the integrity of the motor map is necessary for voluntary control of the corresponding body parts. Studies in the first half of the 20th century on many species by Clinton Woolsey and on humans undergoing surgery by Wilder Penfield demonstrated that the general topographic organization of the rostral bank of the central sulcus is conserved across many species (Figure 34–16). One important observation was that the motor map is not an exact point-to-point reproduction of the body's anatomical form. Instead, the most finely controlled body parts, such as the fingers, face, and mouth, are represented by disproportionately large areas, reflecting the larger number of neurons needed for fine motor control.

Today the best-studied regions of the map are those parts controlling the arm and hand and reveal far more complexity than conveyed in the classic diagrams shown in Figure 34–16A,B. First, neurons controlling the muscles of the digits, hand, and distal arm tend to be concentrated within a central zone, whereas those controlling more proximal arm muscles are located in a horseshoe-shaped ring around the central core (Figure 34–16C). Second, stimulation sites overlap extensively, allowing control of muscles acting across different joints; conversely, each muscle can be activated by stimulating many sites dispersed across the arm/hand motor map. Finally, local horizontal axonal connections link different sites across the motor map, likely allowing coordination of activity across the map during the formation of motor commands.

Some Neurons in the Primary Motor Cortex Project Directly to Spinal Motor Neurons

As already noted, while many corticospinal axons in primates terminate only on spinal interneurons, others also synapse directly onto spinal motor neurons. These corticomotoneuronal (CM) cells are found only in the most caudal part of M1 that lies within the anterior bank of the central sulcus. There is extensive overlap in the distribution of the CM cells that project to the spinal motor neuron pools innervating different muscles (Figure 34–17A).

CM cells are very rare or absent in nonprimate species and become a progressively larger component of the corticospinal tract in primate phylogeny from prosimians to monkeys, great apes, and humans. In monkeys, more CM cells project to the motor pools for muscles of the digits, hand, and wrist than to

A Macaque monkey

B Human

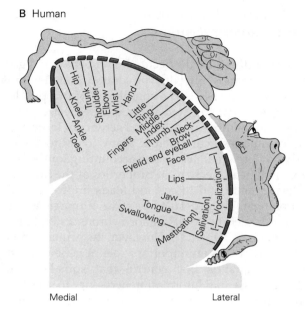

Medial   Lateral

C Forelimb representation

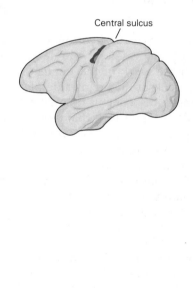

Central sulcus

Figure 34–16 The motor cortex contains a topographic map of motor output to different parts of the body.

A. Studies by Clinton Woolsey and colleagues confirmed that the representation of different body parts in the monkey follows an orderly plan. Motor output to the foot and leg is medial, whereas the arm, face, and mouth areas are more lateral. The areas of cortex controlling the foot, hand, and mouth are much larger than the regions controlling other parts of the body.

B. Wilder Penfield and colleagues showed that the human motor cortex motor map has the same general mediolateral organization as in the monkey. However, the areas controlling the hand and mouth are even larger than in monkeys, whereas the area controlling the foot is much smaller. Penfield

emphasized that this cartoon illustrated the relative size of the representation of each body part in the motor map; he did not claim that each body part was controlled by a single separate part of the motor map.

C. The arm motor map in monkeys has a concentric, horseshoe-shaped organization. Neurons that control the distal arm (digits and wrist) are concentrated in a central core (**pale green**) surrounded by neurons that control the proximal arm (elbow and shoulder; **dark green**). The neuron populations that control the distal and proximal parts of the arm overlap extensively in a zone of proximal-distal co-facilitation (**intermediate green**). (Reproduced, with permission, from Park et al. 2001. Copyright © 2001 Society for Neuroscience.)

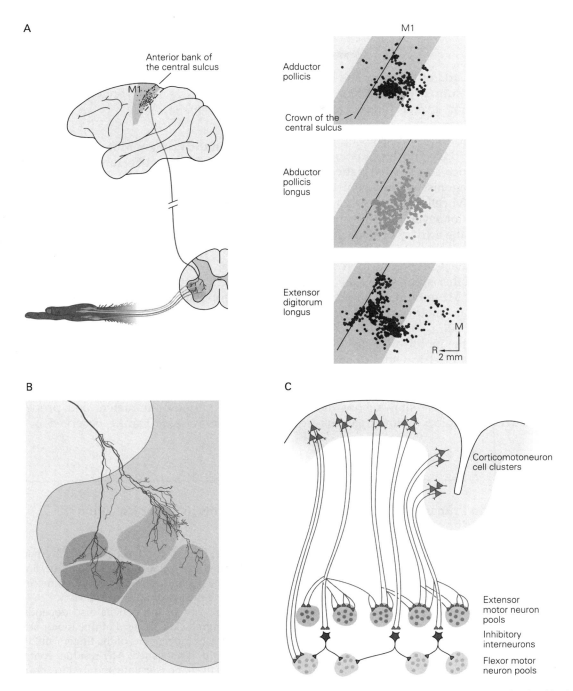

Figure 34–17 Corticomotoneuronal cells activate complex muscle patterns through divergent connections with spinal motor neurons that innervate different arm muscles.

A. Corticomotoneuronal (CM) cells, which project monosynaptically to spinal motor neurons, are located almost exclusively within the anterior bank of the central sulcus in the caudal part of the primary motor cortex (**M1**). The CM cells that control a single hand muscle are widely distributed throughout the arm motor map, and there is extensive overlap of the distribution of neurons projecting to different hand muscles. The distributions of the cell bodies of CM cells that project to the spinal motor pools that innervate the adductor pollicis, abductor pollicis longus, and extensor digitorum communis (shown on the right) illustrate this pattern of wide distribution and extensive overlap of CM cells projecting to different muscles. (Abbreviations: **M**, medial; **R**, rostral.) (Reproduced, with permission, from Rathelot and Strick 2006.)

B. A single CM axon terminal is shown arborized in the ventral horn of one segment of the spinal cord. It forms synapses with the spinal motor neuron pools of four different intrinsic hand muscles (**yellow** and **blue** zones), as well as with surrounding interneuronal networks. Each axon has several such terminal arborizations distributed along several spinal segments. (Reproduced, with permission, from Shinoda, Yokota, and Futami 1981.)

C. Different colonies of CM cells in the primary motor cortex terminate on different combinations of spinal interneuron networks and spinal motor pools, thus activating different combinations of agonist and antagonist muscles. Many other corticospinal axons terminate only on spinal interneurons (not shown). The figure shows CM projections largely onto extensor motor neuron pools. Flexor motor pools receive similar complex projections (not shown). (Adapted, with permission, from Cheney, Fetz, and Palmer 1985.)

those for more proximal parts of the arm. The terminal of a CM cell axon often branches and terminates on spinal motor neurons for several different agonist muscles and can also influence the contractile activity of still more muscles through synapses on spinal interneurons (Figure 34–17B,C). This termination pattern is organized to produce coordinated patterns of activity in a *muscle field* of agonist and antagonist muscles. Most frequently, a CM cell axon directly excites the spinal motor neurons for several agonist muscles and indirectly suppresses the activity of some antagonist muscles through spinal inhibitory interneurons (Figure 34–17C). The fact that CM cells are more prominent in humans than in other species may be one of the reasons why lesions of M1 in humans have a more profound effect on voluntary motor control compared to other mammals (Box 34–3).

The complexity of the motor map in M1—as revealed by short trains of electrical stimuli and anatomical and neurophysiological studies of direct and indirect M1 descending outputs targeting single muscles and small muscle groups—shows how motor commands from M1 to the spinal motor apparatus are able to control movements of every part of the body, with special focus on the fingers, hand, arm, face, and mouth in primates.

Activity in the Primary Motor Cortex Reflects Many Spatial and Temporal Features of Motor Output

As already noted, a given action such as reaching for an object can be described on many levels, ranging from the hand's spatial trajectory and velocity to its joint-centered causal forces and muscle activity (Figure 34–1A). Representational models assume that the motor system directly plans and controls specific parameters of movement. They predict that different neural populations encode the intended movement in a parameter space (ie, hand or joint motion or joint muscular torque) and perform the transformations between them. Dynamical models predict that neural circuits control movements through changes in their activation state from its current state to the desired final state. As their activity changes across time, correlates of various parameters and properties of the intended movement can be observed in the activity

---

Box 34–3 Lesions in Primary Motor Cortex Lead to Impairments in Motor Execution

The effects of primary motor cortex (M1) lesions differ across species. Large lesions in cats do not cause paralysis; the animals can move and walk on a flat open surface. However, they have severe difficulties using visual information to navigate within a complex environment, avoid obstacles, or climb the rungs of a ladder. In cats, the pyramidal tract neurons in M1 are much more strongly activated when an animal must modify its normal stepping motion to clear an obstacle under visual guidance than during normal unimpeded locomotion over a flat, featureless surface (Chapter 33).

Large M1 lesions in monkeys have more drastic consequences, including initial paralysis and usually the permanent loss of independent movements of the thumb and fingers. Monkeys nevertheless recover some ability to make clumsy movements of the hands and arms and to walk and climb.

More focal lesions of M1 typically result in muscle weakness, slowing and imprecision of movements, and discoordination of multi-joint motions, perhaps as a result of selective perturbations of the control circuitry for specific muscles or muscle groups. Lesions limited to part of the motor map, such as the contralateral arm,

leg, or face, lead to paralysis of that body part. There is diminished use of the affected body part, and movements of the distal extremities are much more affected than those of the proximal arm and trunk.

The severity of the deficits also depends on the level of required skill. Control of fine motor skills, such as independent movements of the fingers and hand and precision grip, is abolished. Any residual control of the fingers and the hand is usually reduced to clumsy, claw-like, synchronous flexion and extension motions of all fingers, not unlike the unskilled grasps of young infants. Remaining motor functions, such as postural activity, locomotion, reaching, and grasping objects with the whole hand, are often clumsy.

In humans, large motor cortex lesions are particularly devastating, resulting in severe motor deficits or complete paralysis of affected body parts, usually with limited potential for recovery. This presumably reflects the increased importance in humans of descending signals from M1 onto spinal interneuronal circuits and spinal motor neurons and a diminished capacity of other cortical and subcortical motor structures to compensate for the loss of those descending M1 signals.

of single neurons and neural populations. However, the activity of most neurons reflects a combination of parameters that does not correspond to any identifiable parameter in any specific coordinate framework.

Despite their different assumptions, both perspectives suggest that one can infer the possible contribution of different neurons and different neural structures to motor control by studying how their activity correlates with different parameters of movements. The activity of M1 neurons has been intensively studied since the 1960s to try to reveal, for instance, whether M1 generates a high-level signal about the hand motion or a lower-level kinetic signal more related to the causal forces and muscle activity.

Knowledge about the nature of the control signals generated by M1 also helps to clarify the role of other motor structures, notably the spinal cord. If M1 encodes specific information about muscle activity patterns, less computational processing would be necessary at the spinal level. In contrast, if M1 mainly encodes higher-level information about the intended movement, the spinal cord would have to perform the processes that convert this global signal into detailed patterns of muscle activity.

However, one of the major experimental challenges in identifying how M1 controls movement is the fact that virtually all movement-related parameters are intercorrelated through the laws of motion. As a consequence, a particular muscular force (kinetics) will cause a specific motion (kinematics) given an initial condition (posture, movement) of the body. As a result, if one recorded neural activity while a monkey makes reaching movements in different directions, a neuron that theoretically signals the spatial direction of movement will also inevitably show a correlation with the direction of causal forces. Likewise, the contractile activity of a muscle will co-vary systematically with the spatial direction of movement even though it is clearly generating the causal forces. Unless the task design adequately dissociates these different classes of parameters, it will yield ambiguous information about the functional role of each neuron.

Edward Evarts was the first to examine this issue in the 1960s, in pioneering single-neuron recordings in monkeys while they made simple flexion/extension movements of the wrist. Using a system of pulleys and weights, he applied a load to the wrist of the monkey that pulled the wrist in either the direction of flexion or extension in different trials. This required the monkey to alter the level of wrist muscle activity to compensate for the load while making the movements. As a result, the kinematics (direction and amplitude) of wrist movements remained constant, but the kinetics (forces and muscle activity) changed with the load.

Using a microelectrode, he located single neurons in the M1 motor map that modulated their activity when the monkey made movements of the wrist without the external load. In some neurons, their discharge increased during wrist flexion (*preferred movement direction*) and was suppressed during extension, whereas others displayed the opposite pattern. This movement-related activity typically began 50 to 150 ms before the onset of agonist muscle activity, supporting a causal link between M1 neural activity and movement. When a load was applied, many M1 neurons increased their activity when the load resisted movement in their preferred direction and decreased activity when the load assisted the movement (Figure 34–18). These changes in neural activity paralleled the changes in muscle activity required to compensate for the external load.

Subsequent studies have confirmed that the activity of many M1 neurons varies systematically with the magnitude of muscle force output. This is best shown in tasks in which monkeys generate isometric forces against immovable objects that prevent movement. The activity of many M1 neurons, including CM cells, varies with the direction and level of static isometric output forces generated across a single joint, such as the wrist or elbow, as well as during precise pinches using the thumb and index finger (Figure 34–19A). At least over part of the tested range, these responses vary linearly with the level of static force.

Most natural behaviors involve multi-joint, multi-muscle actions. For instance, reaching movements of the arm in different directions requires different patterns of coordinated motions at the shoulder and elbow. Proximal limb muscle activity during reaching shows a roughly cosine pattern of activity with maximal activity in a specific movement direction, its preferred movement direction, that gradually diminishes as the angle between the desired direction of reach and the muscle's preferred direction increases (Figure 34–20A). Like the proximal arm muscles, single neurons related to shoulder and elbow movements respond in a continuously graded fashion during movements in different reach directions centered on a preferred direction of maximal activity (Figure 34–20B). Different neurons have different preferred directions that cover the entire directional continuum around the circle, and during any given movement, neurons with a wide range of preferred directions discharge at different rates.

As Ed Evarts had shown in single-joint tasks, much of the M1 activity during reaching is closely related to the causal kinetics. For instance, in monkeys trained to make reaching movements in eight directions while compensating for external loads that pulled the arm in different directions, the reach-related activity of both

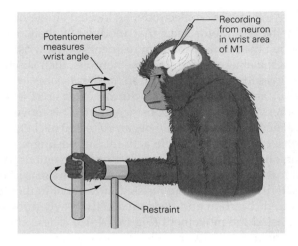

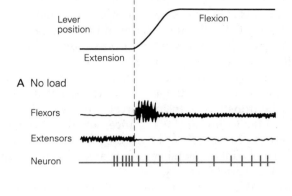

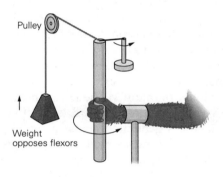

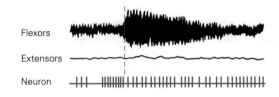

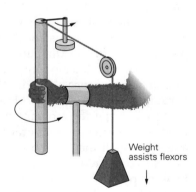

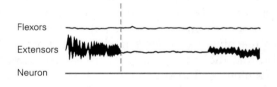

Figure 34–18 Activity of a motor cortex neuron correlates with changes in the direction and amplitude of muscle forces during wrist movements. The records are from an M1 neuron with an axon that projected down the pyramidal tract. The monkey flexes its wrist under three load conditions. When no load is applied to the wrist, the neuron fires before and during flexion (**A**). When a load opposing flexion is applied, the activity of the flexor muscles and the neuron increases

(**B**). When a load assisting wrist flexion is applied, the flexor muscles and neuron fall silent (**C**). In all three conditions, the wrist displacement is the same, but the neural activity changes as the loads and compensatory muscle activity change. Thus, the activity of this motor cortex neuron is better related to the direction and level of forces and to muscle activity exerted during the movement than to the direction of wrist displacement. (Adapted from Evarts 1968.)

proximal-arm muscles and many M1 neurons changed systematically with the direction of the external loads and the corresponding corrective forces that the monkeys had to generate for each reach direction. Both muscle and neural activity increased when the load

resisted movements in their preferred directions and decreased when the loads assisted those movements. In addition, when a monkey uses its whole arm to exert constant isometric force levels in different directions at the hand, the activity of many M1 neurons varies

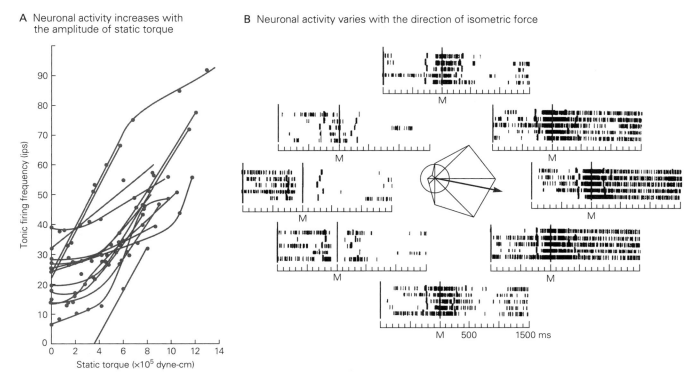

A Neuronal activity increases with the amplitude of static torque

B Neuronal activity varies with the direction of isometric force

Figure 34–19 Activity in many primary motor cortex neurons correlates with the level and direction of force exerted in an isometric action.

A. The activity of many primary motor cortex neurons increases with the amplitude of static torque generated across a single joint. The plot shows the tonic firing rates of several different corticomotoneuronal cells at different levels of static torque exerted in the direction of wrist extension. Other motor cortex neurons show increasing activity with torque exerted in the direction of wrist flexion, and so would show response functions with the opposite slope (not shown). (Reproduced, with permission, from Fetz and Cheney 1980.)

B. When a monkey uses its whole arm to push on an immovable handle in its hand, the activity of some primary motor

cortex neurons varies with the direction of isometric forces. Each of the eight raster plots shows the activity of the same primary motor cortex neuron during five repeated force ramps in one direction. Each row shows the pattern of spikes during a single trial of the task. The position of each raster of activity corresponds to the direction in which the monkey is generating isometric forces on the handle. The onset of the force ramp is indicated by the vertical line labeled **M**. The **thick ticks** on the left of that line in each row indicate when the target appeared on a computer monitor, telling the monkey the direction in which it should push on the handle. The central polar plot illustrates the directional tuning function of the neuron as a function of the direction of isometric forces. (Reproduced, with permission, from Sergio and Kalaska 2003.)

systematically with force direction, and the directional tuning curves for isometric force resemble those for activity during reaching movements (Figure 34–19B).

The complex and nonlinear properties of multisegmented limbs present a major control problem for the motor system. For instance, one can make reaching movements with similar hand trajectories but different arm geometries that require changes in the causal joint-centered torques and muscle activity. In one experiment, when monkeys made horizontal reaching movements along the same planar spatial hand trajectories while holding the arm in different spatial orientations (ie, elbow raised versus lowered), the activity of proximal-arm muscles and many M1 neurons showed corresponding changes in the strength and directional

tuning of their reach-related activity. This indicates that the M1 neurons generate signals that take into account the changes in intrinsic limb biomechanics during the reaching movements.

Similarly, arm movements toward or away from the body require much larger angular motion at the shoulder and elbow joints compared to movements to the right or left. In contrast, muscular torques tend to be larger for movements to the right and left. Both of these factors influence the amount of muscle activity required to move the limb, which can be quantified by a single term, joint muscular power (joint angular velocity multiplied by net muscular torque about that joint). With the limb in the horizontal plane, joint power is greatest for movements away from the body and slightly to the left,

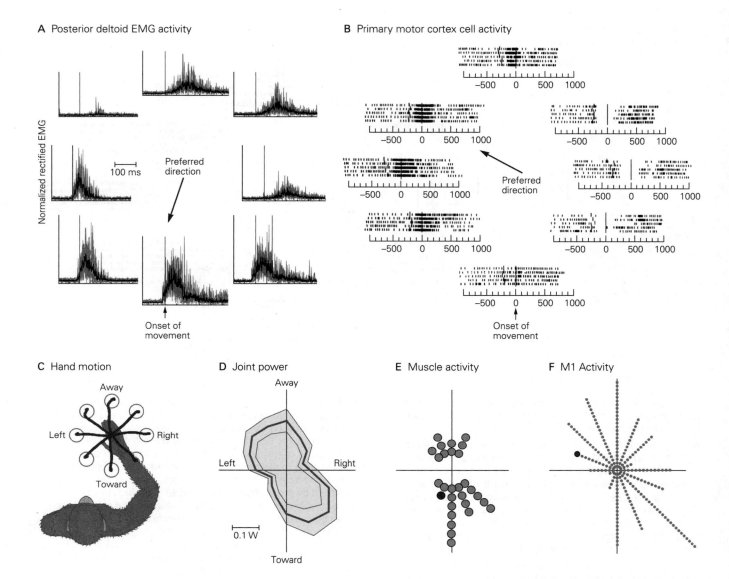

Figure 34–20 Limb muscles and primary motor cortical neurons are broadly tuned to the direction of reaching.

A. Plots show the activity of posterior deltoid of the right arm, a shoulder extensor, during arm movements in eight directions (see panel C) (central panel displays mean hand trajectories). The muscle is initially maximally active for movements at 270° (towards the body, preferred direction = 250°) and diminishes for movements in other directions. **Black lines** denote the mean activity of the muscle across multiple trials, and data are aligned on movement onset (**vertical thin line**). (Abbreviation: EMG, electromyography.)

B. Raster plots show the firing pattern of a single primary motor cortex neuron during whole-arm movements in eight directions. The neuron discharges at the maximal rate for movements near 135° and 180° and at lesser intensities for movements in other directions. The cell's lowest firing rate is for movements opposite the cell's preferred direction. Each row of **thin tics** in each raster plot represents the activity in a single trial, aligned at the time of movement onset (time 0); **thick tics**, time of target appearance. (Reproduced, with permission,

from Georgopoulos et al. 1982. Copyright © 1982 Society for Neuroscience.)

C. Hand trajectories when reaching from a central position in the horizontal plane.

D. Peak joint power (joint muscular torques multiplied by joint velocity) for movements performed in different spatial directions (shoulder and elbow power added together). A large amount of power is required to reach away from the body and to the upper left and to reach toward the body and to the lower right. (Right X-axis is at 0°.)

E. Preferred directions of proximal-limb muscles tend to be for movements that require greater muscular power, reflecting the obvious link between muscle use and the physical requirements of the motor task. Each **dot** represents an individual muscle binned into 22.5° sectors; the **blue dot** represents the preferred direction of the muscle displayed in panel A.

F. Distribution of preferred directions of neurons in primary motor cortex (**M1**). Each **dot** represents an individual neuron, and the **blue dot** represents the preferred direction of the neuron displayed in panel B. (Adapted, with permission, from Scott et al. 2001.)

and toward the body and to the right (Figure 34–20C,D). This bias in the physics of limb movement leads to a bias in the preferred directions of shoulder and elbow muscles, which tend to be maximally active in these same directions (Figure 34–20E). Correspondingly, the distribution of preferred directions of neurons in M1 also parallels this bias, with neurons tending to have preferred directions either away and slightly to the left or toward and to the right (Figure 34–20F). Thus, the physics of the limb dictates the pattern of muscle activity needed to generate movement, and this in turn is reflected in the pattern of neural activity in M1.

The impact of limb physics on M1 activity extends to the level of muscle-related signals. The activity of some single M1 neurons, including CM cells, can be correlated with specific components of the contraction patterns of different muscles during such diverse tasks as isometric force generation, precision pinching of objects between the thumb and index finger, and complex reaching and grasping actions (Figure 34–21). These findings highlight how M1 contributes to the specification of muscle activity patterns for motor actions, including onset times and magnitudes. Nevertheless, the final pattern of muscle activity will only be generated by the spinal motor neurons since they alone take into account the additional influence of other descending supraspinal inputs and local spinal interneuronal processes.

All the studies described so far related the activity of single M1 neurons to motor output. However, voluntary motor control is implemented by the simultaneous coordinated activity of many neurons throughout the motor system. Their activity is noisy, varying stochastically between repetitions of the same movement. Furthermore, their broad symmetrical movement-related tuning curves introduce a high level of uncertainty as to what the limb should do in response to the ambiguous signal generated by each neuron.

A simple computational approach was developed to extract a unique signal about each reaching movement by pooling the heterogeneous single-neuron activity of the recorded M1 population. The activity of each neuron is represented by a vector pointing in its preferred direction; the length of the vector varies as a function of its mean discharge rate during reaches in each direction. This vector notation implies that an increase in the activity of a given M1 neuron evokes changes in activity in the spinal motor apparatus and muscles that causes the arm to move along a path corresponding to the neuron's task-related preferred direction; the strength of that single-neuron influence varies systematically with the difference between the neuron's preferred direction and the desired movement (Chapter 39, Figure 39–6). When the reach-related

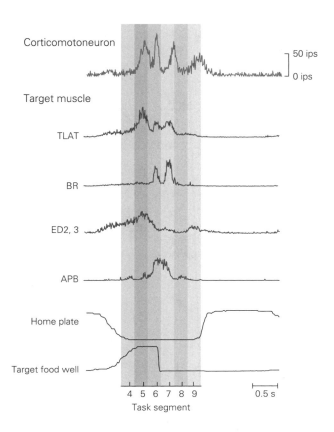

Figure 34–21 The activity of some primary motor cortex neurons can be correlated with particular patterns of muscle activity. Bursts of activity in a single corticomotoneuron during a reach-and-grasp movement to retrieve food pellets from a small well are correlated with bursts of contractile activity in several of its target muscles at different times during the movement. (Abbreviations: **APB**, abductor pollicis brevis; **BR**, brachioradialis; **ED2, 3**, extensor digitorum 2, 3; **ips**, impulses per second; **TLAT**, lateral triceps.) (Reproduced, with permission, from Griffin et al. 2008.)

activity of about 250 M1 neurons was represented by variable-length vectors for each of the eight reach directions and summed, the direction of the net resultant *population vectors* varied systematically with the actual reach directions (Figure 34–22A).

The novel insights of this analysis were that the control of a given reach movement involves coordinated changes in the activity of M1 neurons distributed throughout the M1 arm motor map and that their pooled activity clearly distinguishes the unique identity of each of the reach actions generated by the eight different distributed patterns of population activity. Subsequent studies demonstrated that "instantaneous" population vectors extracted from the pooled activity of large populations of M1 neurons during sequential 20-ms time bins from the start to the end of movement predicted the continually changing trajectory of the

A

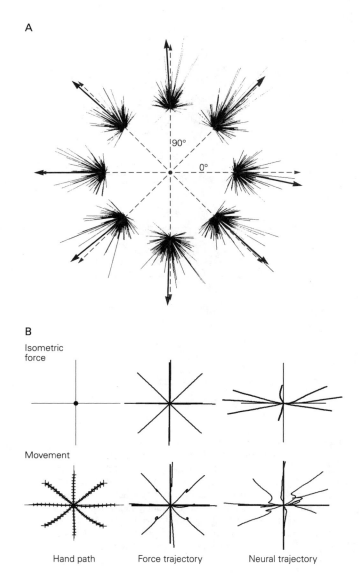

B

Isometric
force

Movement

Hand path   Force trajectory   Neural trajectory

Figure 34–22 Population codes relate M1 activity to different properties of movement.

A. The eight single-neuron vector clusters (**thin black lines**) and the population vectors (**blue arrows**) represent the activity of the same population of cells during reaching movements in eight different directions. Each single-neuron vector points in the neuron's preferred movement direction, and its length is proportional to the discharge of the neuron during that movement. The population vectors were calculated by vectorial addition of all the single-cell vectors in each cluster; **dashed arrows** represent the direction of movement of the arm. (Reproduced, with permission, from Georgopoulos et al. 1983.)

B. Comparison of hand kinematics and kinetics and neural population activity in an isometric task and when moving a handle with a large mass. Force and neural trajectories were generated by linking sequences of 20-ms output force vectors or neural population vectors tip-to-tail for each direction of force or movement output. (Reproduced, with permission, from Sergio et al. 2005.)

arm motions 100 to 150 ms into the future while monkeys made reaching movements or traced spirals on a computer monitor. This showed that the simple vector notation could be used to extract from the activity of populations of neurons a signal about intended motor output even on a moment-to-moment basis. These findings were anticipated by a prescient study in 1970 by Donald Humphrey and colleagues, who showed that the appropriately summed activity of three to five M1 neurons was better correlated to the temporal patterns of motor output during single-joint movements than was the signal of any of the single neurons.

Subsequent studies used the population-vector decoder algorithm to provide further insight into neural processing in M1. In one study, the activity of proximal arm–related M1 neurons was recorded while monkeys performed two tasks (Figure 34–22B). In the first task, they generated isometric force ramps in eight different spatial directions uniformly distributed at 45° intervals in a horizontal plane against a rigid handle that they held in their hand, without arm movements. A 20-ms population-vector decoder was used to extract the net directional bias of the pooled activity of many M1 neurons, and the result showed that these pooled signals varied systematically with the direction of output forces throughout the duration of the force-ramp generation, even though there were no movements. However, unlike the actual uniformly distributed directions of the forces generated by the monkey at the hand, the decoded population-vector signals were skewed toward the x-axis. This showed that the M1 activity reflected the nonlinear relationship between causal shoulder muscle torques and measured isometric forces at the hand resulting from the complex biomechanical properties of the arm (see Figure 34–20).

In the second task, the monkeys made reaching movements of the arm in the same eight directions to move a heavy handle. This required an initial accelerative force in the direction of movement and then a transient reversal of the direction of forces to decelerate the movement of their arm and the mass as it approached the target. The decoded M1 population-vector signals in this task varied dramatically through time. They were directed initially toward the target but then transiently reversed just before the peak of hand velocity. This showed once again that the M1 activity was more closely correlated with the time course of causal forces generating the reaching movements, including their transient directional reversal, than to the uninterrupted motion of the hand toward the target. They also found that correlates of the forces to generate reaching were strongest in M1, weaker in PMd, and largely absent in PE/MIP. This indicated that, unlike M1, reach-related neurons in area PE/MIP generated a reliable signal about stable arm postures and the kinematics of arm movements independent of the underlying causal forces and muscle activity.

Finally, one study has shown that reliable signals about the time-varying activity of proximal-arm muscles during reaching movements can be extracted from the activity of a population of simultaneously recorded M1 neurons. Another study found that the pooled activity of M1 neurons that fire selectively in connection with either shoulder or elbow movements can predict the changes in onset times and levels of contractile activity of the shoulder or elbow muscles during reaches in different directions.

These studies showed that the pooled activity of many M1 neurons is a rich and reliable source of signals about different time-varying attributes of whole-arm movements. This provided an important conceptual foundation for the development of more sophisticated decoder algorithms in brain–machine interfaces that make use of the movement-related information available in the simultaneous activity of many M1 neurons to allow subjects to control the actions of neuroprosthetic devices by covert modulations of M1 neuron activity without overt limb movements (Chapter 39).

Primary Motor Cortical Activity Also Reflects Higher-Order Features of Movement

Activity in M1 is not correlated only with causal forces and muscle activity. Many studies, beginning with those of Ed Evarts, that have attempted to dissociate kinematic from kinetic properties of motor outputs have found that the activity of some M1 neurons varies with the direction of movement but is only weakly influenced or not influenced at all by changes in output forces. Such neurons appear to preferentially signal the kinematic aspects of limb motion.

Changes in behavioral task can influence the relationship between M1 activity and motor output. One study has highlighted how contextual changes in an isometric force task altered the coding of force magnitude by M1 neurons. Either the order of forces or the range of expected forces results in changes in the activity in M1. They suggested that M1 neurons could dynamically adjust their relationship to output forces to optimize precision of control as a function of the range of forces that would be encountered in a given context. Another study found that many CM neurons may discharge intensely when monkeys performed precisely controlled force tasks with low force levels but are relatively inactive when the monkeys generate powerful contractions of the same muscles to make

brisk, back-and-forth movements of the handle. Likewise, a study demonstrated that CM cells in M1 could be very active when monkeys generated a precision-pinch grip of the thumb and index finger with relatively low force output, but were much less active or nearly silent when the animals generated much larger forces with a power grip involving the entire hand.

Still another study has shown that some M1 neurons that respond to loads applied to the limb during postural control can lose this load sensitivity as soon as the monkey makes a reaching movement to another spatial target, and vice versa. That is, those neurons can reflect output forces during postural control, but reflect only kinematics during movement. This change in the cell's response occurs quite abruptly, about 150 ms before the onset of movement. Importantly, any neurons that are sensitive to loads during both posture and movement will retain the same motor field across behaviors; that is, if the neuron responds only to shoulder flexor loads during postural control, it will respond only to shoulder flexor loads during reaching.

Even a simple change in the metrics of limb movement can have a large influence on M1 activity. In a study of monkeys making slow or fast reaching movements in different directions from a central target to peripheral targets, proximal limb muscles displayed relatively simple scaling of their activity patterns, reflecting increased forces for faster and longer reaches. In contrast, M1 neurons displayed a broad range of changes in their activity patterns that rarely paralleled the pattern of changes observed for muscles.

Activity in neurons can also correlate with higher-level features of movement such as the nature of an upcoming motor action. This was demonstrated in a study in which monkeys were trained to make wrist movements to three targets in a row starting from one extreme, stopping at a central position, and then finishing at the other extreme. Visual cues instructed the monkeys when to make each movement. Because the task used a predictable sequence of wrist movements, the monkeys knew before the visual cues appeared what would be the next direction of movement. While many M1 neurons signaled the current wrist posture or the direction of each movement while they were being performed, some M1 neurons reliably signaled the next movement in the sequence before the visual cue appeared. Many subsequent studies have confirmed that M1 neurons can signal impending intended movements, although these planning-like signals are not as prominent in M1 as in premotor cortical areas.

In summary, neural recording studies have revealed a diverse range of response properties within and across movement-related cortical areas, with stronger correlations to causal movement kinetics in M1 and to higher-order motor parameters in premotor and parietal cortex. However, these experimental findings have not yet led to a single unifying hypothesis about how cortical motor circuits control voluntary movements. Part of this uncertainty may result from inadequacies in experimental task design.

Representational motor-control models have interpreted these complex results as evidence of the transformations between different levels of representation of intended movements performed by neural populations distributed across different cortical motor areas. In contrast, nonrepresentational motor-control models such as optimal feedback control argue that these same results can only be interpreted as evidence of when and where neural correlates of different motor output parameters emerge in the dynamical activity distributed across cortical motor areas but do not shed much insight into the underlying neural computations. This illustrates the experimental challenges still confronting researchers as they try to reverse engineer the cortical motor circuitry to reveal its internal computational organization.

Sensory Feedback Is Transmitted Rapidly to the Primary Motor Cortex and Other Cortical Regions

Postcentral and posterior parietal cortex provide much of the sensory information related to the position and motion of the body and the location of spatial goals that is important in voluntary motor control, although the cerebellum is likely another important source (Chapter 37).

The type of afferent information transmitted to M1 differs between the proximal and distal portions of the limb. Afferent input from cutaneous and muscle sensory neurons is equally prevalent for hand-related neurons, reflecting the importance for both sources of sensory feedback when grasping and manipulating objects with the hand. Muscle afferents provide the major source of feedback from the proximal limb. Information from muscles is more prevalent in the rostral M1, whereas cutaneous input is more common in the caudal M1. Muscle afferent feedback to M1 is surprisingly rapid as it takes as little as 20 ms for M1 neurons to respond following a mechanical disturbance to the limb. Analogous to reaching, neural activity is broadly tuned to the direction of the mechanical disturbance.

Sensory feedback supports our ability to make rapid goal-directed corrections for motor errors that arise during movement planning and execution or are caused by unexpected disturbances of the limb. When a perturbing mechanical load is applied to the limb, the motor system generates a multipeaked compensatory electromyographic response, beginning with a short-latency stretch response (20–40 ms after the perturbation), followed by a long-latency response (50–100 ms) and then a so-called "voluntary" response (≥100 ms). The short latency of the initial response indicates that it is generated at the spinal level. The response is relatively small and stereotyped, and its intensity scales with the magnitude of the applied load. In contrast, motor corrections beginning in the long-latency epoch (50–100 ms) are modulated by a broad range of factors necessary to attain a behavioral goal, including the physics of the limb and environment, the presence of obstacles in the environment, the urgency of the goal, and properties of the target, including alternate goals. These context-dependent features suggest the long-latency feedback epoch is an adaptive process in which the control policy (ie, feedback gains) is adjusted based on the behavioral goal, as predicted by the optimal feedback control model.

The ability of the motor system to rapidly generate these goal-directed long-latency motor responses is supported by a transcortical feedback pathway. Neural activity across frontoparietal circuits responds rapidly to mechanical disturbances to a limb, and the pattern of activity across the cortex depends on the behavioral context. Perturbation-related activity is observed in all cortical regions beginning at approximately 20 ms after the disturbance even if the monkey is distracted by watching a movie and does not have to respond to the disturbance (Figure 34–23A,B). If the monkey is actively maintaining its hand at a spatial goal, there is an immediate increase in the neural response in parietal area PE following the disturbance, followed shortly thereafter by changes in activity in other cortical regions (Figure 34–23A,B). If the disturbance is a cue that instructs the monkey to move to another spatial target, then M1 activity reflects the need for a more vigorous response if the disturbance knocks the hand away from the target compared to knocking the hand into the target (Figure 34–23C). In contrast, perturbation-related activity in PE remains similar regardless of target location.

The Primary Motor Cortex Is Dynamic and Adaptable

One of the most remarkable properties of the brain is the adaptability of its circuitry to changes in the environment—the capacity to learn from experience and to store the acquired knowledge as memories. When human subjects practice a motor skill, performance improves.

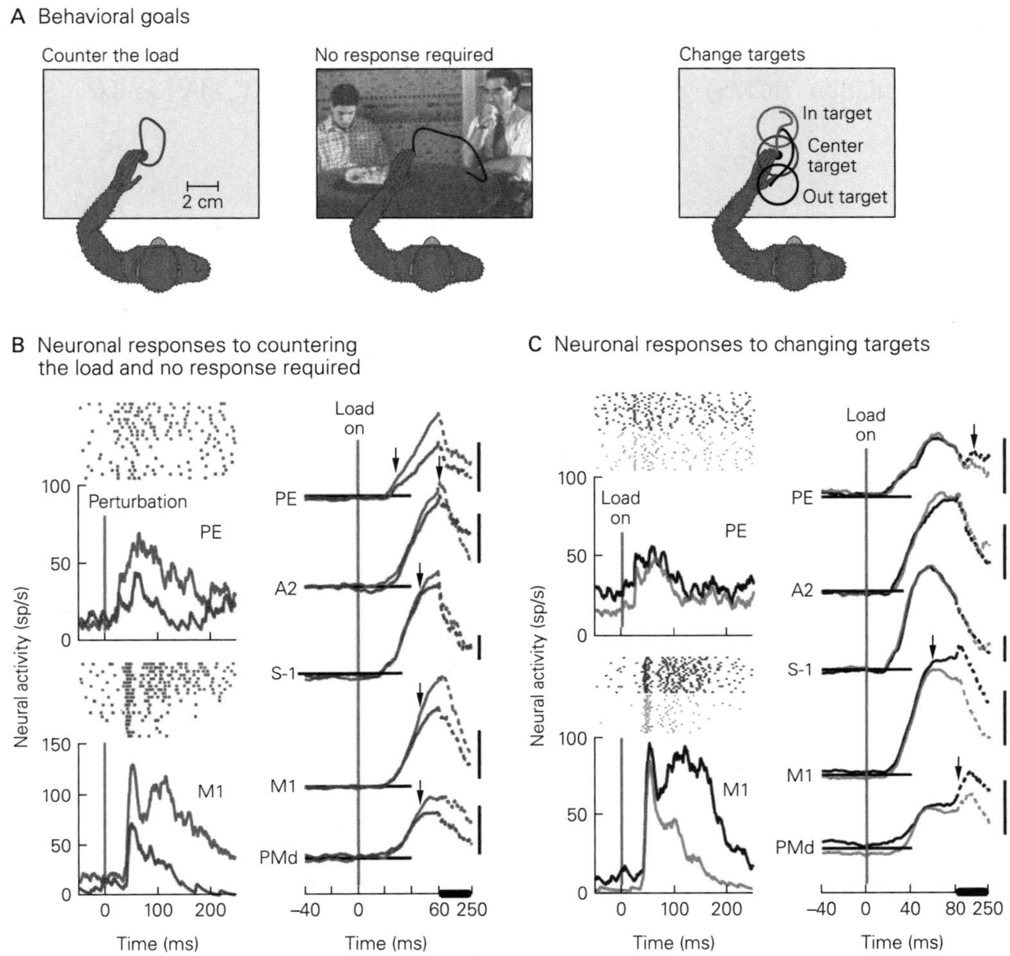

A Behavioral goals

B Neuronal responses to countering the load and no response required

C Neuronal responses to changing targets

Figure 34–23 Changes in behavioral goals alter rapid sensory feedback to parietal and frontal motor cortices. (Reproduced, with permission, from Omrani et al. 2016. Part A photo is from the film American Pie and is reproduced, with permission, from Universal Studios. © 1999, Universal Pictures, All Rights Reserved.)

A. In the experiment described here, the responses of cortical regions to mechanical loads randomly applied to the arm are compared. In the *left* panel, motor corrections return the hand to the spatial goal following the disturbance (**green hand trajectory**). In the *middle* panel, the monkey watches a movie and does not have to respond to the disturbance, leading to the hand remaining to the right following the disturbance (**red hand trajectory**). In the *right* panel, the monkey places its hand at a central start target, and one of two other targets is also presented. The disturbance applied to the limb is a cue for the monkey to move to this second target with its position being either in the direction of the disturbance (**cyan** "in target" trajectory) or away from the disturbance (**blue** "out target" trajectory).

B. *Left:* Response of a neuron in PE and in M1 when a mechanical load was applied to the limb and the monkey had to counter the load and return the hand to a spatial target (**green**) or was not required to respond to the disturbance (**red**). *Right:*

Population signals in each cortical region in response to perturbations. Note how all cortical areas show an increase in activity approximately 20 ms after the applied load. **Arrows** denote when activity was different when the monkey had to respond to the disturbance (**green curve**) as compared to not being required to respond to the disturbance (**red curve**). Note that PE is the first to show a difference in activity between the two conditions. Other cortical areas show changes at 40 ms or later. A2 is a subregion of S-I. (For B and C: Vertical scale bars, 20/spikes/s; Activity between 60–250 ms (**thick horizontal line**) compressed for visualization purposes.)

C. *Left:* Responses of single neurons in PE and M1 when a mechanical load was a cue and instructed the monkey to move to another target. The disturbance either pushed the hand toward the target (**cyan**) or away from it (**blue**). *Right:* Population signals based on perturbation-related activity in each cortical region for the "in target" and "out target" conditions. The initial responses are similar for both "in target" and "out target" disturbances across all cortical areas, and **arrows** denote when there is a difference in activity between conditions. M1 is the first to display an increase in activity for the "out target" disturbance just prior to changes in muscle activity moving the hand to the spatial target.

Motor experience can also modify the motor map. In monkeys trained to use precise movements of the thumb, index finger, and wrist to extract treats from a small well, the area of the motor map in which intracortical microstimulation (ICMS) could evoke

movements at these joints was larger than before training (Figure 34–24). If a monkey did not practice the task for a lengthy period, its skill level decreased, as did the cortical area from which the trained movements could be elicited by ICMS. Similar modifications of the

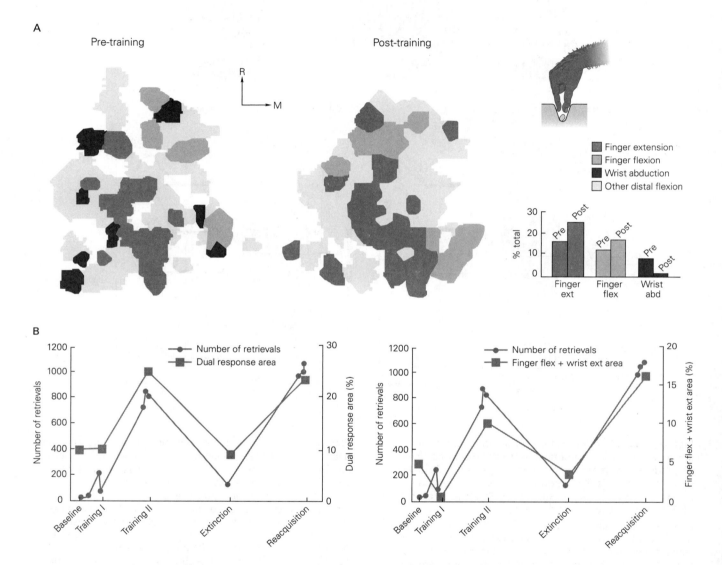

Figure 34–24 Learning a motor skill changes the organization of the M1 motor map. (Reproduced, with permission, from Nudo et al. 1996. Copyright © 1996 Society for Neuroscience.)

A. Motor maps for the hand in a monkey before and after training on retrieval of treats from a small well. Before training, areas of the motor map that generate index finger and wrist movements occupy less than half of a monkey's motor map. After training, the area from which the trained movements can be evoked by intracortical microstimulation expands substantially. The area of the map from which one could elicit individuated movements such as finger extension and flexion has expanded considerably, while the areas controlling wrist abduction, which this monkey used less in the

new skill, became less prominent. (Abbreviations: **M,** medial; **R,** rostral.)

B. The areas of the motor output map parallel the level of performance (number of successful pellet retrievals) during acquisition of the motor skill and extinction (due to lack of practice). Two areas were tested: a "dual response" area (*left plot*), from which any combination of finger and wrist motions could be evoked, and an area from which the specific combination of finger flexion and wrist extension could be evoked (*right plot*). Both areas increased as the monkey's skill improved with practice and decreased as the monkey's skill was extinguished through lack of practice. These data are from a different monkey than the one in part **A** but trained for same task.

cortical representation of practiced actions in humans have been demonstrated by functional imaging and transcranial magnetic stimulation.

At least some of the processes contributing to these changes to the motor map are local to M1 itself. One of the mechanisms contributing to the cortical reorganization underlying improved reach-to-grasp performance in rodents involves changes in synaptic strength similar to long-term potentiation and depression within the local horizontal connections linking different parts of the arm motor map. It has been shown that spike-triggered ICMS could cause specific alterations to the M1 motor output map even without specific training. For instance, one study first identified two different cortical sites (A and B) that caused contractions of different muscles (muscle A and muscle B, respectively) when electrically stimulated. They then recorded the activity of a neuron at site A; whenever that neuron fired, they stimulated site B. Within a day or two of this ICMS conditioning at site B, electrical stimulation of site A was able to cause simultaneous contractions of both muscles A and B. The change likely resulted from a spike-timing dependent increase in synaptic strength that was limited to the horizontal cortical projection from site A to site B. Electromyographic responses elicited by ICMS at a third site that did not receive similar conditioning did not change, confirming that the effect was not generalized.

Motor adaptation to visual or mechanical disturbances has been studied extensively in human subjects (Chapter 30). Neural-recording studies have demonstrated that these alterations lead to changes in the activity of M1 neurons in monkeys as the animals adapt to the perturbations. For instance, when monkeys make reaching movements in a predictable external force field that pushes on the arm in a direction perpendicular to the direction of movement, their initially curved reach trajectories get straighter. As this adaptation evolves, large increases gradually arise in the activity of M1 cells whose preferred directional tuning is opposite to the applied force field. The magnitude of such adaptation-dependent changes in activity diminishes progressively as the angle between the force direction and cell preferred direction increases, following a cosine-like function. This shows that the adaptive changes were specific to the neurons that would make the greatest contribution to compensate for the external force field.

Another example of selective changes in M1 activity during motor learning comes from a visuomotor learning study in which visual feedback from a computer monitor is rotated 90° clockwise such that movements of a monkey's arm to the right result in downward movement of the cursor. Initially, the monkeys make arm movements in the original direction aimed at the visual target location, with corrections made online after movement onset. However, with practice, the monkeys begin to move in a new direction rotated counterclockwise to the visual target so that the cursor moves directly to the target. When training occurs for only one direction, learning generalizes poorly to other directions, suggesting that the adaptive changes occur only in neurons that evoke the adapted movement. The tuning curves of neurons with preferred directions near the learned direction were altered during training, whereas neurons with other preferred directions were not affected by the training. This confirmed that the adaptation was local, consistent with the findings of the force-field adaptation study, and explained why adaptation to the visuomotor rotation in one direction generalized poorly to other directions.

Motor-error signals in the precentral cortex also play an important role in trial-by-trial motor adaptation based on feedback learning. In one study with monkeys, an adjustable prism was used to displace the apparent location of the reach target in the environment. Visual feedback of the target and arm were blocked during the reaching movements, leading to systematic errors in touching the target. The monkeys were allowed to see visual feedback of the position of the hand relative to the target for a brief period of time at the end of movement (Figure 34–25). Activity in M1 and PMd during that brief period of visual feedback after movement reflected the direction of reach end-point errors and could be involved in adapting reaching movements to correct these errors. To test that hypothesis, ICMS was then used in M1 and PMd to simulate those error responses and showed that the monkeys began to make adaptive changes in their reaching movements to compensate for the simulated errors even though no reaching error was actually made.

Some motor skills are relatively easy to learn, such as compensation for a visuomotor rotation. Others, however, are very difficult to learn. Recent studies examined this discrepancy by first measuring the activity of a population of M1 neurons as the monkey moved a cursor on a computer screen using a brain–machine interface and a neural activity decoder. This population-level mapping between M1 activity and cursor motion was then altered by changing the association between the directional tuning of each neuron and cursor motion in the decoder. When the altered decoder mapping retained the normal co-modulation structure of neural activity, as would be the case for instance if the mapping between the activity of all

Figure 34–25 (right) Error signals in the primary motor cortex drive adaptation. After a movement is complete, M1 activity reflects the error between the spatial target and final hand position. (Reproduced, with permission, from Inoue, Uchimura, and Kitazawa 2016. Copyright © 2016 Elsevier Inc.)

A. Monkeys made reaching movements to spatial targets on a touch screen. On each trial, adjustable prism goggles shifted the viewed position of the spatial target by a variable amount during the movement, while a shutter blocked vision of the monkey's hand and the target. Feedback of the final hand position was only provided for 300 ms after contact with the touch screen at the end of movement.

B. Top: Discharge response of a typical M1 neuron. Raster plots and spike-timing histograms are aligned with the initial screen contact (touch).

1. Distribution of reach endpoint errors (**black dots**) where the origin represents the center of the target. **Diameters of green circles** denote the firing rate of the neuron during each movement (**green bar** in B); the firing rates were unrelated to the subsequent endpoint error. The numbers in each quadrant indicate the summed spike activity during movements that ended in the corresponding quadrant; they are all nearly equal.

2. Same as in part B except **purple circles** denote firing rate 100 to 200 ms after movement while the monkey can see its hand while touching the screen (**purple bar** in part B). The circles and spike counts show that the firing rate is greatest for endpoint errors down and to the left relative to the position of the target (0,0), revealing that that the neural activity during this postmovement period is strongly modulated by visual feedback of reach endpoint error.

A Experiment

B Activity of an M1 neuron

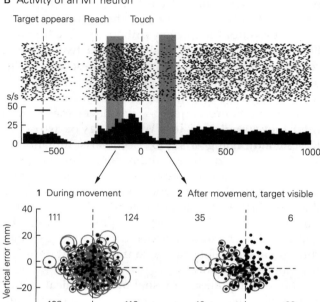

neurons and cursor motions was rotated clockwise by 45°, the monkeys showed significant adaptation to the perturbation within a few hundred trials during a single recording session. In contrast, when the perturbation required the monkeys to learn a more complex "unnatural" remapping, for instance, random clockwise and counterclockwise rotations of the apparent directional tuning of neurons by different amounts, the monkeys showed little ability to recover proficient cursor control over several hundred trials in a single recording session. Importantly, another study found that monkeys could eventually master an "unnatural" change in an M1 neural activity decoder mapping if they could practice with the same altered decoder over several days, indicating that they could learn a new neural co-modulation structure if allowed enough experience with it. These studies reinforce how neural circuits in these cortical motor regions are critical for motor skill learning.

The studies just described used brain–machine interfaces and neural decoders to explore how single neurons and neural populations contribute to motor skill learning. This technology promises to be an increasingly important research tool for developing new insights into the neural mechanisms of voluntary motor control and motor skill learning (Chapter 39).

Highlights

1. Voluntary motor behavior implements an individual's intentional choice or decision to move within, and to interact physically with objects in, the environment. A hallmark of human motor action is the breadth of skills we possess and, when highly practiced, the ease and automaticity of these actions.

2. Voluntary motor control has long been separated into two stages—planning and execution—that

can be dissociated in time. Neural recording studies have found correlates of these two stages differentially distributed across many movement-related cortical areas.

3. The overall computational problem that the motor system must resolve to control voluntary movement is to convert sensory information about the current state of the world and the body into plans for action and ultimately into patterns of muscle activity that generate the causal forces required to execute the desired movement(s), while avoiding or correcting for errors.

4. Representational models of voluntary motor control such as the sensorimotor coordinate transformation hypothesis assume that the motor system directly plans and controls specific features or parameters of intended movements. Single neurons and neural populations express those parameters in their activity and perform definable computations to effect the transformations between the controlled movement parameters in corresponding coordinate frameworks.

5. Dynamical systems models of voluntary motor control, in contrast, assume that motor circuits find empirical solutions for the computations underlying the planning and execution of movements by evolutionary and individual adaptive processes. One recent theory, optimal feedback control, proposes that planning and execution of voluntary movements involve three functional processes, namely, state estimation, task selection, and a control policy. Single neurons and neural populations contribute to voluntary motor control by participating in the computations underlying these three processes.

6. Distributed frontoparietal circuits in cerebral cortex play a pivotal role in voluntary control. There are substantial reciprocal axonal interconnections between frontal and parietal cortical regions, partially segregated based on body part (eg, hand, arm, eye). Frontal motor and parietal cortical regions both directly influence spinal processing through the corticospinal tract and indirectly through brain stem descending pathways.

7. Posterior parietal cortex plays a prominent role in identifying potential goals and objects in the environment, state estimation of the body, and sensory guidance of motor actions. Important sources of sensory signals are transmitted from visual cortex through the dorsal visual pathway and from primary somatosensory cortex. Behavioral goals and objects are represented in many parietal subregions, but how they are represented (relative to the orientation of the eye, head, or arm) varies across subregions. The presence of multiple representations provides a rich basis for defining the movement-relevant properties and the locations of objects in the world and relative to the body that can be used to select and guide movement.

8. Premotor and prefrontal cortices play a prominent role in task selection and motor planning. The dorsal and ventral premotor regions are often implicated when external sensory information plays a dominant role in selecting motor actions. In contrast, more medial premotor regions, such as the supplementary and cingulate motor areas, may play a more dominant role when internal desires are more critical in selecting and initiating a motor action. However, this dichotomy is not absolute, and multiple premotor and prefrontal cortical areas all contribute to the control of voluntary behavior in a broad range of contexts and conditions.

9. Primary motor cortex in primates has a representation of the entire body along its mediolateral axis, with larger cortical territories associated with the hand and face relative to other body parts. This cortical region also provides a large component of the corticospinal tract and has projections to both interneurons and alpha motor neurons in the spinal cord.

10. Neural activity that reflects the causal forces and the spatiotemporal features of muscle activity necessary to move the limb is particularly prominent in the primary motor cortex and can be rapidly altered to correct movement errors or to compensate for displacements of the limb away from the desired movement if the limb is perturbed. However, neural activity in primary motor cortex can also show more complex properties, reflecting changes based on the behavioral context, performance goals and constraints, and features such as movement kinematics. These properties of primary motor cortex activity may reflect the formation of a task-specific control policy within the motor system.

11. Although parietal, premotor, and primary motor cortical regions play prominent roles in state estimation, motor planning, and motor execution, respectively, they are not uniquely responsible for any one aspect; they are instead distributed to some degree across most or all of these cortical regions.

12. The cortical motor system is adaptive and can undergo changes in its functional architecture to adapt to long-term changes in the physical

properties of the world and the body, as well as acquire, retain, and recall new motor skills.

13. New technologies such as large-scale multi-neuron recording and imaging methods, enhanced multi-neuron activity decoding algorithms, and optogenetic control of the activity of specific neural populations will lead to deeper insights into the functional architecture of cortical motor circuits.

Stephen H. Scott
John F. Kalaska

Selected Reading

Battaglia-Mayer A, Babicola L, Satta E. 2016. Parieto-frontal gradients and domains underlying eye and hand operations in the action space. Neuroscience 334:76–92.

Cisek P, Kalaska JF. 2010. Neural mechanisms for interacting with a world full of action choices. Annu Rev Neurosci 33:269–298.

Dum RP, Strick PL. 2002. Motor areas in the frontal lobe of the primate. Physiol Behav 77:677–682.

Hikosaka O, Isoda M. 2010. Switching from automatic to controlled behaviour: cortico-basal ganglia mechanism. Trends Cogn Sci 14:154–161.

Lemon RN. 2008 Descending pathways in motor control. Ann Rev Neurosci 31:195–218.

Passingham RE, Bengtsson SL, Lau HC. 2010. Medial frontal cortex: from self-generated action to reflection on one's own performance. Trends Cogn Sci 14:16–21.

Scott SH. 2004. Optimal feedback control and the neural basis of volitional motor control. Nat Rev Neurosci 5:532–546.

References

Batista AP, Buneo CA, Snyder LH, Andersen RA. 1999. Reach plans in eye-centered coordinates. Science 285:257–260.

Cheney PD, Fetz EE, Palmer SS. 1985. Patterns of facilitation and suppression of antagonist forelimb muscles from motor cortex sites in the awake monkey. J Neurophysiol 53:805–820.

Cherian A, Krucoff MO, Miller LE. 2011. Motor cortical prediction of EMG: evidence that a kinetic brain-machine interface may be robust across altered movement dynamics. J Neurophysiol 106:564–575.

Churchland MM, Shenoy KV. 2006. Temporal complexity and heterogeneity of single-neuron activity in premotor and motor cortex. J Neurophysiol 97:4235–4257.

Cisek P, Crammond DJ, Kalaska JF. 2003. Neural activity in primary motor and dorsal premotor cortex in reaching tasks with the contralateral versus ipsilateral arm. J Neurophysiol 89:922–942.

Cisek P, Kalaska JF. 2004. Neural correlates of mental rehearsal in dorsal premotor cortex. Nature 431:993–996.

Crammond DJ, Kalaska JF. 2000. Prior information in motor and premotor cortex: activity in the delay period and effect on pre-movement activity. J Neurophysiol 84:986–1005.

Duhamel JR, Colby CL, Goldberg ME. 1998. Ventral intraparietal area of the macaque: congruent visual and somatic response properties. J Neurophysiol 79:126–136.

Evarts EV. 1968. Relation of pyramidal tract activity to force exerted during voluntary movement. J Neurophysiol 31:14–27.

Evarts E, Tanji J. 1976. Reflex and intended responses in motor cortex pyramidal tract neurons of monkey. J Neurophysiol 39:1069–1080.

Fetz EE, Cheney PD. 1980. Postspike facilitation of forelimb muscle activity by primate corticomotoneuronal cells. J Neurophysiol 44:751–772.

Ganguly K, Carmena JM. 2009. Emergence of a stable cortical map for neuroprosthetic control. PLoS Biol. 7:e1000153.

Georgopoulos AP, Caminiti R, Kalaska JF, Massey JT. 1983. Spatial coding of movement: a hypothesis concerning the coding of movement direction by motor cortical populations. Exp Brain Res 49(Suppl 7):327–336.

Georgopoulos AP, Kalaska, JF, Caminiti R, Massey JT. 1982. On the relations between the direction of two-dimensional arm movements and cell discharge in primate motor cortex. J Neurosci 2:1527–1537.

Georgopoulos AP, Kettner RE, Schwartz AB. 1988. Primate motor cortex and free arm movements to visual targets in three-dimensional space. II. Coding of the direction of movement by a neuronal population. J Neurosci 8:2928–2937.

Goodale MA, Milner AD. 1992. Separate visual pathways for perception and action. Trends Neurosci 15:20–25.

Griffin DM, Hudson HM, Belhaj-Saïf A, McKiernan BJ, Cheney PD. 2008. Do corticomotoneuronal cells predict target muscle EMG activity? J Neurophysiol 99:1169–1186.

Heming EA, Lillicrap TP, Omrani M, Herter TM, Pruszynski JA, Scott SH. 2016. Primary motor cortex neurons classified in a postural task predict muscle activation patterns in a reaching task. J Neurophysiol 115:2021–2032.

Hepp-Reymond MC, Kirkpatrick-Tanner M, Gabernet L, Qi Hx, Weber B. 1999. Context-dependent force coding in motor and premotor cortical areas. Exp Brain Res 128:123–133.

Humphrey DR, Tanji J. 1991. What features of voluntary motor control are encoded in the neuronal discharge of different cortical areas? In: DR Humphrey, H-J Freund (eds). *Motor Control: Concepts and Issues*, pp 413–443. New York: Wiley.

Hwang EJ, Bailey PM, Andersen RA. 2013. Volitional control of neural activity relies on the natural motor repertoire. Curr Biol 23:353–361.

Inoue M, Uchimura M, Kitazawa S. 2016. Error signals in motor cortices drive adaptation in reaching. Neuron 90:1114–1126.

Kalaska JF, Cohen DA, Hyde ML, Prud'Homme M. 1989. A comparison of movement direction-related versus load direction-related activity in primate motor cortex, using a two-dimensional reaching task. J Neurosci 9:2080–2102.

Kaufman MT, Churchland MM, Ryu SI, Shenoy KV. 2014. Cortical activity in the null space: permitting preparation without movement. Nat Neurosci 17:440–448.

Klaes C, Westendorff S, Chakrabarti S, Gail A. 2011. Choosing goals, not rules: deciding among rule-based action plans. Neuron 70:536–548.

Kurtzer I, Herter TM, Scott SH. 2005. Random change in cortical load representation suggests distinct control of posture and movement. Nat Neurosci 8:498–504.

Maravita A, Iriki A. 2004. Tools for the body (schema). Trends Cogn Sci 8:79–86.

Moritz CT, Perlmutter SI, Fetz EE. 2008. Direct control of paralysed muscles by cortical neurons. Nature 456:639–642.

Muir RB, Lemon RN. 1983. Corticospinal neurons with a special role in precision grip. Brain Res 261:312–316.

Murata A, Wen W, Asama H. 2016. The body and objects represented in the ventral stream of the parieto-premotor network. Neurosci Res 104:4–15.

Nachev P, Kennard C, Husain M. 2008. Functional role of the supplementary and pre-supplementary motor areas. Nat Rev Neurosci 9:856–869.

Nudo RJ, Milliken GW, Jenkins WM, Merzenich MM. 1996. Use-dependent alterations of movement representations in primary motor cortex of adult squirrel monkeys. J Neurosci 16:785–807.

Omrani M, Murnaghan CD, Pruszynski JA, Scott SH. 2016. Distributed task-specific processing of somatosensory feedback for voluntary motor control. eLife 5:e13141.

Park MC, Belhaj-Saïf A, Gordon M, Cheney PD. 2001. Consistent features in the forelimb representation of primary motor cortex in rhesus macaques. J Neurosci 21:2784–2792.

Paz R, Vaadia E. 2004. Learning-induced improvement in encoding and decoding of specific movement directions by neurons in the primary motor cortex. PLoS Biol 2:E45.

Pruszynski JA, Kurtzer I, Nashed JY, Omrani M, Brouwer B, Scott SH. 2011. Primary motor cortex underlies multi-joint integration for fast feedback control. Nature 478: 387–390.

Rathelot JA, Strick PL. 2006. Muscle representation in the macaque motor cortex: an anatomical perspective. Proc Natl Acad Sci U S A 103:8257–8262.

Rizzolatti G, Camarda R, Fogassi L, Gentilucci M, Luppino G, Matelli M. 1988. Functional organization of inferior area 6 in the macaque monkey. II. Area F5 and the control of distal movement. Exp Brain Res 71:491–507.

Rizzolatti G, Fadiga L, Gallese V, Fogassi L. 1996. Premotor cortex and the recognition of motor actions. Cogn Brain Res 3:131–141.

Romo R, Hernández A, Zainos A. 2004. Neuronal correlates of a perceptual decision in ventral premotor cortex. Neuron 41:165–173.

Rozzi S, Calzavara R, Belmalih A, et al. 2006. Cortical connections of the inferior parietal cortical convexity of the macaque monkey. Cereb Cortex 16:1389–1417.

Sadtler PT, Quick KM, Golub MD, et al. 2014. Neural constraints on learning. Nature 512:423–426.

Schaffelhofer S, Scherberger H. 2016. Object vision to hand action in macaque parietal, premotor, and motor cortices. eLife 5:e15278.

Schwartz AB. 1994. Direct cortical representation of drawing. Science 265:540–542.

Scott SH, Cluff T, Lowrey CR, Takei T. 2015. Feedback control during voluntary motor actions. Curr Opin Neurobiol 33:85–94.

Scott SH, Gribble P, Graham K, Cabel DW. 2001. Dissociation between hand motion and population vectors from neural activity in motor cortex. Nature 413:161–165.

Sergio LE, Hamel-Pâquet C, Kalaska JF. 2005. Motor cortex neural correlates of output kinematics and kinetics during isometric-force and arm-reaching tasks. J Neurophysiol 94:2353–2378.

Sergio LE, Kalaska JF. 2003. Systematic changes in motor cortex cell activity with arm posture during directional isometric force generation. J Neurophysiol 89:212–228.

Shenoy KV, Sahani M, Churchland M. 2013. Cortical control of arm movements: a dynamical systems perspective. Annu Rev Neurosci 36:337–359.

Shima K, Tanji J. 1998. Role of cingulate motor area cells in voluntary movement selection based on reward. Science 282:1335–1338.

Shinoda Y, Yokota J, Futami T. 1981. Divergent projections of individual corticospinal axons to motoneurons of multiple muscles in the monkey. Neurosci Lett 23:7–12.

Sommer MA, Wurtz RH. 2008. Brain circuits for the internal monitoring of movements. Ann Rev Neurosci 31:317–338.

Strick PL. 1983. The influence of motor preparation on the response of cerebellar neurons to limb displacements. J Neurosci 3:2007–2020.

Tanji J. 2001. Sequential organization of multiple movements: involvement of cortical motor areas. Ann Rev Neurosci 24:631–651.

Thach WT. 1978. Correlation of neural discharge with pattern and force of muscular activity, joint position, and direction of intended next arm movement in motor cortex and cerebellum. J Neurophysiol 41:654–676.

Wallis JD, Miller EK. 2003. From rule to response: neuronal processes in the premotor and prefrontal cortex. J Neurophysiol 90:1790–1806.

35

The Control of Gaze

IN PRECEDING CHAPTERS, WE LEARNED about the motor systems that control the movements of the body in space. In this and the next chapter, we consider the motor systems that control our gaze, balance, and posture as we move through the world around us. In examining these motor systems, we will focus on three biological challenges that these systems resolve: How do we visually explore our environment quickly and efficiently? How do we compensate for planned and unplanned movements of the head? How do we stay upright?

In this chapter, we describe the oculomotor system and how it uses visual information to guide eye movements. It is one of the simplest motor systems, requiring the coordination of only the 12 evolutionarily old muscles that move the two eyes. In humans and other primates, the primary objective of the oculomotor system is to control the position of the fovea, the central point in the retina that has the highest density of photoreceptors and thus the sharpest vision. The fovea is less than 1 mm in diameter and covers less than 1% of the visual field. When we want to examine an object, we must move its image onto the fovea (Chapter 22).

The Eye Is Moved by the Six Extraocular Muscles

Eye Movements Rotate the Eye in the Orbit

To a good approximation, the eye is a sphere that sits in a socket, the orbit. Eye movements are simply rotations of the eye in the orbit. The eye's orientation can be defined by three axes of rotation—horizontal, vertical, and torsional—that intersect at the center of the eyeball, and eye movements are described as rotations around these axes. Horizontal and vertical eye movements change the line of sight by redirecting the fovea; torsional eye movements rotate the eye around the line of sight but do not change where the eyes are looking.

Horizontal rotation of the eye away from the nose is called *abduction*, and rotation toward the nose is *adduction* (Figure 35–1A). Vertical movements are referred to as *elevation* (upward rotation) and *depression* (downward rotation). Finally, torsional movements include

intorsion (rotation of the top of the cornea toward the nose) and *extorsion* (rotation away from the nose).

Most eye movements are conjugate; that is, both eyes move in the same direction. These eye movements are called *version* movements. For example, during gaze to the right, the right eye abducts and the left eye adducts. Similarly, if the right eye extorts, the left eye intorts. When you change your gaze from far to near, the eyes move in opposite directions—both eyes adduct. These movements are called *vergence* movements.

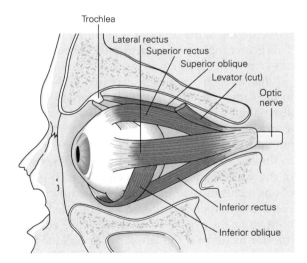

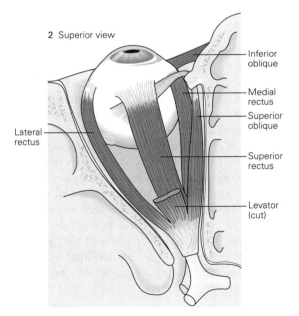

Figure 35–1 The different actions of eye movements and the muscles that control them.

A. View of the left eye and the three dimensions of eye movement.

B. 1. Lateral view of the left eye with the orbital wall cut away. Each rectus muscle inserts in front of the equator of the globe so that contraction rotates the cornea toward the muscle. Conversely, the

oblique muscles insert behind the equator, and contraction rotates the cornea away from the insertion. The superior oblique tendon passes through the trochlea, a bony pulley on the nasal side of the orbit, before it inserts on the globe. The levator muscle of the upper eyelid raises the lid. **2.** Superior view of the left eye with the roof of the orbit and the levator muscle cut away. The superior rectus passes over the superior oblique and inserts in front of it on the globe.

The Six Extraocular Muscles Form Three Agonist–Antagonist Pairs

Each eye is rotated by six extraocular muscles arranged in three agonist–antagonist pairs (Figure 35–1B). The four rectus muscles (lateral, medial, superior, and inferior) share a common origin, the annulus of Zinn, at the apex of the orbit. They insert on the surface of the eye, or sclera, anterior to the center of the eye, so the superior rectus elevates the eye and the inferior rectus depresses it. The origin of the inferior oblique muscle is on the medial wall of the orbit; the superior oblique muscle's tendon passes through the trochlea, or pulley, before inserting on the globe, so that its effective origin is also on the anteromedial wall of the orbit. The oblique muscles insert posterior to the center of the eye, so the superior oblique depresses the eye and the inferior oblique elevates it.

Each muscle has a dual insertion. The part of the muscle farthest from the eye inserts on a soft-tissue pulley through which the rest of the muscle passes on its way to the eye. When the extraocular muscles contract, they not only rotate the eye but also change their pulling directions as a result of these pulleys.

The actions of the extraocular muscles are determined by their geometry and by the position of the eye in the orbit. The medial and lateral recti rotate the eye horizontally; the medial rectus adducts, whereas the lateral rectus abducts. The superior and inferior recti and the obliques rotate the eye both vertically and torsionally. The superior rectus and inferior oblique elevate the eye, and the inferior rectus and superior oblique depress it. The superior rectus and superior oblique intort the eye, whereas the inferior rectus and inferior oblique extort it.

The superior and inferior recti and the obliques are often called the cyclovertical muscles because they produce both vertical and torsional eye rotation. The relative amounts of each rotation depend on eye position. The superior and inferior recti exert their maximal vertical action when the eye is abducted, that is, when the line of sight is parallel to the muscles' pulling directions, while the oblique muscles exert their maximal vertical action when the eye is adducted (Figure 35–2).

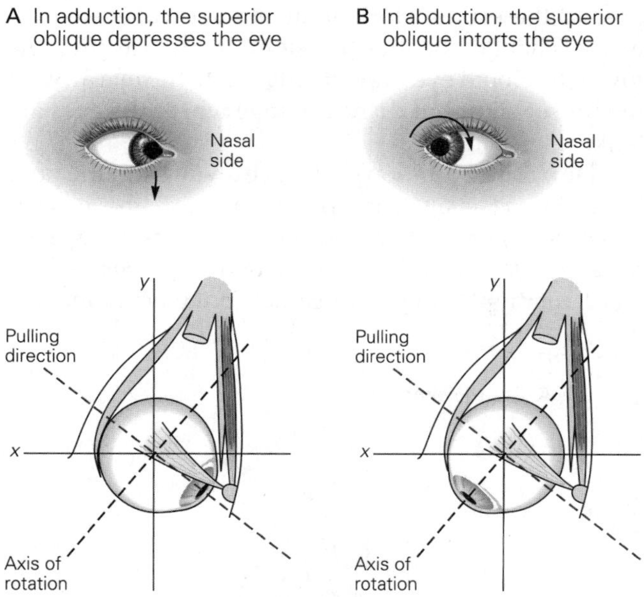

Movements of the Two Eyes Are Coordinated

Humans and other frontal-eyed animals have binocular vision—the fields of vision of the two eyes overlap. This facilitates stereopsis, the ability to perceive a visual scene in three dimensions, as well as depth perception. At the same time, binocular vision requires

A In adduction, the superior oblique depresses the eye

B In abduction, the superior oblique intorts the eye

Figure 35–2 The effect of orbital position on the action of the superior oblique muscle.

A. When the eye is adducted (looking toward the nose), contraction of the superior oblique depresses the eye.

B. When the eye is abducted (looking away from the nose), contraction of the superior oblique intorts the eye.

precise coordination of the movements of the two eyes so that both foveae are always directed at the target of interest. For most eye movements, both eyes must move by the same amount and in the same direction. This is accomplished, in large part, through the pairing of eye muscles in the two eyes.

Just as each eye muscle is paired with its antagonist in the same orbit (eg, the medial and lateral recti), it is also paired with the muscle that moves the opposite eye in the same direction. For example, coupling of the left lateral rectus and right medial rectus moves both eyes to the left during a leftward saccade. The orientations of the vertical muscles are such that each pair consists of one rectus muscle and one oblique muscle. For example, the left superior rectus and the right inferior oblique both move the eyes upward in left gaze, while the right inferior rectus and the left superior oblique both move the eyes downward in right gaze (Table 35–1).

The Extraocular Muscles Are Controlled by Three Cranial Nerves

The extraocular muscles are innervated by groups of motor neurons whose cell bodies are clustered in the three oculomotor nuclei in the brain stem (Figure 35–3).

Table 35–1 Vertical Muscle Action in Adduction and Abduction

Muscle	Action in adduction	Action in abduction
Superior rectus	Intorsion	Elevation
Inferior rectus	Extorsion	Depression
Superior oblique	Depression	Intorsion
Inferior oblique	Elevation	Extorsion

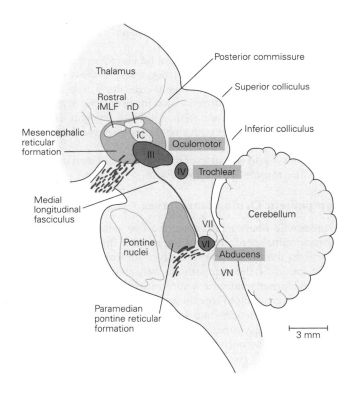

Figure 35–3 The oculomotor nuclei in the brain stem. The nuclei are shown in a parasagittal section through the thalamus, pons, midbrain, and cerebellum of a rhesus monkey. The oculomotor nucleus (cranial nerve III) lies in the midbrain at the level of the mesencephalic reticular formation; the trochlear nucleus (nerve IV) is slightly caudal; and the abducens nucleus (nerve VI) lies in the pons at the level of the paramedian pontine reticular formation, adjacent to the fasciculus of the facial nerve (VII). Compare with Figure 40–5. (Abbreviations: **iC,** interstitial nucleus of Cajal; **iMLF,** interstitial nucleus of the medial longitudinal fasciculus; **nD,** nucleus of Darkshevich; **VN,** vestibular nuclei.) (Adapted from Henn, Hepp, and Büttner-Ennever 1982.)

The lateral rectus is innervated by the abducens nerve (cranial nerve VI), whose nucleus lies in the pons in the floor of the fourth ventricle. The superior oblique muscle is innervated by the trochlear nerve (cranial nerve IV), whose nucleus is located in the contralateral midbrain at the level of the inferior colliculus. (The trochlear nerve gets its name from the trochlea, the bony pulley through which the superior oblique muscle travels.)

All the other extraocular muscles—the medial, inferior, and superior recti, and the inferior oblique—are innervated by the oculomotor nerve (cranial nerve III), whose nucleus lies in the midbrain at the level of the superior colliculus. Superior rectus axons cross the midline and join the contralateral oculomotor nerve. Thus, both superior rectus and superior oblique motor neurons innervate their respective muscles on the opposite side. The oculomotor nerve also contains fibers that innervate the levator muscle of the upper eyelid. Cell bodies of axons innervating both eyelids are located in the central caudal nucleus, a single midline structure within the oculomotor complex. Finally, traveling with the oculomotor nerve are parasympathetic fibers that innervate the iris sphincter muscle, which constricts the pupil, and the ciliary muscles that adjust the curvature of the lens to focus the eye during vergence movements from far to near, the process of accomodation.

The pupil and eyelid also have sympathetic innervation, which originates in the intermediolateral cell column of the ipsilateral upper thoracic spinal cord. Fibers of these neurons synapse on cells in the superior cervical ganglion in the upper neck. Axons of these postganglionic cells travel along the carotid artery to the cavernous sinus and then into the orbit. Sympathetic pupillary fibers innervate the iris dilator muscle, causing the pupil to dilate and thus providing the pupillary component of the so-called "fight or flight" response. Sympathetic fibers also innervate Müller's muscle, a secondary elevator of the upper eyelid. The sympathetic control of pupillary dilatation and lid elevation is responsible for the "wide-eyed" look of excitement and sympathetic overload.

The best way to understand the actions of the extraocular muscles is to consider the eye movements that remain after a lesion of a specific nerve (Box 35–1).

The force generated by an extraocular muscle is determined both by the firing rate of the motor neurons and the number of motor units recruited. Like the motor units for skeletal muscle (Chapter 31), eye motor units are recruited in a fixed sequence. For example, as the eye moves laterally, the number of active abducens neurons and their individual firing rates both increase, thereby increasing the strength of lateral rectus contraction.

Box 35–1 Extraocular Muscle or Nerve Lesions

Patients with lesions of the extraocular muscles or their nerves complain of double vision (diplopia) because the images of the object of gaze no longer fall on the corresponding retinal locations in both eyes. Lesions of each nerve produce characteristic symptoms that depend on which extraocular muscles are affected. In general, double vision increases when the patient tries to look in the direction of the weak muscle.

Abducens Nerve

A lesion of the abducens nerve (VI) causes weakness of the lateral rectus. When the lesion is complete, the eye cannot abduct beyond the midline, such that a horizontal diplopia increases when the subject looks in the direction of the affected eye.

Trochlear Nerve

A left trochlear nerve (IV) lesion affects both torsional and vertical eye movements by weakening the superior oblique muscle (Figure 35–4). Vertical misalignment in superior oblique paresis is also affected by the position of the head. A tilt to one side, such that the ear moves toward the shoulder, induces a small torsion of the eye in the opposite direction, known as ocular counter-roll. For example, when the head tilts to the left, the left eye is ordinarily intorted by the left superior rectus and left superior oblique, while the right eye is extorted by the right inferior rectus and right inferior oblique. In the left eye, the elevation action of the superior rectus is canceled by the depression action of the superior oblique, so the eye only rotates about the line of sight. When the head tilts to the right, the inferior oblique and inferior rectus extort the left eye and the superior oblique and the superior rectus relax.

With paresis of the left superior oblique, the elevating action of the superior rectus is unopposed when the head tilts to the left such that the left eye moves further upward. In contrast, tilting the head to the right relaxes

the superior rectus and superior oblique (Figure 35–4D). Thus, patients with trochlear nerve lesions often prefer to keep their heads tilted away from the affected eye because this reduces the misalignment and can eliminate diplopia.

Oculomotor Nerve

A lesion of the oculomotor nerve (III) has complex effects because this nerve innervates multiple muscles. A complete lesion spares only the lateral rectus and superior oblique muscles. Thus, the paretic eye is typically deviated downward and abducted at rest and cannot move medially or upward. Downward movement is also affected because the inferior rectus muscle is weak; because the eye is abducted, the primary action of the intact superior oblique is intorsion rather than depression.

Because the fibers that control lid elevation, accommodation, and pupillary constriction travel in the oculomotor nerve, damage to this nerve also results in drooping of the eyelid (ptosis), blurred vision for near objects, and pupillary dilation (mydriasis). Although sympathetic innervation is still intact with an oculomotor nerve lesion, the ptosis is essentially complete, since Müller's muscle contributes less to elevation of the upper eyelid than does the levator muscle of the upper eyelid.

Sympathetic Oculomotor Nerves

Sympathetic fibers innervating the eye arise from the thoracic spinal cord, traverse the apex of the lung, and ascend to the eye on the outside of the carotid artery. Interruption of the sympathetic pathways to the eye leads to Horner syndrome, which includes a partial ipsilateral ptosis owing to weakness of Müller's muscle and a relative constriction (miosis) of the ipsilateral pupil. The pupillary asymmetry is most pronounced in low light because the normal pupil is able to dilate but the pupil affected by Horner syndrome is not.

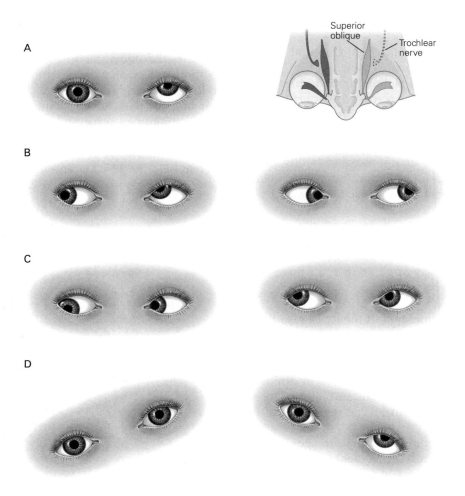

Figure 35–4 Effect of a left trochlear nerve palsy. The trochlear nerve innervates the superior oblique muscle, which inserts behind the equator of the eye. It depresses the eye when it is adducted and intorts the eye when it is abducted.

A. Hypertropia, a permanent upward deviation of the eye, can be seen when a patient is looking straight ahead. The right eye is in the center of the orbit, but the affected left eye is slightly above the right eye.

B. The hypertropia is worse when the eye is adducted because the unopposed inferior oblique pushes the eye higher (*left*). The condition is improved when the eye is abducted (*right*) because the superior oblique contributes less to depression than to intorsion.

C. When the patient looks to the right, the hypertropia is worse on downward gaze (*left*) than it is on upward gaze (*right*).

D. The hypertropia is improved by head tilt to the right (*left*) and worsened by tilt to the left (*right*). The ocular counter-rolling reflex induces intorsion of the left eye on leftward head tilt and extorsion of the eye on rightward head tilt (Chapter 27). With leftward head tilt, intorsion requires increased activity of the superior rectus, whose elevating activity is unopposed by the weak superior oblique, causing increased hypertropia. With rightward head tilt and extorsion of the left eye, the unopposed superior rectus muscle is less active, and the hypertropia decreases.

Six Neuronal Control Systems Keep the Eyes on Target

The oculomotor nuclei are the final common targets for all types of eye movements generated by higher brain networks. Hermann Helmholtz and other 19th-century psychophysicists appreciated that analysis of eye movements was essential for understanding visual perception, but they assumed that all eye movements were smooth. In 1890, Edwin Landott discovered that during reading the eyes do not move smoothly along a line of text but make fast intermittent movements called saccades (French, jerks), each followed by a short pause.

By 1902, Raymond Dodge outlined five distinct types of eye movement that direct the fovea to a visual target and keep it there. All of these eye movements share an effector pathway originating in the three oculomotor nuclei in the brain stem.

- Saccadic eye movements shift the fovea rapidly to a new visual target.
- Smooth-pursuit movements keep the image of a moving target on the fovea.
- Vergence movements move the eyes in opposite directions so that the image of an object of interest is positioned on both foveae regardless of its distance.
- Vestibulo-ocular reflexes stabilize images on the retina during brief head movements.
- Optokinetic movements stabilize images during sustained head rotation or translation.

A sixth system, the fixation system, holds the eye stationary during intent gaze when the head is not moving by actively suppressing eye movement. The optokinetic and vestibular systems are discussed in Chapter 27. We consider the other four systems here.

An Active Fixation System Holds the Fovea on a Stationary Target

Vision is most accurate when the eyes are still. The gaze system actively prevents the eyes from moving when we examine an object of interest. It is not as active in suppressing movement when we are doing something that does not require vision, such as mental arithmetic. Patients with disorders of the fixation system—for example, patients with irrepressible saccadic eye movements (opsoclonus)—have poor vision not because their visual acuity is deficient but because they cannot hold their eyes still enough for the visual system to work correctly.

The Saccadic System Points the Fovea Toward Objects of Interest

Our eyes explore the world in a series of very quick saccades that move the fovea from one fixation point to another (Chapter 25) (Figure 35–5). Saccades allow us to scan the environment quickly and to read. Highly stereotyped, they have a standard waveform with a single smooth increase and decrease of eye velocity. Saccades are also extremely fast, occurring within a fraction of a second at angular speeds up to 900° per second (Figure 35–6A). The velocity of a saccade is determined only by its size. We can voluntarily change

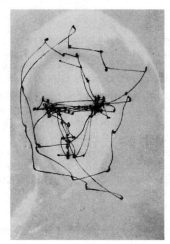

Figure 35–5 Eye movements track the outline of an object of attention. An observer looks at a picture of a woman for 1 minute. The resulting eye positions are then superimposed on the picture. As shown here, the observer concentrated on certain features of the face, lingering over the woman's eyes and mouth (*fixations*) and spending less time over intermediate positions. The rapid movements between fixation points are *saccades.* (Reproduced, with permission, from Yarbus 1967.)

A Saccade

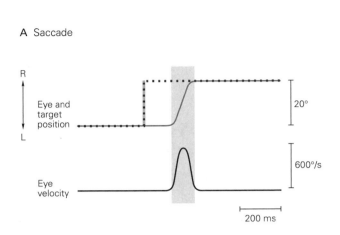

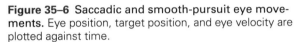

B Smooth pursuit

— Eye position   ▨ Saccade
• • • Target position   ▨ Smooth
■■■ Target moving   pursuit

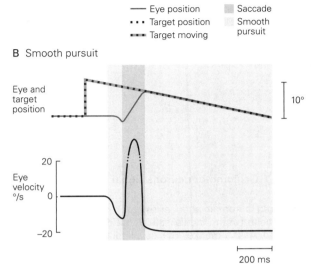

Figure 35–6 Saccadic and smooth-pursuit eye movements. Eye position, target position, and eye velocity are plotted against time.

A. *The human saccade.* At the beginning of the plot, the eye is on the target (the traces representing eye and target positions are superimposed). Suddenly, the target jumps to the right, and within 200 ms, the eye moves to bring the target back to the fovea. Note the smooth, symmetric velocity profile. Because eye movements are rotations of the eye in the orbit, they are described by the angle of rotation. Similarly, objects in the visual field are described by the angle of arc they subtend at the eye. Viewed at arm's length, a thumb subtends an angle of approximately 1°. A saccade from one edge of the thumb to the other therefore traverses 1° of arc. (Abbreviations: **L**, left; **R**, right.)

B. *Human smooth pursuit.* In this example, the subject is asked to make a saccade to a target that jumps away from the center of gaze and then slowly moves back to center. The first movement seen in the position and velocity traces is a smooth-pursuit movement in the same direction as the target movement. The eye briefly moves *away* from the target before a saccade is initiated because the latency of the pursuit system is shorter than that of the saccade system. The smooth-pursuit system is activated by the target moving back toward the center of gaze, the saccade adjusts the eye's position to catch the target, and thereafter, smooth pursuit keeps the eye on the target. The recording of saccade velocity is clipped so that the movement can be shown on the scale of the pursuit movement, an order of magnitude slower than the saccade.

the amplitude and direction of saccades but not their speed, although fatigue, drugs, or pathological states can slow saccades.

Ordinarily, there is no time for visual feedback to modify the course of a saccade as it is being made; instead, corrections to the direction and/or amplitude of movement are made over the course of successive saccades. Accurate saccades can be made not only to visual targets but also to sounds, tactile stimuli, memories of locations in space, and even verbal commands (eg, "look left").

When a saccade is made, the activity of neurons in higher brain centers that control gaze specify only a desired change in eye position (eg, 20° to the right of current gaze, usually based on a target location in the visual field). For the eye movement to be made, this location signal must be transformed into signals for the eye muscles that execute the desired velocity and change in eye position. We can illustrate how the gaze system generates eye movements by considering the activity of an oculomotor neuron during a saccade (Figure 35–7A). To move the eye quickly to a new position in the orbit and keep it there, two passive forces

must be overcome: the elastic force of the orbital tissues, which tends to restore the eye to a central position, and a velocity-dependent viscous force that opposes rapid movement. Thus, the motor signal for an eye movement must include both a position component to counter the elastic force and a velocity component to overcome orbital viscosity and move the eye quickly to the new position.

This eye position and velocity information are coded by the discharge frequencies of oculomotor neurons. When a saccade is made, the firing rate of a neuron rises rapidly as eye velocity increases; this is called the *saccadic pulse* (Figure 35–7B). The frequency of this pulse determines the speed of the saccade, whereas the length of the pulse controls the duration of the saccade and thus its amplitude. When the saccade is completed and the eye has reached its goal, there must be a new level of tonic input to the eye muscles that is appropriate for the elastic restoring force at that orbital position. This difference in the tonic firing rate between before and after the saccade is called the *saccadic step* (Figure 35–7B). If the size of the step is not properly matched to

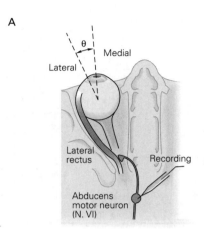

Figure 35–7 Oculomotor neurons signal eye position and velocity.

A. The record is from an abducens neuron of a monkey. When the eye is positioned in the medial side of the orbit, the cell is silent (**position** θ_0). As the monkey makes a lateral saccade, there is a burst of firing (**D1**), but in the new position (θ_1), the eye is still too far medial for the cell to discharge continually. During the next saccade, there is a burst (**D2**), and at the new position (θ_2), there is a tonic position-related discharge. Before and during the next saccade (**D3**), there is again a pulse of activity and a higher tonic discharge when the eye is at the new position (θ_3). When the eye makes a medial movement, there is a period of silence during the saccade (**D4**) even though the eye ends up at a position (θ_4) associated with a tonic discharge. (Adapted from Fuchs and Luschei 1970.)

B. Saccades are associated with a step of activity, which signals the change in eye position, and a pulse of activity, which signals eye velocity. The neural activity corresponding to eye position and velocity is illustrated both as a train of individual spikes and as an estimate of the instantaneous firing rate (spikes per second).

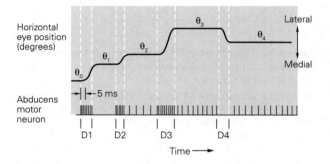

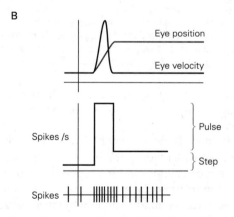

the pulse, then the eye drifts away from the target after the saccade. As described later, the pulse and step are generated by different brain stem structures.

The Motor Circuits for Saccades Lie in the Brain Stem

Horizontal Saccades Are Generated in the Pontine Reticular Formation

The neuronal signal for horizontal saccades originates in the paramedian pontine reticular formation, adjacent to the abducens nucleus to which it projects (Figure 35–8A). The paramedian pontine reticular formation contains a family of *burst neurons* that gives rise to the saccadic pulse. These cells fire at a high frequency just before and during ipsiversive saccades (toward the same side as the discharging neurons), and their activity resembles the pulse component of oculomotor neuron discharge (Figure 35–7B).

There are several types of burst neurons (Figure 35–8B). Medium-lead excitatory burst neurons make direct excitatory connections to motor neurons and interneurons in the ipsilateral abducens nucleus.

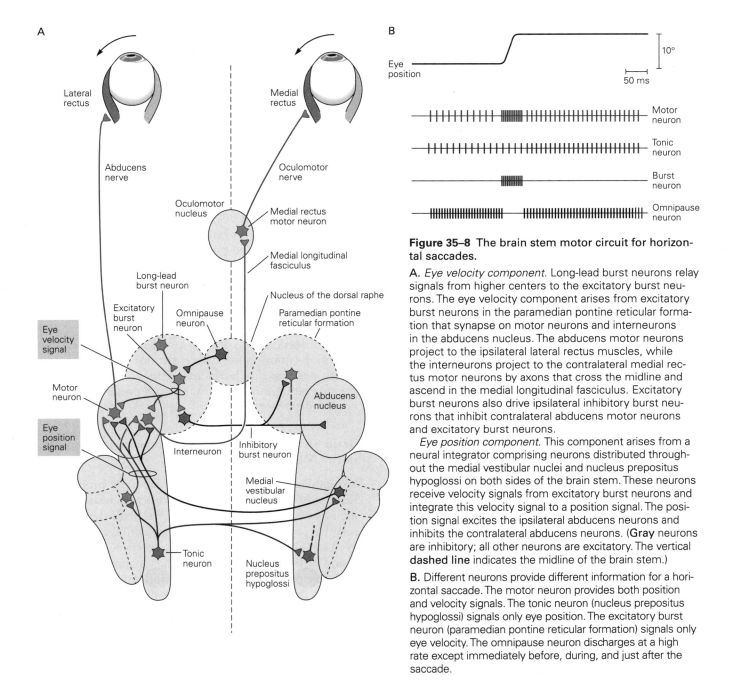

Figure 35–8 The brain stem motor circuit for horizontal saccades.

A. *Eye velocity component.* Long-lead burst neurons relay signals from higher centers to the excitatory burst neurons. The eye velocity component arises from excitatory burst neurons in the paramedian pontine reticular formation that synapse on motor neurons and interneurons in the abducens nucleus. The abducens motor neurons project to the ipsilateral lateral rectus muscles, while the interneurons project to the contralateral medial rectus motor neurons by axons that cross the midline and ascend in the medial longitudinal fasciculus. Excitatory burst neurons also drive ipsilateral inhibitory burst neurons that inhibit contralateral abducens motor neurons and excitatory burst neurons.

Eye position component. This component arises from a neural integrator comprising neurons distributed throughout the medial vestibular nuclei and nucleus prepositus hypoglossi on both sides of the brain stem. These neurons receive velocity signals from excitatory burst neurons and integrate this velocity signal to a position signal. The position signal excites the ipsilateral abducens neurons and inhibits the contralateral abducens neurons. (**Gray** neurons are inhibitory; all other neurons are excitatory. The vertical **dashed line** indicates the midline of the brain stem.)

B. Different neurons provide different information for a horizontal saccade. The motor neuron provides both position and velocity signals. The tonic neuron (nucleus prepositus hypoglossi) signals only eye position. The excitatory burst neuron (paramedian pontine reticular formation) signals only eye velocity. The omnipause neuron discharges at a high rate except immediately before, during, and just after the saccade.

Long-lead burst neurons drive the medium-lead burst cells and receive excitatory input from higher centers. Inhibitory burst neurons suppress the activity of contralateral abducens neurons and contralateral excitatory burst neurons and are themselves excited by medium-lead burst neurons.

A second class of pontine cells, *omnipause neurons,* fires continuously except around the time of a saccade; firing ceases shortly before and during all saccades (Figure 35–8B). Omnipause neurons are located in the nucleus of the dorsal raphe in the midline (Figure 35–8A). They are GABAergic (γ-aminobutyric acid) inhibitory neurons that project to contralateral pontine and mesencephalic burst neurons. Electrical stimulation of omnipause neurons arrests a saccade, which resumes when the stimulation stops. Making a saccade requires simultaneous excitation of burst neurons and inhibition of omnipause cells; this provides the system with additional stability, such that unwanted saccades are infrequent.

If the motor neurons received signals from only the burst cells, the eyes would drift back to the starting position after a saccade, because there would be no new position signal to hold the eyes against elastic restorative forces. The appropriate tonic innervation is required to keep the eye at the new orbital position. This tonic position signal, the saccadic step, can be generated from the velocity burst signal by the neural equivalent of the mathematical process of integration. Velocity can be computed by differentiating position with respect to time; conversely, position can be computed by integrating velocity with respect to time.

For horizontal eye movements, neural integration of the velocity signal is performed by the medial vestibular nucleus and nucleus prepositus hypoglossi (Figure 35–8A) in conjunction with the flocculus of the cerebellum. As expected, animals with lesions of these areas make normal horizontal saccades, but the eyes drift back to a middle position after a saccade. Moreover, integration of the horizontal saccadic burst requires coordination of the bilateral nuclei prepositi hypoglossi and medial vestibular nuclei through commissural connections. Thus, a midline lesion of these connections also causes failure of the neural integrator.

Medium-lead burst neurons in the paramedian pontine reticular formation and neurons of the medial vestibular nucleus and nucleus prepositus hypoglossi project to the ipsilateral abducens nucleus and deliver respectively the pulse and step components of the motor signal. Two populations of neurons in the abducens nucleus receive this signal. One is a group of motor neurons that innervate the ipsilateral lateral rectus muscle. The second group consists of interneurons whose axons cross the midline and ascend in the medial longitudinal fasciculus to the motor neurons for the contralateral medial rectus, which lie in the oculomotor nucleus (Figure 35–8A).

Thus, medial rectus motor neurons do not receive the pulse and step signals directly. This arrangement allows for precise coordination of corresponding movements of both eyes during horizontal saccades and other conjugate eye movements. The susceptibility of the medial longitudinal fasciculus to strokes and multiple sclerosis make it clinically important.

Several cerebellar structures play an important role in the calibration of the saccade motor signal. First, the oculomotor portion of the dorsal vermis, acting through the caudal fastigial nucleus, controls the duration of the pulse and thus the accuracy of the saccade. The fastigial nucleus increases saccade velocity at the beginning of contraversive saccades and contributes to braking ipsiversive saccades to end the saccade. Second, the flocculus and paraflocculus of the vestibulocerebellum calibrate the neural integrator to ensure that the step is properly matched to the pulse, in order to hold the eyes at the new position after each saccade.

Vertical Saccades Are Generated in the Mesencephalic Reticular Formation

The burst neurons responsible for vertical saccades are found in the rostral interstitial nucleus of the medial longitudinal fasciculus in the mesencephalic reticular formation (Figure 35–3). Vertical and torsional neural integration are performed in the nearby interstitial nucleus of Cajal. The pontine and mesencephalic systems participate together in the generation of oblique saccades, which have both horizontal and vertical components.

Purely vertical saccades require activity on both sides of the mesencephalic reticular formation, and communication between the two sides occurs via the posterior commissure. There are not separate omnipause neurons for horizontal and vertical saccades; pontine omnipause cells inhibit both pontine and mesencephalic burst neurons.

Brain Stem Lesions Result in Characteristic Deficits in Eye Movements

We can now understand how different brain stem lesions cause characteristic syndromes. Lesions that include the paramedian pontine reticular formation result in paralysis of ipsiversive horizontal gaze of both eyes but spare contraversive and vertical saccades. A lesion of the abducens nucleus has a similar effect, as both abducens motor neurons and interneurons are affected. Lesions that include the midbrain gaze centers cause paralysis of vertical gaze. Certain neurological disorders cause degeneration of burst neurons and impair their function, leading to a progressive slowing of saccades.

Lesions of the medial longitudinal fasciculus disconnect the medial rectus motor neurons from the abducens interneurons (Figure 35–8A). Thus, during conjugate horizontal eye movements, such as saccades and pursuit, the abducting eye moves normally but adduction of the other eye is impeded. Despite this paralysis in version movements, the medial rectus typically acts normally in vergence movements because the motor neurons for vergence lie in the midbrain, as will be discussed later. This syndrome, called an *internuclear ophthalmoplegia*, is a consequence of a brain stem stroke or demyelinating diseases such as multiple sclerosis.

A lesion of the cerebellar fastigial nucleus causes ipsiversive saccades to overshoot their targets (*hypermetric* saccades), due to failure of normal termination of the saccadic burst. Contraversive saccades undershoot their targets (*hypometric* saccades). Correspondingly, damage to the oculomotor vermis disinhibits the fastigial nucleus and causes hypometric ipsiversive saccades. This may be due to an additional failure to compensate for the position-dependent passive forces of the orbital tissues.

Saccades Are Controlled by the Cerebral Cortex Through the Superior Colliculus

The pontine and mesencephalic burst circuits provide the motor signals necessary to drive the extraocular muscles for saccades. However, among higher mammals, eye movements are ultimately driven by cognitive behavior. The decision of when and where to make a saccade that is behaviorally important is usually made in the cerebral cortex. A network of cortical and subcortical areas controls the saccadic system through the superior colliculus (Figure 35–9).

The Superior Colliculus Integrates Visual and Motor Information into Oculomotor Signals for the Brain Stem

The superior colliculus in the midbrain is a major visuomotor integration region, the mammalian homolog of the optic tectum in nonmammalian vertebrates. It can be divided into two functional regions: the superficial layers and the intermediate and deep layers.

The three superficial layers receive both direct input from the retina and a projection from the striate

Figure 35–9 Cortical pathways for saccades.

A. In the monkey, the saccade generator in the brain stem receives a command from the superior colliculus. That command is relayed through the pontine and mesencephalic burst circuits, providing the motor signals that drive the extraocular muscles for saccades. The colliculus receives direct excitatory projections from the frontal eye fields and the lateral intraparietal area (**LIP**) and an inhibitory projection from the substantia nigra. The substantia nigra is suppressed by the caudate nucleus, which in turn is excited by the frontal eye fields. Thus, the frontal eye fields directly excite the colliculus and indirectly release it from suppression by the substantia nigra by exciting the caudate nucleus, which inhibits the substantia nigra. The oculomotor vermis (**OMV**) of the cerebellum, acting through the fastigial nucleus (**FN**), calibrates the burst to keep saccades accurate.

B. This lateral scan of a human brain shows areas of cortex activated during saccades. (Adapted from Curtis and Connolly 2008.)

A Monkey

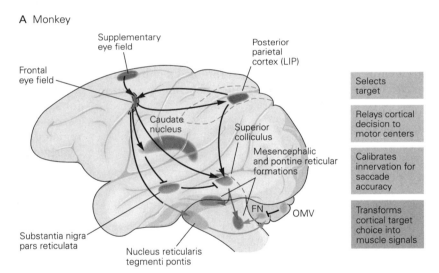

B Human

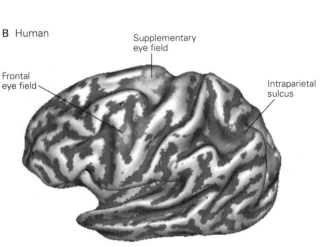

cortex representing the entire contralateral visual hemifield. Neurons in the superficial layers respond to visual stimuli. In monkeys, the responses of half of these vision-related neurons are quantitatively enhanced when an animal prepares to make a saccade to a stimulus in the cell's receptive field. This enhancement is specific for saccades. If the monkey attends to the stimulus without making a saccade to it—for example, by making a hand movement in response to a brightness change—the neuron's response is not augmented. Neurons in the superficial layers of the superior colliculus are functionally arranged in a retinotopic map of the visual field in which representation of the visual field closest to the fovea occupies the largest area (Figure 35–10).

Neuronal activity in the two intermediate and deep layers is primarily related to oculomotor actions. The movement-related neurons in these layers receive visual information from the prestriate, middle temporal, and parietal cortices and motor information from the frontal eye field. The intermediate and deep layers also contain somatotopic, tonotopic, and retinotopic maps of sensory inputs, all in register with one another. For example, the image of a bird will excite a vision-related neuron, whereas the bird's chirp will excite an adjacent audition-related neuron, and both

will excite a bimodal neuron. Polymodal spatial maps enable us to shift our eyes toward auditory or somatosensory stimuli as well as visual ones.

Much of the early research describing the sensory responsiveness of neurons in the intermediate layer was done in anesthetized animals. To understand how the brain generates movement, however, the activity of neurons needs to be studied in alert, active animals. Edward Evarts pioneered this approach in studies of the skeletomotor system, after which it was extended to the oculomotor system.

One of the earliest cellular studies in active animals revealed that individual movement-related neurons in the superior colliculus selectively discharge before saccades of specific amplitudes and directions, just as individual vision-related neurons in the superior colliculus respond to stimuli at specific distances and directions from the fovea (Figure 35–11A). The movement-related neurons form a map of potential eye movements that is in register with the visuotopic and tonotopic arrays of sensory inputs, so that the neurons that control eye movements to a particular target are found in the same region as the cells excited by the sounds and image of that target. Each movement-related neuron in the superior colliculus has a *movement field*, a region of the visual field that is the target for saccades controlled by that neuron. There is a map of movement fields in the intermediate layers that is in register with the map of visual receptive fields in the overlying superficial layers. Each movement neuron discharges before a saccade to the center of the overlying visual receptive field. A map of saccades evoked by electrical stimulation of the intermediate layers resembles the visual map.

Movement fields are large, so each superior colliculus cell fires before a wide range of saccades, although each cell fires most intensely before saccades of a specific direction and amplitude. A large population of cells is thus active before each saccade, and eye movement is encoded by the entire ensemble of these broadly tuned cells. Because each cell makes only a small contribution to the direction and amplitude of the movement, any variability or noise in the discharge of a given cell is minimized. Similar population coding is found in many sensory systems (Chapter 17) and the skeletal motor system (Chapter 34).

Activity in the superficial and intermediate layers of the superior colliculus can occur independently: Sensory activity in the superficial layers does not always lead to motor output, and motor output can occur without sensory activity in the superficial layers. In fact, the neurons in the superficial layers do not provide a large projection directly to the intermediate

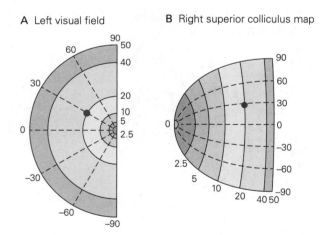

Figure 35–10 Neurons in the superior colliculus are organized in a retinotopic map.

A. Map of the left visual field in polar coordinates. **Dashed lines** represent the angle and **solid lines** the eccentricity.

B. Spatial map of neurons in the superior colliculus represented in polar coordinates of the visual field. In the nucleus, more neurons represent the part of the visual field close to the fovea and fewer neurons represent the periphery. For example, a stimulus appearing at 20° eccentricity and 30° elevation in the visual field (**red dot**) will excite neurons at the location of the red dot on the collicular map. (Reproduced, with permission, from Quaia et al. 1998.)

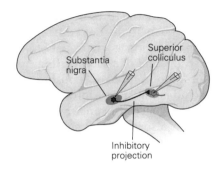

A Superior colliculus neuron

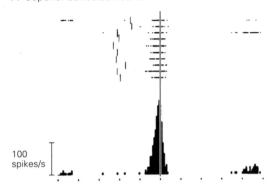

100
spikes/s

B Substantia nigra neuron

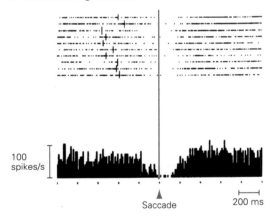

100
spikes/s

Saccade   200 ms

Figure 35–11 Neurons in the superior colliculus and substantia nigra are active around the time of a saccade.
(Reproduced, with permission, from Hikosaka and Wurtz 1989.)
A. A neuron recorded from the region in the superior colliculus from which the neuron in B could be excited antidromically fires in a burst immediately before the saccade. Raster plots of activity in successive trials of the same task are summed to form the histogram below. The small vertical lines in the raster indicate target appearance. The trials are aligned at the beginning of the saccade (**blue line**).

B. A neuron in the substantia nigra pars reticulata is tonically active, becomes quiet just before the saccade, and resumes activity after the saccade. This type of neuron inhibits neurons in the intermediate layers of the superior colliculus.

layers. Instead, their axons terminate on neurons in the pulvinar and lateral posterior nuclei of the thalamus, which relay the signals from the superficial layers of the superior colliculus to cortical regions that project back to the intermediate layers.

Lesions of a small part of the colliculus affect the latency, accuracy, and velocity of saccades. Destruction of the entire colliculus renders a monkey unable to make any contraversive saccades, although with time, this ability is recovered.

The Rostral Superior Colliculus Facilitates Visual Fixation

The most rostral portion of the superior colliculus receives inputs from the fovea and the foveal representation in primary visual cortex (V1). Neurons in the intermediate layers in this region discharge strongly during active visual fixation and before small saccades to the contralateral visual field. Because the neurons are active during visual fixation, this area of the superior colliculus is often called the fixation zone.

Neurons here inhibit the movement-related neurons in the more caudal parts of the colliculus and also project directly to the nucleus of the dorsal raphe, where they inhibit saccade generation by exciting the omnipause neurons. With lesions in the fixation zone, an animal is more likely to make saccades to distracting stimuli.

The Basal Ganglia and Two Regions of Cerebral Cortex Control the Superior Colliculus

The superior colliculus receives a powerful GABAergic inhibitory projection from neurons in the substantia nigra, which fire spontaneously with high frequency. This discharge is suppressed at the time of voluntary eye movements to the contralateral visual field (Figure 35–11B) by inhibitory input from neurons in the caudate nucleus, which fire before saccades to the contralateral visual field.

The superior colliculus is controlled by two regions of the cerebral cortex that have overlapping but distinct functions: the lateral intraparietal area of the posterior parietal cortex (part of Brodmann's area 7) and the frontal eye field (part of Brodmann's area 8). Each of these areas contributes to the generation of saccades and the control of visual attention.

Perception of attended objects in the visual field is better than perception of unattended objects, as measured either by a subject's reaction time to an object suddenly appearing in the visual field or by the subject's ability to perceive a stimulus that is just barely noticeable. Saccadic eye movements and visual attention are closely intertwined (Figure 35–5).

The lateral intraparietal area in the monkey is important in the generation of both visual attention and saccades. The role of this area in the processing of eye movements is best illustrated by a memory-guided saccade. To demonstrate this saccade, a monkey first fixates a spot of light. An object (the stimulus) appears in the receptive field of a neuron and then disappears; then the spot of light is extinguished. After a delay, the monkey must make a saccade to the former location of the vanished object. Neurons in the lateral intraparietal area respond from the moment the object appears and continue firing after the object has vanished and throughout the delay until the saccade begins (Figure 35–12A), but their activity can be also dissociated from saccade planning. If the monkey is planning a saccade to a target outside the receptive field of a neuron and a distractor appears in the field during the delay period, the neuron responds as vigorously to the distractor as it does to the target of a saccade (Figure 35–12B).

Lesioning of a monkey's posterior parietal cortex, which includes the lateral intraparietal area, increases the latency of saccades and reduces their accuracy. Such a lesion also produces selective neglect: A monkey with a unilateral parietal lesion preferentially attends to stimuli in the ipsilateral visual hemifield. In humans as well, parietal lesions—especially right parietal lesions—initially cause dramatic attentional deficits. Patients act as if the objects in the neglected field do not exist, and they have difficulty making eye movements into that field (Chapter 59).

Patients with Balint syndrome, which is usually the result of bilateral lesions of the posterior parietal

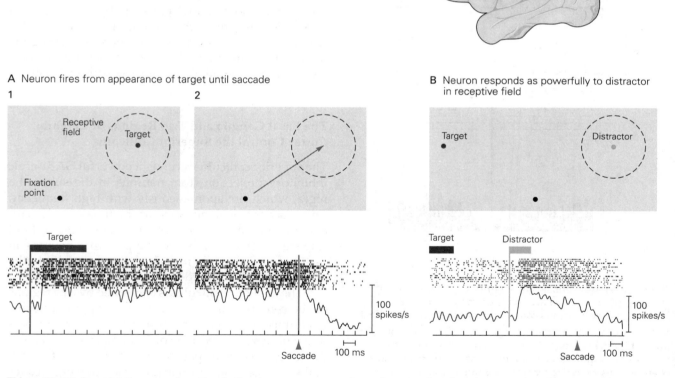

Figure 35–12 A parietal neuron is active before memory-guided saccades. Traces are aligned at events indicated by vertical lines. (Adapted, with permission, from Powell and Goldberg 2000.)

A. The monkey plans a saccade from a fixation point to a target in the receptive field of a neuron in the lateral intraparietal cortex. The neuron responds to the appearance of the target (1).

It continues to fire after the target has disappeared but before the signal to make the saccade and stops firing after the onset of the saccade (2).

B. The monkey plans a saccade to a target outside the receptive field. The neuron responds initially to a distractor in the receptive field as strongly as it did to the target of a saccade.

and prestriate cortex, tend to see and describe only one object at a time in their visual environment. These patients make few saccades, as if they are unable to shift the focus of their attention from the fovea, and can therefore describe only a foveal target. Even after these patients have recovered from most of their deficits, their saccades are delayed and inaccurate.

Compared to the neurons in the parietal cortex, neurons in the frontal eye field are more closely associated with saccades. Three different types of neurons in the frontal eye field discharge before saccades.

Visual neurons respond to visual stimuli, and half of these neurons respond more vigorously to stimuli that are the targets of saccades (Figure 35–13A). Activity in these cells is not enhanced when an animal responds to the stimulus without making a saccade to it. Likewise, these cells are not activated before saccades that are made without visual targets; monkeys can be trained to make saccades of a specific direction and amplitude in total darkness.

Movement-related neurons fire before and during saccades to their movement fields. Unlike the movement-related cells in the superior colliculus, which fire before all saccades, movement-related neurons of the frontal eye field fire only before saccades that are relevant to the monkey's behavior (Figure 35–13B). These neurons, especially those whose receptive fields lie in the visual periphery, project more strongly to the superior colliculus than do the visual neurons.

Visuomovement neurons have both visual and movement-related activity and discharge most strongly before visually guided saccades. Electrical stimulation of the frontal eye field evokes saccades to the movement fields of the stimulated cells. Bilateral stimulation of the frontal eye field evokes vertical saccades.

Movement-related neurons in frontal eye field control the superior colliculus through two pathways. They excite the superior colliculus directly and they release it from the inhibitory influence of the substantia nigra by exciting the caudate nucleus, which in turn inhibits the nigra (Figure 35–9A). The frontal eye field also projects to the pontine and mesencephalic reticular formations, although not directly to the burst cells.

Two other cortical regions besides LIP that have inputs to the frontal eye field are thought to be important in the cognitive aspects of saccades. The supplementary eye field at the most rostral part of the supplementary motor area contains neurons that encode spatial information other than the direction of the desired eye movement. For example, a neuron in the left supplementary eye field that ordinarily fires before rightward eye movements will fire before a leftward saccade if that saccade is to the right side of the

target. The dorsolateral prefrontal cortex has neurons that discharge when a monkey makes a saccade to a remembered target. The activity commences with the appearance of the stimulus and continues throughout the interval during which the monkey must remember the location of the target.

We can now understand the effects of lesions of these regions on the generation of saccades. Lesions of the superior colliculus in monkeys produce only transient damage to the saccade system because the projection from the frontal eye field to the brain stem remains intact. Animals can likewise recover from cortical lesions if the superior colliculus is intact. However, when both the frontal eye field and the colliculus are damaged, the ability to make saccades is permanently compromised. The predominant effect of a parietal lesion is an attentional deficit. After recovery, however, the system can function normally because the frontal eye field signals are sufficient to suppress the substantia nigra and stimulate the colliculus.

Damage to the frontal eye field alone causes more subtle deficits. Lesions of the frontal eye field in monkeys cause transient contralateral neglect and paresis of contraversive gaze, which recover rapidly. The latter deficit may reflect the loss of frontal eye field control of the substantia nigra; this loss of control means that the constant inhibitory input from the substantia nigra to the colliculus does not get suppressed, and the colliculus is unable to generate any saccades. Eventually the system adapts, and the colliculus responds to the remaining parietal signal. After recovery, the animals have no trouble producing saccades to targets in the visual field but have great difficulty with memory-guided saccades. Bilateral lesions of both the frontal eye fields and the superior colliculus render monkeys unable to make saccades at all.

Humans with lesions of the frontal cortex have difficulty suppressing unwanted saccades to attended stimuli. This is easily shown by asking subjects to make an eye movement away from a stimulus, the "anti-saccades task." For example, if a stimulus appears on the left, the subject should make a saccade of the same size to the right. To do this, the subject must attend to the stimulus, without turning the eyes toward it, and use its location to calculate the desired saccade to the opposite direction. Patients with frontal lesions have great difficulty suppressing the unwanted saccade to the stimulus.

As we have seen, neurons in the lateral intraparietal area of monkeys are active when the animal attends to a visual stimulus whether or not the animal makes a saccade to the stimulus. In the absence of frontal eye field signals, this undifferentiated signal is the only one to reach the superior colliculus. In humans, the failure to suppress a saccade is therefore to be expected

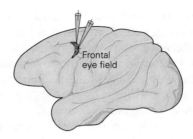

A Visual neuron responds to the stimulus and not to movement

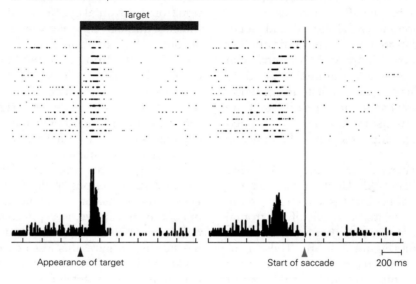

Appearance of target   Start of saccade   200 ms

B Movement-related neuron responds before movement but not to stimulus

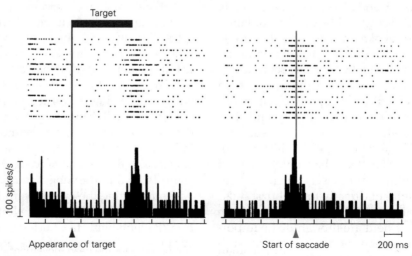

Appearance of target   Start of saccade   200 ms

Figure 35–13 Visual and movement-related neurons in the frontal eye field. (Reproduced, with permission, from Bruce and Goldberg 1985.)

A. Activity of a visual neuron in the frontal eye field as a monkey makes a saccade to a target in its visual field. Raster plots of activity in successive trials of the same task are summed to form the histogram below. In the record on the left, the individual trials are aligned at the appearance of the stimulus. A burst

of firing is closely time-locked to the stimulus. In the record on the right, the trials are aligned at the beginning of the saccade. Activity is not well aligned with the beginning of the saccade and stops before the saccade itself commences.

B. Activity of a movement-related neuron in the frontal eye field. The records of each trial are aligned as in part A. The cell does not respond to appearance of the saccade target (*left*) but is active at the time of the saccade (*right*).

if the superior colliculus responds to a parietal signal that generates attention to the stimulus without the frontal-nigral control that normally prevents saccades in response to parietal signals.

The Control of Saccades Can Be Modified by Experience

Quantitative study of the neural control of movement is possible because the discharge rate of a motor neuron has a predictable effect on a movement. For example, a certain frequency of firing in the abducens motor neuron has a predictable effect on eye position and velocity.

This relationship can change if disease damages an oculomotor nerve or causes an eye muscle to become weak, although the brain can compensate to some degree for such changes. Guntram Kommerell described a case that dramatically illustrates this point. A diabetic patient had an acute partial abducens nerve lesion affecting one eye and a retinal hemorrhage in the other. Because of the poor vision in the eye with a normal abducens nerve, he ordinarily used the eye with the newly weakened lateral rectus muscle. After a few days, the eye recovered the ability to make fairly accurate eye movements. When the weak eye was patched and the subject attempted to make a saccade with the visually poor eye, the saccade overshot the target. This implies that in order to compensate for the weakness of the visually normal eye the brain increased the neural signal to both eyes, resulting in too large a signal to the eye with normal motor input. This change in the motor response depends on the fastigial nucleus and vermis of the cerebellum (Figure 35–9A) and results from the visual system signaling that the preceding eye movement was inaccurate.

Some Rapid Gaze Shifts Require Coordinated Head and Eye Movements

So far, we have described how the eyes are moved when the head is still. When we look around, however, our head is moving as well. Head and eye movements must be coordinated to direct the fovea to a target.

Because the head has a much greater inertia than the eyes, a small shift in gaze drives the fovea to its target before the head begins to move. A small gaze shift usually consists of a saccade followed by a small head movement during which the vestibulo-ocular reflex moves the eyes back to the center of the orbit in the new head position (Figure 35–14). For larger gaze shifts, the eyes and the head move simultaneously in the same direction. Because the vestibulo-ocular reflex ordinarily moves the eyes in the direction opposite that of the head, the reflex must be temporarily suppressed.

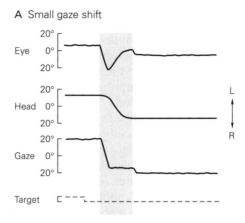

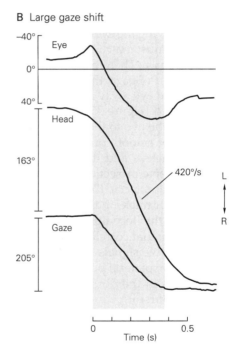

Figure 35–14 Directing the fovea to an object when the head is moving requires coordinated head and eye movements.

A. For a small gaze shift, the eye and head move in sequence. The eye begins to move 300 ms after the target appears. Near the end of the eye movement, the head begins to move as well. The eye then rotates back to the center of the orbit to compensate for the head movement. The gaze record is the sum of eye and head movements. (Abbreviations: **L**, left; **R**, right.) (Reproduced, with permission, from Zee 1977.)

B. For a large gaze shift, the eye and head move in the same direction simultaneously. Near the end of the gaze shift, the vestibulo-ocular reflex returns, the eye begins to compensate for head movement as in part **A**, and gaze becomes still. (Reproduced, with permission, from Laurutis and Robinson 1986.)

The Smooth-Pursuit System Keeps Moving Targets on the Fovea

The smooth-pursuit system holds the image of a moving target on the fovea by calculating how fast the target is moving and moving the eyes at the same speed. Smooth-pursuit movements have a maximum angular velocity of approximately 100° per second, much slower than saccades. Drugs, fatigue, alcohol, and even distraction degrade the quality of these movements.

Smooth pursuit and saccades have very different central control systems. This is best seen when a target jumps away from the center of gaze and then slowly moves back toward it. A smooth-pursuit movement is initiated first because the smooth-pursuit system has a shorter latency and responds to target motion on the peripheral retina as well as on the fovea. The task of the smooth-pursuit system differs from that of the saccade system. Instead of driving the eyes as rapidly as possible to a point in space, it must match the velocity of the eyes to that of a target in space. Therefore, as the target moves back toward the center of gaze, the smooth-pursuit system briefly moves the eye away from the target before the saccade is initiated (Figure 35–6B). The subsequent saccade then brings the eye to the target. Neurons that signal eye velocity for smooth pursuit are found in the medial vestibular nucleus and the nucleus prepositus hypoglossi. They receive projections from the flocculus of the cerebellum and project to the abducens nucleus as well as the oculomotor nuclei in the midbrain.

Neurons in both the flocculus and vermis transmit an eye-velocity signal that correlates with smooth pursuit. These areas receive signals from the cerebral cortex relayed by the dorsolateral pontine nucleus (Figure 35–15). Thus, lesions in the dorsolateral pons disrupt ipsiversive smooth pursuit.

There are two major cortical inputs to the smooth-pursuit system in monkeys. One arises from motion-sensitive regions in the superior temporal sulcus and the middle temporal and medial superior temporal areas. The other arises from the frontal eye field.

The middle temporal and medial superior temporal areas were named because of their position in sulcus-free cortex of the owl monkey, a New World monkey. In humans and Old World monkeys, these areas lie in the superior temporal sulcus, at the junction between the occipital and parietal lobes. Neurons in both the middle temporal and medial superior temporal areas calculate the velocity of the target. When the eye accelerates to match the target's speed, the rate of the target's motion across the retina decreases. As the speed of the retinal image decreases, neurons in the middle temporal area, whose activity signals retinal-image motion, stop firing, even though the target continues to move in space. Neurons in the medial superior temporal area continue to fire even if the target disappears briefly. These neurons have access to a process that adds the speeds of the moving eye and the target moving on the retina to compute the speed of the target in space.

Lesions of either the middle temporal or medial superior temporal area disrupt the ability of a subject to respond to targets moving in regions of the visual field represented in the damaged cortical area. Lesions of the latter area also diminish smooth-pursuit movements toward the side of the lesion, no matter where the target lies on the retina.

The two motion-selective areas provide the sensory information to guide pursuit movements but may not be able to initiate them. Electrical stimulation of either area does not initiate smooth pursuit but can affect pursuit movement, accelerating ipsiversive pursuit and slowing contraversive pursuit. The frontal eye field may

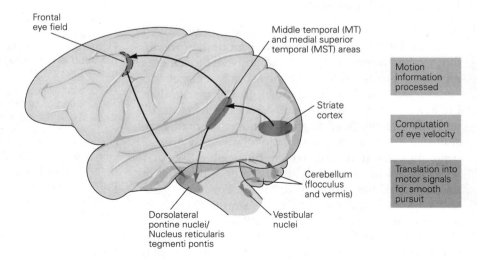

Figure 35–15 Cortical pathways for smooth-pursuit eye movements in the monkey. The cerebral cortex processes information about motion in the visual field and sends it to the oculomotor neurons via the dorsolateral pontine nuclei, the vermis and flocculus of the cerebellum, and the vestibular nuclei. The initiation signal for smooth pursuit may originate in part from the frontal eye field.

Frontal eye field

Middle temporal (MT) and medial superior temporal (MST) areas

Striate cortex

Cerebellum (flocculus and vermis)

Vestibular nuclei

Dorsolateral pontine nuclei/ Nucleus reticularis tegmenti pontis

Motion information processed

Computation of eye velocity

Translation into motor signals for smooth pursuit

be more important for initiating pursuit. This area has neurons that fire in association with ipsiversive smooth pursuit. Electrical stimulation of the frontal eye field initiates ipsiversive pursuit, whereas lesions of this area diminish but do not eliminate smooth pursuit.

In humans, disruption of the pursuit pathway anywhere along its course, including lesions at the level of cortical, cerebellar, and brain stem areas, prevents adequate smooth-pursuit eye movements. Instead, moving targets are tracked using a combination of defective smooth-pursuit movements (the velocity is less than that of the target) and small saccades. Patients with brain stem and cerebellar lesions cannot pursue targets moving toward the side of the lesion.

Patients with parietal deficits that include the motion-sensitive areas have two different types of deficit. The first is a directional deficit that resembles that of monkeys with lesions of the medial superior temporal area: targets moving toward the side of the lesion cannot be tracked. The second is a retinotopic deficit that resembles the deficit of monkeys

with lesions of the middle temporal area: There is an impairment of smooth pursuit of a stimulus limited to the visual hemifield opposite the lesion, regardless of the direction of motion.

The Vergence System Aligns the Eyes to Look at Targets at Different Depths

The smooth-pursuit and saccade systems produce conjugate eye movements: Both eyes move in the same direction and at the same speed. In contrast, the vergence system produces disconjugate movements of the eyes. When we look at an object that is close to us, our eyes *converge* or rotate toward each other; when we look at an object that is farther away, they *diverge* or rotate away from each other (Figure 35–16). These disconjugate movements ensure that the image of the object falls on the foveae of both retinas. Whereas the visual system uses slight differences in left and right retinal positions, or *retinal disparity*, to create a

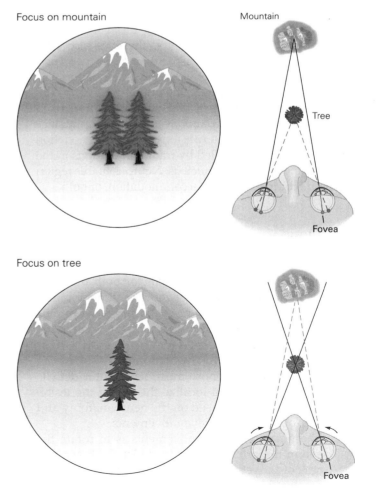

Figure 35–16 Vergence movements. When the eyes focus on a distant mountain, images of the mountain lie on the foveae, while those of the tree in the forefront occupy different retinal positions, yielding the percept of a double image. When the viewer looks instead at the tree, the vergence system must rotate each eye inward. Now the tree's image occupies similar positions on both foveae and is seen as one object, but the mountain's images occupy different locations on the retinas and appear double. (Reproduced, with permission, from F.A. Miles.)

sense of depth, vergence movements eliminate retinal disparity at the fovea.

Vergence is a function of the horizontal rectus muscles only, because the two eyes are horizontally, not vertically, displaced. Convergence of the eyes for near-field viewing is accomplished by simultaneously increasing the tone of the medial recti muscles and decreasing the tone of the lateral recti muscles to converge the eyes. Conversely, distance viewing is accomplished by reducing the tone of the medial rectus and increasing the tone of the lateral rectus.

At any given time, the entire visual field is not in focus on the retina. When we look at something nearby, distant objects are blurred. When we look at something far away, near objects are blurred. When we wish to focus on an object in a closer plane in the visual field, the oculomotor system contracts the ciliary muscle, thereby changing the radius of curvature of the lens. This process is called *accommodation*. With age, accommodation declines owing to increased rigidity of the lens; reading glasses are then needed to focus images at short distances.

Accommodation and vergence are linked. Accommodation is elicited by the blurring of an image, and whenever accommodation occurs, the eyes also converge. Conversely, retinal disparity induces vergence, and whenever the eyes converge, accommodation also takes place. At the same time, the pupils transiently constrict to increase the depth of field of the focus. The linked phenomena of accommodation, vergence, and pupillary constriction comprise the *near response*. Accommodation and vergence are controlled by midbrain neurons in the region of the oculomotor nucleus. Neurons in this region discharge during vergence, accommodation, or both.

Highlights

1. The oculomotor system provides a valuable window into the nervous system for both the clinician and the scientist. Patients with oculomotor deficits may experience alarming symptoms such as double vision that quickly send them to seek medical help. A physician with a thorough knowledge of the oculomotor system can describe and diagnose most oculomotor deficits at the bedside and localize the site of the lesion within the brain based on the neuroanatomy and neurophysiology of eye movements.

2. The purpose of eye movements is to rotate the eye in the orbit in order to direct the fovea, the area of the retina with best acuity, to the point of greatest interest in the visual scene and then to keep the image steady.

3. Six muscles work together to move each eye. These eye muscles are yoked in three pairs. The lateral rectus abducts the eye horizontally, and the medial rectus adducts it. The cyclovertical eye muscles move the eye both vertically and torsionally.

4. Motor neurons for the extraocular muscles lie in three brainstem nuclei. The abducens nucleus in the pons contains the neurons for the lateral rectus. The other oculomotor neurons are in the midbrain: The trochlear nucleus contains superior oblique neurons, and the oculomotor nucleus has the motor neurons for the medial, superior, and inferior rectus muscles and the inferior oblique muscle. Neurons that constrict the pupil and those that elevate the eyelid also lie in the oculomotor nucleus.

5. There are six different types of eye movements, with different control systems: (1) Saccades shift the fovea rapidly to a new visual target. (2) Smooth-pursuit movements keep the image of a moving object on the fovea. (3) Vergence movements rotate the eyes in opposite directions so that the image of an object of interest is positioned on both foveae regardless of its distance. (4) Vestibulo-ocular reflexes hold images still on the retina during brief, rapid head movements. (5) Optokinetic movements hold images stationary during sustained or slow head movements. (6) Fixation is an active process that keeps the eye still during intent gaze when the head is not moving.

6. The firing pattern of eye muscle neurons combines independent signals that code eye position and velocity. The neurons that generate the velocity signal for horizontal saccades lie in the paramedian pontine reticular formation, and this velocity signal is integrated in the medial vestibular nucleus and nucleus prepositus hypoglossi to provide the position signal.

7. The mesencephalic reticular formation provides the position and velocity signals for vertical and torsional eye movements as well as vergence eye movements.

8. Presaccadic burst neurons in the superior colliculus project a desired displacement signal to the reticular formation. These neurons are inhibited by a GABAergic projection from the substantia nigra and excited by projections from the frontal eye field and the posterior parietal cortex. A motor signal from the frontal eye field excites the caudate nucleus, which then inhibits the substantia nigra, allowing a saccade to occur.

9. The posterior parietal cortex projects an attentional signal to the superior colliculus that does not distinguish between attention and movement.

10. Most large gaze shifts involve head movements as well as eye movements. Because the eye moves faster than the head, it typically reaches the target first. The vestibulo-ocular reflex maintains the eye on target by driving the eye with a velocity opposite to that of the head movement.

11. The cerebellum calibrates eye movements based on visual feedback and mediates the learning process that keeps them accurate over time.

12. Smooth pursuit is driven by a network that includes the medial vestibular nucleus, the flocculus of the cerebellum, the dorsolateral pontine nucleus, and two motion-selective areas that are found in the superior temporal sulcus of some monkeys—the middle temporal and medial superior temporal areas. Homologous areas in the human brain are located at the parieto-occipital junction. The pursuit area of the frontal eye fields initiates smooth-pursuit movements.

13. Although the motor programming of eye movements is well understood, the great bulk of physiological research in this field was done with a monkey making a directed saccade to a spot of light. The neural mechanisms underlying the free choice of saccade targets as we explore the visual world are poorly understood. This question, lying at the intersection of cognition and motor control, is one of the great unknowns in neuroscience and will be at the center of oculomotor research in the future.

Michael E. Goldberg
Mark F. Walker

Selected Reading

Bisley JW, Goldberg ME. 2010. Attention, intention, and priority in the parietal lobe. Annu Rev Neurosci 33:1–21.

Krauzlis RJ, Goffart L, Hafed ZM. 2017. Neuronal control of fixation and fixational eye movements. Philos Trans R Soc Lond B Biol Sci 372:20160205.

Leigh RJ, Zee DS. 2015. *The Neurology of Eye Movements*, 6th ed. Philadelphia: FA Davis.

Lisberger SG. Visual guidance of smooth-pursuit eye movements: sensation, action, and what happens in between. Neuron 2010:477–491.

Sparks D. 2002. The brainstem control of saccadic eye movements. Nat Rev Neurosci 3:952–964.

Wurtz RH, Joiner WM, Berman RA. 2011. Neuronal mechanisms for visual stability: progress and problems. Philos Trans R Soc Lond B Biol Sci 366:492–503.

Yarbus AL. 1967. *Eye Movements and Vision*. New York: Plenum.

References

Andersen RA, Asanuma C, Essick G, Siegel RM. 1990. Corticocortical connections of anatomically and physiologically defined subdivisions within the inferior parietal lobule. J Comp Neurol 296:65–113.

Andersen RA, Cui H. 2009. Intention, action planning, and decision making in parietal-frontal circuits. Neuron 63:568–583.

Barnes GR. 2008. Cognitive processes involved in smooth pursuit eye movements. Brain Cogn 68:309–326.

Bisley JW, Goldberg ME. 2006. Neural correlates of attention and distractibility in the lateral intraparietal area. J Neurophysiol 95:1696–1717.

Bruce CJ, Goldberg ME. 1985. Primate frontal eye fields. I. Single neurons discharging before saccades. J Neurophysiol 53:603–635.

Büttner-Ennever JA, Büttner U, Cohen B, Baumgartner G. 1982. Vertical gaze paralysis and the rostral interstitial nucleus of the medial longitudinal fasciculus. Brain 105:125–149.

Büttner-Ennever JA, Cohen B, Pause M, Fries W. 1988. Raphe nucleus of the pons containing omnipause neurons of the oculomotor system in the monkey, and its homologue in man. J Comp Neurol 267:307–321.

Cannon SC, Robinson DA. 1987. Loss of the neural integrator of the oculomotor system from brain stem lesions in monkey. J Neurophysiol 57:1383–1409.

Cohen B, Henn V. 1972. Unit activity in the pontine reticular formation associated with eye movements. Brain Res 46:403–410.

Colby CL, Duhamel J-R, Goldberg ME. 1996. Visual, presaccadic and cognitive activation of single neurons in monkey lateral intraparietal area. J Neurophysiol 76:2841–2852.

Cumming BG, Judge SJ. 1986. Disparity-induced and blur-induced convergence eye movement and accommodation in the monkey. J Neurophysiol 55:896–914.

Curtis CE, Connolly JD. 2008. Saccade preparation signals in the human frontal and parietal cortices. J Neurophysiol 99:133–145.

Dash S, Thier P. 2014. Cerebellum-dependent motor learning: lessons from adaptation of eye movements in primates. Prog Brain Res 210:121–155.

Demer JL, Miller JM, Poukens V, Vinters HV, Glasgow BJ. 1995. Evidence for fibromuscular pulleys of the recti extraocular muscles. Invest Ophthalmol Vis Sci 36:1125–1136.

Duhamel J-R, Colby CL, Goldberg ME. 1992. The updating of the representation of visual space in parietal cortex by intended eye movements. Science 255:90–92.

Dürsteler MR, Wurtz RH, Newsome WT. 1987. Directional pursuit deficits following lesions of the foveal representation within the superior temporal sulcus of the macaque monkey. J Neurophysiol 57:1262–1287.

Fuchs AF, Luschei ES. 1970. Firing patterns of abducens neurons of alert monkeys in relationship to horizontal eye movement. J Neurophysiol 33:382–392.

Funahashi S, Bruce CJ, Goldman-Rakic PS. 1989. Mnemonic coding of visual space in the monkey's dorsolateral prefrontal cortex. J Neurophysiol 61:331–349.

Gottlieb JP, MacAvoy MG, Bruce CJ. 1994. Neural responses related to smooth-pursuit eye movements and their correspondence with electrically elicited smooth eye movements in the primate frontal eye field. J Neurophysiol 74:1634–1653.

Hécaen J, de Ajuriaguerra J. 1954. Balint's syndrome (psychic paralysis of visual fixation). Brain 77:373–400.

Henn V, Hepp K, Büttner-Ennever JA. 1982. The primate oculomotor system. II. Premotor system. A synthesis of anatomical, physiological, and clinical data. Hum Neurobiol 12:87–95.

Highstein SM, Baker R. 1978. Excitatory termination of abducens internuclear neurons on medial rectus motoneurons: relationship to syndrome of internuclear ophthalmoplegia. J Neurophysiol 41:1647–1661.

Hikosaka O, Wurtz RH. 1983. Visual and oculomotor functions of monkey substantia nigra pars reticulata. IV. Relation of substantia nigra to superior colliculus. J Neurophysiol 49:1285–1301.

Hikosaka O, Sakamoto M, Usui S. 1989. Functional properties of monkey caudate neurons. I. Activities related to saccadic eye movements. J Neurophysiol 61:780–798.

Hikosaka O, Wurtz RH. 1989. The basal ganglia. Rev Oculomotor Res 3: 257–281.

Huk A, Dougherty R, Heeger D. 2002. Retinotopy and functional subdivision of human areas MT and MST. J Neurosci 22:7195–7205.

Keller EL. 1974. Participation of medial pontine reticular formation in eye movement generation in monkey. J Neurophysiol 37:316–332.

Laurutis VP, Robinson DA. 1986. The vestibular reflex during human saccadic eye movements. J Physiol (Lond) 373: 209–233.

Luschei ES, Fuchs AF. 1972. Activity of brain stem neurons during eye movements of alert monkeys. J Neurophysiol 35:445–461.

Lynch JC, Graybiel AM, Lobeck LJ. 1985. The differential projection of two cytoarchitectonic subregions of the inferior parietal lobule of macaque upon the deep layers of the superior colliculus. J Comp Neurol 235:241–254.

McFarland JL, Fuchs AF. 1992. Discharge patterns in nucleus prepositus hypoglossi and adjacent medial vestibular nucleus during horizontal eye movement in behaving macaques. J Neurophysiol 68:319–332.

Munoz DP, Wurtz RH. 1993. Fixation cells in monkey superior colliculus. I. Characteristics of cell discharge. J Neurophysiol 70:559–575.

Mustari MJ, Fuchs AF, Wallman J. 1988. Response properties of dorsolateral pontine units during smooth pursuit in the rhesus macaque. J Neurophysiol 60:664–686.

Newsome WT, Wurtz RH, Komatsu H. 1988. Relation of cortical areas MT and MST to pursuit eye movements. II. Differentiation of retinal from extraretinal inputs. J Neurophysiol 60:604–620.

Olson CR, Gettner SN. 1995. Object-centered direction selectivity in the macaque supplementary eye field. Science 269:985–988.

Powell KD, Goldberg ME. 2000. Response of neurons in the lateral intraparietal area to a distractor flashed during the delay period of a memory-guided saccade. J Neurophysiol 84:301–310.

Quaia C, Aizawa H, Optican LM, Wurtz RH. 1998. Reversible inactivation of monkey superior colliculus. II. Maps of saccadic deficits. J Neurophysiol 79:2097–2110.

Quaia C, Lefevre P, Optican LM. 1999. Model of the control of saccades by superior colliculus and cerebellum. J Neurophysiol 82:999–1018.

Ramat S, Leigh RJ, Zee DS, Optican LM. 2007. What clinical disorders tell us about the neural control of saccadic eye movements. Brain 130:10–35.

Rao HM, Mayo JP, Sommer MA. 2016. Circuits for presaccadic visual remapping. J Neurophysiol 116:2624–2636.

Raybourn MS, Keller EL. 1977. Colliculo-reticular organization in primate oculomotor system. J Neurophysiol 269:985–988.

Robinson DA. 1970. Oculomotor unit behavior in the monkey. J Neurophysiol 33:393–404.

Schall JD. 1995. Neural basis of saccade target selection. Rev Neurosci 6:63–85.

Schiller PH, Koerner F. 1971. Discharge characteristics of single units in superior colliculus of the alert rhesus monkey. J Neurophysiol 34:920–936.

Schiller PH, True SD, Conway JL. 1980. Deficits in eye movements following frontal eye field and superior colliculus ablations. J Neurophysiol 44:1175–1189.

Scudder C, Kaneko C, Fuchs A. 2002. The brainstem burst generator for saccadic eye movements—a modern synthesis. Exp Brain Res 142:439–462.

Segraves MA, Goldberg ME. 1987. Functional properties of corticotectal neurons in the monkey's frontal eye field. J Neurophysiol 58:1387–1419.

Silver MA, Kastner S. 2009. Topographic maps in human frontal and parietal cortex. Trends Cogn Sci (Regul Ed) 13: 488–495.

Strupp M, Kremmyda O, Adamczyk C, et al. 2014. Central ocular motor disorders, including gaze palsy and nystagmus. J Neurol 261:S542–S558.

Stuphorn V, Brown JW, Schall JD. 2010. Role of supplementary eye field in saccade initiation: executive, not direct, control. J Neurophysiol 103:801–816.

Takagi M, Zee DS, Tamargo RJ. 1998. Effects of lesions of the oculomotor vermis on eye movements in primate: saccades. J Neurophysiol 80:1911–1931.

von Noorden GK, Campos EC. 2002. *Binocular Vision and Ocular Motility: Theory and Management of Strabismus*, 6th ed. St. Louis, MO: Mosby.

Wurtz RH, Goldberg ME. 1972. Activity of superior colliculus in behaving monkey. III. Cells discharging before eye movements. J Neurophysiol 35:575–586.

Xu-Wilson M, Chen-Harris H, Zee DS, Shadmehr R. 2009. Cerebellar contributions to adaptive control of saccades in humans. J Neurosci 29:12930–12939.

Zee DS. 1977. Disorders of eye-head coordination. In B Brooks, FJ Bajandas (Eds.), *Eye Movements*, Plenum Press, New York, 1977, pp. 9–40.

36

Posture

THE CONTROL OF POSTURE INVOLVES TWO INTER-RELATED GOALS, equilibrium (balance) and orientation, crucial for most tasks of daily living. Balance control maintains the body in stable equilibrium to avoid falls. Orientation aligns the body segments with respect to each other and to the world, such as maintaining the head vertical. Both balance and orientation use several different types of control: automatic postural responses, anticipatory postural adjustments, postural sway in stance, sensory integration for a body schema, orientation to vertical, and dynamic stability during gait.

To appreciate the complexity of maintaining balance and orientation, imagine that you are waiting tables on a tour boat. You have a tray full of drinks to be delivered to a table on the other side of the rolling deck. Even as your mind is occupied with remembering customer orders, unconscious but complex sensorimotor processes for controlling postural orientation and balance allow you to move about in an efficient and coordinated manner without falling. As you cross the rolling deck, your brain rapidly integrates and interprets sensory information and adjusts motor output to maintain your balance and the upright orientation of your head and trunk, as well as stabilize the arm supporting the tray of full glasses. Sudden unexpected motions of the boat evoke automatic postural responses that prevent falls. Before you reach out to place a glass on the table, your nervous system makes anticipatory postural adjustments to maintain your balance.

Somatosensory, vestibular, and visual information are integrated to provide a coherent sense of the position and velocity of the body in space with respect to the support surface, gravity, and visual environment. Since the surface is unstable and vision is not providing earth-stable information, your dependence on vestibular information is greater than usual. Your head is kept stable while your trunk motions and walking pattern adjust for disequilibrium caused by the moving surface. You notice that both your voluntary tasks and your balance control deteriorate when trying to attend to both goals.

Equilibrium and Orientation Underlie Posture Control

Postural equilibrium refers to the ability to actively stabilize the upper body by resisting external forces acting on the body. Although the dominant external force affecting equilibrium on earth is gravity, other inertial forces and external perturbations must also be resisted. Depending on the particular task or behavior, different sets of muscles are activated in response to or in anticipation of disturbance to equilibrium.

Postural orientation refers to the ability to actively align body segments, such as the trunk and head, with respect to each other and to the environment. Depending on the particular task or behavior, body segments may be aligned with respect to gravitational vertical, visual vertical, or the support surface. For example, when skiing downhill, the head may be oriented to gravitational and inertial vertical, but not to the visual or support surface references that are inclined.

The biomechanical requirements of postural control depend on anatomy and postural orientation and thus vary among species. Nevertheless, in a variety of species, the control mechanisms for postural equilibrium and orientation have many common features. The sensorimotor mechanisms for postural control are quite similar in humans and quadrupedal mammals even though their habitual stance is different.

Postural Equilibrium Controls the Body's Center of Mass

With many segments linked by joints, the body is mechanically unstable. To maintain balance, the nervous system must control the position and motion of the body's *center of mass* as well as the body's rotation about it. The center of mass is a point that represents the average position of the body's total mass. In the standing adult, for example, the center of mass is located about 2 cm in front of the second lumbar vertebra; in a young child, it is higher. The location of the center of mass is not fixed but depends on postural orientation. For example, when you flex at the hips while standing, the center of mass moves from a location inside the body to a position outside the body.

Although gravity pulls on all body segments, the net effect on body equilibrium acts through the body's center of mass. The force due to gravity is opposed by the forces between the feet and the ground. Each point on the surface will generate a force on the foot. All the forces acting between the foot and the ground can be summed to yield a single force vector termed the *ground reaction force*. This origin of the ground reaction force vector on the surface is the point at which the rotational effect of all the forces on the feet are balanced and is termed the *center of pressure* (Box 36–1).

Maintaining balance while standing requires keeping the downward projection of the center of mass within the base of support, an imaginary area defined by those parts of the body in contact with the environment. For example, the two feet or one foot of a standing human define a *base of support* (Box 36–1). However, when a standing person leans against a wall or is supported by crutches, the base of support extends from the ground under the feet to the contact point between the body and the wall or crutches. Because the body is always in motion, even during stable stance, the body's center of mass continually moves about with respect to the base of support. Postural instability is determined by how fast the center of mass is accelerating toward and beyond the boundary of its base of support and how close the downward projection of the body's center of mass is to the boundary.

Upright stance requires two actions: (1) maintaining support against gravity by keeping the center of mass at some height and joints stable and (2) maintaining balance by controlling the trajectory of the center of mass in the horizontal plane. Balance and antigravity support are controlled separately by the nervous system and may be differentially affected in certain pathological conditions. For example, antigravity support can be excessive when spasticity is present after a stroke or insufficient in the hypotonia of cerebral palsy, although balance control may be preserved. Alternatively, in vestibular disorders, antigravity support can be normal, although balance control is disordered.

The center of pressure (CoP) is defined as the origin of the *ground reaction force* vector on the support surface. For the body to be in static equilibrium, that is, to remain motionless, the force caused by gravity and the ground reaction force must be equal and opposite, and the CoP must be directly under the center of mass (CoM) (Figure 36–1A). Misalignment of the CoM and CoP causes motion of the CoM. For example, if the CoM projection onto the base of support is to the right of the CoP, the body will sway to the right until the CoP moves to the right to move the CoM back over the base of support.

However, standing is never truly static. While the body is in motion (postural sway), CoM and CoP are not aligned and dynamic equilibrium must maintain balance (Figure 36–1B). In fact, when the body is unsupported, CoP and CoM are continually in motion and are rarely aligned, although when averaged over time during quiet stance, they are coincident. The sway of the body during quiet stance can be described by the trajectory of either CoM or CoP over time, such as sway path, area, velocity, and frequency.

In more dynamic situations like walking, running, turning, and jumping, stability can be achieved even when the CoM briefly goes outside the base of support. For example, when standing on one leg or on a narrow beam, momentum from rotating the hips, arms, and other body parts or movement of the CoP can be used to change the direction of the ground reaction force to return the body CoM over its base of support to maintain stability (Figure 36–1B). If the CoM is outside the base of support and heading away from it, subjects may need to take a step or grab a stable object to change the base of support and avoid a fall.

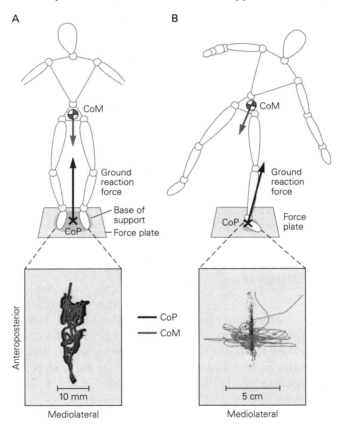

Figure 36–1 The center of mass is controlled by moving the center of pressure.

A. The force caused by gravity passes through the center of mass (**CoM**) in the trunk. The surface exerts an upward force against each foot, such that the ground reaction force vector originates at the center of pressure (**CoP**) on the support surface. **Below:** Even when the feet remain in place, the CoP (**blue displacement**) and CoM (**gold displacement**) are always in motion as we sway. During normal standing on two feet, the projection of the CoM of the body remains within the base of support (**light blue**

rectangle around the feet in contact with the ground) for equilibrium. The base of support of the standing human is defined by contact of the feet on the support surface.

B. In a dynamic situation, such as standing on one leg on a narrow beam, equilibrium can be maintained even when the body CoM displacements (**gold displacements**) go outside of the base of support for brief periods. Strategies such as counter-rotation of the lower and upper body can tilt the ground reaction force so that it accelerates the body CoM back over its base of support. (Adapted, with permission, from Otten 1999. Permission conveyed through Copyright Clearance Center, Inc.)

Antigravity support, or "postural tone," is provided by the tonic activation of muscles that generate force against the ground to keep the trunk and limbs extended and the center of mass at the appropriate height. In humans, much of the support against gravity is provided by passive bone-on-bone forces in joints such as the knees, which can be fully extended during stance, and in stretched ligaments such as those at the front of the hips. Nevertheless, antigravity support in humans also requires active tonic muscle contraction, for example, in ankle, trunk, and neck extensors. Postural tone, however, should not be considered a static state of muscle activation, as can be seen in pathologies such as decerebrate rigidity or the rigidity of parkinsonism. Normal postural tone is constantly changing, as a "wave" or "reed in the wind," to accommodate changes in postural alignment, voluntary movements, and task requirements.

Postural tone is not sufficient, however, for maintaining balance. Both bipeds and quadrupeds are inherently unstable, and their bodies sway during quiet stance. Actively contracted muscles exhibit a spring-like stiffness that helps to resist body sway, but muscle stiffness alone is insufficient for maintaining balance. Even stiffening of the limbs through muscle co-contraction is not sufficient for balance control. Instead, complex patterns of muscle activation produce direction-specific forces to control the body's center of mass. Body sway caused by even subtle movements, such as the motion of the chest during breathing, is actively counteracted by alterations in postural tone.

Postural Orientation Anticipates Disturbances to Balance

Postural orientation is the manner in which body parts are aligned with respect to each other and to the environment. Animals arrange their bodies to accomplish specific tasks efficiently. Although this postural orientation interacts with balance control, the two systems can act independently. For example, soccer goalies may orient their body to intercept a ball by sacrificing the goal of maintaining balance. In contrast, a patient with Parkinson disease or thoracic kyphosis may use an inefficient, flexed postural alignment to maintain effective control of balance while standing.

The energy needed to maintain body position over a period of time can influence postural orientation. In humans, for example, the upright orientation of the trunk with respect to gravity minimizes the forces and thus the energy required to hold the body's center of mass over the base of support. Task requirements also affect postural orientation. For some tasks, it is important to stabilize the arrangement of the body in space, whereas for others, it is necessary to stabilize one body part with respect to another. When walking while carrying a full glass, for example, it is important to stabilize the hand against gravity to prevent spillage. In contrast, when walking while reading a cell phone, the hand must be stabilized with respect to the head and eyes to maintain visual acuity.

Subjects may adopt a particular postural orientation to optimize the accuracy of sensory signals regarding body motion. For example, when standing and walking inside a ship, in which the surface and visual references may be unstable, information about earth vertical is derived primarily from vestibular inputs. A person often aligns his head with respect to gravitational vertical when balancing on an unstable surface because the perception of vertical is most accurate when the head is upright and stable.

Anticipatory alterations of habitual body orientation can minimize the effect of a possible disturbance. For example, people often lean in the direction of an anticipated external force, or they flex their knees, widen their stance, and extend their arms when anticipating that stability will be compromised.

Postural Responses and Anticipatory Postural Adjustments Use Stereotyped Strategies and Synergies

When a sudden disturbance causes the body to sway, various postural motor strategies are used to maintain the center of mass within the base of support. In one strategy, the base of support remains fixed relative to the support surface: While the feet remain in place, the body rotates about the ankles back to the upright position (Figure 36–2A). In other strategies, the base of support is moved or enlarged, for example, by taking a step or by grabbing a support with the hand (Figure 36–2B).

Older views of motor control focused on trunk and proximal limb muscles as the main postural effectors. Recent behavioral studies show that any group of muscles from the neck and trunk, legs and arms, or feet and hands can act as postural muscles depending on the body parts in contact with the environment and the biomechanical requirements of equilibrium.

When studying the posture control system, scientists disrupt balance in a controlled manner to determine the subject's automatic postural response. This response is described by the ground reaction force vector, the motion of the center of pressure, and movements of parts of the body. The electrical activity

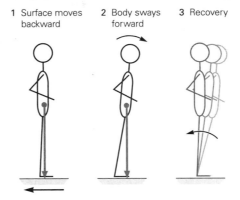

A Bringing center of mass back over base of support

1 Surface moves backward   2 Body sways forward   3 Recovery

Figure 36–2 Automatic postural responses keep the center of mass within the base of support.

A. One postural strategy for regaining balance is to bring the center of mass back to its origin over the base of support. When the platform on which a subject is standing is suddenly moved backward, the body sways forward and the projection of the center of mass moves toward the toes. During recovery, the body actively exerts force into the surface about the ankles, bringing the center of mass back to the original position over the feet.

B. An alternative postural strategy enlarges the base of support to keep the center of mass within the base. A disturbance causes the subject to sway forward and the center of mass moves toward the boundary of the base of support (**blue area** on the ground). The base can be enlarged in two ways: taking a step and placing the foot in front of the center of mass to decelerate the body's motion, or grabbing a support and thereby extending the base to include the contact point between the hand and support.

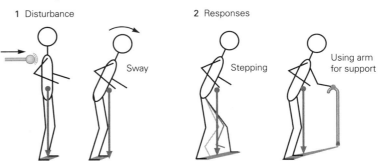

B Extending base of support to capture center of mass

1 Disturbance   2 Responses

Sway   Stepping   Using arm for support

of many muscles is recorded by electromyography (EMG), which reflects the firing of alpha motor neurons that innervate skeletal muscle and thus provides a window into the nervous system's output for balance control. The combination of all these measurements allows investigators to infer the active neural processes underlying balance control.

Automatic Postural Responses Compensate for Sudden Disturbances

An automatic postural response to a sudden disturbance is not a simple stretch reflex but rather the synergistic activation of a group of muscles in a characteristic sequence with the goal of maintaining equilibrium. That is, the recruitment of a muscle for a postural response serves the requirements of equilibrium and is not a reflexive change in the muscle's length caused by the disturbance. For example, when the surface under a person is rotated in the toes-up direction, the ankle extensor (gastrocnemius) is lengthened and a small stretch reflex may occur. However, the postural response for balance recruits the antagonist ankle flexor

(tibialis anterior), which itself is shortened by the surface rotation, while suppressing the stretch response in the gastrocnemius. In contrast, when the platform is moved backward, the gastrocnemius is again lengthened but now it is recruited for the postural response, as evidenced by a second burst of EMG activity after the stretch reflex. Thus, the initial change in length of a muscle induced by perturbation does not determine whether that muscle is recruited for postural control, and stretch reflexes are not the basis for postural control. In fact, monosynaptic stretch reflexes are too weak to move the body center of mass effectively, and very often, the postural muscles activated to recover equilibrium have not been stretched.

Automatic postural responses to sudden disturbances have characteristic temporal and spatial features. A postural response in muscles must be recruited rapidly following the onset of a disturbance. Sudden movement of the support surface under a standing cat evokes EMG activity within 40 to 60 ms (Figure 36–3). Humans have longer latencies of postural response (90–120 ms in the ankle muscles); the increased delay is attributed to the larger body size of humans and thus

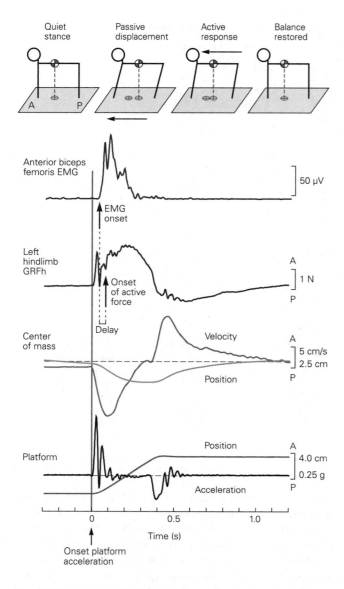

Quiet stance   Passive displacement   Active response   Balance restored

Anterior biceps femoris EMG

50 µV

EMG onset

Left hindlimb GRFh

A

1 N

Onset of active force

P

Delay

Center of mass

Velocity

A

5 cm/s
2.5 cm

Position

P

Platform

Position

A

4.0 cm

0.25 g

P

Acceleration

0   0.5   1.0

Time (s)

Onset platform acceleration

Figure 36–3 Automatic postural responses have stereotypical temporal characteristics. Electromyographic (**EMG**) activity has a characteristic latency. Anterior motion of the platform evokes an EMG response in the hip extensor muscle (anterior biceps femoris) of a cat approximately 40 ms after the onset of platform acceleration (100 ms in a human). This latency is stereotyped and repeatable across subjects and is approximately four times as long as that of the monosynaptic stretch reflex. As the platform moves, the paws are carried forward and the trunk remains behind owing to inertia, causing the center of mass to move backward with increasing velocity with respect to the platform. The velocity of the center of mass peaks and then decreases as the horizontal component of the ground reaction force (**GRFh**) increases following muscle activation. The delay of approximately 30 ms between the onset of EMG activity and the onset of the active response reflects excitation–contraction coupling and musculoskeletal compliance. The automatic postural response extends the hind limb, propelling the trunk forward and restoring the position of the center of mass with respect to the paws. (Abbreviations: **A**, anterior; **P**, posterior.) (Data from J. Macpherson.)

the greater signal conduction distances from sensory receptors to the central nervous system and thence to leg muscles. The latency of automatic postural responses is shorter than voluntary reaction time but longer than the monosynaptic stretch reflex.

Postural responses involving a change in support base, such as stepping, have even longer latencies than those that occur when the feet remain in place. The longer time presumably affords greater flexibility in the commands transmitted by long loops through the cortex; for example, the choice of foot to begin the step, the direction of the step, and the path of the step around obstacles.

Activation of postural muscles results in contraction and the development of force in the muscles, leading to torque (rotational force) at the joints. The net result is an active response, the ground reaction force (Box 36–1), that restores the center of mass to its original position over the base of support (Figure 36–3). The delay between EMG activation and the active response, approximately 30 ms in the cat and 50 ms in humans, reflects the excitation–contraction coupling time of each muscle as well as the compliance of the musculoskeletal system.

The amplitude of EMG activity in a particular muscle depends on both the speed and direction of postural disturbance. The amplitude increases as the speed of a movable platform under a standing human or cat increases, and it varies in a monotonic fashion as the direction of platform motion is varied systematically. Each muscle responds to a limited set of perturbation directions with a characteristic tuning curve (Figure 36–4).

Although individual muscles have unique directional tuning curves, muscles are not activated independently but instead are activated together in *synergies*, with characteristic time delays. The muscles within a synergy receive a common command signal during postural responses. In this way, the many muscles of the body are controlled by just a few signals, reducing the time needed to compute the appropriate postural response (Box 36–2).

The set of muscles recruited in a postural response to a disturbance depends on the body's initial stance. The same disturbance elicits very different postural responses in someone standing unaided, standing while grasping a stable support, or crouching on all four limbs. For example, forward sway activates muscles in the back of the legs and trunk during upright free stance. When the subject is holding onto a stable support, muscles of the arms rather than those of the legs are activated first. When the subject is crouched on toes and fingers, like a cat, muscles in the front of the legs and in the arms are activated (Figure 36–6A).

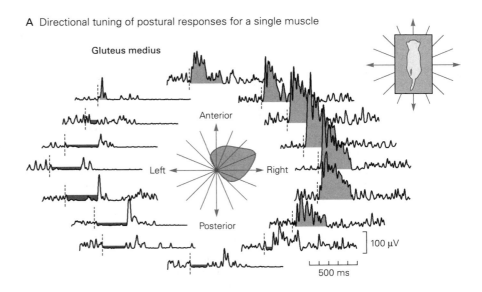

A Directional tuning of postural responses for a single muscle

B Each muscle has unique directional tuning

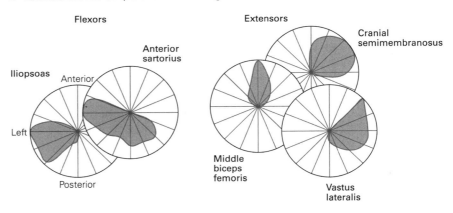

Figure 36–4 Automatic postural responses have stereotypical directional characteristics. (Adapted from Macpherson 1988.)

A. The gluteus medius muscle in the cat, a hip extensor and abductor, responds to a range of directions of motion in the horizontal plane. The electromyographic (EMG) records shown here are from a cat standing on a platform that was moved in the horizontal plane in each of 16 evenly spaced directions. The gluteus medius muscle of the left hind limb was activated by motion in several directions (**pink**) and inhibited in the remaining directions (**gray**). The **dashed vertical lines** indicate the onset of platform acceleration. In the center is a polar plot of the amplitude of EMG activity versus the direction of motion during the automatic postural response; it represents a directional tuning curve for the muscle. EMG amplitude was computed from the area under the curve during the first 80 ms of the response.

B. Every muscle has a characteristic directional tuning curve that differs from that of other muscles, even if they have similar actions. The middle biceps femoris and cranial semimembranosus, for example, are both extensors of the hip.

Because postural responses are influenced by recent experience, they adapt only gradually to new biomechanical conditions. When forward sway is induced by backward motion of a platform on which a subject is standing, the posterior muscles of the ankle, knee, and hip are activated in sequence beginning 90 ms after the platform starts moving. This postural response, the *ankle strategy*, restores balance primarily by rotating the body about the ankle joints. However, when forward sway is induced by backward motion of a narrow beam, it is impossible to use surface torque alone to recover equilibrium and the anterior muscles of the hip and trunk are activated. This postural response, the *hip strategy*, restores the body's center of mass by bending forward at the hip joints and counter-rotating at the ankles (Figure 36–6B).

When a subject moves from standing on a wide platform to a narrow beam, she or he persists in using the ankle strategy in the first few trials. This strategy does not work when standing on the beam, and the subject falls. Over several trials, the subject will gradually switch to using the hip strategy. Similarly, moving from the beam back to the platform requires several trials to adapt the postural response back to the ankle strategy (Figure 36–6C).

Although sensory stimulation changes immediately after subjects move from the beam to the floor, the postural response adjusts gradually as it is tuned for optimal behavior by trial and error. If postural responses were simple reflexes, they would change immediately upon a change in sensory drive. Trial-to-trial changes in postural behavior generally occur at the subconscious level (implicit learning) and involve updating of the body schema and internal model of the world within the right parietal cortex. This body schema is dynamic, as it is constantly updated based on experience.

Postural responses not only improve with practice, but the improvements are retained, a sign of motor

Box 36–2 Synergistic Activation of Muscles

Coordinated movements require precise control of the many joints and muscles in the body. Maintaining control is biomechanically complex, in part because different combinations of joint rotations and muscle activations can achieve the same goal. Such redundancy confers great flexibility, for example, in modifying stepping patterns to negotiate obstacles in our path, but comes at the cost of increased complexity in the brain's computation of movement trajectories and forces.

Many factors must be included in the computation of movement commands, including the effect of external forces such as gravity and the forces that one body segment exerts on another during motion. All these factors come into play when the brain computes postural responses to sudden disturbances, but with the added constraint of a time limit on computation: Responses must occur within a certain time or balance will be lost.

It has long been believed that the brain simplifies the control of movement by grouping control variables, for example, activating several muscles together. Using mathematical techniques that parse complex data into a small number of components, one can determine that only four synergies are needed to account for the vast majority of

activation patterns of 13 muscles of the human leg and trunk during automatic postural responses to many directions of platform motion (Figure 36–5). Activation of each synergy produces a unique direction of force against the ground, suggesting that postural control is based on task-related variables such as the force between foot and ground rather than the contraction force of individual muscles.

Like the arrangement of notes in a musical chord, each muscle synergy specifies the timing and amplitude of activation for a particular muscle together with others. Just as one note belongs to several different chords, each muscle belongs to more than one synergy. When several chords are played simultaneously, the chord structure is no longer evident in the multitude of notes. Similarly, when several synergies are activated concurrently, the observed muscle pattern gives the appearance of unstructured complexity, but a particular muscle's activation is the result of the systematic addition of synergy commands. Concurrent activation of synergies simplifies the neural command signals for movement as only a few central commands are required instead of a separate command for each muscle, while allowing flexibility and adaptability to postural control.

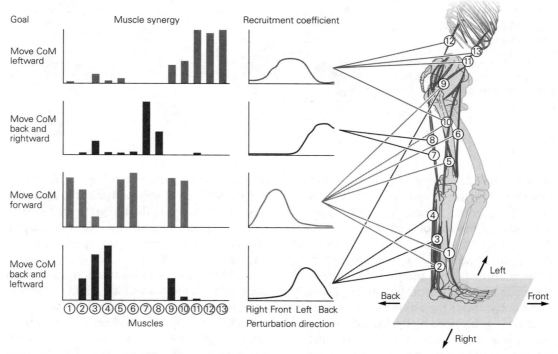

Figure 36–5 Postural commands activate synergies rather than individual muscles. Synergistic activation of several muscles allows movement goals to be translated into specific muscle activity patterns. Each muscle synergy activates a group of muscles in a fixed proportion (**colored bars**) to produce the mechanical output needed to achieve a postural goal. The height of each bar represents the relative amount of activation, or weighting, for each muscle (1–13). Each synergy is activated more or less at particular times during a behavior driven by central commands and sensory drive (recruitment function). For example, different postural synergies are activated for different directions of falling. (Abbreviation: **CoM**, center of mass.) (Reproduced, with permission, from L. Ting.)

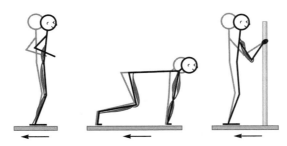

A Stance determines postural response

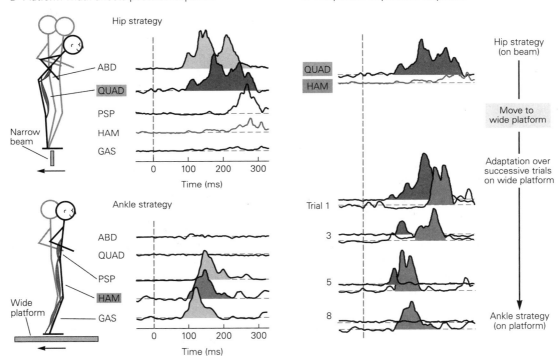

Figure 36–6 Automatic postural responses change with biomechanical conditions.

A. The backward movement of a platform activates different groups of muscles depending on initial stance. **Gray stick figures** show initial positions (upright unsupported, quadrupedal, or upright supported). The muscles activated in each postural response are shown in **red**. (Adapted, with permission, from Dunbar et al. 1986.)

B. When a subject stands on a narrow beam that is abruptly moved backward, the anterior muscles—abdominals (**ABD**) and quadriceps (**QUAD**)—are recruited to flex the trunk and extend the ankles, moving the hips backward (the hip strategy). When the subject instead stands on a wide platform that is moved backward, his posterior muscles—paraspinals (**PSP**), hamstrings (**HAM**), and gastrocnemius (**GAS**)—are activated

to bring the body back to the erect position by rotating at the ankles (the ankle strategy). Muscles representative of different postural responses are highlighted in color. **Dashed vertical lines** in the plots indicate onset of platform (or beam) acceleration.

C. Postural strategy adapts after the subject moves from the narrow beam onto the wide platform. On the beam, the quadriceps are activated and the hamstrings are silent; after adaptation to the wide platform, the reverse is observed. The transition from quadriceps to hamstrings activation occurs over a series of trials; the activity in the quadriceps gradually decreases in amplitude, whereas the hamstrings are activated earlier and earlier until, by trial 8, quadriceps activity disappears altogether. Ankle and trunk muscles show similar patterns of adaptation. (B. and C, adapted, with permission, from Horak and Nashner 1986.)

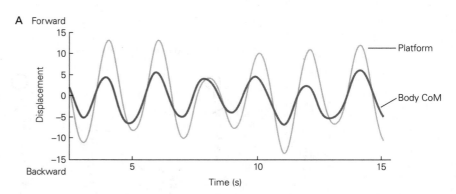

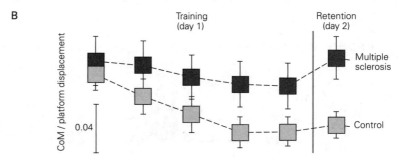

Figure 36–7 Postural responses can be learned and retained with practice.

A. Displacement of body center mass (**CoM, gold oscillation**) in response to forward and backward platform oscillations of varying amplitudes (**gray**) as a healthy subject learns to reduce postural instability.

B. Displacement of the body CoM by forward–backward surface oscillations is reduced across training sessions on day 1, and this improvement is retained on day 2 in healthy control subjects. People with multiple sclerosis also learn to reduce CoM displacements but do not retain this improvement the next day. The mean and standard error of group changes in gain (CoM/surface displacement) are compared. (Adapted, with permission, from Gera et al. 2016.)

learning. For example, when subjects practice standing on an oscillating surface, they gradually learn to decrease the extent of the displacement of their center of mass, and much of this improvement is retained the next day (Figure 36–7). Patients with neurological disorders, such as multiple sclerosis or Parkinson disease, who have significantly impaired postural responses can often learn to improve their postural control with practice, although they may need more practice than normal to retain the improvements (Figure 36–7).

Anticipatory Postural Adjustments Compensate for Voluntary Movement

Voluntary movements can also destabilize postural orientation and equilibrium. For example, rapidly lifting the arms forward while standing produces forces that extend the hips, flex the knees, and dorsiflex the ankles, moving the body's center of mass forward relative to the feet. The nervous system has advance knowledge of the effects of voluntary movement on postural alignment and stability and activates anticipatory postural adjustments, often in advance of the primary movement (Figure 36–8A).

Anticipatory postural adjustments are specific to biomechanical conditions. When a freely standing subject rapidly pulls on a handle fixed to the wall, the leg muscles (gastrocnemius and hamstrings) are activated before the arm muscles (Figure 36–8B). When the

subject performs the same pull while his shoulders are propped against a rigid bar, no anticipatory leg muscle activity occurs because the nervous system relies on the support of the bar to prevent the body from moving forward. When the handle is pulled in response to an external cue, the arm muscles are activated faster in the supported condition than in the freestanding condition. Thus, voluntary arm muscle activation is normally delayed when the task requires active postural stability.

Another common preparatory postural adjustment occurs when one begins to walk. The center of mass is accelerated forward and laterally by the unweighting of one leg. This postural adjustment appears to be independent of the stepping program that underlies ongoing locomotion (Chapter 33). Similarly, a forward shift of the center of mass precedes the act of standing on the toes. A subject is unable to remain standing on his toes if he simply activates the calf muscles without moving his center of mass forward; he rises onto his toes only momentarily before gravity restores a flat-footed stance. Moving the center of mass forward over the toes before activating the calf muscles aligns it over the anticipated base of support and thus stabilizes the toe stance.

Postural equilibrium during voluntary movement requires control not only of the position and motion of the body's center of mass but also of the angular momentum about the center of mass. A diver can

A Ankle force precedes pulling force during voluntary arm pull

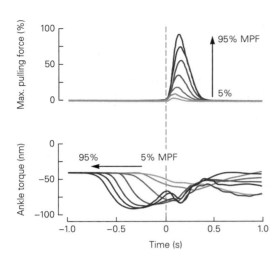

B Postural muscles are recruited only when needed

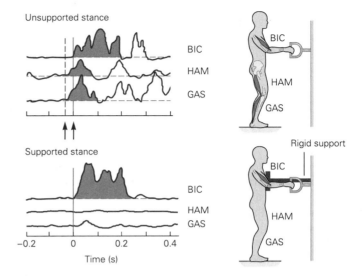

C Center of mass position is controlled during walking by foot placement

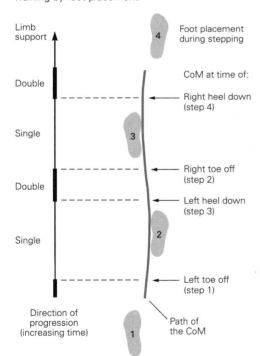

Figure 36–8 Anticipatory postural adjustments precede voluntary movement.

A. The postural component of a voluntary arm pull increases in amplitude and lead time as the pulling force increases. In this experiment, subjects were asked to pull on a handle attached to the wall by a wire. Subjects stood on a force plate and, at a signal, pulled rapidly on the handle to reach a specified peak force varying between 5% and 95% of maximum pulling force. Each pull was preceded by leg muscle activation that produced a rotational force, or torque, about the ankle joints. The larger the pulling force, the larger and earlier was the ankle torque. Traces are aligned at the onset of the pulling force on the handle at time zero. (Abbreviation: **MPF**, maximum pulling force.) (Adapted, with permission, from Lee, Michaels, and Pai 1990.)

B. Postural adjustments accompany voluntary movement only when needed. As in part **A**, subjects were asked to pull on a handle fixed to a wall. Electromyogram traces are aligned at time zero, the onset of activity in the arm muscle (biceps brachii, **BIC**). During unsupported stance, the leg muscles—hamstrings (**HAM**) and gastrocnemius (**GAS**)—are activated prior to the arm muscle to prevent the body from rotating forward during the arm pull. The **red arrow** shows the onset of leg gastrocnemius activation, the **gray arrow** that of the arm biceps brachii. When the subject was supported by a rigid bar at the shoulder, the anticipatory leg muscle activity was not necessary because the body could not rotate forward. Arm activation was earlier when anticipatory postural muscle activity was not needed. Shaded areas indicate anticipatory postural responses (**red**) and the initial arm muscle activation (**brown**). (Adapted, with permission, from Cordo and Nashner 1982.)

C. During walking, the trajectory of the center of mass (CoM) is controlled by foot placement. The body's center of mass is between the feet, moving forward and from side to side as the subject walks forward. When the body is supported by only one leg (single support phase), the CoM is outside the base of support and moves toward the lifting limb. People do not fall while walking because the placement of the foot on the next step decelerates the CoM and propels it back toward the midline. (Adapted from MacKinnon and Winter 1993.)

perform elaborate rolls and twists of the body about the center of mass while airborne, although the trajectory of his center of mass is fixed once he leaves the board. During voluntary movements, postural adjustments control the body's angular momentum by anticipating rotational forces.

Posture Control Is Integrated With Locomotion

During walking and running, the body is in a constant state of falling as the center of mass moves forward and laterally toward the leg that is in the swing phase (Figure 36–8C). During walking, the center of mass is within the base of support only when both feet are on the ground, the double stance phase, which is only one-third of a gait cycle. When one foot is supporting the body, the center of mass moves forward in front of the foot, always medial to the base of support.

Falling is prevented during walking and running by moving the base of support forward and laterally under the falling center of mass. Postural equilibrium during gait relies on the placement of each step to control the speed and trajectory of the center of mass. The nervous system plans foot placement several steps in advance using visual information about the terrain and surrounding environment.

The main postural challenge during walking is controlling the center of mass of the upper body over the moving legs, especially in the lateral direction. Excessive lateral displacement of the trunk and excessive lateral foot placement variability are signs of postural instability during locomotion. Patients with abnormal postural stability during gait may nevertheless exhibit normal automatic and anticipatory postural adjustments, postural sway in stance under different sensory conditions, and orientation to vertical, suggesting that postural control and gait have different nervous system circuits.

Somatosensory, Vestibular, and Visual Information Must Be Integrated and Interpreted to Maintain Posture

Because information about motion from any one sensory system may be ambiguous, multiple modalities must be integrated in postural centers to determine what orientation and motion of the body are appropriate. The influence of any one modality on the postural control system varies according to the task and biomechanical conditions.

According to prevailing theory, sensory modalities are integrated to form an internal representation of the body housed within the parietal cortex that the nervous system uses to plan and execute motor behaviors. Over time, this internal representation must adapt to changes associated with early development, aging, and injury.

Somatosensory Signals Are Important for Timing and Direction of Automatic Postural Responses

Many types of somatosensory fibers trigger and shape the automatic postural response. The largest fibers, those in group I (12–20 μm in diameter), appear to be essential for normal response latencies. The longer latency, slower rise time, and lower amplitude of the EMG response following destruction of the group I fibers reflect a loss of acceleration information encoded by muscle spindle primary receptors (Figure 36–9A). The largest and most rapidly conducting sensory fibers are the Ia afferents from muscle spindles and Ib afferents from Golgi tendon organs, as well as some fibers from cutaneous mechanoreceptors (Chapter 18). Group I fibers provide rapid information about the biomechanics of the body including responses to muscle stretch, muscle force, and directionally specific pressure on the foot soles. However, group II fibers from muscle spindles and cutaneous receptors may also play a role in shaping automatic postural responses. Although they may be too slow to generate the earliest part of the response, they likely encode center of mass velocity and position.

Both proprioceptive and cutaneous inputs provide cues about postural orientation. During upright stance, for example, muscles lengthen and shorten as the body sways under the force of gravity, generating proprioceptive signals related to load, muscle length, and velocity of stretch. Joint receptors may detect compressive forces on the joints, whereas cutaneous receptors in the sole of the foot respond to motion of the center of pressure and to changes in ground reaction force angle as the body sways. Pressure receptors near the kidneys are sensitive to gravity (somatic graviception) and are used by the nervous system to help detect upright or tilted postures. All of these signals contribute to the neural map of the position of body segments with respect to each other and the platform surface and may contribute to the neural computation of center of mass motion.

The large-diameter, fast somatosensory fibers from muscle spindles are critical for maintaining balance during stance. When these axons die, as occurs in some forms of peripheral neuropathy, automatic postural responses to movement of a platform are delayed, retarding the ground reaction force. As a

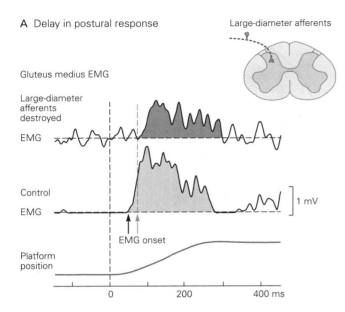

A Delay in postural response

Gluteus medius EMG

Large-diameter afferents

Large-diameter afferents destroyed EMG

Control EMG

]1 mV

EMG onset

Platform position

0   200   400 ms

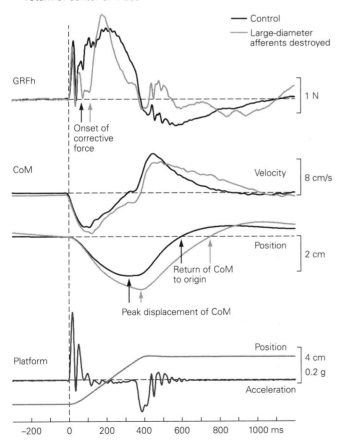

B Delay in development of force at the ground and return of center of mass

— Control
— Large-diameter afferents destroyed

GRFh

]1 N

Onset of corrective force

CoM

Velocity   ]8 cm/s

Position   ]2 cm

Return of CoM to origin

Peak displacement of CoM

Platform

Position   ]4 cm
           0.2 g
Acceleration

−200   0   200   400   600   800   1000 ms

result, the center of mass moves faster and farther from the initial position and takes longer to return (Figure 36–9). Because it is more likely that the center of mass will move outside the base of support, balance is precarious and a fall may occur. Accordingly, individuals with large-fiber peripheral neuropathy in the legs experience ataxia and difficulties with balance.

Vestibular Information Is Important for Balance on Unstable Surfaces and During Head Movements

The otolithic organs of the vestibular apparatus provide information about the direction of gravity, whereas the semicircular canals measure the velocity of head rotation (Chapter 27). Vestibular information thus informs the nervous system about how much the body is tilted with respect to gravity as well as whether it is swaying forward, backward, or sideways.

Somatosensory and vestibular information about the gravitational angle of the body is combined to orient the body with respect to gravity and other inertial forces. To maintain balance while riding a bike in a circular path at high speed, for example, the body and bike must be oriented with respect to a combination of gravitational and centripetal forces (Figure 36–10A).

Unlike somatosensory inputs, vestibular signals are not essential for the normal timing of balance reactions. Instead, they influence the directional tuning of a postural response by providing information about the orientation of the body relative to gravity. In humans and experimental animals lacking functional vestibular afferent pathways, the postural response to *angular* motion or tilt of the support surface is opposite to the

Figure 36–9 (Left) Loss of large-diameter somatosensory fibers delays automatic postural responses. Electromyograms (**EMGs**) of postural responses to horizontal motion of a moveable platform were recorded in a cat before and after destruction of the large-diameter (group I) somatosensory fibers throughout the body by vitamin B$_6$ intoxication. Motor neurons and muscle strength are not affected by the loss of the somatosensory fibers, but afferent information about muscle length and force is diminished. (Reproduced, with permission, from J. Macpherson.)

A. The postural response in the gluteus medius evoked by horizontal motion of the support platform is significantly delayed after destruction of group I fibers. This delay of approximately 20 ms induces ataxia and difficulty in maintaining balance.

B. Destruction of group I fibers delays activation of the hind limb. This delay slows the restoration of the center of mass (**CoM**) and the recovery of balance following platform displacement. The delay in onset of the horizontal component of the ground reaction force (**GRFh**) results in a greater peak displacement of the CoM and a delay in its return to its origin relative to the paws.

A Orienting to gravito-inertial force

B Orienting to rotating visual field

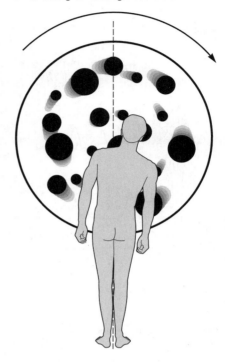

Figure 36–10 The postural system orients the body to various external reference frames.

A. When traveling at high speed along a curved path, a cyclist orients to the gravito-inertial force (angle **A**), the vector sum of the force caused by gravity and the centripetal force caused by acceleration along the curved path. (Used, with permission, from Joseph Daniel, Story Arts Media, LLC. Previously published in McMahon and Bonner 1983.)

B. The postural system can interpret rightward rotation of objects occupying a large region of the visual field as the body tilting to the left. In compensation for this illusion of motion, the subject tilts to the right, adopting a new postural vertical orientation that is driven by the visual system. The **red dashed line** indicates gravitational vertical. (Adapted, with permission, from Brandt, Paulus, and Straube 1986.)

normal response. Rather than resisting the tilt, subjects lacking vestibular signals do the opposite and accentuate the tilt through their own muscular activity. In contrast, the response to horizontal translation motion of a platform has the appropriate directional tuning and latency, even in the acute stage prior to vestibular compensation.

Why does the absence of vestibular signals cause difficulty with tilt but not with linear motion? The answer lies in how the nervous system determines the direction of vertical. Gravity is the main force that causes the body to fall. As the support surface tilts, healthy subjects orient to gravity using vestibular information to remain upright. In contrast, subjects without vestibular function use somatosensory inputs to orient themselves to the support surface and consequently fall downhill as the surface tilts. During linear motion, however, gravitational and surface vertical are collinear, and somatosensory signals are sufficient to compute the correct postural response. Although visual inputs also provide a vertical reference, visual

processing is too slow to participate in the automatic postural response to rapid tilt, especially soon after the loss of vestibular function.

Without vestibular information, the response to linear motion of the support surface is larger than normal (*hypermetria*), leading to overbalancing and instability. Hypermetria is a major cause of ataxia when vestibular information is lost. Vestibular hypermetria may result from reduced cerebellar inhibition of the motor system, for the loss of vestibular inputs reduces the drive to the inhibitory Purkinje cells.

Humans and cats are quite ataxic immediately after loss of the vestibular apparatus. The head and trunk show marked instability, stance and gait are broad-based, and walking follows a weaving path with frequent falling. Instability is especially great on turning the head, probably because trunk motion cannot be distinguished from head motion using somatosensory information alone. Cats and humans lacking vestibular inputs produce motor output that results in them actively pushing themselves toward the side of a voluntary head turn, likely because somatosensory inputs that encode trunk and head motion are misinterpreted in the absence of vestibular inputs. The postural system erroneously senses that the body is falling to the side away from the head tilt and generates a response in the opposite direction, resulting in imbalance.

Immediately following vestibular loss, neck muscles are abnormally activated during ordinary movements and often the head and trunk are moved together as a unit. After several months, routine movement becomes more normal through vestibular compensation, which may involve greater reliance on the remaining sensory information. However, more challenging tasks are hampered by a residual hypermetria, stiffness in head–trunk control, and instability, especially when visual and somatosensory information is unavailable for postural orientation. Vestibular information is critical for balance when visual information is reduced and the support surface is not stable, for example, at night on a sandy beach or on a boat deck.

Visual Inputs Provide the Postural System With Orientation and Motion Information

Vision reduces body sway when standing still and provides stabilizing cues, especially when a new balancing task is attempted or balance is precarious. Skaters and dancers maintain stability while spinning by fixing their gaze on a point in the visual field. However, visual processing is too slow to significantly affect the postural response to a sudden and unexpected disturbance of balance. Vision does play an important role in anticipatory postural adjustments during voluntary movements, such as planning where to place the feet when walking over obstacles.

Vision can have a powerful influence on postural orientation, evident when watching a movie scene filmed from the perspective of a moving viewer and projected on a large screen. Simulated rides in a roller coaster or airplane can induce strong sensations of motion along with activation of postural muscles. An illusion of movement is induced when sufficiently large regions of the visual field are stimulated, as when a large disk in front of a standing subject is rotated. The subject responds to this illusion by tilting his body; clockwise rotation of the visual field is interpreted by the postural system as the body falling to the left, to which the subject compensates by leaning to the right (Figure 36–10B). The rate and direction of optic flow—the flow of images across the retina as people move about—provide clues about body orientation and movement.

Information From a Single Sensory Modality Can Be Ambiguous

Any one sensory modality alone may provide ambiguous information about postural orientation and body motion. The visual system, for example, cannot distinguish self-motion from object motion. We have all experienced the fleeting sensation while sitting in a stationary vehicle of not knowing whether we are moving or the adjacent vehicle is moving.

Vestibular information can also be ambiguous for two reasons. First, vestibular receptors are located in the head and therefore provide information about acceleration of the head but not about the rest of the body. The postural control system cannot use vestibular information alone to distinguish between the head tilting on a stationary trunk and the whole body tilting by rotation at the ankles, both of which activate the semicircular canals and otolith organs. Additional information from somatosensory receptors is required to resolve this ambiguity. The otolith organs also cannot distinguish between acceleration owing to gravity and linear acceleration of the head. Tilting to the left, for example, can produce the same otolithic stimulation as acceleration of the body to the right (Figure 36–11).

Studies suggest that there are neural circuits that can disambiguate the head-tilt component of a linear acceleration by using a combination of canal and otolith inputs. Output from this circuit may allow the postural system to determine the orientation of gravity relative to the head regardless of head position and motion. The distinction between tilt and linear motion is especially important while standing on an unstable or a tilting surface.

Somatosensory inputs may also provide ambiguous information about body orientation and motion. When we stand upright, mechanoreceptors in the soles of our feet and proprioceptors in muscles and joints signal the motion of our body relative to the support surface. But somatosensory inputs alone cannot distinguish between body and surface motion, for example, whether ankle flexion stems from forward body sway or tilting of the surface. Our common experience is that the ground beneath us is stable and that somatosensory inputs reflect movements of the body's center of mass as we sway. But surfaces may move relative to the earth, such as a boat's deck, or may be pliant under our weight, like a soft or spongy surface. Therefore, somatosensory information must be integrated with vestibular and visual inputs to give the nervous system an accurate picture of the stability and inclination of the support surface and of our body's relationship to earth vertical.

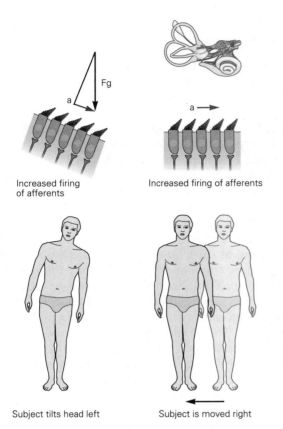

Increased firing
of afferents

Increased firing of afferents

Subject tilts head left

Subject is moved right

Figure 36–11 Vestibular inputs regarding body posture and motion can be ambiguous. The postural system cannot distinguish between tilt and linear acceleration of the body based on otolith inputs alone. The mechanoreceptors of the vestibular system are hair bundles that bend in response to shearing forces, thus changing the firing rate of the tonically active sensory afferents. The same shearing force can result from tilting of the head (*left*), which exposes the hair cells to a portion of the acceleration (**a**) owing to gravity (**Fg**), or from horizontal linear acceleration of the body (*right*).

The Postural Control System Uses a Body Schema That Incorporates Internal Models for Balance

Because of the mechanical complexity of the body, with its many skeletal segments and muscles, the nervous system requires a detailed representation of the body and its interaction with the environment. To execute the simple movement of raising your hand and touching your nose with your index finger while your eyes are closed, your nervous system must know the characteristics (length, mass, and connections) of each segment of the arm, the shoulder, and head as well as the orientation of your arm with respect to the gravity vector and your nose. Thus, information from multiple sensory systems is integrated into a central representation of the body, often called the body schema.

The body schema for postural control, as developed by Viktor Gurfinkel, is not simply a sensory map like the somatotopic representation of the skin in primary sensory cortex. Instead, it incorporates internal models of the body's relationship with the environment. The body schema is used to compute appropriate anticipatory and automatic postural reactions to maintain balance and postural orientation. A simplified example of such an internal model is one in which the body is represented as a single segment hinged at the foot (Figure 36–12A). The internal model generates an estimate of the orientation of the foot in space, which also serves as an estimate of the orientation of the support surface, a variable that cannot be directly sensed.

Henry Head, a neurologist working in the early part of the 20th century, described the body schema as a dynamic system in which both spatial and temporal features are continually updated, a concept that remains current. To allow adequate planning of movement strategies, the body schema must incorporate not only the relationship of body segments to space and to each other but also the mass and inertia of each segment and an estimate of the external forces acting on the body including gravity.

The body schema integrates sensory information from the somatosensory, vestibular, and visual systems to orient the body to vertical. Even in the dark, people can accurately reorient a projected line to a vertical position (visual vertical) and they can reorient themselves to vertical when sitting on a tilting swing (gravitational vertical). Visual vertical and gravitational vertical are independent of each other. Patients with asymmetrical vestibular function show abnormal visual vertical but normal gravitational vertical, whereas patients with hemi-neglect from stroke show abnormal gravitational vertical but normal visual vertical.

Another component of the body schema is a model of the sensory information expected as a result of a movement. Disorientation or motion sickness may result when the actual sensory information received by the nervous system does not match the expected sensory information, as in the microgravity environment of space flight. With continued exposure to the new environment, however, the model is gradually updated until expected and actual sensory information agree and the person is no longer spatially disoriented.

The internal model for balance control must be continually updated, both in the short term, as we use experience to improve our balance strategies, and in the long term, as we age and our bodies change in shape and size. One way the body schema is updated is by changing the relative sensitivity or weighting of each sensory system.

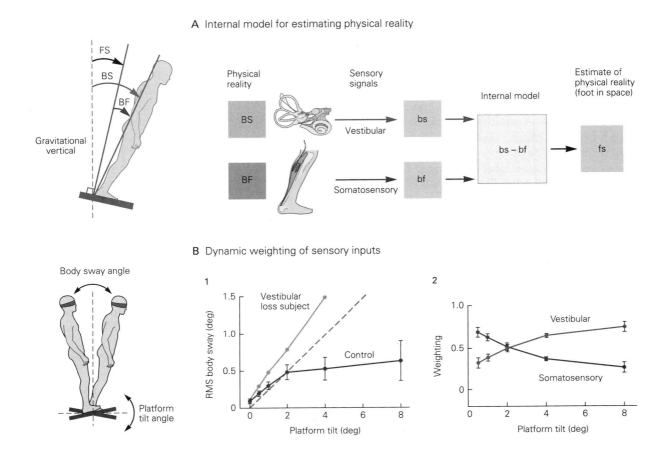

Figure 36–12 Many types of sensory signals are integrated and weighted in an internal model that optimizes balance and orientation. (Adapted from Peterka 2002.)

A. The simple example of a person standing on a tilted surface illustrates how the nervous system might estimate physical variables that are not sensed directly. The physical variables are body tilt with respect to earth vertical or body-in-space (**BS**), and body angle relative to the foot (**BF**). The angle of the foot in space (**FS**) is simply the difference BS − BF. The neural estimate of body in space (**bs**) comes from vestibular and other receptors that detect tilt of the body relative to gravity. The neural estimate of body angle to foot (**bf**) comes from somatosensory signals related to ankle joint angle. The internal model for estimating physical reality, bs − bf, produces a neural estimate of the foot in space (**fs**). Such estimates of the physical world are continually updated based on experience.

B. Sensory information is weighted dynamically to maintain balance and orientation under varying conditions. The figure illustrates findings from an experiment in which human subjects stood blindfolded on a platform that slowly rotated continuously in the toes-up or toes-down direction at amplitudes of up to 8° (peak to peak). **1.** Comparison of body sway during surface oscillations in a subject with loss of vestibular function and a group of control subjects. Body-sway angle is measured relative to gravitational vertical during platform tilt and expressed as root mean square (**RMS**) sway in degrees. The **dashed line** represents equal platform and body sway; for example, for a platform tilt of 4°, an equal amount of body sway is 1° RMS. In control subjects, the body and platform sway are equal for small platform tilts up to 2°, suggesting that people normally use somatosensory signals to remain perpendicular to the platform (minimizing changes in ankle angle). With larger platform tilts, body sway does not increase much beyond 0.5° RMS. In contrast, subjects with vestibular loss sway even more than the platform (1.5° RMS of body tilt at 4° of platform tilt) and cannot remain standing at platform tilts above 4°. Thus, when both vestibular and visual signals are absent, a person attempts to maintain his position only relative to the support surface and has difficulty maintaining balance as that surface moves. **2.** In control subjects, as platform tilt increases, the influence of somatosensory input decreases with increasing platform tilt while the influence of vestibular input increases. At larger tilt angles, the greater influence of vestibular input minimizes the degree of body sway away from gravitational vertical.

Control of Posture Is Task Dependent

The senses and muscles used to control posture vary, depending on task constraints and requirements. For example, when vestibular and somatosensory information is altered while working on a space station, vision is used to orient the body to tasks, and the goal of postural equilibrium changes from preventing falls due to gravity to preventing unintended collision with objects due to inertia. A healthy nervous system very quickly adapts to changing tasks, goals, and environments by modifying its relative dependence upon different sensory information and by using different sets of muscles to optimize achieving the goals of both posture control and voluntary movements.

Task Requirements Determine the Role of Each Sensory System in Postural Equilibrium and Orientation

The postural control system must be able to change the weighting of different sensory modalities to accommodate changes in the environment and movement goals. Subjects standing on a firm stable surface tend to rely primarily on somatosensory information for postural orientation. When the support surface is unstable, subjects depend more on vestibular and visual information. However, even when the support surface is not stable, light touch with a fingertip on a stable object is more effective than vision in maintaining postural orientation and balance. Vestibular information is particularly critical when visual and somatosensory information is ambiguous or absent, such as when skiing downhill or walking below deck on a ship.

The changeable weighting of individual sensory modalities was demonstrated in an experiment in which subjects were blindfolded and asked to stand quietly on a surface with a tilt that slowly oscillated by varying amounts, up to 8° in magnitude. For tilts of less than 2°, all subjects sway with the platform, suggesting that they use somatosensory information to orient their body to the support surface (Figure 36–12B). At larger tilts, healthy subjects attenuate their sway and orient their posture more with respect to gravitational vertical than to the surface, as they rely more on vestibular information so they stop increasing body sway. Thus, relative sensory weighting changes in control subjects such that somatosensory weight is highest with a stable platform and vestibular weight is highest when standing on an unstable surface, such as with large surface tilts (Figure 36–12B2). In contrast, patients who have lost vestibular function persist in swaying along with the platform and subsequently fall

during large surface tilts. This behavior is consistent with the patients' inappropriate automatic postural response to platform tilts.

Studies such as these suggest that when people are standing on moving or unstable surfaces, the weighting of vestibular and visual information increases, whereas that of somatosensory information decreases. Any sensory modality may dominate at a particular time, depending on the conditions of postural support and the specific motor behavior to be performed.

Control of Posture Is Distributed in the Nervous System

Postural orientation and balance are achieved through the dynamic and context-dependent interplay among all levels of the central nervous system, from the spinal cord to cerebral cortex. The major areas of the brain involved in postural control are shown in Figure 36–13. Signals from specific areas in all lobes of the cerebral cortex converge and are integrated to determine appropriate outputs from motor cortical areas to subcortical structures. The basal ganglia, cerebellum, and pedunculopontine nucleus then send outputs to the brain stem. Ultimately, inputs from these varied sources result in activation of the reticulospinal and vestibulospinal pathways, which descend to the spinal cord where they contact interneurons and spinal motor neurons for postural control.

Afferent inputs from visual, vestibular, and somatosensory sources are integrated along the neuraxis, including the vestibular nuclei and right parietal cortex, to inform the internal model of body orientation and balance. This internal model is continually updated by the cerebellum based on error signals between expected and actual sensory feedback following motor commands.

Spinal Cord Circuits Are Sufficient for Maintaining Antigravity Support but Not Balance

Adult cats with complete spinal transection at the thoracic level can, with experience, support the weight of their hindquarters with fairly normal hind limb and trunk postural orientation, but they have little control of balance. These animals do not exhibit normal postural responses in their hind limbs when the support surface moves. Their response to horizontal motion consists of small, random, and highly variable bursts of activity in extensor muscles, and postural activity in flexor muscles is completely absent. Active balance is absent despite the fact that extensors and flexors can

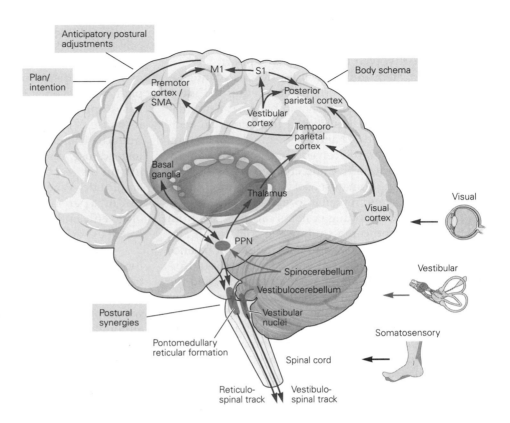

Figure 36–13 Many parts of the nervous system control posture. Areas of frontal, parietal, temporal, and occipital cortex, as well as the basal ganglia, cerebellum, and pendunculopontine nucleus (PPN), provide inputs to the recticulospinal and vestibulospinal pathways descending to spinal motor neurons. Afferent inputs from the visual, vestibular, and somatosensory systems are integrated in the brainstem and cortex to update the body schema and inform future postural commands. (Abbreviations: **M1**, primary motor cortex; **S1**, primary somatosensory cortex; **SMA**, supplementary motor area.) (Adapted from Beristain 2016.)

be recruited for other movements such as stepping on a treadmill, suggesting that unlike locomotion, postural muscle activation requires supraspinal control.

An adult cat with a spinal transection can stand independently for only short periods of time and within a narrow range of stability; head turns in particular cause the animal to lose balance. What stability there is likely results from the broad base of support afforded by quadrupedal stance, the stiffness of the tonically contracting hind limb extensors that support the weight of the hindquarters, and active compensation by forelimbs that continue to produce postural responses. Humans with spinal cord injuries have various amounts of antigravity muscle tonus but lack automatic postural responses below the level of the lesion. These results emphasize that antigravity support and balance control are distinct mechanisms and that the control of balance requires the involvement of supraspinal circuits.

The Brain Stem and Cerebellum Integrate Sensory Signals for Posture

If spinal circuits alone are not capable of producing automatic postural responses, what supraspinal centers are responsible for these responses? Although the answer to this question remains unknown, good candidates include the brain stem and cerebellum, which are highly interconnected and work together to modulate the descending commands to spinal motor centers of the limbs and trunk. These regions have the input–output structure that would be expected of centers for postural control.

Muscle synergies for automatic postural responses may be organized in the brain stem, perhaps the reticular formation. However, adaptation of postural synergies to changes in the environment and task demands may require the cerebellum.

Two regions of the cerebellum influence orientation and balance: the vestibulocerebellum (nodulus, uvula, and fastigial nucleus) and the spinocerebellum (anterior lobe and interpositus nucleus). These regions are interconnected with the vestibular nuclei and reticular formation of the pons and medulla (see Figure 37–4). Lesions of the brain stem and vestibulocerebellum produce a variety of deficits in head and trunk control including a tendency to tilt from vertical, even with eyes open, suggesting a deficit in the internal representation of postural orientation. Lesions of the spinocerebellum result in excessive postural sway that is worse with the eyes closed, ataxia during walking, and hypermetric postural responses, suggesting deficits in balance corrections. Certain regions in the pons and medulla facilitate or depress extensor tonus and could be involved in antigravity support.

The brain stem and cerebellum are sites of integration of sensory inputs, perhaps generating the internal model of body orientation and balance. Vestibular and visual inputs are distributed to brain stem centers (Chapters 25 and 27) and the vestibulocerebellum. The spinocerebellum receives signals from rapidly conducting proprioceptive and cutaneous fibers. More slowly conducting somatosensory fibers project to the vestibular nuclei and reticular formation.

Two major descending systems carry signals from the brain stem and cerebellum to the spinal cord and could trigger automatic postural responses for balance and orientation. The medial and lateral vestibulospinal tracts originate from the vestibular nuclei, and the medial and lateral reticulospinal tracts originate from the reticular formation of the pons and medulla (see Figure 37–5). Lesions of these tracts result in profound ataxia and postural instability. In contrast, lesions of the corticospinal and rubrospinal tracts have minimal effect on balance even though they produce profound disturbance of voluntary limb movements.

The Spinocerebellum and Basal Ganglia Are Important in Adaptation of Posture

Patients with spinocerebellar disorders such as alcoholic anterior-lobe syndrome and basal ganglia deficits such as Parkinson disease experience postural difficulties. Studies suggest that the spinocerebellum and basal ganglia play complementary roles in adapting postural responses to changing conditions.

The spinocerebellum is where the amplitude of postural responses is adapted based on experience. The basal ganglia are important for quickly adjusting the postural set when conditions suddenly change. Both the spinocerebellum and the basal ganglia regulate muscle tone and force for voluntary postural adjustments. They are not necessary, however, for triggering or constructing the basic postural patterns.

Patients with disorders of the spinocerebellum have difficulty modifying the magnitude of balance adjustments with practice, over the course of repeated trials, but can readily adapt postural responses immediately after a change in conditions based on sensory feedback. For example, a patient standing on a movable platform scales the size of postural responses appropriately when platform velocity is increased with each trial. These postural adjustments rely on velocity information, which is encoded by somatosensory inputs at the beginning of platform movement.

In contrast, patients with cerebellar disorders cannot scale the size of postural responses based on feedforward control using the anticipated amplitude of postural displacements. Because the amplitude of platform movement is not known until the platform has stopped moving, well after the initial postural response is complete, a subject cannot use feedback from the trial at hand to guide the response but must instead use his experience from previous trials to inform his response in a subsequent trial of the same amplitude. Whereas a healthy subject does this quite readily, a patient with spinocerebellar disorders is unable to efficiently adapt his postural responses based on recent experience (Figure 36–14A).

A healthy subject standing on a moveable platform is able to scale muscle activity during sudden backward motion of the platform to counteract the forward sway induced by the perturbation. A subject with spinocerebellar disease always overresponds, although the timing of muscle activation is normal (Figure 36–14B). As a result, this individual returns beyond the upright position and oscillates back and forth. Reminiscent of the hypermetria observed immediately after labyrinthectomy, cerebellar hypermetria may also result from loss of Purkinje cell inhibition on spinal motor centers.

A patient with Parkinson disease can, with sufficient practice, gradually modify his postural responses but has difficulty changing responses when conditions change suddenly. Such postural inflexibility is seen when initial posture changes. For example, when a normal subject switches from standing upright to sitting on a stool on a movable platform, the pattern of his automatic postural response to backward movement of the platform changes immediately. Because leg muscle activity is no longer necessary after the switch from standing to sitting, this component ceases to be recruited.

In contrast, a patient with Parkinson disease employs the same muscle activation pattern for both sitting and standing (Figure 36–15). L-DOPA replacement therapy does not improve the patient's ability to switch postural set. With repetition of trials in the seated posture, however, the leg muscle activity eventually disappears, showing that enough experience permits adaptation of postural responses. A patient with Parkinson disease also has difficulty when instructed to increase or decrease the magnitude of a postural response, a difficulty that is consistent with the inability to change cognitive sets quickly.

Patients with Parkinson disease have problems with postural tone and force generation in addition to

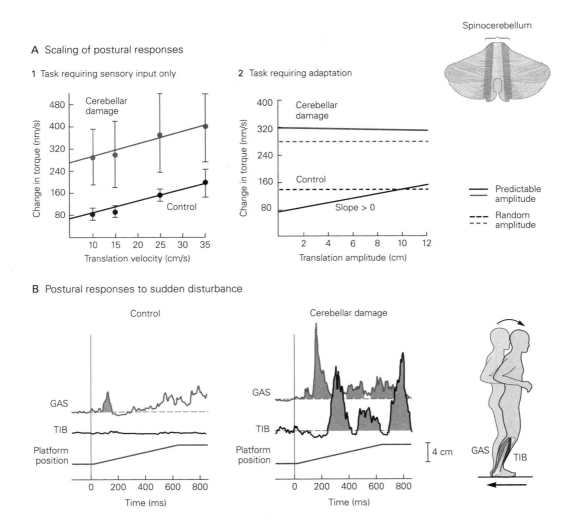

A Scaling of postural responses

Figure 36–14 The spinocerebellum has a role in adapting postural responses to changing conditions and in scaling postural responses to anticipated postural disturbances. The spinocerebellum is important for adapting postural responses based on experience. Patients with a spinocerebellar disorder are able to use immediate sensory input, but not experience, to adjust automatic postural responses. (Adapted, with permission, from Horak and Diener 1994.)

A. 1. In this experiment, subjects stand on a platform that is moved horizontally; the velocity is increased on each trial. Maintaining balance requires scaling responses to the velocity of the platform using sensory feedback. The adjustments in a subject with a spinocerebellar disorder have the same regression coefficient (slope) as those of a control subject, even though in each trial the responses are larger and more variable than those of the control subject. **2.** When subjects are required to anticipate and adapt to platform translation, the postural

adjustments in the spinocerebellar subject are compromised. When translation amplitude is random, responses are large, as if the subject expects a large translation. When trials with the same amplitude are repeated, a control subject learns to predict the amplitude of the disturbance and adjust his response. In contrast, a spinocerebellar subject shows no improvement in performance; he cannot use his experience in one trial to adjust his responses in subsequent trials. All responses are large, as if the subject always expects the large translation.

B. In this experiment, subjects stand on a platform that is moved backward (6 cm amplitude at 10 cm/s). In a control subject, the onset of movement evokes a small burst of activity in the gastrocnemius (**GAS**), an ankle extensor. In a subject with damage to the anterior lobe of the cerebellum, the muscle responses are overly large, with bursts of activity alternating between the gastrocnemius and its antagonist, the tibialis anterior (**TIB**).

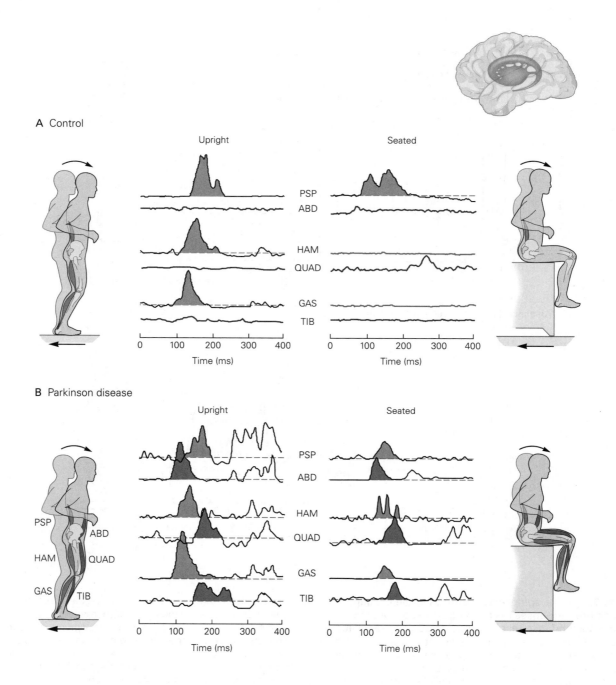

Figure 36–15 The basal ganglia are important for adapting postural responses to a sudden change in initial conditions. (Adapted, with permission, from Horak, Nutt, and Nashner 1992.)

A. When a normal subject switches from upright stance to sitting, he immediately modifies his response to backward movement of the support platform. The postural response to movement while seated does not involve the leg muscles—the gastrocnemius (**GAS**) and hamstrings (**HAM**)—but does

activate the paraspinal muscles (**PSP**) and with shorter latency than in the response to movement while standing. (Abbreviations: **ABD**, abdominals; **QUAD**, quadriceps; **TIB**, tibialis anterior.)

B. A patient with Parkinson disease does not suppress the leg muscle response in the first trial after switching from standing to sitting. The postural response of this subject is similar for both initial positions: Antagonist muscles (**purple**) are activated along with agonists (**pink**).

an inability to adapt to changing conditions. The disease's bradykinesia (slowness of movement) is reflected in slow development of force in postural responses, and its rigidity is manifested in co-contraction. L-DOPA replacement greatly improves a patient's ability to generate not only forceful voluntary movements but also the accompanying postural adjustments, such as rising onto the toes and gait. However, neither the automatic postural response to an unexpected disturbance nor postural adaptation is improved by L-DOPA, suggesting that these functions involve the nondopaminergic pathways affected by Parkinson disease.

Cerebral Cortex Centers Contribute to Postural Control

Several areas of cerebral cortex influence postural orientation and equilibrium, including both anticipatory and automatic postural responses. Most voluntary movements, which are initiated in the cerebral cortex, require postural adjustments that must be integrated with the primary goal of the movement in both timing and amplitude. Where this integration occurs is not clear.

The cerebral cortex is more involved in anticipatory postural adjustments than in automatic postural reactions. However, recent electroencephalographic studies show that areas of cerebral cortex are activated by anticipation of a postural disturbance before an automatic postural response is initiated. This finding is consistent with the idea that the cortex optimizes balance control as part of motor planning.

The supplementary motor area and temporoparietal cortex have both been implicated in postural control. The supplementary motor area, anterior to the motor cortex, is likely involved with anticipatory postural adjustments that accompany voluntary movements. The temporoparietal cortex appears to integrate sensory information and may comprise internal models for perception of body verticality. Lesions of insular cortex can impair perception of the visual vertical, whereas lesions of superior parietal cortex impair perception of the postural vertical, and either of these defects may impair balance when standing on an unstable support.

Sensorimotor cortex receives somatosensory inputs signaling balance disturbances and postural responses. However, this region is not essential for automatic postural adjustments. Lesioning the motor cortex in cats impairs lifting of the forelimb in response to a light touch during stance but does not abolish the accompanying postural adjustment in the contralateral forelimb. Although the sensorimotor cortex is not

responsible for postural adjustments, it may have a role in the process.

Behavioral studies, too, have implicated cortical processes in postural control. Control of posture, like control of voluntary movement, requires attention. When subjects must press a button following a visual or auditory cue while also maintaining balance, their reaction time increases with the difficulty of the task (balancing on one foot versus sitting, for example). Moreover, when subjects try to perform a cognitive task while actively maintaining posture, the performance of either or both can degrade. For example, when a subject is asked to count backward by threes while standing on one foot, both the cognitive task and postural adjustment deteriorate. The timing of automatic postural responses to unexpected disturbances is little affected by cognitive interference.

Balance control is also influenced by emotional state, thus implicating the limbic system in posture control. Fear of falling, for example, can increase postural tone and stiffness, reduce sway area, increase sway velocity, and alter balancing strategies in response to disturbances.

Finally, balance control is also influenced by attentional ability and demands, thus implicating the frontoparietal attention network. There is evidence of competition for central processing resources in dual task conditions, where a person must maintain balance and perform a concurrent cognitive task. Both postural control and cognitive performance may be impaired in dual-task conditions as compared to single-task conditions assessing either postural or cognitive performance in isolation. As cognitive demands increase, responses to postural perturbations are smaller in amplitude and occur at longer latency. However, when necessary, healthy individuals prioritize postural control over the cognitive task and demonstrate decreasing cognitive performance as postural demands increase. In contrast, individuals with nervous system disorders such as Parkinson disease may not prioritize postural control in dual-task situations and may be at increased risk for falls in dual-task situations.

Although the roles of specific areas of cerebral cortex in postural control are largely undefined, there is no doubt that the cortex is important for learning new, complex postural strategies. The cortex must be involved in the amazing improvement in balance and postural orientation of athletes and dancers who use cognitive information and advice from coaches. In fact, the cerebral cortex is involved in postural control each time we consciously maintain our balance while walking across a slippery floor, standing on a moving bus, or waiting tables on a rocking ship.

Highlights

1. The two goals of posture are balance and orientation. Balance control maintains the body in stable equilibrium to avoid falls. Postural orientation aligns the body segments with respect to each other and to the world, such as maintaining the head vertical.

2. A sudden displacement of the body center of mass while standing triggers ankle, hip, and/or stepping strategies to return the center of mass within the base of foot support.

3. Postural responses are fast and automatic, but adapt quickly to changes in environmental context, intention, and conditions. Postural responses can also be improved with practice.

4. Activation of centrally organized muscle synergies is used to control balance. This synergy organization simplifies the neural control so only a few central commands are required, instead of a separate command for each muscle, while allowing flexibility and adaptability for postural control.

5. Somatosensory, vestibular, and visual sensory modalities are integrated to form an internal representation of the body that the nervous system uses for postural orientation and balance control. Somatosensory signals trigger the fastest, largest postural responses and are most critical for control of postural sway in standing. Vestibular signals are particularly critical when standing on an unstable surface, when it is difficult to use somatosensory information for postural orientation. Visual inputs provide spatial orientation and motion information.

6. The body center of mass is often outside of the base of foot support during walking and running so balance is provided by adjusting foot placement and lateral trunk stability to control the center of mass with respect to the changing base of support.

7. The vestibulocerebellum and the spinocerebellum are interconnected with the vestibular nuclei and reticular formation of the brainstem for control of balance and postural orientation.

8. The basal ganglia are important for control of axial postural tone, adapting postural response strategies based on initial conditions, and anticipatory postural control. The cerebellum is important for adapting the magnitude of balance responses with practice, over the course of repeated trials, and for scaling the size of postural responses.

9. Posture control involves many brain areas from the brainstem to the frontal cortex, but the specific circuits involved in different types of posture control (automatic postural responses, anticipatory postural adjustments, body sway in stance, sensory integration for a body schema and verticality) have yet to be determined.

Fay B. Horak
Gammon M. Earhart

Suggested Reading

Chiba R, Takakusaki K, Ota J, Yozu A, Haga N. 2016. Human upright posture control models based on multisensory inputs; in fast and slow dynamics. Neurosci Res 104:96–104.

Cullen KE. 2016. Physiology of central pathways. In: JM Furman, T Lempert (eds). *Handbook of Clinical Neurology*, vol 137, pp 17–40. New York: Elsevier.

Dietz V. 1992. Human neuronal control of automatic functional movements—interaction between central programs and afferent input. Physiol Rev 72:33–69.

Horak FB. 2006. Postural orientation and equilibrium: what do we need to know about neural control of balance to prevent falls? Age Ageing 35(Suppl 2):ii7–ii11.

Horak FB, Macpherson JM. 1996. Postural orientation and equilibrium. In: LB Rowell, JT Shepherd (eds). *Handbook of Physiology, Section 12, Exercise: Regulation and Integration of Multiple Systems*, pp. 255–292. New York: Oxford Univ. Press.

Macpherson JM, Deliagina TG, Orlovsky GN. 1997. Control of body orientation and equilibrium in vertebrates. In: PSG Stein, S Grillner, AI Selverston, DG Stuart (eds). *Neurons Networks and Motor Behavior*, pp. 257–267. Cambridge, MA: MIT Press.

Massion J. 1994. Postural control system. Curr Opin Neurobiol 4:877–887.

Woollacott M, Shumway-Cook A. 2002. Attention and the control of posture and gait: a review of an emerging area of research. Gait Posture 16:1–14.

References

Beristain X. 2016. Gait. In: Salardini A, Biller J (eds). *The Hospital Neurology Book*. Beijing, China: McGraw-Hill Education.

Brandt T, Paulus W, Straube A. 1986. Vision and posture. In: W Bles, T Brandt (eds). *Disorders of Posture and Gait*, pp. 157–175. Amsterdam: Elsevier.

Cavallari P, Bolzoni F, Burttinit C, Esposti R. 2016. The organization and control of intra-limb anticipatory postural adjustments and their role in movement performance. Front Hum Neurosci 10:525.

Cordo PJ, Nashner LM. 1982. Properties of postural adjustments associated with rapid arm movements. J Neurophysiol 47:287–302.

Darriot J, Mohsen J, Cullen K. 2015. Rapid adaptation of multisensory integration in vestibular pathways. Front Syst Neurosci 16:1–5.

De Havas J, Gomi H, Haggard P. 2017. Experimental investigations of control principles of involuntary movement: a comprehensive review of the Kohnstamm phenomenon. Exp Brain Res 235:1953–1997.

Dunbar DC, Horak FB, Macpherson JM, Rushmer DS. 1986. Neural control of quadrupedal and bipedal stance: implications for the evolution of erect posture. Am J Phys Anthropol 69:93–105.

Gera G, Fling BW, Van Ooteghem K, Cameron M, Frank JS, Horak FB. 2016. Postural motor learning deficits in people with MS in spatial but not temporal control of center of mass. Neurorehabil Neural Repair 30:722–730.

Gurfinkel VS, Levick YS. 1991. Perceptual and automatic aspects of the postural body scheme. In: J Paillard (ed). *Brain and Space*, pp. 147–162. Oxford: Oxford Univ. Press.

Hof AL, Curtze C. 2016. A stricter condition for standing balance after unexpected perturbations. J Biomech 49:580–585.

Horak FB, Diener HC. 1994. Cerebellar control of postural scaling and central set in stance. J Neurophysiol 72:479–493.

Horak FB, Nashner LM. 1986. Central programming of postural movements: adaptation to altered support-surface configurations. J Neurophysiol 55:1369–1381.

Horak FB, Nutt J, Nashner LM. 1992. Postural inflexibility in parkinsonian subjects. J Neurol Sci 111:46–58.

Inglis JT, Horak FB, Shupert CL, Jones-Rycewicz C. 1994. The importance of somatosensory information in triggering and scaling automatic postural responses in humans. Exp Brain Res 101:159–164.

Jacobs JV, Horak FB. 2007. Cortical control of postural responses. J Neural Transm 114:1339–1348.

Jahn K, Deutschländer A, Stephan T, Strupp M, Wiesmann M, Brandt T. 2004. Brain activation patterns during imagined stance and locomotion in functional magnetic resonance imaging. Neuroimage 22:1722–1731.

Lee WA, Michaels CF, Pai YC. 1990. The organization of torque and EMG activity during bilateral handle pulls by standing humans. Exp Brain Res 82:304–314.

MacKinnon CD, Winter DA. 1993. Control of whole body balance in the frontal plane during human walking. J Biomech 26:633–644.

Macpherson JM. 1988. Strategies that simplify the control of quadrupedal stance. 2. Electromyographic activity. J Neurophysiol 60:218–231.

Macpherson JM, Everaert DG, Stapley PJ, Ting LH. 2007. Bilateral vestibular loss in cats leads to active destabilization of balance during pitch and roll rotations of the support surface. J Neurophysiol 97:4357–4367.

Macpherson JM, Fung J. 1999. Weight support and balance during perturbed stance in the chronic spinal cat. J Neurophysiol 82:3066–3081.

Maki BE, McIlroy WE. 1997. The role of limb movements in maintaining upright stance: the "change-in-support" strategy. Phys Ther 77:488–507.

Maurer C, Mergner T, Peterka RJ. 2006. Multisensory control of human upright stance. Exp Brain Res 171:231–250.

McMahon TA, Bonner JT. 1983. *On Size and Life*. New York: W.H. Freeman.

Mittelstaedt H. 1998. Origin and processing of postural information. Neurosci Biobehav Rev 22:473–478.

Otten W. 1999. Balancing on a narrow ridge: biomechanics and control. Phil Trans R Soc Lond B 354:869–875.

Peterka RJ. 2002. Sensorimotor integration in human postural control. J Neurophysiol 88:1097–1118.

Peterson DS, Horak FB. 2016. Neural control of walking in people with Parkinsonism. Physiology 3:95–107.

Rousseaux M, Honore J, Saj A. 2014. Body representations and brain damage. Neurophysiol Clin 44:59–67.

Stapley PJ, Ting LH, Kuifu C, Everaert DG, Macpherson JM. 2006. Bilateral vestibular loss leads to active destabilization of balance during voluntary head turns in the standing cat. J Neurophysiol 95:3783–3797.

Takakusaki K. 2017. Functional neuroanatomy for posture and gait control. J Mov Disord 10:1–17.

Ting LH, Chiel HJ, Trumbower RD, et al. 2015. Neuromechanical principles underlying movement modularity and their implications for rehabilitation. Neuron 86:38–54.

37

The Cerebellum

THE CEREBELLUM CONSTITUTES ONLY 10% of the total volume of the brain but contains more than one-half of its neurons. The cerebellar cortex comprises a series of highly regular, repeating units, each of which contains the same basic microcircuit. Different regions of the cerebellum receive projections from distinct brain and spinal structures and then project back to the brain. The similarity of the architecture and physiology in all regions of the cerebellum implies that different regions of the cerebellum perform similar computational operations on different inputs.

The symptoms of cerebellar damage in humans and experimental animals provide compelling evidence that the cerebellum participates in the control of movement. The symptoms, in addition to being diagnostic for clinicians, thus help define the possible roles of the cerebellum in controlling behavior.

Several fundamental principles define our understanding of the physiological function of the cerebellum. First, the cerebellum acts in advance of sensory feedback arising from movement, thus providing feedforward control of muscular contractions. Second, to achieve such control, the cerebellum relies on internal models of the body to process and compare sensory inputs with copies of motor commands. Third, the cerebellum plays a special role in motor and perceptual timing. Fourth, the cerebellum is critical for adapting and learning motor skills. Finally, the primate cerebellum has extensive connectivity to nonmotor areas of the cerebral cortex, suggesting it performs similar functions in the performance and learning of motor and nonmotor behaviors.

Damage of the Cerebellum Causes Distinctive Symptoms and Signs

Damage Results in Characteristic Abnormalities of Movement and Posture

Disorders that involve the cerebellum typically disrupt normal movement patterns, demonstrating the cerebellum's critical role in movement. Patients describe a loss of the automatic, unconscious nature of most movements. In the early 20th century, Gordon Holmes recorded the self-report of a man with a lesion of his right cerebellar hemisphere: "movements of my left arm are done subconsciously, but I have to think out each movement of the right arm. I come to a dead stop in turning and have to think before I start again."

This has been interpreted as an interruption in the automatic level of processing by cerebellar inputs and outputs. With a malfunctioning cerebellum, it seems that the cerebral cortex needs to play a more active role in programming the details of motor actions. Importantly, individuals with cerebellar damage do not experience the paralysis that can be associated with cerebral cortical damage. Instead, they show characteristic abnormalities in voluntary movement, walking, and posture that have provided important clues about cerebellar function.

The most prominent symptom of cerebellar disorders is *ataxia*, or lack of coordination of movement. Ataxia is a generic term used to describe the collective motor features associated with cerebellar damage. People with cerebellar disorders make movements that qualitatively appear jerky, irregular, and highly variable. *Limb ataxia* during reaching is characterized by curved hand paths that are *dysmetric* in that they over- or undershoot the intended target and oscillate (Figure 37–1A). Patients often break a movement down into components, presumably in an effort to simplify control of multi-joint movements (*decomposition of movement*). Yet this may not be effective. For example, patients often have difficulty holding the shoulder steady while moving the elbow, a deficit thought to be due to poor predictions of how the movement at the elbow mechanically affects the shoulder (Figure 37–1B). If prediction fails, then patients are forced to try to steady the shoulder using time-delayed feedback, which is less effective.

At the end of reaching movements, there can be marked oscillation as the hand approaches the target. This *action* (or *intention*) *tremor* is the result of a series of erroneous, overshooting attempts to correct the movement. It largely disappears when the eyes are shut, suggesting that it is driven by time-delayed visual feedback of the movement. Finally, patients show abnormalities in the rate and regularity of repeated movements, a sign referred to as *dysdiadochokinesia* (Greek, impaired alternating movement) that can be readily demonstrated when a patient attempts to perform rapid alternating movements (Figure 37–1C).

People with cerebellar damage also exhibit *gait ataxia* and poor balance. When walking, they take steps that are irregularly timed and placed. They have difficulty shifting their weight from one foot to the other, which can lead to falling. The trunk oscillates when they are unsupported in sitting, standing, and during walking, particularly as they start, stop, or turn. A wide stepping pattern with feet spread apart is common and is thought to be a compensatory measure to improve stability.

Other signs that are commonly observed with cerebellar dysfunction can also occur with damage to other brain regions. People with cerebellar damage often have slurred speech with irregular timing (dysarthria); repetitive to-and-fro movements of the eyes with a slow and fast phase (nystagmus); and reduced resistance to passive limb displacements (hypotonia), which is thought to be related to so-called "pendular reflexes" often observed in cerebellar patients. In patients with cerebellar disease, the leg may oscillate like a pendulum many times after a knee jerk produced by a tap on the patellar tendon with a reflex hammer, instead of coming to rest immediately.

Damage Affects Specific Sensory and Cognitive Abilities

It is now known that cerebellar damage affects proprioceptive abilities (the sense of limb position and movement), but only during active movement. Proprioceptive acuity—the sense of the position and movement of the limbs—is normally more precise for active movements than for passive movements. Cerebellar patients show normal proprioceptive acuity when they have to judge which of two passive movements is larger. However, their proprioceptive acuity is worse than that of healthy individuals when they move a limb actively. One interpretation of these findings is that the cerebellum normally helps to predict how active movements will unfold, which would be important for movement coordination and for perceiving where the limbs are during active movements.

Damage to the cerebellum also affects cognitive processes, although these deficits are less obvious compared to the pronounced disturbances of sensory–motor function. Some of the earliest studies implicating the cerebellum in a range of cognitive tasks involved functional imaging to study the brain activity during behavior in healthy individuals. For example, in a study using positron emission tomography to image the brain activity of subjects during silent reading, reading aloud, and speech, areas of the cerebellum involved in the control of mouth movements were more active when subjects read aloud than when they read silently. Surprisingly, however, cerebellar activation was more pronounced in a task with greater cognitive load, when subjects were asked to name a verb associated with a noun; a subject might respond with "bark" if he or she saw the word "dog." Compared with simply reading aloud, the word-association task produced a pronounced increase in activity within the right lateral cerebellum. Consistent with this finding, a patient with damage in the right cerebellum could not learn a word-association task.

By now, many studies have revealed clear deficits in executive function, visual spatial cognition, language, and emotional processing after cerebellar damage. There appears to be some regional specificity within the cerebellum for different types of cognitive function. Damage to the midline cerebellum or *vermis* seems to be related to emotional or affective dysregulation, likely due to its interconnectivity with limbic structures. Damage to the right cerebellar hemisphere is related to language and verbal dysfunction, presumably because this hemisphere is interconnected with the left cerebral cortical hemisphere. Likewise, damage to the left cerebellar hemisphere is related to visuospatial dysfunction, probably because this hemisphere is interconnected with the right cerebral cortical hemisphere. Additionally, studies that examine cognitive dysfunction produce variable results; patients perform normally in one study but not another. Some studies show that cognitive deficits are most pronounced when patients are tested shortly after damage to the cerebellum and that compensations at the level of cerebral cortex might gradually make up for

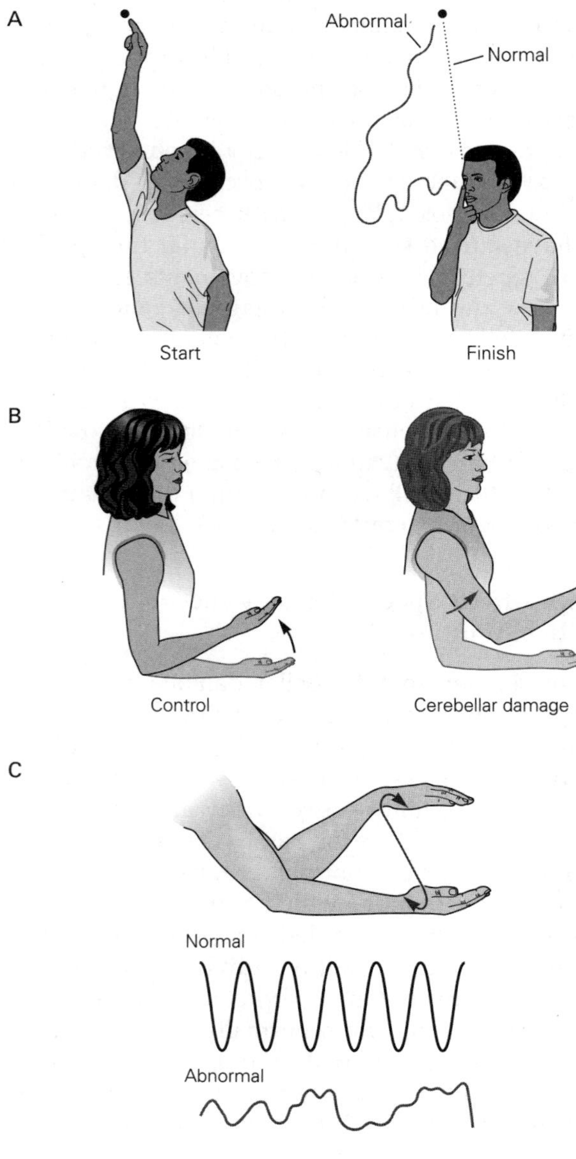

Figure 37–1 (Left) Typical defects observed in cerebellar diseases.

A. A cerebellar patient moving his arm from a raised position to touch the tip of his nose exhibits inaccuracy in range and direction (dysmetria) and moves his shoulder and elbow separately (decomposition of movement). Tremor increases as the finger approaches the nose.

B. Failure of compensation for interaction torques can account for cerebellar ataxia. Subjects flex their elbows while keeping their shoulder stable. In both the control subject and the cerebellar patient, the net elbow torque is large because the elbow is moved. In the control subject, there is relatively little net shoulder torque because the interaction torques are automatically cancelled by muscle torques. In the cerebellar patient, this compensation fails; the muscle torques are present but are inappropriate to cancel the interaction torques. As a result, the patient cannot flex her elbow without causing a large perturbation of her shoulder position. (Adapted, with permission, from Bastian, Zackowski, and Thach 2000.)

C. A subject was asked to alternately pronate and supinate the forearm while flexing and extending at the elbow as rapidly as possible. Position traces of the hand and forearm show the normal pattern of alternating movements and the irregular pattern (dysdiadochokinesia) typical of cerebellar disorder.

cerebellar loss of function. However, cognitive deficits may be more robust and long lasting when cerebellar damage is acquired in childhood.

Thus, cognitive deficits arising from cerebellar damage sometimes can be difficult to characterize. What is clear is that the motor dysfunction after cerebellar loss is more obvious than cognitive dysfunction. It may be that cortical regions of motor control are less able to compensate for losses of cerebellar motor control compared to cortical compensation for impairment of cerebellar computations involved in cognitive processes.

The Cerebellum Indirectly Controls Movement Through Other Brain Structures

Understanding the anatomy of the cerebellum and how it interacts with different brain structures is vital to understanding its function. In this section, we consider the general anatomy of the cerebellum as well as its inputs and outputs.

The Cerebellum Is a Large Subcortical Brain Structure

The cerebellum occupies most of the posterior cranial fossa. It is composed of an outer mantle of gray matter (the cerebellar cortex), internal white matter, and three pairs of deep nuclei: the fastigial nucleus, the interposed nucleus (itself composed of the emboliform and globose nuclei), and the dentate nucleus (Figure 37–2A). The surface of the cerebellum is highly convoluted, with many parallel folds or folia (Latin, leaves).

Two deep transverse fissures divide the cerebellum into three lobes. The primary fissure on the dorsal surface separates the anterior and posterior lobes, which together form the body of the cerebellum (Figure 37–2A). The posterolateral fissure on the ventral surface separates the body of the cerebellum from the smaller flocculonodular lobe (Figure 37–2B). Each lobe extends across the cerebellum from the midline to the most lateral tip. In the orthogonal, anterior-posterior direction, two longitudinal furrows separate three regions: the midline vermis (Latin, worm) and the two cerebellar hemispheres, each split into intermediate and lateral regions (Figure 37–2D).

The cerebellum is connected to the dorsal aspect of the brain stem by three symmetrical pairs of peduncles: the inferior cerebellar peduncle (also called the restiform body), the middle cerebellar peduncle (or brachium pontis), and the superior cerebellar peduncle (or brachium conjunctivum). Most of the output axons of the cerebellum arise from the deep nuclei and project through the superior cerebellar peduncle to other brain areas. The main exception is a group of Purkinje cells in the flocculonodular lobe that project to vestibular nuclei in the brain stem.

The Cerebellum Connects With the Cerebral Cortex Through Recurrent Loops

Many parts of the cerebellum form recurrent loops with the cerebral cortex. The cerebral cortex projects to the lateral cerebellum through relays in the pontine nuclei. In turn, the lateral cerebellum projects back to the cerebral cortex through relays in the thalamus. Peter Strick and his colleagues used viruses for transneuronal tracing in nonhuman primates to show that this recurrent circuit is organized as a series of parallel closed loops, where a given part of the cerebellum connects reciprocally with a specific part of the cerebral cortex (Figure 37–3A). Through these reciprocal connections, the cerebellum interacts with vast regions of the neocortex, including substantial connections to motor, prefrontal, and posterior parietal regions. More recently, Strick's group also demonstrated disynaptic connections between the cerebellum and basal ganglia in nonhuman primates.

The resting state connectivity between the cerebellum and cerebral cortex in humans was studied using fMRI scans of 1,000 subjects. Correlations in activity in different regions of the brain were assessed at low frequencies, measured by blood flow while subjects were at rest. They found that different regions of the cerebellum are functionally connected with cerebral cortical regions across the entire cerebral cortex (Figure 37–3C). Taken together, these studies demonstrate the vast impact the cerebellum could have on many aspects of brain function.

Different Movements Are Controlled by Functional Longitudinal Zones

The cerebellum can be broadly divided into three areas that have distinctive roles in different kinds of movements: the vestibulocerebellum, spinocerebellum, and cerebrocerebellum (Figure 37–4).

The *vestibulocerebellum* consists of the flocculonodular lobe and is the most primitive part of the cerebellum. It receives vestibular and visual inputs, projects to the vestibular nuclei in the brain stem, and participates in balance, other vestibular reflexes, and eye movements. It receives information from the semicircular canals and the otolith organs, which sense the head's motion and its position relative to gravity. Most of this

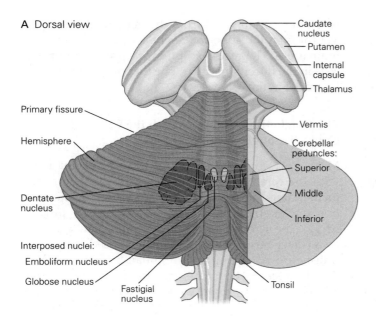

A   Dorsal view

Caudate nucleus
Putamen
Internal capsule
Thalamus
Primary fissure
Hemisphere
Vermis
Cerebellar peduncles:
Superior
Middle
Inferior
Dentate nucleus
Interposed nuclei:
Emboliform nucleus
Globose nucleus
Fastigial nucleus
Tonsil

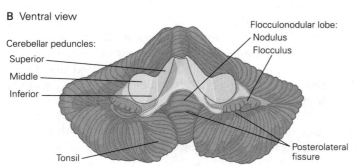

B   Ventral view

Cerebellar peduncles:
Superior
Middle
Inferior
Flocculonodular lobe:
Nodulus
Flocculus
Tonsil
Posterolateral fissure

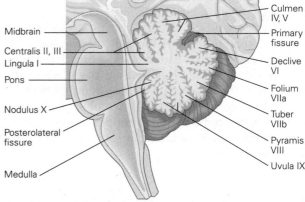

C   Midsagittal section

Midbrain
Centralis II, III
Lingula I
Pons
Nodulus X
Posterolateral fissure
Medulla
Culmen IV, V
Primary fissure
Declive VI
Folium VIIa
Tuber VIIb
Pyramis VIII
Uvula IX

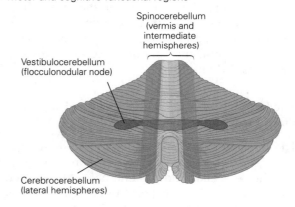

D   Motor and cognitive functional regions

Spinocerebellum (vermis and intermediate hemispheres)
Vestibulocerebellum (flocculonodular node)
Cerebrocerebellum (lateral hemispheres)

Figure 37–2 Gross features of the cerebellum. (Adapted, with permission, from Nieuwenhuys, Voogd, and van Huijzen 1988.)

A. Part of the right hemisphere has been cut away to reveal the underlying cerebellar peduncles.

B. The cerebellum is shown detached from the brain stem.

C. A midsagittal section through the brain stem and cerebellum shows the branching structure of the cerebellum. The cerebellar lobules are labeled with their Latin names and Larsell's Roman numeral designations. (Reproduced, with permission, from Larsell and Jansen 1972.)

D. Functional regions of the cerebellum.

vestibular input arises from the vestibular nuclei in the brain stem. The vestibulocerebellum also receives visual input, from both the pretectal nuclei that lie deep in the midbrain beneath the superior colliculus and the primary and secondary visual cortex through the pontine and pretectal nuclei.

The vestibulocerebellum is unique in that its output bypasses the deep cerebellar nuclei and proceeds directly to the vestibular nuclei in the brain stem. Purkinje cells in the midline parts of the vestibulocerebellum project to the lateral vestibular nucleus to modulate the lateral and medial vestibulospinal tracts, which predominantly control axial muscles and limb extensors to assure balance

during stance and gait (Figure 37–5A). Disruption of these projections through lesions or disease impairs equilibrium.

The most striking deficits following lesions of the lateral vestibulocerebellum are in smooth-pursuit eye movement toward the side of the lesion. A patient with a lesion of the left lateral vestibulocerebellum can smoothly track a target that is moving to the right, but only poorly tracks motion to the left, using saccades predominantly (Figure 37–6A). These patients can have normal vestibulo-ocular reflex responses to head rotations but cannot suppress the reflex by fixating an object that rotates with the head (Figure 37–6B). These deficits occur commonly if the lateral

A Cortical-cerebellar circuit

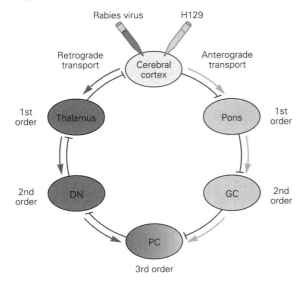

B Cortical-cerebellar connections in the monkey

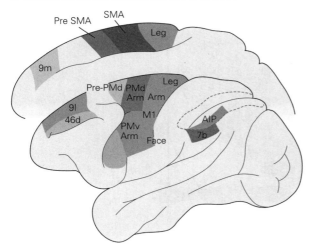

C Cortical-cerebellar connections in the human

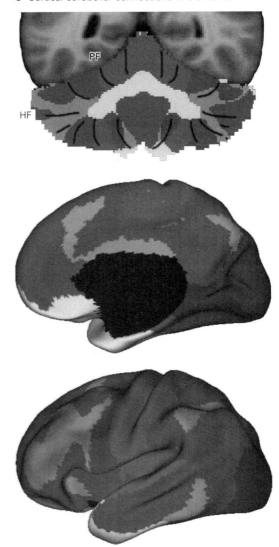

Figure 37–3 The cerebellum connects to many areas of cerebral cortex. (Parts A and B adapted, with permission, from Bostan, Dum, and Strick 2013. Copyright © 2013 Elsevier Ltd. part C adapted, with permission, from Buckner et al. 2011. Copyright © 2011 American Physiological Society.)

A. The cortical-cerebellar circuit in monkeys was traced with fluorescence-labeled transsynaptic viruses that can move in an anterograde or retrograde direction. Injection into the cerebral cortex of a retrograde virus, such as rabies virus, will label neurons that project to it and, by crossing synapses, can label second- and possibly higher-order neurons in a pathway. These are shown here in **red** as first-order (thalamus), second-order (deep nucleus), and third-order neurons (Purkinje cells). Injection into the cerebral cortex of an anterograde virus, such as the H129 strain of herpes simplex virus, will label neurons that are targets of the cerebral cortex. These are shown here in **yellow** as first-order (pons), second-order (granule cells), and third-order neurons (Purkinje cells). (Abbreviations: **DN**, dentate

nuclei; **GC**, granule cell; **H129**, strain of herpes simplex virus; **PC**, Purkinje cell rabies virus.)

B. Areas of the cerebral cortex connected to the cerebellum. The numbers refer to cytoarchitectonic areas. (Abbreviations: **AIP**, anterior intraparietal area; **M1**, face, arm, and leg areas of the primary motor cortex; **PMd arm**, arm area of the dorsal premotor area; **PMv arm**, arm area of the ventral premotor area; **PrePMd**, predorsal premotor area; **PreSMA**, presupplementary motor area; **SMA arm**, arm area of the supplementary motor area.)

C. Color-coded coronal section of the human cerebellum (**top**) and lateral and medial views of the human cerebral cortex (**bottom**) created from resting state functional connectivity maps (based on functional magnetic resonance imaging scans of 1,000 subjects). Colors correspond to cerebellar and cerebral areas that are connected. Note that the cerebellum is functionally connected with nearly all cerebral areas. (Abbreviations: **HF**, horizontal fissure; **PF**, primary fissure.)

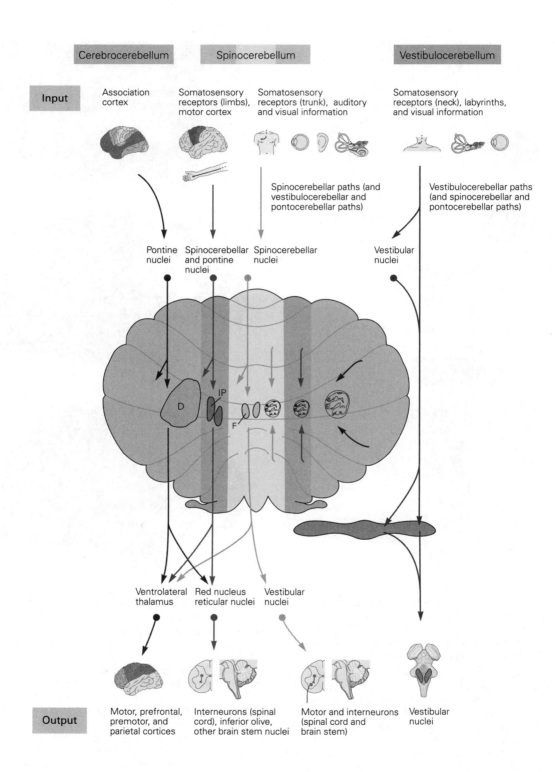

Figure 37–4 The three functional regions of the cerebellum have **different inputs and different output targets.** The cerebellum is shown unfolded, and **arrows** indicate the inputs and outputs of the different functional areas. The body maps in the deep nuclei are based on anatomical tracing and single-cell recordings in nonhuman primates. (Abbreviations: **D,** dentate nucleus; **F,** fastigial nucleus; **IP,** interposed nucleus.) (Adapted, with permission, from Brooks and Thach 1981).

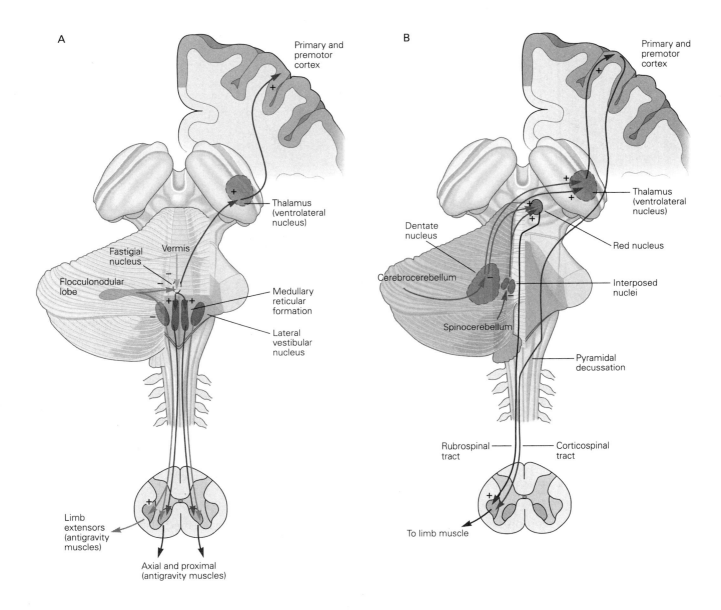

Figure 37–5 Input and output pathways of the cerebellum.

A. Nuclei in the vestibulocerebellum and the vermis control proximal muscles and limb extensors. The vestibulocerebellum (flocculonodular lobe) receives input from the vestibular labyrinth and projects directly to the vestibular nuclei. The vermis receives input from the neck and trunk, the vestibular labyrinth, and retinal and extraocular muscles. Its output is focused on the ventromedial descending systems of the brain stem, mainly the reticulospinal and vestibulospinal tracts and the corticospinal fibers acting on medial motor neurons. The

oculomotor connections of the vestibular nuclei have been omitted for clarity.

B. Nuclei in the intermediate and lateral parts of the cerebellar hemispheres control limb and axial muscles. The intermediate part of each hemisphere (spinocerebellum) receives sensory information from the limbs and controls the dorsolateral descending systems (rubrospinal and corticospinal tracts) acting on the ipsilateral limbs. The lateral area of each hemisphere (cerebrocerebellum) receives cortical input via the pontine nuclei and influences the motor and premotor cortices via the ventrolateral nucleus of the thalamus, and directly influences the red nucleus.

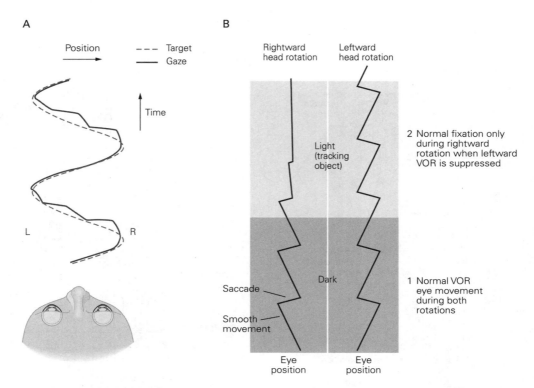

Figure 37–6 Lesions in the vestibulocerebellum have large effects on smooth-pursuit eye movements.

A. Sinusoidal target motion is tracked with smooth-pursuit eye movements as the target moves from left (**L**) to right (**R**). With a lesion of the left vestibulocerebellum, smooth pursuit is punctuated by saccades when the target moves from right to left.

B. In the same patient, responses to vestibular stimulation are normal, whereas object fixation is disrupted during leftward rotation. The traces on the left and right show the eye movements evoked by rightward and leftward head rotation experienced in separate sessions. In each session the patient sat in a chair that rotated continuously in one direction, first in the dark then in the light while fixating on a target that moves along with him. (1) In the dark, the eyes show a normal vestibulo-ocular reflex (**VOR**) during rotation in both directions: The eyes move smoothly in the direction opposite to the head's rotation, then reset with saccades in the direction of head rotation. (2) In the light, the eye position during rightward head rotation is normal: Fixation on the target is excellent and the vestibulo-ocular reflex is suppressed. During leftward head rotation, however, the subject is unable to fixate on the object and the vestibulo-ocular reflex cannot be suppressed.

vestibulocerebellum is compressed by an acoustic neuroma, a benign tumor that grows on the eighth cranial nerve as it courses directly beneath the lateral vestibulocerebellum.

The *spinocerebellum* is composed of the vermis and intermediate parts of the cerebellar hemispheres (Figure 37–4). It is so named because it receives extensive input from the spinal cord via the dorsal and ventral spinocerebellar tracts. These pathways convey information about touch, pressure, and limb position as well as the spiking activity of spinal interneurons. Thus, these inputs provide the cerebellum with varied information about the changing state of the organism and its environment.

The vermis receives visual, auditory, and vestibular input as well as somatic sensory input from the head and proximal parts of the body. It projects by way of the fastigial nucleus to cortical and brain stem regions that give rise to the medial descending systems controlling proximal muscles of the body and limbs (Figure 37–5A). The vermis governs posture and locomotion as well as eye movements. For example, lesions of the oculomotor region of the vermis cause saccadic eye movements that overshoot their target, much as patients with cerebellar damage make arm movements that overshoot their target.

The adjacent intermediate parts of the hemispheres also receive somatosensory input from the limbs. Neurons here project to the interposed nucleus, which provides inputs to lateral corticospinal and rubrospinal systems on the contralateral side of the brain and controls the more distal muscles of the limbs and digits (Figure 37–5B). Because corticospinal and rubrospinal systems cross the midline as they descend to the spinal cord, cerebellar lesions disrupt ipsilateral limb movements.

The *cerebrocerebellum* comprises the lateral parts of the hemispheres (Figure 37–4). These areas are phylogenetically the most recent and are much larger relative to the rest of the cerebellum in humans and apes than in monkeys and cats. Almost all of the inputs to and outputs from this region involve connections with the cerebral cortex. The output is transmitted through the dentate nucleus, which projects via the thalamus to contralateral motor, premotor, parietal, and prefrontal cortices. The dentate nucleus also projects to the contralateral red nucleus. The lateral hemispheres have many functions but seem to participate most extensively in planning and executing movement. They also have a role in cognitive functions unconnected with motor planning, such as visuospatial and language processes. There is now some correlative evidence implicating the cerebellar hemispheres in aspects of schizophrenia (Chapter 60), dystonia (Chapter 38), and autism (Chapter 62).

Two important principles of cerebellar function have emerged from recordings of the action potentials of single neurons in the cerebellar cortex and deep cerebellar nuclei during arm movements, along with controlled, temporary inactivation of specific cerebellar regions.

First, neurons in these areas discharge vigorously in relation to voluntary movements. Cerebellar output is related to the direction and speed of movement. The deep nuclei are organized into somatotopic maps of different limbs and joints, as in the motor cortex, although the organization of the cerebellar cortex has been characterized as "fractured somatotopy" with multiple disconnected and partial maps. Moreover, the interval between the onset of modulation of the firing of cerebellar neurons and movement is remarkably similar to that for neurons in the motor cortex. This result emphasizes the cerebellum's participation in recurrent circuits that operate synchronously with the cerebral cortex.

Second, the cerebellum provides feedforward control of muscle contractions to regulate the timing of movements. Rather than awaiting sensory feedback, cerebellar output anticipates the muscular contractions that will be needed to bring a movement smoothly, accurately, and quickly to its desired endpoint. Failure of these mechanisms causes the intention tremor of cerebellar disorders. For example, a rapid single-joint movement is initiated by the contraction of an agonist muscle and terminated by an appropriately timed contraction of the antagonist. The contraction of the antagonist starts early in the movement, well before there has been time for sensory feedback to reach the brain, and therefore must be programmed as part of

the movement. When the dentate and interposed nuclei are experimentally inactivated, however, contraction of the antagonist muscle is delayed until the limb has overshot its target. The programmed anticipatory contraction of the antagonist in normal movements is replaced by a correction driven by sensory feedback. This correction is itself dysmetric and results in another error, necessitating a new adjustment (Figure 37–7).

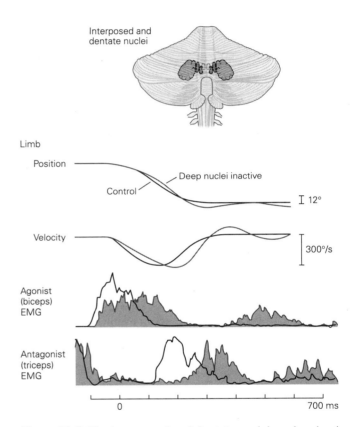

Figure 37–7 The interposed and dentate nuclei are involved in the precise timing of agonist and antagonist activation during rapid movements. The interposed (medial) and dentate (lateral) nuclei are highlighted in the drawing of the cerebellum. The records of limb movement show how a monkey normally makes a rapid elbow flexion limb movement and attempts to make the same movement when the interposed and dentate nuclei are inactivated by cooling. The electromyographic (**EMG**) traces show limb position and velocity and EMG responses of the biceps and triceps muscles. When the deep nuclei are inactivated, activation of the agonist (biceps) becomes slower and more prolonged. Activation of the antagonist (triceps), which is needed to stop the movement at the correct location, is likewise delayed and protracted so that the initial movement overshoots its appropriate extent. Delays in successive phases of the movement produce oscillations similar to the terminal tremor seen in patients with cerebellar damage.

The Cerebellar Cortex Comprises Repeating Functional Units Having the Same Basic Microcircuit

The cellular organization of the microcircuit in the cerebellar cortex is striking, and one of the premises of cerebellar research has been that the details of the microcircuit are an important clue to how the cerebellum works. In this section, we describe three major features of the microcircuit.

The Cerebellar Cortex Is Organized Into Three Functionally Specialized Layers

The three layers of the cerebellar cortex contain distinct kinds of neurons and are functionally specialized (Figure 37–8).

The deepest, or *granular layer*, is the input layer. It contains a vast number of granule cells, estimated at 100 billion, which appear in histological sections as small, densely packed, darkly stained nuclei. The granular layer also contains a few larger Golgi cells and, in some cerebellar regions, a smattering of other neurons such as cells of Lugaro, unipolar brush cells, and chandelier cells. The mossy fibers, one of the two principal afferent inputs to the cerebellum, terminate in this layer. The bulbous terminals of the mossy fibers excite granule cells and Golgi neurons in synaptic complexes called *cerebellar glomeruli* (Figure 37–8). As we will see later when discussing recurrent circuits in the cerebellum, Golgi cells inhibit granule cells.

The middle or *Purkinje cell layer* is the output layer of the cerebellar cortex. This layer consists of a single sheet of Purkinje cell bodies, each 50 to 80 μm in diameter. The fan-like dendritic trees of Purkinje cells extend upward into the molecular layer where they receive inputs from the second major type of afferent to the cerebellum, the climbing fibers, as well as from granule cells and inhibitory interneurons. Purkinje cell axons conduct the entire output of the cerebellar cortex, projecting to the deep nuclei in the underlying white matter or to the vestibular nuclei in the brain stem, where they release the inhibitory transmitter GABA (γ-aminobutyric acid).

The outermost or *molecular layer* contains the spatially polarized dendrites of Purkinje cells, which extend approximately 1 to 3 mm in the anterior-posterior direction but occupy only a very narrow territory in the medial-lateral direction. The molecular layer contains the cell bodies and dendrites of two types of "molecular layer interneurons," the stellate and basket cells, both of which inhibit Purkinje cells. It also contains the axons of the granule cells, called the *parallel fibers* because they run parallel to the long axis of the folia (Figure 37–8). Parallel fibers run perpendicular to the dendritic trees of the Purkinje cells and thus have the potential to form a few synapses with each of a large number of Purkinje cells.

The Climbing-Fiber and Mossy-Fiber Afferent Systems Encode and Process Information Differently

The two main types of afferent fibers in the cerebellum, the mossy fibers and climbing fibers, probably mediate different functions. Both form excitatory synapses with neurons in the deep cerebellar nuclei and in the cerebellar cortex. However, they terminate in different layers of the cerebellar cortex, affect Purkinje cells through very different patterns of synaptic convergence and divergence, and produce different electrical events in the Purkinje cells.

Climbing fibers originate in the inferior olivary nucleus in the brain stem and convey sensory information to the cerebellum from both the periphery and the cerebral cortex. The climbing fiber is so named because each one wraps around the proximal dendrites of a Purkinje neuron like a vine on a tree, making numerous synaptic contacts (Figure 37–9). Each Purkinje neuron receives synaptic input from only a single climbing fiber, but each climbing fiber contacts 1 to 10 Purkinje cells that are arranged topographically along a parasagittal strip in the cerebellar cortex. Indeed, the axons from clusters of related olivary neurons terminate in thin parasagittal strips that extend across several folia, and the Purkinje cells from one strip converge on a common group of neurons in the deep nuclei.

Climbing fibers have an unusually powerful influence on the electrical activity of Purkinje cells. Each action potential in a climbing fiber generates a protracted, voltage-gated Ca^{2+} conductance in the soma and dendrites of the postsynaptic Purkinje cell. This results in prolonged depolarization that produces an electrical event called a "complex spike": an initial large-amplitude action potential followed by a high-frequency burst of smaller-amplitude action potentials (Figure 37–9). Whether these smaller spikes are transmitted down the Purkinje cell's axon is not clear. In awake animals, complex spikes occur spontaneously at low rates, usually around one per second. Specific sensory or motor events cause one or two complex spikes that occur at precise times in relation to those events.

Mossy fibers originate from cell bodies in the spinal cord and brain stem. They carry sensory information from the periphery as well as both sensory information and corollary discharges that report the current movement command (Chapter 30) from the cerebral cortex

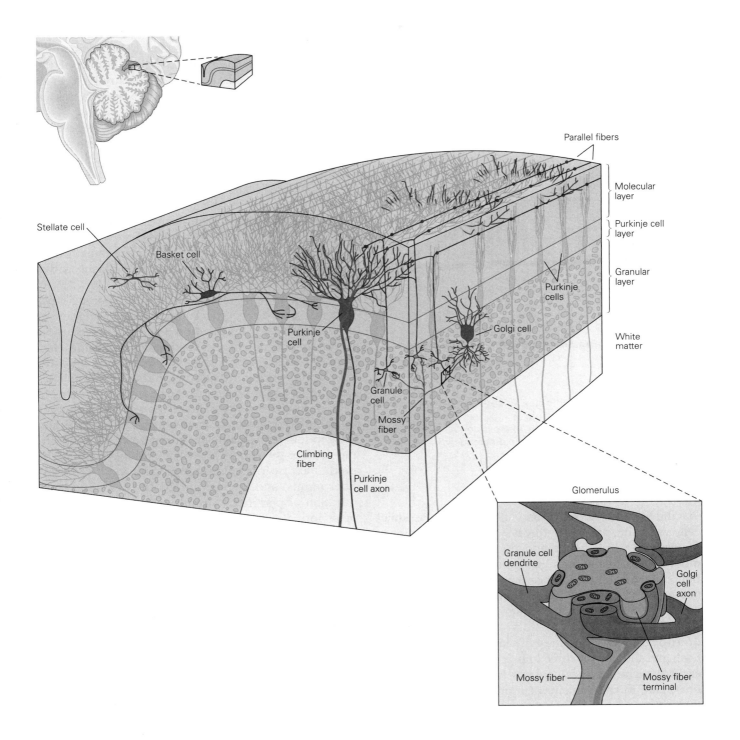

Figure 37–8 The cerebellar cortex contains five main types of neurons organized into three layers. A vertical section of a single cerebellar folium illustrates the general organization of the cerebellar cortex. The detail of a cerebellar glomerulus in the granular layer is also shown. A glomerulus is the synaptic complex formed by the bulbous axon terminal of a mossy fiber and the dendrites of several Golgi and granule cells. Mitochondria are present in all of the structures in the glomerulus, consistent with their high metabolic activity.

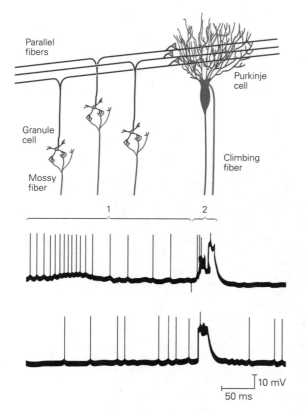

Parallel
fibers

Purkinje
cell

Granule
cell

Climbing
fiber

Mossy
fiber

1   2

10 mV

50 ms

Figure 37–9 Simple and complex spikes recorded intracellularly from a cerebellar Purkinje cell. Simple spikes are produced by mossy-fiber input (**1**), whereas complex spikes are evoked by climbing-fiber synapses (**2**). (Reproduced, with permission, from Martinez, Crill, and Kennedy 1971.)

via the pontine nuclei. Mossy fibers affect Purkinje cells via multisynaptic pathways that have intriguing patterns of convergence and divergence. Individual mossy fibers, acting through granule cells and parallel fibers, have a tiny influence on Purkinje cell output, but collectively, the whole population of mossy fibers has massive effects on cerebellar output.

Mossy fibers form excitatory synapses on the dendrites of granule cells in the granular layer (Figure 37–8). Each granule cell has three to five short dendrites, and each dendrite receives contacts from a single mossy fiber. Due to this paucity of inputs, the spatial integration by a granule cell of its different mossy fiber synapses is not extensive; however, the cell can be the site of convergence of mossy fibers from multiple sensory modalities and motor corollary discharge. The next synaptic relay, between the granule cell axons and Purkinje cells, distributes information with very wide divergence and convergence. The parallel fibers allow each mossy fiber to influence a large number

of Purkinje cells, and each Purkinje cell is contacted potentially by axons from somewhere between 200,000 and 1 million granule cells. Importantly, in response to changing conditions there seems to be tremendous potential for adaptation of cerebellar output at the synapses between parallel fibers and Purkinje cells. It appears that only a small fraction of these synapses are active at any given time.

Parallel fibers produce brief, small excitatory potentials in Purkinje cells (Figure 37–9). These potentials converge in the cell body and spread to the initial segment of the axon where they generate conventional action potentials called "simple spikes" that propagate down the axon. In awake animals, Purkinje cells emit a steady stream of simple spikes, with spontaneous firing rates as high as 100 per second even when an animal is sitting quietly. Purkinje cells fire at rates as high as several hundred spikes per second during active eye, arm, and face movements.

The climbing-fiber and mossy-fiber/parallel-fiber systems seem to be specialized for transmission of different kinds of information. Climbing fibers cause complex spikes that seem specialized for event detection. Although complex spikes occur only infrequently, synchronous firing in multiple climbing fibers enables them to signal important events. Synchrony seems to arise partly because signaling between many neurons in the inferior olivary nucleus occurs electrotonically (at gap-junction channels). In contrast, the high firing rates of the simple spikes in Purkinje cells can be modulated up or down in a graded way by mossy-fiber inputs, and thereby encode the magnitude and duration of peripheral stimuli or centrally generated behaviors.

The Cerebellar Microcircuit Architecture Suggests a Canonical Computation

The cerebellar microcircuit is replicated many times across the surface of the cerebellar cortex. This repeating architecture and pattern of convergence and divergence has led to the suggestion that since every such module has the same architecture and pattern of convergence and divergence, the cerebellar cortex performs the same basic "canonical" computation on all of its inputs, and that it potentially transforms cerebellar inputs in a similar way for all cerebellar output systems. Inspection of a diagram of the cerebellar microcircuit (Figure 37–10) reveals a number of different computational components. One general feature is the existence of parallel excitatory and inhibitory pathways to the Purkinje cells or deep cerebellar nuclei. The other general feature is the prevalence of recurrent loops.

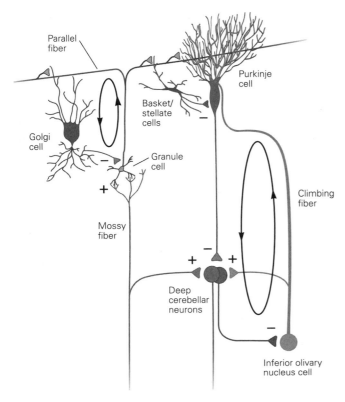

Figure 37–10 Synaptic organization of the cerebellar microcircuit. Excitation and inhibition converge both in the cerebellar cortex and in the deep nuclei. Recurrent loops involve Golgi cells within the cerebellar cortex and the inferior olive outside the cerebellum. (Adapted, with permission, from Raymond, Lisberger, and Mauk 1996. Copyright © 1996 AAAS.)

Parallel Feedforward Excitatory and Inhibitory Pathways

The excitatory inputs relayed from mossy fibers to granule cells to Purkinje cells work in parallel with feedforward inhibitory inputs through the two molecular layer interneurons, the stellate and basket cells. Both of these interneurons receive inputs from parallel fibers and inhibit Purkinje cells, but they have quite different architectures.

The short axons of stellate cells contact the nearby dendrites of Purkinje cells. Thus, a stellate cell acts locally in the sense that it and the Purkinje cell it inhibits are excited by the same parallel fibers. In contrast, a basket cell acts more widely. Its axon runs perpendicular to the parallel fibers (Figure 37–8) and creates flanks of inhibition on Purkinje cells that receive input from parallel fibers other than those that excite the basket cell. Stellate cells affect Purkinje cells via synapses that are on distal dendrites, whereas basket cells make powerful synapses on the cell body of Purkinje cells and seem to be positioned for a powerful influence on

Purkinje cell simple spiking. Remarkably, even 60 years after the architecture of the cerebellar microcircuit was described, the functional role of molecular layer interneurons remains a mystery.

Convergence of excitatory and inhibitory pathways is a predominant feature also in the deep cerebellar nuclei. Here, inhibitory inputs from Purkinje cells converge with excitatory inputs from axon collaterals of mossy and climbing fibers (Figure 37–10). Thus, a mossy fiber affects target neurons in the deep nuclei in two ways: directly by excitatory synapses and indirectly by pathways through the cerebellar cortex and the inhibitory Purkinje cells. Neurons of the deep cerebellar nuclei are active spontaneously even in the absence of synaptic inputs, so the inhibitory output of the Purkinje cells both modulates this intrinsic activity and sculpts the excitatory signals transmitted from mossy fibers to the deep nuclei. In almost all parts of the cerebellum, collaterals from climbing fibers to the deep cerebellar nuclei create the opportunity for a similar interaction of excitatory and inhibitory inputs.

Recurrent Loops

An important recurrent loop is contained entirely within the cerebellar cortex and employs Golgi cells to sculpt the activity of the granule cells, the input elements in the cerebellar cortex. Golgi cells receive a few large excitatory inputs from mossy fibers, many smaller excitatory inputs from parallel fibers, and inhibitory inputs from neighboring Golgi cells. The GABAergic terminals from Golgi cells inhibit granule cells (Figure 37–10) and thereby regulate the activity of granule cells and the signals conveyed by the parallel fibers. This loop is evidence that important processing may occur within the granular layer. It may shorten the duration of bursts in granule cells, limiting the magnitude of the excitatory response of granule cells to their mossy fiber inputs, or could ensure that the granule cells respond only when a certain number of their mossy fiber inputs are active.

A second recurrent loop provides Purkinje cells with a way to regulate their own climbing fiber inputs (Figure 37–10). Purkinje cells inhibit GABAergic inhibitory neurons in the deep cerebellar nuclei that project to the inferior olive. When the simple-spike firing of a group of Purkinje cells decreases, the activity of these inhibitory interneurons increases, leading to decreases in the excitability of neurons in the inferior olive. The decreased excitability of the inferior olive reduces both the probability of action potentials in climbing fibers that project to the original group of Purkinje cells and the duration of each burst of climbing fiber action potentials. In the section on cerebellar learning, we will see how this recurrent loop could allow the cerebellar cortex to control the inputs that cause adaptive changes in the synapses on its Purkinje cells.

The Cerebellum Is Hypothesized to Perform Several General Computational Functions

We know that the cerebellum is important for motor control and some nonmotor functions. Even though we do not yet know how the cerebellar circuit controls these functions, we are able to identity aspects of the control that seem to be particularly "cerebellar." These include reliable feedforward control, internal control of timing, integration of sensory inputs with corollary discharge, and state estimation through internal models.

The Cerebellum Contributes to Feedforward Sensorimotor Control

Sensory feedback is by its nature delayed. Therefore, when a movement is initiated there is a period of time before any useful sensory feedback is received about the movement. We saw earlier that cerebellar damage causes movement disorders that appear to result from out-of-date sensory feedback. If so, it is reasonable to assume that the cerebellum regulates and coordinates movement by preprogramming and coordinating commands for muscular contraction prior to the arrival of useful sensory feedback. The cerebellar output anticipates the muscular contractions that will be needed to bring a movement smoothly, accurately, and quickly to the desired endpoint, and uses sensory feedback mainly to monitor and improve its own performance.

Like neurons in the motor cortex, cerebellar neurons are activated before movement. Still, lesion studies and the symptoms in human motor disorders imply that the cerebellum and motor cortex play very different roles in movement. Lesions of the cerebellum disrupt the accuracy and coordination of voluntary movement, while lesions of the cerebral cortex largely prevent movement.

In addition, the pattern of cerebellar activity, not simply the rate of activity, conveys information for movement control. This is illustrated in mouse models of cerebellar disease. Deletion of certain ion channels produces excessive variability of Purkinje cell simple-spike firing patterns, which seems to lead to ataxia. This suggests that the regularity of cerebellar activity must be closely regulated to achieve normal movement.

The Cerebellum Incorporates an Internal Model of the Motor Apparatus

To program the correct muscle contractions for a smooth, accurate arm movement, the cerebellum needs to have some information about the physical configuration of the arm. Thus, it needs to create and maintain what are called "internal models" of the motor apparatus (Chapter 30). Internal models allow the cerebellum to perform a computation that helps the brain make good estimates of the exact muscle forces needed to move an arm in a desired manner.

An accurate *inverse dynamic* model of the arm, for example, can process sensory data about the current posture of the arm and automatically generate a sequence of properly timed and scaled commands to move the hand to a new desired position. An accurate *forward dynamic* model does the opposite: It processes a copy of a motor command and makes a prediction about the upcoming kinematics (ie, position and speed) of the arm movement. Recordings of the output of the cerebellum have provided evidence compatible with the idea that the cerebellum contains both types of models and that they are used to program both arm and eye movements.

One reason that the cerebellum may need these types of models for motor control is because of the complexities associated with moving linked segments of the body. Consider the mechanics of making a simple arm movement. Because of the mechanics of the arm and the momentum it develops when moving, movement of the forearm alone causes inertial forces that passively move the upper arm. If a subject wants to flex or extend the elbow without simultaneously moving the shoulder, then muscles acting at the shoulder must contract to prevent its movement. These stabilizing contractions of the shoulder joint occur almost perfectly in healthy subjects but not in patients with cerebellar damage, who experience difficulty controlling the inertial interactions among multiple segments of a limb (Figure 37–1B). As a result, patients exhibit greater inaccuracy of multi-joint versus single-joint movements.

In conclusion, the cerebellum uses internal models to allow it to preprogram a sequence of muscle contractions that will generate smooth, accurate movement. It also anticipates the forces that result from the mechanical properties of a moving limb. We do not yet know what these internal models look like in terms of the activity of cerebellar neurons, the circuits that operate as internal models, or how the cerebellar output is transformed into muscle forces. However, given that the properties of the limbs change throughout life, we can be confident that the cerebellum's learning capabilities are involved in adapting these internal models to help generate the most proficient movements.

The Cerebellum Integrates Sensory Inputs and Corollary Discharge

Sensory signals converge in the cerebellum with motor signals that are called a corollary discharge (or efference copy) because they report commands that are being sent to motor nerves at the same time. For example, some neurons in the dorsal spinocerebellar tract relay inputs from sensory afferents in the spinal cord and transmit sensory signals to the cerebellum. In contrast, the neurons in the spinal cord that give rise to the axons in the ventral spinocerebellar tract receive the same afferent and descending inputs as do spinal motor neurons, and they transmit the final motor command back to the cerebellum. The interaction of sensory signals and corollary discharge allows comparison of the plans for a movement with the sensory consequences. This comparison occurs to some degree at Purkinje cells, but we now know that at least some granule cells receive converging sensory and corollary discharge inputs and could perform the comparison.

Internal models and corollary discharge together provide one possible explanation of the role of the cerebellum in movement. To be able to program accurate movements the cerebellum must be able to estimate the state of the motor system through sensory feedback and knowledge of prior motor activity. Next, it must combine information on the state of the motor system with the goals of the next movement and use internal models of the effector to help create commands for muscle forces that will generate an accurate and efficient movement. During the movement, the cerebellum must monitor movement performance through sensory feedback. Current thinking is that much of this is done by an internal model that converts corollary discharge into predictions of the sensory feedback. The cerebellum then compares real and predicted sensory feedback to determine a sensory prediction error and uses the sensory prediction error to guide corrective movements and learning.

Using a paradigm that required monkeys to ignore the sensory signals caused by their own movement, Kathy Cullen and colleagues have identified a neural correlate of a sensory prediction error in the deep cerebellar nuclei. Specifically, they studied the vestibular sensory signals that result from an animal's active head movements. They showed that the brain attenuates or even eliminates the vestibular sensory signals caused by one's own active head movement in order to better detect unpredictable vestibular signals due to the environment. However, when the head is effectively made heavier by adding resistance via a mechanical device, the vestibular sensory signals no longer match the predicted sensory signals that normally would attenuate the vestibular input. They showed that the cerebellum adjusts its predictions of the vestibular input to account for the changes in head movement caused by resistance due to the mechanical device. After some practice, the predicted and actual self-generated sensory inputs again match, and neurons in the deep cerebellar nucleus return to being unresponsive to vestibular inputs. Cerebellum-dependent learning is described in detail later in this chapter.

The Cerebellum Contributes to Timing Control

The cerebellum seems to have a role in movement timing that goes well beyond its role in regulating the timing of contractions in different muscles (Figure 37–7). When patients with cerebellar lesions attempt to make regular tapping movements with their hands or fingers, the rhythm is irregular and the motions vary in duration and force.

Based on a theoretical model of how tapping movements are generated, Richard Ivry and Steven Keele inferred that medial cerebellar lesions interfere only with accurate execution of the response, whereas lateral cerebellar lesions interfere with the timing of serial events. Such timing defects are not limited to motor events. They also affect the patient's ability to judge elapsed time in purely mental or cognitive tasks, as in the ability to distinguish whether one tone is longer or shorter than another or whether the speed of one moving object is greater or less than that of another. We will see in our discussion of motor learning that the cerebellum is critical for learning the timing of motor acts.

The Cerebellum Participates in Motor Skill Learning

In the early 1970s, on the basis of mathematical modeling of cerebellar function and the cerebellar microcircuit, David Marr and James Albus independently suggested that the cerebellum might be involved in learning motor skills. Along with Masao Ito, they proposed that the climbing-fiber input to Purkinje cells causes changes at the synapses that relay mossy fiber input signals from parallel fibers to Purkinje cells. According to their theory, the synaptic plasticity would lead to changes in simple-spike firing, and these changes would cause behavioral learning. Subsequent experimental evidence has supported and extended this theory of cerebellar motor learning.

Climbing-Fiber Activity Changes the Synaptic Efficacy of Parallel Fibers

Climbing fibers can selectively induce *long-term depression* in the synapses between parallel fibers and Purkinje cells that are activated concurrently with the climbing fibers. Many studies in brain slices and cultured Purkinje cells have found that concurrent stimulation of climbing fibers and parallel fibers depresses the Purkinje cell responses to subsequent stimulation of the same parallel fibers. The depression is selective for the parallel fibers that were activated in conjunction with the climbing-fiber input

and does not appear in synapses from parallel fibers that had not been stimulated along with climbing fibers (Figure 37–11A). The resulting depression can last for minutes to hours.

Many studies in a variety of motor learning systems have recorded activity in Purkinje cells that is consistent with the predictions of the cerebellar learning theory. For example, if an unexpected resistance is applied to a well-practiced arm movement, extra muscle tension will be required to move. Climbing fiber activity can signal error until the unexpected resistance is learned. They presumably depress the synaptic

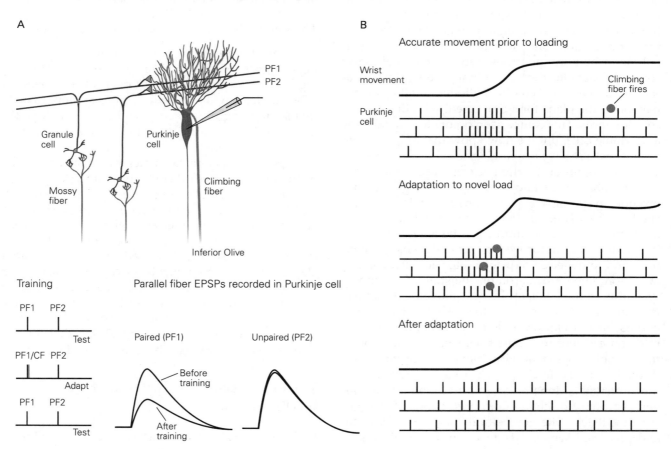

Figure 37–11 Long-term depression of the synaptic input from parallel fibers to Purkinje cells is one plausible mechanism for cerebellar learning.

A. Two different groups of parallel fibers and the presynaptic climbing fibers are electrically stimulated in vitro. Repeated stimulation of one set of parallel fibers (**PF1**) at the same time as the climbing fibers produces a long-term reduction in the responses of those parallel fibers to later stimulation. The responses of a second set of parallel fibers (**PF2**) are not depressed because they are not stimulated simultaneously with the presynaptic climbing fibers. (Abbreviations: **CF**, climbing fiber; **EPSP**, excitatory postsynaptic potential.) (Adapted from Ito et al. 1982.)

B. Top: An accurate wrist movement by a monkey is accompanied by a burst of simple spikes in a Purkinje cell, followed later by discharge of a single climbing fiber in one trial. **Middle:** When the monkey must make the same movement against a novel resistance (adaptation), climbing-fiber activity occurs during movement in every trial and the movement itself overshoots the target. **Bottom:** After adaptation, the frequency of simple spikes during movement is quite attenuated, and the climbing fiber is not active during movement or later. This is the sequence of events expected if long-term depression in the cerebellar cortex plays a role in learning. Climbing fiber activity is usually low (1/s) but increases during adaptation to a novel load. (Adapted, with permission, from Gilbert and Thach 1977.)

strength of parallel fibers involved in generating those errors, namely those that drove Purkinje simple-spike firing at the time of the climbing-fiber activity (Figure 37–11B). With successive movements, the parallel-fiber inputs conveying the flawed central command are increasingly suppressed, a more appropriate pattern of simple-spike activity emerges, and eventually movement errors disappear, along with the climbing-fiber error signal. Although this kind of result is consistent with the theory of cerebellar learning, it stops short of proving that the neural and behavioral learning was caused by long-term depression of the synapses from parallel fibers onto Purkinje cells.

The Cerebellum Is Necessary for Motor Learning in Several Different Movement Systems

The cerebellum is involved in learning a wide variety of movements, ranging from limb and eye movements to walking. In each movement system, motor learning operates to improve the feedforward control of movement. Errors render motor control transiently dependent on sensory feedback, and motor learning restores the ideal situation where performance is accurate without relying on sensory feedback.

Adaptation of limb movements that rely on eye–hand coordination can be demonstrated by having people wear prisms that deflect the light path sideways. When a person plays darts while wearing prism goggles that displace the entire visual field to the left, the initial dart throw lands to the left side of the target by an amount proportional to the strength of the prisms. The subject gradually adapts to the distortion through practice; within 10 to 30 throws, the darts land on target (Figure 37–12). When the prisms are removed, the adaptation persists, and the darts hit to the right of the target by roughly the same distance as the initial prism-induced error. Patients with a damaged cerebellar cortex or inferior olive are severely impaired or unable to adapt at all in this test.

Classical conditioning of the eye-blink response also depends on an intact cerebellum. In this form of associative learning, a puff of air is directed at the cornea, causing the eye to blink at the end of a neutral stimulus such as a tone. If the tone and the puff are paired repeatedly with a fixed duration of the tone, then the brain learns the tone's predictive power and the tone alone is sufficient to cause a blink. Michael Mauk and his colleagues have shown that the brain also can learn about the timing of the stimulus so that the eye blink occurs at the right time. It is even possible to learn to blink at different times in response to tones of different frequencies.

All forms of conjugate eye movement require the cerebellum for correct performance, and each form is subject to motor learning that involves the cerebellum. For example, the vestibulo-ocular reflex normally keeps the eyes fixed on a target when the head is rotated (Chapter 27). Motion of the head in one direction is sensed by the vestibular labyrinth, which initiates eye movements in the opposite direction to prevent visual images from slipping across the retina. When humans and experimental animals wear glasses that change the size of a visual scene, the vestibulo-ocular reflex initially fails to keep images stable on the retina because the amplitude of the reflex is inappropriate to the new conditions. After the glasses have been worn continuously for several days, however, the size of the reflex becomes progressively reduced (for miniaturizing glasses) or increased (for magnifying glasses) (Figure 37–13A). These changes are required to prevent images from slipping across the retina because magnified (or miniaturized) images also move faster (or slower). The performance of the baseline vestibulo-ocular reflex does not depend heavily on the cerebellum, but its adaptation does and can be blocked in experimental animals by lesions of the lateral part of the vestibulocerebellum called the flocular complex.

Saccadic eye movements also depend on the integrity of Purkinje cells in the oculomotor vermis in lobules V, VI, and VII of the vermis (Figure 37–2C). These cells discharge prior to and during saccades, and lesions of the vermis cause saccades to become hypermetric, much as we see in the arm movements of cerebellar patients. The outputs from neurons of the vermis concerned with saccades are transmitted through a very small region of the caudal fastigial nucleus to the saccade generator in the reticular formation.

The same Purkinje cells participate in a form of motor learning called saccadic adaptation. This adaptation is demonstrated by having a monkey fixate on a target straight ahead and then displaying a new target at an eccentric location. During the saccade to the new target, the experimenter moves the new target to a more eccentric location. Initially, the subject needs to make a second saccade to fixate on the target. Gradually, over several hundred trials, the first saccade grows in amplitude so that it brings the eye directly to the final location of the target (Figure 37–13B). Recordings during saccadic adaptation have revealed that climbing fiber inputs to the Purkinje cells in the oculomotor vermis signal saccadic errors during learning, and the simple-spike firing rate of the same cells adapts gradually along with the monkey's eye movements. Thus, the oculomotor vermis is a likely site for motor learning of the amplitude of saccadic eye movements. The story

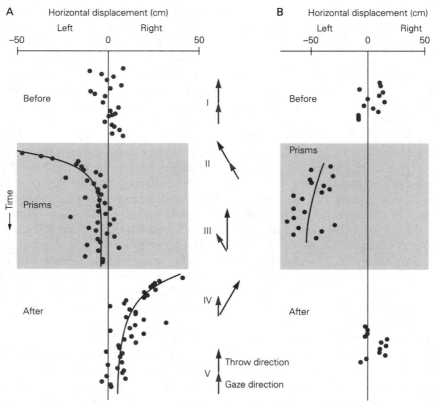

Figure 37–12 Adjustment of eye–hand coordination to a change in optical conditions. The subject wears prism goggles that bend the optic path to her right. She must look to her left along the bent light path to see the target directly ahead. (Adapted, with permission, from Martin et al. 1996).

A. Without prisms, the subject throws with good accuracy (**I**). The first hit after the prisms have been put in place is displaced left of center because the hand throws where the eyes are directed. Thereafter, hits trend rightward toward the target, away from where the eyes are looking (**II**). After removal of the prisms, the subject fixes her gaze in the center of the target; the first throw hits to the right of center, away from where the eyes are directed. Thereafter, hits trend toward the target (**III**).

Immediately after removing the prisms, the subject directs her gaze toward the target; her adapted throw is to the right of the direction of gaze and to the right of the target (**IV**). After recovery from adaptation, she again looks at and throws toward the target (**V**). Data during and after prism use have been fit with exponential curves. Gaze and throw directions are indicated by the **blue** and **brown arrows**, respectively, on the *right*. The inferred gaze direction assumes that the subject is fixating the target.

B. Adaptation fails in a patient with unilateral infarctions in the territory of the posterior inferior cerebellar artery that affect the inferior cerebellar peduncle and inferior lateral posterior cerebellar cortex.

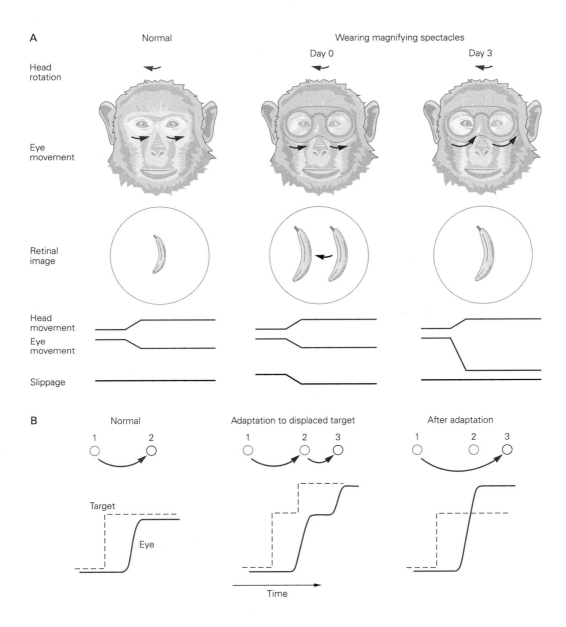

Figure 37–13 Cerebellar learning in the vestibulo-ocular reflex and in saccadic eye movements.

A. Motor learning in the vestibulo-ocular reflex of a monkey wearing magnifying spectacles. The columns show normal conditions before learning, the situation when the monkey first dons the spectacles (day 0), and after complete adaptation (day 3). Eye movements are normally equal and opposite to head turns, and the banana stays stable in the retina during head turns. With the spectacles on, the banana appears larger; when the head turns, the vestibulo-ocular reflex is too small and the banana's image slips across the retina. After adaptation, the eye movements are large enough that the image of the banana

again remains stable on the retina during head turns. (Adapted, with permission, from Lisberger 1988.)

B. Motor learning in saccadic eye movements. The columns show saccades under normal conditions, on the first adaptation trial, and after full adaptation. Normally, the saccade responds to a change in target position by bringing the eye almost perfectly to the new target position. During adaptation, the target moves to a new position during the initial saccade, requiring a second saccade to bring the eye to the new, final target position. After adaptation, the original target position evokes a larger saccade that is appropriate to bring the eye to the new target position, even though the target does not move.

is very similar for smooth-pursuit eye movements, except that the relevant part of the cerebellum is the floccular complex, using the same Purkinje cells that participate in adaptation of the vestibulo-ocular reflex.

Finally, learning of new walking patterns has been studied in cerebellar patients using a split-belt treadmill that requires one leg to move faster than the other. Cerebellar damage does not impair the ability to use feedback to immediately change the walking pattern when the two belt speeds differ: Patients can lengthen the time that they stand on the slower tread-mill belt and shorten the time that they stand on the faster treadmill belt. However, cerebellar patients cannot learn over hundreds of steps to make their walking pattern symmetric, whereas healthy individuals can (see Figure 30–14).

Learning Occurs at Several Sites in the Cerebellum

We know now that there are many sites of synaptic and cellular plasticity in the cerebellar microcircuit. Almost every synapse that has been studied undergoes either potentiation or depression, and the theory of cerebellar learning has been broadened accordingly. Detailed analyses of the role of cerebellar circuits in motor learning have been conducted in several motor systems: adaptation of multiple kinds of eye movements, classical conditioning of the eye blink, and motor learning in arm movements.

In today's broadened theory of cerebellar learning, learning occurs not only in the cerebellar cortex, as postulated by Marr, Albus, and Ito, but also in the deep cerebellar nuclei (Figure 37–14). Our understanding of learning in the cerebellar cortex is based partly on long-term depression of the synapses from parallel fibers to Purkinje cells, but many other synapses are characterized by plasticity, and they also probably participate. Available evidence is still compatible with the long-standing idea that inputs from climbing fibers provide the primary instructive signals that lead to changes in synaptic strength within the cerebellar cortex, but now there is room for the possibility of other instructive signals as well. Learning probably results from coordinated synaptic plasticity at multiple sites rather than from changes at a single site.

Studies of classical conditioning of the eye blink and adaptation of the vestibulo-ocular reflex provide strong evidence that learning occurs in both the cerebellar cortex and the deep cerebellar nuclei. Further, considerable evidence suggests that learning may occur first in the cerebellar cortex and then be transferred to the deep cerebellar nuclei. At least for eye

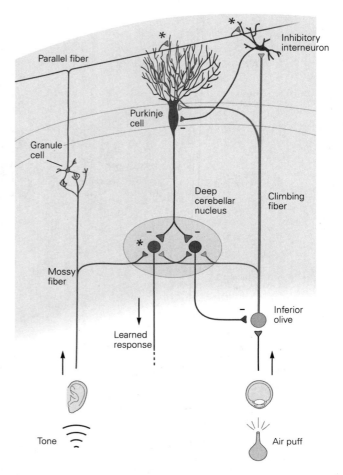

Figure 37–14 Learning in the cerebellar microcircuit can occur in the cerebellar cortex and the deep cerebellar nuclei. The diagram is based on classical conditioning of blinking, which is driven by pairing a tone (so-called conditioned stimulus carried by mossy fibers) and an air puff (the unconditioned stimulus carried by climbing fibers). Learning occurs at the parallel fiber–Purkinje cell synapses when the climbing fiber and parallel fibers are active together. Learning also occurs at the mossy-fiber synapse onto the deep cerebellar nuclei. (Sites of learning are denoted by **asterisks**.) While this example diagrams a classical conditioning paradigm, plasticity occurs at the same sites during adaptation of the vestibulo-ocular reflex when head turns are associated with image motion on the retina (Chapter 27). (Adapted, with permission, from Carey and Lisberger 2002.)

blink conditioning, the cerebellar cortex may play a special role in learning timing.

As discussed earlier, the cerebellum makes use of internal models to ensure smooth and accurate movement in advance of any guidance by sensory feedback. Synaptic changes that lead to circuit learning could be the mechanisms that create and maintain accurate internal models. One important function of learning in the cerebellum may be the continuous tuning of internal models. Cerebellar internal models may use sensory feedback to adjust synaptic and cellular function so that motor commands produce movements that are rapid, accurate, and smooth. Thus, the cerebellum appears to be the learning machine envisioned by the earliest investigators, but its learning capabilities may be greater and more widely dispersed than originally imagined and may affect all cerebellar contributions to behavior.

Highlights

1. The cerebellum plays a critical role in movement. Damage to the cerebellum leads to profound movement incoordination called ataxia, which affects all movements ranging from eye and limb movements to balance and walking. Cerebellar damage also leads to some sensory deficits but only during active movement.

2. The cerebellum also plays a role in cognitive and emotional behavior. Deficits in these domains are less immediately obvious after cerebellar damage but appear with formalized testing. There is probably a common mechanism for deficits across both motor and nonmotor domains, but the mechanism is not yet understood.

3. The cerebellum acts through its connections to other brain structures. Its inputs come indirectly from wide regions of the cerebral cortex, as well as from the brainstem and spinal cord. Cerebellar outputs project to the vestibular nuclei, the brainstem reticular formation, and the red nucleus and via the thalamus to wide regions of the cerebral cortex.

4. Reciprocal connections between the cerebellum and the cerebral cortex include sensory and motor cortices as well as wide regions of the parietal and prefrontal cortices. Cerebrocerebellar connections are organized as a series of parallel, closed, recurrent loops, where a given region of the cerebral cortex makes both efferent and afferent connections with a given part of the cerebellum.

5. The circuit of the cerebellar cortex is highly stereotyped, suggesting a common computational mechanism for its interactions with other brain regions. It includes an input granular layer where mossy fibers synapse on granule cells and Golgi cells provide inhibitory feedback; an inhibitory Purkinje cell layer, with the sole output neurons of the cerebellar cortex; and a molecular layer where Purkinje cell dendrites and inhibitory interneurons receive inputs from the parallel fibers that emerge from the axons of granule cells.

6. The climbing-fiber and mossy-fiber inputs to the cerebellum are very different anatomically. Each Purkinje cell receives many synaptic contacts from a single climbing fiber but can be influenced via granule cells by a huge number of mossy fibers. Climbing fibers fire at very low frequencies and cause unitary "complex spikes" in Purkinje cells. Mossy fibers cause "simple spikes" that can discharge at very high rates. It is thought that the interplay between these inputs is essential for learning.

7. Theories of cerebellar motor control emphasize several general principles. The cerebellum is important for generating reliable feedforward action before there has been time for useful sensory feedback to occur. It plays a key role in the internal control of timing. The cerebellum relies on computations that combine sensory inputs with corollary discharge reporting the movement that was commanded. Internal models of the motor effector organs and the world allow the cerebellum to estimate the state of the motor system and guide accurate feedforward actions.

8. Learning and adaptation of movement are fundamental functions of the cerebellum. Cerebellar learning requires feedback about movement errors and updates movement on a trial-by-trial basis. There are many sites of synaptic plasticity in the cerebellum, and current evidence for motor learning systems supports at least two sites of learning in the cerebellum. One site involves long-term depression of the synapses from parallel fibers to Purkinje cells, guided by errors signaled by climbing-fiber inputs. The other site is in the deep cerebellar nuclei. It is likely that the same learning mechanism is used for cognitive and emotional processing.

Amy J. Bastian
Stephen G. Lisberger

Selected Reading

Bodranghien F, Bastian A, Casali C, et al. 2016. Consensus paper: revisiting the symptoms and signs of cerebellar syndrome. Cerebellum 15:369–391.

Bostan AC, Dum RP, Strick PL. 2013. Cerebellar networks with the cerebral cortex and basal ganglia. Trends Cogn Sci 17:241–254.

Boyden ES, Katoh A, Raymond JL. 2004. Cerebellum-dependent learning: the role of multiple plasticity mechanisms. Annu Rev Neurosci 27:581–609.

Ito M. 1984. *The Cerebellum and Neural Control*. New York: Raven.

Raymond JL, Lisberger SG, Mauk MD. 1996. The cerebellum: a neuronal learning machine? Science 272:1126–1131.

Stoodley CJ, Schmahmann JD. 2010. Evidence for topographic organization in the cerebellum of motor control versus cognitive and affective processing. Cortex 46:831–844.

References

Adamaszek M, D'Agata F, Ferrucci R, et al. 2017. Consensus paper: cerebellum and emotion. Cerebellum 16:552–576.

Adrian ED. 1943. Afferent areas in the cerebellum connected with the limbs. Brain 66:289–315.

Albus JS. 1971. A theory of cerebellar function. Math Biosci 10:25–61.

Arshavsky YI, Berkenblit MB, Fukson OI, Gelfand IM, Orlovsky GN. 1972. Recordings of neurones of the dorsal spinocerebellar tract during evoked locomotion. Brain Res 43:272–275.

Arshavsky YI, Berkenblit MB, Fukson OI, Gelfand IM, Orlovsky GN. 1972. Origin of modulation in neurones of the ventral spinocerebellar tract during locomotion. Brain Res 43:276–279.

Bastian AJ, Martin TA, Keating JG, Thach WT. 1996. Cerebellar ataxia: abnormal control of interaction torques across multiple joints. J Neurophysiol 176:492–509.

Bastian AJ, Zackowski KM, Thach WT. 2000. Cerebellar ataxia: torque deficiency or torque mismatch between joints? J Neurophys 83:3019–3030.

Bhanpuri NH, Okamura AM, Bastian AJ. 2013. Predictive modeling by the cerebellum improves proprioception. J Neurosci 33:14301–14306.

Brooks VB, Thach WT. 1981. Cerebellar control of posture and movement. In *Handbook of Physiology, The Nervous System*, Sect. I, Vol. 2, ed. V. B. Brooks, pp. 877–46. Bethesda: Am Physiol Soc.

Brooks JX, Carriot J, Cullen KE. 2015. Learning to expect the unexpected: rapid updating in primate cerebellum during voluntary self-motion. Nat Neurosci 18:1310–1317.

Buckner RL, Krienen FM, Castellanos A, Diaz JC, Yeo BT. 2011 The organization of the human cerebellum estimated by intrinsic functional connectivity. J Neurophysiol 106:2322–2345.

Courchesne E, Yeung-Courchesne R, Press GA, Hesselink JR, Jernigan TL. 1988. Hypoplasia of cerebellar vermal lobules VI and VII in autism. N Engl J Med 318:1349–1354.

Carey MR, Lisberger SG. 2002. Embarrassed but not depressed: some eye opening lessons for cerebellar learning. Neuron 35: 223–226.

Eccles JC, Ito M, Szentagothai J. 1967. *The Cerebellum as a Neuronal Machine*. New York: Springer.

Fiez JA, Petersen SE, Cheney MK, Raichle ME. 1992. Impaired non-motor learning and error detection associated with cerebellar damage. Brain 115:155–178.

Flament D, Hore J. 1986. Movement and electromyographic disorders associated with cerebellar dysmetria. J Neurophysiol 55:1221–1233.

Gao Z, van Beugen BJ, De Zeeuw CI. 2012. Distributed synergistic plasticity and cerebellar learning. Nat Rev Neurosci 13:619–635.

Ghasia FF, Meng H, Angelaki DE. 2008. Neural correlates of forward and inverse models for eye movements: evidence from three-dimensional kinematics. J Neurosci 28:5082–5087.

Gilbert PFC, Thach WT. 1977. Purkinje cell activity during motor learning. Brain Res 128:309–328.

Groenewegen HJ, Voogd J. 1977. The parasagittal zonation within the olivocerebellar projection. I. Climbing fiber distribution in the vermis of cat cerebellum. J Comp Neurol 174:417–488.

Heck DH, Thach WT, Keating JG. 2007. On-beam synchrony in the cerebellum as the mechanism for the timing and coordination of movement. Proc Natl Acad Sci U S A 104:7658–7663.

Hore J, Flament D. 1986. Evidence that a disordered servo-like mechanism contributes to tremor in movements during cerebellar dysfunction. J Neurophysiol 56:123–136.

Ito M, Sakurai M, Tongroach P. 1982. Climbing fibre induced depression of both mossy fibre responsiveness and glutamate sensitivity of cerebellar Purkinje cells. J Physiol Lond 324:113–134.

Ivry RB, Keele SW. 1989. Timing functions of the cerebellum. J Cogn Neurosci 1:136–152.

Jansen J, Brodal A (eds). 1954. *Aspects of Cerebellar Anatomy*. Oslo: Grundt Tanum.

Kelly RM, Strick PL. 2003. Cerebellar loops with motor cortex and prefrontal cortex of a nonhuman primate. J Neurosci 23:8432–8444.

Kim SG, Ugurbil K, Strick PL. 1994. Activation of a cerebellar output nucleus during cognitive processing. Science 265:949–951.

Larsell O, Jansen J. 1972. *The Comparative Anatomy and Histology of the Cerebellum: The Human Cerebellum, Cerebellar Connection and Cerebellar Cortex*, pp. 111–119. Minneapolis, MN: Univ. of Minnesota Press.

Lisberger SG. 1988. The neural basis for motor learning in the vestibulo-ocular reflex in monkeys. Trends in Neurosci 11:147–152.

Lisberger SG. 1994. Neural basis for motor learning in the vestibulo-ocular reflex of primates. III. Computational and behavioral analysis of the sites of learning. J Neurophysiol 72:974–998.

Lisberger SG, Fuchs AF. 1978. Role of primate flocculus during rapid behavioral modification of vestibulo-ocular reflex. I. Purkinje cell activity during visually guided horizontal smooth-pursuit eye movements and passive head rotation. J Neurophysiol 41:733–763.

Marr D. 1969. A theory of cerebellar cortex. J Physiol 202:437–470.

Martin TA, Keating JG, Goodkin HP, Bastian AJ, Thach WT. 1996. Throwing while looking through prisms. I. Focal olivocerebellar lesions impair adaptation. Brain 119:1183–1198.

Martinez FE, Crill WE, Kennedy TT. 1971. Electrogenesis of cerebellar Purkinje cell responses in cats. J Neurophysiol 34:348–356.

McCormick DA, Thompson RF. 1984. Cerebellum: essential involvement in the classically conditioned eyelid response. Science 223:296–299.

Medina JF, Lisberger SG. 2008. Links from complex spikes to local plasticity and motor learning in the cerebellum of awake-behaving monkeys. Nat Neurosci 11:1185–1192.

Nieuwenhuys R, Voogd J, van Huijzen C. 1981. The human central nervous system: a synopsis and atlas. Springer.

Nieuwenhuys R, Voogd J, van Huijzen Chr. 1988. The Human Central Nervous System: A Synopsis and Atlas, 3rd rev. ed. Berlin: Springer.

Ohyama T, Nores WL, Medina JF, Riusech FA, Mauk MD. 2006. Learning-induced plasticity in deep cerebellar nucleus. J Neurosci 26:12656–12663.

Pasalar S, Roitman AV, Durfee WK, Ebner TJ. 2006. Force field effects on cerebellar Purkinje cell discharge with implications for internal models. Nat Neurosci 9:1404–1411.

Robinson DA. 1976. Adaptive gain control of vestibuloocular reflex by the cerebellum. J Neurophysiol 39:954–969.

Strata P, Montarolo PG. 1982. Functional aspects of the inferior olive. Arch Ital Biol 120:321–329.

Strick PL, Dum RP, Fiez JA. 2009. Cerebellum and nonmotor function. Annu Rev Neurosci 32:413–434.

Thach WT. 1968. Discharge of Purkinje and cerebellar nuclear neurons during rapidly alternating arm movements in the monkey. J Neurophysiol 31:785–797.

Tseng YW, Diedrichsen J, Krakauer JW, Shadmehr R, Bastian AJ. 2007. Sensory prediction errors drive cerebellum-dependent adaptation of reaching. J Neurophysiol 98:54–62.

Yeo CH, Hardiman MJ, Glickstein M. 1984. Discrete lesions of the cerebellar cortex abolish the classically conditioned nictitating membrane response of the rabbit. Behav Brain Res 13:261–266.

38

The Basal Ganglia

Attention Deficit Hyperactivity Disorder and Tourette Syndrome May Also Be Characterized by Intrusions of Nonselected Options

Obsessive-Compulsive Disorder Reflects the Presence of Pathologically Dominant Options

Addictions Are Associated With Disorders of Reinforcement Mechanisms and Habitual Goals

Highlights

THE TRADITIONAL VIEW THAT THE BASAL ganglia play a role in movement arises primarily because diseases of the basal ganglia, such as Parkinson and Huntington disease, are associated with prominent disturbances of movement, and from the belief that basal ganglia neurons send their output exclusively to the motor cortex by way of the thalamus. However, we now know that the basal ganglia also project to wide areas of the brain stem and via the thalamus to nonmotor areas of the cerebral cortex and limbic system, thereby providing a mechanism whereby they contribute to a wide variety of cognitive, motivational, and affective operations. This understanding also explains why diseases of the basal ganglia are frequently associated with complex cognitive, motivational, and affective dysfunction in addition to the better-known motor disturbances.

This chapter provides a perspective on the fundamental contributions of the basal ganglia (Figure 38–1) to overall brain function. Recent advances in the fields of artificial neural networks and robotics emphasize that behavioral function is an emergent property of signal processing in physically connected networks (Chapter 5). Thus, how components of networks are connected and how their input signals are transformed into output signals impose important constraints on final behavioral outputs. We first describe the principal anatomical and physiological features of the basal ganglia network and consider the constraints these might impose on their function. We consider the extent to which the basal ganglia have been conserved during vertebrate brain evolution and, based on these insights, review evidence suggesting that the basal ganglia's normal functions are to select between incompatible behaviors and to mediate reinforcement learning. We conclude by examining important insights into how the system can malfunction in some of the major diseases involving the basal ganglia.

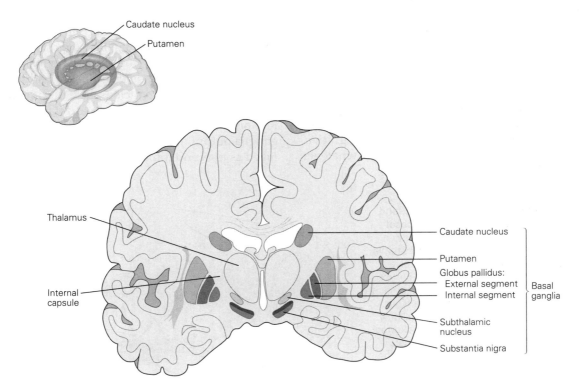

Figure 38–1 The basal ganglia and surrounding structures. The nuclei of the basal ganglia are identified on the right in this coronal section of a human brain. (Adapted from Nieuwenhuys, Voogd, and van Huijzen 1981.)

The Basal Ganglia Network Consists of Three Principal Input Nuclei, Two Main Output Nuclei, and One Intrinsic Nucleus

The striatum (a collective term for the caudate nucleus and putamen; see Figure 38–1), subthalamic nucleus, and substantia nigra pars compacta/ventral tegmental area are the three major input nuclei of the basal ganglia, receiving signals directly and indirectly from structures distributed throughout the neuraxis (Figure 38–2).

The Striatum, Subthalamic Nucleus, and Substantia Nigra Pars Compacta/Ventral Tegmental Area Are the Three Principal Input Nuclei of the Basal Ganglia

The striatum is the largest nucleus of the basal ganglia. It receives direct input from most regions of the cerebral cortex and limbic structures, including the amygdala and hippocampus. Important input from sensorimotor and motivational regions of the brain stem is relayed indirectly via the thalamus. In rodents, the number of contacts received in the striatum from the cerebral cortex and thalamus are approximately equivalent. Finally, important modulatory input to the striatum comes from the substantia nigra pars compacta (dopamine), midbrain raphe (serotonin), and pedunculopontine nucleus (acetylcholine).

The striatum is subdivided functionally on the basis of the organization of input connections, principally the topographically organized afferents from the cerebral cortex. Limbic, associative, and sensorimotor territories are generally recognized along a ventromedial-dorsolateral continuum. This diversity of input shows that the basal ganglia receive signals from brain regions involved in different motivational, emotional, cognitive, and sensorimotor processes, implying that whatever

the basal ganglia are doing, they are doing it for a wide range of brain processes.

An additional architectural feature of the striatum suggests that the basal ganglia are performing more or less the same operations on their inputs from functionally diverse afferent structures. Specifically, within each of the striatum's functional territories, the cellular architecture is remarkably similar. In all regions, inhibitory γ-aminobutyric acid (GABA)-ergic medium spiny neurons are the principal cell type (>90% of all neurons). In addition, in all functionally defined regions, the medium spiny neurons are separated into two populations according to the relative expression of neuroactive peptides (substance P and dynorphin versus enkephalin) or the expression of D_1 and D_2 dopamine receptors, which are thought to positively and negatively modulate cyclic adenosine monophosphate signaling in these neurons. These populations contribute differentially to different efferent projections of the striatum. In addition to these long-range inhibitory connections to other basal ganglia nuclei, medium spiny neurons also send local collaterals to adjacent cells. Colocalized GABAergic and peptidergic neurotransmission provides local mutually inhibitory and excitatory influences. The remaining 5% to 10% of neurons in the striatum are purely GABAergic and cholinergic interneurons, which can be distinguished according to neurochemical, electrophysiological, and in some cases morphological characteristics. The fact that this local cellular architecture is present in all functional regions suggests that neurons in the striatum are applying the same or similar computations on functionally diverse afferent pathways.

The subthalamic nucleus has traditionally been considered an important internal relay in the "indirect output pathway" from the striatum to the basal ganglia output nuclei (see below). It is now also recognized as a second important input nucleus of the basal

Figure 38–2 The principal input, intrinsic, and output connections of the mammalian basal ganglia. The main input nuclei are the striatum (**STR**), subthalamic nucleus (**STN**), and substantia nigra pars compacta (not shown). They receive input directly from the thalamus, cerebral cortex, and limbic structures (amygdala and hippocampus). The main output nuclei are the substantia nigra pars reticulata (**SNr**) and internal globus pallidus/entopeduncular nucleus (not shown). The external globus pallidus (**GP**) is classified as an intrinsic nucleus as most of its connections are with other basal ganglia nuclei. Structures are shown on a sagittal schematic of the rodent brain. **Red** and **dark gray arrows** denote excitatory and inhibitory connections, respectively.

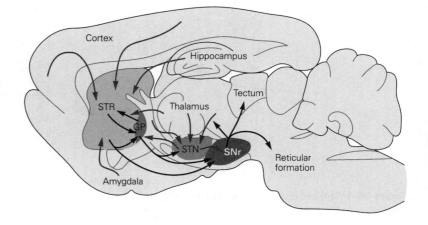

ganglia. Topographically organized inputs derive not only from large parts of frontal cortex, but also from various thalamic and brain stem structures. The subthalamic nucleus is the only component of the basal ganglia that has excitatory (glutamatergic) output connections. These project to both output nuclei and to the intrinsic external globus pallidus.

The substantia nigra pars compacta/ventral tegmental area contain an important population of dopaminergic neurons. These neurons represent the third major input station of the basal ganglia and give rise to the nigrostriatal and mesolimbic/mesocortical dopamine projections. They receive significant afferent connections from other basal ganglia nuclei (the striatum, globus pallidus, and subthalamus), but also from many structures in the brain stem (eg, superior colliculus, rostromedial tegmental region, raphe nuclei, pedunculopontine nucleus, and parabrachial area). Other afferent connections are from the frontal cortex and the amygdala. This pattern of connectivity is important because it suggests the most important direct influence over the dopaminergic neurons arises from evolutionarily ancient parts of the brain (see below).

Individual dopaminergic neurons have highly branching axons that project into extensive regions of not only the other basal ganglia nuclei but also external structures (eg, frontal cortex, septal area, amygdala, habenula). This suggests that their important modulatory signals are widely broadcast throughout targeted structures. The highest concentration of dopaminergic terminals is found in the striatum, where synaptic and nonsynaptic contacts are formed with both medium spiny cells and interneurons. The existence of nonsynaptic contacts gives rise to what has been called *volume transmission*. This occurs when neurotransmitters diffuse through the brain's extracellular fluid from release points that may be remote from targeted cells. Consequently, volume transmission typically has a longer time course than synaptic neurotransmission. Deployment of volume transmission in targeted structures is further evidence for the idea that the effects of dopamine in targeted structures are widely broadcast and spatially imprecise. Variable proportions of GABAergic neurons (substantia nigra and the ventral tegmental area) and glutamatergic neurons (ventral tegmental area) contribute to local processing in these structures.

The Substantia Nigra Pars Reticulata and the Internal Globus Pallidus Are the Two Principal Output Nuclei of the Basal Ganglia

The internal globus pallidus/entopeduncular nucleus is one of the two principal output nuclei. It receives inputs from other basal ganglia nuclei and projects to external targets in the thalamus and brain stem. GABAergic input from the striatum and external globus pallidus are inhibitory, while input from the subthalamic nucleus is glutamatergic and excitatory. Neurons of the internal globus pallidus are themselves GABAergic and have high levels of tonic activity. Under normal circumstances, this imposes powerful inhibitory effects on targets in the thalamus, lateral habenula, and brain stem.

The substantia nigra pars reticulata is the second principal output nucleus. It also receives afferents from other basal ganglia nuclei and provides efferent connections to the thalamus and brain stem. Inhibitory (GABAergic) inputs come from the striatum and globus pallidus (external) and excitatory input from the subthalamus. Pars reticulata neurons are also GABAergic and impose strong inhibitory control over parts of the thalamus and brain stem, including the superior colliculus, pedunculopontine nucleus, and parts of the midbrain and medullary reticular formation.

The External Globus Pallidus Is Mostly an Intrinsic Structure of the Basal Ganglia

Most connections of the globus pallidus are with other basal ganglia nuclei, including inhibitory (GABAergic) input from the striatum and excitatory (glutamatergic) input from the subthalamus, and the globus pallidus provides inhibitory efferent connections to all the basal ganglia's input and output nuclei. This pattern of connections suggests that that the external globus pallidus is an essential regulator of internal basal ganglia activity.

Having described the core components of the basal ganglia, we will now consider in more detail how they are connected, first with each other and then with external structures in the brain.

The Internal Circuitry of the Basal Ganglia Regulates How the Components Interact

The Traditional Model of the Basal Ganglia Emphasizes Direct and Indirect Pathways

An influential interpretation of the intrinsic circuitry of the basal ganglia was proposed in the late 1980s by Roger Albin and colleagues (Figure 38–3A). In their scheme, signals originating in the cerebral cortex are distributed to two populations of medium spiny output neurons in the striatum.

Neurons containing substance P and a preponderance of D_1 dopamine receptors make direct inhibitory

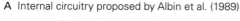

A Internal circuitry proposed by Albin et al. (1989)   B Contemporary view of internal circuitry

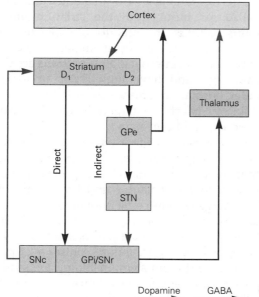
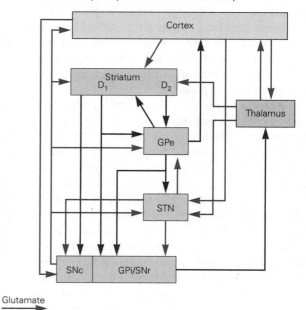

Dopamine   GABA   Glutamate

Figure 38–3 Intrinsic connections within the basal ganglia.

A. The influential proposal by Roger Albin and colleagues (1989) is presented, where output of the basal ganglia is determined by the balance between a *direct pathway* from the striatum to the output nuclei (internal globus pallidus [**Gpi**] and substantial nigra pars reticulata [**SNr**]), which promotes behavior, and an *indirect pathway* from the striatum to the output nuclei via relays in the external globus pallidus (**GPe**) and subthalamic

nucleus (**STN**), which suppresses behavior. The balance between the direct and indirect projections was thought to be regulated by afferent dopaminergic signals from substantia nigra pars compacta (**SNc**) acting on differentially distributed D$_1$ and D$_2$ dopamine receptors.

B. More recent anatomical investigations have revealed a rather more complex organization where the transformations of basal ganglia inputs that generate outputs are less easy to predict.

contact with the basal ganglia output nuclei—the direct pathway. In contrast, striatal neurons containing enkephalin and expressing mainly D$_2$ dopamine receptors make excitatory contact with the output nuclei via relays in the globus pallidus and subthalamus—the indirect pathway. Basal ganglia output was thought to reflect a cortically determined balance between these inhibitory and excitatory projections terminating on the two output structures (the internal globus pallidus and substantia nigra pars reticulata). In this model, a behavior would be promoted when the direct pathway was dominant and inhibited when the indirect pathway was dominant.

Detailed Anatomical Analyses Reveal a More Complex Organization

Recent anatomical observations show that the internal circuitry of the basal ganglia is more complex than originally envisaged (Figure 38–3B). The main findings have been that: (1) medium spiny neurons of the direct pathway also provide collateral input to the globus pallidus; (2) globus pallidus neurons also make direct

contact with the output nuclei in addition to the traditional indirect connections to the subthalamus—often with branching collaterals to all three structures; (3) the globus pallidus also projects back to the striatum and to structures outside the basal ganglia; (4) the subthalamic nucleus also projects back to the external globus pallidus, in addition to the feedforward connections to the two basal ganglia output nuclei; and (5) major inputs to the subthalamic nucleus originate from both cortical and subcortical structures external to the basal ganglia. Consequently, the subthalamus is now considered a major input structure of the basal ganglia (see above), rather than a simple relay in the intrinsic indirect projection. A modern appreciation of this complex organization within the basal ganglia suggests it is no longer possible to intuit how a particular input might be transformed by the basal ganglia to generate a specific output. For this reason, computational modeling of the internal circuitry of the basal ganglia has become increasingly important.

Although the overall pattern of intrinsic circuitry is complex (Figure 38–3B), connections between components of the basal ganglia are topographically ordered throughout. Some of these projections are comparatively focused (eg, the striatonigral projection), while others are more diffuse (eg, the subthalamonigral projection). Significant reductions in the comparative numbers of neurons in afferent structures, the striatum, and the output nuclei suggest a dramatic compression of information as it is processed within the basal ganglia.

Basal Ganglia Connections With External Structures Are Characterized by Reentrant Loops

Inputs Define Functional Territories in the Basal Ganglia

The functional status of inputs to the striatum from the cerebral cortex, limbic structures, and thalamus provides the rationale for classifying functional territories within the basal ganglia nuclei (limbic, associative, and sensorimotor). However, the manner in which the afferent projections make contact with neurons of the basal ganglia nuclei suggests important functional differences. For example, axons arriving in the striatum from the cerebral cortex and central lateral thalamic nucleus appear to make few contacts with many striatal neurons. In contrast, inputs from other regions, principally the parafascicular thalamic nucleus, have axons that make many contacts with fewer individual striatal neurons. Afferent connections to the subthalamic nucleus, at least from cerebral cortex, are also topographically organized according to the limbic, associative, and sensorimotor classification. However, there is no evidence of the same kind of precise topographical input from external structures to SNc and VTA dopamine neurons in the ventral midbrain.

Output Neurons Project to the External Structures That Provide Input

Basal ganglia output neurons project to regions of the thalamus (the intralaminar and ventromedial nuclei) that project back to basal ganglia input nuclei as well as to those regions of cortex that provided the original inputs to the striatum. Similarly, outputs from the basal ganglia to the brain stem tend to target those regions that provide input to the striatum via the thalamic midline and intralaminar nuclei. Importantly, projections from the basal ganglia output nuclei to the thalamus and brain stem are also topographically ordered.

Finally, many output projections of the basal ganglia are extensively collateralized, thereby simultaneously contacting targets in the thalamus, midbrain, and hindbrain. An example of the functional consequences of this organization is that a subset of neurons in the substantia nigra pars reticulata associated with oral behavior can simultaneously influence the activity in the specific regions of the thalamus/cortex, midbrain, and hindbrain that interact during the production of oral behavior.

Reentrant Loops Are a Cardinal Principle of Basal Ganglia Circuitry

Spatial topographies associated with input projections, intrinsic connections, and outputs of the basal ganglia provided the basis for the influential organizational principle suggested by Garrett Alexander and colleagues in 1989. Connections between the cerebral cortex and basal ganglia can be viewed as a series of reentrant parallel projecting, partly segregated, cortico-striato-nigro-thalamo-cortical loops or channels (Figure 38–4). Thus, an important component of the projections from different functional areas of cerebral cortex (eg, limbic, associative, sensorimotor) makes exclusive contact with specific regions of the basal ganglia input nuclei. This regional separation is maintained in forward projections throughout the internal circuitry. Focused output signals from functional territories represented in the basal ganglia output nuclei are returned, via appropriate thalamic relays, to the cortical regions providing the original input signals.

The concept of parallel projecting reentrant loops through the basal ganglia has been extended to their connections with sensorimotor and motivational structures in the brain stem, including the superior colliculus, periaqueductal gray, pedunculopontine, and parabrachial nuclei. This implies that the reentrant loop architecture through the basal ganglia must have predated the evolutionary expansion of the cerebral cortex. An important difference is that for the cortical loops the thalamic relay is on the output side of the loop, whereas for the subcortical loops, the thalamic relay is on the input side (Figure 38–5). Further work will be required to test whether projections from different brain stem structures, as they pass through the thalamic and basal ganglia relays, are functionally distinct channels.

In summary, the partially segregated reentrant loop organization is one of the dominant features characterizing the connections between the basal ganglia and external structures. This pattern of connections provides important clues as to the role played by the basal ganglia nuclei in overall brain function. However, at

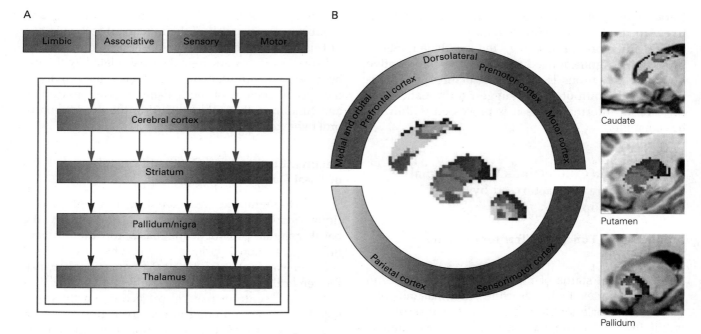

Figure 38–4 Connections between the basal ganglia and cerebral cortex.

A. The connections between the cerebral cortex and basal ganglia can be viewed as a series of parallel projecting, largely segregated loops or channels. Functional territories represented at the level of cerebral cortex are maintained throughout the basal ganglia nuclei and thalamic relays. However, for each loop, the relay points in the cortex, basal ganglia, and thalamus offer opportunities for activity inside the loop to be modified by signals from outside the loop. **Red** and **dark gray arrows** represent excitatory and inhibitory connections, respectively.

B. Spatially segregated rostral-caudal gradient of human frontal cortical connectivity in caudate, putamen, and pallidum. The color-coded ring denotes regions of cerebral cortex in the sagittal plane. (Reproduced, with permission, from Draganski et al. 2008. Copyright © 2008 Society for Neuroscience.)

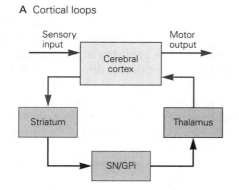

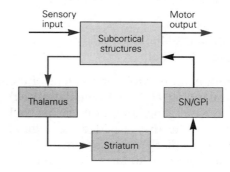

Figure 38–5 Cortical and subcortical sensorimotor loops through the basal ganglia.

A. For cortical loops, the position of the thalamic relay is on the return arm of the loop.

B. In the case of all subcortical loops, the position of the thalamic relay is on the input side of the loop. **Red** indicates predominantly excitatory regions and connections, while **dark gray** indicates inhibitory regions and connections. (Abbreviations: **SN/GPi**, substantia nigra/globus pallidus; **Thal**, thalamus.)

this point, it is important not to think of the reentrant loop architecture as comprising a series of independent and isolated functional channels. At each node or relay point in the loop (eg, in the cortex, the input nuclei, the output nuclei, and the thalamus), there is the opportunity for information flow within the loop to be modified by information from outside the loop (see the section on reinforcement learning below).

At the beginning of this chapter, we stated that behavior is an emergent property of signal processing within a neural network. Having specified the systems-level network of the basal ganglia, we now consider the signals that are being processed within this system.

Physiological Signals Provide Further Clues to Function in the Basal Ganglia

The Striatum and Subthalamic Nucleus Receive Signals Mainly from the Cerebral Cortex, Thalamus, and Ventral Midbrain

Signals received by the striatum from the cerebral cortex and thalamus are conveyed by excitatory glutamatergic neurotransmission. These fast, phasically active excitatory inputs are mediated predominantly by α-amino-3-hydroxy-5-methyl-4-isoxazolepropionic acid (AMPA) and kainate receptors when the medium spiny neurons are near resting potential; N-methyl-D-aspartate (NMDA) receptors play a greater role when the neurons are depolarized. Glutamatergic inputs from both cerebral cortex and thalamus also impinge on striatal interneurons.

It is important to appreciate that these signals come from external structures that are simultaneously generating a wide range of behavioral options. Since these options could not all be expressed at the same time, these inputs to the basal ganglia are thought to be in competition with each other. Another important signal to the striatum is an efference copy of the output activity from the external structures that generate behavioral responses. For example, the sensorimotor territories of the dorsolateral striatum receive collateral fibers from motor cortex axons that send signals to the spinal cord.

The effects of dopaminergic inputs from the ventral midbrain on striatal neuronal activity are complicated, with many conflicting results. In part, this is due to the problem of evoking normal patterns of input activity in slice and anaesthetized preparations. However, recent developments in optogenetic technology in alert, active animals have enabled investigators both to record and manipulate dopamine signals to

the striatum in a temporally controlled manner. Consequently, current evidence suggests dopamine can increase signal-to-noise ratios in the striatum, enhancing the effects of strong external inputs while suppressing weak ones. There is further evidence that dopamine can increase the excitability of medium spiny neurons in the direct pathways while at the same time decreasing the excitability of those in the indirect pathway.

Finally, dopamine input is necessary for both long-term potentiation and long-term depression of glutamatergic inputs to striatal medium spiny neurons from both cortex and thalamus. This latter point is of great significance for the role played by the basal ganglia in reinforcement learning (see below). Dopamine can also influence the activity of GABAergic and cholinergic interneurons. Although anatomically significant, much less is known about the role(s) of serotoninergic inputs to the basal ganglia.

The main external sources of input to the striatum also provide parallel inputs to the subthalamic nucleus. The subthalamus therefore receives phasic excitatory (glutamatergic) signals from the cerebral cortex, thalamus, and brain stem. Following cortical activation, short-latency excitatory effects in the subthalamus are thought to be mediated via these "hyperdirect" connections, whereas longer-latency suppressive effects are more likely to come from indirect inhibitory inputs from other basal ganglia nuclei, principally the external globus pallidus. The subthalamus receives short-latency excitatory sensory input from the brain stem (eg, the superior colliculus); it is also influenced by dopaminergic, serotonergic, and cholinergic modulatory inputs.

Ventral Midbrain Dopamine Neurons Receive Input From External Structures and Other Basal Ganglia Nuclei

Afferent signals to the dopaminergic neurons in the ventral midbrain come from a wide variety of autonomic, sensory, and motor areas and operate over a range of time scales. For example, laterally located neurons in the substantia nigra receive short-latency excitatory inputs from cortical and subcortical sensorimotor regions, while more medially positioned neurons receive both short-latency sensory signals and autonomic-related inputs from the hypothalamus over longer time scales.

Important inhibitory control over dopaminergic neurons is exercised by GABAergic neurons, both local and distant from areas like the rostromedial tegmentum. However, the densest inputs to the dopaminergic neurons are inhibitory inputs from the striatum and globus pallidus and excitatory signals from the sub-thalamic nucleus. The midbrain raphe nuclei provide important modulatory serotonergic input, while both the pedunculopontine nucleus and lateral dorsal tegmental nucleus provide cholinergic and glutamatergic inputs. An important functional question concerning the wide range of afferent signals to dopaminergic neurons is whether dopamine performs a highly integrative role or performs an essential function that is accessed by numerous different systems at different times.

Disinhibition Is the Final Expression of Basal Ganglia Output

The basal ganglia exercise influence over external structures by the fundamental processes of inhibition and disinhibition (Figure 38–6). GABAergic neurons in the basal ganglia output nuclei typically have high tonic firing rates (40–80 Hz). This activity ensures that target regions of the thalamus and brain stem are maintained under a tight and constant inhibitory control.

Focused excitatory inputs from external structures to the striatum can impose focused suppression (mediated via direct pathway GABAergic inhibitory connections) on subpopulations of output nuclei neurons. This focused reduction of inhibitory output effectively releases or disinhibits targeted regions in the thalamus (eg, ventromedial nucleus) and brain stem (eg, superior colliculus) from normal inhibitory control. This sudden release from tonic inhibition allows activity in the targeted region to influence behavioral output, which in the case of the midbrain superior colliculus is to elicit saccadic eye movements.

The patterns of signaling within the basal ganglia architecture provide important insights into what the overall functional properties of these networks might be (see below). Further constraints on the likely core functions of the basal ganglia also become apparent when considering the evolutionary history of the vertebrate brain.

Throughout Vertebrate Evolution, the Basal Ganglia Have Been Highly Conserved

Detailed comparisons between the mammalian basal ganglia and those found in phylogenetically ancient vertebrates (eg, the lamprey) have found striking similarities in their individual components, internal organization, inputs from external structures (the cortex/pallium and thalamus), and the efferent projections of their output nuclei. For example, both direct and

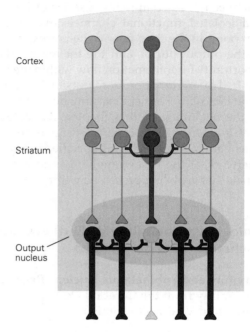

Selective disinhibition

Cortex

Striatum

Output nucleus

Figure 38–6 The diagram illustrates the principle of selection operating at the level of the basal ganglia output nuclei. Throughout the figure, the relative levels of activity within the competing channels are represented by the thickness of projections, and for clarity, the indirect pathway and the return connections of the loops via the thalamus have been omitted. One of the competing inputs to the striatum (the middle one) is more active than its competitors. Relative activities in the direct inhibitory pathways (shown here) differentially suppress activity in the different channels within the output nuclei. Because output nuclei neurons are also inhibitory and tonically active, the selected channel will be the one with the strongest inhibitory input from the striatum. Tonic inhibitory output is maintained on the nonselected channels. This selective disinhibitory mechanism operating at the level of the output nuclei means that selection will be an emergent property of the entire reentrant network. Disinhibition of selected external targets will allow them to direct movement, while nonselected targets remain inhibited and unable to influence behavior. **Red**, excitatory; **gray**, inhibitory.

indirect pathways from striatal medium spiny neurons have been observed in the lamprey. Similarly, tonically active GABAergic output neurons are present in the lamprey internal globus pallidus and substantia nigra pars reticulata. The neurotransmitters and membrane properties of basal ganglia neurons are also remarkably similar in evolutionarily ancient and modern species.

This high degree of morphological and neurochemical conservation implies that the architecture and operation of basal ganglia circuits have been retained for more than 500 million years. The basal ganglia are therefore an essential component of brain architecture that is shared by all vertebrate species. Bearing in mind that a function emerges from specific patterns of signals being processed in specific neural networks, the conservation of basal ganglia architecture across vertebrate species places an additional important constraint on their overall function. Whatever computational problems the basal ganglia evolved to solve in evolutionarily ancient species, the same problems are likely to have remained unchanged and to confront all vertebrate species, including humans.

Thus far, we have identified features of basal ganglia morphology, connectional architecture, signal processing, and evolution that provide potential insights as to the role of the basal ganglia in overall brain function. Thus, proposed functions must be consistent with the predominant looped architecture that connects external structures with the basal ganglia and with an internal circuitry that is shared across the limbic, associative, and sensorimotor territories of the basal ganglia nuclei, and they must be shared by all vertebrate species. With these constraints in mind, we now consider functional properties that could be supported by the basal ganglia.

Action Selection Is a Recurring Theme in Basal Ganglia Research

Despite numerous suggestions that the basal ganglia are involved in a wide range of functions, including perception, learning, memory, attention, many aspects of motor function, and even analgesia and seizure suppression, evidence is accumulating that these nuclei have an underlying role in a variety of selection processes. Thus, throughout the prodigious literature on the basal ganglia there are recurring references to the involvement of these nuclei in the essential brain functions of action selection and reinforcement learning. In this and the next section we will evaluate the extent to which these core processes are consistent with the functional constraints identified above.

All Vertebrates Face the Challenge of Choosing One Behavior From Several Competing Options

Vertebrates are multifunctional organisms: They have to maintain energy and fluid balances, defend against harm, and engage in reproductive activities. Different areas of the brain operate in parallel to deliver these essential functions but must share limited motor resources. Sherrington's "final common motor path" means it is impossible to talk and drink at the same time. Thus, a fundamental selection problem,

continually faced by all vertebrates, is determining which functional system should be allowed to direct behavioral output at any point in time. This is a problem that has not changed materially over the course of 500 million years of evolutionary history. What has changed over this time are the behavioral options that have evolved in different species to implement the core functions of survival and reproduction. Consequently, there has to be a system in the vertebrate brain that can adjudicate between the motivational systems that simultaneously compete for behavioral expression.

A similar selection problem also arises within vertebrate multimodal sensory systems. The visual, auditory, olfactory, and tactile systems are continually faced with multiple external stimuli, each one of which could drive a movement incompatible with one specified by others (eg, orienting/approach, avoidance/escape). It is therefore imperative to select a stimulus that will become the focus of attention and direct movement. The problem is which stimulus should be given access to the motor systems at any one time. Selective attention provides an effective solution to this problem, making it an essential feature of vertebrate brain function.

In summary, despite great evolutionary changes in the range, power, and sophistication of the sensory, motivational, cognitive, and motor systems that compete for behavioral expression in different species, the fundamental computational problems of selection have remained unaltered. And, if the basal ganglia provide a generic solution to the problems of selection, a high degree of structural and functional conservation within vertebrate brain evolution would be expected.

Selection Is Required for Motivational, Affective, Cognitive, and Sensorimotor Processing

In his *Principles of Psychology* (1890), William James observed, "Selection is the very keel on which our mental ship is built." In this statement, he is telling us that the neural systems of motivation, emotion, cognition, perception, and motor performance, at some stage, need to consult a mechanism that can select between parallel processed but incompatible options (Figure 38–7). It is therefore significant that intrinsic circuits in the basal ganglia nuclei are similar across the limbic, associative, and sensorimotor territories.

Such repetition within the basal ganglia circuitry suggests that the same or similar computational processes are applied to inputs from very different functional origins. This duplicated circuitry would therefore be in a position to resolve competitions between high-level motivational goals in the limbic territories;

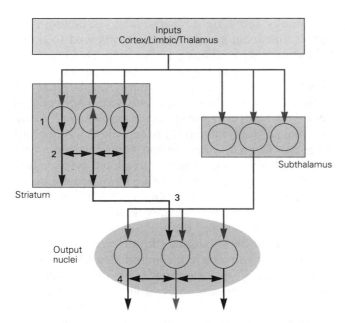

Figure 38–7 Cooperative mechanisms in the basal ganglia that would promote selection.

1. Because cortical and some thalamic inputs make comparatively few contacts with individual striatal neurons, a large population of sufficiently synchronized excitatory inputs is required to depolarize the membrane of a medium spiny neuron to an "up" state sufficient for it to fire action potentials. This mechanism can be seen as an input filter to exclude weak or less biologically significant competitors. The **internal arrows** in striatal neurons denote "up" (**red**) and "down" (**gray**) states.

2. Local GABAergic and peptidergic inhibitory collaterals between striatal spiny neurons and longer-range inhibitory effects of interneurons should cause highly activated striatal elements to suppress activity in adjacent more weakly activated channels.

3. The combination of focused inhibition from the striatum with the more diffuse excitation from the subthalamus would both decrease the activity in selected channels and increase activity in nonselected channels in the basal ganglia output nuclei. The output from just one of the striatal and subthalamic neurons has been illustrated to make this point.

4. Local inhibitory collaterals between output nuclei neurons should further sharpen the difference between inhibited and noninhibited channels.

competitions between incompatible cognitive representations in the central associative territories; and competitions between incompatible sensory and motor options resolved in the lateral sensorimotor regions.

The Neural Architecture of the Basal Ganglia Is Configured to Make Selections

At various times during the past 40 years, and more so recently, it has been argued that the principle function

of the basal ganglia is to select between competing and incompatible behavioral options. It has now been recognized that many aspects of the basal ganglia architecture are consistent with this view (Figure 38–6). The parallel loops originating from and returning to diverse cortical and subcortical functional systems can be viewed as the basic substrate for selection.

The phasic excitatory input signals from the cerebral cortex and thalamus to the different functional territories of the striatum can be seen to carry signals representing the behavioral options competing for expression. To ensure that all options could in principle be assessed against all others, there needs to be a "common currency." This term refers to the parameter according to which qualitatively different functional options can be compared for the purpose of selection. This parameter would be represented by the relative magnitudes of input signals to the striatum, thereby providing each competitor with a measure of relative biological importance or salience. In principle, it should be necessary only for the basal ganglia to appreciate which option is most salient in terms of the common currency.

Processing within the parallel projecting internal architecture (Figure 38–6) would ensure that channels associated with the most salient input activity would cause focused inhibition at the level of the output nuclei (the winning options), while at the same time maintaining or increasing the tonic inhibitory activity in output channels returning to regions specifying weaker (losing) options. Experiments that have recorded neural activity in basal ganglia output nuclei in active animals describe populations of task-sensitive neurons whose activity is reduced or paused prior to movement (the winning option). Conversely, there is a separate, often larger population whose high level of tonic activity is further increased or at least maintained (the losing options). The returning signals within the disinhibited channel(s) are necessary to permit the structures providing the strongest motivational, cognitive, or sensorimotor inputs to access the shared motor resources. Importantly, the maintained or increased levels of inhibitory efferent signals within nonselected channels would prevent the output of nonselected target structures from distorting the selected option's input to the motor system. Thus, this model of the basal ganglia works by keeping all potential behavioral options under tight inhibitory control and selectively removing the inhibition from the option proving the most salient input.

A central-selection control architecture, similar to the systems-level architecture of the basal ganglia just described, was used successfully to select actions for an autonomous mobile robot. Subsequently, it was confirmed that a biologically constrained computer simulation of basal ganglia architecture could do likewise. This work with artificial agents is important because it confirms that selection is indeed an emergent property of systems-level basal ganglia circuitry. The next question is: If the overall architecture can select, are there mechanisms within the basal ganglia that would support or facilitate this function?

Intrinsic Mechanisms in the Basal Ganglia Promote Selection

At each of the major relay points within each of the reentrant loops passing though the basal ganglia (external structures, input nuclei, intrinsic nuclei, output nuclei, and the thalamus), signals flowing within the parallel channels can be subjected to influences originating outside the loop. The selection model outlined above requires features within the internal circuitry of the basal ganglia that permit different channels to interact competitively with each other. Several of these can be identified (Figure 38–7). Together, these mechanisms can be viewed as a cooperative sequence of processes, each of which would facilitate the overall goal of selection. In addition, there is substantial evidence that the relative activity of direct and indirect striatal projection pathways is critical for action selection. The traditional and widely accepted view is that the relative activity in the direct and indirect pathways determines whether or not an animal will perform a particular movement. For example, recent optogenetic stimulation of direct pathway neurons leads to more movement, while optogenetic stimulation of indirect pathway neurons leads to less movement. However, an alternative view for which there is increasing evidence is that simultaneous activity in both pathways is critical for the process of selecting what to do. Here, the idea is that the direct pathway conveys signals representing the most salient options, while the indirect pathway is important for inhibiting the competing weaker options. The latter idea is consistent with the now repeated observations that both projection pathways are concurrently active during movement initiation and that specific patterns of activity in each pathway are associated with different movements.

Selection Function of the Basal Ganglia Questioned

Despite the wide appeal of the selection hypothesis of basal ganglia function, it is by no means universally accepted. Indeed, arguments against it have been voiced based on different studies, the results of which are considered incompatible with the selection model. For example, it has been reported that lesions or suppression of neural activity in motor territories of the internal globus pallidus failed to alter the reaction time between a sensory cue and the triggered movement.

These results could indicate that the basal ganglia are mainly involved in selecting and executing actions that are self-paced, or memory-driven, rather than cue-driven. However, a possibility not considered by these studies is that for well-practiced tasks it is likely that the sensory regions of the basal ganglia will be the most important. This is because such tasks can be performed under stimulus–response habitual control, where selection of the stimulus that triggers the response would be the critical selection. Thus, a failure to disrupt sensory cue selection in such tasks following experimental disruption of the relevant sensory region of basal ganglia would be far stronger evidence against the selection model.

Another recent study claims that in some tasks action choice is already clear in cortical activity even before it reaches the basal ganglia and that the basal ganglia activity is mainly related to reinforcing the commitment to perform the action. This study, and many others like it, base their claims on recordings from afferent structures showing that the neurons are coding the selected stimulus/action/motor program prior to relevant neural responses recorded from within the basal ganglia. An alternative interpretation of these data would be that recordings from all afferent structures that provide competing inputs to the basal ganglia will have shorter latencies than related signals recorded from within the basal ganglia. If in these experiments afferent recordings were from the structure proving the most salient of the competing inputs, then it will be coding for the ultimately selected option *before* it has been selected by basal ganglia.

Other findings are that recordings in the basal ganglia correlate with metrics of movement (eg, speed) and that dopamine signals in the striatum can affect the probability and also the vigor of movement. It is sometimes argued that these results are more indicative of the basal ganglia helping to commit to movement and determining the parameters of movement rather than simply selecting what to do. At least two alternative views could explain why recorded activity in the basal ganglia correlates with movement metrics. First, as mentioned above, one of the significant inputs to the striatum is an efference copy of signals relayed to the motor plant. It would be strange if these signals did not contain information about movement metrics. Second, at this point, it is probably wise to recognize that actions are multidimensional and, as they are learned, require selections about not only *what* to do but also *where* to do it, *when* to do it, and *how* to do it.

The fact that correlates of these various properties of action can be recorded within the basal ganglia nuclei should not necessarily be surprising. Recent studies suggest that *what* and *where* options may arrive to the basal ganglia via glutamatergic input, for example, from the cortex, while *when* options may be modulated by dopaminergic inputs. One of the reasons we know that actions comprise these different dimensions is that each of them can be independently manipulated by reinforcement learning. It is to that topic, which is likely to be an inherent property of a selection architecture, that we now turn.

Reinforcement Learning Is an Inherent Property of a Selection Architecture

The basal ganglia have long been associated with fundamental processes of reinforcement learning. In his famous Law of Effect, first published in his book *Animal Intelligence* (1911), Edward Lee Thorndike proposed that "any act which in a given situation produces satisfaction becomes associated with that situation so that when the situation recurs the act is more likely than before to recur also." Using contemporary language, Thorndike is stating that in a given context an action that has been associated with reward is more likely to be selected in the future when the same or similar contexts are encountered.

Stated in this way, reinforcement learning can be seen as a process for biasing action selection; consequently, it would be expected to operate by modulating activity in the mechanism(s) responsible for selection. How would a reinforcer (reward or punishment) bias selection in the basal ganglia architecture described above? Theoretically, competition between the options represented in the reentrant loops could be biased by sensitizing a reward-related loop at any of its relay points (cortex, input nuclei, globus pallidus, output nuclei, and thalamus). Here, we present just two examples where there is good evidence that reward operating at different nodes within the basal ganglia's reentrant loop circuitry can bias selection (Figure 38–8).

Intrinsic Reinforcement Is Mediated by Phasic Dopamine Signaling Within the Basal Ganglia Nuclei

The popular view of reinforcement in the basal ganglia is that action selection is biased by a dopamine teaching signal that adjusts the sensitivity of intrinsic circuitry so that responses to inputs associated with unpredicted rewards are enhanced (Figure 38–8A). In this model, therefore, the process of reinforcement learning is intrinsic to the basal ganglia nuclei. However, as we have seen above, dopaminergic neurons have highly divergent axons that terminate in wide areas of targeted nuclei. Add to this the problem of volume transmission and the fact that dopaminergic neurons often respond together as a population to relevant events and the problem of how to reinforce only those elements associated with reward or punishment immediately becomes apparent.

It is thought that this issue is addressed by invoking the concept of a decaying eligibility trace. That is, spiking activity in the population of neurons associated with an action that leads to reward is thought to alter the state specifically of those neurons, making them receptive to later widely broadcast reward-related reinforcement signals. There is evidence that this process operates within the basal ganglia. Thus, in most contemporary models, competing behavioral options are represented by specific neurons, the activity of which can be reinforced by phasic increases or decreases in afferent dopamine signals.

Because behavioral experiments have established that unpredicted reward rather than reward per se is critical for learning, the phasic response properties of dopaminergic neurons have captured the imagination of both the biological and computational neuroscience communities. The powerful combination of biological experimentation and computational analyses now indicates clearly that the phasic activity of midbrain dopaminergic neurons provides a teaching signal for reinforcement learning.

While recording from dopaminergic neurons in the ventral midbrain, most studies presented subjects (usually monkeys) either with rewards or neutral stimuli that predicted rewards. The results of these experiments showed that the phasic dopamine responses evoked by unexpected rewards, or the onset of stimuli that predict them, had short response latencies (~100 ms from stimulus onset) and short durations (again ~100 ms). The magnitude of these responses was shown to be influenced by a range of factors including the size, reliability, and extent the reward would be delayed. Importantly, when a neutral stimulus predicted reward (as in traditional Pavlovian conditioning), the phasic dopamine response transferred from the reward to the predicting stimulus. Alternatively, if a reward was predicted but not delivered, the dopaminergic neurons paused briefly at the time the reward would have been delivered. A particularly exciting

A Intrinsic reinforcement

B Extrinsic reinforcement

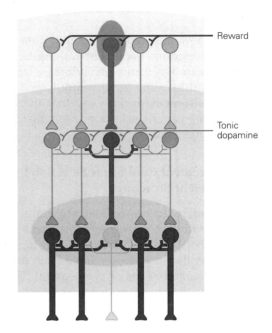

Figure 38–8 Two separate reinforcement mechanisms can bias selection within the reentrant parallel-loop architecture of the basal ganglia. (Return connections of the loops via the thalamus have been omitted for clarity.)

A. Intrinsic reinforcement (**red oval**) involves the selective sensitization of corticostriatal neurotransmission (indicated by the relative thickness of striatal projection neurons in different channels). Transmission in recently active (selected) channel(s) is reinforced by the combined phasic release of dopamine and glutamate evoked by an unpredicted biologically salient sensory event (eg, reward). Nonactive channels lack the eligibility trace required for dopamine reinforcement at the synapse.

The resulting selective plasticity would cause reinforced versions of recent behavioral output to be preferentially reselected, thereby establishing an association between action and outcome.

B. Recent investigations demonstrate that an association with reward (**red oval**) can potentiate processing in structures providing afferent signals to the striatum. Insofar as selection by the basal ganglia is determined in part by the relative strength of inputs to the striatum (the common currency), reward-related modulation of afferent signals would effectively bias selection to favor reward-related inputs. Again, the thickness of projections in the figure denotes relative levels of activation.

finding was that these responses resembled the reward prediction error term in a machine learning reinforcement algorithm. It was therefore widely concluded that phasic dopamine responses could be operating as the brain's teaching signal in reinforcement learning.

With the advent of optogenetic methodology, it has now been established beyond reasonable doubt that phasic dopamine responses can signal both positive and negative reward prediction errors and that these responses correspondingly increase and decrease the probability that prior behavior will be selected. It is thought that phasic dopamine acts by strengthening inputs onto direct pathway neurons in the striatum and weakening inputs onto indirect pathway neurons. Consequently, there is evidence that direct pathway activity can lead animals to do more of a certain action, while indirect pathway activity would lead animals not to do an action.

However, the roles of these pathways may be more complex than this simple dichotomy. In accordance with the different action dimensions outlined above (*what, where, when,* and *how*), activity in both the direct and indirect pathways can reinforce or discourage faster or slower movements, depending on which movements lead to reward in that context. Furthermore, the effects of optogenetic self-stimulation of these pathways on action reinforcement seem to be different between the associative (dorsomedial) and sensorimotor (dorsolateral) domains of the striatum. This could be consistent with different dopamine signals observed in the ventral tegmental area, which projects more medially in the striatum, compared to those in the substantia nigra pars compacta, which projects more laterally. The latter has a higher proportion of dopaminergic neurons that respond to stimulus salience and preferentially respond when the animal initiates self-paced movements (eg, pressing a lever for food whenever it wants, rather than when a sensory cue is presented).

Nonetheless, a wealth of experimental data indicates that phasic dopamine-evoked neural plasticity in the basal ganglia can bias future behavioral selections according to the value of the predicted outcome. This conclusion is consistent with the view that the basal ganglia operate as a generic selection mechanism that can support reinforcement learning.

Extrinsic Reinforcement Could Bias Selection by Operating in Afferent Structures

A second, less widely acknowledged mechanism for biasing selection within the reentrant loop architecture is by modulation of the input salience of competing behavioral options that previously have been associated with a reinforcer—reward or punishment (Figure 38–8B). Since the relative magnitudes of input saliences in competing channels are the common currency by which competing options are judged, reinforcement-induced boosting of a particular channel's input to a selection mechanism would increase the probability of that option being selected in the future.

Evidence in the literature indicates that when a particular stimulus is associated with reward, its representation is enhanced in many of the afferent structures projecting to the basal ganglia. The origin of the reinforcement signals that modulate processing in the input structures is currently unknown. However, the pretuning of basal ganglia inputs by association with reward implies that options associated with high-value outcomes would have a correspondingly higher probability of being selected. Continual updating of input saliences by reward and punishment would bias selections in such a way that the acquisition of reward (or avoidance of punishment) would be maximized in the long term. Finally, it is probably the reward-related tuning of afferent input to the ventral midbrain that enables dopaminergic neurons to accurately report reward prediction errors.

In summary, it is likely that reinforcement learning will be an additional inherent property of a selection architecture. The synaptic relay points at various locations around the parallel reentrant loop architecture provide ample opportunity for the activity in specific loops to be modulated by reward and punishment. There is now good evidence that selection bias can be achieved by reward via a mechanism involving the widespread release of dopamine within the basal ganglia. Reinforcement selectivity is likely to be achieved via some form of eligibility mechanism. A second

possibility is that the relative salience of behavioral options can be modulated by reward and punishment acting directly within the structures that provide input to the basal ganglia.

Behavioral Selection in the Basal Ganglia Is Under Goal-Directed and Habitual Control

Over the past decades, it has become apparent that actions can be learned and then selected based on goal-directed or habitual control. Initially, as we learn to perform particular actions to obtain specific outcomes, these actions are goal-directed, and their performance is highly sensitive to changes in the expected value of the outcome or to changes in the contingency between the action and the outcome. With repetition and consolidation, actions can become not only more efficient but also more automatic, controlled by a stimulus–response type circuit.

In the case of habits, performance becomes less sensitive to changes in the outcome value or changes in contingency between action and the outcome, but rather is controlled by the salience of antecedent stimuli or contexts. Interestingly, shifts from goal-directed to habitual behaviors can be produced not only by extended training, but also by different schedules of reinforcement. Thus, the formation of habits is favored when rewards are delivered according to random time intervals, while goal-directed control is favored when rewards are delivered after a random number of actions.

Different cortical-basal ganglia loops seem to support the learning and performance of goal-directed actions versus habits. The acquisition of goal-directed actions appears to rely on the associative cortical-basal ganglia circuit involving the dorsomedial or associative striatum, the prelimbic cortex, the mediodorsal thalamus, the orbitofrontal cortex, and the amygdala. On the other hand, the formation of habits depends upon circuits coursing through the dorsolateral or sensorimotor striatum, infralimbic cortex, and the central amygdala.

It has been shown that since these two fundamental modes of behavioral control operate through different reentrant loops it has been possible to cause shifts between them through specific manipulations within the basal ganglia. Thus, damage to or inactivation of the associative territories effectively blocks goal-directed control while leaving automatic habitual control relatively unimpaired. Conversely, disruption of the sensorimotor basal ganglia causes habitual performance to switch back to goal-directed control.

Finally, efficient habits, where known stimuli or circumstances trigger a particular response, are very helpful in everyday life such as tying one's shoelaces or locking the front door. However, we also encounter circumstances that cause us to reevaluate our actions. Shifting between goal-directed actions and habits allows us to act flexibly in the environment, and inability to do so may underlie distorted behaviors observed in addiction and other behavioral and neurological disorders of the basal ganglia. It is to this topic that we now turn.

Diseases of the Basal Ganglia May Involve Disorders of Selection

The focus of this chapter has been how the functional architecture of the basal ganglia and their evolutionary history have determined their role in overall brain function. One of the motivations for this exercise that we all have is an intrinsic scientific interest in trying to understand something we currently do not. However, there is another important reason to better understand how the basal ganglia operate. In humans, basal ganglia dysfunction is associated with numerous debilitating conditions including Parkinson disease, Huntington disease, Tourette syndrome, schizophrenia, attention

deficit disorder, obsessive-compulsive disorder, and many addictions. Numerous studies have attempted to shed light on how basal ganglia dysfunction leads to the symptoms that characterize these disorders. This effort can only be helped if we have a better understanding of what a complicated system like the basal ganglia is trying to do when it is operating normally.

A Selection Mechanism Is Likely to Be Vulnerable to Several Potential Malfunctions

Thus far, we have considered the theoretical and empirical evidence supporting the idea that the looped circuitry of the basal ganglia acts as a generic selection mechanism within which reinforcement learning operates to maximize reward and minimize punishment. If action selection and reinforcement learning are the normal functions of the basal ganglia, it should be possible to interpret many of the human basal ganglia–related disorders in terms of selection or reinforcement malfunctions.

Normal selection requires that the selected option is disinhibited at the level of basal ganglia output, while inhibition of nonselected or losing options is maintained or increased (Figure 38–9A). An obvious failure in such a system would be if none of the options were able to achieve sufficient disinhibition to reach a

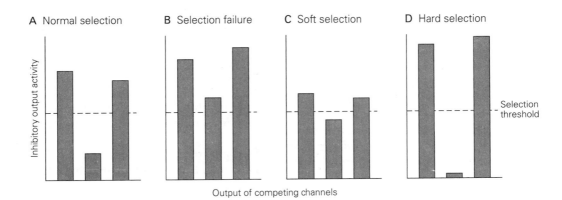

Figure 38–9 Potential disorders of behavior selection.

A. Normal selection within the basal ganglia is characterized by a reduction in inhibition of selected channels below a proposed selection threshold (central channel) while maintaining or increasing inhibition of nonselected channels (left and right channels). Consequently, the disinhibited target structure is able to initiate the action it controls, while the nonselected targets are maintained under inhibitory control.

B. Insufficient reduction in tonic inhibition of all channels means no target structure would be sufficiently disinhibited. This circumstance could explain the akinesia in Parkinson disease.

C. A failure to adequately disinhibit the selected channel or suppress disinhibitory activity in competing channels would cause current selections to be vulnerable to interruption. This disorder could account for the inability to maintain a train of thought and easy distraction by nonattended events in schizophrenia and attention deficit hyperactivity disorder.

D. One channel may become pathologically dominant either through abnormal disinhibition of the selected channel or excessive tonic inhibition of competing channels. This would make the relevant option easy to select and highly resistant to interruption. Hard selections may explain obsessive-compulsive disorder and addictive behaviors.

critical selection threshold (Figure 38–9B). However, a further important point when thinking about selection malfunctions is to appreciate that output inhibition and disinhibition are likely to be continuously variable rather than discrete on/off states. In that case, the difference between the disinhibited and inhibited channels would determine how "hard" or "soft" the selection is. When the difference is large (Figure 38–9D), competing options are likely to find the current selection is resistant to interruption—a larger than normal input salience would be required to cause the system to switch selections. Conversely, when the difference is small (Figure 38–9C), it would be comparatively easy for a competing option to initiate a selection switch.

Support for these ideas comes from behavioral observations showing that at the beginning of task learning there is frequently easy switching between strategies. However, as the task becomes well learned, the system becomes increasingly resistant to alternative strategies. Appreciation of the concepts of hard and soft selection could therefore play an important role when thinking about how a selection mechanism might become dysfunctional in the context of basal ganglia diseases.

Parkinson Disease Can Be Viewed in Part as a Failure to Select Sensorimotor Options

The cardinal symptoms of Parkinson disease are akinesia (difficulties in initiating movement), bradykinesia (initiated movements are slow), and rigidity (stiffness and resistance to passive movement). Tremor is often but not always present. The principal neurological deficit responsible for the motor symptoms of Parkinson disease is thought to be the progressive degeneration of dopaminergic neurotransmission in the basal ganglia.

A consequence of this loss of dopamine is increased tonic and oscillatory activity in the recordings from basal ganglia output nuclei. Since the output of the basal ganglia is GABAergic and inhibitory, in Parkinson disease, targeted structures are receiving high and uneven levels of inhibitory input. This condition impairs the normal selective (disinhibitory) function of the basal ganglia; movements are difficult to select and, when possible, are slow to execute.

Parkinson disease is, however, more nuanced than this. Over much of this progressive condition, the loss of dopaminergic transmission differentially affects the sensorimotor territories of the basal ganglia, leaving the limbic and associative territories comparatively unaffected. As discussed in the section on goal-directed and habitual control, the sensorimotor territories of the basal ganglia

play an essential role in selecting habitual actions. Perhaps, therefore, it is not surprising that many of the motor features of Parkinson disease can be interpreted in terms of a loss of automatic habits. While patients can do things, they are trapped in the slower, serial, and voluntary mode of goal-directed control. In the future, it will be interesting to see if subtle losses of habitual control can be detected before clinical symptoms appear, thereby acting as an early marker for the condition.

Huntington Disease May Reflect a Functional Imbalance Between the Direct and Indirect Pathways

Huntington disease is a genetically transmitted disorder, the initial symptoms of which are subtle changes in mood, personality, cognition, and physical skills. The abnormal movements are characterized by jerky, random, and uncontrollable movements called chorea. The disease is associated with neuronal degeneration. Damage in the early stage is most evident in the striatal medium spiny neurons, but later spreads to other regions of the nervous system.

Observations that neuronal degeneration is evident in limbic, associative, and sensorimotor territories of the striatum would explain why the disease is characterized by disturbances of affect, cognition, and sensorimotor function. Also noteworthy is that the most vulnerable neurons are those in the striatum that project to the external globus pallidus (the indirect pathway) rather than the neurons that project directly to the basal ganglia output nuclei. At the level of the output nuclei, this disturbance would tip the balance in favor of the striatal projection responsible for disinhibition. Consequently, the symptoms of Huntington disease could reflect interference with expression of the selected affective, cognitive, and sensorimotor behaviors by competitors not being sufficiently suppressed.

Schizophrenia May Be Associated With a General Failure to Suppress Nonselected Options

Schizophrenic psychosis is a condition in which there are also disturbances of affect, cognition, and sensorimotor function. Typical symptoms include delusions (false beliefs not based in reality), hallucinations (hearing or seeing things that do not exist), disorganized thinking (inferred from disorganized speech), and abnormal motor behavior (unpredictable agitation, stereotypy, and failure to concentrate on the matter in hand). The disease is progressive, and in later stages, negative symptoms characterized by flattened affect, social withdrawal, absence of thought, and reduced motor behavior become evident (Chapter 60).

Understanding the neurobiological basis of schizophrenia has been complicated by many inconsistent experimental procedures, high variability in symptoms, the side effects of medications, substance abuse, and variability in response to treatments. There is, however, a consistent link between schizophrenia and the basal ganglia insofar as a major class of antipsychotic drugs acts to suppress dopaminergic neurotransmission. In terms of simple regional density of axon terminals and postsynaptic dopamine receptors, dopaminergic transmission within the basal ganglia is likely to be influenced most profoundly by dopamine-related pharmacological therapies. Moreover, there is evidence that dopamine dysregulation in the basal ganglia is intrinsic to the pathology of schizophrenia rather than a medication side effect; predates the psychosis; and is a risk factor for the illness. The implication here is that schizophrenia is associated with a net excess of dopaminergic transmission in the basal ganglia.

So how might dysregulation of this form distort the normal functions of selection and reinforcement? First, the observation that schizophrenia is characterized by disturbances of affect, cognition, and sensorimotor behavior again suggests that the neurobiological substrate will be present in each of the basal ganglia's functional territories. Second, a recurrent theme is that with the positive symptoms there seems to be too much of everything—intense emotional intrusions, too many ideas out of control, spontaneous sensory experiences, too many distracting stimuli, and unpredictable motor agitation. One way of unifying this confusing array of symptoms is to assume that they represent a similar basic fault playing out in different functional territories of the basal ganglia. Here, the basic fault could be a failure on the part of the mechanism responsible for suppressing the impact of competing but nonselected options. Consequently, in all functional territories, the currently selected option would be pathologically vulnerable to interruption (Figure 38–9C).

Attention Deficit Hyperactivity Disorder and Tourette Syndrome May Also Be Characterized by Intrusions of Nonselected Options

Further examples of hyperactive conditions that have been linked to basal ganglia dysfunction may also be due to a faulty selection mechanism where the system in each case is vulnerable to intrusions. Attention deficit hyperactivity disorder (ADHD), like schizophrenia, could in part be the result of a failure in the mechanism responsible for suppressing nonselected sensory options, thereby making it difficult to maintain a focus of attention. Alternatively, the impulsive aspects of

the condition could reflect a malfunction in the neural systems that generate behavioral options based on the value of likely consequences. In this situation, options driven by immediately desired sensory events would take precedence over competing representations of disadvantageous longer-term consequences.

In the case of Tourette syndrome, converging evidence indicates that the involuntary behavioral intrusions (verbal and motor tics) are associated with aberrant activity in the cortical–basal ganglia–thalamic loops. In animal models, similar motor tics can be evoked by blocking inhibitory neurotransmission in local areas of the sensorimotor striatum. Were the disease state also to cause a similar failure of inhibition or inappropriate excitation in parts of the striatum not engaged by the current selection, disruptive motor intrusions might be expected. Furthermore, were the locus of the excessive excitation to remain constant and the motor characteristics of the intrusion to be repeated, it is likely that the mechanism for establishing automatic habits would be engaged, thereby further enhancing the automatic involuntary nature of the intrusion.

Obsessive-Compulsive Disorder Reflects the Presence of Pathologically Dominant Options

Persons with obsessive-compulsive disorder compulsively repeat specific actions (hand washing, counting things, checking things) or have particular thoughts repeatedly come to mind uninvited (obsessions). Studies using functional neuroimaging when the symptoms are present consistently report abnormal activation at various locations within the cortical-striatum-thalamus-cortical loops.

In terms of a selection mechanism dysfunction, the symptoms of obsessive-compulsive disorder would be expected when, for whatever reason, the input salience of relevant functional channels would be abnormally dominant, thereby making it difficult for competing options to interrupt or cause behavioral or attentional switching (hard selection). The fact that the obsessional and compulsive options are dominant behaviors that have been learned suggests that the fault responsible for obsessive-compulsive disorder may lie with the reinforcement mechanism capable of adjusting input salience. Of course, such a fault could be of genetic and/or environmental origin.

Addictions Are Associated With Disorders of Reinforcement Mechanisms and Habitual Goals

Addiction to drugs and other behaviors (eg, gambling, sex, eating) represents a dramatic dysregulation of motivational selections. This is caused by an exaggerated salience of addiction-related stimuli, binge indulgence, and withdrawal anxiety. When addictions are being acquired, changes in dopaminergic and opioid peptide transmission in the basal ganglia have been reported.

Insofar as these transmission systems have been linked with fundamental mechanisms of reinforcement, it might be expected that the selective reinforcement of addiction-related stimuli would lead to observed increases in the ability of these stimuli to capture behavior. Alternatively, the increases in negative emotional states and stress-like responses experienced during withdrawal have been associated with reductions in dopamine function. In the limbic territories of the basal ganglia, such reductions are typically associated with negative reinforcement.

A final point to note is that if addiction-associated stimuli can automatically trigger the motivation/goal to indulge (ie, an automatic stimulus–goal association), a similar kind of mechanism may be operating in the limbic territories as is currently assigned to stimulus–response habits in the sensorimotor territories. Thus, if in the case of drug addiction the goal of drug acquisition may be correctly described as a stimulus-driven habit, the practicalities of obtaining the drug can be highly goal directed (eg, robbing a convenience store, phoning the dealer) and not at all habitual.

From the above sections, it can be seen that interpreting disorders of the basal ganglia in terms of dysregulations of selection and reinforcement does not require implausible intellectual contortions. Indeed, this could be regarded as further support for the view that the systems-level function of the basal ganglia is to operate as a generic selection mechanism. Moreover, having an overriding conceptual framework based on potential disorders of normal function has an important advantage for guiding future research. Instead of fishing in the brains of patients and animal models for clues of what might have gone wrong, one is hunting within a specified network for a malfunction that would be expected to produce the observed disorder.

Highlights

1. The basal ganglia are an interconnected group of nuclei located at the base of the forebrain and midbrain. There are three major input structures (the striatum, the subthalamic nucleus, and the dopamine cells of substantia nigra) and two major output structures (the internal globus pallidus and substantia nigra pars reticulata).

2. Input structures receive projections from most regions of the cerebral cortex, limbic system, and brain stem, many via relays in the thalamus. Inputs to the striatum and subthalamus are topographically organized.

3. The spatial topography is maintained throughout the intrinsic basal ganglia connections, as well as in projections back to the cortex, limbic system, and brain stem structures. Thus, an essential feature of systems-level basal ganglia architecture can be viewed as a series of reentrant loops.

4. The striatum was thought to be connected to the output nuclei via direct and indirect pathways. However, recent anatomical evidence suggests a more complex internal architecture.

5. Phasic excitatory input to the basal ganglia is mediated by the neurotransmitter glutamate. Tonic inhibitory output from the basal ganglia is mediated by the neurotransmitter GABA. The reentrant loops keep afferent structures under strong inhibitory control. For any task, the tonic inhibitory firing of some output neurons pauses, while for others, it is maintained or increased.

6. Basal ganglia architecture appeared at the outset of vertebrate evolution and has been highly conserved throughout. This suggests that the computational problems they solve are likely to be problems faced by all vertebrate species.

7. The internal microarchitecture of the intrinsic basal ganglia nuclei is largely the same throughout their motivational, affective, cognitive, and sensorimotor territories. This suggests that the same basal ganglia algorithm is applied to all general classes of brain function.

8. A recurring theme within the basal ganglia literature is their involvement in action selection and reinforcement learning.

9. The selection hypothesis is supported by the following: (1) Selection is a generic problem faced by all vertebrates. (2) A selection algorithm common to all basal ganglia territories could resolve competitions between incompatible motivational, affective, cognitive, and sensorimotor options. (3) Many intrinsic processes could support a selection function. (4) Selective removal of output inhibition within a multiple reentrant looped architecture is necessarily a selection process. (5) Computational models of basal ganglia architecture effectively select the actions of multifunctional robots.

10. Abundant evidence indicates that the basal ganglia are an essential substrate for reinforcement

learning where selections are biased by the valence/value of past outcomes.

11. The multidimensional aspects of action (what, where, when, and how to do something) can be independently modified by reinforcement learning. It will be important to determine whether these different aspects of action are learned within the same or different functional territories of the basal ganglia.

12. Recent optogenetic investigations have confirmed that phasic dopamine signaling can act as a training signal for reinforcement learning.

13. Within the reentrant looped architecture, future selections can be biased not only within the basal ganglia by dopamine but also at synapses in external afferent structures and the thalamic relays.

14. Reinforcement learning can bias selections on the basis of outcome value (goal-directed), or by operating on an acquired automatic stimulus–response association (habit). Goal-directed and habitual selections are made in different functional territories of the basal ganglia.

15. Insofar as diseases of the basal ganglia in humans can be interpreted as selection malfunctions, additional support is provided for the idea that the basal ganglia operate as a generic selection module.

Peter Redgrave
Rui M. Costa

Suggested Reading

Cui G, Jun SB, Jin X, et al. 2013. Concurrent activation of striatal direct and indirect pathways during action initiation. Nature 494:238–242.

da Silva JA, Tecuapetla F, Paixão V, Costa RM. 2018. Dopamine neuron activity before action initiation gates and invigorates future movements. Nature 554:244–248.

Grillner S, Robertson B, Stephenson-Jones M. 2013. The evolutionary origin of the vertebrate basal ganglia and its role in action selection. J Physiol 591:5425–5431.

Hikosaka O, Ghazizadeh A, Griggs W, Amita H. 2018. Parallel basal ganglia circuits for decision making. J Neural Transm (Vienna) 125:515–529.

Kravitz AV, Freeze BS, Parker PR, et al. 2010. Regulation of parkinsonian motor behaviours by optogenetic control of basal ganglia circuitry. Nature 466:622–626.

Redgrave P, Prescott T, Gurney KN. 1999. The basal ganglia: a vertebrate solution to the selection problem? Neuroscience 89:1009–1023.

Redgrave P, Rodriguez M, Smith Y, et al. 2010. Goal-directed and habitual control in the basal ganglia: implications for Parkinson's disease. Nat Rev Neurosci 11:760–772.

Saunders A, Oldenburg IA, Berezovskii VK, et al. 2015. A direct GABAergic output from the basal ganglia to frontal cortex. Nature 521:85–89.

Yin HH, Knowlton BJ. 2006. The role of the basal ganglia in habit formation. Nat Rev Neurosci 7:464–476.

Yttri EA, Dudman JT. 2016. Opponent and bidirectional control of movement velocity in the basal ganglia. Nature 533:402–406.

References

Albin RL, Mink JW. 2006. Recent advances in Tourette syndrome research. Trends Neurosci 29:175–182.

Albin RL, Young AB, Penney JB. 1989. The functional anatomy of basal ganglia disorders. Trends Neurosci 12:366–375.

Alexander GE, Crutcher MD, Delong MR. 1990. Functional architecture of basal ganglia circuits: neural substrates of parallel processing. Trends Neurosci 13:226–271.

Arbuthnott GW, Wickens J. 2007. Space, time and dopamine. Trends Neurosci 30:62–69.

Carmona S, Proal E, Hoekzema EA, et al. 2009. Ventro-striatal reductions underpin symptoms of hyperactivity and impulsivity in attention-deficit/hyperactivity disorder. Biol Psychiatry 66:972–977.

Chevalier G, Deniau JM. 1990. Disinhibition as a basic process in the expression of striatal functions. Trends Neurosci 13:277–281.

DeLong MR, Wichmann T. 2007. Circuits and circuit disorders of the basal ganglia. Arch Neurol 64:20–24.

Deniau JM, Mailly P, Maurice N, Charpier S. 2007. The pars reticulata of the substantia nigra: a window to basal ganglia output. In: JM Tepper. ED Abercrombie, JP Bolam (eds). *Gaba and the Basal Ganglia: From Molecules to Systems.* Prog Brain Res 160:151–172.

Desmurget M, Turner RS. 2010. Motor sequences and the basal ganglia: kinematics, not habits. J Neurosci 30:7685–7690.

Draganski B, Kherif F, Klöppel S, et al. 2008. Evidence for segregated and integrative connectivity patterns in the human basal ganglia. J Neurosci 28:7138–7152.

Fan D, Rossi MA, Yin HH. 2012. Mechanisms of action selection and timing in substantia nigra neurons. J Neurosci 32:5534–5548.

Gerfen CR, Surmeier DJ. 2011. Modulation of striatal projection systems by dopamine. Ann Rev Neurosci 34:441–466.

Gerfen CR, Wilson CJ. 1996. The basal ganglia. In: LW Swanson, A Bjorklund, T Hokfelt (eds). *Handbook of Chemical Neuroanatomy, Vol 12: Integrated Systems of the CNS, Part III*, pp. 371–468. Amsterdam: Elsevier.

Graybiel AM. 2008. Habits, rituals, and the evaluative brain. Ann Rev Neurosci 31:359–387.

Hikosaka O. 2007. Basal ganglia mechanisms of reward-oriented eye movement. Ann NY Acad Sci 1104:229–249.

Howes OD, Kapur S. 2009. The dopamine hypothesis of schizophrenia: version III—the final common pathway. Schizophr Bull 353:549–562.

Humphries MD, Stewart RD, Gurney KN. 2006. A physiologically plausible model of action selection and oscillatory activity in the basal ganglia. J Neurosci 26:12921–12942.

Kelly RM, Strick PL. 2004. Macro-architecture of basal ganglia loops with the cerebral cortex: use of rabies virus to reveal multisynaptic circuits. Prog Brain Res 143:449–459.

Klaus A, Martins GJ, Paixao VB, Zhou P, Paninski L, Costa RM. 2017. The spatiotemporal organization of the striatum encodes action space. Neuron 95:1171–1180.

Koob GF, Volkow ND. 2016. Neurobiology of addiction: a neurocircuitry analysis. Lancet Psychiatry 38:760–773.

MacDonald AW, Schulz SC. 2009. What we know: findings that every theory of schizophrenia should explain. Schizophr Bull 3:493–508.

Matsuda W, Furuta T, Nakamura KC, et al. 2009. Single nigrostriatal dopaminergic neurons form widely spread and highly dense axonal arborizations in the neostriatum. J Neurosci 29:444–453.

Matsumoto M, Takada M. 2013. Distinct representations of cognitive and motivational signals in midbrain dopamine neurons. Neuron 79:1–14.

McHaffie JG, Stanford TR, Stein BE, Coizet V, Redgrave P. 2005. Subcortical loops through the basal ganglia. Trends Neurosci 28:401–407.

Mink JW. 1996. The basal ganglia: focused selection and inhibition of competing motor programs. Prog Neurobiol 50:381–425.

Minski M. 1986. *The Society of Mind*. London: Heinemann Ltd.

Nambu A. 2011. Somatotopic organization of the primate basal ganglia. Front Neuroanat 5:26.

Nambu A, Tokuno H, Takada M. 2002. Functional significance of the cortico-subthalamo-pallidal 'hyperdirect' pathway. Neurosci Res 43:111–117.

Nasser HM, Calu DJ, Schoenbaum G, Sharpe MJ. 2017. The dopamine prediction error: contributions to associative models of reward learning. Front Psychol 8:244.

Nieuwenhuys R, Voogd J, van Huijzen C. 1981. *The Human Central Nervous System: A Synopsis and Atlas*, 2nd ed. Berlin: Springer.

Piron C, Kase D, Topalidou M, et al. 2016. The globus pallidus pars interna in goal-oriented and routine behaviors: resolving a long-standing paradox. Mov Disord 31:1146–1154.

Plotkin JL, Surmeier DJ. 2015. Corticostriatal synaptic adaptations in Huntington's disease. Curr Opin Neurobiol 33:53–62.

Redgrave P, Gurney KN. 2006. The short-latency dopamine signal: a role in discovering novel actions? Nat Rev Neurosci 7:967–975.

Reiner AJ. 2010. The conservative evolution of the vertebrate basal ganglia. In: H Steiner, KY Tseng (eds). *Handbook of Basal Ganglia Structure and Function*, pp. 29–62. Burlington, MA: Academic Press

Reiner A, Jiao Y, DelMar N, Laverghetta AV, Lei WL. 2003. Differential morphology of pyramidal tract-type and intratelencephalically projecting-type corticostriatal neurons and their intrastriatal terminals in rats. J Comp Neurol 457:420–440.

Schultz W. 2007. Multiple dopamine functions at different time courses. Annu Rev Neurosci 30:259–288.

Silberberg G, Bolam JP. 2015. Local and afferent synaptic pathways in the striatal microcircuitry. Curr Opin Neurobiol 33:182–187.

Smith Y, Galvan A, Ellender TJ, et al. 2014. The thalamostriatal system in normal and diseased states. Front Syst Neurosci 8:5.

Surmeier DJ, Plotkin J, Shen W. 2009. Dopamine and synaptic plasticity in dorsal striatal circuits controlling action selection. Curr Opin Neurobiol 19:621–628.

Tecuapetla F, Jin X, Lima SQ, Costa RM. 2016. Complementary contributions of striatal projection pathways to action initiation and execution. Cell 166:703–715.

Thorndike EL. 1911. *Animal Intelligence*. New York: Macmillan.

van den Heuvel OA, van Wingen G, Soriano-Mas C, et al. 2016. Brain circuitry of compulsivity. Eur Neuropsychopharmacol 26:810–827.

Watabe-Uchida M, Zhu LS, Ogawa SK, Vamanrao A, Uchida N. 2012. Whole-brain mapping of direct inputs to midbrain dopamine neurons. Neuron 74:858–873.

Yael D, Vinner E, Bar-Gad I. 2015. Pathophysiology of tic disorders. Mov Disord 30:1171–1178.

Yin HH, Knowlton BJ. 2006. The role of the basal ganglia in habit formation. Nat Rev Neurosci 7:464–476.

39

Brain–Machine Interfaces

UNDERSTANDING THE NORMAL FUNCTION of the nervous system is central to understanding dysfunction caused by disease or injury and designing therapies. Such treatments include pharmacological agents, surgical interventions, and, increasingly, electronic medical devices. These medical devices fill an important gap between largely molecularly targeted and systemic medications and largely anatomically targeted and focal surgical lesions.

In this chapter, we focus on medical devices that measure or alter electrophysiological activity at the level of populations of neurons. These devices are referred to as brain–machine interfaces (BMIs), brain–computer interfaces, or neural prostheses. We use the term BMI to refer to all such devices because there is no standard distinction among them. BMIs can be organized into four broad categories: those that restore lost sensory capabilities, those that restore lost motor capabilities, those that regulate pathological neural activity, and those that restore lost brain processing capabilities.

BMIs can help people perform "activities of daily living," such as feeding oneself, physically dressing and grooming oneself, maintaining continence, and walking. A type of BMI that we will discuss extensively in this chapter converts electrical activity from neurons in the brain into signals that control prosthetic devices to help people with paralysis. By understanding how neuroscience and neuroengineering work together to create current BMIs, we can more clearly envision how many neurological diseases and injuries can be treated with medical devices.

BMIs Measure and Modulate Neural Activity to Help Restore Lost Capabilities

Cochlear Implants and Retinal Prostheses Can Restore Lost Sensory Capabilities

One of the earliest and most widely used BMIs is the cochlear implant. People with profound deafness can benefit from restoration of even some audition. Since the 1970s, several hundred thousand people who have a peripheral cause of deafness that leaves the cochlear nerve and central auditory pathways intact have received cochlear implants. These systems have restored considerable hearing and spoken language, even to children with congenital deafness who have learned to perceive speech using cochlear implants.

Cochlear implants operate by capturing sounds with a microphone that resides outside the skin and sending these signals to a receiver surgically implanted under the skin near the ear. After conversion (encoding) to appropriate spatial-temporal signal patterns, these signals electrically stimulate spiral ganglion cells in the cochlear modiolus (Chapter 26). In turn, signals from the activated cochlear cells are transmitted through the auditory nerve to the brain stem and higher auditory areas where, ideally, the neural signals are interpreted as the sounds captured by the microphone.

Another example of a BMI is a retinal prosthesis. Blindness can be caused by diseases such as retinitis pigmentosa, an inherited retinal degenerative disease. At present, there is no cure and no approved medical therapy to slow or reverse the disease. Retinal prostheses currently enable patients to recognize large letters and locate the position of objects. They operate by capturing images with a camera and sending these signals to a receiver positioned within the eye. After conversion to appropriate spatial-temporal patterns, these electrical signals stimulate retinal ganglion cells in the retina through dozens of electrodes. In turn, these cells send their signals through the optic nerve to the thalamus and higher visual areas where, ideally, the afferent signals are interpreted as the image captured by the camera.

Motor and Communication BMIs Can Restore Lost Motor Capabilities

BMIs are also being developed to assist paralyzed people and amputees by restoring lost motor and communication function. This is the central topic of this chapter. First, electrical neural activity in one or more brain areas is measured using penetrating multielectrode arrays placed, for example, in the arm and hand region of the primary motor cortex, dorsal and ventral premotor cortex, and/or intraparietal cortex (particularly the parietal reach region and medial intraparietal area) (Figure 39–1).

Second, an arm movement is attempted but cannot be made in the case of people with paralysis. Action potentials and *local field potentials* are measured during these attempts. With 100 electrodes placed in the primary motor cortex and another 100 in the dorsal premotor cortex, for example, action potentials from approximately 200 neurons and local field potentials from 200 electrodes are measured. Local field potentials are lower-frequency signals recorded on the same electrodes as the action potentials and believed to arise from local synaptic currents of many neurons near the electrode tips. Together, these neural signals contain considerable information about how the person wishes to move her arm.

Third, the relationship between neural activity and attempted movements is characterized. This relationship makes it possible to predict the desired movement from new neural activity, a statistical procedure we refer to as *neural decoding*. Fourth, the BMI is then operated in its normal mode where neural activity is measured in real time and desired movements are decoded from the neural activity by a computer. The decoded movements can be used to guide prosthetic devices, such as a cursor on a computer screen or a robotic arm. It is also possible to electrically stimulate muscles in a paralyzed limb to enact the decoded movements, a procedure known as *functional electrical stimulation*. Many other prosthetic devices can be envisioned as we increasingly interact with the world around us electronically (eg, smart phones, automobiles, and everyday objects that are embedded with electronics so that they can send and receive data—known as the "internet of things").

Finally, because the person can see the prosthetic device, she can alter her neural activity by thinking different thoughts on a moment-by-moment basis so as to guide the prosthetic device more accurately. This closed-loop feedback control system can make use of nonvisual sensory modalities as well, including delivering pressure and position information from electronic sensors wrapped on or embedded in a prosthetic arm. Such surrogate sensory information can be transformed into electrical stimulation patterns that are delivered to proprioceptive and somatosensory cortex.

The BMIs described above include motor and communication BMIs. Motor BMIs aim to provide natural control of a robotic limb or a paralyzed limb. In the case of upper-limb prostheses, this involves the

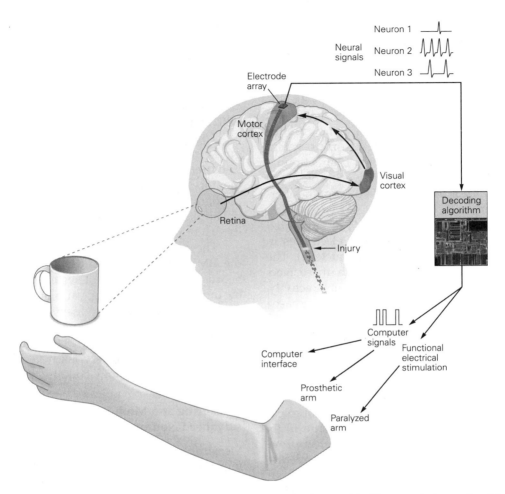

Figure 39–1 Concept of motor and communication brain–machine interfaces. One or more electrode arrays are implanted in brain regions such as the primary motor cortex, dorsal and ventral premotor cortex, or intraparietal cortex. They record action potentials from tens to hundreds of neurons and local field potentials. The recorded neural activity is then converted by a decoding algorithm into (1) computer commands for controlling a computer interface or a prosthetic (robotic) arm, or (2) stimulation patterns for functional electrical stimulation of muscles in a paralyzed arm.

precise movement of the arm along a desired path and with a desired speed profile. Such control is indeed an ambitious ultimate goal, but even intermediate steps toward this goal could improve quality of life by restoring some lost motor function and improving the patient's ability to carry out "activities of daily living." For example, numerous people with tetraplegia could benefit from being able to feed themselves.

Communication BMIs are designed to provide a fast and accurate interface with a plethora of electronic devices. The ability to move a computer cursor around an on-screen keyboard allows a patient to type commands for computers, smart phones, voice synthesizers, smart homes, and the "internet of things." Ideally, communication BMIs would allow for a communication rate at which most people speak or type.

Such BMIs would benefit people with amyotrophic lateral sclerosis (ALS), who often become "locked in" and unable to communicate with the outside world through any movements. Communication BMIs would also benefit people with other neurodegenerative diseases that severely compromise the quality of movement and speech, as well as those with upper spinal cord injury. The ability to reliably type several words per minute is a meaningful improvement in quality of life for many patients.

Motor and communication BMIs build on basic neuroscientific research in voluntary movement (Chapter 34). The design and development of BMIs have so far depended on laboratory animal research, largely with nonhuman primates; recently, however, pilot clinical trials with humans with paralysis have begun.

Pathological Neural Activity Can Be Regulated by Deep Brain Stimulation and Antiseizure BMIs

BMIs have been developed to help people with disorders involving pathological neural activity in the brain, such as Parkinson disease and epilepsy. People with Parkinson disease benefit by having hand and arm tremor reduced. At present, there is no cure for Parkinson disease, and many people become resistant to pharmacological treatments. A deep brain stimulator (DBS) can help these people by delivering electrical pulses to targeted areas in the brain to disrupt the aberrant neural activity.

DBS is controlled by a neurostimulator implanted in the chest, with wires to stimulating electrodes in deep brain nuclei (eg, the subthalamic nucleus). The nuclei are continuously stimulated with these electrodes in order to alter the aberrant neural activity. This method can often greatly reduce Parkinson disease–related tremor for years. A DBS applied to different brain areas can also help people with essential tremor, dystonia, chronic pain, major depression, and obsessive-compulsive disorder.

Millions of people experiencing epileptic seizures are currently treated with antiseizure medications or neurosurgery, both of which often result in incomplete or impermanent seizure reduction. Antiseizure BMIs have shown considerable promise for further improving quality of life. These fully implanted BMIs operate by continuously monitoring neural activity in a brain region determined to be involved with seizures. They identify unusual activity that is predictive of seizure onset and then respond within milliseconds to disrupt this activity by electrically stimulating the same or a different brain region. This closed-loop response can be fast enough that seizure symptoms are not felt and seizures do not occur.

Replacement Part BMIs Can Restore Lost Brain Processing Capabilities

BMIs are capable of restoring more than lost sensory or motor capabilities. They are, in principle, capable of restoring internal brain processing. Of the four categories of BMIs, this is the most futuristic. An example is a "replacement part" BMI. The central idea is that if enough is known about the function of a brain region, and if this region is damaged by disease or injury, then it may be possible to replace this brain region.

Once the normal input activity to a brain region is measured (see next section), the function of the lost brain region could then be modeled in electronic hardware and software, and the output from this substitute

processing center would then be delivered to the next brain region as though no injury had occurred. This would involve, for example, reading out neural activity with electrodes, mimicking the brain region's computational functions with low-power microelectronic circuits, and then writing in electrical neural activity with stimulating electrodes.

This procedure might also be used to initiate and guide neural plasticity. A replacement part BMI that is currently being investigated focuses on restoring memory by replacing parts of the hippocampus that are damaged due to injury or disease. Another potential application would be to restore the lost functionality of a brain region damaged by stroke.

These systems represent the natural evolution of the BMI concept, a so-called "platform technology" because a large number of systems can be envisioned by mixing and matching various write-in, computational, and read-out components. The number of neurological diseases and injuries that BMIs should be able to help address ought to increase as our understanding of the functions of the nervous system and the sophistication of the technology continue to grow.

Measuring and Modulating Neural Activity Rely on Advanced Neurotechnology

Measuring and modulating neural activity involves four broad areas of electronic technologies applied to the nervous system (so-called neurotechnology). The first area is the type of neural sensor; artificial neural sensors are designed with different levels of invasiveness and spatial resolution (Figure 39–2). Sensors that are external to the body, such as an *electroencephalogram* (EEG) cap, have been used extensively in recent decades. The EEG measures signals from many small metal disks (electrodes) applied to the surface of the scalp across the head. Each electrode detects average activity from a large number of neurons beneath it.

More recently, implantable electrode-array techniques, such as subdural *electrocorticography* (ECoG) and finely spaced micro-ECoG electrodes, have been used. Since ECoG electrodes are on the surface of the brain and are thus much closer to neurons than EEG electrodes, ECoG has higher spatial and temporal resolution and thus provides more information with which to control BMIs.

Most recently, arrays of *penetrating intracortical electrodes*, which we focus on in this chapter, have been used. The intracortical electrode arrays are made of silicon or other materials and coated with biocompatible materials. The arrays are implanted on the surface of the brain, with the electrode tips penetrating 1 to 2 mm

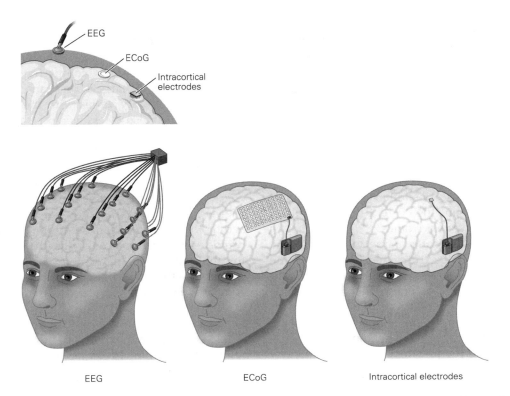

Figure 39–2 Brain–machine interfaces use different types of neural sensors. Electrical neural signals can be measured with various techniques ranging from electroencephalography (**EEG**) electrodes on the surface of the skin, to electrocorticography (**ECoG**) electrodes on the surface of the brain, to intracortical electrodes implanted in the outer 1 to 2 mm of cortex. The signals that can be measured range from the average of many neurons, to averages across fewer neurons, and finally to action potentials from individual neurons. (Adapted, with permission, from Blabe et al. 2015.)

into the cortex. They have the ability to record action potentials from individual neurons, as well as local field potentials from small clusters of neurons near each electrode tip. The electrodes are able to record high-fidelity signals because they are inserted into the brain, bringing the electrode tips within micrometers of neurons. This is beneficial for BMI performance because individual neurons are the fundamental information-encoding units in the nervous system, and action potentials are the fundamental units of the digital code that carries information from the input to the output region of a neuron. Moreover, intracortical electrodes can deliver electrical microstimulation to either disrupt neural activity (eg, DBS) or write in surrogate information (eg, proprioceptive or somatosensory information).

The second area of neurotechnology is scaling up the number of neurons measured at the same time. While one neuron contains some information about a person's intended movement, tens to hundreds of neurons are needed to move a BMI more naturally, and even more neurons are needed to approach naturalistic levels of motor function. Although it is possible to

place electrode arrays in many areas across the brain, thereby gaining more information from multiple areas, a key challenge is to measure activity from thousands of neurons within each individual brain area. Many efforts are underway to achieve this goal, including use of electrode arrays with many tiny shafts, each with hundreds of electrode contacts along its length; many tiny electrodes that are not physically wired together, but are instead inserted into the brain as stand-alone islands that transmit data outside of the head and receive power wirelessly; and optical imaging technologies that can capture the activity of hundreds or more neurons by detecting how each neuron's fluorescence changes over time.

The third area is low-power electronics for signal acquisition, wireless data communications, and wireless powering. In contrast to the BMI systems described above, which implant a passive electrode array in which each electrode is wired to the outside world by a connector passing through the skin, future BMIs will be fully implanted like DBS systems. Electronic circuits are needed to amplify neural signals, digitize them, process them (eg, to detect when an action potential occurred or to estimate local field potential power), and transmit this information to a nearby receiver incorporated into a prosthetic arm, for example. Power consumption must be minimized for two reasons. First, the more power is consumed, the more power a battery or a wireless charging system would need to provide. Batteries would therefore need to be larger and replaced more often, and delivering power wirelessly is challenging. Second, using power generates heat, and the brain can only tolerate a small temperature increase before there are deleterious effects. These trade-offs are similar to those of smart phones, which represent the current best technology available for low-power electronics.

The final area is so-called supervisory systems. Software running on electronic hardware is at the heart of BMIs. Some software implements the mathematical operations of the neural decoding, while other software must tend to aspects of the BMI's overall operation. For example, the supervisory software should monitor whether or not a person wishes to use the prosthesis (eg, if the person is sleeping); if neural signals have changed, thereby requiring recalibration of the decoder; and overall BMI performance and safety.

Having discussed the range of different BMIs and neurotechnologies being developed, in the rest of this chapter we focus on motor and communication BMIs. We first describe different types of decoding algorithms and how they work. We then describe recent progress in BMI development toward assisting paralyzed people and amputees. Next, we consider how sensory feedback can improve BMI performance and how BMIs can be used as an experimental paradigm to address basic scientific questions about brain function. Finally, we conclude with a cautionary note about ethical issues that can arise with BMIs.

BMIs Leverage the Activity of Many Neurons to Decode Movements

Various aspects of movement—including position, velocity, acceleration, and force—are encoded in the activity of neurons throughout the motor system (Chapter 34). Even though our understanding of movement encoding in the motor system is incomplete, there is usually a reliable relationship between aspects of movement and neural activity. This reliable relationship allows us to estimate the desired movement from neural activity, a key component of a BMI.

To study movement encoding, one typically considers the activity of an individual neuron across repeated movements (referred to as "trials") to the same target. The activity of the neuron can be averaged across many trials to create a spike histogram for each target (Figure 39–3A). By comparing the spike histograms for different targets, one can characterize how the neuron's activity varies with the movement produced. One can also assess using the spike histograms whether the neuron is more involved in movement preparation or movement execution.

In contrast, estimating a subject's desired movement from neural activity (referred to as movement *decoding*) needs to be performed on an individual trial while the neural activity is being recorded. The activity of a single neuron cannot unambiguously provide such information. Thus, the BMI must monitor the activity of many neurons on a single trial (Figure 39–3B) rather than one neuron on many trials. A desired movement can be decoded from the neural activity associated with either preparation or execution of the movement. Whereas preparation activity is related to the movement goal execution activity is related to the moment-by-moment details of movement (Chapter 34).

Millions of neurons across multiple brain areas work together to produce a movement as simple as reaching for a cup. Yet in many BMIs, desired movements can be decoded reasonably accurately from the activity of dozens of neurons recorded from a single brain area. Although this may seem surprising, the fact is that the motor system has a great deal of redundancy—many neurons carry similar information about a desired movement (Chapter 34). This is reasonable because millions of neurons are involved in controlling the contractions of dozens of muscles. Thus, most of the neurons in regions of dorsal premotor cortex and primary motor cortex controlling arm movement are informative about most arm movements.

When decoding a movement, the activity of one neuron provides only incomplete information about the movement, whereas the activity of many neurons can provide substantially more accurate information about the movement. This is true for activity associated with both movement preparation and execution. There are two reasons why using multiple neurons is helpful for decoding. First, a typical neuron alone cannot unambiguously determine the intended movement direction. Consider a neuron whose activity (during either preparation or execution) is related to movement direction via a cosine function, known as a *tuning curve* (Figure 39–4A). If this neuron fires at 30 spikes per second, the intended movement direction could be either 120° or 240°. However, by recording from a second neuron whose tuning curve is different from that of the first neuron, the movement direction can be

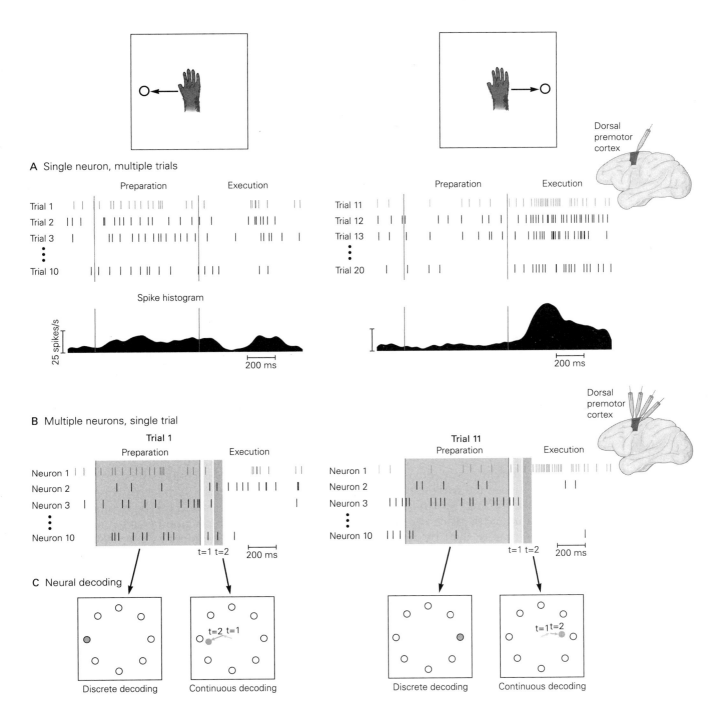

Figure 39–3 Movement encoding uses the activity of individual neurons averaged across experimental trials, whereas movement decoding uses the activity of many neurons on individual experimental trials.

A. Activity of one neuron recorded in the dorsal premotor cortex of a monkey preparing and executing leftward arm movements (*left*) and rightward arm movements (*right*). Characterizing the movement encoding of a neuron involves determining how the activity of the neuron on repeated leftward or rightward movements (each row of spike trains) relates to aspects of arm movement. **Below** is the spike histogram for this neuron for leftward and rightward movements, obtained by averaging neural activity across trials. This neuron shows a greater level of preparation activity for leftward movements and a greater level of execution activity for rightward movements. Many neurons in the dorsal premotor cortex and primary motor cortex show movement-related activity in both the preparation and execution epochs like the neuron shown.

B. Neural activity for many neurons recorded in the dorsal premotor cortex for one leftward movement (*left*) and one rightward movement (*right*). The spike trains for neuron 1 correspond to those shown in part **A**. Spike counts are taken during the preparation epoch, typically in a large time bin of 100 ms or longer to estimate movement goal. In contrast, spike counts are taken during the execution epoch typically in many smaller time bins, each lasting tens of milliseconds. Using such short time bins provides the temporal resolution needed to estimate the moment-by-moment details of the movement.

C. Neural decoding involves extracting movement information from many neurons on a single experimental trial. In the subject's workspace, there are eight possible targets (**circles**). Discrete decoding (see Figure 39–5) extracts the target location; the estimated target is filled in with **gray**. In contrast, continuous decoding (see Figure 39–6) extracts the moment-by-moment details of the movement; the **orange dot** represents the estimated position at one moment in time.

A One neuron

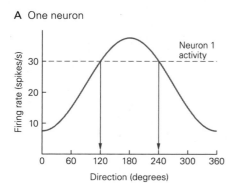

B Two neurons (noiseless)

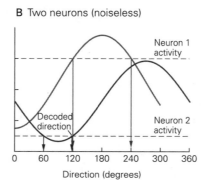

C Two neurons (noisy)

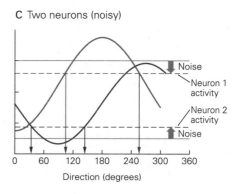

Figure 39–4 More than one neuron is needed for accurate movement decoding.

A. The tuning curve of one neuron defines how the neuron's activity varies with movement direction. If this neuron shows activity of 30 spikes/s, it could correspond to movement in the 120° or 240° direction.

B. A second neuron (**green**) with a different tuning curve shows activity of 5 spikes/s, which could correspond to

movement in the 60° or 120° direction. The only movement direction consistent with the activity of both neurons is 120°, which is determined to be the decoded direction.

C. Because neural activity is "noisy" (represented as a vertical displacement of the dashed lines), it is usually not possible to conclusively determine the movement direction from the activity of two neurons. Here, no one movement direction is consistent with the activity of both neurons.

more accurately determined. If the second neuron fires at 5 spikes per second, corresponding to a movement in either the 60° or 120° direction, the only movement direction that is consistent between the two neurons is 120° (Figure 39–4B). Thus, by recording from these two neurons simultaneously, the intended reach direction can be determined more accurately than by recording from one neuron. (However, two neurons do not necessarily provide a perfect estimate of the intended reach direction due to noise, as described next.)

The second reason why decoding a movement from the activity of several neurons gives greater accuracy is because a neuron's activity level usually varies across repeated movements in the same direction. This variability is typically referred to as spiking "noise." Let us say that due to spiking noise the first neuron fires at slightly less than 30 spikes per second and the second neuron fires at slightly more than 5 spikes per second (Figure 39–4C). Under these conditions, no single movement direction is consistent with the activity level of both neurons. Instead, a compromise must be made between the two neurons to determine a movement direction that is as consistent as possible with their activities. By extending this concept to more than two neurons, the movement direction can be decoded even more accurately as the number of neurons increases.

Decoding Algorithms Estimate Intended Movements From Neural Activity

Movement decoders are a central component of BMIs. There are two types of BMI decoders: discrete

and continuous (Figure 39–3C). A *discrete decoder* estimates one of several possible movement goals. Each of these movement goals could correspond to a letter on a keyboard. A discrete decoder solves a classification problem in statistics and can be applied to either preparation activity or execution activity. A *continuous decoder* estimates the moment-by-moment details of a movement trajectory. This is important, for example, for reaching around obstacles or turning a steering wheel. A continuous decoder solves a regression problem in statistics and is usually applied to execution activity rather than preparation activity because the moment-by-moment details of a movement can be more accurately estimated from execution activity (Chapter 34).

Motor BMIs must produce movement trajectories as accurately as possible to achieve the desired movement and typically use a continuous decoder to do this. In contrast, communication BMIs are concerned with enabling the individual to transmit information as rapidly as possible. Thus, the speed and accuracy with which movement goals (or keys on a keyboard) can be selected are of primary importance. Communication BMIs can use a discrete decoder to directly select a desired key on a keyboard or a continuous decoder to continuously guide the cursor to the desired key, where only the key eventually struck actually contributes to information conveyance. This seemingly subtle distinction has implications that influence the type of neural activity required and therefore the brain area that is targeted, as well as the type of decoder that is used.

Neural decoding involves two phases: calibration and ongoing use. In the calibration phase, the relationship between neural activity and movement is characterized by a statistical model. This can be achieved by recording neural activity while a paralyzed person attempts to move, imagines moving, or passively observes movements of a computer cursor or robotic limb. Once the relationship has been defined, the statistical model can then be used to decode new observed neural activity (ongoing use phase). The goal during the ongoing use phase is to find the movement that is most consistent with the observed neural activity (Figure 39–4B,C).

Discrete Decoders Estimate Movement Goals

We first define a population activity space, where each axis represents the firing rate of one neuron. On each trial (ie, movement repetition), we can measure the firing rate of each neuron during a specified period, and together they yield one point in the population activity space. Across many trials, involving multiple movement goals, there will be a scatter of points in the population activity space. If the neural activity is related to the movement goal, then the points will be separated in the population activity space according to the movement goal (Figure 39–5A). During the calibration phase, *decision boundaries* that partition the population

activity space into different regions are determined by a statistical model. Each region corresponds to one movement goal.

During the ongoing use phase, we measure new neural activity for which the movement goal is unknown (Figure 39–5B). The decoded movement goal is determined by the region in which the neural activity lies. For example, if the neural activity lies within the region corresponding to the leftward target, then the discrete decoder would guess that the subject intended to move to the leftward target on that trial. It is possible that the subject intended to move to the rightward target, even though the recorded activity lies within the region corresponding to the leftward target. In this case, the discrete decoder would incorrectly estimate the subject's intended movement goal. Decoding accuracy typically increases with an increasing number of simultaneously recorded neurons.

Continuous Decoders Estimate Moment-by-Moment Details of Movements

Arm position, velocity, acceleration, force, and other aspects of arm movement can be decoded using the methods described here with varying levels of accuracy. For concreteness, we will discuss decoding movement velocity because it is one of the quantities most strongly reflected in the activity of motor cortical

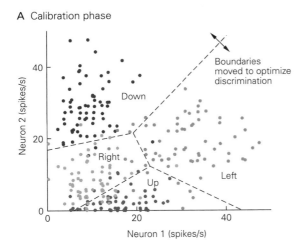

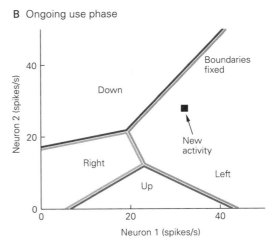

Figure 39–5 Discrete decoding.

A. Calibration phase. A population activity space is shown for two neurons, where each axis represents the firing rate of one neuron. On each trial (ie, movement repetition), the activity of the two neurons together defines one point in the population activity space. Each point is colored by the movement goal, which is known during the calibration phase. Decision boundaries (**dashed lines**) are determined by a statistical model to optimize discrimination among the movement goals. The

decision boundaries define a region in the population activity space for each movement goal.

B. Ongoing use phase. During this phase, the decision boundaries are fixed. If we record new neural activity (**square**) for which the movement goal is unknown, the movement goal is determined by the region in which the neural activity lies. In this case, the neural activity lies in the region corresponding to the leftward target, so the decoder would guess that the subject intended to move to the leftward target.

neurons and is the starting point for the design of most BMI systems.

Consider a population of neurons whose level of activity indicates the movement velocity (ie, speed and direction). During the calibration phase, a "pushing vector" is determined for each neuron (Figure 39–6A). A pushing vector indicates how a neuron's activity influences movement velocity. Various continuous decoding algorithms differ in how they determine the pushing vectors. One of the earliest decoding algorithms, the population vector algorithm (PVA), assigns each neuron's pushing vector to point along the neuron's preferred direction (see Figure 34–22A). A neuron's preferred direction is defined as the direction of movement for which the neuron shows the highest level of activity (ie, peak of curves in Figure 39–4). Much of the pioneering work on BMIs used the PVA. However, the PVA does not take into account the properties of the spiking noise (ie, its variance and covariance across neurons), which influences the accuracy of the decoded movements. A more accurate decoder, the optimal linear estimator (OLE), incorporates the properties of the spiking noise to determine the pushing vectors.

During the ongoing use phase, the pushing vectors are each scaled by the number of spikes emitted by the corresponding neuron at each time step (Figure 39–6B). At each time step, the decoded movement is the vector sum of the scaled pushing vectors across all neurons. The decoded movement represents a change in position during one time step (ie, velocity). The BMI cursor (or limb) position (Figure 39–6C) is then updated according to the decoded movement.

To further improve decoding accuracy, the estimation of velocity at each time step should take into account not only current neural activity (as illustrated in Figure 39–6), but also neural activity in the recent past. The rationale is that movement velocity (and other kinematic variables) changes gradually over time, and so neural activity in the recent past should be informative about the movement velocity. This can be achieved by temporally smoothing the neural activity before applying a PVA or OLE or by using a Kalman filter to define a statistical model describing how movement velocity (or other kinematic variables) changes smoothly over time. With a Kalman filter, the estimated velocity is a combination of the scaled pushing vectors at the current time step (as in Figure 39–6B) and the estimated velocity at the previous time step. Indeed, continuous decoding algorithms that take into account neural activity in the recent past have been shown to provide higher decoding accuracy than those that do not. The Kalman filter and its extensions are widely used in BMIs and among the most accurate continuous decoding algorithms available.

Increases in Performance and Capabilities of Motor and Communication BMIs Enable Clinical Translation

Patients with paralysis wish to perform activities of daily living. For people with ALS or upper spinal cord injury who are unable to speak or to move their arms, the most desired tasks are often the ability to communicate, to move a prosthetic (robotic) arm, or to move the paralyzed arm by stimulating the musculature. Having described how neural signals can be read out from motor areas of the brain and how these electrical signals can be decoded to arrive at BMI control signals, we now describe recent progress toward restoring these abilities.

The majority of laboratory studies are carried out in able-bodied nonhuman primates, although paralysis is sometimes transiently induced in important control experiments. Three types of experimental paradigms are in broad use, differing in the exact way in which arm behavior is instructed and visual feedback is provided during BMI calibration and ongoing use. Setting

Figure 39–6 (Opposite) Continuous decoding.

A. During the calibration phase, a pushing vector is determined for each of 97 neurons. Each vector represents one neuron and indicates how one spike from that neuron drives a change in position per time step (ie, velocity). Thus, the units of the plot are millimeters per spike during one time step. Different neurons can have pushing vectors of different magnitudes and directions.

B. During ongoing use, spikes are recorded from the same neurons as in panel **A** during movement execution. At each time step, the new length of an arrow is obtained by starting with its previous length in panel **A** and scaling it by the number of spikes produced by the neuron of the same color during that time step. If a neuron does not fire, there is no arrow for that neuron during that time step. The decoded movement (**black arrow**) is the vector sum of the scaled pushing vectors, representing a change in position during one time step (ie, velocity). For a given neuron, the direction of its scaled pushing vectors is the same across all time steps. However, the magnitudes of the scaled pushing vectors can change from one time step to the next depending on the level of activity of that neuron.

C. The decoded movements from panel **B** are used to update the position of a computer cursor (**orange dot**), robotic limb, or paralyzed limb at each time step.

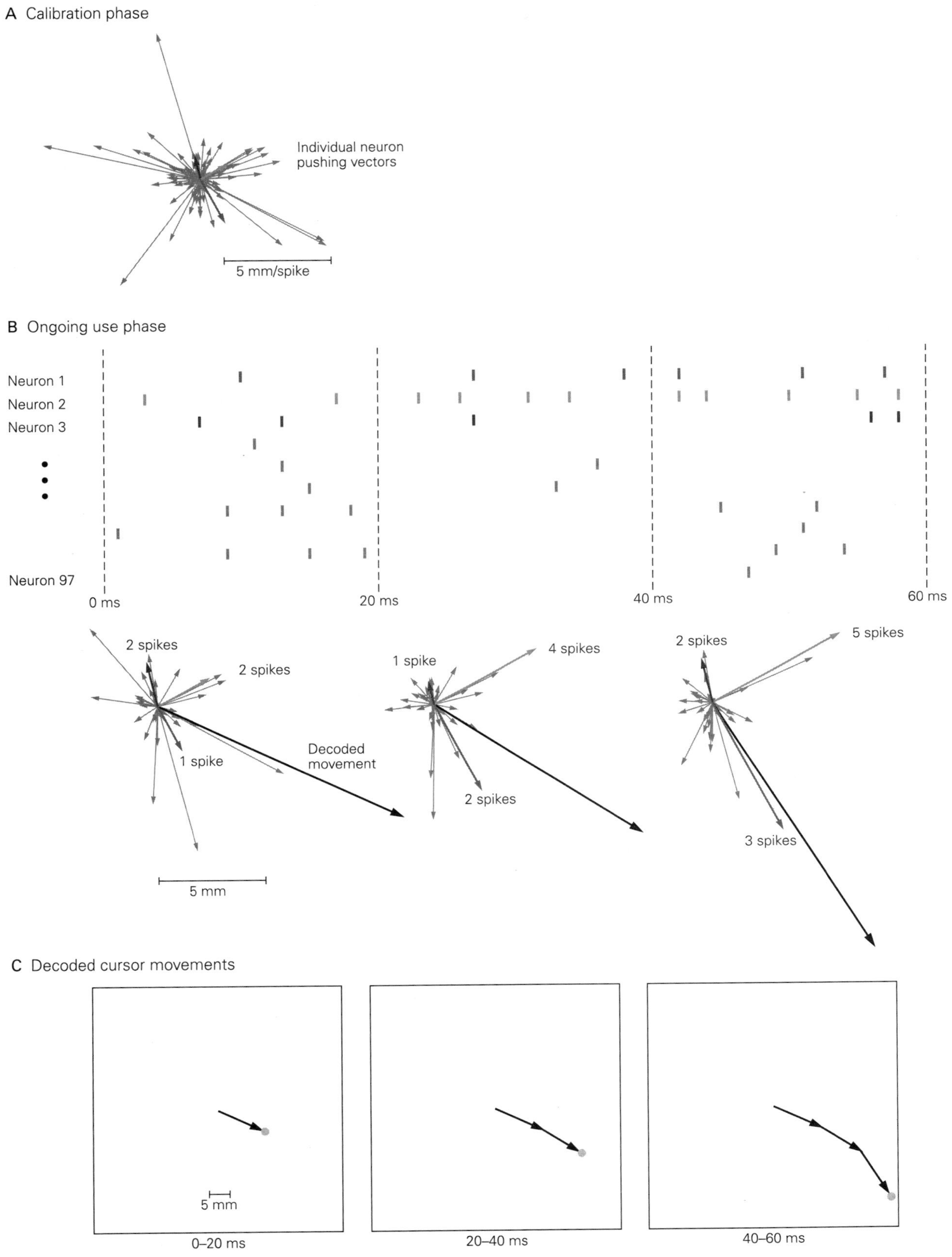

A Calibration phase

Individual neuron
pushing vectors

5 mm/spike

B Ongoing use phase

Neuron 1
Neuron 2
Neuron 3

Neuron 97

0 ms   20 ms   40 ms   60 ms

2 spikes

2 spikes

1 spike

Decoded
movement

1 spike   4 spikes

2 spikes

2 spikes   5 spikes

3 spikes

5 mm

C Decoded cursor movements

5 mm

0–20 ms   20–40 ms   40–60 ms

these differences aside, we focus below on how BMIs function and perform. We also highlight recent pilot clinical trials with people with paralysis.

Subjects Can Type Messages Using Communication BMIs

To investigate how quickly and accurately a communication BMI employing a discrete decoder and preparation activity can operate, monkeys were trained to fixate and touch central targets and prepare to reach to a peripheral target that could appear at one of several different locations on a computer screen. Spikes were recorded using electrodes implanted in the premotor cortex. The number of spikes occurring during a particular time window during the preparation epoch was used to predict where the monkey was preparing to reach (Figure 39–7A). If the decoded target matched the peripheral target, a liquid reward was provided to indicate a successful trial.

By varying the duration of the period in which spike counts are taken and the number of possible targets, it was possible to assess the speed and accuracy

of target selections (Figure 39–7B). Decoding accuracy tended to increase with the period in which spike counts are taken because spiking noise is more easily averaged out in longer periods.

An important metric for efficient communication is information transfer rate (ITR), which measures how much information can be conveyed per unit time. A basic unit of information is a bit, which is specified by a binary value (0 or 1). For example, with three bits of information, one can specify which of $2^3 = 8$ possible targets or keys to press. Thus, the metric for ITR is bits per second (bps). ITR increases with the period in which spike counts are taken, then declines. The reason is that ITR takes into account both how accurately and how quickly each target is selected. Beyond some point of diminishing returns of a longer period, accuracy fails to increase rapidly enough to overcome the slowdown in target-selection rate accompanying a longer period.

Overall performance (ITR) increases with the number of possible targets, despite a decrease in decoding accuracy, because each correct target selection conveys more information. Fast and accurate communication

A Experimental setup

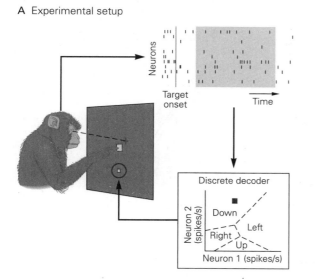

B Single-trial decoding accuracy decreases and ITR increases as more target locations are used

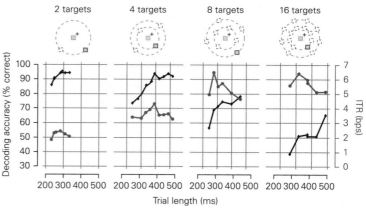

Figure 39–7 A communication brain–machine interface can control a computer cursor using a discrete decoder based on neural activity during the preparation epoch.

A. After a monkey touched a central target (**large yellow square**) and fixated a central point (**red +**), a peripheral target (**small yellow square**) appeared and the monkey prepared to reach to it. Spike counts were taken during the preparation epoch and fed into a discrete decoder. The duration of the period in which spike counts are taken (ie, width of **light blue** shading) affects decoding performance and information transfer rate (ITR) (see panel **B**). Based on the spike counts

(**blue square**), the discrete decoder guessed the target the monkey was preparing to reach to.

B. Decoding accuracy (**black**) and information transfer rate (ITR, **bits/s**; **red**) are shown for different trial lengths and numbers of targets. Trial length was equal to the duration of the period in which spike counts were taken (varied during the experiment) plus 190 ms (fixed during the experiment). The latter provided time for visual information of the peripheral target to reach the premotor cortex (150 ms), plus the time to decode the target location from neural activity and render the decoded target location on the screen (40 ms). (Adapted, with permission, from Santhanam et al. 2006.)

has been demonstrated in BMIs with this design based on a discrete decoder applied to preparatory activity. The ITR of this BMI is approximately 6.5 bps, which corresponds to approximately two to three targets per second with greater than 90% accuracy.

Recent studies have also investigated how quickly and accurately a communication BMI employing a continuous decoder and execution activity can operate. Two different types of continuous decoders were evaluated: a standard Kalman filter decoding movement velocity (V-KF) and a recalibrated feedback intention-trained Kalman filter (ReFIT-KF). The V-KF was calibrated using the neural activity recorded during actual arm movements (ie, open-loop control). The ReFIT-KF incorporated the closed-loop nature of BMIs into decoder calibration by assuming that the user desired to move the cursor straight to the target at each time step.

To assess performance, both types of decoders were used in closed-loop BMI control (Figure 39–8A). Monkeys were required to move a computer cursor from a central location to eight peripheral locations and back. A gold standard for performance evaluation was established by having the monkeys also perform the same task using arm movements. The ReFIT-KF outperformed the V-KF in several ways: Cursor movements using ReFIT-KF were straighter, producing less movement away from a straight line to the target; cursor movements were faster, approaching the speed of arm movements (Figure 39–8B); and there were fewer (potentially frustrating) long trials.

Given its performance benefits, the ReFIT-KF is being used in clinical trials by people with paralysis (Figure 39–8C). Spiking activity was recorded using a 96-channel electrode array implanted in the hand control area of the left motor cortex. Signals were filtered to extract action potentials and high-frequency local field potentials, which were decoded to provide "point-and-click" control of the BMI-controlled cursor. The subject was seated in front of a computer monitor and was asked, "How did you encourage your sons to practice music?" By attempting to move her right hand, the computer cursor moved across the screen and stopped over the desired letter. By attempting to squeeze her left hand, the letter beneath the cursor was selected, much like clicking a mouse button.

BMI performance in the clinical trials was assessed by measuring the number of intended characters subjects were able to type (Figure 39–8D). Subjects were able to demonstrate that the letters they typed were intended by using the delete key to erase occasional mistakes. These clinical tests showed that it is possible to type at a rate of many words per minute using a BMI.

Subjects Can Reach and Grasp Objects Using BMI-Directed Prosthetic Arms

Patients with paralysis would like to pick up objects, feed themselves, and generally interact physically with the world. Motor BMIs with prosthetic limbs aim to restore this lost motor functionality. As before, neural activity is decoded from the brain but is now routed to a robotic arm where the wrist is moved in three dimensions (x, y, and z) and the hand is moved in an additional dimension (grip angle, ranging from an open hand to a closed hand).

In one test of a robotic arm, a patient with paralysis was able to use her neural activity to direct the robotic arm to reach out, grab a bottle of liquid, and bring it to her mouth (Figure 39–9). The three-dimensional reaches and gripping were slower and less accurate than natural arm and hand movements. Importantly, this demonstrated that the same BMI paradigm originally developed with animals, including measuring and decoding signals from motor cortex, works in people even years after the onset of neural degeneration or the time of neural injury.

BMI devices directing prosthetic arms and hands are now able to do more than just control three-dimensional movement or open and close the hand. They can also orient the hand and grasp, manipulate, and carry objects. A person with paralysis was able to move a prosthetic limb with 10 degrees of freedom to grasp objects of different shapes and sizes and move them from one place to another (Figure 39–10). Completion times for grasping and moving objects were considerably slower than natural arm movements, but the results are encouraging. These studies illustrate the existing capabilities of prosthetic arms and also the potential for even greater capabilities in the future.

Subjects Can Reach and Grasp Objects Using BMI-Directed Stimulation of Paralyzed Arms

An alternative to using a robotic arm is to restore lost motor function to the biological arm. Arm paralysis results from the loss of neural signaling from the spinal cord and brain, but the muscles themselves are often still intact and can be made to contract by electrical stimulation. This capacity underlies functional electrical stimulation (FES), which sends electrical signals via internal or external electrodes to a set of muscle groups. By shaping and timing the electrical signals sent to the different muscle groups, FES is able to move the arm and hand in a coordinated fashion to pick up objects.

Laboratory studies in monkeys have demonstrated that this basic approach is viable in principle.

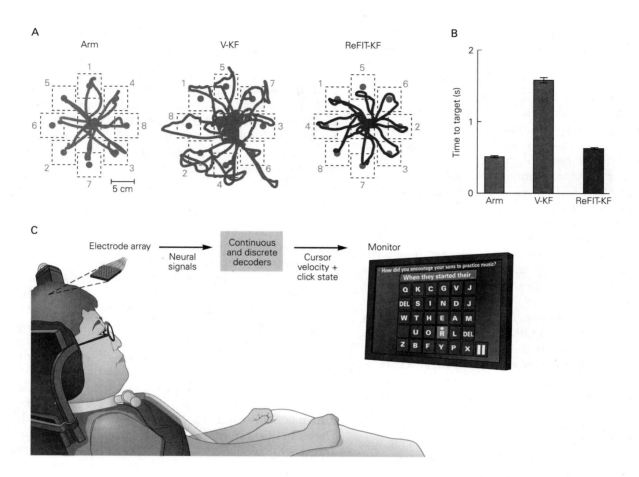

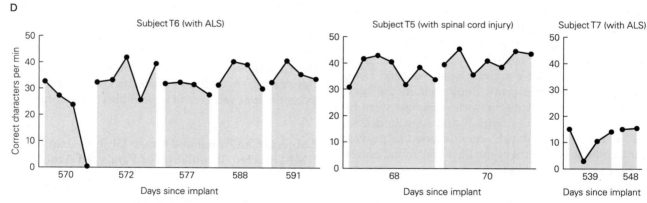

Figure 39–8 A communication brain–machine interface (BMI) can control a computer cursor using a continuous decoder based on neural activity during the execution epoch.

A. Comparison of cursor control by a monkey using its arm, a standard decoder that estimates velocity (BMI with Kalman filter decoding movement velocity [V-KF]), and a feedback intention-trained decoder (BMI with recalibrated feedback intention-trained Kalman filter [ReFIT-KF]). Traces show cursor movements to and from targets alternating in the sequence indicated by the numbers shown. Traces are continuous for the duration of all reaches. (Adapted, with permission, from Gilja et al. 2012.)

B. Time required to move the cursor between the central location and a peripheral location on successful trials (mean ± standard error of the mean). (Adapted, with permission, from Gilja et al. 2012.)

C. Pilot clinical trial participant T6 (53-year-old female with amyotrophic later sclerosis [ALS]) using a BMI to type the answer to a question. (Adapted, with permission, from Pandarinath et al. 2017.)

D. Performance in a typing task for three clinical trial participants. Performance can be sustained across days or even years after array implantation. (Adapted, with permission, from Pandarinath et al. 2017.)

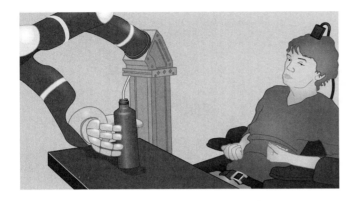

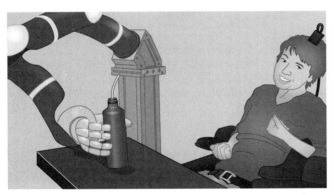

Figure 39–9 A subject with paralysis drinks from a bottle using a robotic arm controlled by a motor brain–machine interface using a continuous decoder. Three sequential images from the first successful trial show the subject using the robotic arm to grasp the bottle, bring it to her mouth and drink coffee through a straw, and place the bottle back on the table. (Adapted from Hochberg et al. 2012.)

It is implemented by calibrating a continuous decoder to predict the intended activity of each of several of the muscles, transiently paralyzed with a nerve block. These predictions are then used to control the intensity of stimulation of the same paralyzed muscles, which in turn controls motor outputs such as a grip angle and force. This process in effect bypasses the spinal cord and restores some semblance of voluntary control

of the paralyzed arm and hand. Similar results have recently been demonstrated in patients with paralysis using either externally applied or fully implanted state-of-the-art FES electrodes. Intracortically recorded signals from motor cortex were decoded to restore movement via FES in a person with upper spinal cord injury (Figure 39–11). The subject was able to achieve control of different wrist and hand motions, including finger movements, and perform various activities of daily living.

Subjects Can Use Sensory Feedback Delivered by Cortical Stimulation During BMI Control

During arm movements, we rely on multiple sources of sensory feedback to guide the arm along a desired path or to a desired goal. These sources include visual, proprioceptive, and somatosensory feedback. However, in most current BMI systems, the user receives only visual feedback about the movements of the computer cursor or robotic limb. In patients with normal motor output pathways but lacking proprioception, arm movements are substantially less accurate than in healthy individuals, both in terms of movement direction and extent. Furthermore, in tests of BMI cursor control in healthy nonhuman primate subjects, the arm continues to provide proprioceptive feedback even though arm movements are not required to move the cursor. BMI cursor control is more accurate when the arm is passively moved together with the BMI cursor along the same path, rather than along a different path. This demonstrates the importance of "correct" proprioceptive feedback. Based on these two lines of evidence, it is perhaps not surprising that BMI-directed movements relying solely on visual feedback are slower and less accurate than normal arm movements. This has motivated recent attempts to demonstrate how providing surrogate (ie, artificial) proprioceptive or somatosensory feedback can improve BMI performance.

Several studies have attempted to write in sensory information by stimulating the brain using cortical electrical microstimulation. Laboratory animals can discriminate current pulses of different frequencies and amplitudes, and this ability can be utilized to provide proprioceptive or somatosensory information in BMIs by using different pulse frequencies to encode different physical locations (akin to proprioception) or different textures (akin to somatic sensation). Electrical microstimulation in the primary somatosensory cortex can be used by nonhuman primates to control a cursor on a moment-by-moment basis without vision. In these subjects, the use of electrical microstimulation

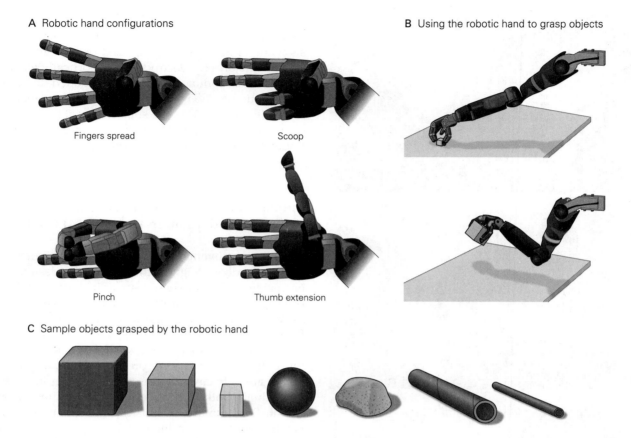

A Robotic hand configurations

Fingers spread

Scoop

Pinch

Thumb extension

B Using the robotic hand to grasp objects

C Sample objects grasped by the robotic hand

Figure 39–10 A motor brain–machine interface (BMI) can control a prosthetic arm with 10 degrees of freedom.

A. Examples of different hand configurations directed by the BMI. The 10 degrees of freedom are three-dimensional arm translation, three-dimensional wrist orientation, and four-dimensional hand shaping.

B. A subject uses the prosthetic arm to pick up an object and move it.

C. Objects of different shapes and sizes are used to test the generalization ability of the BMI. (Adapted from Wodlinger et al. 2015.)

and visual feedback together led to more accurate movements than either type of sensory feedback alone.

Furthermore, electrical microstimulation in the primary somatosensory cortex can also be used to provide tactile information. Nonhuman primates moved a BMI-directed cursor under visual feedback to hit different visual targets, each of which elicited a different stimulation frequency. Subjects learned to use differences in the stimulation feedback to distinguish the rewarded target from the unrewarded targets. This demonstrates that electrical microstimulation can also be used to provide somatosensory feedback during BMI control.

Finally, surrogate somatosensory information was delivered via electrical microstimulation to a person with paralysis and compromised sensory afferents. The person reported naturalistic sensations at different locations of his hand and fingers corresponding to different locations of stimulation in the primary somatosensory cortex.

BMIs Can Be Used to Advance Basic Neuroscience

BMIs are becoming an increasingly important experimental tool for addressing basic scientific questions about brain function. For example, cochlear implants have provided insight into how the brain processes sounds and speech, how the development of these mechanisms is shaped by language acquisition, and how neural plasticity allows the brain to interpret a few channels of stimulation carrying impoverished auditory information. Similarly, motor and communication BMIs are helping to elucidate the neural mechanisms underlying sensorimotor control. Such scientific findings can then be used to refine the design of BMIs.

The key benefit of BMIs for basic science is that they can simplify the brain's input and output interface with the outside world, without simplifying the complexities of brain processing that one wishes to study.

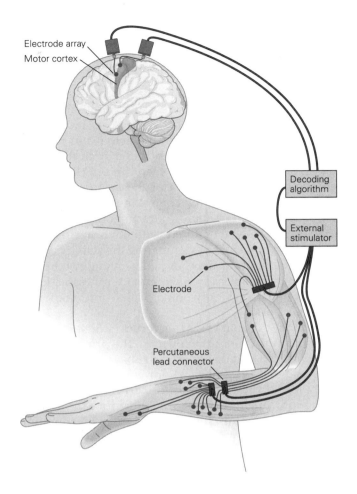

Figure 39–11 A motor brain–machine interface (BMI) can control the muscles of a paralyzed arm using a continuous decoder and functional electrical stimulation. Neural activity recorded in the motor cortex is decoded into command signals that control the stimulation of deltoid, pectoralis major, biceps, triceps, forearm, and hand muscles. This enables cortical control of whole-arm movements and grasping. Muscle stimulation is performed through percutaneous intramuscular fine-wire electrodes. (Adapted, with permission, from Ajiboye et al. 2017. Copyright © 2017 Elsevier Ltd.)

To illustrate this point, consider the output interface of the brain for controlling arm movements. Thousands of neurons from the motor cortex and other brain areas send signals down the spinal cord and to the arm, where they activate muscles that move the arm. Understanding how the brain controls arm movement is challenging because one can typically record from only a small fraction of the output neurons that send signals down the spinal cord, the relationship between the activity of the output neurons and arm movements is unknown, and the arm has nonlinear dynamics that are difficult to measure. Furthermore, it is usually difficult to determine which recorded neurons are output neurons.

One way to ease this difficulty is to use a BMI. Because of the way a BMI is constructed, only those neurons that are recorded can directly affect the movement of the cursor or robotic limb. Neurons throughout the brain are still involved, but they can influence the cursor movements only indirectly through the recorded neurons. Thus, in contrast to arm and eye movement studies, one can record from the entire set of output neurons in a BMI, and BMI-directed movements can be causally attributed to specific changes in the activity of the recorded neurons. Furthermore, the mapping between the activity of the recorded neurons and cursor movement is defined by the experimenter, so it is fully known. This mapping can be defined to be simple and can be easily altered by the experimenter during an experiment. In essence, a BMI defines a simplified sensorimotor loop, whose components are more concretely defined and more easily manipulated than for arm or eye movements.

These advantages of BMIs allow for studies of brain function that are currently difficult to perform using arm or eye movements. For example, one class of studies involves using BMIs to study how the brain learns. The BMI mapping defines which population activity patterns will allow the subject to successfully move the BMI-directed cursor to hit visual targets. By defining the BMI mapping appropriately, the experimenter can challenge the subject's brain to produce novel neural activity patterns.

A recent study explored what types of activity patterns are easier and more difficult for the brain to generate. They found that it was easier for subjects to learn new associations between existing activity patterns and cursor movements than to generate novel activity patterns. This finding has implications for our ability to learn everyday skills. A second class of studies involves asking how the activity of neurons that directly control movement differ from those that do not directly control movement. In a BMI, one can choose to use only a subset of the recorded neurons (the output neurons) for controlling movements. At the same time, other neurons (the nonoutput neurons) can be passively monitored without being used for controlling movements. Comparing the activity of output and nonoutput neurons can provide insight into how a network of neurons internally processes information and relays only some of that information to other networks.

Using this paradigm, a recent study recorded neural activity simultaneously in the primary cortex and striatum and designated a subset of the M1 neurons as the output neurons for controlling the BMI. They found that, during BMI learning, M1 neurons that were most relevant for behavior (the output neurons) preferentially increased their coordination with the striatum, which is known to play an important role during natural behavior (Chapter 38). Identifying output versus nonoutput neurons in a study using arm or eye movements would be challenging.

BMIs Raise New Neuroethics Considerations

A growing number of biomedical ethics considerations centered on the brain have arisen from the dramatic expansion in our understanding of neuroscience and our capabilities with neurotechnology. These advances are driven by society's curiosity about the functioning of the brain, the least-well understood organ in the body, as well as the desire to address the massive unmet need of those suffering from neurological disease and injury. The use of BMIs raises new ethical questions for four principal reasons.

First, recording high-fidelity signals (ie, spike trains) involves risk, including the risks associated with initial implantation of the electrodes as well as possible biological (immunological or infectious) responses during the lifetime of the electrodes and the associated implanted electronics. Electrodes implanted for long periods currently have functional lifetimes on the order of many months to a few years, during which

time glial scar tissue can form around the electrodes and electrode materials can fail. Efforts to increase the functional lifetime of electrodes range from nanoscale flexible electrodes made with new materials to mitigating immunological responses, as is done with cardiac stents.

For these reasons, patients considering receiving implanted recording technologies will need to evaluate the risks and benefits of a BMI, as is the case for all medical interventions. It is important for patients to have options, as each person has personal preferences involving willingness to undergo surgery, desire for functional restoration and outcome, and cosmesis—be it while deliberating cancer treatment or BMI treatment. BMIs based on different neural sensors (Figure 39–2) have different risks and benefits.

Second, because BMIs can read out movement information from the brain at fine temporal resolution, it seems plausible that they will be able to read out more personal and private types of information as well. Future neuroethics questions that may arise as the technology becomes more sophisticated include whether it is acceptable, even with patient consent, to read out memories that may otherwise be lost to Alzheimer disease; promote long-term memory consolidation by recording fleeting short-term memories and playing them back directly into the brain; read out subconscious fears or emotional states to assist desensitization psychotherapy; or read out potential intended movements, including speech, that would not naturally be enacted.

Third, intracortical write-in BMIs, similar to DBS systems currently used to reduce tremor, may one day evoke naturalistic spatial-temporal activity patterns across large populations of neurons. In the extreme it may not be possible for a person to distinguish self-produced and volitional neural activity patterns from artificial or surrogate patterns. Although there are numerous therapeutic and beneficial reasons for embracing this technology, such as reducing tremor or averting an epileptic seizure, more dubious uses can be envisioned such as commandeering a person's motor, sensory, decision making, or emotional valence circuits.

Finally, ethical questions also involve the limits within which BMIs should operate. Current BMIs focus on restoring lost function, but it is possible for BMIs to be made to enhance function beyond natural levels. This is as familiar as prescribing a pair of glasses that confer better than normal vision, or overprescribing a pain medication, which can cause euphoria that is often addictive. Should BMIs be allowed, if and when it becomes technically possible, to move a robotic arm faster and more accurately than a native arm? Should continuous neural recordings from BMIs, covering hours, days, or weeks, be saved for future analysis, and are the security and privacy issues the same or different from personal genomics data? Should BMIs with preset content be available for purchase, for example, to skip a grade of mathematics in high school? Should an able-bodied person be able to elect to receive an implanted motor BMI? While the safe and ethical limits of such sensory, motor, and cognitive BMI treatments might seem readily apparent, society continues to wrestle with these same questions concerning other currently available medical treatments. These include steroids that enhance musculature, energy drinks (eg, caffeine) that enhance alertness, and elective plastic surgery that alters appearance.

Although many of these ideas and questions may appear far-fetched at present, as mechanisms of brain function and dysfunction continue to be revealed, BMI systems could build on these discoveries and create even more daunting ethical quandaries. But equally important is the immediate need to help people suffering from profound neurological disease and injury through restorative BMIs. In order to achieve the right balance, it is imperative that physicians, scientists, and engineers proceed in close conversation and partnership with ethicists, government oversight agencies, and patient advocacy groups.

Highlights

1. Brain–machine interfaces (BMIs) are medical devices that read out and/or alter electrophysiological activity at the level of populations of neurons. BMIs can help to restore lost sensory, motor, or brain processing capabilities, as well as regulate pathological neural activity.

2. BMIs can help to restore lost sensory capabilities by stimulating neurons to convey sensory information to the brain. Examples include cochlear implants to restore audition or retinal prostheses to restore vision.

3. BMIs can help to restore lost motor capabilities by measuring the activity from many individual neurons, converting this neural information into control signals, and guiding a paralyzed limb, robotic limb, or computer cursor.

4. Whereas motor BMIs aim to provide control of a robotic limb or paralyzed limb, communication BMIs aim to provide a fast and accurate interface with a computer or other electronic devices.

5. BMIs can help to regulate pathological neural activity by measuring neural activity, processing the neural activity, and subsequently stimulating neurons. Examples include deep brain stimulators and antiseizure systems.

6. Neural signals can be measured using different technologies, including electroencephalography, electrocorticography, and intracortical electrodes. Intracortical electrodes record the activity of neurons near the electrode tip and can also be used to deliver electrical stimulation.

7. To study movement encoding, one usually considers the activity of an individual neuron across many experimental trials. In contrast, for movement decoding, one needs to consider the activity of many neurons across an individual experimental trial.

8. A discrete decoder estimates one of several possible movement goals from neural population activity. In contrast, a continuous decoder estimates the moment-by-moment details of a movement from neural population activity.

9. The field is making substantial progress in increasing the performance of BMIs, measured in terms of the speed and accuracy of the estimated movements. It is now possible to move a computer cursor in a way that approaches the speed and accuracy of arm movements.

10. In addition to controlling computer cursors, BMIs can also guide a robotic limb or a paralyzed limb using functional electrical stimulation. Developments from preclinical experiments with able-bodied, nonhuman primates have subsequently been tested in clinical trials with paralyzed people.

11. Future advances of BMI will depend, in part, on developments in neurotechnology. These include advances in hardware (eg, neural sensors and low-power electronics), software (eg, supervisory systems), and statistical methods (eg, decoding algorithms).

12. An important direction for improving BMI performance is to provide the user with additional forms of sensory feedback in addition to visual feedback. An area of current investigation uses stimulation of neurons to provide surrogate sensory feedback, representing somatosensation and proprioception, during ongoing use.

13. Beyond helping paralyzed patients and amputees, BMI is being increasingly used as a tool for understanding brain function. BMIs simplify the brain's input and output interfaces and allow the experimenter to define a causal relationship between neural activity and movement.

14. BMIs raise new neuroethics questions, which need to be considered together with the benefits provided by BMIs to people with injury or disease.

Krishna V. Shenoy
Byron M. Yu

Selected Reading

Andersen RA, Hwang EJ, Mulliken GH. 2010. Cognitive neural prosthetics. Annu Rev Psychol 61:169–190.

Donoghue JP, Nurmikko A, Black M, Hochberg LR. 2007. Assistive technology and robotic control using motor cortex ensemble-based neural interface systems in humans with tetraplegia. J Physiol 579:603–611.

Fetz EE. 2007. Volitional control of neural activity: implications for brain-computer interfaces. J Physiol 579:571–579.

Green AM, Kalaska JF. 2011. Learning to move machines with the mind. Trends Neurosci 34:61–75.

Hatsopoulos NG, Donoghue JP. 2009. The science of neural interface systems. Annu Rev Neurosci 32:249–266.

Kao JC, Stavisky SD, Sussillo D, Nuyujukian P, Shenoy KV. 2014. Information systems opportunities in brain-machine interface decoders. Proc IEEE 102:666–682.

Nicolelis MAL, Lebedev MA. 2009. Principles of neural ensemble physiology underlying the operation of brain-machine interfaces. Nat Rev Neurosci 10:530–540.

Schwartz AB. 2016. Movement: how the brain communicates with the world. Cell 164:1122–1135.

Shenoy KV, Carmena JM. 2014. Combining decoder design and neural adaptation in brain-machine interfaces. Neuron 84:665–680.

References

Aflalo T, Kellis S, Klaes C, et al. 2015. Decoding motor imagery from the posterior parietal cortex of a tetraplegic human. Science 348:906–910.

Ajiboye AB, Willett FR, Young DR, et al. 2017. Restoration of reaching and grasping movements through brain-controlled muscle stimulation in a person with tetraplegia: a proof-of-concept demonstration. Lancet 389:1821–1830.

Anumanchipalli GK, Chartier J, Chang EF. 2019. Speech synthesis from neural decoding of spoken sentences. Nature 568:493–498.

Blabe CH, Gilja V, Chestek CA, Shenoy KV, Anderson KD, Henderson JM. 2015. Assessment of brain-machine interfaces from the perspective of people with paralysis. J Neural Eng 12:043002.

Bouton CE, Shaikhouni A, Annetta NV, et al. 2016. Restoring cortical control of functional movement in a human with quadriplegia. Nature 533:247–250.

Carmena JM, Lebedev MA, Crist RE, et al. 2003. Learning to control a brain-machine interface for reaching and grasping by primates. PLoS Biol 1:E42.

Chapin JK, Moxon KA, Markowitz RS, Nicolelis MA. 1999. Real-time control of a robot arm using simultaneously recorded neurons in the motor cortex. Nat Neurosci 2:664–670.

Collinger JL, Wodlinger B, Downey JE, et al. 2013. High-performance neuroprosthetic control by an individual with tetraplegia. Lancet 381:557–564.

Dadarlat MC, O'Dohert JE, Sabes PN. 2015. A learning-based approach to artificial sensory feedback leads to optimal integration. Nat Neurosci 18:138–144.

Ethier C, Oby ER, Bauman MJ, Miller LE. 2012. Restoration of grasp following paralysis through brain-controlled stimulation of muscles. Nature 485:368–371.

Fetz EE. 1969. Operant conditioning of cortical unit activity. Science 163:955–958.

Flesher SN, Collinger JL, Foldes ST, et al. 2016. Intracortical microstimulation of human somatosensory cortex. Sci Transl Med 8:361ra141.

Ganguly K, Carmena JM. 2009. Emergence of a stable cortical map for neuroprosthetic control. PLoS Biol 7:e1000153.

Gilja V, Nuyujukian P, Chestek CA, et al. 2012. A high-performance neural prosthesis enabled by control algorithm design. Nat Neurosci 15:1752–1757.

Gilja V, Pandarinath C, Blabe CH, et al. 2015. Clinical translation of a high-performance neural prosthesis. Nat Med 21:1142–1145.

Golub MD, Chase SM, Batista AP, Yu BM. 2016. Brain-computer interfaces for dissecting cognitive processes underlying sensorimotor control. Curr Opin Neurobiol 37:53–58.

Hochberg LR, Bacher D, Jarosiewicz B, et al. 2012. Reach and grasp by people with tetraplegia using a neurally controlled robotic arm. Nature 485:372–375.

Hochberg LR, Serruya MD, Friehs GM, et al. 2006. Neuronal ensemble control of prosthetic devices by a human with tetraplegia. Nature 442:164–171.

Humphrey DR, Schmidt EM, Thompson WD. 1970. Predicting measures of motor performance from multiple cortical spike trains. Science 170:758–762.

Jackson A, Mavoori J, Fetz EE. 2006. Long-term motor cortex plasticity induced by an electronic neural implant. Nature 444:56–60.

Jarosiewicz B, Sarma AA, Bacher D, et al. 2015. Virtual typing by people with tetraplegia using a self-calibrating intracortical brain-computer interface. Sci Transl Med 7:313ra179.

Kennedy PR, Bakay RA. 1998. Restoration of neural output from a paralyzed patient by a direct brain connection. Neuroreport 9:1707–1711.

Kim SP, Simeral JD, Hochberg LR, Donoghue JP, Black MJ. 2008. Neural control of computer cursor velocity by decoding motor cortical spiking activity in humans with tetraplegia. J Neural Eng 5:455–476.

Koralek AC, Costa RM, Carmena JM. 2013. Temporally precise cell-specific coherence develops in corticostriatal networks during learning. Neuron 79:865–872.

McFarland DJ, Sarnacki WA, Wolpaw JR. 2010. Electroencephalographic (EEG) control of three-dimensional movement. J Neural Eng 7:036007.

Moritz CT, Perlmutter SI, Fetz EE. 2008. Direct control of paralysed muscles by cortical neurons. Nature 456:639–642.

Musallam S, Corneil BD, Greger B, Scherberger H, Andersen RA. 2004. Cognitive control signals for neural prosthetics. Science 305:258–262.

O'Doherty JE, Lebedev MA, Ifft PJ, et al. 2011. Active tactile exploration using a brain-machine-brain interface. Nature 479:228–231.

Pandarinath C, Nuyujukian P, Blabe CH, et al. 2017. High performance communication by people with paralysis using an intracortical brain-computer interface. eLife 6:e18554.

Sadtler PT, Quick KM, Golub MD, et al. 2014. Neural constraints on learning. Nature 512:423–426.

Santhanam G, Ryu SI, Yu BM, Afshar A, Shenoy KV. 2006. A high-performance brain-computer interface. Nature 442:195–198.

Schalk G, Miller KJ, Anderson NR, et al. 2008. Two-dimensional movement control using electrocorticographic signals in humans. J Neural Eng 5:75–84.

Serruya MD, Hatsopoulos NG, Paninski L, Fellows MR, Donoghue JP. 2002. Instant neural control of a movement signal. Nature 416:141–142.

Shenoy KV, Meeker D, Cao S, et al. 2003. Neural prosthetic control signals from plan activity. Neuroreport 14:591–596.

Stavisky SD, Willett FR, Wilson GH, Murphy BA, Rezaii P, Avansino DT, et al. 2019. Neural ensemble dynamics in dorsal motor cortex during speech in people with paralysis. eLife;8:e46015.

Suminski AJ, Tkach DC, Fagg AH, Hatsopoulos NG. 2010. Incorporating feedback from multiple sensory modalities enhances brain-machine interface control. J Neurosci 30:16777–16787.

Taylor DM, Tillery SIH, Schwartz AB. 2002. Direct cortical control of 3d neuroprosthetic devices. Science 296:1829–1832.

Velliste M, Perel S, Spalding MC, Whitford AS, Schwartz AB. 2008. Cortical control of a prosthetic arm for self-feeding. Nature 453:1098–1101.

Wessberg J, Stambaugh CR, Kralik JD, et al. 2000. Real-time prediction of hand trajectory by ensembles of cortical neurons in primates. Nature 408:361–365.

Wodlinger B, Downey JE, Tyler-Kabara EC, Schwartz AB, Boninger ML, Collinger JL. 2015. Ten-dimensional anthropomorphic arm control in a human brain-machine interface: difficulties, solutions, and limitations. J Neural Eng 12:016011.

Part VI

VI The Biology of Emotion, Motivation, and Homeostasis

EMOTIONAL AND HOMEOSTATIC BEHAVIORS ALL INVOLVE the coordination of one or more somatic, autonomic, hormonal, or cognitive processes. Subcortical brain regions concerned with a range of functions—including feeding, drinking, heart rate, breathing, temperature regulation, sleep, sex, and facial expressions—play a critical role in this coordination. Subcortical brain regions are bidirectionally connected with cortical brain areas, providing a means for reperesentations of internal state variables (eg, visceral information) to influence cognitive operations, such as subjective feelings, decision-making, and attention, and for cognitive functions to regulate or extinguish neural representations in subcortical brain areas that help coordinate behavior reflecting emotional states.

Our consideration of these systems begins with the brain stem, a structure critical for wakefulness and conscious attention on the one hand and sleep on the other. The significance of this small region of the brain—located between the spinal cord and the diencephalon—is disproportionate to its size. Damage to the brain stem can profoundly affect motor and sensory processes because it contains all of the ascending tracts that bring sensory information from the surface of the body to the cerebral cortex and all of the descending tracts from the cerebral cortex that deliver motor commands to the spinal cord. Finally, the brain stem contains neurons that control respiration and heartbeat as well as nuclei that give rise to most of the cranial nerves that innervate the head and neck.

Six neurochemical modulatory systems in the brain stem modulate sensory, motor, and arousal systems. The dopaminergic pathways that connect the midbrain to the limbic system and cortex are particularly important, because they are involved in processing stimuli and events in relation to reinforcement expectation, and therefore contribute to motivational state and learning. Addictive drugs such as nicotine, alcohol, opiates, and cocaine are thought to produce their actions by co-opting the same neural pathways that positively reinforce behaviors essential for survival. Other modulatory transmitters regulate sleep and wakefulness, in part by controlling information flow between the thalamus and cortex. Disorders of electrical excitation in corticothalamic circuits can result in seizures and epilepsy.

Rostral to the brain stem lies the hypothalamus, which functions to maintain the stability of the internal environment by keeping physiological variables within the limits favorable to vital bodily processes. Homeostatic processes in the nervous system have profound consequences for behavior that have intrigued many of the founders of modern physiology, including Claude Bernard, Walter B. Cannon, and Walter Hess. Neurons controlling the internal environment are concentrated in the hypothalamus, a small area of the diencephalon that comprises less than 1% of the total brain volume. The hypothalamus, with closely linked structures in the brain stem and limbic system, acts directly on the internal environment, through its control of the endocrine system and autonomic nervous system, to achieve goal-directed behavior. It acts indirectly through its connections to higher brain regions to modulate emotional and motivational states. In addition to influencing motivated behaviors, the hypothalamus, together with the brain stem below and the cerebral cortex above, maintains a general state of arousal, which ranges from excitement and vigilance to drowsiness and stupor.

The neurobiological investigation of emotion has relied on experiments that define emotions in terms of specific measures ranging from subjective reports of feelings in humans, to approach or defensive behaviors, to physiological responses such as autonomic reactivity. Charles Darwin observed in his seminal book *The Expression of the Emotions in Man and Animals* that many emotions are conserved across species, making clear the relevance of studying emotions by using animal models to probe neural mechansisms. In experimental frameworks, emotional states are thereby considered to be central brain states that can cause coordinated behavioral, physiological, and cognitive responses across species.

In recent years, much work on emotion has focused on the amygdala, which can orchestrate different responses via its connections to the cortex, hypothalamus, and brain stem. Lesions of the amygdala in humans impair fear learning and expression, as well as fear recognition in others, due to decreased allocation of attention to features of faces that communicate fear. Symptoms in a variety of psychiatric disorders—ranging from addiction to anxiety to social deficits—likely involve amygdala dysfunction. However, the amygdala is only one component of a larger set of brain regions that includes parts of the hypothalamus, the brain stem, and cortical areas also responsible for coordinating emotional responses. In particular, the medial and ventral prefrontal cortex and amygdala are closely interconnected. Dynamic processing within and between these structures likely subserves many functions beyond coordinated emotional behavior, including extinction, the cognitive regulation of emotional states, interactions between social and emotional domains, and the influence of the amgydalar representations on decision-making and subjective feelings.

Part Editors: C. Daniel Salzman and John D. Koester

Part VI

40

The Brain Stem

IN PRIMITIVE VERTEBRATES—REPTILES, amphibians, and fish—the forebrain is only a small part of the brain and is devoted mainly to olfactory processing and to the integration of autonomic and endocrine function with the basic behaviors necessary for survival. These basic behaviors include feeding, drinking, sexual reproduction, sleep, and emergency responses. Although we are accustomed to thinking that the forebrain orchestrates most human behaviors, many complex responses, such as feeding—the coordination of chewing, licking, and swallowing—are actually made up of relatively simple, stereotypic motor responses governed by ensembles of neurons in the brain stem.

The importance of this pattern of organization in human behavior is clear from observing infants born without a forebrain (hydranencephaly). Hydranencephalic infants are surprisingly difficult to distinguish from normal babies. They cry, smile, suckle, and move their eyes, face, arms, and legs. As these sad cases illustrate, the brain stem can organize virtually all of the behavior of the newborn.

In this chapter, we describe the functional anatomy of the brain stem, particularly the cranial nerves, as well as the ensembles of local circuit neurons that organize the simple behaviors of the face and head. Finally, we consider the modulatory functions of nuclei in the brain stem that adjust the sensitivity of sensory, motor, and arousal systems.

The brain stem is the rostral continuation of the spinal cord, and its motor and sensory components are similar in structure to those of the spinal cord. But the portions of the brain stem that control the cranial nerves are much more complex than the corresponding parts of the spinal cord that control the spinal nerves because cranial nerves mediate more complex behaviors. The core of the brain stem, the *reticular formation*, is homologous to the intermediate gray matter of the spinal cord but is also more complex. Like the spinal cord, the reticular formation contains ensembles of local-circuit interneurons that generate motor and autonomic patterns and coordinate reflexes and simple behaviors. In addition, the brain stem contains glutamatergic and GABAergic circuitry that regulates arousal, wake–sleep cycles, breathing, and other vital functions, as well as monoaminergic modulatory neurons that act to optimize the functions of the nervous system.

The Cranial Nerves Are Homologous to the Spinal Nerves

Because the spinal nerves reach only as high as the first cervical vertebra, the cranial nerves provide the somatic and visceral, sensory and motor innervation for the head. Two cranial nerves, the glossopharyngeal and vagus nerves, also supply visceral sensory and motor innervation of the neck, chest, and most of the abdominal organs with the exception of the pelvis. In addition, some cranial nerves are associated with specialized functions, such as vision or hearing, that go beyond the sensory and motor plan of the spinal cord.

Assessment of the cranial nerves is an important part of the neurological examination because abnormalities of function can pinpoint a site in the brain stem that has been damaged. Therefore, it is important to know the origins of the cranial nerves, their intracranial course, and where they exit from the skull.

The cranial nerves are traditionally numbered I through XII in rostrocaudal sequence. Cranial nerves I and II enter at the base of the forebrain. The other cranial nerves arise from the brain stem at characteristic locations (Figure 40–1). All but one exit from the ventral surface of the brain stem (Figure 40–2).

The exception is the trochlear (IV) nerve, which leaves the midbrain from its dorsal surface just behind the inferior colliculus and wraps around the lateral surface of the brain stem to join the other cranial nerves concerned with eye movements. The cranial nerves with sensory functions (V, VII, VIII, IX, and X) have associated sensory ganglia that operate much as dorsal root ganglia do for spinal nerves. These ganglia are located along the course of individual nerves as they enter the skull.

The olfactory (I) nerve, which is associated with the forebrain, is described in detail in Chapter 29; the optic (II) nerve, which is associated with the diencephalon, is described in Chapters 21 and 22. The spinal accessory (XI) nerve can be considered a cranial nerve anatomically but actually is a spinal nerve originating from the higher cervical motor rootlets. It runs up into the skull before exiting through the jugular foramen to innervate the trapezius and sternocleidomastoid muscles in the neck.

Cranial Nerves Mediate the Sensory and Motor Functions of the Face and Head and the Autonomic Functions of the Body

Three ocular motor nerves control movements of the eyes. The *abducens (VI) nerve* has the simplest action; it contracts the lateral rectus muscle to move the globe laterally. The *trochlear (IV) nerve* also innervates a single muscle, the superior oblique, which both depresses the eye and rotates it inward depending on the eye's position. The *oculomotor (III) nerve* supplies all of the other muscles of the orbit, including the retractor of the lid. It also provides the parasympathetic innervation responsible for pupillary constriction in response to light and accommodation of the lens for near vision. The ocular motor system is considered in detail in Chapter 35.

The *trigeminal (V) nerve* is a mixed nerve (containing both sensory and motor axons) that leaves the brain stem in two roots. The motor root innervates the muscles of mastication (the masseter, temporalis, and pterygoids) and a few muscles of the palate (tensor veli palatini), inner ear (tensor tympani), and upper neck (mylohyoid and anterior belly of the digastric muscle). The sensory fibers arise from neurons in the trigeminal ganglion, located at the floor of the skull in the middle cranial fossa.

Three branches emerge from the trigeminal ganglion. The *ophthalmic division* (V_1) runs with the ocular motor nerves through the superior orbital fissure (Figure 40–2A) to innervate the orbit, nose, and forehead and scalp back to the vertex of the skull (Figure 40–3). Some fibers from this division also innervate

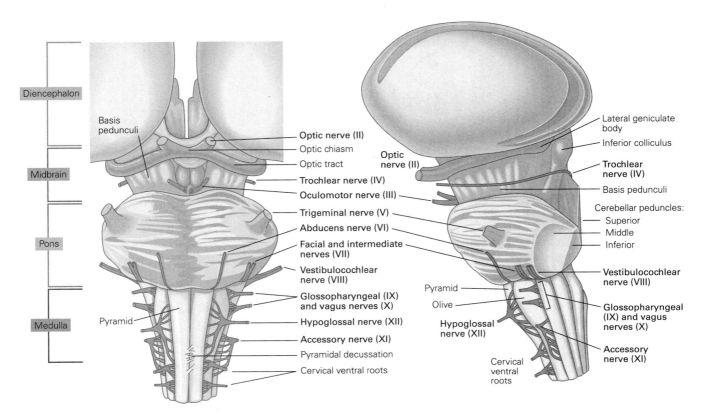

Figure 40–1 The origins of cranial nerves in the brain stem (ventral and lateral views). The olfactory (I) nerve is not shown because it terminates in the olfactory bulb in the forebrain. All of the cranial nerves except one emerge from the ventral surface of the brain; the trochlear (**IV**) nerve originates from the dorsal surface of the midbrain.

the meninges and blood vessels of the anterior and middle intracranial fossas. The *maxillary division* (V_2) runs through the round foramen of the sphenoid bone to innervate the skin over the cheek and the upper portion of the oral cavity. The *mandibular division* (V_3), which also contains the motor axons of the trigeminal nerve, leaves the skull through the oval foramen of the sphenoid bone. It innervates the skin over the jaw, the area above the ear, and the lower part of the oral cavity, including the tongue.

Complete trigeminal sensory loss results in numbness of the entire face and the inside of the mouth. One-sided trigeminal motor weakness does not cause much weakness of jaw closure because the muscles of mastication on either side are sufficient to close the jaw. Nevertheless, the jaw tends to deviate toward the side of the lesion when the mouth is opened because the internal pterygoid muscle on the opposite side, when unopposed, pulls the jaw toward the weak side.

The *facial (VII) nerve* is also a mixed nerve. Its motor root innervates the muscles of facial expression as well as the stapedius muscle in the inner ear, stylohyoid muscle, and posterior belly of the digastric muscle in the upper neck. The sensory root runs as a separate bundle, the intermediate nerve, through the internal auditory canal and arises from neurons in the geniculate ganglion, located near the middle ear. Distal to the geniculate ganglion, the sensory fibers diverge from the motor branch. Some innervate skin of the external auditory canal while others form the chorda tympani, which joins the lingual nerve and conveys taste sensation from the anterior two-thirds of the tongue. The *autonomic component* of the facial nerve includes parasympathetic fibers that travel through the motor root to the sphenopalatine and submandibular ganglia, which innervate lacrimal and salivary glands (except the parotid gland) and the cerebral vasculature.

The facial nerve may suffer isolated injury in Bell palsy, a common complication of certain viral infections. Early on, the patient may complain mainly of the face pulling toward the unaffected side because of the weakness of the muscles on the side of the lesion. Later, the ipsilateral corner of the mouth droops, food falls out of the mouth, and the eyelids no longer close on that side. Loss of blinking may result in drying

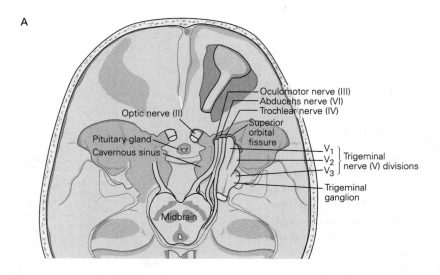

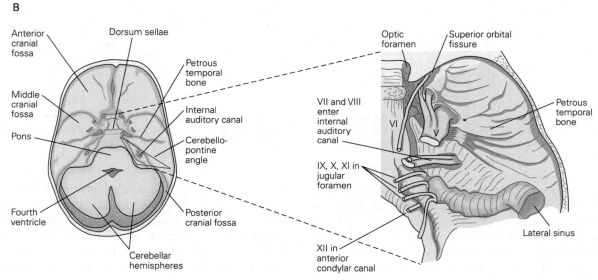

Figure 40–2 The cranial nerves exit the skull in groups.

A. Cranial nerves II, III, IV, V, and VI exit the skull near the pituitary fossa. The optic (II) nerve enters the optic foramen, but the oculomotor (III), trochlear (IV), and abducens (VI) nerves, and the first division of the trigeminal (V) nerve leave through the superior orbital fissure. The second and third divisions of the trigeminal nerve exit through the round and oval foramina, respectively.

B. In the posterior fossa, the facial (VII) and vestibulocochlear (VIII) nerves exit through the internal auditory canal, whereas the glossopharyngeal (IX), vagus (X), and accessory (XI) nerves leave through the jugular foramen. The hypoglossal nerve (XII) has its own foramen.

and injury to the cornea. The patient may complain that sound has a booming quality in the ipsilateral ear because the stapedius muscle fails to tense the ossicles in response to a loud sound (the stapedial reflex). Taste may also be lost on the anterior two-thirds of the tongue on the ipsilateral side. If the Bell palsy is caused by a herpes zoster infection of the geniculate ganglion, small blisters may form in the outer ear canal, the ganglion's cutaneous sensory field.

The *vestibulocochlear (VIII) nerve* contains two main bundles of sensory axons from two ganglia. Fibers from the vestibular ganglion relay sensation of angular and linear acceleration from the semicircular canals, utricle, and saccule in the inner ear. Fibers from the cochlear ganglion relay information from the cochlea concerning sound. A vestibular schwannoma, one of the most common intracranial tumors, may form along the vestibular component of cranial nerve VIII

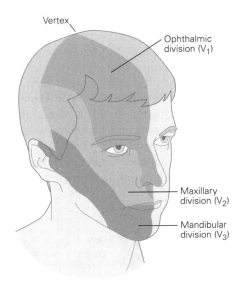

Vertex

Ophthalmic
division (V₁)

Maxillary
division (V₂)

Mandibular
division (V₃)

Figure 40–3 The three sensory divisions of the trigeminal (V) nerve innervate the face and scalp. The V_2 and V_3 divisions also innervate the upper and lower parts of the oral cavity, including the tongue. The C2 cervical root innervates the back of the head. The area around the ear is innervated by branches of the VII and X nerve.

as it runs within the internal auditory meatus. Most patients complain only about hearing loss, as the brain is usually able to adapt to the gradual loss of vestibular input from one side.

The *glossopharyngeal (IX) nerve* and *vagus (X) nerve* are mixed nerves and provide parasympathetic autonomic input to thoracic and visceral organs. These closely related nerves transmit sensory information from the pharynx and upper airway as well as taste from the posterior third of the tongue and oral cavity. The glossopharyngeal nerve transmits visceral information from the neck (for example, information on blood oxygen and carbon dioxide from the carotid body, and arterial pressure from the carotid sinus), whereas the vagus nerve transmits visceral information from the thoracic and abdominal organs except for the distal colon and pelvic organs. Both nerves include parasympathetic motor fibers. The glossopharyngeal nerve provides parasympathetic control of the parotid salivary gland, whereas the vagus nerve innervates the rest of the internal organs of the neck, thorax, and abdomen. The glossopharyngeal nerve innervates only one muscle of the palate, the stylopharyngeus, which raises and dilates the pharynx. The remaining striated muscles of the larynx and pharynx are under control of the vagus nerve.

The vagal sensory neurons innervate the length of the gastrointestinal tract and thus are able to regulate multiple postprandial functions. One excellent example is the role of vagal afferents in regulating food intake following a meal. Cholecystokinin (CCK) is an endogenous peptide secreted by duodenal enteroendocrine cells during meals, which helps to induce satiety. CCK acts (at least in part) via action on vagal afferents in the gut, stimulating a feeling of fullness. Exogenous electrical stimulation of the vagus nerve is now being used clinically to treat a wide variety of conditions including obesity, intractable epilepsy, and even depression. However, the neuroanatomic and molecular mechanisms underlying these effects remain poorly understood. Similarly, bariatric surgery remains one of the most widely used and effective strategies to combat obesity. Some studies have suggested that surgical alterations in the responsiveness of vagal afferents to gut signals may contribute to the sustained weight loss following these surgeries.

Because many of the functions of nerves IX and X are bilateral and partially overlapping, unilateral injury of nerve IX may be difficult to detect. Patients with unilateral cranial nerve X injury are hoarse, because one vocal cord is paralyzed, and they may have some difficulty swallowing. Examination of the oropharynx shows weakness and numbness of the palate on one side.

The *spinal accessory (XI) nerve* is purely motor and originates from motor neurons in the upper cervical spinal cord. It innervates the trapezius and sternocleidomastoid muscles on the same side of the body. Because the mechanical effect of the sternocleidomastoid is to turn the head toward the opposite side, an injury of the left nerve causes weakness in turning the head to the right. A lesion of the cerebral cortex on the left will cause weakness of voluntary muscles on the entire right side of the body except for the sternocleidomastoid; instead, the ipsilateral sternocleidomastoid will be weak (because the left cerebral cortex is concerned with muscles that interact with the right side of the world, and the left sternocleidomastoid turns the head to the right).

The *hypoglossal (XII) nerve* is also purely motor, innervating the muscles of the tongue. When the nerve is injured, for example during surgery for head and neck cancer, the tongue atrophies on that side. The muscle fibers exhibit twitches of muscle fascicles (fasciculations), which may be seen clearly through the thin mucosa of the tongue.

Cranial Nerves Leave the Skull in Groups and Often Are Injured Together

In assessing dysfunction of the cranial nerves, it is important to determine whether the injury is within the brain or further along the course of the nerve. As cranial nerves leave the skull in groups through specific foramina, damage at these locations can affect several nerves.

The cranial nerves concerned with orbital sensation and movement of the eyes—the oculomotor, trochlear, and abducens nerves, as well as the ophthalmic division of the trigeminal nerve—are gathered together in the *cavernous sinus*, along the lateral margins of the sella turcica, and then exit the skull through the *superior orbital fissure* adjacent to the optic foramen (Figure 40–2A). Tumors in this region, such as those arising from the pituitary gland, often make their presence known first by pressure on these nerves or the adjacent optic chiasm.

Cranial nerves VII and VIII exit the brain stem at the *cerebellopontine angle*, the lateral corner of the brain stem at the juncture of the pons, medulla, and cerebellum (Figure 40–2B), and then leave the skull through the internal auditory meatus. A common tumor of the cerebellopontine angle is the vestibular schwannoma (sometimes erroneously called an "acoustic neuroma"), which derives from Schwann cells in the vestibular component of nerve VIII. A large tumor of the cerebellopontine angle may not only impair the function of nerves VII and VIII but may also press on nerve V near its site of emergence from the middle cerebellar peduncle, causing facial numbness, or compress the cerebellum or its peduncles on the same side, causing ipsilateral clumsiness.

The lower cranial nerves (IX, X, and XI) exit through the *jugular foramen* (Figure 40–2B) and are vulnerable to compression by tumors at that site. Nerve XII leaves the skull through its own (hypoglossal) foramen and is generally not affected by tumors located in the adjacent jugular foramen, unless the tumor becomes quite large. If a tumor involves nerves IX and X, but nerve XI is spared, it is generally within or near the brain stem rather than near the jugular foramen.

The Organization of the Cranial Nerve Nuclei Follows the Same Basic Plan as the Sensory and Motor Areas of the Spinal Cord

Cranial nerve nuclei are organized in rostrocaudal columns that are homologous to the sensory and motor laminae of the spinal cord (Chapters 18 and 31). This pattern is best understood from the developmental plan of the caudal neural tube that gives rise to the brain stem and spinal cord.

The transverse axis of the embryonic caudal neural tube is subdivided into alar (dorsal) and basal (ventral)

plates by the sulcus limitans, a longitudinal groove along the lateral walls of the central canal, fourth ventricle, and cerebral aqueduct (Figure 40–4). The alar plate forms the sensory components of the dorsal horn of the spinal cord, whereas the basal plate forms the motor components of the ventral horn. The intermediate gray matter is made up primarily of the interneurons that coordinate spinal reflexes and motor responses.

The brain stem shares this basic plan. As the central canal of the spinal cord opens into the fourth ventricle, the walls of the neural tube are splayed outward

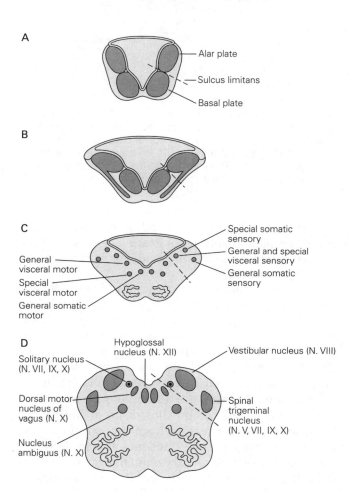

Figure 40–4 The developmental plan of the brain stem is the same general plan as that of the spinal cord.

A. The neural tube is divided into a dorsal sensory portion (the alar plate) and a ventral motor portion (the basal plate) by a longitudinal groove, the sulcus limitans.

B–D. During development, the sensory and motor cell groups migrate into their adult positions but largely retain their relative locations. In maturity (part **D**), the sulcus limitans (**dashed line**) is still recognizable in the walls of the fourth ventricle and the cerebral aqueduct, demarcating the border between dorsal sensory (**orange**) and ventral motor (**green**) structures. The section in part D is from the rostral medulla.

so that the dorsal sensory structures (derived from the alar plate) are displaced laterally, whereas the ventral motor structures (derived from the basal plate) remain more medial. The nuclei of the brain stem are divided into *general nuclei*, which serve functions similar to those of the spinal cord laminae, and *special nuclei*, which serve functions unique to the head, such as hearing, balance, taste, and control of the musculature related to the jaw, face, oropharynx, and larynx.

Embryonic Cranial Nerve Nuclei Have a Segmental Organization

Although the columns of sensory and motor nuclei in the adult hindbrain are organized rostrocaudally, the arrangement of neurons at each level derives from a strikingly segmental pattern in the early embryo. Before neurons appear, the future hindbrain region of the neural plate becomes subdivided into a series of eight segments of approximately equal size, known as *rhombomeres* (Figure 40–5A).

Each rhombomere develops a similar set of differentiated neurons, as if the developing hindbrain is made up of series of modules. Pairs of rhombomeres are associated with specific sets of muscles derived from the embryonic branchial arches (eg, rhombomeres 2 and 3 with the muscles of mastication and 4 and 5 with the muscles of facial expression) (Figure 40–5A). The even-numbered rhombomeres differentiate ahead of the odd-numbered ones. Rhombomeres 2, 4, and 6 form the branchial motor nuclei of the trigeminal, facial, and glossopharyngeal nerves, respectively. Later, rhombomeres 3, 5, and 7 contribute motor neurons to these nuclei, again respectively; in each case, the axons of individual motor neurons from odd-numbered rhombomeres extend rostrally as they join those of their even-numbered neighbors.

At this developmental stage, each of these nuclei is composed of homologous neurons derived from two adjacent segments. This early transverse segmental organization changes later in development, as rhombomere boundaries disappear and the dorsolateral migration of the cell bodies aligns the cells into rostrocaudal columns. Ultimately, some somatic and parasympathetic motor neurons migrate into the ventrolateral tegmentum; for example, the migration of the facial motor neurons of rhombomere 4 around the abducens nucleus generates the internal genu of the facial nerve (Figure 40–5A). Furthermore, neural crest cells from each rhombomere migrate into the corresponding branchial arches where they provide sensory and autonomic ganglion cells, as well as positional cues for the development of the arch muscles.

Adult Cranial Nerve Nuclei Have a Columnar Organization

Overall, the brain stem nuclei on each side are organized in six rostrocaudal columns, three of sensory nuclei and three of motor nuclei (Figure 40–6). These are considered later, in dorsolateral to ventromedial sequence. Although the columns are discontinuous along the rostrocaudal axis of the brain stem, nuclei with similar functions (sensory or motor, somatic or visceral) have similar dorsolateral-ventromedial positions at each level of the brain stem.

Within each motor nucleus, motor neurons for an individual muscle are also arranged in a cigar-shaped longitudinal column. Thus, each motor nucleus in cross section forms a mosaic map of the territory that is innervated. For example, in a cross section through the facial nucleus, the clusters of neurons that innervate the different facial muscles form a topographic map of the face.

General Somatic Sensory Column

The general somatic sensory column occupies the most lateral region of the alar plate and includes the trigeminal sensory nuclei (N. V). The *spinal trigeminal nucleus* is a continuation of the dorsal-most laminae of the spinal dorsal horn (Figure 40–5A) and is sometimes called the medullary dorsal horn. Along its outer surface lies the spinal trigeminal tract, a direct continuation of Lissauer's tract of the spinal cord (Chapter 20), thus allowing some cervical sensory fibers to reach the trigeminal nuclei and some trigeminal sensory axons to reach the dorsal horn in upper cervical segments. This arrangement allows dorsal horn sensory neurons to have a range of inputs that are much broader than that of individual spinal or trigeminal segments and ensures the integration of trigeminal and upper cervical sensory maps.

The spinal trigeminal nucleus receives sensory axons from the trigeminal ganglion (N. V) and from all cranial nerve sensory ganglia concerned with pain and temperature in the head, including geniculate ganglion (N. VII) neurons that relay information from the external auditory meatus, petrosal ganglion (N. IX) cells that convey information from the posterior part of the palate and tonsillar fossa, and nodose ganglion (N. X) axons that relay information from the posterior wall of the pharynx. The spinal trigeminal nucleus thus represents the entire oral cavity as well as the surface of the face.

The somatotopic organization of the afferent fibers in the spinal trigeminal nucleus is inverted: The forehead

A

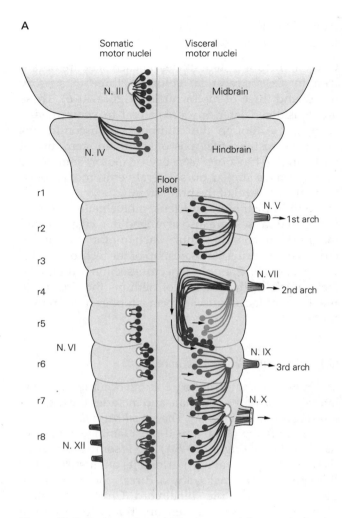

B

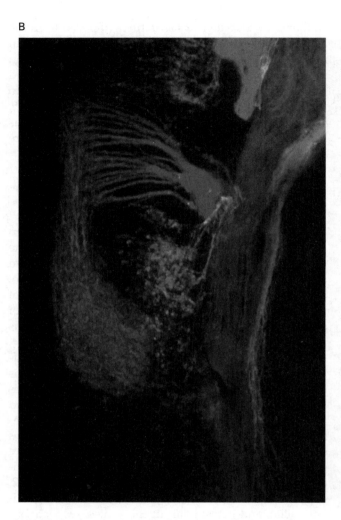

Figure 40–5 Embryonic cranial nerve nuclei are organized segmentally.

A. In the developing hindbrain (seen here from the ventral side), special and general visceral motor neurons (represented on the right side of the brain stem) form in each hindbrain segment (rhombomere) except rhombomere 1 (**r1**). Each special visceral motor nucleus comprises neurons in two rhombomeres: the trigeminal motor nucleus is formed by neurons in r2 and r3, the facial nucleus by neurons in r4 and r5, the glossopharyngeal nucleus by neurons in r6 and r7, and the motor nuclei of the vagus by neurons in r7 and r8. Axons of neurons in each of these nuclei course laterally within the brain, leaving the brain through exit points in the lateral neuroepithelium (of r2, r4, r6, and r7) and running together outside the brain to form the respective cranial motor nerves (V, VII, IX, X). The trigeminal (**V**) nerve innervates muscles in the 1st branchial arch, the facial (**VII**) nerve innervates muscles in the 2nd branchial arch, and the glossopharyngeal (**IX**) nerve innervates muscles in the 3rd branchial arch.

All of the visceral motor neurons (various shades of **green**, represented on the right side of the brain stem) develop initially next to the floor plate at the ventral midline; after extending their axons toward their respective exit points, the cell bodies then migrate laterally (**arrows**). Exceptions are the facial motor neurons formed in r4 (**red**); the cell bodies, after extending their axons toward the exit point, migrate caudally to the axial level

of r6 before migrating laterally. General visceral (parasympathetic) motor neurons associated with nerve VII (**light green**) take a more conventional course (see panel **B**).

General somatic motor nuclei (**various shades of blue**, represented on the left side of the brain stem) are formed in r1 (trochlear nucleus), r5 and r6 (abducens nucleus), and r8 (hypoglossal nucleus). The cell bodies of these neurons remain close to their place of birth, next to the floor plate. The axons of abducens and hypoglossal neurons exit the brain directly ventrally, without coursing laterally. The axons of trochlear neurons (**light blue**) extend laterally and dorsally within the brain until, caudal to the inferior colliculus, they turn medially, decussate just behind the inferior colliculus, and exit near the midline of the opposite side.

B. The brain stem of a mouse embryo in which fluorescent dyes label different populations of cranial nerve VII motor neurons. A red-fluorescing dye fills the cell bodies of facial motor neurons via retrograde transport from the motor root of the facial nerve. These neurons develop initially in r4 and then migrate posteriorly, alongside the floor plate, to r6 (see **red** neurons in part **A**). A green-fluorescing dye fills the cell bodies of general visceral motor neurons in r5 (see **light green** neurons in part **A**) via retrograde transport from the root of the intermediate nerve (sensory and preganglionic general visceral motor axons). (Micrograph reproduced, with permission, from Dr. Ian McKay.)

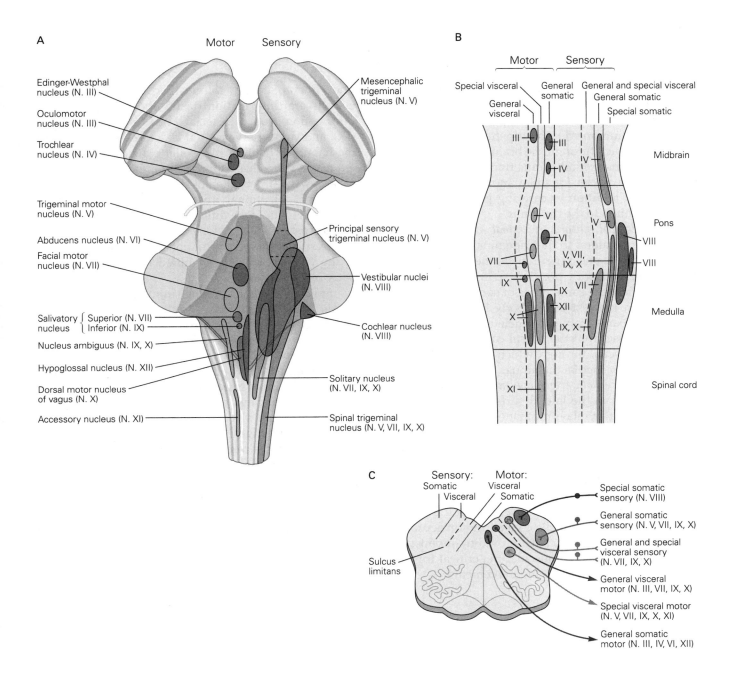

Figure 40–6 Adult cranial nerve nuclei are organized in six functional columns on the rostrocaudal axis of the brain stem.

A. This dorsal view of the human brain stem shows the location of the cranial nerve sensory nuclei (*right*) and motor nuclei (*left*).

B. A schematic view of the functional organization of the cranial nerve nuclei makes it clearer that they form motor and sensory columns.

C. The medial-lateral arrangement of the cranial nerve nuclei is shown in a cross section at the level of the medulla (compare with Figure 40–4D).

is represented ventrally and the oral region dorsally, with the tongue extending medially toward the taste region of the nucleus of the solitary tract, with which it shares some afferent information concerning food texture and temperature. Axons from the spinal trigeminal nucleus descend on the same side of the brain stem into the cervical spinal cord, where they cross the midline in the anterior commissure with spinothalamic axons and join the opposite spinothalamic tract. (For this reason, upper cervical spinal cord injury may cause facial numbness.) The trigeminothalamic axons then ascend back through the brain stem in close association with the spinothalamic tract, providing inputs to brain stem nuclei for reflex motor and autonomic responses in addition to carrying pain and temperature information to the thalamus.

The *principal sensory trigeminal nucleus* lies in the mid-pons just lateral to the trigeminal motor nucleus. It receives the axons of neurons in the trigeminal ganglion concerned with position sense and fine touch discrimination, the same types of sensory information carried from the rest of the body by the dorsal columns. The axons from this nucleus are bundled just medial to those from the dorsal column nuclei in the medial lemniscus, through which they ascend to the ventroposterior medial thalamus.

The *mesencephalic trigeminal nucleus*, located at the midbrain level in the lateral surface of the periaqueductal gray matter, relays mechanosensory information from the muscles of mastication and the periodontal ligaments. The large cells of this nucleus are not central neurons but primary sensory ganglion cells that derive from the neural crest and, unlike their relatives in the trigeminal ganglion, migrate into the brain during development. The central branches of the axons of these pseudo-unipolar cells contact motor neurons in the trigeminal motor nucleus, providing monosynaptic feedback to the jaw musculature, critical for rapid and precise control of chewing movements.

Special Somatic Sensory Column

The special somatic sensory column has inputs from the acoustic and vestibular nerves and develops from the intermediate region of the alar plate. The *cochlear nuclei* (N. VIII), which lie at the lateral margin of the brain stem at the pontomedullary junction, receive afferent fibers from the spiral ganglion of the cochlea. The output of the cochlear nuclei is relayed through the pons to the superior olivary and trapezoid nuclei and bilaterally on to the inferior colliculus (Chapter 28). The *vestibular nuclei* (N. VIII) are more complex. They include four distinct cell groups that relay information from the vestibular ganglion to various motor sites in the brain stem, cerebellum, and spinal cord concerned with maintaining balance and coordination of eye and head movements (Chapter 27).

Visceral Sensory Column

The visceral sensory column is concerned with special visceral information (taste) and general visceral information from the facial (VII), glossopharyngeal (IX), and vagus nerves (X). It is derived from the most medial tier of neurons in the alar plate. All of the afferent axons from these sources terminate in the *nucleus of the solitary tract*. The solitary tract is analogous to the spinal trigeminal tract or Lissauer's tract, bundling afferents from different cranial nerves as they course rostrocaudally along the length of the nucleus. As a result, sensory information from different regions of the viscera produces a unified map of the internal body in the nucleus.

Special visceral afferents from the anterior two-thirds of the tongue travel to the nucleus of the solitary tract through the chorda tympani branch of the facial nerve, whereas those from the posterior parts of the tongue and oral cavity arrive through the glossopharyngeal and vagus nerves. These afferents terminate in roughly somatotopic fashion in the anterior third of the nucleus of the solitary tract (or solitary nucleus). General visceral afferents are relayed through the glossopharyngeal and vagus nerves. Those from the rest of the gastrointestinal tract (down to the transverse colon) terminate in the middle portion of the solitary nucleus in topographic order, whereas those from the cardiovascular and respiratory systems terminate in the caudal and lateral portions.

The solitary nucleus projects directly to parasympathetic and sympathetic preganglionic motor neurons in the medulla and spinal cord that mediate various autonomic reflexes, as well as to parts of the reticular formation that coordinate autonomic and respiratory responses. Most ascending projections from the solitary nucleus that carry information from the viscera to the forebrain are relayed through the parabrachial nucleus in the pons, although some reach the forebrain directly. Together, the solitary and parabrachial nuclei supply visceral sensory information to the hypothalamus, basal forebrain, amygdala, thalamus, and cerebral cortex.

General Visceral Motor Column

All motor neurons initially develop adjacent to the floor plate, a longitudinal strip of non-neuronal cells at the ventral midline of the neural tube (Chapter 45). Neurons fated to become the three types of brain stem motor neurons migrate dorsolaterally, settling in three distinct rostrocaudal columns. The neurons that form the general visceral motor column take up a position along the most lateral region of the basal plate, just medial to the sulcus limitans. During development, the parasympathetic motor neurons destined to join the superior salivatory nucleus (part of the facial nerve) and nucleus ambiguus (part of the vagus nerve) migrate ventrolaterally, leaving behind axons that ascend medially before turning laterally to exit the brain stem, in a course similar to the facial motor neurons.

The *Edinger-Westphal nucleus* (N. III) lies in the midline separating the somatic oculomotor neurons just below the floor of the cerebral aqueduct. It contains preganglionic neurons that control pupillary constriction and lens accommodation through the ciliary ganglion.

The *superior salivatory nucleus* (N. VII) lies just dorsal to the facial motor nucleus and comprises parasympathetic preganglionic neurons that innervate the sublingual and submandibular salivary glands and the lacrimal glands and intracranial circulation through the sphenopalatine and submandibular parasympathetic ganglia.

Parasympathetic preganglionic neurons associated with the gastrointestinal tract form a column at the level of the medulla just dorsal to the hypoglossal nucleus and ventral to the nucleus of the solitary tract. At the most rostral end of this column is the *inferior salivatory nucleus* (N. IX) comprising the preganglionic neurons that innervate the parotid gland through the otic ganglion. The rest of this column constitutes the *dorsal motor vagal nucleus* (N. X). Most of the preganglionic neurons in this nucleus innervate the gastrointestinal tract below the diaphragm; a few are cardiomotor neurons.

The *nucleus ambiguus* (N. X) runs the rostrocaudal length of the ventrolateral medulla and contains parasympathetic preganglionic neurons that innervate thoracic organs, including the esophagus, heart, and respiratory system, as well as special visceral motor neurons that innervate the striated muscle of the larynx and pharynx, and neurons that generate respiratory motor patterns (see later in chapter). The parasympathetic preganglionic neurons are organized in topographic fashion, with the esophagus represented most rostrally and dorsally.

Special Visceral Motor Column

The special visceral motor column includes motor nuclei that innervate muscles derived from the branchial (pharyngeal) arches. Because these arches are homologous to the gills in fish, the muscles are considered special visceral muscles, even though they are striated. During development, these cell groups migrate to an intermediate position in the basal plate and are eventually located ventrolaterally in the tegmentum.

The *trigeminal motor nucleus* (N. V) lies at midpontine levels and innervates the muscles of mastication. Nearby in separate clusters are located the *accessory trigeminal nuclei* that innervate the tensor tympani, tensor veli palatini, and mylohyoid muscles, and the anterior belly of the digastric muscle.

The *facial motor nucleus* (N. VII) lies caudal to the trigeminal motor nucleus at the level of the caudal pons and innervates the muscles of facial expression. During development, facial motor neurons migrate medially and rostrally around the medial margin of the abducens nucleus before turning laterally, ventrally, and caudally toward their definitive position at the pontomedullary junction (Figure 40–5A). This sinuous course that the axons leave behind forms the *internal genu of the facial nerve*. The adjacent *accessory facial motor nuclei* innervate the stylohyoid and stapedius muscles and the posterior belly of the digastric muscle.

The nucleus ambiguus contains branchial motor neurons with axons that run in the glossopharyngeal and vagus nerves. These neurons innervate the striated muscles of the larynx and pharynx. During development, these motor neurons migrate into the ventrolateral medulla, and as a consequence, their axons run dorsomedially toward the dorsal motor vagal nucleus, then turn sharply within the medulla to exit laterally, similar to the course of the facial motor axons.

General Somatic Motor Column

The neurons of the somatic motor column migrate the least during development, remaining close to the ventral midline. The *oculomotor nucleus* (N. III) lies at the midbrain level; it consists of five rostrocaudal columns of motor neurons innervating the medial, superior, and inferior rectus muscles, the inferior oblique muscle, and the levator of the eyelids. The motor neurons for the medial and inferior rectus and inferior oblique muscles are on the side of the brain stem from which the nerve exits, whereas those for the superior rectus are on the opposite side. The levator motor neurons are bilateral.

The *trochlear nucleus* (N. IV), which innervates the trochlear muscle, lies at the midbrain/rostral pontine level on the side of the brain stem opposite from which the nerve exits. The *abducens nucleus* (N. VI), which innervates the lateral rectus muscle, is located at the midpontine level. The *hypoglossal nucleus* (N. XII) in the medulla consists of several columns of neurons, each of which innervates a single muscle of the tongue.

The Organization of the Brain Stem Differs From the Spinal Cord in Three Important Ways

One major difference between the organization of the brain stem and that of the spinal cord is that many long ascending and descending sensory tracts that run along the outside of the spinal cord are incorporated within the interior of the brain stem. Thus, the ascending sensory tracts (the medial lemniscus and spinothalamic tract) run through the reticular formation of the brain stem, as do the auditory, vestibular, and visceral sensory pathways.

A second major difference is that in the brain stem, the cerebellum and its associated pathways form additional structures that are superimposed on the basic plan of the spinal cord. Fibers of the cerebellar tracts and nuclei are bundled with those of the pyramidal and extrapyramidal motor systems to form a large ventral portion of the brain stem. Thus, from the midbrain to the medulla, the brain stem is divided into a dorsal portion, the tegmentum, which follows the basic segmental plan of the spinal cord, and a ventral portion, which contains the structures associated with the cerebellum and the descending motor pathways. At the level of the midbrain, the ventral (motor) portion includes the cerebral peduncles, substantia nigra, and red nuclei. The base of the pons includes the pontine nuclei, corticospinal tract, and middle cerebellar peduncle. In the medulla, the ventral motor structures include the pyramidal tracts and inferior olivary nuclei.

A third major difference is that, although the hindbrain is segmented into rhombomeres during development, there is no clear repeating pattern in the adult brain. In contrast, the spinal cord is not segmented during development, but the final pattern consists of repeating segments. The prominent ladder-like arrays of ventral root axons and dorsal root ganglia suggest that segmentation is imposed by a polarizing effect of the adjacent body segments, or somites into which they migrate—in each somite, the rostral part attracts axonal growth cones and neural crest cells, whereas the caudal part is repulsive. In the head, such patterning is lacking as the cranial mesoderm is not segmented into somites but rather develops under the influence of the rhombomeres.

Neuronal Ensembles in the Brain Stem Reticular Formation Coordinate Reflexes and Simple Behaviors Necessary for Homeostasis and Survival

In the 19th century, Charles Darwin pointed out in his book *The Expression of the Emotions in Man and Animals* that the muscles of facial expression are activated in similar patterns in all mammals during similar emotional situations (fear, anger, disgust, happiness). He hypothesized that the patterns of facial expression must be deeply embedded in the organization of the brain stem. We now recognize that a wide range of reflexes and simple, repetitive, coordinated behaviors, such as facial emotional expression, breathing, and eating, are controlled by neurons in the brain stem reticular formation called *pattern generators*, which produce stereotyped innate responses. Impairment of cranial nerve reflexes and motor patterns in patients with neurological disease can indicate the precise site of brain stem damage.

Cranial Nerve Reflexes Involve Mono- and Polysynaptic Brain Stem Relays

The responses of the pupils to light (*pupillary light reflexes*) are determined by the balance between sympathetic tone in the pupillodilator muscles and parasympathetic tone in the pupilloconstrictor muscles of the iris. Sympathetic tone is maintained by postganglionic neurons in the superior cervical ganglion, which in turn are innervated by preganglionic neurons in the first and second thoracic spinal segments. Parasympathetic tone is supplied by postganglionic ciliary ganglion cells under the control of preganglionic neurons in the Edinger-Westphal nucleus and adjacent areas of the midbrain.

Light impinging on the retina activates a special class of retinal ganglion cells that act as brightness detectors. These cells receive inputs from photopigment-containing rod and cone cells, but they also have their own photopigment, melanopsin, which allows them to respond to light even when the rods and cones have degenerated. These cells send their axons through the optic nerve, chiasm, and tract to the olivary pretectal nucleus, where they terminate on neurons whose axons project to preganglionic neurons in the Edinger-Westphal nucleus (Figure 40–7). Thus, injury to the dorsal midbrain in the region of the posterior commissure can prevent pupillary light responses (midposition, fixed pupils), whereas injury to the oculomotor nerve eliminates parasympathetic tone to that pupil (fixed and dilated pupil). The melanopsin-containing

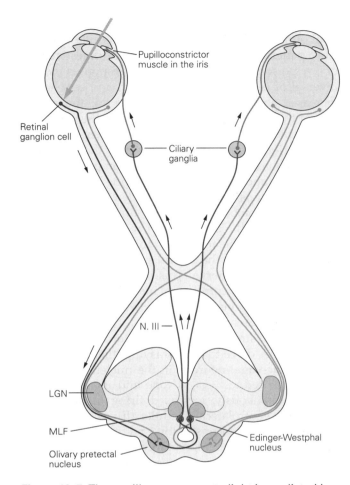

Figure 40–7 The pupillary response to light is mediated by parasympathetic innervation of the iris. Retinal ganglion cells that contain the photopigment melanopsin act as luminance detectors, sending their axons through the optic tract to the olivary pretectal nucleus, at the junction of the midbrain and the thalamus. Neurons in this nucleus project through the posterior commissure to parasympathetic preganglionic neurons in and around the Edinger-Westphal nucleus. The axons of the preganglionic cells exit with the oculomotor (III) nerve and contact ciliary ganglion cells, which control the pupilloconstrictor muscle in the iris. (Abbreviations: **LGN**, lateral geniculate nucleus; **MLF**, medial longitudinal fasciculus.)

retinal ganglion cells also project to the suprachiasmatic nucleus of the hypothalamus, where they entrain circadian rhythms to the day–night cycle (Chapter 44).

Vestibulo-ocular reflexes stabilize the image on the retina during head movement by rotating the eyeballs counter to the rotation of the head. These reflexes are activated by pathways from the vestibular ganglion and nerve to the medial, superior, and lateral vestibular nuclei, and from there to neurons in the reticular formation and ocular motor nuclei that coordinate eye movements. The reflex movements are seen most

clearly in comatose patients, in whom turning the head will elicit counter-rotational movements of the eyes (so-called doll's eye movements). Damage to these pathways in the pons impairs these movements.

The *corneal reflex* involves closure of both eyelids as well as upward turning of the eyes (Bell phenomenon) when the cornea is gently stimulated (eg, with a wisp of cotton). The sensory axons from the first division of the trigeminal nerve terminate in the spinal trigeminal nucleus, which relays the sensory signals to pattern generator neurons in the reticular formation adjacent to the facial motor nucleus. The pattern generator neurons provide bilateral inputs to the motor neurons that protect the cornea from damage by causing the orbicularis oculi muscle to close the eyelid and the oculomotor nuclei to roll the eyes upward and back in the orbit. Because the output of the pattern generator is bilateral, damage along the sensory pathway prevents the reflex in both eyes, whereas damage to the facial nerve prevents closure on the same side only.

The *stapedial reflex* contracts the stapedius muscle in response to a loud sound, thus damping movement of the ossicles. The sensory pathway is through the cochlear nerve and nucleus to the reticular formation adjacent to the facial motor nucleus and from there to the stapedial motor neurons, which run in the facial nerve. As described earlier, in patients with injury to the facial nerve (eg, Bell palsy), the stapedial reflex is impaired, and the patient complains that sounds in that ear have a "booming" quality (hyperacusis).

A variety of gastrointestinal reflexes are controlled by multisynaptic brain stem relays. For example, the tasting of food causes neurons in the solitary nucleus that project to the reticular formation adjacent to the motor facial and dorsal motor vagal nuclei to stimulate the preganglionic salivary neurons. The contact of food in the mouth can also elicit gastric contractions and acid secretion, presumably through inputs from the solitary nucleus directly to parasympathetic preganglionic gastric neurons in the dorsal motor vagal nucleus. In patients who have had Bell palsy, the damaged VII nerve parasympathetic axons may regrow aberrantly so that salivary axons reach the lacrimal gland in error, causing tasty food to initiate reflex tearing (crocodile tears).

The *gag reflex* protects the airway in response to stimulation of the posterior oropharynx. The afferent sensory fibers in the glossopharyngeal and vagus nerves terminate in the spinal trigeminal nucleus, whose axons project to the reticular formation adjacent to the nucleus ambiguus. Branchial motor neurons in the nucleus ambiguus innervate the posterior pharyngeal muscles, resulting in elevation of the palate, constriction of pharyngeal muscles (to expel the offending stimulus), and closure of the airway. Loss of the gag reflex on one side of the throat indicates injury to the medulla or to cranial nerve X on that side (cranial nerve IX has such a small territory of sensory and motor innervations in the pharynx that transection of this nerve does not cause a noticeable deficit).

Pattern Generators Coordinate More Complex Stereotypic Behaviors

As Darwin proposed, pools of pattern generator neurons in the reticular formation adjacent to the facial nucleus control facial emotional expression through stereotypic patterns of contraction of facial muscles simultaneously on the two sides of the face. Pattern generator neurons on each side of the brain stem project to the facial motor neurons on both sides of the brain, so that spontaneous facial expressions are virtually always symmetric. Even patients who have had major strokes in the cerebral hemispheres and cannot voluntarily move the contralateral orofacial muscles still tend to smile symmetrically when they hear a joke and can raise their eyebrows symmetrically, both of which are initiated by facial pattern generators.

Similarly, orofacial movements involved in eating are produced by pattern generator neurons in the reticular formation near the cranial motor nuclei that mediate the behaviors. Licking movements are organized in the reticular formation near the hypoglossal nucleus, chewing movements near the trigeminal motor nucleus, sucking movements near the facial and ambiguus nuclei, and swallowing near the nucleus ambiguus. Not surprisingly, neurons in these reticular areas are closely interconnected with each other and receive inputs from the part of the nucleus of the solitary tract concerned with taste and from the part of the spinal trigeminal nucleus concerned with tongue and oral sensation, as well as from neurons in the adjacent reticular formation that respond to more complex combinations of taste, texture, and temperature of food. As a result, even a decerebrate rat is able to make appropriate choices of which foods to swallow and which to reject.

Vomiting is another example of a coordinated response mediated by pattern generator neurons. Toxic substances in the blood stream can be detected by nerve cells in the area postrema, a small region adjacent to the nucleus of the solitary tract along the floor of the fourth ventricle. Unlike most of the brain, which is protected by a blood–brain barrier, the area postrema contains fenestrated capillaries that allow its neurons to sample the contents of the blood stream.

These neurons, when they detect a toxin, activate a pool of neurons in the ventrolateral medulla that control a pattern of responses that clears the digestive tract of any poisonous substances. These responses include reversal of peristalsis in the stomach and esophagus, increased abdominal muscle contraction, and activation of the same motor patterns used in the gag reflex to clear the oropharynx of unwanted material.

A variety of responses organized by the brain stem require coordination of cranial motor patterns with autonomic and sometimes endocrine responses. A good example is the *baroreceptor reflex*, which ensures an adequate blood flow to the brain (Chapter 41). The nucleus of the solitary tract receives information about stretch of the aortic arch through the vagus (X) nerve and stretch of the carotid sinus through the glossopharyngeal (IX) nerve. This information is relayed to neurons in the ventrolateral medulla that produce a coordinated response that protects the brain against a fall in blood pressure.

Reduced stretch of the aortic arch and carotid sinus reduces drive to the parasympathetic preganglionic cardiac-vagal neurons in the nucleus ambiguus, resulting in reduced vagal tone and increased heart rate. Simultaneously, increased firing of neurons in the rostral ventrolateral medulla drives sympathetic preganglionic vasoconstrictor and cardioaccelerator neurons. This combination of increased cardiac output and increased vascular resistance elevates blood pressure. Meanwhile, other neurons in the ventrolateral medulla increase the firing of hypothalamic neurons that secrete vasopressin from their terminals in the posterior pituitary gland. Vasopressin also has a direct vasoconstrictor effect, and it maintains blood volume by reducing water excretion through the kidney.

Control of Breathing Provides an Example of How Pattern Generators Are Integrated Into More Complex Behaviors

One of the most important functions of the brain stem is control of breathing. The brain stem automatically generates breathing movements beginning in utero at 11 to 13 weeks of gestation in humans, and continues nonstop from birth until death. This behavior does not require any conscious effort, and in fact, it is rare for us to even think about the need to breathe. The primary purpose of breathing is to ventilate the lungs to control blood levels of oxygen, carbon dioxide, and hydrogen ions (pH). (These are often measured together clinically and referred to as "blood gases.") Breathing movements involve contraction of the diaphragm, activated by the phrenic nerve. The diaphragm is assisted

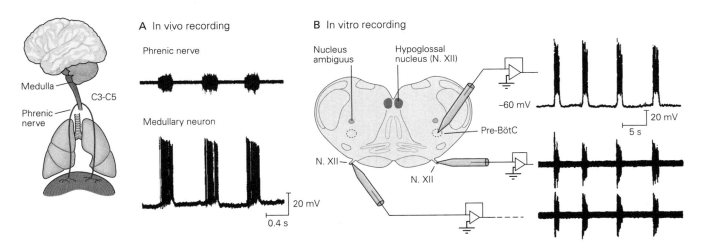

Figure 40–8 Rhythmic breathing is generated within the medulla.

A. Rhythmic activity in the phrenic motor nerve of a guinea pig causes contraction of the diaphragm. The firing of the phrenic nerve is phase-locked to bursts of firing by neurons in the medulla. The activity in a single medullary neuron, recorded intracellularly, is shown. (Reproduced, with permission, from Richerson and Getting 1987. Copyright © 1987. Published by Elsevier B.V.)

B. Similar rhythmic firing can be recorded in vitro from accessory respiratory nerves, such as the hypoglossal (XII) nerve. The minimal tissue necessary to support this rhythm is a slice about 0.5 mm thick at the level of the rostral medulla. Neurons in the pre-Bötzinger complex (**Pre-BötC**) near the nucleus ambiguus fire bursts that are phase-locked to the motor rhythm. (Reproduced, with permission, from Smith et al. 1991. Copyright © 1991 AAAS.)

when necessary by accessory muscles of respiration, including the intercostal muscles, pharyngeal muscles (to change airway diameter), some neck muscles (which help expand the chest), the tongue protruder muscles (to open the airway), and even some facial muscles (which flare the nares).

Respiratory activity can be generated by the medulla even when it is isolated from the rest of the nervous system. Many medullary neurons have patterns of firing that correlate with inspiration or expiration (Figure 40–8A). Some have more refined patterns, such as firing only during early inspiration or late inspiration. These respiratory neurons are concentrated in two regions, the dorsal and ventral respiratory groups.

The *dorsal respiratory group* is located bilaterally in and around the ventrolateral part of the nucleus of the solitary tract. Neurons in this group receive respiratory sensory input, including afferents from stretch receptors in the lungs and peripheral chemoreceptors, and participate in such reflex actions as limitation of lung inflation at high volume (the Hering-Breuer reflex) and the ventilatory response to low oxygen (*hypoxia*). The *ventral respiratory group*, a column of neurons in and around the nucleus ambiguus, coordinates respiratory motor output. Some of these neurons are motor neurons with axons that leave the brain through the vagus nerve and innervate accessory muscles of respiration or premotor neurons that innervate the phrenic

motor nucleus, whereas others form a pattern generator, the *pre-Bötzinger complex*, that generates respiratory rhythm.

The intrinsic rhythmicity of the pre-Bötzinger complex is so resilient that, even in a transverse brain slice from the rostral medulla, neurons in the pre-Bötzinger complex are able independently to generate a respiratory rhythm that can be recorded in the rootlets of the hypoglossal (XII) nerve that emerge from the ventral surface of the slices (Figure 40–8B). Acute destruction of this cell group in an intact animal results in inability to maintain a normal respiratory rhythm.

The most important inputs to the respiratory pattern generator come from chemoreceptors that sense oxygen and carbon dioxide. Under normal conditions, ventilation is primarily regulated by the levels of CO_2 rather than O_2 (Figure 40–9A). However, breathing is also strongly stimulated if O_2 becomes sufficiently low, such as at high altitude or in people with lung disease. The peripheral chemoreceptors in the carotid and aortic bodies normally respond primarily to a decrease in blood oxygen, but during hypoxia, they also become more sensitive to elevated levels of CO_2 (*hypercapnia*). Afferent fibers from the carotid sinus nerve travel in the glossopharyngeal nerve and activate neurons in the dorsal respiratory group.

The response to hypercapnia is largely driven by *central chemoreceptors* in the brain stem that sense the

Figure 40–9 Respiratory motor output is regulated by carbon dioxide in the blood.

A. Lung ventilation (determined by the rate and depth of breathing) in humans is steeply dependent on the partial pressure of carbon dioxide (PCO_2) at normal levels of the partial pressure of oxygen (PO_2) (>100 mm Hg). When PO_2 drops to very low values (<50 mm Hg), breathing is stimulated directly and also becomes more sensitive to an increase in PCO_2 (seen here as an increase in the slope of the curves for alveolar PO_2 of 37 and 47 mm Hg). (Reproduced, with permission, from Nielsen and Smith 1952.)

B. Central chemoreceptors in the medulla control ventilatory motor output to maintain normal blood CO_2. The firing rate of serotonergic neurons within the raphe nuclei of the medulla increases when elevated PCO_2 causes a pH decrease. The records shown here are from in vitro recordings of a neuron in the raphe nuclei of a rat at two different levels of pH (7.4, control, and 7.2, acidosis). (Reproduced, with permission, from Wang et al. 2002.)

C. Serotonergic neurons are closely associated with large arteries in the ventral medulla where they can monitor local changes in PCO_2. Two images of the same transverse section of the rat medulla show blood vessels after injection of a red fluorescent dye into the arterial system (*left*) and green antibody staining for tryptophan hydroxylase, the enzyme that synthesizes serotonin (*right*). The basilar artery (**B**) is on the ventral surface of the medulla between the pyramidal tracts (**P**). (Reproduced, with permission, from Bradley et al. 2002. Copyright © 2002 Springer Nature.)

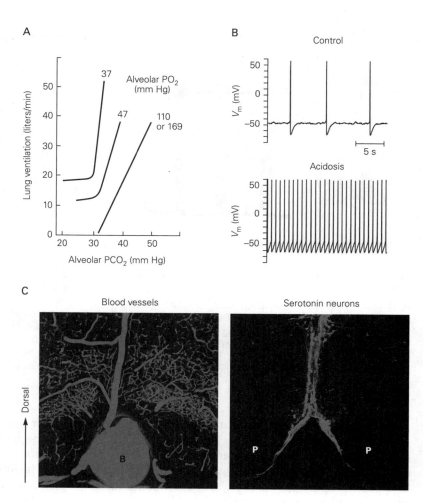

accompanying decrease in pH. The most sensitive area for this is along the ventral surface of the medulla lateral to the pyramidal tract. This region contains at least two sets of neurons that respond to elevated CO_2. Glutamatergic neurons in the retrotrapezoid nucleus in the rostral ventrolateral medulla, near the facial motor nucleus, are highly sensitive to CO_2 levels. Absence of these neurons, due to a mutation in the phox2b transcription factor required for their development, causes congenital central hypoventilation syndrome, in which there is failure to breathe adequately, particularly during sleep. In addition, serotonergic neurons in the rostral ventrolateral medulla, like retrotrapezoid neurons, lie along penetrating arteries and are sensitive to acidosis (Figure 40–9B,C). Genetic deletion of these neurons reduces the ventilatory response to hypercapnia, especially during sleep. Recent studies demonstrate that a serotonin 5-HT_{2A} agonist can restore arousal responses to CO_2, suggesting that the serotonergic neurons play a modulatory role, increasing the sensitivity of the CO_2 reflexes during hypercapnia, and that this may be especially important during sleep.

The motor pattern generated by the respiratory system is remarkably stable in healthy people, but a variety of diseases can alter these patterns. One of the most common and easily recognized patterns is Cheyne-Stokes respiration, which is characterized by repeated cycles of gradually increasing then decreasing ventilation, alternating with cessation of breathing (apnea). This periodic breathing is seen, for example, in congenital central hypoventilation syndrome, where the central neurons are not sufficiently sensitive to rising CO_2, particularly during sleep. By the time they begin to respond, CO_2 levels may already be quite high. This causes hyperventilation, which reduces CO_2 levels below the threshold where breathing is required. The result is a period of apnea, until the CO_2 levels again become quite high (Figure 40–10).

A similar pattern is seen in people who have cardiac or pulmonary disease that increases the time it takes for the change in alveolar CO_2 to register with the medulla. Cheyne-Stokes respiration often occurs in hospitalized patients with marginal cardiac or respiratory reserve when they fall asleep, thus reducing

Figure 40–10 Respiratory motor patterns can become unstable during sleep.

A. Sleep apnea (cessation of breathing) is a common problem that often goes undetected. The records here show blood oxygen saturation (**SaO₂**) and CO_2 partial pressure (**PCO₂**) during sleep in a healthy person and a patient with obstructive sleep apnea. In the healthy person, SaO_2 remains near 100%, and PCO_2 remains near 40 mm Hg during both rapid eye movement (**REM**) and non-REM sleep. In the patient with sleep apnea, reduced muscle tone (hypotonia) during sleep leads to collapse of the upper airway, resulting in obstruction and apnea. Repetitive apnea at the rate of approximately once per minute causes the patient's SaO_2 to fall repetitively and dramatically. (The inset shows a period of approximately 80 seconds on an expanded scale. Ventilation [**V**] begins at the nadir of the SaO_2 and again ceases when the blood oxygen increases.) During non-REM sleep, the patient's PCO_2 increases to near 60 mm Hg. During REM sleep, the SaO_2 and PCO_2 become even more abnormal, as worsening airway hypotonia causes greater obstruction. Many people with sleep apnea wake up repeatedly during the night because of the apnea, but the arousals are too brief for them to be aware that their sleep is interrupted. (Adapted, with permission, from Grunstein and Sullivan 1990.)

B. Breathing in most normal individuals becomes unstable during sleep at high altitudes. The upper trace shows an example of a Cheyne-Stokes breathing pattern in a healthy person, during the first night after arriving at an altitude of 17,700 feet, where the low partial pressure of oxygen in the air reduces the blood SaO_2 to approximately 75% to 80%. Repeated cycles of waxing and waning ventilation are separated by periods of apnea. Administration of supplemental oxygen results in a rapid return to a normal respiratory pattern. This abnormal pattern disappears in most people after they have acclimated to the altitude. (Reproduced, with permission, from Lahiri et al. 1984.)

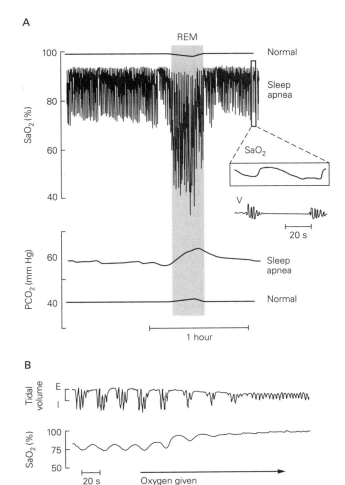

other behavioral drives for respiration. Although not dangerous in itself, it can indicate that there is a serious underlying cardiorespiratory problem that needs to be corrected.

Other inputs to the respiratory pattern generator come from the circuitry mediating particular behaviors, as breathing must be coordinated with many motor actions that share the same muscles. To accomplish this coordination, respiratory neurons in the medulla receive input from neuronal networks concerned with vocalization, swallowing, sniffing, vomiting, and pain. For example, the ventral respiratory group is connected with a part of the parabrachial complex in the pons termed the *pontine respiratory group* or *pneumotaxic center*. These pontine neurons coordinate breathing with behaviors such as chewing and swallowing. They can cause holding of the breath at full inspiration (called *apneusis*), which is required during eating and drinking. The reserve of air in the lungs permits a cough, if necessary, to expel any food or drink that may enter

the airway. Other neurons in the intertrigeminal zone, between the motor and principal sensory trigeminal nuclei, receive facial and upper airway sensory inputs and project to the ventrolateral medulla to temporarily stop breathing to protect against accidental inspiration of dust or water.

Voluntary motor pathways can take over the control of breathing during talking, eating, singing, swimming, or playing a wind instrument. Descending inputs cause hyperventilation at the onset of exercise, in anticipation of an increase in oxygen demand. In fact, this leads to a sustained drop in blood CO_2 during exercise—the opposite of what would be expected for a negative feedback control system. Other descending inputs from the limbic system produce hyperventilation in connection with pain or anxiety and, in some people, may be responsible for causing spontaneous panic attacks, characterized by hyperventilation and a feeling of suffocation. These various descending inputs allow efficient integration of breathing with other behaviors, but they ultimately must yield to the need to maintain blood gas homeostasis, as even a small increase in CO_2 produces severe air hunger or *dyspnea*. Thus, the respiratory control system is a fascinating example of a brain stem pattern generator that must be sufficiently stable to ensure survival yet flexible enough to accommodate a wide variety of behaviors.

Monoaminergic Neurons in the Brain Stem Modulate Sensory, Motor, Autonomic, and Behavioral Functions

In addition to containing the primary sensory and motor nuclei of the cranial nerves and the reflex and pattern generator mechanisms that control basic behaviors, the brain stem also contains a set of modulatory cell groups. In a groundbreaking series of experiments in the 1970s, Hans Kuypers used the newly discovered method of retrograde transport of axonal tracers to identify the cell groups in the brain stem and diencephalon that contribute to modulation of spinal cord sensory and motor systems and those that send inputs directly to the cerebral cortex. To a surprising extent, these two sets of experiments, starting at opposite ends of the neuraxis, identified a common substrate whose role it is to modulate circuitry at other levels of the nervous system, almost as if it were an "autonomic system" for the brain.

These cell groups have direct connections to the forebrain, brain stem, and spinal cord that regulate the overall level of function of their targets. Like the way serotonergic neurons set the overall sensitivity of CO_2 reflexes, brain stem monoaminergic modulatory systems adjust the overall responsiveness of a wide variety of sensory systems by means of projections to sensory neurons in the spinal cord and brain stem, including nociceptive systems. Descending projections from these modulatory systems also control motor tone, which is critical for adjusting posture and gait as well as initiating finer movements. Ascending inputs to the forebrain control overall arousal as well as responses to rewarding situations. While these modulatory systems are not sufficient to accomplish motor, sensory, or cognitive tasks on their own, their ability to adjust the responsiveness of these systems plays an enormously influential role in overall behavior.

Many Modulatory Systems Use Monoamines as Neurotransmitters

The monoaminergic systems use decarboxylated derivatives of the cyclic amino acids tyrosine, tryptophan, and histidine as neurotransmitters. They were among the first in the brain to be identified and mapped due to the property that some of them possess to fluoresce when exposed to formaldehyde. In the 1960s, Dahlstrom and Fuxe used this property to identify serotonergic, noradrenergic, and dopaminergic cell groups in the brain stem. In the 1970s, with the development of immunohistochemical methods able to map the enzymes that synthesize monoamines, other investigators mapped neurons containing epinephrine and histamine.

The cell groups of these modulatory systems in general were unlike earlier identified nuclei in the brain. Rather than forming compact clusters of cell bodies, the monoaminergic cell groups tended to form columns that extended longitudinally through the brain stem and hypothalamus (see Figure 40–6). The monoamine systems were therefore designated with letters and numbers, to avoid confusion with other systems of nomenclature for the brain (Figure 40–11).

The first cell groups identified by Dahlstrom and Fuxe were simply identified alphabetically as the "A" cell groups, and then numbered sequentially from caudal to rostral. It was later determined that the A1–A7 cell groups produce norepinephrine and the A8–A14 groups produce dopamine. The A1, A3, and A5 designations were applied to neurons located in the ventrolateral corner of the medullary and pontine tegmentum (the A3 group was quite small and the term is no longer used), while the A2, A4, A6, and A7 names were applied to cell groups located more dorsally, similar to the columns of motor neurons in the brain stem (Figure 40–11A). The A1 and A2 groups, located among the neurons of the nucleus ambiguus and the nucleus of the solitary tract (respectively), are mainly concerned with autonomic functions. Together, they modulate hypothalamic and brain stem systems that regulate the autonomic nervous system. The noradrenergic A4–A7 cell groups have widespread influence over sensory and motor systems, ranging from the cerebral cortex to the spinal cord, and provide important modulation of arousal and wakefulness.

The dopaminergic systems (Figure 40–11E) include the A8–A10 cell groups, located in the midbrain in and near the substantia nigra, that modulate motor systems as well as forebrain mechanisms of reward and motivation. The A11 and A13 dopaminergic neurons, in the dorsal hypothalamus, provide input to sensory, motor, and autonomic systems in the brain stem and spinal cord. The A12, A14, and A15 neurons have a neuroendocrine role, including release of dopamine as a pituitary release-inhibiting hormone for prolactin secretion. The A16 cell group modulates olfactory inputs, and the A17 neurons in the retina modulate vision.

A Norepinephrine/Epinephrine

—— Noradrenergic projections
—— Adrenergic projections

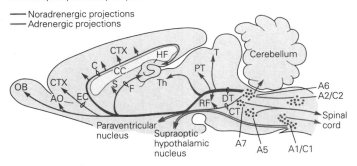

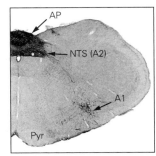

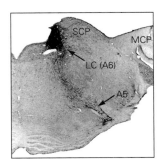

B Histamine

—— Histaminergic innervation

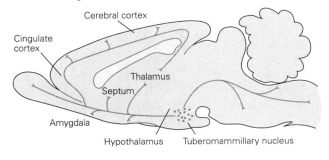

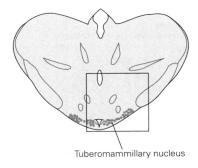

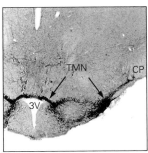

Tuberomammillary nucleus

C Serotonin

—— Serotonergic innervation

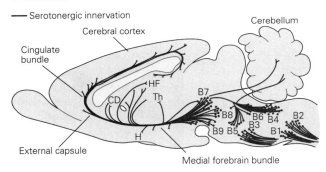

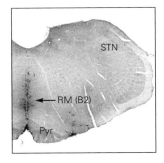

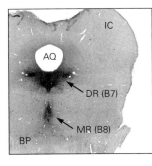

Figure 40–11 Locations and projections of monoaminergic and cholinergic neurons in the rat brain. (Abbreviations: **3V,** third ventricle; **AC,** anterior commissure; **AP,** area postrema; **AQ,** Sylvian aqueduct; **ARC,** arcuate nucleus; **BM,** nucleus basalis of Meynert; **BP,** brachium pontis; **CD,** caudate; **CP,** cerebral peduncle; **DBh,** horizontal limb of the diagonal band; **DR,** dorsal raphe; **FX,** fornix; **IC,** inferior colliculus; **LC,** locus ceruleus; **LDT,** laterodorsal tegmental nucleus; **MCP,** middle cerebellar peduncle; **MGN,** medial geniculate nucleus; **MR,** median raphe; **MS,** medial septum; **MTT,** mammillothalamic tract; **NTS,** nucleus tractus solitarius; **OC,** optic chiasm; **PPT,** pedunculopontine tegmental nucleus; **PUT,** putamen; **Pyr,** pyramidal tract; **RM,** raphe magnus; **SC,** superior colliculus; **SCP,** superior cerebellar peduncle; **SN,** substantia nigra; **STN,** spinal trigeminal nucleus; **TMN,** tuberomammillary nucleus; **VTA,** ventral tegmental area.)

A. Noradrenergic neurons (A groups) and adrenergic neurons (C groups) are located in the medulla and pons. The A2 and C2 groups in the dorsal medulla are part of the nucleus of the solitary tract. The A1 and C1 groups in the ventral medulla are located near the nucleus ambiguus. Both groups project to

the hypothalamus; some C1 neurons project to sympathetic preganglionic neurons in the spinal cord and control cardiovascular and endocrine functions. The A5, A6 (locus ceruleus), and A7 cell groups in the pons project to the spinal cord and modulate autonomic reflexes and pain sensation. The locus ceruleus also projects rostrally to the forebrain and plays an important role in arousal and attention.

B. All histaminergic neurons are located in the posterior lateral hypothalamus, mostly within the tuberomammillary nucleus. These neurons project to virtually every part of the neuraxis and play a major role in arousal.

C. Serotonergic neurons (B groups) are found within the medulla, pons, and midbrain, mostly near the midline in the raphe nuclei. Those within the medulla (the B1–B4 groups corresponding to the raphe magnus, raphe obscurus, and raphe pallidus) project throughout the medulla and spinal cord and modulate afferent pain signals, thermoregulation, cardiovascular control, and breathing. Those within the pons and midbrain (the B5–B9 groups in the raphe pontis, median raphe, and dorsal raphe) project throughout the forebrain and contribute to arousal, mood, and cognition.

D Acetylcholine

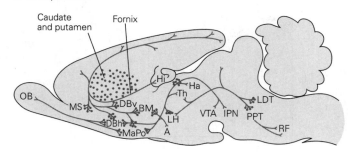
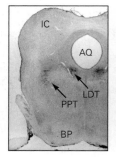
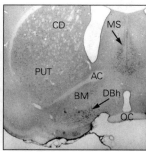

E Dopamine

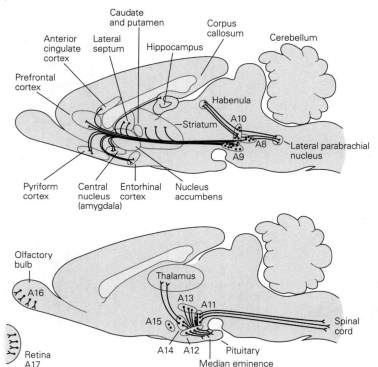
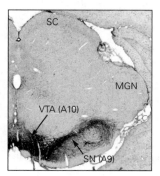
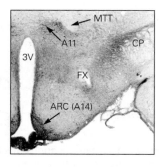

Figure 40–11 (*Continued*) **D.** Cholinergic neurons (sometimes called Ch groups) are located in the pons, midbrain, and basal forebrain. Those in the pons and midbrain (mesopontine groups) are divided into a ventrolateral cluster (pedunculopontine nucleus) and the dorsomedial cluster (laterodorsal tegmental nucleus). The mesopontine cholinergic neurons project to the brain stem reticular formation and the thalamus. Those in the basal forebrain are divided into the medial septum, the nuclei of the vertical and horizontal limbs of the diagonal band, and the nucleus basalis of Meynert. These neurons project throughout the cerebral cortex, hippocampus, and amygdala. Both groups play an important role in arousal, and the basal forebrain groups are also involved in more selective attention.

E. Dopaminergic neurons are located in the midbrain and hypothalamus. The dopaminergic cell groups were originally included with the noradrenergic cell groups and are still labeled as

A groups (A8–A17). The A8 group is in the midbrain dorsally adjacent to the substantia nigra. The A9 cell group constitutes the substantia nigra pars compacta. These two groups of neurons project to the striatum and play an important role in initiation of movement. The A10 group is located in the ventral tegmental area just medial to the substantia nigra. These cells project to the frontal and temporal cortex and limbic structures of the basal forebrain and play a role in emotion and memory. The A11 and A13 cell groups in the zona incerta of the hypothalamus project to the lower brain stem and spinal cord and regulate sympathetic preganglionic neurons. The A12, A14, and A15 cell groups are components of the neuroendocrine system. Some of them inhibit release of prolactin into the hypophysial portal circulation, and others control gonadotrophin secretion. Dopaminergic neurons are also found in the olfactory bulb (A16) and the retina (A17).

The B cell groups, which had a slightly different color of fluorescence, were found to produce serotonin. They are associated with the midline raphe cell groups in the pons and medulla (Figure 40–11C). The B1–B4 cell groups in the medulla mainly provide descending modulation of sensory, motor, and autonomic neurons in the brain stem and spinal cord. The B5–B7 neurons in the pons mainly provide serotonergic innervation of the thalamus, hypothalamus, and cerebral cortex. The functions of serotonin in modulating these targets can be quite complex to decipher, mainly because there are at least 14 different serotonin receptors, and different ones can be expressed by different cell types in a target area.

A few years after the A and B cell groups were named, immunohistochemical studies demonstrated that some medullary neurons have the enzymes to make dopamine and norepinephrine but do not fluoresce. These neurons, named cell groups C1–C3, were found to process these other catecholamines to adrenalin, or epinephrine. They are closely related to the A1–A3 cell groups in the medulla (Figure 40–11A).

Histaminergic cell groups are mainly found in the tuberomammillary nucleus and adjacent areas of the posterior hypothalamus (near the mammillary body) and are named E1–E5 (Figure 40–11B). They are the sole source of histaminergic actions in the entire brain, from the cerebral cortex to the spinal cord, and are involved in a variety of arousal responses.

Although cholinergic neurons are not, strictly speaking, monoaminergic, some of them also participate in modulatory systems, and these have been numbered Ch1–Ch6 (Figure 40–11D). This classification system did not include the many other cholinergic neurons in the nervous system, such as motor neurons or striatal interneurons, and is not used much anymore. Rather, scientists refer to the cholinergic neurons by their location, eg, the pedunculopontine (Ch6) and laterodorsal tegmental (Ch5) neurons in the pons, which project widely from the cerebral cortex to the medulla, and the basal forebrain (Ch1–Ch4) groups, which project to the cerebral cortex, hippocampus, and amygdala.

Monoaminergic Neurons Share Many Cellular Properties

Neurons that use monoamines as neurotransmitters have many similar electrophysiological properties. For example, most continue to fire spontaneous action potentials in a highly regular pattern when isolated from their synaptic inputs in brain slice preparations. Their action potentials typically are followed by a slow membrane depolarization that leads to the next spike

(Figure 40–12). The spontaneous regular firing pattern of monoaminergic neurons is regulated by intrinsic pacemaker currents (Chapter 10). Tonic firing in vivo may be important for ensuring continuous delivery of monoamines to targets. For example, the basal ganglia depend on continuous exposure to dopamine from the neurons of the substantia nigra to facilitate motor responses.

The properties of monoaminergic neurons are suited to their unique and widespread modulatory roles in brain function. Indeed, some axon terminals of monoaminergic cells do not even form conventional synaptic connections, instead releasing neurotransmitter diffusely to many targets at once. Most

A Firing pattern of a locus ceruleus neuron

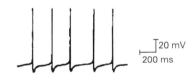

B Firing of a locus ceruleus neuron across wake-sleep

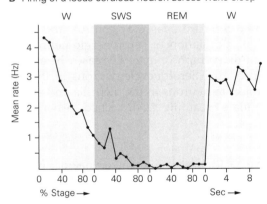

Figure 40–12 Monoaminergic neurons have similar firing patterns across the wake–sleep cycle.

A. When monoaminergic neurons are isolated from synaptic input, they fire spontaneously at a regular rate. This recording is from a noradrenergic neuron in the locus ceruleus. Action potentials are followed by a characteristic afterhyperpolarization followed by a slow depolarization to the next spike, producing a pacemaker-like activity (Chapter 10). Serotonergic and histaminergic neurons exhibit similar spontaneous activity.

B. All three monoaminergic cell types show similar patterns of firing across the wake–sleep cycle. The plot shows that a locus ceruleus neuron in a rat fires fastest when the animal is awake (**W**), slows down as wakefulness wanes and during slow-wave sleep (**SWS**), and almost completely ceases to fire during rapid eye movement (**REM**) sleep. (Adapted, with permission from Aston-Jones and Bloom 1981. © Society for Neuroscience.)

monoaminergic neurotransmission occurs by means of metabotropic synaptic actions through G protein–coupled receptors. Many monoaminergic neurons co-release neuropeptides, which have slow effects through other G protein–coupled receptors. Thus, although some monoaminergic synaptic actions involve fast synaptic mechanisms (Chapter 13), many involve slower metabotropic and neuromodulatory pathways as well (Chapter 14).

Autonomic Regulation and Breathing Are Modulated by Monoaminergic Pathways

Neurons in the adrenergic C1 group in the rostral ventrolateral medulla play a key role in maintaining resting vascular tone as well as adjusting vasomotor tone necessitated by various behaviors. For example, an upright posture disinhibits neurons in the rostral ventrolateral medulla that directly innervate the sympathetic preganglionic vasomotor neurons, thus increasing vasomotor tone to prevent a drop in blood pressure (the baroreceptor reflex). Neurons in the noradrenergic A5 group in the pons inhibit the sympathetic preganglionic neurons and play a role in depressor reflexes (eg, the fall in blood pressure in response to deep pain).

Serotonin regulates many different autonomic functions including gastrointestinal peristalsis, thermoregulation, cardiovascular control, and breathing. Electrical stimulation of serotonergic neurons within the medullary raphe nuclei increases heart rate and blood pressure. Serotonergic neurons in the medulla also project to neurons in the medulla and spinal cord that regulate breathing, as described earlier.

The role of serotonergic neurons as CO_2 receptors may explain why defects in the serotonergic system have been linked to sudden infant death syndrome (SIDS) (Figure 40–13A). SIDS is the leading cause of postneonatal mortality in the Western world, responsible for six infant deaths every day in the United States. A widely held theory holds that some SIDS cases are due to defective CO_2 chemoreception, breathing, and arousal. A relatively high number of serotonergic neurons are found in the raphe nuclei of infants who die of SIDS, but these have an immature morphology, and they are associated with relatively low serotonin levels and low serotonergic receptor densities.

A plausible neurobiological mechanism for SIDS is that a defect in development of serotonergic neurons leads to reduced ability to detect a rise in partial pressure of CO_2 when airflow is obstructed during sleep, thus blunting the normal protective response, which includes arousal and increased ventilation (Figure 40–13C). Infants sleeping face down would be unable to arouse sufficiently to change position when bedding blocks the airway. The Back to Sleep campaign, which encourages parents to place infants on their backs when put down to sleep, has reduced the incidence of SIDS by 50%.

Pain Perception Is Modulated by Monoamine Antinociceptive Pathways

Although pain is necessary for an animal to minimize injury, continued pain following an injury may be maladaptive (eg, if the pain prevents vigorous escape from

Figure 40–13 (Opposite) Serotonergic neurons have a role in the response to a rise in CO_2 levels as well as sudden infant death syndrome.

A. Serotonergic neurons in the medulla are central respiratory chemoreceptors that are thought to stimulate breathing in response to an increase in arterial blood PCO_2 (partial pressure of CO_2). The dendrites of these neurons wrap around large arteries and are stimulated by an increase in PCO_2 (see Figure 40–9C). They project to and excite motor neurons in the medulla and spinal cord that control breathing.

B. Serotonergic neurons in the midbrain are also PCO_2 sensors. Shown here is the increase in firing rate of a serotonergic neuron from the dorsal raphe nucleus in response to an increase in PCO_2 (monitored by the resultant decrease in external pH). This increase in firing rate may sensitize ascending arousal pathways from the parabrachial nucleus, which also receives input from other CO_2 sensory pathways. This important response prevents suffocation during sleep when the airway is obstructed. (Reproduced, with permission, from Richerson 2004. Copyright © 2004 Springer Nature.)

C. Sudden infant death syndrome (SIDS).

1. *Triple risk hypothesis of SIDS.* Infants are at risk to die from SIDS when three conditions coincide. First, the infant must be vulnerable because of an underlying abnormality of the brain stem, such as a genetic predisposition or an environmental insult (eg, exposure to cigarette smoke). Second, the baby must be in the stage of development (usually 2–6 months of age) when it may be difficult to change position to escape suffocation. Third, there also must be an exogenous stressor (eg, lying face down in a pillow). (Reproduced, with permission from, Filiano and Kinney 1994. © 1994 S. Karger AG.)

2. *Proposed mechanism of SIDS.* The combination of abnormal serotonergic neurons (eg, from exposure to cigarette smoke) and postnatal immaturity of neurons involved in respiratory control leads to the inability to respond effectively to airway obstruction (eg, from lying face down in a crib). The infant then does not wake up and turn its head or breathe faster, either of which would correct the problem. As a result, blood oxygenation decreases severely (hypoxia) while blood CO_2 rises (hypercapnia).

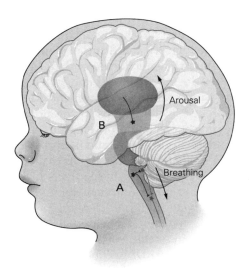

A Serotonergic neurons

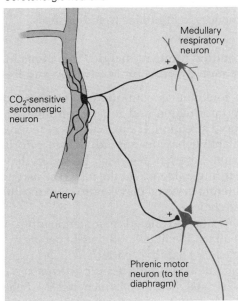

Medullary respiratory neuron

CO_2-sensitive serotonergic neuron

Artery

Phrenic motor neuron (to the diaphragm)

B

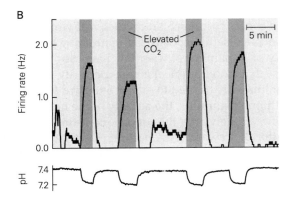

C Sudden Infant Death Syndrome

1 Triple risk hypothesis

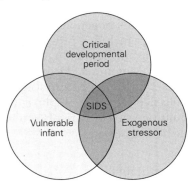

Critical developmental period

SIDS

Vulnerable infant

Exogenous stressor

2 Proposed mechanism

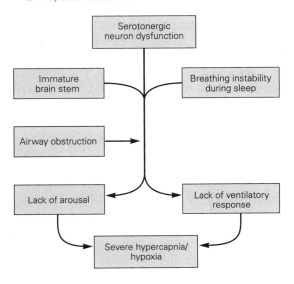

Serotonergic neuron dysfunction

Immature brain stem

Breathing instability during sleep

Airway obstruction

Lack of arousal

Lack of ventilatory response

Severe hypercapnia/hypoxia

a predator). The monoaminergic systems include important descending projections to the dorsal horn of the spinal cord that modulate pain perception (Chapter 20).

The noradrenergic inputs to the spinal cord originate from the pontine cell groups A5–A7, with the locus ceruleus (A6) providing most of the input to the dorsal horn. Similarly, the serotonergic raphe nuclei in the medulla, particularly the nucleus raphe magnus, project to the dorsal horn where they modulate the processing of information about noxious stimuli. Direct application of serotonin to dorsal horn neurons inhibits their response to noxious stimuli, and intrathecal administration of serotonin attenuates the defensive withdrawal of the paw evoked by noxious stimuli. In addition, intrathecal administration of antagonists of serotonin receptors blocks the pain inhibition evoked by stimulation of the raphe nuclei.

Insight into the role of serotonin in pain processing has been used in treating migraine headaches. In particular, the triptan agonists of 5-HT_{1D} receptors have been found to be therapeutically effective. One of the possible mechanisms of action of this family of tryptamine-based drugs includes presynaptic inhibition of pain afferents from the meninges, preventing sensitization of central neurons. Drugs that block monoamine reuptake, including both traditional antidepressants and selective serotonin reuptake inhibitors, are effective in limiting pain in patients with chronic pain and migraine headaches.

Motor Activity Is Facilitated by Monoaminergic Pathways

The dopaminergic system is critical for normal motor performance. A massive projection ascends from the substantia nigra pars compacta to the striatum, where dopaminergic fibers act on striatal neuron receptors to release inhibition of motor responses (Chapter 38).

Patients with Parkinson disease in whom midbrain dopaminergic neurons have degenerated have trouble initiating movement and difficulty sustaining movements. Such patients speak softly, write with small letters, and take small steps. Conversely, drugs that facilitate dopaminergic transmission in the striatum can result in unintended behaviors, ranging from motor tics (small muscle twitches), to chorea (large-scale, jerky limb movements), to complex cognitive behaviors (such as compulsive gambling or sexual activity).

As first shown by Sten Grillner, serotonergic neurons play an important role in modulating motor programs. Drugs that activate serotonin receptors can induce hyperactivity, myoclonus, tremor, and rigidity, all of which are part of the "serotonin syndrome."

Increases in the firing of raphe neurons have been observed in animals during repetitive motor activities such as feeding, grooming, locomotion, and deep breathing. Conversely, the firing of both serotonergic raphe and noradrenergic locus ceruleus neurons practically ceases during the atonia and lack of movement that occur during rapid eye movement (REM) sleep.

Noradrenergic cell groups in the pons also send extensive projections to motor cell groups. This modulatory input acts on presynaptic β- and α_1-adrenergic receptors to facilitate excitatory inputs to motor neurons (Chapter 31). The sum of these effects is to facilitate motor neuron responses in stereotypic and repetitive behaviors such as rhythmic chewing, swimming, or locomotion. Conversely, increased β-adrenergic activation during stress can exaggerate motor responses and produce tremor. Drugs that block β-adrenergic receptors are used clinically to reduce certain types of tremor and are often taken by musicians prior to performances to minimize tremulousness.

Ascending Monoaminergic Projections Modulate Forebrain Systems for Motivation and Reward

The forebrain is continuously bombarded with sensory information and must determine which stimuli deserve attention. It must also decide which of many available behaviors should receive priority, based in part on experience—which behaviors have achieved rewarding outcomes in the past. The ascending monoaminergic systems play key roles in modulating all of these choices.

As noted earlier, dopaminergic inputs to the striatum adjust the likelihood that a specific motor pattern or even a cognitive pattern will be expressed. Low dopamine levels reduce output from the direct pathway striatal neurons (which release behaviors) and increase activity of indirect pathway striatal neurons (which inhibit behavior). Dopamine also has been linked to reward-based learning. Rewards are objects or events for which an animal will work (Chapter 42) and are useful in positively reinforcing behavior. Activity of dopaminergic neurons increases when a reward (such as food or juice) is unexpectedly given. But after animals are trained to expect a reward following a conditioned stimulus, the activity of the neurons increases immediately after the conditioned stimulus rather than after the reward. This pattern of activity indicates that dopaminergic neurons provide a reward-prediction error signal, an important element in reinforcement learning. The importance of dopamine in learning is also supported by observations that lesions of dopaminergic systems prevent reward-based learning.

The same dopaminergic pathways that are important for reward and learning are involved in addiction to many drugs of abuse (Chapter 43)

Noradrenergic neurons of the locus ceruleus play an important role in attention. These neurons have a low baseline level of activity in drowsy monkeys. In alert, attentive monkeys the cells have two firing patterns. In the *phasic mode*, the baseline activity of the neurons is low to moderate, but there are bursts of firing just before the monkey responds to stimuli to which it has been attentive. This pattern of activity is thought to facilitate selective attention to a stimulus that is about to initiate a behavior. In contrast, in the *tonic mode*, the baseline level of activity is elevated and does not change in response to external stimuli. This mode of firing may promote the search for a new behavioral and attentional goal when the current task is no longer rewarding (Figure 40–14).

Many monoaminergic neurons also participate in regulating overall arousal (Figure 40–15). The noradrenergic locus ceruleus, serotonergic dorsal and median raphe nuclei, dopaminergic A10 neurons, and histaminergic tuberomammillary neurons innervate the thalamus, hypothalamus, basal forebrain, and cerebral cortex. All of these systems have the property of firing fastest during wakefulness, slowing down during slow wave (or non-REM) sleep, and grinding to a halt during REM sleep.

Stimulation of noradrenergic neurons in the locus ceruleus or histaminergic cells in the tuberomammillary nucleus increases electroencephalogram (EEG) arousal, indicating that these systems play an important role in cortical and behavioral arousal. However, lesions restricted to one or even a combination of monoaminergic cell groups do not cause profound loss of wakefulness, suggesting that the various cell groups

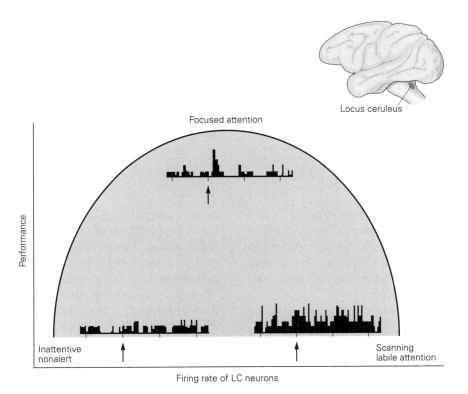

Figure 40–14 Locus ceruleus (LC) neurons exhibit different patterns of activity with different levels of attentiveness and task performance. The inverted U curve shows the relationship between a monkey's performance on a target detection task and the level of locus ceruleus activity. Histograms show the responses of LC neurons to presentation of the target during different levels of task performance. Performance is poor at low levels of LC activity because the animals are not alert. Performance is optimal when baseline activity is moderate and phasic activation follows presentation of the target. Performance is also poor when baseline activity is high because the higher baseline is incompatible with focusing on the assigned task. The tonic mode (with high baseline activity) might be optimal for tasks (or contexts) that require behavioral flexibility instead of focused attention. If so, the LC could regulate the balance between focused and flexible behavior. (Adapted, with permission, from Aston-Jones, Rajkowski, and Cohen 1999. Copyright © 1999 Society of Biological Psychiatry. Published by Elsevier Inc.)

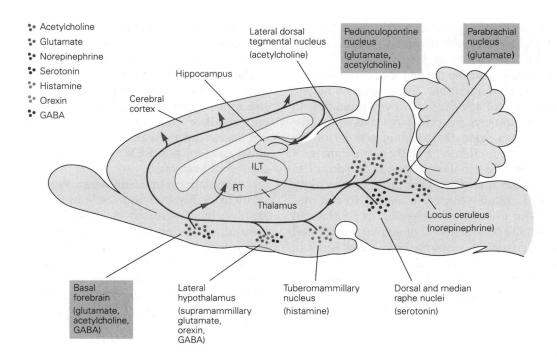

Figure 40–15 Major cell groups in the ascending arousal system. Neurons using the neurotransmitters norepinephrine, serotonin, dopamine, histamine, and acetylcholine have widespread forebrain projections. Although they all contribute to arousal by modulating various brain functions, ablation of any one of these cell groups has little effect on the waking state, suggesting that none of them are essential for maintaining a waking state. On the other hand, extensive damage to glutamatergic neurons in the parabrachial and pedunculopontine nuclei or to the GABAergic, glutamatergic, and cholinergic neurons in the basal forebrain (**orange boxes**) can cause a profound and prolonged coma. Thus, the parabrachial–pedunculopontine–basal forebrain–cortical pathway appears to be the only one that is essential to maintaining a waking state. (Abbreviations: **GABA**, γ-aminobutyric acid; **ILT**, intralaminar thalamic nuclei; **LC**, locus ceruleus; **RT**, reticular nucleus of the thalamus.)

probably have overlapping and at least partly redundant roles in sleep/wake regulation. The monoaminergic pathways modulate specific cellular properties of postsynaptic neurons in the thalamus and cerebral cortex, enhancing alertness and interaction with environmental stimuli.

Monoaminergic and Cholinergic Neurons Maintain Arousal by Modulating Forebrain Neurons

The monoaminergic and cholinergic neurons induce arousal by activating cortical neurons both directly and indirectly. They do this in part by modulating the activity of neurons in the brain stem, hypothalamus, basal forebrain, and thalamus that activate the cerebral cortex.

Both noradrenergic and serotonergic neurons innervate the parabrachial complex, a glutamatergic cell group that is critical for maintaining a waking forebrain. Noradrenergic inputs also activate histaminergic and orexin neurons in the lateral hypothalamus

as well as cholinergic and GABAergic neurons in the basal forebrain, all of which project directly to the cerebral cortex. The parabrachial, histaminergic, orexin, and cholinergic basal forebrain neurons all excite cortical pyramidal cells, whereas the GABAergic basal forebrain neurons inhibit cortical inhibitory interneurons, thus disinhibiting the cortical pyramidal cells. The net effect of these inputs is to make the cortical pyramidal neurons more responsive to incoming sensory and cognitive inputs.

Parabrachial, noradrenergic, serotonergic, histaminergic, and cholinergic inputs also innervate the thalamus and modulate its ability to transmit sensory information to the cerebral cortex. Thalamic relay neurons fire in rhythmic bursts during sleep (Chapter 44) but fire single spikes related to incoming sensory stimuli during wakefulness. The firing pattern of thalamic and cortical neurons changes from burst mode to single-spike mode when the cells are depolarized following application of acetylcholine, norepinephrine,

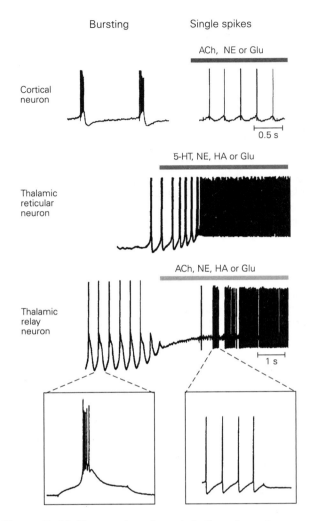

Figure 40–16 Monoaminergic and cholinergic systems modulate the activity of thalamic and cortical neurons to maintain arousal. The firing patterns of cortical and thalamic neurons are converted from burst mode to single-spike mode by the action of acetylcholine or monoamines. Recordings are from neurons in brain slices. Thalamic and cortical neurons have limited ability to convey information when firing in rhythmic bursts. However, when in single-spike mode, their firing activity reflects the inputs they receive. Therefore, the monoaminergic and cholinergic arousal systems keep open the lines of communication necessary for cortical information processing. (Reproduced, with permission, from Steriade, McCormick, and Sejnowski 1993. Copyright © 1993 AAAS.)

serotonin, or histamine (Figure 40–16). Thus, the monoaminergic neurons that participate in the ascending arousal system regulate cortical activity in part by altering the firing of thalamic neurons.

Many pharmacological agents that target monoaminergic and cholinergic systems influence arousal. For example, antihistamines cause drowsiness, serotonin reuptake blockers decrease the amount of REM

sleep, and nicotine is a powerful stimulant. In addition, arousal is induced by amphetamines, cocaine, and other drugs that block dopamine reuptake; mice lacking the dopamine transporter are insensitive to such drugs.

Patients with Parkinson disease, who lose dopaminergic neurons in the substantia nigra, also lose noradrenergic neurons in the locus ceruleus and tend to be abnormally sleepy during the day. Some drugs used to treat Parkinson disease activate the D_2 dopamine receptor on presynaptic terminals of the remaining dopaminergic arousal neurons, which results in presynaptic inhibition, thus reducing dopamine release. As a result, although these drugs may make the movement disorder better (through their effects on postsynaptic D_2 receptors on neurons in the striatum), the inhibitory effect on remaining dopaminergic cells in the arousal system may exacerbate daytime sleepiness.

Highlights

1. The plan for the brain stem and the cranial nerves unfolds early in development, as neurons assemble into clusters that come, in time, to assume their functional organization. Building on the basic plan of the spinal cord, motor and sensory neurons concerned with the face, head, neck, and internal viscera form into discrete nuclei with specific functions and territories of innervation.

2. Neurons in the reticular formation surrounding these cranial nerve nuclei develop into ensembles of neurons that can generate patterns of autonomic and motor responses that subserve simple, stereotyped, coordinated functions, ranging from facial expression to feeding and breathing. These behavior patterns are sufficiently complex and flexible to represent the entire behavioral repertory of a newborn baby.

3. As the forebrain develops and exerts its control over these brain stem pattern generators, a variety of more complex responses and ultimately volitional control of behavior evolve.

4. Even a skilled actor, however, finds it difficult to produce the facial expressions associated with specific emotions unless he recreates the emotional states internally, thereby triggering the prepatterned facial expressions associated with those feeling states. Thus, some of the most complex human emotions and behaviors are played out unconsciously by means of stereotypic patterns of motor and autonomic responses in the brain stem.

5. The brain stem also contains a series of cell groups that have long-ranging and diffuse projections. Their targets range from the cognitive and behavioral systems in the cerebral cortex, to hypothalamic and brain stem autonomic control areas, to sensory and motor control systems in the spinal cord. Many of the neurons that participate in these modulatory systems, which set the tone for more specific sensory, motor, behavioral, and autonomic outputs, use monoamines as neuromodulators.

6. As a result of the diffuseness of these modulatory pathways and the multiplicity of receptors that they employ, a large portion of all central nervous system–active drugs act on these pathways. Unfortunately, many of the off-target effects of these drugs are due to the diffuseness of these pathways and their use of the same neurotransmitters and receptors at multiple locations. A challenge for the future of central nervous system pharmacology will be to develop drugs more highly selective for the targeted functions that require modulation.

Clifford B. Saper
Joel K. Elmquist

Selected Reading

Feldman JL, Del Negro CA, Gray PA. 2013. Understanding the rhythm of breathing: so near and yet so far. Ann Rev Physiol 75:423–452.

Gautron L, Elmquist JK, Williams KW. 2015. Neural control of energy balance: translating circuits to therapies. Cell 161:133–145.

Guyenet PG, Bayliss DA. 2015. Control of breathing and CO_2 homeostasis. Neuron 87:94–961.

Hodges MR, Richerson GB. 2010. Medullary serotonin neurons and their roles in central respiratory chemoreception. Respir Physiol Neurobiol 173:256–263.

Llorca-Torralba M, Borges G, Neto F, Mico JA, Berrocoso E. 2016. Noradrenergic locus coeruleus pathways in pain modulation. Neuroscience 338:93–113.

Plum F, Posner JB, Saper CB, Schiff ND. 2007. *Plum and Posner's Diagnosis of Stupor and Coma*, 4th ed. Philadelphia: Davis.

Saper CB. 2002. The central autonomic nervous system: conscious visceral perception and autonomic pattern generation. Annu Rev Neurosci 25:433–469.

Saper CB, Stornetta RL. 2014. Central autonomic system. In: G Paxinos (ed). *The Rat Nervous System*, 4th ed., pp. 627–671. San Diego: Elsevier.

Schultz W. 2016. Dopamine reward prediction-error signalling: a two-component response. Nat Rev Neurosci 17:183–195.

Sohn JW, Elmquist JK, Williams KW. 2013. Neuronal circuits that regulate feeding behaviour and metabolism. Trends Neurosci 36:504–512.

References

Aston-Jones G, Cohen JD. 2005. An integrative theory of locus coeruleus-norepinephrine function: adaptive gain and optimal performance. Annu Rev Neurosci 28:403–450.

Aston-Jones G, Rajkowski J, Cohen J. 1999. Role of locus coeruleus in attention and behavioral flexibility. Biol Psychiatry 46:1309–1320.

Aston-Jones G, Bloom FE. 1981. Activity of norepinephrine-containing locus ceruleus neurons in behaving rats anticipates fluctuations in the sleep-wake cycle. J Neurosci 1:876–886.

Bieger D, Hopkins DA. 1987. Viscerotropic representation of the upper alimentary tract in the medulla oblongata in the rat: the nucleus ambiguus. J Comp Neurol 262:546–562.

Blessing WW, Li Y-W. 1989. Inhibitory vasomotor neurons in the caudal ventrolateral region of the medulla oblongata. Prog Brain Res 81:83–97.

Bouret S, Richmond BJ. 2015. Sensitivity of locus coeruleus neurons to reward value for goal-directed attention. J Neurosci 35:4005–4014.

Bradley SR, Pieribone VA, Wang W, Severson CA, Jacobs RA, Richerson GB. 2002. Chemosensitive serotonergic neurons are closely associated with large medullary arteries. Nat Neurosci 5:401–402.

Bruinstroop E, Cano G, Vanderhorst VG, et al. 2012. Spinal projections of the A5, A6 (locus coeruleus), and A7 noradrenergic cell groups in rats. J Comp Neurol 520:1985–2001.

Chang RB, Strochlic DE, Williams EK, Umans BD, Liberles SD. 2015. Vagal sensory neuron subtypes that differentially control breathing. Cell 161:622–633.

Filiano JJ, Kinney HC. 1994. A perspective on neuropathologic findings in victims of the sudden infant death syndrome: the triple-risk model. Biol Neonate 65:194–197.

Gray PA, Janczewski WA, Mellen N, McCrimmon DR, Feldman JL. 2001. Normal breathing requires pre-Bötzinger complex neurokinin-1 receptor-expressing neurons. Nat Neurosci 4:927–930.

Grunstein RR, Sullivan CE. 1990. Neural control of respiration during sleep. In: MJ Thorpy (ed). *Handbook of Sleep Disorders*. New York: Marcel Dekker.

Jenny AB, Saper CB. 1987. Organization of the facial nucleus and cortico-facial projection in the monkey: a reconsideration of the upper motor neuron facial palsy. Neurol 37:930–939.

Lahiri S, Maret K, Sherpa M, Peters R Jr. 1984. Sleep and periodic breathing at high altitude: Sherpa natives vs. sojourners. In: J West, S lahiri (eds). *High Altitude and Man*, pp. 73–90. Bethesda: American Physiological Society.

Morecraft RJ, Louie JL, Herrick JL, Stilwell-Morecraft KS. 2001. Cortical innervation of the facial nucleus in the non-human primate: a new interpretation of the effects of stroke and related subtotal brain trauma on the muscles of facial expression. Brain 124(Pt 1):176–208.

Mulkey DK, Stornetta RL, Weston MC, et al. 2004. Respiratory control by ventral surface chemoreceptor neurons in rats. Nat Neurosci 7:1360–1369.

Nielson M, Smith H. 1952. Studies on the regulation of respiration in acute hypoxia; with an appendix on respiratory control during prolonged hypoxia. Acta Physiol Scan 24:293–313.

Richerson GB. 2004. Serotonergic neurons as carbon dioxide sensors that maintain pH homeostasis. Nat Rev Neurosci 5:449–461.

Richerson GB, Getting PA. 1987. Maintenance of complex neural function during perfusion of the mammalian brain. Brain Res 409:128–132.

Rinaman L, Card JP, Schwaber JS, Miselis RR. 1989. Ultrastructural demonstration of a gastric monosynaptic vagal circuit in the nucleus of the solitary tract in rat. J Neurosci 9:1985–1996.

Smith JC, Ellenberger HH, Ballanyi K, Richter DW, Feldman JL. 1991. Pre-Bötzinger complex: a brain stem region that may generate respiratory rhythm in mammals. Science 254:726–729.

Steriade M, McCormick DA, Sejnowski TJ. 1993. Thalamocortical oscillations in the sleeping and aroused brain. Science 262:679–685.

Wang W, Bradley SR, Richerson GB. 2002. Quantification of the response of rat medullary raphe neurons to independent change in pH and PCO_2. J Physiol 540:951–970.

Williams EK, Chang RB, Strochlic DE, Umans BE, Lowell BB, Liberles SD. 2016. Sensory neurons that detect stretch and nutrients in the digestive system. Cell 166:209–221.

41

The Hypothalamus: Autonomic, Hormonal, and Behavioral Control of Survival

T HE SURVIVAL OF AN INDIVIDUAL requires tight control of body temperature, water balance, and blood pressure, together with sufficient food intake and appropriate regulation of sleep/wakefulness cycles. Survival of a species requires that individuals be fertile, mate, and nurture their offspring, and that aggression toward others be appropriate and adaptive. Neurons in the hypothalamus control all of these key survival activities.

As we shall learn in this chapter, the hypothalamus together with interconnected areas of the brain responds to bodily and emotional challenges by recruiting appropriate behavioral and physiological responses. Coordination of these activities ensures constancy of the internal environment, a process known as homeostasis. The hypothalamus acts on three major systems: the autonomic motor system, the neuroendocrine system, and neural pathways that mediate motivated behavior.

The autonomic motor system is distinct from the somatic motor system, which controls skeletal muscle. Whereas somatic motor neurons regulate contractions of striated muscles (Chapter 31), autonomic motor neurons regulate blood vessels, the heart, the skin, and visceral organs through synapses upon smooth and cardiac muscle cells, upon glands cells that serve endocrine and exocrine functions, and upon metabolic targets such as adipocytes. The neuroendocrine system works differently, by secreting several peptide hormones from the pituitary, the "master gland," located just beneath the hypothalamus. These pituitary hormones control water retention by the kidney, parturition, lactation, somatic growth, gamete development, and also the release of nonpeptide hormones from three downstream glands—the gonads, adrenal cortex, and thyroid.

Although largely involuntary, autonomic and neuroendocrine responses are tightly integrated with voluntary behavior executed by the somatic motor system. Running, climbing, and lifting exemplify voluntary actions that have metabolic, cardiovascular, and thermoregulatory consequences. These needs are automatically met by the autonomic and neuroendocrine systems through changes in cardiorespiratory drive, cardiac output, regional blood flow, heat dissipation, and fuel mobilization. Such compensatory changes are implemented primarily by feedforward central commands, supplemented by reflexes activated by sensory feedback. Similarly, emotional states evoke autonomic and neuroendocrine responses. Feelings of fear, anger, happiness, and sadness have characteristic autonomic and hormonal manifestations.

In this chapter, we first explore the concept of homeostasis and the general means by which it is achieved. We then discuss the anatomical and functional organization of the hypothalamus and its two "involuntary" motor arms—the autonomic and neuroendocrine systems. After that, we focus in depth on

three classic examples of hypothalamic homeostatic control—regulation of body temperature, of water balance and its related deficiency drive, thirst, and of energy balance and its drive, hunger. We conclude by examining sexually dimorphic regions of the hypothalamus and their role in regulating sexual behavior, aggression, and parenting. Additional discussion of sleep cycles and regulation of circadian rhythms can be found in Chapter 44.

Homeostasis Keeps Physiological Parameters Within a Narrow Range and Is Essential for Survival

In the mid-19th century, the French physiologist and founder of experimental medicine Claude Bernard drew attention to the stability of the body's internal environment over a broad range of behavioral states and external conditions. "The internal environment (le milieu interior)," he wrote, "is a necessary condition for a free life." Building on this idea, in the 1930s, the American physiologist Walter B. Cannon introduced the concept of homeostasis to describe the mechanisms that maintain the constancy of composition of the bodily fluids, body temperature, blood pressure, and other physiological variables—all of which are necessary for survival.

Homeostatic mechanisms are highly adaptive because they greatly extend the range of conditions that can be tolerated. For example, during exercise, many parameters can increase dramatically—cardiac output by 4- to 5-fold, oxygen and fuel consumption by 5- to 10-fold, and heat production to a similar degree. In the absence of compensatory responses, blood pressure would increase in proportion to cardiac output, rupturing blood vessels; circulating fuels would fall to critically low levels, starving cells of energy; and hyperthermia would denature cell proteins. Indeed, the capacity of homeostasis is remarkable, making it possible for animals to survive at high latitudes where seasonal temperatures can fluctuate by 70°C and for humans to run 251 km in the sands of the extremely hot Sahara Desert (Marathon des Sables). Homeostatic mechanisms greatly extend the range of habitats, activities, and traumas that can be survived.

Homeostasis requires negative sensory feedback from the body. The concept of feedback loops evolved from the discovery of sensors that detect critical physiological variables and then couple them with behavioral, autonomic, and neuroendocrine motor outputs. Drawing upon the engineering principle of negative feedback control, this led to the concept that physiological "set points" help control key parameters like body temperature, blood osmolarity, blood pressure, and body fat content.

Set point models are appealing because thermostats are so effective in maintaining room temperature at a targeted set point and, by analogy, physiological variables like body temperature are likewise tightly controlled. In such models, a "set point" exists for a

A   Set point model

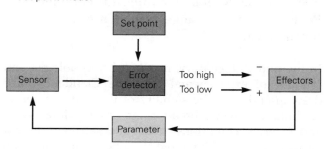

B   Settling point model

C   Combined model: settling point of multiple afferent/efferent loops

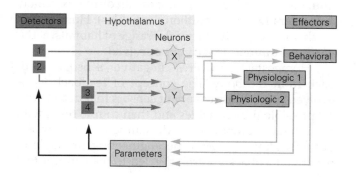

parameter, 37°C in the case of body temperature, and at any given moment, the real level of the parameter is assessed and compared with the targeted set point through feedback and error detection (Figure 41–1A). Any deviation above or below triggers counteracting corrective responses—if too hot, cutaneous vasodilatation, sweating, and a dip in the pool; if too cold, vasoconstriction, thermogenesis, shivering, and the donning of a sweater. For regulation of body temperature, the set point and detection of error were historically seen as emergent properties of neurons in the preoptic area of the hypothalamus (POA).

Over time, the set point model required revision because intensive investigation failed to uncover any molecular or neuronal bases for encoding set points and performing error detection. In addition, "set point–like" regulation can, in principle, be achieved without a set point, feedback, or error detection—the so-called "settling point" model (Figure 41–1B). Consider the changing level of a lake. When rainfall is excessive, its level rises; the rivers draining the lake rise and their flow increases. The converse is true when rainfall is low. The changing flow of the rivers draining the lake thus maintains its level near a settling point without requiring an idealized set point, feedback, or error detection. While aspects of the settling point model have appeal, it too is incomplete because homeostatic processes clearly receive important feedback

Figure 41–1 (Left) Set points, settling points, and homeostasis.

A. The set point view was inspired by engineering principles. As with a thermostat, constancy is achieved by providing feedback on the existing level of a parameter, determining how it compares to an idealized set point, and then instituting corrective measures to return the parameter to the set point. While popular for many years, it has fallen out of favor as years of research have failed to uncover molecular and neural bases for encoding set points and performing error detection.

B. The settling point model was inspired by observations that many systems achieve constancy in the absence of any feedback or error detection. In this example, the level of outflow of water from a lake is proportional to the depth of the lake. When it rains, the increase or decrease in the level of the lake causes more or less water to flow out of the lake. The level of the lake remains fairly constant without a set point or error detection. A related example is regulation of body weight. Increased food intake leads to increased body weight. As body weight increases, the energy cost of carrying and sustaining that increased weight goes up. Because of this, body weight too should have its settling point. (Reproduced, with permission, from Speakman et al. 2011.)

C. In this model the concepts of feedback in part **A** and settling point in part **B** are combined. The apparent set point is in fact the settling point, an emergent property of multiple feedback-informed afferent/efferent loops.

regarding disturbances, and this feedback produces vital responses that hasten recovery. As we shall see, temperature, osmolarity, and body fat are directly or indirectly "sensed," and this affects the activity of neurons in the hypothalamus that generate counteracting responses.

Most physiologists today have adopted a "distributed settling point" model that incorporates strong feedback control of multiple sensory/effector loops (Figure 41–1C). With body temperature, for example, there is no single specific set point and no location in the brain where a single set point is encoded and error detection takes place; in short, there is no thermostat. Instead, there are multiple temperature detectors located in different sites (skin, core, and brain), and each is coupled through neuronal pathways that traverse the preoptic area on their way to different body temperature effectors (cutaneous blood vessels, sweat glands, brown fat metabolism, shivering and behavioral pathways). When engaged, each of these effectors impact body temperature. The apparent set point for body temperature is in fact the emergent settling point that results from the combined activities of the multiple feedback-informed afferent/efferent loops. As we will see later, this nuanced model also applies to regulation of blood pressure, blood osmolarity, and body fat.

The Hypothalamus Coordinates Homeostatic Regulation

The hypothalamus integrates the status of physiological parameters with outputs to behavioral, autonomic, and neuroendocrine motor systems and thereby regulates six vital physiological functions (Table 41–1). The hypothalamus lies at the base of the brain immediately above the pituitary gland (Figure 41–2). It is bounded anteriorly (rostrally) by the diagonal band of Broca; dorsally by the anterior commissure, the bed nuclei of the stria terminalis, the zona incerta, and thalamus; and posteriorly (caudally) by the ventral tegmental area and interpeduncular nucleus.

The Hypothalamus Is Commonly Divided Into Three Rostrocaudal Regions

Regions of the hypothalamus are named according to their location and appearance in Nissl-stained sections. The hypothalamus is divided, rostral to caudal, into three regions. (1) The *preoptic hypothalamus* lies above the optic chiasm and contains neurons that control water balance and thirst, temperature, sleep, sexual behavior, and circadian rhythms. (2) The *tuberal*

Table 41–1 The Hypothalamus Integrates Behavioral (Somatomotor), Autonomic, and Neuroendocrine Responses Involved in Six Vital Functions

1. *Blood pressure and electrolyte composition.* The hypothalamus regulates thirst, salt appetite, and drinking behavior; autonomic control of vasomotor tone; and the release of hormones such as vasopressin (via the paraventricular nucleus).
2. *Energy metabolism.* The hypothalamus regulates hunger and feeding behavior, the autonomic control of digestion, and the release of hormones such as glucocorticoids, growth hormone, and thyroid-stimulating hormone (via the arcuate and paraventricular nuclei).
3. *Reproductive (sexual and parental) behaviors.* The hypothalamus controls autonomic modulation of the reproductive organs and endocrine regulation of the gonads (via the medial preoptic, ventromedial, and ventral premammillary nuclei).
4. *Body temperature.* The hypothalamus influences thermoregulatory behavior (seeking a warmer or cooler environment), controls autonomic body heat conservation/loss mechanisms, and controls secretion of hormones that influence metabolic rate (via the preoptic region).
5. *Defensive behavior.* The hypothalamus regulates the stress response and fight-or-flight response to threats in the environment such as predators (via the paraventricular, anterior hypothalamic, and dorsal premammillary nuclei, and the lateral hypothalamic area).
6. *Sleep–wake cycle.* The hypothalamus regulates the sleep–wake cycle (via a circadian clock in the suprachiasmatic nucleus) and levels of arousal when awake (via the lateral hypothalamic area and tuberomammillary nucleus).

hypothalamus lies above the pituitary and contains neurons controlling pituitary hormone secretion, autonomic outflow, and various behaviors including hunger, sexual behavior, and aggression. (3) The *posterior hypothalamus* includes the posterior and mammillary nuclei, as well as histaminergic neurons in the tuberomammillary nucleus that affect arousal. The functions of other neurons in the posterior hypothalamus areas are less well defined.

The *lateral hypothalamic area* (LHA) spans from the middle to the caudal hypothalamus. It is linked more closely to reward pathways and arousal than to maintenance of homeostasis and specific survival behaviors. Indeed, it is heavily connected with the nucleus accumbens and ventral tegmental area, two areas involved in reward (Chapter 43), and contains neurons that project extensively throughout the cortex. Lastly, LHA neurons expressing the neuropeptide orexin (hypocretin) play a critical role in stabilizing wakefulness (Chapter 44).

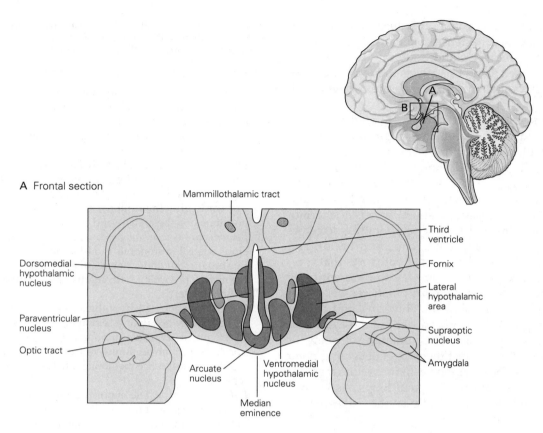

A Frontal section

Mammillothalamic tract

Third ventricle

Dorsomedial hypothalamic nucleus

Fornix

Lateral hypothalamic area

Paraventricular nucleus

Supraoptic nucleus

Optic tract

Amygdala

Arcuate nucleus

Ventromedial hypothalamic nucleus

Median eminence

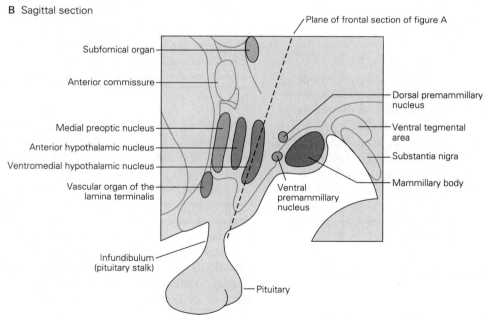

B Sagittal section

Plane of frontal section of figure A

Subfornical organ

Anterior commissure

Dorsal premammillary nucleus

Ventral tegmental area

Medial preoptic nucleus

Substantia nigra

Anterior hypothalamic nucleus

Ventromedial hypothalamic nucleus

Mammillary body

Vascular organ of the lamina terminalis

Ventral premammillary nucleus

Infundibulum (pituitary stalk)

Pituitary

Figure 41–2 The structure of the hypothalamus.

A. Frontal view of the hypothalamus (section along plane A shown in the sagittal view of the brain, upper right). The third ventricle is in the midline; the paraventricular, dorsomedial, and arcuate nuclei adjacent to the ventricle form the neuroendocrine motor zone and periventricular region at this level. The ventromedial nucleus is part of the medial column of hypothalamic nuclei, and the lateral hypothalamic area is the lateral zone component represented in the part of the hypothalamus shown here.

B. Sagittal (rostrocaudal) view of the medial column of hypothalamic nuclei, showing the adjacent (caudal) substantia nigra and ventral tegmental area of the midbrain. The functional significance of key hypothalamic nuclei is summarized in Table 41–1.

Modality-Specific Hypothalamic Neurons Link Interoceptive Sensory Feedback With Outputs That Control Adaptive Behaviors and Physiological Responses

General principles of hypothalamic function have emerged over several decades. Neurons in the periphery and brain respond when parameters under homeostatic control are disturbed. Such neurons can respond directly to the stimulus or indirectly to changes in hormones and other factors that track the regulated parameter. This sensory information is then relayed to functionally appropriate regulatory neurons in a particular site (or sites) within the hypothalamus. Once the information is integrated by hypothalamic neurons, the results are then conveyed downstream to motor circuits that control specific behaviors and physiological responses. The result is a coordinated corrective response (eg, warmth-seeking plus heat production and retention, thirst plus water retention by the kidney, or hunger plus decreased energy expenditure).

Our understanding of the functions of hypothalamic neurons has been refined recently using optogenetic and chemogenetic techniques in active animals. By selectively activating subsets of hypothalamic neurons, one can evoke specific behaviors and physiological responses, even when the need is totally absent. Key regulatory neurons for body temperature are located in the median preoptic nucleus (MnPO). Water balance is regulated by neurons in three sites—the MnPO, the vascular organ of the lamina terminalis (OVLT), and the subfornical organ (SFO)—and energy balance by neurons in the arcuate nucleus (Figure 41–2A).

Modality-Specific Hypothalamic Neurons Also Receive Descending Feedforward Input Regarding Anticipated Homeostatic Challenges

In addition to input from sensory signals that provide important feedback regarding the status of the body, key regulatory neurons in the hypothalamus receive "top-down" feedforward inputs from neurons that anticipate future homeostatic challenges. For example, when food-deprived animals detect cues that predict the availability of food, there is a rapid drop in the firing of hunger-promoting neurons in the arcuate nucleus even before food is ingested. Such top-down feedforward control prepares the body for anticipated homeostatic challenges. In addition, such rapid regulation, by countering an aversive state represented by high activity in deficiency-driven neurons, could be important for motivating

deficiency-based behaviors such as thirst and hunger (discussed below).

Next, we examine two effectors arms of the hypothalamus—the autonomic motor system and the neuroendocrine system.

The Autonomic System Links the Brain to Physiological Responses

Although the autonomic motor system implements many of the physiological responses initiated by the hypothalamus, the autonomic system is also regulated by circuits in the brain stem and spinal cord (Chapter 40). As a consequence, autonomic functions vary in their dependence on the hypothalamus. For example, micturition is largely independent of the hypothalamus, while blood pressure regulation depends heavily on circuits in the brain stem but can also be modulated by the hypothalamus. In contrast, thermogenesis by brown adipose tissue is largely subservient to the hypothalamus.

Visceral Motor Neurons in the Autonomic System Are Organized Into Ganglia

Unlike the somatic motor system, in which motor neurons are located in the ventral spinal cord and brain stem, the cell bodies of autonomic motor neurons are found in enlargements of peripheral nerves called ganglia.[1] The autonomic motor neurons innervate secretory epithelial cells in glands, smooth and cardiac muscle, and adipose tissue.

Efforts to understand the principles of organization of autonomic ganglia began in 1880 in England with the work of Walter Gaskell and were later continued by John N. Langley. They stimulated autonomic nerves and observed the responses of end organs (eg, vasoconstriction, piloerection, sweating, pupillary constriction). They used nicotine to block signals from individual ganglia to test interactions between ganglia. Langley proposed that specific chemical substances must be released by preganglionic neurons of the autonomic ganglia and that these substances act by binding to receptors on the postganglionic neurons, which target the end organs. These ideas set the stage for the later investigations of chemical synaptic transmission.

---

[1]The peripheral nerves also have sensory ganglia, located on the dorsal roots of the spinal cord and on five of the cranial nerves: trigeminal (V), facial (VII), vestibulocochlear (VIII), glossopharyngeal (IX), and vagus (X) (Chapter 40).

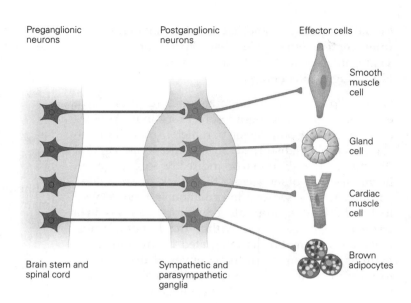

Figure 41–3 Distinct cell types in peripheral autonomic pathways selectively control target cells with different phenotypes. Autonomic motor neurons lie outside the central nervous system in ganglia controlled by preganglionic neurons in the spinal cord and brain stem. These downstream neurons within parasympathetic and sympathetic ganglia regulate three types of effector cells: smooth muscle, gland cells, and cardiac muscle. Additionally, downstream neurons found only in sympathetic ganglia selectively control brown adipocytes and immune cells in lymphoid tissue. This figure illustrates the three basic cell types—preganglionic neurons, downstream ganglionic neurons, and different target effector cells—that control function.

Langley also distinguished the autonomic and somatic motor systems and in so doing, created much of our current nomenclature.

The autonomic system is divided into three divisions: sympathetic, parasympathetic, and enteric. All neurons in sympathetic and parasympathetic ganglia are controlled by *preganglionic neurons* whose cell bodies are located in the spinal cord and brain stem. The preganglionic neurons synthesize and release the neurotransmitter acetylcholine (ACh), which acts on nicotinic ACh receptors on *postganglionic neurons*, producing fast excitatory postsynaptic potentials and initiating action potentials that propagate to synapses with effector cells in *end organs* (Figure 41–3). The sympathetic and parasympathetic systems are differentiated by five criteria:

1. The segmental organization of their preganglionic neurons in the spinal cord and brain stem
2. The peripheral locations of their ganglia
3. The types and locations of end organs they innervate
4. The effects they produce on end organs
5. The neurotransmitters employed by their postganglionic neurons

Preganglionic Neurons Are Localized in Three Regions Along the Brain Stem and Spinal Cord

The parasympathetic pathways arise from a cranial nerve zone in the brain stem and a second zone in sacral segments of the spinal cord (Figure 41–4). These parasympathetic zones surround a sympathetic zone that extends continuously in thoracic and lumbar segments of the cord.

The cranial parasympathetic pathways arise from preganglionic neurons in the general visceral motor nuclei of four cranial nerves: the oculomotor (N. III) in the midbrain and the facial (N. VII), glossopharyngeal (N. IX), and vagus (N. X) in the medulla. The cranial parasympathetic nuclei are described in Chapter 40 together with the mixed cranial nerves (eg, the facial, glossopharyngeal, and vagus). The spinal parasympathetic pathway originates in preganglionic neurons in sacral segments S2–S4. Their cell bodies are located in intermediate regions of the gray matter, and their axons project in peripheral nerves through the ventral roots.

The sympathetic preganglionic cell column extends between the cervical and lumbosacral enlargements of the spinal cord, corresponding to the first thoracic segment and third lumbar segment (Figure 41–4). Most of the cell bodies of sympathetic preganglionic neurons are located in the intermediolateral cell column; others are found in the central autonomic area surrounding the central canal and in a band connecting the central area with the intermediolateral cell column. The axons of preganglionic sympathetic neurons project from the spinal cord through the nearest ventral root and then run with small connecting nerves known as rami communicantes before terminating on postganglionic cells in the paravertebral sympathetic chain (Figure 41–5).

Sympathetic Ganglia Project to Many Targets Throughout the Body

The sympathetic motor system regulates systemic physiological parameters such as blood pressure and body temperature by influencing target cells within virtually every tissue throughout the body (Figure 41–4).

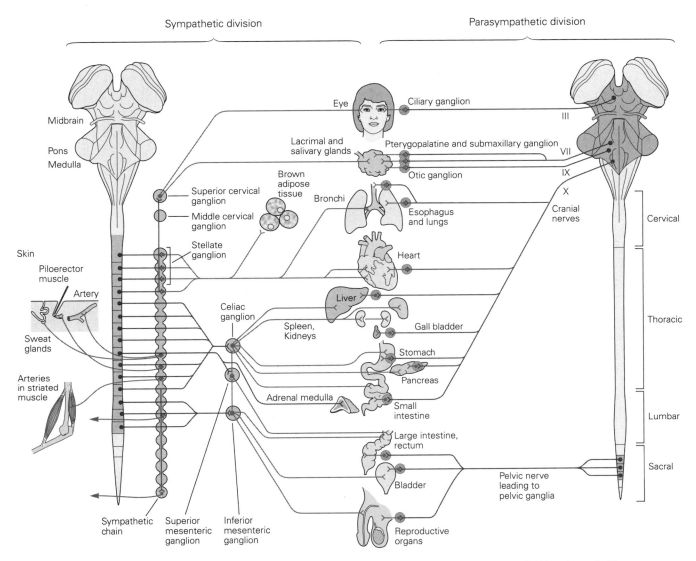

Sympathetic division

Parasympathetic division

Figure 41–4 Sympathetic and parasympathetic divisions of the autonomic motor system. The sympathetic ganglia lie close to the spinal column and supply virtually every tissue in the body. Some tissues, such as skeletal muscle, are regulated only indirectly through their arterial blood supply. The parasympathetic ganglia are located near their targets, which do not include the skin or skeletal muscle.

This regulation depends on synaptic input from the spinal cord and from supraspinal structures that control the activity of the preganglionic neurons.

Important groups of supraspinal neurons that excite preganglionic sympathetic activity are located in the rostral ventrolateral medulla, the raphe pallidus in the brain stem, and the paraventricular nucleus in the hypothalamus. Preganglionic neurons integrate these descending inputs along with local segmental sensory inputs and form synapses with neurons in paravertebral and prevertebral sympathetic ganglia (Figure 41–5).

Ganglionic neurons in turn form synapses with a variety of end organs, including blood vessels, heart, bronchial airways, piloerector muscles, brown fat, and salivary and sweat glands. Sympathetic neurons also regulate immune function through projections to primary lymphoid tissue in the bone marrow and thymus and to secondary lymphoid cells in the spleen. A subset of preganglionic neurons synapse on chromaffin cells in the medulla of the adrenal gland (Figure 41–5), which secrete epinephrine (adrenaline) and norepinephrine (noradrenaline) into the circulation as hormones to act on distant targets.

The paravertebral and prevertebral sympathetic ganglia differ in both location and organization. Paravertebral ganglia are distributed segmentally,

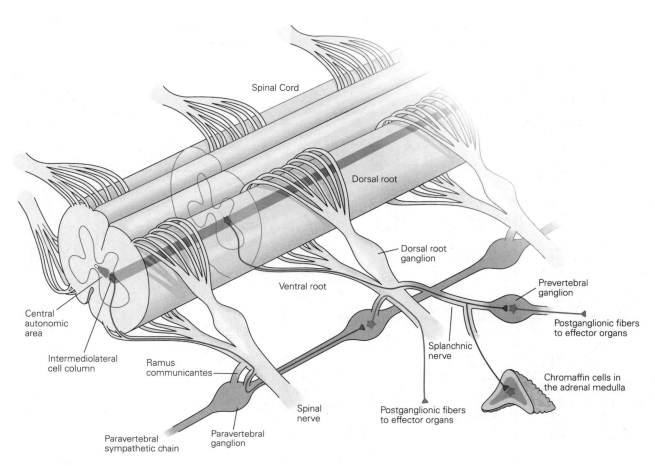

Figure 41–5 The sympathetic outflow is organized into groups of paravertebral and prevertebral ganglia. The axons of preganglionic cells in the spinal cord reach postganglionic neurons by way of ventral roots and the paravertebral sympathetic chain. The axons either form synapses on postganglionic neurons in paravertebral ganglia or project out of the chain into splanchnic nerves. Preganglionic axons in the splanchnic nerves form synapses with postganglionic neurons in prevertebral ganglia and with chromaffin cells in the adrenal medulla.

extending bilaterally as two chains from the first cervical segment to the last sacral segment. The chains lie lateral to the vertebral column at its ventral margin and generally contain one ganglion per segment (Figures 41–4 and 41–5). Two important exceptions are the superior cervical and stellate ganglia. The superior cervical ganglion is a coalescence of several cervical ganglia and supplies sympathetic innervation to the entire head, including the cerebral vasculature. The stellate ganglion, which innervates the heart and lungs, is a coalescence of ganglia from lower cervical segments and the first thoracic segment. These sympathetic pathways have an orderly somatotopic relation to one another from their segmental origin in preganglionic neurons to their terminus in peripheral targets.

The prevertebral ganglia are midline structures that lie close to the arteries for which they are named (Figures 41–4 and 41–5). In addition to sending sympathetic signals to visceral organs in the abdomen and pelvis, these ganglia also receive sensory feedback from their end organs.

Parasympathetic Ganglia Innervate Single Organs

In contrast to sympathetic ganglia, which regulate many targets and lie some distance from their targets close to the spinal cord, parasympathetic ganglia generally innervate single end organs and lie near to or within the end organs they regulate (Figure 41–4). In addition, the parasympathetic system does not influence lymphoid tissue, skin, or skeletal muscle except in the head, where it regulates vascular beds in the jaw, lip, and tongue.

The cranial and sacral parasympathetic ganglia innervate different targets. The cranial outflow includes four ganglia in the head (Chapter 40). The oculomotor

(III) nerve projects to the ciliary ganglion, which controls pupillary size and focus by innervating the iris and ciliary muscles. The facial (VII) nerve and a small component of the glossopharyngeal (IX) nerve project to the pterygopalatine (or sphenopalatine) ganglion, which promotes production of tears by the lacrimal glands and mucus by the nasal and palatine glands. Cranial nerve IX and a small component of nerve VII project to the otic ganglion. Its postganglionic neurons innervate the parotid, the largest salivary gland. Nerve VII also projects to the submandibular ganglion, which controls secretion of saliva by the submaxillary and sublingual glands.

The vagus (X) nerve projects broadly to parasympathetic ganglia in the heart, lungs, liver, gallbladder, and pancreas. It also projects to the stomach, small intestine, and more rostral segments of the gastrointestinal tract. The caudal parasympathetic outflow supplies the large intestine, rectum, bladder, and reproductive organs.

The Enteric Ganglia Regulate the Gastrointestinal Tract

The entire gastrointestinal tract, from the esophagus to the rectum—and including the pancreas and gallbladder—is controlled by the system of enteric ganglia. This system, by far the largest and most complex division of the autonomic nervous system, contains as many as 100 million neurons.

The enteric system has been studied most extensively in the small intestine of the guinea pig. Its activity is coordinated by two interconnected plexuses, small islands of interconnected neurons. The myenteric plexus controls smooth muscle movements of the gastrointestinal tract; the submucous plexus controls mucosal function (Figure 41–6). Working together, this distributed network of ganglia coordinates the orderly peristaltic propulsion of gastrointestinal contents and controls the secretions of the stomach and intestines and other components of digestion. In addition, the enteric system regulates local blood flow and also immune function in Peyer's patches. The enteric system is modulated by external inputs from sympathetic prevertebral ganglia and from parasympathetic components of the vagus nerve.

Unlike the sympathetic and parasympathetic divisions of the autonomic system, the enteric plexus contains interneurons and sensory neurons in addition to motor neurons. This intrinsic neural circuitry can maintain the basic functions of the gut even after the splanchnic sympathetic and vagal parasympathetic pathways are cut. Through splanchnic nerves and the afferent portion of the vagus nerve, the gastrointestinal tract also sends sensory information about the physiological status of the tract to the spinal cord and brain stem.

Acetylcholine and Norepinephrine Are the Principal Transmitters of Autonomic Motor Neurons

All preganglionic neurons in the sympathetic and parasympathetic systems use ACh as their excitatory neurotransmitter, activating ionotropic nicotinic ACh receptors on ganglionic neurons. These receptors resemble those at the neuromuscular junction in having nonselective cation pores, but they are encoded by different genes.

Activation of the ganglionic neurons triggers action potentials that propagate to postganglionic synapses with end organs in the periphery. At these end organ synapses, parasympathetic neurons release ACh, which activates muscarinic G protein–coupled receptors; sympathetic neurons release norepinephrine, which activates α- and β-adrenergic G protein–coupled receptors. The postsynaptic action can be either excitatory or inhibitory, depending on the type of target cell and its receptors (Table 41–2). Notable exceptions to this organization are the sympathetic postganglionic neurons that control sweat glands. They assume a cholinergic phenotype after birth.

In addition to acting on different receptors in different postsynaptic cells, one transmitter can activate different receptor types in the same postsynaptic cell. This principle was first discovered in sympathetic ganglia where ACh activates both nicotinic and muscarinic postsynaptic receptors to produce both a fast and slow excitatory postsynaptic potential (Figure 41–7A and Chapter 14). In some cases, one transmitter can activate both a postsynaptic receptor as well as a receptor on the presynaptic terminals from which the transmitter was released. Such presynaptic responses can cause either presynaptic inhibition or presynaptic facilitation (Figure 41–7B and Chapter 15). This specialization of synaptic transmission in sympathetic and parasympathetic neurons leads to functional diversity in the regulation of end organ function.

Cholinergic and adrenergic synaptic transmission in the peripheral autonomic motor system is often modulated by the co-release of various neuropeptides, nitric oxide, or adenosine triphosphate, which by activating multiple receptor types further contribute to functional diversity (Table 41–2 and Figure 41–7C). The motor responses elicited in end organs depend on the identity of the postganglionic neurotransmitters and the pre- and postsynaptic receptors

A Cross section of intestinal wall

B Layers of wall

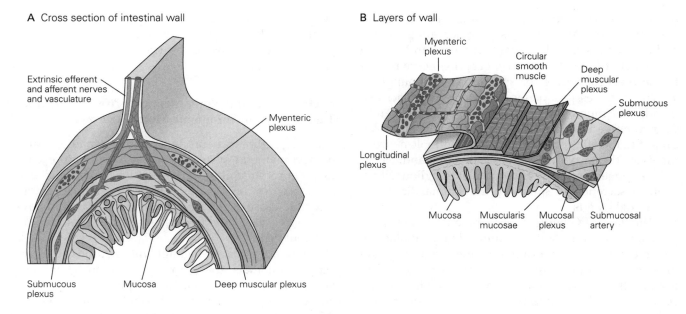

C Laminar distribution of neurons within the intestinal wall

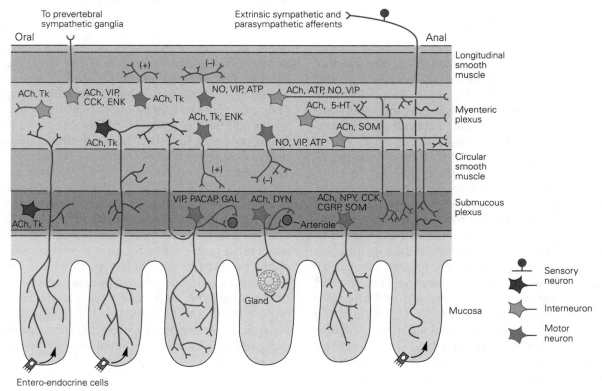

Figure 41–6 Organization of the enteric plexuses in the guinea pig. The myenteric plexus and submucous plexus lie between the layers of intestinal wall (**A** and **B**). At least 14 types of neurons have been identified within the enteric system based on morphology, chemical coding, and functional properties (**C**). Four sets of motor neurons provide excitatory (+) and inhibitory (–) inputs to two smooth muscle layers. Three additional groups of motor neurons control secretions from the mucosa and produce vasodilation. The network also includes two major classes of intrinsic sensory neurons. (Abbreviations: **ACh**, acetylcholine; **ATP**, adenosine triphosphate; **CCK**, cholecystokinin; **CGRP**, calcitonin gene-related polypeptide; **DYN**, dynorphin; **ENK**, enkephalin; **GAL**, galanin; **NO**, nitric oxide; **NPY**, neuropeptide Y; **PACAP**, pituitary adenylate cyclase-activating peptide; **SOM**, somatostatin; **Tk**, tachykinin; **VIP**, vasoactive intestinal peptide; **5-HT**, serotonin.) (Parts A and B adapted, with permission, from Furness and Costa 1980; part C reproduced, with permission, from Furness et al. 2004. Copyright © 2004 Elsevier Ltd.)

Table 41–2 Autonomic Neurotransmitters and Their Receptors

Transmitter	Receptor	Responses
Norepinephrine	α_1	Stimulates smooth muscle contraction in arteries, urethra, gastrointestinal tract, iris (pupillary dilation), uterine contractions during pregnancy, ejaculation; glycogenolysis in liver; glandular secretion (salivary glands, lacrimal glands).
	α_2	Presynaptic inhibition of transmitter release from sympathetic and parasympathetic nerve terminals; stimulates contraction in some arterial smooth muscle.
	β_1	Increases heart rate and strength of contraction.
	β_2	Relaxes smooth muscle in airways and gastrointestinal tract; stimulates glycogenolysis in liver.
	β_3	Stimulates lipolysis in white adipocytes and thermogenesis in brown adipocytes; inhibits bladder contraction.
Acetylcholine	Nicotinic	Fast EPSP in autonomic ganglion cells.
	Muscarinic: M_1, M_2, M_3	Glandular secretion; ocular circular muscle (pupillary constriction); ciliary muscles (focus of lens); stimulates endothelial production of NO and vasodilation; slows EPSPs in sympathetic neurons; slows heart rate; presynaptic inhibition at cholinergic nerve terminals; bladder contraction; salivary gland secretion.
Neuropeptide Y	Y_1, Y_2	Stimulates arterial contraction and potentiates responses mediated by α_1-adrenergic receptors; presynaptic inhibition of transmitter release from some postganglionic sympathetic nerve terminals.
NO	Diffuses through membranes; often acts to stimulate intracellular soluble guanylate cyclase	Vasodilation, penile erection, urethral relaxation.
Vasoactive intestinal peptide	VIPAC1, VIPAC2	Glandular secretion and dilation of blood vessels supplying glands.
ATP	P_{2X}, P_{2Y}	Fast and slow excitation of smooth muscle in bladder, vas deferens, and arteries.

ATP, adenosine triphosphate; EPSP, excitatory postsynaptic potential; NO, nitric oxide.

at the postganglionic synapse. For example, ACh and vasoactive intestinal peptide (VIP) are frequently co-released from neurons that control glandular secretion (Figure 41–7C). In salivary glands, the two transmitters act directly to evoke secretion. In addition, VIP causes dilation of the blood vessels supplying the gland. Because cotransmitters can be released in varying proportions that depend on the frequency of presynaptic firing, different patterns of activity can regulate the volume of secretions, their protein and water content, and their viscosity. This regulation operates both through a direct effect on the gland cells and through indirect effects on the glandular blood flow that provides the water contained in secretions.

Understanding the pharmacology of these receptors and the second-messenger signaling pathways they control is important in the treatment of numerous medical conditions, including hypertension, heart failure, asthma, emphysema, allergies, sexual dysfunction, and incontinence.

Autonomic Responses Involve Cooperation Between the Autonomic Divisions

To survive, animals and humans must have "fight-or-flight" responses in order to stand and fight a predator or run away and live to see another day. Walter Cannon, in addition to introducing the concept of

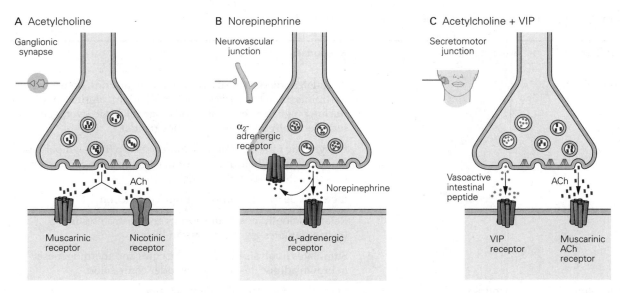

Figure 41–7 Synaptic transmission in the peripheral autonomic system.

A. In sympathetic ganglia, acetylcholine (**ACh**) can activate both nicotinic and muscarinic receptors to produce fast and slow postsynaptic potentials, respectively.

B. At neurovascular junctions, norepinephrine can simultaneously activate postsynaptic α_1-adrenergic receptors to produce

vasoconstriction and presynaptic α_2-adrenergic receptors to inhibit further transmitter release.

C. Cotransmission involves the co-activation of more than one type of receptor by more than one transmitter. Parasympathetic postganglionic nerve terminals in the salivary glands release both ACh and vasoactive intestinal peptide (**VIP**) to control secretion. At some autonomic synapses with end organs, three or more receptor types are activated.

homeostasis, also appreciated that this fight-or-flight response is a critical sympathetic function.

Two important ideas underlie this insight. First, the sympathetic and parasympathetic systems play complementary, even antagonistic, roles; the sympathetic system promotes arousal, defense, and escape, whereas the parasympathetic system promotes eating and procreation. Second, actions of the sympathetic system are relatively diffuse; they influence all parts of the body and once turned on can persist for some time. These ideas are behind the popular notion of the "adrenaline rush" produced by excitement, as by a roller coaster ride.

We now know that extreme sympathetic responses such as "fight-or-flight" can have long-term pathological consequences resulting in the syndrome known as post-traumatic stress disorder (Chapter 61). This disorder was first recognized in soldiers during World War I, when it was referred to as "shell shock." A variety of life-threatening experiences, ranging from sexual abuse and domestic violence to aircraft disasters, can also induce post-traumatic stress disorder, which affects millions of people in the United States alone.

Because the fight-or-flight model assumes antagonistic roles for the sympathetic and parasympathetic systems, Cannon's model led to an overemphasis on

the extremes of autonomic behavior. Actually, during everyday life, the different divisions of the autonomic system are tightly integrated. In addition, we now know that the sympathetic system is less diffusely organized than first envisioned by Cannon. Even within the sympathetic division, subsets of neurons control specific targets, and these pathways can be activated independently.

As in the somatic motor system, reflexes in the autonomic motor system are elicited through sensory pathways and are hierarchically organized. An important feature of this organization is that it allows for coordination between the different divisions of the autonomic system. The interplay between different systems in simple autonomic behaviors is analogous to the role of antagonist muscles in locomotion. To walk, one must alternately contract antagonist muscles that flex and extend a joint. Similarly, the sympathetic and parasympathetic systems are often partners in the regulation of end organs. In most cases, ranging from the simplest reflexes to more complex behaviors, all three peripheral divisions of the autonomic system work together. We illustrate this organization with two examples: control of the bladder (micturition reflex) and regulation of blood pressure.

Bladder Control

The micturition reflex is an example of a physiological cycle resulting from coordination between sympathetic and parasympathetic systems. In this cycle, the bladder is emptied by the parasympathetic pathway, which contracts the bladder and relaxes the urethra. The sympathetic system allows the bladder to fill by stimulating the urethra and inhibiting the parasympathetic pathway, thus inhibiting the reflex for bladder emptying. The sensory feedback required for this behavior is integrated with the motor outflow at both spinal and supraspinal levels (Figure 41–8).

Spinal components of the reflex are most influential during the storage phase of the micturition cycle, when sympathetic and somatic motor effects predominate. When the bladder is full, its distension triggers a sensory signal sufficient to activate the pontine micturition center (PMC). Descending signals from the PMC then increase parasympathetic outflow. Somatic control of the external urinary sphincter, which consists of striated muscle, contributes to both phases of the micturition cycle and is a voluntary behavior that originates through forebrain mechanisms (Figure 41–8). Patients with spinal cord injuries at the cervical or thoracic levels retain the reflex but not voluntary control of urination, because the connections between the bladder and the pons are severed.

Blood Pressure Regulation

The baroreceptor reflex is one of the simplest mechanisms for regulating blood pressure and further illustrates coordinated homeostatic control by antagonist sympathetic and parasympathetic pathways. It prevents orthostatic hypotension and fainting by compensating for rapid hydrostatic effects produced by changes in posture. When a recumbent person stands up, the sudden elevation of the head above the heart causes a transient decrease of cerebral blood pressure that is rapidly sensed by baroreceptors in the carotid sinus in the neck (Figure 41–9). Other important pressure sensors are located in the aortic arch and in the pulmonary circulation.

When neurons in the ventrolateral medulla detect the decrease in afferent baroreceptor activity produced by low blood pressure, they produce a reflexive suppression of parasympathetic activity to the heart and stimulation of sympathetic activity to the heart and vascular system. These changes in autonomic tone restore blood pressure by increasing heart rate, the strength of cardiac contractions, and the overall vascular resistance to blood flow through arterial vasoconstriction.

Under the converse condition of elevated arterial pressure, the increase in baroreceptor activity enhances parasympathetic inhibition of the heart and decreases sympathetic stimulation of cardiac function and vascular resistance. In general, the parasympathetic component of the baroreceptor reflex has a more rapid onset and is briefer than the sympathetic component. Consequently, parasympathetic activity is critical for the rapid response of baroreceptor reflexes but less important than sympathetic activity for long-term blood pressure regulation.

Visceral Sensory Information Is Relayed to the Brain Stem and Higher Brain Structures

Visceral sensory information reaches the brain mainly through two cranial nerves (IX and X), which end in caudal segments of the nucleus of the solitary tract (NTS), and through the abdominal splanchnic nerves, which end in the spinal cord (Chapter 40). The splanchnic information is transmitted to the brain through the spinothalamic tract (Chapter 4), which branches out along the way and also sends afferents to the NTS and lateral parabrachial nucleus.

The NTS relays sensory information in two different directions. First, it projects to networks in the brain stem and spinal cord that control and coordinate autonomic reflexes (as we saw for the baroreceptor reflex). In this way, visceral sensory signals relayed through the NTS regulate vagal motor control of the heart and gastrointestinal tract directly. Some neurons in the NTS project to neurons in the ventrolateral medullary reticular formation that control blood pressure by differentially regulating blood flow in particular vascular beds (Figure 41–9). Second, the NTS sends ascending projections to the forebrain, relaying visceral information to higher structures (Figure 41–10A). These higher structures, including the hypothalamus, use this information to coordinate autonomic, neuroendocrine, and behavioral responses.

Visceral sensory information is relayed from the NTS to the forebrain via direct and indirect projections (Figure 41–10A). The major indirect pathway involves the lateral parabrachial nucleus, which receives afferents from the NTS and sends efferents to higher structures, including the amygdala, hypothalamus, bed nucleus of the stria terminalis, insular cortex, and infralimbic/prelimbic cortex. The direct projections from the NTS target many of these same forebrain sites. The rostral NTS is an important part of the afferent taste pathway (Chapter 29). Information

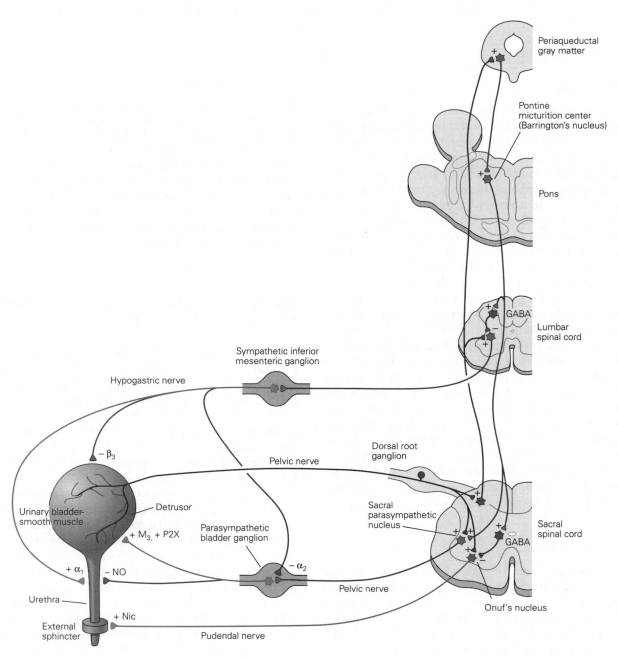

Figure 41–8 The micturition reflex requires interplay between the parasympathetic and sympathetic divisions of the autonomic system. (Adapted from DeGroat, Booth, and Yoshimura 1993.)

When bladder volume is low, urinary outflow is inhibited because activity in the sympathetic pathway is greater than activity in the parasympathetic pathway. Mild distension of the detrusor (storage portion of the bladder) initiates a low level of sensory activity, which reflexively activates spinal preganglionic neurons. The resulting low level of preganglionic activity is effectively transmitted and amplified by the sympathetic inferior mesenteric ganglion but filtered out by the parasympathetic bladder ganglion because of differences in patterns of synaptic convergence in the two ganglia. The resulting predominance of sympathetic tone keeps the detrusor relaxed and the urethra constricted. Sympathetic postganglionic fibers also reduce parasympathetic activity by inhibiting preganglionic release of acetylcholine. In addition to their effects on the autonomic outflow, the sensory signals are sufficient to keep the external urinary sphincter closed.

When filling causes the bladder to reach a critical volume, the associated increase in sensory activity reaches a threshold that allows impulses to pass through the pontine micturition center (Barrington's nucleus). Descending activity from this nucleus then further excites the parasympathetic outflow. The resulting increase in parasympathetic preganglionic firing promotes summation of fast excitatory postsynaptic potentials and initiation of postsynaptic action potentials in the bladder ganglion as it switches to its "on" state. During the emptying process, descending pathways also inhibit the sympathetic and somatic outflows through inhibitory spinal interneurons. Inhibition of somatic motor neurons in Onuf's nucleus causes relaxation and opening of the external sphincter. In this figure, the sacral spinal cord is enlarged relative to the other slices.

(Abbreviations: α_1, alpha-1 adrenergic receptor, α_2, alpha-2 adrenergic receptor, β_3, beta-3 adrenergic receptor, **GABA**, γ-aminobutyric acid; **M$_3$**, muscarinic ACh receptor 3; **nic**, nicotinic receptor; **NO**, nitric oxide; **P2X**, purinergic receptor.)

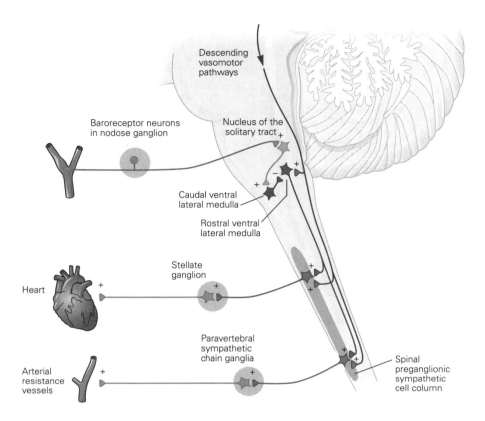

Figure 41–9 The baroreceptor reflex behaves as a negative feedback loop with gain. Arterial blood pressure is sensed by baroreceptors, a type of stretch receptor neuron, in the carotid sinus near the base of the brain. After integration in the medulla, this information provides negative feedback control of the cardiovascular system. The sympathetic component of the circuit includes outputs that stimulate the heart's pumping capacity (cardiac output) by increasing heart rate and the strength of contractions. In addition, sympathetic stimulation causes arteries to contract, which raises the hydraulic resistance to blood flow. Together, the effects of increased cardiac output and increased vascular resistance raise mean arterial blood pressure. Inhibitory projections from the caudal to the rostral ventral lateral medulla create negative feedback so that an increase in blood pressure inhibits sympathetic activity, whereas a decrease raises sympathetic activity. Although omitted for simplicity, parasympathetic neurons in the cardiac ganglion also contribute to the reflex by creating an inhibitory cardiac input that is functionally antagonistic to the sympathetic pathway (Figure 41–10). During baroreceptor reflexes, parasympathetic activity within the heart is therefore increased by hypertension and reduced by hypotension.

in this pathway is relayed via the medial parabrachial nucleus to the taste area of insular cortex.

Central Control of Autonomic Function Can Involve the Periaqueductal Gray, Medial Prefrontal Cortex, and Amygdala

The periaqueductal gray, which surrounds the cerebral aqueduct in the midbrain, receives inputs from most parts of the central autonomic network and projects to the medullary reticular formation to initiate integrated behavioral and autonomic responses. For example, in the defensive "fight-or-flight" response, the periaqueductal gray helps redirect blood flow from the digestive system to the hind limbs, thus enhancing running (Figure 41–10B).

The medial prefrontal cerebral cortex is a visceral sensory-motor region. It includes two functional areas that interact with each other: the rostral insular cortex and the rostromedial tip of the cingulate gyrus (also referred to as the infralimbic and prelimbic areas). Stimulation here can produce a variety of autonomic effects including contractions of the stomach and changes in blood pressure. These visceral sensory and motor areas of cortex send descending projections to the parts of the central autonomic network in the brain stem discussed above.

Finally, visceral regions of cortex, along with many subcortical parts of the central autonomic

A Afferent pathways

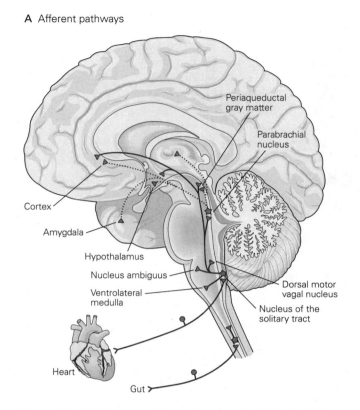

B Efferent pathways

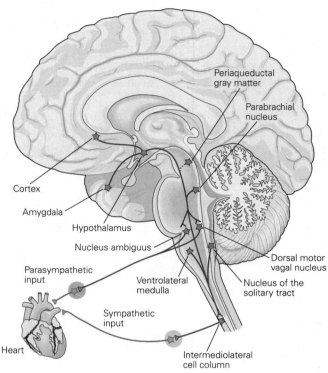

Figure 41–10 The central autonomic network. Nearly all of the cell groups illustrated here are interconnected with one another, forming the central autonomic network.

A. Visceral information (**solid lines**) is distributed to the brain from the nucleus of the solitary tract and from ascending spinal pathways activated by the splanchnic nerves (from the gut, for example). The nucleus of the solitary tract distributes this information to preganglionic parasympathetic neurons (the dorsal motor vagal nucleus and nucleus ambiguus), to regions of the ventrolateral medulla that coordinate autonomic and respiratory reflexes, and to more rostral parts of the central autonomic network in the pons (parabrachial nucleus), midbrain (periaqueductal gray), and forebrain. The parabrachial nucleus also projects to many of the more rostral components of the central autonomic network, including visceral and gustatory nuclei of the thalamus (**dotted lines**).

Other pathways from the spinal cord (not shown) also transmit visceral information to many parts of the central autonomic network, including the nucleus of the solitary tract, parabrachial nucleus, periaqueductal gray, hypothalamus, amygdala, and cortex. The spinal cord also projects to the main somatosensory nucleus of the thalamus (ventral posterolateral nucleus).

B. All of the efferent pathways shown here (except perhaps for the periaqueductal gray) project directly to autonomic preganglionic neurons. In the hypothalamus, the descending division of the paraventricular nucleus and three cell clusters in the lateral zone project heavily to both parasympathetic and sympathetic preganglionic neurons. Other pathways (not shown) arise from certain monoaminergic cell groups in the brain stem, including noradrenergic neurons in the A5 region and serotonergic neurons in the raphe nuclei.

network, interact with the amygdala. Complex pathways between certain amygdalar cell groups underlie certain conditioned emotional responses—learned associations between specific stimuli and behaviors with accompanying autonomic responses. When a rat learns that a mild electric shock follows an auditory cue, the auditory cue alone comes to produce the elevated heart rate and freezing that was originally elicited by the shock alone (Chapters 42 and 53). Such learned responses are prevented by selective lesions of the amygdalar region, which projects to the

hypothalamus and lower brain stem parts of the central autonomic network.

The Neuroendocrine System Links the Brain to Physiological Responses Through Hormones

Another effector arm of the hypothalamus is the neuroendocrine system, which controls secretion of hormones by the pituitary gland. The pituitary has two functionally and anatomically distinct subdivisions, the anterior and posterior pituitary. The posterior pituitary is an extension of the brain and contains hormone-secreting axon terminals of hypothalamic neurons. These terminals secrete vasopressin or oxytocin directly into the systemic circulation. The anterior pituitary, on the other hand, is entirely nonneuronal and is composed of five types of endocrine cells. Hormone secretion from these cells is controlled by stimulatory and inhibitory factors released by hypothalamic neurons into a specialized circulatory system that carries blood from the base of the brain (median eminence) to the anterior pituitary.

Hypothalamic Axon Terminals in the Posterior Pituitary Release Oxytocin and Vasopressin Directly Into the Blood

Large neurons in the paraventricular and supraoptic nuclei form the magnocellular component of the neuroendocrine motor system of the hypothalamus (Figure 41–11). The magnocellular neurons send their axons through the hypothalamo-hypophysial tract to the posterior pituitary, or *neurohypophysis* (Figure 41–12). Approximately one-half of these neurons synthesize and secrete vasopressin (the antidiuretic hormone) into the general circulation; the other half synthesize and secrete oxytocin, a structurally similar hormone. Both hormones circulate to organs, where vasopressin controls blood pressure and water reabsorption by the kidney and oxytocin controls uterine smooth muscle and milk release.

Vasopressin and oxytocin are nine-amino acid peptide hormones. Like other peptide hormones, they are synthesized in the cell body as larger prohormones (Chapter 16) and then cleaved within Golgi transport vesicles before traveling down the axon to release sites in the posterior pituitary. The genes for these peptides have similar sequences and probably arose by duplication.

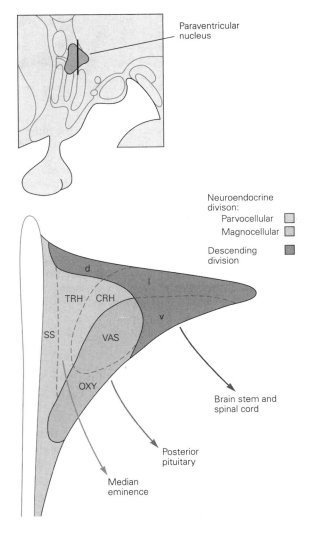

Figure 41–11 The paraventricular nucleus in the hypothalamus is a microcosm of neuroendocrine, autonomic, and sensory-motor integration. The three structural-functional divisions of the paraventricular nucleus are shown. The *magnocellular neuroendocrine division* comprises two distinct although partly interdigitated pools of neurons that normally release vasopressin (**VAS**) or oxytocin (**OXY**). Their axons course through the internal zone of the median eminence and terminate in the posterior pituitary. Two other populations of magnocellular vasopressin and oxytocin neurons lie in the supraoptic nucleus along the base of the brain.

The *parvocellular neuroendocrine division* includes three major, separate (although partly interdigitated) pools of neurons that control anterior pituitary hormone secretion. Their axons end in the external zone of the median eminence, where they release their peptide neurotransmitters—somatostatin (**SS**), growth hormone-inhibiting hormone (**GIH**), thyrotropin-releasing hormone (**TRH**), or corticotropin-releasing hormone (**CRH**)—into the hypophysial portal veins.

The *descending division* has three parts—dorsal (**d**), lateral (**l**), and ventral (**v**)—each comprising topographically organized conventional neurons that project to the brain stem and spinal cord. Their axons terminate in many parts of the central autonomic network in the brain stem (Figure 41–10), the marginal zone (lamina I) of the dorsal horn of the spinal cord and spinal trigeminal nucleus, and a number of regions in the brain stem reticular formation and periaqueductal gray matter. The descending division modulates autonomic outflow (and inflow), the inflow of nociceptive information, and eating and drinking behaviors. Appropriate integration of magnocellular neuroendocrine, parvocellular neuroendocrine, autonomic, and behavioral responses is mediated primarily by external inputs rather than by interneurons or extensive recurrent axon collaterals of projection neurons. Circulating steroid and thyroid hormones also produce selective effects on particular types of neurons in the paraventricular nucleus.

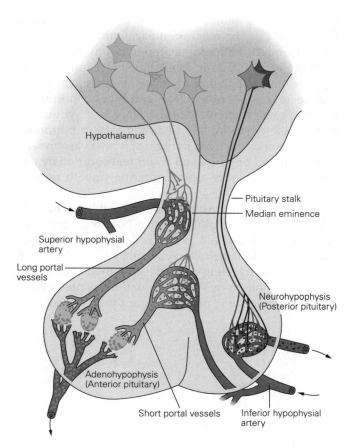

Figure 41–12 The hypothalamus controls the pituitary gland both directly and indirectly through neuroendocrine neurons. Neurons in the magnocellular neuroendocrine system (**blue**) send their axons directly to the posterior pituitary (neurohypophysis) where they release the peptides vasopressin and oxytocin into the general circulation. Neurons in the parvocellular neuroendocrine system (**yellow**) send their axons to the hypophysial portal system in the median eminence and pituitary stalk. Portal veins transport hypothalamic hormones (peptides and dopamine) to the anterior pituitary (adenohypophysis) where they increase the release of hormones from five classic types of endocrine cells (Figure 41–11). The output of neuroendocrine neurons is regulated in large part by inputs from other regions of the brain. (Adapted from Reichlin 1978, and Gay 1972.)

Labels in figure: Hypothalamus; Pituitary stalk; Median eminence; Superior hypophysial artery; Long portal vessels; Neurohypophysis (Posterior pituitary); Adenohypophysis (Anterior pituitary); Short portal vessels; Inferior hypophysial artery

Endocrine Cells in the Anterior Pituitary Secrete Hormones in Response to Specific Factors Released by Hypothalamic Neurons

In the 1950s, Geoffrey Harris proposed that the anterior pituitary, or *adenohypophysis*, is regulated indirectly by the hypothalamus. He showed that the hypophysial portal veins, which carry blood from the hypothalamic median eminence to the anterior pituitary, transport factors released from hypothalamic neurons that control anterior pituitary hormone secretion (Figure 41–12). In the 1970s, Andrew Schally, Roger Guillemin, and Wylie Vale determined the structure of a group of hypothalamic peptide hormones that control pituitary hormone secretion from the five classic endocrine cell types in the anterior pituitary. These hormones, which are released into the median eminence by hypothalamic neurons, fall into two classes: releasing hormones and release-inhibiting hormones. Only one anterior pituitary hormone, prolactin, is under predominantly inhibitory control (mediated by dopamine).

The *parvocellular neuroendocrine motor zone* of the hypothalamus is centered along the wall of the third ventricle (Figure 41–2A) and contains neurons that project to and release their hormones into the median eminence. The parvocellular neurons releasing *gonadotropin-releasing hormone* (GnRH) are atypical in that they are scattered in a continuum extending from the medial septum through to the mediobasal hypothalamus. They are controlled by upstream neurons that release *kisspeptin*. The remaining parvocellular neuroendocrine neurons lie within the paraventricular and arcuate nuclei and the short periventricular region between them (Figures 41–2 and 41–11).

Distinct pools of neurons in and around the paraventricular nucleus release *corticotropin-releasing hormone* (CRH), *thyrotropin-releasing hormone* (TRH), or *somatostatin* (or growth hormone release-inhibiting hormone) (Figure 41–11). The CRH neurons control the release of anterior pituitary adrenocorticotropic hormone (ACTH), which in turn controls the release of cortisol (glucocorticoids) from the adrenal cortex. Thus, this pool of CRH neurons is the "final common pathway" for all centrally mediated glucocorticoid stress hormonal responses. The arcuate nucleus contains two pools of parvocellular neuroendocrine neurons. One group releases *growth hormone–releasing hormone* (GHRH) and the other dopamine, which inhibits prolactin secretion. Some of the dopaminergic neurons are distributed dorsally as far as the paraventricular nucleus.

The axons of all these parvocellular neuroendocrine neurons travel in the hypothalamo-hypophysial tract and end in the specialized proximal end of the pituitary stalk, the median eminence (Figure 41–12). There, in a region of capillary loops in the external zone of the median eminence, the axon terminals release the various hypophysiotropic factors. While the median eminence is within the brain, it is considered outside the blood–brain barrier. This is due to the fenestrated nature of the median eminence

Table 41–3 Hypothalamic Substances That Release or Inhibit the Release of Anterior Pituitary Hormones

Hypothalamic substance	Anterior pituitary hormone
Releasing:	
Thyrotropin-releasing hormone (TRH)	Thyrotropin (TSH), prolactin (PRL)
Corticotropin-releasing hormone (CRH)	Adrenocorticotropin (ACTH), β-lipotropin
Gonadotropin-releasing hormone (GnRH)	Luteinizing hormone (LH), follicle-stimulating hormone (FSH)
Growth hormone–releasing hormone (GHRH or GRH)	Growth hormone (GH)
Inhibiting:	
Prolactin release-inhibiting hormone (PIH), dopamine	Prolactin
Growth hormone release-inhibiting hormone (GIH or GHRIH; somatostatin)	Growth hormone, thyrotropin

capillaries, which allow diffusion of the hypophysiotropic factors into the portal circulation. The median eminence capillary loops are the proximal end of the hypophysial portal system of veins that carry the factors to the anterior pituitary, where they act on cognate receptors on the five types of endocrine cells (Figure 41–12 and Table 41–3).

Dedicated Hypothalamic Systems Control Specific Homeostatic Parameters

Body Temperature Is Controlled by Neurons in the Median Preoptic Nucleus

Body Temperature Reflects the Balance Between Heat Production and Loss

The body generates heat through all of its exothermic biochemical reactions and ion fluxes. These processes can be greatly increased above a baseline level, the resting metabolic rate, by exercise and shivering (both of which increase skeletal muscle heat production), by the digestion and assimilation of food (the so-called thermic effect of food), and by sympathetic stimulation of thermogenic activity in brown adipose tissue (Box 41–1).

The body loses heat by radiation, convection, conduction (if immersed in cool water), and endothermic evaporation of either sweat from the skin or moisture from the respiratory tract (a process augmented in some species by panting). The defensive reaction to cold, in addition to producing heat, involves sympathetically mediated cutaneous vasoconstriction and piloerection (goose bumps). By sending less blood to the skin, vasoconstriction conserves core temperature. Piloerection helps insulate the skin by creating a layer of motionless air near the skin's surface. Conversely, defenses against overheating include inhibition of sympathetic pathways that activate cutaneous vasodilation and brown adipose tissue. Voluntary behavioral responses like taking a swim or putting on a sweater play a particularly important role in thermoregulation.

Box 41–1 Brown Adipose Tissue, Bioenergetics, and Sympathetically Driven Thermogenesis

Brown adipose tissue is a remarkably specialized heat-producing tissue that is especially abundant in newborns and small mammals, but is also found in adult humans. It has a rich blood supply for delivery of fuel and oxygen and for removal of heat and is densely innervated by postganglionic sympathetic nerves. Brown adipocytes, the producers of heat, are found in concentrated deposits in and around the core and also as isolated cells within larger white adipose tissue depots.

Sympathetic stimulation of β-adrenergic receptors activates uncoupling protein-1 (UCP1), a mitochondrial proton transport protein that is unique to brown adipocytes. When activated, UCP1 "leaks" protons across the mitochondrial inner membrane into the mitochondrial matrix, down the proton electrochemical gradient. This uncouples mitochondrial respiration from adenosine triphosphate (ADP) availability, greatly increasing fuel oxidation and, importantly, the production of heat.

Exercise and shivering, on the other hand, increase heat production by using adenosine triphosphate (ATP) to perform work. The resulting increase in ADP activates proton transport into mitochondria via ATP synthase, which then increases coupled mitochondrial respiration, fuel oxidation, and ultimately the production of heat.

They usually begin before the onset of physiologic responses. Like thirst/drinking and hunger/eating, activities generated in response to cold or hot challenges are motivated behaviors.

Body Temperature Is Detected at Multiple Sites

Core temperature is held relatively constant. At the shell, on the other hand, temperature fluctuates extensively because the shell is adjacent to the external environment, it has a high surface-to-mass ratio (in the case of limbs favoring heat loss over heat production), and thermal challenges dramatically affect its supply of warm blood (decreased when heat needs to be conserved; increased when heat needs to be lost).

Most primary afferents that detect temperature have their cell bodies in the spinal dorsal root ganglia. Neurons that detect noxious temperatures are part of the pain pathway (Chapter 20). Their function is to limit local tissue damage by promoting withdrawal as opposed to regulating body temperature. Neurons that respond to innocuous temperatures are often called thermoreceptors. Some thermoreceptor neurons have their endings in the skin, just below the epidermis, and these respond to shell temperature. They are predominantly, but not entirely, cold-responsive. Other thermoreceptor neurons have their endings in and around the large organs and respond to core temperature. They are also largely, but not entirely, cold-responsive. The fibers for the deep-tissue thermoreceptor neurons travel in the splanchnic nerves and, like the thermoreceptor neurons in the shell, have their cell bodies in the dorsal root ganglia. In addition, some also travel in the vagal afferent nerve. Finally, there are warm-sensing neurons in the hypothalamic medial preoptic area.

The molecular sensors utilized by thermoreceptor neurons for detecting changes in temperature are a subset of excitatory transient receptor potential (TRP) channels. Different TRP channels respond to different ranges of temperatures (Chapter 20). Recent studies have implicated specific TRP channel types in various forms of sensing innocuous temperature in the three sites mentioned above: TRMP8 channels mediate cold-sensing by shell thermoreceptor neurons, and TRPM2 channels mediate warm-sensing both by somatosensory thermoreceptor neurons and by neurons in the hypothalamic preoptic area.

Multiple Thermoreceptor/Thermoeffector Loops Control Temperature

Involuntary thermal regulation is controlled by a multisensor, multieffector thermoregulatory system.

Thermal information from the shell and the viscera ascends via primary afferents whose cell bodies are in the dorsal root ganglia. They project to second-order neurons in the dorsal horn of the spinal cord. These neurons project via the spinothalamic tract to the lateral parabrachial nucleus where neurons relay cold-sensing and warm-sensing information to neurons in the hypothalamic MnPO. Activation of the cold-sensing or warm-sensing afferent pathways induces appropriate physiological responses aimed at increasing or decreasing body temperature.

The neurons in the MnPO that indirectly respond to cold and warmth send efferent signals via relays through the medial preoptic area, dorsomedial hypothalamus, and raphe pallidus in the ventral medulla, and from there on to the sympathetic preganglionic neurons in the intermediolateral nucleus of the spinal cord. These latter neurons excite postganglionic sympathetic neurons that project to blood vessels, sweat glands, and arrector muscles of hair follicles to control cutaneous blood flow, sweating, and piloerection, respectively, as well as to brown adipose tissue to control thermogenesis. In addition, cold causes shivering when gamma motor neurons in the ventral horn of the spinal cord are activated by excitatory neurons in the raphe pallidus (Chapter 32). The resultant contraction of intrafusal muscle fibers within muscle spindles activates IA afferents from the spindles to alpha motor neurons. This proprioceptive feedback increases activity of alpha motor neurons, as well as their propensity to undergo rhythmic bursts of activity, causing increased muscle tone and frank shivering.

The neural pathways controlling voluntary thermoregulatory behaviors involve the same thermoreceptor pathways. Stimulation of warm-sensitive neurons in and around the MnPO evokes dramatic cold-seeking behavior, decreases heat production, and increases heat loss. The conscious perception of body temperature relies upon the same first-order thermoreceptor neurons, but the afferent pathway diverges to activate second-order neurons in the dorsal horn, which project directly or indirectly to neurons in the ventromedial nucleus of the thalamus. These thalamic neurons project to the insular cortex.

From the above discussion, it is clear that there is neither a set point for body temperature nor a "thermostat" that maintains it at 37°C. Instead, as mentioned earlier, an apparent set point for body temperature emerges as a settling point controlled by multiple sensory-motor feedback loops containing thermoreceptors and thermoeffectors. It is a major achievement of evolution that this multicomponent afferent/efferent system is so effective in keeping the temperature of the body core remarkably constant.

Dysregulation of Circuits Controlling Temperature Causes Fever

In the past, when the set point view of thermoregulation was dominant, fever was thought to be caused by raising the body temperature set point—a view that still persists in major medical textbooks. Based on the advances described above, fever is now thought to arise through modulation of the afferent/efferent loops, particularly as they traverse the hypothalamic preoptic area. Prostaglandin E2, generated by the action of inflammatory cytokines on endothelial cells in the preoptic area, inhibits warm-activated GABAergic neurons in the MnPO, thus disinhibiting the effector pathways that promote cutaneous vasoconstriction, brown adipose tissue thermogenesis, and shivering. Nonsteroidal anti-inflammatory drugs, such as aspirin, ibuprofen, and acetaminophen, reduce fever by inhibiting hypothalamic generation of prostaglandin E2.

Water Balance and the Related Thirst Drive Are Controlled by Neurons in the Vascular Organ of the Lamina Terminalis, Median Preoptic Nucleus, and Subfornical Organ

Changes in Blood Osmolarity Cause Cells to Shrink or Swell

Driven by osmosis, water moves freely across cell membranes. This has a number of important consequences. First, because of its large size, the intracellular compartment contains two-thirds of the body's water. Second, if blood osmolarity changes from its normal value (~290 mOsm/kg)—because water is gained by drinking or lost by renal excretion and by sweating, or because solutes have been added by eating (or by drinking, eg, sea water—1000 mOsm/kg)—water will move and the osmolarity of all compartments will equilibrate, including the intracellular one.

Because the intracellular content of osmotically active molecules is relatively fixed over the short term, increases in blood osmolarity cause cells to shrink, and conversely, decreases cause cells to swell. This is particularly dangerous for the brain because it is encased by the rigid skull. With extreme hyperosmolarity (too little water), the brain shrinks, pulling away from the skull and tearing blood vessels. With hypo-osmolarity (too much water), the brain swells, causing cerebral edema, seizures, and coma. To prevent such incidents, the brain acts to maintain normal osmolarity. It does

this by detecting changes in osmolarity and then regulating the motivation to drink (thirst) and the kidney's capacity to excrete water.

Osmolarity Is Affected When Water Is Lost or Gained and When Food Is Ingested

Water is gained by drinking and, to a small degree, by the oxidation of fuel (fuel + $O_2 \rightarrow CO_2 + H_2O$). It is lost in a number of ways—by breathing (dry air in, humidified air out), via the gastrointestinal tract (especially when diarrhea is present), by sweating, and by urination. Eating also increases blood osmolarity by moving water from the blood to the gut to aid digestion and by adding solutes to the bloodstream as food is broken down and absorbed.

Because of these effects, there are significant interactions between neural systems that control hunger and thirst. For example, eating is such a significant osmotic challenge that dehydration and its associated hyperosmolarity strongly suppress hunger (dehydration-induced anorexia). Conversely, the act of eating itself, even in an individual with normal water content, rapidly stimulates thirst so as to mitigate the anticipated, eating-induced increase in osmolarity.

Vasopressin Released From the Posterior Pituitary Regulates Renal Water Excretion

The ability of the kidney to excrete water is tightly controlled by vasopressin. When it is absent, humans can excrete up to approximately 900 mL/h of urine, and when it is at maximal levels, humans excrete as little as approximately 15 mL/h. Vasopressin decreases water excretion by increasing its reabsorption from urine by the kidney.

Osmolarity Is Detected by Osmoreceptor Neurons

The brain maintains water balance by monitoring sensory input from osmoreceptors—sensory neurons that respond to osmolarity—which reflects the body's state of hydration. Osmoreceptor neurons are found in the periphery and on neurons in and around the hypothalamus. The central osmoreceptors monitor systemic osmolarity, while the peripheral osmoreceptors monitor osmolarity in and around the gut and related structures.

Peripheral Osmoreceptors Allow Changes in Systemic Osmolarity to Be Anticipated

Sensory information about peripheral osmolarity enables the brain to make preemptive changes in thirst and vasopressin secretion that anticipate and mitigate future shifts in systemic osmolarity, such as the decrease that occurs with drinking or the increase that occurs with eating. Such regulation serves to prevent overshooting normal osmolarity, which would otherwise occur when previously ingested water, which has not yet affected systemic osmolarity, is slowly absorbed from the gut. Indeed, when a dehydrated, hyperosmolar individual ingests water, thirst and vasopressin secretion rapidly decrease, well before systemic osmolarity falls. The identity of peripheral osmoreceptors is unknown.

Central Osmoreceptors and the Afferent/Efferent Circuits Control Water Balance

Three nuclei in the lamina terminalis, which forms the anterior wall of the third ventricle, play a key role in detecting and in responding to disturbances in systemic osmolarity (Figure 41–13). From ventral to dorsal, they are the OVLT, the MnPO, and the SFO. The OVLT and SFO are circumventricular organs, and like the previously discussed median eminence, they lie outside the blood–brain barrier. Because of this, neurons in these two nuclei can rapidly detect changes in blood osmolarity as well as blood-borne circulating factors that are unable to cross the blood–brain barrier (an important example being angiotensin II).

Consistent with this arrangement, osmoreceptor neurons in the OVLT and SFO make extensive connections to neurons in the MnPO. While MnPO neurons themselves do not directly sense osmolarity, they are indirectly responsive to osmolarity via relays from the OVLT and SFO. While all neurons in the OVLT and SFO appear to be dedicated to the regulation of water balance, some neurons in the MnPO are involved in regulation of body temperature (as noted earlier), cardiovascular function, and sleep. Regulation of water balance or body temperature is carried out by subsets of modality-specific neurons in the MnPO.

Neurons from all three lamina terminalis nuclei send dense excitatory projections to secretory vasopressin neurons in the paraventricular hypothalamic nucleus (PVH) and supraoptic nucleus. As described below, these three lamina terminalis nuclei are also able to cause thirst.

Central osmoreceptors, and probably also peripheral osmoreceptors, detect changes in osmolarity by

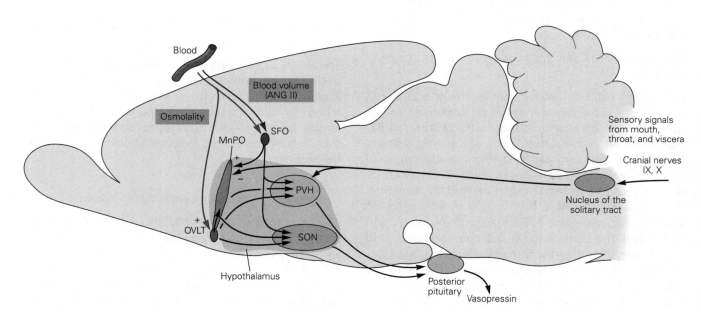

Figure 41–13 Neural and endocrine components combine to regulate fluid balance. The circuitry is shown in a sagittal section through the rat brain. Information from baroreceptors in the circulatory system and from sensory receptors in the mouth, throat, and viscera is conveyed to the nucleus of the solitary tract and neighboring structures in the caudal brain stem through the glossopharyngeal (**IX**) and vagal (**X**) nerves. The hormone angiotensin II (**ANG II**) provides the brain with an additional signal concerning low blood volume. Circulating angiotensin II is sensed by receptors in the subfornical organ (**SFO**); SFO neurons project to the median preoptic area (**MnPO**), paraventricular nucleus of the hypothalamus (**PVH**), supraoptic nucleus (**SON**), and the vascular organ of the lamina terminalis (**OVLT**). The osmolality of the blood is sensed by receptors in and near the OVLT that project to the MnPO, PVH, and SON. Neurosecretory cells in the PVH and SON nuclei trigger release of vasopressin from the posterior pituitary, thus decreasing water excretion by the kidney. (Adapted, with permission, from Swanson 2000.)

responding to changes in cell volume. Shrinking or swelling, which increases or decreases, respectively, their cation permeability, causes increases or decreases in firing rate.

Decreased intravascular volume—for example, that due to acute blood loss—also potently stimulates thirst and vasopressin secretion. Low blood volume is detected by the kidney, which increases its secretion of renin. Renin is a protease that converts circulating angiotensinogen to *angiotensin I* (ANG I). ANG I is then further cleaved by angiotensin-converting enzyme in the lung, generating *angiotensin II* (ANG II). ANG II excites SFO neurons that directly, and also via a relay in the MnPO, drive vasopressin neurons and presumptive thirst neurons.

Thirst Is Controlled by Neurons in the OVLT, MnPO, and SFO

As with vasopressin secretion, all three lamina terminalis structures participate in generating the motivational state of thirst, the desire to seek and ingest water. Lesion of all three structures completely blocks dehydration- and ANG II–induced thirst, as well as secretion of vasopressin. Electrical stimulation of these structures, on the other hand, elicits drinking. Activation of excitatory glutamatergic neurons in the SFO and MnPO induces intense drinking in an otherwise water-satiated mouse within seconds.

Thus, excitatory neurons in the SFO and MnPO, and likely also in the OVLT, have a remarkable capacity to induce thirst. Importantly, the behavior induced is specific—only water drinking occurs. Notably, the excitatory SFO neurons driving this behavior are the same subset of SFO neurons that are activated by dehydration and that express ANG II receptors. The downstream pathway by which these neurons stimulate thirst is presently unknown.

The activity of both the thirst neurons in the SFO and vasopressin neurons in the SON and PVH decreases or increases rapidly in response to sensory cues, such as drinking or eating, respectively, that anticipate future homeostatic disturbances. This rapid regulation occurs independent of any changes in systemic osmolarity and is therefore independent of feedback; hence, it is an example of feedforward control. The likely function of this feedforward regulation is to anticipate disturbances, institute preemptive corrective actions, and thus greatly reduce or eliminate their impact.

In summary, years of research have led to a clear model. Dehydration (water deficiency) increases the activity of neurons in the SFO and OVLT, and in the MnPO via relays from the SFO and OVLT, and this increase in activity enhances thirst and vasopressin secretion. As we shall see, a similar general system, but with different neural structures, controls caloric deficiency–based regulation of hunger and energy metabolism.

Energy Balance and the Related Hunger Drive Are Controlled by Neurons in the Arcuate Nucleus

As with temperature and water balance, energy balance is regulated by feedback signals from the body that modulate activity of key hypothalamic neurons, which then initiate adaptive changes in both physiology and behavior. Regulation of energy balance differs, however, in important ways.

First, the feedback signals monitored are numerous and, in many cases, only very indirectly related to the key parameter, energy balance. Examples of this feedback include neural and hormonal signals from the gut, leptin from adipocytes, insulin from pancreatic beta cells, and metabolite levels in the blood. This is in striking contrast to the single, directly sensed signals monitored for thermoregulation and water balance. Second, energy can be stored as fat. The amount of energy that can be accumulated is remarkably high, so high that the energy needs of a starving person can be met for more than a month. Heat and water, in contrast, are not stored. Thus, organisms have an "energy buffer" that allows survival during prolonged deficiency.

Third, since storage has benefits, regulation of energy balance, as opposed to temperature and water balance, is asymmetric in that low energy stores are defended against very aggressively, while high stores are defended against only very weakly—hence the high prevalence of obesity in societies with calorically dense, palatable food. Fourth, energy storage can be a liability—when excessive, it promotes diseases such as obesity, diabetes, heart disease, and cancer. Finally, in circuits regulating energy balance, neuropeptides play a remarkably important role.

Fat Is Stored When Intake of Energy Exceeds Expenditure

Consistent with the first law of thermodynamics, the calories that are stored as fat equal the number of calories ingested minus calories expended. While there is only one way to gain energy—by eating—there are many ways to expend energy.

Most energy is expended by biochemical reactions that are required for basic life functions. As these processes are constantly in operation, such "obligatory energy expenditure" is fixed and not regulated. Two other types of energy expenditure, however, are dramatically different; one is voluntary physical activity, while the other is involuntary, resulting from sympathetic stimulation of brown adipose tissue and shivering. Sympathetically controlled energy expenditure, often referred to as adaptive thermogenesis, is controlled by the brain. Its function is to respond to perturbations in temperature and in energy stores.

The Intake and Expenditure of Energy Are Usually Matched

For most individuals, body fat stores are relatively constant over time. Thus, calories ingested roughly equal calories expended. A simple calculation demonstrates this point. An average middle-aged person expends 3,392 kcal per day (kcal corresponds to the common term "calorie"). Over the course of a year, this typical person gains 350 g of fat (which equates to 9 kcal of fat per day). Thus, on average, 9 kcal extra must have been consumed per day to account for this gain. This is the amount of energy found in 4% of a typical candy bar or expended by walking about 150 meters. Thus, the mismatch between intake and expenditure (9 kcal) is tiny—only 0.27% of total energy expenditure.

Such close matching is the result of powerful homeostatic mechanisms that use feedback from the body to regulate intake and expenditure. As is true for regulation of temperature and osmolarity, the constancy of body weight, and the close matching between intake and expenditure, is unrelated to any specific "set point." Instead, this remarkable control is the emergent settling point of multiple afferent/efferent feedback loops.

Obesity Is Caused by Genes and Recent Lifestyle Changes

Dysregulation of the above-mentioned afferent/efferent feedback loops results in obesity. While some cases of obesity are due to known mutations in genes required for homeostatic regulation, most are of undetermined cause. Of these, many are likely due to multiple mutations, many of which are uncharacterized. Because the incidence of obesity in Western societies has increased greatly in recent years, too fast to be due to new mutations, changes in diet and physical activity must also play an important role. Homeostatic systems that evolved to achieve energy balance in hunter-gatherers are likely overwhelmed by abundant, palatable, calorically dense food.

But even in our modern obesogenic environment, there are still large variations in fat stores: Only 41% to 70% of interindividual variation in fat stores can be attributed to genetic factors. Thus, genetic predisposition and environment together cause obesity. Of

interest, many of the predisposing genetic loci identified to date involve genes that affect brain function.

Multiple Afferent Signals Control Appetite

The major afferent signals affecting energy balance can be divided into two major categories. (1) Short-term signals from cells that line the gastrointestinal tract report the status of food in the gut. All but one of these signals increase with eating and function to terminate meals; the exception is ghrelin, which increases with fasting and stimulates hunger. (2) Longer-term signals report the status of energy reserves (ie, fat stores). These include the pancreatic hormone insulin and the adipocyte hormone leptin, both of which are released in proportion to fat stores. Their levels, especially that of leptin, inform the brain whether fat stores are adequate (Figure 41–14A).

Signals From the Gut Trigger Meal Termination. During eating, as food enters the stomach and intestine, physical distention increases firing of stretch-sensitive vagal afferents. In addition, chemodetection of food by intestinal endocrine cells stimulates secretion of hormones such as cholecystokinin (CCK), glucagon-like peptide-1 (GLP-1), and peptide YY (PYY). These responses have three primary functions.

First, they cause contraction of the pyloric sphincter, a valve between the stomach and intestine. This limits further passage of food, preventing the small intestine from being overloaded. Second, the intestinal hormones stimulate secretion of bile and enzymes into the intestinal lumen to aid digestion. Third, the vagal afferents and the intestinal hormones decrease subsequent food intake, bringing about meal termination (satiation). The intestinal hormones accomplish this primarily by stimulating local vagal afferent terminals, which in turn excite neurons in the caudal region of the NTS.

Two of these hormones, GLP-1 and PYY, may also directly stimulate neurons in the brain. The activated neurons in the NTS project directly, or via a relay in the lateral parabrachial nucleus, to neurons in the forebrain, including the amygdala and hypothalamus. Neurons in the lateral parabrachial nucleus that express calcitonin gene-related polypeptide (CGRP) are one such important relay involved in satiation. These circuits then bring about meal termination.

When food is absorbed, the increase in blood glucose stimulates β-cells to release insulin and the hormone amylin. Amylin then excites neurons in the area postrema (a circumventricular organ outside the blood–brain barrier, located just above the NTS).

Circulating amylin increases within minutes following a meal, decreasing subsequent food intake.

Ghrelin is released by endocrine cells in the stomach. Unlike the factors described above, its secretion is high before eating and falls during the meal. It may play a role in meal initiation. Indeed, ghrelin is the only known systemic factor that increases hunger and thus eating. It excites neurons in a number of sites, including agouti-related peptide neurons in the arcuate nucleus (see below). The physiological significance of ghrelin is unclear because deletion of its gene does not appear to affect hunger.

Blood Glucose and Insulin Affect Appetite. Glucose is sensed by neurons in the periphery, hindbrain, and hypothalamus. Although glucose sensing does not appear to play a role in the day-to-day regulation of energy balance, the detection of and response to dangerously low blood glucose levels (*glucopenia*) is an important function of the brain. Two adaptive responses are initiated: (1) intense glucoprivic hunger, due at least in part to indirect activation of agouti-related peptide neurons, and (2) secretion of glucagon, epinephrine, and corticosteroids, which stimulate hepatic glucose production. The hormonal responses are caused by increases in sympathetic outflow as well as activation of the CRH pathway associated with stress (Chapter 61). Amino acids also can be sensed and, consequently, regulate energy balance and dietary choice—the latter to ensure ingestion of sufficient quantity and quality of protein.

Insulin, on the other hand, is thought to signal an increase in fat stores. Insulin's primary function is to control blood glucose, the stimulus for its secretion. Insulin lowers blood glucose by driving it into muscle and fat cells and by decreasing its production by the liver. As fat stores increase, its ability to do this is decreased (a phenomenon called insulin resistance). Thus, higher fat stores increase basal and meal-stimulated insulin secretion in an effort to overcome resistance and normalize glucose. The fat store–mediated increase in insulin levels inhibits neurons in the hypothalamus, especially the arcuate nucleus, which is thought to decrease hunger.

The Fat Cell Hormone Leptin Signals the Brain About Fat Stores and Affects Hunger and Energy Expenditure. In 1949, scientists at the Jackson Laboratory in Maine noted the appearance of "some very plump young mice." This obesity was due to a genetic mutation, which they named *obese* (*ob*). Sixteen years later, they identified another obesity mutation, *diabetes* (*db*). The extreme obesity of *ob/ob* and *db/db* mice results from intense hyperphagia and reduced brown fat thermogenesis. Based on a series of parabiosis experiments, Douglas Coleman proposed that *ob/ob* mice lack a circulating satiety factor and that *db/db* mice lack its receptor.

In a tour-de-force positional cloning effort led by Jeffrey Friedman and Rudolph Leibel, the *ob* gene was localized on a small region on chromosome 6. Friedman and his lab then homed in on and identified the *ob* gene. Renamed the *leptin* gene, it encodes leptin, a 167-amino acid protein secreted by adipocytes in proportion to the size of fat stores. Treating *ob/ob* mice with leptin cures their obesity. The *db* gene was identified a few years later, and as predicted by Coleman, it turned out to encode leptin's receptor and was found to be expressed by neurons in the hypothalamus. It is an interleukin-6–type class I cytokine receptor that produces its antiobesity effects by activating the JAK2/STAT3 signaling pathway.

Much has been learned subsequently about leptin. First, humans with a genetic deficiency of leptin or its receptor, like the mutant mice, are massively obese; hence, leptin's function is highly conserved, and such mutations are extremely rare. Second, humans with common forms of obesity have very high circulating levels of leptin, a by-product of their increased fat stores. This finding initially led to the view, later questioned, that "garden variety" obesity is caused by resistance to leptin action. Third, starvation, which reduces fat stores, drastically decreases leptin levels. This reduction is of interest because fasting causes many adaptive responses that are also seen in mice and humans that lack leptin: hunger, low energy expenditure, decreased fertility, and other altered neuroendocrine responses. Indeed, restoration of normal leptin levels in fasted individuals reverses or ameliorates many of fasting's effects. Thus, leptin's primary function is to signal, when its levels are low, that fat stores are inadequate.

These low leptin levels then bring about key adaptive responses such as increased hunger, decreased sympathetically mediated thermogenesis (to conserve limited fuel stores), decreased fertility (to prevent pregnancy when its demands cannot be met), and others. According to this view, leptin's dynamic range for signaling extends from the very low levels seen with fasting, signaling that fat stores are too low, to the levels found in well-fed, nonobese individuals, signaling that fat stores are sufficient. Levels above this may produce some effect to restrain obesity, but this effect, if present, is remarkably weak. Thus, the defense of energy balance is asymmetric—strong against low stores, weak against high stores. A corollary of this is that obese individuals do not have leptin resistance;

A Maintenance of energy balance

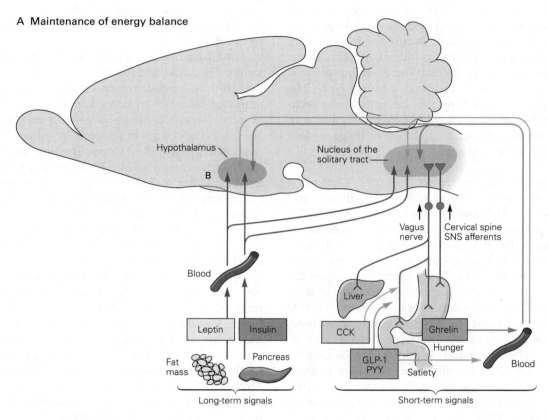

B

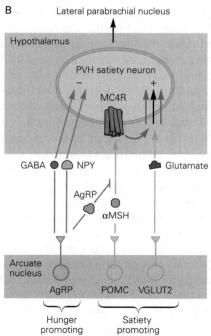

rather, they simply have leptin levels that exceed the maximally effective concentration.

POMC, AgRP, and MC4R Neurons Are Key Nodes in the Afferent/Efferent Loop

Neuron-specific manipulation technologies have revealed two antagonistic populations of neurons in the arcuate nucleus that control energy balance: one expresses agouti-related peptide (AgRP) and the other the precursor polypeptide proopiomelanocortin (POMC) (Figure 41–14B). POMC neurons decrease hunger and stimulate sympathetically driven energy expenditure; AgRP neurons do the opposite. POMC neurons release the processed peptide α-melanocyte-stimulating hormone (αMSH), which activates the melanocortin-4 receptor (MC4R), a G protein–coupled receptor.

The downstream MC4R-expressing neurons that control hunger lie within the PVH. When these MC4R neurons are excited by αMSH released from POMC afferents, hunger is decreased. The PVH-MC4R "satiety neurons" are glutamatergic; they decrease hunger via their excitatory projections to the lateral parabrachial nucleus.

The MC4R-expressing neurons that control energy expenditure are sympathetic preganglionic neurons in the spinal cord. POMC neurons project to these sites, in addition to the PVH, increasing sympathetically driven energy expenditure.

The AgRP neurons increase hunger in part by opposing the actions of POMC neurons (Figure 41–14B). They release three factors: AgRP, an inverse agonist of MC4R, and neuropeptide Y and γ-aminobutyric acid (GABA), two inhibitory transmitters. The AgRP neurons project to and inhibit the PVH-MC4R satiety neurons and directly inhibit POMC neurons. In addition, different subsets of arcuate AgRP neurons project to other sites, including the lateral hypothalamus and the bed nucleus of the stria terminalis. These sites, when inhibited by AgRP inputs, can also stimulate hunger.

A third group of neurons in the arcuate nucleus express VGLUT2, release glutamate, and act in parallel with POMC neurons to induce satiety (Figure 41–14B). Like POMC neurons, and opposite to AgRP neurons, they excite the PVH-MC4R satiety neurons. αMSH/MC4R signaling in PVH-MC4R neurons causes satiety by two mechanisms: by directly activating the PVH-MC4R satiety neurons and by upregulating excitatory transmission from the VGLUT2 neurons to the PVH-MC4R neurons via postsynaptic facilitation.

The importance of the POMC, AgRP, and PVH-MC4R satiety neurons in regulating food intake is supported by a number of compelling findings. First, fasting activates AgRP neurons and inhibits POMC neurons, while feeding or leptin treatment does the opposite. The downstream PVH-MC4R satiety neurons are inhibited by fasting and excited by feeding. Second, genetic deficiency of the POMC protein or MC4R causes massive obesity. Third, genetic ablation of AgRP neurons in mice causes starvation, while stimulation of AgRP neurons rapidly brings about extreme hyperphagia, even in mice that are calorically replete and otherwise sated. Finally, several findings implicate the PVH-MC4R satiety neurons as an important

Figure 41–14 (Opposite) Neural and endocrine components combine to regulate energy balance.

A. *Short-term signals.* During meals, cholecystokinin (CCK) from the intestinal tract stimulates sensory fibers of the vagus nerve, thus promoting satiation (meal termination). Glucagon-like peptide-1 (GLP-1) and peptide YY (PYY), also released by the intestinal tract, appear to work on both sensory fibers of the vagus and neurons in the brain. The vagal sensory fibers, along with sympathetic fibers from the gut and orosensory information, converge in the nucleus of the solitary tract (NTS). Prior to mealtime, release of ghrelin from the stomach peaks, providing a blood-borne signal to neurons in the brain. Whereas CCK promotes satiety, ghrelin promotes eating.

Long-term signals. Leptin and insulin are among the humoral signals that inform the brain about the status of the fat stores. Leptin is produced in fat-storing cells, whereas insulin is produced in the pancreas. Both hormones are sensed by receptors in the arcuate nucleus of the hypothalamus as well as by receptors in the NTS. Leptin and insulin reduce food intake and increase energy expenditure. (Abbreviation: **SNS**, sympathetic nervous system.)

B. Neurons in the arcuate nucleus that synthesize agouti-related peptide (**AgRP**), proopiomelanocortin (**POMC**), and vesicular glutamate transporter 2 (**VGLUT2**) project to the paraventricular nucleus of the hypothalamus (**PVH**) where they control hunger and satiety. Satiety-promoting POMC neurons release the processed POMC peptide, α-melanocyte stimulating hormone (αMSH), which binds to melanocortin-4 receptors (**MC4R**) on neurons in the PVH. Activation of these neurons causes satiety. In contrast, the hunger-promoting AgRP neurons release two inhibitory transmitters, γ-aminobutyric acid (**GABA**) and neuropeptide Y (**NPY**), and the MC4R antagonist AgRP. Their combined effect is to inhibit the MC4R-expressing neurons, causing hunger. The MC4R-expressing neurons also receive direct excitatory input from another population of arcuate neurons, VGLUT2 neurons, which also promote satiety. The binding of αMSH to MC4R causes satiety by two mechanisms: by directly activating the PVH-MC4R neurons and by upregulating excitatory transmission from the arcuate VGLUT2 neurons to the PVH-MC4R neurons (**blue arrows**). Finally, PVH-MC4R neurons project to the lateral parabrachial nucleus where they promote satiety.

downstream target of AgRP and POMC neurons. Most notable are the development of marked hyperphagia and obesity in mice genetically engineered to lack MC4R neurons in the PVH and the induction of intense feeding following optogenetic stimulation of AgRP terminals in the PVH, which inhibits the satiety neurons.

Surprisingly, environmental cues that predict future ingestion of food induce inhibition of AgRP neurons. Indeed, in fasted mice, which have high AgRP neuron activity, presentation of food alone without ingestion decreases AgRP neuron firing. This is roughly analogous to rapid, feedforward inhibition of vasopressin secretion and thirst neuron activity (mentioned previously). Thus, in addition to receiving strong bottom-up feedback signals from the body, the hunger-promoting AgRP neurons also receive top-down feedforward information from the environment. The function of this input is not yet clear, but it could serve as an anticipatory signal to limit future ingestion of excessive calories, or as discussed below, it could serve as a reward-related signal to motivate feeding.

Finally, the complete pathway accounting for regulation of hunger by the AgRP and POMC neuron → PVH pathway is presently unknown. It is likely that, via relays through a number of synapses, it affects neuronal activity in pathways controlling reward as well as perception. This is the case because, in the fasted state, food and cues predictive of food are both more rewarding and much more likely to become the focus of attention. How specificity for a given goal—in this case food—is retained as neural information flows from the highly specific deficiency-regulated homeostatic neurons in the hypothalamus to the "nonspecific" reward and perception pathways in the accumbens and cortex is one of the great mysteries of motivated behaviors such as hunger and thirst. The solution could provide clues for disorders of motivated behavior like drug addiction.

Psychological Concepts Are Used to Explain Motivational Drives Such as Hunger

In a simplified stimulus–response view of behavior, one might assume that neural detection of water or energy deficiency (the stimulus) is hardwired to motor pathways for drinking or eating (the response), and thus analogous to the knee-jerk stretch reflex (Chapter 3). However, this cannot be the case because the responses that can be employed to obtain food, all motivated by the deficiency stimulus, are remarkably varied and complex—to such a degree that they could not be hardwired. Indeed, animals can complete an infinite

number of complex operant learning tasks to obtain water or food rewards.

The challenge to understanding motivational drive is to devise a model that accounts for the ability of deprivation states to induce behavior that is remarkably varied and complex, while remaining completely specific for one goal. Two compelling theories are relevant. According to *incentive motivation theory*, deficiency increases the reward value of food and water. *Drive reduction theory* posits that deficiency generates an aversive state, the resolution of which is thought to motivate behavior. Notably, these two views are not mutually exclusive and may in some ways be two sides of the same coin.

Incentive Motivation: Fasting Increases the Reward Value of Food. The incentive motivation theory is the work of theorists over many years, most recently refined by Frederick Toates and Kent Berridge. Consider eating and the hunger drive. Briefly, food is viewed as inherently rewarding. Through learned associations, cues and tasks related to obtaining food also become rewarding; in this way, varied and complex behavioral responses are learned (Figure 41–15A).

The theory posits that the deprivation state increases the reward value of food and of the related cues and tasks (ie, their incentive salience). Thus, during fasting, the reward value is increased, and all food and cues and tasks related to food are extremely rewarding. After a meal, the reward value is decreased, and only the most inherently palatable foods, for example, ice cream, are still sufficiently rewarding to be eaten. The task of neuroscientists is to determine how deprivation increases reward value. Experimental activation of AgRP neurons in an otherwise sated mouse dramatically increases the reward value of food—remarkably, the reward value of food is increased to the same extremely high level seen with fasting.

Drive Reduction: Activity of AgRP Hunger Neurons Can Be Aversive. As we know from personal experience, the behavioral states created by dehydration and caloric deficiency, namely thirst and hunger, are unpleasant. It was originally proposed many years ago that reduction of these states, which relieves this discomfort, is rewarding and hence motivates drinking and feeding. Recently, Scott Sternson's group has provided compelling new support for a modified version of this model (Figure 41–15B). Using a behavioral conditioning paradigm such as the place preference test, they discovered that optogenetic activation of AgRP neurons in sated mice was aversive. When the same mice were then studied in a food-deprived state (which is associated with

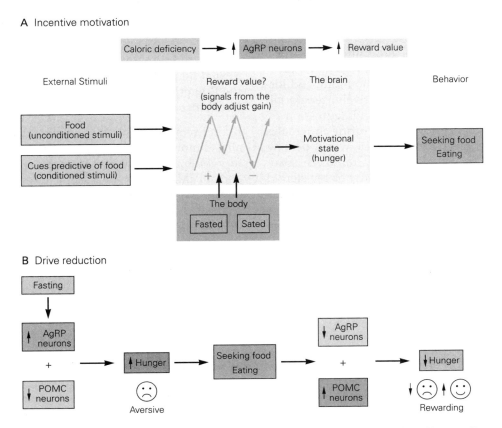

Figure 41–15 Two theories of how fasting promotes eating.

A. *Incentive motivation.* Food is inherently rewarding, and different foods have different reward values (lettuce–low, ice cream–high). Through learned associations, cues that predict food become rewarding. The fed (sated) versus fasted state sets the gain determining how rewarding food and food cues are. Fasting greatly increases while the sated state decreases reward value.

B. *Drive reduction.* The feeling of hunger is aversive. Eating reduces this aversive state. Consistent with this theory, experimental stimulation of hunger-promoting agouti-related peptide (**AgRP**) neurons is aversive, while stimulation of satiety-promoting paraventricular nucleus of the hypothalamus melanocortin-4 receptors (**MC4R**) neurons (which are downstream of the AgRP and proopiomelanocortin [**POMC**] neurons) in an otherwise hungry mouse creates a pleasant feeling.

increased AgRP neuron activity), they engaged in behaviors that in the earlier conditioning had lowered AgRP neuron activity—in short, they acted as if motivated to turn off the AgRP neuron–induced aversive state. Similar results were obtained with thirst neurons in the SFO.

In further support of this view, optogenetic activation of downstream PVH-MC4R satiety neurons in calorically deprived mice, but not in sated mice, is emotionally positive (ie, the mice like it). Thus, causing satiety when otherwise hungry is pleasant. In total, these findings provide strong evidence for the view that homeostatic deficiency is unpleasant, that the aversive state is caused by activation of deficiency-responsive homeostatic neurons, and that when afflicted by the deficiency-induced aversive state animals engage in behaviors that they associate with relief.

This model provides an explanation for why dieting is so difficult. It generates an aversive, unpleasant state that can only be relieved by eating. Finally, the rapid reduction in AgRP neuron activity in response to sensory cues that predict food, and the alleviation of the aversive state that this should cause, could function as a rewarding "teaching signal" that motivates pursuit of the goal (food).

Sexually Dimorphic Regions in the Hypothalamus Control Sex, Aggression, and Parenting

Now we turn to behaviors that are not homeostatic, but are controlled by the hypothalamus, involve integration between sensory cues and signals from the body (ie, gonadal steroids), and are critical for survival of the species.

Males and females differ in their sexual, aggressive, and parenting behaviors. These differences are especially notable in animals, for example in mice, where they are clearly hardwired (ie, require no prior training). These differences include mounting and lordosis by males and females, respectively; territorial-related behaviors such as marking and aggression by males; and the tendency toward nurturing in females versus aggressive behavior in males when dealing with the young. The latent capacity for these sexually dimorphic behaviors is the product of sex steroid action on the brain during embryogenesis (Chapter 51). Full actualization of adult sex-specific behaviors also requires adult levels of gonadal steroids. Sex chromosome–specific genes, other than *Sry*, which causes male sex determination, as well as genes that are imprinted in a sexually dimorphic way, also subtly modulate sex-specific behaviors independent of gonadal steroids. Ultimately, the behaviors themselves are triggered by cues from the environment, such as pheromones.

Two regions of the hypothalamus are critically involved in the control of these behaviors, the POA and the ventral lateral aspect of the ventromedial hypothalamic nucleus (vlVMH). Both sites are sexually dimorphic: The POA contains more neurons in males, and the vlVMH contains more progesterone-expressing neurons in females. These sites are heavily interconnected, and they receive strong input from two other sexually dimorphic areas outside the hypothalamus: the medial division of the posteromedial bed nucleus of the stria terminalis (BNSTmpm) and medial amygdala (MeA).

Sexual Behavior and Aggression Are Controlled by the Preoptic Hypothalamic Area and a Subarea of the Ventromedial Hypothalamic Nucleus

Brain lesion studies have demonstrated that the sexually dimorphic brain regions—the accessory olfactory bulb, BNSTmpm, MeA, and particularly POA and vlVMH—play important roles in sex-specific behaviors. Neurons in these regions are highly interconnected, are downstream of pathways involved in detecting and responding to pheromones (BNSTmpm and MeA), express sex hormone receptors, and, with the exception of neurons in the vlVMH, also express aromatase (Figure 41–16). Neurons in both the POA and vlVMH send strong projections to the lateral periaqueductal gray area, which is thought to mediate and coordinate the motor and autonomic aspects of sexual and aggressive behavior.

The vlVMH plays a critical role in controlling sexually dimorphic behaviors. Firing rates of vlVMH neurons in male mice increase during mating or periods of aggression toward a male intruder. Stimulation of these neurons triggers intense attack behavior toward intruder males and toward atypical targets for male aggression such as castrated males, females, or even rubber gloves! Silencing these neurons eliminates aggressive behavior toward male intruders.

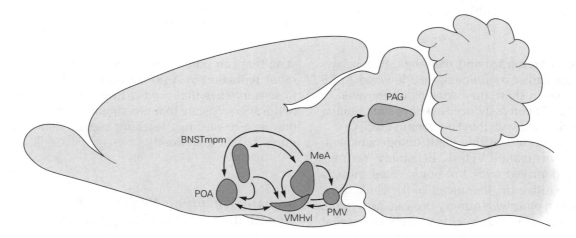

Figure 41–16 Sexually dimorphic neural structures comprise highly interconnected behavioral circuits. Hypothalamic and amygdalar nuclei that regulate sexually dimorphic behaviors are extensively interconnected. These areas process pheromonal information, and subsets of adult neurons within each of these regions express sex hormone receptors; neurons within some of these regions (**blue**) also express aromatase. (Abbreviations: **BNSTmpm**, medial division of the posteromedial bed nucleus of the stria terminalis; **MeA**, medial amygdala; **PAG**, periaqueductal gray; **PMV**, ventral premammillary nucleus; **POA**, preoptic hypothalamus; **VMHvl**, ventrolateral component of the ventromedial hypothalamus.) (Reproduced, with permission, from Yang and Shah 2014. Copyright © 2014 Elsevier Inc.)

Furthermore, stimulation of a subset of vlVMH neurons that express estrogen receptor evokes either sexual behavior (mounting) or aggression, depending on the number of neurons activated and their degree of activation: Lower levels of activation induce mounting, whereas higher levels induce aggression. Consistent with these results, genetic ablation of related progesterone receptor–expressing vlVMH neurons causes loss of both sexual behavior and aggression in males and loss of sexual behavior in females. Thus, it is clear that sex neurons in the vlVMH-expressing steroid receptors play a critical role in driving sexual behavior in males and females and aggression in males.

Parental Behavior Is Controlled by the Preoptic Hypothalamic Area

Nurturing parental behavior is key to the survival of one's offspring. Male rodents demonstrate strikingly different patterns of behavior. Males can be nurturing or hostile to offspring, even to the point of infanticide, depending on whether they view the offspring as their own or those of another male. Female mice, on the other hand, are generally nurturing.

Social interaction between mouse pups and appropriately receptive adult female and male mice, but not hostile adult male mice, induces activity in subsets of galanin-expressing neurons in the POA. These offspring-activated neurons are largely distinct from POA neurons activated by mating. Genetic ablation of galanin-expressing POA neurons prevents nurturing parental behavior, even to the point of inducing uncharacteristic aggression in females toward their offspring. On the other hand, stimulation of these galanin-positive neurons in males, which are normally extremely hostile to unrelated pups, decreases aggression and induces nurturing pup grooming. Thus, neurons in the POA, in addition to controlling sexual behavior itself, also play a role in ensuring survival of the fruit of sexual behavior.

Highlights

1. The hypothalamus and the autonomic and neuroendocrine motor systems coordinate and control body homeostasis by inducing adaptive behaviors; by controlling glands, smooth muscle, cardiac muscle, and adipocytes; and by releasing hormones from the pituitary gland.

2. Homeostatic control of body temperature, fluid and electrolyte balance, and blood pressure allows organisms to function under harsh environmental conditions.

3. Feedback loops that sense temperature, osmolarity, blood pressure, and body fat are essential for homeostatic control. The combined action of multiple feedback-informed sensory-afferent/efferent-effector control loops results in emergent settling points.

4. Modality-specific hypothalamic neurons link specific interoceptive sensory feedback with outputs that control adaptive behaviors and physiologic responses. In addition to feedback, these modality-specific neurons also receive feedforward information regarding future anticipated homeostatic challenges.

5. The autonomic motor system contains neurons located in sympathetic, parasympathetic, and enteric ganglia located near the spinal column or embedded within peripheral targets. Functional subsets of autonomic neurons selectively innervate effector tissues comprised of smooth muscle, cardiac muscle, glandular epithelia, and adipocytes.

6. Sympathetic neurons are activated in response to exercise and stress. Parasympathetic and sympathetic neurons generally have antagonistic functions, but often act in concert. The enteric system coordinates peristaltic contractions of the gastrointestinal tract with mucosal function and local blood flow.

7. Preganglionic neurons that control the sympathetic and parasympathetic outflows are located in the spinal cord and brain stem.

8. Acetylcholine, norepinephrine, and neuropeptide cotransmitters act as synaptic signaling molecules in the autonomic motor system. Excitatory fast synaptic transmission in autonomic ganglia is mediated by acetylcholine acting on nicotinic receptors. G protein–coupled receptors in ganglia mediate additional pre- and postsynaptic excitatory and inhibitory effects. G protein–coupled receptors mediate transmitter actions at autonomic neuroeffector junctions.

9. The neuroendocrine system links the hypothalamus, via the pituitary gland, to various physiologic responses in the body. The posterior pituitary contains hypothalamic axon terminals that release two neurohormones into the blood: Vasopressin stimulates water reabsorption by the kidney, while oxytocin controls uterine contraction and milk release. The anterior pituitary contains endocrine cells that secrete hormones in response to factors released by hypothalamic neurons. These anterior pituitary hormones control the thyroid gland, glucocorticoid secretion

by the adrenal cortex, sex steroid secretion by the gonads, lactation, and linear growth.

10. Body temperature is detected in multiple sites including the periphery, in and around major organs, and in the brain. Constancy of body temperature is maintained by multiple thermo-receptor-afferent/thermoeffector-efferent control loops.

11. Some neurons in the lamina terminalis are activated by both dehydration and loss of intravascular volume. Key parameters sensed in these deficiency states include osmolarity and angiotensin II levels, respectively. When these neurons are activated, they cause thirst and release of vasopressin from the posterior pituitary. Vasopressin release is also rapidly regulated in a feedforward fashion by cues that anticipate future disturbances in osmolarity.

12. Energy balance involves short-term and long-term feedback signals. Short-term signals from the gut mediate satiation, which terminates meals. CCK, released by intestinal endocrine cells, plays a key role in satiation. A key long-term signal is leptin, which is secreted by adipocytes in proportion to the amount of fat stores. When fat stores are low, the consequent low levels of leptin signal the brain to induce a hunger state and to decrease energy expenditure, resulting in replenished fat stores.

13. Leptin is more effective in defending against low fat stores than in resisting obesity.

14. POMC-, AgRP-, and MC4R-expressing neurons in the hypothalamus are key nodes in the afferent/efferent loop controlling energy balance. Neurons that signal satiety are activated by satiety-promoting POMC neurons and inhibited by hunger-promoting AgRP neurons.

15. How specificity for a given goal (eg, food) is retained as neural information flows from the highly specific deficiency-regulated homeostatic neurons in the hypothalamus to the "nonspecific" reward and perception pathways in the accumbens and cortex is one of the great mysteries of motivated behaviors such as hunger and thirst. Solving this could provide clues for disorders of motivated behavior like drug addiction.

16. Leptin regulates hunger and energy expenditure in part by activating POMC neurons and inhibiting AgRP neurons. Hunger-promoting AgRP neurons are also rapidly regulated in a feedforward fashion by cues that anticipate future changes in energy balance.

17. Motivational drives such as hunger have been explained by two mechanisms: The deficiency state (starvation) increases the reward value of food, or deficiency generates an aversive state, the resolution of which motivates behavior.

18. Sexually dimorphic regions in the hypothalamus control sexual behavior and aggression. Neural activity in the sexually dimorphic preoptic area controls parental behavior. Full actualization of adult sex-specific behaviors also requires adult levels of gonadal steroids.

Bradford B. Lowell
Larry W. Swanson
John P. Horn

Selected Reading

Andermann ML, Lowell BB. 2017. Towards a wiring diagram understanding of appetite control. Neuron 95:757–778.

Berridge KC. 2004. Motivation concepts in behavioral neuroscience. Physiol Behav 81:179–209.

Bourque CW. 2008. Central mechanisms of osmosensation and systemic osmoregulation. Nat Rev Neurosci 9:519–531.

Clarke IJ. 2015. Hypothalamus as an endocrine organ. Compr Physiol 5:217–253.

Dulac C, O'Connell LA, Wu Z. 2014. Neural control of maternal and paternal behaviors. Science 345:765–770.

Guyenet PG. 2006. The sympathetic control of blood pressure. Nat Rev Neurosci 7:335–346.

Jænig, W. 2006. *The Integrative Action of the Autonomic Nervous System.* Cambridge, England: Cambridge Univ. Press.

Leib DE, Zimmerman CA, Knight ZA. 2016. Thirst. Curr Biol 26:R1260–R1265.

Morrison SF. 2016. Central control of body temperature. F1000Res 5:F1000.

Morton GJ, Meek TH, Schwartz MW. 2014. Neurobiology of food intake in health and disease. Nat Rev Neurosci 15:367–378.

Romanovsky AA. 2007. Thermoregulation: some concepts have changed. Functional architecture of the thermoregulatory system. Am J Physiol Regul Integr Comp Physiol 292:R37–R46.

Rosenbaum M, Leibel RL. 2014. 20 years of leptin: role of leptin in energy homeostasis in humans. J Endocrinol 223:T83–T96.

Yang CF, Shah NM. 2014. Representing sex in the brain, one module at a time. Neuron 82:261–278.

References

Ahima RS, Prabakaran D, Mantzoros C, et al. 1996. Role of leptin in the neuroendocrine response to fasting. Nature 382:250–252.

Aponte Y, Atasoy D, Sternson SM. 2011. AGRP neurons are sufficient to orchestrate feeding behavior rapidly and without training. Nat Neurosci 14:351–355.

Balthasar N, Dalgaard LT, Lee CE, et al. 2005. Divergence of melanocortin pathways in the control of food intake and energy expenditure. Cell 123:493–505.

Berthoud HR, Neuhuber WL. 2000. Functional and chemical anatomy of the afferent vagal system. Auton Neurosci 85:1–17.

Betley JN, Xu S, Cao ZF, et al. 2015. Neurons for hunger and thirst transmit a negative-valence teaching signal. Nature 521:180–185.

Brookes SJ, Spencer NJ, Costa M, Zagorodnyuk VP. 2013. Extrinsic primary afferent signalling in the gut. Nat Rev Gastroenterol Hepatol 10:286–296.

Burnstock G. 2013. Cotransmission in the autonomic nervous system. Handb Clin Neurol 117:23–35.

Campos CA, Bowen AJ, Schwartz MW, Palmiter RD. 2016. Parabrachial CGRP neurons control meal termination. Cell Metab 23:811–820.

Chambers AP, Sandoval DA, Seeley RJ. 2013. Integration of satiety signals by the central nervous system. Curr Biol 23:R379–R388.

Chen Y, Lin YC, Kuo TW, Knight ZA. 2015. Sensory detection of food rapidly modulates arcuate feeding circuits. Cell 160:829–841.

Coleman DL. 2010. A historical perspective on leptin. Nat Med 16:1097–1099.

DeGroat WC, Booth AM, Yoshimura N. 1993. Neurophysiology of micturition and its modification in animal models of human disease. In: CA Maggi (ed). *Nervous Control of the Urogenital System*, pp. 227–348. Chur, Switzerland: Harwood Academic Publishers.

Fenselau H, Campbell JN, Verstegen AM, et al. 2017. A rapidly acting glutamatergic ARC→PVH satiety circuit postsynaptically regulated by α-MSH. Nat Neurosci 20: 42–51.

Furness JB. 2012. The enteric nervous system and neurogastroenterology. Nat Rev Gastroenterol Hepatol 9: 286–294.

Furness JB, Costa M. 1980. Types of nerves in the enteric nervous system. Neurosci 5:1–20.

Furness JB, Jones C, Nurgali K, Clerc N. 2004. Intrinsic primary afferent neurons and nerve circuits within the intestine. Prog Neurobiol 72:143–164.

Garfield AS, Li C, Madara JC, et al. 2015. A neural basis for melanocortin-4 receptor-regulated appetite. Nat Neurosci 18:863–871.

Gay VL. 1972. The hypothalamus: physiology and clinical use of releasing factors. Fertil Steril 23:50–63.

Gibbins IL. 1995. Chemical neuroanatomy of sympathetic ganglia. In: E McLachlan (ed). *Autonomic Ganglia*, pp. 73–122. Luxembourg: Harwood Academic Publishers.

Huszar D, Lynch CA, Fairchild-Huntress V, et al. 1997. Targeted disruption of the melanocortin-4 receptor results in obesity in mice. Cell 88:131–141.

Ingalls AM, Dickie MM, Snell GD. 1950. Obese, a new mutation in the house mouse. J Hered 41:317–318.

Krashes MJ, Koda S, Ye C, et al. 2011. Rapid, reversible activation of AgRP neurons drives feeding behavior in mice. J Clin Invest 121:1424–1428.

Krashes MJ, Lowell BB, Garfield AS. 2016. Melanocortin-4 receptor-regulated energy homeostasis. Nat Neurosci 19:206–219.

Lechner SG, Markworth S, Poole K, et al. 2011. The molecular and cellular identity of peripheral osmoreceptors. Neuron 69:332–344.

Lee H, Kim DW, Remedios R, et al. 2014. Scalable control of mounting and attack by Esr1+ neurons in the ventromedial hypothalamus. Nature 509:627–632.

Lin D, Boyle MP, Dollar P, et al. 2011. Functional identification of an aggression locus in the mouse hypothalamus. Nature 470:221–226.

Locke AE, Kahali B, Berndt SI, et al. 2015. Genetic studies of body mass index yield new insights for obesity biology. Nature 518:197–206.

Lowell BB, Spiegelman BM. 2000. Towards a molecular understanding of adaptive thermogenesis. Nature 404:652–660.

Luquet S, Perez FA, Hnasko TS, Palmiter RD. 2005. NPY/AgRP neurons are essential for feeding in adult mice but can be ablated in neonates. Science 310:683–685.

Mandelblat-Cerf Y, Kim A, Burgess CR, et al. 2017. Bidirectional anticipation of future osmotic challenges by vasopressin neurons. Neuron 93:57–65.

Mandelblat-Cerf Y, Ramesh RN, Burgess CR, et al. 2015. Arcuate hypothalamic AgRP and putative POMC neurons show opposite changes in spiking across multiple timescales. Elife 4:e07122.

McKinley MJ, Yao ST, Uschakov A, McAllen RM, Rundgren M, Martelli D. 2015. The median preoptic nucleus: front and centre for the regulation of body fluid, sodium, temperature, sleep and cardiovascular homeostasis. Acta Physiol (Oxf) 214:8–32.

Mountjoy KG, Robbins LS, Mortrud MT, Cone RD. 1992. The cloning of a family of genes that encode the melanocortin receptors. Science 257:1248–1251.

Myers MG Jr, Leibel RL, Seeley RJ, Schwartz MW. 2010. Obesity and leptin resistance: distinguishing cause from effect. Trends Endocrinol Metab 21:643–651.

Nakamura K, Morrison SF. 2011. Central efferent pathways for cold-defensive and febrile shivering. J Physiol 589:3641–3658.

Oka Y, Ye M, Zuker CS. 2015. Thirst driving and suppressing signals encoded by distinct neural populations in the brain. Nature 520:349–352.

Powley TL, Phillips RJ. 2004. Gastric satiation is volumetric, intestinal satiation is nutritive. Physiol Behav 82:69–74.

Reichlin S. 1978. The hypothalamus: introduction. Res Publ Assoc Res Nerv Ment Dis 56:1–14.

Romanovsky AA. 2014. Skin temperature: its role in thermoregulation. Acta Physiol (Oxf) 210:498–507.

Rossi J, Balthasar N, Olson D, et al. 2011. Melanocortin-4 receptors expressed by cholinergic neurons regulate energy balance and glucose homeostasis. Cell Metab 13:195–204.

Saper CB. 2002. The central autonomic nervous system: conscious visceral perception and autonomic pattern generation. Annu Rev Neurosci 25:433–469.

Saper CB, Romanovsky AA, Scammell TE. 2012. Neural circuitry engaged by prostaglandins during the sickness syndrome. Nat Neurosci 15:1088–1095.

Shah BP, Vong L, Olson DP, et al. 2014. MC4R-expressing glutamatergic neurons in the paraventricular hypothalamus regulate feeding and are synaptically connected to the parabrachial nucleus. Proc Natl Acad Sci U S A 111:13193–13198.

Song K, Wang H, Kamm GB, et al. 2016. The TRPM2 channel is a hypothalamic heat sensor that limits fever and can drive hypothermia. Science 353:1393–1398.

Speakman JR, Levitsky DA, Allison DB, et al. 2011. Set points, settling points and some alternative models: theoretical options to understand how genes and environments combine to regulate body adiposity. Dis Model Mech 4:733–745.

Stricker EM, Hoffmann ML. 2007. Presystemic signals in the control of thirst, salt appetite, and vasopressin secretion. Physiol Behav 91:404–412.

Swanson LW. 2000. Cerebral hemisphere regulation of motivated behavior. Brain Res 886:113–164.

Tan CH, McNaughton PA. 2016. The TRPM2 ion channel is required for sensitivity to warmth. Nature 536:460–463.

Tan CL, Cooke EK, Leib DE, et al. 2016. Warm-sensitive neurons that control body temperature. Cell 167:47–59 e15.

Tanaka M, Owens NC, Nagashima K, Kanosue K, McAllen RM. 2006. Reflex activation of rat fusimotor neurons by body surface cooling, and its dependence on the medullary raphe. J Physiol 572:569–583.

Toates F 1986. Motivational Systems. New York: Cambridge University Press.

Williams EK, Chang RB, Strochlic DE, Umans BD, Lowell BB, Liberles SD. 2016. Sensory neurons that detect stretch and nutrients in the digestive system. Cell 166:209–221.

Wong LC, Wang L, D'Amour JA, et al. 2016. Effective modulation of male aggression through lateral septum to medial hypothalamus projection. Curr Biol 26:593–604.

Wu Z, Autry AE, Bergan JF, Watabe-Uchida M, Dulac CG. 2014. Galanin neurons in the medial preoptic area govern parental behaviour. Nature 509:325–330.

Yang CF, Chiang MC, Gray DC, et al. 2013. Sexually dimorphic neurons in the ventromedial hypothalamus govern mating in both sexes and aggression in males. Cell 153:896–909.

Zhang Y, Proenca R, Maffei M, Barone M, Leopold L, Friedman JM. 1994. Positional cloning of the mouse obese gene and its human homologue. Nature 372:425–432.

Zimmerman CA, Lin YC, Leib DE, et al. 2016. Thirst neurons anticipate the homeostatic consequences of eating and drinking. Nature 537:680–684.

42

Emotion

ELATION, COMPASSION, SADNESS, FEAR, and anger are commonly considered examples of emotions. These states have an enormous impact on our behavior and well-being. But what exactly is an emotion? Distinguishing different emotion states is difficult and requires an account of the environmentally or internally generated challenge an organism faces as well as its physiological responses. For example, before we can conclude that a rat is in a state of fear, we need to know that the rat is evaluating a specific threatening stimulus (a predator in its environment) and is mounting an adaptive response, such as high arousal and freezing.

Emotions are often represented along two dimensions: valence (ie, pleasantness to unpleasantness) and intensity (ie, low to high arousal), called "core affect" in many psychological theories. However, emotions can also be grouped into categories, such as categories of basic emotions (happiness, fear, anger, disgust, sadness) and categories of more complex emotions that help regulate social or moral behaviors (eg, shame, guilt, embarrassment, pride, jealousy). There is considerable debate about whether all the categories that are in common usage (like the ones just mentioned) will correspond to scientifically useful categories in a future neuroscience of emotion.

Within experimental contexts, the term *emotion* is used in several different ways, often related to the ways in which emotion is measured (Box 42–1). In everyday conversation, most people use the term "emotion" synonymously with "conscious experience of emotion" or "feeling," and most psychological studies in humans have focused on this sense of "emotion" as well. Most research in animals has focused instead on specific behavioral or physiological responses, in good part because it is impossible to obtain verbal reports in animal studies. Yet emotions have been conserved throughout the evolution of species, as Charles Darwin first observed in his seminal book, *The Expression of the Emotions in Man and Animals.* The empirical approach we describe in this chapter thus considers emotions as central brain states that can be studied in humans as well as many other animals, provided that we distinguish between emotions and feelings.

Emotion states typically cause a broad range of physiological responses that occur when the brain

Box 42–1 Ways of Measuring Emotion

Measures Commonly Used in Humans

Psychophysiology. Psychophysiology uses several measures to assay the physiological parameters associated with emotional states. These measures include autonomic responses (Chapter 41) as well as some somatic responses. The most commonly used measure is the galvanic skin response (also known as the skin conductance response), a measure of sympathetic autonomic arousal derived from the sweatiness of the palms of the hands. Other measures include heart rate, heart rate variability, blood pressure, respiration, pupil dilation, facial electromyography (EMG), and the startle response (see below). Some of these measures mostly correlate with basic dimensions of emotion, such as valence (eg, the

Table 42–1 Common Questionnaires Used to Assess Fear in Human Emotion Studies

Questionnaire	Type of fear questions
Fear Survey Schedule II	Probes an individual's level of fear across a range of different objects and situations that commonly evoke fear
Fear of Negative Evaluation Scale	Measures fear of being evaluated negatively by others
Social Avoidance and Distress Scale	Measures fear of social situations
Anxiety Sensitivity Index	Measures fear of experiencing different bodily sensations and feelings
Beck Anxiety Inventory	Measures fear and panic-related symptoms experienced over the prior week
Albany Panic and Phobia Questionnaire	Has the subject estimate the amount of fear they would experience in different situations
Fear Questionnaire	Measures the degree of avoidance due to fear
PANAS-X Fear (general)	Measures how much, in general, a person feels fear-related affective states
PANAS-X Fear (moment)	Measures how much, during the present moment, a person feels fear-related affective states

PANAS, Positive and Negative Affect Schedule.

detects certain environmental situations. These physiological responses are relatively automatic, yet depend on context, and occur within the brain as well as throughout the body. In the brain, they involve changes in arousal levels and in cognitive functions such as attention, memory processing, and decision making. Somatic responses involve endocrine, autonomic, and musculoskeletal systems (Chapter 41). In sum, emotions are neurobiological states that cause coordinated behavioral and cognitive responses triggered by the brain. This can occur when an individual detects a significant stimulus (positively or negatively charged) or has a specific thought or memory that leads to an endogenously generated emotion state.

Some stimuli—objects, animals, or situations—trigger emotions without the organism having to learn anything about those stimuli. Such stimuli have innately reinforcing qualities and are called unconditioned stimuli; examples are a painful shock or a disgusting taste. However, the vast majority of stimuli acquire their emotional significance through associative learning.

When an individual detects an emotionally significant stimulus, three physiological systems are engaged: the endocrine glands, the autonomic motor system, and the musculoskeletal system (Figure 42–1). The endocrine system is responsible for the secretion and regulation of hormones into the bloodstream that affect bodily tissues and the brain. The autonomic system mediates changes in the various physiological control systems of the body: the cardiovascular system, the visceral organs, and the tissues in the body cavity (Chapter 41). The skeletal motor system mediates

magnitude of the startle response) or arousal (eg, the galvanic skin response), whereas others (eg, facial EMG) can provide more fine-grained information about emotions. Facial expression has been used extensively but has no simple relationship to specific emotions.

Subjective ratings. Subjective ratings are often used in human studies and include categorical and continuous ratings (Table 42–1). These ratings can range along emotion dimensions, such as valence (pleasantness/unpleasantness), or the intensity of specific emotions. Subjective ratings necessarily depend on culture-specific words and concepts for emotions.

Experience sampling. Psychologists use experience sampling to quantify the emotions that people actually experience in everyday life. Participants might have their cell phone sound an alarm every few hours, and they then have to stop whatever activity they are doing and fill out a brief questionnaire about what they are feeling at the moment. In this way, a plot of the data can characterize how people's emotions change throughout the day or over longer periods. It turns out that we are actually fairly good at predicting what emotion people will feel next, from knowing how they currently feel.

Hormonal measures. Hormonal responses to emotional states are typically slower than psychophysiological measures. Emotion researchers measure a variety of hormones to assay emotional states over these lengthy periods. Relatively undifferentiated arousal responses are used to evaluate stress. The stress hormone cortisol (Chapter 61) is easily measured from people's saliva.

Specific experimental probes. Several specific behavioral and physiological assays are used to probe emotions with specific stimuli. These assays generally fall within the field of psychophysiology. A common measure is the amplitude of a subject's eyeblink (or other startle reflexes) when a loud sound is presented. This is potentiated when the subject is in a negatively valenced emotional state. Potentiation of the startle reflex is often used to assay the level of anxiety in people, and the same measure has also been validated in animals.

Measures Commonly Used in Nonhuman Animals

Innate behavioral responses. Animals often exhibit stereotyped behaviors as a consequence of certain emotional states. Observing and scoring the behavior is one method of measuring emotional behaviors. Such behaviors can include approaching a stimulus that is rewarding or that promises reward in the future (a positively valenced emotional state), as well as avoiding or defending against threatening stimuli (a negatively valenced emotional state). In addition, analysis of facial expressions can be utilized in many animal model systems, and has even been used for mice.

Psychophysiology and specific experimental probes. As in the case of humans, animal studies can use several psychophysiological measures (eg, heart rate, respiratory rate, galvanic skin response, pupil diameter, startle). In addition, specific behavioral assays have been developed in animals, often derived from initial observations of their innate behavioral responses. Behaviors such as freezing, attacking, exploring, approaching, and hiding can be measured in response to well-controlled experimental stimuli that are designed to induce certain emotional states. The correspondence between human and animal behaviors, which Charles Darwin originally noted in his 1872 book *The Expression of the Emotions in Man and Animals*, provides powerful animal models for investigating human emotions and their pathology.

overt behaviors such as freezing, fight-or-flight, and particular facial expressions. Together, these three systems control the physiological expression of emotion states in the body.

We begin this chapter with a discussion of the historical antecedents of modern research on the neuroscience of emotion. We then describe the neural circuits and cellular mechanisms that underlie the most thoroughly studied emotion, fear, and in so doing, we will focus on the amygdala. However, it is important to note that there does not appear to be any single brain structure that participates in only one emotion. For instance, the amygdala, which has been known to participate in negatively valenced emotions, also plays a central role in positively valenced emotions: Distinct populations of neurons within the amygdala process positively valenced versus negatively valenced stimuli. We briefly review how emotion states can be changed through extinction and regulation and how emotion interacts with other cognitive processes. We conclude with a survey on the relevance of emotion research to understanding psychiatric disorders.

The Modern Search for the Neural Circuitry of Emotion Began in the Late 19th Century

The modern attempt to understand emotions began in 1890 when William James, the founder of American psychology, asked: What is the nature of fear? Do we run from the bear because we are afraid, or are we afraid because we run? James proposed that the conscious

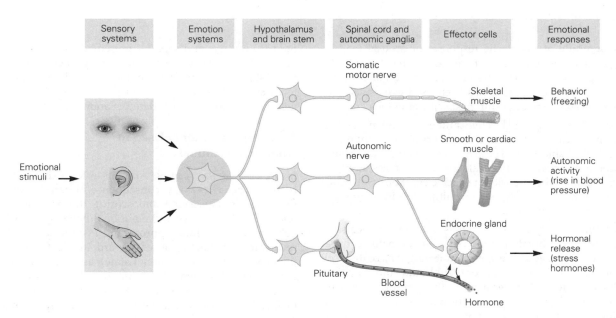

Figure 42–1 Neural control of emotional responses to external stimuli. External stimuli processed by sensory systems converge on "emotion systems" (eg, the amygdala). If the stimuli are emotionally salient, the emotion systems are activated, and their outputs are relayed to hypothalamic and brain stem regions that control physiological responses, including skeletomuscular action, autonomic nervous system activity, and hormonal release. The figure shows some responses associated with fear. It omits many of the complexities of emotion (eg, the effects of emotion states on cognition).

feeling of fear is a consequence of the bodily changes that occur during the act of running away—we feel afraid because we run. James's *peripheral feedback theory* drew on the knowledge of the brain at the time, namely, that the cortex had areas devoted to movement and sensation (Figure 42–2). Little was known at that time about specific areas of the brain responsible for emotion and feeling, but James's view is still debated to this day.

At the turn of the 20th century, researchers found that animals were still capable of emotional responses after the complete removal of the cerebral hemispheres, demonstrating that some aspects of emotion are mediated by subcortical regions. The fact that electrical stimulation of the hypothalamus could elicit autonomic responses similar to those that occur during emotional responses in an intact animal suggested to Walter B. Cannon that the hypothalamus might be a key region in the control of fight-or-flight responses and other emotions.

In the 1920s, Cannon showed that transection of the brain above the level of the hypothalamus (by means of a cut that separates the cortex, thalamus, and anterior hypothalamus from the posterior hypothalamus and lower brain areas) left an animal still capable of showing rage. By contrast, a transection below the hypothalamus, which left only the brain stem and

the spinal cord, eliminated the coordinated reactions of natural rage. This clearly implicated the hypothalamus in organizing emotional reactions. Cannon called the hypothalamically mediated reactions "sham rage" because these animals lacked input from cortical areas, which he assumed were critical for the emotional experience of "real" rage (Figure 42–3).

Cannon and his student Phillip Bard proposed an influential theory of emotion centered on the hypothalamus and thalamus. According to their theory, sensory information processed in the thalamus is sent both to the hypothalamus and to the cerebral cortex. The projections to the hypothalamus were thought to produce emotional responses (through connections to the brain stem and spinal cord), while the projections to the cerebral cortex were thought to produce conscious feelings (Figure 42–2). This theory implied that the hypothalamus is responsible for the brain's evaluation of the emotional significance of external stimuli and that emotional reactions depend on this appraisal.

In 1937, James Papez extended the Cannon-Bard theory. Like Cannon and Bard, Papez proposed that sensory information from the thalamus is sent to both the hypothalamus and the cerebral cortex. The descending connections to the brain stem and spinal cord give rise to emotional responses, and the ascending connections to the cerebral cortex give rise to feelings. But Papez

Peripheral feedback theory

Central theory

Papez circuit

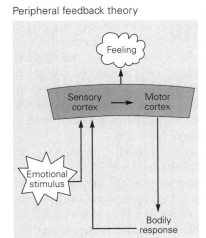
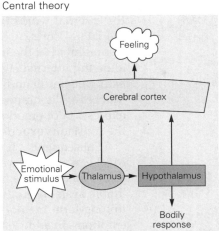
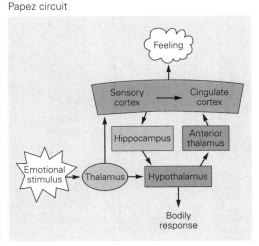

Figure 42–2 Early theories of the emotional brain. (Adapted, with permission, from LeDoux 1996.)

William James's peripheral feedback theory. James proposed that information about emotionally competent stimuli is processed in sensory systems and transmitted to the motor cortex to produce responses in the body. Feedback signals to the cortex convey sensory information about the body responses. The cortical processing of this sensory feedback is the "feeling," according to James.

The Cannon-Bard central theory. Walter Cannon and Philip Bard proposed that emotions are explained by processes within the central nervous system. In their model, sensory information

is transmitted to the thalamus where it is then relayed to both the hypothalamus and the cerebral cortex. The hypothalamus evaluates the emotional qualities of the stimulus, and its descending connections to the brain stem and spinal cord give rise to somatic responses, while the thalamocortical pathways give rise to conscious feelings.

The Papez circuit. James Papez refined the Cannon-Bard theory by adding additional anatomical specificity. He proposed that the cingulate cortex is the cortical region that receives hypothalamic output in the creation of feelings. The outputs of the hypothalamus reach the cingulate via the anterior thalamus, and the outputs of the cingulate reach the hypothalamus via the hippocampus.

went on to expand the neural circuitry of feelings considerably beyond the Cannon-Bard theory by interposing a new set of structures between the hypothalamus and the cerebral cortex. He argued that signals from the hypothalamus go first to the anterior thalamus and then to the cingulate cortex, where signals from the hypothalamus and sensory cortex converge. This convergence accounts for the conscious experience of feeling in Papez's theory. The sensory cortex then projects to both the cingulate cortex and the hippocampus, which in turn makes connections with the mammillary bodies of the hypothalamus, thus completing the loop (Figure 42–2).

The hypothalamus is currently receiving intense interest in studies of emotion in animals, particularly in experiments using optogenetics to manipulate the activity of precise cell populations. These studies have shown that specific populations in the mouse ventromedial hypothalamus are necessary and sufficient for defensive emotion states. Thus the hypothalamus does not merely orchestrate emotional behaviors, but is part of the neural circuitry that constitutes the emotion state itself. The role of the hypothalamus in emotion is much less studied in humans, in part because functional

magnetic resonance imaging (fMRI) does not have the spatial resolution to investigate specific hypothalamic nuclei, let alone neuronal subpopulations within them.

In the late 1930s, Heinrich Klüver and Paul Bucy removed the temporal lobes of monkeys bilaterally, thus lesioning all temporal cortex as well as subcortical structures like the amygdala and hippocampus, and found a variety of psychological disturbances, including alterations in feeding habits (the monkeys put inedible objects in their mouth) and sexual behavior (they attempted to have sex with inappropriate partners, like members of other species). In addition, the monkeys had a striking lack of concern for previously feared objects (eg, humans and snakes). This remarkable set of findings came to be known as the Klüver-Bucy syndrome and already suggested that the amygdala might be important for emotion (although it was not the only structure lesioned in these experiments).

Building on the Cannon-Bard and Papez models and the findings of Klüver and Bucy, Paul MacLean suggested in 1950 that emotion is the product of the "visceral brain." According to MacLean, the visceral brain includes the various cortical areas that had long been referred to as the limbic lobe, so named by Paul

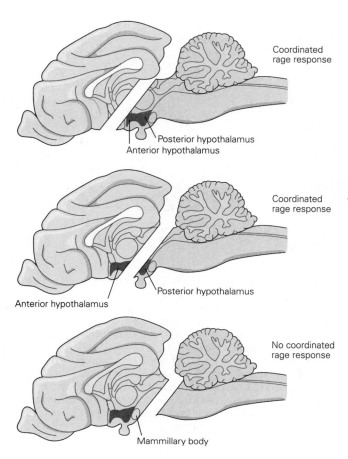

Figure 42–3 Sham rage. An animal exhibits sham rage following transection of the forebrain and the disconnection of everything above the transection (**top**) or transection at the level of the anterior hypothalamus and the disconnection of everything above it (**middle**). Only isolated elements of rage can be elicited if the posterior hypothalamus also is disconnected (**bottom**). This work derives from historical lesion studies in animals. More recent work suggests a more complex picture, in which the hypothalamus is intimately involved in creating the emotion state itself, not merely its behavioral expression.

Broca because these areas form a rim (Latin *limbus*) in the medial wall of the hemispheres. The visceral brain was later renamed the *limbic system*. The limbic system includes the various cortical areas that make up Broca's limbic lobe (especially medial areas of the temporal and frontal lobes) and the subcortical regions connected with these cortical areas, such as the amygdala and hypothalamus (Figure 42–4).

MacLean intended his theory to be an elaboration of Papez's ideas. Indeed, many areas of MacLean's limbic system are parts of the Papez circuit. However, MacLean did not share Papez's view that the cingulate cortex was the seat of feelings. Instead, he thought

of the hippocampus as the part of the brain where the external world (represented in sensory regions of the lateral cortex) converged with the internal world (represented in the medial cortex and hypothalamus), allowing internal signals to give emotional weight to external stimuli and thereby to conscious feelings. For MacLean, the hippocampus was involved both in the expression of emotional responses in the body and in the conscious experience of feelings.

Subsequent findings raised problems for MacLean's limbic system theory. In 1957, it was found that damage to the hippocampus, the keystone of the limbic system, produced deficits in converting short- to long-term memory, a function that is distinct from emotions. In addition, animals with damage to the hippocampus are able to express emotions, and humans with hippocampal lesions appear to express and feel emotions normally. In general, damage to areas of the limbic system did not have the expected effects on emotional behavior.

Several of MacLean's other ideas on emotion are nevertheless still relevant. MacLean thought that emotional responses are essential for survival and therefore involve relatively primitive circuits that have been conserved in evolution, an idea already proposed by Charles Darwin almost a century earlier. This notion is key to an evolutionary perspective of emotion. It is now clear that emotions are processed by many subcortical and cortical regions and that the limbic system is by no means the primary system for emotion. Nonetheless, one component of the original limbic system, the amygdala, has received the most attention in studies of both humans and animals. Today, the role of the amygdala in learned fear is probably the best worked-out example of emotion processing in a specific brain structure, and therefore, we consider it next.

The Amygdala Has Been Implicated in Both Learned and Innate Fear

In Pavlovian fear conditioning, an association is learned between an unconditioned stimulus (US) (eg, electric shock) and a conditioned stimulus (CS) (eg, a tone) that predicts the US. For example, if an animal is presented with an emotionally neutral CS (a tone) for several seconds and then shocked during the final second of the CS, especially if this pairing of tone and shock is repeated several times, presentation of the tone alone will elicit defensive freezing and associated changes in autonomic and endocrine activity. In addition, many defensive reflexes, such as eyeblink and startle, will be facilitated by the tone alone.

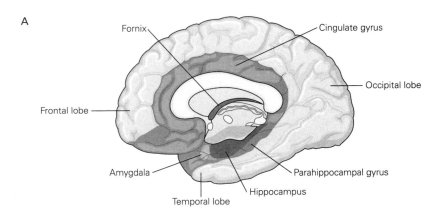

A

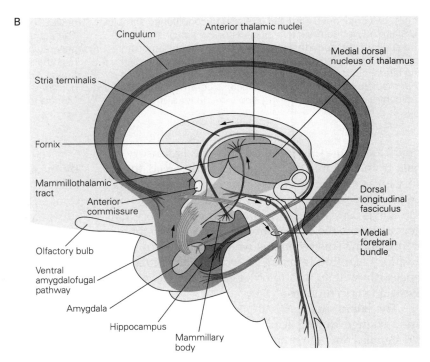

B

Figure 42–4 The limbic system consists of the limbic lobe and deep-lying structures. (Adapted, with permission, from Nieuwenhuys et al. 1988.)

A. This medial view of the brain shows the prefrontal limbic cortex and the limbic lobe. The limbic lobe consists of primitive cortical tissue (**blue**) that encircles the upper brain stem as well as underlying cortical structures (hippocampus and amygdala).

B. Interconnections of the deep-lying structures included in the limbic system. The **arrows** indicate the predominant direction of neural activity in each tract, although these tracts are typically bidirectional.

Research in many laboratories has established that the amygdala is necessary for Pavlovian fear conditioning: Animals with amygdala damage fail to learn the association between the CS and the US and thus do not express fear when the CS is later presented alone.

The amygdala consists of approximately 12 nuclei, but the lateral and central nuclei are especially important in fear conditioning (Figure 42–5). Damage to either nucleus, but not to other regions, prevents fear conditioning. The lateral nucleus of the amygdala receives most sensory input (but the medial nucleus receives olfactory input), including sensory information about the CS (eg, a tone) from both the thalamus and the cortex. The cellular and molecular mechanisms within the amygdala that underlie learned fear, especially in the lateral nucleus, have been elucidated in great detail. The findings support the view that the lateral nucleus is a site of memory storage in fear conditioning. Neurons in the central nucleus, by contrast, mediate outputs to brain stem areas involved in the control of defensive behaviors and associated autonomic and humoral responses (Chapter 41). The lateral and central nuclei are connected by way of several local circuits within the amygdala, including connections with the basal and intercalated masses. The actual circuitry for Pavlovian learning is thus considerably more complex than what is indicated by Figure 42–5, involving multiple relays among amygdala regions.

Sensory inputs reach the lateral nucleus from the thalamus both directly and indirectly. Much as

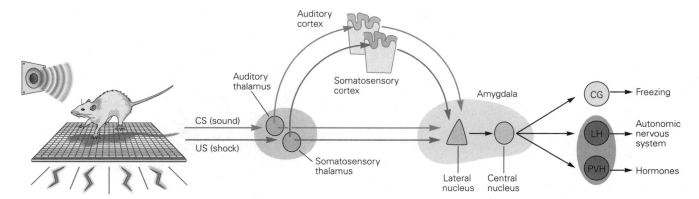

Figure 42–5 Neural circuits engaged during fear conditioning. The conditioned stimulus (**CS**) and unconditioned stimulus (**US**) are relayed to the lateral nucleus of the amygdala from the auditory and somatosensory regions of the thalamus and cerebral cortex. Convergence of the CS and US pathways in the lateral nucleus is believed to underlie the synaptic changes that mediate learning. The lateral nucleus communicates with the central nucleus both directly and through intra-amygdala pathways (not shown) involving the basal and intercalated nuclei. The central nucleus relays these signals to regions that control various motor responses, including the central gray region (**CG**), which controls freezing behavior; the lateral hypothalamus (**LH**), which controls autonomic responses; and the paraventricular hypothalamus (**PVH**), which controls stress hormone secretion by the pituitary–adrenal axis. (Adapted from Medina et al. 2002.)

predicted by the Cannon-Bard hypothesis, sensory signals from thalamic relay nuclei are conveyed to sensory areas of cerebral cortex. As a result, the amygdala and cortex are activated simultaneously. However, the amygdala is able to respond to an auditory danger cue before the cortex can fully process the stimulus information. This scheme is well worked out only for auditory fear conditioning in rodents, and it remains unclear how it applies to other cases, such as visually evoked fear in humans.

The lateral nucleus is thought to be a site of synaptic change during fear conditioning. The CS and US signals converge on neurons in the lateral nucleus; when the CS and US are paired, the effectiveness of the CS in eliciting action potentials is enhanced. This basic mechanism for a form of associative learning is similar to cellular mechanisms that underlie declarative memory in the hippocampus as well (Chapter 54). In particular, the synaptic plasticity found in the hippocampus has also been demonstrated in specific central amygdala circuits. The central amygdala thus does not simply drive motor outputs but is also part of the circuitry through which fear associations are formed and stored, very likely by transmitting information about the CS and US from the lateral nucleus. Neural plasticity likely also occurs in the basal and accessory basal nuclei during fear learning. As with the hypothalamus, recent work in rodents using tools such as optogenetics to manipulate specific subpopulations of amygdala neurons has begun to dissect this circuitry in further detail.

The emotional charge of a stimulus is evaluated by the amygdala together with other brain structures, such as the prefrontal cortex. If this system detects danger, it orchestrates the expression of behavioral and physiological responses by way of connections from the central amygdala and parts of prefrontal cortex to the hypothalamus and brain stem. For example, freezing behavior is mediated by connections from the central nucleus to the ventral periaqueductal gray region. In addition, the basal and accessory basal nuclei of the amygdala send projections to many parts of the cerebral cortex, including the prefrontal, rhinal, and sensory cortices; these pathways provide a means for neural representations in the amygdala to influence cognitive functions. For example, through its widespread projections to cortical areas, the amygdala can modulate attention, perception, memory, and decision making. Its connections with the modulatory dopaminergic, noradrenergic, serotonergic, and cholinergic nuclei that project to cortical areas also influence cognitive processing (Chapter 40). Given these very widespread connections and functional effects, the amygdala is well situated to implement one of the key features of an emotion: its coordinated and multicomponent responses.

The Amygdala Has Been Implicated in Innate Fear in Animals

Although the majority of stimuli acquire their emotional significance through learning, especially in humans, many animals also rely on innate (unconditioned) signals in the detection of threats, mates, food, and so forth. For example, rodents exhibit freezing and other defensive behaviors when fox urine is detected. Recent studies have made considerable progress in uncovering the circuits underlying this innate fear.

In mammals, sensory signals of unconditioned threats involving predator or conspecific odors are transmitted from the vomeronasal component of the olfactory system (Chapter 29) to the medial amygdala. This stands in contrast to auditory and visual threats, which as noted above are processed via the lateral amygdala. Outputs of the medial amygdala reach the ventromedial hypothalamus, which connects with the premammillary hypothalamic nucleus. In contrast to learned fear, which depends on the ventral periaqueductal gray region, unconditioned fear responses depend on inputs from the hypothalamus to the dorsal periaqueductal gray region. There are other subcortical systems specialized for processing specific innate threats; for instance, the mouse superior colliculus is involved in detecting aerial predators, such as a hawk flying overhead.

It is difficult to study unconditioned emotional responses in humans because the possibility of learning begins right at birth and cannot be experimentally controlled, and because there appear to be large individual differences. For instance, it is thought that threat-related stimuli such as snakes and spiders may be innately fear-inducing stimuli for those people with phobias toward these animals but not for people who keep them as pets. These large individual differences, and the relative roles of innate and learned fear, are important topics for understanding psychiatric illnesses such as anxiety disorders.

The Amygdala Is Important for Fear in Humans

The basic findings from animal studies regarding the role of the amygdala in emotion have been confirmed in studies of humans. Patients with damage to the amygdala fail to show fear conditioning when presented with a neutral CS paired with a US (electric shock or loud noise). In normal human subjects, activity in the amygdala increases during CS–US pairing, as measured with fMRI.

Studies of rare human patients with bilateral amygdala lesions have led to the surprising finding of a dissociation in fear reactions to exteroceptive and interoceptive stimuli (Figure 42–6). Not only do such patients fail to show any autonomic fear reactions to exteroceptive stimuli, to either the CS or the US, but they also appear to lack any conscious experience of fear, as evidenced either from behavioral observation or through subjective verbal report on a questionnaire. In one study, such a patient was confronted with snakes

and spiders in an exotic pet store, with monsters in a haunted house, and with autobiographical recollections of highly traumatic personal events (eg, being threatened with death by another person). In none of these instances was there any evidence of fear, and the patient reported feeling no fear at all (even though the patient was able to feel other emotions). These findings argue that the amygdala is necessary for the induction and experience of fear in humans.

By striking contrast, the very same patients with amygdala lesions report intense panic when they are made to feel as though they are suffocating (an interoceptive fear cue, achieved by inhaling carbon dioxide, which lowers blood pH). The dissociation of fear reactions to exteroceptive and interoceptive stimuli supports the idea that there are multiple fear systems in the human brain and that the amygdala cannot be the only structure essential for all forms of fear. Ongoing work is providing more insight, such as mapping out the specific amygdala nuclei that are damaged in these patients and which nuclei are responsible for what types of deficits. This level of resolution is standard in animal studies of the amygdala but has been difficult to achieve in humans, since the amygdala lesions cannot be made experimentally but instead must rely on rare patients that reflect accidents of nature. Equally important, there are theoretical frameworks for how to subdivide the different types of fear. For example, fear can be mapped onto a dimension of threat imminence, which may cover a range from threats that are very far away (perhaps evoking mild anxiety, and engaging monitoring and attention), to threats that are more proximal (evoking fear, and engaging responses such as freezing), to threats that are about to cause death (evoking panic, and engaging defensive behaviors). Eventually, we will need to have a more fine-grained mapping between brain systems and varieties of emotion that incorporates all of these details.

Certain forms of fear learning are relatively unique to humans. For example, simply telling a human subject that the CS may be followed by a shock is enough to allow the CS to elicit fear responses. The CS elicits characteristic autonomic responses even though it was never associated with the delivery of the shock. Humans can also be conditioned by allowing them to observe someone else being conditioned—the observer learns to fear the CS even though the CS or US was never directly presented to the observing subject. Some other animals also are able to learn fear through such observational learning, although this seems to be more rare than is the case in humans. One form of learning that is ubiquitous in humans appears to be unique to our species: active pedagogy, whereby another person teaches somebody that a stimulus is dangerous. While learning what to avoid

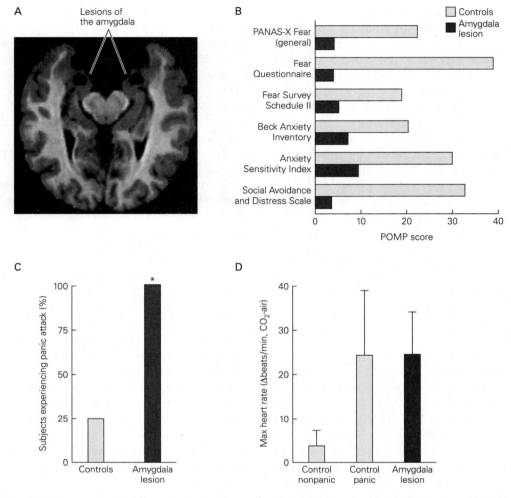

Figure 42–6 In humans, the amygdala is necessary for fear responses to external, but not internal, stimuli.

A. Magnetic resonance imaging scan of a subject's brain with bilateral amygdala lesions. Lesions were relatively restricted to the entire amygdala, a very rare lesion in humans.

B. The subject with bilateral amygdala lesions, S.M., did not report feeling fear for any of the questionnaire-based measures normally used to assess fear and anxiety (percentage of maximum score possible [**POMP**]). This was consistent with other findings: She did not exhibit fear when watching horror movies, when confronted with large spiders and snakes, or when visiting a haunted house during Halloween. These findings show that the human amygdala is necessary for inducing fear in response to these external stimuli. (Abbreviation: **PANAS**, Positive and Negative Affect Schedule.)

C. By contrast, a study of S.M. and two other subjects with bilateral amygdala lesions found that they exhibited strong panic when given an internal stimulus. They were asked to inhale carbon dioxide (CO_2), which produces a feeling of suffocation. This caused all three patients with amygdala lesions and 3 out of 12 of the control subjects with intact amygdalae to experience panic attacks.

D. Change from baseline in maximum heart rate during CO_2 inhalation relative to air trials. Both the amygdala lesion patients (n = 2) and the control subjects who panicked (n = 3) had higher increases in heart rate than the control subjects who did not panic (n = 9). (Mean ± standard error of the mean.) (Adapted, with permission from Feinstein et al 2011, 2013.)

and what to approach in the world is a large part of development in the young of all species, active teaching about the significance of stimuli has not been found in any species other than humans so far (learning through passive observation is more common).

The emotional learning and memory capacities of the human amygdala fall into the category of *implicit*

learning and memory, which includes forms of memory such as the unconscious recall of perceptual and motor skills (Chapter 53). In situations of danger, however, the hippocampus and other components of the medial temporal lobe system that participate in *explicit learning* and memory (the conscious recall of people, places, and things) will be recruited as well and will encode aspects of the learning episode. As a result, the learned indicators of danger can also be recalled consciously, at least in humans and probably in some other species as well.

Studies of patients with bilateral damage to the amygdala or hippocampus illustrate the separate contributions of these structures to implicit and explicit memory for emotional events, respectively. Patients with damage to the amygdala show no conditioned skin-conductance responses to a CS (suggesting no implicit emotional learning) but have normal declarative memory of the conditioning experience (indicating intact explicit learning). By contrast, patients with hippocampal damage show normal conditioned skin-conductance responses to the CS (suggesting intact implicit emotional learning) but have no conscious memory of the conditioning experience (indicating impaired explicit learning).

Amygdala function is altered in a number of psychiatric disorders in humans, especially disorders of fear and anxiety (Chapter 61). In addition, the amygdala plays an important role in processing cues related to addictive drugs (Chapter 43). In all of these cases, the amygdala is but one component of a distributed neural network that includes other cortical and subcortical regions. For instance, declarative memory for highly emotional events involves interactions between the amygdala and hippocampus; motivational consequences of Pavlovian conditioning involve interactions between the amygdala and the ventral striatum; and learning that a previously dangerous stimulus is now safe involves interactions between the amygdala and the prefrontal cortex. An important future direction will be to go beyond examining each component in isolation in order to better understand how emotions are processed by complex multicomponent networks of brain regions. This level of analysis is common in studies of human emotion using fMRI (see below).

The Amygdala's Role Extends to Positive Emotions

Although most work on the neural basis of emotion during the past half century has focused on aversive responses, especially fear, other studies have shown that the amygdala is also involved in positive emotions, in particular the processing of rewards. In monkeys and rodents, the amygdala participates in associating neutral stimuli with rewards (appetitive Pavlovian conditioning), just as it participates in associating neutral stimuli with punishments, and there appear to be distinct populations of neurons that encode rewards and punishments in the amygdala. This is broadly similar to findings from the rodent hypothalamus, where

neurons involved in defense and in mating are also close together and only modern molecular techniques can test their independent roles.

Studies in nonhuman primates and rodents have investigated a suggestion first made by Larry Weiskrantz that the amygdala represents stimulus reward as well as punishment. For example, in a recent study, monkeys were trained to associate abstract visual images with rewarding or aversive USs. The meaning was then reversed (eg, by pairing an aversive outcome with a visual image that had previously been associated with a reward). In this way, it was possible to distinguish the role of the amygdala in representing visual information from its role in representing the reinforcement (a rewarding or aversive stimulus) predicted by a visual image. Changes in the type of reinforcement associated with an image modulated neural activity in the amygdala, and the modulation occurred rapidly enough to account for behavioral learning.

Subsequent studies using modern molecular and genetic techniques have demonstrated that distinct circuitry within the amygdala mediates a neural representation of rewarding USs, as well as rewarding experiences. The activation of a neural representation of an appetitive US in the amygdala is sufficient to induce innate valenced physiological responses as well as appetitive learning. Moreover, reactivation of neurons activated earlier by an enjoyable experience appears to be sufficient to elicit positive emotions. These findings are consistent with a growing number of functional imaging studies in humans that have shown that the amygdala is involved in emotions quite broadly. For example, the human amygdala is activated when subjects observe pictures of stimuli associated with food, sex, and money or when people make decisions based on the reward value of stimuli.

Emotional Responses Can Be Updated Through Extinction and Regulation

Once conditioned fear has been learned, it can be extinguished by later experiencing that the CS is now safe, for instance, by repeatedly presenting the CS without any US pairing. The circuitry underlying fear extinction has been studied in detail as it is highly relevant to psychiatric illnesses such as post-traumatic stress disorder (PTSD). Projections from the prefrontal cortex to the amygdala are required to override the conditioned fear in the amygdala. While conditioned fear responses decline during extinction, they are never completely erased, as demonstrated by the phenomenon of reinstatement, where fear can suddenly reappear.

Cognitive therapies for changing emotion states have also been studied, primarily in humans. For instance, a focused effort to increase or decrease the intensity of an emotion like fear has some effect on the emotion state. Indeed, neuroimaging studies have found that people can, to some degree, change their amygdala activation to fear-inducing stimuli just by how they think about those stimuli. Emotion regulation is a complex phenomenon, since there are multiple strategies for changing the emotion, ranging from just suppressing the motor behaviors to better control over how we evaluate a situation. These multiple sources of emotion regulation, especially in humans, highlight the fact that emotions must often be adjusted in keeping with complex social norms.

Emotion Can Influence Cognitive Processes

As evidenced in the above examples, emotion interacts with many other aspects of cognition, including memory, decision making, and attention. We discussed above an example of nondeclarative emotional memory, Pavlovian fear conditioning, but emotions can also influence declarative memory. Projections from the amygdala to the hippocampus can influence how learning is encoded and consolidated into long-term declarative memory. This accounts for why we remember best those events in our lives that are the most emotional, such as weddings and funerals.

Emotion has complex effects on decision making, as one might expect, since the subjective evaluation of such variables as risk, effort, and value is modulated by emotion. For instance, different choices with the same objective risk can elicit different behavioral decisions depending on whether they are framed as a win or a loss. For example, subjects typically prefer a sure gain of $5 to a 50% chance of winning $10, but prefer a 50% chance of losing $10 to a sure loss of $5. Interestingly, fMRI studies have revealed that such framing modulates amygdala activation. There is greater amygdala activation in the "win" frame when subjects choose a sure amount over a risky gamble, and greater amygdala activation in the "loss" frame when subjects choose the gamble over the sure amount. Thus, value representations in the amygdala are not rigidly associated with stimuli but are modulated by context-dependent evaluation.

Because emotionally relevant stimuli are highly salient to an organism's self-interest, they typically capture attention. For instance, people tend to orient toward, and look at, emotionally relevant visual stimuli, even when those stimuli are presented under conditions where they cannot be consciously perceived. One intriguing finding is that patients with bilateral amygdala lesions are impaired not only in their experience and expression of fear, as described above, but also in their recognition of fear in other people. One such patient, a woman called S.M., was selectively impaired in recognizing fear from facial expressions. This impairment in turn appears to result from a more basic impairment in allocating visual attention to those regions of the face that normally signal fear. S.M. does not spontaneously fixate on the eye region of the face when she looks at facial expressions and therefore does not process detailed visual information from wide eyes that would normally contribute to the recognition of fear when one is looking at a fearful face (Figure 42–7).

These findings suggest an important role for the amygdala in attention and highlight the possibility that apparently specific deficits for certain emotions (like fear) might arise from more basic attentional or motivational effects. There is ongoing debate about the precise role of the human amygdala in attentional aspects of emotion processing: Some studies argue that it comes into play even for nonconscious threat-related stimuli and in a very automatic fashion; other studies argue that the amygdala requires more elaborated and conscious processing once attention has already been allocated. Single-neuron recordings from the human amygdala support the latter view, whereas some fMRI studies support the former view. All of the findings from human lesion studies will need to be more finely dissected; some recent work on patients who have damage only to specific amygdala subnuclei is yielding further insights.

Many Other Brain Areas Contribute to Emotional Processing

As seen in the case of conditioned and unconditioned fear, the amygdala contributes to emotional processing as part of a larger circuit, or set of circuits, that includes regions of the hypothalamus and brain stem, eg, the periaqueductal gray region in the brain stem. Cortical areas are also important components of this circuit.

A number of human studies have implicated the ventral region of the anterior cingulate cortex, the insular cortex, and the ventromedial prefrontal cortex in various aspects of emotional processing. The medial prefrontal cortex and amygdala are closely connected with one another, and neurons in these brain regions show complex responses that encode information about many emotional and cognitive variables. These findings contribute to an emerging picture of a dynamic neural substrate for emotion states: Individual states are not the outcome of a single structure or specific neurons, but are more

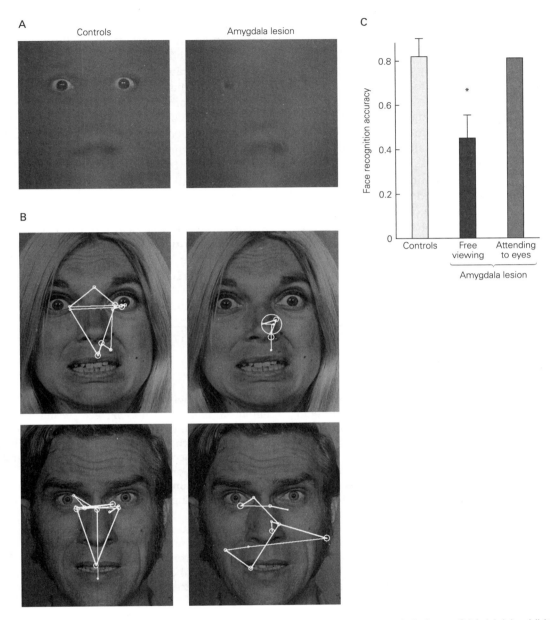

Figure 42–7 Bilateral amygdala lesions impair the recognition of fear in the facial expressions of others. This impairment may be due to abnormal processing of information from the face. (Reproduced, with permission, from Adolphs et al. 2005.)

A. S.M. made significantly less use of information from the eye region of faces when judging emotion. These images show the regions of the face from which control subjects (*left*) or S.M. (*right*) were able to recognize fear. The results were obtained by showing subjects many trials with only small parts of the face revealed. All those trials in which subjects were able to recognize fear could then be summed to produce an image like this, which shows the regions of the face that viewers make use of in order to discriminate fearful from happy faces (these particular parts of the face allow viewers to tell apart fearful from happy faces, whereas other parts do not help with this discrimination).

B. While looking at whole faces, S.M. (*right*) exhibited abnormal face gaze (indicated by **white lines**), making far fewer fixations to the eyes than did controls (*left*). This shows that S.M. failed to attend to and hence process visual information from the eye region. This deficit was observed across all emotions, but was most important for fear recognition because wide eyes normally predict fear.

C. S.M. showed poor ability to recognize fear when freely observing whole faces (**free viewing**), but her performance improved remarkably when instructed to look at the eyes (**attending to eyes**). This result shows that the role of the amygdala in processing fearful expressions involves directing attention onto features that are particularly significant (the eyes), rather than the downstream process of interpreting the sensory input.

flexibly assembled over a distributed population of multifunction neurons.

Some emotions are associated with social interaction and range from empathy and pride to embarrassment and guilt. Like the primary emotions such as fear, pleasure, or sadness, these social emotions produce various bodily changes and behaviors and can be experienced consciously as distinct feelings. This class of emotions may depend especially on cortical regions in the prefrontal cortex.

Studies of patients with neurological disease and focal brain lesions have advanced the understanding of the neural circuitry of emotions (Box 42–2). For example, damage to some sectors of the prefrontal cortex markedly impairs social emotions and related feelings. In addition, these patients show marked changes in social behavior that resemble the behavior of patients with developmental sociopathic personalities. Patients with damage to some sectors of the prefrontal cortex are unable to hold jobs, cannot maintain stable social relationships, are prone to violate social conventions, and cannot maintain financial independence. It is common for family ties and friendships to break after the onset of this condition. Recent studies reveal that, under controlled experimental conditions, the moral judgments of these patients can also be flawed.

Patients with medial and ventral frontal lobe damage, unlike patients with more dorsal or lateral frontal lobe damage, do not have motor defects such as limb paralysis or speech defects and thus may appear at first to be neurologically normal. Their perceptual abilities, attention, learning, recall, language, and motor abilities often show no signs of disturbance. Some patients have IQ scores in the superior range.

Box 42–2 Lesion Studies of Emotion

Examination of patients with focal lesions complements neuroimaging studies of the neural correlates of emotions. In addition to studies of the amygdala, lesion studies have provided insights into the role of several other brain regions in processing emotions.

One of the most famous set of studies harks back to the accident of Phineas Gage, who in 1848 suffered an injury to his ventromedial prefrontal cortex. Gage was working on constructing a railway in Vermont and was tamping gunpowder into a hole with a long metal rod, called a tamping iron. By accident, he struck a spark in the rock and the gunpowder exploded, shooting the tamping iron straight through his head.

Amazingly, Gage lived for many years after this horrible accident, but he was a changed person with notable changes in his social and emotional behavior. This was the first evidence that parts of the prefrontal cortex played a role in emotions. Since Gage, several patients with damage centered on the ventromedial prefrontal cortex have been described. These patients have poor insight and decision-making abilities and tend to have blunted or unusual emotional responses, especially for social emotions.

Unlike normal individuals, patients with these frontal lesions do not exhibit changes in heart rate or degree of palm sweating when shown pictures that have emotional content, although they can describe the pictures flawlessly. Likewise, patients with frontal lesions do not show skin conductance changes, a sign of sympathetic activation, during the period that precedes making risky and disadvantageous decisions, suggesting that their emotional memory is not engaged during that critical period. Also unlike normal subjects, these patients fail in tasks in which they have to make a decision under conditions of uncertainty, and in which reward and punishment are important factors.

Several brain regions are also more specifically involved in feelings. Damage to the right somatosensory cortex (primary and secondary somatosensory cortices and insula) impairs social feelings such as empathy. Consistent with this finding, patients with lesions in the right somatosensory cortex fail to guess accurately the feelings behind the facial expressions of other individuals. This ability to read faces is not impaired in patients with comparable lesions of the *left* somatosensory cortex, indicating that the right cerebral hemisphere is dominant in the processing of at least some feelings. Body sensations such as pain and itch remain intact, as do feelings of basic emotions such as fear, joy, and sadness.

On the other hand, damage to the human insular cortex, especially on the left, can suspend addictive behaviors, such as smoking. This suggests that the insular cortices play a role in associating external cues with internal states such as pleasure and desire. Interestingly, complete bilateral damage to the human insular cortices, as caused by herpes simplex encephalitis, does not eliminate emotional feelings or body sensations, suggesting that the somatosensory cortices and subcortical nuclei in the hypothalamus and brain stem are also involved in generating feeling states.

For these reasons, they sometimes attempt to return to their work and social activities after their initial recovery from brain damage. Only when they start to interact with others are their defects noticed.

In the prefrontal cortex, the ventromedial sector is particularly important for such interactions. In most patients with impaired social emotions, this sector is damaged bilaterally, although damage restricted to the right side can be sufficient to cause impairments. The critical region encompasses Brodmann's areas 12, 11, 10, 25, and 32, which receive extensive projections from the dorsolateral and dorsomedial sectors of the prefrontal cortex. Some of these areas project extensively to subcortical areas related to emotions: the amygdala, the hypothalamus, and the periaqueductal gray region in the brain stem.

Interestingly, when asked about punishment, reward, or responsibility, adult patients with damage to the ventromedial prefrontal cortex often respond as if they still have the basic knowledge of the rules, but their actions indicate that they fail to use them in real-life situations. This dissociation suggests that their behavioral defects are not caused by a loss of factual knowledge but rather by impairment of the brain's assignment of motivational value to factors that normally exert control over behavior. In some respects, this dissociation is similar to the dissociation between explicit and implicit emotional learning vis-à-vis the hippocampus and the amygdala. An interesting hypothesis arising from these dissociations is that one might find greater deficits following lesions to emotion-related structures, like the amygdala or ventromedial prefrontal cortex, in other species, or in children, in whom explicit behavioral control has not yet evolved or developed to the degree that it has in adults. There is some support for this idea: Lesions to these structures early in life can result in more severe deficits in emotional and social behaviors than if the lesions are sustained in adulthood (a pattern opposite to that of most other lesions, which show better recovery of function the earlier the onset). These findings also suggest hypotheses for neural dysfunction that may contribute to the emotional difficulties seen in developmental psychiatric disorders, such as autism.

The above lesion studies have been complemented by controlled experimental studies using fMRI, which provide further insight into mechanisms. Functional imaging of value-based decision making in healthy human subjects shows that the ventromedial prefrontal cortex is activated during the period before making a choice. That same region is activated also just by the administration of punishment and reward, supporting the notion that the emotional significance of anticipated punishments and rewards is computed as part of the mechanism that guides this kind of decision making. Punishment and reward are frequently featured in experiments involving economic and moral decisions, and such decision making prominently involves many of the same structures that are also involved in processing emotions.

The prefrontal cortex, especially areas in the ventromedial sector, operates in parallel with the amygdala. During an emotional response, ventromedial areas govern the attention accorded to certain stimuli, influence the content retrieved from memory, and help shape mental plans for responding to the triggering stimulus. By influencing attention, both the amygdala and the ventromedial prefrontal cortex are also likely to alter cognitive processes, for example, by speeding up or slowing down the flow of sensory representations (Chapter 17).

Functional Neuroimaging Is Contributing to Our Understanding of Emotion in Humans

Neuroimaging studies of emotions typically use fMRI. These studies have contributed to our understanding of emotion in three important ways. First, they have begun to dissociate and experimentally manipulate specific aspects of emotion, such as feelings, value, or concepts of emotions. These studies are beginning to show how all these different aspects can be coordinated by activity in different brain regions.

Second, fMRI studies on emotion have been accumulating at an ever-increasing pace, and much of the data from such studies are now widely available. This provides the opportunity for meta-analyses of many studies, avoiding the limitations that may be inherent in any one study in isolation. For instance, some meta-analyses have confirmed the role of the ventromedial prefrontal cortex in representing value for many different kinds of stimuli, including food and money. Other meta-analyses have suggested that specific basic emotions (eg, fear, anger, or happiness) activate a widely distributed and overlapping set of brain regions, confirming the view that no brain structure is responsible for a single emotion.

Finally, fMRI studies have begun to use novel methods in their analyses. For example, the pattern of activation seen across many voxels in a brain region, rather than the mean level of activation in that region, is used to train powerful machine-learning algorithms to classify emotion states. This approach is demonstrating that it is possible to decode specific emotion states from distributed patterns of brain activation.

Functional Imaging Has Identified Neural Correlates of Feelings

Conscious experiences of an emotion are generally referred to as feelings. Evidence for the neural correlates of feelings comes primarily from functional imaging studies of humans and from neuropsychological testing of patients with specific brain lesions. A main challenge for these studies is in dissociating the conscious experience of the emotion from other aspects of the emotion, such as the elicitation of physiological responses, since these tend to occur contemporaneously. Another challenge is how to connect such studies with studies of emotion in animals, where we have no agreed-upon dependent measures to assay what they consciously experience.

One early functional imaging study used positron emission tomography to test the idea that feelings are correlated with activity in those cortical and subcortical somatosensory regions that specifically receive inputs related to the internal environment—the viscera, endocrine glands, and musculoskeletal system. Healthy subjects were asked to recall personal episodes and to attempt to reexperience as closely as possible the emotions that accompanied those events. Activity changed in many regions known to represent and regulate body states, such as the insular cortex, secondary somatosensory cortex (S-II), cingulate cortex, hypothalamus, and upper brain stem. These results support the idea that at least a part of the neural substrate for feelings involves brain regions that regulate and represent bodily states, a finding that bears some resemblance to the hypothesis of William James mentioned earlier, that feelings are based on an awareness of bodily reactions.

The importance of both cortical and subcortical structures in processing feelings is also borne out by more recent fMRI studies. One such study examined the feeling of fear induced by anticipation of electrical shock (Figure 42–8). In this study, subjects lay in the scanner while they saw a game on a video screen in which a virtual predator (a red dot) gets closer to the subject. Once the predator caught them, they could receive a painful electric shock to the hand. The anxiety produced when the predator was some distance away was associated with activation of the medial prefrontal cortex; as the predator closed in, the periaqueductal gray became activated, and this was correlated with reports by the subjects of a feeling of dread. This finding supports a role for the medial prefrontal cortex in planning and anticipation related to a distant threat and a role for the periaqueductal gray in mounting the defensive responses required for coping with an immediate threat.

Another brain region of interest in relation to feelings is the subgenual sector of the anterior cingulate cortex (Brodmann's area 25), which has been found in neuroimaging studies to be activated when subjects are experiencing sadness. This region is of special interest because it is also differentially activated in patients with bipolar depression, and it appears thinned in structural MRI scans of patients with chronic depression. Direct electrical stimulation of this brain region (deep brain stimulation) can dramatically improve the mood of some patients with severe depression.

Emotion Is Related to Homeostasis

While it seems clear that no brain region is specialized for any specific emotion, it is even doubtful that there are any brain regions specialized for emotions in general. It may be that all brain regions involved in emotions also carry out other functions. In fact, those nonemotional functions may give us clues about how emotions evolved and, indeed, may be the basic building blocks through which emotion states are assembled.

For example, sectors of the human insular cortex that are activated during recall of feelings are also activated during the conscious sensation of pain and temperature. The insular cortex receives homeostatic information (about temperature and pain, changes in blood pH, carbon dioxide, and oxygen) through pathways that originate in peripheral nerve fibers. These afferent fibers include, for example, the C and Aδ fibers that form synapses with neurons in lamina I of the posterior horn of the spinal cord or the pars caudalis of the trigeminal nerve nucleus in the brain stem. The pathways from lamina I and the trigeminal nucleus project to brain stem nuclei (nucleus of the solitary tract and parabrachial nucleus) and from there to the thalamus and on to the insular cortex. The identification of this functional system is further support for the idea that signals in the afferent somatosensory pathways play a role in the processing of feelings.

Moreover, in patients with pure autonomic failure, a disease in which visceral afferent information is severely compromised, functional imaging studies reveal a blunting of emotional processes *and* attenuation of activity in the somatosensory areas that contribute to feelings. Like other feelings, social feelings engage the insular cortices and the primary and secondary somatosensory cortices (S-I and S-II), as has been found in functional neuroimaging experiments evaluating empathy for pain and, separately, admiration and compassion.

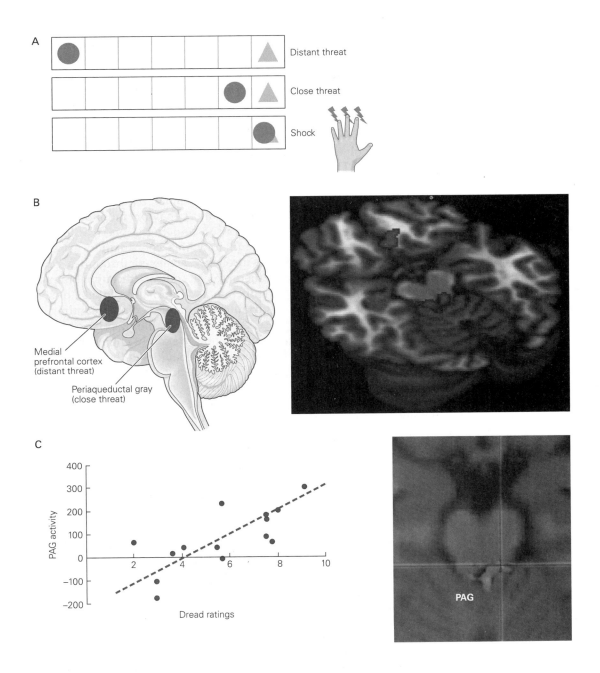

Figure 42–8 Both cortical and subcortical regions come into play during emotion states. Results are from a functional magnetic resonance imaging study in which a subject lies in the scanner while watching a virtual predator (**red dot**) move around on the screen and get closer to a subject (**blue triangle**, representing the actual research participant). (Reproduced, with permission, from Mobbs et al. 2007. Copyright © 2007 AAAS.)

A. Once the predator catches up to the subject, there is a chance that a real and painful electric shock will be delivered to the hand.

B. When the predator gets closer to the subject, activity in the prefrontal cortex and periaqueductal gray matter increases. Notably, this pattern of neural activation shifts such that a distant predator elicits greater activity in the medial prefrontal cortex, whereas a predator close by elicits more activity in the periaqueductal gray.

C. Activation of the periaqueductal gray (**PAG**) is correlated with the subjective sense of dread measured by ratings that subjects gave while in the scanner.

Using these data as support, some influential modern theories build on William James's original hypothesis and propose that the feeling of all emotions is grounded in the brain's representation of bodily homeostasis. As in the case of the amygdala's role in both positive and negative emotions, the insula's role in processing both interoceptive and emotional information is still compatible with the possibility that these processes are distinct. That is, different populations of neurons within these structures may be involved in processing different emotions. Therefore, fMRI may not provide the level of resolution needed to elucidate distinct yet anatomically intermingled neuronal populations, and cellular techniques in animal models may be required.

Although most neuroscience research thus far has focused on negatively valenced emotions, the neural circuitry for positively valenced emotions is being elucidated in studies in both humans and animals. These studies consistently implicate the medial prefrontal cortex in computing the subjective value of rewards, as well as the nucleus accumbens and other nuclei of the basal ganglia in processing the hedonic component (or pleasure) of positive emotions. A growing number of functional imaging studies in humans—especially in the fields of neuroeconomics and social neuroscience—links the role of these structures in emotion processing to their role in value-based decision making and social behavior.

Highlights

1. In the overall physiology of regulating the body and behavior of organisms, emotion states carry out functions intermediate to those of the simpler processes of reflexes and homeostatic regulation, on the one hand, and those of cognitive processes and deliberate behavior on the other. Emotions are more flexible, context-dependent, and controlled than are simple reflexes, but less flexible, context-dependent, and controlled than deliberate behavior. Emotions evolved to produce behavior in response to recurring environmental and internal challenges that are too varied for reflexes, but sufficiently stereotyped that they do not require the full flexibility of cognition.

2. Emotion states need to be carefully distinguished from the conscious experience of emotion (feelings) and also from the concepts and words that we have in everyday language to describe emotions. For example, a hissing cat's behavior is caused by an emotion state, but whether the cat

consciously feels afraid is unclear. The cat probably has no concept, and certainly no words, with which to think about the emotion. Human subjects who recognize fear while observing a facial expression are attributing fear to another person and are thinking about a particular emotion, but are not themselves necessarily in a state of fear or experiencing fear. It is a major challenge in designing experiments, especially in humans, to independently control and manipulate these different components of emotion.

3. Emotions coordinate integrated changes in many organismal parameters, including effects on somatic behavior, autonomic and endocrine responses, and cognition. We do not yet understand how this coordination arises, although it is probably achieved through a combination of hierarchical control (through brain regions that function as "command centers" of sorts) and distributed dynamics. Understanding how this is accomplished in biological organisms will also inform how we might engineer robots that exhibit emotional behaviors in the future.

4. Different specific emotions can be thought of categorically (eg, happiness, fear, anger) or dimensionally (in terms of arousal and valence or other dimensional frameworks). It is likely that many of the categories for which we have words in a particular language (like the examples just given) will need to be revised once we have a more scientific understanding. New analytic methods applied to data acquired using fMRI, including methods that take into account the spatial and temporal patterns of brain activity and utilize powerful machine-learning algorithms, may provide new insights into how the brain mediates a broad range of emotions.

5. In humans, emotions can be regulated by several mechanisms. Thus, we have some control over how we feel and some control over how we express emotional behaviors, for instance, through facial expressions. Nonhuman animals do not have this same level of control, and so their emotional behaviors will generally be honest signals of their emotion state, whereas humans often engage in strategic deception.

6. Fear is probably the emotion whose neurobiology is best understood. It depends on the amygdala, in both animals and humans. However, some data suggest that certain types of fear, such as the panic of suffocating induced by inhaling carbon dioxide, are independent of the amygdala. Indeed, we now know that the amygdala is part of a distributed

brain system, and therefore, many other brain regions also participate in processing fear. Increasingly, modern studies use sophisticated genetic and cellular techniques to image and to causally manipulate brain function, allowing us to understand the necessary and sufficient roles of multiple brain structures in mediating different emotional behaviors.

7. The ventral and medial prefrontal cortex is intimately involved in emotion and connected with the amygdala. Social emotions, reward representations, and emotion regulation and extinction all involve specific sectors of prefrontal cortex. This region of the brain, together with the insula, may also be the most important for our conscious experience of emotions, an aspect of emotion that remains the most challenging to study.

C. Daniel Salzman
Ralph Adolphs

Selected Reading

Amaral DG and Adolphs R (eds). 2016. *Living Without an Amygdala*. New York: Guilford Press.

Anderson, DJ, Adolphs R. 2018. *The Neuroscience of Emotion in People and Animals: A New Synthesis*. Princeton University Press.

Bechara A, Tranel D, Damasio H, Adolphs R, Rockland C, Damasio AR. 1995. A double dissociation of conditioning and declarative knowledge relative to the amygdala and hippocampus in humans. Science 269:1115–1118.

Craig AD. 2002. How do you feel? Interoception: the sense of the physiological condition of the body. Nat Rev Neurosci 3:655–666.

Damasio AR. 1994. *Descartes's Error: Emotion, Reason, and the Human Brain*. New York: Penguin Books.

Darwin, C. 1872/1965. *The Expression of the Emotions in Man and Animals*. Chicago: Univ of Chicago Press.

Dolan RJ. 2002. Emotion, cognition, and behavior. Science 298:1191–1194.

Feinstein JS, Adolphs R, Damasio A, Tranel D. 2011. The human amygdala and the induction and experience of fear. Curr Biol 21:34–38.

Feinstein JS, Buzza C, Hurlemann R, et al. 2013. Fear and panic in humans with bilateral amygdala damage. Nat Neurosci 16:270–272.

Feldman Barrett L, Adolphs R, Marsella S, Martinez AM, Pollack SD. 2019. Emotional expressions reconsidered: challenges to inferring emotion from human facial movements. Psychol Sci Public Interest 20:1–68.

McGaugh JL. 2003. *Memory and Emotions: The Making of Lasting Memories*. New York: Columbia Univ Press.

Salzman CD, Fusi S. 2010. Emotion, cognition, and mental state representation in amygdala and prefrontal cortex. Ann Rev Neurosci 33:173–202.

Thornton MA, Tamir DI. 2017. Mental models accurately predict emotion transitions. Proc Natl Acad Sci U S A 114:5982–5987.

Whalen PJ, Phelps EA. 2009. *The Human Amygdala*. New York: Guilford Press.

References

Adolphs R, Gosselin F, Buchanan T, Tranel D, Schyns P, Damasio A. 2005. A mechanism for impaired fear recognition in amygdala damage. Nature 433:68–72.

Anderson SW, Bechara A, Damasio H, Tranel D, Damasio AR. 1999. Impairment of social and moral behavior related to early damage in human prefrontal cortex. Nat Neurosci 2:1032–1037.

Berridge KC, Kringelbach ML. 2013. Neuroscience of affect: brain mechanisms of pleasure and displeasure. Curr Opin Neurobiol 23:294–303.

Cahill L, McGaugh JL. 1998. Mechanisms of emotional arousal and lasting declarative memory. Trends Neurosci 21:294–299.

Clithero JA, Rangel A. 2014. Informatic parcellation of the network involved in the computation of subjective value. Soc Cogn and Affect Neurosci 9:1289–1302.

Damasio AR, Grabowski TJ, Bechara A, et al. 2000. Feeling emotions: subcortical and cortical brain activity during the experience of self-generated emotions. Nat Neurosci 3:1049–1056.

Damasio H, Grabowski T, Frank R, Galaburda AM, Damasio AR. 1994. The return of Phineas Gage: clues about the brain from the skull of a famous patient. Science 264:1102–1105.

De Martino B, Kumaran D, Seymour B, Dolan RJ. 2006. Frames, biases, and rational decision-making in the human brain. Science 313:684–687.

Gore F, Schwartz EC, Brangers BC, et al. 2015. Neural representations of unconditioned stimuli in basolateral amygdala mediate innate and learned responses. Cell 162:132–145.

Holland PC, Gallagher M. 2004. Amygdala-frontal interactions and reward expectancy. Curr Opin Neurobiol 14:148–155.

Jin J, Gottfried JA, Mohanty A. 2015. Human amygdala represents the complete spectrum of subjective valence. J Neurosci 35:15145–15156.

LeDoux, JE. 1996. *The Emotional Brain*. 1996. New York: Simon & Schuster.

LeDoux JE. 2000. Emotion circuits in the brain. Annu Rev Neurosci 23:155–184.

Lin D, Boyle MP, Dollar P, Lee H, Perona P, Anderson DJ. 2011. Functional identification of an aggression locus in the mouse hypothalamus. Nature 470:221–226.

MacLean PD. 1990. *The Triune Brain in Evolution*. New York: Plenum.

Mayberg HS, Lozano AM, Voon V, et al. 2005. Deep brain stimulation for treatment-resistant depression. Neuron 45:651–660.

Medina JF, Repa CJ, Mauk MD, LeDoux JE. 2002. Parallels between cerebellum- and amygdala-dependent conditioning. Nat Rev Neurosci 3:122–131.

Mobbs D, Petrovic P, Marchant JL, et al. 2007. When fear is near: threat imminence elicits prefrontal-periaqueductal gray shifts in humans. Science 317:1079–1083.

Nieuwenhuys R, Voogd J, van Huijzen Chr. 1988. *The Human Central Nervous System: A Synopsis and Atlas*, 3rd ed. Berlin: Springer-Verlag.

Ochsner KN, Gross JJ. 2005. The cognitive control of emotions. Trends Cogn Sci 9:242–249.

Paton JJ, Belova MA, Morrison SE, Salzman CD. 2006. The primate amygdala represents the positive and negative value of visual stimuli during learning. Nature 439:865–870.

Pessoa L, Adolphs R. 2010. Emotion processing and the amygdala: from a "low road" to "many roads" of evaluating biological significance. Nat Neurosci 11:773–782.

Phelps EA. 2006. Emotion and cognition: insights from studies of the human amygdala. Annu Rev Psychol 57:27–53.

Rauch SL, Shin LM, Phelps EA. 2006. Neurocircuitry models of posttraumatic stress disorder and extinction: human neuroimaging research—past, present, and future. Biol Psychiat 60:376–382.

Redondo RL, Kim J, Arons AL, Ramirez S, Liu X, Tonegawa S. 2014. Bidirectional switch of the valence associated with a hippocampal contextual memory engram. Nature 513:426–430.

Saez A, Rigotti M, Ostojic S, Fusi S, Salzman CD. 2015 Abstract context representations in primate amygdala and prefrontal cortex. Neuron 87:869–881.

Weiskrantz L. 1956. Behavioral changes associated with ablation of the amygdaloid complex in monkeys. J Comp Physiol Psychol 49:381–391.

43

Motivation, Reward, and Addictive States

Motivational States Influence Goal-Directed Behavior

ONE DAY A CHEETAH, TAKING REFUGE from the mid-day sun in the shade of a tree, views a distant antelope with apparent indifference. Later in the afternoon, the sighting of the antelope provokes immediate orienting and stalking behavior. The stimulus is the same, but the behavioral responses are very different. What has changed is the motivational state of the animal.

Motivational states influence attentiveness, goal selection, investment of effort in the pursuit of goals, and responsiveness to stimuli. They thus drive approach, avoidance, and action selection. This chapter focuses on the neurobiological basis of motivational states related to rewards and the manner in which reward-related brain circuits are implicated in mechanisms underlying drug addiction.

Both Internal and External Stimuli Contribute to Motivational States

Motivational states reflect one's desires, and desires can be influenced by physiological status as well as by stimuli that predict future rewarding and aversive events. Motivational states thus depend on both internal and external variables. Internal variables include physiological signals concerning hunger or thirst, as well as variables related to the circadian clock. For example, the frequency and duration of foraging vary with the time of day, the time since an animal has last eaten, and whether, if female, she is lactating.

Other internal variables are related to cognitive processes. In the game of blackjack, for instance, being dealt the same card in different hands can cause a player to go bust or make 21, leading to very different emotional responses and adjustments in subsequent decision making and action selection. The differential meaning of the same stimulus (a particular card) is made possible by the cognitive understanding of the rules of the game of blackjack. The cognitive understanding of a rule is an internal variable. Similarly, different social situations often elicit distinct behavioral responses to the same stimulus, such as whether one chugs wine at a college party or sips it at a formal dinner.

External variables also influence motivational states. These variables include *rewarding incentive stimuli.* For example, when a dehydrated cheetah comes across a watering hole during a search for antelopes, the sight of the water may serve as an incentive stimulus, tipping the balance between hunger and thirst and driving the animal to interrupt its quest for food to drink. However, an internal variable—the state of the cheetah's hydration—can also lead to a different reward value being assigned to the same sensory stimulus, the watering hole. Even innately rewarding stimuli, such as a sweet tastant that normally elicits pleasure, can in some circumstances become unpleasurable. Chocolate cake may be innately rewarding to chocolate lovers, but satiation to the chocolate—which involves modulation of an internal variable—can decrease the reward value of this stimulus and thereby affect motivational state.

Rewards Can Meet Both Regulatory and Nonregulatory Needs on Short and Long Timescales

Feeding, drinking, and thermoregulatory behaviors and their underlying motivational states typically arise in response to (or anticipation of) a physiological imbalance. In these cases, actions acquire rewards in a relatively short timescale. In contrast, some motivational states serve biological imperatives other than short-term physiological homeostasis. More complex long-term goals, such as finding and sustaining a love partner or achieving an educational or professional goal, require goal-oriented actions on longer timescales. Nonregulatory motivational states may resemble those arising from physiological signals, but motivated behaviors often involve sequences of actions in which not every action is immediately rewarded (except in the sense of making progress toward a longer-term goal).

In general, incentive stimuli, even stimuli that only signal progress toward a longer-term goal, can influence motivational states so that complex behavioral sequences are completed. A simple example of this concept is when a cheetah must stalk, chase, run down, and kill an antelope, and then drag the carcass to a refuge before beginning to feed. Of course, even the complexity of the actions involved in foraging and feeding is far simpler than the steps required of a student motivated to achieve a graduate degree and develop an academic career. Motivational states must be sustained across challenging circumstances in order to achieve such goals.

The Brain's Reward Circuitry Provides a Biological Substrate for Goal Selection

Rewards are objects, stimuli, or activities that have positive value. Rewards can incite an animal to switch from one behavior to another or to resist interruption of ongoing action. For example, a rat that encounters a seed while scouting the environment may cease exploring in order to eat the seed or carry it to a safer place; while nibbling the seed, the rat will resist the efforts of another rat to steal the food from its paws. If seeds are made available only at a particular location and time, the rat will go to that location as the expected moment of reward delivery approaches.

Much contemporary work in neuroscience is directed at elucidating the neural systems that process different types of rewards. These systems must link the initial sensory representation of a reward to different behaviors that respond to physiological needs and environmental challenges and opportunities. Pathologies such as addiction can highjack these reward systems, resulting in maladaptive behavior (discussed in the latter part of this chapter).

Goal-directed behaviors entail the assessment of risks, costs, and benefits. Straying from the herd may offer an antelope better opportunities for foraging but only at the risk of becoming an easier target for a lurking cheetah. Attacking the venturesome antelope offers the cheetah an easier promise of a meal but at the risk that energetic and hydromineral resources will be depleted needlessly if the antelope gets away. Thus, the neural mechanisms responsible for goal selection must weigh the costs and benefits of different behaviors that might attain a specific goal.

In 1954, James Olds and Peter Milner reported their work on the neural pathways responsible for reward-related behaviors. These classic studies employed electrical brain stimulation as a goal. Rats and other vertebrates ranging from goldfish to humans will work for electrical stimulation of certain brain regions. The avidity and persistence of this self-stimulation are remarkable. Rats will cross electrified grids, run uphill while leaping over hurdles, or press a lever for hours on end in order to trigger the electrical stimulation. The phenomenon that leads the animal to work for self-stimulation is called *brain stimulation reward* (Figure 43–1A). Brain stimulation, therefore, elicits a motivational state, a strong drive to perform an action (eg, lever pressing) that will deliver further stimulation.

Although brain stimulation reward is an artificial goal, it mimics some of the properties of natural goal objects. For example, brain stimulation can compete with, summate with, or substitute for other reward-predictive stimuli to induce motivational states that

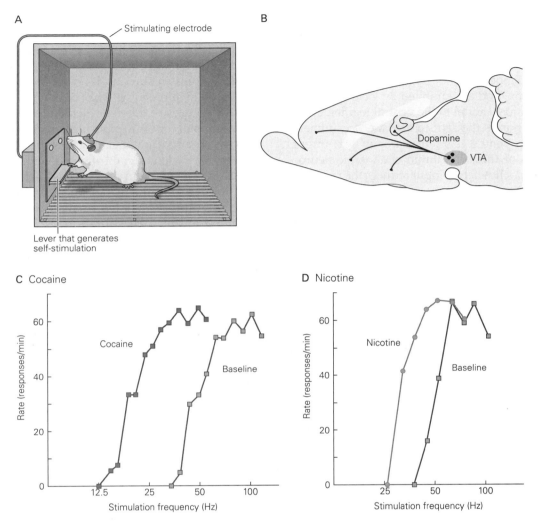

Figure 43–1 Intracranial self-stimulation recruits reward circuitry and dopaminergic neural pathways.

A. Classic testing apparatus for self-stimulation experiments. In this example, an electrode is implanted in a brain region of a rodent. Lever pressing by the rodent triggers electrical stimulation of that brain area.

B. Brain structures that produce self-stimulation behavior typically activate dopaminergic pathways emanating from the ventral tegmental area (**VTA**), among other pathways.

C–D. Cocaine and nicotine affect the rate of electrical self-stimulation. The rate at which the animal presses the stimulation lever increases with increases in the frequency of the self-stimulation current. In the presence of the drugs, animals press the lever at lower stimulation frequencies, indicating that the drugs augment the effects of the electrical stimulation.

drive goal-directed behaviors. The circuitry that mediates brain stimulation reward is broadly distributed. Rewarding effects can be produced by electrical stimulation of sites at all levels of the brain, from the olfactory bulb to the nucleus of the solitary tract.

Particularly effective sites lie along the course of the medial forebrain bundle and along longitudinally oriented fiber bundles coursing near the midline of the brain stem. Stimulation of either of these pathways results in activation of dopaminergic neurons in the ventral tegmental area of the midbrain. These neurons project to several areas of the brain, including the nucleus accumbens (the major component of the ventral striatum), the ventromedial portion of the head of the caudate nucleus (in the dorsal striatum), the basal forebrain, and regions of the prefrontal cortex (Figure 43–1B).

Activation of dopaminergic neurons in the ventral tegmental area plays a crucial role in brain stimulation reward. The effects of this activation are strengthened by increases in dopaminergic synaptic transmission and weakened by decreases. These dopaminergic neurons are excited by glutamatergic cells in the prefrontal cortex and amygdala as well as from cholinergic cells in the laterodorsal tegmental and pedunculopontine nuclei of the hindbrain, and are inhibited by local GABAergic cells within or just caudal to the ventral tegmental area. Brain stimulation is thought to activate dopaminergic neurons in the ventral tegmental area in part through the activation of these hindbrain cholinergic neurons. Blockade of this cholinergic input reduces the rewarding effects of the electrical stimulation. While most attention has focused on dopamine pathways in mediating brain stimulation reward, it is important to emphasize the involvement of non-dopaminergic pathways as well.

The strength of brain stimulation reward is indicated by the finding that starving rats provided with brief daily access to food will forego eating to press a lever for brain stimulation. The heedless pursuit of an artificial goal to the detriment of a biological need is one of many parallels between self-stimulation and drug abuse. Indeed, drugs of abuse augment the rewarding effects of activation of dopaminergic pathways with brain stimulation (Figure 43–1C,D). Lower frequencies of stimulating currents accompanied by cocaine or nicotine administration—both of which enhance dopaminergic neurotransmission through different mechanisms—produce a rate of lever pressing equivalent to that obtained during self-stimulation at higher stimulating currents in the absence of these drugs. These results indicate that cocaine and nicotine amplify the effects of neuronal activation elicited by microstimulation.

Dopamine May Act as a Learning Signal

An earlier view of the function of dopamine was that it conveyed "hedonic signals" in the brain and that, in humans, it was directly responsible for subjective pleasure. From this point of view, addiction would reflect the habitual choice of short-term pleasure despite a host of long-term life problems that emerge. In fact, however, new research indicates that the hedonic principle cannot easily explain the persistence of drug use by addicted persons as negative consequences mount.

The effects of dopamine have proven to be far more complex than was first thought. Dopamine can be released by aversive as well as by rewarding stimuli, and the short latency component of a dopamine neuron's response may not be related to the rewarding or aversive qualities of a stimulus at all. Moreover, rodents lacking dopamine—rats in which dopamine is depleted by 6-hydroxydopamine and mice genetically

engineered so that they cannot produce dopamine—continue to exhibit hedonic responses to sucrose. Dopamine delivery itself is not currently considered to produce hedonic qualities. Instead, the degree to which a particular sensory stimulus is rewarding is thought to be processed by a broad network of brain areas, spanning sensory cortices of different modalities, association cortex, prefrontal cortex (in particular, orbitofrontal regions), and many subcortical areas such as the amygdala, hippocampus, nucleus accumbens, and ventral pallidum.

Many of the brain areas whose activity is modulated by reward anticipation or receipt receive dopaminergic input. What information do dopaminergic neurons transmit to these brain areas? Wolfram Schultz and his colleagues discovered that dopaminergic neurons often have a complex and changing pattern of responses to rewards during learning. In one experiment, Schultz trained monkeys to expect juice at a fixed interval after a visual or auditory cue. Before the monkeys learned the predictive cues, the appearance of the juice was unexpected and produced a transient increase in firing above basal levels by ventral tegmental area dopaminergic neurons. As the monkeys learned that certain cues predict the juice, the timing of the firing changed. The neurons no longer fired in response to presentation of the juice—the reward—but earlier, in response to a predictive visual or auditory cue. If a cue was presented but the reward was withheld, firing paused at the time the reward would have been presented. In contrast, if a reward exceeded expectation or was unexpected, because it appeared without a prior cue, firing was enhanced (Figure 43–2).

These observations suggest that dopamine release in the forebrain serves not as a pleasure signal but as a *prediction error* signal. A burst of dopamine would signify a reward or reward-related stimulus that had not been predicted; pauses would signify that the predicted reward is less than expected or absent. If a reward is just as expected based on environmental cues, dopaminergic neurons would maintain their tonic (baseline) firing rates. Alterations in dopamine release are thus thought to modify future responses to stimuli to maximize the likelihood of obtaining rewards and to minimize fruitless pursuits. For natural rewards, like the sweet juice consumed by the monkeys in Schultz's experiments, once the environmental cues for a reward are learned, dopaminergic neuron firing returns toward baseline levels. Schultz has interpreted this to mean that as long as nothing changes in the environment, there is nothing more to learn and therefore no need to alter behavioral responses.

Experiments using functional magnetic resonance imaging in humans have provided further evidence

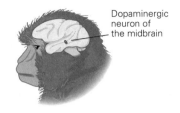

Unexpected reward

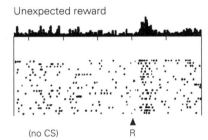

(no CS)   R

Predicted reward

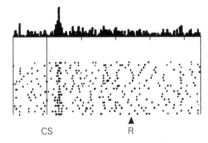

CS   R

Reward predicted but does not occur

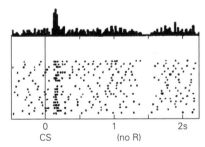

0   1   2s
CS   (no R)

Figure 43–2 Dopaminergic neurons report an error in reward prediction. Graphs show firing rates recorded from midbrain dopaminergic neurons in awake, active monkeys. **Top:** A drop of sweet liquid is delivered without warning to a monkey. The unexpected reward (**R**) elicits a response in the neurons. The reward can thus be construed as a positive error in reward prediction. **Middle:** The monkey has been trained that a conditioned stimulus (**CS**) predicts a reward. In this record, the reward occurs according to the prediction and does not elicit a response in the neurons because there is no error in the prediction of reward. The neurons are activated by the first appearance of a predicting stimulus but not by the reward. **Bottom:** A conditioned stimulus predicts a reward that fails to occur. The dopaminergic neurons show a decrease in firing at the time the reward would have occurred. (Reproduced, with permission, from Schultz, Dayan, and Montague 1997. Copyright © 1997 AAAS.)

that dopaminergic agonists and antagonists modulate reward learning and the blood oxygen level–dependent (BOLD) signal in the nucleus accumbens. However, in some experiments, mice that lack a dopamine synthesis gene can still learn where to find a sugar or cocaine reward, suggesting that dopamine is not required for all forms of reward learning. In addition, rodents who receive amphetamines to elevate presynaptic dopamine levels over a more extended time interval exhibit enhanced "wanting" behavior (ie, increased responding in the presence of a Pavlovian cue predicting sucrose reward).

These considerations have led some investigators to suggest that dopamine has a broader role than simply driving reinforcement learning by providing prediction-error signals. Indeed, several recent studies have demonstrated considerable variation in the response properties of different subpopulations of midbrain dopaminergic neurons. Some neurons are activated by both rewarding and aversive stimuli, while others are activated preferentially by one of the two types of stimuli, and still others show opposite responses (activated by rewards and suppressed by aversive stimuli). There is some evidence that these neuronal differences are related to differences in afferent inputs and efferent projections between subpopulations of dopaminergic neurons. Understanding the precise role of this complex mixture of dopamine signals—in learning, in driving goal-directed behavior, and especially in more complex forms of learning that involve longer timescales of sequences of actions to acquire distant rewards—remains an active area of investigation.

Unlike natural rewards, addictive drugs cause dopamine release in the reward circuitry no matter how often they are consumed, and the magnitude of this release is often greater than that seen with natural rewards—dopamine is released even when the drug no longer produces subjective pleasure. To the brain, consumption of addictive drugs might always signal "better than expected" and in this way would continue to influence behavior to maximize drug seeking and drug taking. If this idea is correct, it might explain why drug seeking and consumption become compulsive and why the life of the addicted person becomes focused increasingly on drug taking at the expense of all other pursuits.

Drug Addiction Is a Pathological Reward State

Drug addiction is a chronic and sometimes fatal syndrome characterized by compulsive drug seeking and consumption despite serious negative consequences such as medical illness and inability to function in the family, workplace, or society. Many drug addicts are aware of the destructive nature of their addiction but are unable to alter their addictive behavior despite numerous attempts at treatment.

An interesting feature of drug addiction is that only a minute fraction of all chemical substances can cause the syndrome. These so-called drugs of abuse do not share a common chemical structure, and they produce their effects by binding to different protein targets in the brain. Rather, these diverse substances can each cause a similar behavioral syndrome of addiction because their actions converge on the brain circuits that control reward and motivation (Figure 43–3).

Advances in understanding these actions have come about in large part based on studies of laboratory animals that self-administer the same drugs that cause addiction in humans. In fact, when animals are given free and unlimited access to these drugs, a subset will lose control over drug consumption—which becomes increasingly involuntary—at the expense of eating and sleeping, and some will even die by overdose. Drug self-administration and other animal models of addiction (Box 43–1) have made it possible to study both the

neural circuitry through which drugs of abuse act to produce their initial rewarding effects and the molecular and cellular adaptations that drugs induce in this circuitry after repeated exposures cause an addiction-like syndrome. Over the past decade, these studies in animals, together with brain imaging studies in human addicts, have provided an increasingly complete view of the addiction process.

All Drugs of Abuse Target Neurotransmitter Receptors, Transporters, or Ion Channels

A great deal is known about the initial interactions of addictive drugs with the nervous system. Virtually all of the proteins with which such drugs interact have been cloned and characterized (Table 43–1).

Each class of drug of abuse produces a different range of acute behavioral effects, consistent with the fact that each class acts on different targets and that these targets have distinct patterns of expression throughout the nervous system and peripheral tissues. Cocaine and other psychostimulants are activating and can cause cardiac side effects because their targets

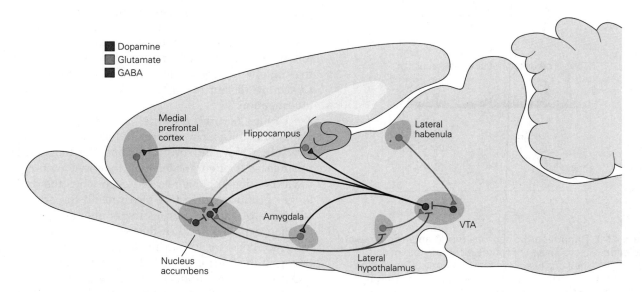

Figure 43–3 Brain reward circuits. A schematic drawing of the major dopaminergic, glutamatergic, and γ-aminobutyric acid (GABA)-ergic connections to and from the ventral tegmental area (**VTA**) and nucleus accumbens (NAc) in the rodent brain. The primary reward circuit includes dopaminergic projections from the VTA to the NAc. The VTA projections release dopamine in response to reward-related stimuli (and in some cases aversion-related stimuli). There are also GABAergic projections from the NAc to the VTA, with some in a direct pathway innervating the VTA and some in an indirect pathway innervating the VTA via intervening GABAergic neurons in the ventral pallidum (not

shown). The NAc also contains numerous types of interneurons. The NAc receives dense innervation from glutamatergic monosynaptic circuits from the medial prefrontal cortex, hippocampus, lateral habenula, and amygdala, among other regions. The VTA receives such inputs from the amygdala and prefrontal cortex and from several brain stem nuclei that use the transmitter acetylcholine (not shown). It also receives the peptidergic terminals of neurons in the lateral hypothalamus as well as other inputs. These various inputs control aspects of reward-related perception and memory. (Adapted from Russo and Nestler 2013.)

Box 43–1 Animal Models of Drug Addiction

Several animal models have played an important role in understanding how addictive drugs produce reward acutely and an addiction-like syndrome after repeated exposures.

Drug Self-administration

The reinforcing effects of a drug can be demonstrated in experiments in which animals perform a task (eg, press a lever) to receive an intravenous drug injection. In addition to studying acquisition of this behavior, scientists assess how hard an animal will work to deliver the drug by use of progressive ratio procedures, where each successive drug dose requires an increasing number of lever presses.

Animals reach a so-called break point when they stop self-administering the drug. After weeks or months of withdrawal from or extinction of drug self-administration, animals display a relapse-like behavior: They will press the stimulation lever, which no longer delivers the drug, in response to a test dose of the drug, cues associated previously with the drug (a light or tone), or stress. These various self-administration behaviors are considered the best-validated models of human addiction.

Conditioned Place Preference

Animals learn to associate a particular environment with passive exposure to drugs. For example, a rodent will spend more time on the side of a box where it was given cocaine than on the side where it received saline. This paradigm offers an indirect measure of potency of a drug reward and demonstrates the strong cue-conditioned effects of addictive drugs.

Locomotor Sensitization

All drugs of abuse stimulate locomotion in rodents upon initial drug exposure, with increasing locomotor activation seen after repeated drug doses. Since the neural circuitry that mediates locomotor responses to drugs of abuse partly overlaps with the circuitry that mediates reward and addiction, locomotor sensitization provides a model with which to study plasticity in this circuitry during a course of chronic drug exposure.

Intracranial Self-Stimulation

Animals will work (eg, press a lever) to deliver electrical current into parts of the brain's reward circuitry (see Figure 43–1). Drugs of abuse reduce the stimulation threshold for such self-stimulation, meaning that in the presence of drug animals will work for stimulation frequencies that have no effect under control conditions.

(monoamine transporters) are expressed in peripheral nerves that innervate the heart. In contrast, opiates are sedatives and potent analgesics because their targets (opioid receptors) are expressed in sleep and pain centers.

Nevertheless, all drugs of abuse acutely induce reward and reinforcement, and this shared action reflects the fact that the drugs, despite their very different initial targets, induce some common functional effects on the brain's reward circuitry (Figure 43–4). The best established of these shared initial effects is increased dopaminergic neurotransmission in the nucleus accumbens, albeit via different mechanisms. For example, cocaine produces this effect by blocking dopamine reuptake transporters located on the terminals of the ventral tegmental neurons, whereas opiates activate ventral tegmental area dopamine neuron cell bodies by inhibiting nearby GABAergic interneurons.

Opiates also produce reward through dopamine-independent actions (eg, by activating opioid receptors on nucleus accumbens neurons themselves). All other drugs of abuse act through a combination of dopamine-dependent and -independent mechanisms (eg, activation of endogenous opioid and cannabinoid signaling) to produce some of the same functional effects on nucleus accumbens neurons. Importantly, by increasing dopaminergic neurotransmission, all such drugs also produce some of the same functional effects mediated by activation of dopamine receptors on the many other projection targets of ventral tegmental dopamine neurons (Figure 43–3), actions that are also instrumental in reward and in initiating some of the deleterious actions of repeated drug exposure.

Repeated Exposure to a Drug of Abuse Induces Lasting Behavioral Adaptations

The acute rewarding actions of drugs of abuse do not account for addiction. Rather, addiction is mediated by the brain's adaptations to the repeated exposure to such acute actions. Two main questions in the field

Table 43-1 Major Classes of Addictive Drugs

Class	Source	Molecular target	Examples
Opiates	Opium poppy	μ opioid receptor (agonist)[1]	Morphine, methadone, oxycodone, heroin, many others
Psychomotor stimulants	Coca leaf Synthetic[2] Synthetic	Dopamine transporter (antagonist)[3]	Cocaine Amphetamines Methamphetamine
Cannabinoids	Cannabis	CB1 cannabinoid receptors (agonist)	Marijuana
Nicotine	Tobacco	Nicotinic acetylcholine receptor (agonist)	Tobacco
Ethyl alcohol	Fermentation	$GABA_A$ receptor (agonist), NMDA-type glutamate receptor (antagonist), and multiple other targets	Various beverages
Phencyclidine-like drugs	Synthetic	NMDA-type glutamate receptor (antagonist)	Phencyclidine (PCP, angel dust)
Sedative/hypnotics	Synthetic	$GABA_A$ receptor (positive allosteric modulator)	Barbiturates, benzodiazepines
Inhalants	Varied	Unknown	Glues, gasoline, nitrous oxide, others

[1]The signaling pathways induced by μ receptor activation differ between opiates, differences that might be related to different addiction liabilities. Additionally, most opiates activate the δ opioid receptor, although the action at μ receptors is most important for reward and addiction.
[2]The original synthesis of amphetamine was based on the natural plant product ephedrine.
[3]While cocaine is an antagonist of the transporter, amphetamine and methamphetamine act differently: They are substrates for the transporter and, once in the nerve terminal cytoplasm, act to stimulate dopamine release.
GABA, γ-aminobutyric acid; **NMDA**, N-methyl-D-aspartate.
Note: Caffeine can produce mild physical dependence but does not result in compulsive use. Some illegal drugs that are abused can be harmful but do not generally produce addiction; these include the hallucinogens lysergic acid diethylamide (LSD), mescaline, psilocybin, and 3,4-methylenedioxymethamphetamine (MDMA), popularly known as ecstasy.

remain: What specific adaptations mediate the behavioral syndrome of addiction, and why are some individuals more likely to become addicted?

We know—in both animals and humans—that roughly 50% of the risk for addiction across all drugs of abuse is genetic, but the specific genes that confer risk remain largely unknown. As for most other common chronic conditions, the genetic risk for addiction is highly complex, reflecting the combined actions of hundreds of genetic variations, each of which has a very small effect. The other 50% of the risk, while incompletely understood, involves a host of environmental factors including early life stress, stress throughout life, and peer pressure.

Historically, the adaptations induced by repeated drug exposure have been described by a series of pharmacological terms. *Tolerance* refers to the diminishing effect of a drug after repeated administration at the same dose or to the need for an increase in dose to produce the same effect. *Sensitization*, also known as reverse tolerance, occurs when repeated administration of the same drug dose elicits escalating effects. *Dependence* is defined as an adaptive state that develops

in response to repeated drug administration and is unmasked during *withdrawal*, which occurs when drug taking stops. The symptoms of withdrawal vary from drug to drug and include effects opposite to a drug's acute actions. Tolerance, sensitization, and dependence/withdrawal are seen with many drugs that are not addicting. For instance, two drugs used to treat hypertension, the β-adrenergic antagonist propranolol and the $α_2$-adrenergic agonist clonidine, produce strong dependence as evidenced by severe hypertension upon their sudden withdrawal.

Drugs of abuse are unique in causing tolerance, sensitization, and dependence/withdrawal in reward- and motivation-related behaviors, and these behaviors contribute to the syndrome of addiction. Reward tolerance, which can be viewed as homeostatic suppression of endogenous reward mechanisms in response to repeated drug exposure, is one factor leading to escalating patterns of drug use. Motivational dependence, which is manifested as negative emotional (eg, depression- and anxiety-like) symptoms seen during early drug withdrawal and also mediated by

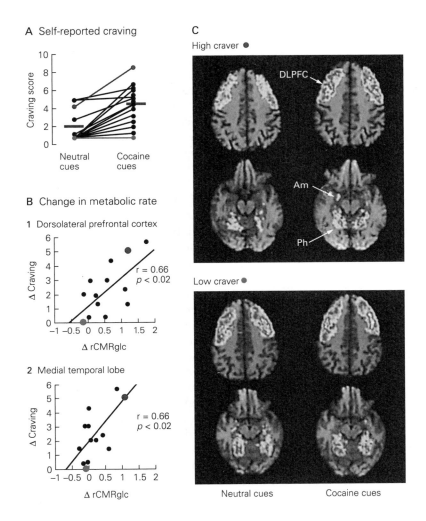

Figure 43–4 Positron emission tomography (PET) imaging reveals neural correlates of cue-induced cocaine craving. (Adapted, with permission, from Grant et al. 1996.)

A. Subjects were shown neutral or cocaine-related cues and asked, "How do you rate your craving or urge for cocaine on a scale of 1–10?" The mean craving score (**horizontal bar**) is significantly higher for exposure to cocaine-related cues than for exposure to neutral stimuli, even though the magnitude of the response across individuals varies considerably. Two subjects, identified by **red** and **blue dots**, represent high-level and low-level craving, respectively.

B. Changes in self-reported craving are correlated with changes in metabolic rate in the dorsolateral prefrontal cortex and medial temporal lobe during exposure to cocaine-related cues. The abscissa plots the difference in metabolic rate between the two sessions (activity with cocaine cues minus activity with neutral cues). Metabolic rate is measured as the regional cerebral metabolic rate for glucose (**rCMRglc**). The ordinate

plots the difference between the average of the responses to the question, "Do you have a craving or urge for cocaine?" in separate sessions with neutral and cocaine-related cues. (Each session lasted 30 minutes, and in each session, the question was asked three times.)

C. When subjects report a craving for cocaine, metabolic activity increases in the dorsolateral prefrontal cortex (**DLPFC**) and in two medial temporal lobe structures, the amygdala (**Am**) and parahippocampal gyrus (**Ph**). Pseudocolored PET images of metabolic activity are spatially aligned with high-resolution structural magnetic resonance images. Metabolic rate markedly increased in the amygdala and parahippocampal gyrus in one subject who reported a large increase in craving during presentation of cocaine-related cues (**red dots** in parts **A** and **B**). This effect is not evident in a subject who reported no increase in craving while exposed to the cocaine-related cues (**blue dots** in parts **A** and **B**). Metabolic activity outside the dorsolateral prefrontal cortex and medial temporal lobe is not shown.

suppressed endogenous reward mechanisms, is a leading factor in driving the return to drug use, or *relapse*. Reward sensitization, which typically occurs after longer withdrawal periods, can trigger relapse in response to exposure to the drug itself or to drug-associated cues (eg, being with people or in a place where drug was previously used).

Interestingly, a given drug can produce all of these adaptations—tolerance, sensitization, and dependence—simultaneously, due to different acute effects of the drug; this phenomenon emphasizes the involvement of multiple cell types and circuits in mediating a drug's global actions. The key challenge for neuroscientists is to identify the changes in specific types of neurons and glia—and in their consequent contributions to circuit function—that are induced by repeated drug exposure and that mediate the behavioral features that define a state of addiction.

Lasting Molecular Adaptations Are Induced in Brain Reward Regions by Repeated Drug Exposure

An extensive literature shows that repeated exposure to a drug of abuse in animal models alters the levels of many neurotransmitters and neurotrophic factors, their receptors and intracellular signaling pathways, and transcriptional regulatory proteins throughout the brain's reward circuitry. Most of these changes cannot be studied in living patients—only a small number of neurotransmitters and receptors can be assessed in patients with brain imaging—although studies of postmortem human brain tissue are being used increasingly to validate findings from animal models. Most of the reported research has focused on the ventral tegmental area and nucleus accumbens, although an increasing number of studies are examining other parts of the reward circuitry.

The most robust experimental findings are available for psychostimulants and opiates, probably because the changes induced by these drugs are larger in magnitude than those of other drugs of abuse. This likely reflects the greater inherent addictiveness of psychostimulants and opiates: With equivalent exposures, a larger fraction of people will become addicted to these drugs as compared with other classes of abused substances. Nevertheless, given the dominant public health consequences of alcohol, nicotine, and marijuana addictions, more attention should be given to these drugs.

Below, we summarize this large literature by focusing on a small number of drug-induced adaptations that have been linked causally to specific behavioral features of addiction in animal models. As will be clear in the next section, the present research focus is on relating these and many other molecular changes to synaptic and circuit adaptations also implicated in addiction.

Upregulation of the cAMP-CREB Pathway

Several drugs of abuse activate G_i protein–linked receptors, such as the D_2 dopamine receptor; the μ, δ, and κ opioid receptors; and the CB1 cannabinoid receptor. This means that, to a certain extent, many drugs of abuse will activate G_i protein–linked signaling pathways, with effects such as inhibition of adenylyl cyclase (Chapter 14), in the nucleus accumbens and other target neurons.

Work over the past two decades has established that, after repeated exposure, the affected neurons adapt to this sustained suppression of the cyclic adenosine monophosphate (cAMP) pathway by upregulating it, including induction of certain isoforms of adenylyl cyclase and protein kinase A. Repeated drug exposure likewise induces upregulation of the transcription factor CREB, which is normally activated by the cAMP pathway. Such upregulation of the cAMP-CREB pathway can be seen as a molecular mechanism of tolerance and dependence: It restores normal activity of these pathways despite the presence of a drug (tolerance and dependence), and when the drug is removed, the upregulated pathway is unopposed, causing abnormally high activity of the pathway (withdrawal) (Figure 43–5). Indeed, upregulation of the cAMP-CREB pathway in nucleus accumbens neurons has been shown to mediate both reward tolerance and motivational dependence and withdrawal in animal models.

Induction of ΔFosB

ΔFosB is a member of the Fos family of transcription factors. It is a truncated product of the *FosB* gene generated through alternative splicing. In contrast to all other members of the Fos family, which are induced rapidly and transiently in response to many perturbations in neural activity or cell signaling, ΔFosB is induced only slightly by initial presentation of stimuli. However, with repeated drug exposure, ΔFosB accumulates in neurons because of its unusual stability, unique among all Fos family proteins.

This phenomenon occurs within neurons of the nucleus accumbens and several other brain reward areas after repeated exposure to virtually any drug of abuse, including cocaine and other psychomotor stimulants, opiates, nicotine, ethanol, cannabinoids, and phencyclidine. Recent studies involving the selective

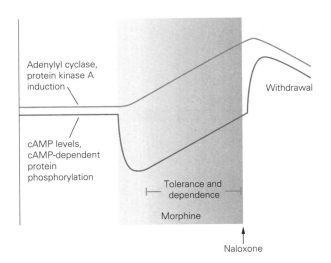

Figure 43–5 Upregulation of the cAMP-CREB pathway is a molecular mechanism underlying drug tolerance and dependence. Morphine or other μ opioid receptor agonists acutely inhibit the functional activity of the cyclic adenosine monophosphate (cAMP) pathway in brain reward neurons, as indicated, for example, by cellular levels of cAMP or protein kinase A (PKA)–dependent phosphorylation of substrates such as CREB. With continued drug exposure (**shading**), functional activity of the cAMP-CREB pathway is gradually upregulated and increases far above control levels upon removal of the drug (eg, by administration of the μ opioid receptor antagonist naloxone). These changes in the functional state of the cAMP-CREB pathway are mediated via the induction of adenylyl cyclase and PKA and activation of PKA substrates such as CREB in response to repeated drug administration. Induction of these proteins accounts for the gradual recovery in the functional activity of the cAMP-CREB pathway seen during chronic drug exposure (tolerance and dependence) and for the elevated activity of the cAMP-CREB pathway seen upon removal of the drug (withdrawal). First demonstrated for opiate drugs, similar regulation is seen in response to several other types of drugs of abuse. (Reproduced, with permission, from Nestler et al. 2020.)

expression or knockdown of ΔFosB in the nucleus accumbens of adult mice have provided direct evidence that induction of ΔFosB mediates reward sensitization, including increased drug self-administration and relapse. This is yet another example of a common adaptation to drugs of abuse that contributes to aspects of addiction shared across numerous drugs of abuse.

CREB and ΔFosB are two of many transcription factors implicated in drug addiction. Ongoing research is focused on characterizing the chromatin regulatory mechanisms through which these factors cooperate to regulate the expression of specific genes in the affected neurons and glia. Work is also underway to understand how these target genes drive their associated behavioral abnormalities via altered expression of proteins involved in synaptic, cell, and circuit function (Figure 43–6).

Lasting Cellular and Circuit Adaptations Mediate Aspects of the Drug-Addicted State

Repeated exposure to a drug of abuse can alter a neural circuit in two major ways. One mechanism, referred to as whole-cell or homeostatic plasticity, involves altering the intrinsic excitability of a nerve cell that will ultimately alter functioning of the larger circuit of which it is a part. It is easy to imagine how whole-cell plasticity in neurons within the brain's reward circuitry might mediate aspects of reward tolerance, sensitization, and dependence and withdrawal.

The other mechanism is synaptic plasticity, where connections between particular neurons are either strengthened or weakened. These synapse-specific adaptations could mediate the features of addiction that involve maladaptive memories, such as memories of the association of drug exposure with a host of environmental cues. This pathological learning and memory can increasingly focus an individual on the drug at the expense of natural rewards. Most attention in the field to date has concentrated on synaptic plasticity.

Synaptic Plasticity

As discussed elsewhere in this book, two major forms of synaptic plasticity have been described at glutamatergic synapses: *long-term depression* (LTD) and *long-term potentiation* (LTP). Over the past two decades, the molecular basis of both adaptations has been established, with distinct mechanisms underlying each of several distinct subtypes of LTD and LTP that occur throughout the nervous system. We now know that several types of drugs of abuse, in particular psychomotor stimulants and opiates, cause LTD- and LTP-like changes at particular classes of glutamatergic synapses in the brain's reward circuitry, with most work to date focused on the ventral tegmental area and nucleus accumbens.

Changes in the nucleus accumbens show interesting time-dependent adaptations as a function of drug withdrawal. At early withdrawal points (hours to days), glutamatergic synapses on neurons of the nucleus accumbens display LTD-like changes, which evolve into LTP-like changes after longer periods of withdrawal (weeks to months). Drug-induced LTD- and LTP-like adaptations in the nucleus accumbens involve morphological changes similar to those in other brain regions (mostly the hippocampus and cerebral cortex) where LTD and LTP occur in association with morphological changes in individual dendritic spines. During early withdrawal, LTD-like responses occur coincidently with increased numbers of

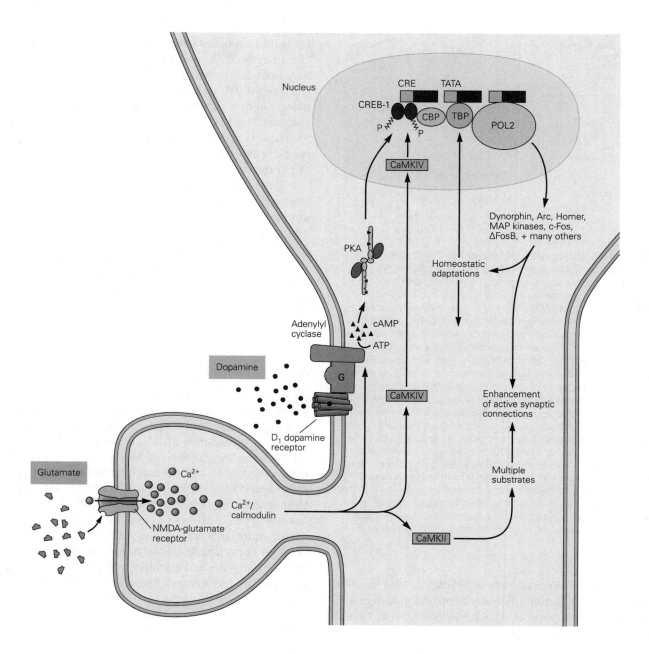

Figure 43–6 Dopamine- and glutamate-activated intracellular signaling pathways implicated in drug addiction. NMDA-type glutamate receptors permit Ca^{2+} entry, which binds calmodulin. The Ca^{2+}/calmodulin complex activates two types of Ca^{2+}/calmodulin-dependent protein kinases, CaMKII in the cytoplasm and CaMKIV in the cell nucleus. Certain dopamine receptors activate a stimulatory G protein that in turn activates adenylyl cyclase to produce cyclic adenosine monophosphate (**cAMP**). The cAMP-dependent protein kinase A (**PKA**) catalytic subunit can enter the nucleus. Once activated in the nucleus, both PKA and CaMKIV phosphorylate and thus activate cAMP response element binding protein (**CREB**). CREB recruits CREB-binding protein (**CBP**) and many other chromatin regulatory proteins and thereby activates the RNA polymerase II–dependent transcription of many genes, giving rise to proteins that can alter cellular function. Arc and Homer are localized in synaptic regions; mitogen-activated protein (**MAP**) kinases are protein kinases that control numerous cellular processes; Fos and ΔFosB are transcription factors; and dynorphin is a type of endogenous opioid peptide. These proteins are thought to contribute both to homeostatic responses to excessive dopamine stimulation and to the morphological and functional changes in synapses associated with memory formation. (Abbreviations: **ATP**, adenosine triphosphate; **NMDA**, *N*-methyl-ᴅ-aspartate; **POL 2**, RNA polymerase 2; **TBP**, TATA binding protein.)

immature, thin dendritic spines, whereas during later withdrawal, LTP-like responses occur coincidently with increased numbers of mature, mushroom-shaped spines. These findings suggest that repeated drug use weakens certain glutamatergic synapses with nucleus accumbens neurons via the induction of so-called *silent synapses* (Chapter 54), with a subset of these synapses strengthening during prolonged withdrawal.

These advances now define several ongoing lines of investigation. We need to understand which particular glutamatergic connections are affected and how those changes contribute to behavioral features of addiction. We need to define the molecular basis of this time-dependent synaptic plasticity, which is mediated in part through transcriptional mechanisms and altered expression levels of a host of proteins, including glutamate receptors, postsynaptic density proteins, proteins that regulate the actin cytoskeleton, and so on (Figure 43–6). In addition, we need to examine drug-induced glutamatergic synaptic plasticity at the several other reward-related brain regions that become corrupted in an addicted state, beyond the ventral tegmental area and nucleus accumbens. Finally, we need to understand how repeated exposure to drugs of abuse also corrupts inhibitory GABAergic synaptic transmission throughout this circuitry.

Whole-Cell Plasticity

As with synaptic plasticity, most examples of drug-induced whole-cell plasticity involve the ventral tegmental area and nucleus accumbens. For example, repeated cocaine exposure increases the intrinsic excitability of nucleus accumbens neurons, which contributes to reward tolerance. This adaptation is due in part to a decrease in expression of specific types of K^+ channels mediated by CREB, thus linking molecular-transcriptional adaptations to altered neural activity and an addiction-related behavioral abnormality. Repeated opiate exposure also increases the intrinsic excitability of dopaminergic neurons in the ventral tegmental area, but in a manner that impedes dopaminergic transmission to the nucleus accumbens. As with repeated cocaine exposure, this adaptation too is mediated by suppression of certain K^+ channels and contributes to reward tolerance.

Circuit Plasticity

Advanced tools are making it possible for the first time to track the activity of specific nerve cell types in the brain in awake, active animals and to experimentally manipulate the activity of those cells and study the behavioral consequences (Chapter 5). This is enabling scientists to define the precise ensembles of neurons within a given brain region that are affected by drug exposure over the life cycle of addiction—from initial drug exposure to compulsive drug consumption to withdrawal and relapse—and to provide causal evidence for the involvement of those neurons and the microcircuits within which they function. This work is beginning to define the distinct roles that various glutamatergic projections to the nucleus accumbens—from the prefrontal cortex, hippocampus, amygdala, and thalamus—play in controlling different cell types in the nucleus accumbens and the broader reward circuitry and in producing distinct addiction-related behavioral abnormalities.

While we have focused exclusively in this chapter on the effects of acute and chronic actions of drugs of abuse on the neural control of behavior, we realize that this is an oversimplification. As discussed in Chapter 7, neuronal function is intricately controlled by a host of nonneural cells in the brain, including astroglia, microglia, oligodendrocytes, and endothelial cells. There is growing evidence that each of these cell types is affected both directly and indirectly by drugs of abuse and that these nonneuronal actions also affect the long-term behavioral consequences of drug exposure. Integrating such actions with the neuronal effects of drugs of abuse will be required to achieve a comprehensive understanding of addiction.

Natural Addictions Share Biological Mechanisms With Drug Addictions

As previously indicated, the brain's reward circuitry evolved to motivate individuals to pursue natural rewards such as food, sex, and social interactions. Just as drug-addicted individuals display compulsive consumption of drugs of abuse, some people exhibit compulsive consumption of nondrug rewards (eg, compulsive overeating, shopping, gambling, video gaming, and sex), with behavioral consequences very similar to those observed in drug addiction. An interesting question for the field is whether these so-called "natural addictions" are mediated by some of the same molecular, cellular, and circuit adaptations that underlie drug addiction.

It is possible that these normal pleasurable behaviors excessively activate reward mechanisms in certain individuals who are particularly susceptible due to genetic or nongenetic factors. As with drugs, such activation may result in profound alterations in motivation that promote the repetition of initially rewarding behavior, despite the impact of negative consequences associated with the resulting compulsive behavior. It is far more difficult to study the neurobiological basis of natural addictions because of limitations in animal models (imagine a mouse model of compulsive shopping!), although progress is being made in developing such paradigms. In any event, brain imaging studies in humans support the notion that addictions to both drugs and behavioral rewards are associated with similar dysregulation of the brain's reward circuitry (Figure 43–3).

Highlights

1. Motivational states drive behaviors that either seek rewards or defend against or avoid aversive stimuli. Motivational states themselves are determined by a variety of internal and external variables. Internal variables include both physiological states and cognitive states. External variables include stimuli that possess innately rewarding or aversive properties, although the motivational significance of these properties may be modified by internal variables.

2. Rewards are desirable objects, stimuli, or actions. Rewards tend to elicit motivational states that drive approach behaviors. Rewards can meet regulatory needs on a short timescale, but can also result from complex sequences of behavior that achieve a long-term goal.

3. Key components of reward-related circuitry in the brain include dopaminergic neurons and brain areas targeted by dopaminergic neurons, such as the nucleus accumbens, ventral pallidum, amygdala, hippocampus, and parts of the prefrontal cortex. However, dopamine itself does not account for hedonic experiences.

4. Many dopaminergic neurons exhibit physiological response properties that suggest they communicate a prediction-error signal, with enhanced activity occurring when something better than expected occurs. This type of signal could play a critical role in different forms of reinforcement learning, learning that links stimuli or actions to rewards. However, recent studies have revealed more response heterogeneity in dopaminergic neurons than previously appreciated, including responses to aversive stimuli. This heterogeneity and its complex effects on neural circuit function remain active areas of investigation.

5. Drug addiction can be defined as the compulsive seeking and taking of a drug despite negative consequences to one's physical health or occupational and social functioning. The risk for addiction is roughly 50% genetic, with many hundreds of genes, each of which contributes a very small effect to this heritability. Important nongenetic risk factors include a history of adverse life events.

6. Drugs of abuse compose only a very small fraction of known chemical compounds. These drugs are chemically diverse, with each type acting initially on a distinct protein target. Nevertheless, the drugs can induce a common behavioral syndrome because their actions at these targets converge in producing similar functional effects on midbrain dopaminergic neurons or their projection regions such as the nucleus accumbens.

7. Addiction requires repeated exposure to a drug of abuse. Such repeated exposure is often accompanied by tolerance, sensitization, and dependence/withdrawal. While many nonabused drugs can produce tolerance and dependence/withdrawal, drugs of abuse are unique in their ability to produce these adaptations as well as sensitization in motivational and reward states.

8. The adaptations underlying drug addiction are mediated in part through lasting changes in gene expression, which result in altered intrinsic activity of neurons as well as structural and functional alterations in their synaptic contacts within the brain's reward circuitry.

9. An important goal of current research is to understand how a myriad of molecular changes summate to underlie specific changes in neural and synaptic function. Likewise, it will be important to understand how these neural and synaptic changes combine to alter the functioning of the brain's larger reward-related circuitry, so as to mediate specific behavioral abnormalities that define an addicted state.

10. This delineation of molecular, cellular, and circuit mechanisms of addiction will require increased attention to the specific cell types (neuronal and nonneuronal) in which certain drug-induced adaptations occur and to the specific microcircuits within the reward pathways affected by those adaptations.

11. A subset of individuals show addiction-like behavioral abnormalities to nondrug rewards, such as food, gambling, and sex. Evidence suggests that such so-called natural addictions are mediated by the same brain circuitry involved in drug addiction, with some common molecular and cellular abnormalities implicated as well.

12. These considerations highlight the need to learn more about the precise molecular, cellular, and

circuit bases of drug addiction. Nonetheless, our evolving understanding of the brain's reward circuitry and how individual synapses and cells in that circuitry are altered by drug exposure in a way that corrupts circuit function and usurps normal systems of reward and associative memory provides a compelling notion of what happens in the addicted brain.

Eric J. Nestler
C. Daniel Salzman

Selected Reading

Berridge KC, Robinson TE. 2016. Liking, wanting, incentive-sensitization theory of addiction. Am Psychologist 71: 670–679.

Di Chiara G. 1998. A motivational learning hypothesis of the role of mesolimbic dopamine in compulsive drug use. J Psychopharmacol 12:54–67.

Hyman SE, Malenka RC, Nestler EJ. 2006. Neural mechanisms of addiction: the role of reward-related learning and memory. Annu Rev Neurosci 29:565–598.

Olds J, Milner PM. 1954. Positive reinforcement produced by electrical stimulation of septal area and other regions of rat brain. J Comp Physiol Psych 47:419–427.

Schultz W. 2015. Neuronal reward and decision signals: from theories to data. Physiol Rev 95:853–951.

Wise RA, Koob GF. 2014. The development and maintenance of drug addiction. Neuropsychopharmacology 39:254–262.

References

Bevilacqua L, Goldman D. 2013. Genetics of impulsive behavior. Philos Trans R Soc Lond B Biol Sci 368:20120380.

Calipari ES, Bagot RC, Purushothaman I, et al. 2016. In vivo imaging identifies temporal signature of D1 and D2 medium spiny neurons in cocaine reward. Proc Natl Acad Sci U S A 113:2726–2731.

Carlezon WA Jr, Chartoff EH. 2007. Intracranial self-stimulation (ICSS) in rodents to study the neurobiology of motivation. Nat Protoc 2:2987–2995.

Dong Y. 2016. Silent synapse-based circuitry remodeling in drug addiction. Int J Neuropsychopharmacol 19:pyv136.

Everitt BJ, Belin D, Economidou D, Pelloux Y, Dalley JW, Robbins TW. 2008. Review. Neural mechanisms underlying the vulnerability to develop compulsive drug-seeking habits and addiction. Philos Trans R Soc Lond B Biol Sci 363:3125–3135.

Gipson CD, Kupchik YM, Kalivas PW. 2014. Rapid, transient synaptic plasticity in addiction. Neuropharmacology 76 Pt B:276–286.

Goldstein RZ, Volkow ND. 2011. Dysfunction of the prefrontal cortex in addiction: neuroimaging findings and clinical implications. Nat Rev Neurosci 12:652–669.

Grant S, London ED, Newlin DB, et al. 1996. Activation of memory circuits during cue-elicited cocaine cravings. Proc Natl Acad Sci U S A 93:12040–12045.

Loweth JA, Tseng KY, Wolf ME. 2014. Adaptations in AMPA receptor transmission in the nucleus accumbens contributing to incubation of cocaine craving. Neuropharmacology 76 Pt B:287–300.

Lüscher C, Malenka RC. 2011. Drug-evoked synaptic plasticity in addiction: from molecular changes to circuit remodeling. Neuron 69:650–663.

Matsumoto M, Hikosaka O. 2009. Two types of dopamine neuron distinctly convey positive and negative motivational signals. Nature 459:837–841.

Nestler EJ, Kenny PJ, Russo SJ, Schaefer A. 2020. *Molecular Neuropharmacology: A Foundation for Clinical Neuroscience*, 4th ed. New York: McGraw-Hill.

Pessiglione M, Seymour B, Flandin G, Dolan RJ, Frith CD. 2006. Dopamine-dependent prediction errors underpin reward-seeking behavior in humans. Nature 442: 1042–1045.

Polter AM, Kauer JA. 2014. Stress and VTA synapses: implications for addiction and depression. Eur J Neurosci 39:1179–1188.

Robbins TW, Clark L. 2015. Behavioral addictions. Curr Opin Neurobiol 30:66–72.

Robinson TE, Kolb B. 2004. Structural plasticity associated with exposure to drugs of abuse. Neuropharmacology 47(Suppl 1):33–46.

Robison AJ, Nestler EJ. 2011. Transcriptional and epigenetic mechanisms of addiction. Nat Rev Neurosci 12:623–637.

Russo SJ, Nestler EJ. 2013. The brain reward circuitry in mood disorders. Nat Rev Neurosci 14:609–625.

Schmidt HD, McGinty JF, West AE, Sadri-Vakili G. 2013. Epigenetics and psychostimulant addiction. Cold Spring Harb Perspect Med 3:a012047.

Schultz W, Dayan P, Montague PR. 1997. A neural substrate of prediction and reward. Science 275:1593–1599.

Scofield MD, Kalivas PW. 2014. Astrocytic dysfunction and addiction: consequences of impaired glutamate homeostasis. Neuroscientist 20:610–622.

Stuber GD, Britt JP, Bonci A. 2012. Optogenetic modulation of neural circuits that underlie reward seeking. Biol Psychiatry 71:1061–1067.

Volkow ND, Morales M. 2015. The brain on drugs: from reward to addiction. Cell 162:712–725.

44

Sleep and Wakefulness

S LEEP IS A REMARKABLE STATE. It consumes fully a third of our lives—approximately 25 years in the average lifetime—yet we know little about what happens in the brain during this daily excursion. Perhaps even more surprising, the exact functions of sleep and of dreaming, one of the more noteworthy components of sleep, are still unknown.

Although the psychological content of dreams has been a rich subject of speculation from Plato and Aristotle to Sigmund Freud, we still do not understand whether dreams carry deep personal meaning, as Freud hypothesized, or represent the brain "throwing out its trash," the bits and pieces of daily experience that are not worth retaining, as Francis Crick speculated. One function of sleep may be to allow synaptic remodeling and consolidation of memory traces reflecting the day's experiences, but the role of dreaming in this process remains a subject of intense debate.

When studying sleep and wakefulness, researchers typically use a polysomnogram, which consists of three physiological measures: brain activity measured by an electroencephalogram (EEG) (see Figure 58–1), eye movements recorded by an electro-oculogram (EOG), and muscle tone measured by an electromyogram (EMG) (Figure 44–1B). In clinical polysomnograms, respiration is also measured, as breathing during sleep is disrupted in many patients with sleep disorders.

Figure 44-1 Electrophysiological patterns of wakefulness and sleep.

A. A hypnogram or graph showing the progression of sleep stages over a typical night in a healthy young person. Periods of rapid eye movement (**REM**) sleep alternate with non-REM sleep about every 90 minutes. An individual typically progresses from the awake state into light non-REM sleep (N1) then progressively deeper non-REM sleep (N2, N3), then back to lighter non-REM sleep before the first period of REM sleep occurs (**light blue bars**). As the night progresses, the individual spends less time in the deepest stage of non-REM sleep, and the duration of REM sleep periods increases.

B. The records show the components of the polysomnogram used to distinguish sleep stages. The electro-oculogram (**EOG**) records eye movements from electrodes on either side of the eyes. The electroencephalogram (**EEG**) records cortical field potentials from the scalp; the electromyogram (**EMG**) records muscle fiber firing through the skin. During the awake state, the EOG shows voluntary eye movements, the EEG shows fast low-amplitude activity, and the EMG shows variable muscle tone. Stage N1 sleep is characterized by a slight slowing of EEG frequencies and slow roving eye movements, with less EMG activity; stage N2 is characterized by bursts of 12- to 14-Hz activity called sleep spindles and high-voltage slow waves called K-complexes; stage N3 is dominated by high-voltage slow waves. During REM sleep, the EEG is similar to that of the awake state. Rapid eye movements can be seen on the EOG, but the EMG is so silent that contamination by tiny electrocardiogram signals can sometimes be seen (as in the illustrated case).

A Hypnogram

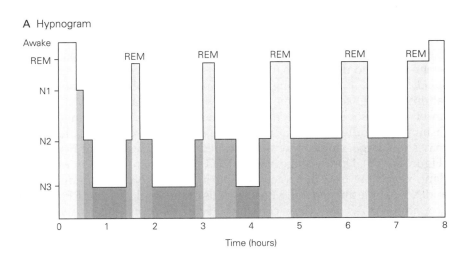

B Components of the polysomnogram

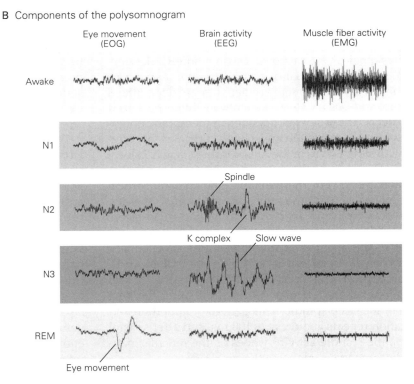

During wakefulness, the EEG is characterized by mainly high-frequency, low-voltage activity, indicative of the unique activity of individual cortical neurons; the EOG shows frequent eye movements; and the EMG shows moderate and variable muscle tone. During quiet wakefulness, with eyes closed, rhythmic EEG waves in the alpha range (8–13 Hz) are common, particularly over the occipital region. For most of the sleep period, the EEG shows slower activity, but periodically during the night, there are shifts into a sleep state with a faster, lower-voltage EEG, loss of muscle tone, and rapid eye movements called rapid eye movement

(REM) sleep. The entire period of slow EEG activity, from light drowsiness to deep sleep, is referred to as non-REM sleep and is divided into three stages, N1 to N3 (Figure 44–1).

Sleep Consists of Alternating Periods of REM Sleep and Non-REM Sleep

As an individual becomes drowsy and transitions into light non-REM sleep (stage N1), the EEG slows and shows waves in the theta range (4–7 Hz) (Figure 44–1B).

Consciousness begins to fade during stage N1, but the individual may still be awakened by minimal stimulation. Stage N2 often contains some slow EEG activity in the theta and delta range (0.5–4 Hz) as well as *sleep spindles*, 10- to 16-Hz waxing and waning EEG oscillations lasting 1 to 2 seconds, typically with a gradual onset and offset so the EEG waves resemble an old-fashioned spindle tapered at both ends. The EEG also may show large, single slow waves called *K-complexes* (Figure 44–1B). During stage N3, the EEG shows abundant, very slow EEG delta activity. During stages N2 and N3, people are generally unconscious of the world around them as the slow cortical activity disrupts information processing. Across all stages of non-REM sleep, eye movements are absent, muscle tone is low, breathing is slow and regular, and body temperature falls.

Slow EEG activity and sleep spindles arise, respectively, from cortico-cortical and cortico-thalamic electrophysiological interactions. During non-REM sleep, the membrane potential of cortical pyramidal neurons fluctuates between *Up states* (when they are depolarized and fire) and *Down states* (when they are hyperpolarized and silent). These slow oscillations in membrane potential, which occur even in an isolated cortical slab, correlate with slow waves in the EEG. During stage N2 sleep, spindles arise from an interaction of neurons in the reticular nucleus of the thalamus and thalamocortical relay neurons. Thalamocortical neurons are generally hyperpolarized and inactive during non-REM sleep, but inhibition from the reticular thalamic neurons can result in the opening of low-threshold Ca^{2+} channels, which drive a burst of Na^+ spikes in the thalamocortical neurons. The thalamocortical neurons then excite and recruit more reticular neurons, initiating the next cycle of the sleep spindle. This pattern of inhibition and excitation repeats about every 100 ms, and after several cycles, the spindle activity wanes as the reticular neurons become less responsive (Figure 44–2).

After about 90 minutes of sleep, people usually enter the stage known as REM sleep, a period in which dreams are often vivid and sometimes bizarre. REM sleep was discovered in 1953 when Eugene Aserinsky and Nathaniel Kleitman observed that across a night of sleep, adults have several episodes of jerky conjugate eye movements, and when awakened from this state, about three-fourths of subjects reported dreams with visual imagery.

Muscle tone is extremely low during REM sleep, owing to inhibition of motor neurons by descending pathways from the brain stem. This paralysis affects nearly all motor neurons except those that support respiration, eye movements, and a few other functions

such as sphincter control. As discussed later in this chapter, this inhibition of motor neurons is crucial as it prevents the physical enactment of dreams.

During REM sleep, the body undergoes many additional physiological changes. Body temperature falls during non-REM sleep, and it can fall further during REM sleep as the generation and retention of heat are minimal. Autonomic regulation is altered such that heart rate and blood pressure can vary wildly. In addition, men experience penile erections and women experience physiological signs of sexual arousal during REM sleep.

Across the night, episodes of non-REM sleep alternate with REM sleep, and each of these sleep cycles takes about 90 minutes. Sleep in a healthy young adult usually begins with a rapid descent into stage N3 non-REM sleep, followed by lighter non-REM sleep and then some REM sleep, and with each cycle, non-REM sleep becomes lighter and the periods of REM sleep become longer (Figure 44–1A). At the end of the sleep period, people often wake spontaneously from an episode of REM sleep.

The Ascending Arousal System Promotes Wakefulness

Modern perspectives of the neural basis of sleep and wakefulness go back about 100 years to the concepts derived by the neurologist and neuropathologist Baron Constantin von Economo. Around World War I, he observed an unusual type of encephalitis, believed to be a viral infection of the brain that specifically attacked the sleep–wake control circuitry. In most cases, patients had "encephalitis lethargica," sleeping 20 or more hours per day. When they awoke, they were generally cogent, but they would stay awake only long enough to eat and then go right back to sleep. This intense sleepiness would persist for many months before improving. But in patients who died during this interval, von Economo found focal damage to the brain, at the junction of the midbrain and diencephalon, leading him to hypothesize that the upper brain stem and posterior hypothalamus contain critical circuitry that activates the forebrain, producing a normal wakeful state.

Other patients afflicted in the same epidemic had just the opposite problem: unrelenting severe insomnia. They would be restless and, despite feeling sleepy, unable to fall asleep. Eventually, they would fall into a fitful sleep for only a few hours each day, but would waken without feeling refreshed. In post mortem examinations of these patients, von Economo found

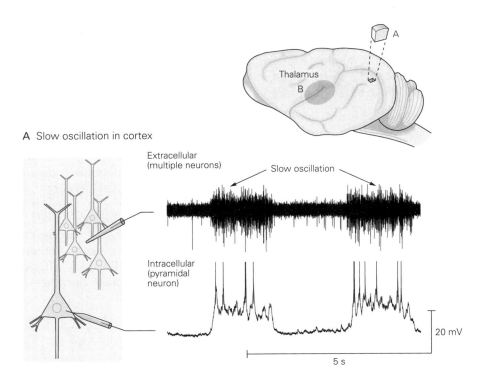

A Slow oscillation in cortex

Extracellular
(multiple neurons)

Slow oscillation

Intracellular
(pyramidal
neuron)

20 mV

5 s

B Spindle wave in thalamus

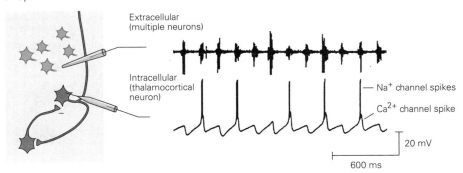

Extracellular
(multiple neurons)

Intracellular
(thalamocortical
neuron)

Na⁺ channel spikes

Ca²⁺ channel spike

20 mV

600 ms

Figure 44–2 Cellular mechanisms of electroencephalogram rhythms during sleep.

A. The slow oscillation that underlies the slow waves of the EEG during non-REM sleep is generated within the cerebral cortex by intrinsic massively recurrent excitatory and inhibitory connections. Slow waves will continue even in an isolated cortical slab. Intracellular recordings from such neurons during slow oscillations show rhythmic down states when the individual neurons are hyperpolarized and do not fire, alternating with up states when the membrane potential is more depolarized and the neurons fire multiple action potentials. This synchronous firing produces waves of dendritic potentials, which appear as slow waves in the EEG. (Data from Dr. David McCormick.)

B. Similarly, in a thalamic slice, recurrent circuitry generates spindle waves. A burst of spikes in reticular nucleus neurons (**gray**) hyperpolarizes thalamocortical relay neurons (**red**) sufficiently to de-inactivate low-threshold (T-type) Ca²⁺ channels. As the hyperpolarization wanes, these Ca²⁺ channels open and the resultant Ca²⁺ current depolarizes the relay neuron, producing a brief burst of Na⁺ channel spikes on top of a Ca²⁺ channel spike plateau.

Meanwhile, as the burst in the relay cell continues, its excitatory output generates a T-type Ca²⁺ channel spike in the reticular neuron, which drives another burst of Na⁺ spikes. The resultant volley of feedback inhibition to the relay cells initiates a new burst cycle. This firing pattern recurs 12 to 14 times per second, and the resulting waves of thalamocortical action potentials reaching the cortex produce sleep spindles in the EEG. The upper trace shows action potentials from a local population of relay cells. The lower trace shows inhibitory postsynaptic potentials and spike bursts from an individual relay cell; on this slow time base, each upstroke in the intracellular record represents a burst of up to six action potentials. As the trace shows, individual relay neurons do not reach spike threshold during every cycle of the spindle wave. As a result, the amplitude of the extracellular spike activity varies from cycle to cycle, depending on which neurons happen to fire and their distances from the extracellular electrode tip. However, each burst of thalamic firing would produce a volley of excitatory postsynaptic potentials in the cortex, resulting in an electroencephalogram wave, time-locked to the thalamic firing. (Reproduced, with permission, from Bal, von Krosigk, and McCormick 1995. Copyright © 1995 The Physiological Society.)

lesions in the anterior hypothalamus. He proposed that neurons in this area are important for inhibiting the brain stem arousal system to allow sleep. Modern studies have shown a system of wake–sleep circuitry in the brain that is remarkably close to von Economo's model.

The Ascending Arousal System in the Brain Stem and Hypothalamus Innervates the Forebrain

The composition of the ascending arousal system has been debated since von Economo's time. In the late 1940s and early 1950s, lesion studies confirmed that damage to the upper midbrain reticular formation could cause coma, whereas electrical stimulation of this region could arouse animals. The location and nature of the wake-promoting neurons were unknown.

In the succeeding decades, it became clear that these lesions damaged the axons of neurons in the upper brain stem that project to the forebrain, including noradrenergic neurons in the locus ceruleus, serotonergic neurons in the dorsal and median raphe, and midbrain dopaminergic neurons (Chapter 40). The axons of other neurons in the posterior hypothalamus, including those producing histamine and orexin, also join this pathway, which splits into two bundles, with some projections innervating the thalamus and others the hypothalamus, basal forebrain, and cerebral cortex (Figure 44–3).

Neurons contributing to all of these ascending pathways fire fastest during the awake state but much slower during sleep, suggesting that they are wake-promoting. However, although many monoamine antagonists cause sleepiness, and lesions of the

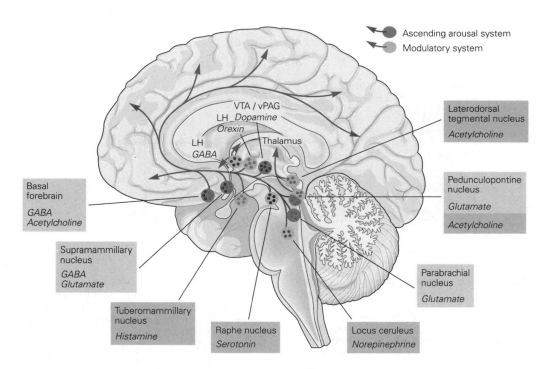

Figure 44–3 The ascending arousal system. The ascending arousal system comprises primarily axons from glutamatergic neurons in the parabrachial and pedunculopontine tegmental nuclei and cholinergic and GABAergic (**dark gray**) neurons in the basal forebrain. Lesions of either the parabrachial and pedunculopontine nuclei or the basal forebrain cause coma. Of somewhat lesser importance are dopaminergic neurons in the ventral tegmental area (**VTA**) and ventral periaqueductal gray (**vPAG**) matter and glutamatergic and GABAergic neurons in the supramammillary nucleus, where lesions can increase sleep by about 20%. In addition, populations of modulatory neurons can strongly promote wakefulness when stimulated, but when damaged cause minimal changes in wake–sleep amounts. These include the

monoaminergic neurons in the noradrenergic locus ceruleus, the serotonergic dorsal and median raphe nuclei, and the histaminergic tuberomammillary nucleus; the cholinergic neurons in the pedunculopontine and lateral dorsal tegmental nuclei; and the orexinergic neurons in the lateral hypothalamus (**LH**). All of these neurons send their axons through the hypothalamus and basal forebrain directly to the cerebral cortex, where their net effect is to increase cortical arousal. Many of the modulatory pathways also activate the thalamus, enabling thalamic transmission of sensory information to the cerebral cortex. GABAergic neurons in the lateral hypothalamus also promote wakefulness by inhibiting neurons in the ventrolateral preoptic area and reticular nucleus of the thalamus that oppose wakefulness.

monoaminergic cell groups impair the ability to stay awake under adverse conditions, such lesions have little lasting effect on the amount or timing of wake or sleep.

Lesions of the orexinergic neurons in the lateral hypothalamus cause narcolepsy, a condition in which sleep–wake states are present in normal amounts but are unstable, as discussed later. In fact, of all of the monoaminergic cell groups that are thought to contribute to arousal, only lesions of the dopaminergic neurons near the dorsal raphe nucleus cause small but long-lasting reductions in arousal, resulting in about a 20% increase in total sleep time. Interestingly, the ability of drugs such as amphetamine or modafinil to promote wakefulness appears to depend upon their ability to block dopamine reuptake, as mice with deletions of the dopamine transporter do not respond to these drugs.

Because lesions of the ascending monoaminergic and orexinergic pathways have little if any effect on the total amount of wakefulness, recent work has emphasized the role of glutamatergic, cholinergic, and GABAergic neurons in maintaining wakefulness. Lesions of glutamatergic neurons in the dorsolateral rostral pons, including the parabrachial nucleus and adjacent pedunculopontine tegmental nucleus, cause a comatose state from which the animals cannot be awakened. Lesions confined to the thalamus impair the content of consciousness but have relatively little effect on wake–sleep cycles. On the other hand, lesions of the posterior lateral hypothalamus cause profound sleepiness, which cannot be accounted for by damage to the orexinergic or histaminergic neurons in this region. Glutamatergic neurons in the supramammillary region, which activate the cortex, and GABAergic neurons in the lateral hypothalamus that inhibit sleep-promoting circuits, may account for this arousal effect. Finally, large bilateral lesions of the basal forebrain also can produce coma, similar to lesions of the dorsolateral pons. Optogenetic or chemogenetic activation of cholinergic, GABAergic, or glutamatergic neurons in the basal forebrain indicates that neurons of all three types may produce arousal.

Thus, the current view of the ascending arousal system is that the crucial components are glutamatergic neurons in the dorsolateral pons, supramammillary hypothalamus, and basal forebrain; cholinergic neurons in the dorsolateral pons and basal forebrain; and GABAergic neurons in the lateral hypothalamus and basal forebrain. These are likely to be augmented by modulatory pathways containing orexin and monoamines, needed to allow full and sustained wakefulness particularly under adverse conditions (Figure 44–3).

Damage to the Ascending Arousal System Causes Coma

Consciousness depends upon the activity of the cerebral hemispheres during the awake state. Hence, loss of consciousness occurs when there is injury to the ascending arousal system or to both cerebral hemispheres, or there is a severe metabolic derangement (eg, low blood sugar, inadequate oxygenation, various forms of drug intoxication) that affects both the arousal system and its cortical targets. A patient who cannot be awakened, even by vigorous stimulation, is said to be in a coma. Those who can be partially awakened by such stimuli are said to be stuporous or obtunded.

The clinical approach to a comatose or obtunded patient is first to determine if there is injury to the ascending arousal system. Because of the proximity of the arousal pathways to those that control eye movement and pupillary responses, as well as respiration and some motor responses (Chapter 40), clinicians examine these brain stem functions carefully. If these functions are intact, it is likely that the problem is due to a metabolic condition, which can be assessed by various blood and spinal fluid tests. In addition, a computed tomographic (CT) scan of the brain is needed to look for pathology affecting both cerebral hemispheres (eg, a large tumor or blood clot).

Circuits Composed of Mutually Inhibitory Neurons Control Transitions From Wake to Sleep and From Non-REM to REM Sleep

In contrast to coma, sleep is a temporary, reversible loss of consciousness produced by specific brain circuitry that inhibits the ascending arousal system. Neurons in the ventrolateral preoptic nucleus contain the inhibitory neurotransmitters GABA and galanin and project extensively to most parts of the ascending arousal system. These preoptic neurons fire slowest during wakefulness, increase their firing as animals fall asleep, and fire fastest during deep sleep after a period of sleep deprivation.

Similarly, GABAergic neurons in the nearby median preoptic nucleus also promote sleep and project to some components of the arousal system. Lesions of these preoptic neurons result in fragmented sleep and cause animals to lose as much as half of their total sleep. Of clinical relevance, elderly people often have fragmented sleep, and those with the most fragmented sleep show the greatest loss of the sleep-promoting ventrolateral preoptic galanin neurons in postmortem examination. In addition, a population of GABAergic neurons in the parafacial zone, a region near the facial nerve as it courses through the brain stem, inhibits the parabrachial nucleus. Lesions of the parafacial zone also result in loss of up to half of total sleep time.

Interestingly, the ventrolateral preoptic neurons receive inhibitory inputs from neurons throughout the arousal system. Mutually inhibitory connections between the ventrolateral preoptic neurons and the arousal system result in a neural circuit with properties similar to an electrical flip-flop switch, in which each side of the circuit turns the other off. Such a circuit produces rapid and full transitions between two states. Although it may sometimes appear that it takes a long time to fall asleep, the actual transitions from wake to sleep, or vice versa, are generally quick, taking only a few seconds to a few minutes. In fact, most animals spend nearly their entire day clearly awake or asleep, with very little time spent in transitions. These rapid transitions are behaviorally adaptive as an animal would be vulnerable in an intermediate, drowsy state. A neural flip-flop switch prevents this situation because when either side of the switch gains advantage over the other, the circuit produces a rapid and complete transition in state (Figure 44–4).

REM sleep is generated by a network of brain stem neurons centered in the pons. The fast EEG rhythms and dream activity of REM sleep are thought to be driven by coordinated activity of glutamatergic neurons in the subceruleus region (ventral to the locus ceruleus in the pons) plus cholinergic and glutamatergic neurons in the parabrachial and pedunculopontine tegmental nucleus that innervate the basal forebrain and thalamus. Other glutamatergic subceruleus neurons produce the paralysis of REM sleep via projections to the ventromedial medulla and spinal cord, where they activate GABAergic and glycinergic neurons that deeply hyperpolarize motor neurons.

The subceruleus area in turn receives input from a population of GABAergic neurons in and just lateral to the periaqueductal gray matter, where the cerebral aqueduct opens into the fourth ventricle. These neurons are most active during wake and non-REM sleep, and they inhibit the subceruleus neurons, preventing entry into REM sleep. Conversely, GABAergic neurons in the subceruleus area also project back to the ventrolateral periaqueductal gray region. The mutual inhibition between the two populations of neurons may form another flip-flop switch that promotes rapid and complete transitions into and out of REM sleep.

Interestingly, the noradrenergic locus ceruleus and serotonergic dorsal raphe nucleus innervate and inhibit the subceruleus region. Thus, REM sleep is often reduced when people take antidepressants that increase brain levels of serotonin or norepinephrine.

In addition, as these monoaminergic neurons are active during wakefulness, they prevent direct transitions from wake into REM sleep (Figure 44–5).

Sleep Is Regulated by Homeostatic and Circadian Drives

The circadian regulation of sleep obeys a 24-hour biological clock (described later), whereas the homeostatic drive for sleep gradually accumulates during the awake state. After a period of sleep deprivation, much of the lost sleep is recovered over the next few nights, which in younger people may involve deeper and longer periods of stage N3 non-REM sleep.

REM sleep is also recovered after REM deprivation; rebound REM sleep can include especially intense dreams, long periods of REM sleep, and occasional breakthrough of REM sleep phenomena into wakefulness, such as dream-like hallucinations or brief paralysis when falling asleep or waking up. Rebound sleep on a weekend is commonly enriched in REM sleep in people who wake up early to an alarm clock on workdays and miss out on the last portion of sleep, which is mainly REM sleep.

The Homeostatic Pressure for Sleep Depends on Humoral Factors

Humoral factors circulating in the brain signal the homeostatic pressure for non-REM sleep. The brain is metabolically quite active during wakefulness and uses ATP, but with sustained periods of wakefulness, ATP is dephosphorylated to adenosine, which acts as a local neuromodulator in the extracellular environment. Adenosine type 1 receptors are inhibitory receptors that are expressed on wake-promoting neurons and in many other parts of the brain, so higher adenosine levels may produce sleepiness by inhibiting these neurons. In addition, adenosine can excite neurons via adenosine type 2a receptors; these receptors are common in the shell of the nucleus accumbens and may cause sleepiness by means of projections to the hypothalamus that activate the ventrolateral preoptic neurons.

The pressure for sleep can be measured by the time it takes an individual to fall asleep if given the opportunity in a comfortable environment. This approach is used by sleep clinicians in the Multiple Sleep Latency Test, in which an individual is given 20-minute intervals to try to fall asleep in a comfortable, quiet bed every 2 hours beginning at 9 am, for five sleep opportunities. An individual who is well rested generally

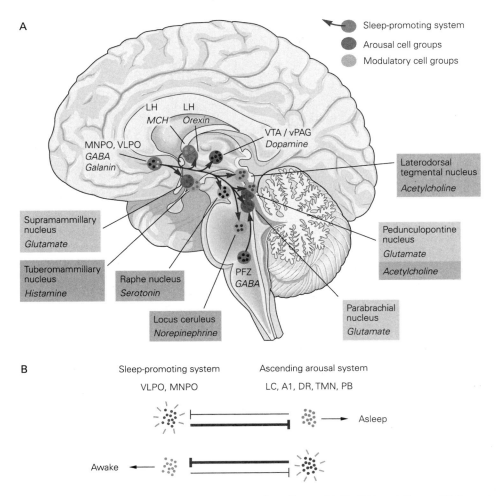

Figure 44–4 Sleep-promoting pathways.

A. The components of the ascending arousal system (Figure 44–3) receive inhibitory, largely GABAergic inputs from sleep-promoting neurons. Neurons in the ventrolateral and median preoptic nuclei (**VLPO, MNPO**) innervate the entire arousal-promoting system, while those in the parafacial zone (**PFZ**) innervate mainly the parabrachial area. Many of the VLPO neurons also contain galanin (GAL), an inhibitory peptide. Neurons in the lateral hypothalamus (**LH**) that release melanin-concentrating hormone (**MCH**) may promote REM sleep by inhibiting both nearby orexinergic neurons as well as neurons in the periaqueductal gray matter that prevent REM sleep

(see Figure 44–5). (Abbreviations: **vPAG**, ventral periaqueductal gray; **VTA**, ventral tegmental area.)

B. The flip-flop switch relationship of the ventrolateral and median preoptic nuclei and the mutually inhibitory components of the ascending arousal system (**LC**, locus ceruleus; **A1**, noradrenergic neurons; **DR**, dorsal raphe; **TMN**, tuberomammillary nucleus; **PB**, parabrachial nucleus). When activated, the sleep-promoting neurons inhibit the components of the ascending arousal system. However, the sleep-promoting cell groups are also inhibited by the arousal system. The net effect is that the individual spends most time fully awake or asleep while minimizing time in transitional states.

takes at least 15 to 20 minutes to fall asleep, but a very sleepy person can easily fall asleep within a few minutes in each nap. Another test of sleep pressure is the Psychomotor Vigilance Task. The subject is told to watch a small lamp and press a button as soon as they see the light turned on. The light then turns on at random times over a 5- to 10-minute test period; sleepy subjects are inattentive and intermittently are slow or completely fail to respond to the light stimulus.

Circadian Rhythms Are Controlled by a Biological Clock in the Suprachiasmatic Nucleus

Circadian rhythms are roughly 24-hour physiological rhythms that synchronize the internal state of an animal with the external daily environment and anticipate various physiologic demands that occur on a daily basis. In humans, circadian wake-promoting signals during the day counterbalance the rising homeostatic

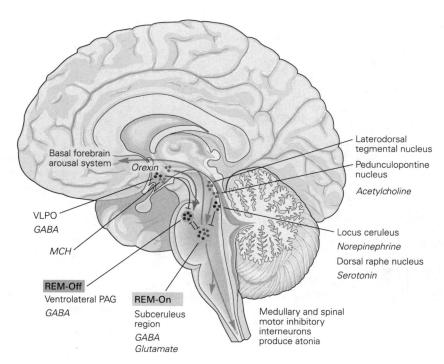

Figure 44–5 **The REM sleep switch.** Brain stem neurons are essential for controlling the transitions between non–rapid eye movement (**REM**) and REM sleep. REM sleep is generated by a population of neurons in the rostral pons, just ventral to the laterodorsal tegmental nucleus and locus ceruleus, in what is called the sublaterodorsal area in rodents and the subceruleus region in humans. These glutamatergic neurons project to other parts of the brain stem, where they initiate the motor and autonomic manifestations of REM sleep, and to the forebrain, where they mediate behavioral and electroencephalographic components of REM sleep. The descending projection activates inhibitory interneurons in the medulla and spinal cord that profoundly hyperpolarize motor neurons and prevent the individual from acting out his or her dreams. These REM-on neurons are inhibited by GABAergic neurons in the ventrolateral periaqueductal gray matter and adjacent pontine reticular formation, while the latter are themselves inhibited by neurons in the REM-on region, thus forming a flip-flop switch (see

Figure 44–4B). These REM-off neurons are under the control of forebrain neurons, including neurons that release the excitatory orexin neuropeptides, neurons in the ventrolateral preoptic nucleus (**VLPO**) that release the inhibitory signaling molecules γ-aminobutyric acid (**GABA**) and galanin, and hypothalamic neurons that release the inhibitory neuropeptide melanin-concentrating hormone (**MCH**). In addition, modulatory neurons in the locus ceruleus and dorsal raphe inhibit the REM generator, whereas cholinergic neurons in the pedunculopontine and laterodorsal tegmental nuclei promote REM sleep. This model explains many clinical observations, such as the fact that cholinergic agonist drugs promote REM sleep, whereas drugs such as antidepressants that increase monoamine levels suppress REM sleep. Loss of orexinergic neurons can cause abrupt onset of REM sleep, whereas loss of REM-on neurons in the sublaterodorsal area abolishes atonia during REM sleep. Thus, individuals with this condition act out their dreams (*REM sleep behavior disorder*).

sleep pressure. The circadian wake-promoting signal dips slightly in the mid-afternoon, when many people take a nap or siesta. Around the habitual bedtime, this circadian waking influence rapidly collapses, the homeostatic drive for sleep is unopposed, and sleep ensues. In the hour or two before the customary waking time, circadian promotion of sleep occurs to ensure an adequate amount of sleep, since homeostatic sleep pressure is low late in the sleep period (Figure 44–6A).

Circadian rhythms are driven by a small group of GABAergic neurons in the suprachiasmatic nucleus located in the hypothalamus just above the optic chiasm. The 24-hour rhythm of activity in this biological

clock is driven by a set of "clock genes," which undergo a transcriptional-translational cycle with an approximately 24-hour period. The positive limb of the loop consists of two proteins, BMAL1 and CLOCK, which dimerize and form a transcription factor that binds to the E-box motif, which is found in the promoter region of hundreds of genes that undergo daily cycles in their expression. Among those genes whose expression is increased by BMAL1 and CLOCK are the *Period* and *Cryptochrome* genes. Their protein products also dimerize, form a complex with casein kinase 1 delta or epsilon, and are translocated to the nucleus of the cell, where they cause BMAL1 and CLOCK to dissociate

Figure 44–6 The circadian drive for wakefulness interacts with the homeostatic drive for sleep to shape wake–sleep cycles.

A. Sleep drive builds up gradually over the course of a long period of wakefulness, whereas the circadian drive for wakefulness varies on a 24-hour cycle, regardless of previous sleep. The peak of this circadian wake cycle occurs in the hours before bed, as the homeostatic sleep drive is rising, whereas the low point occurs in the hours just before the habitual time of awakening, when the homeostatic drive for sleep is ebbing.

B. The 24-hour rhythm in mammalian cells is regulated by a set of proteins that form a transcriptional-translational loop. BMAL1 and CLOCK form a dimer that binds to the E-box motif found on many genes that have circadian rhythms of transcription. Among these are the *Period 1* and *2* genes (*Per1*, *Per2*) and the *Cryptochrome 1* and *2* genes (*Cry1, Cry2*). Their products dimerize and form a complex with casein-1 kinase epsilon or delta (**CK1E/D**). The complex translocates to the nucleus, where it inhibits the dimerization of BMAL1 and CLOCK, causing it to fall off the E-box. This reduces the transcription of *Period* and *Cryptochrome* genes; as PER and CRY proteins are degraded, BMAL1 and CLOCK dimerize once more, and the cycle repeats.

C. Circadian rhythms regulate the timing of sleep and wake. The plot shows the wake cycles (**yellow bars**) of an individual who initially lives under regular lighting conditions for 3 days and then lives in a dimly lit environment with no time cues for 18 days. The individual maintains daily cycles of about 25.2 hours, drifting an entire day over this period. Blind people who cannot relay light signals to the suprachiasmatic nucleus (see Figure 44–7) often live continuously like this, a condition called non–24-hour wake–sleep disorder.

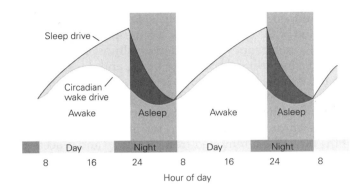

A Wake–sleep cycle with light cues

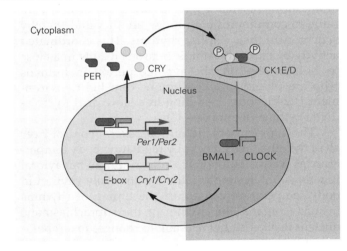

B Genetic clock of the wake–sleep cycle

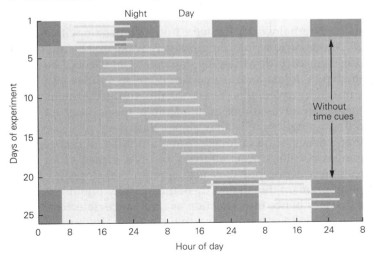

C Wake–sleep cycle without light cues

from the E-box, reducing transcription of many genes, including themselves. This results in a fall in Period and Cryptochrome proteins until BMAL1 and CLOCK can once more dimerize, restarting the cycle. In addition to this core loop, additional genetic side loops modulate the period of the circadian clock (Figure 44–6B).

This daily gene cycle functions in almost all cells in the body, including those in the brain, and it is essential for driving a wide range of circadian rhythms, from secretion of hormones and digestive enzymes to readying the liver for metabolic processing of food and the cardiovascular system for the active period of the day. When removed from the body and placed in a culture dish, most cells in the body rapidly fall out of synchrony, as the individual cellular clock cycles vary between cells by as much as an hour or two from the 24-hour mean. However, when neurons of the suprachiasmatic nucleus are cultured they continue to communicate with one another and thereby synchronize their cellular rhythms. This coordinated activity by suprachiasmatic neurons results in a close to 24-hour rhythm; the average period in humans who are placed in a continuous dim light environment is 24.1 hours, resulting in a slow drift in circadian rhythms (Figure 44–6C).

The suprachiasmatic nucleus exerts control over all of the other body clocks by regulating body temperature as well as autonomic, endocrine, and behavioral functions. Interestingly, although the daily rhythm in body temperature can adjust the timing of rhythms in many organs, the rhythm of the suprachiasmatic nucleus itself is highly resistant to changes in temperature, so its fundamental pacemaking is unaltered. In the end, the circadian timing of the brain and body runs on suprachiasmatic time.

Still, the suprachiasmatic clock must be entrained to the external world. If it were not, each person with a 24.1-hour cycle would progressively wake up six minutes later each day than the day before and would be unable to adjust to seasonal variations in sunrise and sunset. To avoid this situation, the suprachiasmatic nucleus receives direct inputs from a special class of retinal ganglion cells that signal light levels, rather than participating in image formation. Like all retinal ganglion cells, these neurons receive inputs from rods and cones, but they also contain melanopsin, a photopigment that makes them intrinsically photosensitive and hence they function as luminance detectors. In addition to entraining internal circadian rhythms to the ambient light cycle, these cells also regulate other non–image-forming visual functions such as the pupillary light reflex and the feeling of pain that can occur when one looks into bright lights.

Some individuals have unusually short circadian periods due to mutations in clock genes or their regulatory elements. For example, individuals with *familial advanced sleep-phase syndrome* prefer to go to bed early in the evening and cannot sleep past 3 or 4 am. In families with this disorder, mutations in the genes coding for Period or Casein kinase-1 delta result in more rapid cycling of the clock.

Blind people in whom the melanopsin-containing neurons are damaged lack visual input to their suprachiasmatic nucleus, often resulting in *non–24-hour sleep–wake rhythm disorder*. Because most people have an intrinsic cycle longer than 24 hours, circadian rhythms in these individuals drift, becoming a few minutes later each day, so that most of the time they are out of synch with the rest of the world. They lack the ability to entrain to external light–dark conditions because the suprachiasmatic nucleus lacks the crucial resetting signal from the retina. This problem is most common in people who have lost their eyes (eg, due to trauma or infection), but it is not seen in blind people in whom the melanopsin-containing neurons are intact (eg, blindness due to degeneration of rods and cones, or problems with the cornea or lens) and who also retain their pupillary light reflexes.

In contrast to light, which is signaled by the melanopsin neurons, the hormone melatonin signals darkness. Melatonin is made by the pineal gland, and the suprachiasmatic neurons time its release through communication with neurons in the paraventricular nucleus of the hypothalamus that activate sympathetic innervation of the pineal gland. Neurons in the suprachiasmatic nucleus contain melatonin receptors, which reinforce circadian rhythms. Similarly, exogenous melatonin or melatonin agonists can entrain circadian rhythms, promoting sleep by regularizing sleep onset. This treatment approach is particularly useful in entraining circadian rhythms in individuals with non–24-hour sleep–wake rhythm disorder.

Circadian Control of Sleep Depends on Hypothalamic Relays

The suprachiasmatic nucleus is most active during the daily light period in all mammalian species. While humans are diurnal (awake during the day and asleep during the night), nocturnal mammals have the opposite activity cycle. How can such opposite behavioral patterns be set by the suprachiasmatic nucleus if it is most active during the light period?

The answer appears to lie in a series of relays interposed between the suprachiasmatic nucleus and the wake–sleep control circuitry, which give the circadian timing system flexibility in meeting the needs of the individual. The suprachiasmatic neurons are GABAergic, and they send the bulk of their output to an adjacent region called the subparaventricular zone. This area in the anterior hypothalamus contains mostly GABAergic neurons that fire in antiphase to the suprachiasmatic nucleus, ie, are most active at night. The targets of the subparaventricular zone largely overlap those of the suprachiasmatic nucleus, including parts of the paraventricular, dorsomedial, ventromedial, and lateral hypothalamus that regulate various physiological and behavioral systems. Presumably, then, the timing of a particular physiological or behavioral function would depend upon the relationships of these two antiphase circadian inputs to their target neurons.

One crucial target of the subparaventricular zone is the dorsomedial nucleus of the hypothalamus, which regulates a number of circadian behaviors, including the wake–sleep cycle. Lesions of the dorsomedial nucleus severely disrupt circadian rhythms of sleep, feeding, locomotor activity, and corticosteroid secretion. The dorsomedial nucleus is thought to promote wakefulness via GABAergic projections to the ventrolateral preoptic nucleus and glutamatergic projections to the lateral hypothalamus (Figure 44–7).

Sleep Loss Impairs Cognition and Memory

When people are sleepy, they often have impaired vigilance, working memory, judgement, and insight. Some of the attentional problems may be caused by *microsleeps*, brief periods of slower cortical activity. For example, subjects who rarely miss stimuli on the Psychomotor Vigilance Task when well rested may miss over 20% of the visual stimuli when sleepy. In addition to these global lapses in cortical function, sleepiness can also produce *local sleep* with slow EEG waves in focal cortical areas. Executive function is often the first thing to fail with sleepiness, and sleep-deprived people show reduced metabolism and focal slowing in the frontal cortex in EEGs.

While sleepiness impairs cognition, sleep itself helps consolidate memories. When subjects are taught a simple motor task, such as pressing buttons in a predetermined sequence, they become more efficient with practice. Robert Stickgold and colleagues found that if the training is in the morning, and the subjects are tested 12 hours later in the evening (without intervening sleep), they perform at about the same level as when they stopped their training. However, if they are tested the next morning after a night of sleep, they usually perform better than on the day of training. Subjects who are trained in the evening still perform better 12 hours later if they have had a chance to sleep

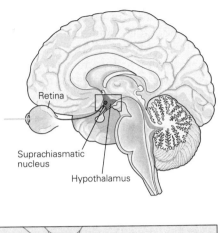

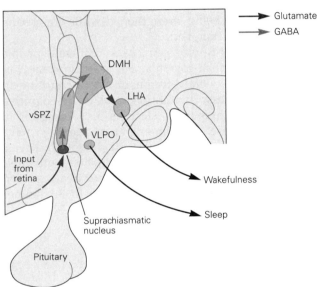

Figure 44–7 Neurons in the suprachiasmatic nucleus provide a master clock for wake–sleep. Retinal inputs signaling light activate the suprachiasmatic nucleus (**upper figure**), which then drives the wake–sleep cycle through a series of relays in the hypothalamus (**lower figure**). The sagittal section through the hypothalamus shows the suprachiasmatic nucleus projecting to neurons in the ventral subparaventricular zone (**vSPZ**), which in turn project to the dorsomedial nucleus of the hypothalamus (**DMH**). The DMH contains glutamatergic neurons that excite orexinergic and glutamatergic neurons in the lateral hypothalamic area (**LHA**), causing wakefulness. GABAergic neurons in the DMH inhibit the ventrolateral preoptic nucleus (**VLPO**), turning off the sleep-promoting system. Animals with DMH lesions fail to show circadian rhythms of wake–sleep and sleep about an hour more per day.

overnight but not if they remained awake. Improvement of certain types of memory consolidation (eg, memory for a visual perception task) is correlated with the amount of REM sleep, while other types (eg, memory for a finger-tapping sequence task) correlate with stage N2 non-REM sleep.

These studies suggest that in each stage of sleep the cerebral cortex undergoes synaptic reorganization to consolidate the memory of specific types of salient information. Conversely, this memory consolidation is lost when subjects are deprived of sleep or have fragmented sleep. A related theory, proposed by Giulio Tononi and Chiara Cirelli, is that rebalancing of synaptic strengths based on recent experience (synaptic homeostasis) occurs during sleep. The size of many excitatory synapses is increased during learning, requiring that some excitatory inputs be reduced to avoid overexciting the target neuron. Tononi and Cirelli found that the size of smaller synapses in motor and sensory cortex is reduced during sleep, resulting in strong inputs being strengthened while competing weaker ones are removed.

Diseases that cause sleep loss or that wake people from sleep can impair cognition. For example, *obstructive sleep apnea* can severely fragment sleep, resulting in daytime sleepiness, inattention, and other cognitive impairments. Fragmented sleep is also common in Alzheimer disease. Alzheimer patients tend to have fewer neurons in the ventrolateral preoptic nucleus, and the extent of neuronal loss correlates with their degree of sleep fragmentation. Whether treating sleep fragmentation can improve cognition in Alzheimer patients remains to be determined.

Sleep Changes With Age

Sleep changes with age in striking and characteristic ways. As every new parent quickly learns, the lengthy sleep time of a newborn is distributed almost randomly throughout the day. Although the EEG rhythms in newborns are not as well formed as those of older children or adults, more than 50% (8–9 hours per day) of that sleep is spent in a state much like REM sleep.

Sleep recordings from a premature infant exhibit an even higher percentage of REM-like sleep, indicating that in utero the fetus spends a large fraction of the day in a brain-activated but movement-inhibited state. As neuronal activity influences the development of functional circuits in the brain (Chapters 48 and 49), it is reasonable to think that the spontaneous activity of the immature brain during sleep facilitates the development of neural circuits.

By approximately 4 months of age, the average baby begins to show diurnal rhythms that are synchronized with day and night, much to the relief of weary parents. The total duration of sleep gradually declines, and by 5 years of age, the child may sleep 11 hours each night plus a nap, and 10 hours of sleep is typical

around age 10. At these early ages, sleep is deep; stage N3 is prominent, with an abundance of delta waves in the EEG. As a result, children are not easily wakened by environmental stimuli.

With age, sleep becomes lighter and more fragmented. The percentage of time spent in stage N3 sleep drops across adulthood, and by the age of 50 to 60, it is not unusual for N3 to fade entirely, especially in men. This shift toward lighter stages of non-REM sleep results in two to three times as many spontaneous awakenings and more easily disrupted sleep. Many sleep disorders, including insomnia and sleep apnea, become more prevalent with age, and insomnia is common, often due to waking in response to neural signals to empty the bladder or due to discomfort from menopausal symptoms or from arthritis and other diseases. Why this change occurs with age is unclear; homeostatic sleep pressure appears normal, but the neural mechanisms for producing deep non-REM sleep may be less effective.

Disruptions in Sleep Circuitry Contribute to Many Sleep Disorders

Insomnia May Be Caused by Incomplete Inhibition of the Arousal System

Insomnia is one of the most common problems in all of medicine, yet the underlying neurobiology remains a mystery. Insomnia is defined as difficulty falling asleep or trouble staying asleep, so that function the next day is impaired. Positron emission tomography studies in patients with chronic insomnia demonstrate unusual activation of brain arousal systems during sleep, and the EEG often shows persistence of high-frequency activity (15–30 Hz) that is usually seen only during wake.

In addition, rats exposed to acute stress show high-frequency EEG activity during sleep, as well as simultaneous activity in neurons of the ventrolateral preoptic nucleus and components of the arousal system, such as the locus ceruleus and histamine neurons. This simultaneous activation can produce a unique state in which the EEG shows slow waves consistent with sleep along with high-frequency activity consistent with the awake state; this may explain why some patients appear asleep on the polysomnogram recording but may feel awake.

Clinically, insomnia is often treated with cognitive behavioral therapy that is aimed at reducing the hyperarousal and improving sleep habits. Some patients may be treated with benzodiazepines and related drugs that potentiate GABA transmission and, therefore, may help reduce activity in arousal-promoting brain regions. Other patients derive benefit from drugs that block the arousal system more directly, such as antihistamines.

Sleep Apnea Fragments Sleep and Impairs Cognition

Sleep apnea is one of the most common sleep disorders, affecting about 5% of adults and children. Patients with *obstructive sleep apnea* have repeated episodes of airway obstruction that force the individual to briefly awaken from sleep to resume breathing. During sleep, muscle tone falls, and in people with small airways, relaxation of airway dilator muscles such as the genioglossus (which normally acts to pull the tongue forward) results in collapse of the airway. This causes a brief period of no air flow, and consequently, blood levels of carbon dioxide rise while oxygen levels fall, activating chemosensory systems in the medulla that increase respiratory effort.

These chemosensory systems also activate neurons in the parabrachial nucleus that promote awakening, which results in a further increase in muscle tone that reopens the airway. These airway obstructions can occur hundreds of times per night, but the arousals are usually so brief that the individual may not remember them in the morning. Many people with obstructive sleep apnea do not feel rested in the morning; they feel sleepy all day and they have difficulty with a wide variety of cognitive tasks, especially those that require vigilance or learning.

Clinicians often treat sleep apnea with a *continuous positive airway pressure (CPAP) device* that delivers mildly pressurized air via the nose to inflate and open the airway during sleep. Sleep apnea can also be treated with upper airway surgery to remove obstructions such as large tonsils, a dental device to move the tongue forward, or weight loss to reduce adipose tissue in the neck. Treated patients often feel more alert and have better cognitive function, although there may be some residual cognitive impairment, possibly due to neuronal injury from repeated episodes of low oxygen saturation or hypoxia (Figure 44–8).

Narcolepsy Is Caused by a Loss of Orexinergic Neurons

Narcolepsy was first described in the late 1800s, but the underlying cause, a deficiency in a single

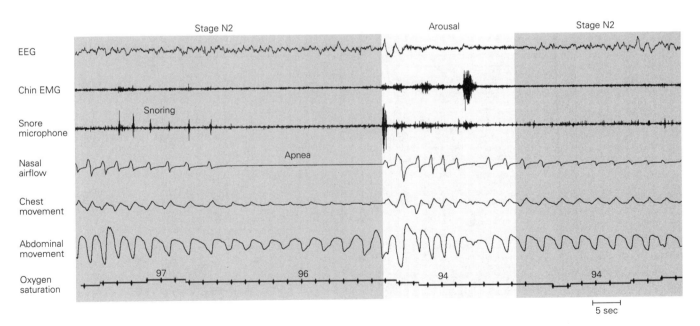

Figure 44–8 An episode of sleep apnea. At the start of this polysomnogram, an individual is in stage N2 sleep. Some snoring is detected, but nasal airflow is good and oxygen saturation is normal. The individual then experiences an obstructive apnea with no nasal airflow; nevertheless, respiratory effort persists (shown by the abdominal movement). The apnea is terminated by a brief awakening (low-voltage fast electroencephalogram [EEG]), accompanied by a loud snore, increased electromyographic activity, intensified respiratory effort, and opening of the airway. Oxygen saturation drops by about 3%, reaching its nadir about 15 seconds after the apnea finishes, as it takes time for blood to get from the lungs to the fingertip where oxygen saturation is measured.

neurotransmitter, has become clear only in the last two decades. Narcolepsy typically begins in the teen years as moderate to severe sleepiness every day, even with ample amounts of sleep at night. People with narcolepsy can easily fall asleep in class, while driving, or during other activities when sleep might be embarrassing or dangerous. Unlike sleep apnea, their sleep is restorative, and they often feel much more alert after a 15- to 20-minute nap.

In addition, in people with narcolepsy, elements of REM sleep often occur during wakefulness. For example, at night, while falling asleep or waking up, an individual with narcolepsy might find himself unable to move (*sleep paralysis*) or may have vivid dream-like hallucinations (*hypnagogic or hypnopompic hallucinations*) superimposed on wakefulness. Even more mysteriously, during the day, when surprised with a good joke or by unexpectedly seeing a friend, a person with narcolepsy can develop *cataplexy*, emotionally triggered muscle weakness that is similar to the paralysis of REM sleep. Mild cataplexy can cause weakness of the face and neck, but when severe, the individual can lose all muscle control, collapse to the ground, and be unable to move for 1 to 2 minutes.

Narcolepsy remained mysterious until the late 1990s when a new family of peptide neurotransmitters,

orexins (also known as hypocretins), was discovered. There are two orexin peptides, derived from the same mRNA precursor, and they are found only in cells in the posterior lateral hypothalamus. It was soon found that loss of orexin signaling in animals or humans could reproduce the entire narcolepsy phenotype. People with narcolepsy show a highly selective loss of more than 90% of their orexin neurons, while other types of hypothalamic neurons are spared. This cell loss is probably due to an autoimmune attack as it is linked to genes that affect immune function and has been triggered by seasonal influenza epidemics and use of a certain influenza vaccine. Recently, researchers discovered that people with narcolepsy have immune cells (T lymphocytes) that target the orexin neuropeptides (Figure 44–9).

Orexinergic neurons promote wakefulness and suppress REM sleep, in part by activating monoaminergic neurons in the locus ceruleus and dorsal raphe as well as REM-off GABAergic neurons in the periaqueductal gray matter, all of which inhibit the REM sleep generating neurons in the pons. Thus, people and animals with loss of orexinergic neurons have great difficulty remaining awake for long periods, and REM sleep is disinhibited, such that REM sleep (or components of REM sleep, such as motor atonia during

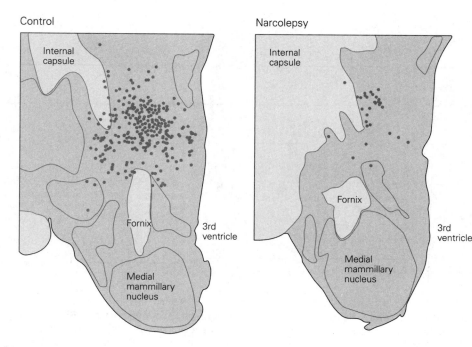

Figure 44–9 Narcolepsy is associated with a loss of hypothalamic neurons that produce the orexin neuropeptides. A dramatic loss of orexinergic neurons (**green dots**) is evident in these drawings of sections through the brain at the level of the mammillary bodies in an individual with narcolepsy (*right*) compared to a normal brain (*left*). (Reproduced, with permission, from Crocker et al. 2005. Copyright © 2005 American Academy of Neurology.)

wakefulness, or cataplexy) breaks through at inappropriate times during the day.

In terms of sleep circuitry, loss of orexinergic neurons can be considered to destabilize both the wake–sleep and the REM/non-REM switches in the brain. Thus, patients with narcolepsy can easily doze off during the day but also spontaneously wake from sleep more frequently at night. The dysregulation of REM sleep is also apparent on the Multiple Sleep Latency Test; healthy individuals almost never experience REM sleep during the day as it is under tight circadian control, but patients with narcolepsy often experience REM sleep during daytime naps.

The absence of orexin signaling also explains the mysterious symptom of cataplexy. Evidence from mice lacking orexins suggests that pleasant experiences turn on neurons in the prefrontal cortex and amygdala that can activate the brain stem pathways that trigger REM sleep paralysis. This influence is normally opposed by the orexin system, so that one may feel slightly "weak with laughter." When orexin signaling is absent, full-blown paralysis can occur.

Narcolepsy is treated with medications and behavioral approaches. Sleepiness can be substantially attenuated with wake-promoting medications such as amphetamine and modafinil. One to two strategically timed naps during the day are often helpful and can improve alertness for a couple of hours. Cataplexy often responds well to antidepressants such as serotonin or norepinephrine reuptake inhibitors, as these drugs strongly suppress REM sleep. Sodium oxybate taken during the night enhances deep sleep, and through an unknown mechanism, it helps consolidate wakefulness and reduce cataplexy during the day.

REM Sleep Behavior Disorder Is Caused by Failure of REM Sleep Paralysis Circuits

REM sleep behavior disorder—the loss of paralysis during REM sleep in some older adults—is the direct opposite of cataplexy. The lack of paralytic inhibition permits the patients to act out their dreams. The individual often calls out and may grab or violently punch or kick; injuries from hitting nearby furniture or the bed partner are not unusual. These dramatic movements typically awaken the patient, who can then recall a dream about fighting off an attacker in a way that closely matches the actual movements.

REM sleep behavior disorder was first identified in 1986 by Mahowald and Schenck. Ten years later, they reported that 40% of their original cohort of 19 patients had developed Parkinson disease or a related neurodegenerative disorder with deposition of alpha-synuclein, such as Lewy body dementia or multiple system atrophy. Subsequent studies have shown that about half of patients with REM sleep behavior disorder develop a synucleinopathy by 12 to 14 years after onset, and nearly all by 25 years. It is now thought that the synucleinopathy begins in the brain stem and early on damages the subceruleus neurons that normally drive REM sleep paralysis. If this relationship is confirmed, the diagnosis of REM sleep behavior disorder may identify individuals with nascent synucleinopathies who could be treated with drugs, not yet developed, that slow the neurodegeneration.

Restless Legs Syndrome and Periodic Limb Movement Disorder Disrupt Sleep

Restless legs syndrome occurs in about 10% of the population and is characterized by an irresistible urge to move the legs, usually accompanied by an annoying internal discomfort like "ants in the pants." This restless sensation typically occurs in the evening and first half of the night and often makes it hard to fall asleep. The sensation is much worse with rest and improves by moving the legs in bed or walking about.

Many people suffering from restless legs syndrome also experience *periodic limb movement disorder*, in which the legs and sometimes arms flex in a stereotyped way every 20 to 40 seconds during non-REM sleep. These leg movements fragment sleep and can produce daytime sleepiness. Iron deficiency is a common cause of restless legs, and treatment with iron can be very helpful. Genome-wide association studies have found genes common to both conditions, but the underlying pathophysiology is not yet understood. Patients with both disorders often improve with low doses of a D_2 dopamine agonist, the antiepileptic drug pregabalin, or an opiate drug.

Non-REM Parasomnias Include Sleepwalking, Sleep Talking, and Night Terrors

Parasomnias are unusual behaviors that occur during either REM or non-REM sleep. Non-REM parasomnias are common in children and include sleepwalking, sleep talking, confusional arousals, bed-wetting, and night terrors. About 15% of young adolescents have some sleepwalking, but this usually fades over time, so only about 1% of adults regularly sleepwalk.

The non-REM parasomnias typically begin with a sudden arousal from stage N3 sleep, which can occur spontaneously or be triggered by a noise or airway obstruction from sleep apnea. These are not full arousals, as for the first minute or two, the EEG still shows the slow EEG delta waves typical of stage N3 sleep even as the child walks, dresses, or eats. Over time, the EEG changes to the pattern typical of wakefulness and eventually the individual wakes up. Sleepwalkers or talkers often have no memory of these events, so that reports from the family are necessary to make the diagnosis. Bed-wetting (enuresis) may also occur during deep non-REM sleep in some children.

Night terrors also occur in stage N3 sleep and are common in children age 2 to 5. The child often sits up and cries as if in great fear, sometimes with dilated pupils and a fast heart rate. During the episode, the child is inconsolable; attempts to calm or wake the child may only cause the screams and fearful behavior to worsen. Like sleep walking, but in contrast to ordinary nightmares, the child usually does not remember the night terror, and the events are typically much more difficult for the parent than the child.

The underlying cause of non-REM parasomnias is unknown. They are usually managed by ensuring adequate sleep to reduce pressure for deep non-REM sleep, reducing stress, and treating underlying sleep disorders such as sleep apnea that might trigger arousals from sleep. Fluid restriction in the evening may help with enuresis. Most children outgrow non-REM parasomnias as their N3 sleep decreases in late adolescence. Drugs that reduce the amount of N3 sleep, such as tricyclic antidepressants, are sometimes used as well. As with REM sleep behavior disorder, people can be seriously hurt sleepwalking if they fall down stairs or trip over furniture, and it is important to make the bedroom layout safe. The amount of time spent in stage N3 sleep is high in individuals with high homeostatic sleep pressure, so getting adequate sleep is also helpful.

Sleep Has Many Functions

Although there has been remarkable progress in our understanding of the brain circuitry that regulates sleep and wakefulness, we still understand relatively little about the actual functions of sleep. For an activity that occupies one-third of the life of humans, and much more in some other species, we have very little understanding of sleep's purposes. Allan Rechtschaffen, who first systematized the stages of sleep (and was an author of this chapter in earlier editions of this book), once said that if sleep did not have a vital function it would be the biggest mistake that evolution ever made. He found that rats would die of overwhelming infection and hypothermia if chronically deprived of sleep. However, the methods for keeping animals

continuously awake were stressful, and it is unclear whether the consequences observed were due to loss of sleep or continuous stress. Indeed, it is not clear if prolonged sleep loss and stress can be dissociated.

One proposed function of sleep suggests that a period of brain inactivity is needed to permit metabolic recovery of the brain. The role of adenosine as a sleep-promoting humoral factor is based on the rundown of adenosine triphosphate (ATP) stores to adenosine during the awake period. Another idea is that sleep may permit the body to reconstitute injured tissue and replenish energy stores, but there is little evidence that sleep deprivation impairs any of these processes.

A recent hypothesis has been raised by the observation that during sleep the extracellular space in the brain expands, thus permitting the cerebrospinal fluid to "clean out" undesirable molecules that should not be allowed to build up extracellularly. The waking brain has very little extracellular space, largely due to ion fluxes to and from neurons during synaptic communication. These fluxes establish an osmotic gradient that drives most fluid in the brain into cells. During sleep, neurons and glia may shrink as that fluid moves back into the extracellular space. Among the molecules that may be washed out from the extracellular space during sleep are beta-amyloid peptides. In mice engineered to produce high levels of human beta-amyloid, sleep deprivation reduces the clearance of beta-amyloid from the extracellular space in the brain, thus accelerating its deposition in the plaques that are characteristic of Alzheimer disease (Chapter 64). Because buildup of beta-amyloid peptide in the brain is thought to be an early step in Alzheimer disease, work is now underway to determine whether poor sleep may predispose people to this disease.

In addition to these biochemical functions, sleep also promotes memory formation. As described earlier, the synaptic homeostasis model suggests that synapses are rebalanced during sleep, although it is not clear why this process would require sleep. A more basic need may be to provide a time for synapses to consolidate new memory traces. During the waking state, experience can modify synaptic strength on the fly by such processes as protein phosphorylation, insertion of premade receptors into the postsynaptic membrane, or translation of mRNA in the dendrites into new protein. But some portion of the synaptic remodeling that underlies memory formation requires nuclear-dependent transcription of new mRNA. Because synaptic sites on dendrites may be a millimeter from the nucleus or even more in some neurons, time is required for messenger molecules that are produced at the synapse to reach the nucleus and alter transcription and then for the resulting mRNA to be transported back to the dendrite where it can result in new protein synthesis. This process may require a time when these messengers are not competing with new incoming signals to complete their work in stabilizing memories.

One thing about sleep is certain: It is required for normal brain function, and inadequate sleep, as defined by an increased tendency to fall asleep during the day, is associated with impaired cognitive function. Medical training programs are now being redesigned to reduce the risk of interns and residents making critical medical decisions while sleep deprived. Similar approaches to school start times, drowsy driving, and other aspects of our society could potentially improve productivity and save many lives.

Highlights

1. Sleep involves distinct changes in the electroencephalogram (EEG), electromyogram (EMG), and electro-oculogram (EOG) that are recorded on a polysomnogram. These changes can be used to divide sleep into rapid eye movement (REM) sleep—during which the EEG is similar to wake, but the body has such low muscle tone that it is essentially paralyzed—and three stages of non-REM sleep (N1–N3), with low to high amounts of slow waves in the EEG.

2. During the night, sleep alternates between periods of non-REM sleep followed by bouts of REM sleep, with the entire cycle taking about 90 minutes. Over the course of a night, non-REM sleep becomes progressively lighter, while REM sleep bouts become longer.

3. The waking state is actively produced by an ascending arousal network. The key neurons required to drive wakefulness are glutamatergic neurons in the parabrachial and pedunculopontine tegmental nuclei, dopaminergic neurons in the midbrain, glutamatergic neurons in the supramammillary nucleus, and GABAergic and cholinergic neurons in the basal forebrain that directly innervate the cerebral cortex. Modulatory cell groups, using mainly monoamines such as norepinephrine, serotonin, and histamine as neurotransmitters, can drive arousal under appropriate conditions, but unlike the main pathways, lesions of these cell groups do not impair baseline wakefulness.

4. During sleep, the ascending arousal system is inhibited by GABAergic neurons in the ventrolateral preoptic nuclei and the parafacial zone.

Conversely, during wake, the ventrolateral preoptic neurons are inhibited by neurons in the ascending arousal system. These mutually antagonistic pathways produce a neural circuit resembling an electrical flip-flop switch, which favors rapid and complete transitions between sleep and wakefulness. Similarly, populations of mutually inhibitory neurons in the caudal midbrain and pons govern transitions between REM and non-REM sleep. Monoamine neurotransmitters, such as serotonin and norepinephrine, also act on these switching neurons and prevent transitions into REM sleep during wakefulness. Orexin neurons in the lateral hypothalamus activate REM sleep-suppressing neurons, preventing transitions from wake into REM sleep.

5. Sleep is regulated by a homeostatic drive to sleep, so that the longer one is awake, the more intense the drive, and the more sleep is required to satisfy the need to sleep. There is also a circadian influence on sleep that inhibits sleep during the day but promotes it at night, especially during the latter part of the night, when homeostatic sleep drive wanes. The circadian cycle is synchronized with the outside world by light signals from the retina to the brain's master circadian clock in the suprachiasmatic nucleus. The suprachiasmatic nucleus then activates hypothalamic pathways that regulate wake–sleep states, as well as many other behaviors, hormonal cycles, and physiological adjustments.

6. Sleep needs change throughout development, from about 16 hours per day in a newborn to about 8 hours per day in a healthy young adult. However, sleep-promoting mechanisms weaken with aging, and so individuals over 70 years old have more fragmented sleep and sleep about an hour less per day.

7. Sleep apnea is a condition in which the airway collapses due to reduced muscle tone during sleep. This impaired breathing causes frequent awakenings and can impair cognition. Restoring airway patency with continuous positive airway pressure (CPAP) can overcome this problem.

8. Insomnia may be caused by hyperactivation of the arousal system, and it is best treated with cognitive behavioral therapy.

9. Narcolepsy is caused by selective loss of the orexin (also called hypocretin) neurons in the hypothalamus. The orexin neuropeptides normally promote wake and regulate REM sleep, and loss of orexin signaling results in chronic daytime sleepiness and poor control of REM sleep.

Specifically, people with narcolepsy may quickly transition into REM sleep after dozing off, and they can have fragments of REM sleep, such as cataplexy and hypnogogic hallucinations, during wake. Narcolepsy is usually treated with medications that promote wake and suppress REM sleep.

10. REM sleep behavior disorder is due to loss of atonia during REM sleep, causing patients to act out their dreams. REM sleep behavior disorder is usually an early manifestation of either Parkinson disease or Lewy body dementia.

11. Restless legs syndrome is a genetically influenced disorder in which people feel that they have to move their legs. This makes them very uncomfortable when awake, and they can have periodic leg movements during sleep that disrupt sleep.

12. Sleepwalking and related parasomnias usually occur in young children during deep (stage N3) non-REM sleep. They are best managed by ensuring adequate, good-quality sleep.

13. Sleep loss impairs the ability to maintain sustained attention and clouds judgment. The reasons for this are not understood. Theories about the brain requiring down time to recharge its metabolic status or to allow it to flush out unwanted products from the extracellular space have received attention, but it is unclear whether this accounts for the penalty paid due to lack of sleep. One attractive theory for the function of sleep is that it may be required for synaptic remodeling that is necessary for certain types of learning.

Clifford B. Saper
Thomas E. Scammell

Selected Reading

Buhr ED, Takahashi JS. 2013. Molecular components of the mammalian circadian clock. Handb Exp Pharmacol 217:3–27.

Kryger MH, Roth T, Dement WC. 2017. *Principles and Practice of Sleep Medicine*, 6th ed. Philadelphia: Elsevier.

Saper CB. 2013. The central circadian timing system. Curr Opin Neurobiol 23:747–751.

Saper CB, Fuller PM, Pedersen NP, Lu J, Scammell TE. 2010. Sleep state switches. Neuron 68:1023–1042.

Scammell TE, Arrigoni E, Lipton JO. 2017. Neural circuitry of wakefulness and sleep. Neuron 93:747–765.

References

Achermann P, Borbely AA. 2003. Mathematical models of sleep regulation. Front Biosci 8:683–693.

Anaclet C, Pedersen NP, Ferrari LL, et al. 2015. Basal forebrain control of wakefulness and cortical rhythms. Nat Commun 6:8744.

Aschoff J. 1965. Circadian rhythms in man. Science 148:1427–1432.

Aserinsky E, Kleitman N. 1953. Regularly occurring periods of eye motility and concomitant phenomena during sleep. Science 118:273–274.

Bal T, von Krosigk M, McCormick DA. 1995. Synaptic and membrane mechanisms underlying synchronized oscillations in the ferret lateral geniculate nucleus *in vitro*. J Physiol 483:641–663.

Boeve BF. 2013. Idiopathic REM sleep behaviour disorder in the development of Parkinson's disease. Lancet Neurol 12:469–482.

Buyssee DJ, Germain A, Hall M, Monk TH, Nofzinger EA. 2011. A neurobiological model of insomnia. Drug Discov Today Dis Models 8:124–137.

Chemelli RM, Willie JT, Sinton CM, et al. 1999. Narcolepsy in orexin knockout mice: molecular genetics of sleep regulation. Cell 98:437–451.

Crocker A, Espana RA, Papadopoulou M, et al. 2005. Concomitant loss of dynorphin, NARP, and orexin in narcolepsy. Neurology 65:1184–1188.

Dement W, Kleitman N. 1957. Cyclic variations in EEG during sleep and their relation to eye movements, body motility, and dreaming. Electroencephalogr Clin Neurophysiol Suppl 9:673–690.

de Vivo L, Bellesi M, Marshall W, et al. 2017. Ultrastructural evidence for synaptic scaling across the wake/sleep cycle. Science 355:507–510.

Dijk DJ, Czeisler CA. 1995. Contribution of circadian pacemaker and sleep homeostat to sleep propensity, sleep structure, electroencephalographic slow waves, and sleep spindle activity in humans. J Neurosci 15:3526–3538.

Fuller PM, Sherman D, Pedersen NP, Saper CB, Lu J. 2011. Reassessment of the structural basis of the ascending arousal system. J Comp Neurol 519:933–956.

Lim AS, Ellison BA, Wang JL, et al. 2014. Sleep is related to neuron numbers in the ventrolateral preoptic/intermediate nucleus in older adults with and without Alzheimer's disease. Brain 137:2847–2861.

Lin L, Faraco J, Li R, et al. 1999. The sleep disorder canine narcolepsy is caused by a mutation in the hypocretin (orexin) receptor 2 gene. Cell 98:365–376.

Lu J, Sherman D, Devor M, Saper CB. 2006. A putative flip-flop switch for control of REM sleep. Nature 41:589–594.

Mahowald MW, Schenck CH. 2005. Insights from studying human sleep disorders. Nature 437:1279–1285.

McCormick DA, Bal T. 1997. Sleep and arousal: thalamocortical mechanisms. Annu Rev Neurosci 20:185–215.

Moruzzi G, Magoun HW. 1949. Brain stem reticular formation and activation of the EEG. Electroencephalogr Clin Neurophysiol Suppl 1:455–473.

Peyron C, Faraco J, Rogers W, et al. 2000. A mutation in a case of early onset narcolepsy and a generalized absence of hypocretin peptides in human narcoleptic brains. Nat Med 6:991–997.

Saper CB, Scammell TE, Lu J. 2005. Hypothalamic regulation of sleep and circadian rhythms. Nature 437:1257–1263.

Stickgold R. 2005. Sleep-dependent memory consolidation. Nature 437:1272–1278.

Tononi G, Cirelli C. 2014. Sleep and the price of plasticity: from synaptic and cellular homeostasis to memory consolidation and integration. Neuron 81:12–34.

Xie L, Kang H, Xu Q, et al. 2013. Sleep drives metabolite clearance from the adult brain. Science 342:373–377.

Xu M, Chung S, Zhang S, et al. 2015 Basal forebrain circuit for sleep-wake control. Nat Neurosci 18:1641–1647.

Part VII

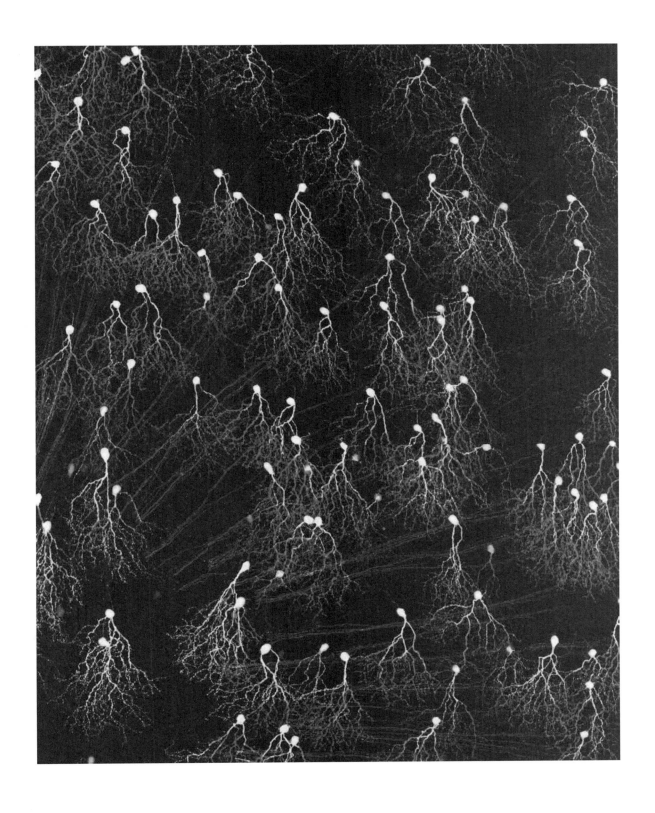

VII Development and the Emergence of Behavior

THE INNUMERABLE BEHAVIORS controlled by the mature nervous system—our thoughts, perceptions, decisions, emotions, and actions—depend on precise patterns of synaptic connectivity among the billions of neurons in our brain and spinal cord. These connections form during embryonic and early postnatal life but can then be remodeled throughout life. In this section, we describe how the nervous system develops and matures.

The history of developmental neurobiology is long and illustrious. Nearly 150 years ago, Santiago Ramón y Cajal undertook a comprehensive series of anatomical studies on the structure and organization of the nervous system and then set out to probe its development. The only method available to him was light microscopic analysis of fixed tissue, but from his observations, he deduced many developmental principles that are still recognized as correct. During the first half of the 20th century, other anatomists followed in his footsteps. Progress then accelerated as new methods became available—first electrophysiology and electron microscopy and, more recently, molecular biology, genetics, and live imaging. We now know a great deal about molecules that determine how nerve cells acquire their identities, how they extend axons to target cells, and how these axons choose appropriate synaptic partners once they have arrived at their destinations.

It is useful to divide the numerous steps that compose neural development into three epochs, which are conceptually distinct even though they overlap temporally to some extent. The first, beginning at the earliest stages of embryogenesis, leads to the generation and differentiation of neurons and glia. One can think of this epoch as devoted to producing the components from which neural circuits will be assembled: the hardware. These steps depend on the expression of particular genes at particular times and places. Some of the molecules that control these spatial and temporal patterns are transcription factors that act at the level of DNA to regulate gene expression. They act within the differentiating cells and are therefore called cell-autonomous factors. Other factors, called cell non-autonomous, include cell surface and secreted molecules that arise from other cells.

They act by binding to receptors on the differentiating cells and generating signals that regulate the activity of the cell-autonomous transcriptional programs. The interaction of these intrinsic and extrinsic factors is critical for the proper differentiation of each nerve cell.

A second epoch encompasses the steps by which neurons wire up: the migration of their somata to appropriate places, the guidance of axons to their targets, and the formation of synaptic connections. The complexity of the wiring problem is staggering—axons of many neuronal types must navigate, often over long distances, and then choose among a hundred or more potential synaptic partners. Nonetheless, progress has been encouraging. A major factor has been the ability to address the problem through the analysis of simple and genetically accessible organisms such as the fruit fly *Drosophila* and the nematode worm *Caenorhabditis elegans*. It turns out that many of the key molecules that control the formation of the nervous system are conserved in organisms separated by millions of years of evolution. Thus, despite the great diversity of animal forms, the developmental programs that govern body plan and neural connectivity are conserved throughout phylogeny.

In the third epoch, the genetically determined patterns of connectivity (the hardware) are molded by activity and experience (the software). Unfortunately for investigators, these steps in mammals are shared to a very limited degree with invertebrates and lower vertebrates. A newly hatched bird or fly is not remarkably different in its behavioral repertoire from its adult self, but no one could say that about a person. This is largely because our nervous system is something of a rough draft at birth. The hardwired circuits that lay out its basic plan are modified over a prolonged postnatal period by experience, acting via neural activity. In this way, the experience of each individual can leave indelible imprints on his or her nervous system and the cognitive abilities of the brain can be enhanced by learning. These processes act in all mammals, and neuroscientists now use mice to probe the mechanisms that underlie them—but they are especially prominent and prolonged in humans. It may be that the prolonged period during which experience can sculpt the human nervous system is the most important single factor in making its capabilities unique among all species.

As our understanding of development increases, it is increasingly informing neurology and psychiatry. Many genes that regulate the first two epochs have now been implicated as susceptibility factors for, or even causes of, some neurodegenerative and behavioral disorders. Thus, studies of neural development are beginning to provide insights into the etiology of neurological diseases and to suggest rational strategies for restoring neural connections and function after disease or traumatic injury. More recently, as we learn about the cellular and molecular changes that underlie experience-dependent remodeling, we can hope to understand how, for example, the plasticity that is so evident during early life can be recruited in adults to improve rehabilitative therapy after injury, stroke, or neurodegenerative disease. Moreover, there is increasing reason to believe that some behavioral disorders, such as autism or schizophrenia, may result in part from defects in the experience-dependent tuning of neural circuits during early postnatal life.

Part VII summarizes these epochs in a sequential manner. Beginning with the early stages of neural development, we concentrate on the factors that control the diversity and survival of nerve cells, guide axons, and regulate the formation of synapses. We then explain how interactions with the environment, both social and physical, modify or consolidate the neural connections formed during early development. Finally, we examine ways in which developmental processes can be harnessed in adults and how factors such as steroid hormones mold the brain, affecting sexual and gender identity. The last steps—changes that occur as the brain ages—are covered in Section IX (Chapter 64).

Part Editor: Joshua R. Sanes

Part VII

45

Patterning the Nervous System

A VAST ARRAY OF NEURONS AND GLIAL CELLS is produced during development of the vertebrate nervous system. Different types of neurons develop in discrete anatomical positions, acquire varied morphological forms, and establish connections with specific populations of target cells. Their diversity is far greater than that of cells in any other organ of the body. The retina, for example, has dozens of types of interneurons, and the spinal cord has more than a hundred types of motor neurons. At present, the true number of neuronal types in the mammalian central nervous system remains unknown, but it is surely more than a thousand. The number of glial types is even less clear; unexpected heterogeneity is being discovered in what was thought, until recently, to be rather homogeneous classes of astrocytes and oligodendrocytes.

The diversity of neuronal types underlies the impressive computational properties of the mammalian nervous system. Yet, as we describe in this chapter and those that follow, the developmental principles that drive the differentiation of the nervous system are begged and borrowed from those used to direct the development in other tissues. In one sense, the development of the nervous system merely represents an elaborate example of the basic challenge that pervades all of developmental biology: how to convert a single cell, the fertilized egg, into the highly differentiated cell types that characterize the mature organism. Only at later stages, as the neurons form complex circuits and experience modifies their connections, do principles of neural development diverge from those in other organs.

Early developmental principles are conserved not only among tissues but also across species and phyla. Indeed, much of what we know about the cellular and molecular bases of neural development in vertebrates comes from genetic studies of so-called simple organisms, most notably the fruit fly *Drosophila melanogaster* and the worm *Caenorhabditis elegans*. Nevertheless, because a main goal of studying neural development is to explain how the assembly of the nervous system underlies both human behavior and brain disorders, our description of the rules and principles of nervous system development focus primarily on vertebrate organisms.

The Neural Tube Arises From the Ectoderm

The vertebrate embryo arises from the fertilized egg. Cell divisions initially form a ball of cells, called the morula, which then hollows out to form the blastula. Next, infoldings and growth generate the gastrula, a structure with polarity (dorsal-ventral and anterior-posterior) and three layers of cells—the endoderm, mesoderm, and ectoderm (Figure 45–1A).

The *endoderm* is the innermost germ layer that later gives rise to the gut, as well as to the lungs, pancreas, and liver. The *mesoderm* is the middle germ layer that gives rise to muscle, connective tissues, and much of the vascular system. The *ectoderm* is the outermost layer. Most of the ectoderm gives rise to the skin, but a narrow central strip flattens out to become the *neural plate* (Figure 45–1B). It is from the neural plate that the central and peripheral nervous systems arise.

Soon after the neural plate forms, it begins to invaginate, forming the *neural groove*. The folds then deepen and eventually separate from the rest of the ectoderm to form the *neural tube*, through a process called neurulation (Figure 45–1C,D). The caudal region of the neural tube gives rise to the spinal cord, whereas the rostral region becomes the brain. As the neural tube closes, cells at its junction with the overlying ectoderm are set aside to become the neural crest, which eventually gives rise to the autonomic and sensory nervous systems, as well as several nonneural cell types (Figure 45–1E).

Secreted Signals Promote Neural Cell Fate

As with other organs, the emergence of the nervous system is the culmination of a complex molecular program that involves the tightly orchestrated expression of specific genes. For the nervous system, the first step is the formation of the neural plate from a restricted region of the ectoderm. This step reflects the outcome of an early choice that ectodermal cells have to make: whether to become neural or epidermal cells. This decision has been the subject of intense study for nearly 100 years.

Much of this work has focused on a search for signals that control the fate of ectodermal cells. We now know that two major classes of proteins work together to promote the differentiation of an ectodermal cell into a neural cell. The first are *inductive factors*, signaling molecules that are secreted by nearby cells. Some of these factors are freely diffusible and exert their actions at a distance, but others are tethered to the cell surface and act locally. The second are surface receptors that enable cells to respond to inductive factors. Activation of these receptors triggers the expression of genes encoding intracellular proteins—transcription factors, enzymes, and cytoskeletal proteins—that push ectodermal cells along the pathway to becoming neural cells.

The ability of a cell to respond to inductive signals, termed its *competence*, depends on the exact repertoire of receptors, transduction molecules, and transcription factors that it expresses. Thus, a cell's fate is determined not only by the signals to which it is exposed—a consequence of when and where it finds itself in the embryo—but also by the profile of genes it expresses as a consequence of its prior developmental history. We will see in subsequent chapters that the interaction of localized inductive signals and intrinsic cell responsiveness is evident at virtually every step throughout neural development.

Development of the Neural Plate Is Induced by Signals From the Organizer Region

The discovery that specific signals are responsible for triggering the formation of the neural plate was the first major advance in understanding the mechanisms that pattern the nervous system. In 1924, Hans Spemann and Hilde Mangold made the remarkable observation that the differentiation of the neural plate from uncommitted ectoderm depends on signals secreted by a specialized group of cells they called the *organizer region*.

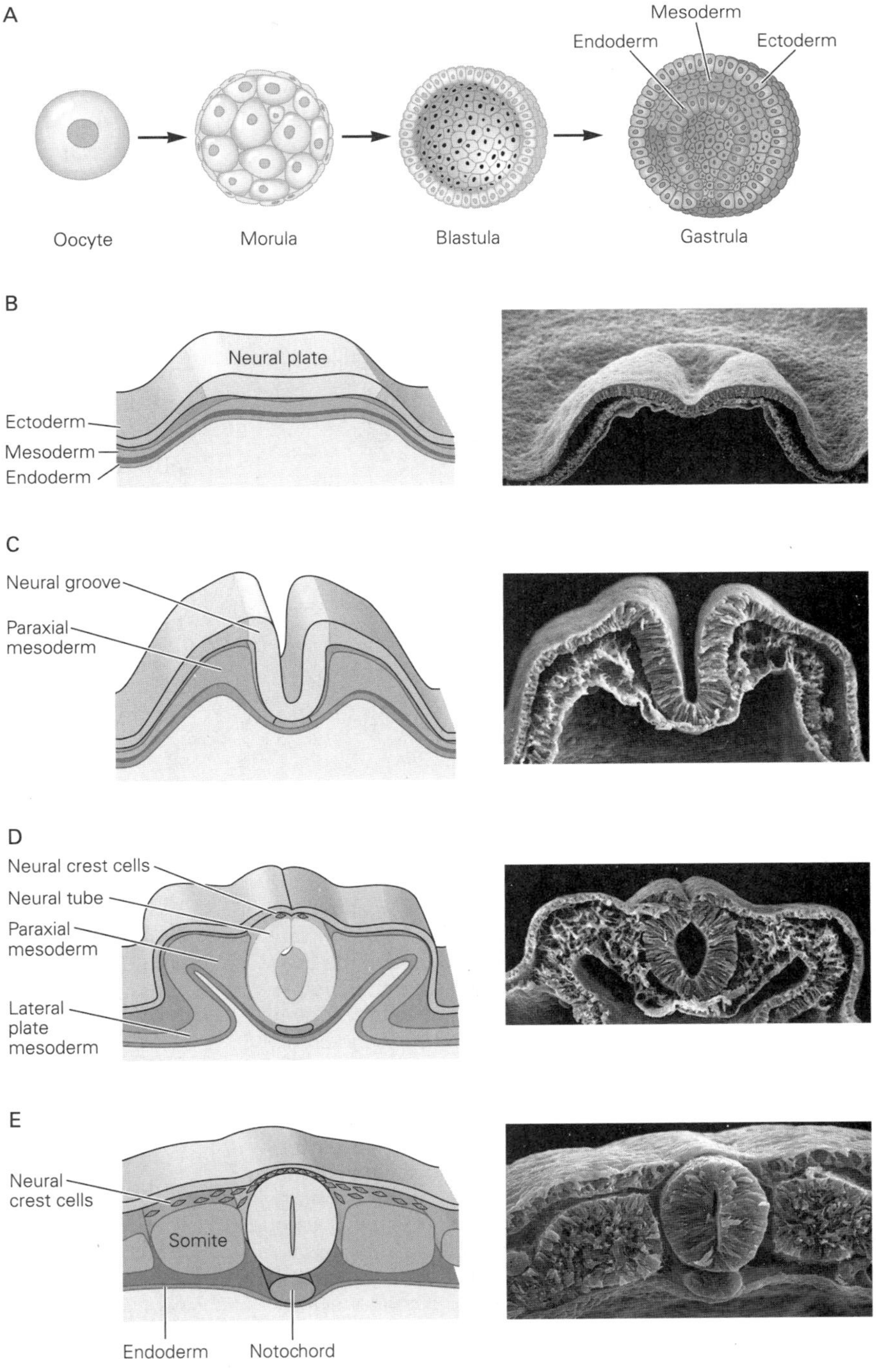

Figure 45-1 The neural plate folds to form the neural tube. (Scanning electron micrographs of chick neural tube reproduced, with permission, from G. Schoenwolf.)

A. Following fertilization of the egg by sperm, cell divisions give rise successively to the morula, blastula, and gastrula. Three germ cell layers—the ectoderm, mesoderm, and endoderm—form during gastrulation.

B. A strip of ectoderm becomes the neural plate, the precursor of the central and peripheral nervous systems.

C. The neural plate buckles at its midline to form the neural groove.

D. Closure of the dorsal neural folds forms the neural tube.

E. The neural tube lies over the notochord and is flanked by somites, an ovoid group of mesodermal cells that give rise to muscle and cartilage. Cells at the junction between the neural tube and overlying ectoderm are set aside to become the neural crest.

Their experiments involved transplanting small pieces of tissue from one amphibian embryo to another. Most telling were transplantations of the dorsal lip of the blastopore, which is destined to form the dorsal mesoderm, from its normal dorsal position to the ventral side of a host embryo. The dorsal lip lies underneath the dorsal ectoderm, a region that normally gives rise to dorsal epidermis, including the neural plate (Figure 45-2). They grafted the tissue from a pigmented embryo into unpigmented host, allowing them to distinguish the position and fate of donor and host cells.

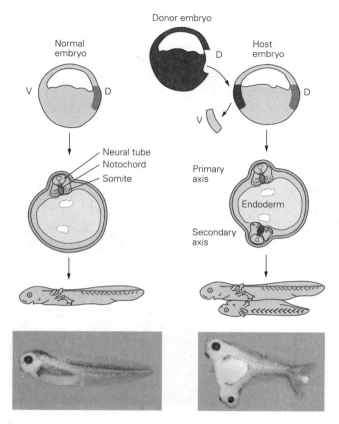

Figure 45–2 Signals from the organizer region induce a second neural tube. (Micrographs reproduced, with permission, from Eduardo de Robertis.)

Left: In the normal frog, embryo cells from the organizer region (the dorsal blastopore lip) populate the notochord, floor plate, and somites. *Right:* Spemann and Mangold grafted the dorsal blastopore lip from an early gastrula stage embryo into a region of a host embryo that normally gives rise to the ventral epidermis. Signals from grafted cells induce a second embryonic axis, which includes a virtually complete neural tube. The donor tissue was from a pigmented embryo, whereas the host tissue was unpigmented, permitting the fate of grafted cells to be monitored by their color. Grafted cells themselves contribute only to the notochord, floor plate, and somites of the host embryo. As the embryo matures, the secondary neural tube develops into a complete nervous system. In the *Xenopus* embryo shown in the micrograph, the second neural axis was induced by injection of an antagonist of bone morphogenetic protein (BMP), in effect substituting for the organizer signal (Figure 45–3). The primary neural axis is also apparent. (Abbreviations: **D**, dorsal; **V**, ventral.)

Spemann and Mangold found that transplanted cells from the dorsal lip of the blastopore followed their normal developmental program, generating midline mesoderm tissue such as the somites and notochord. But the transplanted cells also caused a striking change in the fate of the neighboring ventral ectodermal cells

of the host embryo. Host ectodermal cells were induced to form a virtually complete copy of the nervous system (Figure 45–2). They therefore called the donor tissue the *organizer*. Spemann and Mangold went on to show that the dorsal lip of the blastopore was the only tissue that possessed this "organizing" effect.

These pioneering studies also demonstrated that "induction" plays a critical role in neural development. Induction is a process by which cells of one tissue direct the development of neighboring cells at a region where the two come into proximity. The importance is that it provides a mechanism by which signals from one tissue can lead to subdivision of a second tissue. In this case, the mesoderm induces one part of the ectoderm to become the neural plate, and eventually the nervous system, while the remainder goes on to become epithelium, and eventually skin. The new juxtaposition thereby formed could, in principle, set the stage for a cascade of subsequent inductions and subdivisions. Indeed, we will see that many aspects of neural tube patterning are now known to depend on signals secreted by local organizing centers through actions similar in principle to that of the classical organizer region.

Neural Induction Is Mediated by Peptide Growth Factors and Their Inhibitors

For decades after Spemann and Mangold's pioneering studies, identification of the neural inducer constituted a Holy Grail of developmental biology. The search was marked by little success until the 1980s, when the advent of molecular biology and the availability of better markers of early neural tissue led to breakthroughs in our understanding of neural induction and its chemical mediators.

The first advance came from a simple finding: When the early ectoderm is dissociated into single isolated cells, effectively preventing cell-to-cell signaling, the cells readily acquire neural properties in the absence of added factors (Figure 45–3A). The surprising implication of this finding was that the "default" fate of ectodermal cells is neural differentiation and that this fate is prevented by signaling among ectodermal cells. In this model, the long sought-after "inducer" is actually a "de-repressor": It prevents ectoderm from repressing neural fate.

These ideas immediately raised two further questions. What ectodermal signal represses neural differentiation, and what does organizer tissue provide to overcome the effects of the repressor? Studies of neural induction in frogs and chicks have now provided answers to these questions.

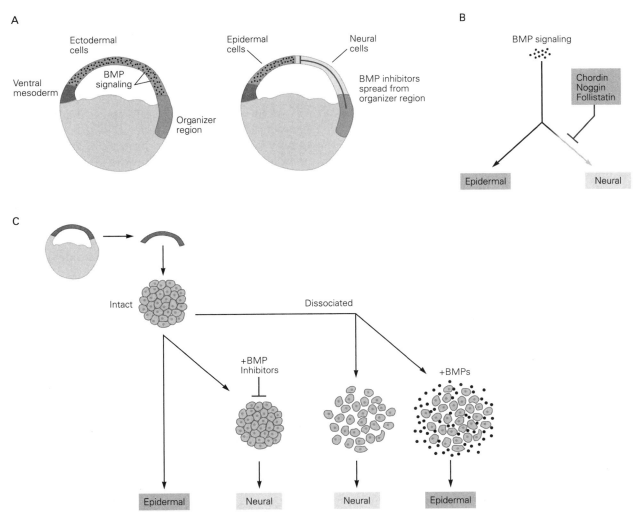

Figure 45–3 Inhibition of bone morphogenetic protein (BMP) signaling initiates neural induction.

A. In *Xenopus* frog embryos, signals from the organizer region (**red line**) spread through the ectoderm to induce neural tissue. Ectodermal tissue that is beyond the range of organizer signals gives rise to epidermis.

B. BMP inhibitors secreted from the organizer region (including noggin, follistatin, and chordin) bind to BMPs and block the

ability of ectodermal cells to acquire an epidermal fate, thus promoting neural character.

C. Ectodermal cells acquire neural or epidermal character depending on the presence or absence of BMP signaling. When ectodermal cell aggregates are exposed to BMP signaling, they differentiate into epidermal tissue. When BMP signaling is blocked, either by dissociating ectodermal tissue into single cells or by addition of BMP inhibitors to ectodermal cell aggregates, the cells differentiate into neural tissue.

In the absence of signals from the organizer, ectodermal cells synthesize and secrete *bone morphogenetic proteins* (BMPs), members of a large family of transforming growth factor β (TGFβ)-related proteins. The BMPs, acting through serine/threonine kinase class receptors on ectodermal cells, suppress the potential for neural differentiation and promote epidermal differentiation (Figure 43–3B). Key evidence for the role of BMPs as

neural repressors came from experiments in which a truncated version of a BMP receptor, which blocks BMP signaling, was found to trigger the differentiation of neural tissue in the *Xenopus* frog embryo. Conversely, exposure of ectodermal cells to BMP signaling promoted differentiation as epidermal cells (Figure 45–3C).

The identification of BMPs as suppressors of neuronal differentiation in turn suggested that the organizer might induce neural differentiation in ecto-dermal cells by secreting factors that antagonize BMP signaling. Direct support for this idea came from the finding that cells of the organizer region express many secreted proteins that act as BMP antagonists. These proteins include noggin, chordin, follistatin, and even some variant BMP proteins. Each of these proteins has the ability to induce ectodermal cells to differentiate into neural tissue (Figure 45–3B). Thus, there is no single neural inducer. In fact, multiple classes of proteins are required for induction, as shown by the later finding that the exposure of ectodermal cells to fibroblast growth factors (FGFs) is also a necessary step in neural differentiation.

Together, these studies provided a molecular explanation of the cellular phenomenon of neural induction. Although many details of the pathway remain to be clarified and some mechanistic differences among species remain perplexing, a key chapter in neural development has now been brought to a satisfying conclusion nearly a century after the organizer was discovered by Spemann and Mangold.

Rostrocaudal Patterning of the Neural Tube Involves Signaling Gradients and Secondary Organizing Centers

As soon as cells of the neural plate have been induced, they begin to acquire regional characteristics that mark the first steps in dividing the nervous system into regions such as forebrain, midbrain, hindbrain, and spinal cord. The subdivision is directed by a series of secreted inductive factors and follows the same basic principles of neural induction. Neural plate cells in different regions of the neural tube respond to these inductive signals by expressing distinct transcription factors that gradually constrain the developmental potential of cells in each local domain. In this way, neurons in different positions acquire functional differences. Signaling occurs along both the rostrocaudal and the dorsoventral axes of the neural tube. We begin by describing rostrocaudal patterning and then return to dorsoventral patterning.

The Neural Tube Becomes Regionalized Early in Development

After the neural tube forms, cells divide rapidly, but rates of proliferation are not uniform. Individual regions of the neural epithelium expand at different rates and begin to form the specialized parts of the

mature central nervous system. Differences in the rate of proliferation of cells in rostral regions of the neural tube result in the formation of three brain vesicles: the forebrain (or prosencephalic) vesicle, the midbrain (or mesencephalic) vesicle, and the hindbrain (or rhombencephalic) vesicle (Figure 45–4A).

At this early three-vesicle stage, the neural tube flexes twice: once at the *cervical flexure,* at the junction of the spinal cord and hindbrain, and once at the *cephalic flexure,* at the junction of the hindbrain and midbrain. A third flexure, the *pontine flexure,* forms later, and later still, the cervical flexure straightens out and becomes indistinct (Figure 45–4D). The cephalic flexure remains prominent throughout development, and its persistence is the reason why the orientation of the longitudinal axis of the forebrain deviates from that of the brain stem and spinal cord.

As the neural tube develops, two of the primary embryonic vesicles divide further, thus forming five vesicles (Figure 45–4B,C). The forebrain vesicle divides to form the telencephalon, which will give rise to the cortex, hippocampus, and basal ganglia, and the diencephalon, which will give rise to the thalamus, hypothalamus, and retina. The mesencephalon, which does not divide further, gives rise to the inferior and superior colliculi and other midbrain structures. The hindbrain vesicle divides to form the metencephalon, which will give rise to the pons and cerebellum, and the myelencephalon, which will give rise to the medulla. Together with the spinal cord, these divisions make up the major functional regions of the mature central nervous system (see Chapter 4). The progressive subdivision of the neural tube into these functional domains is regulated by a variety of secreted signals.

Signals From the Mesoderm and Endoderm Define the Rostrocaudal Pattern of the Neural Plate

It was originally believed that the organizer, as defined by Spemann and Mangold, was uniform in character, and therefore induced an initially uniform neural plate. Subsequent studies showed, however, that the organizer is regionally specialized and secretes factors that initiate the rostrocaudal patterning of the neural plate almost as soon as induction commences. One important class of factors comprises the Wnt proteins (an acronym based on their founding family members, the *Drosophila* Wingless protein and the mammalian Int1 proto-oncogene protein). Others include retinoic acid and FGFs. They are produced by mesodermal cells of the organizer as well as nearby paraxial mesoderm.

The net level of Wnt signaling activity is low at rostral levels of the neural plate and increases

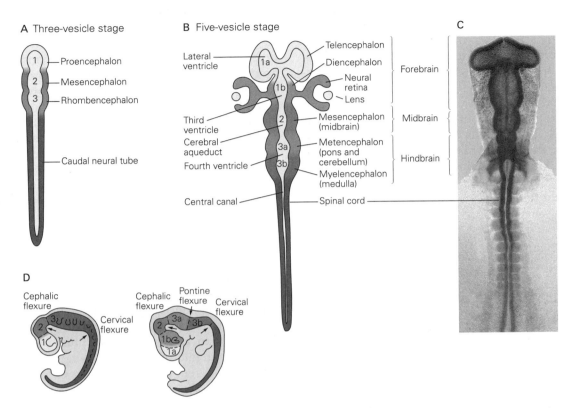

Figure 45–4 Sequential stages of neural tube development.

A. At early stages of neural tube development, there are three brain vesicles, which will form the prosencephalon (forebrain), mesencephalon (midbrain), and rhombencephalon (hindbrain).

B. Further division within the prosencephalon and rhombencephalon generate additional vesicles. The prosencephalon splits to form the telencephalon and diencephalon, and the

rhombencephalon splits to form the metencephalon and the myelencephalon.

C. Top-down view of the neural tube of a chick embryo at the five-vesicle stage. (Reproduced, with permission, from G. Schoenwolf.)

D. The neural tube bends at borders between vesicles, forming the cephalic, pontine, and cervical flexures.

progressively in the caudal direction. This activity gradient arises because the mesoderm adjacent to the caudal region of the neural plate expresses high levels of Wnt. Sharpening this gradient, tissue that underlies the rostral region of the neural plate is a source of secreted proteins that inhibit Wnt signaling, much as BMP inhibitors attenuate BMP signaling at an earlier stage. Thus, cells at progressively more caudal positions along the neural plate are exposed to increasing levels of Wnt activity and acquire a more caudal regional character, spanning the entire range from forebrain to midbrain to hindbrain and finally to spinal cord (Figure 45–5A). These results suggest that an anterior character is the "default" state for neural tissue, with signals such as Wnt imposing a posterior character. Indeed, when ectodermal cells are induced to become neural by application of BMP inhibitors, they differentiate into cells characteristic of anterior structures.

Signals From Organizing Centers Within the Neural Tube Pattern the Forebrain, Midbrain, and Hindbrain

The early influence of mesodermal and endodermal tissues on rostrocaudal neural pattern is further refined by signals from specialized cell groups in the neural tube itself. One that has been studied in particular detail is called the *isthmic organizer*, which forms at the boundary of the hindbrain and midbrain (Figure 45–5B). The isthmic organizer serves a key role in patterning these two domains of the neural tube as well as in specifying the neuronal types within them. Dopaminergic neurons of the substantia nigra and ventral tegmental area are generated in the midbrain, just rostral to the isthmic organizer, whereas serotonergic neurons of the raphe nuclei are generated just caudal to the isthmic organizer, within the hindbrain. As an illustration of how these secondary neural signaling centers impose neural pattern, we describe the origin and signaling activities of the isthmic organizer.

The rostrocaudal positional character of the neural plate stems from the expression of homeodomain transcription factors, the homeodomain being a section of the protein that binds to a specific DNA sequence in regulatory regions of genes, leading to changes in the gene's transcription. Cells in forebrain and midbrain domains of the neural plate express Otx2, whereas cells in the hindbrain domain express Gbx2, both of which are homeodomain transcription factors. The point of transition between Otx2 and Gbx2 expression

is located at the midbrain–hindbrain boundary (Figure 45–5B) and marks the position at which the isthmic organizer will emerge after neural tube closure. At this boundary, other transcription factors are expressed, notably En1 (an Engrailed class transcription factor).

These transcription factors in turn control the expression of two signaling factors, Wnt1 and FGF8, by cells of the isthmic organizer. Wnt1 is involved in the proliferation of cells in the midbrain–hindbrain domain and in the maintenance of FGF8 expression. The spread of FGF8 from the isthmic organizer into the midbrain domain marked by Otx2 expression induces differentiation of dopaminergic neurons, whereas its spread into the hindbrain domain marked by Gbx2 expression triggers the differentiation of serotonergic neurons (Figure 45–5C).

The roles of FGF8 and Wnt1 in signaling from the *isthmic organizer* illustrate an important economy in early neural patterning. The early actions of inductive signals impose discrete domains of transcription factor expression, and these transcriptional domains then allow cells to interpret the actions of the same secreted factor in different ways, producing different neuronal subtypes. In this way, a relatively small number of secreted factors—FGFs, BMPs, hedgehog proteins, Wnt proteins, and retinoic acid—are used in different regions and at different times to program the vast diversity of neuronal cell types generated within the central and peripheral nervous systems.

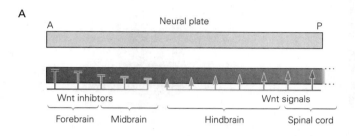

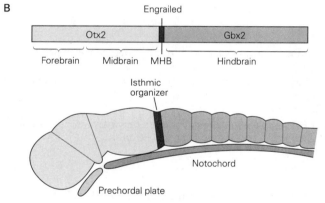

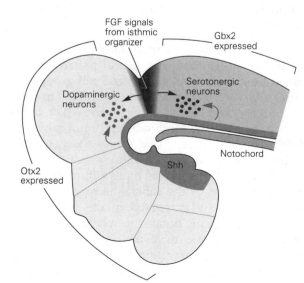

Figure 45–5 (Left) Early anteroposterior patterning signals establish distinct transcription factor domains and define the position of the midbrain–hindbrain boundary region.

A. The anteroposterior pattern of the neural plate is established by exposure of neural cells to a gradient of Wnt signals. Anterior (**A**) regions of the neural plate are exposed to Wnt inhibitors secreted from the endoderm and thus perceive only low levels of Wnt activity. Progressively more posterior (**P**) regions of the neural plate are exposed to high levels of Wnt signaling from the paraxial mesoderm and to lower levels of Wnt inhibitors.

B. In response to this Wnt signaling gradient and other signals, cells in anterior and posterior regions of the neural plate begin to express different transcription factors: Otx2 at anterior levels and Gbx2 at more posterior levels. The intersection of these two transcription factor domains marks the region of the midbrain–hindbrain boundary (**MHB**), where Engrailed transcription factors are expressed. The neural tube then forms segments anterior and posterior to the MHB.

C. Fibroblast growth factor (**FGF**) signals from the isthmic organizer act in concert with sonic hedgehog (**Shh**) signals from the ventral midline to specify the identity and position of dopaminergic and serotonergic neurons. The distinct fates of these two classes of neurons result from the expression of Otx2 in the midbrain and Gbx2 in the hindbrain.

(Adapted, with permission, from Wurst and Bally-Cuif 2001. Copyright © 2001 Springer Nature.)

Figure 45–6 Local signaling centers in the developing neural tube. This side view of the neural tube at a later stage shows the positions of three key signaling centers that pattern the neural tube along the anterior-posterior axis: the anterior neural ridge, the zona limitans intrathalamica (**ZLI**) at the boundary between the rostral and caudal forebrain (diencephalon), and the isthmic organizer, the boundary of the midbrain and hindbrain. The ZLI is a source of sonic hedgehog, and the isthmic organizer and anterior neural ridge are sources of fibroblast growth factor (see Figure 45–5).

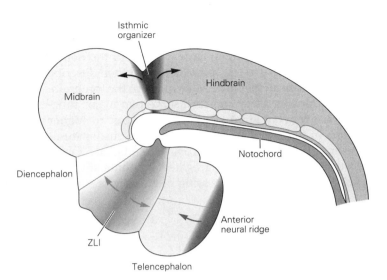

Other cell groups serve similar roles in subdividing the neural tube into domains. For example, at the very rostral margin of the neural tube, a specialized group of cells, called the *anterior neural ridge*, secretes FGF that patterns the telencephalon (Figure 45–6). More caudally is a restricted region called the *zona limitans intrathalamica*, which appears as a pair of horn-like spurs within the diencephalon. Zona limitans intrathalamica cells secrete the protein sonic hedgehog (Shh), which patterns nearby cells that give rise to the nuclei of the thalamus. FGFs and Shh are described in detail below in the context of their prominent role in patterning the cortex and spinal cord, respectively.

Repressive Interactions Divide the Hindbrain Into Segments

An important next step in patterning the neural tube along the rostrocaudal axis is the subdivision of the forebrain and hindbrain into segments, compartmental units that are arrayed along the rostrocaudal axis. These units are called *prosomeres* in the forebrain and *rhombomeres* in the hindbrain.

We use the formation of rhombomeres 3 and 4 (of 7 total) to illustrate the mechanisms leading to segmentation (Figure 45–7). An initial morphogen gradient leads to expression of two distinct transcription factors in this region—*krox20* in what will become rhombomere 3, endowing these cells with a rhombomere 3 identity, and *hoxb1* in what will become rhombomere 4, endowing these cells with a rhombomere 4 identity. Cells near the border express both factors and therefore have an uncertain identity. However, these two factors inhibit each other's expression, so eventually the identity of each cell is fixed.

The problem is that some cells are trapped within the wrong rhombomere. This intermingling is rectified in several ways, one of which is a second inhibitory interaction, this one of a markedly different type. Krox20 and Hoxb1 induce the expression of cell surface recognition and signaling molecules called EphA4 and ephrinB3, respectively. These two proteins bind to each other, leading to transmission of a repulsive signal that separates the cells. We will see below that this repulsion is also key to later decisions that axons make as they grow to their targets. In the hindbrain, before neurons form, it sharpens the borders between rhombomeres. More broadly, rhombomere segregation provides another example of a general theme in neural development: that inductive or adhesive interactions combine with repressive or inhibitory ones to pattern the nervous system.

Dorsoventral Patterning of the Neural Tube Involves Similar Mechanisms at Different Rostrocaudal Levels

As the neural epithelium acquires its rostrocaudal character, cells located at different positions along its dorsoventral axis also begin to acquire distinct identities. Together, patterning along the rostrocaudal and dorsoventral axes divides the neural tube into a three-dimensional grid of molecularly distinct cell types, leading eventually to generation of the various neuronal and glial cell types that distinguish one part of the nervous system from another.

In contrast to the diversity of signals and organizing centers responsible for rostrocaudal patterning of developing neurons, there is a striking consistency in the strategies and principles that establish dorsoventral pattern. We focus initially on the mechanisms of dorsoventral patterning at caudal levels of the neural tube that give rise to the spinal cord and then describe how similar strategies are used to pattern the forebrain.

Neurons in the spinal cord serve two major functions. They relay cutaneous sensory input to higher centers in the brain, and they transform sensory input into motor output. The neuronal circuits that mediate these functions are segregated anatomically. Circuits involved in the processing of cutaneous sensory information are located in the dorsal half of the spinal cord, whereas those involved in the control of motor output are mainly located in the ventral half of the spinal cord.

The neurons that form these circuits are generated at different positions along the dorsoventral axis of the spinal cord in a patterning process that begins with the establishment of distinct progenitor cell types. Motor neurons are generated close to the ventral midline, and most of the interneuron classes that control motor output are generated just dorsal to the position at which motor neurons appear (Figure 45–8). The dorsal half of the neural tube generates projection neurons and local circuit interneurons that process incoming sensory information.

How are the position and identity of spinal neurons established? The dorsoventral patterning of the neural tube is initiated by signals from mesodermal and ectodermal cells that lie close to the ventral and dorsal poles of the neural tube and is perpetuated by signals from two midline neural organizing centers. Ventral patterning signals are initially provided by the notochord, a mesodermal cell group that lies immediately under the ventral neural tube (Figure 45–1). This signaling activity is transferred to the floor plate, a specialized glial cell group that sits at the ventral midline of the neural tube itself. Similarly, dorsal signals are provided initially by cells of the epidermal ectoderm that span the dorsal midline of the neural tube, and subsequently by the roof plate, a glial cell group embedded at the dorsal midline of the neural tube (Figure 45–8).

Thus, neural patterning is initiated through a process of *homogenetic* induction, in which like begets like: Notochord signals induce the floor plate, which

A Induction of segmental identity

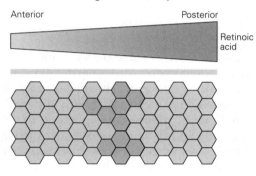

B Establishment of mutually exclusive expression

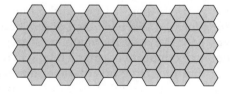

C Cell segregation and border sharpening

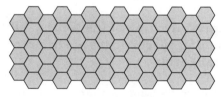

D

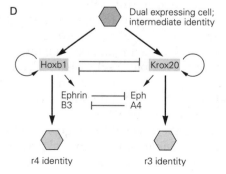

Figure 45–7 (Left) **Repressive interactions divide the hindbrain into rhombomeres.** The sharp border between hindbrain rhombomeres 3 and 4 forms in several steps. (Adapted, with permission, from Addison and Wilkinson 2016. Copyright © 2016 Elsevier Inc.)

A. A gradient of retinoic acid upregulates expression of *hoxb1* in anterior cells (**blue**) and *krox20* in posterior cells (**yellow**), with some cells at the prospective border expressing both genes (**green**).

B. Hoxb1 expression and Krox20 expression become mutually exclusive, thus endowing each cell with a unique molecular identity.

C. Cells trapped in the wrong domain migrate to sharpen the border.

D. Inhibitory interactions underlying border formation. Hoxb1 and Krox20 repress each other's expression in individual cells, so a modest imbalance in level leads to exclusive expression of one factor. Krox20 then upregulates EphA4 in r3 cells, whereas ephrinB3 is upregulated in r4 cells. EphA4 and ephrinB3 repel each other, driving migration of isolated cells and sharpening the segment border.

Figure 45–8 Distinct precursor populations form along the dorsoventral axis of the developing spinal cord.

A. The neural plate is generated from ectodermal cells that overlie the notochord (**N**) and the future somites (**S**). It is flanked by the epidermal ectoderm. (See also Figure 45–1)

B. The neural plate folds dorsally at its midline to form the neural fold. Floor plate cells (**blue**) differentiate at the ventral midline of the neural tube.

C. The neural tube forms by fusion of the dorsal tips of the neural folds. Roof plate cells form at the dorsal midline of the neural tube. Neural crest cells migrate from the neural tube into and past the somites before populating the sensory and sympathetic ganglia.

D. Distinct classes of neurons are generated at different dorsoventral positions in the embryonic spinal cord. Ventral interneurons (**V0–V3**) and motor neurons (**MN**) differentiate from progenitor domains in the ventral spinal cord. Six classes of early dorsal interneurons (**D1–D6**) develop in the dorsal half of the spinal cord. (Adapted from Goulding et al. 2002.)

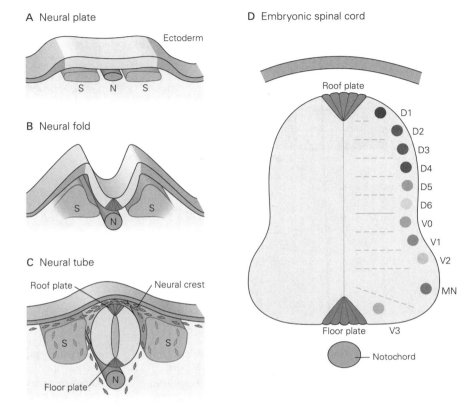

induces ventral neurons, and signals from ectoderm induce the roof plate, which induces dorsal neurons. This strategy ensures that inductive signals are positioned appropriately to control neural cell fate and pattern over a prolonged period of development, as tissues grow and cells move.

The Ventral Neural Tube Is Patterned by Sonic Hedgehog Protein Secreted from the Notochord and Floor Plate

Within the ventral half of the neural tube, the identity and position of developing motor neurons and local interneurons depend on the inductive activity of the Shh protein, which is secreted by the notochord and subsequently by the floor plate. Shh is a member of a family of secreted proteins related to the *Drosophila* hedgehog protein, which had been discovered earlier and shown to control many aspects of embryonic development.

Shh signaling is necessary for the induction of each of the neuronal classes generated in the ventral half of the spinal cord. How can a single inductive signal specify the fate of at least half a dozen neuronal classes? The answer lies in the ability of Shh to act as a morphogen—a signal that can direct different cell fates at different concentration thresholds. The secretion of Shh from the notochord and floor plate establishes a ventral-to-dorsal gradient of Shh protein activity in the ventral neural tube, such that progenitor cells occupying different dorsoventral positions within the neural epithelium are exposed to small (two- to three-fold) differences in ambient Shh signaling activity. Different levels of Shh signaling activity direct progenitor cells in different ventral domains to differentiation as motor neurons and interneurons (Figure 45–9A).

These findings raise two additional questions. How is the spread of Shh protein within the ventral neural epithelium controlled in such a precise manner? And how are small differences in Shh signaling activity converted into all-or-none decisions about the identity of progenitor cells in the ventral neural tube?

Active Shh protein is synthesized from a larger precursor protein, cleaved through an unusual autocatalytic process that involves a serine protease-like activity resident within the carboxy terminus of the precursor protein. Cleavage generates an amino

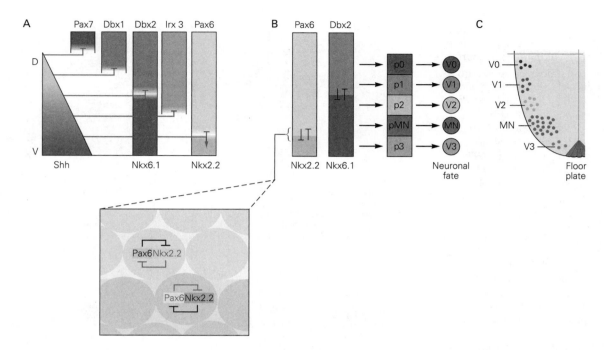

Figure 45–9 A sonic hedgehog signaling gradient controls neuronal identity and pattern in the ventral spinal cord.

A. A ventral-to-dorsal (**V–D**) gradient of sonic hedgehog (**Shh**) signaling establishes dorsoventral domains of homeodomain protein expression in progenitor cells within the ventral half of the neural tube. At each concentration, a different homeodomain transcription factor (Pax7, Dbx1, Dbx2, Irx3, or Pax6) is repressed, with Pax7 the most sensitive and Pax6 the least sensitive to repression. Other homeodomain transcription factors (Nkx6.1 and Nkx2.2) are induced at different Shh levels. The homeodomain proteins that abut a common progenitor domain boundary have similar Shh concentration thresholds for repression and activation. Graded Shh signaling generates a

corresponding gradient of Gli transcription factor activity (not shown).

B. Cross-repression between transcription factors induced or repressed by Shh/Gli signaling specifies different neuronal classes. For example, Pax6 and Nkx2.2, and Dbx2 and Nkx6.1, act in a cell-autonomous manner to repress each other's expression (inset), conferring cell identity to progenitor cells in an unambiguous manner. The sequential influence of graded Shh and Gli signaling, together with homeodomain transcriptional cross-repression, establishes five cardinal progenitor domains.

C. The postmitotic neurons that emerge from these domains give rise to the five major classes of ventral neurons: the interneurons **V0–V3** and motor neurons (**MN**).

terminal protein fragment that possesses all of the signaling activity of Shh. During cleavage, the active amino terminal fragment is modified covalently by the addition of a cholesterol molecule. Following Shh secretion, this lipophilic anchor tethers most of the protein to the surface of notochord and floor plate cells. Nevertheless, a small fraction of the anchored protein is released from the cell surface and transferred from cell to cell within the ventral neural epithelium. In reality, the molecular machinery that ensures the formation of a long-distance gradient of extracellular Shh protein is more complex, involving specialized transmembrane proteins that promote the release of Shh from the floor plate, as well as proteins that regulate Shh protein transfer between cells.

How does the gradient of Shh protein within the ventral neural tube direct progenitor cells along different pathways of differentiation? Shh signaling is

initiated by its interaction with a transmembrane receptor complex that consists of a ligand-binding subunit called *patched* and a signal-transducing subunit called *smoothened* (named for the corresponding *Drosophila* genes). The binding of Shh to patched relieves its inhibition of smoothened and so activates an intracellular signaling pathway that involves several protein kinase enzymes, transport proteins, and most important, the Gli proteins, a class of zinc finger transcription factors.

In the absence of Shh, the Gli proteins are proteolytically processed into transcriptional repressors that prevent the activation of Shh target genes. Activation of the Shh signaling pathway inhibits this proteolytic processing, with the result that transcriptional activator forms of Gli predominate, thus directing the expression of Shh target genes. In this way, an extracellular gradient of Shh protein is converted into a nuclear gradient of Gli activator proteins. The ratio of

Gli repressor and activator proteins at different dorsoventral positions determines which target genes are activated.

What genes are activated by Shh-Gli signaling, and how do they participate in the specification of ventral neuronal subtypes? The major Gli targets are genes encoding yet more transcription factors. One major class of Gli targets encodes homeodomain proteins, transcription factors that contain a conserved DNA-binding motif termed a *homeobox*. A second major class of target genes encodes proteins with a basic helix-loop-helix DNA-binding motif. Some homeodomain and basic helix-loop-helix proteins are repressed and others activated by Shh signaling, each at a particular concentration threshold. In this way, cells in the ventral neural tube are allocated to one of five cardinal progenitor domains, each marked by its own transcription factor profile (Figure 45–9B,C).

The transcription factors that define adjacent progenitor domains repress each other's expression. Thus, although a cell may initially express several transcription factors that could direct the cell along different pathways of differentiation, a minor imbalance in the starting concentration of the two factors is rapidly amplified through repression, and only one of these proteins is stably expressed. This winner-take-all strategy of transcriptional repression sharpens the boundaries of progenitor domains and ensures that an initial gradient of Shh and Gli activity will resolve itself into clear distinctions in transcription factor profile. The transcription factors that specify a ventral progenitor domain then direct the expression of downstream genes that commit progenitor cells to a particular postmitotic neuronal identity. Thus, studies of Shh signaling have not only revealed the logic of ventral neuronal patterning but also demonstrated that the fate of a neuron is determined in part by the actions of transcriptional repressors rather than activators. This principle operates in many other tissues and organisms.

Although originally studied in the context of neural development, defects in Shh signaling have now been implicated in a wide variety of human diseases. Mutations in human Shh pathway genes result in defects in the development of ventral forebrain structures (holoprosencephaly), as well as neurological defects such as spina bifida, limb deformities, and certain cancers.

The Dorsal Neural Tube Is Patterned by Bone Morphogenetic Proteins

A signaling strategy based on graded morphogen levels activating sets of transcriptional programs has also been found to determine the patterning of cell types in the dorsal spinal cord. The differentiation of roof plate cells at the dorsal midline of the neural tube is triggered by BMP signals from epidermal cells that initially border the neural plate and later flank the dorsal neural tube.

After the neural tube has closed, roof plate cells themselves begin to express BMP as well as Wnt proteins. Wnt proteins promote the proliferation of progenitor cells in the dorsal neural tube. BMP proteins induce the differentiation of neural crest cells at the very dorsal margin of the neural tube and later induce generation of diverse populations of sensory relay neurons that settle in the dorsal spinal cord.

Dorsoventral Patterning Mechanisms Are Conserved Along the Rostrocaudal Extent of the Neural Tube

The strategies used to establish dorsoventral pattern in the spinal cord also control cell identity and pattern along the dorsoventral axis of the hindbrain and midbrain, as well as throughout much of the forebrain.

In the mesencephalic region of the neural tube, Shh signals from the floor plate act in concert with the rostrocaudal patterning signals discussed earlier to specify dopaminergic neurons of the substantia nigra and ventral tegmental area as well as serotonergic neurons of the raphe nuclei (see Figure 45–5C). In the forebrain, Shh signals from the ventral midline and BMP signals from the dorsal midline act in combination to establish different regional domains. Shh signaling from the ventral midline sets up early progenitor domains that later produce neurons of the basal ganglia and some cortical interneurons, whereas BMP signaling from the dorsal midline is involved in establishing early neocortical character.

Local Signals Determine Functional Subclasses of Neurons

To this point, we have seen how a uniform group of neural precursor cells, the neural plate, is progressively partitioned into discrete rostrocaudal and dorsoventral domains within the neural tube, largely by morphogen-dependent differential expression of different sets of transcriptional regulators. The next question is how cells within these domains go on to generate the extraordinary diversity of neuronal types that characterize the vertebrate central nervous system. We address that question by focusing on development of the motor neuron.

Motor neurons can be distinguished from all other classes of neurons in the central nervous system by the simple fact that they have axons that extend into the periphery. Viewed in this light, motor neurons represent a coherent and distinct class. But motor neuron types can be distinguished by their position within the central nervous system as well as by the target cells they innervate. The primary job of most motor neurons is to innervate skeletal muscles, of which there are approximately 600 in a typical mammal. From this, it follows that there must be an equal number of motor neuron types.

In this section, we discuss the developmental mechanisms that direct the differentiation of these different functional subclasses. The details of motor neuron development are also important for understanding the basis of neurological disorders that affect these neurons, including spinal muscular atrophy and amyotrophic lateral sclerosis (Lou Gehrig disease). In both diseases, some motor neuron types are highly vulnerable whereas others are relatively resilient. Similar principles drive the diversification of other neuronal classes into distinct types.

Rostrocaudal Position Is a Major Determinant of Motor Neuron Subtype

Motor neurons are generated along much of the rostrocaudal axis of the neural tube, from the midbrain to the spinal cord. Distinct motor neuron types develop at each rostrocaudal level (Figure 45–10), suggesting that one goal of the patterning signals that establish rostrocaudal positional identity within the neural tube is to make motor neurons different.

One major class of genes involved in specifying motor neuron types is the *Hox* gene family. Their name reflects the fact that they were the first transcription factors found that contain a homeodomain, a DNA binding domain now known to be present in many transcription factors that regulate developmental processes in organisms as diverse as yeast, plants, and mammals. For example, the *Otx* and *Gbx* genes discussed above contain homeodomains. The mammalian *Hox* gene family is especially large, containing 39 genes organized in four chromosomal clusters. These genes derive from an ancestral *Hox* complex that also gave rise to the *HOM-C* gene complex in *Drosophila*, where they were initially discovered and analyzed (Figure 45–11).

Members of the vertebrate *Hox* gene family are expressed in overlapping domains along the rostrocaudal axis of the developing midbrain, hindbrain, and spinal cord. As in *Drosophila*, the position of an individual *Hox* gene within its cluster predicts its rostrocaudal domain of expression within the neural tube. In most but not all cases, *Hox* genes located at more 3′ positions within the chromosomal cluster are expressed in more rostral domains, within the midbrain and hindbrain, whereas genes at more 5′ positions are expressed in progressively more caudal positions within the spinal cord (Figures 45–10 and 45–11). This spatial array of *Hox* gene expression determines many aspects of neuronal diversity.

Genetic studies, mostly in mice, have revealed how *Hox* genes control motor neuron identity in the hindbrain and spinal cord. We saw above that Hox genes contribute to formation of the rhombomeres, the fundamental cellular building blocks of the hindbrain. Later, the same genes help to determine the identity of motor neurons within rhombomeres. For example, *Hoxb1* is expressed at high levels in rhombomere 4, the domain that gives rise to facial motor neurons, but is absent from rhombomere 2, the domain that gives rise to trigeminal motor neurons (Figure 45–10).

In the mouse, mutations that eliminate the activity of *Hoxb1* change the fate of cells in rhombomere 4; there is a switch in the identity and connectivity of the motor neurons that emerge from this domain. In the absence of *Hoxb1* function, cells in rhombomere 4 generate motor neurons that innervate trigeminal rather than facial targets, that is, the motor neuron subtype normally generated within rhombomere 2 (Figure 45–12). Many additional studies have confirmed the general principle that motor neuron identity in the hindbrain is controlled by the spatial distribution of *Hox* gene expression.

The control of spinal motor neuron identity is more complicated. Spinal motor neurons are clustered within longitudinal columns that occupy discrete segmental positions, in register with their peripheral targets. Motor neurons that innervate forelimb and hindlimb muscles are found in the lateral motor columns at cervical and lumbar levels of the spinal cord, respectively. In contrast, motor neurons that innervate sympathetic neuronal targets are found within the preganglionic motor column at thoracic levels of the spinal cord. Within the lateral motor columns, motor neurons that innervate a single limb muscle are clustered together into discrete groups, termed *motor pools*. Because each limb in higher vertebrates contains more than 50 different muscle groups, a corresponding number of motor pools are required.

The identity of motor neurons in the spinal cord is controlled by the coordinate activity of *Hox* genes found at more 5′ positions within the chromosomal *Hox* clusters. For example, the spatial domains of expression and activity of Hox6 and Hox9 proteins establish the identities of motor neurons in the

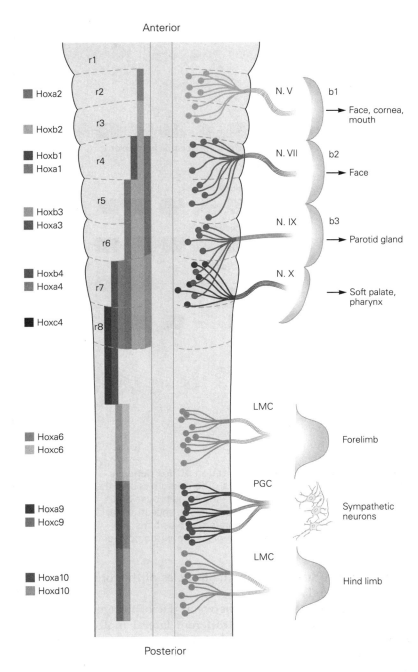

Figure 45–10 The anteroposterior profile of *Hox* gene expression determines the subtype of motor neurons in the hindbrain and spinal cord. Different Hox proteins are expressed in discrete but partially overlapping rostrocaudal domains of the hindbrain and spinal cord. The position of *Hox* genes on the four mammalian chromosomal clusters roughly corresponds to their domain of expression along the anteroposterior axis of the neural tube.

At hindbrain levels, motor neurons sending axons into cranial nerves V (trigeminal), VII (facial), IX (glossopharyngeal), and X (vagus) are depicted. These cranial motor nerves project to peripheral targets in the branchial arches **b1–b3**. The hindbrain rhombomeres (**r1–r8**) and Hox profiles are shown on the left.

At spinal levels, motor neurons that send axons to the forelimb and hind limb are contained within the lateral motor columns (**LMC**), located at brachial and lumbar levels of the spinal cord, respectively. Preganglionic autonomic motor neurons (**PGC**) destined to innervate sympathetic ganglion targets are generated at thoracic levels. (Adapted, with permission, from Kiecker and Lumsden 2005. Copyright © 2005 Springer Nature.)

brachial lateral motor column and the preganglionic motor column. Hox6 proteins specify brachial lateral motor column identity, whereas Hox9 proteins specify preganglionic motor column identity. Motor neurons at the boundary of the forelimb and thoracic regions acquire an unambiguous columnar identity because the Hox6 and Hox9 proteins are mutually repressive (Figure 45–13A), similar to the transcriptional cross-repression that occurs in the dorsoventral patterning of the spinal cord.

Local Signals and Transcriptional Circuits Further Diversify Motor Neuron Subtypes

How do motor neurons within the lateral motor columns develop more refined identities, directing their axons to specific limb muscles? Once again, *Hox* genes control this stage of motor neuron diversification. We illustrate this function of Hox proteins by considering the pathway that generates the distinct divisional and pool identities of neurons within the brachial lateral

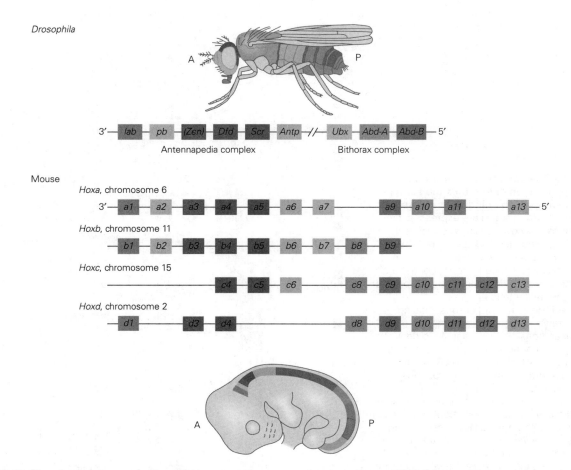

Figure 45–11 The clustered organization of *Hox* genes is conserved from flies to vertebrates. The diagram shows the chromosomal arrangement of *Hox* genes in the mouse and *HOM-C* genes in *Drosophila*. Insects have one ancestral *Hox* gene cluster, whereas higher vertebrates such as birds and mammals have four duplicate *Hox* gene clusters. The position of a given *Hox* or *HOM-C* gene on the chromosomal cluster is typically related to the position on the anteroposterior body axis where the gene is expressed. (Adapted, with permission, from Wolpert et al. 1998. Permission conveyed through Copyright Clearance Center, Inc.)

motor column that innervate the muscles of the fore-limb (Figure 45–13A).

Repressive interactions between Hox proteins expressed by the neurons in different lateral motor columns ensure that neurons that populate different motor pools express distinct profiles of Hox protein expression. These Hox profiles direct the expression of downstream transcription factors as well as the axonal surface receptors that enable motor axons to respond to local cues within the limb that guide them to specific muscle targets. For example, the expression of Hox6 proteins activates a retinoic acid signaling pathway that directs the expression of two homeodo-main transcription factors, Is11 and Lhx1. These fac-tors in turn assign motor neurons to two divisional classes and determine the pattern of expression of the ephrin receptors that guide motor axons in the limb. The axons of motor neurons in these two divisions project into the ventral and dorsal halves of the limb mesenchyme under the control of ephrin signaling (Figure 45–14).

Not all motor neuron columns are determined by Hox protein activity, however. The median motor col-umn is generated at all segmental levels of the spinal cord in register with axial muscles. Development of median motor column cells is controlled by Wnt4/5 signals secreted from the ventral midline of the spi-nal cord and by the expression of the homeodomain proteins Lhx3 and Lhx4, which render neurons in this column immune to the segmental patterning actions of Hox proteins.

Thus, in both the hindbrain and spinal cord, the point-to-point connectivity of motor neurons with specific muscles emerges through tightly orchestrated programs of homeodomain protein expression and activity. In vertebrates, these genes have evolved to

Wild type

Hoxb1 mutant

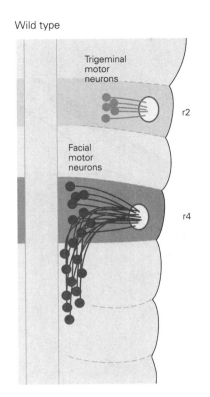
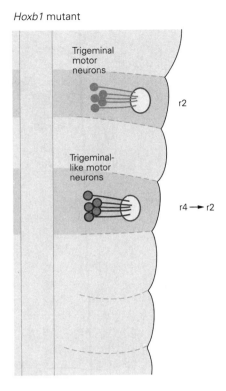

Figure 45–12 The mouse *Hoxb1* gene controls the identity and projection of hindbrain motor neurons. *Hoxb1* is normally expressed at highest levels by cells in rhombomere r4. In wild-type mice, trigeminal motor neurons are generated in rhombomere r2, and their cell bodies migrate laterally before projecting their axons out of the hindbrain at the r2 level. In contrast, the cell bodies of facial motor neurons generated in rhombomere r4 migrate caudally yet project their axons out of the hindbrain at the r4 level. In mouse *Hoxb1* mutants, motor neurons generated in rhombomere r4 migrate laterally instead of caudally, acquiring the features of r2 level trigeminal motor neurons. **Ellipses** indicate axonal exit points. (Adapted, with permission, from Struder et al. 1996. Copyright © 1996 Springer Nature.)

direct neuron subtype and connectivity as well the basic body plan.

The Developing Forebrain Is Patterned by Intrinsic and Extrinsic Influences

Neurons in the mammalian forebrain form circuits that mediate emotional behaviors, perception, and cognition and participate in the storage and retrieval of memories. Much like the hindbrain, the embryonic forebrain is initially divided along its rostrocaudal axis into transversely organized domains called *prosomeres*. Prosomeres 1 to 3 develop into the caudal part of the diencephalon, from which the thalamus emerges. Prosomeres 4 to 6 give rise to the rostral diencephalon and telencephalon. The ventral region of the rostral diencephalon gives rise to the hypothalamus and basal ganglia, whereas the telencephalon gives rise to the neocortex and hippocampus.

Inductive Signals and Transcription Factor Gradients Establish Regional Differentiation

Finally, we turn to the patterning of the neocortex itself, asking whether the developmental mechanisms and principles that govern the development of other

regions of the central nervous system also control the emergence of cortical areas specialized for particular sensory, motor, and cognitive functions.

From the time of Brodmann's classical anatomical description at the beginning of the 20th century, we have known that the cerebral cortex is subdivided into many different areas. Recent studies of cortical development have begun to provide insight into the signaling mechanisms that establish somatosensory, auditory, and visual areas.

There is now evidence for the existence of a cortical "protomap," a basic plan in which different cortical areas are established early in development before inputs from other brain regions can influence development. This view is supported by studies of transcription factor expression in the developing neocortex. Two homeodomain transcription factors, Pax6 and Emx2, are expressed in complementary anteroposterior gradients in the ventricular zone of the developing neocortex—high levels of Pax6 at anterior levels and high levels of Emx2 at posterior levels. These early patterns are established in part by a local rostral source of FGF signals, which promote Pax6 and repress Emx2 expression (Figure 45–15A). As is the case in the hindbrain, the distinct spatial domains of expression of Pax6 and Emx2 are sharpened by cross-repressive interactions between the two transcription factors.

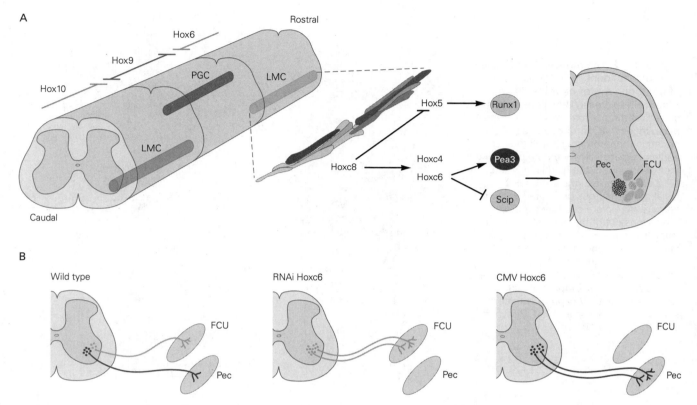

Figure 45–13 Hox proteins control the identity of neurons in motor columns and pools. (Adapted, with permission, from Dasen et al. 2005.)

A. Hox6, Hox9, and Hox10 proteins are expressed in motor neurons at distinct rostrocaudal levels of the spinal cord and direct motor neuron identity and peripheral target connectivity. Hox6 activities control the identity of cells in the brachial lateral motor column (**LMC**), Hox9 controls the identity of cells in the preganglionic column (**PGC**), and Hox10 controls the identity of cells in the lumbar column (**LMC**). Cross-repressive interactions between Hox6, Hox9, and Hox10 proteins refine Hox profiles, and Hox activator functions define LMC and PGC identities. A more complex Hox transcriptional network controls motor pool identity and connectivity. Hox genes determine the rostrocaudal position of motor pools within the LMC. Hoxc8 is required in caudal LMC neurons to generate the motor pools

for the pectoralis (**Pec**) and flexor carpi ulnaris (**FCU**) muscles; these neurons express the transcription factors Pea3 and Scip, respectively. The patterns of Hox expression in the Pec and FCU pools are established through a transcriptional network that appears to be driven largely by Hox cross-repressive interactions.

B. Changing the Hox code within motor pools changes the pattern of muscle connectivity. Alterations in the profile of Hox6 expression determine the expression of Pea3 and Scip and control the projection of motor axons to the Pec or FCU muscles. RNA interference (**RNAi**) knock-down of Hox6 suppresses innervation of the Pec muscle so that motor axons innervate the FCU muscle only. Ectopic expression of Hoxc6 driven by a cytomegalovirus (**CMV**) promoter represses connectivity with FCU, so that motor axons innervate only the Pec muscle.

The spatial distribution of Pax6 and Emx2 helps to establish the initial regional pattern of the neocortex. In mice lacking Emx2 activity, there is an expansion of rostral neocortex—the motor and somatosensory areas—at the expense of the more caudal auditory and visual areas. Conversely, in mice lacking Pax6 activity, visual and auditory areas are expanded at the expense of motor and somatosensory areas (Figure 45–15B).

Thus, as in the spinal cord, hindbrain, and midbrain, early neocortical patterns are established through the interplay between local inductive signals and gradients of transcription factor expression. How

these gradients specify discrete functional areas in the neocortex remains unclear. Unlike segmentation in the hindbrain, where transcription factors precisely specify rhombomeres, transcriptional markers of individual neocortical areas have not yet been identified.

Afferent Inputs Also Contribute to Regionalization

In the adult neocortex, different functional areas can be distinguished by differences in the layering pattern of neurons—the cytoarchitecture of the areas—and by their neuronal connections. One striking instance of

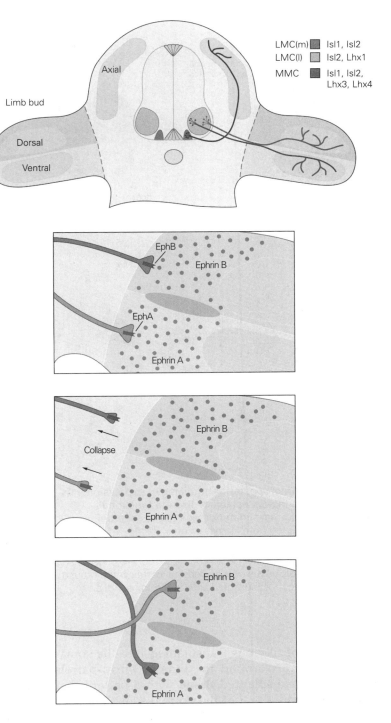

Figure 45–14 The axons of lateral motor column neurons are guided into the limb by ephrin class tyrosine kinase receptors. Motor neurons in the medial and lateral divisions of the lateral motor column (**LMC**) project axons into the ventral and dorsal halves of the limb mesenchyme, respectively. The profile of expression of LIM class homeodomain proteins regulates this dorsoventral projection. The LIM homeodomain protein Isl1 expressed by medial LMC neurons directs a high level of expression of EphB receptors, such that as the axons of these cells enter the limb, they are prevented from projecting dorsally by the high level of repellant ephrin B ligands expressed by cells of the dorsal limb mesenchyme. These axons therefore project into the ventral limb mesenchyme. Conversely, the LIM homeodomain protein Lhx1 expressed by lateral LMC neurons directs a high level of expression of EphA receptors, such that as the axons of these cells enter the limb, they are prevented from projecting ventrally by the high level of repellant ephrin A ligands expressed by cells of the ventral limb mesenchyme. These axons therefore project into the dorsal limb mesenchyme. Eph and ephrin signaling is discussed in greater detail in Chapter 47. (Abbreviation: **MMC**, medial motor column.)

regional distinctiveness in cell pattern is a grid-like array of neurons and glial cells termed "barrels" in the primary somatosensory cortex of rodents. Each cortical barrel receives somatosensory information from a single whisker on the snout, and the regular array of cortical barrels reflects the somatotopic organization of afferent information from the body surface,

culminating in the projection of thalamic efferents to specific cortical barrels (Figure 45–16A).

Cortical barrels are evident soon after birth, and their development depends on a critical period of afferent input from the periphery; their formation is disrupted if the whisker field in the skin is eliminated during this critical period. Strikingly, if prospective

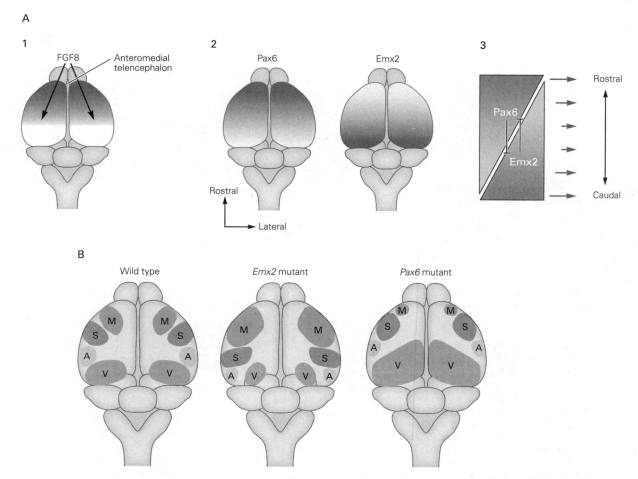

Figure 45–15 Anteroposterior gradients of expression of transcription factors establish discrete functional areas along the anteroposterior axis of the developing forebrain. (Adapted from Hamasaki et al. 2004.)

A. (1) FGF8 signals from the anteromedial telencephalon establish the rostrocaudal pattern of the cerebral cortex. (2) A top-down view of the developing cerebral cortex in the mouse shows inverse rostrocaudal gradients of the transcription factors Pax6 and Emx2. (3) These two transcription factors mutually repress each other's expression.

B. Different functional areas develop at different rostrocaudal positions. Motor areas develop in the anterior region (**M**) and visual areas in more posterior regions (**V**). Genetic elimination of Emx2 function results in expansion of the motor areas and contraction in auditory (**A**) and visual areas. Conversely, elimination of Pax6 function results in an expansion of the visual areas and a contraction of motor and auditory areas. (Abbreviation: **S**, somatosensory areas.)

visual cortical tissue is transplanted into the somatosensory cortex around the time of birth, barrels form in the transplanted tissue with a pattern that closely resembles that of the normal somatosensory barrel field (Figure 45–16B). Together, these findings demonstrate that afferent input superimposes aspects of neocortical patterning on the basic features of the protomap.

The nature of the input to different cortical areas influences neural function as well as cytoarchitecture. This can be shown by monitoring physiological and behavioral responses after rerouting afferent pathways of one sensory modality to a region of neocortex that normally processes a different modality. In animals in which retinal inputs are rerouted into the auditory pathway, the primary auditory cortex contains a systematic representation of visual space rather than of sound frequency (Figure 45–17). When these animals are trained to discriminate a visual from an auditory cue, they perceive a cue as visual when the rewired auditory cortex is activated by vision.

Thus, brain pathways and neocortical regions are established through genetic programs during early development but later depend on afferent inputs for their specialized anatomical, physiological, and behavioral functions.

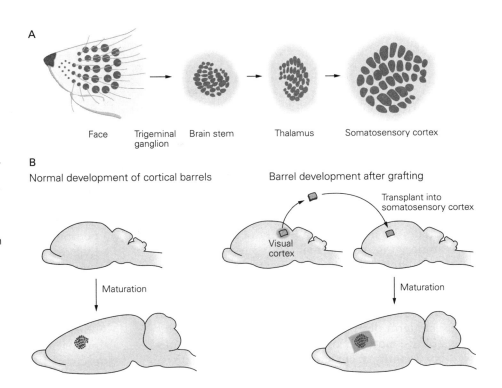

Figure 45–16 Sensory input regulates the organization of "barrels" in the developing somatosensory cortex in rodents. (Adapted from Schlaggar and O'Leary 1991.)

A. The barrel area of the rodent somatosensory cortex forms a somatotopic representation of the rows of whiskers on the animal's snout. Similar representations of the whisker field are present upstream—in the brain stem and in the thalamic nuclei that relay somatosensory inputs from the face to the cortex.

B. A barrel-like cellular organization is induced in developing visual cortex tissue that was grafted at an early postnatal stage into the somatosensory cortex.

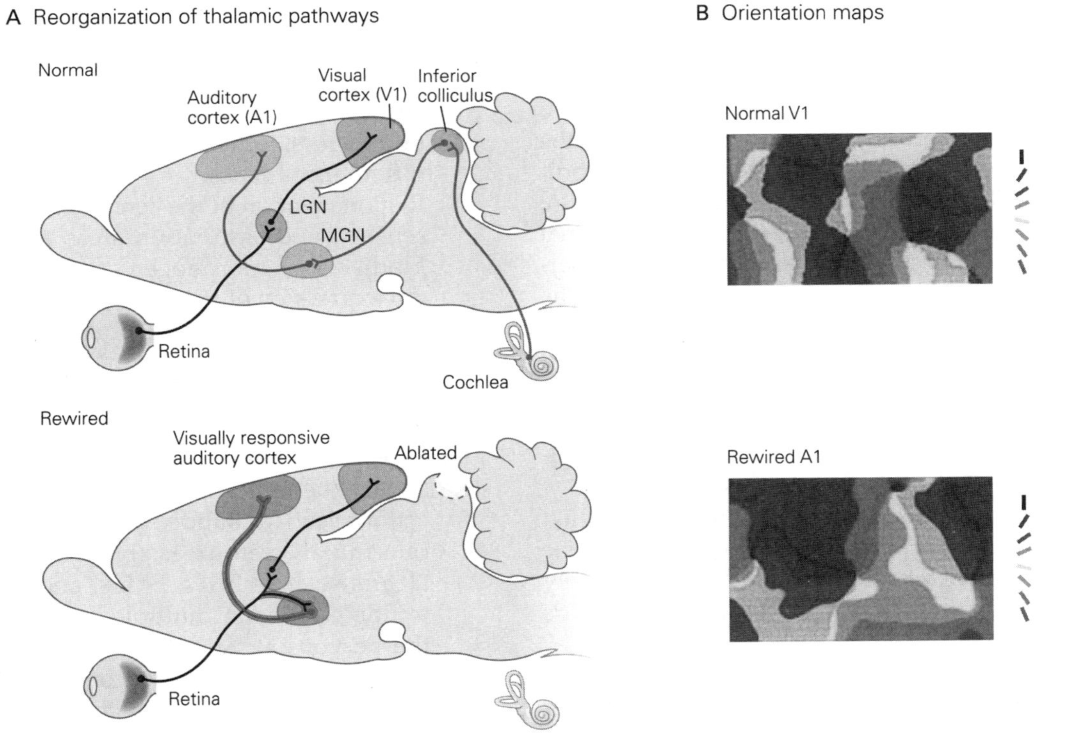

Figure 45–17 Rerouting thalamocortical input can recruit cortical areas for new sensory functions. (Adapted, with permission, from Sharma, Angelucci, and Sur 2000. Copyright © 2000 Springer Nature.)

A. The visual pathway consists of afferent fibers from the retina that innervate the lateral geniculate nucleus (**LGN**) and superior colliculus. Axons from the LGN project to the primary visual cortex (**V1**). The auditory pathway projects from the cochlear nucleus (not shown) to the inferior colliculus, and then to the medial geniculate nucleus (**MGN**) and on to the primary auditory cortex (**A1**). Ablating the inferior colliculus in neonatal ferrets causes retinal afferents to innervate the MGN. As a consequence, the auditory cortex is reprogrammed to process visual information.

B. Visual orientation maps similar to those seen in normal V1 cortex are observed in rewired A1 auditory cortex of ferrets using optical imaging of intrinsic signals. The different colors represent different receptor field orientations (see bars at right). The pattern of activity in rewired A1 resembles that of normal V1.

Highlights

1. The early vertebrate embryo consists of three layers of cells—ectoderm, mesoderm, and endoderm. The entire nervous system arises from the ectoderm, and more specifically from a central strip of ectoderm called the neural plate.

2. Formation of neural plate within the ectoderm occurs by a process called induction, in which underlying mesodermal cells secrete soluble factors that induce a neural program of gene expression in neighboring ectodermal cells. Induction involves a "de-repression" mechanism in which mesoderm-derived soluble factors prevent ectoderm-derived bone morphogenetic proteins (BMPs; members of the transforming growth factor β family) from suppressing the neural fate.

3. Following induction, the neural plate invaginates from the ectoderm to form a neural tube. The tube gives rise to the central nervous system, while cells at the border between neural tube and ectoderm form neural crest, which migrates through the embryo to form the sensory and autonomic ganglia of the peripheral nervous system.

4. As soon as the neural tube forms, it begins to become regionalized. Regionalization along the anterior-posterior axis leads to a series of subdivisions. The anterior region becomes the brain, and the posterior region becomes the spinal cord. Divisions of the prospective brain generate the forebrain, midbrain, and hindbrain. The forebrain divides further to form the telencephalon, from which cortex, hippocampus, and basal ganglia arise; and the diencephalon, which gives rise to thalamus, hypothalamus, and retina. The hindbrain divides to form the pons and cerebellum anteriorly and the medulla posteriorly.

5. Anterior-posterior patterning is established by gradients of Wnt signaling, which arise from selective production of Wnts posteriorly and selective production of Wnt inhibitors anteriorly.

6. Subdivisions along the anteroposterior axis are established by groups of cells called organizing centers at defined positions within the neural tube. The organizing centers secrete factors that pattern neighboring regions of the neural tube and specify neuronal types within them. For example, the isthmic organizer at the boundary of the hindbrain and midbrain secretes Wnts and fibroblast growth factors (FGFs). They act differentially in anterior and posterior regions because the earlier patterning events led to expression of different transcription factors by cells in these regions.

7. Still later, further subdivisions form segments called prosomeres in the forebrain and rhombomeres in the hindbrain, with differential expression of transcription factors leading to generation of distinct neural types in each.

8. In both the hindbrain and the spinal cord, motor neurons acquire distinct properties according to their anterior-posterior position, differentiating into the groups that innervate distinct muscles. Differential expression of transcription factors called Hox proteins is particularly important in diversification of motor neurons. They act with other transcription and soluble factors to divide motor neurons into columns and pools, with each pool destined to innervate a specific muscle.

9. The neural tube is also patterned along the dorsoventral axis. Similar to anterior-posterior regionalization, patterning results from gradients of morphogens. The most important are sonic hedgehog (Shh), which forms a ventral high–dorsal low gradient, and BMPs, which form a dorsal high–ventral low gradient. Different levels of Shh and BMPs induce different transcription factors, which in turn lead to generation of different cell types.

10. Regionalization of the cerebral cortex into motor, sensory, and association areas also begins with gradients of morphogens that induce differential expression of transcription factors, leading to establishment of a "protomap" of area identity. Interactions among areas along with input from subcortical regions refine the protomap to form definitive cortical areas.

11. Several general principles explain many aspects of early neural development: (a) Inductive interactions lead to subdivision of a uniform set of cells into discrete areas. (b) A small set of soluble factors such as FGFs, BMPs, and Wnts are used multiple times at multiple stages to regionalize the nervous system. (c) Varying levels of these factors lead to expression of different transcription factors, which in turn generate different neural cell types. (d) Repressive interactions between cells expressing different transcription factors sharpen boundaries along both anteroposterior and dorsoventral axes.

12. Until recently, studies on early stages of neural development have been restricted to experimental animals. Recent advances now enable neuroscientists to recapitulate some of these processes using cultured human cells. It should therefore

soon be possible to learn whether there are critical early differences between humans and other species that contribute to the complexity of the human brain and to human brain disorders.

Joshua R. Sanes
Thomas M. Jessell

Selected Reading

Anderson C, Stern CD. 2016. Organizers in development. Curr Top Dev Biol 117:435–454.

Catela C, Shin MM, Dasen JS. 2015. Assembly and function of spinal circuits for motor control. Annu Rev Cell Dev Biol 31:669–698.

Dessaud E, McMahon AP, Briscoe J. 2008. Pattern formation in the vertebrate neural tube: a sonic hedgehog morphogen-regulated transcriptional network. Development 135: 2489–2503.

Goulding M. 2009. Circuits controlling vertebrate locomotion: moving in a new direction. Nat Rev Neurosci 10:507–518.

Hamburger V. 1988. *The Heritage of Experimental Embryology. Hans Spemann and the Organizer.* New York: Oxford Univ. Press.

Kiecker C, Lumsden A. 2012. The role of organizers in patterning the nervous system. Annu Rev Neurosci 35:347–367.

Ozair MZ, Kintner C, Brivanlou AH. 2013. Neural induction and early patterning in vertebrates. Wiley Interdiscip Rev Dev Biol 2:479–498.

Rakic P. 2002. Evolving concepts of cortical radial and areal specification. Prog Brain Res 136:265–280.

Sur M, Rubenstein JL. 2005. Patterning and plasticity of the cerebral cortex. Science 310:805–810.

References

Addison M, Wilkinson DG. 2016. Segment identity and cell segregation in the vertebrate hindbrain. Curr Top Dev Biol 117:581–596.

Bell E, Wingate RJ, Lumsden A. 1999. Homeotic transformation of rhombomere identity after localized Hoxb1 misexpression. Science 284:2168–2171.

Cholfin JA, Rubenstein JL. 2007. Patterning of frontal cortex subdivisions by Fgf17. Proc Natl Acad Sci U S A 104:7652–7657.

Dasen JS. 2017. Master or servant? Emerging roles for motor neuron subtypes in the construction and evolution of locomotor circuits. Curr Opin Neurobiol 42:25–32.

Dasen JS, Tice BC, Brenner-Morton S, Jessell TM. 2005. A Hox regulatory network establishes motor neuron pool identity and target-muscle connectivity. Cell 123:477–491.

Goulding M, Lanuza G, Sapir T, Narayan S. 2002. The formation of sensorimotor circuits. Curr Opin Neurobiol 12:505–515.

Hamasaki T, Leingartner A, Ringstedt T, O'Leary DD. 2004. EMX2 regulates sizes and positioning of the primary sensory and motor areas in neocortex by direct specification of cortical progenitors. Neuron 43:359–372.

Horng S, Sur M. 2006. Visual activity and cortical rewiring: activity-dependent plasticity of cortical networks. Prog Brain Res 157:3–11.

Ille F, Atanasoski S, Falkm S, et al. 2007. Wnt/BMP signal integration regulates the balance between proliferation and differentiation of neuroepithelial cells in the dorsal spinal cord. Dev Biol 304:394–408.

Kiecker C, Lumsden A. 2005. Compartments and their boundaries in vertebrate brain development. Nat Rev Neurosci 6:553–564.

Levine AJ, Brivanlou AH. 2007. Proposal of a model of mammalian neural induction. Dev Biol 308:247–256.

Lim Y, Golden JA. 2007. Patterning the developing diencephalon. Brain Res Rev 53:17–26.

Liu A, Niswander LA. 2005. Bone morphogenetic protein signalling and vertebrate nervous system development. Nat Rev Neurosci 6:945–954.

Lupo G, Harris WA, Lewis KE. 2006. Mechanisms of ventral patterning in the vertebrate nervous system. Nat Rev Neurosci 7:103–114.

Mallamaci A, Stoykova A. 2006. Gene networks controlling early cerebral cortex arealization. Eur J Neurosci 23:847–856.

Nordstrom U, Maier E, Jessell TM, Edlund T. 2006. An early role for WNT signaling in specifying neural patterns of Cdx and Hox gene expression and motor neuron subtype identity. PLoS Biol 4:1438–1452.

Rash BG, Grove EA. 2006. Area and layer patterning in the developing cerebral cortex. Curr Opin Neurobiol 16:25–34.

Schlaggar BL, O'Leary DDM. 1991. Potential of visual cortex to develop an array of functional units unique to somatosensory cortex. Science 252:1556–1560.

Sharma J, Angelucci A, Sur M. 2000. Induction of visual orientation modules in auditory cortex. Nature 404:841–847.

Song MR, Pfaff SL. 2005. Hox genes: the instructors working at motor pools. Cell 123:363–365.

Stamataki D, Ulloa F, Tsoni SV, Mynett A, Briscoe J. 2005. A gradient of Gli activity mediates graded Sonic Hedgehog signaling in the neural tube. Genes Dev 19:626–641.

Struder M, Lumsden A, Ariza-McNaughton L, Bradley A, Krumlauf R. 1996. Altered segmental identity and abnormal migration of motor neurons in mice lacking *Hoxb-1*. Nature 384:630–634.

von Melchner L, Pallas SL, Sur M. 2000. Visual behaviour mediated by retinal projections directed to the auditory pathway. Nature 404:871–876.

Wolpert L, Beddington R, Brockes J, Jessell TM, Lawrence PA, Meyerowitz E. 1998. *Principles of Development.* New York: Oxford Univ Press.

Wolpert L, Smith J, Jessell T, Lawrence P, Robertson E, Meyerowitz E. 2006. *Principles of Development,* 3rd ed. New York: Oxford Univ. Press.

Wurst W, Bally-Cuif L. 2001. Neural plate patterning: upstream and downstream of the isthmic organizer. Nat Rev Neurosci 2:99–108.

46

Differentiation and Survival of Nerve Cells

IN THE PRECEDING CHAPTER, WE DESCRIBED how local inductive signals pattern the neural tube and establish the early regional subdivisions of the nervous system—the spinal cord, hindbrain, midbrain, and forebrain. Here, we turn to the issue of how progenitor cells within these regions differentiate into neurons and glial cells, the two major cell types of the nervous system. The mature brain comprises billions of nerve cells and a similar number of glial cells arranged in complex patterns, yet its precursor, the neural plate, initially contains only a few hundred cells arranged in a simple columnar epithelium. From this observation alone, it should be apparent that the generation of neural cells and their delivery to appropriate sites must be carefully regulated.

We begin by discussing some of the molecules that specify neuronal and glial cell fates. The basic mechanisms of neurogenesis endow cells with common neuronal properties, features that are largely independent of the region of the nervous system in which they are generated or the specific functions they perform. We also describe mechanisms by which developing neurons become specialized, for example by acquiring the machinery to synthesize specific neurotransmitters.

We next discuss how neurons are delivered from their sites of origin to their final destinations. A common theme is that neurons are frequently "born"—that is, become postmitotic—far from where they end up, for example, in the layers of the cerebral cortex or the ganglia of the peripheral nervous system. Such distances necessitate elaborate migratory mechanisms, which differ among neuronal types.

After the identity and functional properties of the neuron have begun to emerge, additional developmental processes determine whether the neuron will live or die. Remarkably, approximately half of the neurons generated in the mammalian nervous system are lost through programmed cell death. We examine the factors that regulate the survival of neurons and the possible benefits of widespread neuronal loss. Finally, we describe a core biochemical pathway in nerve cells destined for elimination.

The Proliferation of Neural Progenitor Cells Involves Symmetric and Asymmetric Cell Divisions

Histologists in the late 19th century showed that neural epithelial cells close to the ventricular lumen of the embryonic brain exhibit features of mitosis. We now know that the proliferative zones surrounding the ventricles are the major sites for the production of neural cells in the central nervous system. Moreover, newborn cells in the proliferative zones often become committed to neuronal or glial fates before migrating from these zones.

At early stages of embryonic development, most progenitor cells in the ventricular zone of the neural tube proliferate rapidly. Many of these early neural progenitors have the properties of stem cells: They can generate additional copies of themselves, a process called *self-renewal*, and also give rise to differentiated neurons and glial cells. In a later chapter, we will describe the more recent discovery that stem cells resembling those of embryos also exist in the adult brain and may be harnessed for therapeutic purposes (Chapter 50).

As with other types of stem cells, neural progenitor cells undergo stereotyped programs of cell division. One mode of cell division is symmetric: Neural stem cells divide to produce two stem cells, and in this way expand the population of proliferative progenitor cells. This mode predominates at the earliest times, as the neuroepithelium expands. A second mode is asymmetric: The progenitor produces one differentiated daughter and another daughter that retains its stem cell–like properties. This mode retains but does not amplify the stem cell population. A third mode leads to production of two differentiated daughters. In this symmetric mode, the stem cell population is depleted. All three modes have been found in the embryonic cerebral cortex in vivo and in cortical cells grown in tissue culture (Figure 46–1).

The incidence of symmetric and asymmetric cell division is influenced by signals in the local environment of the dividing cell, making it possible to control the probability of self-renewal or differentiation. Environmental factors can influence the outcome of progenitor cell divisions in two fundamental ways. They can act in an "instructive" manner, biasing the outcome of the division process and causing the stem cell to adopt one fate at the expense of others. Or they can act in a "selective" manner, permitting the survival and maturation of only certain cell progeny.

Radial Glial Cells Serve as Neural Progenitors and Structural Scaffolds

Radial glial cells are the earliest morphologically distinguishable cell type to appear within the primitive neural epithelium. Their cell bodies are located in the ventricular zone, and their long process extends to the pial surface. As the brain thickens, the processes of radial glial cells remain attached to the ventricular and pial surfaces. After the generation of neurons is complete, many radial glial cells differentiate into astrocytes. The elongated shape of the radial glial cell places it in a favorable position to serve as a scaffold for the migration of neurons that emerge from the ventricular zone (Figure 46–2).

The ventricular zone was once thought to contain two major cell types: radial glial cells and a set of neuroepithelial progenitors that serve as the primary source of neurons. More recently, this classical view has changed dramatically. Once symmetric divisions of stem cells have expanded the neuroepithelium, these cells give rise to radial glial cells. The radial glial cells serve as progenitor cells that generate both neurons and astrocytes in addition to their role in neuronal migration (Figure 46–2). Labeling of radial glial cells with fluorescent dyes or viruses shows that their clonal progeny include both neuronal and radial glial cells. These findings indicate that radial glial cells are able to undergo both asymmetric and self-renewing cell division and serve as a major source of postmitotic neurons as well as astrocytes.

The Generation of Neurons and Glial Cells Is Regulated by Delta-Notch Signaling and Basic Helix-Loop-Helix Transcription Factors

How do radial glial cells make the decision to self-renew, generate neurons, or give rise to mature astrocytes? The answer to this question involves an evolutionarily conserved signaling system.

In flies and vertebrates, neural fate is regulated by a cell-surface signaling system, comprised of the

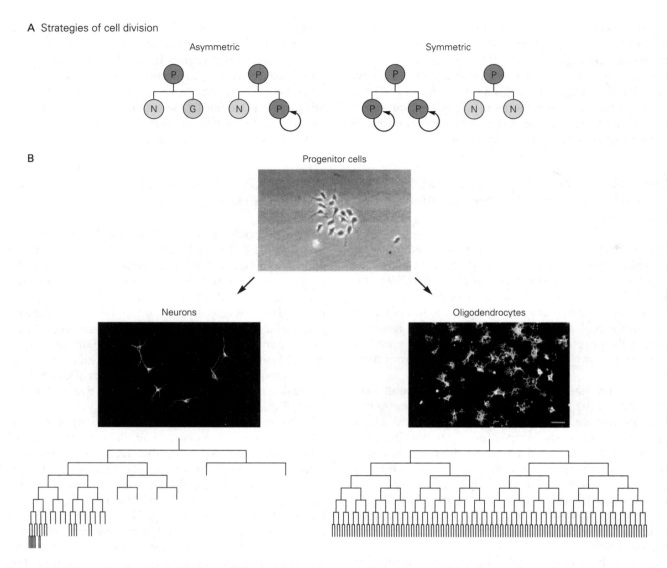

A Strategies of cell division

Asymmetric

Symmetric

B

Progenitor cells

Neurons

Oligodendrocytes

Figure 46–1 Neural progenitor cells have different modes of division.

A. Asymmetric and symmetric modes of cell division. A progenitor cell (**P**) can undergo asymmetric division to generate a neuron (**N**) and a glial cell (**G**), or a neuron and another progenitor. This mode of division contributes to the generation of neurons at early stages of development and of glial cells at later stages, typical of many regions of the central nervous system. Progenitor cells can also undergo symmetric division to generate two additional progenitor cells or two postmitotic neurons.

B. Time-lapse cinematography captures the divisions and differentiation of isolated cortical progenitor cells in the rodent. Lineage diagrams illustrate cells that undergo predominantly asymmetric division, giving rise to neurons, or symmetric division, giving rise to oligodendrocytes. (Adapted, with permission, from Qian et al. 1998. Permission conveyed through Copyright Clearance Center, Inc.)

transmembrane ligand Delta and its receptor Notch. This signaling system was revealed in genetic studies in *Drosophila*. Neurons emerge from within a larger cluster of ectodermal cells, called a *proneural region*, all of which have the potential to generate neurons. Yet within the proneural region, only certain cells form neurons; the others become epidermal support cells.

Delta and Notch are initially expressed at similar levels by all proneural cells (Figure 46–3A). With time, however, Notch activity is enhanced in one cell and suppressed in its neighbor. The cell in which Notch activity is highest loses the potential to form a neuron and acquires an alternative fate. The binding of Delta to Notch results in proteolytic cleavage of the Notch cytoplasmic domain, which then enters the nucleus.

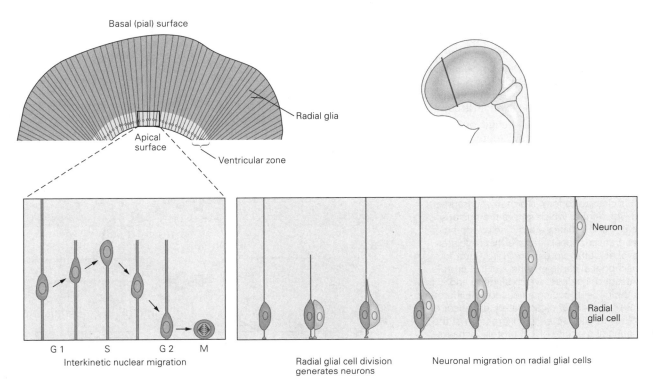

Figure 46–2 Radial glial cells serve as precursors to neurons in the central nervous system and also provide a scaffold for radial neuronal migration. The nuclei of progenitor cells in the ventricular zone of the developing cerebral cortex migrate along the apical-basal axis as they progress through the cell cycle. *Left:* During the G1 phase, the nuclei rise from the inner (apical) surface of the ventricular zone. During the S phase, they reside in the outer (basal) third of the ventricular zone. During the G2 phase, they migrate apically, and mitosis (**M**) occurs when the nuclei reach the ventricular surface. *Right:* During cell division, radial glial cells give rise to postmitotic neurons that migrate away from the ventricular zone using radial glial cells as a guide.

There, it functions as a transcription factor, regulating the activity of a cascade of other transcription factors of the basic helix-loop-helix (bHLH) family. The bHLH transcription factors suppress the ability of the cell to become a neuron and reduce the level of expression of the ligand Delta (Figure 46–3B,C).

The initial difference in Notch levels between cells may be small and in some cases stochastic (random). Through this feedback pathway, however, these initial minor differences are amplified to generate all-or-none differences in the status of Notch activation and, consequently, the fates of the two cells. This basic logic of Delta-Notch and bHLH signaling has been conserved in vertebrate and invertebrate neural tissues.

How does Notch signaling regulate neuronal and glial production in mammals? At early stages in the development of the mammalian cortex, Notch signaling promotes the generation of radial glial cells by activating members of the Hes family of bHLH transcriptional repressors. Two of these proteins, Hes1 and Hes5, appear to maintain radial glial cell character by activating the expression of an ErbB class tyrosine kinase receptor for neuregulin, a secreted signal that promotes radial glial cell identity. The Notch ligand Delta1 as well as neuregulin are expressed by newly generated cortical neurons; thus, the radial glial cells depend on feedback signals from their neuronal progeny for continued production.

At later stages of cortical development, Notch signaling continues to activate Hes proteins, but a change in the intracellular response pathway results in astrocyte differentiation. At this stage, the Hes proteins work by activating a transcription factor, STAT3, which recruits the serine-threonine kinase JAK2, a potent inducer of astrocyte differentiation. STAT3 also activates expression of astrocyte-specific genes such as the glial-fibrillary acidic protein (GFAP).

The generation of oligodendrocytes, the second major class of glial cells in the central nervous system, follows many of the principles that control neuron

Figure 46–3 Delta binds the receptor Notch and determines neuronal fate.

A. At the onset of the interaction between two cells, Delta engages the receptor Notch. Delta and Notch are expressed at similar levels in each cell, and thus their initial signaling strength is equal.

B. A small imbalance in the strength of Delta-Notch signaling breaks the symmetry of the interaction. In this example, the left cell provides a slightly greater Delta signal, thus activating Notch signaling in the right cell to a greater extent. On binding by Delta, the cytoplasmic domain of Notch is cleaved to form a proteolytic fragment called Notch-Intra, which enters the nucleus of the cell and initiates a basic helix-loop-helix (bHLH) transcriptional cascade that regulates the level of Delta expression. Notch-Intra forms a transcriptional complex with a bHLH protein, suppressor of hairless, which binds to and activates the gene encoding a second bHLH protein, enhancer of split. Once activated, enhancer of split binds to and represses expression of the gene encoding a third bHLH protein, achaete-scute. Achaete-scute activity promotes expression of Delta. Thus, by repressing achaete-scute, enhancer of split decreases transcriptional activation of the Delta gene and production of Delta protein. This diminishes the ability of the cell on the right to activate Notch signaling in the left cell.

C. Once the level of Notch signaling in the left cell has been reduced, suppressor of hairless no longer activates enhancer of split, and the level of expression of achaete-scute increases, resulting in enhanced expression of Delta and further activation of Notch signaling in the right cell. In this way, a small initial imbalance in Delta-Notch signaling is rapidly amplified into a marked asymmetry in the level of Notch activation in the two cells. In the mammalian central nervous system, cells with high levels of Notch activation are diverted from neuronal fates, whereas cells with low levels of Notch activation become neurons.

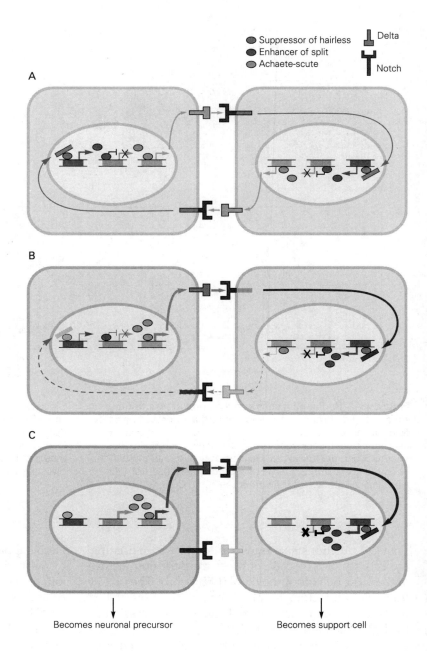

and astrocyte production (Figure 46–4). Notch signaling regulates the expression of two bHLH transcription factors, Olig1 and Olig2, which have essential roles in the production of embryonic and postnatal oligodendrocytes.

Additional mechanisms exist to ensure that the effects of Notch signals are avoided in cells destined to become neurons. One involves a cytoplasmic protein called Numb. The key role of Numb in neurogenesis was first shown in *Drosophila*, where

it determines the neuronal fate of daughter cells of asymmetrically dividing progenitors. In the mammalian cortex, Numb is preferentially localized in neuronal daughters and antagonizes Notch signaling. Loss of Numb activity causes progenitor cells to proliferate extensively. The inhibition of Notch signaling results in the expression of several proneural bHLH transcription factors, notably Mash1, neurogenin-1, and neurogenin-2. Neurogenins promote neuronal production by activating downstream

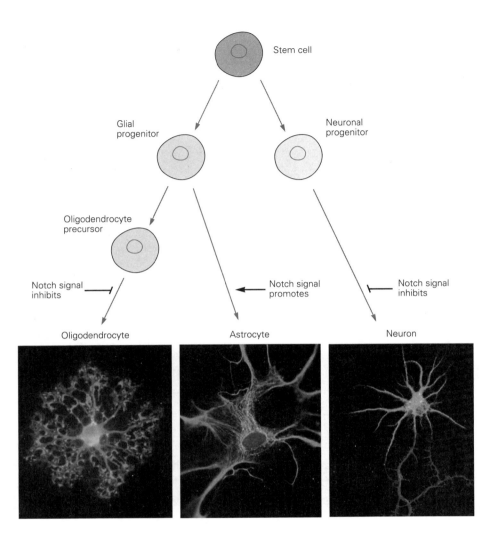

Figure 46–4 Notch signaling regulates the fate of cells in the developing cerebral cortex. Notch signaling has several roles in cell differentiation in the developing cerebral cortex. Activation of Notch signaling in glial progenitor cells results in differentiation of the cells as astrocytes and inhibits differentiation as oligodendrocytes (left pathway). Notch signaling also inhibits progenitor cells from differentiating into neurons (right pathway). (Photo of oligodendrocyte reproduced, with permission, from David H. Rowitch; photo of astrocyte reproduced, with permission, from SAASTA on behalf of photographers Edward Nyatia and Dirk Michael Lang; photo of neuron reproduced, with permission, from Masatoshi Takeichi.)

bHLH proteins such as neuroD, and they block the formation of astrocytes by inhibiting JAK and STAT signaling.

Although Delta-Notch signaling and bHLH transcription factor activators lie at the heart of the decision to produce neurons or glial cells, several additional transcriptional pathways augment this core molecular program. One important transcription factor, REST/NRSF, represses the expression of neuronal genes in neural progenitors and glial cells. REST/NRSF is rapidly degraded as neurons differentiate, permitting the expression of neurogenic bHLH factors and other neuronal genes. Homeodomain transcription factors of the SoxB class also play an important role in maintaining neural progenitors by blocking neurogenic bHLH protein activity. The differentiation of neurons therefore requires the avoidance of REST/NRSF and SoxB protein activity.

The Layers of the Cerebral Cortex Are Established by Sequential Addition of Newborn Neurons

The ventricular zone in the most anterior portion of the mammalian neural tube gives rise to the cerebral cortex in a series of steps. Cells from the ventricular zone, which is on the apical edge of the neuroepithelium, initially migrate basally to form a subventricular zone, which houses a set of progenitor cells with a more restricted set of fates. Next to form is an intermediate zone, through which newly formed neurons migrate, and a preplate, which houses the earliest-born neurons. Additional neurons migrate to form a layer called the cortical plate, which lies within the preplate. The cortical plate thereby divides the preplate into an apical subplate and a basal marginal zone (Figure 46–5A).

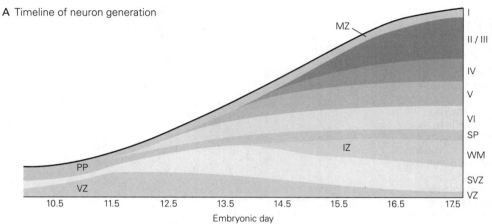

A Timeline of neuron generation

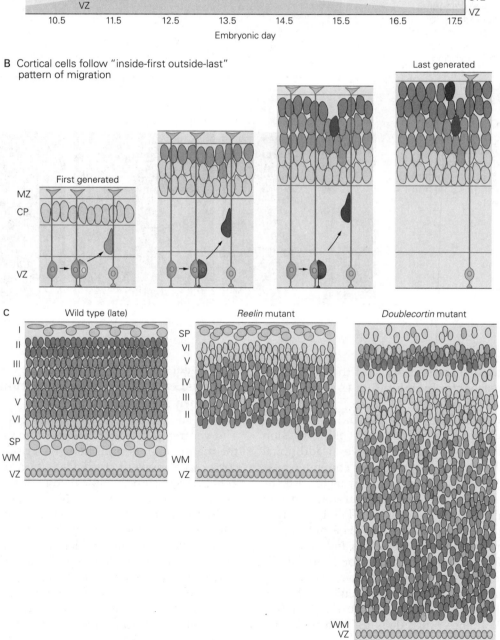

B Cortical cells follow "inside-first outside-last" pattern of migration

C

Once within the cortical plate, neurons become organized into well-defined layers. The layer in which a neuron settles is correlated precisely with the neuron's *birthday*, a term that refers to the time at which a dividing precursor cell undergoes its final round of cell division and gives rise to a postmitotic neuron. Cells that migrate from the ventricular and subventricular zones and leave the cell cycle at early stages give rise to neurons that settle in the deepest layers of the cortex. Cells that exit the cell cycle at progressively later stages migrate over longer distances and pass earlier-born neurons before settling in more superficial layers of the cortex. Thus, the layering of neurons in the cerebral cortex follows an inside-first, outside-last rule (Figure 46–5B).

Neurons Migrate Long Distances From Their Site of Origin to Their Final Position

The migration of neurons from the cortical ventricular zone to the cortical plate follows a process termed *radial migration*. In this mode, the neurons move along the long unbranched processes of radial glial cells to reach their destinations. In contrast, interneurons enter the cortex from subcortical sites by a process called *tangential migration*. We discuss these modes in turn and then describe a third migratory strategy, *free migration*, which predominates in the peripheral nervous system.

Excitatory Cortical Neurons Migrate Radially Along Glial Guides

Classical anatomical studies of cortical development in the 1970s provided evidence that neurons generated in the ventricular zone migrate to their settling position along a pathway of radial glial fibers. Radial glial cells serve as the primary scaffold for radial neuronal migration. Their cell bodies are located close to

the ventricular surface and give rise to elongated fibers that span the width of the developing cerebral wall. Each radial glial cell has one basal end-foot at the apical surface of the ventricular zone and processes that terminate in multiple end-feet at the pial surface (Figure 46–6). Radial glial scaffolds are especially important in the development of the primate cortex, where neurons are required to migrate over long distances as the cortex expands. A single radial glial cell scaffold can support the migration of up to 30 generations of cortical neurons before eventually differentiating into an astrocyte.

What forces and molecules power neuronal migration on radial glial cells? After a neuron leaves the cell cycle, its leading process wraps around the shaft of the radial glial cell and its nucleus translocates within the cytoplasm of the leading process. Although the leading process of the migrating neuron extends slowly and steadily, the nucleus moves in an intermittent, stepwise manner because of complex rearrangements of the cytoskeleton. A microtubular lattice forms a cage around the nucleus; movement of the nucleus depends on a centrosome-like structure, termed a *basal body*, from which a system of microtubules projects into the leading process, providing tracks along which the nucleus moves (Figure 46–7A).

Neuronal migration along radial glia also involves adhesive interactions between cells. Adhesive receptors such as integrins promote neuronal extension on radial glial cells. The migration of neurons along glial fibers is nevertheless different from the extension of axons driven by growth cones (Chapter 47). In neuronal migration, the leading process is devoid of the structured actin filaments that typify growth cones and more closely resembles an extending dendrite, an inference made first by Santiago Ramón y Cajal.

Disruption in the migratory and settling programs of cortical neurons underlies much human cortical pathology (Figure 46–5C). For example, in lissencephaly

Figure 46–5 (Opposite) The migration of neurons within the embryonic cerebral cortex leads to layered cortical organization. (Adapted from Olsen and Walsh 2002.)

A. This temporal sequence of neurogenesis is for the mouse cerebral cortex. Neurons begin to accumulate in the cortical plate during the last 5 days of embryonic development. Within the cortical plate, neurons populate the deep layers before settling in the superficial layers. (Abbreviations: IZ, intermediate zone; MZ, marginal zone; PP, preplate; SP, subplate; SVZ, subventricular zone; VZ, ventricular zone; WM, white matter.)

B. During normal cortical development, neurons use radial glial cells as migratory scaffolds as they enter the cortical plate.

As they approach the pial surface, neurons stop migrating and detach from radial glial cells. This orderly inside-out pattern of neuronal migration results in the formation of six neuronal layers in the mature cerebral cortex, arranged between the white matter and subplate. (Abbreviation: CP, cortical plate.)

C. In the mouse mutant *reeler*, which lacks functional reelin protein, the layering of neurons in the cortical plate is severely disrupted and partially inverted. In addition, the entire cortical plate develops beneath the subplate. In *doublecortin* mutants, the cortex is thickened, neurons lose their characteristic layered identity, and some layers contain fewer neurons. A similar disruption is observed in *Lis1* mutants, which underlies certain forms of human lissencephaly.

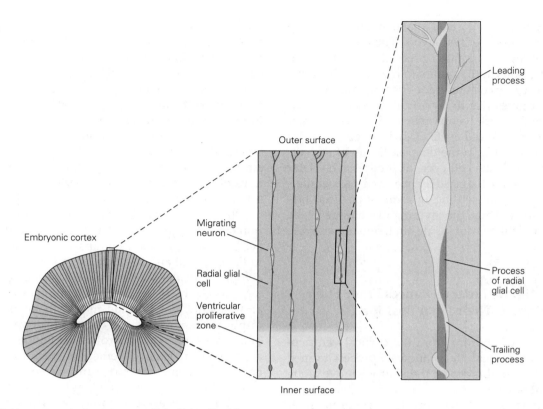

Figure 46–6 Neurons migrate along radial glial cells. After their generation from radial glial cells, newly generated neurons in the embryonic cerebral cortex extend a leading process that wraps around the shaft of the radial glial cell, thus using the radial glial cells as scaffolds during their migration from the ventricular zone to the pial surface of cortex.

(Greek, smooth brain, referring to the characteristic smoothing of the cortical surface in patients with the disorder), neurons leave the ventricular zone but fail to complete their migration into the cortical plate. As a result, the mature cortex is typically reduced from six to four neuronal layers, and the arrangement of neurons in each remaining layer is disordered. Occasionally, lissencephaly is accompanied by the presence of an additional group of neurons in the subcortical white matter. Patients with lissencephalies from mutations in the *Lis1* and *doublecortin* genes often suffer severe intellectual disability and intractable epilepsy. The Lis1 and doublecortin proteins have been localized to microtubules, suggesting that they are involved in microtubule-dependent nuclear movement, although their precise functions in neuronal migration remain unclear.

Mutations that disrupt the reelin signaling pathway disrupt the final stage of neuronal migration through the cortical subplate. The reelin protein is secreted from the Cajal-Retzius cells, a class of neurons found in the preplate and marginal zone. Signals from these cells are crucial for the migration of cortical neurons. In mice lacking functional reelin, neurons fail to detach from their radial glial scaffolds and pile up underneath the cortical plate, disobeying the inside-out migratory rule. As a consequence, the normal layering of cell types is partially inverted and the marginal zone is lost. Reelin acts through cell-surface receptors that include the ApoE receptor 2 and the very-low-density lipoprotein receptor. The binding of reelin to these receptors activates an intracellular protein, Dab1, which transduces reelin signals. Not surprisingly, the loss of proteins that transduce reelin signals produces similar migratory phenotypes.

Cortical Interneurons Arise Subcortically and Migrate Tangentially to Cortex

Progenitor cells in the cortical ventricular zone were initially believed to give rise to all cortical neurons. However, as better molecular labels for distinct neuronal types became available, it was found that interneurons arise in the ventricular zone of subcortical structures. Most of them originate in regions of the ventral telencephalon called the ganglionic eminences (Figure 46–8).

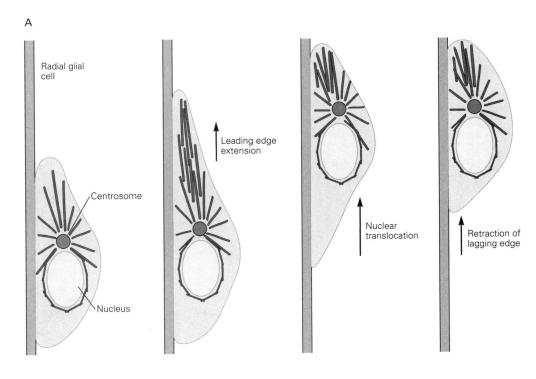

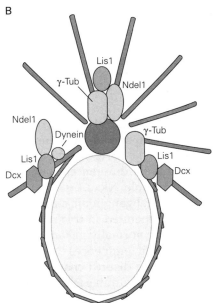

Figure 46–7 Cytoskeletal proteins power the migration of neurons along radial glia.

A. The microtubular cytoskeleton has an important role in neuronal migration. Microtubules envelop the nucleus in a cage-like structure. Migration along radial glial cells involves elongation of the leading process of the neuron in the direction of movement, under the control of attractive and repellant extracellular guidance cues. These cues regulate the phosphorylation status of the microtubule-associated proteins Ndel1 and Lis1 (two components of the dynein motor complex) and of doublecortin (**Dcx**), which together stabilize the microtubule cytoskeleton. (Adapted from Gleeson and Walsh 2000.)

B. Microtubules are attached to the centrosome by a series of proteins that are targets for disruption in neuronal migration disorders.

The medial and central eminences generate most cortical interneurons, which migrate dorsally from their sites of origin to enter the cortex. Some enter through the intermediate zone, while others enter through the marginal zone (Figure 46–5A). Once they reach particular anterior-posterior and mediolateral positions, they switch to a radial mode of migration to travel the final distance to appropriate layers. Distinct populations of neurons generated in the ganglionic eminences migrate at different times and through different routes, contributing to the diversity of the interneuronal population. Precise relationships between time and place of origin, migratory route, and ultimate fate remain to be determined.

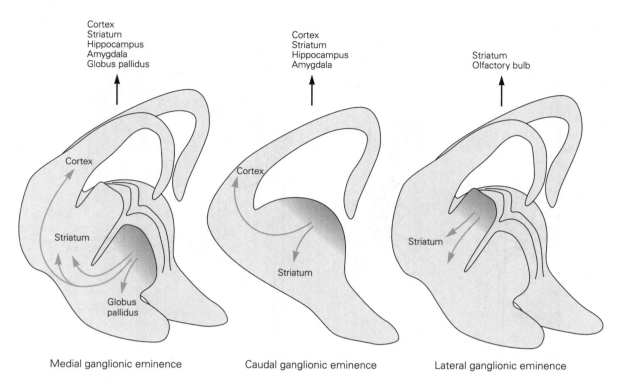

Figure 46–8 Forebrain interneurons are generated in the ventral telencephalon and migrate tangentially to the cerebral cortex. Neurons generated in the ganglionic eminences migrate to and settle in many regions of the forebrain, where they differentiate into interneurons. Cortical interneurons arise from medial and caudal ganglionic eminences. Other cells generated in these regions migrate in other directions, populating the hippocampus, striatum, globus pallidus, and amygdala with interneurons. The lateral ganglionic eminence generates cells that migrate to the striatum and the olfactory bulb. Cells migrating to the bulb use neighboring migrating cells as substrates for migration, a process called chain migration. (Adapted, with permission, from Bandler, Mayer, and Fishell 2017. Copyright © 2017 Elsevier Ltd.)

Nonetheless, it is now clear that cortical neurons originate from two sources: excitatory neurons from the cortical ventricular zone and interneurons from the ganglionic eminences.

Interneurons in other forebrain structures also arise from the ganglionic eminences, as well as a few other subcortical sites such as the preoptic area. Cells migrating caudally from the medial and caudal eminences populate the hippocampus, while cells migrating ventrolaterally from these regions populate the basal ganglia. In contrast, neurons generated in the lateral ganglionic eminence migrate rostrally and contribute the periglomerular and granule interneurons of the olfactory bulb. In this rostral migratory stream, neurons use neighboring neurons as substrates for migration (chain migration). In the adult brain, neurons that follow the rostral migratory stream originate instead in the subventricular zone of the striatum.

Transcription factors control the character of ganglionic eminence neurons. The homeodomain proteins Dlx1 and Dlx2 are expressed by cells in the ganglionic eminences. In mice lacking Dlx1 and Dlx2 activity, the resultant perturbation of neuronal migration leads to a profound reduction in the number of GABAergic interneurons in the cortex. Other transcription factors are responsible for differences among ganglionic eminences. For example, Nkx2.1 is selectively expressed by cells in the medial ganglionic eminence. In its absence, interneurons generated in this region take on characteristics of those normally generated in the lateral and caudal ganglionic eminences. Yet other transcription factors specify the distinct characteristics of subpopulations of neurons within each ganglionic eminence.

One of the main features that these transcription factors specify is the migratory path that the newborn interneurons take. A host of soluble and cell surface factors produced by cells in and near the ganglionic eminences provide repulsive cues that lead to expulsion of cells from the ventricular zone, so-called motogenic (movement-promoting) cues that speed their migration and attractive cues that direct them to their targets. These factors include slits, semaphorins, and ephrins, all of which we will encounter in Chapter 47 as molecules that guide axons to their targets.

Neural Crest Cell Migration in the Peripheral Nervous System Does Not Rely on Scaffolding

The peripheral nervous system derives from neural crest stem cells, a small group of neuroepithelial cells at the boundary of the neural tube and epidermal ectoderm. Soon after their induction, neural crest cells are transformed from epithelial to mesenchymal cells and begin to detach from the neural tube. They then migrate to many sites throughout the body (Figure 46–9). Neural crest cell migration does not rely on scaffolding (ie, radial glial cells or preexisting axon tracts) and thus is called free migration. This form of neuronal migration requires significant cytoarchitectural and cell adhesive changes and differs from most of the migratory events in the central nervous system.

Neural crest migration is promoted and guided by several families of secreted factors. For example, bone morphogenetic proteins (BMPs), which are critical for neural crest induction at an earlier stage (Chapter 45), are required for neural crest migration at later stages. Exposure of neural epithelial cells to BMPs triggers molecular changes that convert epithelial cells to a mesenchymal state, causing them to delaminate from the neural tube and migrate into the periphery. BMPs trigger changes in neural crest cells by inducing expression of transcription factors, notably the zinc finger proteins snail, slug, and twist, which have a conserved role in promoting epithelial-to-mesenchymal transitions. These transcription factors direct expression of proteins that regulate the properties of the cytoskeleton as well as enzymes that degrade extracellular matrix proteins. These enzymes give neural crest cells the ability to break down the basement membrane surrounding the epithelium of the neural tube, permitting them to embark on their migratory journey into the periphery.

As neural crest cells begin to delaminate, their expression of cell adhesion molecules changes. Alterations in expression of adhesive proteins, notably cadherins, permit neural crest cells to loosen their adhesive contacts with neural tube cells and begin the delamination process. Neural crest cells also begin to express integrins, receptors for extracellular matrix proteins such as laminins and collagens that are found along peripheral migratory paths.

The first structures encountered by migrating neural crest cells are somites, epithelial cells that later give rise to muscle and cartilage. Neural crest cells pass through the anterior half of each somite but avoid the posterior half (Figure 46–9A). The rostral channeling of migratory neural crest cells is imposed by ephrin B proteins, which are concentrated in the posterior half

of each somite. Ephrins provide a repellant signal that interacts with EphB class tyrosine kinase receptors on neural crest cells to prevent their invasion. Neural crest cells that remain within the anterior sclerotome of the somite differentiate into sensory neurons of the dorsal root ganglia; those that migrate around the dorsal region of the somite approach the skin and give rise to melanocytes.

Differentiation of the neural crest into its various derivatives depends on complex interactions between the distinct cues that cells receive along their journey and intrinsic predispositions that vary along the rostrocaudal axis. Development of sensory neurons is initiated at the time the cells emigrate from the neural tube. The cells are exposed to signals from the dorsal neural tube and somites that induce expression of neurogenin, a transcription factor of the bHLH family, which in turn promotes a sensory fate. Subsequent influences diversify the neurons into multiple sensory types, such as nociceptive and proprioceptive neurons (Figure 46–10). In contrast, those neural crest cells that follow a more medial and ventral migratory path are exposed to BMPs secreted from the dorsal aorta. They express the bHLH factor Mash1, which leads to their differentiation into sympathetic neurons.

Structural and Molecular Innovations Underlie the Expansion of the Human Cerebral Cortex

No mice or monkeys are reading this book. This is in large part because the human brain is different from that of even our closest relatives, both qualitatively and quantitatively. Yet most studies of mammalian neural development have been carried out on mice, whose brain contains approximately 1,000-fold fewer neurons than those of the human brain and 100-fold fewer than the best-studied nonhuman primate, the rhesus macaque. Recently, however, new methods are making it possible to elucidate some of the molecular and structural features that lead to the expansion of the human brain and particularly the human cerebral cortex.

Classical anatomical studies made clear that the primate cortex has not only a far larger size and thickness than that of rodents but also more discrete areas and more layers (Figure 46–11A). In addition, the packing density of neurons is higher in primates than in mice, so the difference in neuronal number is greater than would be expected from size alone. One main contributor to the expansion in primates is a large pool of neuronal progenitors. Many of these progenitors are a second type of radial glial cell, called the outer radial

A Migratory paths

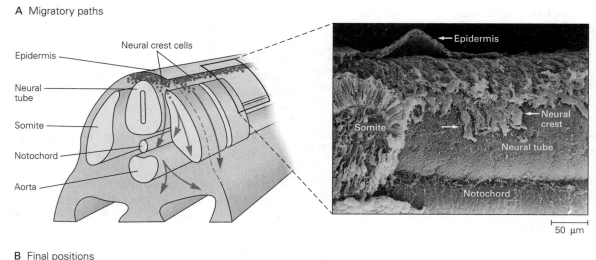

B Final positions

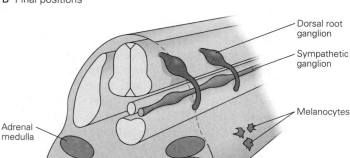

Figure 46–9 Neural crest cell migration in the peripheral nervous system.

A. A cross section through the middle part of the trunk of a chick embryo shows the main pathways of neural crest cells. Some cells migrate along a superficial pathway, just beneath the ectoderm, and differentiate into pigment cells of the skin. Others migrate along a deeper pathway that takes them through the somites, where they coalesce to form dorsal root

sensory ganglia. Still others migrate between the neural tube and somites, past the dorsal aorta. These cells differentiate into sympathetic ganglia and adrenal medulla. The scanning electron micrograph shows neural crest cells migrating away from the dorsal surface of the neural tube of a chick embryo. (Micrograph reproduced, with permission, from K. Tosney.)

B. Neural crest cells reach their final settling positions where they complete differentiation.

glial cell to distinguish it from the canonical or inner radial glia described above. Outer radial glia, unlike inner radial glia, lack contact with the ventricular surface and exhibit molecular differences from inner radial glia. However, they are capable of generating neurons and serving as a migratory guide. The massive increase in their number in primates, and particularly humans, provides a partial explanation for the increase in the number of neurons in the human cerebral cortex.

How can human-specific developmental features be analyzed experimentally? New methods of molecular analysis are making it possible to compare the proteins, transcripts, and genes of humans with those of our close relatives, resulting in the discovery of intriguing specializations. However, hypotheses

derived from these findings are difficult to test: Most of the developmental studies we describe in these chapters cannot be performed on humans, and even nonhuman primates are difficult subjects for developmental analysis. A possible solution is the recently devised "organoid" culture system.

Cells from adult skin can be reprogrammed to become multipotential progenitors called induced pluripotent stem cells (iPSCs) by methods that we will discuss in Chapter 50. When placed into culture under carefully controlled conditions and allowed to expand in three dimensions (quite unlike conventional two-dimensional cultures), they proliferate and self-organize into structures that resemble the developing forebrain and exhibit species-specific features

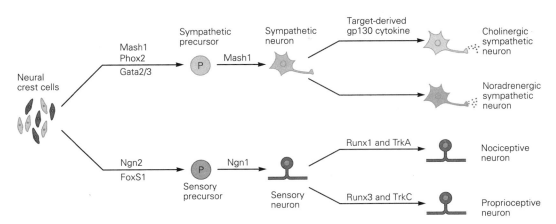

Figure 46–10 Neural crest cells differentiate into sympathetic and sensory neurons. The neuronal fates of trunk neural crest cells are controlled by transcription factor expression. Expression of the basic helix-loop-helix (bHLH) protein Mash1 directs neural crest cells along a sympathetic neuronal pathway. Sympathetic neurons can acquire noradrenergic or cholinergic transmitter phenotypes depending on the target cells they innervate and the level of gp130 cytokine signaling (see Figure 46–13). Two bHLH proteins, neurogenin-1 and -2, direct neural crest cells along a sensory neuronal pathway. Sensory neurons that express the transcription factor Runx1 and the tyrosine kinase receptor TrkA become nociceptors; those that express Runx3 and TrkC become proprioceptors. (Abbreviations: **Ngn-1**, neurogenin-1; **Ngn-2**, neurogenin-2.)

(Figure 46–11B). Most notably, organoids from human cells contain a bilayered, large subventricular zone with numerous outer radial glia, whereas organoids from mouse cells contain a smaller subventricular zone containing predominately conventional or inner radial glia. These organoids can be used to elucidate the development of at least some early aspects of human cortical development.

Additional applications abound. One is to obtain iPSCs from patients with brain disorders. Organoids derived from such patients have features that may lead to cortical malformations such as lissencephaly (see Figure 46–5). The hope is that these organoids can be used to elucidate disease mechanisms and eventually test therapies. A second application is to compare organoids derived from chimpanzee and human iPSCs. This comparison provides a novel means of investigating the most recent evolutionary innovations that separate us from our closest living relatives.

Intrinsic Programs and Extrinsic Factors Determine the Neurotransmitter Phenotypes of Neurons

Neurons continue to develop after they have migrated to their final position, and no aspect of their later differentiation is more important than the choice of chemical neurotransmitter. Neurons that populate the brain use two major neurotransmitters: The amino acid L-glutamate is the major excitatory transmitter, whereas γ-aminobutyric acid (GABA) is the major inhibitory transmitter. Some spinal cord neurons use another amino acid, glycine, as their inhibitory transmitter. In the peripheral nervous system, sensory neurons use glutamate, motor neurons use acetylcholine, and autonomic neurons use acetylcholine or norepinephrine. Smaller numbers of neurons use other transmitters, such as serotonin and dopamine. The choice of neurotransmitter determines which postsynaptic cells a neuron can talk to and what it can say.

Neurotransmitter Choice Is a Core Component of Transcriptional Programs of Neuronal Differentiation

Distinct molecular programs are used to establish neurotransmitter phenotype in different brain regions and neuronal classes. We shall illustrate the general strategy for assignment of amino acid neurotransmitter phenotypes by focusing on neurons in the cerebral cortex and cerebellum.

The cerebral cortex contains glutamatergic pyramidal neurons that are generated within the cortical plate and rely on the bHLH factors neurogenin-1 and neurogenin-2 for their differentiation. In contrast, as discussed earlier in the chapter (see Figure 46–8), most GABAergic inhibitory interneurons migrate into the cortex from the ganglionic eminences; their inhibitory transmitter character is specified by the bHLH protein

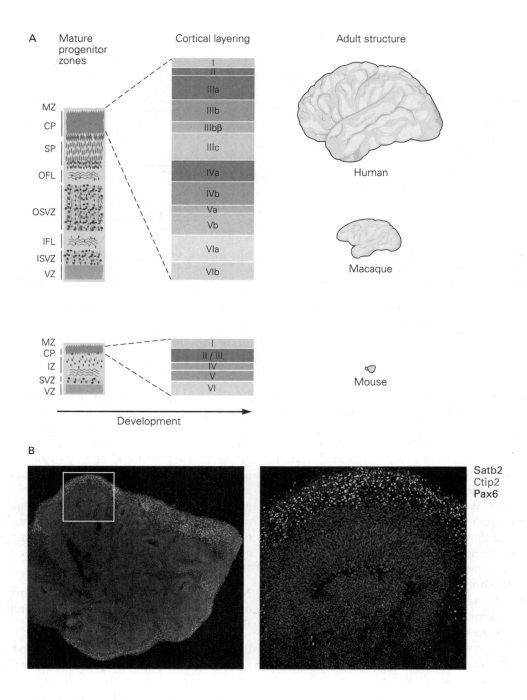

Figure 46–11 Expansion of the proliferative zones contributes to cortical specialization in humans and other primates.

A. The size of the neuroepithelium is initially small in both rodents and humans, but their relative size differs dramatically as development proceeds, owing to increased self-renewal rates and larger numbers of progenitors in humans. The primate subventricular zone is greatly enlarged compared to the mouse and becomes subdivided into inner and outer regions, which contain large populations of radial glial cells, both of which generate neurons. In mice, nearly all radial glial cells are of the inner type. (Abbreviations: **CP**, cortical plate; **IFL**, inner

fiber layer; **ISVZ**, inner subventricular zone; **IZ**, intermediate zone; **MZ**, marginal zone; **OFL**, outer fiberlayer; **OSVZ**, outer subventricular zone; **SP**, subplate; **SVZ**, subventricular zone; **VZ**, ventricular zone.) (Adapted, with permission, from Giandomenico and Lancaster 2017.)

B. Section through an organoid generated from human induced pluripotent stem cells. The area between the white lines is enlarged in the micrograph on the right. The section was stained with antibodies to transcription factors (Satb2, Ctip2, and Pax6) selectively expressed in specific layers in human cortex, demonstrating that a layered cortical structure develops in the organoid. (Micrographs reproduced, with permission, from P. Arlotta.)

Mash1 (Figure 46–12A) as well as by the Dlx1 and Dlx2 proteins.

Similarly, the cerebellum contains several different classes of inhibitory neurons (Purkinje, Golgi, basket, and stellate neurons) and two major classes of excitatory neurons (granule neurons and large cerebellar nucleus neurons). These inhibitory and excitatory neurons have different origins; GABAergic neurons derive from the ventricular zone, whereas glutamatergic neurons migrate into the cerebellum from the rhombic lip. The generation of GABAergic and glutamatergic neurons is controlled by two different bHLH transcription factors, Ptf1a for inhibitory and Math-1 for excitatory neurons

(Figure 46–12B). These bHLH factors are expressed by neuroepithelial cells but not by mature neurons, implying that differentiation into glutamatergic and GABAergic neurons is initiated prior to neuronal generation.

Transcriptional programs also determine the transmitter phenotype in the peripheral nervous system. For example, BMPs promote noradrenergic neuronal differentiation by inducing the expression of a variety of transcription factors that include the bHLH protein Mash1, the homeodomain protein Phox2, and the zinc finger protein Gata2. In contrast, Runx proteins are determinants of the glutamatergic phenotype of sensory neurons (Figure 46–10).

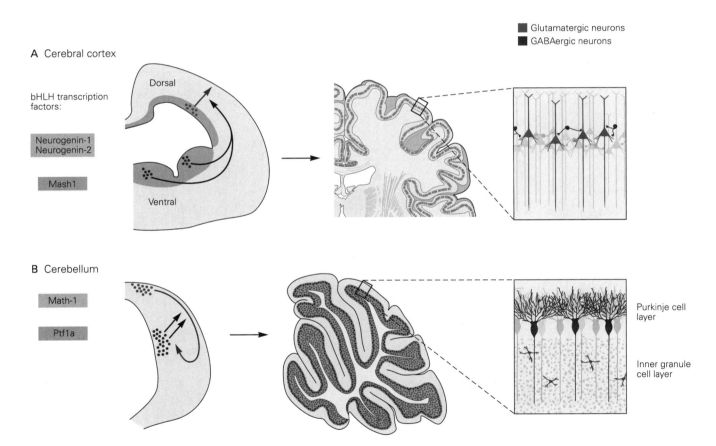

Figure 46–12 The neurotransmitter phenotype of central neurons is controlled by basic helix-loop-helix transcription factors.

A. GABAergic and glutamatergic neurons in the cerebral cortex are generated in different proliferative zones and are specified by different basic helix-loop-helix (**bHLH**) transcription factors. Glutamatergic pyramidal neurons derive from the cortical ventricular zone, and their differentiation depends on the activities of neurogenin-1 and -2. The differentiation of GABAergic interneurons in the ganglionic eminences of the ventral telencephalon depends on the bHLH protein Mash1. These neurons

migrate dorsally to supply the cerebral cortex with most of its inhibitory interneurons.

B. GABAergic and glutamatergic neurons in the developing cerebellum also derive from different proliferative zones and are specified by different bHLH transcription factors. Glutamatergic granule cells migrate into the cerebellum from the rhombic lip, settle in the inner granular layer, and are specified by the bHLH protein Math-1. GABAergic Purkinje neurons migrate from the deep cerebellar proliferative zone, settle in the Purkinje cell layer, and are specified by the bHLH protein Ptf1a.

Signals From Synaptic Inputs and Targets Can Influence the Transmitter Phenotypes of Neurons

Because neurotransmitter phenotype is a core neuronal property, it was long thought that transmitter properties were fixed at the earliest stage of neuronal differentiation. This view was challenged by studies showing that the migratory pathway of a neural crest cell exposes the cell to environmental signals that have a critical role in determining its transmitter phenotype.

Most sympathetic neurons use norepinephrine as their primary transmitter. However, those that innervate the exocrine sweat glands in the footpads use acetylcholine, and even these neurons express norepinephrine when they first innervate the sweat glands of the skin. Only after their axons have contacted the sweat glands do they stop synthesizing norepinephrine and start producing acetylcholine.

When the sweat glands from the footpad of a newborn rat are transplanted into a region that is normally innervated by noradrenergic sympathetic neurons, the synaptic neurons acquire cholinergic transmitter properties, indicating that cells of the sweat gland secrete factors that induce cholinergic properties in sympathetic neurons.

Several secreted factors trigger the switch from a noradrenergic to cholinergic phenotype in sympathetic neurons. The sweat gland secretes a cocktail of interleukin-6–like cytokines, notably cardiotrophin-1, leukemia inhibitory factor, and ciliary neurotrophic factor. Several aspects of neuronal metabolism that are linked to transmitter synthesis and release are controlled by these factors. The neurons stop producing the large dense-core granules characteristic of noradrenergic neurons and start making the small electron-translucent vesicles typical of cholinergic neurons (Figure 46–13).

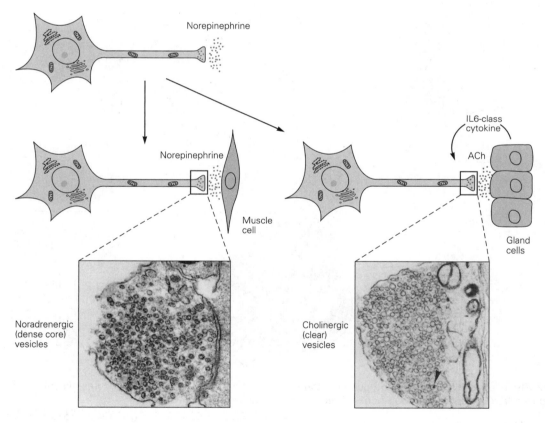

Figure 46–13 The target of sympathetic neurons determines neurotransmitter phenotype. Sympathetic neurons are initially specified with a noradrenergic transmitter phenotype. Most sympathetic neurons, including those that innervate cardiac muscle cells, retain this transmitter phenotype, and their terminals are packed with the dense-core vesicles in which norepinephrine is stored. But the sympathetic neurons that innervate sweat gland targets are induced to switch to a cholinergic transmitter phenotype; their terminals become filled with the small clear vesicles in which acetylcholine (**ACh**) is stored. Sweat gland cells direct the switch by secreting members of the interleukin cytokine family. Several members of this family, including leukemia inhibitory factor and ciliary neurotrophic factor, are potent inducers of cholinergic phenotype in sympathetic neurons grown in cell culture. (Abbreviation: **IL6**, interleukin-6.) (Micrographs reproduced, with permission, from S. Landis.)

More recently, evidence has accumulated that the transmitter phenotype of central neurons can also be influenced by signals including hormones and electrical activity. When the spontaneous activity of embryonic amphibian neurons is increased, some motor neurons can be respecified to synthesize and use the inhibitory neurotransmitter GABA instead of or in addition to acetylcholine. Conversely, when activity is decreased, some inhibitory neurons switch to using the excitatory neurotransmitter glutamate along with or instead of GABA. Postsynaptic partners typically express new receptors that correspond to the transmitter being released onto them. These switches occur without overall respecification of the neuron and are best viewed as homeostatic responses aimed at keeping the overall activity of the system in a narrow range.

Although such transmitter switches in central neurons are likely to occur only rarely under natural conditions, activity-dependent neurotransmitter plasticity may be a more common phenomenon in the adult nervous system. For example, changes in the light cycle where rodents are housed can lead to reciprocal changes in the numbers of neurons that use dopamine and somatostatin as neuromodulators in areas of the brain responsible for maintaining a circadian rhythm. In this and other cases, neurotransmitter switching has measurable consequences on the behavior of the animal, suggesting that this process, along with less drastic synaptic changes discussed in Chapter 49, are employed by the brain as responses to novel environments.

The Survival of a Neuron Is Regulated by Neurotrophic Signals From the Neuron's Target

One of the more surprising findings in developmental neuroscience is that a large fraction of the neurons generated in the embryonic nervous system end up dying later in embryonic development. Equally surprising, we now know that the potential for cell death is preprogrammed in most animal cells, including neurons. Thus, decisions about life and death are aspects of a neuron's fate.

The Neurotrophic Factor Hypothesis Was Confirmed by the Discovery of Nerve Growth Factor

The target of a neuron is a key source of factors essential for the neuron's survival. The critical role of target cells in neuronal survival was discovered in studies of the dorsal root ganglia.

In the 1930s, Samuel Detwiler and Viktor Hamburger discovered that the number of sensory neurons in embryos is increased by transplantation of an additional limb bud into the target field and decreased if the limb target is removed. At the time, these findings were thought to reflect an influence of the limb on the proliferation and subsequent differentiation of sensory neuron precursors. In the 1940s, however, Rita Levi-Montalcini made the startling observation that the death of neurons is not simply a consequence of pathology or experimental manipulation, but rather occurs during the normal program of embryonic development. Levi-Montalcini and Hamburger went on to show that removal of a limb leads to the excessive death of sensory neurons rather than a decrease in their production.

These early discoveries on the life and death of sensory neurons were quickly extended to neurons in the central nervous system. Hamburger found that approximately half of all motor neurons generated in the spinal cord die during embryonic development. Moreover, in experiments similar to those performed on sensory ganglia, Hamburger discovered that motor neuron death could be increased by removing a limb and reduced by adding an additional limb (Figure 46–14A,B). These findings indicate that signals from target cells are critical for the survival of neurons within the central as well as peripheral nervous system. In some cases, manipulating synaptic activity affects the extent of death, perhaps by modulating the types or amount of signals that the target cell produces (Figure 46–14C). We now know that the phenomenon of neuronal overproduction, followed by a phase of neuronal death, occurs in most regions of the vertebrate nervous system.

The early discoveries of Levi-Montalcini and Hamburger laid the foundations for the *neurotrophic factor hypothesis*. The core of this hypothesis is that cells at or near the target of a neuron secrete small amounts of an essential nutrient or trophic factor and that the uptake of this factor by nerve terminals is needed for the survival of the neuron (Figure 46–15). This hypothesis was dramatically confirmed in the 1970s when Levi-Montalcini and Stanley Cohen purified the protein we now know as nerve growth factor (NGF) and showed that this protein is made by target cells and supports the survival of sensory and sympathetic neurons in vitro. Moreover, neutralizing antibodies directed against NGF were found to cause a profound loss of sympathetic and sensory neurons in vivo.

Neurotrophins Are the Best-Studied Neurotrophic Factors

The discovery of NGF prompted a search for additional neurotrophic factors. Today, we know of over a

Figure 46–14 The survival of motor neurons depends on signals provided by their muscle targets. The role of the muscle target in motor neuron survival was demonstrated by Viktor Hamburger in a classic series of experiments performed on the chick embryo. (Adapted from Purves and Lichtman 1985.)

A. A limb bud was removed from a 2.5-day-old chick embryo soon after the arrival of motor nerves. A section of the lumbar spinal cord 1 week later reveals few surviving motor neurons on the deprived side of the spinal cord. The number of motor neurons on the contralateral side with an intact limb is normal.

B. An extra limb bud was grafted adjacent to a host limb prior to the normal period of motor neuron death. A section of the lumbar spinal cord 2 weeks later shows an increased number of limb motor neurons on the side with the extra limb.

C. Blockade of nerve-muscle activity with the toxin curare, which blocks acetylcholine receptors, rescues many motor neurons that would otherwise die. Curare may act by enhancing the release of trophic factors from inactive muscle.

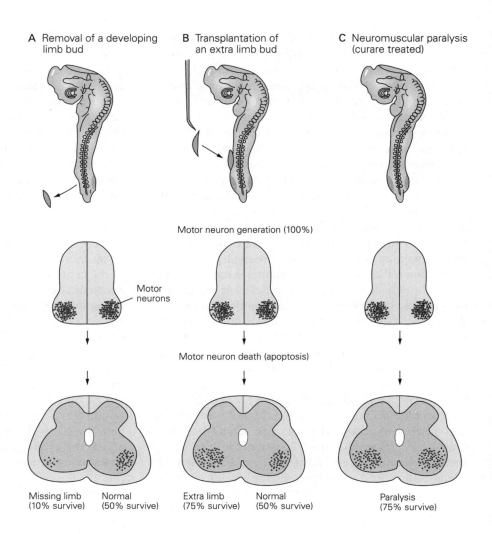

A Removal of a developing limb bud

B Transplantation of an extra limb bud

C Neuromuscular paralysis (curare treated)

Motor neuron generation (100%)

Motor neurons

Motor neuron death (apoptosis)

Missing limb (10% survive)   Normal (50% survive)

Extra limb (75% survive)   Normal (50% survive)

Paralysis (75% survive)

dozen secreted factors that promote neuronal survival. The best-studied are related to NGF and are called the neurotrophin family.

There are four main neurotrophins: NGF itself, brain-derived neurotrophic factor (BDNF), and neurotrophins-3 and -4 (NT-3 and NT-4). Other classes of proteins that promote neuronal survival include members of the transforming growth factor β family, the interleukin-6–related cytokines, fibroblast growth factors, and even certain inductive signals we encountered earlier (BMPs and hedgehogs). Other neurotrophic factors, notably members of the glial cell line–derived neurotrophic factor (GDNF) family, are responsible for the survival of different types of sensory and sympathetic neurons (Figure 46–16).

Neurotrophins interact with two major classes of receptors, the Trk receptors and p75. Neurotrophins promote cell survival through activation of Trk receptors. The Trk family comprises three membrane-spanning

tyrosine kinases named TrkA, TrkB, and TrkC, each of which exists as a dimer (Figure 46–17).

Much is now known about the intracellular signaling pathways activated by binding of neurotrophins to Trks. As with other tyrosine kinase receptors, the binding of neurotrophins to Trk receptors leads to dimerization of the Trk proteins. Dimerization results in phosphorylation of specific tyrosine residues in the activation loop of the kinase domain. This phosphorylation leads to a conformational change in the receptor and to phosphorylation of tyrosine residues that serve as docking sites for adaptor proteins. The adaptors then trigger production of second messengers that both promote the survival of neurons and trigger their maturation. These divergent biological responses involve different intracellular signaling pathways: neuronal differentiation largely via the mitogen-activated protein kinase (MAPK) enzymatic pathways and survival largely via the phosphatidylinositol-3 kinase pathway (Figure 46–18).

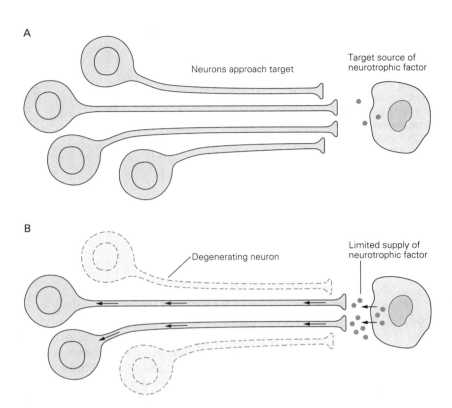

Figure 46–15 The neurotrophic factor hypothesis.

A. Neurons extend their axons to target cells, which secrete low levels of neurotrophic factors. (For simplicity, only one target cell is shown.) The neurotrophic factor binds to specific receptors and is internalized and transported to the cell body, where it promotes neuronal survival.

B. Neurons that fail to receive adequate amounts of neurotrophic factor die through a program of cell death termed apoptosis.

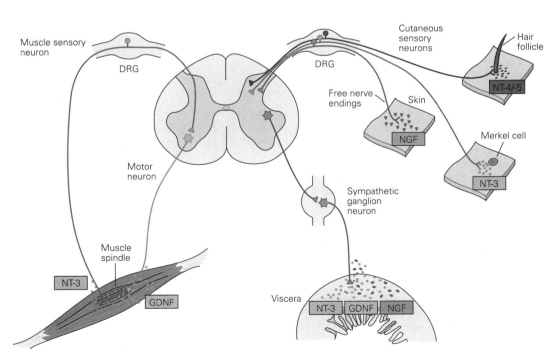

Figure 46–16 Particular neurotrophic factors promote the survival of distinct populations of dorsal root ganglion neurons. Proprioceptive sensory neurons that innervate muscle spindles depend on neurotrophin-3 (**NT-3**); nociceptive neurons that innervate skin depend on nerve growth factor (**NGF**) and neurturin; mechanoreceptive neurons that innervate

Merkel cells depend on NT-3; and those that innervate hair follicles depend on neurotrophin-4 and -5 (**NT-4/-5**) and brain-derived neurotrophic factor. Motor neurons depend on glial cell line–derived neurotrophic factor (**GDNF**) and other factors. Sympathetic neurons depend on NGF, NT-3, and GDNF. (Adapted from Reichardt and Fariñas 1997.)

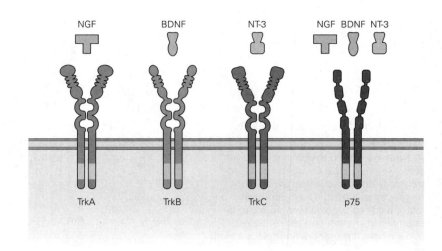

Figure 46–17 Neurotrophins and their receptors. Each of the three main neurotrophins interacts with a different transmembrane tyrosine kinase receptor (**Trk**). In addition, all three neurotrophins can bind to the low-affinity neurotrophin receptor p75. (Abbreviations: **BDNF**, brain-derived neurotrophic factor; **NGF**, nerve growth factor; **NT-3**, neurotrophin-3.) A fourth neurotrophin, NT-4, is not shown. (Adapted from Reichardt and Fariñas 1997.)

In contrast to the specificity of Trk receptor interactions, all neurotrophins bind the receptor p75 (Figure 46–17). In some cases, p75 works along with Trk receptors, tuning the affinity and specificity of Trks for their neurotrophin ligands and thereby contributing to neuronal survival. However, p75 leads a double life. It can also bind unprocessed precursors of neurotrophins, called proneurotrophins, and it can associate with other membrane receptors called sortilins. Binding of proneurotrophins to the p75/sortilin complex promotes neuronal death. Receptor p75 is a member of the tumor necrosis factor (TNF) receptor family and promotes cell death by activating proteases of the caspase family, which we discuss below.

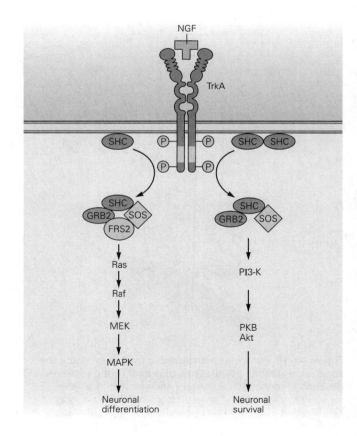

Figure 46–18 Binding of nerve growth factor to the TrkA receptor activates alternative intracellular signaling pathways. The binding of nerve growth factor (**NGF**) induces dimerization of the TrkA receptor, which triggers its phosphorylation at many different residues. Phosphorylation of TrkA results in the recruitment of the adaptor proteins SHC, GRB2, and SOS. The additional recruitment of FRS2 to this complex (*left*) activates a Ras kinase signaling pathway that promotes neuronal differentiation. In the absence of FRS2 (*right*), the complex activates a phosphatidylinositol-3 kinase (**PI3-K**) pathway that promotes neuronal survival. (Abbreviations: **Akt/PKB**, protein kinase B; **MAPK**, mitogen-activated protein kinase; **MEK**, mitogen-activated/ERK kinase; **P**, phosphate.)

Neurotrophin signaling is relayed from the axon terminal to the cell body of the neuron through a process that involves internalization of a complex of neurotrophin bound to Trk receptors. The retrograde transport of this complex occurs in a class of endocytotic vesicles called signaling endosomes. The transport of these vesicles brings activated Trk receptors into cellular compartments able to activate signaling pathways and transcriptional programs essential for neuronal survival, maturation, and synaptic differentiation.

The picture is more complex for neurons in the central nervous system. The survival of motor neurons, for example, is not dependent on a single neurotrophic factor. Instead, different classes of motor neurons require neurotrophins, GDNF, and interleukin-6–like proteins expressed by muscles or peripheral glial cells. The survival of these neuronal classes depends on the exposure of axons to local neurotrophic factors.

Neurotrophic Factors Suppress a Latent Cell Death Program

Neurotrophic factors were once believed to promote the survival of neural cells by stimulating their metabolism in beneficial ways, hence their name. It is now evident, however, that neurotrophic factors suppress a latent death program present in all cells of the body, including neurons.

This biochemical pathway can be considered a suicide program. Once it is activated, cells die by apoptosis (Greek, falling away): They round up, form blebs, condense their chromatin, and fragment their nuclei. Apoptotic cell deaths are distinguishable from necrosis, which typically results from acute traumatic injury and involves rapid lysis of cell membranes without activation of the cell death program.

The first clue that deprivation of neurotrophic factors kills neurons by unleashing an active biochemical program emerged from studies that assessed neuronal survival after inhibition of RNA and protein synthesis. Exposure of sympathetic neurons to protein synthesis inhibitors was found to prevent the death of sympathetic neurons triggered by removal of NGF. These results sparked the idea that neurons have the ability to synthesize proteins that are lethal and that NGF prevents their synthesis, thereby suppressing an endogenous cell death program.

Key insights into the biochemical nature of the endogenous cell death program emerged from genetic studies of the nematode *Caenorhabditis elegans*. During the development of *C. elegans*, a precise number of cells is generated and a fixed number of these cells die—the same number from embryo to embryo. The findings prompted a screen for genes that block or enhance cell death, which led to the identification of the cell death (*ced*) genes. Two of these genes, *ced-3* and *ced-4*, are needed for the death of neurons; in their absence, every one of the cells destined to die instead survives. A third gene, *ced-9*, is needed for survival and works by antagonizing the activities of *ced-3* and *ced-4* (Figure 46–19). Thus, in the absence of *ced-9*, many additional cells die, even though these deaths still depend on *ced-3* and *ced-4* activity.

The cell death pathway in *C. elegans* has been conserved in mammals. Similar proteins and pathways control the apoptotic death of central and peripheral neurons, indeed of all developing cells. The worm *ced-9* gene encodes a protein that is related to members of the mammalian Bcl-2 family, which protect

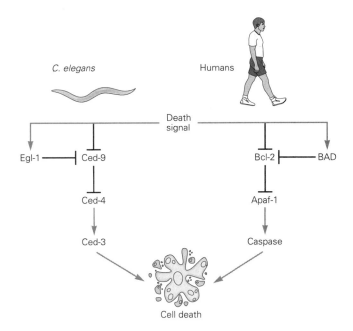

Figure 46–19 Neurons and other cells express a conserved death program. Different cellular insults trigger a genetic cascade that involves a series of death effector genes. These death genes and pathways have been conserved in the evolution of species from worms to humans. The core death pathway activates a set of proteolytic enzymes, the caspases. Caspases cleave many downstream and essential protein substrates (see Figure 46–20), resulting in the death of cells by a process termed *apoptosis*. Genetic analysis of the worm *Caenorhabditis elegans* indicates that the Ced-9 protein acts upstream and inhibits the activity of Ced-4 and Ced-3, two proteins that promote cell death. Many vertebrate homologs of Ced-9, the Bcl-2 family of proteins, have been identified. Some of these proteins, such as Bcl-2 itself, inhibit cell death, but others promote cell death by antagonizing the actions of Bcl-2. The Bcl-2 class proteins act upstream of Apaf-1 (a vertebrate homolog of Ced-4) and the caspases (vertebrate homologs of Ced-3).

lymphocytes and other cells from apoptotic death. The worm *ced-3* gene encodes a protein closely related to a class of mammalian cysteine proteases called caspases. The worm *ced-4* gene encodes a protein that is functionally related to a mammalian protein called apoptosis activating factor-1 (Apaf-1).

The mammalian apoptotic cell death pathway works in a way that resembles the worm pathway (Figure 46–20). The morphological and histochemical changes that accompany the apoptosis of mammalian cells result from the activation of caspases, which cleave specific aspartic acid residues within cellular

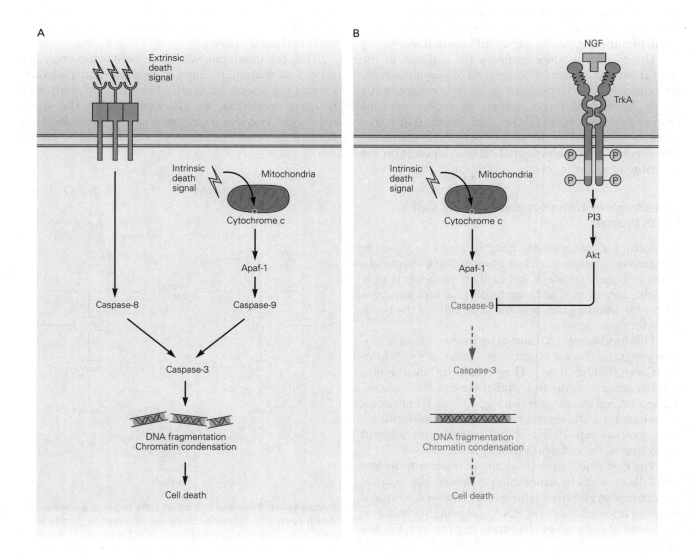

Figure 46–20 Neurotrophic factors suppress caspase activation and cell death. (Adapted from Jesenberger and Jentsch 2002.)

A. Two types of pathways trigger cell death: extrinsic activation of surface membrane death receptors and intrinsic activation of a mitochondrial pathway. Both pathways result in activation of caspases such as caspase-8 and caspase-9, which initiate a proteolytic cleavage cascade that converges at the level of caspase-3 activation. Cleavage of the caspase precursor removes the caspase prodomain and produces a proteolytically active enzyme conformation.

The extrinsic pathway involves activation of death receptors by ligands such as tumor necrosis factor receptor 1 or Fas/CD95. The intrinsic pathway involves stress-induced signals such as DNA damage that initiate the release of cytochrome c from the mitochondrial intermembrane space. Cytochrome c binds to Apaf-1 and recruits and activates caspase-9.

B. Binding of neurotrophins to Trk receptors recruits the PI3 kinase pathway and Akt and suppresses the cell death pathway by inhibiting caspase-9. This pathway is inhibited in developing neurons by neurotrophic factors, explaining why withdrawal of these factors leads to apoptosis. (Abbreviations: **NGF**, nerve growth factor; **P**, phosphate.)

proteins. Two classes of caspases regulate apoptotic death: the initiator and effector caspases. Initiator caspases (caspase-8, -9, and -10) cleave and activate effector caspases. Effector caspases (caspase-3 and -7) cleave other protein substrates, thus triggering the apoptotic process. Perhaps 1% of all proteins in the cell serve as substrates for effector caspases. Their cleavage contributes to neuronal apoptosis through many pathways: by activation of proteolytic cascades, inactivation of repair, DNA cleavage, mitochondrial permeabilization, and initiation of phagocytosis.

The survival of mammalian neurons depends on the balance between antiapoptotic and proapoptotic members of the Bcl-2 family of proteins. Some Bcl-2 proteins such as BAX and BAK increase the permeability of the mitochondrial outer membranes, causing the release of proapoptotic proteins such as cytochrome c into the cytosol. The release of cytochrome c induces Apaf-1 to bind and activate caspase-9, leading to the cleavage and activation of effector caspases. The binding of neurotrophic factors to their tyrosine kinase receptors is thought to lead to the phosphorylation of protein substrates that promote Bcl-2–like activities (Figure 46–20B). Thus, withdrawal of neurotrophic factors from neurons changes the balance from antiapoptotic to proapoptotic members of the Bcl-2 family, which triggers the neuron's demise.

The caspase cell death program can also be activated by many cellular insults, including DNA damage and anoxia. The activation of cell-surface death receptors such as Fas by extracellular ligands results in the activation of caspase-8 or -10 as well as the recruitment of death effector proteins such as FADD. Recruitment of an initiator caspase to the Fas-FADD complex then leads to activation of effector caspases. Because many neurodegenerative disorders result in apoptotic death, pharmacological strategies to inhibit caspases are under investigation.

Highlights

1. Stem cells near the ventricular surface of the neural tube divide to expand the neuroepithelium. Further divisions then generate the neurons and glia of the central nervous system as well as radial glia.

2. Processes of radial glia extend from the ventricular to the pial surface. Radial glial cells continue dividing to form neurons and astrocytes. In the cortex, they also serve as a scaffold on which newborn excitatory neurons migrate to appropriate layers.

3. The choice between neuronal and glial fate is determined by signals from ligands of the Delta family to receptors of the Notch family on neighboring cells. Initially, cells express both Notch and Delta. Activation of Notch leads to a glial fate, downregulating Delta, which in turn attenuates Notch activity on the neighbors, promoting their differentiation into neurons.

4. As cortical principal (excitatory) neurons migrate along radial glia, they form cortical layers in an inside-out sequence (layer 6 forms before layer 5, and so on). Disruptions of migration are among the causes of intellectual disability and epilepsy.

5. Unlike excitatory neurons, forebrain interneurons arise subcortically in ganglionic eminences and then migrate tangentially into the cortex, basal ganglia, and other forebrain structures.

6. Neural crest cells migrate from their source at the dorsal tip of the neural tube through somites and mesenchyme to form sensory and autonomic neurons and glia, as well as several nonneural cell types.

7. For principal neurons, interneurons, and peripheral neurons, intrinsic differences and cues encountered along the migratory path interact to induce expression of distinct combinations of transcription factors. The transcriptional programs then lead to diversification of the developing neurons into multiple classes and types.

8. The greater complexity of the primate, and particularly the human brain compared to those of lower mammals is due in part to a larger pool of neuronal progenitors, including a second type of radial glial cell.

9. A recent advance in the ability to study the human brain is the discovery that complex neuronal ensembles called cerebral organoids can be generated from stem cells. Although they fail to acquire characteristics of the mature cortex, they enable analysis of some aspects of early brain development and its disorders and may be useful in testing possible therapeutics.

10. The neurotransmitters that neurons use are determined as part of the transcriptional program that endows each neuronal type with its defining characteristics. However, extrinsic factors, including patterns of electric activity and hormonal milieu, can lead to transmitter switching in some cases.

11. The nervous system generates up to twice as many neurons as survive in adulthood. The excess is eliminated by a cell death program that is conserved from invertebrates to humans.

12. Trophic factors play a crucial role in determining which neurons within a population live or die. They control survival by holding the cell death program in check. In some cases, neurons appear to compete for a limited supply of neurotrophic factors; the cell death program is activated in those that lose the competition.

13. Multiple trophic factors are produced in the body, with each controlling the fate of only some neuronal types. The best-studied, called neurotrophins (nerve growth factor, brain-derived neurotrophic factor, neurotrophin-3, and neurotrophin-4), bind to and activate kinases called Trk receptors.

Joshua R. Sanes
Thomas M. Jessell

Selected Reading

Di Lullo E, Kriegstein AR. 2017. The use of brain organoids to investigate neural development and disease. Nat Rev Neurosci 18:573–584.

Gleeson JG, Walsh CA. 2000. Neuronal migration disorders: from genetic diseases to developmental mechanisms. Trends Neurosci 23:352–359.

Lodato S, Arlotta P. 2015. Generating neuronal diversity in the mammalian cerebral cortex. Annu Rev Cell Dev Biol 31:699–720.

Spitzer NC. 2017. Neurotransmitter switching in the developing and adult brain. Annu Rev Neurosci 40:1–19.

Wamsley B, Fishell G. 2017. Genetic and activity-dependent mechanisms underlying interneuron diversity. Nat Rev Neurosci 18:299–309.

Wilsch-Bräuninger M, Florio M, Huttner WB. 2016. Neocortex expansion in development and evolution—from cell biology to single genes. Curr Opin Neurobiol 39:122–132.

References

Anderson DJ. 1997. Cellular and molecular biology of neural crest cell lineage determination. Trends Genet 13:276–280.

Bandler RC, Mayer C, Fishell G. 2017. Cortical interneuron specification: the juncture of genes, time and geometry. Curr Opin Neurobiol 42:17–24.

Bershteyn M, Nowakowski TJ, Pollen AA, et al. 2017. Human iPSC-derived cerebral organoids model cellular features of lissencephaly and reveal prolonged mitosis of outer radial glia. Cell Stem Cell 20:435–449.

Costa RO, Perestrelo T, Almeida RD. 2018. PROneurotrophins and CONSequences. Mol Neurobiol 55:2934–2951.

Detwiler SR. 1936. Neuroembryology: An Experimental Study. New York: Macmillan.

Doupe AJ, Landis SC, Patterson PH. 1985. Environmental influences in the development of neural crest derivatives: glucocorticoids, growth factors, and chromaffin cell plasticity. J Neurosci 5:2119–2142.

Duband JL. 2006. Neural crest delamination and migration: integrating regulations of cell interactions, locomotion, survival and fate. Adv Exp Med Biol 589:45–77.

Florio M, Borrell V, Huttner WB. 2017. Human-specific genomic signatures of neocortical expansion. Curr Opin Neurobiol 42:33–44.

Furshpan EJ, Potter DD, Landis SC. 1982. On the transmitter repertoire of sympathetic neurons in culture. Harvey Lect 76:149–191.

Giandomenico SL, Lancaster MA. 2017. Probing human brain evolution and development in organoids. Curr Opin Cell Biol 44:36–43.

Gray GE, Sanes JR. 1992. Lineage of radial glia in the chicken optic tectum. Development 114:271–283.

Guo J, Anton ES. 2014. Decision making during interneuron migration in the developing cerebral cortex. Trends Cell Biol 24:342–351.

Hamburger V. 1975. Cell death in the development of the lateral motor column of the chick embryo. J Comp Neurol 160:535–546.

Hamburger V, Levi-Montalcini R. 1949. Proliferation differentiation and degeneration in the spinal ganglia of the chick embryo under normal and experimental conditions. J Exp Zool 111:457–501.

Hoshino M. 2006. Molecular machinery governing GABAergic neuron specification in the cerebellum. Cerebellum 5: 193–198.

Howard MJ. 2005. Mechanisms and perspectives on differentiation of autonomic neurons. Dev Biol 277:271–286.

Jesenberger V, Jentsch S. 2002. Deadly encounter: ubiquitin meets apoptosis. Nat Rev Mol Cell Biol 3:112–121.

Lancaster MA, Renner M, Martin CA, et al. 2013. Cerebral organoids model human brain development and microcephaly. Nature 501:373–379.

Landis SC. 1980. Developmental changes in the neurotransmitter properties of dissociated sympathetic neurons: a cytochemical study of the effects of medium. Dev Biol 77:349–361.

Le Douarin NM. 1998. Cell line segregation during peripheral nervous system ontogeny. Science 231:1515–1522.

Nowakowski TJ, Pollen AA, Sandoval-Espinosa C, Kriegstein AR. 2016. Transformation of the radial glia scaffold demarcates two stages of human cerebral cortex development. Neuron 91:1219–1227.

Olson EC, Walsh CA. 2002. Smooth, rough and upside-down neocortical development. Curr Opin Genet Dev 12:320–327.

Oppenheim RW. 1981. Neuronal cell death and some related regressive phenomena during neurogenesis: a selective historical review and progress report. In: WM Cowan (ed). *Studies in Developmental Neurobiology: Essays in Honor of Viktor Hamburger*, pp. 74–133. New York: Oxford Univ. Press.

Purves D, Lichtman JW. 1985. *Principles of Neural Development.* Sunderland, MA: Sinauer.

Qian X, Goderie SK, Shen Q, Stern JH, Temple S. 1998. Intrinsic programs of patterned cell lineages in isolated vertebrate CNS ventricular zone cells. Development 125:3143–3152.

Reichardt LF. 2006. Neurotrophin-regulated signaling pathways. Philos Trans R Soc Lond B Biol Sci 361:1545–1564.

Reichardt LF, Fariñas I. 1997. Neurotrophic factors and their receptors: roles in neuronal development and function. In: MW Cowan, TM Jessell, L Zipursky (eds). *Molecular Approaches to Neural Development*, pp. 220–263. New York: Oxford Univ. Press.

Sánchez-Alcañiz JA, Haege S, Mueller W. 2011. Cxcr7 controls neuronal migration by regulating chemokine responsiveness. Neuron 69:77–90.

Shah NM, Groves AK, Anderson DJ. 1996. Alternative neural crest cell fates are instructively promoted by TGF beta superfamily members. Cell 85:331–343.

Sun Y, Nadal-Vicens M, Misono S, et al. 2001. Neurogenin promotes neurogenesis and inhibits glial differentiation by independent mechanisms. Cell 104:365–376.

Wang Y, Li G, Stanco A, et al. 2011. CXCR4 and CXCR7 have distinct functions in regulating interneuron migration. Neuron 69:61–76.

Zeng H, Sanes JR. 2017. Neuronal cell-type classification: challenges, opportunities and the path forward. Nat Rev Neurosci 18:530–546.

47

The Growth and Guidance of Axons

I N THE TWO PRECEDING CHAPTERS, we saw how
neurons are generated in appropriate numbers, at
correct times, and in the right places. These early
developmental steps set the stage for later events that
direct neurons to form functional connections with target cells. To form connections, neurons have to extend
long processes—axons and dendrites—which permit connectivity with postsynaptic cells and synaptic
input from other neurons. In this chapter, we examine
how neurons elaborate axons and dendrites and how
axons are guided to their targets.

We begin the chapter by discussing how certain
neuronal processes become axons and others dendrites. We then describe the growing axon, which
may have to travel a long distance and ignore many
inappropriate neuronal partners before terminating in
just the right region and recognizing its correct synaptic targets. We consider the strategies by which the
axon overcomes these challenges. Finally, we illustrate
general features of axonal guidance by describing the
development of two well-studied axonal pathways:
one that conveys visual information from the retina to
the brain and another that conveys cutaneous sensory
information from the spinal cord to the brain.

Differences Between Axons and Dendrites
Emerge Early in Development

The processes of neurons vary enormously in their
length, thickness, branching pattern, and molecular
architecture. Nonetheless, most neuronal processes fit
into one of two functional categories: axons and dendrites. More than a century ago, Santiago Ramón y
Cajal hypothesized that this distinction underlies the
ability of neurons to transmit information in a particular direction, an idea he formalized as the law of
dynamic polarization. Cajal wrote that "the transmission of the nerve impulse is always from the dendritic
branches and the cell body to the axon." In the decades
before electrophysiological methods were up to the
task, this law provided a means of analyzing neural
circuits histologically. Although exceptions have been
found, Ramón y Cajal's law remains a basic principle

that relates structure to function in the nervous system and highlights the importance of knowing how neurons acquire their polarized form.

Progress in understanding how neuronal polarization occurs comes in large part from studies of neurons taken from the rodent brain and grown in tissue culture. Hippocampal neurons grown in isolation develop processes reminiscent of those seen in vivo: a single, long, cylindrical axon and several shorter, tapered dendrites (Figure 47–1A). As cytoskeletal and synaptic proteins are differentially targeted to these components, axons and dendrites acquire distinctive molecular profiles. For example, a particular form of the Tau protein is localized in axons and the MAP2 protein in dendrites (Figure 47–1B)

Cultured neurons are especially useful for developmental studies because they initially show no obvious sign of polarization and acquire their specialized features gradually in a stereotyped sequence of cellular steps. This sequence begins with extension of several short processes, each equivalent to the others. Soon thereafter, one process is established as an axon and the remaining processes acquire dendritic features (Figure 47–1A).

How does this occur? Cytoskeletal proteins that maintain elongated processes and drive growth are central to this process. If the actin filaments in an early neurite are destabilized, the cytoskeleton becomes reconfigured in a way that commits the neurite to becoming the axon; secondarily, the remaining neurites react by becoming dendrites. If the nascent axon is removed, one of the remaining neurites quickly assumes an axonal character. This sequence suggests that axonal specification is a key event in neuronal polarization and that signals from newly formed axons both suppress the generation of additional axons and promote dendrite formation.

The nature of the axonally derived signal that represses other axons is not known, but some insight into signals that control cytoskeletal arrangements has come from the study of a group of proteins encoded by the Par complex genes. As first shown in the nematode worm Caenorhabditis elegans, Par proteins are involved in diverse aspects of cytoskeletal reorganization, including the polarization of neuronal processes. Mammalian forebrain neurons lacking Par3, Par4, Par6, or relatives of Par1 grow multiple processes that are intermediate in length between axons and dendrites and bear markers of both processes (Figure 47–1B).

Although neurons grown in culture are similar to those in the brain, they are deprived of key extrinsic cues and signals. Cultured neurons become randomly arranged with respect to each other, whereas in many regions of the developing brain, neurons line up in rows, with their dendrites pointing in the same direction (Figure 47–2A). As the neurons migrate to their destinations (Chapter 46), axons and dendrites often grow as extensions of their trailing and leading processes, respectively. This difference in vivo and in vitro implies that extrinsic signals regulate the polarization machinery. In the developing brain, the local release of semaphorins and other axonal guidance factors, discussed later in the chapter, may help to orient dendrites (Figure 47–2C). The job of the Par protein complex is to link these extracellular signals to the cellular machinery that rearranges the cytoskeleton, a process achieved in part through the regulation of proteins that modify actin or tubulin function. In fact, both the Tau protein in axons and the MAP2 protein in dendrites associate with and affect microtubules. Cytoskeletal differences also contribute to other mechanisms that amplify distinctions between axons and dendrites, such as polarized trafficking of molecules and generation of a specialized initial segment in axons.

If local signals are needed to polarize neurons in the brain, how is polarity established in the uniform environment of a tissue culture? One possible explanation is that minor variations in the intensity of signaling within a neuron, or in signals from its immediate environment, will activate Par proteins in one small domain of the neuron, triggering the nearest cell process to become an axon. If, by happenstance, one process grows slightly faster than its neighbors or encounters an environment that speeds neurite extension (Figure 47–2B), its chances of becoming an axon increase markedly. Presumably, this proto-axonal process emits signals that decrease the chance of other processes following suit, forcing them to become dendrites.

Dendrites Are Patterned by Intrinsic and Extrinsic Factors

Once polarization occurs, dendrites grow and mature, acquiring the structural features that distinguish them from axons. Nascent dendrites form branched arbors, with their branches generally being more numerous and closer to the cell body than those of axons. In addition, small protrusions called spines extend from the distal branches of many dendrites. Finally, some dendritic branches are retracted or "pruned" to give the arbor its final and definitive shape (Figure 47–3).

Although the core features of dendrite formation are common to many neurons, there is striking variation in their number, shape, and branching pattern among neuronal types. Indeed, the shape of dendritic arbors is one of the main ways in which neurons can

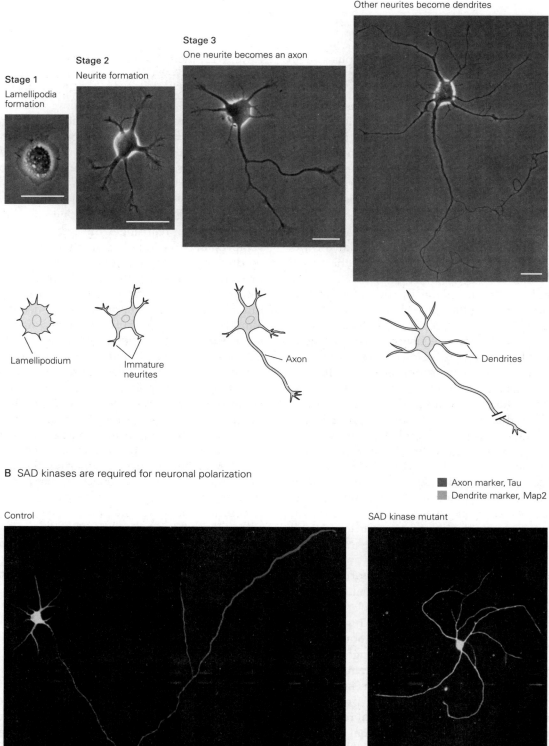

A Developmental stages of a neuron grown in culture

Stage 1
Lamellipodia formation

Stage 2
Neurite formation

Stage 3
One neurite becomes an axon

Stage 4
Other neurites become dendrites

Lamellipodium

Immature neurites

Axon

Dendrites

B SAD kinases are required for neuronal polarization

■ Axon marker, Tau
■ Dendrite marker, Map2

Control

SAD kinase mutant

Figure 47–1 The differentiation of axons and dendrites marks the emergence of neuronal polarity.

A. Four stages in the polarization of a hippocampal neuron grown in tissue culture. (Adapted, with permission, from Kaech and Banker 2006. Copyright © 2007 Springer Nature.)

B. Hippocampal neurons grown in culture possess multiple short, thick dendrites that are enriched in the microtubule-associated protein MAP2. They also possess a single long axon that is marked by a dephosphorylated form of the microtubule-associated protein tau (*left*). A cultured neuron isolated from a mutant mouse lacks expression of a *Par* family gene (SAD kinase). The neuron generates neurites that express both tau and MAP2, markers of axons and dendrites, respectively. The length and diameter of these neurites are intermediate in size between those of axons and dendrites (*right*). (Reproduced, with permission, from Kishi et al. 2005.)

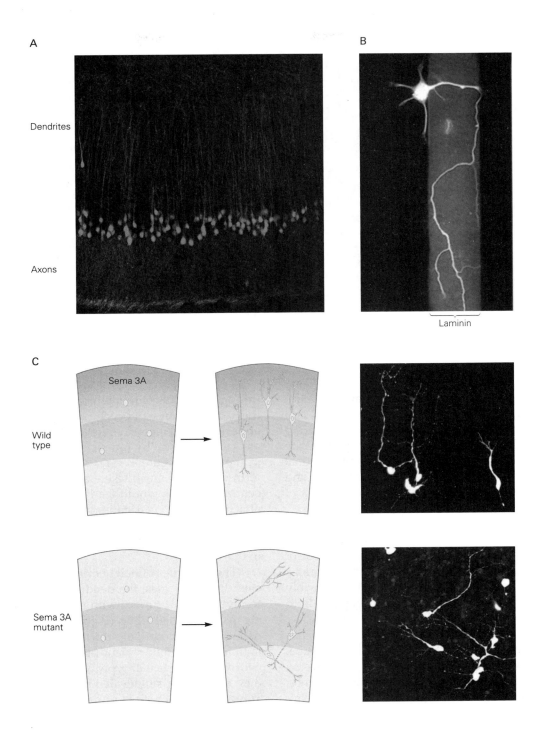

Figure 47–2 Extracellular factors determine whether neu-ronal processes become axons or dendrites.

A. Cortical pyramidal neurons in vivo display a common axonal and dendritic orientation.

B. Neurons growing on laminin acquire polarity. When a cortical neuron extends a process from a less attractive substrate onto laminin, the process grows faster and usually becomes an axon. (Image reproduced, with permission, from Paul Letourneau.)

C. In the developing neocortex, semaphorin-3A (**Sema 3A**) is secreted by cells near the pial surface. Semaphorin-3A is an attractant for growing dendrites, helping to establish neuronal polarity and orientation. The parallel orientation of cortical pyramidal neurons is disrupted in mutant mice lacking func-tional semaphorin-3A. (Reproduced, with permission, from Polleux, Morrow, and Ghosh 2000. Copyright © 2000 Springer Nature.)

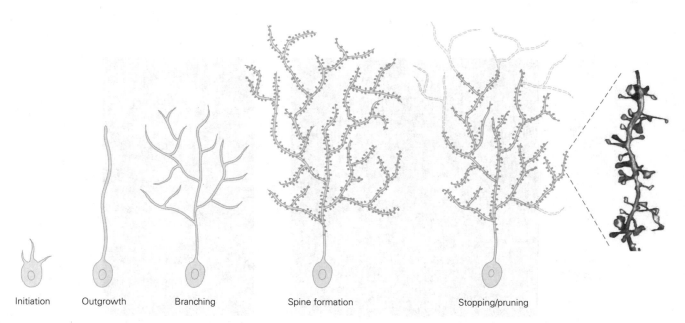

Initiation   Outgrowth   Branching   Spine formation   Stopping/pruning

Figure 47–3 Dendritic branching develops in a series of steps. The outgrowth of dendrites involves the formation of elaborate branches from which spines develop. Certain branches and spines are later pruned to achieve the mature pattern of dendrite arborization. (Image of spines at right reproduced, with permission, from Stefan W. Hell.)

be classified. Cerebellar Purkinje cells can be distinguished from granule cells, spinal motor neurons, and hippocampal pyramidal neurons simply by looking at the pattern of their dendrites. These variations are critical for the distinct functions of different neuronal types. For example, the size of a dendritic arbor and the density of its branches are main determinants of the number of synapses it receives.

How is dendritic pattern established? Neurons must have intrinsic information about their shape because the patterns in tissue culture are strikingly reminiscent of those in vivo (Figure 47–4). The transcriptional programs that specify neuronal subtype (Chapter 46) presumably also encode information about neuronal shape. In both invertebrates and vertebrates, some transcription factors are selectively expressed by specific neuronal types and appear to be devoted to controlling the size, shape, and complexity of their dendritic arbors. They do so by coordinating the expression of downstream genes, including those encoding components of the cytoskeletal apparatus and membrane proteins that mediate interactions with neighboring cells.

A second mechanism for establishing the pattern of dendritic arbors is the recognition of one dendrite by others of the same cell. In some neurons, dendrites are spaced evenly with respect to each other, an arrangement that allows them to sample inputs efficiently without major gaps or clumps (Figure 47–5A). In many cases, this process, called self-avoidance, occurs

through a mechanism in which branches belonging to the same neuron repel each other. Several cell-surface adhesion molecules have now been found that mediate self-avoidance by interacting in a way that results in repulsion (Figure 47–5D). Although it seems counterintuitive that an adhesive interaction between adjacent membranes would lead to repulsion rather than attachment, the consequences of most intercellular interactions are determined by the signaling they initiate rather than by adhesion per se, as we will see later in this chapter.

The dendrites of neighboring neurons also provide cues. In many cases, the dendrites of a particular neuron type cover a surface with minimal overlap, a spacing pattern called *tiling* (Figure 47–5B). The tiling of dendrites is conceptually related to self-avoidance, but in tiling, the inhibitory dendritic interactions are among neurons of a particular type, whereas in self-avoidance, they are among sibling dendrites of a single neuron. Tiling allows each class of neuron to receive information from the entire surface or area it innervates. Tiling of a region by the dendrites of one class of neuron also avoids the confusion that could arise if the dendrites of many different neurons occupied the same area.

A particularly interesting situation is one in which dendrites engage in self-avoidance but synapse on the dendrites of other cells of the same type. In this situation, dendrites face the challenging task of distinguishing nominally identical dendrites from dendrites of

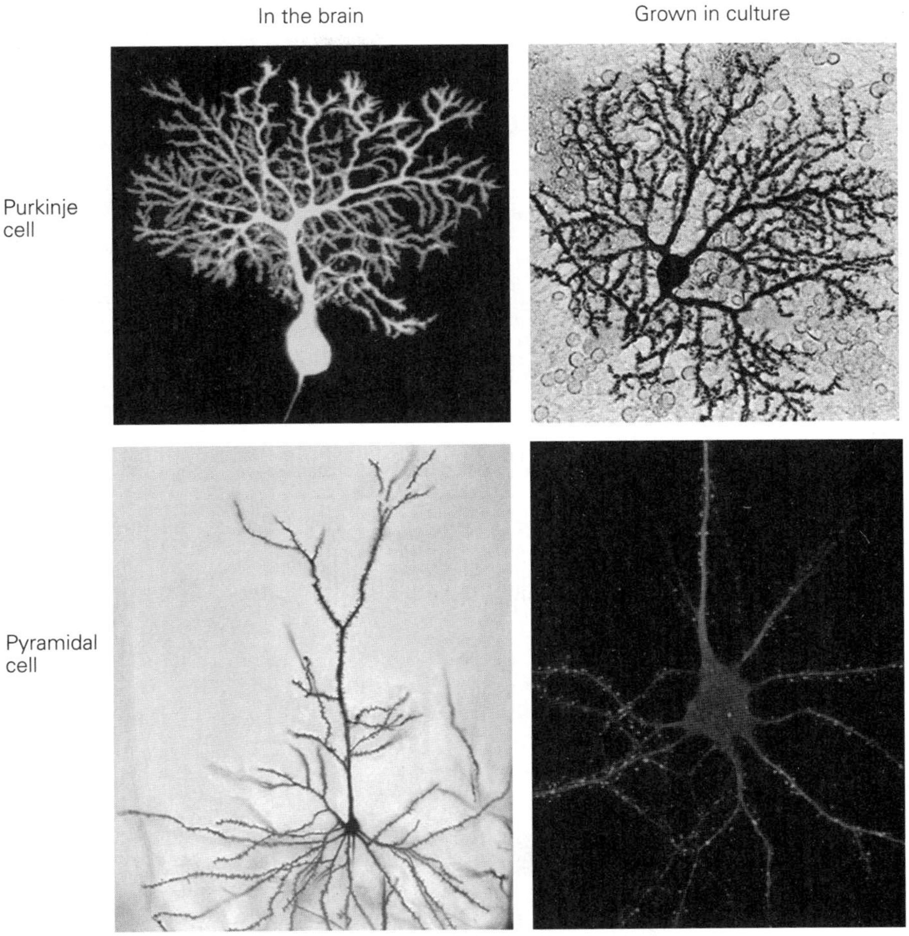

In the brain   Grown in culture

Purkinje cell

Pyramidal cell

Figure 47–4 The morphologies of neurons are preserved in dissociated cell culture. Cerebellar Purkinje neurons and hippocampal pyramidal neurons have distinctive patterns of dendritic branching. These basic patterns are recapitulated when these two classes of neurons are isolated and grown in dissociated cell culture. (Image upper left: Dr. David Becker; upper right reproduced, with permission, from Yoshio Hirabayashi; lower left reproduced, with permission, from Terry E. Robinson; lower right reproduced, with permission, from Kelsey Martin.)

nominally identical cells (Figure 47–5C). Two groups of molecules have been identified that mediate this self-/non–self-discrimination: clustered protocadherins in mammals and DS-CAMs in *Drosophila*. Although they are unrelated structurally, they share several features (Figure 47–5D).

First, both are encoded by large, complex genes that generate large numbers of isoforms. *Drosophila* Dscam1 encodes around 38,000 distinct proteins through alternative splicing, and the clustered protocadherins encode around 60 proteins that can assemble into thousands of distinct multimers. Second, nearly all of the isoforms bind homophilically; for example, protocadherin γa1 on the surface of one dendrite binds well to protocadherin γa1 on a neighboring membrane, but poorly if at all to other isoforms. Third, in ways that remain incompletely understood, each neuron within a population expresses a random subset of all possible Dscam1 or protocadherin isoforms. Given the large number of isoforms, it is unlikely that individual neurons express identical sets of isoforms on their cell surface. The upshot is that

dendrites of each neuron in a population bind homophilically to sibling dendrites, leading to repulsion and self-avoidance, whereas they bind poorly to dendrites of neighboring neurons, enabling other recognition systems to foster synaptogenesis.

Together, the mechanisms we have described, and many others, establish an overall arborization pattern through a combination of intrinsic and extracellular mechanisms. For dendrites, the extrinsic patterning signals determine neuronal morphology. For axons, which we consider next, the signals guide the axons to their targets.

The Growth Cone Is a Sensory Transducer and a Motor Structure

Once an axon forms, it begins to grow toward its synaptic target. The key neuronal element responsible for axonal growth is a specialized structure at the tip of the axon called the *growth cone*. Both axons and dendrites

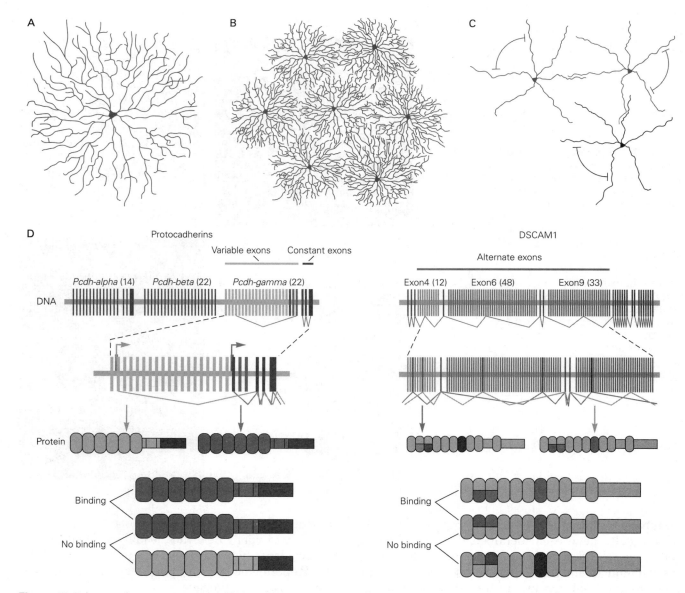

Figure 47–5 Interactions among dendritic branches pattern dendritic arbors.

A. Self-avoidance among sibling dendrites leads to even spacing of branches, minimizing gaps and clumps. In retinal starburst amacrine cells, self-avoidance fails when gamma protocadherins are lost.

B. Tiling of dendrites is conceptually similar to self-avoidance but applies to groups of neurons. It ensures that neighboring neurons of a single type cover territory efficiently.

C. Self-/non-self–discrimination allows sibling dendrites to avoid each other while interacting freely with dendrites of other neurons of the same type.

D. Generation of numerous adhesion molecules from a single genomic complex by promoter choice at the mouse clustered protocadherin (**Pcdh**) locus (*left*) and by alternative splicing at the *Drosophila* DSCAM1 locus (*right*).

use growth cones for elongation, but those linked to axons have been studied more intensively.

Ramón y Cajal discovered the growth cone and had the key insight that it was responsible for axonal pathfinding. With static images alone for inspiration (Figure 47–6A), he envisioned the growth cone to be "endowed with exquisite chemical sensitivity, rapid ameboid movements and a certain motive force, thanks to which it is able to proceed forward and overcome obstacles met in its way . . . until it reaches its destination."

Many studies over the past century have confirmed Ramón y Cajal's intuition. We now know

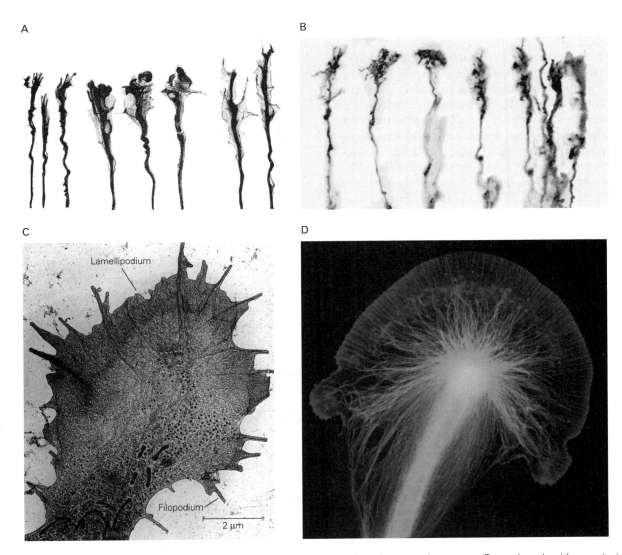

Figure 47–6 Neuronal growth cones.

A. Drawings of growth cones by Santiago Ramón y Cajal, who discovered these cellular structures and inferred their function.

B. Growth cones visualized in dye-labeled retinal ganglion neurons in the mouse. Note the similarities with Cajal's drawings. (Reproduced, with permission, from Carol Mason and Pierre Godement.)

C. The three main domains of the growth cone—filopodia, lamellipodia, and a central core—are shown by whole-mount

scanning electron microscopy. (Reproduced, with permission, from Bridgman and Dailey 1989. Permission conveyed through Copyright Clearance Center, Inc.)

D. The growth cone of a neuron from *Aplysia* in which actin and tubulin have been visualized. Actin (**purple**) is concentrated in lamellipodia and filopodia, whereas tubulin and microtubules (**aquamarine**) are concentrated in the central core. (Reproduced, with permission, from Paul Forscher and Dylan Burnette.)

that the growth cone is both a sensory structure that receives directional cues from the environment and a motor structure whose activity drives axon elongation. Ramón y Cajal also pondered "what mysterious forces precede the appearance of these processes . . . promote their growth and ramification . . . and finally establish those protoplasmic kisses . . . which seem to constitute the final ecstasy of an epic love story." In more modern

and prosaic terms, we now know that the growth cone guides the axon by transducing positive and negative cues into signals that regulate the cytoskeleton, thereby determining the course and rate of axonal growth toward its targets, where it will form synapses.

Growth cones have three main compartments. Their *central core* is rich in microtubules, mitochondria, and other organelles. Long slender extensions called *filopodia* project from the body of the growth cone. Between the filopodia lie *lamellipodia*, which are also motile and give the growth cone its characteristic ruffled appearance (Figure 47–6C,D).

Growth cones sense environmental signals through their filopodia: rod-like, actin-rich, membrane-limited structures that are highly motile. Their surface membranes bear receptors for the molecules that serve as directional cues for the axon. Their length—tens of micrometers in some cases—permits the filopodia to sample environments far in advance of the central core of the growth core. Their rapid movements permit them to compile a detailed inventory of the environment, and their flexibility permits them to navigate around cells and other obstacles.

When filopodia encounter signals in the environment, the growth cone is stimulated to advance, retract, or turn. Several motors power these orienting behaviors. One source of power is the movement of actin along myosin, an interaction similar to the one that powers the contraction of skeletal muscle fibers, although the actin and myosin of neurons are different from those in muscle. The assembly of actin monomers into polymeric filaments also contributes a propulsive force for filopodial extension. As the actin filaments are constantly depolymerized at the base of filopodia, the balance of polymerization and depolymerization enables the filopodia to move forward without becoming longer. Depolymerization slows during periods of growth cone advance, leading to greater net forward motion. The movement of membranes along the substrate provides yet another source of forward motion.

The contribution of each type of molecular motor to the advance of the growth cone is likely to vary from one situation to another. Nevertheless, the final step involves the flow of microtubules from the central core of the growth cone into the newly extended tip, thus moving the growth cone ahead and leaving in its wake a new segment of axon. New lamellipodia and filopodia form in the advancing growth cone and the cycle repeats (Figure 47–7).

Accurate pathfinding can occur only if the growth cone's motor action is linked to its sensory function. Therefore, it is crucial that the recognition proteins on the filopodia are signal-inducing receptors and not merely binding moieties that mediate adhesion. The binding of a ligand to its receptor affects growth in diverse ways. In some cases, it engages the cytoskeleton directly, through the intracellular domain of receptors (Figure 47–7). Integrin receptors couple to actin in growth cones when they bind molecules associated with the surface of adjoining cells or the extracellular matrix, thereby influencing motility.

Of equal if not greater importance is the ability of ligand binding to stimulate the formation, accumulation, and even breakdown of soluble intracellular molecules that function as second messengers. These second messengers affect the organization of the cytoskeleton, and in this way regulate the direction and rate of movement of the growth cone.

One important second messenger is calcium. The calcium concentration in growth cones is regulated by the activation of receptors on filopodia, and this affects the organization of the cytoskeleton, which in turn modulates motility. Growth cone motility is optimal within a narrow range of calcium concentrations, called a *set point*. Activation of filopodia on one side of the growth cone leads to a concentration gradient of calcium across the growth cone, providing a possible basis for changes in the direction of growth.

Other second messengers that link receptors and motor molecules include cyclic nucleotides, which modulate the activity of enzymes such as protein kinases, protein phosphatases, and rho-family guanosine triphosphatases (GTPases). In turn, these messengers and enzymes regulate the activity of proteins that regulate the polymerization and depolymerization of actin filaments, thereby promoting or inhibiting axonal extension.

The critical role of intracellular signals in growth cone motility and orientation can be demonstrated using embryonic neurons grown in culture. Application of growth factors to one side of a growth cone activates receptors locally and leads to extension and turning of the growth cone toward the source of the signal. In essence, the factor attracts the growth cone. Yet when cyclic adenosine monophosphate (cAMP) levels in the neuron are decreased, the same stimulus acts as a repellent and the growth cone turns away from the signal (Figure 47–8A). Other repulsive factors can become attractive when levels of the second messenger cyclic guanosine 3',5'-monophosphate (cGMP) are raised.

Recently, another mechanism for coupling guidance molecules to growth cone behavior has come to light. It was long thought that all neuronal protein synthesis occurs in the cell body, but we now know that growth cones (as well as some dendrites) contain the machinery for protein synthesis, including a subset of messenger RNAs. Initial evidence that these molecules play an important role came from experiments in which axons were severed from their parent cell body. The growth cones continued to advance for a few hours; they could be stimulated to turn toward or away from local depots of guidance molecules, and these behaviors were abolished by inhibitors of protein

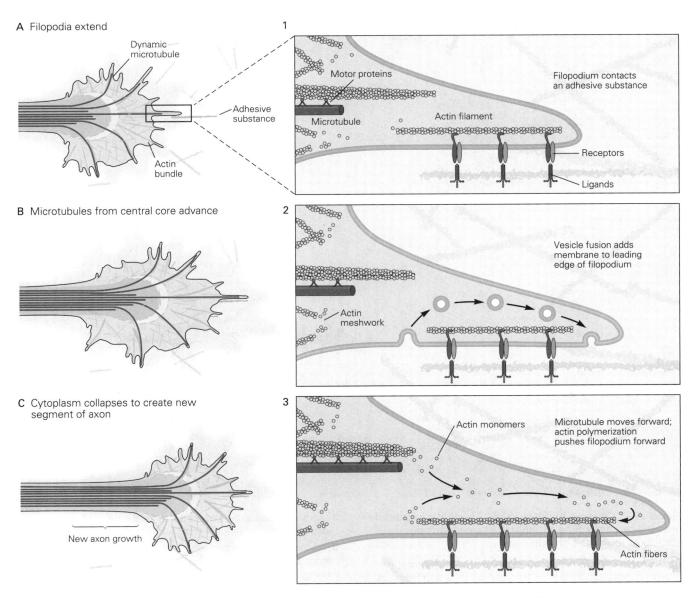

Figure 47–7 The growth cone advances under the control of cellular motors. (Adapted, with permission, from Heidemann 1996. Copyright © 1996 Academic Press Inc.)

A. A filopodium contacts an adhesive cue and contracts, thus pulling the growth cone forward (**1**). Actin filaments assemble at the leading edge of a filopodium and disassemble at the trailing edge, interacting with myosin along the way (**2**). Actin polymerization pushes the filopodium forward (**3**). Force generated by the retrograde flow of actin pushes the filopodium forward. Exocytosis adds membrane to the leading edge of the

filopodium and supplies new adhesion receptors to maintain traction. Membrane is recovered at the back of the filopodium. The actin polymer is linked to adhesion molecules on the plasma membrane.

B. The combined action of these motors creates an actin-depleted space that is filled by the advance of microtubules from the central core.

C. Individual microtubules condense to form a thick bundle, and the cytoplasm collapses around them to create a new segment of axonal shaft.

synthesis. The local protein synthesis is regulated by second messengers produced in response to activation of guidance receptors on the growth cones (Figure 47–8). This mechanism leads to synthesis of new motor

proteins precisely when and where they are needed. Thus, the growth cone has many strategies and mechanisms for integrating molecular signals to direct the axon in specific directions.

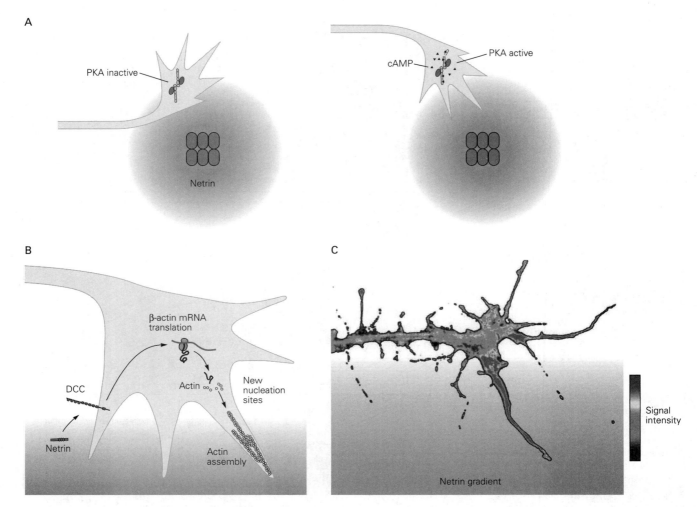

Figure 47–8 Changes in the level of intracellular regulatory proteins can determine whether the same extrinsic cue attracts or repels the growth cone.

A. The state of protein kinase A (**PKA**) activity can alter the growth cone's response to an extracellular orienting factor, in this instance, the protein netrin. When PKA activity and intracellular cyclic adenosine monophosphate (**cAMP**) levels are low, the growth cone is repelled by netrin. When PKA activity is high, the resulting elevation in intracellular cAMP causes

the growth cone to be attracted to a local source of netrin. (Adapted, with permission, from Ming et al. 1997.)

B. Netrin activation of growth cone receptors (deleted in colon cancer, **DCC**) leads to local synthesis of actin, which leads to turning.

C. Immunohistochemical analysis of a growth cone showing local synthesis of actin in response to local application of netrin. (Reproduced, with permission, from Christine Holt. Adapted, with permission, from Leung et al. 2006.)

Molecular Cues Guide Axons to Their Targets

For much of the 20th century, a debate raged between advocates of two very different views of how growth cones navigate embryonic terrains to reach their targets. A molecular view of axonal guidance was first articulated at the turn of the 20th century by the physiologist J. N. Langley. But by the 1930s, many eminent biologists, including Paul Weiss, believed that axonal outgrowth was essentially random and that appropriate connections persisted largely because of

productive, matching patterns of electrical activity in the axon and its target cell.

In our molecular age, Weiss's ideas may seem simplistic, but they were not unreasonable at the time. In tissue culture, axons grow preferentially along mechanical discontinuities (scratches and bumps on a cover slip), and embryonic nerve trunks often align themselves with solid supports (blood vessels or cartilage). It seemed logical to Weiss that mechanical guidance, called *stereotropism*, could account for axonal patterning. Today, we are quite comfortable with the idea that electrical signals can be used to change the way current flows in a computer without the need to resolder connections. Likewise, patterns of activity and experience can strengthen or weaken neural connections without requiring the formation of new axonal pathways. Then why not consider that congruent activity, called *resonance* by Weiss, is involved in establishing appropriate connections?

Today, few scientists believe that stereotaxis or resonance is a crucial force in initial patterning of neuronal circuits. The tipping point that shifted opinion in favor of the molecular view was an experiment performed with frogs and other amphibia in the 1940s by Roger Sperry (ironically, a student of Weiss). Sperry manipulated the information carried from the eye to the brain by the axons of retinal ganglion cells. These axons terminate in their target areas—the lateral geniculate body in the thalamus and the superior colliculus (called optic tectum in lower vertebrates) in the midbrain—in such a way that an orderly retinotopic map of the visual field is created.

Because of the optics of the eye, the visual image on the retina is an inversion of the visual field. The retinal ganglion cells reinvert the image by the pattern in which their axons terminate in the optic tectum, the main visual center in the brain of frogs (Figure 47–9A). If the optic nerve is cut, the animal is blinded. In lower vertebrates, cut retinal axons can reestablish projections to the tectum, whereupon vision is restored. This is not the case in mammals, as we will discuss in Chapter 50.

Sperry's key experiment was to sever the optic nerve in a frog and then rotate the eye in its socket by 180° before regeneration of the nerve. Remarkably, the frog exhibited orderly responses to visual input, but the behavior was wrong. When the frog was presented with a fly on the ground, it jumped up, and when offered a fly above its head, it struck downward (Figure 47–9B). Importantly, the animal never learned to correct its mistakes. Sperry suggested—and later verified with anatomical and physiological methods—that the retinal axons had reinnervated their original tectal targets, even though these connections provided the brain with erroneous spatial information that led to aberrant behavior. The inference of these experiments was that recognition between axons and their targets relied on molecular matching rather than functional validation and refinement of random connections.

But Weiss's ideas are by no means obsolete. Indeed, we now recognize that the activity of neural circuits can play a crucial role in shaping connectivity. The current view is that molecular matching predominates during embryonic development and that activity and experience modify circuits after they have formed. In this chapter and the next, we describe the molecular cues that guide the formation of neural connections, and then in Chapter 49, we examine the role of activity and experience in the fine-tuning of synaptic connections.

Sperry's conjecture, often called the *chemospecificity hypothesis*, prompted developmental neurobiologists to search for axonal and synaptic "recognition molecules." Success was limited for the first few decades, in part because these molecules are present in small amounts and on discrete subsets of neurons and there were no effective methods for isolating rare molecules from complex tissues. Eventually advances in biochemical and molecular-biological methods made this task more feasible, and many proteins involved in the guidance of axons to their targets have now been discovered. These proteins typically consist of paired ligands and receptors: The ligands are presented by cells along the pathway an axon follows and the receptors by the growth cone itself.

In the most general terms, guidance cues can be presented on cell surfaces, in the extracellular matrix, or in soluble form. As described above (Figure 47–8), they interact with receptors embedded in the growth cone membrane to promote or inhibit outgrowth of the axon. Most receptors have an extracellular domain that selectively binds the cognate ligand and an intracellular domain that couples to the cytoskeleton, either directly or through intermediates such as second messengers. The ligands can speed or slow growth. Ligands presented to one side of the growth cone can result in local activation or inhibition, leading to turning. In this way, the local distribution of environmental cues determines the pathway of the advancing growth cone.

As a result of these recent discoveries, axon guidance—a process that appeared mysterious years ago—can now be viewed as the orderly consequence of protein–protein interactions that instruct the growth cone to grow, turn, branch, or stop (Figures 47–10, 11). This limited set of instructions is sufficient, when presented with spatial precision, to choreograph growth cone behaviors with exquisite subtlety. Axonal guidance can therefore be explained by describing how and where ligands are presented and how the growth cone integrates this information to generate an orderly response. In the rest of the chapter, we illustrate lessons learned by describing the journeys of two types of axons: those of retinal ganglion neurons and those of a particular class of sensory relay neurons in the spinal cord.

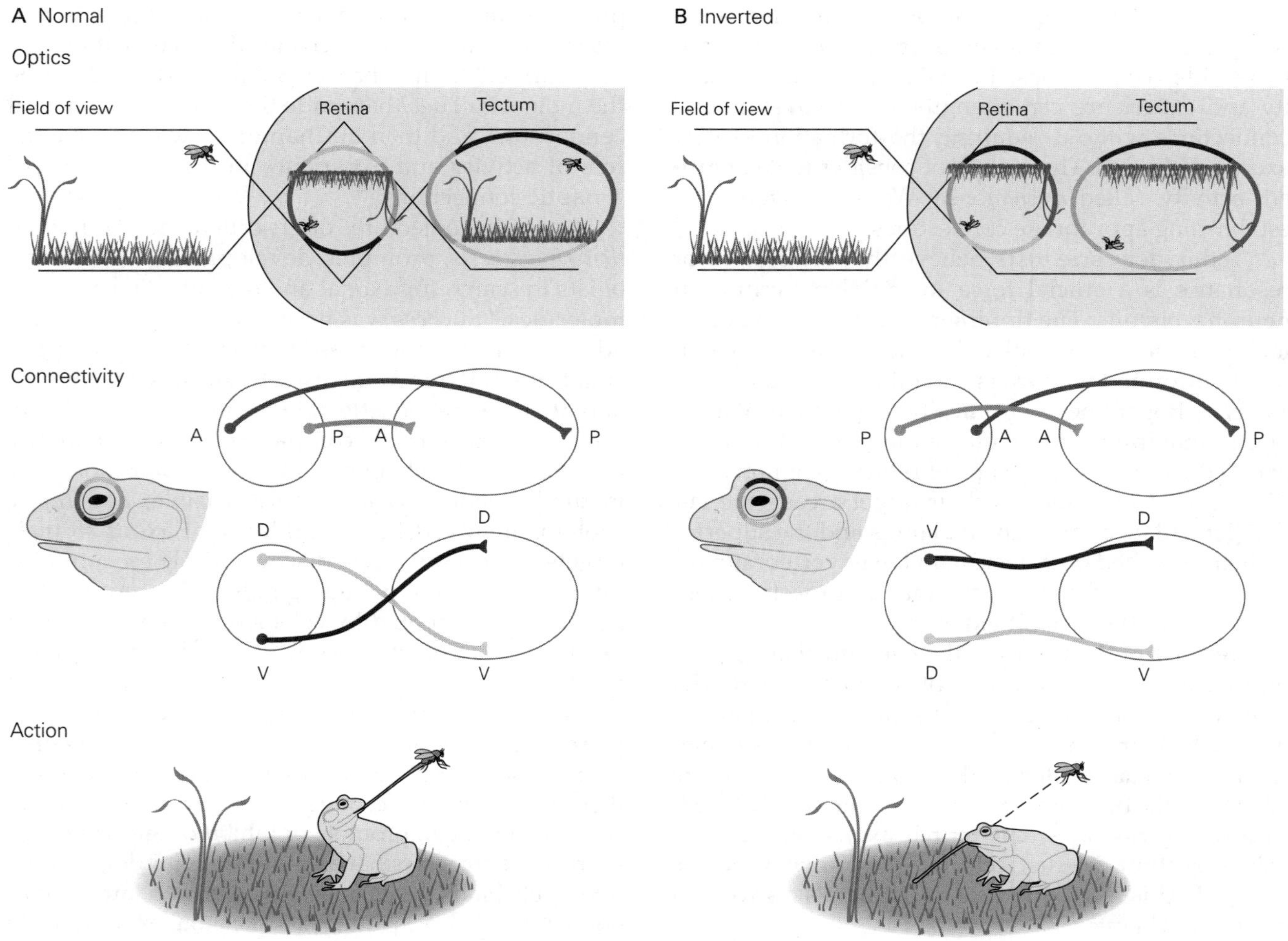

Figure 47–9 Roger Sperry's classical experiments on regeneration in the visual system provided evidence for chemoaffinity in the wiring of connections.

A. In the visual system of the frog, the lens projects an inverted visual image onto the retina and the optic nerve then transfers the image, with an additional inversion, to the optic tectum. The spatial arrangement of retinal inputs to the tectum allows for this transfer. Neurons in the anterior retina project axons to the posterior tectum, while neurons in the posterior retina project to the anterior tectum. Similarly, neurons in the dorsal retina project to the ventral tectum, and neurons in the ventral retina project to the dorsal tectum. As a result, visually guided behaviors (here catching a fly) are accurate. (Abbreviations: **A,** anterior; **D,** dorsal; **P,** posterior; **V,** ventral.)

B. If the optic nerve is cut and the eye is surgically rotated in its socket before the nerve regenerates, visually guided behavior is aberrant. When a fly is presented overhead, the frog perceives it as below, and vice versa. The inversion of behavioral reflexes results from the connection of regenerating retinal axons to their original targets, even though these connections now transfer an inverted, inappropriate map of the world into the brain.

The Growth of Retinal Ganglion Axons Is Oriented in a Series of Discrete Steps

Sperry's experiment implied the existence of axon guidance cues but did not reveal where they were or how they worked. For a time, one prominent view was that recognition occurred mostly at or near the target and that mechanical forces or long-range chemotactic factors sufficed to get axons to the vicinity of the target.

We now know that axons reach distant targets in a series of discrete steps, making frequent decisions at closely spaced intervals along their route. To illustrate this point, we shall trace in greater detail the path that Sperry was trying to understand, that of a retinal axon growing to the optic tectum.

1 Extracellular matrix adhesion

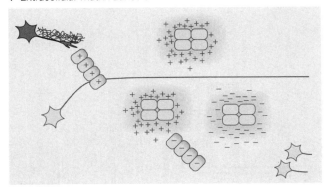

2 Cell surface adhesion

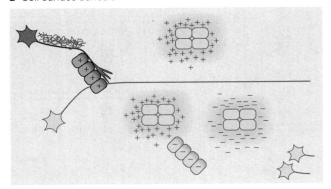

3 Fasciculation

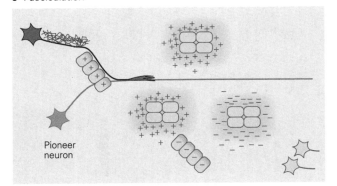

4 Chemoattraction

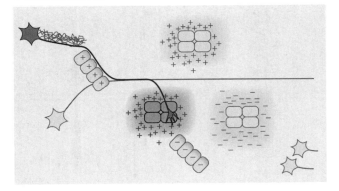

5 Contact inhibition

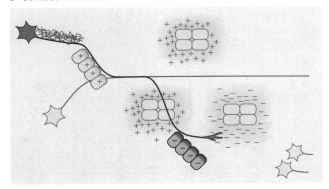

6 Chemorepulsion

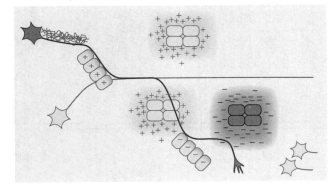

7 Collateral branching

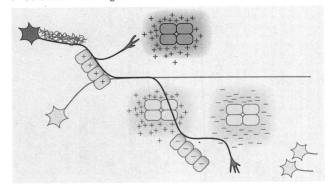

8 Terminal branching

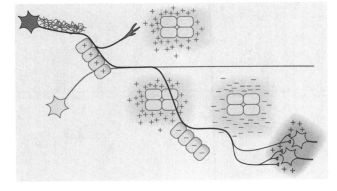

Figure 47–10 Extracellular cues use a variety of mechanisms to guide growth cones. The axon can interact with growth-promoting molecules in the extracellular matrix (**1**). It can interact with adhesive cell-surface molecules on neural cells (**2**). The growing axon can encounter another axon from a "pioneer" neuron and track along it, a process termed *fasciculation* (**3**). Soluble chemical signals can attract the growing axon to its cellular source (**4**). Intermediate target cells that express cell-surface repellent cues can cause the axon to turn away (**5**). Soluble chemical signals can repel the growing axon (**6**). Extracellular signals also lead to formation of collaterals from axon shafts (**7**) or branching of the growing axon (**8**).

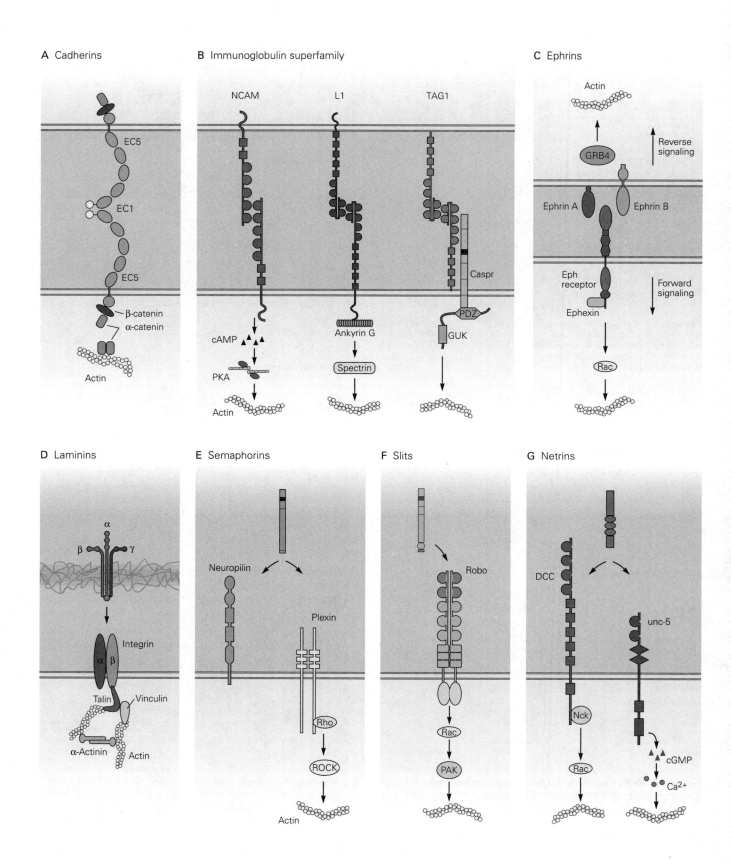

Growth Cones Diverge at the Optic Chiasm

The first task of the axon of a retinal ganglion cell is to leave the retina. As it enters the optic fiber layer, it extends along the basal lamina and glial end-feet at the retina's edge. The growth of the axon is oriented from the outset, indicating that it can read directional cues in the environment. As it approaches the center of the retina, it comes under the influence of attractants emanating from the optic nerve head (the junction of the optic nerve with the retina proper), which guide it into the optic stalk. It then follows the optic nerve toward the brain (Figure 47–12).

The first axons to travel this route follow the cells of the optic stalk, the rudiment of the neural tube that connects the retina to the diencephalon from which it arose. These "pioneer" axons then serve as scaffolds for later-arriving axons, which are able to extend accurately simply by following their predecessors (see "fasciculation" in Figure 47–10). Once they reach the optic chiasm, however, the retinal axons must make a choice. Axons that arise from neurons in the nasal hemiretina of each eye cross the chiasm and proceed to the opposite side of the brain, whereas those from the temporal half are deflected as they reach the chiasm and so stay on the same side of the brain (Figure 47–13A).

This divergence in trajectory reflects the differential responses of axons from the nasal and temporal hemiretinas to guidance cues presented by midline chiasm cells. Some retinal axons contact and traverse chiasm cells, whereas others are inhibited by these cells and deflected away, thus remaining on the ipsilateral side. One of the key molecules presented by chiasm cells is a membrane-bound repellent of the ephrin-B family (Figure 47–13B), which also figures in later steps of retinal ganglion cell axon guidance.

The fraction of temporal retinal axons that project ipsilaterally varies among species: few in lower vertebrates, some in rodents, and many in humans. These differences reflect placement of the eyes. In many animals, the eyes point to the sides and monitor different parts of the visual world, so that information from the two eyes need not be combined. In humans, both eyes look forward and sample largely overlapping regions of the visual world, so coordination of visual input is essential.

After crossing the optic chiasm, retinal axons assemble in the optic tract along the ventral surface of the diencephalon. Axons then leave the tract at different points. In most vertebrate species, the tectum of the midbrain (called the superior colliculus in mammals) is the major target of retinal axons, but a small number of axons project to the lateral geniculate nucleus of the thalamus. In humans, however, most axons project to the lateral geniculate, a sizable number reach the

Figure 47–11 (Opposite) Diverse molecular families control the growth and guidance of developing axons.

A. A large family of classical cadherins promote cell and axonal adhesion, primarily through homophilic interactions between cadherin molecules on adjacent neurons. Adhesive interactions are mediated through interactions of the extracellular EC1 domains. Cadherins transduce adhesive interactions though their cytoplasmic interactions with catenins, which link cadherins to the actin cytoskeleton.

B. A diverse array of immunoglobulin superfamily proteins are expressed in the nervous system and mediate adhesive interactions. The three examples shown here, NCAM, L1, and TAG1, can bind both homophilically and heterophilically to promote axon outgrowth and adhesion. These proteins contain both immunoglobulin domains (**circles**) and fibronectin type III domains (**squares**). Homophilic interactions typically involve amino terminal immunoglobulin domains. Different immunoglobulin adhesion molecules interact with the cytoskeleton via diverse cytoplasmic mediators, only a few of which are shown here.

C. Different ephrin proteins bind to Eph class tyrosine kinase receptors. Class A ephrins are linked to the surface membrane through a glycosyl phosphatidylinositol tether, whereas class B ephrins are transmembrane proteins. Class A ephrins typically bind class A Eph kinases, and class B ephrins typically bind

class B Eph kinases. Forward Eph signaling usually elicits repellant or inhibitory responses in receptive cells, whereas reverse ephrin signaling can elicit adhesive or inhibitory responses. Ephrin-Eph signaling involves many different cytoplasmic mediators.

D. Laminin proteins are components of the extracellular matrix and promote cell adhesion and axon extension through interactions with integrin receptors. Integrins mediate adhesion and axon growth through interactions with the cytoskeleton via many intermediary proteins.

E. Semaphorin proteins can promote or inhibit axonal growth through interaction with a diverse array of plexin and neuropilin receptors, which transduce signals via Rho class GTPases and downstream kinases.

F. Slit proteins typically mediate repellant responses through interaction with Robo class receptors, which influence axonal growth via intermediary GTPases such as Rac.

G. The secreted or extracellular matrix–associated netrin proteins mediate both chemoattractant and chemorepellent responses. Attractant responses are mediated through interaction with **DCC** (deleted in colorectal cancer) receptors, whereas repellent responses involve interactions with DCC and unc-5 coreceptors. Netrin receptors signal via GTPases and cyclic guanosine monophosphate (cGMP) cascades.

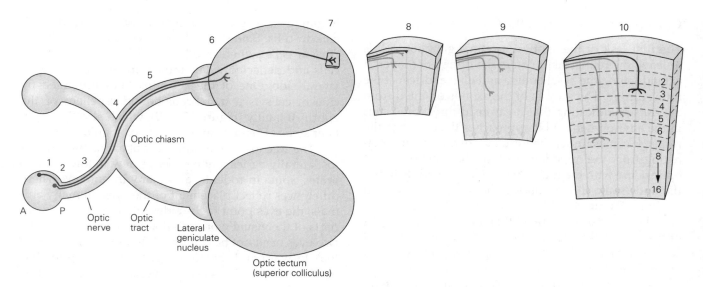

Figure 47–12 The axons of retinal ganglion cells grow to the optic tectum in discrete steps. Two neurons that carry information from the nasal half of the retina are shown. The axon of one crosses the optic chiasm to reach the contralateral optic tectum. The axon of the other also crosses the optic chiasm but projects to the lateral geniculate nucleus. The numbers indicate important landmarks on the axon's journey. The growing axon is directed toward the optic nerve head (the junction of the nerve with the retina) (1), enters into the optic nerve

(2), extends through the optic nerve (3), swerves to remain ipsilateral (not shown) or crosses to the contralateral side at the optic chiasm (4), extends through the optic tract (5), enters into the optic tectum or lateral geniculate nucleus (not shown) (6), navigates to an appropriate rostrocaudal and dorsoventral position on the tectum (7), turns to enter the neuropil (descends in chicks as shown here; ascends in mammals) (8), stops at an appropriate layer where a rudimentary terminal arbor is formed (9), and finally is remodeled (10). (Abbreviations: **A**, anterior; **P**, posterior.)

colliculus, and small numbers project to the pulvinar, superchiasmatic nucleus, and pretectal nuclei. Within these targets, different retinal axons project to different regions. As Sperry showed, the retinal axons form a precise retinotopic map on the tectal surface. Similar maps form in other areas innervated by retinal axons such as the lateral geniculate nucleus.

Having reached an appropriate position within the tectum, retinal axons need to find an appropriate synaptic partner. To achieve this last leg of their journey, retinal axons turn and dive into the tectal neuropil (Figure 47–12), descending (or, in mammals, ascending) along the surface of radial glial cells, which provide a scaffold for radial axonal growth. Although radial glial cells span the entire extent of the neuroepithelium, each retinal axon confines its synaptic terminals to a single layer. The dendrites of many postsynaptic cells extend through multiple layers and form synapses along their entire length, but retinal inputs are restricted to a small fraction of the target neuron's dendritic tree. These organizational features imply that layer-specific cues arrest axonal elongation and trigger arborization.

The problem of long-distance axon navigation is therefore solved by dividing the journey into short

segments in which intermediate targets guide the axons along the path to their final targets. Some intermediate targets, such as the optic chiasm, are "decision" regions where axons diverge.

Reliance on intermediate targets is an effective solution to the problem of long-distance axonal navigation but is not the only one. In some cases, the first axons reach their targets when the embryo is small and the distance to be covered is short. These "pioneer" axons respond to molecular cues embedded in cells or the extracellular matrix along their way. The first axons to exit the retina fall within this class. Axons that appear later, when distances are longer and obstacles more numerous, can reach their targets by following the pioneers. Yet another guidance mechanism is a molecular gradient. Indeed, as we will see, gradients of cell-surface molecules in the tectum inform axons about their proper termination zone.

Gradients of Ephrins Provide Inhibitory Signals in the Brain

So far, we have seen how retinal axons reach the tectum by responding to a series of discrete directional cues. However, these choices during growth do not account

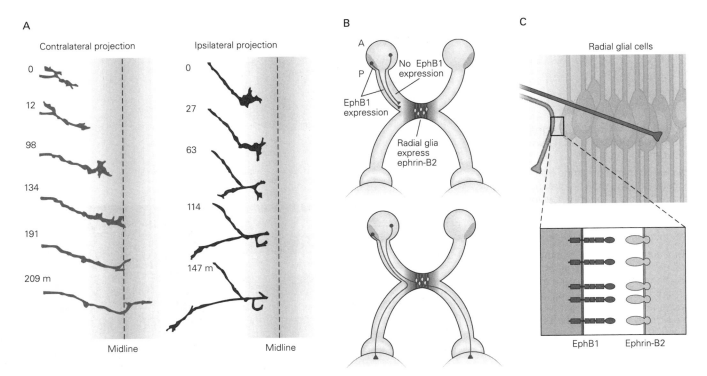

Figure 47–13 Axons of retinal ganglion neurons diverge as they reach the optic chiasm.

A. A time lapse series shows axons approaching the midline. Axons that arise from the nasal hemiretina cross the optic chiasm and project to the contralateral tectum (*left*). In contrast, axons from the temporal hemiretina reach the chiasm but fail to cross and thus project toward the ipsilateral tectum (*right*). (Reproduced, with permission, from Godement, Wang, and Mason 1994.)

B. The axons of neurons from the temporal hemiretina, which express the tyrosine kinase receptor EphB1, encounter ephrin-B2 expressed by midline radial glial cells at the optic chiasm and so are prevented from crossing the midline. The axons of nasal hemiretina neurons, which lack EphB1 receptors, are unaffected by the presence of ephrin-B2 and cross to the contralateral side. (Abbreviations: **A**, anterior; **P**, posterior.)

C. Higher-power view illustrating the trajectories of retinal ganglion cell axons at the chiasm.

for the smoothly graded connections implied by Sperry's analysis of the retinotopic map in the tectum. The quest for the hypothetical "map molecules" became a major focus for developmental neurobiologists, and so we describe it in some detail.

A key breakthrough in the quest for these molecules came with the development of bioassays in which explants from defined portions of the retina were laid on substrates of tectal membrane fragments. The membrane fragments were taken from defined anteroposterior portions of the tectum and arranged in alternating stripes. Axons from the temporal (posterior) hemiretina were found to grow preferentially on membranes from anterior tectum, a preference similar to that exhibited in vivo (Figure 47–14). This preference was found to result from the presence of inhibitory factors in posterior membranes rather than from attractive or adhesive substances in anterior membranes. This observation

was one of the first to demonstrate the role of inhibitory or repellent substances in axon guidance.

This stripe assay permitted the characterization of an inhibitory cue, present in membranes from the posterior but not the anterior tectum. Independently, molecular biologists identified a family of receptor tyrosine kinases, the Eph kinases, and a large family of membrane-associated ligands, the ephrins. Both receptors and ligands are divided into A and B subfamilies. The ephrin-A proteins bind and activate EphA kinases; conversely, ephrin-B proteins bind and activate EphB kinases (Figure 47–11C).

The two lines of research converged when the tectal inhibitory cue was identified as ephrin-A5. We now know that the Eph kinases and ephrins serve many functions in neural and nonneural tissues and that each class of proteins can serve as ligands or receptors, depending on cellular context. In the developing

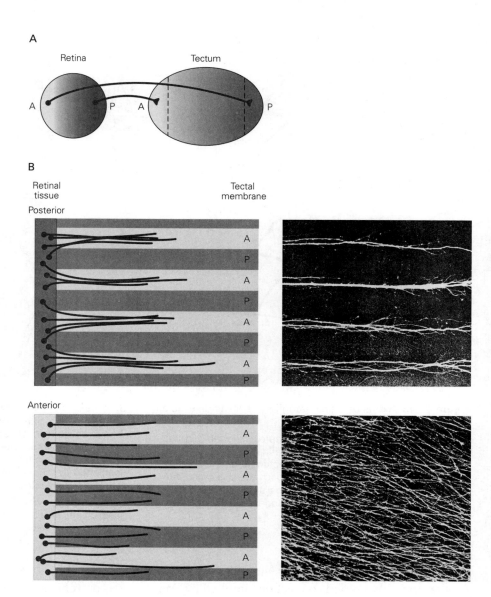

Figure 47–14 Repellent signals guide developing retinal axons in vitro.

A. Retinal ganglion axons from the posterior (temporal) hemiretina project into the anterior developing tectum. Conversely, axons from the anterior (nasal) hemiretina project into the posterior tectum.

B. Fragments of membrane were taken from specified anteroposterior portions of the tectum and arranged in alternating strips. Axons from explants of posterior retina grow selectively on the fragments from anterior tectum. The preferential growth of axons on anterior membrane results from an inhibitory cue in the posterior membrane. In contrast, axons from anterior retina grow on both anterior and posterior tectal membrane fragments. (Abbreviations: **A**, anterior; **P**, posterior.) (Adapted, with permission, from Walter, Henke-Fahle, and Bonhoeffer 1987.)

nervous system, these proteins comprise a major group of repellent signals.

Ephrin–Eph interactions account in large part for formation of the retinotopic map in the tectum. Levels of ephrin-A2 and ephrin-A5 in the tectum as well as levels of the Eph receptors in the retina are graded along the anteroposterior axis. These gradients run in the same direction. Ephrin-A concentrations run from posterior-high to anterior-low in tectum, while Eph A concentrations run from posterior-high to anterior-low in retina (Figure 47–15A). Such counter-gradients account, at least in part, for topographic mapping. Axons from posterior retinal ganglion cells with high levels of EphA receptors are repelled most strongly by the high level of ephrin-A in the posterior tectum and thus are confined

to the anterior tectum. The less sensitive axons from the anterior retina are able to penetrate further into the posterior domain of the tectum. Ephrin-A2 and ephrin-A5 are therefore strong candidates for chemospecificity factors of the type postulated by Sperry.

The crucial role of the interaction of ephrins and Eph kinases in the formation of retinotopic maps has been confirmed in vivo. Overexpression of ephrin-A2 in the developing optic tectum of chick embryos generates small patches of cells in the rostral tectum that are abnormally rich in ephrin-A2. Temporal retinal axons, which normally avoid the ephrin-rich caudal tectum, also avoid these patches in the rostral tectum, and they terminate in abnormal positions. In contrast, nasal retinal axons, which normally grow toward the

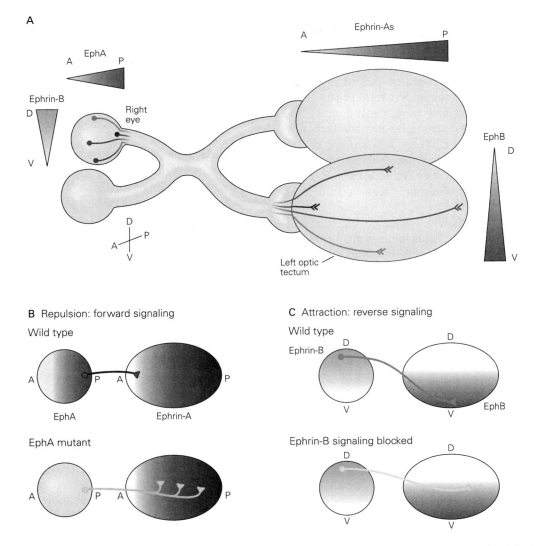

Figure 47–15 The formation of retinotopic maps in vivo depends on ephrin-Eph kinase signaling.

A. In the retina, EphA receptors are expressed in an anteroposterior (A-P) gradient, and ephrin-B is expressed in a dorsoventral (D-V) gradient. In the tectum, ephrin-A receptors are distributed in an anteroposterior gradient and EphB in a dorsoventral gradient.

B. Expression of EphA in retinal axons that derive from neurons in the posterior (temporal) retina directs axon growth to

the anterior tectum through avoidance of ephrin-A proteins. In EphA mutant mice, posterior retinal axons are able to project to a more posterior domain within the tectum.

C. EphB signaling directs the projection of dorsal retinal axons to the ventral tectum. Blocking ephrin-B signaling with soluble EphB protein causes dorsal axons to project to an abnormally dorsal domain within the tectum.

caudal tectum, are not perturbed by encounters with excess ephrin-A.

Conversely, in mice with targeted mutations in the relevant *ephA* or *ephrin-A* genes, some posterior retinal axons terminate in inappropriately posterior tectal regions (Figure 47–15B). Anterior retinal axons, which naturally express low levels of EphA proteins,

project normally in these mutants. In mice lacking both ephrin-A proteins, these deficits are more severe than with either single mutant. Thus, the interaction of ephrin-A with EphA receptors is crucial for the targeting of retinal axons in the tectum. These ephrin/EphA pairs possess the properties of the recognition molecules that Sperry predicted were necessary to direct topographic mapping along the anteroposterior axis of the tectum.

Of course, the retinal map also has a dorsoventral axis. Ephrin/EphB pairs are involved in establishing order along this axis. Just as ephrin-A and EphA are graded along the anteroposterior axis, ephrin-B and EphB are graded along the dorsoventral axis, and manipulation of ephrin-B and EphB levels affects dorsoventral mapping (Figure 47–15C). Thus, at a simple level, the retinotopic map is arranged in rectangular coordinates with ephrin-A/EphA and ephrin-B/EphB labeling the anteroposterior and dorsoventral axes, respectively.

Although this simple view is satisfying, the reality is more complex. First, EphB kinases are expressed in the tectum as well as in the retina, and ephrins-A are expressed in the retina as well as in the tectum. Thus, so-called "cis" interactions (Eph and ephrin on the same cell) as well as "trans" interactions (Eph on growth cone, ephrin on target cell) may be involved. Second, both ligands and receptors are present at multiple points along the optic pathway and play multiple roles. As we have seen, ephrin-B/EphB interactions affect not only dorsoventral mapping but also the decision of an axon to cross to the contralateral side at the optic chiasm. Finally, in developing visual circuits, more precise spatial mapping of retinal inputs is regulated by patterns of neural activity, as discussed in the next two chapters. Nonetheless, we now have the outline of a molecular strategy for the initial formation of topographic projections from the eye to the brain.

Axons From Some Spinal Neurons Are Guided Across the Midline

One of the fundamental features of the central nervous system is the need to coordinate activity on both sides of the body. To accomplish this task, certain axons need to project to the opposite side.

We have seen one example of axonal crossing in the optic chiasm. Another example that has been studied in detail is the axonal crossing of *commissural neurons* that convey sensory information from the spinal cord to the brain at the ventral midline of the spinal cord across the floor plate. After crossing, axons turn abruptly and grow up toward the brain. This simple trajectory raises several questions. How do these axons reach the ventral midline? How do they cross the midline, and after crossing, how do they *ignore* cues that axons on the other side are using to get to the midline? In other words, why do they turn toward the brain instead of crossing back?

Netrins Direct Developing Commissural Axons Across the Midline

Many of the neurons that send axons across the ventral midline are generated in the dorsal half of the spinal cord. The first task for these axons is to reach the ventral midline. Ramón y Cajal considered the possibility that chemotactic factors emitted by targets could attract axons, but this idea lay dormant for nearly a century. We now know that such factors do exist, and one of them, the protein netrin-1, is expressed by cells of the floor plate as well as by progenitors along the ventral midline. When presented in culture, netrin attracts commissural axons; when mice are deprived of netrin-1 function, axons fail to reach the floor plate (Figure 47–16). It may act as both a secreted factor (chemotaxis) and a membrane guidance molecule (haptotaxis) to guide the axons of commissural neurons to the floor plate.

The netrin protein is structurally related to the protein product of *unc-6*, a gene shown to regulate axon guidance in the nematode *Clostridia elegans*. Two other *C. elegans* genes, *unc-5* and *unc-40*, encode receptors for the unc-6 protein. Vertebrate netrin receptors are related to the unc-5 and unc-40 receptors. The unc-5H proteins are homologs of unc-5, and DCC (deleted in colorectal cancer) are related to unc-40 (see Figure 47–11G). These receptors are members of the immunoglobulin superfamily, and their functions have been remarkably conserved throughout animal evolution (Figure 47–17). This conservation supports the use of simple and genetically accessible invertebrates to unravel developmental complexities. In no area has this approach been more fruitful than in the analysis of axon guidance. Dozens of genes that affect this process were first identified and cloned in *Drosophila* and *C. elegans* and then shown to play important and related roles in mammals.

Chemoattractant and Chemorepellent Factors Pattern the Midline

Other signaling systems work with netrins to guide commissural axons. One group consists of bone morphogenetic proteins, which are secreted by the roof plate. They act as repellents, directing commissural axons ventrally as they begin their journey. Additional factors from the floor plate, such as the hedgehog proteins initially involved in patterning the spinal cord (Chapter 45), may collaborate with netrins at a later stage, serving as axonal attractants.

Once commissural axons reach the midline, they find themselves exposed to the highest available levels

A Wild type

B Netrin or DCC mutants

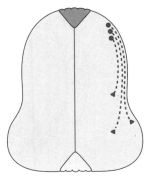

Figure 47–16 Netrin signaling attracts the axons of spinal commissural neurons to the floor plate. (Micrographs reproduced, with permission, from Marc Tessier-Lavigne.)

A. Netrin-1 is generated by floor plate cells and ventral neural progenitors. It attracts the axons of commissural neurons to the floor plate (**FP**) at the ventral midline of the spinal cord.

B. Most commissural axons fail to reach the floor plate when netrin or deleted in colorectal cancer (**DCC**) proteins are eliminated.

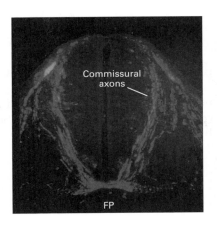

Netrin-1 mutant

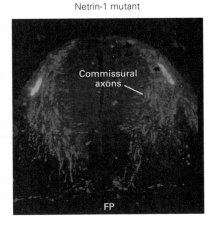

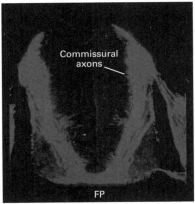

DCC mutant

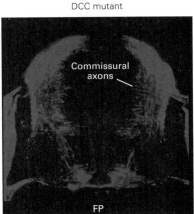

of netrin-1 and sonic hedgehog. Yet this netrin-rich environment does not keep the axons at the midline indefinitely. Instead they cross to the other side of the spinal cord, even while their contralateral counterparts are navigating up the netrin chemoattractant gradient.

This puzzling behavior is explained by the fact that growth cones change their responsiveness to attractive and repellent signals as a consequence of exposure to floor plate signals. This switch illustrates an important property of intermediate targets involved in axon guidance. Factors presented by intermediate targets not only guide the growth of axons but also change the sensitivity of the growth cone, preparing it for the next leg of its journey.

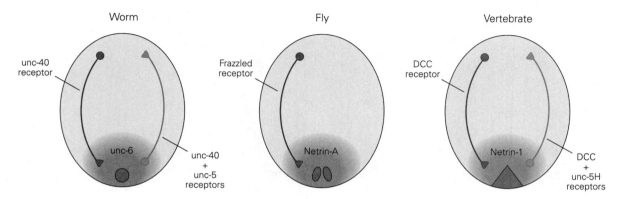

Figure 47–17 The expression and activity of netrins have been conserved throughout evolution. Netrins are secreted by ventral midline cells in worms, flies, and vertebrates and interact with receptors on cells or axons that migrate or extend along the dorsoventral axis. The netrin receptors unc-40 (worm), frazzled (fly), and deleted in colorectal cancer (**DCC**) (vertebrate) mediate netrin's attractant activity, whereas unc-5 class receptors mediate its repellent activity.

Once axons arrive at the floor plate, they become sensitive to Slit, a chemorepellent signal secreted by floor plate cells (Figure 47–18). Before commissural axons reach the floor plate, the Robo proteins that serve as Slit receptors are kept inactive by expression of a related protein, Rig-1. As axons reach the floor plate, levels of Rig-1 on their surface decline, unleashing Robo activity and causing axons to respond to the repellant actions of Slit. This repellent action propels growth cones *down* the Slit gradient into the contralateral side

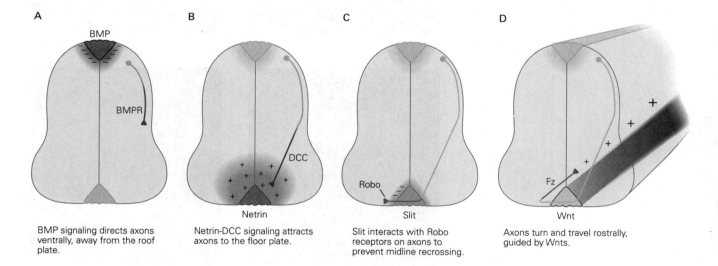

BMP signaling directs axons ventrally, away from the roof plate.

Netrin-DCC signaling attracts axons to the floor plate.

Slit interacts with Robo receptors on axons to prevent midline recrossing.

Axons turn and travel rostrally, guided by Wnts.

Figure 47–18 Guidance cues expressed by roof plate and floor plate cells guide commissural axons in the developing spinal cord.

A. Bone morphogenetic proteins (**BMP**) secreted by roof plate cells interact with BMP receptors (**BMPR**) on commissural axons to direct the axons away from the roof plate.

B. Netrin expressed by floor plate cells attracts deleted in colorectal cancer (**DCC**)-expressing commissural axons to the ventral midline of the spinal cord. Sonic hedgehog has also been implicated in the ventral guidance of commissural axons.

C. Slit proteins secreted by floor plate cells interact with Robo receptors on commissural axons to prevent these axons from recrossing the midline. Prior to crossing, but not after, commissural axons express robo3 (Rig-1) in addition to robo1 and robo2. The Rig-1 protein inactivates the Robo receptors, preventing the axons from responding to the repellent effects of Slits as they approach the ventral midline.

D. After commissural axons cross the midline, Wnt proteins secreted from floor plate cells and distributed in a rostrocaudal gradient interact with frizzled (**Fz**) proteins on the commissural axons, guiding the axons toward the brain.

of the spinal cord. In addition, activated Robo forms a complex with DCC, rendering these Netrin receptors incapable of responding to their ligand. The decreased sensitivity of growth cones to the attractive properties of the floor plate helps to account for the transient influence of floor plate signals on axons.

Finally, once axons have left the floor plate, they turn rostrally toward their eventual synaptic targets in the brain. A rostrocaudal gradient of Wnt proteins expressed by floor plate cells appears to direct axon growth rostrally at the ventral midline (Figure 47–18D). Thus, different cues guide commissural axons during distinct phases of their overall trajectory. This same process is presumably played out for hundreds and even thousands of classes of neurons to establish the mature pattern of brain wiring.

Highlights

1. As neurons extend processes, one generally becomes an axon and the others become dendrites. This process is called polarization. The two types of processes differ in structure and molecular architecture as well as function.

2. Cell types differ markedly in the shape, size, and branching patterns of their dendrites. Type-specific dendritic features arise both from intrinsic differences in transcriptional programs among types and from extrinsic influences on the developing dendrites.

3. Interactions among dendrites are critical for dendritic patterning. Repellent interactions among the dendrites of a single cell, a process called self-avoidance, leads to even coverage of an area, with minimal gaps or clumps. Repellent actions between dendrites of neighboring cells, a process called tiling, minimizes overlap of dendritic fields. In some cases, dendrites avoid other dendrites from their own neuron but interact with dendrites of nominally identical neighboring cells. This process is called self-/non–self-discrimination.

4. Growth cones at the tips of axons serve as both sensory and motor elements to guide axons to their destinations. Cytoskeletal elements of the growth cone, including actin and myosin, propel the growth.

5. Receptors on the growth cone recognize and bind ligands in the environment through which the axon is extending, guiding the growth. These interactions lead to generation of their second messengers that mediate growth, turning and stopping of the growth cone, and branching of the axon.

6. Some growth cones contain protein synthetic machinery including messenger RNAs. In these cases, receptors can promote local synthesis of specific proteins that mediate growth or turning.

7. Ligand–receptor pairs include several key families of molecules including cadherins, Slits and their Robo receptors, semaphorins and their plexin receptors, and ephrins and their Eph kinase receptors.

8. The growth of an axon to a distant target is broken into discrete shorter steps. At each step, molecules on the surface of or secreted by neighboring structures guide the axon. They can also lead to alterations in the growth cone's complement of receptors, allowing it to respond to different sets of cues at the subsequent stage.

9. Roger Sperry proposed a chemospecificity hypothesis to explain the specific growth of axons from different parts of the retina to different parts of the optic tectum (superior colliculus), forming an orderly retinotopic map. The ephrins and their receptors, the Eph kinases, are key molecules that guide map formation. They are graded in expression along the retina and tectum and act in large part by repelling axons from incorrect positions rather than attracting them to correct positions.

10. Both attractive and repellent molecules guide axons across midline structures, a process called decussation. Evolutionarily conserved signals include Slits, netrins, and Wnts. Mutations in genes that encode these ligands and receptors can result in developmental neurological disorders.

Joshua R. Sanes

Selected Reading

Bentley M, Banker G. 2016. The cellular mechanisms that maintain neuronal polarity. Nat Rev Neurosci 17:611–622.

Cang J, Feldheim DA. 2013. Developmental mechanisms of topographic map formation and alignment. Annu Rev Neurosci 36:51–77.

Dong X, Shen K, Bülow HE. 2015. Intrinsic and extrinsic mechanisms of dendritic morphogenesis. Annu Rev Physiol 77:271–300.

Herrera E, Erskine L, Morenilla-Palao C. 2017. Guidance of retinal axons in mammals. Semin Cell Dev Biol pii: S1084-9521.

Jung H, Gkogkas CG, Sonenberg N, Holt CE. 2014. Remote control of gene function by local translation. Cell 157:26–40.

Lai Wing Sun K, Correia JP, Kennedy TE. 2011. Netrins: versatile extracellular cues with diverse functions. Development 138:2153–2169.

Lefebvre JL, Sanes JR, Kay JN. 2015. Development of dendritic form and function. Annu Rev Cell Dev Biol 31:741–777.

Tojima T, Hines JH, Henley JR, Kamiguchi H. 2011. Second messengers and membrane trafficking direct and organize growth cone steering. Nat Rev Neurosci 12:191–203.

Zhang C, Kolodkin AL, Wong RO, James RE. 2017. Establishing wiring specificity in visual system circuits: from the retina to the brain. Annu Rev Neurosci 40:395–424.

Zipursky SL, Grueber WB. 2013. The molecular basis of self-avoidance. Annu Rev Neurosci 36:547–568.

References

Barnes AP, Lilley BN, Pan YA, et al. 2007. LKB1 and SAD kinases define a pathway required for the polarization of cortical neurons. Cell 129:549–563.

Bridgman PC, Dailey ME. 1989. The organization of myosin and actin in rapid frozen nerve growth cones. J Cell Biol 108:95–109.

Campbell DS, Holt CE. 2001. Chemotropic responses of retinal growth cones mediated by rapid local protein synthesis and degradation. Neuron 32:1013–1026.

Fazeli A, Dickinson SL, Hermiston ML, et al. 1997. Phenotype of mice lacking functional Deleted in colorectal cancer (Dcc) gene. Nature 386:796–804.

Forscher P, Smith SJ. 1988. Actions of cytochalasins on the organization of actin filaments and microtubules in a neuronal growth cone. J Cell Biol 107:1505–1516.

Frisen J, Yates PA, McLaughlin T, Friedman GC, O'Leary DD, Barbacid M. 1998. Ephrin-A5 (AL-1/RAGS) is essential for proper retinal axon guidance and topographic mapping in the mammalian visual system. Neuron 20:235–243.

Godement P, Wang LC, Mason CA. 1994. Retinal axon divergence in the optic chiasm: dynamics of growth cone behavior at the midline. J Neurosci 14:7024–7039.

Grueber WB, Jan LY, Jan YN. 2003. Different levels of the homeodomain protein cut regulate distinct dendrite branching patterns of Drosophila multidendritic neurons. Cell 112:805–818.

Harrison RG. 1959. The outgrowth of the nerve fiber as a mode of protoplasmic movement. J Exp Zool 142:5–73.

Heidemann SR. 1996. Cytoplasmic mechanisms of axonal and dendritic growth in neurons. Int Rev Cytol 165:235–296.

Kaech S, Banker G. 2006. Culturing hippocampal neurons. Nat Protoc 1:2406–2415.

Kalil K, Dent EW. 2014. Branch management: mechanisms of axon branching in the developing vertebrate CNS. Nat Rev Neurosci 15:7–18.

Kapfhammer JP, Grunewald BE, Raper JA. 1986. The selective inhibition of growth cone extension by specific neurites in culture. J Neurosci 6:2527–2534.

Keino-Masu K, Hinck L, Leonardo ED, Chan SS, Culotti JG, Tessier-Lavigne M. 1996. Deleted in colorectal cancer (DCC) encodes a netrin receptor. Cell 87:75–85.

Kidd T, Brose K, Mitchell KJ, et al. 1998. Roundabout controls axon crossing of the CNS midline and defines a novel subfamily of evolutionarily conserved guidance receptors. Cell 92:205–215.

Kishi M, Pan YA, Crump JG, Sanes JR. 2005. Mammalian SAD kinases are required for neuronal polarization. Science 307:929–932.

Lefebvre JL, Kostadinov D, Chen WV, Maniatis T, Sanes JR. 2012. Protocadherins mediate dendritic self-avoidance in the mammalian nervous system. Nature 488:517–521.

Letourneau PC. 1979. Cell-substratum adhesion of neurite growth cones, and its role in neurite elongation. Exp Cell Res 124:127–138.

Leung K-M, van Horck FPG, Lin AC, Allison R, Standart N, Holt CE. 2006. Asymmetrical beta-actin mRNA translation in growth cones mediates attractive turning to netrin-1. Nat Neurosci 9:1247–1256.

Ming GL, Song HJ, Berninger B, Holt CE, Tessier-Lavigne M, Poo MM. 1997. cAMP-dependent growth cone guidance by netrin-1. Neuron 19:1225–1235.

Polleux F, Morrow T, Ghosh A. 2000. Semaphorin 3A is a chemoattractant for cortical apical dendrites. Nature 404:567–573.

Serafini T, Colamarino SA, Leonardo ED, et al. 1996. Netrin-1 is required for commissural axon guidance in the developing vertebrate nervous system. Cell 87:1001–1014.

Shigeoka T, Jung H, Jung J, et al. 2016. Dynamic axonal translation in developing and mature visual circuits. Cell 166:181–192.

Sperry RW. 1943. Visuomotor coordination in the newt (Triturus viridescens) after regeneration of the optic nerve. J Compar Neurol 79:33–55.

Sperry RW. 1945. Restoration of vision after crossing of optic nerves and after contralateral transplantation of eye. J Neurophysiol 8:17–28.

Thu CA, Chen WV, Rubinstein R, et al. 2014. Single-cell identity generated by combinatorial homophilic interactions between α, β, and γ protocadherins. Cell 158:1045–1059.

Walter J, Henke-Fahle S, Bonhoeffer F. 1987. Avoidance of posterior tectal membranes by temporal retinal axons. Development 101:909–913.

Wang L, Marquardt T. 2013. What axons tell each other: axon-axon signaling in nerve and circuit assembly. Curr Opin Neurobiol 23:974–982.

Weiss P. 1941. Nerve patterns: the mechanics of nerve growth. Growth 5:163–203. Suppl.

Zhang XH, Poo MM. 2002. Localized synaptic potentiation by BDNF requires local protein synthesis in the developing axon. Neuron 36:675–688.

48

Formation and Elimination of Synapses

S O FAR, WE HAVE EXAMINED THREE STAGES in the development of the mammalian nervous system: the formation and patterning of the neural tube,

the generation and differentiation of neurons and glia, and the growth and guidance of axons. One additional step must occur before the brain becomes functional: the formation of synapses. Only when synapses are formed and functional can the brain go about the business of processing information.

Three key processes drive synapse formation. First, axons make choices among many potential postsynaptic partners. By forming synaptic connections only on particular target cells, neurons assemble functional circuits that can process information. In many cases, synapses are even formed at specific sites on the postsynaptic cell; some types of axons form synapses on dendrites, others on cell bodies, and yet others on axons or nerve terminals. Although cellular and subcellular specificity are evident throughout the brain, the general features of synapse formation can be illustrated with a few well-studied examples.

Second, after cell–cell contacts have formed, the portion of the axon that contacts the target cell differentiates into a presynaptic nerve terminal, and the domain of the target cell contacted by the axon differentiates into a specialized postsynaptic apparatus. Precise coordination of pre- and postsynaptic differentiation depends on interactions between the axon and its target cell. Much of what we know about these interactions comes from studies of the neuromuscular junction, the synapse between motor neurons and skeletal muscle fibers. The simplicity of this synapse made it a favorable system to probe the structural and electrophysiological principles of chemical synapses (Chapter 12), and this simplicity has also helped in the analysis of developing synapses. We will use the neuromuscular synapse to illustrate key features of synaptic development and then apply insights from this peripheral synapse to examine synapses that form in the brain.

Finally, once formed, synapses mature, often undergoing major rearrangements. One striking aspect of the rearrangement is that as some synapses grow and strengthen, many others are eliminated. Like neuronal cell death (Chapter 46), synapse elimination at first glance is a puzzling and seemingly wasteful step in neural development. It is increasingly clear, however, that it plays a key role in refining initial patterns of connectivity. We will discuss the main features of synaptic rearrangement at the neuromuscular junction, where it has been studied intensively, as well as at synapses between neurons, where it also is prominent.

Synapse formation stands at an interesting crossroads in the sequence of events that assemble the nervous system. The initial steps in this process appear to be largely "hardwired" by molecular programs. However, as soon as synapses form, the nervous system begins to function, and the activity of neural circuits plays a critical role in subsequent development. Indeed, the information-processing capacity of the nervous system is refined through its use, most dramatically in early postnatal life but also into adulthood. In this sense, the nervous system continues to develop throughout life. We will consider this interplay of molecular programs and neural activity as we describe synapse formation and rearrangement. This discussion will be a useful prelude to Chapter 49, in which we discuss how genes and the environment—nature and nurture—interact to customize nervous systems early in postnatal life.

Neurons Recognize Specific Synaptic Targets

Once axons reach their designated target areas, they must choose appropriate synaptic partners from the many potential targets within easy reach. Although synapse formation is a highly selective process at both cellular and subcellular levels, few of the molecules that confer synaptic specificity have been identified.

The specificity of synaptic connections is particularly evident when intertwined axons select subsets of target cells. In these cases, axon guidance and selective synapse formation can be distinguished. The first report of such specificity came more than 100 years ago when J. N. Langley, studying the autonomic nervous system, proposed the first version of a chemospecificity hypothesis (see Chapter 46). Langley observed that autonomic preganglionic neurons are generated at distinct rostrocaudal levels of the spinal cord. Their axons enter sympathetic ganglia together but form synapses

with different postsynaptic neurons that innervate distinct targets. Using behavioral assays as a guide, Langley inferred that the axons of preganglionic neurons located in the rostral spinal cord form synapses on ganglion neurons that project their axons to relatively rostral targets such as the eye, whereas neurons that derive from more caudal regions of the spinal cord synapse on ganglion neurons that project to caudal targets such as the ear (Figure 48–1A). He then showed that similar patterns were reestablished after the preganglionic axons were severed and allowed to regenerate, leading him to postulate that some sort of molecular recognition was responsible (Figure 48–1B).

Electrophysiological studies later confirmed Langley's intuition about the specificity of synaptic connections in these ganglia. Moreover, this selectivity is apparent from early stages of innervation, even though specific types of postsynaptic neurons are interspersed within the ganglion. The reestablishment of selectivity in adults after nerve damage shows that specificity does not emerge through peculiarities of embryonic timing or neuronal positioning.

Recognition Molecules Promote Selective Synapse Formation in the Visual System

To illustrate the idea of target specificity in more detail, we will first consider retinal ganglion cells. These neurons differ in their response properties—some ganglion neurons respond to increases in light level (ON cells), others to decreases (OFF cells), others to moving objects, and still others to light of a particular color. The axons of all ganglion cells run through the optic nerve, forming parallel axonal pathways from the retina to the brain.

The response properties of each class of ganglion cell depend on the synaptic inputs they receive from amacrine and bipolar interneurons, which in turn receive synapses from light-sensitive photoreceptors. All of the synapses from bipolar and amacrine cells onto ganglion cell dendrites occur in a narrow zone of the retina called the inner plexiform layer. Axons and dendrites therefore have the daunting task of recognizing their correct partners within a large crowd of inappropriate bystanders.

One important contributor to synaptic matchmaking in the inner plexiform layer is its division into sublayers. The processes of each amacrine and bipolar cell type, as well as the dendrites of each functionally distinct ganglion cell type, branch and synapse in just one or a few of approximately 10 sublayers. For example, the dendrites of ON and OFF cells are restricted to inner and outer portions of the plexiform layer,

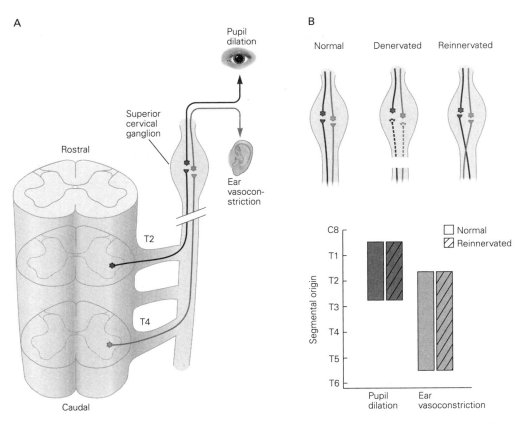

Figure 48–1 Preganglionic motor neurons regenerate selective connections with their sympathetic neuronal targets.

A. Preganglionic motor neurons arise from different levels of the thoracic spinal cord. Axons that arise from rostrally located thoracic neurons innervate superior cervical ganglion neurons that project to rostral targets, including the intrinsic eye muscles. Axons that arise from neurons at caudal levels of the thoracic spinal cord innervate ganglion neurons that project to more

caudal targets, such as the blood vessels of the ear. These two classes of ganglion neurons are intermingled in the ganglion, which suggested to J. N. Langley that preganglionic axons from different thoracic levels selectively form synapses with ganglion neurons that terminate in specific peripheral targets.

B. After nerve damage in adults, similar segment-specific patterns of connectivity form during reinnervation, supporting the notion that synapse formation is selective. (Adapted from Njå and Purves 1977.)

respectively, and therefore receive synapses from different interneurons; particular types of ON and OFF cells have narrower restrictions within these zones (Figure 48–2). This layer-specific arborization of pre- and postsynaptic processes restricts the choice of synaptic partners to which they have ready access. Similar lamina-specific connections are found in many other regions of the brain and spinal cord. For example, in the cerebral cortex, distinct populations of axons confine their dendritic arbors and synapses to just one or two of the six main layers.

Laminar specificity does not, however, completely account for the wiring of the retina. As the number of retinal cell types—currently estimated at around 130 in mice—greatly exceeds the number of plexiform sublayers, the processes of many cell types arborize within

each sublayer. Anatomical and physiological studies have shown that connectivity is specific even within individual sublayers. Moreover, patterns of connectivity appear to be largely, although not entirely, "hardwired," occurring before visual experience has a chance to affect circuitry. Thus, there must be molecules that restrict axons and dendrites to specific sublayers, as well as molecules that distinguish synaptic partners within a sublayer.

One clue to the basis of both laminar and intralaminar synaptic specificity in the retina comes from the finding that specific types of interneurons and ganglion neurons express different classes of recognition molecules of the immunoglobulin and cadherin families (Chapter 47). Thus, the processes of cells that express a particular recognition molecule are confined

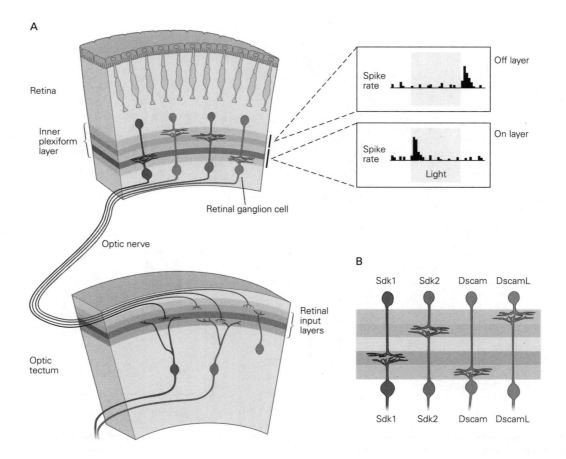

Figure 48–2 Retinal ganglion neurons form layer-specific synapses. (Reproduced, with permission, from Sanes and Yamagata 2009.)

A. The dendrites of retinal ganglion neurons receive input from the processes of retinal interneurons (amacrine and bipolar cells) in the inner plexiform layer, which is subdivided into at least 10 sublaminae. Specific subsets of interneurons and ganglion cells often arborize and synapse in just one layer. These lamina-specific connections determine which aspects of visual stimuli (their onset or offset) activate each type of retinal

ganglion cell. The responses of OFF and ON retinal ganglion cells are shown on the right.

B. Immunoglobulin superfamily adhesion molecules (Sdk1, Sdk2, Dscam, and DscamL) are expressed by different subsets of amacrine and retinal ganglion neurons in the developing chick embryo. Amacrine neurons that express one of these four proteins form synapses with retinal ganglion cells that express the same protein. Manipulating Sdk or Dscam expression alters these patterns of lamina-specific arborization.

to one or a few plexiform sublayers (Figure 48–2B). Many of these proteins promote homophilic interactions; that is, they bind to the same protein on other cell surfaces. The roles of several recognition molecules have now been assessed in chick and mouse retina, either by removing them during development or by implanting them into neurons that do not normally express them. Results of these so-called "loss-of-function" and "gain-of-function" experiments hint at the existence of a complex code of recognition molecules that promotes specific connectivity within a target region. In mice, for example, two cadherins direct bipolar interneurons to appropriate sublayers, while Sidekick 2, a member of the immunoglobulin

superfamily, is required for interneurons to choose among ganglion cells with dendrites in one particular sublayer.

Sensory Receptors Promote Targeting of Olfactory Neurons

A different type of specificity is evident in the olfactory system. Each olfactory sensory neuron in the nasal epithelium expresses just one of approximately 1,000 types of odorant receptors. Neurons expressing one receptor are randomly distributed across a large sector of the epithelium, yet all of their axons converge on the dendrites of just a few target neurons in the olfactory bulb, forming synapse-rich glomeruli (Figure 48–3A). When an individual olfactory receptor is deleted, the axons that normally express the receptor reach the olfactory bulb but fail to converge into specific glomeruli or to terminate on the appropriate postsynaptic cells (Figure 48–3B). Conversely, when neurons are forced to express a different odorant receptor, their axons form glomeruli at a different position within the olfactory bulb (Figure 48–3C).

Together, these experiments suggest that olfactory receptors not only determine a neuron's responsiveness to specific odorants but also help the axon to form appropriate synapses on target neurons. Initially, it was suspected that specific olfactory receptors served not only as odor detectors but also as recognition molecules. More recent studies provide evidence for a different mechanism: that second messengers generated from activation of the olfactory receptors influence the expression of recognition molecules that match olfactory axons with appropriate targets in the olfactory bulb.

The matching occurs in two steps. First, intrinsic differences in the abilities of olfactory receptors to stimulate formation of the second messenger cyclic adenosine monophosphate lead to differential expression of guidance molecules in embryos, generating a coarse matching of olfactory neurons and olfactory bulb targets along the anterior-posterior axis. Second, selective expression of recognition molecules by four groups of olfactory sensory neurons targets them to corresponding domains along the dorsoventral axis of the olfactory bulb.

Thus, an early phase of molecular recognition generates a coarse map of nose-to-brain connectivity by activity-independent mechanisms (Figure 48–4A). Then, postnatally, odorant receptors are activated by odorants, and because of developmental changes in

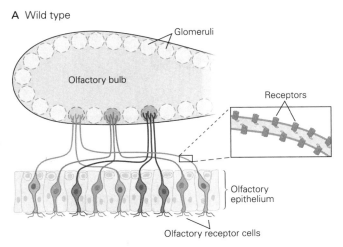

A Wild type

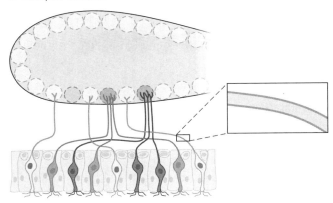

B Receptor deletion

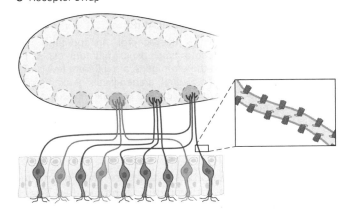

C Receptor swap

Figure 48–3 (Right) Odorant receptors influence the targeting of sensory axons to discrete glomeruli in the olfactory bulb. (Adapted, with permission, from Sanes and Yamagata 2009.)

A. Each olfactory receptor neuron expresses one of approximately 1,000 possible odorant receptors. Neurons expressing the same receptor are distributed sparsely throughout the olfactory epithelium of the nose. The axons of these neurons form synapses with target neurons in a single glomerulus in the olfactory bulb.

B. In mouse mutants in which an odorant receptor gene has been deleted, the sensory neurons that would have expressed the gene send their axons to other glomeruli, in part because these neurons now express other receptors.

C. When one odorant receptor gene replaces another in a set of sensory neurons, their axons project improperly.

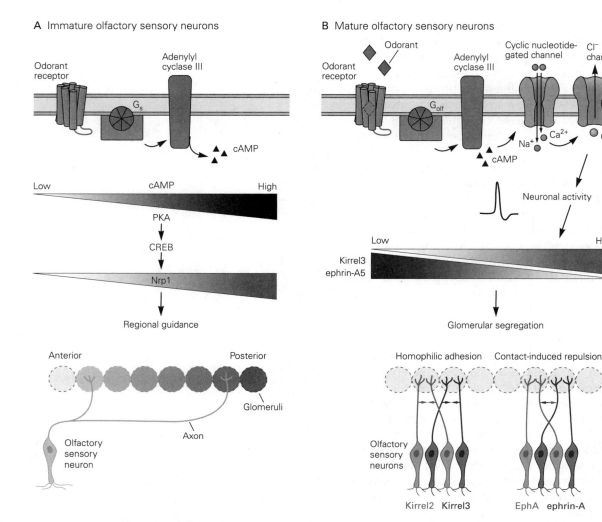

Figure 48–4 Odorant receptors promote specific connections in the olfactory bulb by controlling expression of guidance and recognition molecules. Activation of olfactory receptors in olfactory sensory neurons leads to activation of adenylyl cyclase and production of the second messenger cyclic adenosine monophosphate (**cAMP**).

A. Prenatally, prior to olfaction, the receptors are spontaneously active. Different receptor types exhibit different levels of spontaneous activity and therefore generate different levels of cAMP, which in turn induce distinct, graded levels of axon guidance molecules such as neuropilins and semaphorins. These guidance molecules mediate interactions among axons that

guide them to appropriate regions of the olfactory bulb. (Abbreviations: **CREB**, cAMP response element-binding protein; **Nrp1**, neuropilin1; **PKA**, protein kinase A.)

B. Postnatally, olfactory receptors are activated by odorant molecules. This olfactory activity also generates distinct levels of cAMP in each type of odorant receptor neuron, but now the second messenger acts through ion channels to induce new sets of guidance molecules such as kirrels and ephrins. These molecules mediate interactions that segregate axonal terminals into glomeruli. Thus, successive phases of receptor activity, the first spontaneous and the second evoked by odorants, act together to map olfactory sensory axons of different types onto different glomeruli.

intracellular signaling, this activation leads to induction of a second set of recognition molecules. These molecules lead to convergence of axons onto glomeruli, thus refining the projection by an activity-dependent mechanism (Figure 48–4B). Segregation of axons first to particular regions and then to particular glomeruli occurs via both adhesive and repulsive interactions.

Different Synaptic Inputs Are Directed to Discrete Domains of the Postsynaptic Cell

Nerve terminals not only discriminate among candidate targets but also terminate on a specific portion of the target neuron. In the cerebral cortex and hippocampus, for example, axons arriving in layered structures

Figure 48–5 The axons of inhibitory interneurons in the cerebellum terminate on a distinct region of the cerebellar Purkinje cell. Many neurons form synapses on cerebellar Purkinje neurons, each selecting a distinct domain on the Purkinje cell. The axons of inhibitory basket cells form most of their synapses on the axon hillock and initial segment. Basket cells select these domains by recognizing neurofascin, a cell surface immunoglobulin superfamily adhesion molecule that is anchored to the initial segment of the axon by ankyrin G. When the localization of neurofascin is perturbed, basket cell axons fail to restrict synapse formation to the initial segment. (Adapted from Huang 2006.)

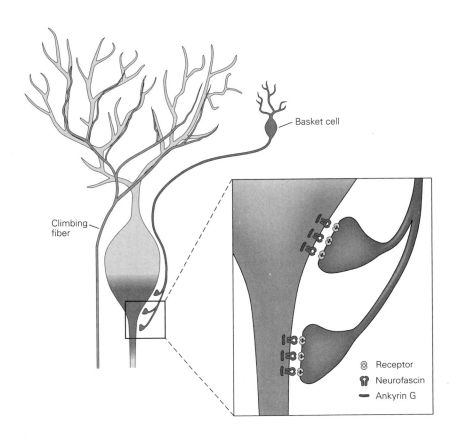

often confine their terminals to one layer, even if the dendritic tree of the postsynaptic cell traverses numerous layers. In the cerebellum, the axons of different types of neurons terminate on distinct domains of the Purkinje neurons. Granule cell axons contact distal dendritic spines, climbing fiber axons contact proximal dendritic shafts, and basket cell axons contact the axon hillock and initial segment (Figure 48–5).

Such specificity presumably relies on molecular cues on the postsynaptic cell surface. For Purkinje neurons of the cerebellum, one such cue is neurofascin, an adhesion molecule of the immunoglobulin superfamily. Neurofascin is present at high levels on the axonal initial segment, thus directing basket cells to form axons selectively on this axonal domain. Adhesion molecules can therefore also serve as recognition molecules for particular domains of a neuron. Since individual neurons can form synapses with several classes of pre- and postsynaptic cells, it follows that each neuronal subtype must express a variety of synaptic recognition molecules.

Neural Activity Sharpens Synaptic Specificity

So far, we have emphasized the role of recognition molecules in the initial formation of synapses. Once synapses form, however, neural activity within the circuit plays a critical role in refining synaptic patterns. For example, as described above, guidance of olfactory neurons to the olfactory bulb includes an initial activity-independent crude mapping followed by an activity-dependent phase in which the projection is refined.

A similar biphasic pattern has been studied in detail in the visual system. Retinal ganglion cells project to the optic tectum (superior colliculus), where interactions between ephrins and Eph kinases result in formation of a crude retinotopic map of retinal axons on the tectal surface (Chapter 47). Activity-dependent processes then sculpt the axonal arbors of retinal ganglion cells. The axons initially form broad diffuse arbors, which gradually become denser but more focused, sharpening the tectal map (Figure 48–6). This refinement is inhibited when the activity of synapses is blocked. The molecular mechanisms of this activity-dependent refinement are largely unknown. As in the olfactory system, an attractive idea is that the level and pattern of neuronal activity regulate the expression of recognition molecules.

These examples from the olfactory and visual systems illustrate a widespread phenomenon: Molecular cues initially control synapse specificity, but once the circuit begins to function, specificity is sharpened

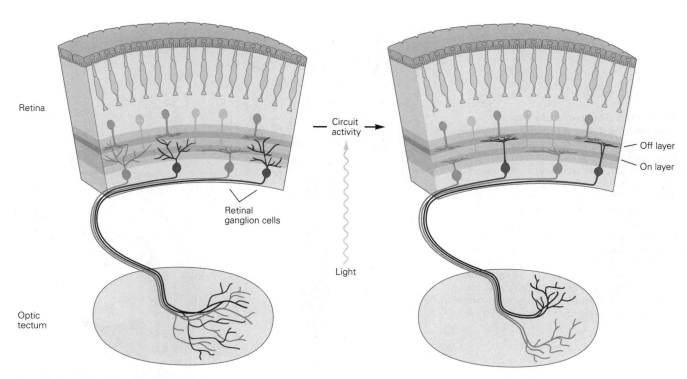

Figure 48–6 Electrical activity refines the specificity of synaptic connections of retinal ganglion cells. Some retinal ganglion cells initially form dendritic arbors that are limited to specific sublaminae in the inner plexiform layer of the retina, whereas others initially form diffuse arbors that are later pruned to form large specific patterns. Similarly, the axonal arbors of retinal ganglion cells initially innervate a large region of their target fields in the superior colliculus. This expansive axonal arbor is then refined so as to concentrate many branches in a small region. Abolishing electrical activity in retinal ganglion cells decreases the remodeling of dendritic and axonal arbors.

through neural activity. In the visual system, sharpening involves loss of synapses. We will return to this process of synapse elimination at the end of this chapter and consider its consequences for behavior in the next chapter.

In a few cases, neural activity promotes specificity in a different way, by turning an inappropriate target into an appropriate one. This mechanism has been most clearly demonstrated in skeletal muscle, where mammalian muscle fibers can be divided into several categories according to their contractile characteristics (Chapter 31). Muscle fibers of particular types express genes for distinctive isoforms of the main contractile proteins, such as myosins and troponins.

Few muscles are composed exclusively of a single type of fiber; most have fibers of all types. Yet the branches of an individual motor axon innervate muscle fibers of a single type, even in "mixed" muscles in which fibers of different types are intermingled (Figure 48–7A). This pattern implies a remarkable degree of synaptic specificity. However, matching does not always come about through recognition in the motor axon of the appropriate type of muscle fiber. The motor axon can also convert the target muscle fiber to an appropriate type. When a muscle is denervated at birth, before the properties of its fibers are fixed, a nerve that normally innervates a slow muscle can be redirected to innervate a muscle destined to become fast, and vice versa. Under these conditions, the contractile properties of the muscle are partially transformed in a direction imposed by the firing properties of the motor nerve (Figure 48–7B,C).

Different patterns of neural activity in fast and slow motor neurons are responsible for the switch in muscle properties. Most strikingly, direct electrical stimulation of a muscle with patterns normally evoked by slow or fast nerves leads to changes that are nearly as dramatic as those produced by cross-innervation (Figure 48–7D). Although activity-based conversion of the type observed at the neuromuscular junction is unlikely to be a major contributor to synaptic specificity in the central nervous system, it is likely that central axons modify the properties of their synaptic targets, contributing to the diversification of neuronal subtypes and refining connectivity imposed by recognition molecules.

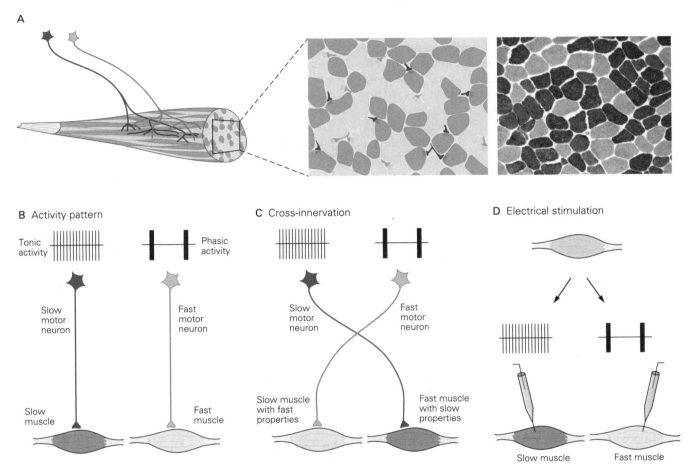

Figure 48–7 The pattern of motor neuron activity can change the biochemical and functional properties of skeletal muscle cells.

A. Muscle fibers have characteristic metabolic, molecular, and electrical properties that identify them as "slow" (tonic) or "fast" (phasic) types. The micrograph on the right shows a section of muscle tissue with histochemical staining for myosin ATPase. The middle sketch shows a section through the muscle, in which motor neurons (**green** and **brown**) form synapses on a single type of muscle fiber. (Photo on right reproduced, with permission, from Arthur P. Hays.)

B. Motor neurons that connect with fast and slow muscle fibers (fast and slow motor neurons) exhibit distinct patterns of electrical activity: steady low-frequency (tonic) firing for slow fibers and intermittent high-frequency bursts (phasic) for fast fibers.

C. Cross-innervation experiments showed that some property of the motor neuron helps to determine whether muscle fibers are fast or slow. Cross-innervation was achieved by surgically rerouting fast axons to slow muscle and vice versa. Although the properties of the motor neurons are little changed, the properties of the muscle change profoundly. For example, fast motor neurons induce fast properties in the slow muscle. (Adapted, with permission, from Salmons and Sreter 1976.)

D. The effects of innervation by fast and slow nerves on muscle are mediated in part by their distinct patterns of activity. Stimulation of a fast muscle in a slow tonic pattern converts the muscle into a slow type. Conversely, fast phasic stimulation of a slow muscle can convert it to a faster type.

Principles of Synaptic Differentiation Are Revealed at the Neuromuscular Junction

The neuromuscular junction comprises three types of cells: a motor neuron, a muscle fiber, and Schwann cells. All three types are highly differentiated in the region of the synapse.

The process of synapse formation is initiated when a motor axon, guided by the multiple factors described in Chapter 47, reaches a developing skeletal muscle and approaches an immature muscle fiber. Contact is made, and the process of synaptic differentiation gets underway. As the growth cone begins its transformation into a nerve terminal, the portion of the muscle surface opposite the nerve terminal begins to acquire its own specializations. As development proceeds, synaptic components are added and structural signs of synaptic differentiation become apparent in the pre- and postsynaptic cells and in the synaptic cleft. Eventually, the neuromuscular junction acquires its mature and complex form (Figure 48–8).

Three general features of neuromuscular junction development have provided clues about the molecular mechanisms that underlie synapse formation. First, nerve and muscle organize each other's differentiation. In principle, the precise apposition of pre- and postsynaptic specializations might be explained by independent programming of nerve and muscle properties. However, in muscle cells cultured alone, acetylcholine (ACh) receptors are generally distributed uniformly on the surface, although some are clustered as in mature postsynaptic membranes. Yet, when motor neurons are added to the cultures, they extend neurites that contact the muscle cells more or less randomly, instead of seeking out the ACh receptor clusters. New receptor clusters appear precisely at the points of contact with the presynaptic neurites, while preexisting uninnervated clusters eventually disperse (Figure 48–9). Thus, factors on or released by motor axons exert a profound influence on the synaptic organization of the muscle cell.

Likewise, muscles signal retrogradely to motor nerve terminals. When motor neurons in culture extend neurites, they assemble and transport synaptic vesicles, some of which form aggregates similar to those found in nerve terminals. When the neurites contact muscle cells, new vesicle clusters form opposite the muscle membrane, and most of the preexisting clusters disperse.

These studies also revealed a second feature of neuromuscular development: that motor neurons and muscle cells can synthesize and arrange most synaptic components without each other's help. Uninnervated myotubes can synthesize functional ACh receptors and gather them into high-density aggregates. Likewise, motor axons can form synaptic vesicles and cluster them into varicosities in the absence of muscle. In fact, vesicles in growth cones can synthesize and release ACh in response to electrical stimulation, before the growth cone has reached its target cells. Thus, the developmental signals that pass between nerve and muscle do not induce wholesale changes in cell properties; rather, they assure that components of the pre- and postsynaptic machinery are organized at the correct time and in the right places. It is useful therefore to think of the intercellular signals that control synaptogenesis as organizers rather than inducers.

A third key feature of neuromuscular junction development is that new synaptic components are added in several distinct steps. The newly formed synapse is not simply a prototype of a fully developed synapse. Although nerve and muscle membrane form close contacts at early stages of synaptogenesis, only later does the synaptic cleft widen and the basal lamina appear. Similarly, ACh receptors accumulate in the postsynaptic membrane before acetylcholinesterase accumulates in the synaptic cleft, and the postsynaptic membrane acquires junctional folds only after the nerve terminal has matured. Several different axons innervate each myotube around the time of birth, but during early postnatal life, all but one axon withdraws.

This elaborate sequence is not orchestrated by the simple act of contact between nerve and muscle. Instead, multiple signals pass between the cells—the nerve sends a signal to the muscle that triggers the first steps in postsynaptic differentiation, at which point the muscle sends a signal that triggers the initial steps of nerve terminal differentiation. The nerve then sends further signals to the muscle, and this interaction continues.

We now consider retrograde (from muscle to nerve) and anterograde (from nerve to muscle) organizers in more detail.

Differentiation of Motor Nerve Terminals Is Organized by Muscle Fibers

Soon after the growth cone of a motor axon contacts a developing myotube, a rudimentary form of neurotransmission begins. The axon releases ACh in vesicular packets, the transmitter binds to receptors, and the myotube responds with depolarization and weak contraction.

The onset of transmission at the new synapse reflects the intrinsic capabilities of each synaptic partner. Nevertheless, these intrinsic capabilities cannot readily explain the marked increase in the rate of transmitter release that occurs after nerve-muscle contact is made, nor can they explain the accumulation of synaptic vesicles and the assembly of active zones in the small portion of the motor axon that contacts the muscle surface. These developmental steps require signals from muscle to nerve.

A clue to the source of these signals came from studies on the reinnervation of adult muscle. Although axotomy leaves muscle fibers denervated and leads to insertion of ACh receptors in nonsynaptic regions, the postsynaptic apparatus remains largely intact. It is still recognizable by its synaptic nuclei, junctional folds, and the ACh receptors, which remain far more densely

A Development stages

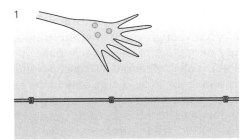

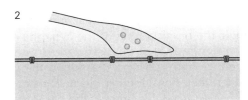

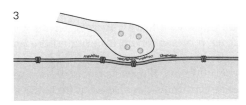

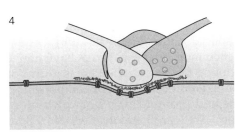

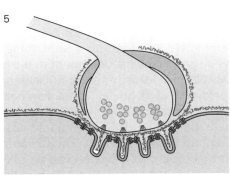

B Mature neuromuscular junction

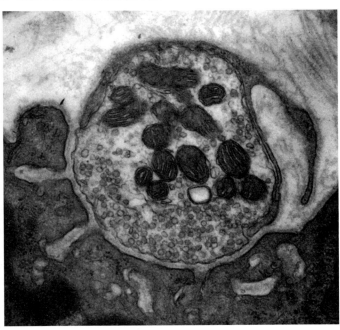

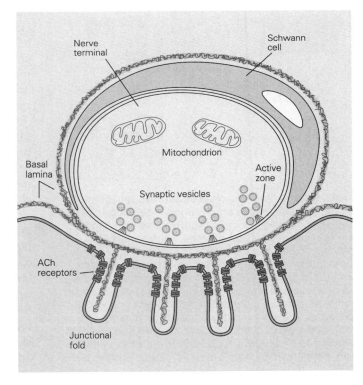

Figure 48–8 The neuromuscular junction develops in sequential stages.

A. A growth cone approaches a newly fused myotube (**1**) and forms a morphologically unspecialized but functional contact (**2**). The nerve terminal accumulates synaptic vesicles and a basal lamina forms in the synaptic cleft (**3**). As the muscle matures, multiple axons converge on a single site (**4**). Finally, all axons but one are eliminated and the surviving terminal matures (**5**). As the synapse matures, acetylcholine (**ACh**) receptors become concentrated in the postsynaptic membrane and depleted from the extrasynaptic membrane. (Adapted, with permission, from Hall and Sanes 1993.)

B. At the mature neuromuscular junction, pre- and postsynaptic membranes are separated by a synaptic cleft that contains basal lamina and extracellular matrix proteins. Vesicles are clustered at presynaptic release sites, transmitter receptors are clustered in the postsynaptic membrane, and nerve terminals are coated by Schwann cell processes. (Micrograph reproduced, with permission, from T. Gillingwater.)

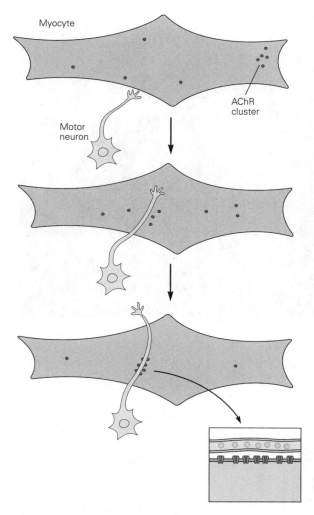

Figure 48–9 Nerve and muscle cells express synaptic components, but synaptic organization requires cell interactions. Acetylcholine receptors (**AChR**) are synthesized by muscle cells cultured without neurons. Many receptors are diffusely distributed, but some form high-density aggregates similar to those found in the postsynaptic membrane of the neuromuscular junction. When neurons first contact muscle, they do not restrict themselves to the receptor-rich aggregates. Instead, new receptor aggregates form at sites of neurite-muscle contact, and many of the preexisting clusters disperse. Similarly, motor axons contain synaptic vesicles that cluster at sites of neurite contact with muscle cells. (Adapted, with permission, from Anderson and Cohen 1977; Lupa, Gordon, and Hall 1990.)

packed in synaptic areas than in extrasynaptic areas of the cell. Damaged peripheral axons regenerate readily (unlike those in the central nervous system) and form new neuromuscular junctions that look and perform much like the original ones.

A century ago, Fernando Tello-Muñóz, a student of Santiago Ramón y Cajal, noted that the new junctions

form at preexisting synaptic sites on the denervated muscle fibers even though the postsynaptic specializations occupy only 0.1% of the muscle fiber surface (Figure 48–10A). Later, electron microscopy showed that specialization in the axon occurs only in the terminals that contact the muscle. For example, active zones form directly opposite the mouths of the postsynaptic junctional folds. This striking example of subcellular specificity implies that motor axons recognize signals associated with the postsynaptic apparatus.

When regenerating axons reach a muscle fiber, they encounter the basal lamina of the synaptic cleft. To explore the significance of this association, muscles were damaged in vivo in a way that killed the muscle fibers but left their basal lamina intact. The necrotic fibers were phagocytized, leaving behind basal lamina sheaths on which synaptic sites were readily recognizable. At the same time that the muscle was damaged, the nerve was cut and allowed to regenerate. Under these conditions, motor axons reinnervated the empty basal lamina sheaths, contacting synaptic sites as precisely as they would have if muscle fibers were present. Moreover, nerve terminals developed at these sites and active zones even formed opposite struts of basal lamina that once lined junctional folds. These observations implied that components of the basal lamina organize presynaptic specialization (Figure 48–10B).

Several such molecular organizers have now been identified. Among the best studied are isoforms of the protein laminin. Laminins are major components of all basal laminae and promote axon outgrowth in many neuronal types. They are heterotrimers of α, β, and γ chains, comprising a family of five α, four β, and three γ chains (Chapter 47). Muscle fibers synthesize multiple laminin isoforms and incorporate them into the basal lamina. Laminin-211, a heterotrimer containing the α2, β1, and γ1 chains, is the major laminin in the basal lamina, and its absence leads to severe muscular dystrophy. In the synaptic cleft, however, isoforms bearing the β2 chain predominate (Figure 48–11A), and nerve terminals fail to differentiate fully in mutant mice that lack the β2 laminin (Figure 48–11B). The β2 laminins appear to act by binding to voltage-sensitive calcium channels that reside in the axon terminal membrane, where they couple activity to transmitter release. Laminins act on the extracellular domain of the channels, whereas the intracellular segment recruits or stabilizes other components of the release apparatus.

The finding that presynaptic differentiation is only partially compromised in the absence of laminins indicated that additional muscle-derived organizers of axonal specialization must exist. Several have now been identified, including members of the fibroblast

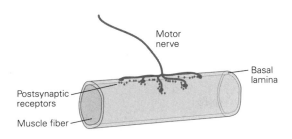

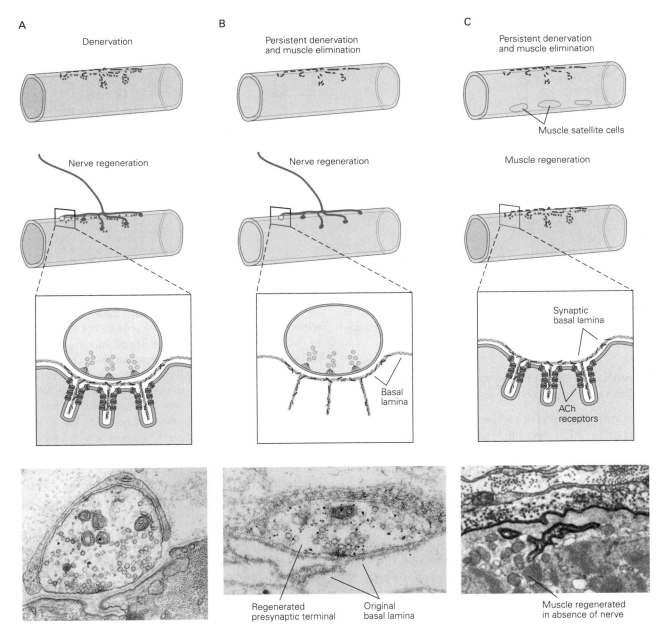

Figure 48–10 Synaptic portions of basal lamina contain proteins that organize developing and regenerating nerve terminals.

A. Damaged motor axons regenerate and form new neuromuscular junctions. Nearly all of the new synapses form at the original synaptic sites. (Micrograph reproduced, with permission, from Glicksman and Sanes 1983.)

B. A strong preference for innervation at original synaptic sites persists even after the muscle fibers have been removed, leaving behind basal lamina "ghosts." Regenerated axons develop synaptic specialization on contact with the original synaptic

sites on the basal lamina. (Micrograph reproduced, with permission, from Glicksman and Sanes 1983.)

C. Following denervation of a skeletal muscle fiber and elimination of mature muscle fibers, muscle satellite cells proliferate and differentiate to form new myofibers. The expression of acetylcholine (**ACh**) receptors on the regenerated myofiber surface is concentrated in the synaptic areas of basal lamina, even when reinnervation is prevented. (Micrograph reproduced, with permission, from Burden, Sargent, and McMahan 1979. © The Rockefeller University Press. Permission conveyed through Copyright Clearance Center, Inc.)

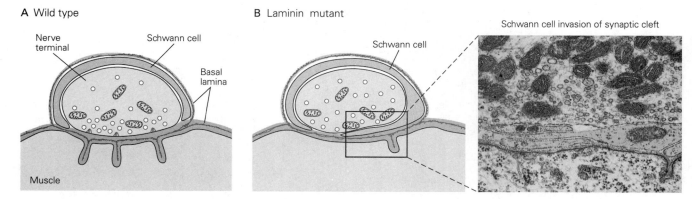

A Wild type

Nerve terminal

Schwann cell

Basal lamina

Muscle

B Laminin mutant

Schwann cell

Schwann cell invasion of synaptic cleft

Figure 48–11 Different laminin isoforms are localized at synaptic and extrasynaptic areas of the basal lamina.

A. Different laminin isoforms are found in synaptic (**brown**) and extrasynaptic (**green**) areas of basal lamina. Isoforms, containing the β2 chain, are concentrated in the synaptic areas.

B. Maturation of neuromuscular junctions is impaired in mice lacking β2 laminins. These mutants have few active zones, and the synaptic cleft is invaded by Schwann cell processes (**blue**). (Micrograph reproduced, with permission, from Noakes et al. 1995.)

growth factor and collagen IV families, as well as a muscle membrane-associated protein, LRP4, that we will soon encounter again in the context of postsynaptic differentiation. Thus target-derived proteins from multiple families collaborate to organize the presynaptic nerve terminal.

Differentiation of the Postsynaptic Muscle Membrane Is Organized by the Motor Nerve

Soon after myoblasts fuse to form myotubes, the genes that encode ACh receptor subunits are activated. Receptor subunits are synthesized, assembled into pentamers in the endoplasmic reticulum, and inserted into the plasma membrane. As noted above, some receptors spontaneously form aggregates, but the majority are distributed throughout the membrane at a low density, approximately 1,000 per μm^2.

Once synapse formation is complete, however, the distribution of the receptors changes drastically. The receptors become concentrated at the synaptic sites of the membrane (to a density up to 10,000 per μm^2) and depleted in the nonsynaptic membrane (reduced to 10 per μm^2 or less). This thousand-fold difference in ACh receptor density occurs within a few tens of micrometers from the edge of the nerve terminal.

Appreciation of the critical role of the nerve in the redistribution of ACh receptors inspired a search for factors that might promote their clustering. This quest led to the discovery of a proteoglycan, agrin. Agrin is synthesized by motor neurons, transported down the axon, released from nerve terminals, and incorporated into the synaptic cleft (Figure 48–12A,B). Some agrin

isoforms are also made by muscle cells, but the neuronal isoforms are about a thousand-fold more active in aggregating ACh receptors.

The phenotype of mutant mice lacking agrin shows that agrin has a central role in the organization of ACh receptors. Agrin mutants have grossly perturbed neuromuscular junctions and die at birth. The number, size, and density of ACh receptor aggregates are severely reduced in these mice (Figure 48–12C). Other components of the postsynaptic apparatus—including cytoskeletal, membrane, and basal lamina proteins—are also reduced. Interestingly, the differentiation of presynaptic elements is also perturbed. However, the defects in the presynaptic element do not result directly from lack of agrin in the motor neuron, but rather indirectly from the failure of the disorganized postsynaptic apparatus to generate signals for presynaptic specialization.

How does agrin work? Agrin's major receptor is a complex of a muscle-specific tyrosine kinase called MuSK (muscle-specific trk-related receptor with a kringle domain) and a coreceptor subunit called LRP4 (Figure 48–12A). MuSK and LRP4 are normally concentrated at synaptic sites in the muscle membrane, and muscles of mutant mice lacking MuSK or LRP4 do not have ACh receptor clusters (Figure 48–12C). Myotubes generated in vitro from these mutants express normal levels of ACh receptors, but these receptors cannot be clustered by agrin. Binding of agrin to the MuSK/LRP4 complex initiates a chain of events that ends in receptor clustering. Key events are agrin-induced activation of MuSK's kinase activity; autophosphorylation of the MuSK intracellular domain; recruitment of

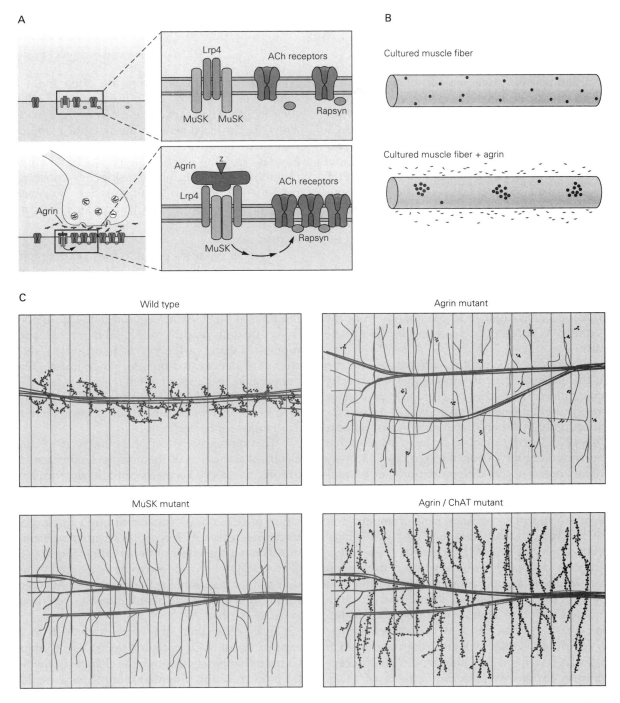

Figure 48–12 Agrin induces aggregation of acetylcholine (ACh) receptors at synaptic sites.

A. Agrin is a large (~400 kDa) extracellular matrix proteoglycan. Alternative splicing includes a "z" exon that confers the ability to cluster ACh receptors. When released by a nerve terminal, agrin binds Lrp4 on the muscle membrane, activating the membrane-associated receptor tyrosine kinase MuSK and triggering an intracellular cascade that results in ACh receptor clustering. Key intracellular signaling molecules are Dok7, Crk, and CrkL. These signal to rapsyn, a cytoplasmic ACh receptor–associated protein, which physically interacts with and clusters the ACh receptors. (Adapted, with permission, from DeChiara et al. 1996.)

B. Few ACh receptor clusters form on myofibers grown in culture under control conditions, but addition of agrin induces ACh receptor clustering. (Adapted, with permission, from Misgeld et al. 2005.)

C. Muscles from wild-type neonatal mice and from three mutant types. Muscles were labeled for ACh receptors (**green**) and motor axons (**brown**). In wild-type mice, ACh receptor clusters have formed under each nerve terminal by birth, whereas in agrin mutants, most clusters have dispersed. ACh receptor clusters are also absent in MuSK, Dok7, and rapsyn mutant mice. When the genes for agrin and **ChAT** (choline acetyltransferase) are mutated, clusters of ACh receptors remain, indicating that agrin works by counteracting receptor dispersion mediated by ACh. All mutant conditions also show axonal abnormalities, reflecting defects in retrograde signaling to the motor axon. (Abbreviation: MuSK, muscle-specific trk-related receptor with a kringle domain.) (Adapted, with permission, from Gautam et al. 1996.)

adaptor proteins Dok-7, Crk, and CrkL; and strengthening of an interaction between a cytoplasmic protein rapsyn and the ACh receptors. Rapsyn may be the final element in the sequence: It binds directly to the ACh receptors and can induce their aggregation in vitro. In mice lacking rapsyn, muscles form normally and ACh receptors accumulate in normal numbers but fail to aggregate at the synaptic sites on the membrane. Accordingly, muscles of mutant mice lacking Dok7 or rapsyn resemble those lacking MuSK or LRP4: They synthesize ACh receptors but do not have ACh receptor clusters.

Thus, an extracellular protein (agrin), transmembrane proteins (MuSK and LRP4), adaptor proteins (Dok-7, Crk, and CrkL), and a cytoskeletal protein (rapsyn) form a chain that links commands from the motor axon to ACh receptor clustering in the muscle membrane.

Nevertheless, postsynaptic differentiation can occur in the absence of agrin signaling. This capacity was apparent in early studies on cultured muscle (see Figure 48–9) and is also seen in vivo: ACh receptor clusters form initially but then disperse in agrin mutants (Figure 48–12C). Clustering also occurs in muscles that lack innervation entirely. Thus, the signaling pathway that initiates postsynaptic differentiation can be activated without agrin, but agrin is required to maintain clustering of ACh receptors.

The role of agrin is perhaps best understood in terms of the requirement that pre- and postsynaptic specializations be perfectly aligned. ACh receptor aggregates persist in uninnervated muscles but disappear in agrin mutant muscles, suggesting that axons sculpt the postsynaptic membrane through the combined action of agrin and a dispersal factor. One major dispersal factor is ACh itself; clustering persists in mutants that lack both agrin and ACh (Figure 48–12C). Thus, agrin may render ACh receptors immune to the declustering effects of ACh. Through a combination of positive and negative factors, the motor neuron ensures that the patches of postsynaptic membrane contacted by axon branches are rich in ACh receptors.

The Nerve Regulates Transcription of Acetylcholine Receptor Genes

Along with redistribution of ACh receptors in the plane of the membrane, the motor nerve orchestrates the transcriptional program responsible for expression of ACh receptor genes in muscle. To understand this aspect of transcriptional control, it is important to appreciate the geometry of the muscle.

Individual muscle fibers are often more than a centimeter long and contain hundreds of nuclei along their length. Most nuclei are far from the synapse, but a few are clustered beneath the synaptic membrane, so that their transcribed and translated products do not have far to go to reach the synapse. In newly formed myotubes, most nuclei express genes encoding ACh receptor subunits. In adult muscles, however, only synaptic nuclei express ACh receptor genes; nonsynaptic nuclei do not. Two processes contribute to this transformation.

First, as synapses begin to form, expression of the ACh receptor subunit genes is increased in synaptic nuclei (Figure 48–13). Signals acting through MuSK are needed for this specialization. Second, around the time of birth, ACh receptor gene expression shuts down in nonsynaptic nuclei. This change reflects a repressive effect of the nerve, as originally shown by studies of denervated muscle. When muscle fibers are denervated, as happens when the motor nerve is damaged, the density of ACh receptors in the postsynaptic membrane increases markedly, a phenomenon termed *denervation supersensitivity*.

This repressive effect of the nerve is mediated by electrical activation of the muscle. Under normal conditions, the nerve keeps the muscle electrically active, and fewer ACh receptors are synthesized in active muscle than in inactive muscle. Indeed, direct stimulation of denervated muscle through implanted electrodes decreases ACh receptor expression, preventing or reversing the effect of denervation (Figure 48–13B). Conversely, when nerve activity is blocked by application of a local anesthetic, the number of ACh receptors throughout the muscle fiber increases, even though the synapse is intact.

In essence, then, the nerve uses ACh to repress expression of ACh receptor genes extrasynaptically. Current that passes through the channel of the receptor leads to an action potential that propagates along the entire muscle fiber. This depolarization opens voltage-dependent Ca^{2+} channels, leading to an influx of Ca^{2+}, which activates a signal transduction cascade that reaches nonsynaptic nuclei and regulates transcription of ACh receptor genes. Thus, the same voltage changes that produce muscle contraction over a period of milliseconds also regulate transcription of ACh receptor genes over a period of days.

The increase in transcription of ACh receptor genes in nuclei beneath the synapse, along with the decrease in nuclei distant from synapses, leads to localization of ACh receptor mRNA and thus preferential synthesis and insertion of ACh receptors near synaptic sites. This local synthesis is reminiscent of that seen

Chapter 48 / Formation and Elimination of Synapses   1197

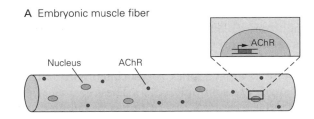

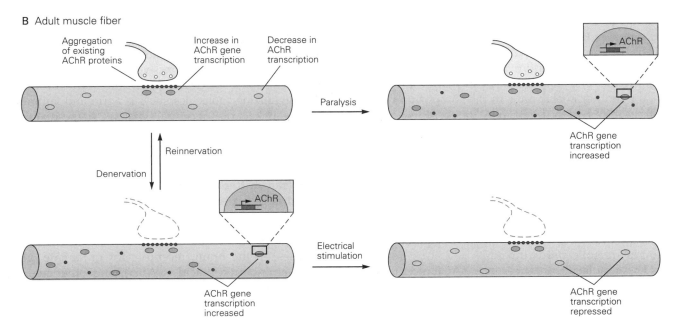

Figure 48–13 Clustering of acetylcholine (ACh) receptors at the neuromuscular junction results from transcriptional regulation and local protein trafficking.

A. ACh receptors (**AChR**) are distributed diffusely on the surface of embryonic myotubes.

B. After the muscle is innervated by a motor axon, the number of receptors in extrasynaptic regions decreases, whereas receptor density at the synapse increases. This reflects the aggregation of preexisting receptors and enhanced expression of ACh receptor genes in nuclei that lie directly beneath the

nerve terminal. In addition, the transcription of receptor genes is repressed in nuclei in extrasynaptic regions. Electrical activity in muscle represses ACh gene expression in nonsynaptic nuclei, leading to a lower density of ACh receptors in these regions. The nuclei at synaptic sites are immune to this repressive effect. Following denervation, ACh receptor gene expression is upregulated in extrasynaptic nuclei, although not to the high level attained by synaptic nuclei. Paralysis mimics the effect of denervation, whereas electrical stimulation of denervated muscle mimics the influence of the nerve and decreases the density of ACh receptors in the extrasynaptic membrane.

at postsynaptic sites on dendritic spines in the brain. Local synthesis in muscle is advantageous since ACh receptors synthesized near the ends of fibers would never reach the synapse without degradation.

Many components of the postsynaptic apparatus are regulated in ways similar to those we have described for ACh receptors—their aggregation depends on agrin and MuSK, and their transcription is enhanced in synaptic nuclei and repressed in extrasynaptic nuclei by electrical activity. Thus, synaptic components have tailor-made regulatory mechanisms, but many of these components are regulated in parallel.

The Neuromuscular Junction Matures in a Series of Steps

The adult neuromuscular junction is dramatically different in its molecular architecture, shape, size, and functional properties from the simple nerve-muscle contact that initiates neurotransmission in the embryo. Maturation of the nerve terminal, the postsynaptic membrane, and the intervening synaptic cleft occurs in a complex series of steps. We illustrate this stepwise synaptic construction with a continued focus on the development of ACh receptors.

As we have seen, ACh receptors aggregate in the plane of the membrane as the neuromuscular junction begins to form, and receptor gene transcription is enhanced in postsynaptic nuclei. A few days later, activity begins to decrease the level of extrasynaptic receptors. These transcriptional changes are soon followed by changes in the stability of the receptors. In embryonic muscle, ACh receptors are turned over rapidly (with a half-life of approximately 1 day) in both synaptic and extrasynaptic regions. In contrast, in adult muscle, the receptors are relatively stable (with a half-life of approximately 2 weeks). The metabolic stabilization of ACh receptors helps concentrate them at synaptic sites and stabilize the postsynaptic apparatus.

Yet another alteration is in the composition of the ACh receptors. In the embryo, ACh receptors are composed of α-, β-, δ-, and γ-subunits. During the first few postnatal days, the γ gene is turned off and a closely related gene called ε is activated. As a result, new ACh receptors inserted in the membrane are composed of α-, β-, δ-, and ε-subunits. This altered subunit composition tunes the receptor in a way that is suited to its mature function. However, although it occurs at the same time as the metabolic stabilization, the two changes are not causally linked.

These molecular changes in the ACh receptors are accompanied by changes in their distribution (Figure 48–14). Soon after birth, junctional folds begin to form in the postsynaptic membrane and ACh receptors become concentrated at the crests of the folds, along with rapsyn, whereas other membrane and cytoskeletal proteins are localized in the depths of the folds. The initial aggregate of ACh receptors appears to have a plaque-like appearance. Perforations that undergo fusion and fission eventually transform the dense plaque into a pretzel shape that follows the branches of the nerve ending. New receptor-associated cytoskeletal proteins are added to the aggregate, presumably to drive the geometric changes. Finally, the postsynaptic membrane enlarges and eventually contains many more ACh receptors than were present in the initial cluster. Each of these changes occurs while the synapse is functional, suggesting that ongoing activity plays an important role in synaptic maturation.

Central Synapses and Neuromuscular Junctions Develop in Similar Ways

Synapses in the central nervous system are structurally and functionally similar to neuromuscular junctions in many ways. Presynaptically, most of the major protein components of synaptic vesicles are identical at both types of synapses. Likewise, the mechanisms of transmitter release differ only quantitatively, not qualitatively. Postsynaptically, neurotransmitter receptors are concentrated beneath the nerve terminal and associated with "clustering" proteins.

These parallels extend to synaptic development. Studies of cultured neurons have shown that the cellular logic of synapse formation is conserved between neuromuscular junctions and central synapses. At both synaptic types, pre- and postsynaptic elements regulate each other's differentiation by organizing synaptic components rather than by inducing their expression, and synapses develop in a progressive series of steps (Figure 48–15). The molecular details differ, however. Neuromuscular organizing molecules such as agrin and laminins do not play key roles at central synapses, suggesting that other synaptic organizers are involved. Recently, some of these organizing molecules have been identified.

Neurotransmitter Receptors Become Localized at Central Synapses

The concentration of neurotransmitter receptors in the postsynaptic membrane is a feature shared by many synapses. In the brain, receptors for glutamate, glycine, γ-aminobutyric acid (GABA), and other neurotransmitters are concentrated in patches of membrane aligned with nerve terminals that contain the corresponding transmitter.

The processes by which these receptors become localized may be similar to those at the neuromuscular junction. In cultures of dissociated hippocampal neurons, for example, both glutamatergic and GABAergic nerve terminals appear to stimulate clustering of appropriate receptors in the postsynaptic membrane. Moreover, nerves can induce expression of genes encoding glutamate receptors in central neurons, much as occurs for ACh receptors in muscle. Finally, electrical activity also regulates expression of neurotransmitter receptors in neurons.

In forming receptor clusters, central neurons face an obvious challenge that myotubes do not: They are contacted by axon terminals from distinct classes of neurons that use different neurotransmitters (Figure 48–16A). Thus, the nerve terminal probably has an instructive role in the clustering of receptors. In cultures of hippocampal neurons, glutamatergic and GABAergic axons terminate on adjacent regions of the same dendrite. Initially, glutamate and GABA receptors are dispersed, but soon, each type becomes selectively clustered beneath terminals that release that neurotransmitter. This observation implies the existence of multiple clustering signals with parallel pathways of signal transduction.

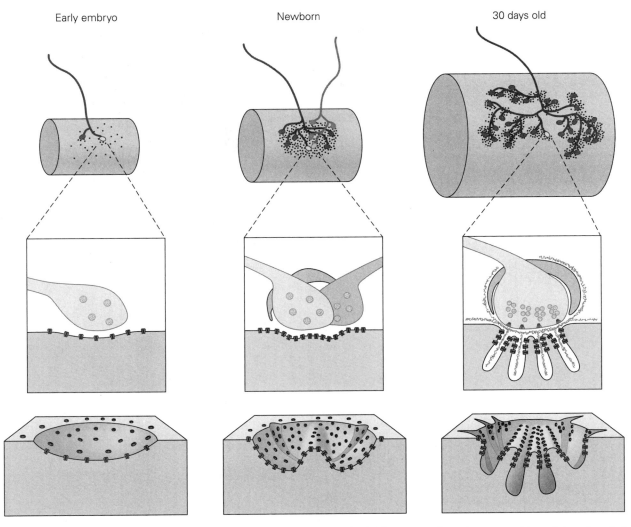

Figure 48–14 The postsynaptic membrane at the neuromuscular junction matures in stages. During early embryogenesis, ACh receptors exist as loose aggregates. Later, these aggregates condense into a plaque-like structure. After birth, the dense cluster opens up as the nerve develops multiple terminals. These axon branches expand in an intercalary fashion as the muscle grows, and the plaque indents to form a gutter, which then invaginates to form folds. Receptors are concentrated at the crests of the folds. (Adapted, with permission, from Sanes and Lichtman 2001.)

At the neuromuscular junction, rapsyn binds to the intracellular domain of ACh receptors and clusters them. Several proteins have been found to play similar roles at central synapses. One, gephyrin, is highly concentrated in the synaptic densities at glycinergic and some GABAergic synapses (Figure 48–16A). Gephyrin is not structurally related to rapsyn but has the same function: It links the receptors to the underlying cytoskeleton. In nonneural cells, glycine receptors cluster when gephyrin is co-expressed; conversely, clusters fail to form at inhibitory synapses in gephyrin-deficient mutant mice (Figure 48–16B). Similarly, a class of proteins that share conserved segments called PDZ domains—the prototypes being PSD-95 or SAP-90—facilitate clustering of N-methyl-D-aspartate (NMDA)-type glutamate receptors and their associated proteins. Other PDZ-containing proteins interact with α-amino-3-hydroxy-5-methylisoxazole-4-propionate acid (AMPA), kainate, and metabotropic types of glutamate receptors.

Synaptic Organizing Molecules Pattern Central Nerve Terminals

Although central synapses and neuromuscular junctions share many features, their synaptic clefts differ

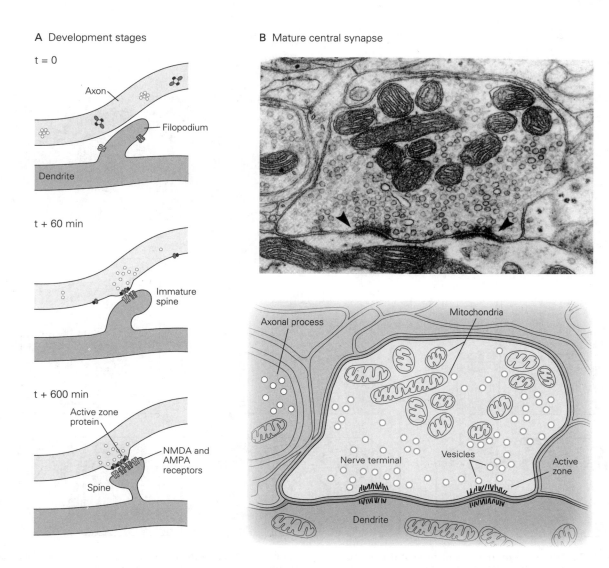

A Development stages

t = 0

Axon

Filopodium

Dendrite

t + 60 min

Immature
spine

t + 600 min

Active zone
protein

NMDA and
AMPA
receptors

Spine

B Mature central synapse

Axonal process

Mitochondria

Nerve terminal

Vesicles

Active
zone

Dendrite

Figure 48–15 Ultrastructure of a synapse in the mammalian
central nervous system.

A. Initial contact between an axon and a filopodium on
a developing dendrite leads to a stable dendritic spine
and an axodendritic synapse. This entire process can
take as little as 60 minutes. (Abbreviations: **AMPA**,

α-amino-3-hydroxy-5-methylisoxazole-4-propionate acid; **NMDA**,
N-methyl-D-aspartate.)

B. In a mature interneuron synapse in the cerebellum, synaptic
vesicles in the nerve terminal are clustered at active zones
(**arrows**) directly opposite receptor-rich patches of postsynaptic
membrane. (Reproduced, with permission, from J.E. Heuser
and T.S. Reese.)

dramatically. Whereas muscle fibers are ensheathed by
a basal lamina that has a distinctive molecular structure at the neuromuscular junction, central neurons
do not have a prominent basal lamina. Instead, formation of central synapses is regulated in large part
by molecules embedded in the pre- and postsynaptic
membranes.

Several interacting pairs of membrane proteins
have now been found that link the pre- and postsynaptic

membranes and also organize synaptic differentiation
as synapses form. Perhaps the best studied are a set of
proteins called neurexins, which are enriched in presynaptic membranes, and their partners, the neuroligins,
which are concentrated in postsynaptic membranes
(Figure 48–17A). There are three neurexin and four
neuroligin genes in the mammalian genome. The ability of neurexins and neuroligins to promote synaptic
differentiation was first revealed by culturing neurons

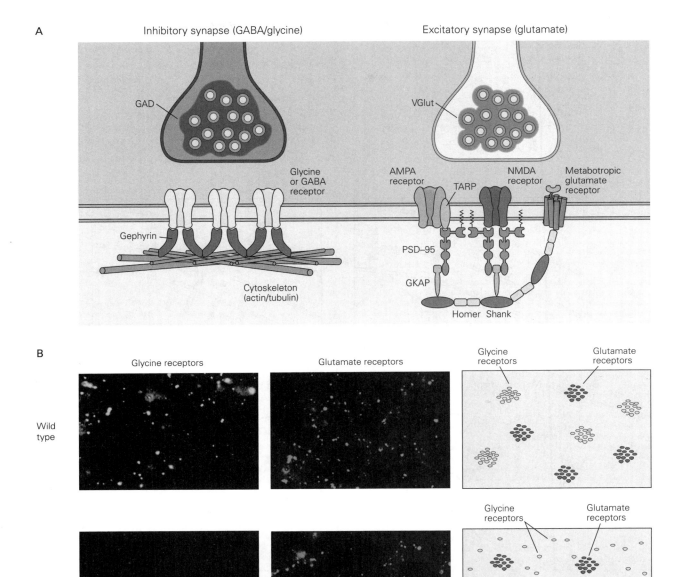

Figure 48–16 Localization of neurotransmitter receptors in central neurons.

A. Glutamate receptors are localized at excitatory synapses, and γ-aminobutyric acid (**GABA**) and glycine receptors are localized at inhibitory synapses. The receptors are linked to the cytoskeleton by adaptor proteins. Glycine receptors are linked to microtubules by gephyrin (*left*), and N-methyl-D-aspartate (**NMDA**)-type glutamate receptors are linked to each other and to the cytoskeleton by PSD-95–related molecules (*right*). The PSD family of molecules contains PDZ domains that interact with a variety of synaptic proteins to assemble signaling complexes. Other PDZ-containing proteins interact with

α-amino-3-hydroxy-5-methylisoxazole-4-propionate acid (**AMPA**)-type and metabotropic glutamate receptors (see Chapter 13). (Abbreviations: **GAD**, glutamate decarboxylase; **GKAP**, Guanylate-kinase-associated protein; **TARP**, transmembrane AMPA receptor regulatory proteins; **VGlut**, vesicular glutamate transporter.)

B. In gephyrin mutant mice, glycine receptors do not cluster at synaptic sites on spinal motor neurons, and the animals show spasticity and hyperreflexia. In the same neurons, glutamate receptor clusters are unaffected. (Adapted, with permission, from Feng et al. 1998.)

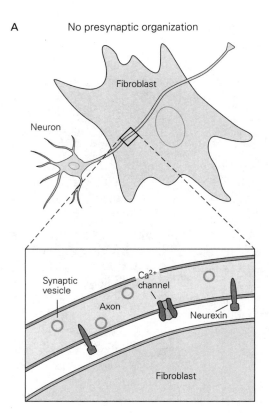

A No presynaptic organization

Fibroblast

Neuron

Synaptic vesicle

Ca²⁺ channel

Axon

Neurexin

Fibroblast

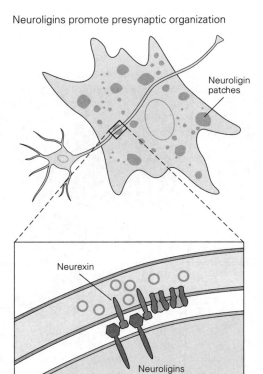

Neuroligins promote presynaptic organization

Neuroligin patches

Neurexin

Neuroligins

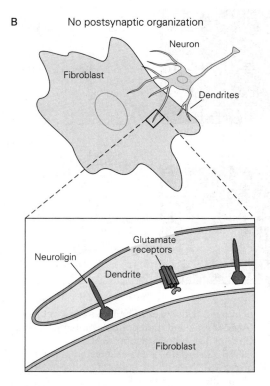

B No postsynaptic organization

Neuron

Fibroblast

Dendrites

Neuroligin

Glutamate receptors

Dendrite

Fibroblast

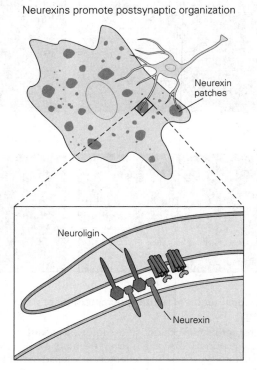

Neurexins promote postsynaptic organization

Neurexin patches

Neuroligin

Neurexin

Figure 48–17 Synaptic organizers such as neurexins and neuroligins promote differentiation of central synapses.

A. When brain neurons are cultured with fibroblast cells that express neuroligin, those segments of the axon that contact these cells form presynaptic specializations, marked by clustered neurexin, Ca²⁺ channels, and synaptic vesicles.

B. Similarly, when neurons are cultured with cells that express neurexin, dendrites that contact these cells accumulate aggregates of glutamate receptors, accompanied by scaffolding molecules (not shown) and clustered neuroligins. Neurons grown with control cells fail to form such pre- and postsynaptic specializations.

with nonneural cells engineered to express one or the other. In culture, synaptic vesicles form clusters at sites of contact with the neuroligin-expressing cells, and they are capable of releasing neurotransmitter when stimulated (Figure 48–17A). Conversely, neurotransmitter receptors in dendrites aggregate at sites that contact nonneural cells engineered to express neurexins (Figure 48–17B). Thus, neurexin–neuroligin interactions facilitate precise apposition of pre- and postsynaptic specializations.

How do neurexins and neuroligins work? Part of the answer is that their carboxy terminal tails bind to PDZ domains in proteins such as PSD-95 (Figure 48–16). Indeed, a remarkable number of proteins in both pre- and postsynaptic membranes have PDZ domain-binding motifs, notably adhesion molecules, neurotransmitter receptors, and ion channels. Moreover, many cytoplasmic proteins that possess PDZ domains are present in nerve terminals and beneath the postsynaptic membrane. Thus, PDZ-containing proteins can serve as scaffolding molecules that link key components on both sides of the synapse. Interactions of proteins such as neurexins and neuroligins may provide a means of coupling the intercellular interactions required for synaptic recognition to the intracellular interactions required to cluster synaptic components within the cell membrane.

Although neurexin–neuroligin interactions promote synaptic differentiation in culture, mice lacking neurexins or neuroligins form synapses in vivo. However, the synapses that form in the mutants are defective, with the nature and severity of the defects varying among synaptic types. Thus, the primary role of these synaptic organizers may be to specify the properties of particular synapses. For example, neuroligin1 is concentrated in the postsynaptic membrane of excitatory synapses, and levels of glutamate receptors are reduced at excitatory synapses in neuroligin1 mutants. Conversely, neuroligin2 is concentrated at inhibitory synapses and plays a critical role in patterning the inhibitory postsynaptic membrane.

Additional complexity in the tuning of central synapses by neurexins arises from the fact that they bind to multiple postsynaptic organizing molecules in addition to the neuroligins (Figure 48–18). Moreover, thousands of neurexin isoforms are generated from each neurexin gene as a result of differences in promoter choice (generating α and β forms) and alternative splicing at multiple sites. Different neurexin isoforms are differentially expressed by neurons and have different affinities for the various neurexin ligands. Neuroligins are also alternatively spliced and differentially expressed and thus are likely to have multiple presynaptic partners.

More recently, other synaptic organizing molecules have been found; they include protein tyrosine phosphatases and leucine-rich repeat proteins as well as members of the fibroblast growth factor (FGF) and

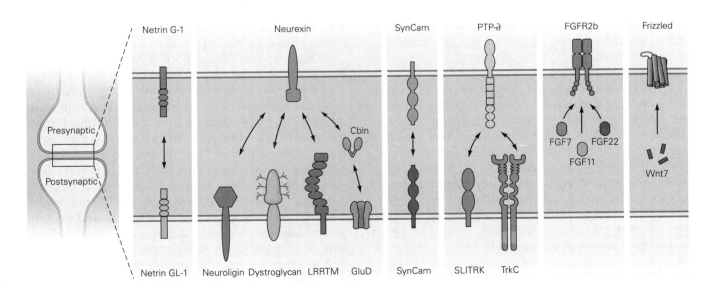

Figure 48–18 Numerous macromolecular complexes link pre- and postsynaptic membranes at central synapses. The figure shows some of the many transsynaptic proteins that interact at synaptic sites. Some bias synapse formation in favor of appropriate partners, whereas others act to regulate the properties of the synapse; some may do both.

Wnt families of secreted morphogens and their receptors (Figure 48–18). They are present at specific subsets of synapses and play distinct roles. For example, similar to neuroligin1 and neuroligin2, FGF22 and FGF7 are localized to and promote differentiation of excitatory and inhibitory synapses, respectively. Some of these organizing proteins may act in parallel with neurexins, while others may act as initial organizers, with neuroligins and neurexins consolidating the synapses at a later time and specifying their particular properties.

Together, these results suggest that central synapses are not patterned by master organizers akin to agrin, MuSK, LRP4, and laminins. Indeed, loss of no single central organizer studied to date is lethal in the manner observed for agrin, MuSK, LRP4, and laminin mutants. Instead, the enormous variety of neuronal and synaptic types in the central nervous system and their wide range of functional properties arise from a multitude of organizers that act combinatorially and in cell type–specific ways. Consistent with this view, genetic variation in many central organizers and synaptic recognition molecules, including neurexins, neuroligins, cadherins, and contactins, has been associated with behavioral perturbations in experimental animals and with behavioral disorders, including autism, in humans (Chapter 62).

Some Synapses Are Eliminated After Birth

In adult mammals, each muscle fiber bears only a single synapse. However, this is not the case in the embryo. At intermediate stages of development, several axons converge on each myotube and form synapses at a common site. Soon after birth, all inputs but one are eliminated.

The process of synapse elimination is not a manifestation of neuronal death. Indeed, it generally occurs long after the period of naturally occurring cell death (Chapter 46). Each motor axon withdraws branches from some muscle fibers but strengthens its connections with others, thus focusing its increasing capacity for transmitter release on a decreasing number of targets. Moreover, axonal elimination is not targeted to defective synapses; all inputs to a neonatal myotube are morphologically and electrically similar, and each can activate the postsynaptic cell (Figure 48–19).

What is the purpose of the transient stage of polyneuronal innervation? One possibility is that it ensures that each muscle fiber is innervated. A second is that it allows all axons to capture an appropriate set of target cells. A third, intriguing idea is that synapse elimination provides a means by which activity can change

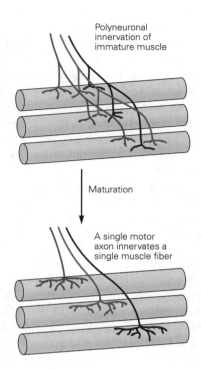

Figure 48–19 Some neuromuscular synapses are eliminated after birth. Early in the development of the neuromuscular junction, each muscle fiber is innervated by several motor axons. After birth, all motor axons but one withdraw from each fiber, and the surviving axon becomes more elaborate. Synapse elimination occurs without any overall loss of axons—axons that "lose" at some muscle fibers "win" at others. Central synapses are also subject to elimination.

the strength of specific synaptic connections. We will explore this idea in Chapter 49.

Like synapse formation, synapse elimination results from intercellular interactions. Every muscle fiber ends up with exactly one input: None have zero, and very few have more than one. It is difficult to imagine how this could occur without feedback from the muscle cell. Moreover, the axons that remain after partial denervation at birth have a larger number of synapses than they did initially. Thus, synapse elimination appears to be a competitive process.

What drives the competition, and what is the reward? There is good evidence that neural activity plays a role: Paralysis of muscle reduces synapse elimination, whereas direct stimulation enhances it. These findings showed that activity was involved but did not reveal how the outcome was determined, because all axons were stimulated or paralyzed together. Because the essence of the competitive process is that some synapses gain territory at the expense of others, differential activity among axons may be a determinant of axon winners and losers. Changing the activity of only a subset of axons in a living animal has been a technical challenge, but genetic approaches have made this possible in mice. In fact, when the activity of one of the inputs to a muscle fiber is decreased, that axon is highly likely to withdraw.

If the more active axon wins the competition, there is a new problem. Because all synapses made by an axon have the same activity pattern, one might predict that the least active axon in the muscle would eventually lose all of its synapses and the most active would retain all of its synapses. Yet this does not happen. Instead, all axons win at some sites and lose at others, so that every axon ends up innervating a substantial number of muscle fibers.

One possible resolution to this paradox is that the outcome of competition may not depend on the number of synaptic potentials from the winning axon at a synapse but rather on the total amount of synaptic input that the axon provides to the muscle—a product of the number of impulses and the amount of transmitter released per impulse. In this case, an axon that loses at several synapses might redistribute its resources (eg, synaptic vesicles) so that the remaining terminals would be strengthened and more likely to win at their synapses. Conversely, an axon that wins many competitions might find itself with insufficient vesicles to generate large synaptic potentials and thus would eventually lose to competitors at some synapses. Accordingly, the number of muscle fibers innervated by individual axons would vary much more among axons than is actually observed.

If activity drives the competition, what is the object of the competition? One idea is that the mechanisms are similar to those that determine whether neurons live or die. The muscle might produce limited amounts of a trophic substance for which the axons compete. As the winner grows, it either deprives the loser of its sustenance or gains enough strength to mount an attack that results in removal of its competitor. Alternatively, the muscle might release a toxic or punitive factor. In these scenarios, although the muscle does contribute a factor in the competition, the outcome is entirely dependent on differences between axons. These differences could be related to activity. The more active axon might be better able to take up trophic factor or resist a toxin. Such positive and negative competitive interactions have been demonstrated at nerve-muscle synapses in culture, although not in vivo.

Nevertheless, the muscle could play a selective role in synapse elimination rather than just providing a broadly distributed signal. For example, the more active axon might trigger a signal from the muscle fiber that strengthens its adhesive interactions with the synaptic cleft, whereas the less active axon might elicit a signal that weakens those interactions.

The complexity of the brain makes direct demonstration of synapse elimination problematic, but electrophysiological evidence from many parts of the central nervous system indicates that synapse elimination is widespread. In autonomic ganglia and cerebellar Purkinje cells, synapse elimination has been documented directly and its rules seem similar to those found at neuromuscular junctions. Individual axons withdraw from some postsynaptic cells while simultaneously increasing the size of the synapses they form with other neurons.

Glial Cells Regulate Both Formation and Elimination of Synapses

Classical studies of synapse formation and maturation focused, logically enough, on the pre- and postsynaptic partners. More recently, however, there has been a growing appreciation of the role played by a third type of cell: the glial cells that cap nerve terminals. Schwann cells are the glia at neuromuscular junctions, and astrocytes are the glia at central synapses. Both have been implicated in synapse formation and maturation.

The most penetrating analyses were performed by the late Ben Barres and his colleagues. They devised methods to culture neurons in defined media and in the complete absence of nonneuronal cells. Using this system, they found that neurons formed few synapses when cultured in isolation but many when astrocytes were present (Figure 48–20). The astrocytes provide multiple signals to neurons. Some, such as thrombospondin, promote postsynaptic maturation, whereas others, such as cholesterol, promote presynaptic maturation.

Another glial type, the microglial cell, also plays critical roles. Microglia are relatives of macrophages and monocytes in other tissues, sharing their ability to eliminate dead cells or debris. Initially thought to be primarily involved in the brain's response to damage, they have now been found to phagocytose synaptic terminals during the period of synapse elimination. True to their phagocytic origins, they use the complex system of complement factors, initially studied in the context of immunity, to target terminals; the targeting is activity dependent, providing a possible mechanism for the activity dependence of synapse elimination (Figure 48–21). An intriguing possibility is that dysregulation of microglial pruning contributes to synaptic loss in neurodegenerative diseases such as Alzheimer disease and schizophrenia (see Chapters 60 and 64).

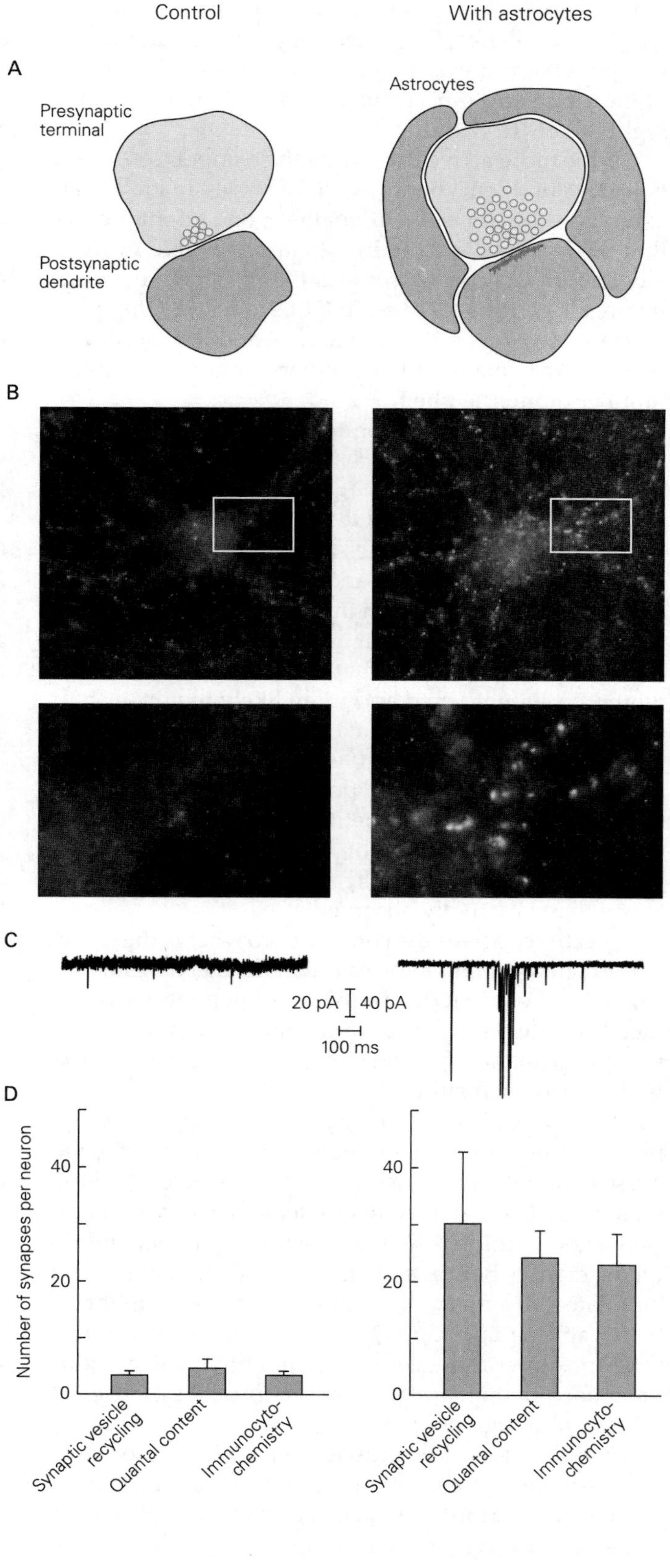

Figure 48–20 Signals from astrocytes promote synapse formation.

A. Astrocytes promote the maturation of both pre- and postsynaptic elements of the synapse.

B. Neurons cultured with astrocytes form more synapses, as assessed by expression of synaptic proteins (**yellow dots**). (Reproduced, with permission, from Ben A. Barres.)

C. Retinal neurons cultured with astrocytes form a greater number of synapses, as shown by increased transmitter release.

D. Synapse formation is enhanced in the presence of astrocytes by three measures.

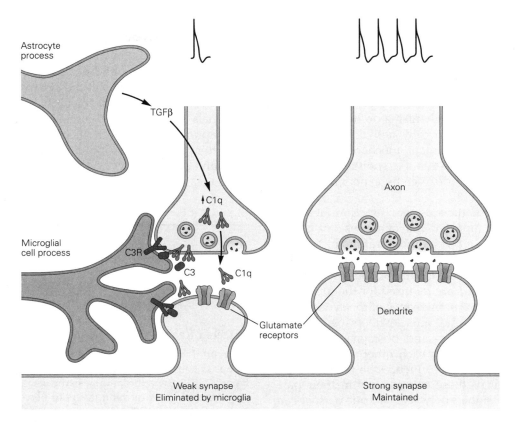

Figure 48–21 Microglia prune synapses, contributing to synapse elimination. Microglia engulf weak synapses. The engulfment is stimulated by complement components such as C1q, which tags the inactive terminal and marks it for removal by a process involving interaction of C3 with the complement receptor C3R on the microglia. Astrocytes play a role by secreting transforming growth factor β (TGFβ), which promotes production of C1q. (Adapted, with permission, from Allen 2014. Permission conveyed through Copyright Clearance Center, Inc.)

The roles of glia in synaptic development are only beginning to be worked out, and the assignments of astrocytes and microglia to synapse formation and elimination are clearly oversimplifications. Both glial types are involved in both processes, and Schwann cells may play both roles at the neuromuscular junction. Moreover, a complex set of signals passes between astrocytes and microglia, and between neurons and glia, all of which contribute to development and are at risk of going awry in brain disorders.

Highlights

1. Elaborate guidance mechanisms bring axons to appropriate target areas, but within those areas they still need to choose synaptic partners, often from among many neuronal types. Multiple mechanisms guide these choices.

2. Matching cell-surface recognition molecules on pre- and postsynaptic partners provide one prevalent mechanism for synaptic specificity. They include members of the cadherin, immunoglobulin, and leucine-rich repeat protein superfamilies. Individual members are selectively expressed by subsets of neurons and exhibit selective binding. Often, the binding is homophilic, biasing connectivity in favor of partners expressing the same molecule.

3. Other mechanisms promoting specificity include selective interactions among axons, the ability of some axons to convert their targets to the appropriate types, and selective elimination of inappropriate contacts.

4. At present, it remains unknown how many molecular species are required to wire up neural circuits in the mammalian brain. At one time, it seemed that molecular complexity might need to approach the complexity of circuits, but it is more likely that a few hundred recognition molecules will suffice, given their combinatorial use, as well as deployment of the same gene at multiple times and in multiple regions.

5. Spatial constraints that enhance specificity include restriction of axons and dendrites to particular laminae within a target region—thereby restricting their choice of partners—and restriction of synapses of particular types to defined domains on the target cell surface.

6. Some specificity mechanisms do not require the partners to be electrically active, but in many cases, activity-dependent mechanisms sharpen specificity. Activity can be spontaneous, early in development, or driven by experience at later stages.

7. The skeletal neuromuscular junction, at which the axon of a motor neuron synapses on a muscle fiber, has been a favored preparation for working out principles of synaptic development. A key finding is that multiple interactions between the synaptic partners are required for the formation, maturation, and maintenance of the synapse.

8. Motor neurons and muscle fibers can express genes encoding pre- and postsynaptic components, respectively, in each other's absence, but they exert profound influences on the levels and distribution of these components in their partners. Thus, signals between synaptic partners are best viewed as organizers rather than inducers.

9. At the neuromuscular junction, a layer of basal lamina occupies the synaptic cleft between the motor nerve terminal and the postsynaptic membrane. Nerve and muscle secrete signaling molecules into the cleft, where they become stabilized and organize differentiation.

10. A key nerve-derived organizer of postsynaptic differentiation is agrin. It acts through the receptors MuSK and LRP4 to cluster acetylcholine receptors and other postsynaptic components beneath the nerve terminal. Nerve-evoked activity also affects postsynaptic differentiation by modulating expression of postsynaptic components. Key muscle-derived organizers of presynaptic differentiation include members of the laminin and fibroblast growth factor families.

11. Central synapses develop in ways similar to those discovered at the neuromuscular junction. Many central synaptic organizers have now been discovered, including neuroligins, neurexins, protein tyrosine phosphatases, leucine-rich repeat proteins, and numerous others.

12. Many of the synapses that form initially in both the peripheral and central nervous systems are subsequently eliminated, generally by competitive, activity-dependent mechanisms. The consequence is that as circuits mature, the number of inputs a neuron receives may decrease dramatically, but the size and strength of the remaining inputs increase even more dramatically.

13. Along with pre- and postsynaptic partners, glial cells play key roles at the synapse. In particular, both astrocytes and microglial cells receive signals from and send signals to developing synaptic partners, with these signals contributing to synapse formation, maturation, maintenance, and elimination.

Joshua R. Sanes

Selected Reading

Allen NJ, Lyons DA. 2018. Glia as architects of central nervous system formation and function. Science 362:181–185.

Baier H. 2013. Synaptic laminae in the visual system: molecular mechanisms forming layers of perception. Annu Rev Cell Dev Biol 29:385–416.

Darabid H, Perez-Gonzalez AP, Robitaille R. 2014. Neuromuscular synaptogenesis: coordinating partners with multiple functions. Nat Rev Neurosci 15:703–718.

Hirano S, Takeichi M. 2012. Cadherins in brain morphogenesis and wiring. Physiol Rev 92:597–634.

Krueger-Burg D, Papadopoulos T, Brose N. 2017. Organizers of inhibitory synapses come of age. Curr Opin Neurobiol 45:66–77.

Nishizumi H, Sakano H. 2015. Developmental regulation of neural map formation in the mouse olfactory system. Dev Neurobiol 75:594–607.

Südhof TC. 2017. Synaptic neurexin complexes: a molecular code for the logic of neural circuits. Cell 171:745–769.

Takahashi H, Craig AM. 2013. Protein tyrosine phosphatases PTPδ, PTPσ, and LAR: presynaptic hubs for synapse organization. Trends Neurosci 36:522–534.

Thion MS, Ginhoux F, Garel S. 2018. Microglia and early brain development: an intimate journey. Science 362:185–189.

Yogev S, Shen K. 2014. Cellular and molecular mechanisms of synaptic specificity. Annu Rev Cell Dev Biol 30:417–437.

References

Allen NJ. 2014. Astrocyte regulation of synaptic behavior. Annu Rev Cell Dev Biol. 30:439–463.

Anderson, MJ, Cohen MW. 1977. Nerve-induced and spontaneous redistribution of acetylcholine receptors on cultured muscle cells. J Physiol 268:757–773.

Ango F, di Cristo G, Higashiyama H, Bennett V, Wu P, Huang ZJ. 2004. Ankyrin-based subcellular gradient of neurofascin, an immunoglobulin family protein, directs GABAergic innervation at Purkinje axon initial segment. Cell 119:257–272.

Buller AJ, Eccles JC, Eccles RM. 1960. Interactions between motoneurons and muscles in respect of the characteristic speeds of their responses. J Physiol 150:417–439.

Burden SJ, Sargent PB, McMahan UJ. 1979. Acetylcholine receptors in regenerating muscle accumulate at original synaptic sites in the absence of the nerve. J Cell Biol 82:412–425.

Christopherson KS, Ullian EM, Stokes CC, et al. 2005. Thrombospondins are astrocyte-secreted proteins that promote CNS synaptogenesis. Cell 120:421–433.

DeChiara TM, Bowen DC, Valenzuela DM, et al. 1996. The receptor tyrosine kinase MuSK is required for neuromuscular junction formation in vivo. Cell 85:501–512.

Duan X, Krishnaswamy A, De la Huerta I, Sanes JR. 2014. Type II cadherins guide assembly of a direction-selective retinal circuit. Cell 158:793–807.

Feng G, Tintrup H, Kirsch J, et al. 1998. Dual requirement for gephyrin in glycine receptor clustering and molybdoenzyme activity. Science 282:1321–1324.

Fox MA, Sanes JR, Borza DB, et al. 2007. Distinct target-derived signals organize formation, maturation, and maintenance of motor nerve terminals. Cell 129:179–193.

Gautam M, Noakes PG, Moscoso L, et al. 1996. Defective neuromuscular synaptogenesis in agrin-deficient mutant mice. Cell 85:525–535.

Glicksman MA, Sanes JR. 1983. Differentiation of motor nerve terminals formed in the absence of muscle fibres. J Neurocytol 12:661–671.

Graf ER, Zhang X, Jin SX, Linhoff MW, Craig AM. 2004. Neurexins induce differentiation of GABA and glutamate postsynaptic specializations via neuroligins. Cell 119:1013–1026.

Hall ZW, Sanes JR. 1993. Synaptic structure and development: the neuromuscular junction. Cell 72:99–121. Suppl.

Huang ZJ. 2006. Subcellular organization of GABAergic synapses: role of ankyrins and L1 cell adhesion molecules. Nat Neurosci 9:163–166.

Imai T, Suzuki M, Sakano H. 2006. Odorant receptor-derived cAMP signals direct axonal targeting. Science 314:657–661.

Krishnaswamy A, Yamagata M, Duan X, Hong YK, Sanes JR. 2015. Sidekick 2 directs formation of a retinal circuit that detects differential motion. Nature 2524:466–470.

Lupa MT, Gordon H, Hall ZW. 1990. A specific effect of muscle cells on the distribution of presynaptic proteins in neurites and its absence in a C2 muscle cell variant. Dev Biol 142:31–43.

Misgeld T, Kummer TT, Lichtman JW, Sanes JR. 2005. Agrin promotes synaptic differentiation by counteracting an inhibitory effect of neurotransmitter. Proc Natl Acad Sci U S A 102:11088–11093.

Nishimune H, Sanes JR, Carlson SS. 2004. A synaptic laminin-calcium channel interaction organizes active zones in motor nerve terminals. Nature 432:580–587.

Nja A, Purves D. 1977. Re-innervation of guinea-pig superior cervical ganglion cells by preganglionic fibres arising from different levels of the spinal cord. J Physiol 272:633–651.

Noakes PG, Gautam M, Mudd J, Sanes JR, Merlie JP. 1995. Aberrant differentiation of neuromuscular junctions in mice lacking s-laminin/laminin beta 2. Nature 374:258–262.

Salmons S, Sreter FA. 1976. Significance of impulse activity in the transformation of skeletal muscle type. Nature 263:30–34.

Sanes JR, Lichtman JW. 2001. Induction, assembly, maturation and maintenance of a postsynaptic apparatus. Nat Rev Neurosci 2:791–805.

Sanes JR, Yamagata M. 2009. Many paths to synaptic specificity. Annu Rev Cell Dev Biol 25:161–195.

Schafer DP, Lehrman EK, Kautzman AG, et al. 2012. Microglia sculpt postnatal neural circuits in an activity and complement-dependent manner. Neuron 74:691–705.

Scheiffele P, Fan J, Choih J, Fetter R, Serafini T. 2000. Neuroligin expressed in nonneuronal cells triggers presynaptic development in contacting axons. Cell 101:657–669.

Serizawa S, Miyamichi K, Takeuchi H, Yamagishi Y, Suzuki M, Sakano H. 2006. A neuronal identity code for the odorant receptor-specific and activity-dependent axon sorting. Cell 127:1057–1069.

Terauchi A, Johnson-Venkatesh EM, Toth AB, Javed D, Sutton MA, Umemori H. 2010. Distinct FGFs promote differentiation of excitatory and inhibitory synapses. Nature 465:783–787.

Uezu A, Kanak DJ, Bradshaw TW, et al. 2016. Identification of an elaborate complex mediating postsynaptic inhibition. Science 353:1123–1129.

Vaughn JE. 1989. Fine structure of synaptogenesis in the vertebrate central nervous system. Synapse 3:255–285.

Yamagata M, Sanes JR. 2012. Expanding the Ig superfamily code for laminar specificity in retina: expression and role of contactins. J Neurosci 32:14402–14414.

Yumoto N, Kim N, Burden SJ. 2012. Lrp4 is a retrograde signal for presynaptic differentiation at neuromuscular synapses. Nature 489:438–442.

49

Experience and the Refinement of Synaptic Connections

THE HUMAN NERVOUS SYSTEM IS FUNCTIONAL at birth—newborn babies can see, hear, breathe, and suckle. However, the capabilities of human infants are quite rudimentary compared to those of other species. Wildebeest calves can stand and run within minutes of birth, and many birds can fly shortly after they hatch from their eggs. In contrast, a human baby cannot lift its head until it is 2 months old, cannot bring food to its mouth until it is 6 months old, and cannot survive without parental care for a decade.

What accounts for the delayed maturation of our motor, perceptual, and cognitive abilities? One main factor is that the embryonic connectivity of the nervous system, discussed in Chapters 45 through 48, is only a "rough draft" of the neural circuits that exist in our adult selves. Embryonic circuits are refined by sensory stimulation—our experiences. This two-part sequence—genetically determined connectivity followed by experience-dependent reorganization—is a common feature of mammalian neural development, but in humans, the second phase is especially prolonged.

At first glance, this delay in human neural development might seem dysfunctional. It does exact a toll, but it also provides an advantage. Because our mental abilities are shaped largely by experience, we gain the ability to custom fit our nervous systems to our individual bodies and unique environments. It has been argued that it is not just the large size of the human brain but also its experience-dependent maturation that makes our mental capabilities superior to those of other species.

The plasticity of the nervous system in response to experience endures throughout life. Nevertheless, periods of heightened susceptibility to modification, known as *sensitive periods*, occur at particular times in development. In some cases, the adverse effects of deprivation or atypical experience during circumscribed periods in early life cannot easily be reversed by providing appropriate experience at a later age. Such periods are referred to as *critical periods*. As we shall see, new discoveries are blurring the distinction between sensitive and critical periods, so we will use the term "critical periods" to refer to both.

Behavioral observations have helped us appreciate critical periods. Imprinting, a form of learning in birds, is one of the most striking illustrations of a lifelong behavior established during a critical period. Just after hatching, birds become indelibly attached, or imprinted, to a prominent moving object in their environment and follow it around. This is typically their mother, but it could be an experimenter who is near the newborn chick. The process of imprinting is important for the protection of the hatchling. Although the attachment is acquired rapidly and persists, imprinting can only occur during a critical period soon after hatching—in some species, only a few hours.

In humans, critical periods are evident in the ways children acquire the capacities to perceive the world around them, learn a language, or form social relationships. A 5-year-old child can quickly and effortlessly learn a second language, whereas a 15-year-old adolescent may become fluent but is likely to speak with an accent, even if he lives to be 90 years old. Likewise, deaf children fitted with a cochlear implant during the first 3 to 4 years of life generally acquire and understand spoken language well, whereas neither production nor understanding may ever be normal following implantation at later ages. Such critical periods demonstrate that experience-dependent neural development is concentrated in, although certainly not confined to, early postnatal life.

We begin this chapter by examining the evidence that early experience shapes a range of human mental capacities, from our ability to make sense of what we see to our ability to engage in appropriate social interactions. The neural basis of these experiential effects

has been analyzed in numerous parts of the brains of experimental animals, including the auditory, somatosensory, motor, and visual systems. Here, rather than surveying multiple systems, we will focus primarily on the visual system because research on this system has provided a particularly rich understanding of how experience shapes neural circuitry. We will see that experience is needed to refine patterns of synaptic connections and to stabilize these patterns once they have formed. Finally, we will consider recent evidence that critical periods in many systems are less restrictive than once thought and, in some cases, can be extended or even "reopened."

Understanding critical periods in childhood and the extent to which they can be reopened in adulthood has many important practical consequences. First, much educational policy is based on the idea that early experience is crucial, so it is important to know exactly when a particular form of enrichment will be optimally beneficial. Second, medical treatment of many childhood conditions, such as congenital cataracts or deafness, is now predicated on the idea that early intervention is imperative if long-lasting deficits are to be avoided. Third, there is increasing suspicion that some behavioral disorders, such as autism, may be caused by impairment of reorganization of neural circuits during critical periods. Finally, the possibility of reopening critical periods in adulthood is leading to new therapeutic approaches to neural insults, such as stroke, that previously were thought to have irreversible consequences.

Development of Human Mental Function Is Influenced by Early Experience

Early Experience Has Lifelong Effects on Social Behaviors

One of the first indications that early social and perceptual experiences have irreversible consequences for human development came from studies of children who had been deprived of these experiences early in life. In rare cases, children abandoned in the wild and later returned to human society have also been studied. As might be expected, these children were socially maladjusted, but surprisingly, the defects proved to persist throughout life.

In the 1940s, the psychoanalyst René Spitz provided more systematic evidence that early interactions with other humans are essential for normal social development. Spitz compared the development of infants raised in a foundling home with the development of infants raised in a nursing home attached to a women's prison. Both institutions were clean and both provided adequate food and medical care. The babies in the prison nursing home were all cared for by their mothers, who, although in prison and away from their families, tended to shower affection on their infants in the limited time allotted to them each day. In contrast, infants in the foundling home were cared for by nurses, each of whom was responsible for several babies. As a result, children in the foundling home had far less contact with other humans than did those in the prison's nursing home.

The two institutions also differed in another respect. In the prison nursing home, the cribs were open, so that the infants could readily watch other activities in the ward; they could see other babies play and observe the staff go about their business. In the foundling home, the bars of the cribs were covered by sheets that prevented the infants from seeing outside. In reality, the babies in the foundling home were living under conditions of severe sensory and social deprivation.

Infants at the two institutions were followed through their early years. At the end of the first 4 months, the infants in the foundling home fared better on several developmental tests than those in the prison nursing home, suggesting that intrinsic factors did not favor the infants in the prison nursing home. But by the end of the first year, the motor and intellectual performance of the children in the foundling home had fallen far below that of children in the prison nursing home. Many of the children in the foundling home had developed a syndrome that Spitz called *hospitalism* and is now sometimes called *anaclitic depression*. These children were withdrawn and displayed little curiosity or gaiety. Moreover, their defects extended beyond emotional and cognitive signs. They were especially prone to infection, implying that the brain exerts complex controls over the immune system as well as behavior. By their second and third years, children in the prison nursing home were similar to children raised in normal families at home—they were agile, had a vocabulary of hundreds of words, and spoke in sentences. In contrast, the development of children in the foundling home was still further delayed—many were unable to walk or to speak more than a few words.

More recent studies of other similarly deprived children have confirmed these conclusions and shown that the defects are long-lasting. Longitudinal studies of orphans who were raised for several years in large impersonal institutions with little or no personal care, then adopted by caring families, have been especially revealing. Despite every effort

of the adoptive parents, many of the children were never able to develop appropriate, caring relationships with family members or peers (Figure 49–1A). More recent imaging studies have revealed defects in brain structure correlated with, and presumably due to, this deprivation (Figure 49–1B).

As compelling as these observations are, it is difficult to derive definitive conclusions from them. An influential set of studies that extended the analysis of social behavior to monkeys was carried out in the 1960s by two psychologists, Harry and Margaret Harlow. The Harlows reared newborn monkeys in isolation for 6 to 12 months, depriving them of contact with their mothers, other monkeys, or people. At the end of this period, the monkeys were physically healthy but behaviorally devastated. They crouched in a corner of their cage and rocked back and forth like autistic children (Figure 49–1C). They did not interact with other monkeys, nor did they fight, play, or show any sexual interest. Thus, a 6-month period of social isolation during the first 18 months of life produced persistent and serious disturbances in behavior. By comparison, isolation of an older animal for a comparable period was found to be without such drastic consequences. These results confirmed, under controlled conditions, the critical influence of early experience on later behavior. For ethical reasons, these studies would not be possible today.

Development of Visual Perception Requires Visual Experience

The dramatic dependence of the brain on experience and the ability of that experience to shape perception is evident in people born with cataracts. Cataracts are opacities of the lens that interfere with the optics of the eye but not directly with the nervous system; they are easily removed surgically. In the 1930s, it became apparent that patients who had congenital binocular cataracts removed after the age of 10 years experienced permanent deficits in visual acuity and had difficulties perceiving shape and form. In contrast, when cataracts that develop in adults are removed decades after they form, normal vision returns immediately.

Likewise, children with *strabismus* (crossed eyes) do not have normal depth perception (*stereopsis*), an ability that requires the two eyes to focus on the same location at the same time. They can acquire this ability if their eyes are aligned surgically during the first few years of life, but not if surgery occurs later in adolescence. As a result of these observations, congenital cataracts are now usually removed, and strabismus is corrected surgically, in early childhood. Over the

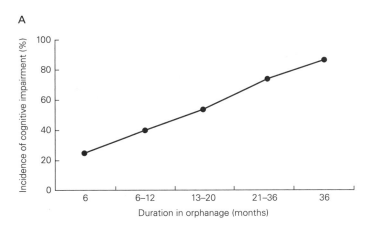

Figure 49–1 Early social deprivation has a profound impact on later brain structure and behavior.

A. Neurocognitive dysfunction is evident in children raised under conditions of social deprivation in orphanages. The incidence of cognitive impairment increases with the duration of stay in the orphanage. (Adapted from Behen et al. 2008.)

B. Diffusion tensor magnetic resonance imaging (MRI) scans show a well-developed and robust uncinate fasciculus (**red region**) in a normal child (*left*), whereas in a socially deprived child (*right*), it is thin and poorly organized. (Reproduced, with permission, from Eluvathingal et al. 2006. Copyright © 2006 by the AAP.)

C. Early social interactions impact later social behavior patterns. Monkeys reared in the presence of their siblings acquire social skills that permit effective interactions in later life (*left*). A monkey reared in isolation never acquires the capacity to interact with others and remains secluded and isolated in later life (*right*). (Source: Harry F. Harlow. Used with permission.)

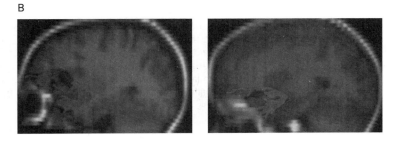

past five decades, researchers have elucidated structural and physiological underpinnings of these critical periods.

Development of Binocular Circuits in the Visual Cortex Depends on Postnatal Activity

Because sensory experience of the world is transformed into patterns of electrical activity in the brain, one might imagine that electrical signals in neural circuits affect the brain's circuitry. But is this true? And if it is true, what changes occur, and how does activity trigger them?

Our most detailed understanding of these links comes from studies of the neural circuits that mediate binocular vision. The key figures in the early phases of

this work were David Hubel and Torsten Wiesel. Following their pioneering studies on the structural and functional organization of the visual cortex in cats and monkeys (Chapter 23), they undertook another set of studies on how experience affects the circuits they had delineated.

Visual Experience Affects the Structure and Function of the Visual Cortex

In one influential study, Hubel and Wiesel raised a monkey from birth to 6 months of age with one eyelid sutured shut, thus depriving the animal of vision in that eye. When the sutures were removed, it became clear that the animal was blind in the deprived eye, a condition called *amblyopia*. They then performed electrophysiological recordings from cells along the visual

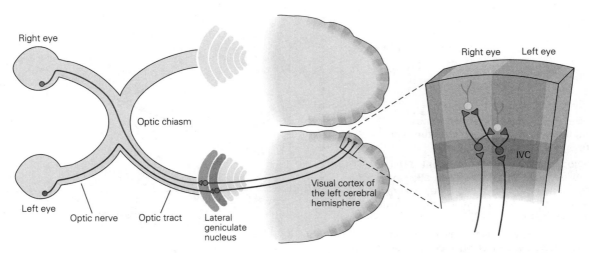

Figure 49–2 Afferent pathways from the two eyes project to discrete columns of neurons in the visual cortex. Retinal ganglion neurons from each eye send axons to separate layers of the lateral geniculate nucleus. The axons of neurons in this nucleus project to neurons in layer IVC of the primary visual cortex, which is organized in alternating sets of ocular dominance columns; each column receives input from only one eye. The axons of the neurons in layer IVC project to neurons in adjacent columns as well as to neurons in the upper and lower layers of the same column. As a result, most neurons in the upper and lower layers of the cortex receive information from both eyes.

pathway to determine where the defect arose (Figure 49–2). They found that retinal ganglion cells in the deprived eye, as well as neurons in the lateral geniculate nucleus that receive input from the deprived eye, responded well to visual stimuli and had essentially normal receptive fields.

In contrast, cells in the visual cortex were fundamentally altered. In the cortex of normal animals, most neurons are responsive to binocular input. In animals that had been monocularly deprived for the first 6 months, most cortical neurons did not respond to signals from the deprived eye (Figure 49–3). The few cortical cells that were responsive were not sufficient for visual perception. Not only had the deprived eye lost its ability to drive most cortical neurons, but little recovery ever occurred: The loss was permanent and irreversible.

Hubel and Wiesel went on to test the effects of visual deprivation imposed for shorter periods and at different ages. They obtained three types of results, depending on the timing and duration of the deprivation. First, monocular deprivation for a few weeks shortly after birth led to loss of cortical responses from the deprived eye that was reversible after the eye had been opened, especially if the opposite eye was then closed to encourage use of the initially deprived eye. Second, monocular deprivation for a few weeks during the next several weeks also resulted in a substantial loss of cortical responsiveness to signals from the deprived eye, but in this case, the effects were irreversible. Finally, deprivation in adults, even for periods of

many months, had no effect on the responses of cortical cells to signals from the deprived eye or on visual perception. These results demonstrated that the cortical connections that control visual perception are established within a critical period of early development.

Are there anatomical correlates of these functional defects? To address this question, we need to recall three basic facts about the anatomy of the visual cortex (Figure 49–2). First, inputs from the two eyes remain segregated in the lateral geniculate nucleus. Second, the geniculate inputs carrying information from the two eyes to the cortex terminate in alternating columns, termed *ocular dominance columns*. Third, lateral geniculate axons terminate on neurons in layer IVC of the primary visual cortex; convergence of input from the two eyes on a common target cell occurs at the next stage of the pathway, in cells above and below layer IVC.

To examine whether the architecture of ocular dominance columns depends on visual experience early in postnatal life, Hubel and Wiesel deprived newborn animals of vision in one eye and then injected a labeled amino acid into the normal eye. The injected label was incorporated into proteins in retinal ganglion cell bodies, transported along the retinal axons to the lateral geniculate nucleus, transferred to geniculate neurons, and then transported to the synaptic terminals of these axons in the primary visual cortex. After closure of one eye, the columnar array of synaptic terminals relaying input from the deprived eye was reduced, whereas the columnar array of terminals relaying input from the normal eye was expanded (Figure 49–4). Thus, sensory deprivation early in life alters the structure of the cerebral cortex.

How are these striking anatomical changes brought about? Does sensory deprivation alter ocular dominance columns after they have been established, or does it interfere with their formation? A columnar organization of the visual cortex is already evident by birth in monkeys, although the mature pattern is not achieved until several weeks after birth (Figure 49–5). Only at this time do the terminals of fibers from the lateral geniculate nucleus become completely segregated in the cortex. Because the inputs are partially but not completely segregated at the time visual deprivation exerts its effects, we can conclude that the deprivation perturbs the ability of the inputs to acquire their mature pattern. We shall return to the question of what leads to the initial, experience-independent phases of segregation in a later section of this chapter.

Patterns of Electrical Activity Organize Binocular Circuits

How does activity lead to maturation of ocular dominance columns? The crucial factor may be the

Figure 49–3 (Right) Responses of neurons in the primary visual cortex of a monkey to visual stimuli. (Adapted from Hubel and Wiesel 1977.)

A. A diagonal bar of light is moved leftward across the visual field, traversing the receptive fields of a binocularly responsive cell in area 17 of visual cortex. Receptive fields measured through the right and left eye are drawn separately. The receptive fields of the two cells are similar in orientation, position, shape, and size, and respond to the same form of stimulus. Recordings (below) show that the cortical neuron responds more effectively to input from the ipsilateral eye. (Abbreviation: **F**, fixation point.)

B. The responses of individual cortical neurons in area 17 can be classified into seven groups. Neurons receiving input only from the contralateral eye (**C**) fall into group 1, whereas neurons that receive input only from the ipsilateral eye (**I**) fall into group 7. Other neurons receive inputs from both eyes, but the input from one eye may influence the neuron much more than the other (groups 2 and 6), or the differences may be slight (groups 3 and 5). Some neurons respond equally to input from both eyes (group 4). According to these criteria, the cortical neuron shown in part A falls into group 6.

C. Responsiveness of neurons in area 17 to stimulation of one or the other eye. **1.** The responses of more than 1,000 neurons in area 17 in the left hemisphere of normal adult and juvenile monkeys. Neurons in layer IV that normally receive only monocular input have been excluded. **2.** The responses of neurons in the left hemisphere of a monkey in which the contralateral (right) eye was closed from the age of 2 weeks to 18 months and then reopened. Most neurons respond only to stimulation of the ipsilateral eye.

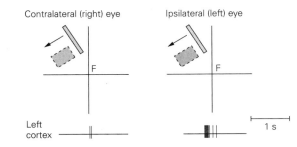

A Movement across the retina

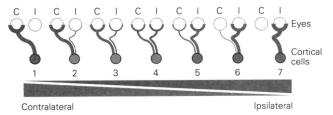

B Variation in responses of single cortical cells

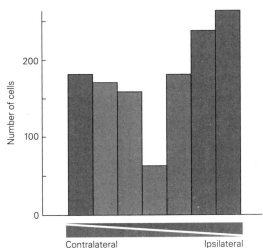

C₁ Normal area 17

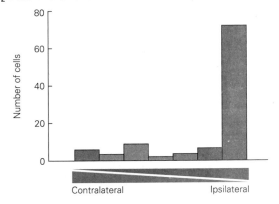

C₂ Area 17 after closure of contralateral eye

A Normal

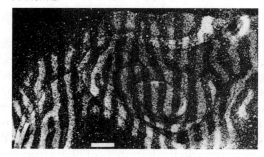

Figure 49–4 Visual deprivation of one eye during a critical period of development reduces the width of the ocular dominance columns for that eye. (Scale bars = 1 mm) (Adapted, with permission, from Hubel, Wiesel, and LeVay 1977.)

A. A tangential section through area 17 of the right hemisphere of a normal adult monkey, 10 days after one eye was injected with a radiolabeled amino acid. Radioactivity is localized in stripes (**white**) in layer IVC of the visual cortex, indicating sites of termination of the axons from the lateral geniculate nucleus that carry input from the injected eye. The alternating unlabeled (**dark**) stripes indicate sites of termination of the axons carrying signals from the uninjected eye. Labeled and unlabeled stripes are of equal width.

B. A comparable section through the visual cortex of an 18-month-old monkey whose right eye had been surgically closed at 2 weeks of age. Label was injected into the left (open) eye. The wider (**white**) stripes are the labeled terminals of afferent axons carrying signals from the open eye; the narrow (**dark**) stripes are terminals of axons with input from the closed eye.

C. A section comparable to that in part B from an 18-month-old animal whose right eye had been shut at 2 weeks. Label was injected into the closed eye, giving rise to narrow (**white**) stripes of labeled axon terminals and wide (**dark**) stripes of unlabeled terminals.

B Deprived: open eye labeled (white)

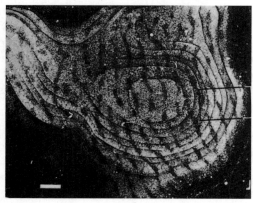

C Deprived: closed eye labeled

differences in the proportion of inputs from each eye that converge on common target cells at birth. If by chance the fibers conveying input from one eye are initially more numerous in one local region of cortex, those axons may have an advantage, leading to further segregation.

How might this occur? An attractive idea, based on a theory first proposed in the 1940s by Donald Hebb, is that synaptic connections are strengthened when pre- and postsynaptic elements are active together. In the case of binocular interactions, neighboring axons from the same eye tend to fire in synchrony because they are activated by the same visual stimulus at any instant.

The synchronization of their firing means that they cooperate in the depolarization and excitation of a target cell. This cooperative action maintains the viability of those synaptic contacts at the expense of the noncooperating synapses.

Cooperative activity could also promote branching of axons and thus create the opportunity for the formation of additional synaptic connections with cells in the target region. At the same time, the strengthening of synaptic contacts made by the axons of one eye will impede the growth of synaptic inputs from the opposite eye. In this sense, fibers from the two eyes may be said to compete for a target cell. Together,

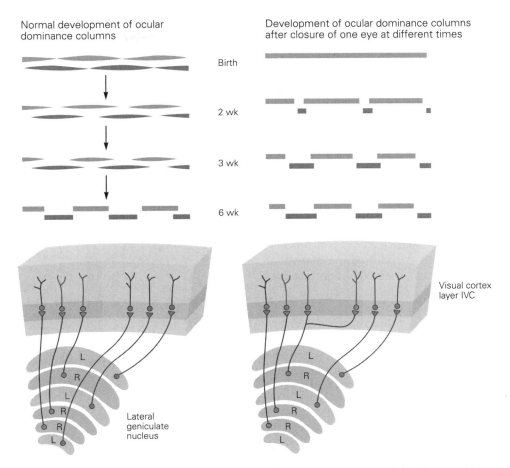

Normal development of ocular dominance columns

Development of ocular dominance columns after closure of one eye at different times

Birth

2 wk

3 wk

6 wk

Visual cortex layer IVC

Lateral geniculate nucleus

Figure 49–5 The effects of eye closure on the formation of ocular dominance columns. The top diagrams show the gradual segregation of the terminals of lateral geniculate afferents in layer IVC of the visual cortex under normal conditions (*left*) and when one eye is deprived of stimulation (*right*). **Blue domains** represent the areas of termination of inputs from one eye, **red domains** those of the other eye. The lengths of the domains represent the density of the terminals at each point along layer IVC. For clarity, the columns are shown here as one above the other, whereas in reality, they are side by side in

the cortex. During normal development, layer IVC is gradually divided into alternating sites of input from each eye. The consequences of depriving sight in one eye depend on the timing of eye closure. Closure at birth leads to dominance by the open eye (red) because at this point little segregation has occurred. Closure at 2, 3, and 6 weeks has a progressively weaker effect on the formation of ocular dominance columns because the columns become more segregated with time. (Abbreviations: L, left; R, right.) (Adapted, with permission, from Hubel, Wiesel, and LeVay 1977.)

cooperation and competition between axons ensure that two populations of afferent fibers will eventually innervate distinct regions of the primary visual cortex with little local overlap.

Competition and cooperation are not simply the outcome of neural activity per se or of differences in absolute levels of activity among axons. Instead, they appear to depend on precise temporal patterns of activity in the competing (or cooperating) axons. The principle was dramatically illustrated by Hubel and Wiesel in a set of studies that examined stereoscopic vision—the perception of depth. The brain normally computes depth perception by comparing the disparity in retinal images between the two eyes. When the

eyes are improperly aligned, this comparison cannot be made and stereoscopy is impossible. Such misalignments occur in children who are "cross-eyed," or strabismic. As noted above, this condition can be surgically repaired, but unless the surgery occurs during the first few years of life, the children forever remain incapable of stereoscopy.

Hubel and Wiesel examined the impact of strabismus on the organization of the visual system in cats. To render cats strabismic, the tendon of an extraocular muscle was severed in kittens. Both eyes remained fully functional but misaligned. Inputs from the two eyes that converged on a binocular cell in the visual cortex now carried information about different stimuli in slightly different parts of the visual field. As a result, cortical cells became monocular, driven by input from one eye or the other but not both (Figure 49–6). Conversely, cortical neurons remained binocularly responsive following binocular visual deprivation, leading to a decrease but not an imbalance in activity arising from the two eyes. These findings suggested to Hubel and Wiesel that disruption of the synchrony of inputs led to competition rather than cooperation, so that cortical cells came to be dominated by one eye, presumably the one that had dominated at the outset.

These physiological studies led investigators to test whether pharmacological blockade of electrical activity in retinal ganglion cells could affect neural connectivity in the visual system. Activity was blocked by injecting both eyes with tetrodotoxin, a toxin that selectively blocks voltage-sensitive Na^+ channels. Signals from the two eyes were generated separately by direct electric stimulation of the bilateral optic nerves. In kittens, ocular dominance columns are not established if activity in retinal ganglion neurons is blocked before the critical period of development. When the two optic nerves were stimulated synchronously, ocular dominance columns still failed to form. Only when the optic nerves were stimulated asynchronously were ocular dominance columns established.

If the development of ocular dominance columns indeed depends on competition between fibers from the two eyes, might it be possible to induce the formation of columns where they normally are not present, simply by establishing competition between two sets of axons? This radical possibility was tested in frogs, where retinal ganglion neurons from each eye project only to the contralateral side of the brain. In normal frogs, afferent fibers from the two eyes do not compete for the same cells, so there is no columnar segregation of afferent inputs. To generate competition, a third eye was transplanted early in larval development into a region of the frog's head near one of the normal eyes. The retinal ganglion neurons of the extra eye extended axons to the contralateral optic tectum. Remarkably, axon terminals from the transplanted and normal eyes segregated, generating a pattern of alternating columns (Figure 49–7).

This finding provided dramatic support for the idea that competition between afferent axons for the same population of target neurons drives their segregation into distinct target territories. The columnar segregation of retinal inputs in the frog brain is dependent on synaptic activity, presumably at the synapses between retinal axons and tectal neurons. Thus, neural activity has powerful roles in fine-tuning visual circuits.

A Alignment of eyes

Normal   Strabismic

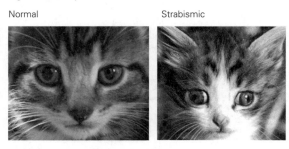

B Ocular dominance columns

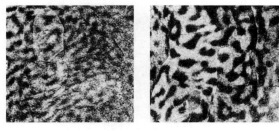

C Ocular dominance preference of V1 cells

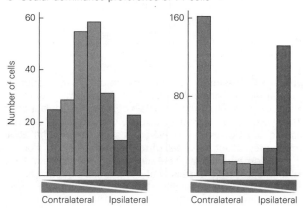

Figure 49–6 Inducing strabismus in kittens impairs the formation of binocular response regions in the primary visual cortex.

A. The eyes of strabismic cats are misaligned. (Photos [*left*] Steve Richardson/Alamy Stock Photo and [*right*] reproduced with permission from Van Sluyters and Levitt 1980.)

B. In strabismic animals, left and right eye domains are more sharply defined, an indication of the paucity of binocular regions. (Reproduced, with permission, from Löwel 1994. Copyright © 1994 Society for Neuroscience.)

C. Strabismic animals have fewer binocularly tuned neurons in the visual cortex. (Reproduced, with permission, from Hubel and Wiesel 1965.)

Figure 49–7 Ocular dominance columns can be experimentally induced in a frog by transplantation of a third eye. (Adapted, with permission, from Constantine-Paton and Law 1978. Copyright © 1978 AAAS.)

A. Three days before the transplant, the right eye was injected with a radiolabeled amino acid. The autoradiograph in a coronal section of the hindbrain shows the entire superficial neuropil of the left optic lobe filled with silver grains, indicating the region occupied by synaptic terminals from the labeled (contralateral) eye.

B. Some time after a third eye was transplanted near the normal right eye, the right eye was injected with a radiolabeled amino acid. The autoradiograph shows that the left optic lobe receives inputs from both the labeled eye and the transplanted eye. The normally continuous synaptic zone of the contralateral eye has become divided into alternating dark and light zones that indicate the sites of inputs from each eye.

A Inputs are normally segregated in the tectum

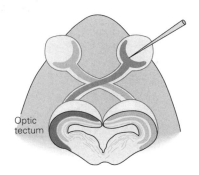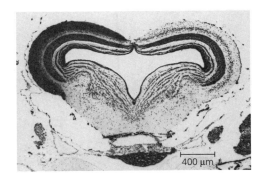

B Transplanted eye induces ocular dominance columns

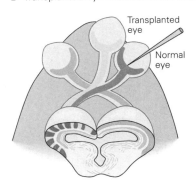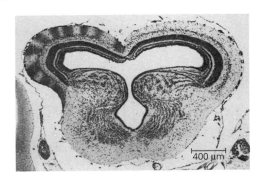

Reorganization of Visual Circuits During a Critical Period Involves Alterations in Synaptic Connections

The pioneering work of Hubel, Wiesel, and their colleagues showed that early experience is required for the emergence of normal structure and function in the visual cortex. However, the cellular and molecular mechanisms that underlie the critical period remained mysterious. In recent years, many investigators have begun addressing these issues. Much of their work has involved the use of mice, because mice are more amenable to mechanistic analysis than the cats and monkeys studied by Hubel, Wiesel, and their disciples.

Cortical Reorganization Depends on Changes in Both Excitation and Inhibition

Unlike cats and monkeys, most of the mouse visual cortex receives only contralateral input and its binocular region is not divided into ocular dominance columns. Nonetheless, the small binocular region contains a mixture of monocularly and binocularly driven neurons, and closure of the contralateral eye during the critical period for ocular dominance markedly shifts the preference of binocular neurons to inputs from the ipsilateral eye (Figure 49–8).

What converts this early loss of input into a permanent alteration of functional capability? One idea is that thalamic axons carrying information from the deprived eye lose their ability to activate cortical neurons. However, although a decrease in efficacy of the thalamocortical synapse may contribute to this effect, this is not the whole story. Each thalamic axon carries input from only one eye (Figure 49–2). Because loss of responsiveness to the deprived eye occurs only if the other eye remains active, one might imagine that the earliest changes would occur at the first site where inputs from the two eyes have the opportunity to interact. Consistent with this idea, the first physiological changes are not observed in layer IV neurons, each of which receives input from only one eye. Rather, they occur in the binocular neurons of layers II/III and V, which receive convergent input from both right eye– and left eye– driven monocular layer IV neurons. This implies that the loss of cortical responsiveness to the deprived eye results from a circuit alteration rather than from a simple loss of input.

Several possible cellular mechanisms have been proposed to account for these changes in circuitry.

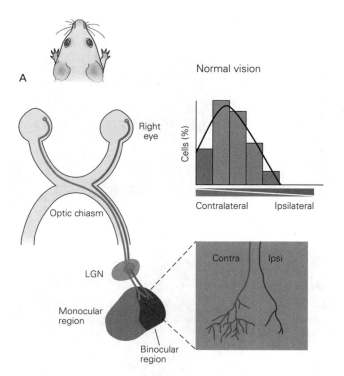

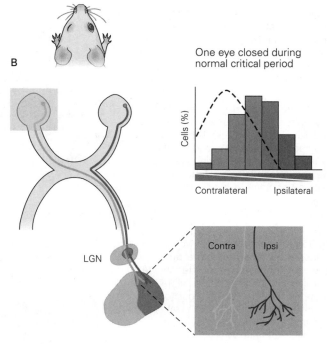

Figure 49–8 A critical period for ocular dominance plasticity is evident in mice. (Adapted, with permission, from Hensch 2005.)

A. The visual cortex in mice contains a small region that receives thalamic (lateral geniculate nucleus [**LGN**]) inputs from both eyes. In this binocular region, most neurons are predominantly responsive to contralateral eye input, fewer respond to binocular inputs, and very few respond to ipsilateral eye input only.

B. When the contralateral eye has been closed during the normal critical period and then reopened, inputs from that eye are underrepresented, and many more neurons respond to binocular or ipsilateral eye input. Eye closure before or after the time of the normal critical period does not elicit the same shift in responsiveness.

First, excitatory synapses within the primary visual cortex may weaken because of the decreased input from the closed eye, perhaps through long-term depression (LTD) (Chapter 53). Second, excitatory synapses carrying input from the open eye may become stronger. Third, the strength of inhibitory synapses may be altered, leading to a net decrease in the level of excitation of cortical neurons by inputs from the closed eye or a net increase in excitation from the open eye. Fourth, neuromodulation within the cortex may tune the circuit in more subtle ways, altering the balance between excitation and inhibition.

Careful analysis of neurons in mouse cortex has provided insight into roles played by some of these mechanisms. During the first few days after closing one eye, responses to input from the closed eye are greatly weakened, with no major effect on inputs from the open eye. The weakening results from a process like LTD or a closely related phenomenon called spike timing–dependent plasticity (STDP). Then, over the following few days, responses to inputs from the open eye become stronger. The increase results from

a combination of synaptic changes called long-term potentiation and homeostatic plasticity. Homeostatic plasticity is a circuit mechanism that endeavors to maintain a steady level of input to neurons. In this case, loss of excitatory drive from the closed eye leads to a compensatory increase in excitatory drive from the open eye.

Further studies demonstrated that inhibitory interneurons have an important role in the timing of the critical period. Maturation of inhibitory input onto visual cortical neurons coincides with the beginning of the critical period. Moreover, manipulations that lead to earlier development of γ-aminobutyric acid (GABA) signaling result in advancing the critical period (Figure 49–9). Conversely, delaying GABA signaling delays the period in which monocular deprivation enhances the preference for ipsilateral eye input (Figure 49–9). Together these results and others suggest that a sufficient level of inhibitory input plays a critical role in "gating" the opening of the critical period, whereas excitatory mechanisms may play a more prominent role in enacting the alterations that occur during the critical period.

Synaptic Structures Are Altered During the Critical Period

Many studies have sought structural changes that correlate with the altered responsiveness of the visual cortex to input from the closed and open eyes. Particular attention has been paid to dendritic spines as potential sites of plasticity.

Spines are small protrusions from the dendrites of many cortical neurons on which excitatory synapses form. They are dynamic structures, and their appearance and loss are thought to reflect the formation and elimination of synapses. Spine motility is especially marked during early postnatal development, and increases in spine dynamics and number have been associated with changes in behavior.

Striking alterations in the motility and number of dendritic spines on neurons in the mouse visual cortex are observed following closure of one eye. Two days after eye closure in young mice, the motility and turnover of dendritic spines on neurons in the visual cortex increases, suggesting that synaptic connections are beginning to rearrange (Figure 49–10). A few days later, the number of spines begins to change; the number of spines on the apical dendrites of pyramidal neurons decreases initially, but after longer periods of deprivation increases again.

These alterations in spine motility and number can be correlated with three known features of the critical period. First, rather than occurring in layer IV, the changes occur primarily in superficial and deep layers of the cortex, where binocular cells lie. Second, they occur only in the portion of the visual cortex that normally receives binocular input. Third, they fail to occur following eye closure in adult mice (Figure 49–10).

Together, these results support a linkage of spine dynamics with critical period plasticity. According to one model, spine motility may result from the imbalance of inputs to binocular neurons from the open and closed eyes, and it may reflect the first stages in synaptic rearrangement. In turn, the loss of spines, and presumably of synapses, corresponds in time and space to the loss of input from the closed eye and may provide a structural basis for the permanence of this loss. The later growth of new spines occurs as or after responsiveness to the open eye increases and may underlie the adaptive rearrangement that permits the cortex to make the best use of the input available to it.

Thalamic Inputs Are Remodeled During the Critical Period

How are local changes in spines related to the large-scale structural changes in ocular dominance columns

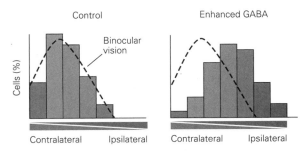

Deprivation before normal critical period

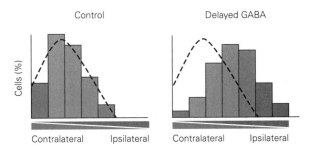

Deprivation after normal critical period

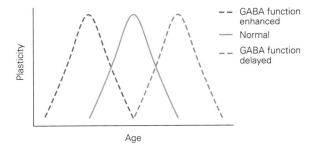

Critical period for plasticity after monocular deprivation

Figure 49–9 The timing of the critical period for ocular dominance plasticity in mice is sensitive to the level of GABAergic neurotransmission. Altering the status of γ-aminobutyric acid (GABA) synthesis and signaling shifts the period in which monocular deprivation can change the response properties of neurons in the visual cortex. Enhancing GABA signaling (through administration of benzodiazepines) shifts the critical period for monocular deprivation to an earlier developmental time. In contrast, delaying GABA signaling (by reducing GABA synthesis genetically and then administering benzodiazepines at a later time) shifts the critical period for monocular deprivation to a later developmental time. (Adapted from Hensch et al. 1998.)

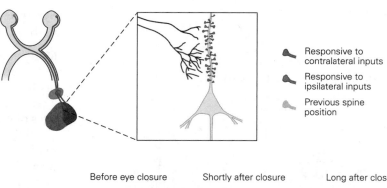

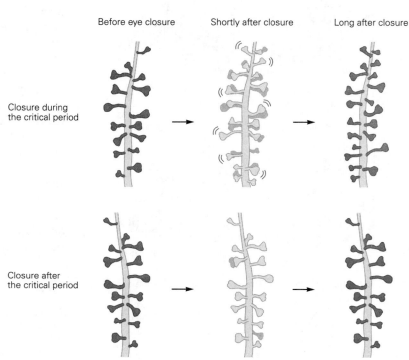

Figure 49–10 The motility of dendritic spines in the mouse visual cortex changes after one eye is closed. The dendrites of pyramidal neurons in the visual cortex have many spines, the density of which remains comparatively constant under normal conditions. Closure of one eye (contralateral in this example) during the critical period for binocular development enhances the motility of dendritic spines and, over time, results in an increase in the proportion of spines that receive synaptic input from the open eye. Similar changes in spine motility are not observed if the eye is closed after the critical period. (Adapted from Oray, Majewska, and Sur 2004.)

shown in Figure 49–4? When developing axons from the lateral geniculate nucleus first reach the cortex, the terminal endings of several neurons overlap extensively. Each fiber extends a few branches over an area of the visual cortex that spans several future ocular dominance columns. As the cortex matures, axons retract some branches, expand others, and even form new branches (Figure 49–11A).

With time, each geniculate neuron becomes connected almost exclusively to a group of neighboring cortical neurons within a single column. The arbors become segregated into columns through the pruning or retraction of certain axons and the sprouting of others. This dual process of axon retraction and sprouting occurs widely throughout the nervous system during development.

What happens after one eye is closed? Axons from a closed eye are at a disadvantage, and a greater

than normal proportion retract. At the same time, axons from the open eye sprout new terminals at sites vacated by fibers that would otherwise convey input from the closed eye (Figure 49–11B). If an animal is deprived of the use of one eye early during the critical period of axonal segregation, the normal processes of axon retraction and outgrowth are perturbed. In contrast, if an animal is deprived of the use of one eye after the ocular dominance columns are almost fully segregated, axons conveying input from the open eye actually sprout collaterals in regions of the cortex that they had vacated earlier (see Figure 49–5).

Initially, it was believed that rearrangements of thalamocortical axons in monocularly deprived animals caused the changes in cortical responsiveness to the open and closed eyes. We now know, through electrophysiological recording and imaging of spines, that physiological changes and synaptic alterations

A Normal development

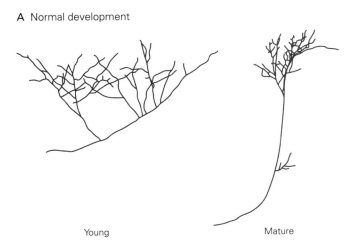

Young   Mature

B Development after eye closure

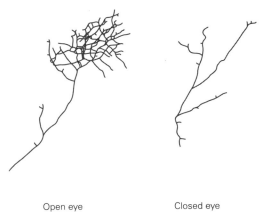

Open eye   Closed eye

Figure 49–11 The branching of thalamocortical fibers in the visual cortex of kittens changes after the closure of an eye. (Adapted, with permission, from Antonini and Stryker 1993. Copyright © 1993 AAAS.)

A. During normal postnatal development, the axons of lateral geniculate nucleus cells branch widely in the visual

cortex. The branching eventually becomes confined to a small region.

B. After one eye is closed, the terminal arbors of neurons in the pathway from that eye are dramatically smaller compared to those of the open eye.

precede the large-scale axonal rearrangements. So rather than causing the physiological changes, axonal remodeling may contribute to making these changes enduring and irreversible. The question then becomes: How do alterations in synaptic structure and function within the cortex lead to alterations in the input?

One idea is that synaptic activity regulates the secretion of neurotrophic factors by cortical neurons. Such factors may then regulate survival of some neurons at the expense of others (Chapter 46) or promote the expansion of some axonal arbors at the expense of others. One such factor, brain-derived neurotrophic factor (BDNF) is synthesized and secreted by cortical neurons, and administering excess BDNF or interfering with its receptor trkB modifies the formation of ocular dominance columns. Nevertheless, interpreting the actions of BDNF is not straightforward. BDNF and trkB signaling affect the cortex in many ways, including enhancing the growth of thalamocortical axons. BDNF can also speed the maturation of inhibitory circuits, which, as noted above, can influence plasticity. It remains unclear whether BDNF is a specific catalyst of the competition that preferentially promotes expansion of some arbors.

Synaptic Stabilization Contributes to Closing the Critical Period

A hallmark of critical periods is that the interval in which experience affects the development of neural

circuits is limited. What brings this period of heightened plasticity to a close?

Since synapses and circuits are labile during critical periods, investigators have sought developmental changes in cortex that could lead to stabilization. One parameter is the state of myelination of axons, which occurs around the time the critical period closes. Formation of myelin creates physical barriers to sprouting and axonal growth. Moreover, as discussed in detail in Chapter 50, myelin contains factors such as Nogo and myelin-associated glycoprotein that actively inhibit growth of axons. In mutant mice lacking Nogo or one of its receptors, NogoR, the critical period remains open into adulthood, suggesting that the appearance of these receptors normally contributes to closing the critical period (Figure 49–12).

Another possible agent of closure is the perineuronal net, a web of glycosaminoglycans that wraps certain classes of inhibitory neurons. These nets form around the time that the critical period closes. Infusion of the enzyme chondroitinase, which digests perineuronal nets, maintains plasticity. Thus, critical periods may close once molecular barriers to synaptic growth and rearrangement come into play.

Additional agents of closure may be intrinsic to the neurons. In Chapter 50, we will see that neuronal growth programs decrease with age, and in Chapter 51, we will describe epigenetic mechanisms that "lock in" experience-dependent patterns of gene expression established in early postnatal life.

Why should there be an end to critical periods? Would it not be advantageous for the brain to maintain its ability to remodel into adulthood? Perhaps not—the ability of our brain to adapt to variations in sensory input, to gradual physical growth (eg, increases in the distance between the eyes affecting binocular correspondence), and to various congenital disorders is a valuable asset. At an extreme, if one eye is lost, it is advantageous to devote all available cortical real estate to the remaining eye. However, one would not want wholesale reorganization, possibly accompanied by loss of skills and memories, if vision through one eye were lost temporarily in adulthood due to disease or injury. So, enhancing plasticity during a critical period may represent an adaptive compromise between flexibility and stability.

Experience-Independent Spontaneous Neural Activity Leads to Early Circuit Refinement

As noted above, the segregation of visual cortex into ocular dominance columns in cats and monkeys begins before the onset of visual experience. What drives this early phase of segregation? One possibility is that axons from the ipsilateral and contralateral eyes bear different molecular labels that lead to their association. A similar mechanism occurs in the formation of the olfactory projection (Chapter 48). However, no such molecule or mechanism has yet been discovered in the visual projection. Instead segregation appears to rely on spontaneous activity, which not only occurs prior to sensory input but also exhibits striking patterning. This mechanism was initially discovered in studies of the lateral geniculate nucleus, whose neurons provide visual input to the visual cortex.

The arbors of retinal ganglion cells from the two eyes are segregated into alternating layers in the lateral geniculate nucleus, much as the projections from this nucleus are segregated in alternating ocular dominance

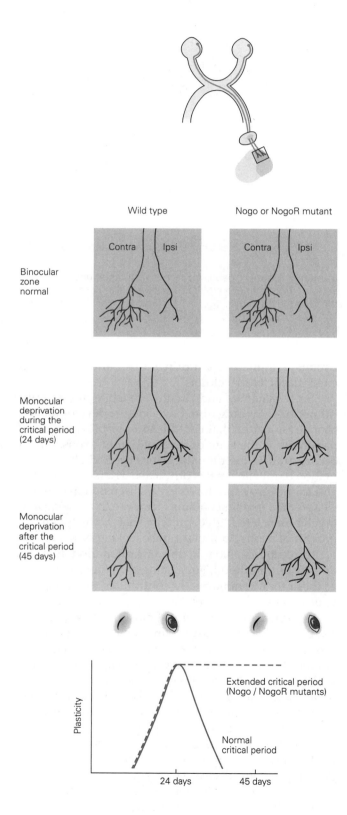

Figure 49–12 (Left) The critical period for monocular deprivation is extended in mice lacking Nogo signaling. The drawings show arborization patterns of thalamocortical axons carrying signals from contralateral and ipsilateral eyes to the binocular zone in visual cortex. Monocular deprivation during the critical period results in a shift in ocular preference in neurons in the binocular zone in both wild-type mice and mice mutant for Nogo or the Nogo receptor (**NogoR**). After the normal critical period (at 45 days), the shift in ocular preference continues in mice with mutant Nogo-A or the Nogo receptor but not in wild-type mice. The plot shows that elimination of Nogo signaling prevents closure of the critical period. (Adapted from McGee et al. 2005.)

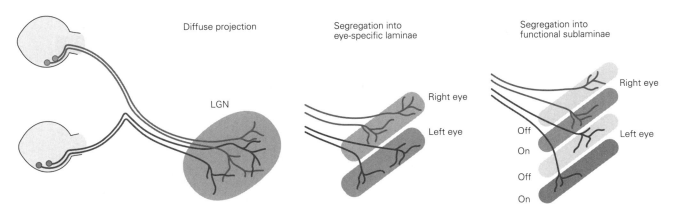

Figure 49–13 The terminals of retinal ganglion cells in the lateral geniculate nucleus (LGN) become segregated during normal development. At early stages of development, the terminals of axons from each eye intermingle, but at later stages, they segregate into separate layers of the nucleus. In some species, axons from one eye even segregate into functionally specialized sublayers (on and off layers in ferrets). (Adapted, with permission, from Sanes and Yamagata 1999.)

columns in the visual cortex (Figure 49–13). In both structures, individual axons at first form terminals in multiple domains (layers in the geniculate nucleus, columns in the cortex). Later, the terminals become segregated by a process of refinement. The refinement involves both growth of terminal arbors in the "appropriate" layer and elimination of terminals from the inappropriate layer (Chapter 48).

As in the cortex, application of tetrodotoxin to the optic nerves disrupts the segregation of the inputs from each eye, indicating that activity is essential for segregation. In contrast to cortex, however, segregation of inputs is complete before the onset of visual experience—prior to birth in monkeys and postnatally but prior to eye opening in mice. Thus, vision cannot drive the neural activity essential for segregation.

It turns out that the axons of retinal ganglion neurons are spontaneously active in utero, well before the eyes open. Neighboring ganglion cells fire in synchronous bursts that last a few seconds, followed by silent periods that may last for minutes. Sampling the activity of retinal ganglion neurons across the entire retina revealed that these bursts propagate across much of the retina in a wave-like manner (Figure 49–14). This pattern of ganglion cell activity appears to be coordinated by excitatory inputs from amacrine cells in the overlying layer of the retina (Chapter 22).

The spontaneous, synchronous firing of a select group of ganglion neurons excites a local group of neurons in the lateral geniculate nucleus. Such synchronized activity appears to strengthen these synapses at the expense of other nearby synapses, perhaps by a Hebbian mechanism similar to that posited for experience-dependent refinement. This does not mean

that visually evoked activity has no role in sculpting the retinogeniculate pathway. At a later stage, other aspects of refinement, such as spatial rearrangement of synapses along the axon, are regulated by visual experience.

The discovery that spontaneous activity can lead to circuit refinement provides a likely explanation for the initial segregation of inputs to the visual cortex. More generally, the parsing of activity-dependent circuit refinement into two phases, the first dependent on spontaneous activity and the second on sensory input, now appears to be a general theme in the development of brain circuits that begin refinement before they have the chance to respond to environmental stimulation.

Activity-Dependent Refinement of Connections Is a General Feature of Brain Circuitry

We have seen that neural activity is critical for segregating axons from the two retinas into distinct layers in the lateral geniculate nucleus and then into distinct columns in the visual cortex. Is this developmental role of activity a special case, or does activity also affect maturation elsewhere in the visual system, and even in other parts of the brain? Studies of many systems show that activity-dependent control of refinement is a general property of neural circuits in the mammalian brain.

Many Aspects of Visual System Development Are Activity-Dependent

One well-studied example of activity-dependent development in the visual system is the sharpening

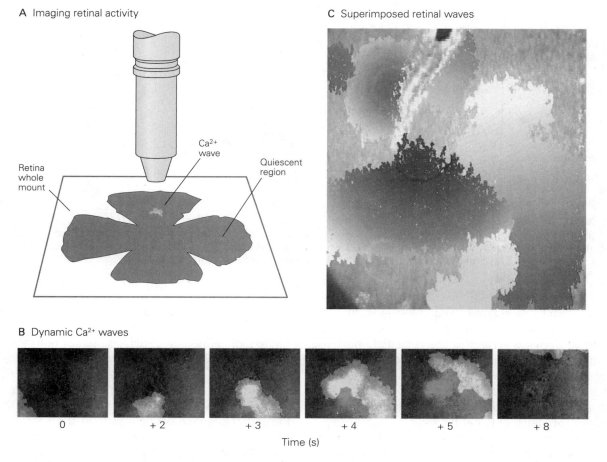

A Imaging retinal activity

Ca²⁺ wave

Quiescent region

Retina whole mount

C Superimposed retinal waves

B Dynamic Ca²⁺ waves

0   + 2   + 3   + 4   + 5   + 8

Time (s)

Figure 49–14 Correlated waves of neural activity in the developing retina.

A. Microscopic visualization of the activity of retinal ganglion neurons in a flat-mounted preparation of mammalian retina. Spontaneous waves of neural activity are visualized by monitoring Ca²⁺ transients (**yellow domain**) after loading of cells with dyes that change their fluorescent emission spectrum in response to changes in intracellular Ca²⁺ concentration.

B. These still images from a movie sequence show the propagation of one Ca²⁺ activity focus (**yellow domain**) across the

retina. Images were taken 1 second apart. Many cells within the activity focus are activated synchronously. (Reproduced, with permission, from Blankenship et al. 2009. Copyright © 2009 Elsevier Inc.)

C. Retinal activity waves recorded over time are superimposed in this image. Discrete waves are indicated in different colors; the origin of a wave is indicated by a darker hue. These waves originate in different retinal foci and spread in distinct, unpredictable directions. (Reproduced, with permission, from Meister et al. 1991. Copyright © 1991 AAAS.)

of the topographic distribution of retinal ganglion cell axons onto their central targets, a topic we introduced in Chapter 47. In vertebrates, molecular cues such as ephrins guide axons from the retina to appropriate sites in the optic tectum (called the superior colliculus in mammals—see Figure 47–11), but they are not sufficient to form the refined visual map.

Histological and physiological studies have found that the map formed initially in the superior colliculus/optic tectum is coarse and that individual retinal ganglion cell axons have large, overlapping arbors. These axonal arbors are later pruned to their mature size, resulting in a more restricted and precise

field of termination. If retinal activity is inhibited, only the initial coarse map forms.

Is it the pattern of activity or activity itself that is important in visual map formation? Put another way, is activity simply a precondition for refinement, or does it have an organizing role, determining exactly which axons win or lose the competition? Many experiments show that the latter idea is closer to the truth.

In one study, the accuracy of the retinotectal map was assessed in fish raised in a tank illuminated only by brief flashes from a strobe light. A control group was raised in a normal laboratory environment. The total light intensity presented to the fish was similar under both conditions, but the resulting pattern was very different. In control fish, the images fell haphazardly on various parts of the retina as the fish swam around their tanks. This input produces local synchronous activity of the sort generated by the waves of spontaneous activity described above—neighboring ganglion cells tend to fire together, but there is little correlation with the firing patterns of distant ganglion cells. In these fish, the map becomes precise. In contrast, stroboscopic illumination synchronously activates nearly all of the ganglion cells, and in these fish, the retinotectal map remains coarse.

Presumably, the tectum determines which retinal axons are near neighbors by judging which ones fire in synchrony, much as activity patterns in the lateral geniculate nucleus or visual cortex determine which axons carry signals from the same eye. This information is then used to refine the topographic map, through mechanisms similar to those in the cortex. When all of the axons fire in synchrony, the tectum cannot determine which axons are neighbors; refinement fails, and the map remains coarse.

Sensory Modalities Are Coordinated During a Critical Period

Our experience of the world is shaped by synthesizing sensory input from multiple modalities. For example, our mental image of where an object is with respect to our body is the same whether we localize it by touch, sound, or sight. For each modality, information is mapped in an orderly way within relevant brain areas, much like the retinotopic maps in the optic tectum and visual cortex. Multimodal localization requires that these maps, which are formed independently during development, be brought into register. This aspect of refinement occurs during critical periods.

Studies on barn owls have provided insight into how auditory and visual maps are coordinated during a critical period. During the day, owls use vision to localize their prey—mice or other small rodents—but at night, they rely on auditory cues, and at dusk, both sensory channels are used. The localization of sound must be precise if owls are to succeed in finding prey, and it is intuitively obvious that the visual and auditory cues for the same location need to be consistent.

Auditory localization in owls, as in people, results from the presence of neurons that vary in their sensitivity to sounds sensed by the two ears. For example, sounds arising from a source to the left arrive slightly sooner at the left ear than at the right ear and are slightly louder in the left ear. These discrepancies help us determine the point in horizontal space from which a sound arises (Chapter 28). Computation of the temporal difference in the arrival of sounds at the two ears is particularly crucial. The difference is only a few tens of microseconds, as expected from calculations based upon the speed of sound and the width of the head. Remarkably, the auditory system is sensitive to these extremely short interaural time differences (ITDs) and can calculate prey position from them (Figure 49–15). Moreover, many auditory neurons in the optic tectum

Figure 49–15 The barn owl uses interaural time differences to localize its prey. Sound waves generated by movements of a mouse are received by the owl's left and right ears. As the prey emits noise, the difference in the time of arrival of auditory stimuli at the two ears—the interaural time difference (ITD)—is used to calculate the precise position of the prey target. (Reproduced, with permission, from Knudsen 2002. Copyright © 2002 Springer Nature.)

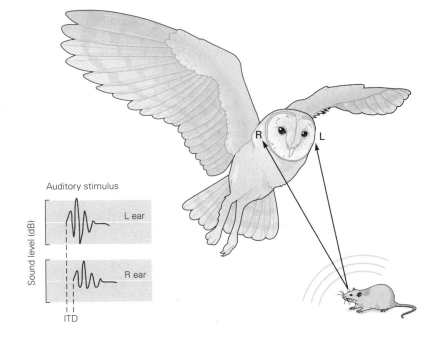

Auditory stimulus

Sound level (dB)

L ear

R ear

ITD

with receptive fields centered on a particular location are also tuned to ITDs that correspond to sounds emitted from that same point in space. The registration is imprecise at early stages but becomes progressively more precise during early adolescence as a consequence of the animal's experience.

Crucial insight into how this registration occurs came from experiments in which prisms were mounted over the eyes of young owls. The prisms shifted the retinal image horizontally so that the visual map in the tectum reflected a world systematically displaced from its "actual" orientation. This change abruptly disrupted the correspondence between visual and auditory receptive fields. Over the next several weeks, however, the ITD to which tectal neurons responded optimally, ie, their auditory receptive field, changed until the visual and auditory maps came back into register (Figure 49–16). Thus, the visual map instructs the auditory map.

Further experiments showed that this reorganization resulted from rewiring of connections between two deeper auditory nuclei (Figure 49–17). When prism goggles were placed on young owls, changes in ITD tuning were fully adaptive in that the animals compensated completely for the effects of the prisms. In contrast, goggles placed on mature owls (older than 7 months of age) had little effect. Thus, reorganization of this auditory projection occurs optimally during a critical juvenile period.

Different Functions and Brain Regions Have Different Critical Periods of Development

Not all brain circuits are stabilized at the same time. Even within the visual cortex, the critical periods for organization of inputs differ among layers in both mice

Head orientation in response to:
○ Auditory stimuli
● Visual stimuli mutations

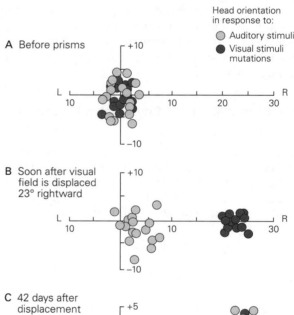

A   Before prisms

B   Soon after visual field is displaced 23° rightward

C   42 days after displacement

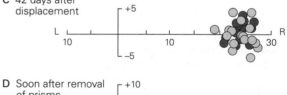

D   Soon after removal of prisms

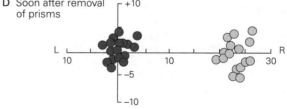

Figure 49–16 (Right) Reorganization of sensory maps in the optic tectum of owls after systematic displacement of the retinal image. The retinal image in adolescent owls can be displaced by prism goggles, which shift images from 5° to 30°. (Adapted, with permission, from Knudsen 2002. Copyright © 2002 Springer Nature.)

A. Before application of the prisms, the visual and auditory neural maps coincide.

B. The prism goggles displace the retinal image by 23°. Consequently, the neural and auditory maps are out of alignment.

C. The two brain maps are once again congruent 42 days after prism application because the auditory map has shifted to realign with the visual map.

D. Soon after the prisms are removed, the visual map reverts to its original position, but the auditory map remains in its shifted position.

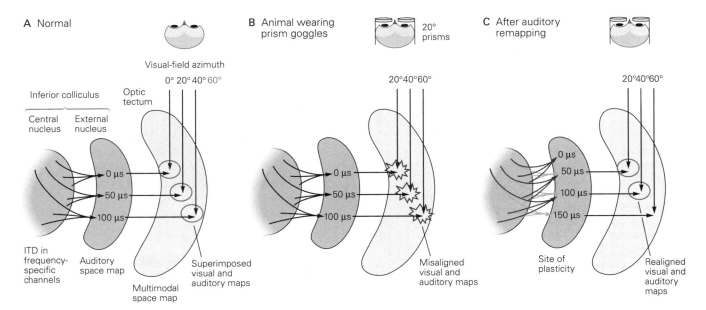

A Normal

B Animal wearing prism goggles

C After auditory remapping

Visual-field azimuth
0° 20° 40° 60°

20° prisms

20° 40° 60°

20° 40° 60°

Inferior colliculus

Optic tectum

Central nucleus   External nucleus

0 μs

50 μs

100 μs

0 μs

50 μs

100 μs

0 μs

50 μs

100 μs

150 μs

ITD in frequency-specific channels

Auditory space map

Superimposed visual and auditory maps

Multimodal space map

Misaligned visual and auditory maps

Site of plasticity

Realigned visual and auditory maps

Figure 49–17 The effect of prism experience on information flow in the midbrain auditory localization pathway in the barn owl. (Adapted from Knudsen 2002.)

A. The auditory pathway in a normal owl. The interaural time difference (**ITD**) is measured and mapped in frequency-specific channels in the brain stem. This information ascends to the inferior colliculus, where a neural map of auditory space is created.

The map is conveyed to the optic tectum where it merges with a map of visual space.

B. After an owl is fitted with prism goggles, the visual and auditory space maps in the optic tectum become misaligned.

C. After reorganization of auditory maps, the visual and auditory maps are once again in alignment.

and monkeys. As an example, the neural connections in layer IVC of the visual cortex of the monkey are not affected by monocular deprivation by the time the animal is 2 months old. In contrast, connections in the upper and lower layers continue to be influenced by sensory experience (or lack of it) for almost the entire first year after birth. Critical periods for other features of the visual system, such as orientation tuning, occur at different developmental stages (Figure 49–18A).

The timing of critical periods also varies between brain regions (Figure 49–18B). The adverse consequences of sensory deprivation for the primary sensory regions of the brain are generally fully realized early in postnatal development. In contrast, social experience can affect the intracortical connections over a much longer period. These differences may explain why certain types of learning are optimal at particular stages of development. For example, certain cognitive capacities—language, music, and mathematics—usually must be acquired well before puberty if they are to develop at all. In addition, insults to the brain at specific early stages of postnatal life may selectively affect the development of certain perceptual abilities and behavior.

Critical Periods Can Be Reopened in Adulthood

By definition, critical periods are limited in time. Nevertheless, they are less sharply defined than originally thought. Extending or reopening critical periods in adulthood could increase brain plasticity and make it possible to facilitate recovery from strokes and other insults that impair discrete regions of the nervous system.

Some of the first evidence for plasticity in the adult cortex came from studies by Merzenich and colleagues on the representation of the fingers of monkeys in the somatosensory cortex. Recordings of neuronal receptive fields in normal adult animals showed that each digit is mapped in an orderly way on the cortical surface, with abrupt discontinuities between areas responding to different digits (Figure 49–19A). Amputation of a digit left the cortical representation of that digit initially unresponsive, but after several months, areas serving the neighboring digits filled in the gap (Figure 49–19B). Much as happens in the visual cortex following monocular deprivation, the somatosensory map was readjusted so that the cortex could devote

A Critical periods for visual function in cats

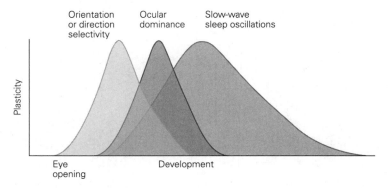

Figure 49–18 The timing of critical periods varies with brain function. (Reproduced, with permission, from Hensch 2005. Copyright © 2005 Springer Nature.)

A. In cats, the critical periods for development of orientation or direction selectivity in visual neurons occur earlier than those for establishment of ocular dominance and slow-wave sleep oscillation.

B. In humans, the timing of periods for development of sensory processing, language, and cognitive functions varies.

B Critical periods for sensory and cognitive skills in humans

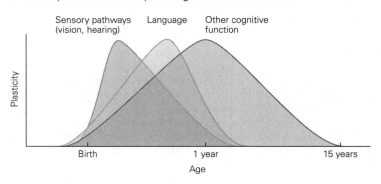

most of its resources to useful inputs. Conversely, when two fingers were sutured together so they received coincident input, a swath of cortex on both sides of the border between the two digit areas eventually became responsive to both areas (Figure 49–19C). This result suggests that, as in the visual system, borders may result from competition and can be blurred when competition declines. What was most surprising was that these effects occurred in adulthood, long after all known critical periods had closed.

In the years since Merzenich's studies, evidence has accumulated that critical periods can be reopened in many systems. We illustrate this principle by returning to two areas in which critical periods have been well mapped, the optic tectum in the owl and the visual cortex in the mouse.

Visual and Auditory Maps Can Be Aligned in Adults

In initial studies of the matching of auditory and visual maps in owls, realignment following displacement of the visual field with prism goggles was largely restricted to an early sensitive period (Figures 49–16 and 49–17). However, three strategies dramatically enhance binaural tuning plasticity in adult owls.

First, when adult owls that had worn goggles as adolescents are refitted with the goggles, the auditory map again shifts to align with the new visual map (Figure 49–20A). In contrast, in adult owls that had not worn the goggles as adolescents, the use of goggles has little effect on the organization of the auditory map. Thus, the events of map rearrangement during the normal critical period must leave a neural trace that permits rearrangement later in life. In fact, in the owls that wore prisms in early life, axons to auditory nuclei that were normally pruned were maintained, providing a structural basis for the reorganization in adulthood.

A second method for inducing late plasticity is to displace the retinal image in small steps by having the owl wear a series of prism spectacles of progressively increasing strength. Under these conditions, adjustment of the auditory map is typically three- to fourfold greater than the response to a single large displacement of the retinal image (Figure 49–20B).

The third technique is to allow owls to hunt live prey. In earlier experiments, animals were housed and fed under standard laboratory conditions. However, when adult prism-wearing owls are allowed to capture live mice under low light conditions for 10 weeks, they exhibit far greater plasticity of binaural tuning than owls fed dead mice (Figure 49–19C), albeit less than that exhibited by juvenile owls that did not hunt. The finding that hunting increases the plasticity of binaural tuning in adult owls dramatically demonstrates that behavioral context affects the ability of the nervous system to reorganize. Whether this effect results from increased sensory information, attention, arousal, motivation, or reward needs to be resolved.

Binocular Circuits Can Be Remodeled in Adults

As the body of observations on monocular deprivation grew, it became apparent that some plasticity persisted beyond the classical critical period in cats, rats, and mice. In mice, for example, modest shifts in ocular dominance occur even when one eye is deprived of vision at 2 or 3 months of age. By 4 months of age, however, monocular deprivation has no detectable effect.

Over the past decade, several interventions have been discovered that enhance the extent of ocular dominance plasticity in young adults and even enable substantial plasticity in older animals. Some are noninvasive: Environmental enrichment, social interaction (via group housing), visual stimulation, and exercise all increase the magnitude and speed of changes that occur following monocular deprivation in adults. A second group of interventions targets mechanisms that appear to affect the timing of the normal critical period. As noted above, treating the cortex with chondroitinase to disrupt perineuronal nets or interference with the inhibitory effects of myelin on axonal growth can both extend and reopen the critical period. Remarkably, transplantation of immature inhibitory interneurons into the visual cortex also reopens the critical period even in 6-month-old mice.

How can we reconcile the strong evidence for critical periods with the newer evidence for reorganization

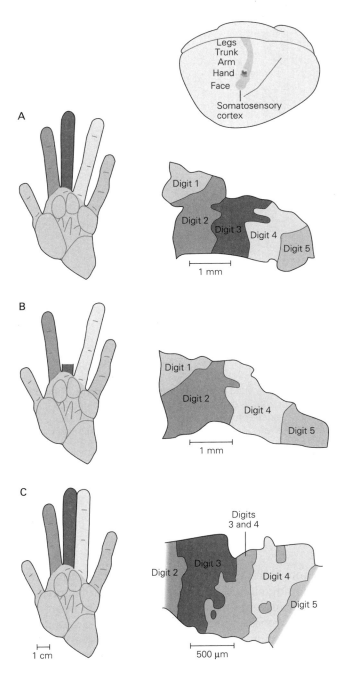

Figure 49–19 (Left) Representation of digits in somatosensory cortex can be remapped in adult monkeys. (Adapted, with permission, from Merzenich et al. 1984 and Allard et al. 1991.)

A. Lightly touching specific spots on the digits (*left*) elicits responses from neurons in somatosensory cortex (*right*), revealing orderly topographic maps of each digit on the cortical surface. Abrupt discontinuities distinguish regions serving adjacent digits.

B. Following amputation of a digit, the cortical region it previously supplied is left unresponsive. Several months later, axons from the adjacent digits (2 and 4) have formed synaptic connections in the unresponsive area.

C. After digits 3 and 4 have been sutured together, they received simultaneous sensory stimulation, and cortical regions at the border between areas representing the digits become responsive to both.

A The effect of early experience

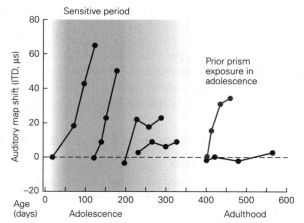

B The effect of incremental change

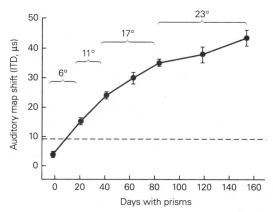

C The effect of hunting

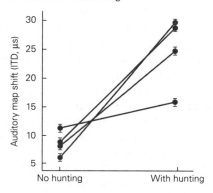

Figure 49–20 Different behavioral conditions have different effects on the realignment of visual and auditory neural maps in the mature barn owl.

A. The remodeling of the auditory maps that results from wearing prism goggles for a brief period during adolescence leaves a neural trace that can be reactivated in the adult. When these birds are fitted with the goggles as adults, the auditory map is still able to realign with the visual map. (Abbreviation: **ITD**, interaural time difference.) (Reproduced, with permission, from Knudsen 2002. Copyright © 2002 Springer Nature.)

B. When an animal is fitted with a series of prisms, each of which produces a small displacement in the visual image, the

auditory map is successfully brought into alignment. The dotted line shows the extent of realignment if the animal is fitted with a 23° prism on day 0. (Reproduced, with permission, from Linkenhoker and Knudsen 2002. Copyright © 2002 Springer Nature.)

C. If an adult owl has the opportunity to hunt live prey while wearing prism goggles, auditory remapping occurs, perhaps because of enhanced motivation to sharpen perception. (Reproduced, with permission, from Bergan et al. 2005. Copyright © 2005 Society for Neuroscience.)

of circuitry in adults? The plasticity observed in adults is modest and slow compared to that seen during the critical period, and its mechanisms differ in some respects from those for earlier deprivation. These differences result from two factors. First, from early postnatal life into adolescence, the molecular environment in the brain is conducive to axonal growth, and cellular mechanisms are optimal for promoting the formation, strengthening, weakening, and elimination of synapses. Under these conditions, circuits can readily change in response to experience. Conversely, in mature circuits, molecular and structural elements

promote stability and impede plasticity. Second, in a developing circuit, no particular pattern of connectivity is firmly entrenched, so there is less to overcome. The connections specified by genetic determinants are less precise, and the connections themselves are relatively weak. The patterns of neural activity that result from experience sharpen and even realign these patterns of connectivity.

In sum, experience during critical periods has a potent effect on circuits because the cellular and molecular conditions are optimal for plasticity and because the instructed pattern of connectivity does not have to compete with a long-existing pattern. These differences help explain the special behavioral, pharmacological, or genetic interventions needed to stimulate plasticity in adults.

Highlights

1. Although the nervous system is malleable throughout life, plasticity is particularly great during restricted intervals in early postnatal life called critical periods. Alterations that occur during these periods are nearly irreversible.

2. Critical periods vary in time among brain areas and tasks. For example, children with strabismus (crossed eyes) will never have good stereoscopic vision unless their eyes are brought into alignment during the first few postnatal years, and people cannot learn a new language without an accent after their early teens.

3. The richest understanding of critical periods comes from studies initiated by Hubel and Wiesel on how input from the two eyes is integrated in the cortex. They deprived one eye of vision for varying periods in young cats or monkeys. In normal animals, most neurons in visual cortex are binocularly responsive, but following monocular deprivation for a brief period in early postnatal life, most cortical cells permanently lost responsiveness to input from the once-closed eye. Responses in the eye itself and the lateral geniculate nucleus were nearly normal, pinpointing the cortex as the site of change. Much longer deprivation in adulthood had little effect.

4. A structural basis for the loss of binocularity was seen in the alternating pattern of ocular dominance columns, within which neurons are dominated by input from one eye or the other. Following monocular deprivation during the critical period, columns representing the open eye expanded at the expense of those representing the closed eye. This form of plasticity may be designed to optimize the use of cortical space for each individual at each time period—for example, subtly shifting binocular interactions as the head grows and the eyes become further apart.

5. The binocular interaction reflects competition between the two sets of inputs, since vision and symmetrical columns are retained following binocular deprivation. Many lines of evidence indicate that the competition depends on patterns of activity arising in the two eyes, with inputs from each eye being more synchronous with each other than with

inputs from the other eye. Postnatally, synchrony is driven by visual experience. Prenatally or prior to eye opening, patterned spontaneous activity in the two eyes accounts for the synchrony.

6. Cellular mechanisms underlying the effects of monocular deprivation have been studied in greatest detail in mice. Following monocular deprivation, input from the closed eye is weakened rapidly by a process akin to long-term depression (LTD). Shortly thereafter, input from the other eye is strengthened, partly by a compensatory mechanism called homeostatic plasticity. Structural remodeling of thalamic axons and cortical dendrites occurs later.

7. Maturation of inhibitory interneurons is a main determinant of when the critical period opens. The end of the critical period is marked by the formation of myelin and proteoglycan-rich perineuronal structures that hamper structural remodeling.

8. Although plasticity of binocular interactions was initially believed to be confined to early postnatal life, it is now apparent that critical periods can be "reopened" to some extent in adults. In some cases, this can be done by altering the animal's environment or the way in which the altered experience is delivered. Critical periods can also be reopened by manipulating some of the factors that normally close them in adolescence.

9. Plasticity in adulthood is modest in magnitude and difficult to trigger compared to early postnatal critical periods. Nonetheless, reopening of critical periods could, if properly controlled, enable reorganization to compensate for losses incurred from injuries, disease, and early maladaptive experience.

10. Critical periods occur during development of numerous systems, such as formation of orderly maps of auditory, somatosensory, and visual input onto relevant sensory cortices. Many of the principles and mechanisms that characterize the plasticity of binocular interactions also regulate these critical periods, including roles of spontaneous and experience-dependent activity, competition, alterations in excitatory and inhibitory synapses, and selective growth and pruning of inputs to achieve appropriate patterns of adult connectivity.

11. The existence of critical periods demonstrates that the brain's ability to remodel declines precipitously in adulthood. This seems disadvantageous but may represent a useful adaptation, allowing each brain to adapt to its environment

as it develops, but then buffering it against excessive change later, perhaps even enabling skills and memories to persist. If this is the case, therapies based on reopening critical periods in adults may come at a cost.

Joshua R. Sanes

Selected Reading

Espinosa JS, Stryker MP. 2012. Development and plasticity of the primary visual cortex. Neuron 75:230–249.

Harlow HF. 1958. The nature of love. Am Psychol 13: 673–685.

Hensch TK, Quinlan EM. 2018. Critical periods in amblyopia. Vis Neurosci 35:E014.

Hübener M, Bonhoeffer T. 2014. Neuronal plasticity: beyond the critical period. Cell 159:727–737.

Knudsen EI. 2002. Instructed learning in the auditory localization pathway of the barn owl. Nature 417:322–328.

Leighton AH, Lohmann C. 2016. The wiring of developing sensory circuits: from patterned spontaneous activity to synaptic plasticity mechanisms. Front Neural Circuits 10:71.

Thompson A, Gribizis A, Chen C, Crair MC. 2017. Activity-dependent development of visual receptive fields. Curr Opin Neurobiol 42:136–143.

Wiesel TN. 1982. Postnatal development of the visual cortex and the influence of environment. Nature 299:583–591.

References

Allard T, Clark SA, Jenkins WM, Merzenich MM. 1991. Reorganization of somatosensory area 3b representations in adult owl monkeys after digital syndactyly. J Neurophysiol 66:1048–1058.

Antonini A, Stryker MP. 1993. Rapid remodeling of axonal arbors in the visual cortex. Science 260:1819–1812.

Behen ME, Helder E, Rothermel R, Solomon K, Chugani HT. 2008. Incidence of specific absolute neurocognitive impairment in globally intact children with histories of early severe deprivation. Child Neuropsychol 14: 453–469.

Bergan JF, Ro P, Ro D, Knudsen EI. 2005. Hunting increases adaptive auditory map plasticity in adult barn owls. J Neurosci 25:9816–9820.

Blankenship A, Ford K, Johnson J, et al. 2009. Synaptic and extrasynaptic factors governing glutamatergic retinal waves. Neuron 62:230–241.

Buonomano DV, Merzenich MM. 1998. Cortical plasticity: from synapses to maps. Annu Rev Neurosci 21:149–186.

Constantine-Paton M, Law MI. 1978. Eye-specific termination bands in tecta of three-eyed frogs. Science 202:639–641.

Davis MF, Figueroa Velez DX, et al. 2015. Inhibitory neuron transplantation into adult visual cortex creates a new critical period that rescues impaired vision. Neuron 86:1055–1066.

Eluvathingal TJ, Chugani HT, Behen ME, et al. 2006. Abnormal brain connectivity in children after early severe socioemotional deprivation: a diffusion tensor imaging study. Pediatrics 117:2093–2100.

Galli L, Maffei L. 1988. Spontaneous impulse activity of rat retinal ganglion cells in prenatal life. Science 242: 90–91.

Hebb DO. 1949. *Organization of Behavior: A Neuropsychological Theory*. New York: Wiley.

Hensch TK. 2005. Critical period plasticity in local cortical circuits. Nat Rev Neurosci 6:877–888.

Hensch TK, Fagiolini M, Mataga N, Stryker MP, Baekkeskov S, Kash SF. 1998. Local GABA circuit control of experience-dependent plasticity in developing visual cortex. Science 282:1504–1508.

Hofer S, Mrsic-Flogel T, Bonhoeffer T, Hubener M. 2009. Experience leaves a lasting structural trace in cortical circuits. Nature 457:313–317.

Hong YK, Park S, Litvina EY, Morales J, Sanes JR, Chen C. 2014. Refinement of the retinogeniculate synapse by bouton clustering. Neuron 84:332–339.

Hubel DH, Wiesel TN. 1965. Binocular interaction in striate cortex of kittens reared with artificial squint. J Neurophysiol 28:1041–1059.

Hubel DH, Wiesel TN. 1977. Ferrier lecture: functional architecture of macaque monkey visual cortex. Proc R Soc Lond B Biol Sci 198:1–59.

Hubel DH, Wiesel TN, LeVay S. 1977. Plasticity of ocular dominance columns in monkey striate cortex. Philos Trans R Soc Lond B Biol Sci 278:377–409.

Khibnik LA, Cho KK, Bear MF. 2010. Relative contribution of feed forward excitatory connections to expression of ocular dominance plasticity in layer 4 of visual cortex. Neuron 66:493–500.

Kral A, Sharma A. 2012. Developmental neuroplasticity after cochlear implantation. Trends Neurosci 35:111–122.

Linkenhoker BA, Knudsen EI. 2002 Incremental training increases the plasticity of the auditory space map in adult barn owls. Nature 419:293–296.

Löwel S. 1994. Ocular dominance column development: strabismus changes the spacing of adjacent columns in cat visual cortex. J Neurosci 14:7451–7468.

McGee AW, Yang Y, Fischer QS, Daw NW, Strittmatter SM. 2005. Experience-driven plasticity of visual cortex limited by myelin and Nogo receptor. Science 309:2222–2226.

Meister M, Wong ROL, Baylor DA, Shatz CJ. 1991. Synchronous bursts of action potentials in ganglion cells of the developing mammalian retina. Science 252:939–943.

Merzenich MM, Nelson RJ, Stryker MP, Cynader MS, Schoppmann A, Zook JM. 1984. Somatosensory cortical map changes following digit amputation in adult monkeys. J Comp Neurol 224:591–605.

Nelson CA 3rd, Zeanah CH, Fox NA, Marshall PJ, Smyke AT, Guthrie D. 2007. Cognitive recovery in socially deprived young children: the Bucharest Early Intervention Project. Science 318:1937–1940.

Oray S, Majewska A, Sur M. 2004. Dendritic spine dynamics are regulated by monocular deprivation and extracellular matrix degradation. Neuron 44:1021–1030.

Pizzorusso T, Medini P, Berardi N, Chierzi S, Fawcett JW, Maffei L. 2002. Reactivation of ocular dominance plasticity in the adult visual cortex. Science 298:1248–1251.

Rakic P. 1981. Development of visual centers in the primate brain depends on binocular competition before birth. Science 214:928–931.

Sanes JR, Yamagata M. 1999. Formation of lamina-specific synaptic connections. Curr Opin Neurobiol 9:79–87.

Shatz CJ, Stryker MP. 1988. Prenatal tetrodotoxin infusion blocks segregation of retino-geniculate afferents. Science 242:87–89.

Van Sluyters RC, Levitt FB. 1980. Experimental strabismus in the kitten. J Neurophysiol 43:686–699.

Zhang J, Ackman JB, Xu HP, Crair MC. 2011. Visual map development depends on the temporal pattern of binocular activity in mice. Nat Neurosci 15:298–307.

50

Repairing the Damaged Brain

FOR MUCH OF ITS HISTORY, NEUROLOGY has been a
discipline of outstanding diagnostic rigor but little therapeutic efficacy. Simply put, neurologists
have been renowned for their ability to localize lesions
with great precision but until recently have had little
to offer in terms of treatment. This situation is now
changing.

Advances in our understanding of the structure,
function, and chemistry of the brain's neurons, glial
cells, and synapses have led to new ideas for treatment. Many of these are now in clinical trials, and
some are already available to patients. Developmental
neuroscience is emerging as a major contributor to this
sea change for three main reasons. First, efforts to preserve or replace neurons lost to damage or disease rely
on recent advances in our understanding of the mechanisms that control the generation and death of nerve
cells in embryos (Chapters 45 and 46). Second, efforts
to improve the regeneration of neural pathways following injury draw heavily on what we have learned
about the growth of axons and the formation of synapses (Chapters 47 and 48). Third, there is increasing
evidence that some devastating brain disorders, such
as autism and schizophrenia, are the result of disturbances in the formation of neural circuits in embryonic
or early postnatal life. Accordingly, studies of normal
development provide an essential foundation for discovering precisely what has gone wrong in disease.

In this chapter, we focus on the first two of these
issues: how neuroscientists hope to augment the limited ability of neurons to recover normal function. We
shall begin by describing how axons degenerate following the separation of the axon and its terminals
from the cell body. The regeneration of severed axons
is robust in the peripheral nervous system of mammals and in the central nervous system of lower vertebrates, but very poor in the central nervous system of

mammals. Many investigators have sought the reasons for these differences in the hope that understanding them will lead to methods for augmenting recovery of the human brain and spinal cord following injury. Indeed, we shall see that several differences in regenerative capacity of mammalian neurons have been discovered, each of which has opened promising new approaches to therapy.

We shall then consider an even more dire consequence of neural injury: the death of neurons. The inability of the adult brain to form new neurons has been a central dogma of neuroscience since the pioneering neuroanatomist Santiago Ramón y Cajal asserted that in the injured central nervous system, "Everything may die, nothing may be regenerated." This pessimistic view dominated neurology for most of the last century despite the fact that Ramón y Cajal added, "It is for the science of the future to change, if possible, this harsh decree." Remarkably, in the past few decades, evidence has accumulated that neurogenesis does occur in certain regions of the adult mammalian brain. This discovery has helped accelerate the pace of research on ways to stimulate neurogenesis and to replace neurons following injury. More than a century later, neuroscientists are finally beginning to reverse Cajal's "harsh decree."

Damage to the Axon Affects Both the Neuron and Neighboring Cells

Because many neurons have very long axons and cell bodies of modest size, most injuries to the central or peripheral nervous system involve damage to axons. Transection of the axon, either by cutting or by crushing, is called *axotomy*, and its consequences are numerous.

Axon Degeneration Is an Active Process

Axotomy divides the axon in two: a proximal segment that remains attached to the cell body and a distal segment that has lost this crucial attachment. Axotomy dooms the distal segment of the axon because energy supplies dwindle during a short-lived latent period. Soon the alterations become irreversible. Synaptic transmission fails at severed nerve terminals, and calcium levels increase within the axon. The calcium activates proteases, initiating a program of cytoskeletal disassembly and degradation, and physical degeneration of the axon ensues. Once the denervation begins, its progression is relatively rapid and inexorably proceeds to completion (Figure 50–1). This degenerative response is the first step in an elaborate constellation

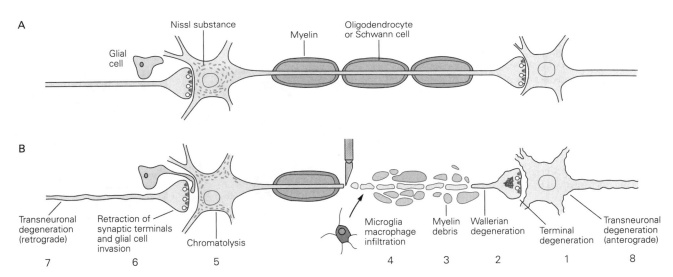

Figure 50–1 Axotomy affects the injured neuron and its synaptic partners.

A. A normal neuron with an intact functional axon wrapped by myelinating cells contacts a postsynaptic neuron. The neuron's cell body is itself a postsynaptic target.

B. After axotomy, the nerve terminals of the injured neuron begin to degenerate (**1**). The distal axonal stump separates from the parental cell body, becomes irregular, and undergoes

Wallerian degeneration (**2**). Myelin begins to fragment (**3**) and the lesion site is invaded by phagocytic cells (**4**). The cell body of the damaged neuron undergoes chromatolysis: The cell body swells and the nucleus moves to an eccentric position (**5**). Synaptic terminals that contact the damaged neuron withdraw and the synaptic site is invaded by glial cell processes (**6**). The injured neuron's inputs (**7**) and targets (**8**) can atrophy and degenerate.

of changes, called *Wallerian degeneration*, that were initially described in 1850 by Augustus Waller.

The degeneration of transected axons was long thought to be a passive process, the consequence of separation from the cell body, where most of the cell's proteins are synthesized. Lacking a source of new protein, the distal stump was thought to simply wither away. But the discovery and analysis in mice of a spontaneously occurring mutation called *Wlds* (Wallerian degeneration slow) challenged this view (Figure 50–2). In *Wlds* mutant mice, the distal stumps of peripheral nerves persist for several weeks after transection, about 10-fold longer than in normal mice. This remarkable finding suggested that degeneration is not a passive consequence of separation from the cell body, but is rather an actively regulated response.

Analysis of the *Wlds* mutant mice led to insights into the nature of this regulation. The mutation led to formation of a mutant form of nicotinamide mononucleotide adenyltransferase 1 (NMNAT1), an enzyme involved in biosynthesis of a metabolic cofactor, nicotinamide adenine dinucleotide (NAD). A related enzyme, NMNAT2, which is normally present in the axon, becomes quite unstable and breaks down rapidly following axotomy, leading to loss of NAD, which is critical for maintenance of energy homeostasis in the axon. Although normal NMNAT1 is confined to the nucleus, the mutant *Wlds* form mislocalizes to the axon, where it substitutes for NMNAT2 to prolong axonal survival. Surprisingly, a main way that both the wild type and *Wlds* forms of NMNAT maintain NAD levels is not by synthesizing it but by inhibiting another protein, SARM1, that breaks down NAD. Thus, loss of SARM protects damaged axons, whereas activation of SARM1 leads to degeneration (Figure 50–3A). Several other proteins modulate this core pathway (Figure 50–3B).

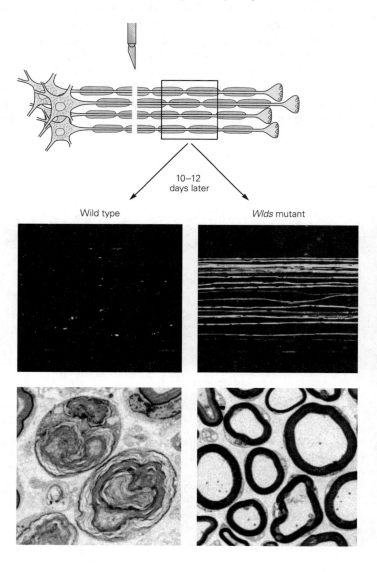

Figure 50–2 Axonal degeneration is delayed in *Wlds* mutant mice. In wild type animals, axons in the distal stump degenerate rapidly after sectioning of a peripheral nerve, as shown by disrupted axonal fragments (**yellow**) and the lack of myelinated axonal profiles at the electron micrographic level. In *Wlds* mutant mice the distal portion of severed axons persists for a long time. (Confocal micrographs reproduced, with permission, from Beirowski et al. 2004. Copyright © 2004 Elsevier B.V.; electron micrographs reproduced, with permission, from Mack TGA, Reiner M, Beirowski B, et al. 2001. Copyright © 2001 Springer Nature.)

10–12 days later

Wild type   *Wlds* mutant

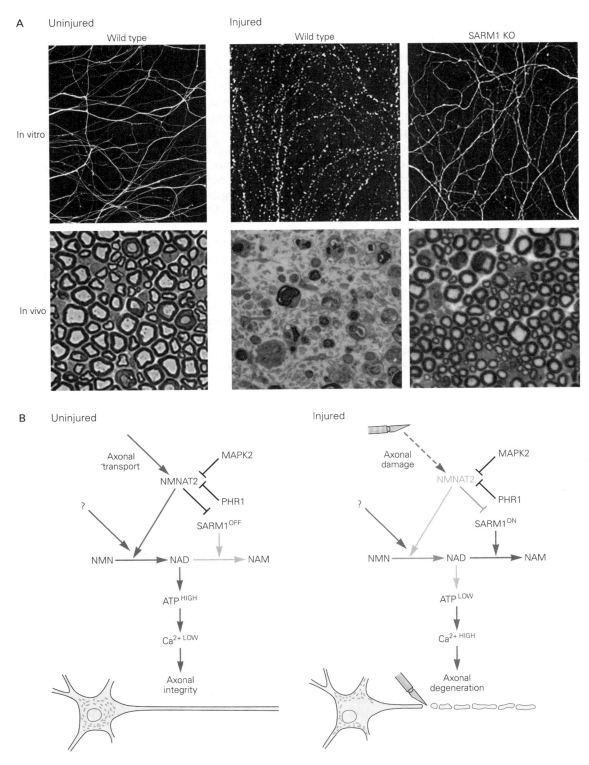

Figure 50–3 A core pathway regulates axon degeneration following axotomy in mice.

A. Damage to neurites in vitro leads to degeneration of the portions separated from the cell body. Likewise, axotomy in vivo leads to Wallerian degeneration, as shown by loss of myelin profiles in the cross section. Both in vitro and in vivo axons are spared if the *SARM1* gene is deleted. (From Gerdts et al. 2013.)

B. NMNAT2, closely related to the mutant *Wlds* protein, is normally present in axons. It can generate nicotinamide adenine dinucleotide (**NAD**) and inhibit SARM1, which degrades NAD. High NAD levels are required for energy metabolism, keeping adenosine triphosphate (**ATP**) levels high and calcium levels low in the axon. Following axotomy, NMNAT2 levels decrease rapidly, disinhibiting SARM1. NAD levels fall, ATP is depleted, calcium levels rise, calcium-dependent proteases are activated, and the axon is degraded. Kinases (MAPK) and a ubiquitin ligase (Phr1) regulate the pathway.

Together, these exciting new discoveries provide an answer to the question of why, following axotomy, the distal stump degenerates while the proximal stump is preserved. The conventional explanation that the distal stump is deprived of nutrients normally delivered from the cell body is incomplete. Instead, a signaling pathway in the axon senses damage and rapidly triggers degeneration. In this scenario, the key element supplied by the axon is NMNAT2. Its breakdown following axotomy disinhibits SARM1 and, perhaps in parallel with activation of factors that stimulate SARM1, triggers the loss of NAD, leading to the energy crisis that results in Wallerian degeneration.

These recent discoveries may be useful in devising treatments for neurological disorders in which axonal degeneration is prominent and generally precedes neuronal death. A fatal disease of motor neurons, amyotrophic lateral sclerosis, falls into this category. Other possibilities include some forms of spinal muscular atrophy, Parkinson disease, and even Alzheimer disease. Axon degeneration that occurs in these diseases, as well as after metabolic, toxic, or inflammatory insults, resembles the degeneration that follows acute trauma and may be regulated in similar ways. Thus, while methods for saving transected distal axons are unlikely to be useful clinically for treating patients who have suffered traumatic injury, the same techniques could be useful in treating neurodegenerative diseases.

Even though the proximal portion of the axon remains attached to the cell body, it too suffers. And in some cases, the neuron itself dies by apoptosis, probably because axotomy isolates the cell body from its supply of target-derived trophic factors. Even when this does not occur, the cell body often undergoes a series of cellular and biochemical changes called the *chromatolytic reaction:* The cell body swells, the nucleus moves to an eccentric position, and the rough endoplasmic reticulum becomes fragmented (Figure 50–1B). Chromatolysis is accompanied by other metabolic changes, including an increase in protein and RNA synthesis as well as a change in the pattern of genes that the neuron expresses. These changes are reversed if regeneration is successful.

Axotomy Leads to Reactive Responses in Nearby Cells

Axotomy sets in motion a cascade of responses in numerous types of neighboring cells. Among the most important responses are those of the glial cells that ensheath the distal nerve segment. One is fragmentation of the myelin sheath, which is then removed by phagocytes. This process is rapid in the peripheral nervous system, where the myelin-producing Schwann cells break the myelin into small fragments and engulf it. Schwann cells, which then divide, secrete factors that recruit macrophages from the blood stream. The macrophages in turn assist the Schwann cells in disposing of debris. Schwann cells also produce growth factors that promote axon regeneration, a point to which we will return later.

In contrast, in the central nervous system, the myelin-forming oligodendrocytes have little or no ability to dispose of myelin, and removal of debris depends on resident phagocytic cells called *microglia*. This difference in cellular properties may help explain the observation that Wallerian degeneration proceeds to completion much more slowly in the central nervous system.

Axotomy also affects both the synaptic inputs to and the synaptic targets of the injured neuron. When axotomy disrupts the major inputs to a cell—as happens in denervated muscle, or to neurons in the lateral geniculate nucleus when the optic nerve is cut—the consequences are severe. Usually the target atrophies and sometimes dies. When targets are only partially denervated, their responses are more limited. In addition, axotomy affects presynaptic neurons. In many instances, synaptic terminals withdraw from the cell body or dendrites of chromatolytic neurons and are replaced by the processes of glial cells—Schwann cells in the periphery and microglia or astrocytes in the central nervous system. This process, called *synaptic stripping*, depresses synaptic activity and can impair functional recovery.

Although the mechanism of synaptic stripping remains unclear, two possibilities have been suggested. One is that postsynaptic injury causes axon terminals to lose their adhesiveness to synaptic sites so that they are subsequently wrapped by glia. The other is that glia initiate the process of synaptic stripping in response to factors released from the injured neuron or to changes in its cell surface. Whatever the trigger, the activation of microglia and astrocytes by axotomy clearly contributes to the stripping process. In addition, biochemically altered astrocytes, called reactive astrocytes, contribute to formation of a *glial scar* near sites of injury.

As a result of these transsynaptic effects, neuronal degeneration can propagate through a circuit in both anterograde and retrograde directions. For example, a denervated neuron that becomes severely atrophic can fail to activate its target, which in turn becomes atrophic. Likewise, when synaptic stripping prevents an afferent neuron from obtaining sufficient sustenance from its target cell, the afferent neuron's inputs are placed at risk. Such chain reactions help to explain how injury in one area in the central nervous system eventually affects regions far from the site of the injury.

Central Axons Regenerate Poorly After Injury

Central and peripheral nerves differ substantially in their ability to regenerate after injury. Peripheral nerves can often be repaired following injury. Although the distal segments of peripheral axons degenerate, connective tissue elements surrounding the distal stump generally survive.

Axonal sprouts grow from the proximal stump, enter the distal stump, and grow along the nerve toward its targets (Figure 50–4). The mechanisms that drive this process are related to those that guide embryonic axons. Chemotropic factors secreted by Schwann cells attract axons to the distal stump, adhesive molecules

within the distal stump promote axon growth along cell membranes and extracellular matrices, and inhibitory molecules in the perineural sheath prevent regenerating axons from going astray.

Once regenerated peripheral axons reach their targets, they are able to form new functional nerve endings. Motor axons form new neuromuscular junctions; autonomic axons successfully reinnervate glands, blood vessels, and viscera; and sensory axons reinnervate muscle spindles. Finally, those axons that lost their myelin sheaths are remyelinated, and chromatolytic cell bodies regain their original appearance. Thus, in all three divisions of the peripheral nervous system—motor, sensory, and autonomic—the effects of axotomy are reversible. Peripheral regeneration is not perfect, however. In the motor system, recovery of strength may be substantial, but recovery of fine movements is usually impaired. Some motor axons never find their targets, some form synapses on inappropriate muscles, and some motor neurons die. Nevertheless, the regenerative capacities in the peripheral nervous system are impressive.

Peripheral nervous system

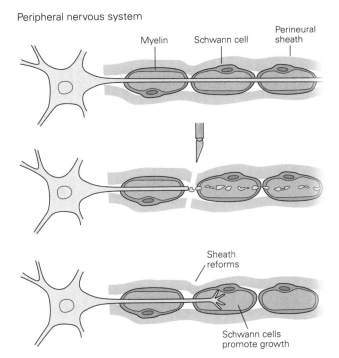

Central nervous system

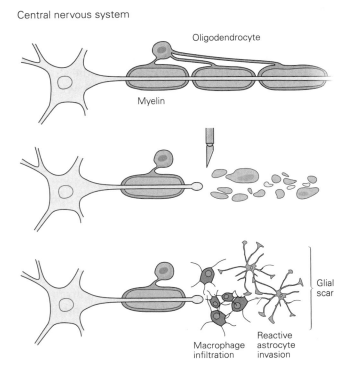

Figure 50–4 Axons in the periphery regenerate better than those in the central nervous system. After sectioning of a peripheral nerve, the perineural sheath reforms rapidly and Schwann cells in the distal stump promote axonal growth by producing trophic and attractant factors and expressing high

levels of adhesive proteins. After sectioning of an axona in the central nervous system, the distal segment disintegrates and myelin fragments. In addition, reactive astrocytes and macrophages are attracted to the lesion site. This complex cellular milieu, termed a *glial scar*, inhibits axonal regeneration.

In contrast, regeneration after injury is poor in the central nervous system (Figure 50–4). The proximal stumps of damaged axons can form short sprouts, but these soon stall and form swollen endings called "retraction bulbs", which fail to progress. Long-distance regeneration is rare. The failure of central regeneration is what led to the long-standing belief that injuries to the brain and spinal cord are largely irreversible and that therapy must be restricted to rehabilitative measures.

For some time, neurobiologists have been seeking the reasons why regenerative capacity in the central and peripheral nervous systems differs so dramatically. The goal of this work has been to identify the crucial barriers to regeneration so that they can be overcome. These studies have begun to bear fruit, and there is now cautious optimism that the injured human brain and spinal cord have a regenerative capacity that can eventually be exploited.

Before discussing these new developments, it is helpful to consider the problem of neural regeneration in a broader biological context. Is it the ability of peripheral axons to regenerate that is unusual, or the inability of central axons to do so? It is in fact the latter. Obviously, central axons grow well during development. More surprisingly, axons in immature mammals can also regenerate following transection in the brain or spinal cord. Moreover, regeneration is robust in the adult central nervous systems of lower vertebrates such as fish and frogs, as exemplified by the studies of Roger Sperry on restoration of vision following damage to the optic nerve (Chapter 47).

So why have mature mammals lost this seemingly important capacity for repair? The answer may lie in what the mammalian brain *can* do peerlessly, which is to remodel its basic wiring diagram in accordance with experience during critical periods in early postnatal life, so that each individual's brain is optimized to deal with the changes and challenges of internal and external worlds (Chapter 49). Once remodeling has occurred, it must be stabilized. Although it is obviously useful to reassign cortical space to one eye if the other is blinded in childhood, we would not want our cortical connections similarly rearranged in response to a brief period of unusual illumination or darkness. Maintaining constancy in the face of small perturbations in connectivity may therefore have the unavoidable consequence of limiting the ability of central connections to regenerate in response to injury. In this view, our limited regenerative capacity is a Faustian bargain in which we have sacrificed recuperative power to ensure the maintenance of precisely wired circuits that underlie our superior intellectual capacity.

Therapeutic Interventions May Promote Regeneration of Injured Central Neurons

In seeking reasons for the poor regeneration of central axons, one critical question is whether it reflects an inability of neurons themselves to grow or an inability of the environment to support axonal growth. This issue was addressed by Albert Aguayo and his colleagues in the early 1980s. They inserted segments of a central nerve trunk into a peripheral nerve, and segments of a peripheral nerve into the brain or spinal cord, to find out how axons would respond when confronted with a novel environment.

As expected, axons in the grafts, which were separated from their somata, promptly degenerated, leaving "distal stumps" containing glia, support cells, and extracellular matrix. What was striking was the behavior of axons near the translocated segments. Spinal axons that regenerated poorly following spinal cord injury grew several centimeters into the peripheral graft (Figure 50–5). Similarly, retinal axons, which regenerated poorly following damage to the optic nerve, grew long distances into a peripheral graft placed in their path.

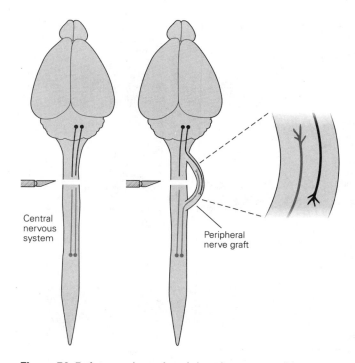

Central nervous system

Peripheral nerve graft

Figure 50–5 A transplanted peripheral nerve provides a favorable environment for the regeneration of central axons. *Left:* After sectioning of the spinal cord, ascending and descending axons fail to cross the lesion site. *Right:* Insertion of a peripheral nerve graft that bypasses the lesion site promotes regeneration of both ascending and descending axons. (Adapted from David and Aguayo 1981.)

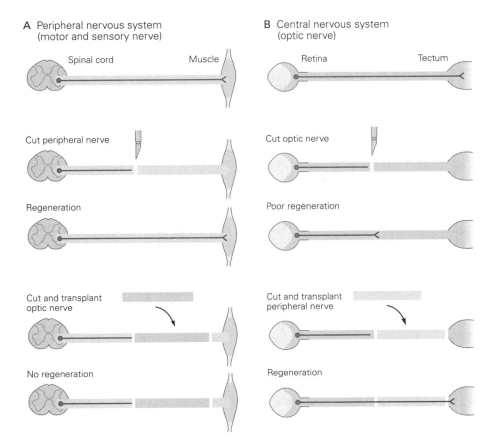

A Peripheral nervous system
(motor and sensory nerve)

Spinal cord   Muscle

Cut peripheral nerve

Regeneration

Cut and transplant
optic nerve

No regeneration

B Central nervous system
(optic nerve)

Retina   Tectum

Cut optic nerve

Poor regeneration

Cut and transplant
peripheral nerve

Regeneration

Figure 50–6 Peripheral and central nerves differ in their ability to support axonal regeneration.

A. In the peripheral nervous system, severed axons regrow past the site of injury. Insertion of a segment of optic nerve into a peripheral nerve suppresses the ability of the peripheral nerve to regenerate.

B. In the central nervous system, severed axons typically fail to regrow past the site of injury. Insertion of a section of peripheral nerve into a central nerve tract promotes regeneration.

Conversely, peripheral axons regenerated well through their own distal nerve trunk, but fared poorly when paired with a severed optic nerve (Figure 50–6).

Aguayo extended these studies to show that axons from multiple regions, including the olfactory bulb, brain stem, and mesencephalon, could all regenerate long distances if provided with a suitable environment. Even an optimal environment cannot fully restore the growth potential of central axons for reasons we will discuss in a later section. Nevertheless, these pioneering experiments focused attention on components of the central environment that inhibit regenerative ability and motivated an intensive search for the molecular culprits.

Environmental Factors Support the Regeneration of Injured Axons

In probing the differences between peripheral and central growth environments, initial searches were influenced by the results of experiments performed by Ramón y Cajal's student Francisco Tello nearly a century before Aguayo's studies. Tello transplanted segments of peripheral nerves into the brains of experimental animals and found that injured central axons grew toward the implants, whereas they barely grew when implants were not available.

This result implied that peripheral cells provide growth-promoting factors to the injured areas, factors normally absent from the brain. Ramón y Cajal reasoned that central nerve pathways lacked "substances able to sustain and invigorate the indolent and scanty growth" similar to those provided by peripheral pathways. Numerous studies over the succeeding century identified constituents of peripheral nerves that are potent promoters of neurite outgrowth. These include components of Schwann cell basal laminae, such as laminin, and cell adhesion molecules of the immunoglobulin superfamily. In addition, cells in denervated distal nerve stumps begin to produce neurotrophins and other trophic molecules of the sort described in Chapter 46. Together, these molecules nourish neurons and guide growing axons in the embryonic nervous system, so it makes sense that they also promote the regrowth of axons. By contrast, central neuronal tissue is a poor source of these molecules, containing little laminin and low levels of trophic molecules. Thus, in the embryo, both central and peripheral nervous systems provide environments that promote axon outgrowth. But only the peripheral environment retains this capacity in adulthood or is able to regain it effectively following injury.

The practical implications of this view are that supplementing the central environment with growth-promoting molecules might improve regeneration. To this end, investigators have infused neurotrophins into areas of injury or inserted fibers rich in extracellular matrix molecules such as laminin to serve as scaffold for axonal growth. In some attempts, Schwann cells themselves, or cells engineered to secrete trophic factors, have been grafted into sites of injury. In many of these cases, injured axons grow more extensively than they do under control conditions. Yet regeneration remains limited, with axons generally failing to extend long distances. More important, functional recovery is minimal.

Components of Myelin Inhibit Neurite Outgrowth

What accounts for such disappointingly limited regeneration? One part of the explanation is that the environment encountered by severed central axons is not only poor in growth-promoting factors but also rich in growth-inhibiting factors, some of which are derived from myelin. In culture, fragments of central but not peripheral myelin potently inhibit neurite outgrowth from co-cultured central or peripheral neurons. Conversely, sprouting of spinal axon collaterals following injury is enhanced in rats treated to prevent myelin formation in the spinal cord (Figure 50–7).

These findings implied that although both central and peripheral environments might contain a supply of growth-promoting elements, central nerves also contain inhibitory components. The fact that myelin inhibits neurite growth may seem peculiar, but not if we consider that myelination normally occurs postnatally, after axon extension is largely complete.

Searches for the inhibitory components of central myelin turned up an embarrassment of riches. Several classes of molecules that occur at higher levels in central myelin compared to peripheral myelin are able to inhibit neurite outgrowth when presented to cultured neurons. The first to be discovered was identified when an antibody generated against myelin proteins proved

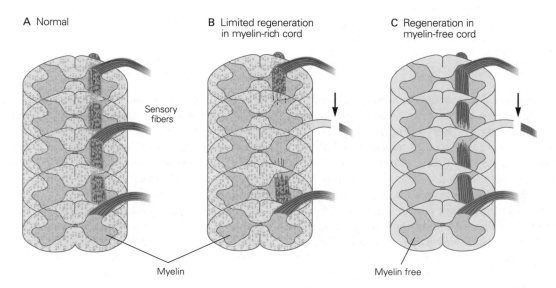

Figure 50–7 Myelin inhibits regeneration of central axons. (Adapted, with permission, from Schwegler, Schwab, and Kapfhammer 1995.)

A. Sensory fibers normally extend rostrally in a myelin-rich spinal cord.

B. Right dorsal root fibers were sectioned in 2-week-old normal rats. Regeneration of the fibers was assessed histochemically

20 days later. The central branches of the sectioned axons degenerated, leaving a portion of the spinal cord denervated. Little regeneration occurred in the myelin-rich cord.

C. Some littermates received local x-irradiation to block myelination. In these animals, sensory fibers that entered the cord through neighboring uninjured roots sprouted new collaterals following denervation.

to be capable of partially neutralizing myelin's ability to inhibit neurite outgrowth. Use of this antibody to isolate the corresponding antigen yielded the protein now called Nogo. Two other proteins, myelin-associated glycoprotein (MAG) and oligodendrocyte-myelin glycoprotein (OMgp), initially isolated as major components of myelin, have also been found to inhibit the growth of some neuronal types.

Intriguingly, Nogo, MAG, and OMgp bind to common membrane receptors, NogoR and PirB (Figure 50–8). NogoR, as well as related receptors such as LINGO that have been implicated in growth

inhibition, all interact with the neurotrophin receptor p75 (Chapter 46). This interaction converts p75 from a growth-promoting to a growth-inhibiting receptor. Perhaps because there are so many growth inhibitory factors and receptors, regeneration of central axons is not greatly enhanced in mutant mice lacking any one of them. However, many of the inhibitory components trigger the same intracellular signaling pathway in which RhoA is activated, thereby stimulating Rho kinase (ROCK); ROCK in turn leads to the collapse of growth cones and blocks actin and tubulin polymerization required for neurite growth. Current studies are

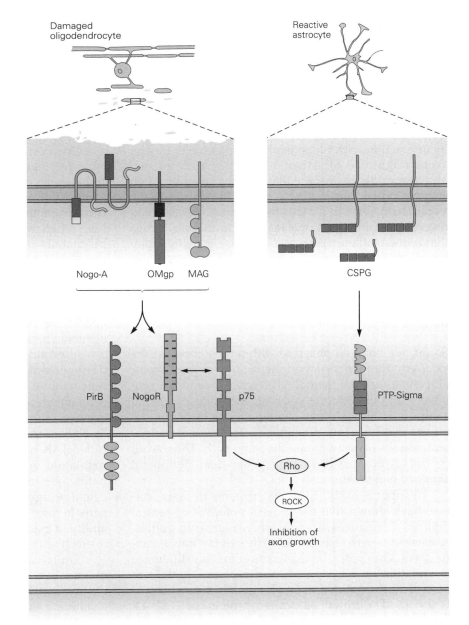

Figure 50–8 Myelin and glial scar components that inhibit regeneration of central axons. (Adapted from Yiu and He 2006.)

Left: Myelin contains the proteins Nogo-A, oligodendrocyte-myelin glycoprotein (**OMgp**), and myelin-associated glycoprotein (**MAG**). All three proteins are exposed when myelin breaks down. They can bind to the receptor protein NogoR, which can associate with the neurotrophin receptor p75, as well as an immunoglobulin-like receptor protein PirB. Inactivation of PirB results in a modest enhancement of corticospinal axon regeneration. *Right:* Chondroitin sulphate proteoglycans (**CSPG**) are major components of the glial scar and are thought to suppress axon regeneration through interaction with the receptor tyrosine phosphatase PTP-sigma, which activates intracellular mediators such as Rho and ROCK.

exploring whether interference with that shared pathway might neutralize the impact of many inhibitors in one fell swoop.

Injury-Induced Scarring Hinders Axonal Regeneration

Myelin debris is not the only source of growth-inhibiting material in the injured brain or spinal cord. As noted earlier, astrocytes become activated and proliferate following injury, acquiring features of reactive astrocytes that generate scar tissue at sites of injury. Scarring is an adaptive response that helps to limit the size of the injury, reestablish the blood-brain barrier, and reduce inflammation.

But the scar itself hinders regeneration in two ways: through mechanical interference with axon growth and through growth-inhibiting effects of proteins produced by cells within the scar. Chief among these inhibitors are a class of chondroitin sulfate proteoglycans (CSPG) that are produced in abundance by reactive astrocytes and directly inhibit axon extension by interaction with tyrosine phosphatase receptors on axons (Figure 50–8). Attention has therefore focused on ways of dissolving the glial scar by infusion of an enzyme called *chondroitinase*, which breaks down the sugar chains on CSPG. This treatment promotes axon regeneration and functional recovery in animals. Drugs that reduce inflammation and decrease scarring, notably prednisolone, are also beneficial if administered shortly after injury, before the scar forms.

An Intrinsic Growth Program Promotes Regeneration

So far, we have emphasized differences between the local environments of peripheral and central axons. However, environmental differences cannot completely account for the poor regeneration of central axons. Even though they can regenerate in peripheral nerves, central axons grow much less well than peripheral axons when navigating the same path. Thus, adult central axons may be less capable than peripheral axons of regeneration.

In support of this idea, experiments in tissue culture have shown that the growth potential of central neurons decreases with age, whereas mature peripheral neurons extend axons robustly in a favorable environment. One potential explanation for this difference is variation in the expression of proteins thought to be critical for optimal axon elongation. One example is the 43 kDa growth-associated protein, or GAP-43. This protein is expressed at high levels in embryonic central and peripheral neurons. In peripheral neurons, the level remains high in maturity and increases even more following axotomy, whereas in central neurons, its expression decreases as development proceeds. Transcription factors required to coordinate axonal growth programs are also expressed at high levels during development, and then are downregulated in maturity.

Is this reduced ability of central axons to regenerate reversible? Hope is provided by two sets of studies. One involves what has been called a "conditioning lesion." Recall that primary sensory neurons in dorsal root ganglia have a bifurcated axon, with a peripheral branch that extends to skin, muscle, or other targets, and a central branch that enters the spinal cord. The peripheral branch regenerates well following injury, whereas the central branch regenerates poorly. However, the central branch will regenerate successfully if the peripheral branch is damaged several days before the central branch is damaged (Figure 50–9). Somehow, prior injury or conditioning lesion activates an axonal growth program.

One component of the growth program responsible for regeneration of the central branch appears to be cyclic adenosine monophosphate (cAMP). This second-messenger molecule activates enzymes that in turn promote neurite outgrowth. Levels of cAMP are high when neurons initially form circuits; they decline postnatally in central but not peripheral neurons. In some instances, increased supplies of cAMP or proteins normally activated by cAMP can promote regeneration of central axons following injury. Accordingly, drugs that increase cAMP levels or activate targets of cAMP are being actively considered as therapeutic agents to be administered following spinal cord injury.

A second group of investigations has manipulated developmentally regulated intrinsic factors to restore regenerative ability in adults. For example, injury sometimes leads to formation of cytokines such as ciliary neurotrophic factors (CNTFs) that promote growth by activating a signaling pathway involving molecules called JAK and STAT that travel to the nucleus and regulate a growth program. In adults, however, the pathway is inhibited by a protein called suppressor of cytokine signaling 3 (SOCS3). Deletion of the *SOCS3* gene in mice relieves the inhibition and augments the ability of cytokines to promote regeneration of injured axons (Figure 50–10A).

Similarly, a signaling pathway involving the kinase mammalian target or rapamycin (mTOR) regulates energy metabolism, promoting an anabolic growth-promoting state required for axon regeneration. However, mTOR is downregulated as central neurons

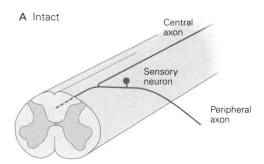

A Intact

Central axon

Sensory neuron

Peripheral axon

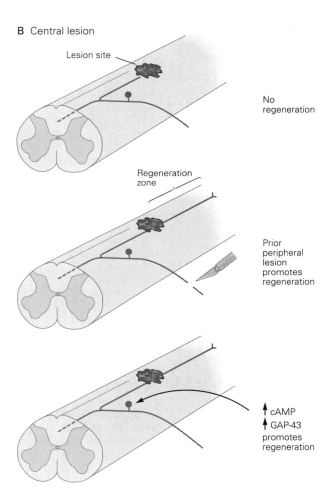

B Central lesion

Lesion site

No regeneration

Regeneration zone

Prior peripheral lesion promotes regeneration

↑ cAMP
↑ GAP-43
promotes regeneration

Figure 50–9 A conditioning lesion promotes regeneration of the central branch of a primary sensory neuron axon. After lesions of the spinal cord, there is little regeneration of the central branch beyond the injury site. However, if the peripheral branch of the axon is sectioned before the central branch is damaged, the latter will grow beyond the lesion site. The impact of such a "conditioning lesion" can be mimicked by elevating levels of cyclic adenosine monophosphate (**cAMP**) or of the growth-associated protein GAP-43 in the peripheral branch.

mature and is further inhibited by a phosphatase called PTEN. Analogous to SOCS3 and JAK/STAT signaling, deletion of the *PTEN* gene in mice promotes axonal regrowth following injury to the optic nerve or spinal cord (Figure 50–10B). Moreover, loss of SOCS3 and PTEN stimulates regeneration significantly more than loss of either one. Although their multiple roles make it unlikely that either SOCS3 or PTEN is a useful target for therapy, the signaling pathways they regulate provide multiple starting points for designing drugs that could augment regeneration.

Formation of New Connections by Intact Axons Can Lead to Recovery of Function Following Injury

So far, we have discussed interventions designed to enhance the limited regenerative capacity of injured central axons. An alternative strategy focuses on the significant, although incomplete, functional recovery that can occur following injury even without appreciable regeneration of cut axons. If the basis for this limited recovery of function can be understood, it may be possible to enhance it.

A rearrangement of existing connections in response to injury may contribute to recovery of function. We have learned that axotomy leads to changes in both the inputs to and the targets of the injured neuron. Although many of these changes are detrimental to function, some are beneficial. In particular, the central nervous system can, following injury, spontaneously undergo adaptive reorganization that helps it regain function. For example, after transection of the descending corticospinal pathway, which occurs with many traumatic injuries of the spinal cord, the cortex can no longer transmit commands to motor neurons below the site of the lesion. Over several weeks, however, intact corticospinal axons rostral to the lesion begin to sprout new terminal branches and form synapses on spinal interneurons whose axons extend around the lesion, thereby forming an intraspinal detour that contributes to limited recovery of function (Figure 50–11).

Similar instances of functional reorganization have been demonstrated in the motor cortex and brain stem. These compensatory responses attest to the latent plasticity of the nervous system. The ability of the nervous system to rewire itself is most vigorous during the critical periods of early postnatal life but can be revived by traumatic events in adulthood (Chapter 49).

How can the rewiring ability of the central nervous system be improved? It is possible that some of the beneficial effects of grafts in experimental animals reflect reorganization of intact axons rather than regeneration of transected axons. As the nervous system's

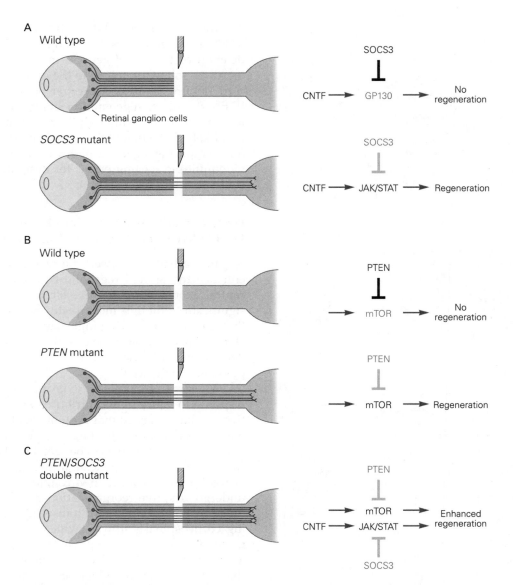

Figure 50–10 Signaling pathways that regulate axon regeneration in the optic nerve.

A. The regeneration of retinal ganglion cell axons in the optic nerve is normally constrained by neuronal expression of several genes. One encodes SOCS3, which blocks the ability of ciliary neurotrophic factor (**CNTF**) to bind its receptor GP130 and thus blocks CNTF from promoting regeneration. In *SOCS3* mutant mice, ambient levels of CNTF are sufficient to improve optic nerve regeneration. Elimination of GP130 as well as SOCS3 blocks the capacity for regeneration. Addition of extra

CNTF enhances the capacity for regeneration in *SOCS3* mutant mice.

B. Another gene encodes PTEN, which blocks signaling through the mammalian target of rapamycin (**mTOR**) pathway, which regulates energy metabolism. Accordingly, regeneration is enhanced in *PTEN* mutant mice.

C. Because *SOCS3* and *PTEN* regulate different growth-promoting signals, mutant mice lacking both genes exhibit greater regenerative ability than either single mutant. (Adapted from Smith et al. 2009.)

plasticity becomes better understood, therapeutic strategies that promote specific changes in circuitry may become possible. Perhaps most promising is an approach in which cellular or molecular interventions that promote growth are combined with behavioral therapies that result in circuit rewiring.

Neurons in the Injured Brain Die but New Ones Can Be Born

The failure to grow a new axon is by no means the worst fate that can befall an injured neuron. For many neurons, axotomy leads to the death of the cell. Efforts

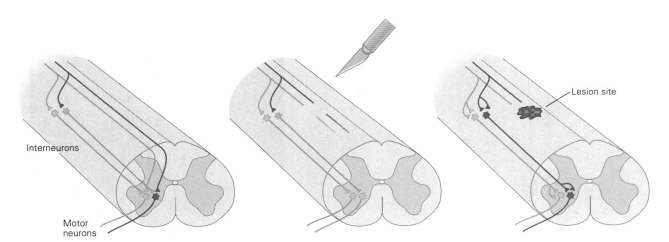

Figure 50–11 Function can be recovered after spinal cord injury through reorganization of spinal circuits. Severed corticospinal axons can reestablish connections with motor neurons by sprouting axon collaterals that innervate propriospinal interneurons whose axons bypass the lesion and contact motor neurons located caudal to the lesion site. (Adapted from Bareyre et al. 2004.)

to improve recovery following injury therefore need to consider survival of neurons and not simply the regrowth of axons. Since neuronal death is a frequent consequence of other neural insults, such as stroke and neurodegenerative disease, improved ways of retaining or replacing neurons would have broad utility.

The loss of cells following injury is not unique to the nervous system, although in other tissues, new cells are often effective at repairing damage. This regenerative capacity is most dramatic in the hematopoietic system, where a few stem cells can repopulate the entire adaptive immune system. In contrast, it has long been believed that the generation of neurons is complete by birth. Because of this, approaches to regeneration have often focused on finding ways to spare neurons that would otherwise die.

This traditional view has changed, prompted initially by Joseph Altman's discovery in the 1960s that neurogenesis continues into adulthood in some parts of the mammalian brain. Since this finding challenged fundamental tenets of prevailing dogma, the idea that new neurons could form in postnatal rodents was met with skepticism for three decades.

Eventually, however, the application of better cell labeling technologies amply supported Altman's conclusion and showed that it also applies to nonhuman primates and even, in a limited way, to humans. We are now confident that new neurons are added to the dentate gyrus of the hippocampus and to the olfactory bulb throughout life, although the rate of addition declines with age. Some of the newborn cells in the dentate gyrus of the adult hippocampus die soon after

they are born and others become glial cells, but a substantial minority differentiate into granule cells that are indistinguishable from those born at embryonic stages (Figure 50–12). New neurons are also added to the adult olfactory bulb. They are generated near the surface of the lateral ventricles, far from the bulb itself, and then migrate to their destination (Figure 50–13). In both cases, the new neurons extend processes, form synapses, and become integrated into functional circuits. Thus, neurons born at embryonic stages are gradually replaced by later-born neurons, so that the total number of neurons in these regions of the brain is maintained.

The properties of neurons born in mature animals are not completely understood, but they appear able to recapitulate many of the properties of neurons that arise in the embryo. When the generation of new neurons in the adult is prevented, certain behaviors mediated by the olfactory bulb and hippocampus are degraded. Conversely, some behavioral alterations are accompanied by alterations in the tempo of adult neurogenesis. Adult neurogenesis can be decreased in animal models of depression and chronic stress, whereas enrichment of the habitat of an animal or an increase in the physical activity of otherwise sedentary rodents can increase the generation of new neurons.

What cells give rise to adult-born neurons? The principle that embryonic neurons and glia arise from multipotential progenitors also applies to neurons born in adults. Stem cells are the source of neurons in the adult as well as the embryo. They are likely derived from radial glia, which also serve as a source of

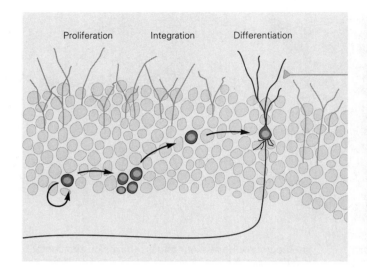

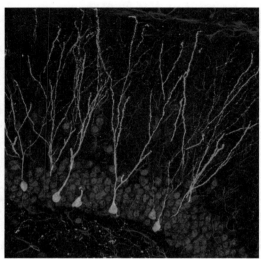

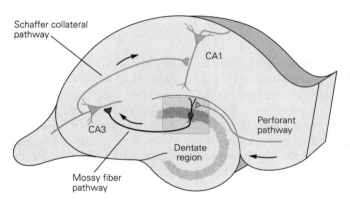

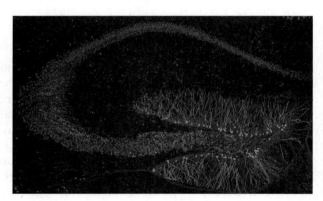

Figure 50–12 Neurons born in the germinal zone of the dentate gyrus in adult rodents are integrated into hippocampal circuits. The diagrams on the left show the pathways of neuronal differentiation and integration into dentate gyrus circuits. The images on the right show newly generated neurons and their dendritic arbors labeled with a virus expressing green fluorescence protein. (Micrographs reproduced, with permission, from F. Gage.)

neurons during embryonic development (Chapter 46). A subset of these cells exit the cell cycle during gestation, become quiescent, and take up residence near the ventricular surface. In adulthood, they are activated, reenter the cell cycle, and give rise to neurons.

Although so far adult neurogenesis has not been directly linked to repair of damaged tissue, its discovery has influenced research on recovery from injury in two important ways. First, the findings that endogenously generated neurons can differentiate and extend processes through the thicket of adult neuropil, and can be integrated into functional circuits, led researchers to test the idea that the same could be true for transplanted neurons or precursors. Second, since neural precursors can be induced to divide and differentiate, strategies designed to augment this innate ability are now being considered, with the goal of producing neurons in large enough numbers to replace those lost to injury or neurodegenerative disease. As we describe below, these ideas have progressed over the past few decades from science fiction to efforts that are tantalizingly close to clinical tests.

Therapeutic Interventions May Retain or Replace Injured Central Neurons

Transplantation of Neurons or Their Progenitors Can Replace Lost Neurons

For many years, neurologists have transplanted developing neurons into experimental animals to see if the new neurons could reverse the effects of injury or disease. These attempts have had promising results in a few cases.

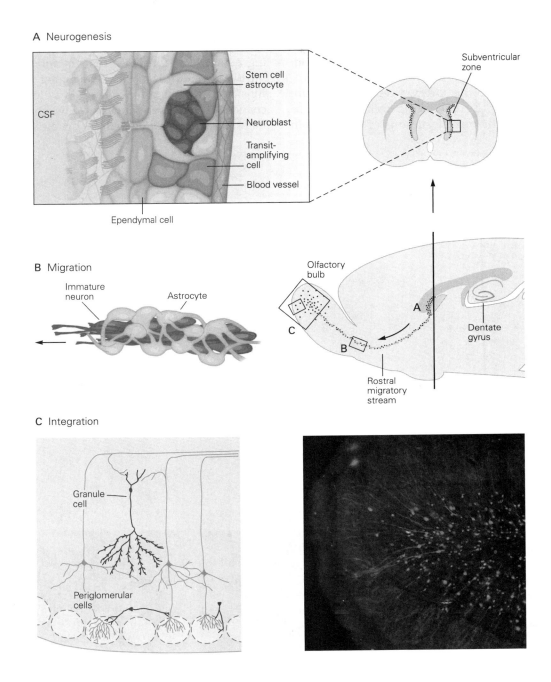

Figure 50–13 The origin and fate of neurons born in the adult ventricular zone. (Adapted from Tavazoie et al. 2008.)

A. Neuroblasts develop in an orderly progression from astrocytic stem cells via a population of cells within a local niche close to blood vessels in the subventricular zone. (Abbreviation: **CSF,** cerebrospinal fluid.)

B. Neuroblasts differentiate into immature neurons that migrate to the olfactory bulb using astrocytes as guides. They crawl along each other in a process called chain migration.

C. On arrival in the olfactory bulb, immature neurons differentiate into granule cells and periglomerular cells, two classes of olfactory bulb interneurons. (Image reproduced, with permission, from A. Mizrahi.)

One is to replace dopaminergic cells that die in Parkinson disease. When transplanted into the striatum, these neurons release dopamine onto their targets without the need to grow long axons or form elaborate synapses (Figure 50–14). Another is to transplant immature inhibitory interneurons from the ganglionic eminences in which they are produced (Chapter 46) to the cortex, where they mature and form synapses. By enhancing inhibition, these neurons attenuate the manifestations of disorders in which insufficient inhibitory drive plays a role, such as epilepsy and anxiety.

Unfortunately, application of these methods to human patients has been fraught with difficulties. One is the difficulty of obtaining and growing developing neurons in sufficient numbers and with sufficient purity. Second, it has been challenging to modify neurons by introducing new genes so as to improve their chances of functioning in a new environment. Third, in many cases, the grafted neurons are already too mature

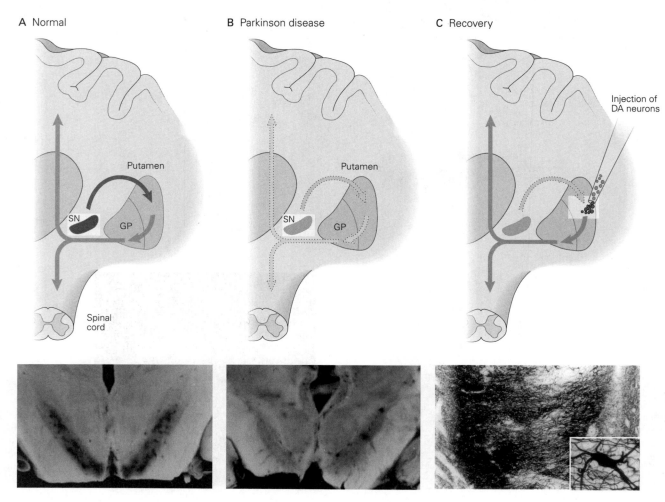

Figure 50–14 Loss of dopaminergic (DA) neurons in Parkinson disease can be treated by grafting embryonic cells into the putamen.

A. In the healthy brain, dopaminergic projections from the substantia nigra (**SN**) innervate the putamen, which in turn activates neurons in the globus pallidus (**GP**). Pallidal outputs to the brain and spinal cord facilitate movement. The image below shows melanin-rich dopaminergic neurons in human substantia nigra.

B. In Parkinson disease, the loss of dopaminergic neurons in the substantia nigra deprives the putamen–globus pallidus

pathways of their drive. The image beneath the diagram shows the virtual absence of melanin-rich dopaminergic neurons in the substantia nigra of an individual with Parkinson disease.

C. Direct injection of embryonic dopaminergic neurons into the putamen reactivates the globus pallidus output pathways. The image below shows tyrosine hydroxylase expression in the cell bodies and axons of embryonic mesencephalic dopaminergic neurons grafted into the putamen of a human patient. (Image reproduced, with permission, from Kordower and Sortwell 2000. Copyright © 2000. Published by Elsevier B.V.)

to differentiate properly or to integrate effectively into functional circuits.

These obstacles can be overcome by transplanting neural precursors into the adult brain where they can go on to differentiate into neurons in a hospitable environment. Several classes of precursors have been transplanted successfully, including neural stem cells and committed precursors. Some initial success has been obtained with embryonic stem (ES) cells. These cells are derived from early blastocyst stage embryos and can give rise to all cells of the body. Because they can divide indefinitely in culture, large numbers of cells can be generated, induced to differentiate, and then engrafted.

More recently, this technology has been enhanced by the molecular reprogramming of skin fibroblast cells to create induced pluripotent stem (iPS) cells

(Figure 50–15). These cells have a distinct advantage over ES cells; embryos are not required for their production, effectively bypassing a minefield of practical, political, and ethical concerns that have hindered research using human ES cells. Another advantage of iPS cells is that they can be generated from an individual patient's own skin cells, neatly avoiding issues of immunological incompatibility. It is also possible to genetically modify the iPS cells in culture by repairing a defective gene before transplantation.

Because ES and iPS cells have the potential to generate any cell type, it is essential that their differentiation be guided along specific pathways in culture before they are transplanted. Methods for generating specific classes of neural precursors, neurons, and glial cells from ES and iPS cells have now been devised (Figure 50–15).

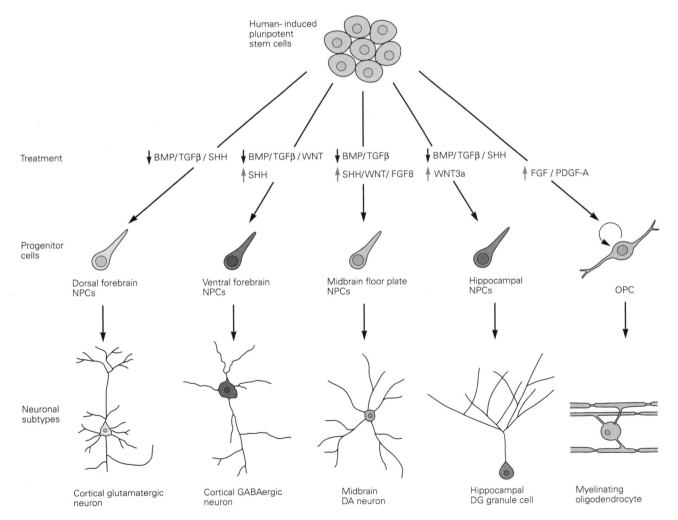

Figure 50–15 Induced pluripotent stem cells can be reprogrammed to generate precursors of many neuronal and glial types. The precursors can then be transplanted into the brain or spinal cord, where cells complete their differentiation and integrate into functional circuits. (Abbreviations: **DA,** dopamine; **DG,** dentate gyrus; **NPC,** neural progenitor cell; **OPC,** oligodendrocyte progenitor cell.) (Adapted, with permission, from Wen et al. 2016. Copyright © 2016 Elsevier Ltd.)

For example, it is possible to generate neurons that possess many or all of the properties of the spinal motor neurons that are lost in amyotrophic lateral sclerosis (Figure 50–16) or to generate the dopaminergic neurons lost from the striatum in Parkinson disease and then to engraft such neurons into the spinal cord or brain.

Although many hurdles need to be overcome, clinical trials using ES and iPS cell-derived neurons are underway. In addition, these cells are being used in chemical screens to identify compounds that counteract the cellular defects that underlie human neurodegenerative disease.

Stimulation of Neurogenesis in Regions of Injury May Contribute to Restoring Function

What if, following injury in adults, endogenous neuronal precursors could be stimulated to produce neurons capable of replacing those that have been lost? Two sets of recent findings suggest that this idea is not so far-fetched.

First, precursors capable of forming neurons in culture have been isolated from many parts of the adult nervous system, including the cerebral cortex and spinal cord, even though neurogenesis in adults is ordinarily confined to the olfactory bulb and hippocampus. This diversion of cell fate led to the idea that neurogenesis in the adult occurs in only a few sites, because only they contain appropriate permissive or stimulatory factors. This hypothesis has spurred a search for such factors, in the hope that they could be used to render a larger range of sites capable of supporting neurogenesis.

Second, in a few cases, the generation of new neurons can be stimulated by traumatic or ischemic injury (akin to stroke), even in areas such as the cerebral cortex

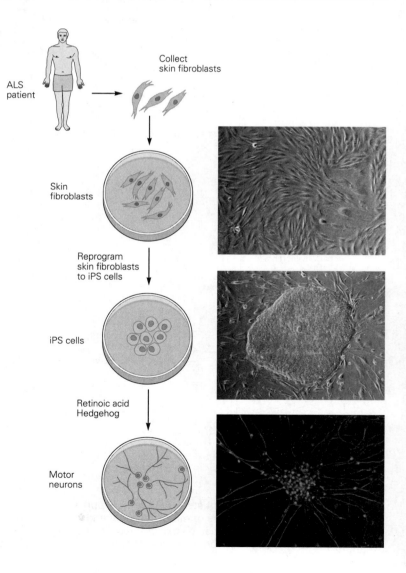

Figure 50–16 Induced pluripotent stem cells derived from an individual with amyotrophic lateral sclerosis (ALS) can differentiate into spinal motor neurons. Fibroblasts from the skin of a patient with ALS were used to generate induced pluripotent stem (iPS) cells, which were then directed to a motor neuron fate (see Figure 50–15). These cells can be used to analyze mechanisms that underlie motor neuron loss in ALS. The images at right show (from top to bottom) cultured fibroblasts, an iPS cell clump, and differentiated motor neurons expressing characteristic nuclear transcription factors (**green**) and axonal proteins (**red**). (Micrographs reproduced, with permission, from C. Henderson, H. Wichterle, G. Croft, and M. Weygandt.)

or spinal cord in which neurogenesis normally fails to occur. The fact that recovery after stroke and injury is poor demonstrates that spontaneous compensatory neurogenesis, if it occurs in humans, is insufficient for tissue repair. However, injury-induced neurogenesis has been enhanced in experimental animals in several ways. In one, administration of growth factors promotes neuronal production from progenitors grown in culture. In another, glial cells that retain the capacity to divide, such as Müller glia in the retina or astrocytes in the cortex, are reprogrammed to differentiate into neurons. If such interventions could be adapted to humans, the range of neurons subject to replacement would be greatly increased.

Transplantation of Nonneuronal Cells or Their Progenitors Can Improve Neuronal Function

Cells other than neurons are lost after brain injury. Among the most profound losses are those of oligodendrocytes, the cells that form the myelin sheath around central axons. The stripping of myelin continues long after traumatic injury and contributes to progressive loss of function of axons that may not have been injured directly.

Although the adult brain and spinal cord are capable of generating new oligodendrocytes and replacing lost myelin, this production is insufficient to restore function in many cases. Since several common neurological diseases, most notably multiple sclerosis, are accompanied by a profound state of demyelination, there is strong interest in providing the nervous system with additional oligodendrocyte precursors in order to augment remyelination.

Neural stem cells, multipotential progenitors, ES cells, and iPS cells can give rise not only to neurons but also to nonneural cells, including oligodendrocytes and their direct precursors. Indeed, at present, human ES cells are being channeled into oligodendrocyte progenitor cells and implanted into injured spinal cords of experimental animals. Transplanted cells that differentiate into oligodendrocytes enhance remyelination and substantially improve the locomotor ability of experimental animals (Figure 50–17).

Restoration of Function Is the Aim of Regenerative Therapies

We need to bear in mind that efforts to replace central neurons or to enhance the regeneration of their axons

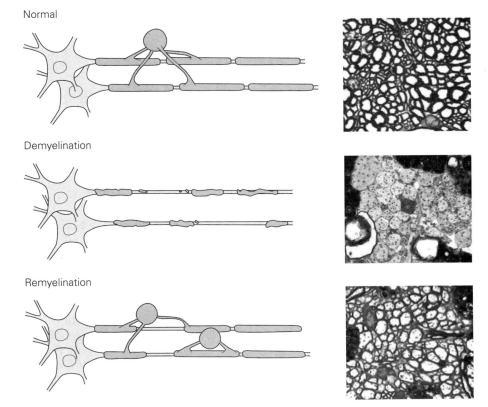

Figure 50–17 Restoration of myelination in the central nervous system by transplanted oligodendrocyte stem cells. In rodents with demyelinated axons, grafts of oligodendrocyte precursor cells can restore myelination to near normal. Sections through central nerve tracts are shown in the images at right. (Adapted, with permission, from Franklin and ffrench-Constant 2008. Copyright © 2008 Springer Nature.)

Normal

Demyelination

Remyelination

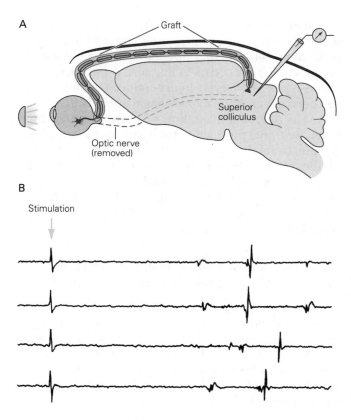

Figure 50–18 Regenerated retinal ganglion axons in the optic nerve can form functional synapses. (Adapted, with permission, from Keirstead et al. 1989. Copyright © 1989 AAAS.)

A. A segment of optic nerve in an adult rat was removed, and a segment of sciatic nerve was grafted in its place. The other end of the sciatic nerve was attached to the superior colliculus. Some retinal ganglion cell axons regenerated through the sciatic nerve and entered the superior colliculus.

B. Once the axons of the retinal ganglion neurons had regenerated, recordings were made from the superior colliculus. Flashes of light delivered to the eye elicited action potentials in collicular neurons, demonstrating that at least some regenerated axons had formed functional synapses.

would be of little use if these axons were unable to form functional synapses with their target cells. The same fundamental questions asked about axon regeneration in adults therefore apply to synaptogenesis: Can it happen, and if not, why not?

It has been difficult to address these questions because axonal regeneration following experimentally induced injury is usually so poor that the axons never reach appropriate target fields. However, several of the studies discussed earlier in this chapter offer hope that synapse formation is possible within the dense adult neuropil. In fact, axon branches that regenerate following injury can form synapses on nearby targets. For example, Aguayo and his colleagues found that retinal axons were able to regrow into the superior colliculus when they were channeled through a peripheral nerve that had been grafted into the optic nerve (Figure 50–18A). Remarkably, some collicular neurons fired action potentials when the eye was illuminated, showing that functional synaptic connections had been reestablished (Figure 50–18B). More recent studies have promoted regeneration of severed axons by enhancing their intrinsic growth programs, as described above, and observed some restoration of function.

Likewise, neurons that arise endogenously or are implanted by investigators can form and receive

synapses. Thus, there is reason to believe that if injured axons can be induced to regenerate, or new neurons supplied to replace lost ones, they will wire up in ways that help restore lost functions and behaviors.

Highlights

1. When axons are transected, the distal segment degenerates, a process called Wallerian degeneration. The proximal segment and cell body also undergo changes, as do the injured neuron's synaptic inputs and targets.

2. It was long thought that Wallerian degeneration was a passive and inevitable consequence of the distal segment being deprived of sustenance from the cell body, but it is not known to be an active, regulated process. Genes called *NMNAT* and *SARM1* are key components of a core signaling pathway that controls the process. Intervention in the pathway can slow or even halt degeneration.

3. Axons can regenerate and form new synapses following injury, but in mammals, regeneration is far more widespread and effective in peripheral axons than in central axons.

4. A key factor in the differential response of peripheral and central axons is that the environment confronting injured central axons is poor at supporting growth. It both lacks nutritive factors present in the pathway of peripheral nerves and contains growth-inhibitory factors absent from peripheral nerves.

5. Structures that inhibit regeneration include myelin fragments that persist following Wallerian degeneration and astrocytes that form glial scars at injury sites. Inhibitory factors in myelin include Nogo and myelin-associated glycoprotein. Inhibitory factors secreted by astrocytes include chondroitin sulfate proteoglycans.

6. Central regeneration is also hindered by intrinsic decreased ability of adult central neurons to grow, due to downregulation of growth programs active during development. Interventions that restore or disinhibit growth pathways, such as JAK/STAT and mTOR signaling, enable regeneration.

7. However, it is important to note that the failure of regeneration following injury may be related to the stabilization of connections that occurs at the end of critical periods. For example, myelination, which occurs largely at the end of a critical period, may have the secondary effect of preventing further, large-scale rearrangement of synaptic connections. Thus, caution will be needed to ensure that treatments aimed at fostering recovery following injury do not end up promoting formation of maladaptive circuits.

8. Another approach for restoring function following damage is to harness the ability of intact axons to form new connections, generating adaptive circuits that can compensate to some extent for those lost to injury.

9. The traditional view that all neurogenesis occurs during or shortly after gestation has now been modified by the discovery that new neurons are born throughout life in a few brain areas. These neurons arise from resident stem cells and can integrate into functional circuits.

10. Cells capable of forming new neurons are also present in many other areas of the brain and spinal cord but remain quiescent. Attempts to activate them by providing growth factors or introducing growth-promoting genes (transcriptional reprogramming) could harness their potential following injury or in neurodegenerative disease.

11. Another approach to neuronal replacement is to implant developing neurons. Although fetal neurons are sometimes used for this purpose in experimental animals, a more useful source may be neurons derived from ES or iPS cells. They can be grown in large quantities, genetically modified if necessary, and treated to differentiate into specific neuronal types. Clinical studies using this approach are now beginning.

Joshua R. Sanes

Selected Reading

Benowitz LI, He Z, Goldberg JL. 2017. Reaching the brain: advances in optic nerve regeneration. Exp Neurol 287:365–373.

Dell'Anno MT, Strittmatter SM. 2017. Rewiring the spinal cord: direct and indirect strategies. Neurosci Lett 652:625–634.

Gerdts J, Summers DW, Milbrandt J, DiAntonio A. 2016. Axon self-destruction: new links among SARM1, MAPKs, and NAD+ metabolism. Neuron 89:449–460.

He Z, Jin Y. 2016. Intrinsic control of axon regeneration. Neuron 90:437–451.

Magnusson JP, Frisén J. 2016. Stars from the darkest night: unlocking the neurogenic potential of astrocytes in different brain regions. Development 143:1075–1086.

McComish SF, Caldwell MA. 2018. Generation of defined neural populations from pluripotent stem cells. Philos Trans R Soc Lond B Biol Sci 373:pii: 20170214.

Zhao C, Deng W, Gage FH. 2008. Mechanisms and functional implications of adult neurogenesis. Cell 132:645–660.

References

Alilain WJ, Horn KP, Hu H, Dick TE, Silver J. 2011. Functional regeneration of respiratory pathways after spinal cord injury. Nature 475:196–200.

Altman J. 1969. Autoradiographic and histological studies of postnatal neurogenesis. IV. Cell proliferation and migration in the anterior forebrain, with special reference to persisting neurogenesis in the olfactory bulb. J Comp Neurol 137:433–457.

Altman J, Das GD. 1965. Autoradiographic and histological evidence of postnatal hippocampal neurogenesis in rats. J Comp Neurol 124:319–335.

Bareyre FM, Kerschensteiner M, Raineteau O, Mettenleiter TC, Weinmann O, Schwab ME. 2004. The injured spinal cord spontaneously forms a new intraspinal circuit in adult rats. Nat Neurosci 7:269–277.

Bei F, Lee HHC, Liu X, et al. 2016. Restoration of visual function by enhancing conduction in regenerated axons. Cell 164:219–232.

Beirowski B, Berek L, Adalbert R, et al. 2004. Quantitative and qualitative analysis of Wallerian degeneration using restricted axonal labelling in YFP-H mice. J Neurosci Methods 134:23–35.

Bradbury EJ, McMahon SB. 2006. Spinal cord repair strategies: why do they work? Nat Rev Neurosci 7:644–653.

Bradbury EJ, Moon LD, Popat RJ, et al. 2002. Chondroitinase ABC promotes functional recovery after spinal cord injury. Nature 416:636–640.

Caroni P, Schwab ME. 1988. Antibody against myelin-associated inhibitor of neurite growth neutralizes nonpermissive substrate properties of CNS white matter. Neuron 1:85–96.

Conforti L, Gilley J, Coleman MP. 2014. Wallerian degeneration: an emerging axon death pathway linking injury and disease. Nat Rev Neurosci 15:394–409.

David S, Aguayo AJ. 1981. Axonal elongation into peripheral nervous system "bridges" after central nervous system injury in adult rats. Science 214:931–933.

Dimos JT, Rodolfa KT, Niakan KK, et al. 2008. Induced pluripotent stem cells generated from patients with ALS can be differentiated into motor neurons. Science 321:1218–1221.

Duan X, Qiao M, Bei F, Kim IJ, He Z, Sanes JR. 2015. Subtype-specific regeneration of retinal ganglion cells following axotomy: effects of osteopontin and mTOR signaling. Neuron 85:1244–1256.

Essuman K, Summers DW, Sasaki Y, Mao X, DiAntonio A, Milbrandt J. 2017. The SARM1 toll/interleukin-1 receptor domain possesses intrinsic NAD+ cleavage activity that promotes pathological axonal degeneration. Neuron 93:1334–1343.

Ferri A, Sanes JR, Coleman MP, Cunningham JM, Kato AC. 2003. Inhibiting axon degeneration and synapse loss attenuates apoptosis and disease progression in a mouse model of motoneuron disease. Curr Biol 13:669–673.

Franklin RJ, ffrench-Constant C. 2008. Remyelination in the CNS: from biology to therapy. Nat Rev Neurosci 9:839–855.

Galtrey CM, Fawcett JW. 2007. The role of chondroitin sulfate proteoglycans in regeneration and plasticity in the central nervous system. Brain Res Rev 54:1–18.

Gerdts J, Brace EJ, Sasaki Y, DiAntonio A, Milbrandt J. 2015. SARM1 activation triggers axon degeneration locally via NAD⁺ destruction. Science 348:453–457.

Gerdts J, Summers DW, Sasaki Y, DiAntonio A, Milbrandt J. 2013. Sarm1-mediated axon degeneration requires both SAM and TIR interactions. J Neurosci 33:13569–13580.

Goldman SA, Kuypers NJ. 2015. How to make an oligodendrocyte. Development 142:3983-3995.

Guo Z, Zhang L, Wu Z, Chen Y, Wang F, Chen G. 2014. In vivo direct reprogramming of reactive glial cells into functional neurons after brain injury and in an Alzheimer's disease model. Cell Stem Cell 14:188–202.

Imayoshi I, Sakamoto M, Ohtsuka T. 2008. Roles of continuous neurogenesis in the structural and functional integrity of the adult forebrain. Nat Neurosci 10:1153–1161.

Jorstad NL, Wilken MS, Grimes WN, et al. 2017. Stimulation of functional neuronal regeneration from Müller glia in adult mice. Nature 548:103–107.

Keirstead HS, Nistor G, Bernal G, et al. 2005. Human embryonic stem cell-derived oligodendrocyte progenitor cell transplants remyelinate and restore locomotion after spinal cord injury. J Neurosci 25:4694–4705.

Keirstead SA, Rasminsky M, Fukuda Y, Carter DA, Aguayo AJ, Vidal-Sanz M. 1989. Electrophysiologic responses in hamster superior colliculus evoked by regenerating retinal axons. Science 246:255–257.

Kordower J, Sortwell C. 2000. Neuropathology of fetal nigra transplants for Parkinson's disease. Prog Brain Res 127:333–344.

Lim DA, Alvarez-Buylla A. 2016. The adult ventricular-subventricular zone (V-SVZ) and olfactory bulb (OB) Neurogenesis. Cold Spring Harb Perspect Biol 8:pii: a018820.

Lois C, Alvarez-Buylla A. 1994. Long-distance neuronal migration in the adult mammalian brain. Science 264:1145–1148.

Mack TGA, Reiner M, Beirowski B, et al. 2001. Wallerian degeneration of injured axons and synapses is delayed by a Ube4b/Nmnat chimeric gene. Nat Neurosci 4:1199–1206.

Magavi SS, Leavitt BR, Macklis JD. 2000. Induction of neurogenesis in the neocortex of adult mice. Nature 405:951–955.

Magnusson JP, Göritz C, Tatarishvili J, et al. 2014. A latent neurogenic program in astrocytes regulated by Notch signaling in the mouse. Science 346:237–241.

Maier IC, Schwab ME. 2006. Sprouting, regeneration and circuit formation in the injured spinal cord: factors and activity. Philos Trans R Soc Lond B Biol Sci 361:1611–1634.

Osterloh JM, Yang J, Rooney TM, et al. 2012. dSarm/Sarm1 is required for activation of an injury-induced axon death pathway. Science 337:481–484.

Schwab ME, Thoenen H. 1985. Dissociated neurons regenerate into sciatic but not optic nerve explants in culture irrespective of neurotrophic factors. J Neurosci 5:2415–2423.

Schwegler G, Schwab ME, Kapfhammer JP. 1995. Increased collateral sprouting of primary afferents in the myelin-free spinal cord. J Neurosci 15:2756–2767.

Smith PD, Sun F, Park KK, et al. 2009. SOCS3 deletion promotes optic nerve regeneration in vivo. Neuron 64:617–623.

Sohur US, Emsley JG, Mitchell BD, Macklis JD. 2006. Adult neurogenesis and cellular brain repair with neural progenitors, precursors and stem cells. Philos Trans R Soc Lond B Biol Sci 361:1477–1497.

Southwell DG, Nicholas CR, Basbaum AI, et al. 2014. Interneurons from embryonic development to cell-based therapy. Science 344:1240622.

Takahashi K, Tanabe K, Ohnuki M, et al. 2007. Induction of pluripotent stem cells from adult human fibroblasts by defined factors. Cell 131:861–872.

Takahashi K, Yamanaka S. 2006. Induction of pluripotent stem cells from mouse embryonic and adult fibroblast cultures by defined factors. Cell 126:663–676.

Tavazoie M, Van der Verken L, Silva-Vargas V, et al. 2008. A specialized vascular niche for adult neural stem cells. Cell Stem Cell 3:279–288.

Thuret S, Moon LD, Gage FH. 2006. Therapeutic interventions after spinal cord injury. Nat Rev Neurosci 7: 628–643.

Torper O, Ottosson DR, Pereira M, et al. 2015. In vivo reprogramming of striatal NG2 glia into functional neurons that integrate into local host circuitry. Cell Rep 12:474–481.

Wen Z, Christian KM, Song H, Ming GL. 2016. Modeling psychiatric disorders with patient-derived iPSCs. Curr Opin Neurobiol 36:118–127.

Wernig M, Zhao JP, Pruszak J, et al. 2008. Neurons derived from reprogrammed fibroblasts functionally integrate into the fetal brain and improve symptoms of rats with Parkinson's disease. Proc Natl Acad Sci U S A 105:5856–5861.

Winkler C, Kirik D, Bjorklund A. 2005. Cell transplantation in Parkinson's disease: how can we make it work? Trends Neurosci 28:86–92.

Yiu G, He Z. 2006. Glial inhibition of CNS axon regeneration. Nat Rev Neurosci 7:617–627.

Zhou FQ, Snider WD. 2006. Intracellular control of developmental and regenerative axon growth. Philos Trans R Soc Lond B Biol Sci 361:1575–1592.

51

Sexual Differentiation of the Nervous System

FEW WORDS ARE MORE LOADED WITH meaning than the word "sex." Sexual activity is a biological imperative and a major human preoccupation.

The physical differences between men and women that underlie partner recognition and reproduction are obvious to all of us, and their developmental origins are well understood. In contrast, our understanding of behavioral differences between the sexes is primitive. In many cases, their very existence remains controversial, and the origins of those that have been clearly demonstrated remain unclear.

In this chapter, we first briefly summarize the embryological basis of sexual differentiation. We then discuss at greater length the behavioral differences between the two sexes, focusing on those differences or dimorphisms for which some neurobiological basis has been found. These dimorphisms include physiological responses (erection, lactation), drives (maternal behavior), and even more complex behaviors (gender identity). In analyzing these dimorphisms, we will discuss three issues.

First, what are the genetic origins of sexual differences? Human males and females have a complement of 23 chromosomal pairs, and only one differs between the sexes. Females have a pair of X chromosomes (and are therefore XX), whereas males have one copy of the X chromosome paired with a Y chromosome (XY). The other 22 chromosome pairs, called *autosomes*, are shared between males and females. We will see that the initial genetic determinants arise from a single gene on the Y chromosome, while later ones arise indirectly from sex-specific patterns of expression imposed upon other genes as development proceeds.

Second, how are sexual differences initiated by the Y chromosome translated into differences between the brains of men and women? We will see that key intermediates are the sex hormones, a set of steroids that includes testosterone and estrogens. These hormones act during embryogenesis as well as postnatally, first organizing the physical development of both genitalia and brain regions, and later activating particular physiological and behavioral responses. Hormonal regulation is especially complex because the nervous system, which is profoundly influenced by sex steroids, also controls their synthesis. This feedback loop may help to explain how the external environment, including social and cultural factors, can ultimately shape sexual dimorphism at a neural level.

Third, what are the crucial neural differences that underlie sexually dimorphic behaviors? Clear physical and molecular differences between the brains of men and women have been found. These differences reflect differences in neural circuitry between the sexes, and in a few cases, these distinctions in connectivity are directly related to behavioral differences. In other cases, however, sexually dimorphic behaviors appear to result from differential usage of the same basic circuits.

Before proceeding, we must define two words that are commonly used in many ways and sometimes confused with each other: *sex* and *gender*. As a descriptor of biological differences between men and women, the word *sex* is used in three ways. First, *anatomical sex* refers to overt differences including the differences in the external genitalia as well as other sexual characteristics such as the distribution of body hair. *Gonadal sex* refers to the presence of male or female gonads, the testes or ovaries. Finally, *chromosomal sex* refers to the distribution of the sex chromosomes between females (XX) and males (XY).

Whereas *sex* is a biological term, *gender* encompasses the collection of social behaviors and mental states that typically differ between males and females. *Gender role* is the set of behaviors and social mannerisms that is typically distributed in a sexually dimorphic fashion within the population. Toy preferences in children as well as distinctive attire are some examples of gender roles that can distinguish males from females. *Gender identity* is the feeling of belonging to the category of the male or female sex. Importantly, gender identity is distinct from *sexual orientation*, the erotic responsiveness displayed toward members of one or the other sex.

Are gender and sexual orientation genetically determined? Or are they social constructs molded by cultural expectations and personal experience? As the examples in this chapter will illustrate, we are still far from untangling the contributions of genes and environment to these complex phenomena. However, our recognition that genes and experience interact to shape

neural circuits gives us a more realistic framework with which to answer this question compared to our predecessors, who were constrained by the simplistic view that genes and experience acted in mutually exclusive ways.

Genes and Hormones Determine Physical Differences Between Males and Females

Chromosomal Sex Directs the Gonadal Differentiation of the Embryo

Sex determination is the embryonic process whereby chromosomal sex directs the differentiation of the gonadal sex of the animal. Surprisingly, this process differs in fundamental ways within the animal kingdom and even among vertebrates. In most mammals, including humans, however, an XY genotype drives differentiation of the embryonic gonad into testes, whereas an XX genotype leads to ovarian differentiation. Hormones produced by the testes and ovaries subsequently direct sexual differentiation of the nervous system and the rest of the body.

It is the presence of the Y chromosome rather than the lack of a second X chromosome that is the crucial determinant of male differentiation. This was first evident in rare individuals born with two or even three X chromosomes in addition to a Y chromosome (XXY or XXXY). These individuals are men who exhibit male-typical traits. In fact, female cells do not have two active X chromosomes. Early in embryogenesis, one of the two X chromosomes in each female cell is chosen at random for inactivation, and the genes on it are rendered transcriptionally silent. Thus, both male and female cells have a single active X chromosome, and male cells also have a Y chromosome.

The sex-determining activity of the Y chromosome is encoded by the gene *SRY* (sex-determining region on Y) whose activity is required for masculinization of the embryonic gonads (Figure 51–1). Inactivation or deletion of *SRY* leads to complete sex reversal: Individuals are chromosomally male (XY) but externally indistinguishable from females. Conversely, in rare instances, *SRY* translocates to another chromosome (to the X chromosome or an autosome) during spermatogenesis. Such sperm can fertilize eggs to produce individuals who are chromosomally female (XX) but externally male. However, such XX sex-reversed men are infertile, as many of the genes required for sperm function are located on the Y chromosome.

How does *SRY* instruct the undifferentiated gonads to develop into testes? The female differentiation

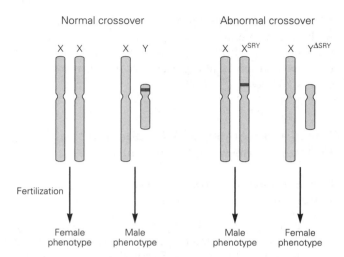

Figure 51–1 The role of the *SRY* gene in sex determination in humans. *SRY*, the sex-determining locus (**blue domain**), resides on the nonhomologous region of the short arm of the Y chromosome. The presence of *SRY* is determinative for male differentiation in many mammals, including primates and most rodents. Normally X- or Y-bearing sperm fertilize an oocyte to generate XX females or XY males, and the resulting phenotypic sex is concordant for the chromosomal sex. Rarely, *SRY* translocates to the X chromosome or an autosome (not shown). In such cases, XX^{SRY} offspring are phenotypically male, whereas XY$^{\Delta SRY}$ offspring (the Δ indicating a gene deletion) are phenotypically female. (Adapted from Wilhelm, Palmer, and Koopman 2007.)

program appears to be the default mode; patterning genes prime the body and gonads to develop along female-specific pathways. The *SRY* gene encodes a transcription factor that regulates expression of genes, some of which prevent execution of the default program and initiate the process of male gonadal differentiation. One of the best-studied targets of the SRY transcription factor is another transcription factor, SOX9, which is required for differentiation of the testes. Thus, SRY initiates a cascade of inductive interactions that ultimately lead to male-specific gonad development.

Gonads Synthesize Hormones That Promote Sexual Differentiation

The chromosomal complement of the embryo directs sexual differentiation of the gonads, and in turn, the gonads determine the sex-specific features of all organs of the body, including the nervous system. They do this by secreting hormones. Gonadal hormones have two major roles. Their developmental role is traditionally referred to as *organizational* because the early effects of hormones on the brain and the rest of the body lead to

major, generally irreversible, aspects of cell and tissue differentiation. Later, some of the same hormones trigger physiological or behavioral responses. These influences, generally termed *activational*, are reversible.

One example of an organizational role of gonadal hormones is seen in the differentiation of structures that connect the gonads to the external genitalia. In males, the Wolffian duct gives rise to the vas deferens, the seminal vesicles, and the epididymis. In females, the Müllerian duct differentiates into the oviduct, the uterus, and the vagina (Figure 51–2). Initially, both female (XX) and male (XY) embryos possess Wolffian and Müllerian ducts. In males, the developing testes secrete a protein hormone, the Müllerian inhibiting substance (MIS), and a steroid hormone, testosterone. MIS leads to a regression of the Müllerian duct, and testosterone induces the Wolffian duct to differentiate into its mature derivatives. In females, the absence of MIS permits the Müllerian duct to differentiate into its adult derivatives, and the absence of circulating testosterone causes the Wolffian duct to resorb. Thus, the Y chromosome overrides a female default program to generate male gonads, which in turn secrete hormones that override a female default program of genital differentiation.

The action of MIS is largely confined to embryos, but steroid hormones exert effects throughout life—that is, they also have activational roles at later stages. All of the steroid hormones derive from cholesterol (Figure 51–3). The sex steroids can be divided into androgens, which generally promote male characteristics, and the estrogens plus progesterone that promote female characteristics. The testes produce mostly the androgen testosterone, while the ovaries produce mostly progesterone and an estrogen, 17-β-estradiol. The menstrual cycle is a good example of the activational function of estrogen and progesterone.

A glance at the metabolic relationships among steroid hormones (Figure 51–3) reveals a surprise. The female hormone progesterone is the precursor of the male hormone testosterone, and testosterone is the direct precursor of the female hormone 17-β-estradiol. Thus, the enzymes that convert one hormone to the other control not only the level of the hormone but also the "sign" (male or female) of the hormonal effect. Aromatase, the enzyme that converts testosterone to estradiol, is present at high levels in the ovaries but not in the testes. Differential expression of aromatase is the reason for sexual dimorphism in circulating testosterone and estrogen. Aromatase is also expressed in various regions of the brain (Figure 51–4A), and many of the effects of testosterone on neurons are thought to occur after its conversion to estrogen. Testosterone

Figure 51–2 Sexual differentiation of the internal genitalia. Embryos of both sexes develop bilateral genital ridges (the gonadal anlagen) that can differentiate into either testes or ovaries; Müllerian ducts, which can differentiate into oviducts, the uterus, and the upper vagina; and Wolffian ducts, which can differentiate into the epididymis, the vas deferens, and the seminal vesicles. In XY embryos, the expression of the *SRY* gene in the genital ridge induces differentiation of this tissue into testes and of the Wolffian ducts into the rest of the male internal genitalia, while the Müllerian ducts are resorbed. In XX embryos, the *absence* of *SRY* permits the genital ridges to develop into ovaries and the Müllerian ducts to differentiate into the rest of the female internal genitalia; in the absence of circulating testosterone, the Wolffian ducts degenerate. (Abbreviation: **MIS**, Müllerian inhibiting substance.) (Adapted, with permission, from Wilhelm, Palmer, and Koopman 2007.)

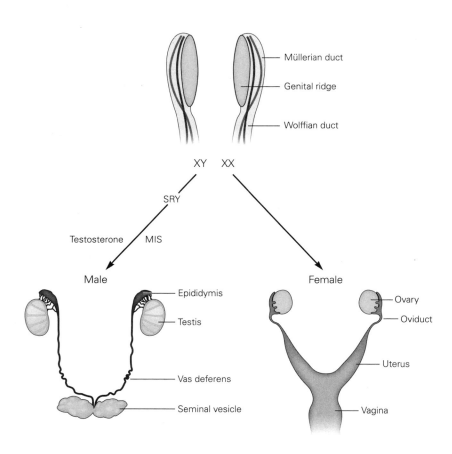

is also converted by the enzyme 5α-reductase into another androgenic steroid, 5α-dihydrotestosterone (DHT), in various target tissues, including the external genitalia. In these tissues, DHT is responsible for induction of secondary male characteristics such as facial and body hair and growth of the prostate. Later in life, DHT is the culprit in male pattern baldness.

Disorders of Steroid Hormone Biosynthesis Affect Sexual Differentiation

As one can imagine, mutations in genes encoding enzymes involved in steroid hormone biosynthesis have far-reaching consequences. The resulting phenotypes dramatically illustrate both the organizational and activational effects of steroid hormones, as well as the difficulty of neatly distinguishing the two. Here, we describe three disorders (Table 51–1).

The first, congenital adrenal hyperplasia (CAH), is a genetic deficiency in the synthesis of corticosteroids by the adrenal glands that results in overproduction of testosterone and related androgens. This condition is autosomal recessive and occurs once in 10,000 to 15,000 live births. In girls born with CAH, excess androgens

lead to some masculinization of the external genitalia, a process called *virilization*. Virilization clearly reflects the organizational roles of steroids. This condition can be diagnosed at birth and resolved by surgical intervention. Treatment with corticosteroids reduces testosterone levels, permitting these females to undergo puberty and become fertile.

A second genetic disorder, 5α-reductase II deficiency, can also affect sexual differentiation. In male fetuses, 5α-reductase II is expressed at high levels in the precursor of the external genitalia, where it converts circulating testosterone into DHT. The high local concentrations of DHT virilize the external genitalia. Clinical 5α-reductase II deficiency is inherited in an autosomal recessive manner, and males present at birth with ambiguous (under-virilized) or overtly feminized external genitalia. In many instances, therefore, chromosomally male patients (XY) with this condition are mistakenly raised as females until puberty, at which time the large increase in circulating testosterone virilizes the body hair, musculature, and, most dramatically, the external genitalia.

The critical role of steroid receptors in controlling sexual differentiation is well illustrated by patients

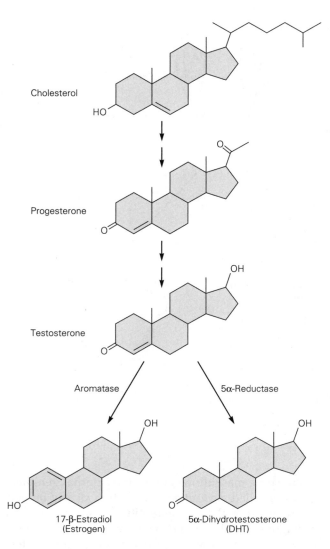

Figure 51–3 Steroid hormone biosynthesis. Cholesterol is the precursor of all steroid hormones and is converted via a series of enzymatic reactions into progesterone and testosterone. Testosterone or related androgens are obligate precursors of all estrogens in the body, a conversion that is catalyzed by aromatase. The expression of 5α-reductase in target tissues converts testosterone into dihydrotestosterone, an androgen.

with a third disorder, complete androgen insensitivity syndrome (CAIS). Testosterone, estrogen, and progesterone are hydrophobic molecules that are able to diffuse across cell membranes, enter the bloodstream, enter cells in many organs, and bind to intracellular ligand-specific receptors. The receptors for these hormones are encoded by distinct but homologous genes.

A single gene encodes a receptor that binds the androgens testosterone and DHT. The androgen receptor binds DHT approximately three-fold more tightly

than testosterone, accounting for the greater potency of DHT. There is also a single receptor for progesterone (progesterone receptor), whereas two genes encode receptors that bind estrogens (estrogen receptors α and β). These steroid hormone receptors are present in many tissues of the body, including the brain (Figure 51–4B).

These receptor proteins are transcription factors that bind specific sites in the genome and modulate transcription of target genes. They contain several signature motifs, including a hormone-binding domain, a DNA-binding domain, and a domain that modulates the transcriptional activity of target genes (Figure 51–5A). Hormones activate the transcriptional activity by binding to the receptor. In the absence of ligand, the receptors bind to protein complexes that sequester them in the cytoplasm. Upon binding of ligand, the receptors dissociate from the complex and enter the nucleus, where they dimerize and bind to specific sequence elements in the promoter and enhancer regions of target genes, modulating their transcription (Figure 51–5B).

Patients with CAIS are chromosomally XY but carry a loss-of-function allele of the X-linked androgen receptor that abolishes cellular responses to testosterone and DHT. Because the pathway of sex determination via *SRY* remains functional, these patients have testes. However, because of deficient androgen signaling, the Wolffian ducts do not develop, the testes fail to descend, and the external genitalia are feminized. In adulthood, most of these patients opt for surgical removal of the testes and hormonal supplementation appropriate for females.

Sexual Differentiation of the Nervous System Generates Sexually Dimorphic Behaviors

Sex-specific behaviors occur because the nervous system differs between males and females. These differences arise from a combination of genetic factors, such as signaling pathways initiated by sex determination, as well as environmental factors, such as social experience. In many cases, both genetic and environmental inputs act through the steroid hormone system to sculpt the nervous system. Many instances of sexual dimorphism have been documented, including differences in the numbers and size of neurons in particular structures, differences in gene expression in various neuronal groups, and differences in the pattern and number of connections. Here, we examine a few cases in which studies in experimental animals have provided insights. In later sections, we ask whether similar

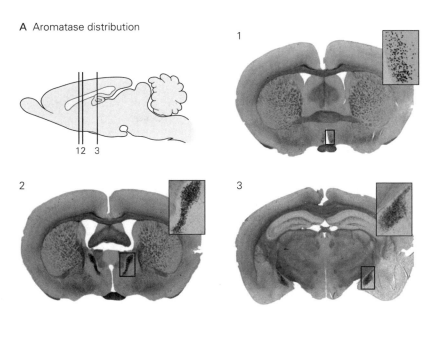

A Aromatase distribution

B Estrogen receptor distribution

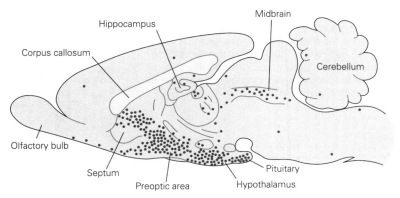

Figure 51–4 Aromatase and estrogen receptors are expressed in specific regions of the brain.

A. The enzyme aromatase catalyzes the conversion of testosterone into estrogen (Figure 51–3) and is expressed in discrete neuronal populations in the brain. The distribution of aromatase-expressing neurons labeled with a reporter protein (**blue**) in transgenic mice is shown here in three coronal planes of the brain: in neurons in the preoptic hypothalamus (**1**), in the bed nucleus of the stria terminalis (BNST) (**2**), and in the medial amygdala (**3**). These areas contain sexually dimorphic neurons that regulate sexual behavior, aggression, and maternal behaviors. (Adapted, with permission, from Wu et al. 2009.)

B. This midsagittal section of an adult rat brain shows binding of estrogen to cells in various hypothalamic regions, including the sexually dimorphic preoptic area. Additional estrogen binding is seen in the septum, hippocampus, pituitary, and midbrain. Other, more lateral areas such as the amygdala (not shown) also contain estrogen receptors.

Table 51–1 Three Clinical Syndromes That Highlight the Role of Androgens in Masculinization in Humans

	Complete androgen insensitivity syndrome (CAIS)	5α-Reductase II deficiency	Congenital adrenal hyperplasia (CAH)
Chromosomal sex	XY	XY	XX
Molecular basis	Nonfunctional androgen receptor, leading to inability to respond to circulating androgens	Nonfunctional 5α-reductase II, leading to deficit in conversion of testosterone to 5α-dihydrotestosterone (DHT) in target tissues	Defect in corticosteroid synthesis, leading to increase in circulating androgens from the adrenals
Gonad	Testis	Testis	Ovary
Wolffian derivatives	Vestigial	Present	Absent
Müllerian derivatives	Absent	Absent	Present
External genitalia			
At birth	Feminized	Variably feminized	Variably virilized
After puberty	Feminized	Masculinized	Feminized
Gender identity	Female	Female or male	Female or male
Sexual partner preference	Male	Female or male	Female or male

A Steroid hormone receptor structure

Figure 51–5 Steroid hormone receptors and their mechanism of action.

A. The canonical receptors for steroid hormones are ligand-activated transcription factors. These receptors have an N-terminal domain, which contains a transcriptional transactivator domain; a central DNA-binding domain; and a C-terminal ligand-binding domain, which may contain an additional transcriptional transactivator domain.

B. Sex steroid hormones are hydrophobic and enter the circulation by diffusing across the plasma membrane of steroidogenic cells in the gonads. They enter target cells in distant tissues such as the brain by passing through the plasma membrane and bind their cognate receptors. The steroid hormone receptor typically exists in a multiprotein complex with chaperone proteins in the cytoplasm of hormone-responsive cells. Ligand-binding promotes dissociation of the receptor from the chaperone complex and translocation into the nucleus. In the nucleus, the receptor is thought to bind to hormone response elements as a homodimer to modulate transcription of target genes. (Adapted from Wierman 2007.)

B Steroid hormone pathway

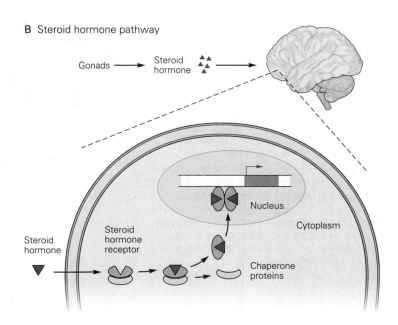

mechanisms underlie sexually dimorphic behaviors in humans.

However, before proceeding, we note that the ways in which chromosomal mechanisms of sex determination are linked to the cellular processes of sexual differentiation in the central nervous system vary widely among species. In insects, sex differences in behavior are independent of hormonal secretion from the gonads, and instead rely exclusively on a sex determination pathway within individual neurons. This mode of sexual differentiation of the brain and behavior is particularly well understood in the fruit fly, where it has been demonstrated that the sex determination cascade initiates expression of a transcription factor, fruitless (Fru), that specifies much of the repertoire of male sexual behaviors (Box 51–1).

Erectile Function Is Controlled by a Sexually Dimorphic Circuit in the Spinal Cord

The lumbar spinal cord of many mammals, including humans, contains a sexually dimorphic motor center, the spinal nucleus of the bulbocavernosus (SNB).

Motor neurons in the SNB innervate the bulbocavernosus muscle, which plays an important part in penile reflexes in males and vaginal movements in females.

In adult rats, the male SNB contains many more motor neurons than the female SNB. In addition, male SNB motor neurons are larger in size and have larger dendritic arbors, with a corresponding increase in the number of synapses they receive. Like the SNB motor neurons, the bulbocavernosus muscle is larger in males than females; it is completely absent in the females of some mammalian species. SNB motor neurons also innervate the levator ani muscle, which is involved in copulatory behavior and is also larger in males than females.

How do these differences arise? Initially, the circuit is not sexually dimorphic. At birth, male and female rats have similar numbers of neurons in the SNB and similar numbers of fibers in the bulbocavernosus and levator ani muscles. In females, however, many motor neurons in the SNB and many fibers in the bulbocavernosus and levator ani muscles die in early postnatal life. Thus, this sexual dimorphism arises not by male-specific generation of cells, but rather by female-specific cell death (Figure 51–7A).

Perinatal injections of testosterone or DHT can rescue a significant number of the dying neurons and muscle fibers in the female rat. Conversely, treatment of male pups with an androgen receptor antagonist increases the number of dying neurons and muscle fibers. So at a deeper level, we see that the dimorphism results from male-specific preservation of motor neurons and muscle fibers that would die in the absence of hormone.

Where does testosterone act to establish this structural dimorphism? Is it primarily a survival factor for the motor neurons, with muscle fibers dying secondarily because they lose their innervation? Or does testosterone act on muscles, which then provide a trophic factor to support the survival of SNB motor neurons? This issue has been examined in rats carrying a mutation of the androgen receptor (*tfm* allele) that reduces binding of ligand to 10% of normal. The receptor resides on the X chromosome, so all males that carry a mutant gene on their one and only X chromosome are feminized and sterile. For female heterozygotes, the situation is more complicated. As described earlier, one of the X chromosomes is randomly inactivated in each XX female.

Female heterozygotes are therefore mosaics: Some cells express a functional androgen receptor allele, others the mutated allele. Each muscle fiber has many nuclei, so most bulbocavernosus muscle fibers in the heterozygous female express functional androgen receptors. Motor neurons have a single nucleus, however, so each neuron is either normal or receptor-deficient. If androgen receptors were required in the neuron, one would expect only receptor-expressing SNB motor neurons to survive, whereas if receptors were required only in muscles, one would expect surviving motor neurons to be a mixture of wild type and mutant.

In fact, the latter situation occurs, indicating that survival of SNB motor neurons does not depend on a neuron-autonomous function of the androgen receptor. Rather, these neurons receive a trophic cue from the androgen-dependent bulbocavernosus and the levator ani muscles (Figure 51–7A). These cues may include the ciliary neurotrophic factor (CNTF) or a related molecule, because mutant male mice lacking a CNTF receptor exhibit a decreased number of SNB motor neurons, typical of females.

Male and female SNB motor neurons also differ in size. Androgens determine the differences in number and size of these neurons in different ways. Studies of *tfm* mutants showed that androgens exert an organizational effect during early postnatal life through a direct effect on muscle. Low levels of androgens during this critical period lead to an irreversible reduction in the number of SNB motor neurons. Later, androgens act directly on SNB motor neurons to increase the extent of their dendritic arbors. A loss of circulating testosterone, such as that occurring after castration, leads to a dramatic pruning of dendritic arbors; injection of supplemental testosterone to a castrated male rat can restore this dendritic branching pattern (Figure 51–7B). This effect persists in adulthood and is reversible, so it can be viewed as an activational influence. Thus, androgens can exert diverse effects, even on a single neuronal type.

Song Production in Birds Is Controlled by Sexually Dimorphic Circuits in the Forebrain

Several species of songbirds learn species-specific vocalizations that are used for courtship rituals and territorial marking (Chapter 55). A set of interconnected brain nuclei controls the learning and production of birdsong (Figure 51–8A). In some songbird species, both sexes sing and the structure of the song circuit is similar in males and females. In other species, such as zebra finches and canaries, males alone sing. In these species, several song-related nuclei are significantly larger in the male than in the female.

The development of sexual dimorphism in song circuitry has been studied in detail in the zebra finch. The robust nucleus of the archistriatum (RA) in the adult male zebra finch contains five times as many neurons as the same nucleus in females. In addition, the afferent projections to RA exhibit a striking sexual dimorphism—only in males does the RA receive input from high vocal centers (HVCs) (Figure 51–8B). These sex differences in cell number and connectivity of RA are not evident until after hatching, when in females a large number of RA neurons die and in males the axons of HVC neurons enter the RA nucleus.

These sexually dimorphic anatomical features are regulated by steroid hormones. When females are supplied with estrogen (or an aromatizable androgen such as testosterone) after hatching, the number of neurons in the RA and the termination pattern in the nucleus are similar to that of the male. However, early hormone administration to young females is not sufficient to masculinize the song nuclei to a size comparable to that of adult males, nor is it sufficient to induce singing in females. To achieve these functions, female birds that receive testosterone or estradiol after hatching must also receive testosterone or dihydrotestosterone (but not estrogen) as adults. Thus, steroids play both organizational and activational roles in this system as well.

Box 51–1 Genetic and Neural Control of Mating Behavior in the Fruit Fly

In the presence of a female fruit fly, the adult male fly engages in a series of essentially stereotyped routines that usually culminate in copulation (Figure 51–6A). This elaborate male courtship ritual is encoded by a cascade of gene transcription within the brain and peripheral sensory organs that masculinizes the underlying neural circuitry.

Sex determination in the fly does not depend on gonadal hormones as it does in vertebrates. Instead, it occurs cell autonomously throughout the body. In other words, sexual differentiation of the brain and the rest of the body is independent of gonadal sex. The male-specific Y chromosome of fruit flies does not bear a sex-determining locus. Instead, sex is determined by the ratio of X chromosome number to autosome number (X:A). A ratio of 1 is determinative for female differentiation, whereas a ratio of 0.5 drives male differentiation.

The X:A ratio sets into motion a cascade of gene transcription and alternative splicing programs that leads to the expression of sex-specific splice forms of two genes, *doublesex* (*dsx*) and *fruitless* (*fru*). The *dsx* gene encodes a transcription factor that is essential for sexual differentiation of the nervous system and the rest of the body, with the sex-specific splice variants responsible for male- and female-typical development.

The *fru* gene encodes a set of putative transcription factors that are generated by multiple promoters and alternative splicing. In males, one particular mRNA (*fru*^{M}) is translated into functional proteins. In female flies, alternative splicing results in the absence of such proteins.

Males carrying a genetically modified *fru* allele that can only be spliced in the female-specific manner (*fru*^{F}) have essentially normal, *dsx*-dependent sexual differentiation. These *fru*^{F} males therefore resemble wild type males externally. However, the loss of Fru^M in these animals abolishes male courtship behavior directed toward females. These data indicate that Fru^M is required for male courtship and copulation.

Conversely, transgenic female flies carrying a *fru*^{M} allele exhibit male mating behavior toward wild type females, indicating that *fru*^{M} is sufficient to inhibit female sexual responses and promote male mating.

Intriguingly, *fru*^{F} males do not court females and, like wild type females, do not reject mating attempts by wild type males or *fru*^{M} females. Similarly, *fru*^{M} females attempt to mate with both *fru*^{M} and wild type females. These data suggest that *fru*^{M} may also specify sexual partner preference, which in the case of wild type males would be directed to females.

In wild type females without *fru*^{M}, the neural pathways are wired such that these flies exhibit sexually receptive behaviors toward males. When groups of *fru*^{F} males (or *fru*^{M} females) are housed together, they court each other vigorously, often forming long chains of flies attempting copulation.

To build the circuitry underlying male courtship rituals, *fru*^{M} appears to initiate cell-autonomous male-typical differentiation of the neurons in which it is expressed. This leads to overt neuroanatomic dimorphism in cell number or projections of many classes of neurons (Figure 51–6B). Some neurons that express *fru*^{M} are not distributed in dimorphic patterns. In these neurons, *fru*^{M} may regulate the expression of particular classes of genes whose products drive a male-specific program of physiology and function.

Are neurons that express *fru*^{M} required for male courtship behavior? When synaptic transmission is genetically blocked in these neurons in adult males, all components of courtship behavior are abolished. Importantly, these males continue to exhibit normal movement, flight, and other behaviors in response to visual and olfactory stimuli. These findings demonstrate that *fru*^{M} appears to be expressed in a neural circuit that is essential for and dedicated to male fly courtship.

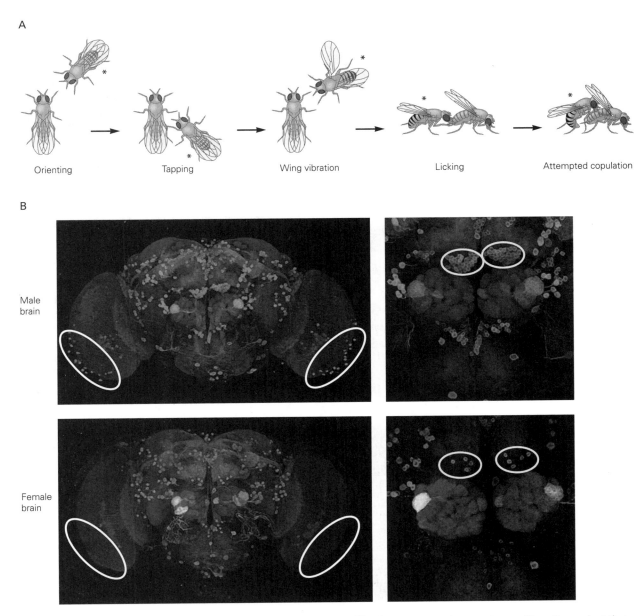

Figure 51–6 Control of male courtship in the fruit fly
Drosophila melanogaster.

A. Male flies (labeled with **asterisk**) engage in a stereotyped
sequence of behavioral routines that culminate in attempted
copulation. The male fly orients toward the female and then
taps her with his forelegs. This is followed by wing extension
in the male and a species-specific pattern of wing vibrations
that is commonly referred to as the fly courtship song. If the
female fly is sexually receptive, she slows down and permits
the male to lick her genitalia. The female then opens her
vaginal plates in order to allow the male to initiate copulation.
All steps in the male mating ritual require the expression of a
sex-specific splice variant of the *fruitless* (*fru*) gene. (Adapted,
with permission, from Greenspan and Ferveur 2000.)

B. The *fru* gene encodes a male-specific splice variant that
is necessary and sufficient to drive most steps in the male
fly courtship ritual. *Fru* expression is visualized using a fluorescent reporter protein (**green**) in transgenic flies. Neuronal
clusters that express *Fru* are present in comparable numbers in the central nervous system of both male and female
flies. However, there are regional sex differences in *Fru*
expression. A cluster of *Fru*-expressing neurons is present
in the male optic lobes (in the area within the **white ellipses**) but absent in the corresponding regions in the female
brain. The two male antennal lobe regions (areas within **yellow ellipses**) contain about 30 neurons each, whereas each
female region has only four to five neurons. (Adapted, with
permission, from Kimura et al. 2005.)

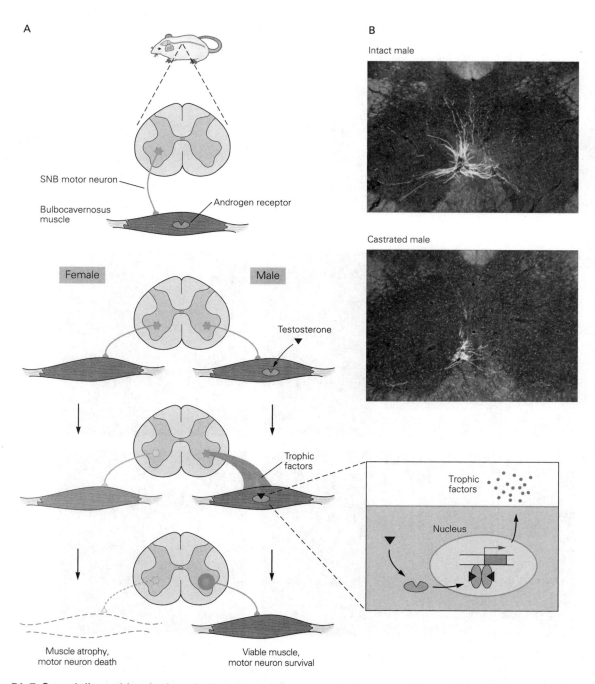

Figure 51–7 Sexual dimorphism in the spinal nucleus of the bulbocavernosus muscle in the rat.

A. The spinal nucleus of the bulbocavernosus (**SNB**) is found in the male lumbar spinal cord but is greatly reduced in the female. The motor neurons of the nucleus are present in both sexes at birth, but the lack of circulating testosterone in females leads to death of the SNB neurons and their target muscles. It is thought that testosterone in the male circulation promotes the survival of the target muscles, which express the androgen receptor. In response to testosterone, the muscles provide trophic support to the innervating SNB neurons. This muscle-derived survival factor is likely to be ciliary neurotrophic factor or a related member of the cytokine family. Thus, testosterone acts on muscle cells to

control the sexual differentiation of SNB neurons. (Reproduced, with permission, from Morris, Jordan, and Breedlove 2004. Copyright © 2004 Springer Nature.)

B. Dendritic branching of SNB neurons is regulated by circulating testosterone in adult male rats. In males, the dendrites arborize extensively within the spinal cord (**upper photo**). The fact that the arbors are pruned in adult castrated male rats (**lower photo**) is evidence that this dendritic branching depends on androgens. The spinal cord is shown in transverse section, and the SNB neurons and their dendrites are labeled by a retrograde tracer injected into target muscles. (From Cooke and Woolley 2005. Reproduced, with permission from D. Sengelaub.)

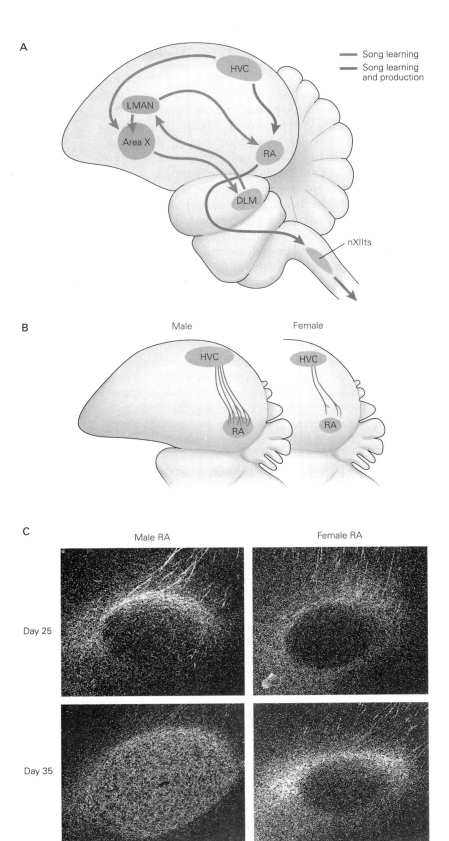

Figure 51–8 Sexual dimorphism in the avian song circuit.

A. Songbirds have a dedicated neural circuit for song production and learning, with distinct components contributing to learning or production. Many of these components are sexually dimorphic in songbirds in which only one sex sings. For example, in zebra finches, only the male sings, and the male high vocal center (**HVC**), robust nucleus of the archistriatum (**RA**), lateral magnocellular nucleus of the anterior neostriatum (**LMAN**), and area X are larger in volume and contain more neurons than the comparable regions in the female. (Abbreviations: **DLM**, medial nucleus of the dorsolateral thalamus; **nXIIts**, hypoglossal nucleus.) (Reproduced, with permission, from Brainard and Doupe 2002. Copyright © 2002 Springer Nature.)

B. In the male, the axons of HVC neurons terminate on neurons in the RA nucleus, whereas in females, the axons terminate in a zone surrounding the nucleus. The sexual dimorphism in cell number and connectivity of these regions is regulated by estrogen. (Reproduced, with permission, from Morris, Jordan, and Breedlove 2004. Copyright © 2004 Springer Nature.)

C. The pattern of termination of the axons of HVC neurons in the RA nucleus varies in males and females at different ages after hatching. (Reproduced, with permission, from Konishi and Akutagawa 1985. Copyright © 1985 Springer Nature.)

Mating Behavior in Mammals Is Controlled by a Sexually Dimorphic Neural Circuit in the Hypothalamus

In many mammalian species, the preoptic region of the hypothalamus and a reciprocally connected region, the bed nucleus of the stria terminalis (BNST), play important roles in sexually dimorphic mating behaviors (Chapter 41; Figure 51–4). In male rodents and monkeys, these areas are activated during mating behavior; surgical lesions that ablate the preoptic region or the BNST result in deficits in male sexual behavior in male rodents and, in the case of preoptic lesions, disinhibit female-type sexual receptivity in males.

Both the preoptic hypothalamus and the BNST are sexually dimorphic, containing more neurons in males compared to females. The sexually dimorphic nucleus of the preoptic area (SDN-POA) also contains significantly more neurons in the male. A male-specific perinatal surge of testosterone promotes survival of neurons in the SDN-POA and BNST, whereas in females, these same cells gradually die off in the early postnatal period. This development is similar to that in the sexually dimorphic nuclei of the rodent spinal cord and the songbird brain, suggesting that androgen control is a common mechanism for production of sex differences in the size of neuronal populations.

Curiously, the ability of brain testosterone to promote the survival of neurons is likely to be exerted via aromatization into estrogen and subsequent activation of the estrogen receptors (see Figures 51–3 and 51–4). How then is the neonatal female brain shielded from the effects of circulating estrogen? In newborn females, there is very little estrogen in the circulation, and the small amount present is easily sequestered by binding to α-fetoprotein, a serum protein. This explains why female mice lacking α-fetoprotein exhibit male-typical behaviors and reduced female-typical sexual receptivity. In this case, then, structural sexual dimorphism does not result from differential effects of androgens and estrogens, but rather from sex differences in the level of hormone available to the target tissue.

Environmental Cues Regulate Sexually Dimorphic Behaviors

Sex-specific behaviors are usually initiated in response to sensory cues in the environment. There are many such cues, and different species use distinct sensory modalities to elicit similar responses. Courtship rituals can be triggered by species-specific vocalizations, visual signals, odors, and even, in the case of weakly electric fish, by electric discharges. Recent genetic and molecular studies have led to significant insight into how sensory experience controls some of these behaviors in rodents. Here, we discuss two examples: the regulation of partner choice by pheromones and the regulation of maternal behavior by experience during infancy.

Pheromones Control Partner Choice in Mice

Many animals rely on their sense of smell to move about, obtain food, and avoid predators. They also rely on pheromones—chemicals that are produced by an animal to affect the behavior of another member of the species. In rodents, pheromones can trigger many sexually dimorphic behaviors, including mate choice and aggression.

Pheromones are detected by neurons in two distinct sensory tissues in the vertebrate nose: the main olfactory epithelium (MOE) and the vomeronasal organ (VNO) (Figure 51–9A). It is thought that sensory neurons in the MOE detect volatile odors, whereas those in the VNO detect nonvolatile chemosensory cues. Removal of the olfactory bulb, the only synaptic target of neurons in the MOE and the VNO, abolishes mating as well as aggression in mice and other rodents. These and other studies indicate an essential role for olfactory stimulation in initiating mating and fighting.

Genetically engineered disruption of pheromone responsiveness in the MOE or VNO reveals that these sensory tissues have a surprisingly complex role in the mating behavior of mice. A functional MOE is essential to trigger male sexual behavior, and an intact VNO is required for sex discrimination and directing the male to mate with females.

Key to these experiments is the fact that olfactory neurons in the MOE and the VNO use different signal transduction cascades to convert olfactory input into electrical responses. The cation channel Trpc2 appears essential for pheromone-evoked signaling in VNO neurons; it is not expressed in MOE neurons, which use a different signal transduction apparatus. Thus, mice lacking the gene *trpc2* have a nonfunctional VNO and an intact MOE. Mating behavior directed to animals of the opposite sex appears unaltered in *trpc2* mutant males as well as females.

However, both male and female *trpc2* mutants often exhibit male sexual behavior with members of either sex. For example, *trpc2* mutant females mate with females in a manner seemingly indistinguishable from wild type males, except of course the females cannot ejaculate. These and other findings suggest that the VNO is used to discriminate among sexual partners.

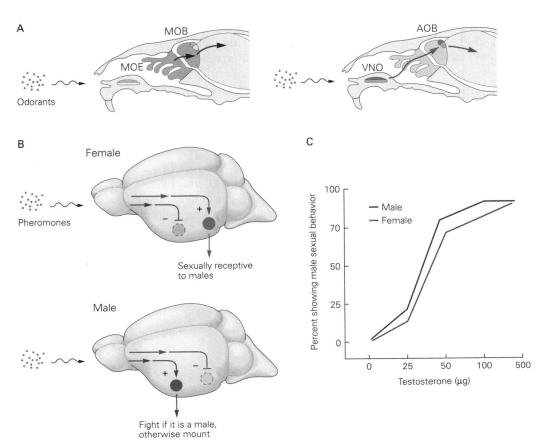

Figure 51–9 Pheromonal and hormonal control of sexually dimorphic behavior in mice.

A. Odorants are detected by sensory neurons in the main olfactory epithelium (**MOE**), which projects to the main olfactory bulb (**MOB**), and by neurons in the vomeronasal organ (**VNO**), which projects to the accessory olfactory bulb (**AOB**). Many of the central connections of the MOE and VNO pathway are anatomically segregated. (Adapted, with permission, from Dulac and Wagner 2006.)

B. Female mice possess the neural circuitry that can activate either male (**blue**) or female (**red**) mating behaviors. In wild type females, pheromones activate female mating behavior and inhibit male-type mating. By contrast, in males, pheromones activate a circuitry that will initiate fights with males and mating

with females. (Adapted, with permission, from Kimchi, Xu, and Dulac 2007.)

C. Testosterone activates male sexual behavior in male and female mice. The data are from a study in which the gonads of male and female mice were surgically removed in adulthood. None of the animals exhibited male sexual behavior with wild type females following surgery. After administration of testosterone, mating behavior was restored in castrated males, and females displayed male sexual behavior. This effect was dose-dependent; at the highest dose, male and female mice exhibited comparable levels of male-type mating behavior toward wild type females. (Adapted, with permission, from Edwards and Burge 1971. Copyright © 1971 Springer Nature.)

When the VNO is inactivated, animals can no longer distinguish between males and females, and mutants therefore exhibit male sexual behavior toward members of both sexes. Similarly, adult wild type females treated with testosterone also exhibit male sexual behavior toward females (Figure 51–9C).

One implication of these studies is that female mice possess the neural circuitry for male sexual behavior (Figure 51–9B). Activation of this neural circuit is inhibited in wild type females by sensory input from

the VNO and by the lack of testosterone. Removal of the VNO or administration of testosterone activates male sexual behavior in females. Male pattern mating behavior has been observed in females of many species, indicating that the findings in mice are likely to be of general relevance. Thus, neural pathways for male sexual behavior appear to be present in both sexes. Similarly, the female-typical behavior of male rats following hypothalamic lesions suggests that the neural pathway for female sexual behavior also exists in the male brain. In such cases, it is the differential regulation of these circuits that underlies the sexually dimorphic expression of male and female sexual behaviors.

Early Experience Modifies Later Maternal Behavior

The preoptic area of the hypothalamus and the BNST are also important for another set of sexually dimorphic behaviors in females. Nursing rodents are good mothers, building a nest for their litter, crouching over the pups to keep them warm, and returning the pups to the nest when they happen to crawl away. Surgical lesioning or experimental stimulation of the preoptic region abolishes or activates these maternal behaviors, respectively.

Studies of these behaviors have shed light on variations among individual females and how these differences exert lifelong effects on behavior of the offspring. Female lab rats exhibit distinct, stable forms of maternal care: Some lick and groom (LG) their pups frequently (high-LG mothers), whereas others lick and groom less frequently (low-LG mothers). Female offspring of high-LG mothers display high-LG activity when they themselves become mothers compared to female offspring of low-LG mothers (Figure 51–10). Moreover, pups of high-LG mothers show less anxiety-like behaviors in stressful conditions than do the pups of low-LG mothers.

These results suggested that levels of licking and grooming behavior and stress responses are genetically determined. However, studies by Michael Meaney and his colleagues provide an alternative explanation. When female rat pups are transferred from

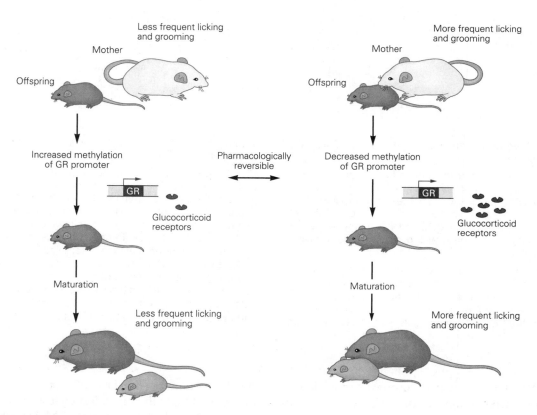

Figure 51–10 Epigenetic regulation of maternal behavior in rats. In a common lab rat strain, different mothers lick and groom their pups at low or high frequencies, resulting in distinct epigenetic modifications at the glucocorticoid receptor (**GR**) promoter. Mothers that lick and groom at high frequency raise progeny with low levels of DNA methylation at the GR promoter, resulting in higher levels of GR expression in the hippocampus. Females raised by these mothers exhibit higher frequencies of licking and grooming behavior with their own pups. Mothers that lick and groom at low frequency raise progeny with high DNA methylation levels at the GR promoter and lower levels of hippocampal GR expression. Females nursed by these mothers subsequently exhibit similar low levels of licking and grooming of their pups. Pharmacological reversal of the epigenetic modifications at the GR promoter results in a corresponding change in both GR expression and maternal behavior. (Adapted from Sapolsky 2004.)

their mother to a foster mother at birth, their maternal behavior and stress responses as adults resemble those of their foster mother rather than those of their biological mother. Thus, experience in infancy can lead to lifelong behavioral patterns. Because these patterns impact maternal behavior, their influence can endure over many generations.

How does brief and early experience lead to such long-lasting changes? One mechanism involves a covalent modification of the genome. Stress responses are coordinated by glucocorticoids acting on glucocorticoid receptors in the hippocampus. Throughout life, tactile stimulation, including grooming, leads to transcriptional activation of the glucocorticoid receptor gene, which ultimately leads to reduced release of hypothalamic hormones that trigger stress responses. Tactile stimulation during early life also regulates the glucocorticoid receptor gene in a second way. A key site in the glucocorticoid receptor gene is methylated by the enzyme DNA methyltransferase, leading to gene inactivation. Initially, gene methylation occurs in all pups, but pups reared by high-LG mothers are selectively demethylated. Thus, in animals reared by high-LG mothers, the effects of adult experience are potentiated. This is an example of epigenetic modification by which genes can be turned on or off more or less permanently. These animals exhibit blunted behavioral responses to stressful stimuli later in life.

What are the biological links between early experience and behavioral variation? A peptide hormone, oxytocin, plays a major role. Classic work showed that oxytocin regulates provision of milk by the mother, which occurs via reflex ejection in response to suckling (milk let-down). Oxytocin is synthesized by neurons in the hypothalamus and released into the general circulation through their projections in the posterior pituitary. It elicits smooth muscle contraction in the mammary gland, resulting in milk ejection. Oxytocin release from the pituitary is controlled by suckling, which provides a sensory stimulus that is conveyed to the hypothalamus by spinal afferent nerves.

Oxytocin and a related polypeptide hormone vasopressin also play important roles in regulating maternal bonding and other social behaviors (Chapter 2). In these cases, experience appears to modulate behaviors by affecting both release of oxytocin and levels of the oxytocin receptor in specific brain areas. In both rats and voles, individual differences in the care females provide their offspring correlate with variations in oxytocin receptor level in specific brain areas. Especially noteworthy is that oxytocin receptor levels in several regions are higher in female offspring reared by high-LG mothers than in female progeny of low

LG-mothers. Thus, sensory stimulation may affect activity of these polypeptide hormone systems, which in turn regulate maternal and other social behaviors.

A Set of Core Mechanisms Underlies Many Sexual Dimorphisms in the Brain and Spinal Cord

In the previous few sections, we described neural circuits that regulate several sexually dimorphic behaviors. Can we discern any common themes?

A variety of sexually dimorphic neural circuits, or wiring diagrams, can in principle generate sex differences in behavior (Figure 51–11). Although it is challenging to trace the chain of causality from genetic factors to dimorphic circuits to sex-specific behaviors, there are a few general possibilities. In one, a neural circuit, from sensory input to motor output, might be unique to one sex. In fact, this alternative is seldom encountered. Most behaviors are shared between the sexes, and even behaviors such as feeding, maternal retrieval of a pup by the scruff of its neck, or biting (during territorial scuffles between males) all call upon similar jaw movements. Consistent with this commonality, it appears that most sexual dimorphisms in behavior arise from sex differences in key neuronal populations within common circuits. The activity and connectivity of these populations alter behavioral output in a male- or female-typical manner.

Estrogen can act not only during development but also in adults to periodically reconfigure presynaptic connectivity within a hypothalamic neural circuit, ensuring that female mice only mate when they are ovulating and fertile. These studies paint a picture of dynamic neural circuits in the female brain: Wiring diagrams are plastic and responsive to hormonal changes across the estrous cycle, which is analogous to the menstrual cycle in humans. Similarly, estrogen also exerts cycle-related effects on dendritic spine plasticity in other brain regions, although the behavioral consequences in these instances are less well understood.

Another recurring theme in the developing brain is that masculinization is controlled by estrogen during the organizational phase. This control has profound enduring effects on social behaviors in adult life. Testosterone (which is aromatized into estrogen) or estrogen treatment of neonatal rodent females masculinizes the brain. As adults, these females are no longer sexually receptive to males, and in fact display male-typical social interactions, albeit at reduced intensity. Providing testosterone to these females, to mimic adult levels of testosterone in males, boosts the intensity of social behaviors, including territorial aggression (the propensity of animals to fight over territory or mates),

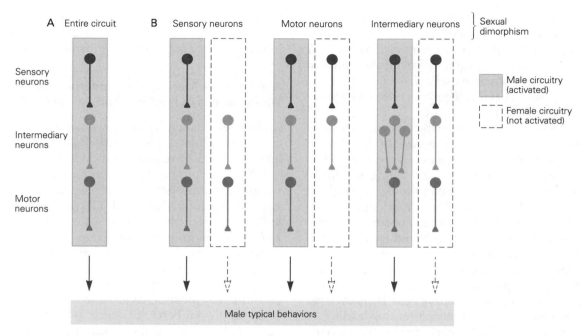

Figure 51–11 Possible circuit configurations that underlie sex differences in behavior. Neural circuit diagrams can be configured to generate sex differences in behaviors. Although it is possible to envision a neural circuit entirely exclusive to one or the other sex, most behaviors are shared between the sexes, and the current consensus is that sex differences in behavior or physiology reflect sexual dimorphisms in key neuronal populations embedded within an otherwise shared neural circuit. Such sexual dimorphisms have been found at the level of sensory neurons, motor neurons (as discussed for spinal nucleus of the bulbocavernosus neurons), or neurons interposed between sensory and motor pathways (such as the BNST and the sexually dimorphic nucleus of the preoptic area).

to male-typical levels. Thus, the perinatal surge of testosterone acts largely via aromatization into estrogens to masculinize the brain, whereas in adult life, both testosterone and estrogen facilitate the display of male-typical social interactions (Figure 51–12A).

These findings imply that male mice lacking androgen receptor exclusively in the nervous system should not only have male genitalia but also exhibit male patterns of social behavior, albeit at reduced intensity. This has in fact been borne out nicely by genetic engineering studies in mice; such mutant male mice indeed appear indistinguishable externally from control males, but they exhibit male-type sexual and aggressive behaviors with diminished intensity. However, there is growing evidence that the developmental control of masculinization of the brain by estrogen has shifted during evolution such that testosterone may be the predominant masculinizing agent in primates, including humans.

How do the actions of the limited number of sex hormones modulate the display of a large array of complex social interactions such as courtship vocalizations (similar to songbirds, many animals, including mice, vocalize as part of their mating ritual), sexual

behavior, marking (the propensity of animals of many species to claim territory with pheromones secreted in bodily fluids), and aggression? As described earlier in this chapter, sex hormones bind to cognate receptors to modulate gene expression in target cells. These steroids are available at different times, amounts, and places in the brain of the two sexes. Accordingly, sex hormone–regulated genes are expressed in sexually dimorphic patterns that are also different for different brain regions. These genes regulate differentiation and adult function of neural circuits along male- or female-typical lines (Figure 51–12B).

Experimental inactivation of such sex hormone–regulated genes reveals that individual genes influence only a subset of the sexually dimorphic social interactions without altering the entire behavioral program of males and females. Thus, an additional emerging theme is that sex hormones control differentiation and function of neural circuits in a modular manner, with different sex hormone–regulated genes acting in distinct neuronal populations to regulate separate aspects of male- or female-typical behaviors. In short, there is no single neuronal population that governs

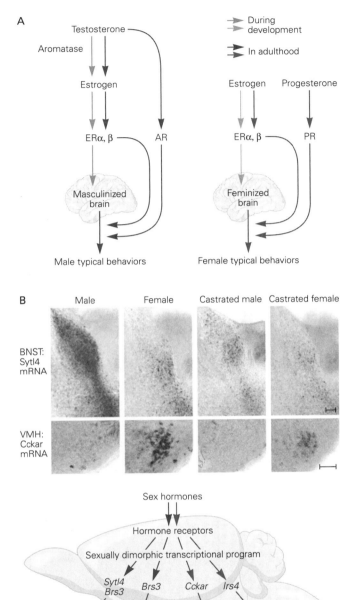

Figure 51–12 Mechanisms whereby sex hormones influence development and function of the nervous system.

A. Masculinization of the nervous systems occurs in at least two distinct steps: a developmental organizational phase largely controlled by estrogen signaling and a postpubertal activational phase controlled by estrogen and testosterone signaling via their cognate hormone receptors to regulate gene expression. (Abbreviations: **AR**, androgen receptor; **ER**, estrogen receptor; **PR**, progesterone receptor.)

B. Histological images show sexually dimorphic expression patterns of *Sytl4* mRNA in the bed nucleus of the stria terminalis (**BNST**) and *Cckar* in the ventromedial hypothalamus (**VMH**) of adult mice. Expression of these genes is clearly different in unmanipulated males and females and dramatically altered upon experimental removal of sex hormones from the circulation following castration in adult life. Both the BNST and VMH regulate mating and aggression in the two sexes.

Current thinking about how sex hormones regulate sex differences in behavior is illustrated in the diagram below. Molecular studies have identified many genes, such as *Sytl4* and *Cckar*, whose expression is sexually dimorphic in the adult brain and controlled by sex hormones. Many such genes, when experimentally mutated in mice via genetic engineering, regulate distinct components of sexually dimorphic behaviors but not the entire repertoire of social interactions. In other words, sex hormones control sexually dimorphic behaviors in a modular genetic manner. (Reproduced, with permission, from Xu et al. 2012.)

gender-typical behaviors; rather, the neural control of distinct behaviors is distributed across multiple different neuronal populations.

This modular control of sexually dimorphic behaviors fits well with our thinking that most circuits are shared between the males and females and that sex differences in behavior arise from key neural populations that alter circuit function in a male- or female-typical fashion. It seems likely that neurons exhibiting

sexually dimorphic molecular or anatomical features represent such key neuronal populations.

The Human Brain Is Sexually Dimorphic

Are sex differences between the brains of male and female mammals also present in humans, and if so, might they be functionally important? Early studies revealed that a few structures are markedly larger in men. These include Onuf's nucleus in the spinal cord, the homolog of the SNB in rodents (Figure 51–7); the BNST, implicated in rodent mating behavior (Figure 51–4); and the interstitial nucleus of the anterior hypothalamus 3 (INAH3), related to the rodent SDN-POA discussed earlier (Figure 51–13).

Advances in high-resolution magnetic resonance imaging (MRI) and histology have uncovered more subtle structural and molecular dimorphisms in the

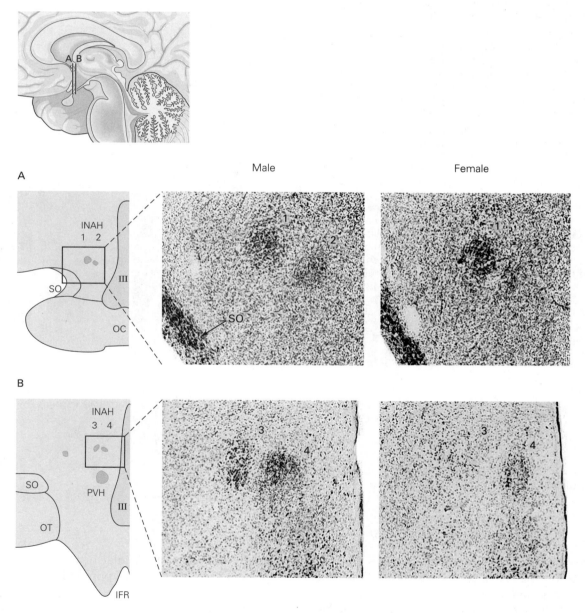

Figure 51–13 Sexual dimorphism in the interstitial nucleus of the anterior hypothalamus (INAH) 3 in the human brain. The human hypothalamus contains four small, discrete neuronal clusters, INAH1 to INAH4. The photomicrographs show these nuclei in adult male and female brains. While INAH1, INAH2, and INAH4 appear similar in men and women, INAH3 is significantly larger in men. The section in part **A** is 0.8 mm anterior to the section in part **B**. (Abbreviations: **IFR**, infundibular recess; **III**, third ventricle; **OC**, optic chiasm; **OT**, optic tract; **PVH**, paraventricular nucleus of the hypothalamus; **SO**, supraoptic nucleus.) (Adapted, with permission, from Gorski 1988.)

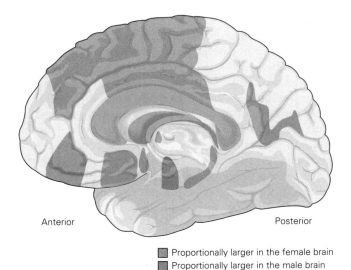

Anterior   Posterior

■ Proportionally larger in the female brain
■ Proportionally larger in the male brain

Figure 51–14 Sexual dimorphism is widespread in the adult human brain. A magnetic resonance imaging study measured the volume of many brain regions in adult men and women. The volume of each region was normalized to the size of the cerebrum for both sexes. Sex differences were significant in many regions, including several cortical areas that likely mediate cognitive functions. (Adapted, with permission, from Cahill 2006. Copyright © 2006 Springer Nature.)

central nervous system. For example, structures such as the fronto-orbital cortex and several gyri—including the precentral, superior frontal, and lingual gyri—occupy a significantly larger volume in adult women compared to a cohort of adult men (Figure 51–14). Moreover, the frontomedial cortex, amygdala, and angular gyrus volumes are larger in men compared to women. Thus, there are likely to be many sexual dimorphisms in the human brain.

Sexual Dimorphisms in Humans May Arise From Hormonal Action or Experience

What remains unclear is how these brain dimorphisms arise and how they relate to behavior. They might arise early from the organizational effects of hormones or later as a result of experience. Sex differences arising before or soon after birth could underlie behavioral differences, whereas those that arise later in life might be results of dimorphic experiences. Answers to these questions are fairly clear in a few cases. For example, studies of the development of neural circuits responsible for penile erection and lactation in rodents translate readily to humans.

Two recent observations suggest that enduring effects of experience on behavior first studied in animals (Figure 51–10) are also relevant to humans. First,

as discussed in Chapter 49, children raised for lengthy periods in orphanages with little individual care have long-lasting defects in a variety of social behaviors. Even years after placement in foster homes, these children have on average lower levels of oxytocin and vasopressin in their serum than children raised with biological parents. Second, people who have suffered abuse as children often grow up to be poor parents. Postmortem studies have shown that adults who had been abused as children exhibited greater promoter methylation of their glucocorticoid receptor genes than adults in control populations. Although these studies are new and require replication, they provide tantalizing hints at the biological mechanisms that underlie the lifelong effects of early parental care.

Dimorphic Structures in the Brain Correlate with Gender Identity and Sexual Orientation

In contrast to progress in mapping the biological bases of some relatively simple sexually dimorphic behaviors in people, differences in sexual partner preference and gender identity remain poorly understood. Little progress has been made in relating sex differences in cognitive functions to structural differences in the brain, in part because the very existence of cognitive differences remains a matter of controversy; if they exist at all, they are small and represent differences in means between highly variable male and female populations. On the other hand, several lines of evidence have connected clear differences in gender identity and sexual orientation to dimorphic structures in the brain.

Early insight into this issue came from observation of people with single-gene mutations that dissociate anatomical sex from gonadal and chromosomal sex, such as CAIS, CAH, and 5α-reductase deficiency (Table 51–1). For example, girls with CAH experience an excess of testosterone during fetal life; the disorder is generally diagnosed at birth and corrected. Nevertheless, the early exposure to androgens is correlated with subsequent changes in gender-related behaviors. On average, girls with CAH tend to have toy preferences and engage in play typical of boys of equivalent age. There is also a small but significant increase in the incidence of homosexual and bisexual orientation in females treated for CAH as children, and a significant proportion of these females also express the desire to live as men, consistent with a change in gender identity. These findings suggest that early organizational effects of steroids affect gender-specific behaviors independent of chromosomal and anatomical sex.

In 5α-reductase II deficiency and CAIS, many of the affected males show completely feminized external genitalia and are mistakenly raised as females until puberty. Thereafter, their histories diverge. In 5α-reductase II deficiency, the symptoms arise from a defect in testosterone processing largely confined to the developing external genitalia. At puberty, the large increase in circulating testosterone virilizes the body hair, musculature, and most dramatically, the external genitalia. At this stage, many but not all patients choose to adopt a male gender. In CAIS, in contrast, defects arise from a body-wide defect in the androgen receptor. These patients commonly seek medical advice after they fail to menstruate at puberty. Concordant with their feminized external phenotype, most CAIS

patients express a female gender identity and a sexual preference for men. They opt for surgical removal of the testes and hormonal supplementation appropriate for females.

What accounts for the different outcomes? Among many possibilities, one is that the dramatic change in behavior in 5α-reductase II patients at puberty results from the effects of testosterone acting on the brain. In CAIS patients, these effects do not occur because androgen receptors are absent from the brain. Clearly, however, this explanation does not rule out social and cultural upbringing as important factors in determining gender identity and sexual orientation.

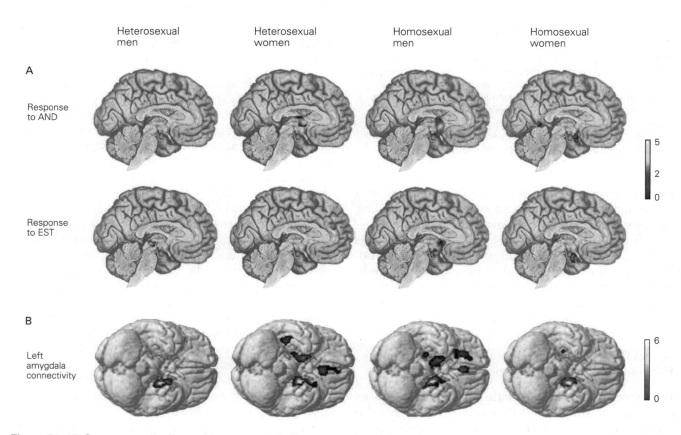

Figure 51–15 Some sexually dimorphic patterns of olfactory activation in the brain correlate with sexual orientation.

A. Positron emission tomography imaging was used to identify brain regions that were activated when subjects sniffed androstadienone (**AND**) or estratetraenol (**EST**) compared to nonodorous air. AND activated several hypothalamic centers in the brains of heterosexual women but not men, whereas EST activated several hypothalamic centers in heterosexual males but not females. Patterns of activation in the hypothalamus of homosexual men were similar to those of heterosexual women in response to AND, whereas similar patterns of activation were found in heterosexual men and homosexual women in response to EST. The color calibration on the right shows the

level of putative neural activity. Because the same brain sections were selected to compare, the figure does not illustrate maximal activation for each condition. (Adapted, with permission, from Berglund, Lindstrom, and Savic 2006; Savic, Berglund, and Lindstrom 2005.)

B. Heterosexual and homosexual subjects were scanned while breathing unscented air, and a measure of covariance was used to estimate connectivity among regions. In heterosexual women and homosexual men, the left amygdala was strongly connected to the right amygdala, whereas connectivity remained local in heterosexual men and homosexual women. (Adapted, with permission, from Savic and Lindstrom 2008.)

A second set of studies probing the biology of sexual orientation assessed responses to pheromones. Pheromone perception in humans is quite different from that of mice and is likely a less important sense. Humans do not have a functional VNO, and most of the genes implicated in pheromone reception in the mouse VNO, such as *trpc2* and those encoding VNO receptors, are absent or nonfunctional in the human genome. To the extent that humans do sense pheromones, they appear to use the main olfactory epithelium and bulb. Chemicals that appear to be human pheromones include androstadienone (AND), an odorous androgenic metabolite, and estratetraenol (EST), an odorous estrogenic metabolite. AND is present at 10-fold higher concentrations in male sweat compared to female sweat, whereas EST is present in the urine of pregnant women. Both compounds can produce sexual arousal—AND in heterosexual women and EST in heterosexual men—even at concentrations so low that there is no conscious olfactory perception.

Brain areas activated by AND and EST have been identified by positron emission tomography (PET) imaging. When AND is presented, certain hypothalamic nuclei are activated in heterosexual women but not heterosexual men, whereas when EST is presented, adjacent regions containing clusters of nuclei are activated in men but not in women (Figure 51–15A). In homosexual men and women, there is a reversal of hypothalamic activation: AND but not EST activates hypothalamic centers in homosexual men, and conversely, EST but not AND activates those areas in lesbian women. Heterosexual and homosexual brains therefore appear to process olfactory sensory information in different ways.

Do sexually dimorphic structures in homosexual brains correlate with anatomical sex or sexual orientation? Imaging studies have provided support for the view that the brains of homosexual men resemble those of heterosexual woman and that the brains of homosexual women resemble those of heterosexual men (Figure 51–15B). Moreover, the volume of the sexually dimorphic BNST is small in male-to-female transsexuals compared to men, whereas female-to-male transsexuals appear to have a larger BNST compared to women (Figure 51–16). It is not clear, however, whether the structural dimorphism in these individuals is a consequence or a cause of gender identity or sexual orientation.

The male mouse counterpart of the human BNST plays a critical role in recognizing the sex of *other* mice and guides subsequent social interactions, such as aggression with males and mating with females. Thus, a region linked to gender identity in the human brain plays an important role in sex recognition in rodents.

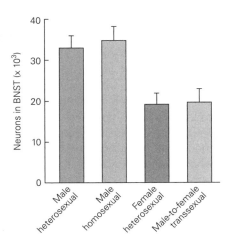

Figure 51–16 Sexual dimorphism in the bed nucleus of the stria terminalis (BNST) in humans. The nucleus has significantly more neurons in men compared to women regardless of male sexual orientation. Similar to women, male-to-female transsexuals have fewer neurons than men. In the one female-to-male transsexual brain available for postmortem analysis (not shown in the bar graph), the number of neurons is well within the normal range for men. (Adapted, with permission, from Kruijver et al. 2000.)

As is the case with the sexual dimorphism of the mouse BNST discussed earlier, hormonal influences are also thought to underlie the dimorphism of the human BNST.

If prenatal influences do lead to dissociation of sex from gender, are those influences genetic? Other than the rare syndromes described earlier, attempts to find genetic bases for sexual orientation or gender identity have not been productive. Claimed genetic contributions are small and claims of associations with specific genomic loci have not been replicated. Thus, while the current weight of evidence favors some contribution of early, even prenatal, factors in these processes, their cause and relative importance remain unknown.

Highlights

1. In humans and many other mammals, the sex determination pathway directs differentiation of the bipotential gonad into testes in males and ovaries in females. The *SRY* gene on the Y chromosome directs the gonad to form testes, whereas the absence of *SRY* enables the gonad to differentiate into ovaries.

2. Sex steroid hormones produced by the gonads—testosterone by testes and estrogens, and progesterone by ovaries—drive sexual differentiation of both the nervous system and the rest of the body.

3. Sex hormones act early during a critical window in development to irreversibly organize neural substrates for behavior in a sexually dimorphic manner, whereas in adult life, these hormones act acutely and reversibly to activate sex-typical physiological and behavioral responses.

4. During the critical window, testes produce a transient surge of testosterone that masculinizes the developing bipotential nervous system. By contrast, the ovaries are quiescent during this period, and it is thought that the absence of sex hormones enables the nervous system in this period to differentiate along a female-typical pathway.

5. Many of the actions of testosterone that masculinize the nervous system occur following its conversion to estrogen locally at the site of action. There is evidence to suggest that in humans and other primates testosterone also acts directly via its cognate hormone receptor to effect masculinization of the neural substrates of behavior.

6. Sex hormones control sexual differentiation of neural pathways by utilizing cellular processes such as apoptosis, neurite extension, and synapse formation that are employed widely during other developmental events.

7. Sex hormones bind to cognate hormone receptors that modulate gene expression. Such genes in turn regulate the cellular processes that result in sex differences in neuronal number, connectivity, and physiology.

8. Many neuronal populations that are sexually dimorphic by morphological and other criteria have been identified in the vertebrate brain over the past few decades. Functional studies show that these regions influence some, but not all, sexually dimorphic behaviors.

9. Recent molecular studies have identified many sex hormone–regulated genes whose expression patterns are sexually dimorphic. These genes as well as the neurons they are expressed in regulate sexually dimorphic social behaviors in a modular manner. In other words, individual genes and the neuronal populations that express them modulate one or a few sexually dimorphic behaviors so that the control of these behaviors is distributed among many different neuronal groups.

10. Such sexually dimorphic neuronal populations are likely embedded within neural circuits found in both sexes, and they are thought to guide behavior along male- or female-typical patterns.

11. Both sensory stimuli and past experience profoundly regulate the display of sexually dimorphic behaviors. In some cases, the influence of past experience can extend across the life span of the animal.

12. Pheromones guide choice of sexual partner in rodents. There is evidence from imaging studies that men and women may also show sexually dimorphic neural responses to male and female pheromones and that these responses can align with sexual orientation; in these cases, however, it is unclear if the neural responses are learned responses based on past experience.

13. There are many sex differences between the brains of men and women, and in some instances, these sex differences align with gender in adult life rather than gender assigned at birth. In these cases, it is not clear whether the sex differences causally reflect gender identity or are a result of it. These issues are difficult to disentangle at present.

Nirao M. Shah
Joshua R. Sanes

Selected Reading

Arnold AP. 2004. Sex chromosomes and brain gender. Nat Rev Neurosci 5:701–708.

Bayless DW, Shah NM. 2016. Genetic dissection of neural circuits underlying sexually dimorphic social behaviours. Philos Trans R Soc Lond B Biol Sci 371:20150109.

Byne W. 2006. Developmental endocrine influences on gender identity: implications for management of disorders of sex development. Mt Sinai J Med 73:950–959.

Cahill L. 2006. Why sex matters for neuroscience. Nat Rev Neurosci 7:477–484.

Curley JP, Jensen CL, Mashoodh R, Champagne FA. 2010. Social influences on neurobiology and behavior: epigenetic effects during development. Psychoneuroendocrinology 36:352–371.

Dulac C, Wagner S. 2006. Genetic analysis of brain circuits underlying pheromone signaling. Annu Rev Genet 40:449–467.

Hines M. 2006. Prenatal testosterone and gender-related behavior. Eur J Endocrinol 155:S115–S121.

Kohl J, Dulac C. 2018. Neural control of parental behaviors. Curr Opin Neurobiol 49:116–122.

Morris JA, Jordan CL, Breedlove SM. 2004. Sexual differentiation of the vertebrate nervous system. Nat Neurosci 7:1034–1039.

Swaab DF. 2004. Sexual differentiation of the human brain: relevance for gender identity, transsexualism and sexual orientation. Gynecol Endocrinol 19:301–312.

Wilhelm D, Palmer S, Koopman P. 2007. Sex determination and gonadal development in mammals. Physiol Rev 87:1–28.

Yang CF, Shah NM. 2014. Representing sex in the brain, one module at a time. Neuron 82:261–278.

References

Bakker J, De Mees C, Douhard Q, et al. 2006. Alpha-fetoprotein protects the developing female mouse brain from masculinization and defeminization by estrogens. Nat Neurosci 9:220–226.

Bayless DW, Yang T, Mason MM, et al. 2019. Limbic neurons shape sex recognition and social behavior in sexually naïve males. Cell 176:1190–1205.

Berglund H, Lindstrom P, Savic I. 2006. Brain response to putative pheromones in lesbian women. Proc Natl Acad Sci U S A 103:8269–8274.

Brainard MS, Doupe AJ. 2002. What songbirds teach us about learning. Nature 417:351–358.

Byne W, Lasco MS, Kemether E, et al. 2000. The interstitial nuclei of the human anterior hypothalamus: an investigation of sexual variation in volume and cell size, number and density. Brain Res 856:254–258.

Cohen-Kettenis PT. 2005. Gender change in 46, XY persons with 5α-reductase-2 deficiency and 17β-hydroxysteroid dehydrogenase-3 deficiency. Arch Sex Behav 34:399–410.

Cooke BM, Woolley CS. 2005. Gonadal hormone modulation of dendrites in the mammalian CNS. J Neurobiol 64: 34–46.

Demir E, Dickson BJ. 2005. *Fruitless* splicing specifies male courtship behavior in *Drosophila*. Cell 121:785–794.

Edwards DA, Burge KG. 1971. Early androgen treatment and male and female sexual behavior in mice. Horm Behav 2:49–58.

Forger NG, de Vries GJ. 2010. Cell death and sexual differentiation of behavior: worms, flies, and mammals. Curr Opin Neurobiol 20:776–783.

Goldstein LA, Kurz EM, Sengelaub DR. 1990. Androgen regulation of dendritic growth and retraction in the development of a sexually dimorphic spinal nucleus. J Neurosci 10:935–946.

Gorski RA. 1988. Hormone-induced sex differences in hypothalamic structure. Bull Tokyo Metropol Inst Neurosci 16 (Suppl 3):67–90.

Gorski RA. 1988. Sexual differentiation of the brain: mechanisms and implications for neuroscience. In: SS Easter Jr, KF Barald, BM Carlson (eds). *From Message to Mind: Directions in Developmental Neurobiology*, pp. 256–271. Sunderland, MA: Sinauer.

Gorski RA, Harlan RE, Jacobsen CD, Shryne JE, Southam AM. 1980. Evidence for the existence of a sexually dimorphic nucleus in the preoptic area of the rat. J Comp Neurol 193:529–539.

Greenspan RJ, Ferveur JF. 2000. Courtship in *Drosophila*. Annu Rev Genet 34:205–232.

Inoue S, Yang R, Tantry A, et al. 2019. Periodic remodeling in a neural circuit governs timing of female sexual behavior. Cell 179:1393–1408.

Juntti SA, Tollkuhn J, Wu MV, et al. 2010. The androgen receptor governs the execution, but not programming, of male sexual and territorial behaviors. Neuron 66:260–272.

Kimchi T, Xu J, Dulac C. 2007. A functional circuit underlying male sexual behavior in the female mouse brain. Nature 448:1009–1014.

Kimura K, Ote M, Tazawa T, Yamamoto D. 2005. *Fruitless* specifies sexually dimorphic neural circuitry in the *Drosophila* brain. Nature 438:229–233.

Kohl J, Babayan BM, Rubinstein ND, et al. 2018. Functional circuit architecture underlying parental behaviour. Nature 556:326–331.

Konishi M, Akutagawa E. 1985. Neuronal growth, atrophy and death in a sexually dimorphic song nucleus in the zebra finch brain. Nature 315:145–147.

Koopman P, Gubbay J, Vivian N, Goodfellow P, Lovell-Badge R. 1991. Male development of chromosomally female mice transgenic for *Sry*. Nature 351:117–121.

Kruijver FP, Zhou JN, Pool CW, Hofman MA, Gooren LJ, Swaab DF. 2000. Male-to-female transsexuals have female neuron numbers in a limbic nucleus. J Clin Endocrinol Metab 85:2034–2041.

Långström N, Rahman Q, Carlström E, Lichtenstein P. 2010. Genetic and environmental effects on same-sex sexual behavior: a population study of twins in Sweden. Arch Sex Behav 39:75–80.

Lee H, Kim DW, Remedios R, et al. 2014. Scalable control of mounting and attack by Esr1+ neurons in the ventromedial hypothalamus. Nature 509:627–632.

LeVay S. 1991. A difference in hypothalamic structure between heterosexual and homosexual men. Science 253:1034–1037.

Leypold BG, Yu CR, Leinders-Zufall T, Kim MM, Zufall F, Axel R. 2002. Altered sexual and social behaviors in *trp2* mutant mice. Proc Natl Acad Sci U S A 99:6376–6381.

Liu YC, Salamone JD, Sachs BD. 1997. Lesions in medial preoptic area and bed nucleus of stria terminalis: differential effects on copulatory behavior and noncontact erection in male rats. J Neurosci 17:5245–5253.

Mandiyan VS, Coats JK, Shah NM. 2005. Deficits in sexual and aggressive behaviors in *Cnga2* mutant mice. Nat Neurosci 8:1660–1662.

Manoli DS, Foss M, Villella A, Taylor BJ, Hall JC, Baker BS. 2005. Male-specific *fruitless* specifies the neural substrates of *Drosophila* courtship behaviour. Nature 436:395–400.

McCarthy MM, Arnold AP. 2011. Reframing sexual differentiation of the brain. Nat Neurosci 14:677–683.

McGowan PO, Sasaki A, D'Alessio AC, et al. 2009. Epigenetic regulation of the glucocorticoid receptor in human brain associates with childhood abuse. Nat Neurosci 12:342–348.

Nottebohm F, Arnold AP. 1976. Sexual dimorphism in vocal control areas of the songbird brain. Science 194:211–213.

Ohno S, Geller LN, Lai EV. 1974. TFM mutation and masculinization versus feminization of the mouse central nervous system. Cell 3:235–242.

Sapolsky RM. 2004. Mothering style and methylation. Nat Neurosci 7:791–792.

Savic I, Berglund H, Gulyas B, Roland P. 2001. Smelling of odorous sex hormone-like compounds causes sex-differentiated hypothalamic activations in humans. Neuron 31:661–668.

Savic I, Berglund H, Lindstrom P. 2005. Brain response to putative pheromones in homosexual men. Proc Natl Acad Sci U S A 102:7356–7361.

Savic I, Lindstrom P. 2008. PET and MRI show differences in cerebral asymmetry and functional connectivity between homo- and heterosexual subjects. Proc Natl Acad Sci U S A 105:9403–9408.

Sekido R, Lovell-Badge R. 2009. Sex determination and *SRY*: down to a wink and a nudge? Trends Genet 25:19–29.

Shah NM, Pisapia DJ, Maniatis S, Mendelsohn MM, Nemes A, Axel R. 2004. Visualizing sexual dimorphism in the brain. Neuron 43:313–319.

Stockinger P, Kvitsiani D, Rotkopf S, Tirian L, Dickson BJ. 2005. Neural circuitry that governs *Drosophila* male courtship behavior. Cell 121:795–807.

Stowers L, Holy TE, Meister M, Dulac C, Koentges G. 2002. Loss of sex discrimination and male-male aggression in mice deficient for TRP2. Science 295:1493–1500.

Unger EK, Burke KJ Jr, Yang CF, Bender KJ, Fuller PM, Shah NM. 2015. Medial amygdalar aromatase neurons regulate aggression in both sexes. Cell Rep 10:453–462.

Weaver IC, Cervoni N, Champagne FA, et al. 2004. Epigenetic programming by maternal behavior. Nat Neurosci 7:847–854.

Wei YC, Wang SR, Jiao ZL, et al. 2018. Medial preoptic area in mice is capable of mediating sexually dimorphic behaviors regardless of gender. Nat Commun 9:279.

Wierman ME. 2007. Sex steroid effects at target tissues: mechanisms of action. Adv Physiol Educ 31:26–33.

Wu MV, Manoli DS, Fraser EJ, et al. 2009. Estrogen masculinizes neural pathways and sex-specific behaviors. Cell 139:61–72.

Wu Z, Autry AE, Bergan JF, Watabe-Uchida M, Dulac CG. 2014. Galanin neurons in the medial preoptic area govern parental behaviour. Nature 509:325–330.

Xu X, Coats JK, Yang CF, et al. 2012. Modular genetic control of sexually dimorphic behaviors. Cell 148:596–607.

Yang CF, Chiang MC, Gray DC, et al. 2013. Sexually dimorphic neurons in the ventromedial hypothalamus govern mating in both sexes and aggression in males. Cell 153:896–909.

Yang T, Yang CF, Chizari MD, et al. 2017. Social control of hypothalamus-mediated male aggression. Neuron 95:955–970.

Zhang J, Webb DM. 2003. Evolutionary deterioration of the vomeronasal pheromone transduction pathway in catarrhine primates. Proc Natl Acad Sci U S A 100:8337–8341.

Zhang TY, Meaney MJ. 2010. Epigenetics and the environmental regulation of the genome and its function. Annu Rev Psychol 61:439–466.

Part VIII

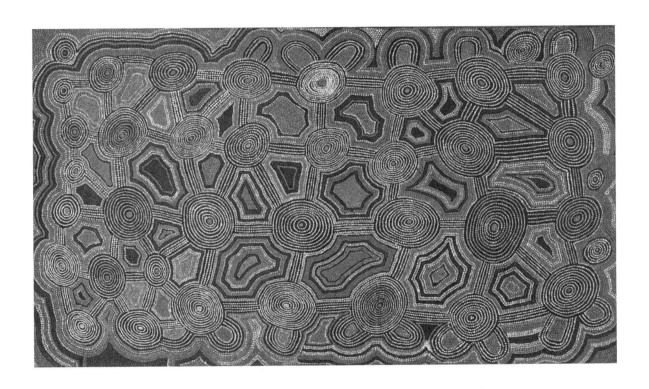

VIII   Learning, Memory, Language and Cognition

MOTOR AND SENSORY FUNCTIONS take up less than one-half of the cerebral cortex in humans. The rest of the cortex is occupied by the association areas, which coordinate events arising in the motor and sensory centers. Three association areas—the prefrontal, parietal-temporal-occipital, and limbic—are involved in cognitive behavior: speaking, thinking, feeling, perceiving, planning skilled movements, learning, memory, decision-making, and consciousness.

Most of the early evidence relating cognitive functions to the association areas came from clinical studies of brain-damaged patients. Thus, the study of language in patients with aphasia yielded important information about how human mental processes are distributed in the two hemispheres of the brain and how they develop. More refined analyses have come from human imaging studies using functional magnetic resonance imaging (fMRI) and other methods.

Deeper insights into the neural circuitry and cellular mechanisms giving rise to cognitive processes have come from electrophysiological recordings and genetic-based manipulations, including cell type–specific gene deletions and cell type–specific optogenetic excitation or inhibition in experimental animals, particularly in rodents. Such studies can evaluate the relative contribution of specific genes, neurons, and synaptic connections to specific types of behavior.

So far in this book, we have considered neural mechanisms associated with basic functions of the brain, including primary sensory perception, movement, and homeostatic control. In this part and the next, we begin to consider the more complex, higher-order brain functions mentioned earlier, the realm of cognitive neural science. The aim of this merger of neurophysiology, anatomy, developmental biology, cell and molecular biology, theory, and cognitive psychology is to ultimately provide an understanding of the neural mechanisms of the mind.

Until the latter part of the 20th century, the study of higher mental function was approached through behavioral observations gleaned from brain-damaged patients and animals with experimental lesions. In the first part of the 20th century, to avoid untestable concepts and hypotheses, psychology became rigidly concerned with behaviors defined strictly in terms of observable stimuli and responses. Orthodox behaviorists thought it unproductive to deal with consciousness, feeling, attention, or even motivation. By concentrating only on observable actions, behaviorists asked: What can an organism do, and how does it do it? Indeed, careful quantitative analysis of stimuli and responses has contributed greatly to our understanding of the acquisition and use of "implicit" knowledge of perceptual and motor skills. However, humans and other higher animals also have "explicit" knowledge of facts and events. They have knowledge of space, rules, and relations—what Edward Tolman termed *cognitive maps*. Animals can choose a newly available route to a goal without ever learning the sensory-response association, and humans can reason deliberatively from what they know to imagine something unknown. Indeed, that is what makes neural science possible—in fact, all of science and the humanities.

Thus, we also need to ask: What does the animal know about the world, and how does it come to know it? How is that knowledge represented in the brain? Does explicit knowledge differ from implicit knowledge? And how can such knowledge be communicated to others and enable us to make rational decisions based on past experience? Much, perhaps most, knowledge is unconscious a great deal of the time. We need to know the nature of the unconscious processes, the systems that mediate them, and their influence on the nature of conscious mental activity. Finally, we need to know about the highest realms of conscious knowledge, the knowledge of oneself as an individual, a thinking and feeling human being.

The modern effort to understand the neural mechanisms of higher mental functions began at the end of the 19th century when Pierre Broca and Carl Wernicke discovered regions of the cerebral cortex responsible for the production and comprehension of language. Throughout the 20th century, studies of patients with brain damage resulting from accidents, war, and disease led to an expansion of knowledge of the roles of specific brain areas responsible for cognitive functions, including attention, intention (planning), reasoning, and learning and memory. However, it was only in the past 20 to 30 years, based in part on new technological approaches, that our understanding of cognitive processes advanced from anatomical localization to an understanding of how neural activity in specific brain regions underlies such processes.

In Part VIII, we explore such questions of cognitive brain science. Chapter 52 introduces basic mechanisms of human learning and memory, focusing on the use of fMRI and behavioral studies to elucidate the role of different brain regions in implicit and explicit memory. In Chapter 53, we discuss the cellular and molecular mechanisms responsible for implicit memory storage, focusing on studies in invertebrates and vertebrates that have elucidated the role of synaptic plasticity in implicit memory storage. In Chapter 54, we expand on the theme of synaptic plasticity, this time for the storage of explicit memory by the hippocampus and related brain regions. We further consider how the synaptic connectivity between the entorhinal cortex and hippocampus enables us to perceive and remember our spatial location in a given environment. Next, in Chapter 55, we focus on the neural mechanisms underlying language, a uniquely human function that enables us to communicate our store of knowledge to others, including brain circuits necessary for speaking and perceiving the spoken word. Finally, in Chapter 56, we examine how the brain enables us to use our knowledge to make rational decisions. Viewed through the lens of decision-making, the distinction between the apparently separate processes of knowledge and know-how can be seen as a unified function, one that provides a basis for understanding how consciousness may emerge from brain activity. Achieving a full understanding of the neural mechanisms that enable us to maintain a rich set of memories of our past experiences over a lifetime, to communicate those memories to others, and to use them to make informed, conscious decisions is perhaps one of the most daunting challenges in all of science.

Part Editors: Eric R. Kandel and Steven A. Siegelbaum

Part VIII

52

Learning and Memory

I N HIS MASTERFUL NOVEL *One Hundred Years of Solitude,* Gabriel Garcia Márquez describes a strange plague that invades a tiny village and robs people of their memories. The villagers first lose personal recollections, then the names and functions of common objects. To combat the plague, one man places written labels on every object in his home. But he soon realizes the futility of this strategy, because the plague eventually destroys even his knowledge of words and letters.

This fictional incident reminds us of how important learning and memory are in everyday life. Learning refers to a change in behavior that results from acquiring knowledge about the world, and memory refers to the processes by which that knowledge is encoded, stored, and later retrieved. Marquez's story challenges us to imagine life without the ability to learn and remember. We would forget people and places we once knew, and no longer be able to use and understand language or execute motor skills we had once learned; we would not recall the happiest or saddest moments of our lives and would even lose our sense of personal identity. Learning and memory are essential to the full functioning and independent survival of people and animals.

In 1861, Pierre Paul Broca discovered that damage to the posterior portion of the left frontal lobe (Broca's area) produces a specific deficit in language. Soon thereafter, it became clear that other mental functions, such as perception and voluntary movement, are also mediated by discrete parts of the brain (Chapter 1). This naturally led to the question: Are there discrete neural systems concerned with memory? If so, is there a "memory center," or is memory processing widely distributed throughout the brain?

Contrary to the prevalent view that cognitive functions are localized in the brain, many students of learning doubted that memory is localized. In fact, until the middle of the 20th century, many psychologists doubted that memory is a discrete function, independent of perception, language, or movement. One reason for the persistent doubt is that memory storage involves many different parts of the brain. We now appreciate, however, that these regions are not all equally important. There are several fundamentally different types of memory, and certain regions of the brain are much more important for encoding some types of memory than for others.

During the past several decades, researchers have made significant progress in the analysis and understanding of learning and memory. In this chapter, we focus on studies of normal human memory behavior, its perturbations following brain lesions due to injury or surgery, and measurements of brain activity during learning and memory recall using functional magnetic resonance imaging (fMRI) and extracellular electrophysiological recordings. These studies have yielded three major insights.

First, there are several forms of learning and memory. Each form of learning and memory has distinctive cognitive and computational properties and is supported by different brain systems. Second, memory involves encoding, storage, retrieval, and consolidation. Finally, imperfections and errors in remembering can provide clues about the nature and function of learning and memory and the fundamental role that memory plays in guiding behavior and planning for the future.

Memory can be classified along two dimensions: (1) the time course of storage and (2) the nature of the information stored. In this chapter, we consider the time course of storage. In the next two chapters, we focus on the cellular, molecular, and circuit-based mechanisms of different forms of learning and memory, based largely on studies of animal models.

Short-Term and Long-Term Memory Involve Different Neural Systems

Short-Term Memory Maintains Transient Representations of Information Relevant to Immediate Goals

When we reflect on the nature of memory, we usually think of the long-term memory that William James referred to as "memory proper" or "secondary memory." That is, we think of memory as "the knowledge of a former state of mind after it has already once dropped from consciousness." This knowledge depends on the formation of a memory trace that is durable, in which the representation persists even when its content has been out of conscious awareness for a long period.

Not all forms of memory, however, constitute "former states of mind." In fact, the ability to store information depends on a form of short-term memory, called working memory, which maintains current, albeit transient, representations of goal-relevant knowledge. In humans, working memory consists of at least two subsystems—one for verbal information and another for visuospatial information. The functioning of these two subsystems is coordinated by a third system called the *executive control processes*. Executive control processes are thought to allocate attentional resources to the verbal and visuospatial subsystems and to monitor, manipulate, and update stored representations.

We use the verbal subsystem when we attempt to keep speech-based (phonological) information in conscious awareness, as when we mentally rehearse a password before entering it. The verbal subsystem consists of two interactive components: a store that represents phonological knowledge and a rehearsal mechanism that keeps these representations active while we need them. Phonological storage depends on posterior parietal cortices, and rehearsal partially depends on articulatory processes in Broca's area.

The visuospatial subsystem of working memory retains mental images of visual objects and of the location of objects in space. The rehearsal of spatial and object information is thought to involve modulation of this information in the parietal, inferior temporal, and occipital cortices by the frontal and premotor cortices.

Single-cell recordings in nonhuman primates indicate that, over a period of seconds, some prefrontal neurons maintain spatial representations, others maintain object representations, and still others represent the integration of spatial and object knowledge. Although neurons concerned with working memory of objects tend to lie in the ventrolateral prefrontal cortex and those concerned with spatial knowledge tend to lie in the dorsolateral prefrontal cortex, all three classes of neurons are found in both prefrontal subregions (Figure 52–1).

Thus, working memory involves activation of representations of information stored in specialized cortical regions that vary based on the content of the information, as well as activation of general control mechanisms in prefrontal cortex. Prefrontal control signals in working memory are further dependent on interaction with the striatum and ascending dopaminergic inputs from the midbrain.

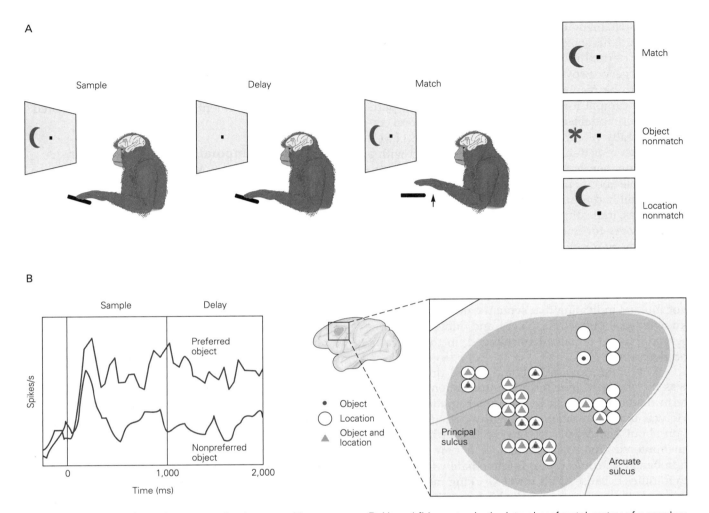

Figure 52–1 The prefrontal cortex maintains a working memory. (Adapted, with permission, from Rainer, Asaad, and Miller 1998.)

A. The role of prefrontal cortex in maintaining information in working memory is often assessed in monkeys using electrophysiological methods in conjunction with a delayed-match-to-sample (DMS) task. In this type of task, each trial begins when the monkey grabs a response lever and visually fixates a small target at the center of a computer screen. An initial visual stimulus (the sample) is briefly presented and must be held in working memory until the next stimulus (the match) appears. In the task illustrated here, the monkey was required to remember the sample ("what") and its location ("where") and release the lever only in response to stimuli that matched on both dimensions.

B. Neural firing rates in the lateral prefrontal cortex of a monkey during the delay period are often maintained above baseline and represent responses to the type of stimulus (what), the location (where), and the integration of the two (what and where). As shown, at left is the activity of a prefrontal neuron in response to a preferred object (to which the neuron responds robustly) and to a nonpreferred object (to which the neuron responds minimally). Activity is robust both when the monkey looks at the preferred object (sample) and during the delay. In the sketch at right, the symbols represent recording sites where neurons maintained each type of information (what, where, and what and where). Typically, several types of neurons were found at one site; hence, many symbols overlap and some symbols indicate more than one neuron.

Information Stored in Short-Term Memory Is Selectively Transferred to Long-Term Memory

In the mid-1950s, startling new evidence about the neural basis of long-term memory emerged from the study of patients who had undergone bilateral removal of the hippocampus and neighboring regions in the medial temporal lobe as treatment for epilepsy. The first and best-studied case was a patient called H.M. studied by the psychologist Brenda Milner and the surgeon William Scoville. (After H.M. died on December 2, 2008, his full name, Henry Molaison, was revealed to the world.)

H.M. had suffered for a number of years from untreatable temporal lobe epilepsy caused by brain damage sustained at age 7 years in a bicycle accident. As an adult, his seizures rendered him unable to work or lead a normal life, and at the age of 27, he underwent surgery. Scoville removed the brain regions thought to be responsible for the seizures, including the hippocampal formation, the amygdala, and parts of the multimodal association area of the temporal cortex bilaterally (Figure 52–2). After the surgery, H.M.'s seizures were better controlled, but he was left with a devastating memory deficit (or amnesia). What was so remarkable about H.M.'s deficit was its specificity.

He still had normal working memory, for seconds or minutes, indicating that the medial temporal lobe is not necessary for transient memory. He also had long-term memory for events that had occurred before the operation. For example, he remembered his name, the job he had held, and childhood events. In addition, he retained a command of language, including his vocabulary, indicating that semantic memory—factual knowledge about people, places, and things—was preserved. His IQ remained unchanged, in the range of bright-normal.

What H.M. now lacked, and lacked dramatically, was the ability to transfer new information into long-term memory, a deficit termed anterograde amnesia. He was unable to retain for lengthy periods information about people, places, or objects that he had just encountered. Asked to remember a new telephone number, H.M. could repeat it immediately for seconds to minutes because of his intact working memory. But when distracted, even briefly, he forgot the number. H.M. could not recognize people he met after surgery, even when he met them again and again. For several years, he saw Milner every month, yet each time she entered the room, he reacted as though he had never seen her before. H.M. is not unique. All patients with extensive bilateral lesions of the limbic association areas of the medial temporal lobe show similar long-term memory deficits.

H.M. is a historic case because his deficit provided the first clear link between memory and the medial temporal lobe, including the hippocampus. Subsequent studies by Larry Squire and others of patients with brain damage more limited to the hippocampus confirmed its central role in memory. The observation that H.M. and others with medial temporal lobe damage had a profound deficit in the formation of new memories while the retrieval of old memories remained largely intact suggested that memories must be transferred over time from the hippocampus and medial temporal lobe to other brain structures. These studies gave rise to four central questions that continue to drive memory research to this day: First,

what is the functional role of the medial temporal lobe memory system? Second, what are the roles of different subregions within this system? Third, how do these subregions work together with other brain circuits to support different forms of memory? Fourth, where are hippocampal-dependent memories ultimately stored?

The Medial Temporal Lobe Is Critical for Episodic Long-Term Memory

A crucial finding about H.M. was that formation of long-term memory was impaired only for certain types of information. H.M. and other patients with damage to the medial temporal lobe were able to form and retain certain types of durable memories just as well as healthy subjects.

For example, H.M. learned to draw the outlines of a star while looking at the star and his hand in a mirror (Figure 52–3). Like healthy subjects learning to remap hand–eye coordination, H.M. initially made many mistakes, but after several days of training, his performance was error-free and comparable to that of healthy subjects. Nevertheless, he did not consciously remember having performed the task.

Long-term memory formation in amnesic patients is not limited to motor skills. These patients retain simple reflexive learning, including habituation, sensitization, and some forms of conditioning (to be discussed later in this chapter). Furthermore, they are able to improve their performance on certain perceptual and conceptual tasks. For example, they do well with a form of memory called priming, in which perception of a word or object or access to the meaning of a word or object is improved by prior exposure. Thus, when shown only the first few letters of previously studied words, a subject with amnesia is able to generate the same number of studied words as normal subjects, even though the amnesic patient has no conscious memory of having recently encountered the words (Figure 52–4).

This pattern of selectively impaired performance in patients with amnesia raised questions about how to classify these different forms of memory: What are the key features that distinguish between memories that survive medial temporal lobe damage and those that do not? Early theories by Squire and colleagues suggested that a critical factor may be conscious awareness—damage to the medial temporal lobe appears to impair forms of memory that can be accessed consciously and can be reported on or expressed in words, while leaving intact forms of memory that cannot. For this reason, memories that depend on the medial temporal

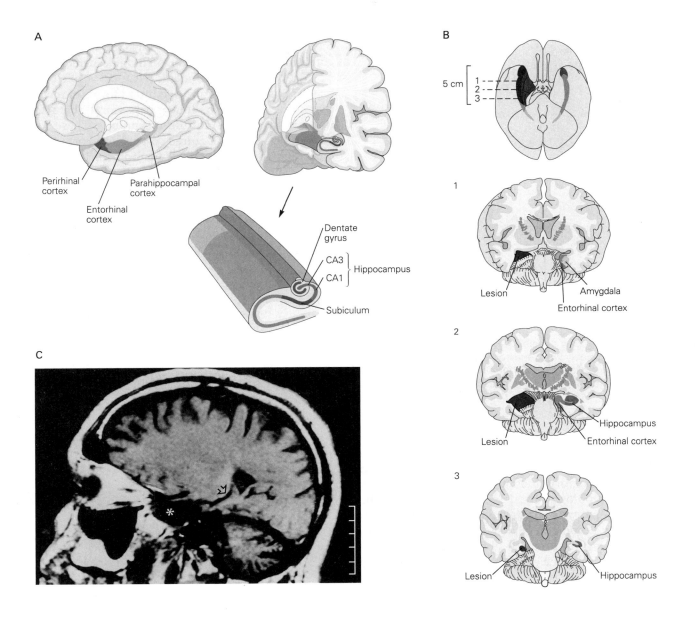

Figure 52–2 The medial temporal lobe and memory storage.

A. The key components of the medial temporal lobe important for memory storage.

B. The areas of temporal lobe resected (**gray shading**) in the patient known as H.M., viewed from the ventral surface of the brain (left hemisphere is on the right side of the image). Surgery was a bilateral, single-stage procedure, but to illustrate the structures that were removed from the left hemisphere (right side of the image), the left hemisphere is shown here intact. The longitudinal extent of the lesion is shown in a ventral view of the brain (top). Cross sections 1 through 3 show

the estimated extent of areas of the brain removed from H.M. (Adapted, with permission, from Corkin et al. 1997.)

C. Magnetic resonance image (MRI) scan of a parasagittal section from the left side of H.M.'s brain. The calibration bar at the right of the panel has 1-cm increments. The **asterisk** in the central area of the scan indicates the resected portion of the anterior temporal lobes. The nearby **arrowhead** points to the remaining portion of the intraventricular portion of the hippocampal formation. Approximately 2 cm of preserved hippocampal formation is visible bilaterally. Note also the substantial degeneration in the enlarged folial spaces of the cerebellum. (Adapted, with permission, from Corkin et al. 1997.)

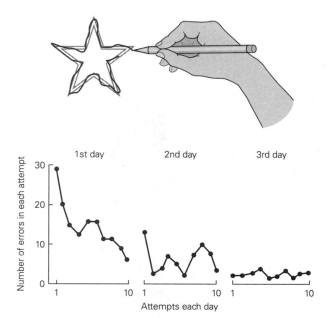

Figure 52–3 The amnesic patient H.M. could learn skilled movements. He was taught to trace between two outlines of a star while viewing his hand in a mirror. The graph plots the number of times, during each attempt, that he strayed outside the outlines as he drew the star. As with healthy subjects, H.M. improved considerably with repeated attempts despite the fact that he had no recollection of performing the task. (Reproduced, with permission, from Blakemore 1977.)

lobe are often referred to as *explicit* (or *declarative*) memory. Explicit memory can be further classified into episodic memory (the memory of personal experiences or autobiographical memory) and semantic memory (memory for facts). *Episodic memory* refers to our ability to remember rich details of moments in time, including information about what happened, when, and where. For example, episodic memory is used to recall that we saw the first flowers of spring yesterday or that we heard Beethoven's "Moonlight Sonata" several months ago. *Semantic memory* is used to recall the meanings of words or concepts, among other facts.

Cognitive psychologists found a similar distinction between different forms of memory in healthy subjects by using tasks that differ in how memories are expressed. One type is a nonconscious form of memory that is evident in the performance of a task. This form of memory is often referred to as *implicit* memory (also referred to as *nondeclarative* or *procedural* memory). Implicit memory is typically manifested in an automatic manner, with little conscious processing on the part of the subject. Different forms give rise to

priming, skill learning, habit memory, and conditioning (Figure 52–5). Explicit memory is considered to be highly flexible; multiple pieces of information can be associated under different circumstances. Implicit memory, however, is tightly connected to the original conditions under which the learning occurred.

The terms "explicit memory" and "implicit memory" are used to describe two broad forms of memory that differ in their hallmark behavioral characteristics and in their neural underpinnings. These forms of memory can be acquired in parallel. For example, one might form an explicit memory of how good a bakery smelled upon entering it yesterday, while at the same time, one might develop an automatic conditioned response of increased salivation upon viewing a picture of the bakery. Moreover, we now believe that these forms of memory, while distinct, normally interact to support behavior, although the precise

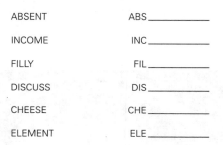

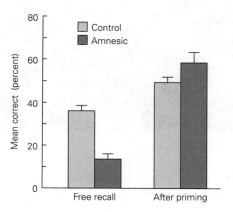

Figure 52–4 Amnesic subjects differ in their ability to recall words under two conditions. Subjects were presented with common words and then asked to recall the words. Amnesic patients did not do well on this test during free recall. However, when subjects were given the first three letters of a word that had been presented and instructed to form the first word that came to mind (word completion), the amnesic subjects performed as well as normal subjects. The baseline guessing rate in the word completion condition for words not previously presented was 9%. (Adapted from Squire 1987.)

Figure 52–5 Long-term memory is commonly classified as either explicit (the memory is reported verbally) or implicit (the memory is expressed through behavior without conscious awareness).

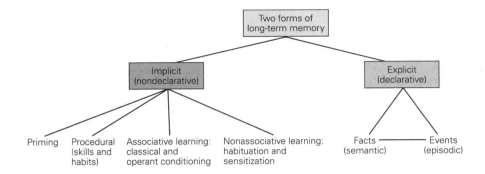

nature and extent of their interactions are a topic of ongoing investigation.

There are also ongoing debates about the role of conscious awareness in memory and about whether it is indeed a necessary feature of memories supported by the medial temporal lobe. These debates are driven by a growing body of work showing that the same medial temporal circuits necessary for explicit memory are also necessary for some forms of implicit memory (as described below). Indeed, although episodic memory is typically assessed by asking subjects to report the content of their memory, it remains unknown whether conscious accessibility is an integral feature of the memories themselves. Nonetheless, the distinction between implicit and explicit memory played an important historical role in differentiating forms of memory and still offers a productive framework for considering the neural bases of memory. Thus, we use the terms "explicit memory" and "implicit memory" here to distinguish these two forms of memory and the classes of subjective experience and behaviors that they are based on. In the following sections, we focus on episodic memory, which has been the target of a great deal of cognitive neuroscience research in both amnesic patients and healthy individuals.

Episodic Memory Processing Involves Encoding, Storage, Retrieval, and Consolidation

Episodic memory has been studied extensively and offers a window into understanding how the brain builds, stores, and retrieves details about episodes in our lives. We now know that the brain does not have a single long-term store of episodic memories. Instead, the storage of any item of knowledge is widely distributed among many brain regions that process different aspects of the content of the memory and can be accessed independently (by visual, verbal, or other sensory clues). Second, episodic memory is mediated by at least four related but distinct types of processing: encoding, storage, consolidation, and retrieval.

Encoding is the process by which new information is initially acquired and processed during the formation of a new memory. The extent of this processing is critically important for determining how well the learned material will be remembered. For a memory to persist and be well remembered, the incoming information must undergo what the psychologists Fergus Craik and Robert Lockhart called "deep" encoding. This is accomplished by attending to the information and associating it with memories that were already established. Memory encoding is also stronger when one is motivated to remember, whether because the information has particular emotional or behavioral relevance (eg, a memory for a particularly delicious meal on an enjoyable first date) or whether the information itself is neutral but is associated with something meaningful (eg, remembering the location of that restaurant).

Storage refers to the neural mechanisms and sites by which the newly acquired information is retained as a lasting memory over time. One of the remarkable features about long-term storage is that it seems to have an almost unlimited capacity. In contrast, working memory storage is very limited; psychologists believe that human working memory can hold only a few pieces of information at any one time.

Consolidation is the process that transforms temporarily stored and still labile information into a more stable form. As we shall learn in the next two chapters, consolidation involves expression of genes and protein synthesis that give rise to structural changes at synapses.

Finally, *retrieval* is the process by which stored information is recalled. It involves bringing back to mind different kinds of information that are stored in different sites. Retrieval of memory is much like perception; it is a constructive process and therefore subject to distortion much as perception is subject to illusions (Box 52–1). When a memory is retrieved, it becomes active again, providing an opportunity for an old memory to be encoded again. Because retrieval is constructive, re-encoding of a retrieved memory can differ from the original memory. For example,

Box 52–1 Episodic Memories Are Subject to Change During Recall

How accurate is episodic memory? This question was explored by the psychologist Frederic Bartlett in a series of studies in the 1930s in which subjects were asked to read stories and then retell them. The recalled stories were shorter and more coherent than the original stories, reflecting reconstruction and condensation of the original.

The subjects were unaware that they were editing the original stories and often felt more certain about the edited parts than about the unedited parts of the retold stories. They were not confabulating; they were merely interpreting the original material so that it made sense on recall.

Observations such as these demonstrate that episodic memory is malleable. Moreover, the fact that people incorporate later edits into their original memories leads us to believe that episodic memory is a constructive

process in the sense that individuals perceive the environment from the standpoint of a specific point in space as well as a specific point in their own history. Much like sensory perception, episodic memory is not a passive recording of the external world but an active process in which incoming bottom-up sensory information is shaped by top-down signals, representing prior experience, along the afferent pathways. Likewise, once information is stored, recall is not an exact copy of the information stored. Past experiences are used in the present as cues that help the brain reconstruct a past event. During recall, we use a variety of cognitive strategies, including comparison, inference, shrewd guessing, and supposition, to generate a memory that seems coherent to us, that is consistent with other memories, and that is consistent with our "memory of the memory."

---

re-encoding can include information from the old memory together with the new context in which it was retrieved. This re-encoding allows memories of separate moments in time to be connected in memory, but it also opens the door to errors in memory, as discussed later in the chapter.

Retrieval of information is most efficient when a retrieval cue reminds individuals of the episodic nature of the events linking the elements of the encoded experience. For example, in a classic behavioral experiment, Craig Barclay and colleagues asked some subjects to encode sentences such as "The man lifted the piano." On a later retrieval test, "something heavy" was a more effective cue for recalling piano than "something with a nice sound." Other subjects, however, encoded the sentence "The man tuned the piano." For them, "something with a nice sound" was a more effective retrieval cue for piano than "something heavy" as it reflected better the initial experience. Retrieval, particularly of explicit memories, also is partially dependent on working memory.

Episodic Memory Involves Interactions Between the Medial Temporal Lobe and Association Cortices

Although studies of amnesic patients during the past few decades have refined our understanding of various types of memory, medial temporal lobe damage affects all four operations of memory—encoding,

storage, consolidation, and retrieval—and thus it is often difficult to discern how the medial temporal lobe contributes to each. fMRI allows us to scan brain activity in the process of building new memories or retrieving existing memories, and thus to identify specific regions that are active during different processes (Chapter 6).

A common method for studying encoding with fMRI is the *subsequent memory paradigm*. In a typical subsequent memory task, a human subject views a series of stimuli (eg, words or pictures) one at a time while being scanned with fMRI, often while engaged in a cover task (eg, determining whether the pictures are in color or black and white). A subject's memory for the stimuli is then tested outside of the scanner, allowing the researchers to sort all the encoding events into those that were later remembered compared to those that were later forgotten. fMRI scans show that remembered items, compared with forgotten items, are associated with greater activity in the hippocampus during encoding. This difference is also evident in simultaneous activity in other parts of the brain, including prefrontal, retrosplenial, and parietal cortices. Often, the activity of these regions covaries on a moment-to-moment basis with the activity in the hippocampus during memory encoding, suggesting that these regions are functionally connected (Figure 52–6).

These fMRI findings, together with findings from patients with amnesia, provide strong support for an

Figure 52–6 In the study illustrated here, neural activity during encoding of visual events (presentation of words) was measured using functional magnetic resonance imaging (fMRI). Subsequently, recall of the studied words was tested, and each word was classified as either remembered or forgotten. The scans taken during encoding were then sorted into two groups: those made during encoding of words that were later remembered and those made during encoding of words that were later forgotten. The activity in regions of the left prefrontal cortex and medial temporal lobe was greater during the encoding of words later remembered than those later forgotten (locations denoted by **white arrows**). At right are the observed fMRI responses in these regions for words later remembered and those later forgotten. (Adapted, with permission, from Wagner et al. 1998.)

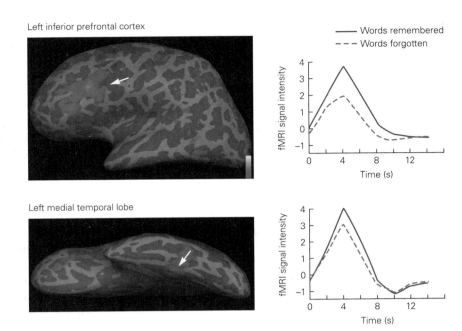

important role of the hippocampus in encoding episodic memories. The fMRI findings also extend the findings from amnesic patients, showing that successful formation of episodic memories depends on interaction between frontoparietal networks and the medial temporal lobe. However, as the medial temporal lobe is a large structure, a key goal is to understand the role of its different subregions. Such information is being provided by higher-resolution fMRI studies that use more powerful brain scanning technologies. These studies reveal that distinct subregions within and outside the hippocampus contribute to different aspects of memory encoding. Thus, whereas some cortical areas surrounding the hippocampus are particularly important for object recognition (perirhinal cortex), others are important for encoding spatial context (parahippocampal cortex). These cortical regions provide strong (but indirect) inputs to the hippocampus proper, which is thought to bind together spatial and object information, forming a unified memory.

Interaction between the medial temporal lobe and widely separated cortical regions is also central in memory consolidation and retrieval. It was initially thought that the hippocampus was not important for retrieval, since patient H.M., whose medial temporal lobe was surgically removed, could still recall childhood memories. In fact, early observations suggested that H.M. could recall many of the experiences of his life up until several years before his operation. These observations of H.M. and other amnesic patients with damage to the medial temporal lobe suggested that old memories must be ultimately stored in various other cortical regions through interaction with the medial temporal lobe. However, even though patients with hippocampal damage like H.M. have some ability to recall older memories, there is evidence that the extent of memory recall may be impaired in these patients. Current thinking suggests that there is a distributed circuit for consolidation and retrieval involving several brain regions, with the hippocampus playing an essential role in the binding of associations during both encoding and retrieval. The cortical regions serve as the long-term repository of the separate elements of information that constitute a memory and in the controlled retrieval and reactivation of the content of the memory itself.

As with studies of encoding, studies of retrieval of episodic knowledge have implicated specific regions of association cortex, frontoparietal networks, and the medial temporal lobe. The retrieval of contextual or event details associated with an episodic memory also involves activity in the hippocampus, with medial temporal lobe retrieval processes facilitating the activation of neocortical representations that were present during encoding.

fMRI scans have a fairly limited time resolution due to the relatively slow time course of changes in blood flow associated with brain activity. To achieve higher temporal resolution of brain activity, researchers can record electrical activity from the human brain using extracellular electrodes. Such recordings are rare and possible only in human patients who are already undergoing brain surgery for medical reasons, such as severe epilepsy, when electrode implantation is used to localize the site of seizure generation. In one study, intracranial electroencephalography (iEEG) signals were measured using subdural electrodes placed in the medial temporal lobe and other areas of cortex. A subject first learned associations between pairs of words and then had to retrieve memories of those associations. The retrieval of memories was associated with neural activity in the hippocampus, coupled with neural activity in temporal association cortex, a region involved in language and multisensory integration. This coupled neural activity was associated with a reactivation of cortical patterns that were initially observed when participants first memorized word pairs. This finding provides a link between the neural activity observed in the hippocampus during initial encoding of a memory and the later coupled activity in the temporal association cortex during retrieval. Related observations of reactivation of encoding patterns during retrieval have been reported in numerous human functional imaging studies, documenting the ubiquity of such effects. As with encoding of episodic memory, retrieval involves a complex interaction between the medial temporal lobe and distributed cortical regions, including frontoparietal networks and other high-level association areas.

Episodic Memory Contributes to Imagination and Goal-Directed Behavior

Memory enables us to use our past experience to predict future events, thus promoting adaptive behavior. Like retrieval of memories, imagination of future events involves construction of details from memory. The first report of a possible connection between memory and imagination came from the case study of patient K.C., as reported by Endel Tulving in 1985. Patient K.C. displayed typical and devastating amnesia as a result of damage to his hippocampus and medial temporal lobe. Similar to patient H.M., he had a complete lack of episodic memory while language and nonepisodic functions were unimpaired. Tulving's studies revealed further that such brain damage was associated with the loss of the ability to imagine events in the future. When asked what he would be doing the next day, K.C. was unable to provide details.

The importance of the hippocampus in imagining future events is also seen with fMRI studies. Such studies examined brain activity of healthy individuals, comparing activity when subjects were asked to remember an event from the past (eg, think of your birthday last year) with activity when they imagined

events in the future (eg, imagine a beach vacation next summer). The subjects were asked to report any vivid details of the event that came to mind. The MRI scans showed a striking overlap in the network of brain regions that were active during memory retrieval and imagination of future events. This network included the hippocampus, prefrontal cortex, posterior cingulate cortex, retrosplenial cortex, and lateral parietal and temporal areas (Figure 52–7).

Further evidence supporting the view that episodic memory and hippocampal function are necessary for planning future behavior comes from a study on human performance of a spatial navigation task using virtual reality simulations. High-resolution fMRI and multivoxel pattern analysis (Chapter 6) showed that activity in the hippocampus was related to simulation of navigation goals. Moreover, hippocampal activity during planning covaried with goal-related activity in prefrontal, medial temporal, and medial parietal cortex (Figure 52–8).

Episodic memory encoding and storage are also influenced by the adaptive value of events. Alison Adcock and colleagues showed that the anticipation of a potential reward can enhance memory by eliciting coordinated activity between the medial temporal lobe and midbrain regions that are rich in dopamine neurons. Reward can also retroactively enhance memories. When human participants navigate a maze for a reward, they have better memory for neutral events that happened right before the reward. The ability to retroactively shape episodic memory based on outcomes is important because the relevance of a specific episode may only become known after the fact. Together with the role of episodic memory in constructing the retrieval of past events and in imagining and simulating future events, the findings on reward support the view that a major function of episodic memory is to guide adaptive behaviors.

The Hippocampus Supports Episodic Memory by Building Relational Associations

In addition to the broad role of the hippocampus in episodic memory, future thinking, and goal-directed behavior, studies of rodents first pointed to a role for the hippocampus in spatial navigation (Chapter 54), findings that were later supported by studies of nonhuman primates and humans. In rodents, single neurons in the hippocampus encode specific spatial information, and lesions of the hippocampus interfere with the animal's memory for spatial location. Functional imaging of the brain in healthy humans shows that activity increases in the right hippocampus when spatial information is recalled and in the left hippocampus when words,

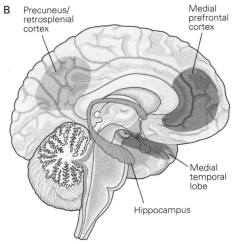

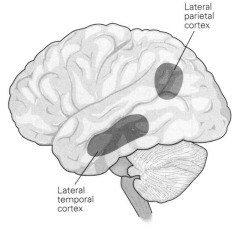

Figure 52–7 Brain regions supporting retrieval of memories for past events and imagination of future events. (Adapted, with permission, from Schacter, Addis, and Buckner 2007.)

A. Subjects were instructed to either remember a personally experienced event in their past or imagine a plausible event in their future while lying inside a functional magnetic resonance imaging scanner. Events are elicited by a cue word (eg, "beach" or "birthday"). Subjective ratings of event phenomenology (eg, vividness and emotionality of the episode) and detailed event descriptions are often obtained in an interview following the scanning in order to confirm that an episodic event was successfully generated.

B. The core brain system that mediates past and future thinking is consistently activated while remembering the past, when envisioning the future, and during related forms of mental simulation. Prominent components of this network include medial prefrontal regions, posterior regions in the medial and lateral parietal cortex (extending into the precuneus and retrosplenial cortex), the lateral temporal cortex, and the medial temporal lobe. Moreover, regions within this core brain system are functionally correlated with each other and with the hippocampus. This core brain system is thought to function adaptively to integrate information about relationships and associations from past experiences to construct mental simulations about possible future events.

objects, or people are recalled. These physiological findings are consistent with the clinical observation that lesions of the right hippocampus differentially give rise to problems with spatial orientation, whereas lesions of the left hippocampus differentially cause deficits in verbal memory.

The fact that the hippocampus supports spatial processing, semantic memory, and episodic memory raises questions about how the hippocampus contributes to such different behaviors. One compelling theory, proposed by Howard Eichenbaum and Neal Cohen, suggests that the hippocampus provides a general mechanism for forming and storing complex multimodal associations. According to this view, the hippocampus binds in memory the separate elements of experiences, encoding events as relational maps of

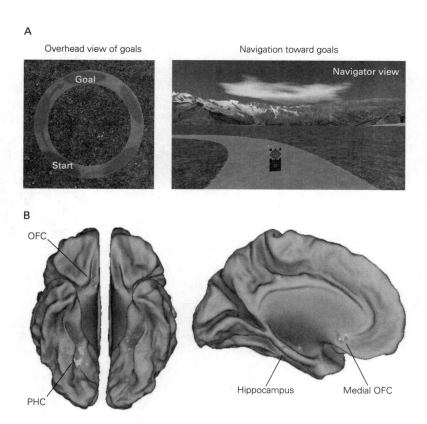

Figure 52–8 Neural circuits supporting memory-based goal-directed navigation. (Reproduced, with permission, from Brown et al. 2016.)

A. Human participants navigate to goals in a virtual reality environment while being scanned with functional magnetic resonance imaging. They first explore the space and learn where goals are located and then are tested on their ability to navigate to specific goals.

B. Navigational planning elicits goal-related activity in a core network including the hippocampus, medial temporal lobe, parahippocampal cortex (**PHC**), and orbitofrontal cortex (**OFC**).

items within spatial and temporal contexts, thus composing a "memory space" that can distinguish distinct episodes, or sequences of events, even when the same (or similar) events occur in different episodes (Figure 52–9). As discussed later in this chapter, the view that the hippocampus encodes relations offers insights into the mechanism by which memories are built and explains why, in some cases, the hippocampus may contribute to memory processes that are not consciously accessible but do encode relations.

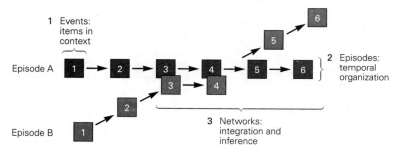

Figure 52–9 The hippocampus supports relational processing underlying episodic memory. A conceptual illustration of a memory space designating three key types of relational processing: events, episodes, and networks. The schematic illustrates processing of two distinct episodes (Episode A and Episode B), which have both distinct and overlapping elements. For example, the episodes might be two distinct visits to an Italian restaurant on separate evenings with the same friend. The evenings are experienced as distinct (different days, different weather, different moods), yet they share some overlap (the company of the same friend at the same restaurant). *Events* (**1**) are defined as items (objects,

behaviors) that are associated with the context in which they occurred (denoted here as events 1 to 6 in each episode, such as the specific table you sat at, the food you ordered, etc.). *Episodes* (**2**) are defined in this view as the temporal organization of these events. While most of the items in each episode are unique, some of them overlap (here, items 3 and 4; in the example, your friend and the restaurant). Relational *networks* (**3**) are formed via associations between events and episodes by way of the overlapping events, supporting the capacity for links between indirectly related events. (Reproduced, with permission, from Eichenbaum and Cohen 2014. Copyright © 2014 Elsevier Inc.)

Implicit Memory Supports a Range of Behaviors in Humans and Animals

Just as there are many ways in which explicit memory guides behavior, there are also many ways in which nonexplicit forms of memory, those without conscious awareness, can influence behavior. Implicit memory refers to forms of knowledge that guide behavior without conscious awareness. Priming, for example, is the automatic influence of exposure to one cue on processing of a later cue.

Priming can be classified as conceptual or perceptual. *Conceptual priming* provides enhanced access to task-relevant semantic knowledge because that knowledge has been used before. It is correlated with decreased activity in left prefrontal regions that subserve initial retrieval of semantic knowledge. In contrast, *perceptual priming* occurs within a specific sensory modality and depends on cortical modules that operate on sensory information about the form and structure of words and objects.

Damage to unimodal sensory regions of cortex impairs modality-specific perceptual priming. For example, one patient with an extensive surgical lesion of the right occipital lobe failed to demonstrate visual priming for words but had normal explicit memory (Figure 52–10). This condition is the reverse of that found in amnesic patients such as H.M., suggesting that the neural mechanisms of priming are distinct from those for explicit memory. The fact that perceptual priming can be intact in patients with amnesia due to medial temporal damage further suggests that it is distinct from explicit memory.

Different Forms of Implicit Memory Involve Different Neural Circuits

Other forms of implicit memory subserve the learning of habits and motor, perceptual, and cognitive skills and the formation and expression of conditioned responses. In general, these forms of implicit memory are characterized by incremental learning, which proceeds gradually with repetition and, in some cases, is driven by reinforcement.

The learning of habits, motor skills, and conditioned responses can take place independently of the medial temporal lobe system. For example, H.M. was able to acquire new visuomotor skills, like the mirror-tracing task (see Figure 52–3). Therefore, early theories posited that these forms of memory generally do not depend on the medial temporal lobe but, rather, depend on the basal ganglia and cerebellum (see Chapters 37 and 38). However, subsequent work

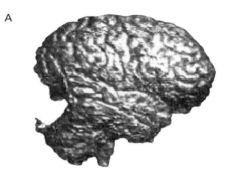

A

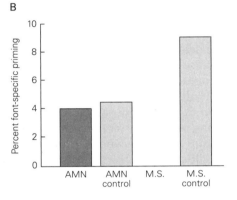

B

Figure 52–10 The right occipital cortex is required for visual priming for words. (Adapted from Vaidya et al. 1998.)

A. Structural magnetic resonance imaging depicts the near-complete removal of the right occipital cortex in a patient, M.S., who suffered pharmacologically intractable epilepsy with a right occipital cortical focus.

B. Font-specific priming is a form of visual priming in which the individual is better able to identify a briefly flashed word when the type font is identical to an earlier presentation, compared to identification when the font is different. Priming is measured as performance when the font is the same minus performance when the font is different. Font-specific priming is intact in amnesic patients (**AMN**) and their controls as well as in the controls for patient M.S., but not in M.S. himself. The patient M.S. has normal explicit memory, even for visual cues (data not shown), but lacks implicit memory for specific properties of visually presented words.

suggested that this is not a general rule and that the medial temporal lobe is required for forms of implicit learning that store relational associations, even when such associations are learned through repetition and appear to take place without conscious awareness.

It is now thought that several kinds of incremental implicit learning involve the medial temporal lobes. For example, Turk-Browne and colleagues investigated implicit learning of regularities between visual cues, called statistical learning. In a typical statistical learning task, human subjects are presented with a stream of sounds or images that follow a structured sequence or "grammar" of repetitions. Learning of the sequence is typically measured by a faster reaction time to repeated compared to nonrepeated sequences. At first glance, it would appear that statistical learning should not involve the medial temporal lobe: The learning is nonverbal, it does not require conscious thought and is therefore implicit, and it is assumed to reflect the accumulated computation of probabilistic relationships across multiple episodes, rather than the specific memory of one episode. Yet fMRI studies show that the hippocampus is active during statistical learning, and damage to the medial temporal lobe has been found to impair performance on this implicit task.

Statistical learning is an example of how learning takes place through repetition. New perceptual, motor, or cognitive abilities are also learned through repetition. With practice, performance becomes more accurate and faster, and these improvements generalize to learning novel information. Skill learning moves from a cognitive stage, where knowledge is represented explicitly and the learner must pay a great deal of attention to performance, to an autonomous stage, where the skill can be executed without much conscious attention. As an example, driving a car initially requires that one be consciously aware of each component of the skill, but after practice, one no longer attends to the individual components.

The learning of sensorimotor skills depends on numerous brain regions that vary with the specific associations being learned. As we learned in Chapter 38, these include the basal ganglia, cerebellum, and neocortex. Dysfunction of the basal ganglia in patients with Parkinson and Huntington disease impairs learning of motor skills. Patients with cerebellar lesions also have difficulties acquiring some motor skills. Functional imaging of healthy individuals during sensorimotor learning shows changes in the activity of the basal ganglia and cerebellum and their connectivity with cortical regions. Danielle Bassett and colleagues have used network-analysis algorithms applied to whole-brain fMRI data to characterize dynamic changes in network functional connectivity that take place during motor skill learning. Finally, skilled behavior can depend on structural changes in motor neocortex, as seen by the expansion of the cortical representation of the fingers in musicians (Chapter 53).

Habits emerge from the repeated association of cues or actions with rewarding outcomes. Habit learning in humans is studied with tasks that involve incremental learning of stimulus–reward associations. In a typical task, subjects perform a series of trials in which they are asked to choose among visual cues and receive trial-by-trial feedback on their choice. The relationship between the cues and the feedback varies probabilistically over the course of the task so that participants must keep updating their responses based on the feedback. Because learning takes place over numerous trials, explicit memory of any one specific trial may not be as useful for successful performance as the gradual accumulation of feedback-driven learning of stimulus–outcome associations.

fMRI studies demonstrate that incremental learning of stimulus–reward associations depends on the striatum, the area of the basal ganglia that receives input from neocortex, and its modulatory dopaminergic inputs. Patients with a loss of striatal dopamine, as occurs in Parkinson disease, are less effective at learning based on trial-by-trial reinforcement. These findings are consistent with other studies that indicate dopamine has an important role in modulating cortico-striatal circuitry for reinforcement learning (see Chapter 38).

At first glance, stimulus-reward learning appears to be precisely the sort of learning that does not depend on the medial temporal lobe: It is implicit rather than explicit, and it occurs gradually rather than through an explicit memory for a single event. Indeed, early theories posited that learning probabilistic stimulus–reward associations does not depend on the medial temporal lobe. However, subsequent work has revealed that the hippocampus does contribute to stimulus–reward learning under some circumstances, such as when the task demands learning of more complex stimulus–stimulus associations (Figure 52–11). The contribution of the hippocampus to implicit learning takes place via interactions with other cortical and subcortical circuits. fMRI studies show functional connectivity between the hippocampus and the striatum in support of learning across a variety of tasks. Interactions between the hippocampus and the striatum are sometimes competitive and sometimes cooperative, depending on the demands of the task.

Implicit Memory Can Be Associative or Nonassociative

Some forms of implicit memory have also been studied in nonhuman animals, and these animal studies have distinguished two types of implicit memory: nonassociative and associative. With nonassociative learning, an animal learns about the properties of a single stimulus. With associative learning, the animal learns about the relationship between two stimuli or between a stimulus and a behavior. We consider the cellular mechanisms of implicit memory in animals in the next chapter.

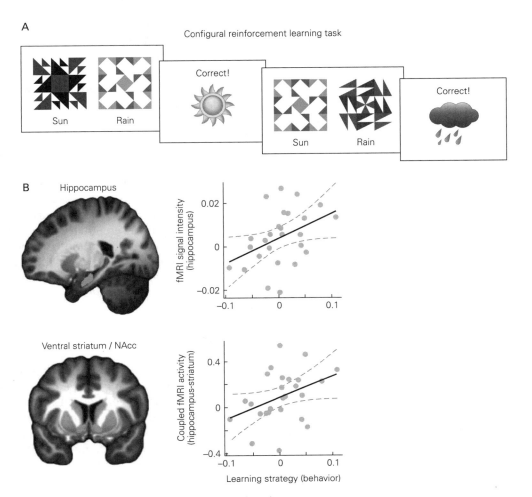

Figure 52–11 Learning stimulus–response associations involves both the striatum and the hippocampus. (Adapted, with permission, from Duncan et al. 2018.)

A. Participants use trial-by-trial reinforcement to learn to predict outcomes (rain or sun) based on cues (colorful shapes). The cues have a probabilistic relation to each weather outcome that the viewer learns by trial and error. The weather can be predicted based either on each individual cue or on the combined presentation of the two cues (their configuration). Reinforcement learning models can discern which strategy each subject uses.

B. The striatum is known to play a critical role in learning to update choices based on reinforcement. When subjects learn

about the configuration, this same task also elicits activity in the hippocampus and increased coupling of activity in the hippocampus and the striatum. Scatter plots show that the extent to which subjects use a configural learning strategy correlates with blood oxygenation level–dependent activity in the hippocampus and with functional coupling between the hippocampus and the striatum. The images show activity in hippocampus and the nucleus accumbens (**NAcc**), a region in the ventral portion of the striatum that responds to rewarding stimuli. (Abbreviation: **fMRI**, functional magnetic resonance imaging.)

Nonassociative learning results when a subject is exposed once or repeatedly to a single type of stimulus. Two forms of nonassociative learning are common in everyday life: habituation and sensitization. Habituation is a decrease in a response that occurs when a benign stimulus is presented repeatedly. For example, most people in the United States are startled when they first hear the sound of a firecracker on Independence Day, but as the day progresses, they become

accustomed to the noise and do not respond. Sensitization (or pseudo-conditioning) is an enhanced response to a wide variety of stimuli after the presentation of an intense or noxious stimulus. For example, an animal will respond more vigorously to a mild tactile stimulus after receiving a painful pinch. Moreover, a sensitizing stimulus can override the effects of habituation, a process called dishabituation. For example, after the startle response to a noise has been reduced by habituation, one can restore the intensity of response to the noise by delivering a strong pinch.

With sensitization and dishabituation, the timing of stimuli is not important because no association between stimuli must be learned. In contrast, with two forms of associative learning, the timing of the stimuli to be associated is critical. Classical conditioning involves learning a relationship between two stimuli, whereas operant conditioning involves learning a relationship between the organism's behavior and the consequences of that behavior.

Classical conditioning was first described in the early 1900s by the Russian physiologist Ivan Pavlov. The essence of classical conditioning is the pairing of two stimuli: a conditioned stimulus and an unconditioned stimulus. The conditioned stimulus (CS), such as a light, a tone, or a touch, is chosen because it produces either no overt response or a weak response usually unrelated to the response that eventually will be learned. The unconditioned stimulus (US), such as food or a shock, is chosen because it normally produces a strong and consistent response (the unconditioned response), such as salivation or withdrawal of a limb. Unconditioned responses are innate; they are produced without learning. Repeated presentation of a CS followed by a US gradually elicits a new or different response called the conditioned response.

One way of explaining conditioning is that repeated pairing of the CS and US causes the CS to become an anticipatory signal for the US. With sufficient experience, an animal will respond to the CS as if it were anticipating the US. For example, if a light is followed repeatedly by the presentation of meat, eventually the sight of the light itself will make the animal salivate. Thus, classical conditioning is one way an animal learns to predict events.

The probability that an established conditioned response will occur decreases if the CS is repeatedly presented without the US. This process is known as extinction. If a light that has been paired with food is later repeatedly presented in the absence of food, it will gradually cease to evoke salivation. Extinction is an important adaptive mechanism; it would be maladaptive for an animal to continue to respond to cues that are no longer meaningful. The available evidence indicates that extinction is not the same as forgetting; instead, something new is learned—the CS now signals that the US will not occur.

For many years, psychologists thought that classical conditioning resulted as long as the CS preceded the US within a critical time interval. According to this view, each time a CS is followed by a US (reinforcing stimulus), a connection is strengthened between the

internal representations of the stimulus and response or between the representations of one stimulus and another. The strength of the connection was thought to depend on the number of pairings of CS and US. A substantial body of evidence now indicates that classical conditioning cannot be adequately explained simply by the fact that two events or stimuli occur one after the other (Figure 52-12). Indeed, it would not be adaptive to depend solely on sequence. Rather, all animals capable of associative conditioning, from snails to humans, remember the salient relationship between associated events. Thus, classical conditioning, and perhaps all forms of associative learning, enables animals to distinguish events that reliably occur together from those that are only randomly associated.

Lesions in several regions of the brain affect classical conditioning. One well-studied example is conditioning of the protective eyeblink reflex, a form of motor learning. A puff of air to the eye naturally causes an eyeblink. A conditioned eyeblink can be established by pairing the puff with a tone that precedes the puff. Studies in rabbits indicate that the conditioned response (an eyeblink in response to a tone) is abolished by a lesion at either of two sites. Damage to the vermis of the cerebellum abolishes the conditioned response but does not affect the unconditioned response (eyeblink in response to a puff of air). Interestingly, neurons in the same area of the cerebellum show learning-dependent increases in activity that closely parallel the development of the conditioned behavior. A lesion in the interpositus nucleus, a deep cerebellar nucleus, also abolishes the conditioned eyeblink. Thus, both the vermis and the deep nuclei of the cerebellum play an important role in conditioning the eyeblink, and perhaps other simple forms of classical conditioning involving skeletal muscle movement.

Another well-studied example is fear conditioning, which depends on the amygdala. In fear conditioning, a neutral cue, such as a tone, is paired with an aversive outcome, such as a shock. This pairing leads to a conditioned fear response in which the neutral tone alone elicits a behavioral reaction, such as freezing. Fear conditioning depends on plasticity in the inputs to and connections between the subnuclei of the amygdala, particularly the basolateral amygdala, as we will discuss in the next chapter.

Operant Conditioning Involves Associating a Specific Behavior With a Reinforcing Event

A second major paradigm of associative learning, discovered by Edgar Thorndike and systematically studied by B. F. Skinner and others, is operant conditioning

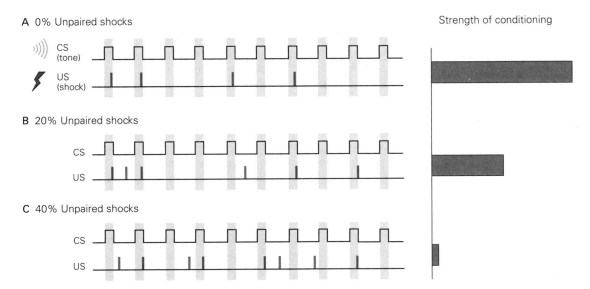

Figure 52–12 Classical conditioning depends on the degree of correlation of two stimuli. In this experiment with rats, a tone (the conditioned stimulus [CS]) was paired with an electric shock (the unconditioned stimulus [US]) in four out of 10 of the trials (**red ticks**). In some trial blocks, the shock was presented without the tone (**green ticks**). Suppression of lever-pressing to obtain food is a sign of freezing, a conditioned defensive response. The degree of conditioning was evaluated by determining how effective the tone alone was in suppressing lever-pressing to obtain food. (Adapted from Rescorla 1968.)

A. Maximal conditioning occurred when the US was presented only with the CS.

B–C. Little or no conditioning occurred when the shock was presented without the tone almost as often as with it (40%). Some conditioning occurred when the shock occurred without the tone 20% of the time.

(also called trial-and-error learning). In a typical laboratory example of operant conditioning, a hungry rat or pigeon is placed in a test chamber in which the animal is rewarded for a specific action. For example, the chamber may have a lever protruding from one wall. Because of previous learning, or through play and random activity, the animal will occasionally press the lever. If the animal promptly receives a positive reinforcer (eg, food) after pressing the lever, it will begin to press the lever more often than the spontaneous rate. The animal can be described as having learned that among its many behaviors (eg, grooming, rearing, and walking) one behavior is followed by food. With this information, the animal is likely to press the lever whenever it is hungry.

If we think of classical conditioning as the formation of a predictive relationship between two stimuli (the CS and the US), operant conditioning can be considered as the formation of a predictive relationship between an action and an outcome. Unlike classical conditioning, which tests the responsiveness of a reflex to a stimulus, operant conditioning tests behavior that occurs either spontaneously or without an identifiable stimulus. Thus, operant behaviors are said to be emitted rather than elicited. In general, actions that are rewarded tend to be repeated, whereas actions followed by aversive, although not necessarily painful, consequences tend not to be repeated. Many experimental psychologists think that this simple idea, called the law of effect, governs much voluntary behavior.

Operant and classical conditioning involve different kinds of association—an association between an action and a reward or between two stimuli, respectively. However, the laws of operant and classical conditioning are quite similar. For example, timing is critical in both. In operant conditioning, the reinforcer usually must closely follow the operant action. If the reinforcer is delayed too long, only weak conditioning occurs. Similarly, classical conditioning is generally poor if the interval between the CS and US is too long or if the US precedes the CS.

Associative Learning Is Constrained by the Biology of the Organism

Animals generally learn to associate stimuli that are relevant to their survival. For example, animals readily learn to avoid certain foods that have been followed by a negative reinforcement (eg, nausea produced by a poison), a phenomenon termed *taste aversion*.

Unlike most other forms of conditioning, taste aversion develops even when the unconditioned response (poison-induced nausea) occurs after a long delay, up to hours after the CS (specific taste). This makes biological sense because the ill effects of infected foods and naturally occurring toxins usually follow ingestion only after some delay. For most species, including humans, taste-aversion conditioning occurs only when certain tastes are associated with illness. Taste aversion develops poorly if a taste is followed by a painful stimulus that does not produce nausea. Also, animals do not develop an aversion to a visual or auditory stimulus that has been paired with nausea.

Errors and Imperfections in Memory Shed Light on Normal Memory Processes

Memory allows us to revisit our personal past; provides access to a vast network of facts, associations, and concepts; and supports learning and adaptive behavior. But memory is not perfect. We often forget events rapidly or gradually, sometimes distort the past, and occasionally remember events that we would prefer to forget. In the 1930s, the British psychologist Frederic Bartlett reported experiments in which people read and tried to remember complex stories. He showed that people often misremember many features of the stories, often distorting information based on their expectations of what should have happened. Forgetting and distortion can provide important insights into the workings of memory.

Memory's imperfections have been classified into seven basic categories, dubbed the "seven sins of memory": transience, absent-mindedness, blocking, misattribution, suggestibility, bias, and persistence. Here, we focus on six of these.

Absent-mindedness results from a lack of attention to immediate experience. Absent-mindedness during encoding is a likely source of common memory failures such as forgetting where one recently placed an object. Absent-mindedness also occurs when we forget to carry out a particular task such as picking up groceries on the way home from the office, even though we initially encoded the relevant information.

Blocking refers to a temporary inability to access information stored in memory. People often have partial awareness of a sought-after word or image but are nonetheless unable to recall the entire word accurately or completely. Sometimes, it feels like a blocked word is on "the tip of the tongue"—we are aware of the initial letter of the word, the number of syllables in it, or a like-sounding word. Determining which information

is correct and which is incorrect requires a great deal of conscious effort.

Absent-mindedness and blocking are sins of omission: At a moment when we need to remember information, it is inaccessible. However, memory is also characterized by sins of commission, situations in which some form of memory is present but wrong.

Misattribution refers to the association of a memory with an incorrect time, place, or person. False recognition, a type of misattribution, occurs when individuals report that they "remember" items or events that never happened. Such false memories have been documented in controlled experiments where people claim to have seen or heard words or objects that had not been presented previously but are similar in meaning or appearance to what was actually presented. Studies using positron emission tomography imaging and fMRI have shown that many brain regions show similar levels of activity during both true and false recognition, which may be one reason why false memories sometimes feel like real ones.

Suggestibility refers to the tendency to incorporate new information into memory, usually as a result of leading questions or suggestions about what may have been experienced. Research using hypnotic suggestion indicates that various kinds of false memories can be implanted in highly suggestible individuals, such as remembering hearing loud noises at night. Studies with young adults have also shown that repeated suggestions about a childhood experience can produce memories of events that never occurred. These findings are important theoretically because they highlight that memory is not simply a "playback" of past experiences (Box 52–1). Despite these important theoretical and practical implications, next to nothing is known about the neural bases of suggestibility.

Bias refers to distortions and unconscious influences on memory that reflect one's general knowledge and beliefs. People often misremember the past to make it consistent with what they presently believe, know, or feel. This idea is consistent with the idea of "predictive coding" supported by studies showing that even low-level neural mechanisms of perception and sensation are shaped by expectations. The specific brain mechanisms by which expectations influence memory are not well understood.

Persistence refers to obsessive memory, constant remembering of information or events that we might want to forget. Neuroimaging studies have illuminated some neurobiological factors that contribute to persistent emotional memories. Some key results implicate the amygdala, the almond-shaped structure near the hippocampus long known to be involved in emotional processing (Chapter 42). Studies indicate that the level of recall of emotional components of a story is correlated with the level of activity in the amygdala during presentation of the story. Related studies implicate the amygdala in the encoding and retrieval of emotionally charged experiences that can repeatedly intrude into consciousness.

Although persistence can be disabling, it also has adaptive value. The persistence of memories of disturbing experiences increases the likelihood that we will recall information about arousing or traumatic events at times when it may be crucial for survival.

Indeed, many memory imperfections may have adaptive value. False memories and suggestibility may both be related to one of the most basic adaptive functions of memory: the integration of experiences separated in time into a network of learned associations. For memory to play an important role in guiding future behavior, it must be flexible so that we can leverage past experiences to make inferences about future events even when the circumstances have changed. Similarly, although the various forms of forgetting (transience, absent-mindedness, and blocking) can be annoying, a memory system that automatically retains every detail of every experience could result in an overwhelming clutter of useless trivia. This is exactly what happened in the fascinating case of Shereshevski, a mnemonist studied by the Russian neuropsychologist Alexander Luria and described in the book *The Mind of a Mnemonist*. Shereshevski was filled with highly detailed memories of his past experiences but was unable to generalize or to think at an abstract level. A healthy memory system does not encode, store, and retrieve all the details of every experience. Thus, transience, absent-mindedness, and blocking allow us to avoid the unfortunate fate of Shereshevski.

Highlights

1. Different forms of learning and memory can be distinguished behaviorally and neurally. Working memory maintains goal-relevant information for short periods. Explicit (or declarative) memory involves two classes of knowledge: episodic memory, which represents personal experiences, and semantic memory, which represents general knowledge and facts. Implicit memory includes forms of perceptual and conceptual priming, as well as the learning of motor and perceptual skills, perceptual regularities, and reinforced habits.

2. Encoding, storage, retrieval, and consolidation of new explicit memories depend on interactions between specific regions within the neocortex and medial temporal lobe and specific hippocampal subregions. The initiation of long-term storage of explicit memory requires the temporal lobe system, as highlighted by studies of amnesic patients such as H.M. Consolidation processes stabilize stored representations, rendering explicit memories less dependent on the medial temporal lobe. Retrieval of explicit memories involves the medial temporal lobe, as well as frontoparietal networks that subserve attention and cognitive control.

3. Multiple processes interact to support memory-guided behavior. Retrieval of episodic memory guides the imagining of future events, which is important for making decisions about future choices and actions. Motivationally significant events are prioritized in memory through the enhancement of encoding, storage, and consolidation processes. Motivation also impacts retrieval, perhaps through different mechanisms of prioritization.

4. Implicit memory emerges automatically in the course of perceiving, thinking, and acting. It tends to be inflexible and expressed in the performance of tasks even without conscious awareness. Implicit memory involves a wide variety of brain regions and circuits, including cortical areas that support the specific perceptual, conceptual, or motor systems recruited to process a stimulus or perform a task, as well as the striatum and the amygdala. Implicit learning that involves the encoding of relational associations additionally involves the hippocampus.

5. Imperfections and errors in remembering provide telltale clues about learning and memory mechanisms. The past can be forgotten or distorted, indicating that memory is not a faithful record of all details of every experience. Retrieved memories are the result of a complex interplay among various brain regions and can be reshaped over time by multiple influences. Various forms of forgetting and distortion tell us much about the flexibility of memory that allows the brain to adapt to the physical and social environment.

Daphna Shohamy
Daniel L. Schacter
Anthony D. Wagner

Suggested Reading

Baddeley AD. 1986. *Working Memory*. Oxford: Oxford Univ. Press.

Eichenbaum H. 2017. Prefrontal-hippocampal interactions in episodic memory. Nat Rev Neurosci 18:547–558.

Eichenbaum H, Cohen NJ. 2001. *From Conditioning to Conscious Recollection: Memory Systems of the Brain*. Oxford: Oxford Univ. Press.

Kamin LJ. 1969. Predictability, surprise, attention, and conditioning. In: BA Campbell, RM Church (eds). *Punishment and Aversive Behavior*, pp. 279–296. New York: Appleton–Century–Crofts.

Kumaran D, Hassabis D, McClelland JL. 2016. What learning systems do intelligent agents need? Complementary learning systems theory updated. Trends Cog Sci 20:512–534.

Milner B, Squire LR, Kandel ER. 1998. Cognitive neuroscience and the study of memory. Neuron 20:445–468.

Schacter DL, Benoit RG, Szpunar KK. 2017. Episodic future thinking: mechanisms and functions. Curr Opin Behav Sci 17:41–50.

Shohamy D, Turk-Browne NB. 2013. Mechanisms for widespread hippocampal involvement in cognition. J Exp Psychol Gen 142:1159–1170.

Tulving E. 1983. *Elements of Episodic Memory*. Oxford: Oxford Univ. Press.

Yonelinas AP, Ranganath C, Ekstrom A, Wiltgen B. 2019. A contextual binding theory of episodic memory: systems consolidation reconsidered. Nat Rev Neurosci 20:364–375.

References

Adcock RA, Thangavel A, Whitfield-Gabrieli S, Knutson B, Gabrieli JD. 2006. Reward motivated learning: mesolimbic activation precedes memory formation. Neuron 50:507–517.

Bartlett FC. 1932. *Remembering: A Study in Experimental and Social Psychology*. Cambridge: Cambridge Univ. Press.

Blakemore C. 1977. *Mechanics of the Mind*. Cambridge: Cambridge Univ. Press.

Brewer JB, Zhao Z, Desmond JE, et al. 1998. Making memories: brain activity that predicts how well visual experience will be remembered. Science 281:1185–1187.

Brown TI, Carr VA, LaRocque KF, et al. 2016. Prospective representation of navigational goals in the human hippocampus. Science 352:1323–1326.

Corkin S. 2002. What's new with the amnesic patient H.M.? Nat Rev Neurosci 3:153–160.

Corkin S, Amaral DG, González RG, et al. 1997. H.M.'s medial temporal lobe lesion: findings from magnetic resonance imaging. J Neurosci 17:3964–3979.

Craik FIM, Lockhart RS. 1972. Levels of processing: a framework for memory research. J Verb Learn Verb Behav 11:671–684.

Duncan K, Doll BB, Daw ND, Shohamy D. 2018. More than the sum of its parts: a role for the hippocampus in configural reinforcement learning. Neuron 98:646–657.

Eichenbaum H, Cohen NJ. 2014. Can we reconcile the declarative memory and spatial navigation views on hippocampal function? Neuron 83:764–770.

Eldridge LL, Knowlton BJ, Furmanski CS, et al. 2000. Remembering episodes: a selective role for the hippocampus during retrieval. Nat Neurosci 3:1149–1152.

Hebb DO. 1966. *A Textbook of Psychology*. Philadelphia: Saunders.

Luria AR. 1968. *The Mind of a Mnemonist*. New York: Basic Books.

Naya Y, Yoshida M, Miyashita Y. 2001. Backward spreading of memory-related signal in the primate temporal cortex. Science 291:661–664.

Nyberg L, Habib R, McIntosh AR, Tulving E. 2000. Reactivation of encoding-related brain activity during memory retrieval. Proc Natl Acad Sci U S A 97:11120–11124.

Pavlov IP. 1927. *Conditioned Reflexes: Investigation of the Physiological Activity of the Cerebral Cortex*. GV Anrep (transl). London: Oxford Univ. Press.

Penfield W. 1958. Functional localization in temporal and deep sylvian areas. Res Publ Assoc Res Nerv Ment Dis 36:210–226.

Petrides M. 1994. Frontal lobes and behavior. Curr Opin Neurobiol 4:207–211.

Poldrack RA, Clark J, Pare-Blagoev EJ, et al. 2001. Interactive memory systems in the human brain. Nature 414:546–550.

Rainer G, Asaad WF, Miller EK. 1998. Memory fields of neurons in the primate prefrontal cortex. Proc Natl Acad Sci U S A 95:15008–15013.

Rescorla RA. 1968. Probability of shock in the presence and absence of CS in fear conditioning. J Comp Physiol Psychol 66:1–5.

Rescorla RA. 1988. Behavioral studies of Pavlovian conditioning. Annu Rev Neurosci 11:329–352.

Schacter DL. 2001. *The Seven Sins of Memory: How the Mind Forgets and Remembers*. Boston and New York: Houghton Mifflin.

Schacter DL, Addis DR. 2007. The cognitive neuroscience of constructive memory: remembering the past and imagining the future. Philos Trans Roy Soc B 362:773–786.

Schacter DL, Addis DR, Buckner RL. 2007. Remembering the past to imagine the future: the prospective brain. Nat Rev Neurosci 8:657–661.

Schacter DL, Guerin SA, St. Jacques PL. 2011. Memory distortion: an adaptive perspective. Trends Cog Sci 15:467–474.

Sestieri C, Shulman GL, Corbetta M. 2017. The contribution of the human posterior parietal cortex to episodic memory. Nat Rev Neurosci 18:183–192.

Shohamy D, Adcock RA. 2010. Dopamine and adaptive memory. Trends Cog Sci 14:464–472.

Skinner BF. 1938. *The Behavior of Organisms: An Experimental Analysis*. New York: Appleton–Century–Crofts.

Squire LR. 1987. *Memory and Brain*. New York: Oxford Univ. Press.

Thorndike EL. 1911. *Animal Intelligence: Experimental Studies*. New York: Macmillan.

Tomita H, Ohbayashi M, Nakahara K, et al. 1999. Top-down signal from prefrontal cortex in executive control of memory retrieval. Nature 401:699–703.

Tulving E, Schacter DL. 1990. Priming and human memory systems. Science 247:301–306.

Uncapher M, Wagner AD. 2009. Posterior parietal cortex and episodic encoding: insights from fMRI subsequent memory effects and dual attention theory. Neurobiol Learn Mem 91:139–154.

Vaidya CJ, Gabrieli JD, Verfaellie M, et al. 1998. Font-specific priming following global amnesia and occipital lobe damage. Neuropsychology 12:183–192.

Vaz AP Inati SK, Brunel N, Zaghloul KA. 2019. Coupled ripple oscillations between the medial temporal lobe and neocortex retrieve human memory. Science 363: 975–978.

Wagner AD. 2002. Cognitive control and episodic memory: contributions from prefrontal cortex. In: LR Squire, DL Schacter (eds). *Neuropsychology of Memory*, 3rd ed., pp. 174–192. New York: Guilford Press.

Wagner AD, Schacter DL, Rotte M, et al. 1998. Building memories: remembering and forgetting of verbal experiences as predicted by brain activity. Science 281:1188–1191.

Wheeler ME, Petersen SE, Buckner RL. 2000. Memory's echo: vivid remembering reactivates sensory-specific cortex. Proc Natl Acad Sci U S A 97:11125–11129.

Wimmer GE, Shohamy D. 2012. Preference by association: how memory mechanisms in the hippocampus bias decisions. Science 338:270–273.

53

Cellular Mechanisms of Implicit Memory Storage and the Biological Basis of Individuality

T HROUGHOUT THIS BOOK WE HAVE EMPHASIZED that all behavior is a function of the brain and that malfunctions of the brain produce characteristic disturbances of behavior. Behavior is also shaped by experience. How does experience act on the neural circuits of the brain to change behavior? How is new information acquired by the brain, and once acquired, how is it stored, retrieved, and remembered?

In the previous chapter, we saw that memory is not a single process but has at least two major forms. Implicit memory operates unconsciously and automatically, as in the memory for conditioned responses, habits, and perceptual and motor skills, whereas explicit memory operates consciously, as in the memory for people, places, and objects. The circuitry for long-term memory storage differs between explicit and implicit memory. Long-term storage of explicit memory begins in the hippocampus and the medial temporal lobe of the neocortex, whereas long-term storage of different types of implicit memory requires a family of neural structures: the neocortex for priming, the striatum for skills and habits, the amygdala for Pavlovian threat conditioning (also known as fear conditioning), the cerebellum for learned motor skills, and certain reflex pathways for nonassociative learning such as habituation and sensitization (Figure 53–1).

Over time, explicit memories are transferred to different regions of the neocortex. In addition, many cognitive, motor, and perceptual skills that we initially store as explicit memory ultimately become so ingrained with practice that they become stored as implicit memory. The transference from explicit to implicit memory and the difference between them is dramatically demonstrated in the case of the English musician and conductor Clive Waring, who in 1985 sustained a viral infection of his brain (herpes encephalitis) that affected

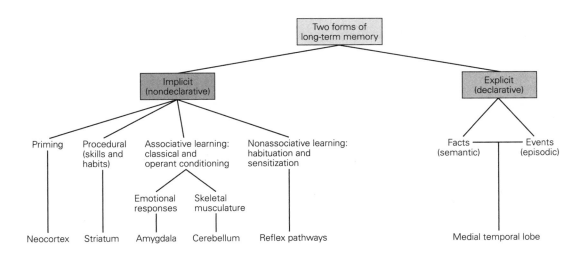

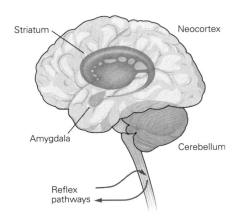

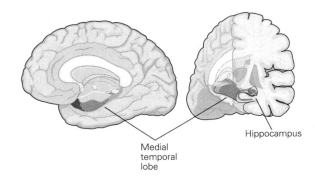

Figure 53–1 Two forms of long-term memory involve different brain systems. Implicit memory involves the neocortex, striatum, amygdala, cerebellum, and, in the simplest cases, the reflex pathways themselves. Explicit memory requires the medial temporal lobe and the hippocampus, as well as certain areas of neocortex (not shown).

the hippocampus and temporal cortex. Waring was left with a devastating loss of memory for events or people he had encountered even a minute or two earlier, but his ability to read music, play the piano, or conduct a chorale was unaffected. Once a performance was completed, however, he could not remember a thing about it.

Similarly, the abstract expressionist painter William de Kooning developed severe disturbances of explicit memory as a result of Alzheimer disease. As the disease progressed and his memory for people, places, and objects deteriorated, he nevertheless continued to produce important and interesting paintings. This aspect of his creative personality was relatively untouched.

In this chapter, we examine the cellular and molecular mechanisms that underlie implicit memory storage in invertebrate and vertebrate animals. We focus on learning about threats (sometimes called fear learning). Implicit memory for motor skills and habits in mammals involving the cerebellum and basal ganglia was considered in Chapters 37 and 38. In the next chapter, we examine the biology of explicit memory in mammals.

Storage of Implicit Memory Involves Changes in the Effectiveness of Synaptic Transmission

Studies of elementary forms of implicit learning—habituation, sensitization, and classical conditioning—provided the conceptual framework for investigating the neural mechanisms of memory storage. Such learning has been analyzed in simple invertebrates and in a variety of vertebrate behaviors, such as the flexion and eye blink reflexes, and also defensive behaviors such as freezing. These simple forms of implicit memory involve changes in the effectiveness of the synaptic pathways that mediate the behavior.

Habituation Results From Presynaptic Depression of Synaptic Transmission

Habituation is the simplest form of implicit learning. It occurs, for example, when an animal learns to ignore a novel stimulus. An animal reacts to a new stimulus with a series of orienting responses. If the stimulus is neither beneficial nor harmful, the animal learns to ignore it after repeated exposure.

The physiological basis of this behavior was first investigated by Charles Sherrington while studying posture and locomotion in cats. Sherrington observed a decrease in the intensity of certain reflexes in response to repeated electrical stimulation of the motor pathways. He suggested that this decrease, which he called *habituation*, is caused by diminished synaptic effectiveness in the stimulated pathways.

Habituation was later investigated at the cellular level by Alden Spencer and Richard Thompson. They found close cellular and behavioral parallels between habituation of a spinal flexion reflex in cats (the withdrawal of a limb from a noxious stimulus) and habituation of more complex human behaviors. They showed that during habituation the strength of the input from local excitatory interneurons onto motor neurons in the spinal cord decreased, whereas the input to the same interneurons from sensory neurons innervating the skin was unchanged.

Because the organization of interneurons in the vertebrate spinal cord is quite complex, it was difficult to analyze further the cellular mechanisms of habituation in the flexion reflex. Progress required a simpler system. The marine mollusk *Aplysia californica*, which has a simple nervous system of about 20,000 central neurons, proved to be an excellent system for studying implicit forms of memory.

Aplysia has a repertoire of defensive reflexes for withdrawing its respiratory gill and siphon, a small fleshy spout above the gill used to expel seawater and waste (Figure 53–2A). These reflexes are similar to the withdrawal reflex of the leg studied by Spencer and Thompson. Mild touching of the siphon elicits reflex withdrawal of both the siphon and gill. With repeated stimulation, these reflexes habituate. As we shall see, these responses can also be dishabituated, sensitized, and classically conditioned.

The neural circuit mediating the gill-withdrawal reflex in *Aplysia* has been studied in detail. Touching the siphon excites a population of mechanoreceptor sensory neurons that innervate the siphon. The release of glutamate from sensory neuron terminals generates fast excitatory postsynaptic potentials (EPSPs) in interneurons and motor cells. The EPSPs from the sensory cells and interneurons summate on motor cells both temporally and spatially, causing them to discharge strongly, thereby producing vigorous withdrawal of the gill. If the siphon is repeatedly touched, however, the monosynaptic EPSPs produced by sensory neurons in both interneurons and motor cells decrease progressively, paralleling the habituation of gill withdrawal. In addition, repeated stimulation also leads to a decrease in the strength of synaptic transmission from the excitatory interneurons to the motor neurons; the net result is that the reflex response diminishes (Figure 53–2B,C).

What reduces the effectiveness of synaptic transmission between the sensory neurons and their postsynaptic cells during repeated stimulation? Quantal analysis (Chapter 15) revealed that the amount of synaptic glutamate released from presynaptic terminals of sensory neurons decreases. That is, fewer synaptic vesicles are released with each action potential in the sensory neuron; the sensitivity of the postsynaptic glutamate receptors does not change. Because the reduction in transmission occurs in the activated pathway itself and does not require another modulatory cell, the reduction is referred to as *homosynaptic depression*. This depression lasts many minutes.

An enduring change in the functional strength of synaptic connections thus constitutes the cellular mechanism mediating short-term habituation. As change of this type occurs at several sites in the gill-withdrawal reflex circuit, *memory is distributed and stored throughout the circuit*. Depression of synaptic transmission by sensory neurons, interneurons, or both is a common mechanism underlying habituation of escape responses of crayfish and cockroaches as well as startle reflexes in vertebrates.

How long can the effectiveness of a synapse change last? In *Aplysia*, a single session of 10 stimuli leads to short-term habituation of the withdrawal reflex lasting minutes. Four sessions separated by periods ranging from several hours to 1 day produce long-term habituation, lasting as long as 3 weeks (Figure 53–3).

Anatomical studies indicate that long-term habituation is caused by a decrease in the number of synaptic contacts between sensory and motor neurons. In naïve animals, 90% of the sensory neurons make physiologically detectable connections with identified motor neurons. In contrast, in animals trained for

A Experimental setup

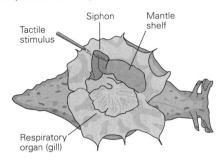

B Gill-withdrawal reflex circuit

C Habituation

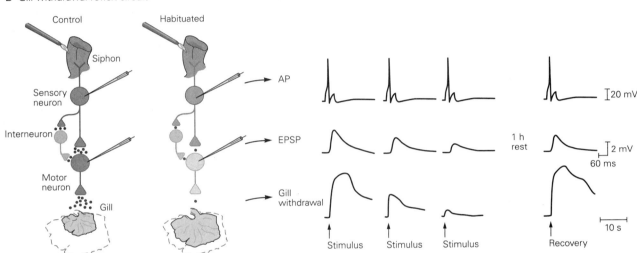

Figure 53–2 Short-term habituation of the gill-withdrawal reflex of the marine snail *Aplysia*.

A. A dorsal view of *Aplysia* illustrates the respiratory organ (gill) and the mantle shelf, which ends in the siphon, a fleshy spout used to expel seawater and waste. Touching the siphon elicits the gill-withdrawal reflex. Repeated stimulation leads to habituation.

B. Simplified diagrams of the gill-withdrawal reflex circuit and sites involved in habituation. Approximately 24 mechanoreceptor neurons in the abdominal ganglion innervate the siphon skin. These sensory cells make excitatory synapses onto a cluster of six motor neurons that innervate the gill, as well as on interneurons that modulate the firing of the motor neurons. (For simplicity, only one of each type of neuron is illustrated here.) Touching the siphon leads to withdrawal of the gill (**dashed**

outline shows original gill size; **solid outline** shows maximal withdrawal).

C. Repeated stimulation of the siphon sensory neuron (**top traces**) leads to a progressive depression of synaptic transmission between the sensory and motor neurons. The size of the motor neuron excitatory postsynaptic potential (**EPSP**) gradually decreases despite no change in the presynaptic action potential (**AP**). In a separate experiment, repeated stimulation of the siphon results in a decrease in gill withdrawal (habituation). One hour after repetitive stimulation, both the EPSP and gill withdrawal have recovered. Habituation involves a decrease in transmitter release at many synaptic sites throughout the reflex circuit. (Adapted, with permission, from Pinsker et al. 1970; Castellucci and Kandel 1974.)

long-term habituation, the incidence of connections is reduced to 30%; the reduction in number of synapses persists for a week and does not fully recover even 3 weeks later (see Figure 53–9). As we shall see, the converse occurs with long-term sensitization, where synaptic transmission is associated with an *increase* in the number of synapses between sensory and motor neurons.

Not all classes of synapses are equally modifiable. In *Aplysia*, the strength of some synapses rarely changes, even with repeated activation. In synapses specifically involved in learning (such as the connections between sensory and motor neurons in the withdrawal reflex circuit), a relatively small amount of training can produce large and enduring changes in synaptic strength.

A Depression of synaptic potentials by long-term habituation

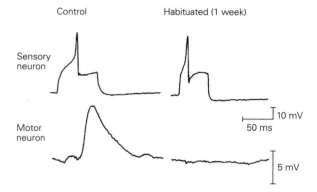

B Inactivation of synaptic connections by long-term habituation

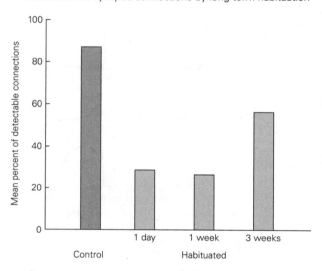

Figure 53–3 Long-term habituation of the gill-withdrawal reflex in *Aplysia*. (Adapted, with permission, from Castellucci, Carew, and Kandel 1978.)

A. Comparison of action potentials in sensory neurons and the postsynaptic potential in motor neurons in an untrained animal (control) and one that has been subjected to long-term habituation. In the habituated animal 1 week after training, no synaptic potential occurs in the motor neuron in response to the sensory neuron action potential.

B. After long-term habituation training, the mean percentage of sensory neurons making physiologically detectable connections with motor neurons is reduced even at 3 weeks.

Sensitization Involves Presynaptic Facilitation of Synaptic Transmission

The ability to recognize and respond to danger is necessary for survival. Not only snails and flies, but all animals, including humans, must distinguish predators from prey and hostile environments from safe ones. Because the ability to respond to threats is a universal requirement of survival, it has been conserved throughout evolution, allowing studies of invertebrates to shed light on neural mechanisms in mammals.

At the beginning of the 20th century, both Freud and Pavlov appreciated that anticipatory defensive responses to danger signals are biologically adaptive, a fact that likely accounts for the profound conservation of this capacity throughout vertebrates and invertebrates. In the laboratory, threat (fear) conditioning is typically studied by presenting a neutral stimulus, such as a tone, prior to the onset of an aversive stimulus, such as electrical shock. The two stimuli become associated such that the tone leads to the elicitation of defensive behaviors that protect against the harmful consequences predicted by the tone. Freud called this "signal anxiety," which prepares the individual for fight or flight when there is even the suggestion of external danger.

When an animal repeatedly encounters a harmless stimulus, its responsiveness to the stimulus habituates, as seen above. In contrast, when the animal confronts a *harmful* stimulus, it typically learns to respond more vigorously to a subsequent presentation of the same stimulus. Presentation of a harmful stimulus can even cause an animal to mount a defensive response to a subsequent *harmless* stimulus. As a result, defensive reflexes for withdrawal and escape become heightened. This enhancement of reflex responses is called *sensitization*.

Like habituation, sensitization can be transient or long lasting. A single shock to the tail of an *Aplysia* produces short-term sensitization of the gill-withdrawal reflex that lasts minutes; five or more shocks to the tail produce sensitization lasting days to weeks. Tail shock is also sufficient to overcome the effects of habituation and enhance a habituated gill-withdrawal reflex, a process termed *dishabituation*.

Sensitization and dishabituation result from an enhancement in synaptic transmission at several connections in the neural circuit of the gill-withdrawal reflex, including the connections made by sensory neurons with motor neurons and interneurons—the same synapses depressed by habituation (Figure 53–4A). Typically, modifiable synapses can be regulated bidirectionally, participate in more than one type of learning, and store more than one type of memory. The bidirectional synaptic changes that underlie habituation and sensitization are the result of different cellular mechanisms. In *Aplysia*, the same synapses that are weakened by habituation through a homosynaptic process can be strengthened by sensitization through a *heterosynaptic* process that depends on modulatory interneurons activated by the harmful stimulus to the tail.

At least three groups of modulatory interneurons are involved in sensitization. The best studied use serotonin as a transmitter (Figure 53–4B). The serotonergic interneurons form synapses on many regions of the sensory neurons, including axo-axonic synapses on the presynaptic terminals of the sensory cells. After a single tail shock, the serotonin released from the interneurons binds to a receptor in the sensory neurons that is coupled to a stimulatory G protein that increases the activity of adenylyl cyclase. This action produces the second messenger cyclic adenosine monophosphate (cAMP), which in turn activates the cAMP-dependent protein kinase (PKA) (Chapter 14). Serotonin also activates a second type of G-protein–coupled receptor that leads to the hydrolysis of phospholipids and the activation of protein kinase C (PKC).

The protein phosphorylation mediated by PKA and PKC enhances the release of transmitter from sensory neurons through at least two mechanisms (Figure 53–4B). In one action, PKA phosphorylates a K^+ channel, causing it to close. This broadens the action potential and thus enhances the duration of Ca^{2+} influx through voltage-gated Ca^{2+} channels, which in turn enhances transmitter release. In a second action, protein phosphorylation through PKC enhances the functioning of the release machinery directly. Presynaptic facilitation in response to release of serotonin by a tail shock lasts for a period of many minutes. Repeated noxious stimuli can strengthen synaptic activity for days (by a mechanism we consider below).

Classical Threat Conditioning Involves Facilitation of Synaptic Transmission

Classical conditioning is a more complex form of learning. Rather than learning about the properties of one stimulus, as in habituation and sensitization, the animal learns to associate one type of stimulus with another. As described in Chapter 52, an initial weak conditioned stimulus (eg, the ringing of a bell) becomes highly effective in producing a response when paired with a strong unconditioned stimulus (eg, presentation of food). In reflexes that can be enhanced by both classical conditioning and sensitization, such as the defensive withdrawal reflexes of *Aplysia*, classical conditioning results in greater and longer-lasting enhancement.

Although aversive classical conditioning is traditionally referred to as fear conditioning, we will use the more neutral term *threat conditioning* to avoid the implication that animals have subjective states comparable

to those that humans experience and label as "fear." This distinction is important because humans can respond to threats behaviorally and physiologically in the absence of any reported feeling of fear. This terminology allows the findings from research on implicit learning in all animals, from the simplest worm to humans, to be interpreted in an objective manner without invoking empirically unverifiable subjective fear states in animals.

For classical conditioning of the *Aplysia* gill-withdrawal reflex, a weak touch to the siphon serves as the conditioned stimulus while a strong shock to the tail serves as the unconditioned stimulus. When the gill-withdrawal reflex is classically conditioned, gill withdrawal in response to siphon stimulation alone is greatly enhanced. This enhancement is even more dramatic than the enhancement produced in an unpaired pathway by tail shock alone (sensitization). In classical conditioning, the timing of the conditioned and unconditioned stimuli is critical. To be effective, the conditioned stimulus (siphon touch) must *precede* (and thus predict) the unconditioned stimulus (tail shock), often within an interval of about 0.5 seconds.

The convergence in individual sensory neurons of the signals initiated by the conditioned and unconditioned stimuli is critical. Alone, a strong shock to the tail (unconditioned stimulus) will excite serotonergic interneurons that form synapses on presynaptic terminals of the siphon sensory neurons, resulting in presynaptic facilitation (Figure 53–5A). However, when the tail shock immediately follows a slight tap on the siphon (conditioned stimulus), the serotonin from the interneurons produces even greater presynaptic facilitation, a process termed *activity-dependent facilitation* (Figure 53–5B).

How does this work? During conditioning, the modulatory interneurons activated by tail shock release serotonin shortly *after* the action potential produced in the siphon sensory neurons by the tap on the siphon. The action potential triggers an influx of Ca^{2+} into the presynaptic terminals of the sensory neurons, and the Ca^{2+} binds to calmodulin, which in turn binds to the enzyme, adenylyl cyclase. This primes the adenylyl cyclase so that it responds more vigorously to the serotonin released following the tail shock. This in turn enhances the production of cAMP, which increases the amount of presynaptic facilitation. If the order of stimuli is reversed so that serotonin release precedes Ca^{2+} influx in the presynaptic sensory terminals, there is no potentiation and no classical conditioning.

Thus, the cellular mechanism of classical conditioning in the monosynaptic pathway of the withdrawal reflex is largely an elaboration of the mechanism of

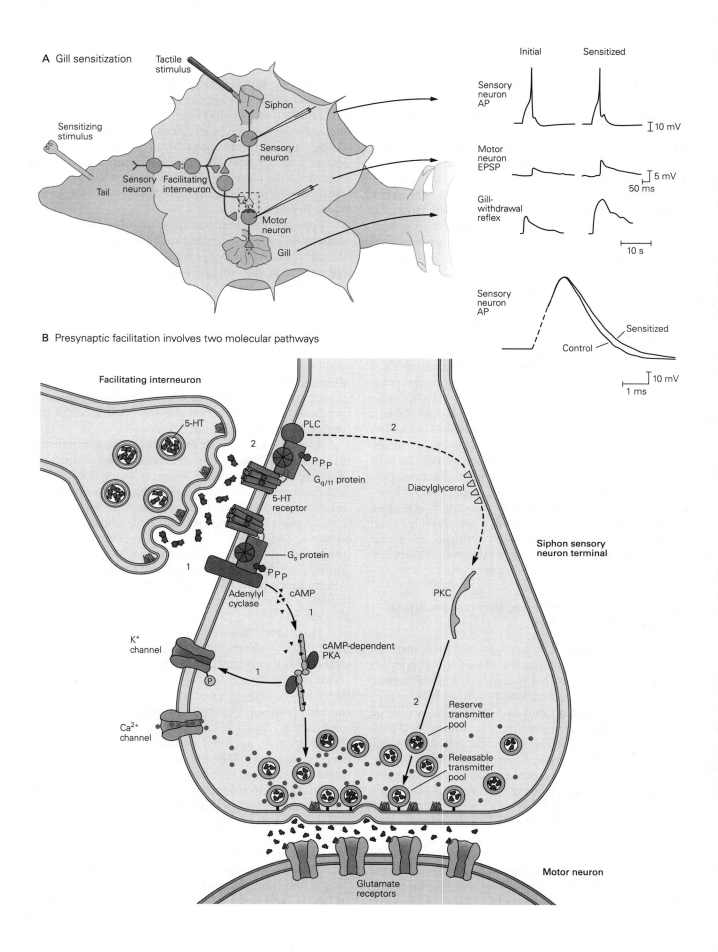

A Gill sensitization

Tactile stimulus

Siphon

Sensory neuron

Sensitizing stimulus

Sensory neuron   Facilitating interneuron

Tail

Motor neuron

Gill

Initial   Sensitized

Sensory neuron AP

$\rfloor$ 10 mV

Motor neuron EPSP

$\rfloor$ 5 mV

50 ms

Gill-withdrawal reflex

10 s

Sensory neuron AP

Sensitized

Control

$\rfloor$ 10 mV

1 ms

B Presynaptic facilitation involves two molecular pathways

Facilitating interneuron

5-HT

PLC

2

PPP

$G_{q/11}$ protein

5-HT receptor

Diacylglycerol

G_s protein

PPP

Adenylyl cyclase

cAMP

1

PKC

Siphon sensory neuron terminal

K^+ channel

cAMP-dependent PKA

P

Ca^{2+} channel

Reserve transmitter pool

Releasable transmitter pool

Glutamate receptors

Motor neuron

sensitization, with the added feature that the adenylyl cyclase serves as a *coincidence detector* in the presynaptic sensory neuron, recognizing the temporal order of the physiological responses to the tail shock (unconditioned stimulus) and the siphon tap (conditioned stimulus).

In addition to the presynaptic component of activity-dependent facilitation, a postsynaptic component is triggered by Ca^{2+} influx into the motor neuron when it is highly excited by the siphon sensory neurons. The properties of this postsynaptic mechanism are similar to those of long-term potentiation of synaptic transmission in the mammalian brain (discussed later in this chapter and in Chapters 13 and 54).

Long-Term Storage of Implicit Memory Involves Synaptic Changes Mediated by the cAMP-PKA-CREB Pathway

Cyclic AMP Signaling Has a Role in Long-Term Sensitization

In all forms of learning, practice makes perfect. Repeated experience converts short-term memory into a long-term form. In *Aplysia*, the form of long-term memory that has been most intensively studied is long-term sensitization. Like the short-term form, long-term sensitization of the gill-withdrawal reflex involves changes in the strength of connections at several synapses. But in addition, it also recruits the growth of new synaptic connections.

Five spaced training sessions (or repeated applications of serotonin) over approximately 1 hour produce long-term sensitization and long-term synaptic facilitation lasting 1 or more days. Spaced training over several days produces sensitization that persists for 1 or more weeks. Long-term sensitization, like the short-term form, requires protein phosphorylation that is dependent on increased levels of cAMP (Figure 53–6).

The conversion of short-term memory into long-term memory, called *consolidation*, requires synthesis of messenger RNAs and proteins in the neurons in the circuit. Thus, activation of specific gene expression is required for long-term memory. The transition from short-term to long-term memory depends on the prolonged rise in cAMP that follows repeated applications of serotonin. The increase in cAMP leads to prolonged activation of PKA, allowing the catalytic subunit of the kinase to translocate into the nucleus of the sensory neurons. It also leads indirectly to activation of a second protein kinase, the mitogen-activated protein kinase (MAPK), a kinase commonly associated with cellular growth (Chapter 14). Within the nucleus, the catalytic subunit of PKA phosphorylates and thereby activates the transcription factor CREB-1 (*c*AMP *r*esponse *e*lement *b*inding protein 1), which binds a promoter element called CRE (*c*AMP *r*ecognition *e*lement) (Figures 53–6 and 53–7).

To turn on gene transcription, phosphorylated CREB-1 recruits a transcriptional coactivator, CREB-binding protein (CBP), to the promoter region. CBP has two important properties that facilitate transcriptional activation: It recruits RNA polymerase II to the

Figure 53–4 (Opposite) Short-term sensitization of the gill-withdrawal reflex in *Aplysia*.

A. Sensitization of the gill-withdrawal reflex is produced by applying a noxious stimulus to another part of the body, such as the tail. A shock to the tail activates tail sensory neurons that excite facilitating (modulatory) interneurons, which form synapses on the cell body and terminals of the mechanoreceptor sensory neurons that innervate the siphon. Through these axo-axonic synapses, the modulatory interneurons enhance transmitter release from the siphon sensory neurons onto their postsynaptic gill motor neurons (presynaptic facilitation), thus enhancing gill withdrawal. Presynaptic facilitation results, in part, from a prolongation of the sensory neuron action potential (**AP; bottom traces**). (Abbreviation: EPSP, excitatory postsynaptic potential.) (Adapted, with permission, from Pinsker et al. 1970; Klein and Kandel 1980.)

B. Presynaptic facilitation in the sensory neuron is thought to occur by means of two biochemical pathways. The diagram shows details of the synaptic complex in the dashed box in part **A**.

Pathway 1: A facilitating interneuron releases serotonin (5-HT), which binds to metabotropic receptors in the sensory neuron terminal. This action engages a G protein (G$_s$), which in turn increases the activity of adenylyl cyclase. The adenylyl cyclase converts adenosine triphosphate to cyclic adenosine monophosphate (**cAMP**), which binds to the regulatory subunit of protein kinase A (**PKA**), thus activating its catalytic subunit. The catalytic subunit phosphorylates certain K$^+$ channels, thereby closing the channels and decreasing the outward K$^+$ current. This prolongs the action potential, thus increasing the influx of Ca^{2+} through voltage-gated Ca^{2+} channels and thereby augmenting transmitter release.

Pathway 2: Serotonin binds to a second class of metabotropic receptor that activates the G$_{q/11}$ class of G protein that enhances the activity of phospholipase C (**PLC**). The PLC activity leads to production of diacylglycerol, which activates protein kinase C (**PKC**). The PKC phosphorylates presynaptic proteins, resulting in the mobilization of vesicles containing glutamate from a reserve pool to a releasable pool at the active zone, thus increasing the efficiency of transmitter release.

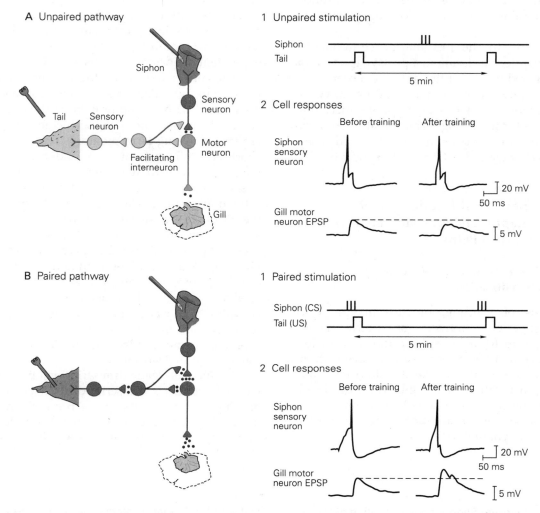

Figure 53–5 Classical conditioning of the gill-withdrawal reflex in *Aplysia*. (Adapted, with permission, from Hawkins et al. 1983.)

A. The siphon is stimulated by a light tap and the tail is shocked, but the two stimuli are not paired in time. The tail shock excites facilitatory interneurons that form synapses on the presynaptic terminals of sensory neurons innervating the mantle shelf and siphon. This is the mechanism of sensitization. **1.** The pattern of unpaired stimulation during training. **2.** Under these conditions, the size of the motor neuron test excitatory postsynaptic potential (**EPSP**) is only weakly facilitated by the tail shock. Often, as in this example, the EPSP actually decreases slightly despite the tail shock because repeated unpaired stimulation of the siphon leads to synaptic depression due to habituation.

B. The tail shock is paired in time with stimulation of the siphon. **1.** The siphon is touched (conditioned stimulus [**CS**]) immediately prior to shocking the tail (unconditioned stimulus [**US**]). As a result, the siphon sensory neurons are primed to be more responsive to input from the facilitatory interneurons in the unconditioned pathway. This is the mechanism of classical conditioning; it selectively amplifies the response of the conditioned pathway. **2.** Recordings of test EPSPs in an identified motor neuron produced by a siphon sensory neuron before training and 1 hour after training. After training with paired sensory input, the EPSP in the siphon motor neuron is considerably greater than either the EPSP before training or the EPSP following unpaired tail shock (shown in part **A2**). This synaptic amplification produces a more vigorous gill withdrawal.

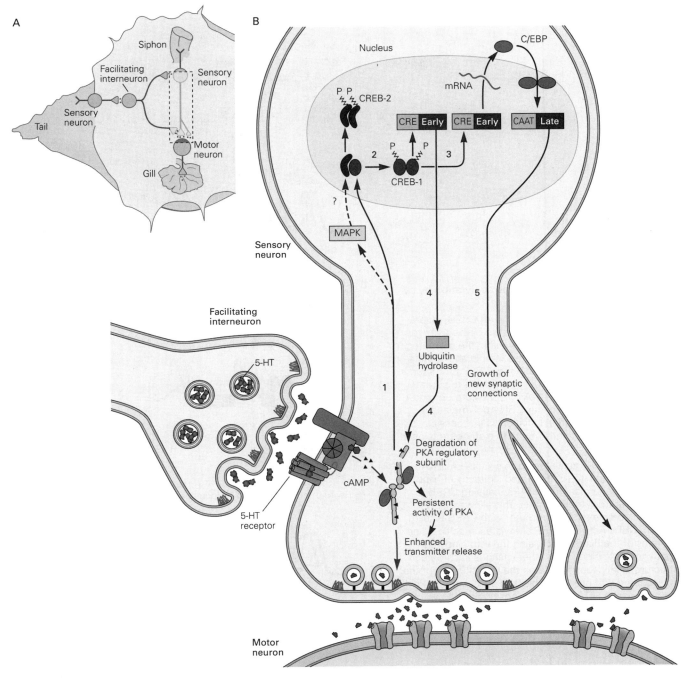

Figure 53–6 Long-term sensitization involves synaptic facilitation and the growth of new synaptic connections.

A. Long-term sensitization of the gill-withdrawal reflex of *Aplysia* involves long-lasting facilitation of transmitter release at the synapses between sensory and motor neurons.

B. Long-term sensitization of the gill-withdrawal reflex leads to persistent activity of protein kinase A (PKA), resulting in the growth of new synaptic connections. Repeated tail shock leads to more pronounced elevation of cyclic adenosine monophosphate (cAMP), producing long-term facilitation (lasting 1 or more days) that outlasts the increase in cAMP and recruits the synthesis of new proteins. This inductive mechanism is initiated by translocation of PKA to the nucleus (pathway 1), where PKA phosphorylates the transcriptional activator cAMP response element binding

protein 1 (CREB-1) (pathway 2). CREB-1 binds cAMP regulatory elements (CRE) located in the upstream region of several cAMP-inducible genes, activating gene transcription (pathway 3). PKA also activates the mitogen-activated protein kinase (MAPK), which phosphorylates the transcriptional repressor cAMP response element binding protein 2 (CREB-2), thus removing its repressive action. One gene activated by CREB-1 encodes a ubiquitin hydrolase, a component of a specific ubiquitin proteasome that leads to the proteolytic cleavage of the regulatory subunit of PKA, resulting in persistent activity of PKA, even after cAMP has returned to its resting level (pathway 4). CREB-1 also activates the expression of the transcription factor C/EBP, which leads to expression of a set of unidentified proteins important for the growth of new synaptic connections (pathway 5).

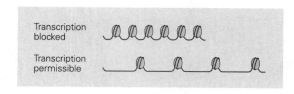

A Basal state

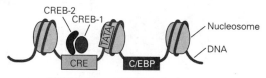

Figure 53–7 Regulation of histone acetylation by serotonin, CREB-1, and CBP.

A. Under basal conditions, the activator CREB-1 (here in complex with CREB-2) occupies the binding site for cAMP recognition element (**CRE**) within the promoter region of its target genes. In the example shown here, CREB-1 binds to the CRE within the C/EBP promoter. In the basal state, CREB-1 binding is not able to activate transcription because the TATA box, the core promoter region responsible for recruiting RNA polymerase II (**Pol II**) during transcription initiation, is inaccessible because the DNA is tightly bound to histone proteins in the nucleosome.

B. Serotonin (**5-HT**) activates protein kinase A (**PKA**), which phosphorylates CREB-1 and indirectly enhances CREB-2 phosphorylation by MAPK, causing CREB-2 to dissociate from the promoter. This allows CREB-1 to form a complex at the promoter with CREB binding protein (**CBP**). Activated CBP acetylates specific lysine residues of the histones, causing them to bind less tightly to DNA. Along with other changes in chromatin structure, acetylation facilitates the repositioning of the nucleosome that previously blocked access of the Pol II complex to the TATA box. This repositioning allows Pol II to be recruited to initiate transcription of the C/EBP gene. (Abbreviation: **TBP**, TATA binding protein.)

B 5-HT produces modifications in chromatin structure, CREB-1 phosphorylation and exclusion of CREB-2

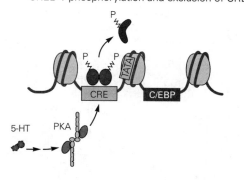

Recruitment of CBP and histone acetylation

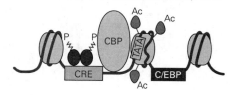

Initiation of transcription by Pol II

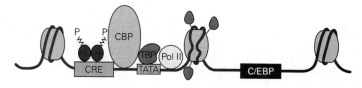

mRNA elongation

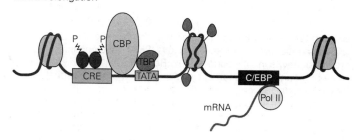

promoter, and it functions as an acetyltransferase, adding acetyl groups to certain lysine residues on its substrate proteins. One of the most important substrates of CBP are DNA-binding histone proteins, which are components of nucleosomes, the fundamental building blocks of chromatin. The histones contain a series of positively charged basic residues that strongly interact with the negatively charged phosphates of DNA. This interaction causes DNA to become tightly wrapped around the nucleosomes, much like string is wrapped around a spool, thereby preventing necessary transcription factors from accessing their gene targets.

The binding of CBP to CREB-1 leads to histone acetylation, which causes a number of important structural and functional changes at the level of the nucleosome. For example, acetylation neutralizes the positive charge of lysine residues in the histone tail domains, decreasing the affinity of histones for DNA. Also, specific classes of transcriptional activators can bind to acetylated histones and facilitate the repositioning of nucleosomes at the promoter region. Together, these and other types of chromatin modifications serve to regulate the accessibility of chromatin to the transcriptional machinery, and thus enhance the ability of a gene to be transcribed. This type of modification of DNA structure is termed *epigenetic* regulation. As we will see in Chapter 54, a mutation in the gene encoding CBP underlies Rubinstein-Taybi syndrome, a disorder associated with mental retardation.

The turning on of transcription by PKA also depends on its ability to indirectly activate the MAPK pathway (Chapter 14). MAPK phosphorylates the transcription factor CREB-2, relieving its inhibitory action on transcription (Figure 53–6B). The combined effects of CREB-1 activation and relief of CREB-2 repression induce a cascade of new gene expression important for learning and memory (Figure 53–7).

The presence of both a repressor (CREB-2) and an activator (CREB-1) of transcription at the first step in long-term facilitation suggests that the threshold for long-term memory storage can be regulated. Indeed, we see in everyday life that the ease with which short-term memory is transferred into long-term memory varies greatly with attention, mood, and social context.

The Role of Noncoding RNAs in the Regulation of Transcription

There are other targets of transcription and chromatin regulation in memory consolidation and reconsolidation besides messenger RNAs. Of particular interest are noncoding RNAs such as microRNAs (miRNAs), PIWI-interacting RNAs (piRNAs), and long noncoding

RNAs. These are also targeted to specific genetic sites, and their expression in turn regulates transcriptional and posttranscriptional mechanisms.

Studies in *Aplysia* show that miRNAs and piRNAs are both regulated by neuronal activity and contribute to long-term facilitation. MicroRNAs are a class of conserved noncoding RNAs, 20 to 23 nucleotides in length, that contribute to transcriptional and posttranscriptional regulation of gene expression through a specific set of RNA–protein machinery. In *Aplysia*, the most abundant and conserved brain species of these miRNAs are present in sensory neurons, where one of them—miRNA-124—normally constrains serotonin-induced synaptic facilitation by inhibiting the translation of CREB-1 mRNA, suppressing levels of CREB-1 protein. Serotonin inhibits the synthesis of miRNA-124, thereby leading to the disinhibition of the translation of CREB-1 mRNA, enabling the initiation of CREB-1–mediated transcription. The piRNAs are 28 to 32 nucleotides in length, slightly longer than miRNAs, and bind to a protein called Piwi. Individual piRNAs promote the methylation of specific DNA sequences, thereby silencing the genes, providing another example of epigenetic regulation. One piRNA, piRNA-F, increases in response to serotonin, which leads to the methylation of the promoter of CREB-2, reducing CREB-2 gene transcription.

Thus, we see here an example of integrative action at the transcriptional level. Serotonin regulates both piRNA and microRNA in a coordinated fashion: Serotonin rapidly decreases levels of miRNA-124 and facilitates the activation of CREB-1, which begins the process of memory consolidation. After a delay, serotonin also increases levels of piRNA-F, resulting in the methylation and silencing of the promoter of the transcription repressor CREB-2. The decrease in CREB-2 increases the duration of action of CREB-1, thereby consolidating a stable form of long-term memory in the sensory neuron (Figure 53–8).

Two of the genes expressed in the wake of CREB-1 activation and the consequential alteration in chromatin structure are important in the early development of long-term facilitation. One is a gene for ubiquitin carboxyterminal hydrolase, the other a gene for a transcription factor, CAAT box enhancer binding protein (C/EBP), a component of a gene cascade necessary for synthesizing proteins needed for the growth of new synaptic connections (Figures 53–6 and 53–7).

The hydrolase facilitates ubiquitin-mediated protein degradation (Chapter 7) and helps enhance activation of PKA. PKA is made up of four subunits; two regulatory subunits inhibit two catalytic subunits (Chapter 14). With long-term training and the

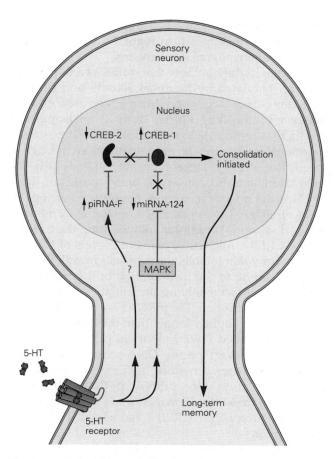

Figure 53–8 Small non-coding RNA molecules contribute to the memory consolidation switch. Long-term facilitation of the sensory to motor neuron synapses is consolidated through the action of two distinct classes of small noncoding RNA molecules. miRNA-124 normally acts to suppress levels of the CREB-1 transcription factor by binding to its mRNA and inhibiting its translation. Serotonin (**5-HT**) downregulates miRNA-124 levels through a mechanism requiring mitogen-activated protein kinase (**MAPK**). This enhances the levels of CREB-1, promoting activation of CREB-1–dependent transcription of gene products necessary for memory consolidation. In a complementary pathway, 5-HT enhances with a delay the synthesis of several piRNAs, including piRNA-F, which bind to the Piwi protein. The piRNA-F/Piwi complex leads to enhanced methylation of the *CREB-2* gene, resulting in long-lasting transcriptional repression of *CREB-2* and decreased levels of CREB-2 protein. Because CREB-2 normally inhibits the action of CREB-1, the increased levels of piRNA-F in response to 5-HT enhance and prolong CREB-1 activity, resulting in more effective memory consolidation.

induction of the hydrolase, approximately 25% of the regulatory subunits are degraded in the sensory neurons. As a result, free catalytic subunits can continue to phosphorylate proteins important for the enhancement of transmitter release and the strengthening of synaptic connections, including CREB-1, long after

cAMP has returned to its resting level (Figure 53–6B). Formation of a constitutively active enzyme is therefore the simplest molecular mechanism for long-term memory. With repeated training, a second-messenger kinase critical for short-term facilitation can remain persistently active for up to 24 hours without requiring a continuous activating signal.

The second and more enduring consequence of CREB-1 activation is the activation of the transcription factor C/EBP. This transcription factor forms both a homodimer with itself and a heterodimer with another transcription factor called *activating factor*. Together, these factors act on downstream genes that trigger the growth of new synaptic connections that support long-term memory.

With long-term sensitization, the number of presynaptic terminals in the sensory neurons in the gill-withdrawal circuit doubles (Figure 53–9). The dendrites of the motor neurons also grow to accommodate the additional synaptic input. Thus, long-term structural changes in both post- and presynaptic cells increase the number of synapses. Long-term habituation, in contrast, leads to *pruning* of synaptic connections, as described above. Long-term disuse of functional connections between sensory and motor neurons reduces the number of terminals of each sensory neuron by one-third (Figure 53–9A).

Long-Term Synaptic Facilitation Is Synapse Specific

A typical pyramidal neuron in the mammalian brain makes 10,000 presynaptic connections with a wide range of target cells. It is therefore generally thought that long-term memory storage should be synapse specific—that is, only those synapses that actively participate in learning should be enhanced. However, the finding that long-term facilitation involves gene expression—which occurs in the nucleus, far removed from a neuron's synapses—raises some fundamental questions regarding information storage.

Is long-term memory storage indeed synapse specific, or do the gene products recruited during long-term memory storage alter the strength of every presynaptic terminal in a neuron? And if long-term memory is synapse specific, what are the cellular mechanisms that enable the products of gene transcription to selectively strengthen just some synapses and not others?

Kelsey Martin and her colleagues addressed these questions for long-term facilitation by using a cell culture system consisting of an isolated *Aplysia* sensory neuron with a bifurcated axon that makes separate synaptic contacts with two motor neurons. The sensory

Figure 53–9 Long-term habituation and sensitization involve structural changes in the presynaptic terminals of sensory neurons.

A. Long-term habituation leads to a loss of synapses, and long-term sensitization leads to an increase in the number of synapses. When measured either 1 day (shown here) or 1 week after training, the number of presynaptic terminals relative to control levels is greater in sensitized animals and less in habituated animals. The drawings below the graph illustrate changes in the number of synaptic contacts. The swellings or varicosities on the sensory neuron processes are called synaptic boutons; they contain all the specialized structures necessary for transmitter release. (Adapted, with permission, from Bailey and Chen 1983. Copyright © 1983 AAAS.)

B. Fluorescence images of a sensory neuron axon contacting a motor neuron in culture before (*left*) and 1 day after (*right*) five brief exposures to serotonin. The resulting increase in varicosities simulates the synaptic changes associated with long-term sensitization. Prior to serotonin application, no presynaptic varicosities are visible in the outlined area (*left*). After serotonin, several new boutons are apparent (**arrows**), some of which contain a fully developed active zone (**asterisk**) or have small immature active zones. Scale bar = 50 μm. (Reproduced, with permission, from Glanzman, Kandel, and Schacher 1990.)

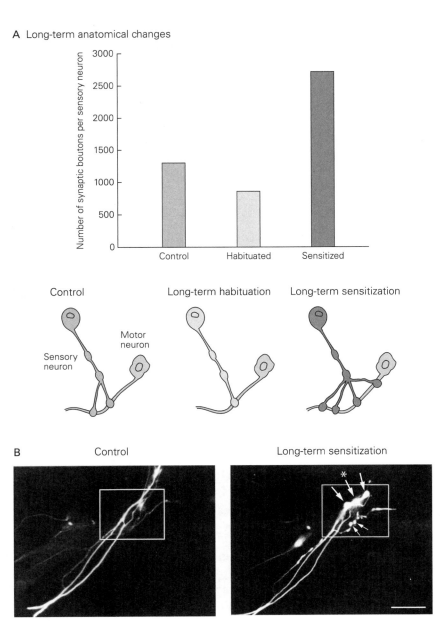

A Long-term anatomical changes

B   Control   Long-term sensitization

neuron terminals on one of the two motor neurons were activated by focal pulses of serotonin, thus mimicking the neural effects of a shock to the tail. When only one pulse of serotonin was applied, those synapses showed short-term facilitation. The synapses on the second motor neuron, which did not receive serotonin, showed no change in synaptic transmission.

When five pulses of serotonin were applied to the same synapses, those synapses displayed both short-term and long-term facilitation, and new synaptic connections were formed with the motor neuron. Although long-term facilitation and synaptic growth require gene transcription and protein synthesis, the synapses that did not receive serotonin showed no enhancement of synaptic transmission (Figure 53–10). Thus, both short-term and long-term synaptic facilitation are synapse specific and manifested only by those synapses that receive the modulatory serotonin signal.

But how are the nuclear products able to enhance transmission at only certain synapses and not others of the same neuron? Are the newly synthesized proteins somehow targeted to only those synapses that receive serotonin? Or are they shipped out to all synapses but used productively for the growth of new synaptic connections only at those synapses that have been marked by at least a single pulse of serotonin?

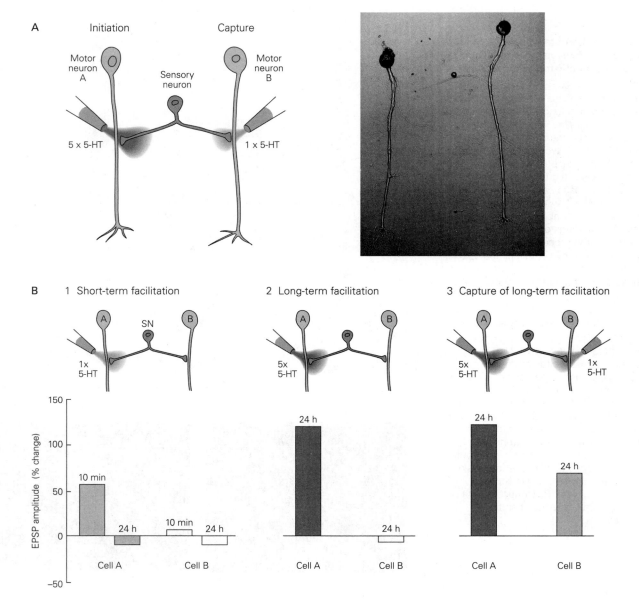

Figure 53–10 The long-term facilitation of synaptic transmission is synapse specific. (Adapted, with permission, from Martin et al. 1997.)

A. The experiment uses a single presynaptic sensory neuron that contacts two postsynaptic motor neurons A and B. The pipette on the left is used to apply five pulses of serotonin (**5-HT**) to a sensory neuron synapse with motor neuron A, initiating long-term facilitation at that synapse. The pipette on the right is used to apply one pulse of 5-HT to a sensory neuron synapse with motor neuron B, allowing this synapse to make use of (capture) new proteins produced in the cell body in response to the five pulses of 5-HT at the synapse with motor neuron A. The image at the right shows the actual appearance of the cells in culture.

B. 1. One pulse of 5-HT applied to the synapse with motor neuron A produces only short-term (10-minute) facilitation of the excitatory postsynaptic potential (**EPSP**) in the neuron. By 24 hours, the EPSP has returned to its normal size. There is no significant change in EPSP size in cell B. **2.** Application of five pulses of 5-HT to the synapses with cell A produces long-term (24-hour) facilitation of the EPSP in that cell but no change in the size of the EPSP in cell B. **3.** When five pulses of 5-HT onto the synapse with cell A are paired with a single pulse of 5-HT onto the synapses with cell B, cell B now displays long-term facilitation and an increase in EPSP size after 24 hours.

To test this question, Martin and her colleagues again selectively applied five pulses of serotonin to the synapses made by the sensory neuron onto one of the motor neurons. This time, however, the synapses with the second motor neuron were simultaneously activated by a single pulse of serotonin (which by itself produces only short-term synaptic facilitation lasting minutes). Under these conditions, the single pulse of serotonin was sufficient to induce long-term facilitation and growth of new synaptic connections at the contacts between the sensory neuron and the second motor neuron. Thus, application of the single pulse of serotonin onto the synapses at the second branch enabled those synapses to use the nuclear products produced in response to the five pulses of serotonin onto the synapses of the first branch, a process called *capture*.

These results suggest that newly synthesized gene products, both mRNAs and proteins, are delivered by fast axonal transport to all the synapses of a neuron but are functional only at synapses that have been marked by previous synaptic activity, that is, by presynaptic release of serotonin. Although one pulse of serotonin at a synapse is insufficient to turn on new gene expression in the cell body, it is sufficient to mark that synapse, allowing it to make use of new proteins generated in the cell body in response to five pulses of serotonin at another synapse. This idea, developed by Martin and her colleagues for *Aplysia* and independently by Frey and Morris for the hippocampus in rodents, is called *synaptic capture* or *synaptic tagging*.

These findings raise the question, what is the nature of the synaptic mark that allows the capture of the gene products for long-term facilitation? When an inhibitor of PKA was applied locally to the synapses receiving the single pulse of serotonin, those synapses could no longer capture the gene products produced in response to the five pulses of serotonin (Figure 53–11). This indicates that local phosphorylation by PKA is required for synaptic capture.

In the early 1980s, Oswald Steward discovered that ribosomes, the machinery for protein synthesis, are present at synapses as well as in the cell body. Martin examined the importance of local protein synthesis in long-term synaptic facilitation by applying a single pulse of serotonin together with an inhibitor of local protein synthesis onto one set of synapses while simultaneously applying five pulses of serotonin to a second set of synapses. Normally, long-term facilitation and synaptic growth would persist for up to 72 hours in response to synaptic capture. In the presence of the local protein synthesis inhibitor, synaptic capture still occurred, producing long-term synaptic facilitation at

the synapses exposed to only one pulse of serotonin. However, the facilitation only lasted 24 hours. After 24 hours, synaptic growth and facilitation at these synapses collapsed, indicating that the maintenance of learning-induced synaptic growth requires new local protein synthesis at the synapse (Figure 53–11B).

Martin and her colleagues thus found that regulation of protein synthesis at the synapse plays a major role in controlling synaptic strength at the sensory-to-motor neuron connection in *Aplysia*. As we shall see in Chapter 54, local protein synthesis is also important for the later phases of long-term potentiation of synaptic strength in the hippocampus.

These findings indicate there are two distinct components of synaptic marking in *Aplysia*. The first component, lasting about 24 hours, initiates long-term synaptic plasticity and synaptic growth, requires transcription and translation in the nucleus, and recruits local PKA activity, but does not require local protein synthesis. The second component, which stabilizes the long-term synaptic change after 72 hours, requires local protein synthesis at the synapse. How might this local protein synthesis be regulated?

Maintaining Long-Term Synaptic Facilitation Requires a Prion-Like Protein Regulator of Local Protein Synthesis

The fact that mRNAs are translated at the synapse in response to marking of that synapse by one pulse of serotonin suggests that these mRNAs may initially be dormant and under the control of a regulator of translation recruited by serotonin. Translation of most mRNAs requires that transcripts contain a long tail of adenosine nucleotides at their 3′ end [poly(A) tail]. Joel Richter had earlier found that in *Xenopus* (frog) oocytes the maternal mRNAs only have a short tail of adenine nucleotides and thus are silent until activated by the cytoplasmic polyadenylation element binding protein (CPEB). CPEB binds to a site on mRNAs and recruits poly(A) polymerase, leading to the elongation of the poly(A) tail.

Kausik Si and his colleagues found that serotonin increases the local synthesis of a novel, neuron-specific isoform of CPEB in *Aplysia* sensory neuron terminals. The induction of CPEB is independent of transcription but requires new protein synthesis. Blocking CPEB locally at an activated synapse blocks the long-term maintenance of synaptic facilitation at the synapse but not its initiation and initial 24-hour maintenance.

How might CPEB stabilize the late phase of long-term facilitation? Most biological molecules have a relatively short half-life (hours to days), whereas memory

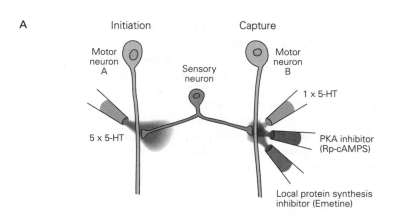

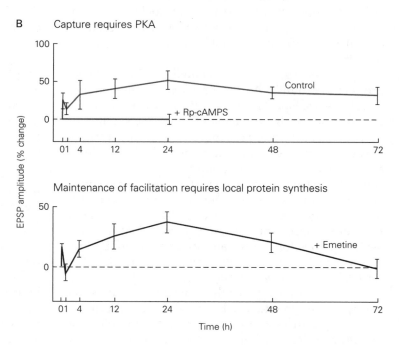

Figure 53–11 Long-term facilitation requires both cyclic adenosine monophosphate (cAMP)-dependent phosphorylation and local protein synthesis. (Adapted, with permission, from Casadio et al. 1999.)

A. Five pulses of serotonin (**5-HT**) are applied to the synapses on motor neuron A, and a single pulse is applied to those of cell B. Inhibitors of protein kinase A (**PKA**; Rp-cAMPS) or local protein synthesis (emetine) are applied to synapses on cell B.

B. Rp-cAMPS blocks the capture of long-term facilitation completely at the synapses on neuron B. Emetine has no effect on the capture of facilitation or the growth of new synaptic connections measured 24 hours after 5-HT application, but by 72 hours, it fully blocks synaptic enhancement. The outgrowth of new synaptic connections is retracted, and long-term facilitation decays after 1 day if capture is not maintained by local protein synthesis. (Abbreviations: **EPSP**, excitatory postsynaptic potential; **Rp-cAMPS**, Rp-diaster-eomer of adenosine cyclic 3′,5′-phosphorothioate.)

lasts days, weeks, or even years. How can learning-induced alterations in the molecular composition of a synapse be maintained for such a long time? Most hypotheses posit some type of self-sustained mechanism that modulates synaptic strength and structure.

Si and his colleagues made the surprising discovery that the neuronal isoform of *Aplysia* CPEB appears to have self-sustaining properties that resemble those of prion proteins. Prions were discovered by Stanley Prusiner, who demonstrated that these proteins were the causative agents of Creutzfeldt-Jakob disease, a devastating neurodegenerative human disease, and mad cow disease. Prion proteins can exist in two forms: a soluble form and an aggregated form that is capable of self-perpetuation. *Aplysia* CPEB also has two conformational states, a soluble form that is inactive and

an aggregated form that is active. This switch depends on an N-terminal domain of CPEB that is rich in glutamine, similar to prion domains in other proteins.

In a naïve synapse, CPEB exists in the soluble, inactive state, and its resting level of expression is low. However, in response to serotonin, the local synthesis of CPEB increases until a threshold concentration is reached that switches CPEB to the aggregated, active state, which is then capable of activating the translation of dormant mRNAs. Once the active state is established, it becomes self-perpetuating by recruiting soluble CPEB to the aggregates, maintaining its ability to activate the translation of dormant mRNAs. Although dormant mRNAs are made in the cell body and distributed throughout the cell, they are translated only at synapses that have active CPEB aggregates.

Whereas conventional prion mechanisms are pathogenic—the aggregated state of most prion proteins causes cell death—the *Aplysia* CPEB is a new form of a prion-like protein, one whose aggregated state plays an important physiological function. The active self-perpetuating form of *Aplysia* CPEB maintains long-term molecular changes in a synapse that are necessary for the persistence of memory storage (Figure 53–12).

Figure 53–12 A self-perpetuating switch for protein synthesis at axon terminals in *Aplysia* maintains long-term synaptic facilitation. Five pulses of serotonin (**5-HT**) set up a signal that goes back to the nucleus to activate synthesis of mRNA. Newly transcribed mRNAs and newly synthesized proteins in the cell body are then sent to all terminals by fast axonal transport. However, only those terminals that have been marked by at least one pulse of serotonin can use the proteins to grow the new synapses needed for long-term facilitation. The marking of a terminal involves two substances: (1) protein kinase A (**PKA**), which is necessary for the immediate synaptic growth initiated by the proteins transported to the terminals, and (2) phosphoinositide 3 kinase (**PI3 kinase**), which initiates the local translation of mRNAs required to maintain synaptic growth and long-term facilitation past 24 hours. Some of the mRNAs at the terminals encode cytoplasmic polyadenylation element binding protein (**CPEB**), a regulator of local protein synthesis. In the basal state, CPEB is thought to exist in a largely inactive conformation as a soluble monomer that cannot bind to mRNAs. Through some as yet unspecified mechanism activated by serotonin and PI3 kinase, some copies of CPEB convert to an active conformation that forms aggregates. The aggregates function like prions in that they are able to recruit monomers to join the aggregate, thereby activating the monomers. The CPEB aggregates bind the cytoplasmic polyadenylation element (**CPE**) site of mRNAs. This binding recruits the poly(A) polymerase machinery and allows poly(A) tails of adenine nucleotides (**A**) to be added to dormant mRNAs. The polyadenylated mRNAs can now be recognized by ribosomes, allowing the translation of these mRNAs to several proteins. For example, in addition to CPEB, this leads to the local synthesis of N-actin and tubulin, which stabilize newly grown synaptic structures. (Model based on Bailey, Kandel, and Si 2004.)

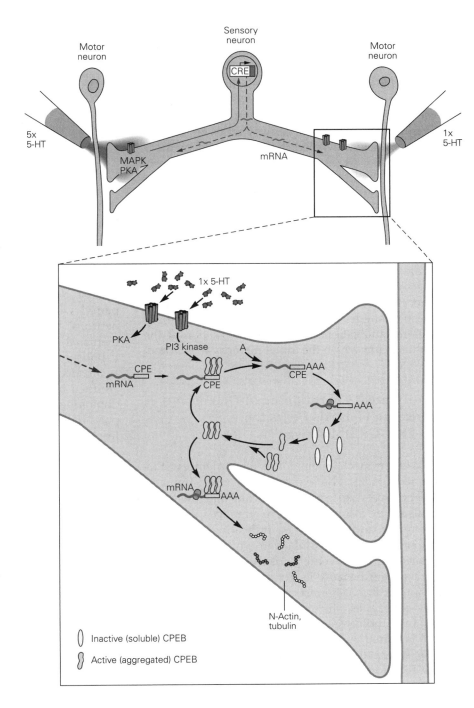

Memory Stored in a Sensory-Motor Synapse Becomes Destabilized Following Retrieval but Can Be Restabilized

A variety of studies in mammals by Karim Nader and others have found that in its early stages long-term memory storage is dynamic and can be disrupted. In particular, a memory trace can become labile after retrieval and require an additional round of consolidation (so-called *reconsolidation*).

Until recently, it was unclear whether the same set of synapses involved in storing a memory are destabilized and restabilized following retrieval or whether, after synaptic reactivation following a memory, a new set of synapses is regulated. This question was examined for retrieval of long-term sensitization of the gill- and siphon-withdrawal reflex in *Aplysia*. These experiments revealed that a retrieved memory becomes labile as a result of ubiquitin-mediated protein degradation and is then reconsolidated by means new protein synthesis.

Does a similar reconsolidation mechanism occur at sensory-motor synapses that have undergone long-term facilitation? Indeed, when a synapse that has undergone long-term facilitation is reactivated by a brief burst of presynaptic action potentials, that synapse becomes destabilized through protein degradation and requires protein synthesis for restabilization. Such results suggest that reconsolidation of memory involves restabilization of synaptic facilitation at the same synapses at which the initial memory was stored.

Classical Threat Conditioning of Defensive Responses in Flies Also Uses the cAMP-PKA-CREB Pathway

Do the cellular mechanisms for implicit memory storage found in *Aplysia* have parallels in other animals? Studies on aversive learning indicate that the same mechanisms are also used to store memory in the fruit fly *Drosophila* and in rodents, indicating conserved mechanisms throughout Metazoan evolution. The fruit fly is particularly convenient for the study of implicit memory storage because its genome is easily manipulated and, as first demonstrated by Seymour Benzer and his colleagues, the fly can be classically conditioned. In a typical classical conditioning paradigm, an odor is paired with repeated electrical shocks to the feet. The extent of learning is then examined by allowing the flies to choose between two arms of a maze, where one arm contains the odor that had been paired with a shock and the other arm contains an unpaired

odor. Following training, a large fraction of wild type flies avoids the arm with the conditioned odor. Several fly mutants have been identified that do not learn to avoid the conditioned odor. These learning-defective mutants have been given imaginatively descriptive names such as *dumb, dunce, rutabaga, amnesiac,* and *PKA-R1.* Of great interest, all of these mutants have defects in the cAMP cascade.

Olfactory conditioning depends on a region of the fly brain called the mushroom bodies. Neurons of the mushroom bodies, called Kenyon cells, receive olfactory input from the antennal lobes, structures similar to the olfactory lobes of the mammalian brain. The Kenyon cells also receive input from dopaminergic neurons that respond to aversive stimuli, such as a foot shock. The dopamine binds to a metabotropic receptor (encoded by the *dumb* gene) that activates a stimulatory G protein and a specific type of Ca^{2+}/calmodulin-dependent adenylyl cyclase (encoded by the *rutabaga* gene), similar to the cyclase involved in classical conditioning in *Aplysia*. The convergent action of dopamine released by the unconditioned stimulus (foot shock) and a rise in intracellular Ca^{2+} triggered by olfactory input leads to the synergistic activation of adenylyl cyclase, producing a large increase in cAMP.

Recent experiments have demonstrated that flies can be classically conditioned when an odorant is paired with direct stimulation of the dopaminergic neurons, bypassing the foot shock. In these experiments, the mammalian P2X receptor (an adenosine triphosphate [ATP]-gated cation channel) is expressed as a transgene in the dopaminergic neurons. The flies are then injected with a caged derivative of ATP. The dopaminergic neurons can then be excited to fire action potentials by shining light on the flies to release ATP from its cage and activate the P2X receptors. When the dopaminergic neurons are activated in this manner in the presence of an odor, the flies undergo aversive conditioning—they learn to avoid the odor. Thus, the unconditioned stimulus activates a dopamine signal that reinforces aversive conditioning, much as serotonin acts as an aversive reinforcement signal for learned defensive responses in *Aplysia*.

A reverse genetic approach has also been used to explore memory formation in *Drosophila*. In these experiments, various transgenes are placed under the control of a promoter that is heat sensitive. The heat sensitivity permits the gene to be turned on at will by elevating the temperature of the chamber housing the flies. This was done in mature animals to minimize any potential effect on the development of the brain. When the catalytic subunit of PKA was blocked by transient expression of an inhibitory transgene, flies were unable to form short-term memory, indicating the importance of the cAMP signal transduction pathway for associative learning and short-term memory in *Drosophila*.

Long-term memory in *Drosophila* requires new protein synthesis just as in *Aplysia* and other animals. Knockout of a CREB activator gene selectively blocks long-term memory without interfering with short-term memory. Conversely, when the gene is overexpressed, a training procedure that ordinarily produces only short-term memory produces long-term memory.

As in *Aplysia*, certain forms of long-term memory in *Drosophila* also involve CPEB and may depend on prion-like behavior in this protein. Male flies learn to suppress their courtship behavior after exposure to unreceptive females. When the N-terminal domain of CPEB is deleted genetically, there is a loss of long-term courtship memory; the male fly fails to recognize the unreceptive female. This N-terminal domain is rich in glutamine residues and corresponds to the glutamine rich prion-like domain of CPEB in *Aplysia*. Thus several molecular mechanisms involved in implicit memory are conserved from *Aplysia* to flies, and as we will see next, this conservation extends to mammals.

Memory of Threat Learning in Mammals Involves the Amygdala

Research over the past several decades has resulted in a detailed understanding of the neural circuits for both innate and learned defensive responses to threats in mammals, often referred to as "fear learning." In particular, as we have noted in Chapter 42, both types of defensive responses crucially involve the amygdala, which participates in the detection and evaluation of a broad range of significant and potentially dangerous environmental stimuli. The amygdala-based defense system quickly learns about new dangers. It can associate a new neutral stimulus (conditioned stimulus) with a known threat (unconditioned stimulus) after a single paired exposure, and this learned association is often retained throughout life.

The amygdala receives information about threats directly from sensory systems. The input nucleus of the amygdala, the lateral nucleus, is the site of convergence for signals from both unconditioned and conditioned stimuli. Both signals are carried by a rapid pathway that goes directly from the thalamus to the amygdala and a slower indirect pathway that projects from the thalamus to sensory areas of neocortex and from there to the amygdala. These parallel pathways both contribute to conditioning (Figure 53–13). The amygdala also receives higher-order cognitive information by means of connections from cortical associational areas, especially medial cortical regions in the frontal and temporal lobes.

During Pavlovian conditioning, the strength of synaptic transmission is modified in the amygdala. In response to a tone, an extracellular electrophysiological signal proportional to the excitatory synaptic response is recorded in the lateral nucleus. Following pairing of the tone with a shock, the electrophysiological response to the tone is enhanced by an increase in synaptic transmission, which depends on the

Figure 53–13 Threat learning engages parallel pathways from the thalamus to the amygdala. The signal for the conditioned stimulus, here a neutral tone, is carried by two pathways from the auditory thalamus to the lateral nucleus of the amygdala: by a direct pathway and by an indirect pathway via the auditory cortex. Similarly, the signal for the unconditioned stimulus, here a shock, is conveyed through parallel nociceptive pathways from the somatosensory part of the thalamus to the lateral nucleus, one a direct pathway and one an indirect pathway via the somatosensory cortex. The lateral nucleus in turn projects to the central nucleus, the output nucleus of the amygdala, which activates neural circuits that increase heart rate, produce other autonomic changes, and elicit defensive behaviors that constitute the defensive state. (Reproduced, with permission, from Kandel 2006.)

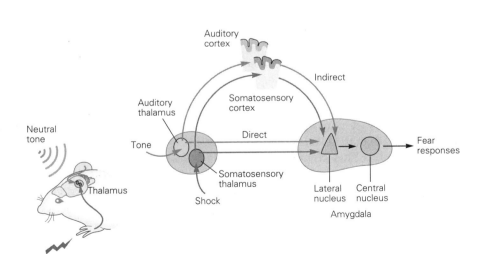

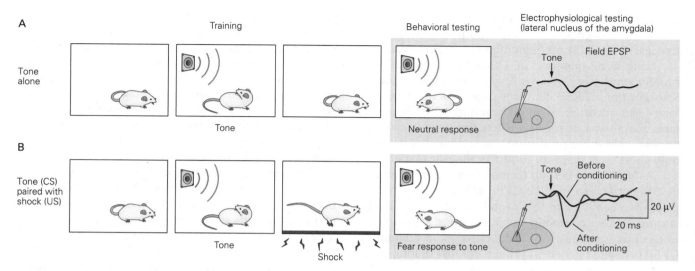

Figure 53–14 Threat learning produces correlated behavioral and electrophysiological changes.

A. An animal ordinarily ignores a neutral tone. The tone produces a small synaptic response in the amygdala recorded by an extracellular field electrode. This field excitatory postsynaptic potential (**field EPSP**) is generated by the small voltage drop between the recording electrode in the amygdala and a second electrode on the exterior of the brain as excitatory synaptic current enters the dendrites of a large population of amygdala neurons.

B. When the tone is presented immediately before a foot shock, the animal learns to associate the tone with the shock. As a result, the tone alone will elicit what the shock previously elicited: It causes the mouse to freeze, an instinctive defense response. After threat conditioning, the electrophysiological response in the lateral nucleus of the amygdala to the tone is greater than the response prior to conditioning. (Abbreviations: **CS**, conditioned stimulus; **US**, unconditioned stimulus.) (Reproduced, with permission, from Rogan et al. 2005.)

convergence of the tone (conditioned stimulus) and the shock (unconditioned stimulus) onto single neurons in the lateral amygdala (Figure 53–14).

It is generally thought that behavioral learning depends on synaptic plasticity. In an effort to understand how such plasticity might occur during learning in the lateral amygdala, researchers have studied *long-term potentiation* (LTP), a cellular model of plasticity. We initially discussed LTP in connection with excitatory synapse function in Chapter 13 and will examine it in detail in Chapter 54 in connection with explicit memory and the hippocampus. In brain slices that include the lateral amygdala, LTP can be induced by high-frequency tetanic stimulation of either the direct or indirect sensory pathways, which produces a long-lasting increase in the excitatory postsynaptic response to these inputs. This change results from a form of homosynaptic plasticity (Figure 53–15).

Long-term potentiation in the lateral nucleus of the amygdala is triggered by Ca^{2+} influx into the postsynaptic neurons in response to strong synaptic activity. The Ca^{2+} entry is mediated by the opening of both N-methyl-D-aspartate (NMDA)-type glutamate receptors and L-type voltage-gated Ca^{2+} channels in the postsynaptic cell. Because NMDA receptors are normally blocked by extracellular Mg^{2+}, they require a large

synaptic input to generate enough postsynaptic depolarization to relieve this blockade (Chapter 13). L-type channels also require a strong depolarization to open. Thus, LTP is only generated in response to coincident synaptic activity. Calcium influx triggers a biochemical cascade that enhances synaptic transmission through both the insertion of additional α-amino-3-hydroxy-5-methyl-4-isoxazolepropionic acid (AMPA)-type glutamate receptors in the postsynaptic membrane and an increase in transmitter release from the presynaptic terminals. As in *Aplysia*, monoamine neurotransmitters, such as norepinephrine and dopamine, released during tetanic stimulation provide a heterosynaptic modulatory signal that contributes to the induction of LTP.

Studies in awake behaving rodents indicate that similar mechanisms contribute to the acquisition of Pavlovian threat conditioning. This form of learning requires postsynaptic NMDA receptors and voltage-gated calcium channels in the lateral amygdala, and it is enhanced by norepinephrine released in lateral amygdala from the locus ceruleus.

In addition, the size of the LTP elicited by electrical stimulation in slices of the amygdala from animals previously trained is less than that found in slices from untrained animals. Because there is an upper limit to the amount by which synapses can be

A Basolateral complex of the amygdala

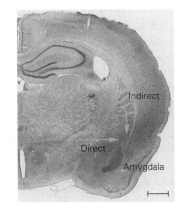
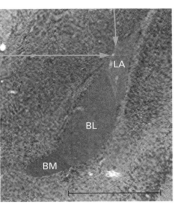

Figure 53–15 Long-term potentiation at synapses in the amygdala may mediate threat conditioning.

A. A coronal brain slice from a mouse shows the position of the amygdala. The enlargement shows three key input nuclei of the amygdala—lateral (**LA**), basolateral (**BL**), and basomedial (**BM**)—which together form the basolateral complex. These nuclei project to the central nucleus, which projects to the hypothalamus and brain stem. (Adapted, with permission, from Maren 1999. Copyright © 1999 Elsevier.)

B. High-frequency tetanic stimulation of the direct or indirect pathway from the thalamus to the lateral nucleus initiates long-term potentiation (**LTP**). The drawing shows the position of the extracellular voltage recording electrode in the lateral nucleus, and the positions of two stimulating electrodes used to activate either the direct pathway or indirect pathway. The plot shows the amplitude of the extracellular field excitatory postsynaptic potential (**EPSP**) in response to stimulation of the indirect cortical pathway during the time course of the experiment. When a pathway is stimulated at a low frequency (once every 30 seconds), the field EPSP is stable. However, when five trains of high-frequency tetanic stimulation are applied (**asterisks**), the response is enhanced for a period of hours. The facilitation depends on protein kinase A (**PKA**) and is compromised when the PKA inhibitor KT5720 is applied (the **bar**). Field EPSPs before and after induction of LTP are also shown. (Adapted, with permission, from Huang and Kandel 1998; Huang, Martin, and Kandel 2000.)

B LTP in the amygdala

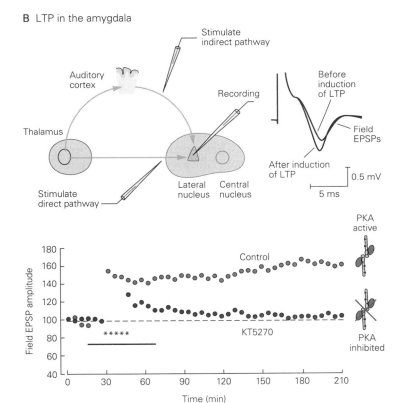

potentiated, this result is taken as evidence that threat conditioning recruits LTP, which precludes further LTP in response to electrical stimulation. Thus, artificially induced LTP and behaviorally induced LTP are closely related.

Two types of genetic experiments also strongly support the idea that an LTP-like phenomenon contributes to the cellular mechanism for storing the memory of a learned threat. First, genetic disruption of the GluN2B (NR2B) subunit of the NMDA receptor interferes both with threat conditioning and the induction of LTP in pathways that transmit the conditioned stimulus signal to the lateral amygdala. Moreover, this mutation affects only learned threats; it does not affect responses to unconditioned threats or routine synaptic transmission. Conversely, overexpression of the GluN2B subunit facilitates learning. Similarly, disruption of CREB signaling, a step downstream from Ca^{2+} influx, interferes with conditioning, whereas enhancement of CREB activity facilitates learning.

Does the LTP important for threat learning involve insertion of new AMPA receptors, as observed in brain slices? To address this question, researchers infected pyramidal neurons in the lateral nucleus with a genetically engineered virus that did not damage the neurons but caused them to express AMPA receptors tagged with a fluorescent label. Threat conditioning led to an increase in insertion of the tagged AMPA receptors into the cell membrane, similar to what is seen during experimentally induced LTP in brain slices. When a different virus was used to express a C-terminal portion of the AMPA receptor that competes with and prevents the insertion of endogenous AMPA receptors, memory for learned threat was substantially reduced, even though the virus infected only 10% to 20% of the neurons in the lateral nucleus. This surprising result suggests that LTP needs to be induced at nearly all activated synapses to effectively support threat learning.

One of the virtues of the Pavlovian paradigm is its amenability to experimental study due to the fact that specific stimuli are transmitted to the amygdala by known pathways. This has allowed experimenters to directly activate conditioned stimulus or unconditioned stimulus pathways, bypassing the normal sensory input. Such studies have provided convincing evidence implicating these pathways to the amygdala in threat learning.

Based on these findings, researchers explored whether threat learning could be induced when they paired an auditory conditioned stimulus (tone) with direct depolarization of lateral amygdala neurons, instead of using an external, pain-eliciting unconditioned shock stimulus to produce the depolarization via the unconditioned stimulus pathway to the lateral amygdala. To accomplish this, they used an optogenetic approach (Chapter 5). They injected a virus into the amygdala to express channelrhodopsin-2, a light-activated excitatory cation channel, in lateral amygdala neurons. Following pairing of the auditory stimulus with a light pulse that depolarized lateral amygdala cells, presentation of the tone alone elicited conditioned freezing. The amount of freezing was greater when norepinephrine was present, which is further evidence that modulatory pathways also have a role in synaptic facilitation in this circuit. Thus, an aversive shock itself is not necessary to induce threat learning. Rather, it is the association of a stimulus with activation of the lateral amygdala that is key.

Other studies demonstrated the possibility of artificially manipulating the amygdala to impair as well as to instantiate, threat memory. They first trained animals to associate a foot shock with optogenetic stimulation of auditory inputs to the amygdala. They then delivered a pattern of optogenetic stimulation that generated *long-term depression* (LTD) of the auditory input to the amygdala, a form of synaptic plasticity in which weak, repetitive stimulation decreases the strength of synaptic transmission. Induction of LTD was able to inactivate the memory of the shock. Then, using a pattern of optical stimulation that produced LTP of the same auditory input, they found that the memory of the shock could be reinstated. The findings that inactivation and reactivation of a memory could be engineered using LTD and LTP strengthened the possible causal link between synaptic strength and behavioral memory storage.

The persistence of the synaptic changes underlying the memory for a threat depend on gene expression and protein synthesis in the amygdala, much like long-term memory in *Aplysia* and *Drosophila*. Thus, cAMP-dependent protein kinase and MAPK activate the transcription factor CREB to initiate gene expression. The importance of CREB is underscored by the finding that different neurons in the lateral amygdala have varying levels of CREB expression prior to threat conditioning. Neurons that express a larger than average amount of CREB are selectively recruited during learning. Conversely, if neurons with a large resting level of CREB are selectively ablated after learning, memory is impaired.

While most of the work on the neural mechanisms of threat conditioning has involved the lateral nucleus of the amygdala, in recent years, evidence has accumulated that plasticity in the central nucleus is also important. The central nucleus receives direct and indirect inputs from the lateral nucleus and forms synaptic connections with neurons in the periaqueductal gray region in the midbrain, which projects to the brain stem to control a number of defensive reactions, including freezing behavior. Within a lateral cell group of the central nucleus, inhibitory cells called PKC delta neurons control the activity of the output neurons in the medial cell group that project to the periaqueductal gray.

Memory for threat conditioning in humans also involves the amygdala. Thus, in humans, damage to the amygdala impairs the implicit memory of threat conditioning but not the explicit memory of having been conditioned. Functional imaging studies have found that the amygdala is activated by threats even when the person is not aware of the presence of the threat because the stimulus was subliminal. Although human studies are limited in their ability to reveal neurobiological details, they demonstrate the relevance of the animal work for human psychopathology.

In summary, Pavlovian threat conditioning has emerged as one of the most useful experimental

Figure 53–16 Training expands representation of inputs from the fingers in the cortex.

A. A monkey was trained for 1 hour per day to perform a task that required repeated use of the tips of fingers 2, 3, and occasionally 4. After training, the portion of area 3b of the somatosensory cortex representing the tips of the stimulated fingers (**dark color**) is substantially greater than normal (measured 3 months prior to training). (Adapted, with permission, from Jenkins et al. 1990.)

B. 1. A human subject trained to do a rapid sequence of finger movements will improve in accuracy and speed after 3 weeks of daily training (10–20 minutes each day). Functional magnetic resonance imaging scans of the primary motor cortex (based on local blood oxygenation level–dependent signals) after training show that the region activated in trained subjects (**orange region**) is larger than the region activated in untrained (controls). The control subjects received no training and performed unlearned finger movements using the same hand as control subjects. The change in cortical representation in trained subjects persisted for several months. (Reproduced, with permission, from Karni et al. 1998. Copyright © 1998 National Academy of Sciences.)

2. The size of the cortical representation of the fifth finger of the left hand is greater in string players than in nonmusicians. The graph plots the dipole strength obtained from magnetoencephalography, a measure of neural activity. The increase is most pronounced in musicians that began musical training before age 13. (Reproduced, with permission, from Elbert et al. 1995. Copyright © 1995 AAAS.)

A Monkey training

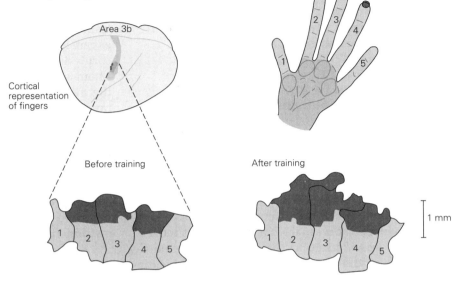

B Human training

1 Acquisition of a motor skill in adulthood

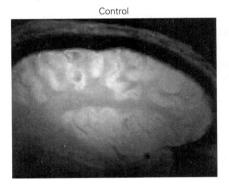

Control

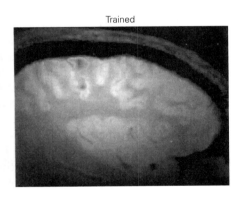

Trained

2 Cortical plasticity in childhood

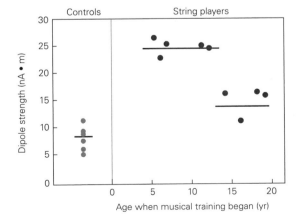

models for studying associative learning and memory in the mammalian brain. In part, this is due to the fact that the behavioral paradigm has been applied successfully across diverse species, from flies to humans, and thus builds upon the earlier progress that invertebrate models have made.

Learning-Induced Changes in the Structure of the Brain Contribute to the Biological Basis of Individuality

To what extent do the anatomical alterations in synapses required for long-term memory storage alter the large-scale functional architecture of the mature brain? The answer is well illustrated by the fact that the maps of the body surface in the primary somatic sensory cortex differ among individuals in a manner that reflects their use of specific sensory pathways. This remarkable finding results from the expansion or retraction of the connections of sensory pathways in the cortex according to the specific experience of the individual (Chapter 49).

The reorganization of afferent inputs as a result of behavior is also evident at lower levels in the brain, specifically at the level of the dorsal column nuclei, which contain the first synapses of the somatic sensory system. Therefore, organizational changes probably occur throughout the somatic afferent pathway.

The process by which experience alters the maps of somatosensory inputs in the cortex is illustrated in an experiment in which adult monkeys were trained to use their middle three fingers at the expense of other fingers to obtain food. After several thousand trials of this behavior, the area of cortex devoted to the middle fingers expanded greatly (Figure 53–16A). Thus, practice may expand synaptic connections by strengthening the effectiveness of existing connections.

The normal development of somatosensory input to cortical neurons may depend on the level of activity in neighboring afferent axons. In one experiment using monkeys, the skin surfaces of two adjacent fingers were surgically connected so that the connected fingers were always used together, thus ensuring that their afferent somatosensory axons were normally coactivated. As a result, the normally sharp discontinuity between the zones in the somatosensory cortex that receive inputs from these digits was abolished. Thus, normal development of the boundaries of representation of adjacent fingers in the cortex may be guided not only genetically but also through experience. Fine tuning of cortical connections may depend on associative mechanisms such as LTP, similar to the

role of cooperative activity in shaping the development of ocular dominance columns in the visual system (Chapter 49).

This plasticity is evident in humans as well. People trained to perform a task with their fingers show an expansion in the fMRI signal in the primary motor cortex during performance of the task (Figure 53–16B). Thomas Elbert explored the hand representation in the motor cortex of string instrument players. These musicians use their left hand for fingering the strings, manipulating the fingers in a highly individuated way. By contrast, the right hand, used for bowing, is used almost like a fist. The representation of the right hand in the cortex of string instrument players is the same as that of nonmusicians. But the representation of the left hand is greater than in nonmusicians and substantially more prominent in players who started to play their instrument prior to age 13 years (Figure 53–16B).

Because each of us is brought up in a somewhat different environment, experiencing different combinations of stimuli and developing motor skills in different ways, each individual's brain is uniquely modified. This distinctive modification of brain architecture, along with a unique genetic makeup, constitutes a biological basis for individuality.

Highlights

1. Many aspects of personality are guided by implicit memory. A great deal of what we experience—what we perceive, think, fantasize—is not directly controlled by conscious thought.

2. In mammals, both innate and learned defensive responses involve the amygdala. The amygdala-based defense system quickly learns about new dangers. It can associate a new neutral (conditioned) stimulus with a known threatening (unconditioned) stimulus on a single paired exposure, and this learned association is often retained throughout life.

3. During Pavlovian conditioning, the strength of synaptic transmission is modified in the lateral amygdala by pairing the conditioned and unconditioned stimuli. As a result, electrophysiological responses of neurons in the lateral amygdala are enhanced and behavioral learning occurs.

4. Many of the molecular mechanisms underlying threat conditioning in invertebrates also contribute to conditioning in mammals.

5. Damage to the human amygdala impairs implicit threat conditioning but does not affect the explicit memory of having been conditioned.

6. Habits are routines that are acquired gradually by repetition and are the result of a distinct form of implicit learning. As with all forms of implicit learning, habits are expressed in action alone, without conscious control, and independent of verbal reports.

7. As these arguments make clear, the empirical study of unconscious psychic processes was severely limited for many years by the lack of suitable experimental methods. Today, however, biology has a wide range of empirical methods that are providing cellular and molecular insights that are expanding our understanding of a wide range of mental activities.

Eric R. Kandel
Joseph LeDoux

Selected Reading

Alberini CM, Kandel ER. 2016. The regulation of transcription in memory consolidation. In: ER Kandel, Y Dudai, MR Mayford (eds). *Learning and Memory*, pp. 157–174. New York: Cold Spring Harbor Laboratory Press.

Bailey CH, Kandel ER, Harris KM. 2015. Structural components of synaptic plasticity and memory consolidation. Cold Spring Harb Perspect Biol 7:a021758.

Busto GU, Cervantes-Sandoval I, Davis RL. 2010. Olfactory learning in *Drosophila*. Physiology (Bethesda) 25:338–346.

Duvarci S, Pare D. 2014. Amygdala microcircuits controlling learned fear. Neuron 82:966–980.

Fanselow MS, Zelikowsky M, Perusini J, Barrera VR, Hersman S. 2014. Isomorphisms between psychological processes and neural mechanisms: from stimulus elements to genetic markers of activity. Neurobiol Learn Mem 108:5–13.

Hawkins RD, Kandel ER, Bailey CH. 2006. Molecular mechanisms of memory storage in *Aplysia*. Biol Bull 210:174–191.

LeDoux JE. 2014. Coming to terms with fear. Proc Natl Acad Sci U S A 111:2871–2878.

LeDoux J. 2015. *Anxious: Using the Brain to Understand and Treat Fear and Anxiety*. New York: Viking.

LeDoux JE. 2019. *The Deep History of Ourselves: The Four-Billion Year History of How We Got Conscious Brains*. New York: Viking.

Nader K. 2016. Reconsolidation and the dynamic nature of memory. In: ER Kandel, Y Dudai, MR Mayford (eds). *Learning and Memory*, pp. 245–260. New York: Cold Spring Harbor Laboratory Press.

Phelps EA. 2006. Emotion and cognition: insights from studies of the human amygdala. Annu Rev Psychol 57:27–53.

Tubon CT Jr, Yin JCP. 2008. CREB responsive transcription and memory formation. In: SM Dudek (ed). *Transcriptional Regulation by Neuronal Activity, Part III*, pp. 377–397. New York: Springer.

References

Bailey CH, Chen MC. 1983. Morphological basis of long-term habituation and sensitization in *Aplysia*. Science 220:91–93.

Bailey CH, Kandel ER, Si K. 2004. The persistence of long-term memory: a molecular approach to self-sustaining changes in learning-induced synaptic growth. Neuron 44:49–57.

Bear MF, Connors BW, Paradiso MA. 2001. *Neuroscience: Exploring the Brain,* 2nd ed. Chicago: Lippincott Williams & Wilkins.

Casadio A, Martin KC, Giustetto M, et al. 1999. A transient, neuron-wide form of CREB-mediated long-term facilitation can be stabilized at specific synapses by local protein synthesis. Cell 99:221–237.

Castellucci VF, Carew TJ, Kandel ER. 1978. Cellular analysis of long-term habituation of the gill-withdrawal reflex in *Aplysia californica*. Science 202:1306–1308.

Castellucci VF, Kandel ER. 1974. A quantal analysis of the synaptic depression underlying habituation of the gill-withdrawal reflex in *Aplysia*. Proc Natl Acad Sci U S A 71:5004–5008.

Claridge-Chang A, Roorda RD, Vrontou E, et al. 2009. Writing memories with light-addressable reinforcement circuitry. Cell 139:405–415.

Ehrlich DE, Josselyn SA. 2016. Plasticity-related genes in brain development and amygdala-dependent learning. Genes Brain Behav 15:125–143.

Eichenbaum H, Cohen NJ. 2001. *From Conditioning to Conscious Recollection: Memory Systems of the Brain*. Oxford: Oxford University Press.

Elbert T, Pantev C, Wienbruch C, Rockstroh B, Taub E. 1995. Increased cortical representation of the fingers of the left hand in string players. Science 270:305–307.

Glanzman DL, Kandel ER, Schacher S. 1990. Target-dependent structural changes accompanying long-term synaptic facilitation in *Aplysia* neurons. Science 249:799–802.

Greco JA, Liberzon I. 2016. Neuroimaging of fear-associated learning. Neuropsychopharmacology 41:320–334.

Gründemann J, Lüthi A. 2015. Ensemble coding in amygdala circuits for associative learning. Curr Opin Neurobiol 35:200–206.

Guan Z, Giustetto M, Lomvardas S, et al. 2002. Integration of long-term–memory-related synaptic plasticity involves bidirectional regulation of gene expression and chromatin structure. Cell 111:483–493.

Hawkins RD, Abrams TW, Carew TJ, Kandel ER. 1983. A cellular mechanism of classical conditioning in *Aplysia*: activity-dependent amplification of presynaptic facilitation. Science 219:400–405.

Hegde AN, Inokuchi K, Pei W, et al. 1997. Ubiquitin C-terminal hydrolase is an immediate-early gene essential for long-term facilitation in *Aplysia*. Cell 89:115–126.

Herry C, Johansen JP. 2014. Encoding of fear learning and memory in distributed neuronal circuits. Nat Neurosci 17:1644–1654.

Huang YY, Kandel ER. 1998. Postsynaptic induction and PKA-dependent expression of LTP in the lateral amygdala. Neuron 21:169–178.

Huang YY, Martin KC, Kandel ER. 2000. Both protein kinase A and mitogen-activated protein kinase are required in the amygdala for the macromolecular synthesis-dependent late phase of long-term potentiation. J Neurosci 20:6317–6325.

Janak PH, Tye KM. 2015. From circuits to behaviour in the amygdala. Nature 517:284–292.

Jenkins WM, Merzenich MM, Ochs MT, Allard T, Guic-Robles E. 1990. Functional reorganization of primary somatosensory cortex in adult owl monkeys after behaviorally controlled tactile stimulation. J Neurophysiol 63:82–104.

Johansen JP, Diaz-Mataix L, Hamanaka H, et al. 2014. Hebbian and neuromodulatory mechanisms interact to trigger associative memory formation. Proc Natl Acad Sci U S A 111:E5584–E5592.

Kandel ER. 2001. The molecular biology of memory storage: a dialogue between genes and synapses. Science 294:1030–1038.

Kandel ER. 2006. *In Search of Memory: The Emergence of a New Science of Mind*. New York: Norton.

Karni A, Meyer G, Rey-Hipolito C, et al. 1998. The acquisition of skilled motor performance: fast and slow experience-driven changes in primary motor cortex. Proc Natl Acad Sci U S A 95:861–868.

Keleman K, Krüttner S, Alenius M, Dickson BJ. 2007. Function of the *Drosophila* CPEB protein Orb2 in long-term courtship memory. Nat Neurosci 10:1587–1593.

Klein M, Kandel ER. 1980. Mechanism of calcium current modulation underlying presynaptic facilitation and behavioral sensitization in *Aplysia*. Proc Natl Acad Sci U S A 77:6912–6916.

Krabbe S, Gründemann J, Lüthi A. 2018. Amygdala inhibitory circuits regulate associative fear conditioning. Biol Psychiatry 83:800–809.

Mahan AL, Ressler KJ. 2011. Fear conditioning, synaptic plasticity and the amygdala: implications for posttraumatic stress disorder. Trends Neurosci 35:24–35.

Maren S. 2017. Synapse-specific encoding of fear memory in the amygdala. Neuron 95:988–990.

Maren S. 1999. Long-term potentiation in the amygdala: a mechanism for emotional learning and memory. Trends Neurosci 22:561–567.

Martin KC, Casadio A, Zhu H, et al. 1997. Synapse-specific, long-term facilitation of *Aplysia* sensory to motor synapses: a function for local protein synthesis in memory storage. Cell 91:927–938.

Nabavi S, Fox R, Proulx CD, Lin JY, Tsien RY, Malinow R. 2014. Engineering a memory with LTD and LTP. Nature 511:348–352.

Pape HC, Pare D. 2010. Plastic synaptic networks of the amygdala for the acquisition, expression, and extinction of conditioned fear. Physiol Rev 90:419–463.

Pavlov IP. 1927. *Conditioned Reflexes: An Investigation of the Physiological Activity of the Cerebral Cortex*. GV Anrep (transl). Oxford: Oxford University Press.

Pinsker H, Kupferman I, Castelucci V, Kandel ER. 1970. Habituation and dishabituation of the gill-withdrawal reflex in *Aplysia*. Science 167:1740–1742.

Rajasethupathy P, Antonov I, Sheridan R, et al. 2012. A role for neuronal piRNAs in the epigenetic control of memory-related synaptic plasticity. Cell 149:693–707.

Rogan MT, Leon KS, Perez DL, Kandel ER. 2005. Distinct neural signatures for safety and danger in the amygdala and striatum of the mouse. Neuron 46:309–320.

Sears RM, Fink AE, Wigestrand MB, Farb CR, de Lecea L, LeDoux JE. 2013. Orexin/hypocretin system modulates amygdala-dependent treat learning trough the locus coeruleus. Proc Natl Acad Sci U S A 110:20260–20265.

Sears RM, Schiff HC, LeDoux JE. 2014. Molecular mechanisms of threat learning in the lateral nucleus of the amygdala. Prog Mol Biol Transl Sci 122:263–304.

Si K, Giustetto M, Etkin A, et al. 2003. A neuronal isoform of CPEB regulates local protein synthesis and stabilizes synapse-specific long-term facilitation in *Aplysia*. Cell 115:893–904.

Si K, Lindquist S, Kandel ER. 2003. A neuronal isoform of the *Aplysia* CPEB has prion-like properties. Cell 115:879–891.

Spencer AW, Thompson RF, Nielson DR Jr. 1966. Response decrement of the flexion reflex in the acute spinal cat and transient restoration by strong stimuli. J Neurophysiol 29:240–252.

Squire LR, Kandel ER. 2008. *Memory: From Mind to Molecules*, 2nd ed. Greenwood Village: Roberts.

Yin JCP, Wallach JS, Del Vecchio M, et al. 1994. Induction of a dominant negative CREB transgene specifically blocks long-term memory in *Drosophila*. Cell 79:49–58.

54

The Hippocampus and the Neural Basis of Explicit Memory Storage

EXPLICIT MEMORY—THE CONSCIOUS recall of information about people, places, objects, and events—is what people commonly think of as memory. Sometimes called *declarative memory*, it binds our mental life together by allowing us to recall at will what we ate for breakfast, where we ate it, and with whom. It allows us to join what we did today with what we did yesterday or the week or month before that.

Two structures in the mammalian brain are particularly critical for encoding and storing explicit memory: the prefrontal cortex and the hippocampus (Chapter 52). The prefrontal cortex mediates working memory, which can be actively maintained for only very short periods and is then rapidly forgotten, such as a password that is remembered only until it is entered. Information in working memory can be stored elsewhere in the brain as long-term memory for periods ranging from days to weeks to years, and throughout a lifetime. Although long-term storage of explicit memory requires the hippocampus, the ultimate storage site for most declarative memory is thought to be the cerebral cortex.

In this chapter, we focus on the cellular, molecular, and network mechanisms of the hippocampus that underlie the long-term storage of explicit memory. Because the hippocampus receives its major input from a region of the cerebral cortex called the entorhinal cortex, an area that processes many forms of sensory input, we also consider how information from the entorhinal cortex is transformed by the hippocampus. In particular, we examine how neural activity in the entorhinal cortex and hippocampus contributes to spatial memory by encoding a representation of an animal's location in its environment.

Explicit Memory in Mammals Involves Synaptic Plasticity in the Hippocampus

Unlike working memory, which is thought to be maintained by ongoing neural activity in the prefrontal cortex (Chapter 52), the long-term storage of information is thought to depend on long-lasting changes in the strength of connections among specific ensembles of neurons (neural assemblies) in the hippocampus that encode particular elements of memory.

The idea that memory storage involves long-lasting structural changes in the brain, first referred to as an "engram" by the German biologist Richard Semon in the early 20th century, dates back to the French philosopher Rene Descartes. In an attempt to locate an engram, the American psychologist Karl Lashley examined the effects of lesions in different regions of the neocortex on the ability of a rat to learn to navigate a maze. Since the performance in the maze seemed to be directly proportional to the size of the lesion, rather than its precise location, Lashley concluded that any memory trace must be distributed throughout the brain. Although it is now generally accepted that storage of an explicit memory is distributed throughout the neocortex, it is also clear that the process of storing memory requires the hippocampus, as demonstrated by the pioneering studies of Brenda Milner on patient H.M. (Chapter 52) and subsequent studies in animals with targeted lesions of the hippocampus. Thus, understanding how the brain stores explicit memory depends on an understanding of how the cortico-hippocampal circuit processes and stores information.

The nature of the basic mechanisms for memory storage was and remains the subject of much speculation and debate among psychologists and neuroscientists. One influential theory was proposed by the Canadian psychologist Donald Hebb, who suggested in 1949 that memory-encoding neural assemblies may be generated when synaptic connections are strengthened based on experience. According to *Hebb's rule*: "When an axon of cell A . . . excites cell B and repeatedly or persistently takes part in firing it, some growth process or metabolic change takes place in one or both cells so that A's efficiency as one of the cells firing B is increased." The key element of Hebb's rule is the requirement for coincidence of pre- and postsynaptic firing, and so the rule has sometimes been rephrased as "Cells that fire together, wire together." A similar Hebbian coincidence principle is thought to be involved

in fine-tuning synaptic connections during the late stages of development (Chapter 49). Hebb's ideas were later refined by the theoretical neuroscientist David Marr, based on a consideration of the hippocampal circuit.

The hippocampus comprises a loop of connections that process multimodal sensory and spatial information from the superficial layers of the nearby entorhinal cortex. This information passes through multiple synapses before arriving at the hippocampal CA1 region, the major output area of the hippocampus. The critical importance of CA1 neurons in learning and memory is seen in the profound memory loss exhibited by patients with lesions in this region alone, an observation supported by numerous animal studies. Information from the entorhinal cortex reaches CA1 neurons along two excitatory pathways, one direct and one indirect.

In the indirect pathway, the axons of neurons in layer II of the entorhinal cortex project through the *perforant pathway* to excite the granule cells of the dentate gyrus (an area considered part of the hippocampus). Next, the axons of the granule cells project in the *mossy fiber pathway* to excite the pyramidal cells in the CA3 region of the hippocampus. Finally, axons of the CA3 neurons project through the *Schaffer collateral pathway* to make excitatory synapses on more proximal regions of the dendrites of the CA1 pyramidal cells (Figure 54–1). (Because of its three successive excitatory synaptic connections, the indirect pathway is often referred to as the *trisynaptic pathway*). Finally, CA1 pyramidal cells project back to the deep layers of entorhinal cortex and forward to the subiculum, another medial temporal lobe structure that connects the hippocampus with a wide diversity of brain regions.

In parallel with the indirect pathway, the entorhinal cortex also projects directly to CA3 and CA1 hippocampal regions. In the direct pathway to CA1, neurons in layer III of the entorhinal cortex send their axons through the *perforant pathway* to form excitatory synapses on the very distal regions of the apical dendrites of CA1 neurons (such projections are also called the *temporoammonic pathway*). Interactions between direct and indirect inputs at each stage of the hippocampal circuit are likely important for memory storage or recall, although the precise nature of these interactions remains to be determined.

In addition to the above pathways that link different stages of the hippocampal circuit, CA3 pyramidal neurons also make strong excitatory connections with one another. This self-excitation through recurrent collaterals is thought to contribute to associative aspects of memory storage and recall. Under pathological conditions, such self-excitation can lead to seizures.

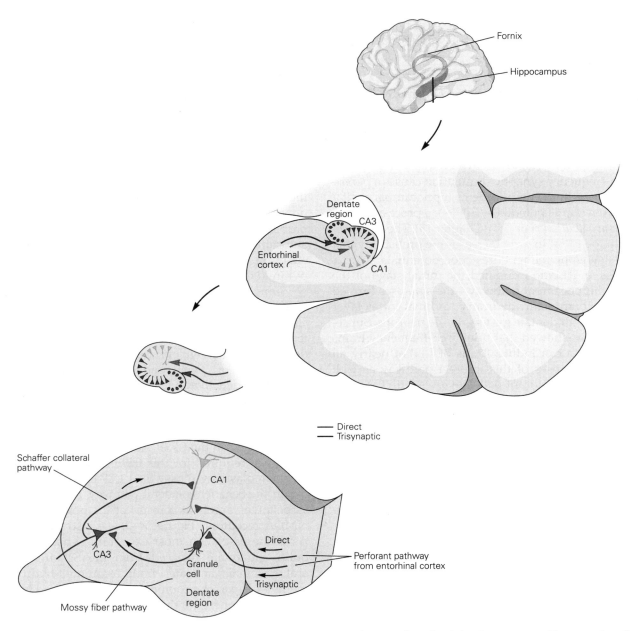

Figure 54–1 The cortico-hippocampal synaptic circuit is important for declarative memory. Information arrives in the hippocampus from the entorhinal cortex through the perforant pathways, which provide both direct and indirect input to pyramidal neurons in area CA1, the major output neurons of the hippocampus. (**Arrows** denote the direction of impulse flow.) The indirect *trisynaptic pathway* has three component connections. Neurons in layer II of the entorhinal cortex send their axons through the perforant path to make excitatory synapses onto the granule cells of the dentate gyrus. The granule cells project through the mossy fiber pathway and make excitatory synapses with the pyramidal cells in area CA3 of the hippocampus. The CA3 cells excite the pyramidal cells in CA1 by means of the Schaffer collateral pathway. In the *direct pathway*, neurons in layer III of the entorhinal cortex project through the perforant path to make excitatory synapses on the distal dendrites of CA3 and CA1 pyramidal neurons without inter vening synapses (shown only for CA1).

Finally, neurons in the relatively small CA2 region, located between CA3 and CA1, receive information from entorhinal cortex layer II through both a direct pathway and an indirect pathway via the dentate gyrus and CA3. The CA2 region also receives strong input from hypothalamic nuclei that release oxytocin and vasopressin, hormones important for social behavior. In turn, CA2 sends a strong output to CA1, providing CA1 with a third source of excitatory input (in addition to the direct and trisynaptic routes from the entorhinal cortex).

Long-Term Potentiation at Distinct Hippocampal Pathways Is Essential for Explicit Memory Storage

How is information stored in the hippocampal circuit to provide a long-lasting memory trace? In 1973, Timothy Bliss and Terje Lømo discovered that a brief period of high-frequency synaptic stimulation causes a persistent increase in the amplitude of hippocampal excitatory postsynaptic potentials (EPSPs), a process termed *long-term potentiation* or LTP (Chapter 13). The enhancement in the EPSP, in turn, increases the probability that the postsynaptic cell will fire action potentials.

Bliss and Lømo examined the initial stage of the indirect hippocampal pathway—the synapses formed by the perforant pathway from entorhinal cortex layer II neurons with dentate gyrus granule neurons. Subsequent studies showed that brief high-frequency trains of stimulation can induce forms of LTP at nearly all excitatory synapses of this indirect pathway as well as at the direct perforant path synapses with CA3 and CA1 neurons (Figure 54–2). LTP can last for days or even weeks when induced in intact animals using implanted electrodes and can last several hours in isolated slices of hippocampus and in hippocampal neurons in cell culture.

Studies in the different hippocampal pathways have shown that LTP at different synapses is not a single process. Rather, it comprises a family of processes that strengthen synaptic transmission at different hippocampal synapses through distinct cellular and molecular mechanisms. Indeed, even at a single synapse, different forms of LTP can be induced by different patterns of synaptic activity, although these distinct processes share many important similarities.

All forms of LTP are induced by synaptic activity in the pathway that is being potentiated—that is, LTP is homosynaptic. In addition, LTP is synapse specific; only those synapses that are activated by the tetanic stimulation are potentiated. However, the various forms of LTP differ in their dependence on specific receptors and ion channels. In addition, different forms of LTP recruit different second-messenger signaling pathways that act at different synaptic sites. Some forms of LTP result from an enhancement of the postsynaptic response to the neurotransmitter glutamate, whereas other forms of LTP result from the enhancement of glutamate release from the presynaptic terminal, and still other forms of LTP engage both the presynaptic and postsynaptic neurons.

The similarities and differences in the mechanisms of different forms of LTP can be seen by comparing LTP at Schaffer collateral, mossy fiber, and direct entorhinal synapses. In all three pathways, synaptic transmission is persistently enhanced in response to a brief tetanic stimulation. However, the contribution of the N-methyl-D-aspartate (NMDA) receptor to the induction of LTP differs in the three pathways. At the Schaffer collateral synapses, the induction of LTP in response to a brief 100-Hz stimulation is completely blocked when the tetanus is applied in the presence of the NMDA receptor antagonist 2-amino-5-phosphonovaleric acid (AP5 or APV). In contrast, APV only partially inhibits the induction of LTP at the direct entorhinal synapses with CA1 neurons and has no effect on LTP at the mossy fiber synapses with CA3 pyramidal neurons (Figure 54–2).

Long-term potentiation in the mossy fiber pathway is largely presynaptic and is triggered by the large Ca^{2+} influx into the presynaptic terminals during the tetanus. The Ca^{2+} influx activates a calcium/calmodulin-dependent adenylyl cyclase, thereby increasing the production of cyclic adenosine monophosphate (cAMP) and activating protein kinase A (PKA; see Chapter 14). This leads to the phosphorylation of presynaptic vesicle proteins that enhance the release of glutamate from the mossy fiber terminals, resulting in an increase in the EPSP. Activity in the postsynaptic cell is not required for this form of LTP. Thus, unlike Hebbian plasticity, mossy fiber LTP is nonassociative.

In the Schaffer collateral pathway, however, LTP is associative, largely as a result of the properties of the NMDA receptors (Figure 54–3; see also Chapter 13). As is the case with most excitatory synapses in the brain, glutamate released from the Schaffer collateral terminals activates both α-amino-3-hydroxy-5-methyl-4-isoxazolepropionic acid (AMPA) and NMDA receptor-channels in the postsynaptic membrane of CA1 pyramidal neurons. However, unlike the AMPA receptors, activation of the NMDA receptors is associative because it requires simultaneous presynaptic and postsynaptic activity. This is because the pore of the NMDA receptor-channel is normally blocked by extracellular Mg^{2+} at typical negative resting potentials, which prevents these channels from conducting ions in response to glutamate. For the NMDA receptor-channel to function efficiently, the postsynaptic membrane must be depolarized sufficiently to expel the bound Mg^{2+} by electrostatic repulsion. In this manner, the NMDA receptor-channel acts as a coincidence detector: It is functional only when (1) the action potentials in the presynaptic neuron release glutamate that binds to the receptor *and* (2) the membrane of the postsynaptic

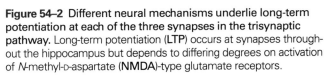

Figure 54–2 Different neural mechanisms underlie long-term potentiation at each of the three synapses in the trisynaptic pathway. Long-term potentiation (**LTP**) occurs at synapses throughout the hippocampus but depends to differing degrees on activation of N-methyl-D-aspartate (**NMDA**)-type glutamate receptors.

A. Tetanic stimulation of the Schaffer collateral fibers (at time 0 in the plot) induces LTP at the synapses between presynaptic CA3 pyramidal neurons and postsynaptic CA1 pyramidal neurons. The plot shows the size of the extracellular field excitatory postsynaptic potential (**fEPSP**) as a percentage of the baseline fEPSP prior to induction of LTP. At these synapses, LTP requires activation of the NMDA receptor-channels in the postsynaptic CA1 neurons as it is completely blocked when the tetanus is delivered in the presence of the NMDA receptor antagonist 2-amino-5-phosphonovaleric acid (**APV**). (Adapted from Morgan and Teyler 2001.)

B. Tetanic stimulation of the direct pathway from entorhinal cortex to CA1 neurons generates LTP of the fEPSP that depends partly on activation of the NMDA receptor-channels and partly on activation of L-type voltage-gated Ca^{2+} channels. It is therefore only partially blocked by APV. Addition of APV and nifedipine, a dihydropyridine that blocks L type channels, is needed to fully inhibit LTP.

C. Tetanic stimulation of the mossy fiber pathway induces LTP at the synapses with the pyramidal cells in the CA3 region. In this experiment, the excitatory postsynaptic current (**EPSC**) was measured under voltage-clamp conditions. This LTP does not require activation of the NMDA receptors and so is not blocked by APV. However, it does require activation of protein kinase A (**PKA**) and so is blocked by the kinase inhibitor H-89. (Reproduced, with permission, from Zalutsky and Nicoll 1990. Copyright © 1990 AAAS.)

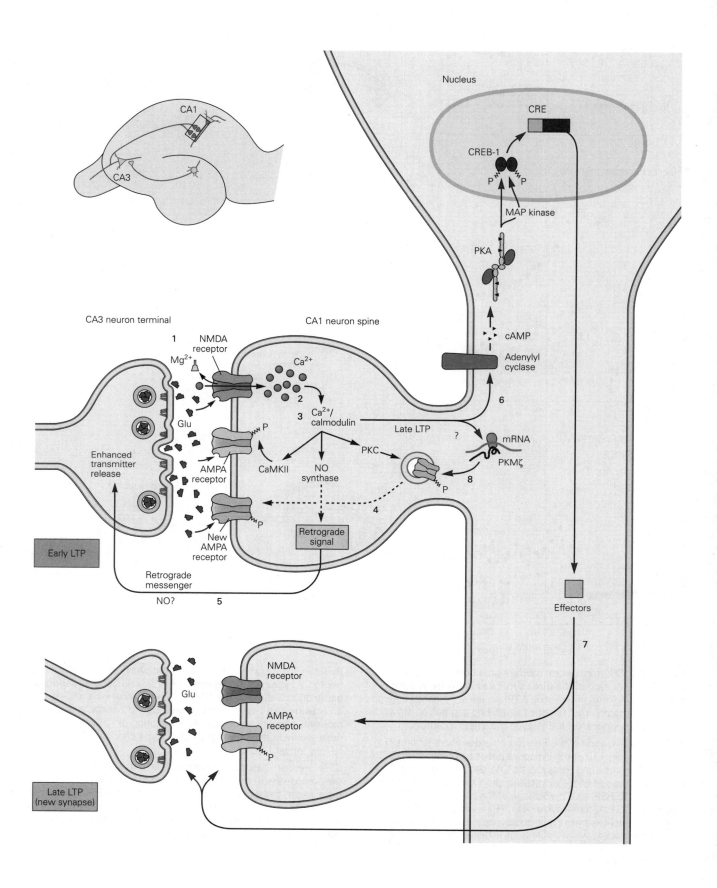

cell is sufficiently depolarized by strong synaptic activity to relieve the Mg^{2+} block. Thus, the NMDA receptor is able to associate presynaptic and postsynaptic activity to recruit plasticity mechanisms that strengthen connections between pairs of cells, fulfilling Hebb's coincidence requirement for synaptic modification.

What are the functional consequences of the activation of NMDA receptors by strong synaptic excitation? Whereas most AMPA receptor-channels conduct only monovalent cations (Na^+ and K^+), the NMDA receptor-channels have a high permeability to Ca^{2+} (Chapter 13). Thus, the opening of these channels leads to a significant increase in the Ca^{2+} concentration in the postsynaptic cell. The increase in intracellular Ca^{2+} activates several downstream signaling pathways—including calcium/calmodulin–dependent protein kinase II (CaMKII), protein kinase C (PKC), and tyrosine kinases—that lead to changes that enhance the magnitude of the EPSP at Schaffer collateral synapses (Figure 54–3).

Different Molecular and Cellular Mechanisms Contribute to the Forms of Expression of Long-Term Potentiation

Neuroscientists often find it useful to distinguish between the *induction* of LTP (the biochemical reactions activated by the tetanic stimulation) and the *expression* of LTP (the long-term changes responsible for enhanced synaptic transmission). The mechanisms for the induction of LTP at the CA3-CA1 synapse are largely postsynaptic. Is the expression of LTP at this synapse caused by an increase in transmitter release, an increased postsynaptic response to a fixed amount of transmitter, or some combination of the two?

A number of lines of experiments suggest that the form of expression of LTP depends on the type of synapse and precise pattern of activity that induces LTP.

In many cases, the expression of LTP in CA1 neurons in response to Ca^{2+} influx through NMDA receptor-channels depends on an increase in the response of the postsynaptic membrane to glutamate. But stronger patterns of stimulation can elicit forms of LTP at the same synapse whose expression depends on presynaptic events that enhance transmitter release.

One of the key pieces of evidence for a postsynaptic contribution to the expression of LTP at Schaffer collateral synapses comes from an examination of so-called "silent synapses." In some recordings from pairs of hippocampal pyramidal neurons, stimulation of an action potential in one neuron fails to elicit a response in the postsynaptic neuron when that neuron is at its resting potential (approximately –70 mV). This result is not surprising, as each hippocampal presynaptic neuron is connected to only a small number of other neurons. What *is* surprising is that in some neuronal pairs that appear unconnected when the postsynaptic membrane is initially at –70 mV, stimulation of the same presynaptic neuron is able to elicit a large excitatory postsynaptic current in the second neuron when the second neuron is depolarized under voltage clamp to +30 mV. In such neuronal pairs, the postsynaptic membrane appears to lack functional AMPA receptors so that the excitatory postsynaptic current (EPSC) is mediated solely by NMDA receptors-channels. As a result, there is no measurable EPSC when the membrane is held at the cell's resting potential (–70 mV) because of the strong Mg^{2+} block of these receptor-channels (the synapse is effectively silent). However, a large EPSC can be generated at +30 mV because the depolarization relieves the block (Figure 54–4).

The key finding from these experiments is seen following the induction of LTP using strong synaptic stimulation. Pairs of neurons initially connected solely by silent synapses now often exhibit large EPSPs at the

Figure 54–3 (Opposite) A model for the induction of long-term potentiation (LTP) at Schaffer collateral synapses. A single high-frequency tetanus induces early LTP. The large depolarization of the postsynaptic membrane (caused by strong activation of the α-amino-3-hydroxy-5-methyl-4-isoxazolepropionic acid [AMPA] receptors) relieves the Mg^{2+} blockade of the N-methyl-D-aspartate (NMDA) receptor-channels (1), allowing Ca^{2+}, Na^+, and K^+ to flow through these channels. The resulting increase of Ca^{2+} in the dendritic spine (2) triggers calcium-dependent kinases (3)—calcium/calmodulin–dependent kinase (CaMKII) and protein kinase C (PKC)—leading to induction of LTP. Second-messenger cascades activated during induction of LTP have two main effects on synaptic transmission. Phosphorylation through activation of protein kinases, including PKC, enhances current through the AMPA receptor-channels, in part by causing insertion of new receptors into the spine synapses

(4). In addition, the postsynaptic cell releases retrograde messengers, such as nitric oxide (NO), that activate protein kinases in the presynaptic terminal to enhance subsequent transmitter release (5). Repeated bouts of tetanic stimulation induce late LTP. The prolonged increase in Ca^{2+} influx recruits adenylyl cyclase (6), which generates cyclic adenosine monophosphate (cAMP) that activates protein kinase A (PKA). This leads to the activation of MAP kinase, which translocates to the nucleus where it phosphorylates CREB-1. CREB-1 in turn activates transcription of targets (containing the CRE promoter) that are thought to lead to the growth of new synaptic connections (7). Repeated stimulation also activates translation of mRNA encoding PKMζ, a constitutively active isoform of PKC (8). This leads to a long-lasting increase in the number of AMPA receptors in the postsynaptic membrane.

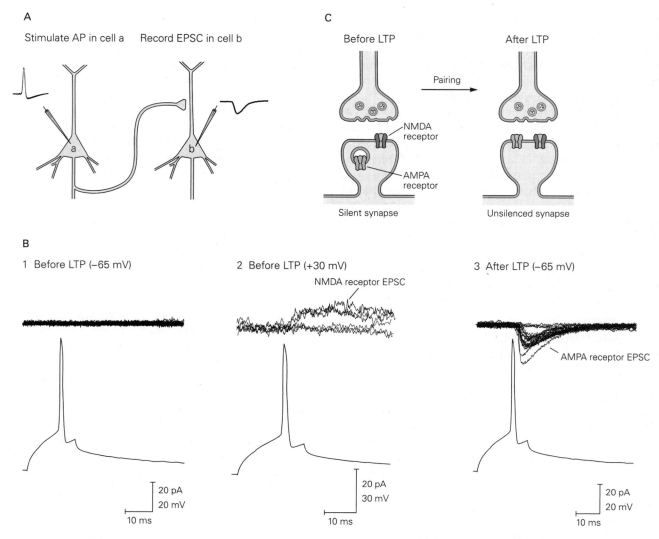

Figure 54–4 Adding α-amino-3-hydroxy-5-methyl-4-isoxazolepropionic acid (**AMPA**) receptors to silent synapses during long-term potentiation (**LTP**).

A. Intracellular recordings are obtained from a pair of hippocampal pyramidal neurons. An action potential (**AP**) is triggered in neuron *a* by a depolarizing current pulse, and the resultant excitatory postsynaptic current (**EPSC**) produced in neuron *b* is recorded under voltage-clamp conditions.

B. Before induction of LTP, there is no EPSC in cell *b* (top traces) in response to an action potential in cell *a* (bottom traces) when the membrane potential of neuron *b* is at its resting value of −65 mV (1). However, when neuron *b* is depolarized by the voltage clamp to +30 mV, the

N-methyl-D-aspartate (**NMDA**) receptors are activated and slow EPSCs characteristic of these receptors are observed (2). LTP is then induced by pairing action potentials in neuron *a* with postsynaptic depolarization in neuron *b* to relieve the Mg^{2+} block of the NMDA receptors. After this pairing, fast EPSCs initiated by activation of AMPA receptors are seen in cell *b* (3). (Reproduced, with permission, from Montgomery, Pavlidis, and Madison 2001. Copyright © 2001 Cell Press.)

C. Mechanism of the unsilencing of silent synapses. Prior to LTP, the dendritic spine contacted by a presynaptic CA3 neuron contains only NMDA receptors. Following induction of LTP, intracellular vesicles containing AMPA receptors fuse with the plasma membrane at the synapse, adding AMPA receptors to the membrane.

negative resting potential, and these EPSPs are mediated by AMPA receptors. The simplest interpretation of this result is that LTP somehow recruits new functional AMPA receptors to the silent synapse membrane, a process Roberto Malinow refers to as "AMPAfication."

How does the induction of LTP increase the response of AMPA receptors? The strong synaptic stimulation used to induce LTP triggers glutamate release at both silent and nonsilent synapses on the same postsynaptic neuron. This leads to the opening of a large number of AMPA receptor-channels at the nonsilent synapses, which in turn produces a large postsynaptic depolarization. The depolarization then propagates throughout the neuron, thus relieving Mg^{2+} block of the NMDA receptor-channels at both the non-silent and silent synapses. At the silent synapses, the Ca^{2+} influx through the NMDA receptor-channels activates a biochemical cascade that ultimately leads to the insertion of clusters of AMPA receptors in the postsynaptic membrane. These newly inserted AMPA receptors are thought to come from a reserve pool stored in endosomal vesicles within dendritic spines, the site of all excitatory input to pyramidal neurons (Chapter 13). Calcium influx through the NMDA receptor-channels elevates spine Ca^{2+} levels, triggering a postsynaptic signaling cascade that leads to phosphorylation of the cytoplasmic tail of the vesicular AMPA receptors by PKC (Chapter 14), leading to their insertion in the postsynaptic membrane (Figure 54–3).

Because the induction of almost all forms of postsynaptic LTP requires Ca^{2+} influx into the postsynaptic cell, the finding that transmitter release is enhanced during some forms of LTP implies that the presynaptic cell must receive a signal from the postsynaptic cell that LTP has been induced. There is now evidence that calcium-activated second messengers in the postsynaptic cell, or perhaps Ca^{2+} itself, cause the postsynaptic cell to release one or more chemical messengers, including the gas nitric oxide, that diffuse to the presynaptic terminals to enhance transmitter release (Figure 54–3 and Chapter 14). Importantly, these diffusible retrograde signals appear to affect only those presynaptic terminals that have been activated by the tetanic stimulation, thereby preserving synapse specificity.

Long-Term Potentiation Has Early and Late Phases

Long-term potentiation has two phases, early and late, that provide a means of regulating the duration

of the enhancement of synaptic transmission. The phase we have focused on up to now lasts for only 1 to 3 hours and is termed early LTP; this phase is typically induced by a single train of 100-Hz tetanic stimulation for 1 second. More prolonged periods of activity (using three or four trains of 100-Hz tetanic stimulation, each lasting 1 second) induce a late phase of LTP that can last 24 hours or even longer. Unlike early LTP, late LTP requires the synthesis of new proteins (Figure 54–5).

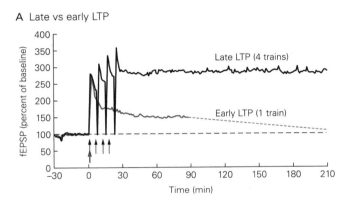

A Late vs early LTP

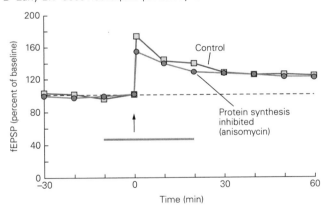

B Early LTP does not require protein synthesis

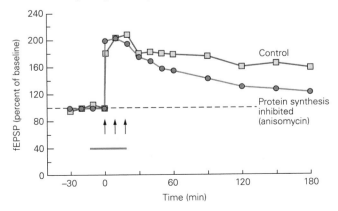

C Late LTP requires protein synthesis

Figure 54–5 (Right) Long-term potentiation (LTP) in the CA1 region of the hippocampus has early and late phases.

A. Early LTP is induced by a single tetanus lasting 1 second at 100 Hz, whereas late LTP is induced by four tetani given 10 minutes apart. Early LTP of the field excitatory postsynaptic potential (fEPSP) lasts only 1 to 2 hours, whereas the late LTP lasts more than 8 hours (only the first 3.5 hours are shown).

B. Early LTP induced by one tetanus is not blocked by anisomycin (**bar**), an inhibitor of protein synthesis.

C. Late LTP, normally induced by three trains of stimulation, is blocked by anisomycin. (Three or four trains can be used to induce late LTP.) (Panels B and C reproduced, with permission, from Huang and Kandel 1994.)

Whereas the early phase of LTP is mediated by changes at existing synapses, late LTP is thought to result from the growth of new synaptic connections between pairs of co-activated neurons.

Although the mechanisms for early LTP in the Schaffer collateral and mossy fiber pathways are quite different, the mechanisms for late LTP in the two pathways appear similar (Figure 54–3). In both pathways, late LTP recruits the cAMP and PKA signaling pathway to activate by phosphorylation the cAMP response element binding protein (CREB) transcription factor, leading to the synthesis of new mRNAs and proteins. Like sensitization of the gill-withdrawal reflex in *Aplysia*, which also involves cAMP, PKA, and CREB (Chapter 53), late LTP in the Schaffer collateral pathway is synapse specific. When two independent sets of synapses in the same postsynaptic CA1 neuron are stimulated using two electrodes spaced some distance apart, the application of four trains of tetanic stimulation to one set of synapses induces late LTP only at the activated synapses; synaptic transmission is not altered at the second set of synapses that were not tetanized.

How can late LTP achieve synapse specificity given that transcription and most translation occurs in the cell body, such that newly synthesized proteins should be available to all synapses of a cell? To explain synapse specificity, Uwe Frey and Richard Morris proposed the synaptic capture hypothesis, in which synapses that are activated during the tetanus are tagged in some way, perhaps by protein phosphorylation, that enables them to make use of ("capture") the newly synthesized proteins. Frey and Morris tested this idea using the two-pathway protocol described above. They delivered four tetani to induce late LTP at one set of synapses with one electrode and delivered a single tetanus to a second set of synapses with the other electrode. Although a single tetanus on its own induces only early LTP, it is able to induce late LTP when delivered within 2–3 hours of the four tetani from the first electrode. This phenomenon is similar to the synapse-specific capture of long-term facilitation at the sensory-motor neuron synapses in *Aplysia* (Chapter 53).

According to Frey and Morris, the single train of tetanic stimulation, although not sufficient to induce new protein synthesis, is sufficient to tag the activated synapses, allowing them to capture the newly synthesized proteins produced in response to the prior delivery of the four trains of tetanic stimulation. The increased synaptic plasticity that this tagging mechanism affords, and its limitation to the period when newly synthesized proteins are around, may explain the recent finding that hippocampal cell assemblies that store memories of events closely spaced in time have a larger number of common neurons than do cell assemblies for events widely separated in time.

How can a few brief trains of synaptic stimulation produce such long-lasting increases in synaptic transmission? One mechanism proposed by John Lisman depends on the unique properties of CaMKII. After a brief exposure to Ca^{2+}, CaMKII can be converted to a calcium-independent state through its autophosphorylation at threonine-286 (Thr286). This ability to become persistently active in response to a transient Ca^{2+} stimulus has led to the suggestion that CaMKII may act as a simple molecular switch that can extend the duration of LTP following its initial activation.

Studies from Todd Sacktor have suggested that longer-lasting changes that maintain late LTP may depend on an atypical isoform of PKC termed PKMζ (PKM zeta). Most isoforms of PKC contain both a regulatory domain and a catalytic domain (Chapter 14). Binding of diacylglycerol, phospholipids, and Ca^{2+} to the regulatory domain relieves inhibitory domain binding to the catalytic domain, allowing PKC to phosphorylate its protein substrates. In contrast, PKMζ lacks a regulatory domain and so is constitutively active.

Levels of PKMζ in the hippocampus are normally low. Tetanic stimulation that induces LTP leads to an increase in synthesis of PKMζ through enhanced translation of its mRNA. Because this mRNA is present in the CA1 neuron dendrites, its translation can rapidly alter synaptic strength. Blockade of PKMζ with a peptide inhibitor during the tetanic stimulation blocks late LTP but not early LTP. If the blocker is applied several hours after LTP induction, the late LTP that had been established will be reversed. This result indicates that the maintenance of late LTP requires the ongoing activity of PKMζ to maintain the increase in AMPA receptors in the postsynaptic membrane (Figure 54–3). A second atypical PKC isoform may substitute for PKMζ under certain conditions, which may explain the surprising finding that genetic deletion of PKMζ has little effect on late LTP.

Constitutively active forms of protein kinases may not be the only mechanism for maintaining long-lasting synaptic changes in the hippocampus. Repeated stimulation may lead to the formation of new synaptic connections, just as long-term facilitation leads to the formation of new synapses during learning in *Aplysia*. In addition, long-lasting synaptic changes likely involve epigenetic changes in chromatin structure. During late LTP, phosphorylated CREB activates gene expression by recruiting the CREB binding protein (CBP), which acts as a histone acetylase, transferring an acetyl group to specific lysine residues on histone proteins, and thereby producing long-lasting changes in gene expression. Mutations in CBP impair late LTP and learning and memory in mice. In humans, de novo mutations in the CBP gene underlie Rubinstein-Taybi syndrome, a developmental disorder associated with intellectual impairment. Other studies implicate a second epigenetic mechanism, DNA methylation, in long-lasting synaptic plasticity and learning and memory.

Spike-Timing-Dependent Plasticity Provides a More Natural Mechanism for Altering Synaptic Strength

Under most circumstances, hippocampal neurons do not produce the high-frequency trains of action potentials typically used to induce LTP experimentally. However, a form of LTP termed spike-timing-dependent plasticity (STDP) can be induced by a more natural pattern of activity in which a single presynaptic stimulus is paired with the firing of a single action potential in the postsynaptic cell at a relatively low frequency (eg, one pair per second over several seconds). However, the presynaptic cell must fire just before the postsynaptic cell. If instead the postsynaptic cell fires just before the EPSP, a long-lasting decrease in the size of the EPSP occurs. Such long-term depression of synaptic transmission represents a distinct form of synaptic plasticity from LTP and is described more fully below. If the postsynaptic action potential occurs more than

a hundred milliseconds before or after the EPSP, the synaptic strength will not change.

The pairing rules of STDP thus follow Hebb's postulate and result in large part from the cooperative properties of the NMDA receptor-channel. If the postsynaptic spike occurs during the EPSP, it is able to relieve the Mg^{2+} blockade of the channel at a time when the NMDA receptor has been activated by the binding of glutamate. This leads to a large influx of Ca^{2+} through the receptor and the induction of STDP. However, if the postsynaptic action potential occurs prior to the presynaptic release of glutamate, any relief from the Mg^{2+} block will occur when the gate of the receptor-channel is closed (because of the absence of glutamate). As a result, there will be only a small influx of Ca^{2+} through the receptor that is insufficient to induce STDP.

Long-Term Potentiation in the Hippocampus Has Properties That Make It Useful as A Mechanism for Memory Storage

NMDA receptor–dependent LTP at the Schaffer collateral pathway and other hippocampal pathways has three properties with direct relevance to learning and memory (Figure 54–6). First, LTP in such pathways requires the near-simultaneous activation of a large number of afferent inputs, a feature called *cooperativity* (Figure 54–6). This requirement stems from the fact

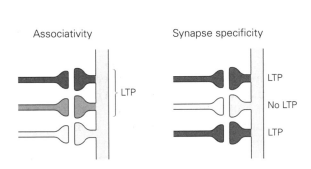

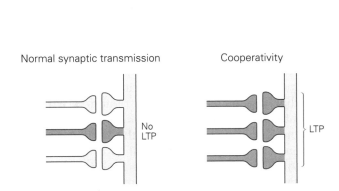

Figure 54–6 Long-term potentiation (LTP) in CA1 pyramidal neurons of the hippocampus shows cooperativity, associativity, and synapse specificity. With normal synaptic transmission, a single action potential in one or a few axons (weak input) leads to a small excitatory postsynaptic potential (EPSP) that is insufficient to expel Mg^{2+} from the N-methyl-D-aspartate (NMDA) receptor-channels and thus cannot induce LTP. This ensures that irrelevant stimuli are not remembered. The near-simultaneous activation of several weak inputs

during strong activation (cooperativity) produces a suprathreshold EPSP that triggers an action potential, resulting in LTP in all pathways. Stimulation of strong and weak inputs together (associativity) causes LTP in both pathways. In this way, a weak input becomes significant when paired with a powerful one. An unstimulated synapse does not undergo LTP despite the strong stimulation of neighboring synapses. This ensures that memory is selectively stored at active synapses (synapse specificity).

that relief of Mg^{2+} block of the NMDA receptor-channel requires a large depolarization, which is achieved only when the postsynaptic cell receives input from a large number of presynaptic cells.

Second, LTP at synapses with NMDA receptor-channels is *associative*. A weak presynaptic input normally does not produce enough postsynaptic depolarization to induce LTP. However, if the weak input is paired with a strong input that produces a suprathreshold depolarization, the resulting large depolarization will propagate to the synapses with weak input, leading to relief of the Mg^{2+} blockade of the NMDA receptors and induction of LTP at those synapses.

Third, NMDA receptor-dependent LTP is *synapse specific*. If a particular synapse is not activated during a period of strong synaptic stimulation, the NMDA receptors at that site will not be able to bind glutamate and thus will not be activated despite the strong postsynaptic depolarization. As a result, that synapse will not undergo LTP.

Each of these three properties—cooperativity, associativity, and synapse specificity—underlies a key requirement of memory storage. Cooperativity ensures that only events of a high degree of significance, those that activate sufficient inputs, will result in memory storage. Associativity, like associative Pavlovian conditioning, allows an event (or conditioned stimulus) that has little significance in and of itself to be endowed with a higher degree of meaning if that event occurs just before or simultaneously with another more significant event (an unconditioned stimulus). In a network with strong recurrent connections, such as CA3, associative LTP enables a pattern of activity in one group of cells to become linked to a distinct pattern of activity in a separate, but partially overlapping, group of synaptically coupled cells. Such linkages of cell assemblies are thought to enable related events to become associated with one another and to be important for storing and recalling large varieties of experiences, as occurs with explicit memory. Finally, synapse specificity ensures that inputs that convey information not related to a particular event will not be strengthened. Synapse specificity is critical when large amounts of information must be stored in one network, because much more information can be stored in a cell through functional alterations at individual synapses than through blanket changes in a property of the cell, such as its excitability.

Spatial Memory Depends on Long-Term Potentiation

Long-term potentiation is an experimentally induced change in synaptic strength produced by strong direct stimulation of neural pathways. Does this or a related form of synaptic plasticity occur physiologically during explicit memory storage? If so, how important is it for explicit memory storage in the hippocampus?

To date, a large number of experimental approaches have shown that inhibiting LTP interferes with spatial memory. In one approach, a mouse is placed in a pool filled with an opaque fluid (the Morris water maze); to escape from the liquid, the mouse must swim to find a platform submerged in the fluid and completely hidden from view. The animal is released at random locations around the pool and initially encounters the platform by chance. However, in subsequent trials, the mouse quickly learns to locate the platform and then remembers its position based on spatial information—distal markings on the walls of the room in which the pool is located. This task requires the hippocampus. In a nonspatial, or cued, version of this test, the platform is raised above the water surface or marked with a flag so that it is visible, permitting the mouse to navigate directly to it using brain pathways that do not require the hippocampus.

When NMDA receptors are blocked by a pharmacological antagonist injected into the hippocampus immediately before an animal is trained to navigate the Morris water maze, the animal cannot remember the location of the hidden platform using spatial information but can find it in the version of the task with the visible marker. These experiments thus suggest that some mechanism involving NMDA receptors in the hippocampus, perhaps LTP, is involved in spatial learning. However, if the NMDA receptor blocker is injected into the hippocampus *after* an animal has learned a spatial memory task, it does not inhibit subsequent memory recall for that task. This is consistent with findings that NMDA receptors are required for the induction, but not the maintenance, of LTP.

More direct evidence correlating memory formation and LTP comes from experiments with mutant mice that have genetic lesions that interfere with LTP. One interesting mutation is produced by the genetic deletion of the NR1 subunit of the NMDA receptor. Neurons lacking this subunit fail to form functional NMDA receptors. Mice with a general deletion of the subunit die soon after birth, indicating the importance of these receptors for neural function. However, it is possible to generate lines of conditional mutant mice in which the NR1 deletion is restricted to CA1 pyramidal neurons and occurs only 1 or 2 weeks after birth (see Chapter 2, Figure 2–8, for a description of how this mouse line is generated). These mice survive into adulthood and show a loss of LTP in the Schaffer collateral pathway. Although this disruption is highly localized, the mutant mice have a serious deficit in spatial memory (Figure 54–7).

In some cases, genetic changes can actually enhance both hippocampal LTP and spatial learning and memory. One of the first examples of such an enhancement comes from studies of a mutant mouse that overexpresses the NR2B subunit of the NMDA receptor. This subunit is normally present at hippocampal synapses in the early stages of development but is downregulated in adults. Receptors that include this subunit allow more Ca^{2+} influx than those without the subunit. In mutant mice that overexpress the NR2B subunit, LTP is enhanced, presumably because of an enhancement in Ca^{2+} influx. Importantly, learning and memory for several different tasks are also enhanced (Figure 54–8).

One concern with gene knockouts or transgene expression is that such mutations might lead to subtle developmental abnormalities. That is, changes in the size of LTP and spatial memory in the mutant animals could be the result of an early developmental alteration in the wiring of the hippocampal circuit rather than a change in the basic mechanisms of LTP. This possibility can be addressed by reversibly turning on and off a transgene that interferes with LTP.

Reversible gene expression has been used to explore the role of CaMKII, whose autophosphorylation properties and function in LTP were discussed earlier in this chapter (see also Chapter 2, Figure 2–9, for a description of the methodology). Mutation of the autophosphorylation Thr286 site to the negatively charged amino acid aspartate mimics the effect of autophosphorylation at Thr286 and converts the CaMKII to a calcium-independent form. Transgenic expression of this dominant mutation of CaMKII (CaMKII-Asp286) results in a systematic shift in the relation between the frequency of a tetanus and the resultant change in synaptic strength during long-term plasticity.

In the transgenic mice, tetanic stimulation at an intermediate frequency of 10 Hz, which normally induces a small amount of LTP, induces long-term depression of synaptic transmission in the Schaffer collateral pathway (Figure 54–9A). In contrast, the transgenic mice showed normal LTP to a 100-Hz tetanus. The defect in synaptic plasticity with 10-Hz stimulation is associated with an inability of the mutant mice to remember spatial tasks. However, the defects in the induction of LTP and in spatial memory can be fully extinguished when the mutant gene is switched off in the adult, showing that the memory defect is not due to a developmental abnormality (Figure 54–9).

These several experiments using restricted knockout and overexpression of the NMDA receptor and regulated overexpression of CaMKII-Asp286 make it clear that the molecular pathways important for LTP at Schaffer collateral synapses are also required for spatial memory. However, such results do not directly show that spatial learning and memory are actually associated with an enhancement in hippocampal synaptic transmission. Mark Bear and his colleagues addressed this question by monitoring the strength of synaptic transmission at the Schaffer collateral synapses in vivo in rats.

Recordings were made of synaptic strength using an array of extracellular electrodes to stimulate the Schaffer collateral inputs and another array to record the extracellular field EPSPs at various locations. Rats were then trained to avoid one side of a box through administration of a foot shock; the field EPSPs were remeasured after training, showing a small but significant increase in the amplitude of synaptic transmission at a subset of the recording electrodes. Does the increase in synaptic transmission during learning result from LTP or some other mechanism? Because the amount of LTP at a given synapse is finite, if learning does indeed recruit an LTP-like process, then the ability to induce LTP by tetanic stimulation after learning should be reduced. Indeed, Bear and his colleagues found that the magnitude of LTP is diminished at those recording sites where the behavioral training produced the greatest enhancement in the field EPSP. This result is similar to findings in the amygdala, where fear learning reduces the magnitude of LTP induced by subsequent tetanic stimulation.

If LTP-like changes take place during memory formation in the hippocampus, such changes would be expected only in a small subset of synapses, namely those that participate in the storage of the particular memory. Different memories probably correspond to different assemblies of cells with strengthened synaptic interconnections. If this is true, however, hippocampal memories should be vulnerable to disruption by manipulations that indiscriminately alter synaptic strength within the network as a whole. To test this idea, investigators induced LTP throughout the dentate gyrus *after* hippocampal-dependent spatial training in the water maze task. This protocol indeed impairs the animal's memory of the goal location in the water maze. Control animals that are given NMDA receptor antagonists after learning but prior to high-frequency stimulation exhibit normal spatial memory. These results indicate that the memory impairment was generated specifically as a consequence of the generation of indiscriminate LTP, which likely disrupts the specific pattern of strong and weak synapses that encode memory of the goal location.

A Action of Cre recombinase is restricted to CA1 region

Wild type

Mutant

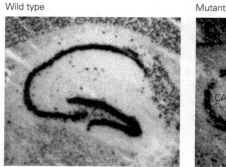
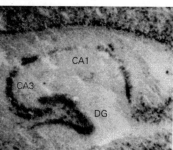

B Long-term potentiation

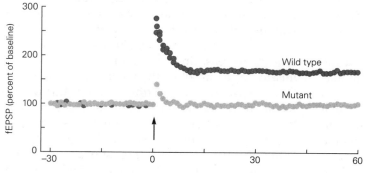
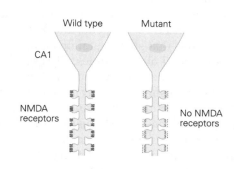

C Morris water maze learning

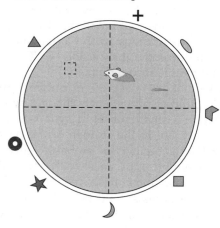
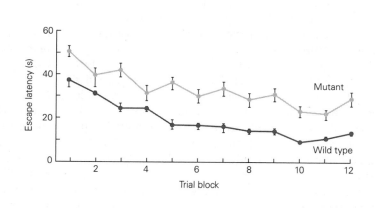

D Probe trial test of memory

Movement patterns

Wild type

Mutant

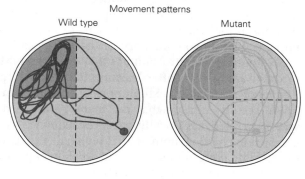
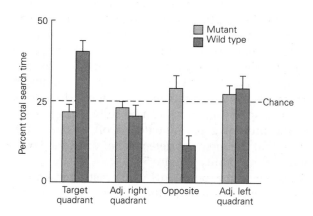

Finally, although most behavioral tests of LTP have used spatial learning tasks to assess memory, studies have also shown that NMDA receptors, and by inference LTP, are necessary for a variety of hippocampal-dependent explicit memories. When NMDA receptors in the CA1 area are blocked, mice are not able to master a nonspatial object recognition task, learn complex odor discrimination, or undergo the social transmission of a food preference, in which an animal learns to accept a novel food by observing a conspecific (another animal of its species) consume that same food. Thus, NMDA receptor–dependent LTP is likely required for many, if not all, forms of explicit memory in the hippocampus (most of which include a spatial recognition element).

Explicit Memory Storage Also Depends on Long-Term Depression of Synaptic Transmission

If synaptic connections could only be enhanced and never attenuated, synaptic transmission might rapidly saturate—the strength of the synaptic connections might reach a point beyond which further enhancement is not possible. Moreover, uniform synaptic strengthening may lead to a loss of memory specificity, with one memory interfering with another. Yet individuals are able to learn, store, and recall new memories throughout a lifetime. This paradox led to the suggestion that neurons must have mechanisms to downregulate synaptic function to counteract LTP.

Such an inhibitory mechanism, termed *long-term depression* (LTD), was first discovered in the cerebellum,

where it is important for motor learning. Since then, LTD has also been characterized at a number of synapses within the hippocampus. Whereas LTP is typically induced by a brief high-frequency tetanus, LTD is induced by prolonged low-frequency synaptic stimulation (Figure 54–10A). As mentioned above, it can also be induced by a spike pairing protocol in which an EPSP is evoked *after* an action potential in the postsynaptic cell. This suggests a corollary to Hebb's learning rule: Active synapses that do not contribute to the firing of a cell are weakened. Like LTP, a number of molecular and synaptic mechanisms are engaged during the induction and expression of LTD.

Surprisingly, many forms of LTD require activation of the same receptors involved in LTP, namely the NMDA receptors (Figure 54–10A). How can activation of a single type of receptor produce both potentiation and depression? A key difference lies in the experimental protocols used to induce LTP or LTD. Compared to the high-frequency stimulation used to induce LTP, the low-frequency tetanus used to induce LTD produces a relatively modest postsynaptic depolarization and thus is much less effective at relieving the Mg^{2+} block of the NMDA receptors. As a result, any increase in Ca^{2+} concentration in the postsynaptic cell is much smaller than the increase observed during induction of LTP and therefore insufficient to activate CaMKII, the enzyme implicated in LTP. Rather, LTD may result from activation of the calcium-dependent phosphatase calcineurin, an enzyme complex that has a higher affinity for Ca^{2+} compared to that of CaMKII (Chapter 14).

Figure 54–7 (Opposite) Long-term potentiation (LTP) and spatial learning and memory are impaired in mice that lack the *N*-methyl-D-aspartate (NMDA) receptor in the CA1 region of the hippocampus. (Reproduced, with permission, from Tsien, Huerta, and Tonegawa 1996.)

A. A line of mice is bred in which the gene encoding the NR1 subunit of the NMDA receptor is selectively deleted in CA1 pyramidal neurons. In situ hybridization is used to detect mRNA for the NR1 subunit in hippocampal slices from wild type and mutant mice that contain two floxed NR1 alleles and express Cre recombinase under the control of the *CaMKIIα* promoter. Note that NR1 mRNA expression (**dark staining**) is greatly reduced in the CA1 region of the hippocampus but not in CA3 and the dentate gyrus (**DG**).

B. LTP at the CA1 Schaffer collateral synapses is abolished in these mice. Field excitatory postsynaptic potentials (**fEPSPs**) were recorded in response to Schaffer collateral stimulation. Tetanic stimulation at 100 Hz for 1 second (**arrow**) caused a large potentiation in wild type mice but failed to induce LTP in the NMDA receptor knockout (mutant) mice.

C. Mice that lack the NMDA receptor in CA1 pyramidal neurons have impaired spatial memory. A platform (**dashed square**) is submerged in an opaque fluid in a circular tank (a Morris water maze). To avoid remaining in the water, the mice have to find the platform using spatial (contextual) cues on the walls surrounding the tank and then climb onto the platform. The graph shows escape latency or the time required by mice to find the hidden platform in successive trials. The mutant mice display a longer escape latency in every block of trials (four trials per day) than do the wild type mice. Also, mutant mice do not reach the optimal performance attained by the control mice after 12 training days, even though they show some improvement with training.

D. After the mice have been trained in the Morris maze, the platform is taken away. In this probe trial, the wild type mice spend a disproportionate amount of time in the quadrant that formerly contained the platform (the target quadrant), indicating that they remember the location of the platform. Mutant mice spend an equal amount of time (25%) in all quadrants; that is, they perform at chance level, indicating deficient memory.

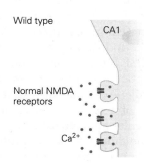

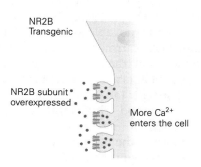

A NMDA-type receptor synaptic current

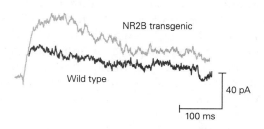

C Morris water maze learning

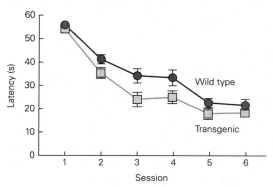

B Long-term potentiation

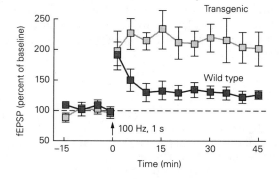

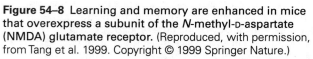

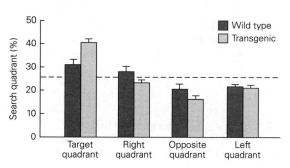

Figure 54–8 Learning and memory are enhanced in mice that overexpress a subunit of the *N*-methyl-D-aspartate (NMDA) glutamate receptor. (Reproduced, with permission, from Tang et al. 1999. Copyright © 1999 Springer Nature.)

A. The amplitude of the current generated by the NMDA receptors in response to a brief pulse of glutamate is enhanced and its time course prolonged in hippocampal neurons obtained from mice that contain a transgene that expresses higher levels of the receptor's NR2B subunit compared to wild type mice.

B. Long-term potentiation produced by tetanic stimulation of the Schaffer collateral synapses is greater in the transgenic

mice than in wild type mice. (Abbreviation: fEPSP, field excitatory postsynaptic potential.)

C. Spatial learning is enhanced in the transgenic mice (**upper plot**). The rate of learning in a Morris water maze (the reduction in time to find the hidden platform, or escape latency) is faster in transgenic mice than in wild type mice. Spatial memory is also enhanced in the transgenic mice (**lower plot**). In the probe trial, the transgenic mice spend more time in the target quadrant, which previously contained the hidden platform, than do wild-type mice.

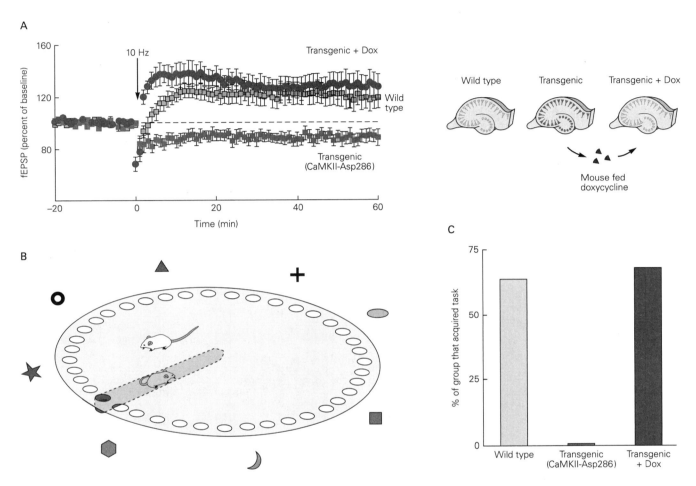

Figure 54–9 Deficits in long-term potentiation (LTP) and spatial memory due to a transgene are reversible. (Reproduced, with permission, from Mayford et al. 1996.)

A. An LTP deficit is seen in hippocampal slices from transgenic mice that overexpress CaMKII-Asp286 kinase, a constitutively active mutant form of CaMKII. Expression of this transgene is driven by a second transgene, the tTA bacterial transcription factor, which is inhibited by the antibiotic doxycycline (**Dox**) (see Chapter 2, Figure 2–9, for a complete description). Four groups of mice were tested: transgenic mice that were fed doxycycline, which blocks expression of the kinase; transgenic mice without doxycycline, in which the kinase is expressed; and wild type mice with and without doxycycline. In wild type mice, a 10-Hz tetanus induces LTP; doxycycline has no effect (data are not shown). In the transgenic mice, the tetanus fails to induce LTP but causes a small synaptic depression. In the transgenic mice that were fed doxycycline, the deficit in LTP is reversed. (Abbreviation: **fEPSP**, field excitatory postsynaptic potential.)

B. The effect of the kinase on spatial memory was tested in a Barnes maze. The maze consists of a platform with 40 holes, one of which leads to an escape tunnel that allows the mouse to exit the platform. The mouse is placed in the center of the platform. Mice do not like open, well-lit spaces and therefore try to escape from the platform by finding the hole that leads to the escape tunnel. The most efficient way of learning and remembering the location of the hole (and the only way of meeting the criteria set for the task by the experimenter) is for the mouse to use distinctive markings on the four walls as spatial cues, thus demonstrating hippocampal spatial memory.

C. Transgenic mice that receive doxycycline perform as well as wild type mice in learning the Barnes maze task (approximately 65% of animals learn the task), whereas transgenic mice without the doxycycline, which thus express CaMKII-Asp286, do not learn the task.

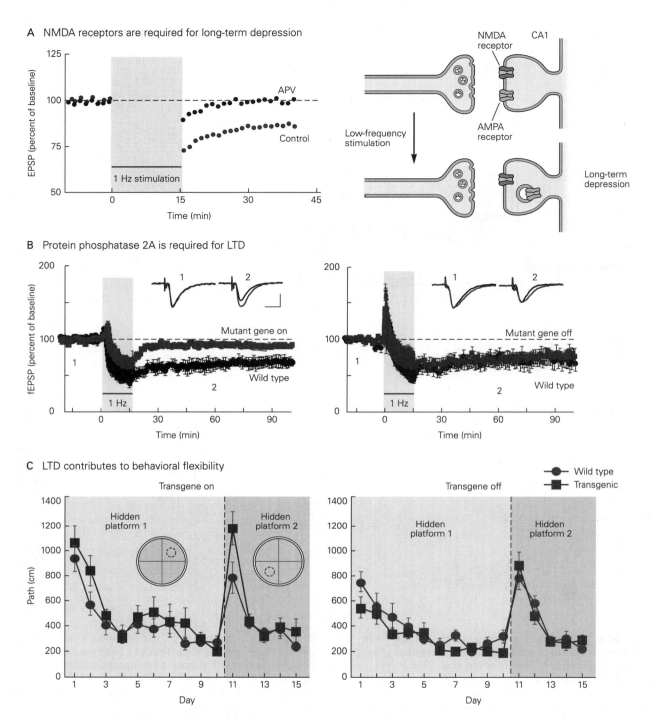

Figure 54–10 Long-term depression of synaptic transmission requires *N*-methyl-D-aspartate (NMDA) receptors and phosphatase activity.

A. Prolonged low-frequency stimulation (1 Hz for 15 minutes) of Schaffer collateral fibers produces a long-term decrease in the size of the field excitatory postsynaptic potential (fEPSP) in the hippocampal CA1 region, a decrease that outlasts the period of stimulation (control). Long-term depression (LTD) occurs when α-amino-3-hydroxy-5-methyl-4-isoxazolepropionic acid (AMPA) receptors are removed from the postsynaptic membrane by endocytosis; it is blocked when the NMDA receptors are blocked by the drug 2-amino-5-phosphonovaleric acid (APV). (Adapted from Dudek and Bear 1992.)

B. LTD requires protein dephosphorylation. The plots compare LTD in the hippocampal CA1 region of wild type mice and transgenic mice that express a protein that inhibits phosphoprotein phosphatase 2A. Transgene expression is under control of the

tTA system. In the absence of doxycycline, the phosphatase inhibitor is expressed, and induction of LTD is inhibited (*left plot*). When expression of the phosphatase inhibitor is turned off by administering doxycycline, a normal-sized LTD is induced (*right plot*).

C. Inhibition of phosphatase 2A reduces behavioral flexibility. Transgenic mice expressing the phosphatase inhibitor learn the location of a submerged platform in the Morris maze at the same rate as wild-type mice (days 1–10). Thus, LTD is not necessary for learning the initial platform location. At the end of day 10, the platform is moved to a new hidden location and the mice are retested (days 11–15). Now the transgenic mice travel significantly longer paths to find the platform on the first day of retesting (day 11), indicating an impaired learning (reduced flexibility). When transgene expression is turned off with doxycycline, the transgenic mice display normal learning on all phases of the test. (Panels B and C reproduced, with permission, from Nicholls et al. 2008.)

Long-term depression may also depend on a surprising metabotropic action of the ionotropic NMDA receptor-channels. Glutamate binding, in addition to opening the receptor pore, is thought to trigger a conformational change in a cytoplasmic domain of the receptor that directly activates a downstream signaling cascade that increases the activity of phosphoprotein phosphatase 1 (PP1). Activation of PP1 or calcineurin eventually leads to changes in protein phosphorylation that promote endocytosis of AMPA receptors, resulting in a decrease in the size of an EPSP.

Distinctly different forms of LTD can be induced through the activation of G protein–coupled metabotropic glutamate receptors. Such forms of LTD depend on activation of mitogen-activated protein (MAP) kinase signaling pathways (Chapter 14) rather than activation of phosphatases. These types of LTD lead to a reduction in synaptic transmission through a decrease in glutamate release from presynaptic terminals as well as through alterations in the trafficking of AMPA receptors in the postsynaptic cells.

Much less is known about the behavioral role of LTD compared to that of LTP, but some insight has come from studies with mice using a transgene that expresses an inhibitor of protein phosphatase. LTD that depends on NMDA receptors is inhibited when the transgene is expressed but is normal when transgene expression is suppressed (Figure 54–10B). Transgene expression does not affect LTP or forms of LTD that involve metabotropic glutamate receptors. Mice that express the transgene show normal learning the first time they are tested in the Morris maze. However, when the same mice are retested after the hidden platform has been moved to a new location, they show a decreased ability to learn the new location and tend to persevere in searching for the platform near the previously learned location (Figure 54–10C). Thus, LTD may be necessary not only to prevent LTP saturation but also to enhance flexibility in memory storage and specificity in memory recall. Studies on fear conditioning suggest that LTD in the amygdala may be important for reversing learned fear.

Memory Is Stored in Cell Assemblies

While the cumulative evidence for a relationship between long-term synaptic plasticity and memory formation is strong, we know less about how specific cellular processes such as LTP enable memory formation. This reflects limitations in our knowledge of how neural circuits operate and how memories might be embedded in them. Theoretical models for memory storage in neural circuits can be traced to Hebb's concept of a cell assembly—a network of neurons that is activated whenever a function is executed; for example, each time a memory is recalled. Cells within an assembly are bound together by excitatory synaptic connections strengthened at the time the memory was formed.

Today, more than half a century later, Hebb's thoughts still form the framework for how the hippocampus mediates the storage and recall of memory, although experimental proof has been difficult to obtain. A proper test requires recording the activity of thousands of neurons simultaneously, in combination with the experimental excitation or inactivation of selected cell groups. Technological advances are now enabling such experiments. By and large, the results obtained so far confirm Hebb's cell assembly model and implicitly point to LTP as the mechanism for their formation.

In a telling study with mice, Susumu Tonegawa and his colleagues tested whether reactivation of neurons that participated in the storage of a specific memory is sufficient to trigger recall of that memory. The researchers first applied an electric shock to an animal as it explored a novel environment. Reexposure of the animal to the same environment a day or more later elicited a freezing response, indicating that the animal associated the environment or context (the conditioned stimulus) with the shock (the unconditioned stimulus). Using a genetic strategy, Tonegawa caused a subset of dentate gyrus granule neurons that were active during the fear conditioning to express the light-activated cation channel channelrhodopsin-2 (Figure 54–11). The conditioned animals were subsequently placed in a novel environment that did not resemble the conditioned environment and so did not elicit a fear response. However, light activation of the subset of granule cells that were active during fear conditioning was able to elicit a strong freezing response, even though the animals were in a nonthreatening environment. This supports the idea that memories are stored in cell assemblies and, more importantly, demonstrates that reactivation of these assemblies is sufficient to induce recall of an experience.

In a complementary experimental approach, a light-activated inhibitory Cl⁻ transporter was expressed in CA1 cells active at the time of fear conditioning. Later, the labeled cells were inactivated and the animals were placed again in the environment in which they received the shock. Under these conditions, the normal freezing behavior (ie, recall of the memory of fear conditioning) was blocked, suggesting that activity in the labeled CA1 cell population was necessary for memory retrieval. Taken together, these findings suggest that reactivation of the specific cell assembly pattern that occurred during encoding is both necessary and sufficient for memory retrieval.

Perhaps the most direct test of the ensemble model is the creation of a false memory. Tonegawa and colleagues expressed channelrhodopsin in cells that were active during exploration of a novel environment (context A), except that no shock was delivered this time. At a later time, the labeled cells were reactivated using light stimulation as the mice explored a second novel environment (context B), this time in combination with an electric shock. When the animals were returned to the neutral context A, they froze, although they had never been shocked in this environment. This result indicates that the reactivation of the original engram of context A when paired with an aversive

experience in context B is able to create a false memory, causing the animals to fear context A. Thus, it is possible to modify the behavioral significance of a neural representation (a pattern of neural firing in response to a given stimulus) by pairing the assembly with a new experience unrelated to the original experience.

Different Aspects of Explicit Memory Are Processed in Different Subregions of the Hippocampus

Explicit memory stores knowledge of facts (semantic memory), places (spatial memory), other individuals (social memory), and events (episodic memory). As discussed above, successful storage and recall of explicit

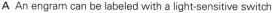

A An engram can be labeled with a light-sensitive switch

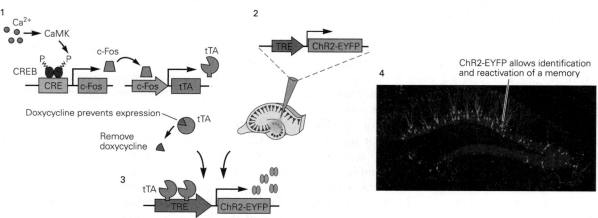

B A memory can be recalled when the engram is activated by light

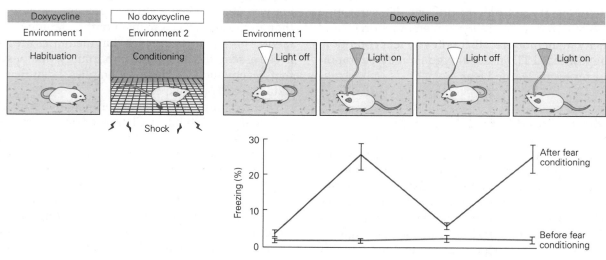

memory requires that patterns of activity be formed within local cell assemblies to avoid mix-up between memories. At the same time, an important psychological feature of hippocampal-dependent memory is that a few cues are usually enough to trigger the recall of a complex memory. How does the hippocampus perform all of these diverse functions? Do its subregions have specialized roles, or is memory a unitary function of the hippocampus? In at least some instances, it has been possible to assign key functions to specific areas of the hippocampus.

The Dentate Gyrus Is Important for Pattern Separation

How does the hippocampus store a different pattern of neural activity in response to every experience that needs to be remembered, including patterns that distinguish between two closely related environments? Contemporary ideas about how neural circuits accomplish this task, often referred to as *pattern separation*, dates to the theoretical work of David Marr in the late 1960s and early 1970s. In a landmark paper on the cerebellum, Marr suggested that the extensive divergence

of mossy-fiber inputs onto an extraordinarily large number of cerebellar granule cells might allow for pattern separation in this system.

This idea of "expansion recoding," in which distinct firing patterns are formed through the projection of a limited number of inputs onto a larger population of synaptic target cells, was later applied by others to the hippocampus. They proposed that hippocampal pattern separation results from the divergence of entorhinal inputs onto a larger number of granule cells in the dentate gyrus. The findings of subsequent experimental studies are broadly in line with these theoretical suggestions: Neural activity patterns recorded in different environments differ more extensively in the dentate gyrus and CA3 than they do one synapse upstream in the entorhinal cortex. The dentate gyrus is also implicated in pattern separation by the fact that lesions or genetic manipulations targeted to this area impair the ability of rats and mice to discriminate between similar locations and contexts.

The dentate gyrus is the site of one of the most unexpected findings in neuroscience, the discovery that the birth of new neurons, or neurogenesis, is not limited to early stages of development. New neurons

Figure 54–11 (Opposite) Stimulating a neuronal assembly associated with a stored memory of fear conditioning elicits fear behavior. (Panels reproduced or redrawn, with permission, from Liu et al. 2012. Copyright © 2012 Springer Nature.)

A. *Experimental protocol.* **1.** Exposure of a mouse to a new environment increases activity in a group of hippocampal neurons (cell assembly) that codes for the environment. The activity increases intracellular Ca^{2+}, which activates a CaM kinase signaling cascade, resulting in phosphorylation of the transcription factor CREB. Phosphorylated CREB increases expression of immediate early genes, including the c-Fos transcription factor. In the *c-fos-tTA* transgenic mouse line, c-Fos binds to the *c-fos* promoter of the transgene and thereby initiates expression of the transcription factor tTA. The antibiotic doxycycline is fed to mice, which binds to and inhibits tTA, until the day of the experiment. **2.** The dentate gyrus of the same transgenic mice was previously injected with an adeno-associated virus that contains a DNA sequence encoding ChR2 fused to the fluorescent marker protein EYFP (ChR2-EYFP). The transcription of this sequence is under control of the TRE promoter, which requires tTA (without doxycycline) for expression. **3.** Exposure of the mice to a novel environment (after removing doxycycline from the feed) leads to expression of tTA and subsequent expression of ChR2-EYFP in a subset of active dentate gyrus neurons. **4.** The ChR2-EYFP remains expressed for several days in the neurons, as seen by the EYFP fluorescence signal in dentate gyrus granule cells in a hippocampal slice. (ChR2-EYPF in **green**, dentate gyrus cell body layer in **blue**.)

B. *Recall of a fear memory.* An optical fiber is implanted above the dentate gyrus. **1.** During fear memory encoding,

mice were first habituated in one environment while being fed doxycycline (which prevents expression of ChR2-EYFP). The mice were then taken off doxycycline and exposed to a new environment for a few minutes. This turns on gene transcription in the assembly of neurons that are active in the new environment, leading to prolonged expression of ChR2-EYFP in these cells. The mice were then given a series of footshocks while in the new environment to induce fear conditioning: The mice learn to associate the new environment with a fearful stimulus. The mice were then returned to their cage and put back on doxycycline. **2.** During fear memory reactivation 5 days after conditioning, mice show a normal defensive freezing behavior when reintroduced to the environment where they received footshocks (not shown). However, when mice are exposed to the environment to which they were initially habituated (no associated foot shock), they normally recognize this as a neutral environment and do not exhibit defensive freezing. However, as the mice explore the neutral environment, delivery of blue light to activate ChR2-expressing neurons in the dentate gyrus causes the mice to freeze. This indicates that activation of the ensemble of ChR2-expressing neurons initially activated in the conditioning environment is sufficient to recall the fear memory associated with that environment. The experimental data show the freezing response in the neutral environment is much greater when light pulses are turned on compared to when the light is off (**red plot**; light delivery indicated in cartoon on top). Delivery of light pulses to an animal that had not undergone fear conditioning does not elicit freezing (**blue plot**).

continue to be born from precursor stem cells throughout adulthood and become incorporated into neural circuits. Nevertheless, adult neurogenesis is limited to granule neurons in two brain regions: inhibitory granule cells in the olfactory bulb and the excitatory granule neurons of the dentate gyrus. Recent experimental findings raise the possibility that newly born granule neurons in the adult are particularly important for pattern separation, even though they represent only a minor fraction of the total number of granule cells. Procedures that stimulate neurogenesis enhance the ability of a mouse to discriminate between closely related environments. Experimental silencing of all dentate gyrus granule neurons except those newly born in the adult does not seem to impair pattern separation, implying that it is the newborn neurons that are most essential to pattern separation. Although some uncertainties remain on the role of neurogenesis in pattern separation and memory encoding, methods that enhance neurogenesis are currently being explored as a means of treating different types of age-related memory loss.

The CA3 Region Is Important for Pattern Completion

A key feature of explicit memory is that a few cues are often sufficient to retrieve a complex stored memory. Marr suggested in a second landmark paper in 1971 that the recurrent excitatory connections of CA3 pyramidal cells might underlie this phenomenon. He proposed that when a memory is encoded, neuronal activity patterns are stored as changes in connections between active CA3 cells. During subsequent retrieval of the memory, the reactivation of a subset of this stored cell assembly would be sufficient to activate the entire original neural ensemble that encodes the memory because of the strong recurrent connections between the cells of the ensemble. This restoration is referred to as *pattern completion*.

The importance of LTP for pattern completion in the CA3 network is seen in studies with mice in which the NMDA glutamate receptor is selectively deleted from the CA3 neurons. These mice experience a selective loss of LTP at the recurrent synapses between CA3 neurons, with no change in LTP at the synapses between mossy fibers and CA3 neurons or at the Schaffer collateral synapses between CA3 and CA1 neurons. Despite this deficit, the mice show normal learning and memory for finding a submerged platform in a water maze using a complete set of spatial cues. However, when the mice are asked to find the platform with fewer spatial cues, their performance is impaired, indicating that LTP at the recurrent

synapses between CA3 neurons is important for pattern completion.

The CA2 Region Encodes Social Memory

Studies comparing neuronal representations in the dentate gyrus and CA3 and CA1 areas have indicated that each region has a unique function in the storage and retrieval of hippocampal memory. Recent evidence suggests that the CA2 region plays a crucial role in social memory, the ability of an individual to recognize and remember other members of its own species (conspecifics). Genetic silencing of CA2 disrupts the ability of a mouse to remember encounters with other mice, but does not impair other forms of hippocampal-dependent memory, including memory of objects and places.

The CA2 region is also unique among hippocampal regions in having very high levels of receptors for the hormones oxytocin and vasopressin, important regulators of social behaviors. Selective stimulation of the vasopressin inputs to CA2 neurons can greatly prolong the duration of a social memory. Social memory also depends on CA1 neurons in the ventral region of the hippocampus, an area linked to emotional behavior, which receives important input from CA2.

A Spatial Map of the External World Is Formed in the Hippocampus

How do hippocampal neurons encode features of the external environment to form a memory of spatial locale, enabling an animal to navigate to a remembered goal? At the end of the 1940s, the cognitive psychologist Edward Tolman proposed that somewhere in the brain there must be representations of one's environment. He referred to these neural representations as cognitive maps. They were thought to form not only an internal map of space but also a mental database in which information is stored in relation to an animal's position in the environment, similar to the GPS coordinates of a photograph.

Tolman did not have the opportunity to determine whether a cognitive map actually existed in the brain, but in 1971, John O'Keefe and John Dostrovsky discovered that many cells in the CA1 and CA3 areas of the rat hippocampus fire selectively when an animal is located at a specific position in a specific environment. They called these cells "place cells" and the spatial location in the environment where the cells preferentially fired "place fields" (Figure 54–12A,B). When the animal enters a new environment, new place fields are formed within minutes and are stable for weeks to months.

Different place cells have different place fields, and collectively, they provide a map of the environment, in the sense that the combination of currently active cells is sufficient to read out precisely where the animal is in the environment. A place-cell map is not egocentric in its organization, like the neural maps for touch or vision on the surface of the cerebral cortex. Rather, it is allocentric (or geocentric); it is fixed with respect to a point in the outside world. Based on these properties, John O'Keefe and Lynn Nadel suggested in 1978 that place cells are part of the cognitive map that Tolman had in mind. The discovery of place cells provided the first evidence for an internal representation of the environment that allows an animal to navigate purposefully around the world.

Entorhinal Cortex Neurons Provide a Distinct Representation of Space

How is the hippocampal spatial map formed? What type of spatial information is carried by afferent connections from the entorhinal cortex to the hippocampal place cells? In 2005, a surprising discovery was made about the spatial representation formed by certain neurons in the medial entorhinal cortex, whose axons provide a major part of the perforant pathway input to the hippocampus. These neurons represent space in a manner very different from the hippocampal place cells. Instead of firing when the animal is in a unique location, like the place cells, these entorhinal neurons, termed *grid cells*, fire whenever the animal is at any of several regularly spaced positions forming a hexagonal grid-like array (Figure 54–12C). When the animal moves about in the environment, different grid cells become activated, such that the activity in the entire population of grid cells always represents the animal's current position.

The grid allows the animal to locate itself within a Cartesian-like external coordinate system that is independent of context, landmarks, or specific markings. A grid cell's firing pattern is expressed in all environments that an animal visits, including during complete darkness. The independence of grid-cell firing from visual input implies that intrinsic networks, as well as self-motion cues, may serve as sources of information to ensure that grid cells are activated systematically throughout the environment. The gridded spatial information conveyed by the entorhinal inputs is then transformed within the hippocampus into unique spatial locations represented by the firing of ensembles of place cells, but how this transformation occurs remains to be determined. Since grid cells were discovered in the medial entorhinal cortex of rats in 2005, they have been identified in mice, bats, monkeys,

and humans. Recordings from flying bats have shown that grid cells and place cells represent locations in three-dimensional space, suggesting the generality of the cortico-hippocampal spatial navigation system. Finally, it has been proposed that grid cells in primates may encode positions in multiple sensory coordinate systems, including eye fixation coordinates.

Grid cells display a characteristic relation between their firing fields and anatomical organization (Figure 54–13). The x,y coordinates of a cell's grid fields—often called the phase of the grid—differ among cells at the same location of the medial entorhinal cortex. The x,y coordinates of two neighboring cells are often as different as those of widely separated grid cells. In contrast, the size of the individual grid fields and the spacing between them generally increase topographically from the dorsal to the ventral part of the medial entorhinal cortex, expanding from a typical grid spacing of 30 to 40 cm at the dorsal pole to several meters in some cells at the ventral pole (Figure 54–13A). The expansion is not linear but step-like, suggesting that the grid-cell network is modular.

Interestingly, a gradual expansion is seen also in the size of the place fields of hippocampal place cells along the dorsal to ventral axis of the hippocampus (Figure 54–13B). This is consistent with the known pattern of synaptic connectivity: Dorsal entorhinal cortex innervates dorsal hippocampus, whereas ventral entorhinal cortex innervates ventral hippocampus. The finding that place fields are larger in the ventral hippocampus is in accord with results suggesting that the dorsal hippocampus is more important for spatial memory, whereas the ventral hippocampus is more important for nonspatial memory, including social memory and emotional behavior.

Grid cells are not the only medial entorhinal cells with projections to the hippocampus. Others include *head direction cells*, which respond primarily to the direction that the animal is facing (Figure 54–14A). Such cells were originally discovered in the presubiculum, another region of the parahippocampal cortex, but they exist also in the medial entorhinal cortex. Many entorhinal head direction cells also have grid-like firing properties. Like grid cells, such head direction cells are active when an animal traverses the vertices of a triangular grid in a two-dimensional environment. However, within each grid field, these cells fire only if the animal is facing a certain direction. Head direction cells and conjunctive grid and head direction cells are thought to provide directional information to the entorhinal spatial map.

Intermingled among grid cells and head direction cells is yet another type of spatially modulated cell, the *border cell* (Figure 54–14B). The firing rate of a border

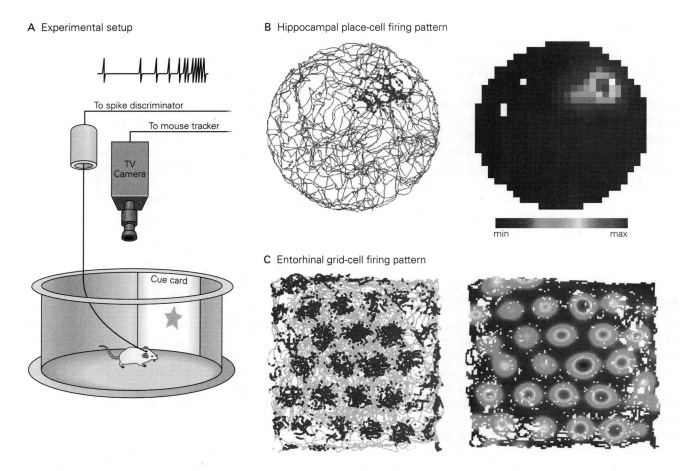

A Experimental setup

To spike discriminator

To mouse tracker

TV Camera

Cue card

B Hippocampal place-cell firing pattern

min   max

C Entorhinal grid-cell firing pattern

Figure 54–12 The firing patterns of cells in the hippocampus and medial entorhinal cortex signal the animal's location in its surroundings.

A. Electrodes implanted in the hippocampus of a mouse are attached to a recording cable, which is connected to an amplifier attached to a computer-based spike-discrimination program. The mouse is placed in an enclosure with an overhead TV camera that transmits to a device that detects the position of the mouse. The enclosure also contains a visual cue to orient the animal. Spikes in individual hippocampal pyramidal neurons ("place cells") are detected by a spike discrimination program. The firing rate of each cell is then plotted as a function of the animal's location in the cylinder. This information is visualized as a two-dimensional activity map for the cell, from which the cell's firing fields can be determined (shown in part **B**). (Adapted, with permission, from Muller, Kubie, and Ranck 1987. Copyright © 1987 Society for Neuroscience.)

B. Location-specific firing of a hippocampal place cell. A rat is running in a cylindrical enclosure similar to the one shown in part **A**. *Left*: The animal's path in the enclosure is shown in **gray**; firing locations of individual spikes are shown for a single place cell as **red dots**. *Right*: The firing rate of the same cell is color-coded (**blue** = low rate, **red** = high rate). In larger environments, place cells usually have more than one firing field but the fields have no apparent spatial relationship.

C. Spatial pattern of firing of an entorhinal grid cell in a rat during 30 minutes of foraging in a square enclosure 220 cm wide. The pattern shows typical periodic grid firing fields. *Left*: The trajectory of the rat is shown in **gray**; individual spike locations are shown as **red dots**. *Right*: Color-coded firing rate map for the grid cell to the left. Color coding as for the place cell in part **B**. (Adapted, with permission, from Stensola et al. 2012.)

cell increases whenever the animal approaches a local border of the environment, such as an edge or a wall. Border cells may help align the phase and orientation of grid cell firing to the local geometry of the environment. A similar role may be played by recently discovered object-vector cells—cells in medial entorhinal cortex that encode the animal's distance and direction relative to salient landmarks. A final entorhinal cell

type is the *speed cell*. Speed cells fire proportionally to the running speed of the animal, irrespective of the animal's location or direction (Figure 54–14C). Together with head direction cells, speed cells can provide grid cells with information about the animal's instantaneous velocity, allowing the ensemble of active grid cells to be updated dynamically in accordance with a moving animal's changing location.

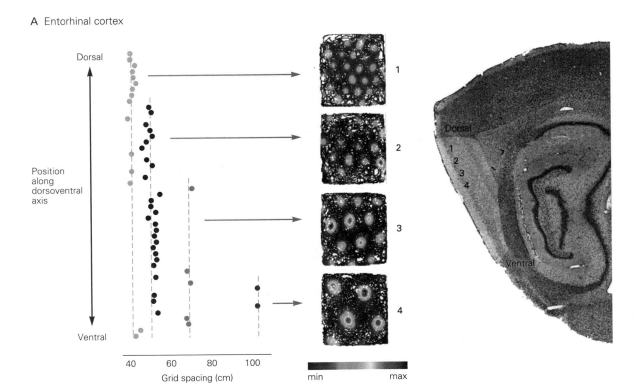

A Entorhinal cortex

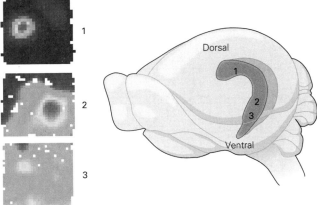

B Hippocampus

Figure 54–13 Grid fields and place fields expand in size as a function of neuronal location along the dorsoventral axis of the entorhinal cortex and hippocampus.

A. Topographical organization of grid scale in the entorhinal cortex. Grid spacing (distance between grid fields) was determined for 49 grid cells (**colored dots**) recorded in the same rat at successive dorsal to ventral levels in the medial entorhinal cortex (**green** area in the sagittal brain section on the right). **Dashed lines** indicate mean grid-spacing values, indicating that grid-spacing falls in one of four discrete modules, with points colored according to module. Firing rate maps for four

of the cells are shown in the middle (similar to those of Figure 54–12C). Recording locations for these cells are indicated by numbers 1 to 4 to the right. (Adapted, with permission, from Stensola et al. 2012.)

B. Place fields from three different locations along the dorsoventral axis of the hippocampus. *Right*: Recording positions (numbers) in the hippocampal formation are shown at right. *Left*: Color-coded maps show the firing fields of each place cell at the recording locations. The field size expands in cells along the dorsoventral axis of the hippocampus. (Reproduced, with permission, from Kjelstrup et al. 2008.)

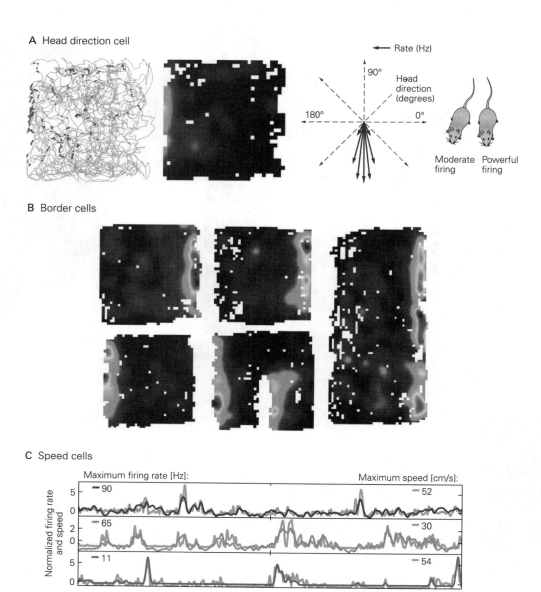

Figure 54–14 The medial entorhinal cortex contains several functional cell types tuned to distinct representations of an animal's navigation.

A. On the left is the trajectory of a rat exploring a 100-cm-wide square enclosure (**red dots** indicate firing locations). A color-coded firing rate map is also shown (color scale as in previous figures). Note that the cell's firing is scattered across the enclosure. The plot on the right shows the same cell's firing rate as a function of head direction, in polar coordinates. The cell fires selectively when the rat faces south, anywhere in the box. (Adapted, with permission, from Sargolini et al. 2006.)

B. Firing rate maps for a representative border cell in enclosures with different geometric shapes (**red** = high rate; **blue** = low rate).

Top row: The firing field map follows the walls when the enclosure is stretched from a square (left and middle maps) to a rectangle (right map). **Bottom row:** The firing field of the same border cell in another environment. Introduction of a discrete wall (**white pixels**, right map) inside the square enclosure causes a new border field to appear to the right of the wall. (Reproduced, with permission, from Solstad et al. 2008.)

C. Speed cells. Traces show normalized firing rate (**colored traces**) and speed (**gray**) for seven representative entorhinal speed cells during 2 minutes of free foraging. Maximum values of firing rate and speed are indicated (left and right, respectively). Note high correspondence between speed and firing rate in these cells. (Reproduced, with permission, from Kropff et al. 2015.)

Taken together, these discoveries point to a network of functionally dedicated cells in the medial entorhinal cortex reminiscent of the feature detectors of the sensory cortices. The functional specificity of each cell type stems from the cell's representation of a specific feature of behavior. In this sense, the entorhinal cell types differ from cells in most other association cortices, which integrate information from many sources in ways that are not straightforward to decode.

What are the key differences between space-coding cells in the hippocampus and the medial entorhinal cortex? A striking property of all entorhinal cell types is the rigidity of their firing patterns. Ensembles of co-localized grid cells maintain the same intrinsic firing pattern regardless of context or environment. When a pair of grid cells has overlapping grid fields in one environment, their grid fields overlap also in other environments. If their grid fields are opposite, or "out of phase," they will be opposite in other environments as well. A similar rigidity is seen in head direction cells and border cells: Cells with similar orientation in one environment have similar orientations in other environments. Speed cells also maintain their unique tuning to running speed across environments. These findings suggest that the medial entorhinal cortex, or modules of this cortical circuit, may operate like a universal map of space that disregards the details of the environment. By doing so, the entorhinal map differs strongly from the place-cell map of the hippocampus.

The firing pattern of a hippocampal place cell is very sensitive to changes in the environment. The place fields of a given place cell in the hippocampus often switch to encode a completely different spatial locale when an animal's environment undergoes a major change, a process referred to as "remapping." Sometimes even minor changes in sensory or motivational inputs are sufficient to elicit remapping. The lack of correlation of hippocampal place maps for different environments (Figure 54–15) is thought to facilitate storage of discrete memories and minimize the risk that one memory will be confused with another, a process termed interference. For an explicit memory system like the hippocampus, with millions of events to be stored, this may be a huge advantage. For accurate and fast representation of an animal's position in space, as occurs in the medial entorhinal cortex, it may instead be beneficial to use a more stereotyped code that is less sensitive to environmental context or nonspatial sensory stimuli.

Place Cells Are Part of the Substrate for Spatial Memory

In addition to representing the animal's current location, place cells are thought to also store the memory of a location in position-related firing patterns that are evoked in the absence of the sensory inputs that originally elicited the firing. For example, as an animal sleeps after running repeated laps along a linear maze,

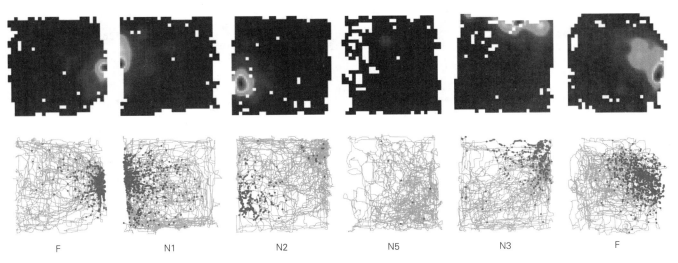

F	N1	N2	N5	N3	F

Figure 54–15 Place cells form independent maps for different environments. Rate maps showing firing patterns of a single hippocampal place cell in different square enclosures, each located in a different room. The rat was tested in one familiar (**F**) and 11 novel (**N**) rooms (recordings only shown for four of the novel rooms). The top row shows firing rate maps, whereas the bottom row shows trajectories of the animal's movement with firing locations in **red**. The cell was active only in some of the rooms (F, N1, N2, N3), where the firing locations were different. When the rat was returned to the familiar room at the end of the experiment, the cell's firing field had a similar location to the initial recording in the familiar room, indicating that a given cell's spatial firing pattern in the same environment is stable. (Adapted, with permission, from Alme et al. 2014.)

place cells spontaneously fire in the same order that they did in the maze, a phenomenon called "replay." Similarly, past trajectories and experiences may influence firing rates at particular locations in the environment. The ability of place cells to represent events and locations experienced in the past likely underlies the ability of the hippocampus to encode complex memories of events.

Once the firing pattern of a population of hippocampal neurons is formed for a given environment, how is it maintained? Because the place cells are the same hippocampal pyramidal neurons that undergo experimental LTP, a natural question is whether LTP is important. This question was addressed in experiments in mice in which LTP was disrupted.

In mice lacking the NR1 subunit of the NMDA receptor, hippocampal pyramidal neurons still fire in place fields despite the fact that LTP is blocked. Thus, this form of LTP is not required for the transformation of spatial sensory information into place fields.

However, the place fields of the mutant mice are larger and fuzzier in outline than those in normal animals. In a second experiment with mutant mice, late LTP and long-term spatial memory were selectively disrupted by expression of a transgene that encodes a protein inhibitor of PKA. In these mice, place fields also form, but the firing patterns of individual cells are stable only for an hour or so (Figure 54–16). Thus, late LTP is required for long-term stabilization of place fields but not their formation.

To what degree do these maps of an animal's surroundings mediate explicit memory? In humans, explicit memory is defined as the conscious recall of facts about people, places, and objects. Although consciousness cannot be studied empirically in the mouse, selective attention, which is required for conscious recall, can be examined.

When mice are presented with different behavioral tasks, the long-term stability of place fields correlates strongly with the degree of attention required to

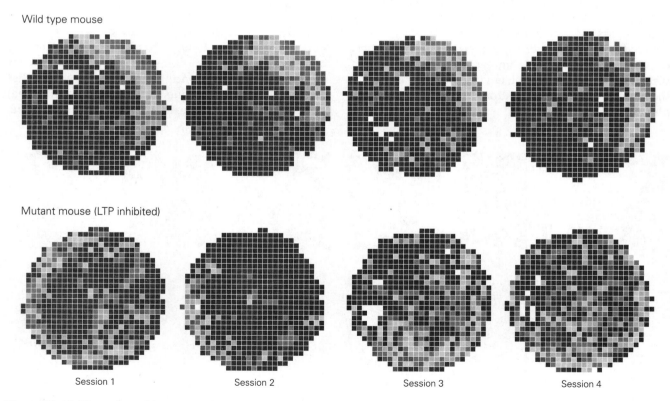

Wild type mouse

Mutant mouse (LTP inhibited)

Session 1   Session 2   Session 3   Session 4

Figure 54–16 Disruption of long-term potentiation (LTP) degrades the stability of place field formation in the hippocampus. Color-coded firing rate maps (see Figure 54–12) show place fields recorded in four successive sessions from a single hippocampal pyramidal neuron in a wild type mouse and from a neuron in a mutant mouse that expresses the persistently active CaMKII (which inhibits the induction of LTP).

Before each recording session, the animal is taken out of the enclosure and sometime later reintroduced into it. In each of the four sessions, the place field for the cell in the wild type animal is stable; the cell fires whenever the animal is in the upper right region of the enclosure. By contrast, the place field for the cell in the mutant mouse is unstable across the four sessions. (Reproduced, with permission, from Rotenberg et al. 1996.)

perform the task. When a mouse does not attend to the space it walks through, place fields form but are unstable after 3 to 6 hours. Animals with unstable place fields are unable to learn a spatial task. However, when a mouse is forced to attend to the space, for example, when trained to run to a specific location, the place fields are stable for days.

How does this attentional mechanism work? Studies in primates have shown the importance of the prefrontal cortex and the modulatory dopaminergic system during attention. Indeed, the formation of stable place fields in mice requires the activation of the dopamine D_1/D_5 type of receptor, which has been shown to enhance the formation of late LTP through production of cAMP and activation of PKA. These results suggest that long-term memory of a place field, rather than being a form of implicit memory that is stored and recalled without conscious effort, requires the animal to attend to its environment, as is the case for explicit memory in humans.

Disorders of Autobiographical Memory Result From Functional Perturbations in the Hippocampus

Our sense of identity is greatly dependent on our store of explicit autobiographical memories and our ability to recognize and navigate through familiar spatial environments. Neurological and psychiatric disorders that disrupt these abilities often occur as a result of changes in neural circuitry and plasticity mechanisms within the hippocampus and related regions in the temporal lobe.

There is now substantial evidence that the devastating memory loss associated with Alzheimer disease is associated with an accumulation of extracellular plaques of the protein fragment β-amyloid (Aβ) and intracellular neurofibrillary tangles of tau, a microtubule associated protein (Chapter 64). However, even before plaques and tangles are apparent, elevated levels of soluble Aβ and tau are thought to disrupt a number of cellular processes, particularly by reducing the magnitude of both early and late LTP at certain synapses. Mouse models of Alzheimer disease also show alterations in hippocampal place cell stability and population-level synchrony, which may contribute to memory loss and spatial disorientation. Changes in grid-cell function have also been observed in electrophysiological recordings in mouse disease models and in humans through functional magnetic resonance imaging studies. Although a number of preclinical studies have shown that agents that decrease

levels of Aβ can rescue synaptic function and memory in rodents, so far these treatments have been less successful in treating patients with Alzheimer disease, perhaps because treatment must be initiated at early stages prior to irreversible synaptic changes.

Altered hippocampal function may also contribute to cognitive problems experienced by individuals with schizophrenia, including disturbances in working memory (Chapter 60). Recent studies using a genetic mouse model of schizophrenia report reduced synchrony between the hippocampus and prefrontal cortex associated with working memory. Furthermore, the place fields of place cells in the hippocampus CA1 region may be overly rigid in this mouse, suggesting that the ability of the hippocampus to distinguish different contexts may be impaired. Finally, a deficit in social memory in these mice has been linked to a reduction of parvalbumin-positive inhibitory neurons in the CA2 region; a similar loss of inhibitory neurons has been observed in postmortem brain tissue from individuals with schizophrenia and bipolar disorder.

Thus, studies of the hippocampus and related temporal lobe structures offer the great promise of providing fundamental insight into how explicit memories are stored and recalled and how functional alterations in these structures may contribute to neuropsychiatric disease. In turn, such insight may aid in the discovery of new treatments for these devastating disorders.

Highlights

1. Explicit memory has both a short-term component, termed working memory, and a long-term component. Both forms depend on the prefrontal cortex and hippocampus.

2. Long-term memory is thought to depend on activity-dependent long-term synaptic plasticity at synapses within the cortico-hippocampal circuit. A brief high-frequency train of tetanic stimulation leads to long-term potentiation (LTP) of excitatory synaptic transmission at each stage of the cortico-hippocampal circuit.

3. LTP at many synapses depends on calcium influx into the postsynaptic cell mediated by the N-methyl-D-aspartate (NMDA) type of glutamate receptor. This receptor acts as a coincidence detector: It requires both glutamate release and strong postsynaptic depolarization to conduct calcium.

4. The expression of LTP depends on either the insertion of the α-amino-3-hydroxy-5-methyl-4-isoxazolepropionic acid (AMPA) type of

glutamate receptors in the postsynaptic membrane or an increase in presynaptic glutamate release, depending on the type of synapse and intensity of tetanic stimulation.

5. LTP has both early and late phases. Early LTP depends on covalent modifications, whereas late LTP depends on new protein synthesis, gene transcription, and growth of new synaptic connections.

6. Pharmacological and genetic manipulations that disrupt LTP often lead to an impairment of long-term memory, indicating that LTP may provide an important cellular mechanism for memory storage.

7. Memories are stored by cell assemblies. LTP may be required for forming event-specific assemblies. Recall of memory may reflect reactivation of the same assemblies that were active during the original event.

8. The hippocampus encodes both spatial and nonspatial signals. Many hippocampal neurons act as place cells, firing action potentials when an animal visits a particular location in its environment.

9. The entorhinal cortex, the area of the cortex that provides most of the input to hippocampus, also encodes both nonspatial and spatial information. The medial portion of entorhinal cortex contains neurons, called grid cells, that fire when an animal crosses the vertices of a hexagonal grid-like lattice of spatial locales. Grid cells are organized into semi-independent semi-topographically organized modules with distinct grid frequencies. The entorhinal map also contains border cells, object-vector cells, head direction cells, and speed cells.

10. Within a grid-cell module, pairs of grid cells maintain firing relationships rigidly across environments and experiences, suggesting that grid cells form a universal map that is expressed similarly in all environments. In contrast, place cells in the hippocampus form maps that are plastic as they are completely uncorrelated between environments.

11. Neuropsychiatric disorders such as Alzheimer disease and schizophrenia have been associated with deficits in hippocampal and entorhinal synaptic function, place-cell properties, and learning and memory. Treatments aimed at restoring such function may yield new therapeutic approaches to disease.

12. Despite their clear differences, implicit (Chapter 53) and explicit memory storage rely on a common logic. Both activity-dependent presynaptic facilitation for storing implicit memory and associative long-term potentiation for storing explicit memory rely on the associative properties of specific proteins: Adenylyl cyclase activation in implicit memory requires neurotransmitter plus intracellular Ca^{2+}, whereas NMDA receptor activation in explicit memory requires glutamate plus postsynaptic depolarization. Such similarities indicate the fundamental importance of associative learning rules for memory storage.

Edvard I. Moser
May-Britt Moser
Steven A. Siegelbaum

Selected Reading

Basu J, Siegelbaum SA. 2015. The corticohippocampal circuit, synaptic plasticity, and memory. Cold Spring Harb Perspect Biol 7:a021733.

Bliss TV, Collingridge GL. 2013. Expression of NMDA receptor-dependent LTP in the hippocampus: bridging the divide. Mol Brain 6:5.

Frey U, Morris RG. 1991. Synaptic tagging and long-term potentiation. Nature 385:533–536.

Hafting T, Fyhn M, Molden S, Moser M-B, Moser EI. 2005. Microstructure of a spatial map in the entorhinal cortex. Nature 436:801–806.

Kessels HW, Malinow R. 2009. Synaptic AMPA receptor plasticity and behavior. Neuron 61:340–350.

Martin SJ, Grimwood PD, Morris RG. 2000. Synaptic plasticity and memory: an evaluation of the hypothesis. Annu Rev Neurosci 23:649–711.

Nicoll RA. 2017. A brief history of long-term potentiation. Neuron 93:281–290.

Rowland DC, Roudi Y, Moser MB, Moser EI. 2016. Ten years of grid cells. Annu Rev Neurosci 39:19–40.

Taube JS. 2007. The head direction signal: origins, and sensory-motor integration. Annu Rev Neurosci 30:181–207.

Tonegawa S, Pignatelli M, Roy DS, Ryan TJ. 2015. Memory engram storage and retrieval. Curr Opin Neurobiol 35:101–109.

References

Abel T, Nguyen PV, Barad M, Deuel TAS, Kandel ER, Bourtchouladze R. 1997. Genetic demonstration of a role for PKA in the late phase of LTP and in hippocampal based long-term memory. Cell 88:615–626.

Alme CB, Miao C, Jezek K, Treves A, Moser EI, Moser M-B. 2014. Place cells in the hippocampus: eleven maps for eleven rooms. Proc Natl Acad Sci U S A 111:18428–18435.

Bliss TVP, Lømo T. 1973. Long-lasting potentiation of synaptic transmission in the dentate gyrus of the anesthetized rabbit following stimulation of the perforant path. J Physiol (Lond) 232:331–356.

Dudek SM, Bear MF. 1992. Homosynaptic long-term depression in area CA1 of hippocampus and effects of N-methyl-D-aspartate receptor blockade. Proc Natl Acad Sci U S A 89:4363–4367.

Fyhn M, Hafting T, Treves A, Moser M-B, Moser EI. 2007. Hippocampal remapping and grid realignment in entorhinal cortex. Nature 446:190–194.

Hebb DO. 1949. *The Organization of Behavior: A Neuropsychological Theory.* New York: Wiley.

Hitti FL, Siegelbaum SA. 2014. The hippocampal CA2 region is essential for social memory. Nature 508:88–92.

Høydal ØA, Skytøen ER, Andersson SO, Moser MB, Moser EI. 2019. Object-vector coding in the medial entorhinal cortex. Nature 568:400–404.

Huang Y-Y, Kandel ER. 1994. Recruitment of long-lasting and protein kinase A-dependent long-term potentiation in the CA1 region of hippocampus requires repeated tetanization. Learn Mem 1:74–82.

Kandel ER. 2001. The molecular biology of memory storage: a dialog between genes and synapses (Nobel Lecture). Biosci Rep 21:565–611.

Kjelstrup KB, Solstad T, Brun VH, et al. 2008. Finite scale of spatial representation in the hippocampus. Science 321:140–143.

Kropff E, Carmichael JE, Moser M-B, Moser EI. 2015. Speed cells in the medial entorhinal cortex. Nature 523:419–424.

Lisman J, Yasuda R, Raghavachari S. 2012. Mechanisms of CaMKII action in long-term potentiation. Nat Rev Neurosci 13:169–182.

Liu X, Ramirez S, Pang PT, et al. 2012. Optogenetic stimulation of a hippocampal engram activates fear memory recall. Nature 484:381–385.

Mayford M, Bach ME, Huang Y-Y, Wang L, Hawkins RD, Kandel ER. 1996. Control of memory formation through regulated expression of a CaMKII transgene. Science 274:1678–1683.

McHugh TJ, Blum KI, Tsien JZ, Tonegawa S, Wilson MA. 1996. Impaired hippocampal representation of space in CA1-specific NMDAR1 knockout mice. Cell 87:1339–1349.

McHugh TJ, Jones MW, Quinn JJ, et al. 2007. Dentate gyrus NMDA receptors mediate rapid pattern separation in the hippocampal network. Science 317:94–99.

Montgomery JM, Pavlidis P, Madison DV. 2001. Pair recordings reveal all-silent synaptic connections and the postsynaptic expression of long-term potentiation. Neuron 29:691–701.

Morgan SL, Teyler TJ. 2001. Electrical stimuli patterned after the theta-rhythm induce multiple forms of LTP. J Neurophysiol 86:1289–1296.

Muller RU, Kubie JL, Ranck JB Jr. 1987. Spatial firing patterns of hippocampal complex-spike cells in a fixed environment. J Neurosci 7:1935–1950.

Nakashiba T, Young JZ, McHugh TJ, Buhl DL, Tonegawa S. 2008. Transgenic inhibition of synaptic transmission reveals role of CA3 output in hippocampal learning. Science 319:1260–1264.

Nakazawa K, Quirk MC, Chitwood RA, et al. 2002. Requirement for hippocampal CA3 NMDA receptors in associative memory recall. Science 297:211–218.

Nicholls RE, Alarcon JM, Malleret G, et al. 2008. Transgenic mice lacking NMDAR-dependent LTD exhibit deficits in behavioral flexibility. Neuron 58:104–117.

O'Keefe J, Dostrovsky J. 1971. The hippocampus as a spatial map: preliminary evidence from unit activity in the freely-moving rat. Brain Res 34:171–175.

O'Keefe J, Nadel L. 1978. *The Hippocampus as a Cognitive Map.* Oxford: Clarendon Press.

Ramirez S, Liu X, Lin PA, et al. 2013. Creating a false memory in the hippocampus. Science 341:387–391.

Rotenberg A, Mayford M, Hawkins RD, Kandel ER, Muller RU. 1996. Mice expressing activated CaMKII lack low frequency LTP and do not form stable place cells in the CA1 region of the hippocampus. Cell 87:1351–1361.

Rumpel S, LeDoux J, Zador A, Malinow R. 2005. Postsynaptic receptor trafficking underlying a form of associative learning. Science 308:83–88.

Sacktor TC. 2011. How does PKMζ maintain long-term memory? Nat Rev Neurosci 12:9–15.

Sargolini F, Fyhn M, Hafting T, et al. 2006. Conjunctive representation of position, direction, and velocity in entorhinal cortex. Science 312:758–762.

Silva AJ, Stevens CF, Tonegawa S, Wang Y. 1992. Deficient hippocampal long-term potentiation in α-calcium-calmodulin kinase II mutant mice. Science 257:201–206.

Solstad T, Boccara CN, Kropff E, Moser M-B, Moser EI. 2008. Representation of geometric borders in the entorhinal cortex. Science 322:1865–1868.

Stensola H, Stensola T, Solstad T, Frøland K, Moser M-B, Moser EI. 2012. The entorhinal grid map is discretized. Nature 492:72–78.

Tang YP, Shimizu E, Dube GR, et al. 1999. Genetic enhancement of learning and memory in mice. Nature 401:63–69.

Taube JS, Muller RU, Ranck JB Jr. 1990. Head-direction cells recorded from the postsubiculum in freely moving rats. I. Description and quantitative analysis. J Neurosci 10:420–435.

Tsien JZ, Huerta PT, Tonegawa S. 1996. The essential role of hippocampal CA1 NMDA receptor-dependent synaptic plasticity in spatial memory. Cell 87:1327–1338.

Whitlock JR, Heynen AJ, Shuler MG, Bear MF. 2006. Learning induces long-term potentiation in the hippocampus. Science 313:1093–1097.

Zalutsky RA, Nicoll RA. 1990. Comparison of two forms of long-term potentiation in single hippocampal neurons. Science 248:1619–1624.

55

Language

LANGUAGE IS UNIQUELY HUMAN and arguably our greatest skill and our highest achievement. Despite its complexity, all typically developing children master it by the age of 3. What causes this universal developmental phenomenon, and why are children so much better at acquiring a new language than adults? What brain systems are involved in mature language processing, and are these systems present at birth? How does brain damage produce the various disorders of language known as the aphasias?

For centuries, these questions about language and the brain have prompted vigorous debate among theorists. In the last decade, however, an explosion of information regarding language has taken us beyond the nature–nurture debates and beyond the standard view that a few specialized brain areas are responsible for language. Two factors have brought about this change.

First, functional brain imaging techniques such as positron emission tomography (PET), functional magnetic resonance imaging (fMRI), electroencephalography (EEG), and magnetoencephalography (MEG) have allowed us to examine activation patterns in the brain while a person carries out language tasks—naming objects or actions, listening to sounds or words, and detecting grammatical anomalies. The results of these studies reveal a far more complex picture than the one first proposed by Carl Wernicke in 1874. Moreover, structural brain imaging techniques, such as diffusion tensor imaging (DTI), tractography, and quantitative magnetic resonance imaging (qMRI), have revealed a network of connections that link specialized language areas in the brain. These discoveries are taking us beyond previous, simpler views of the neural underpinnings of language processing and production that assumed involvement of only a few specific brain areas and connections.

Second, behavioral and brain studies of language acquisition show that infants begin to learn language earlier than previously thought, and in ways that had not been previously envisioned. Well before children produce their first words, they learn the sound patterns underlying the phonetic units, words, and phrase structure of the language they hear. Listening to language alters the infant brain early in development, and early language learning affects the brain for life.

Taken together, these advances are shaping a new view of the functional anatomy of language in the brain as a complex and dynamic network in the adult brain, one in which multiple, spatially distributed brain systems cooperate functionally via long-distance neural fascicles (axon fiber bundles). This mature network arises from the considerable brain structure and function in place at birth and develops in conjunction with powerful innate learning mechanisms responsive to linguistic experience. This new view of language encompasses not only its development and mature state, but also its dissolution when brain damage leads to aphasia.

Humans are not the only species to communicate. Passerine birds attract mates with songs, bees code the distance and direction to nectar by dancing, and monkeys signal a desire for sexual contact or fear at the approach of an enemy with coos and grunts. With language, we accomplish all of the above and more. We use language to provide information and express our emotions, to comment on the past and future, and to create fiction and poetry. Using sounds that have only an arbitrary association with the meanings they convey, we talk about anything and everything. No animal has a communication system that parallels human language either in form or in function. Language is the defining characteristic of humans, and living without it creates a totally different world, as patients with aphasia following a stroke experience so heartbreakingly.

Language Has Many Structural Levels: Phonemes, Morphemes, Words, and Sentences

What distinguishes language from other forms of communication? The key feature is a finite set of distinctive speech sounds or phonemes that can be combined with infinite possibilities. Phonemes are the building blocks of units of significance called morphemes. Each language has a distinctive set of phonemes and rules for combining them into morphemes and words. Words can be combined according to the rules of syntax into an infinite number of sentences.

Understanding language presents an interesting set of puzzles, ones that challenge supercomputers. The advent of virtual personal assistants such as Siri and Alexa, based on machine-learning algorithms, has allowed electronic devices to respond to select kinds of human utterances. However, we are still not conversing with computers. Fundamental advances will need to be made before humans can expect to have a conversation with a machine that resembles a conversation you can have with any 3-year-old. Machine-learning solutions do not accomplish their limited responses by mimicking human brain systems used for language, nor do they learn in the ways that human infants learn. Comparing machine-learning approaches (artificial intelligence) and human approaches is of theoretical and practical interest (Chapter 39) and is a hot topic for future research.

Language presents such a complex puzzle because it involves many functionally interconnected levels, starting at the most basic level with the sounds that distinguish words. For example, in English, the sounds /r/ and /l/ differentiate the words *rock* and *lock*. In Japanese, however, this sound change does not distinguish words because the /r/ and /l/ sounds are used interchangeably. Similarly, Spanish speakers distinguish between the words *pano* and *bano*, whereas English speakers treat the /p/ and /b/ sounds at the beginning of these words as the same sounds. Given that many languages use identical sounds but group them differently, children must discover how sounds are grouped to make meaningful distinctions in their language.

Phonetic units are subphonemic. As we have illustrated above with /r/ and /l/, these two sounds are both phonetic units, but their phonemic status differs in English and Japanese. In English, the two are phonemically distinct, meaning that they change the meaning of a word. In Japanese, /r/ and /l/ belong to the same phonemic category and are not distinct. Phonetic units are distinguished by subtle acoustic variations caused by the shape of the vocal tract called *formant frequencies* (Figure 55–1). The patterns and timing of formant frequencies distinguish words that differ in only one phonetic unit, such as the words *pat* and *bat*. In normal speech, formant changes occur very rapidly, on the order of milliseconds. The auditory system has to track these rapid changes in order for an individual

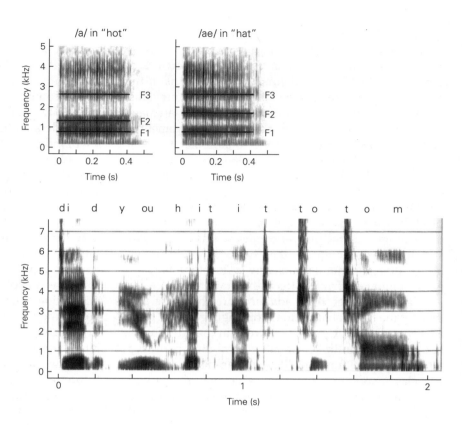

Figure 55–1 Formant frequencies. Formants are systematic variations in the concentration of energy at various sound frequencies and represent resonances of the vocal tract. They are shown here as a function of time in a spectrographic analysis of speech. The formant patterns for two simple vowels (/a/ and /ae/) spoken in isolation are distinguished by differences in formant 2 (F2). Formant patterns for the sentence "Did you hit it to Tom?" spoken slowly and clearly illustrate the rapid changes that underlie normal speech. (Data from Patricia Kuhl.)

to distinguish semantically different sounds and thus understand speech. Whereas in written language, spaces are customarily inserted between words, in speech, there are no acoustic breaks between words. Thus, speech requires a process that can detect words on the basis of something other than sounds bracketed by silence. Computers have a great deal of trouble recognizing words in the normal flow of speech.

Phonotactic rules specify how phonemes can be combined to form words. Both English and Polish use the phonemes /z/ and /b/, for example, but the combination /zb/ is not allowed in English, whereas in Polish, it is common (as in the name *Zbigniew*).

Morphemes are the smallest structural units of a language, best illustrated by prefixes and suffixes. In English, for example, the prefix *un* (meaning *not*) can be added to many adjectives to convey the opposite meaning (eg, *unimportant*). Suffixes often signal the tense or number of a word. For example, in English, we add *s* or *es* to indicate more than one of something (*pot* becomes *pots*, *bug* becomes *bugs*, or box becomes *boxes*). To indicate the tense of a regular verb, we add an ending to the word (eg, *play* can become *plays*, *playing*, and *played*). Irregular verbs do not follow the rule (eg, *go* becomes *went* rather than *goed* and *break* becomes *broke* rather than *breaked*). Every language has a different set of rules for altering the tense and number of a word.

Finally, to create language, words have to be strung together. *Syntax* specifies word and phrase order for a given language. In English, for example, sentences typically conform to a subject-verb-object order (eg, *He eats cake*), whereas in Japanese, it is typically subject-object-verb (eg, *Karewa keeki o tabenzasu*, literally *He cake eats*). Languages have systematic differences in the order of larger elements (noun phrases and verb phrases) of a sentence, and in the order of words within phrases, as illustrated by the difference between English and French noun phrases. In English, adjectives precede the noun (eg, *a very intelligent man*), whereas in French, most follow the noun (eg, *un homme tres intelligent*).

Language Acquisition in Children Follows a Universal Pattern

Regardless of culture, all children initially exhibit universal patterns of speech perception and production that do not depend on the specific language children hear (Figure 55–2). By the end of the first year, infants have learned through exposure to a specific language which phonetic units convey meaning in that language and to recognize likely words, even though they do not yet understand those words. By 12 months of age, infants understand approximately 50 words and have begun to produce speech that resembles the native language. By the age of 3 years, children know approximately 1,000 words (by adulthood 70,000), create long adult-like sentences, and can carry on a conversation. Between 36 and 48 months, children respond to the differences between grammatical and ungrammatical sentences in an adult-like way, although tests using the most complex sentences indicate that the intricacies of grammar are not mastered until late childhood, between 7 and 10 years of age.

In the last half of the 20th century, debate on the nature and acquisition of language was ignited by a highly publicized exchange between a strong learning theorist and a strong nativist. In 1957, the behavioral psychologist B. F. Skinner proposed that language was acquired through learning. In his book *Verbal Behavior*, Skinner argued that language, like all animal behavior, was a learned behavior that developed in children as a function of external reinforcement and careful parental shaping. By Skinner's account, infants learn language as a rat learns to press a bar—through monitoring and management of reward contingencies. The nativist Noam Chomsky, writing a review of *Verbal Behavior*, took a very different position. Chomsky argued that traditional reinforcement learning has little to do with the ability of humans to acquire language. Instead, he proposed that every individual has an innate "language faculty" that includes a universal grammar and a universal phonetics; exposure to a specific language triggers a "selection" process for one language.

More recent studies of language acquisition in infants and children have clearly demonstrated that the kind of learning going on in infancy does not resemble that described by Skinner with its reliance on external shaping and reinforcement. At the same time, a nativist account such as Chomsky's, in which the language the infant hears triggers selection of one of several innate options, also does not capture the process.

The "Universalist" Infant Becomes Linguistically Specialized by Age 1

In the early 1970s, psychologist Peter Eimas showed that infants were especially good at hearing the acoustic changes that distinguish phonetic units in the world's languages. When speech sounds were acoustically varied in small equal steps to form a series ranging from one phonetic unit to another, say from /ba/ to /pa/, Eimas showed that infants could discern very slight acoustic changes at the locations in the series (the "boundary") where adults heard an abrupt change between the two phonetic categories, a phenomenon

called *categorical perception*. Eimas demonstrated that infants could detect these slight acoustic changes at the phonetic boundary between two categories for phonetic units in languages they had never experienced, whereas adults have this ability only for phonetic units in languages in which they are fluent. Japanese people, for example, find it very difficult to hear the acoustic differences between the American English /r/ and /l/ sounds. Both are perceived as Japanese /r/, and as we have seen, Japanese speakers use the two sounds interchangeably when producing words.

Categorical perception was originally thought to occur only in humans, but in 1975, cognitive neuroscientists showed that it exists in nonhuman mammals such as chinchillas and monkeys. Since then, many studies have confirmed this result (as well as identifying species differences between mammals and birds). These studies suggest that the evolution of phonetic units was strongly influenced by preexisting auditory structures and capacities. Infants' ability to hear all possible differences in speech prepares them to learn any language; at birth, they are linguistic "universalists."

Speech production develops simultaneously with speech perception (Figure 55–2). All infants, regardless of culture, produce sounds that are universal. Infants "coo" with vowel-like sounds at 3 months of age and "babble" using consonant–vowel combinations at about 7 months of age. Toward the end of the first year, language-specific patterns of speech production begin to emerge in infants' spontaneous utterances. As children approach the age of 2 years, they begin to mimic the sound patterns of their native language. Chinese toddlers' utterances reflect the pitch, rhythm, and phonetic structure of Mandarin, and the utterances of British toddlers sound distinctly British. Infants develop an ability to imitate the sounds they hear others produce as early as 20 weeks of age. Very early in development, infants begin to master the subtle motor patterns required to produce their "mother tongue." Speech-motor patterns acquired in the earliest stages of language learning persist throughout life and influence the sounds, tempo, and rhythm of a second language learned later.

Right before the onset of first words, infants' abilities to discriminate native and nonnative phonetic units show a dramatic shift. At 6 months of age, infants can discriminate all phonetic units used in all languages, but by the end of the first year, they fail to discriminate phonetic changes that they successfully recognized 6 months earlier. At the same time, infants become significantly more adept at hearing native-language phonetic differences. For example, when American

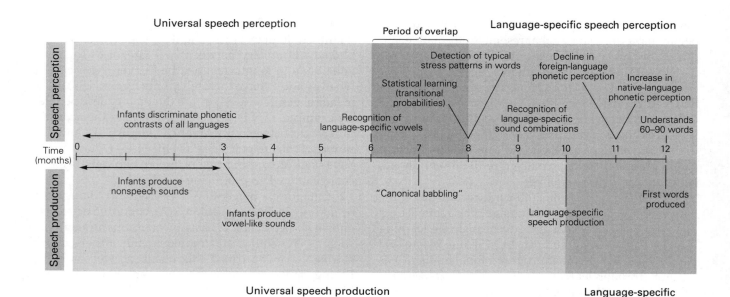

Figure 55–2 Language development progresses through a standard sequence in all children. Speech perception and production in children in various cultures initially follow a language-universal pattern. By the end of the first year of life, language-specific patterns emerge. Speech perception becomes language-specific before speech production. (Adapted, with permission, from Doupe and Kuhl 1999.)

and Japanese infants were tested between 6 and 12 months of age on the discrimination of the American English /r/ and /l/, American infants improved significantly between 8 and 10 months, whereas Japanese infants declined, suggesting that this is a sensitive period for phonetic learning. Moreover, infants' native-language discrimination ability at 7.5 months of age predicts the rate at which known words, sentence complexity, and mean length of utterance grow between 14 and 30 months.

If the second half of the first year is a sensitive period for speech learning, what happens when infants are exposed to a new language during this time? Do they learn? When American infants were exposed to Mandarin Chinese in the laboratory between 9 and 10 months of age, the infants learned if exposure occurred through interaction with a human being; infants exposed to the exact same material through television or audiotape with no live human interaction do not learn (Figure 55–3). When tested, the performance of the group exposed to live speakers was statistically indistinguishable from that of infants raised in Taiwan, China who had listened to Mandarin for 10 months. These results established that, at 9 months of age, the right kind of exposure to a foreign language permits phonetic learning, supporting the view that this is a sensitive period for such learning. The study also demonstrated, however, that social interaction

plays a more significant role in learning than previously thought.

Further work showed that the degree to which infants track the eye movements of the tutor—watching what she is looking at as she names objects in the foreign language—correlates strongly with neural measures of phonetic and word learning after exposure to the new language, again implicating social brain areas in language learning.

An infant's ability to pick up social cues is essential to language learning, but what other skills promote learning during this critical period? Studies suggest that early exposure to speech induces an implicit learning process that increases native-language discrimination and reduces the infant's innate ability to hear distinctions between the phonetic units of all other languages. Infants are sensitive to the statistical properties of the language they hear. Distributional frequency patterns of sounds affect infants' speech learning by 6 months of age. Infants begin to organize speech sounds into categories based on *phonetic prototypes*, the most frequently occurring phonetic units in their language.

Six-month-old infants in the United States and Sweden were tested with prototypical English and Swedish vowels to examine whether infants discriminated acoustic variations in the vowels, like those that occur when different talkers produce them. By 6 months of age, the American and Swedish infants ignored

Language-specific speech perception

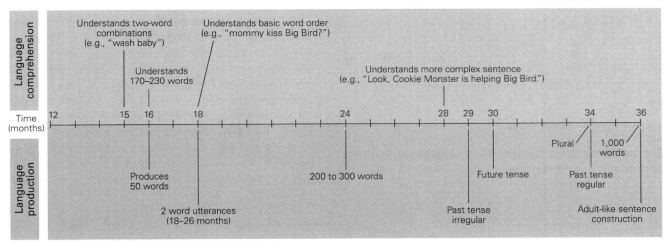

Language-specific speech production

Live exposure

Audiovisual exposure

Figure 55–3 Infants can learn the phonemes of a nonnative language at 9 months of age. Three groups of American infants were exposed for the first time to a new language (Mandarin Chinese) in 12 25-minute sessions between the ages of 9 and 10.5 months. One group interacted with live native speakers of Mandarin; a second group was exposed to the identical material through television; and a third group heard tape recordings only. A control group had similar language sessions but heard only English. Performance on discrimination of Mandarin phonemes was tested in all groups after exposure (age 11 months). (Reproduced, with permission, from Kuhl, Tsao, and Liu 2003.)

Left. Only infants exposed to live Mandarin speakers discriminated the Mandarin phonemes. Infants exposed through TV or tapes showed no learning, and their performance was indistinguishable from that of control infants (who heard only English).

Right. The performance of American infants exposed to live Mandarin speakers was equivalent to that of monolingual Chinese infants of the same age who had experienced Mandarin from birth.

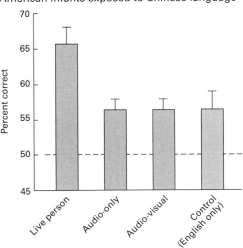
American infants exposed to Chinese language

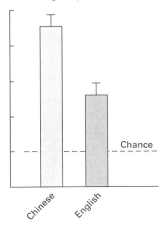
Monolingually raised infants

acoustic variations around native language prototypes but not with nonnative prototypes. Paul Iverson has shown that language experience alters the acoustic features to which speakers of different languages attend and distorts perception around category prototypes. This makes stimuli perceptually more similar to the prototype, which helps explain why 11-month-old Japanese infants fail to discriminate English /r/ and /1/ after experience with Japanese.

The Visual System Is Engaged in Language Production and Perception

Language is ordinarily communicated through an auditory-vocal channel, but deaf individuals communicate through a visual-manual channel. Natural signed languages, such as American Sign Language (Ameslan or ASL), are those invented by the deaf and vary across countries. Deaf infants "babble" with their hands at approximately the same time in development as hearing infants babble orally. Other developmental milestones, such as first words and two-word combinations, also occur on the developmental timetable of hearing infants.

Additional studies indicate that visual information of another kind, the face of the talker, is not only very helpful for communication but also affects the everyday perception of speech. We all experience the benefits of "lip reading" at noisy parties—watching speakers' mouths move helps us understand speech in a noisy environment. The most compelling laboratory demonstration that vision plays a role in everyday speech perception is the illusion that results when discrepant speech information is sent to the visual and auditory modalities. When subjects hear the syllable "ba" while watching a person pronounce "ga" they report hearing an intermediate articulation "da." Such demonstrations support the idea that speech categories are defined both auditorily and visually and that perception is governed by both sight and sound.

Prosodic Cues Are Learned as Early as In Utero

Long before infants recognize that things and events in the world have names, they memorize the global sound patterns typical in their language. Infants learn such prosodic cues as pitch, duration, and loudness changes. In English, for example, a strong/weak pattern of stress is typical—as in the words "BAby," "MOMmy," "TAble," and "BASEball"—whereas in some languages, a weak/strong pattern predominates. Six- and 9-month-old infants given a listening choice between words in English or Dutch show a listening preference for native-language words at the age of 9 months (but not at 6 months).

Prosodic cues can convey both linguistic information (differences in intonation and tone in languages such as Chinese) and paralinguistic information, such as the emotional state of the speaker. Even in utero fetuses learn prosodic cues by listening to their mother's speech. Certain sounds are transmitted through bone conduction to the womb; these are typically intense (above 80 dB), low-frequency sounds (particularly below 300 Hz, but as high as 1,000 Hz with some attenuation). Thus, the prosodic patterns of speech, including voice pitch and the stress and intonation patterns characteristic of a particular language and speaker, are transmitted to the fetus, while the sound patterns that convey phonetic units and words are greatly attenuated. At birth, infants demonstrate having learned this prosodic information by their preference for (1) the language spoken by their mothers during pregnancy, (2) their mother's voice over that of another female, and (3) stories with a distinct tempo and rhythm read out loud by the mother during the last 10 weeks of pregnancy.

Transitional Probabilities Help Distinguish Words in Continuous Speech

Seven- to 8-month-old infants learn to recognize words using the probability that one syllable will follow another. Such transitional probabilities between syllables within a word are high because the sequential order remains fixed. In the word *potato*, for example, the syllable "ta" always follows the syllable "po" (probability of 1.0). Between words, on the other hand, as between "hot" and "po" in the string *hot potato*, are much lower transitional probabilities.

Psychologist Jenny Saffran showed that infants treat phonetic units and syllables with high transitional probabilities as word-like units. In one experiment, infants heard 2-minute strings of pseudo-words, such as *tibudo*, *pabiku*, *golatu*, and *daropi*, without any acoustic breaks between them. They were then tested for recognition of these pseudo-words as well as new ones formed by combining the last syllable of one word with the two initial syllables of another word (such as *tudaro* formed from *golatu* and *daropi*). Infants recognized the original pseudo-words but not the new combinations they had not been previously exposed to, indicating that they used transitional probabilities to identify words.

These forms of learning clearly do not involve Skinnerian reinforcement. Caretakers do not manage the contingencies and gradually shape through reinforcement the statistical analyses performed by infants. Conversely, language learning by infants also does not appear to reflect a process in which innately provided options are chosen based on language experience. Rather, infants learn language implicitly through detailed analysis of the patterns of statistical variation in the natural speech they hear and sophisticated analysis of information provided through social interaction (eg, eye gaze). The learning of these patterns in turn alters perception to favor the native language. In summary, both the statistical properties of language and the social cues provided during language interactions help infants learn. Language evolved to capitalize on the kinds of cues that infants are innately able to recognize. This mirrors the argument that the development of phonetic units was significantly influenced by the features of mammalian hearing, ensuring that infants would find it easy to discriminate phonemes, the fundamental units of meaning in language.

There Is a Critical Period for Language Learning

Children learn language more naturally and efficiently than adults, a paradox given that the cognitive skills of adults are superior. Why should this be the case?

Many consider language acquisition to be an example of a skill that is learned best during a critical period in development. Eric Lenneberg proposed that maturational factors at puberty cause a change in the neural mechanisms that control language acquisition. Evidence supporting this view comes from classic studies of Chinese and Korean immigrants to the United States who had been immersed in English at ages ranging from 3 to 39 years. When asked to identify errors in sentences containing grammatical mistakes, an easy task for native speakers, the responses of second-language learners declined with the age of arrival in the United States. A similar trend emerges when one compares individuals exposed to ASL from birth to those exposed between 5 and 12 years of age. Those exposed from birth were best at identifying errors in ASL, those exposed at age 5 were slightly poorer, and those exposed after the age of 12 years were substantially poorer.

What restricts our ability to learn a new language after puberty? Developmental studies suggest that prior learning plays a role. Learning a native language produces a neural commitment to detection of the acoustic patterns of that language, and this commitment interferes with later learning of a second language. Early exposure to language results in neural circuitry that is "tuned" to detect the phonetic units and prosodic patterns of that language. Neural commitment to native language enhances the ability to detect patterns based on those already learned (eg, phonetic

learning supports word learning) but reduces the ability to detect patterns that do not conform. Learning the motor patterns required to speak a language also results in neural commitment. The motor patterns learned for one language (eg, lip rounding in French) can interfere with those required for pronunciation of a second language (eg, English) and thus can hinder efforts to pronounce the second language without an accent. Early in life, two or more languages can be easily learned because interference effects are minimal until neural patterns are well established.

Neurobiologist Takao Hensch has been working on identifying the chemical switches that open and close neurodevelopmental critical periods in learning, including those in animals and humans. Hensch has found that the neurotransmitter γ-aminobutyric acid (GABA) opens the critical period by inhibiting the firing of excitatory neurons, bringing them into balance with the firing of inhibitory neurons so as to create an excitatory–inhibitory (EI) balance. Studies testing this hypothesis in humans are difficult to conduct, but investigations on the infants of mothers who altered the EI balance of the fetus during pregnancy by taking psychotropic medications (serotonin reuptake inhibitors [SRIs]) for depression support the EI hypothesis. One of fluoxetine's off-target effects is to increase the sensitivity of some GABA receptors to GABA. When compared to infants of depressed mothers who were not exposed prenatally to SRIs and control mothers without depression or SRIs, infants exposed prenatally to SRIs showed an accelerated phonetic learning process, indicating that the well-established timing of the early transition in infants' phonetic perception can be altered.

We do not completely lose the ability later in life to learn a new language, but it is far more difficult. Regardless of the age at which learning begins, second-language learning is improved by a training regimen that mimics critical components of early learning—long periods of listening in a social context (immersion), the use of both auditory and visual information, and exposure to simplified and exaggerated speech resembling "parentese."

The "Parentese" Speaking Style Enhances Language Learning

Everyone agrees that when adults talk to their children they sound unusual. Discovered by linguists and anthropologists in the early 1960s as they listened to languages spoken around the world, "motherese" (or "parentese," as fathers produce it as well) is a special speaking style used when addressing infants and young children. Parentese has a higher pitch, slower tempo, and exaggerated intonation contours, and is easily recognized. Compared to adult-directed speech, the pitch of the voice is increased on average by an octave both in males and in females. Phonetic units are spoken more clearly and are acoustically exaggerated, thus increasing the acoustic distinctiveness of phonetic units. Adults speaking to infants exaggerate just those features of speech that are critical to their native language. For example, when talking to their infants, Chinese mothers exaggerate the four tones in Mandarin that are critical to word meaning.

When given a choice, infants prefer listening to infant-directed rather than adult-directed speech. When infants are allowed to activate recordings of infant-directed or adult-directed speech by turning their head left or right, they will turn in whatever direction is required to turn on infant-directed speech.

Recent research by psychologists Nairan Ramirez-Esparza and Adrian Garcia-Sierra shows that the degree to which parentese is used in language spoken to infants at 11 and 14 months of age at home is strongly correlated with a child's language development by the age of 24 months and remains strongly correlated at the age of 36 months. This relationship holds for both monolingual and bilingual children. However, in bilingual children, early advances in the two languages differ depending on the language spoken in parentese. For example, Spanish-language parentese enhances a child's behavioral and neural responses to Spanish, but not English, and vice versa. Children raised in families in which the amount of language exposure and the use of parentese are low often show deficits in language and literacy by the time they enter school, and these deficits correlate with decreased functional activation in brain areas related to language.

Successful Bilingual Learning Depends on the Age at Which the Second Language Is Learned

How does the brain handle two languages? Behavioral data show that if exposure to two languages begins at birth, children reach the milestones of language at the same age as their monolingual peers—they coo, babble, and produce words at the benchmark ages seen in monolinguals. The idea that bilingual experience produces "confusion" has been debunked by studies that measure "conceptual" vocabulary, that is, word knowledge regardless of the language the child uses to express that knowledge. Older studies measured words in only one of the infants' two languages, and such word counts often showed decreased vocabulary when compared to monolinguals. Conceptual

vocabulary scores show that bilingual children's vocabulary counts meet or exceed those of their monolingual peers.

Exposure to a second language after puberty shows limitations in the degree to which the new language can be learned. Whether subjects are tested on phonological rules, morphological endings, or syntax, the ability to learn a new language appears to decline every 2 years after the age of 7 years, indicating that acquisition of a second language after puberty is quite difficult.

Brain measures on bilingual infants reflect these behavioral data. Psychologist Naja Ferjan Ramirez used MEG to show that activation of the superior temporal area in 11-month-old infants exposed to two languages (English and Spanish) from birth is the same for the sounds of both languages and that brain responses to English sounds are equivalent to those of age-matched monolingual infants for English. Bilingual infants listening to speech also exhibit greater activation in the prefrontal cortex, a region mediating attention, when compared to monolingual infants; this finding is consistent with the fact that bilingual children (adults as well) demonstrate superior cognitive skills related to attention. Arguably, listening to two languages requires multiple shifts in attention to activate one language over another.

If a second language is acquired later in development, the age at which exposure occurs and the degree of eventual proficiency affect how the brain processes both languages. In "late" bilinguals (those who learned a second language after puberty), the second language and native language are processed in spatially separated areas in the language-sensitive left frontal region. In "early" bilinguals (those who acquired both languages as children), the two languages are processed in the same left frontal area.

A New Model for the Neural Basis of Language Has Emerged

Numerous Specialized Cortical Regions Contribute to Language Processing

The classical Wernicke-Geschwind neural model of language was based on the works of Broca (1861), Wernicke (1874), Lichtheim (1885), and Geschwind (1970). In the Wernicke-Geschwind model, acoustic cues contained in spoken words were processed in auditory pathways and relayed to Wernicke's area, where the meaning of a word was conveyed to higher brain structures. The arcuate fasciculus was assumed to

Table 55–1 Differential Diagnosis of the Main Types of Aphasia

Type of aphasia	Speech	Comprehension	Capacity for repetition	Other signs	Region affected
Broca	Nonfluent, effortful	Largely preserved for single words and grammatically simple sentences	Impaired	Right hemiparesis (arm > leg); patient aware of defect and can be depressed	Left posterior frontal cortex and underlying structures
Wernicke	Fluent, abundant, well articulated, melodic	Impaired	Impaired	No motor signs; patient can be anxious, agitated, euphoric, or paranoid	Left posterior superior and middle temporal cortex
Conduction	Fluent with some articulatory defects	Intact or largely preserved	Impaired	Often none; patient can have cortical sensory loss or weakness in right arm	Left superior temporal and supramarginal gyri
Global	Scant, nonfluent	Impaired	Impaired	Right hemiplegia	Massive left perisylvian lesion
Transcortical motor	Nonfluent, explosive	Intact or largely preserved	Intact or largely preserved	Sometimes right-sided weakness	Anterior or superior to Broca's area
Transcortical sensory	Fluent, scant	Impaired	Intact or largely preserved	No motor signs	Posterior or inferior to Wernicke's area

be a unidirectional pathway that brought information from Wernicke's area to Broca's area to enable speech production. Both Wernicke's and Broca's areas interacted with association areas. The Wernicke-Geschwind model formed the basis for a practical classification of the aphasias that clinical neurologists still use today (Table 55–1).

Advancements in basic and clinical neuroscience, the advent of more sophisticated functional brain imaging tools, advanced methods for structural brain imaging, and an increasing number of studies that combine brain and behavioral measures have resulted in the development of a new "dual-stream" model. In the dual-stream model, the processing of language is thought to involve large-scale networks that are composed of different brain areas, each with a specialized function, and the white matter tracts that connect them.

This dual-stream model of language processing is similar to the well-established "what" and "where" dual-stream model of the visual system. The existence of two cortical streams of auditory information processing was first postulated by Josef Rauschecker. Gregory Hickok and David Poeppel further elaborated the dual-stream model, and it has since been even further expanded upon by Angela Friederici as well as others studying the neurobiology of language. Figure 55–4 shows the basic components of the dual-stream model.

Compared to the classic Wernicke-Geschwind model, the dual-stream model comprises a larger number of cortical areas that are more widely distributed in the brain and adds critical connecting bidirectional pathways between specialized brain regions. These improvements in the model for language processing are due to advances in structural brain imaging techniques, such as DTI and diffusion-weighted imaging, which provide quantitative measures on a microscopic scale of the white matter in fascicles that connect various cortical areas and allow for the detailed delineation of neural tracts throughout the brain (tractography).

In the dual-stream model, initial spectrotemporal processing of auditory speech sounds is performed bilaterally in the auditory cortex. This information is then communicated to the posterior superior temporal gyrus bilaterally, where phonological-level processing occurs. Language processing then diverges into a dorsal "sensorimotor stream," which maps sound to articulation, and a ventral "sensory-conceptual" stream, which maps sound to meaning.

The bidirectional dorsal stream connects auditory speech information with motor plans that produce speech. The dorsal stream passes above the lateral ventricles and maps sounds onto articulatory representations, connecting regions of the inferior frontal lobe, premotor cortex, and insula (all involved in speech articulation) to the region that is classically recognized as Wernicke's area. It is considered to comprise two pathways: Dorsal pathway 1 connects the posterior

Figure 55–4 Dual-stream model of language processing. Temporal and spectral analyses of speech signals occur bilaterally in the auditory cortex followed by phonological analysis in the posterior superior temporal gyri (**yellow arrow**). Processing then diverges into two separate pathways: a dorsal stream that maps speech sounds to motor programs and a ventral stream that maps speech sounds to meaning. The dorsal pathway is strongly left hemisphere dominant and has segments that extend to the premotor cortex (dorsal pathway 1) and to the posterior inferior frontal cortex (dorsal pathway 2). The ventral pathway occurs bilaterally and extends to the anterior temporal lobe and the posterior inferior frontal cortex. (Adapted, with permission, from Hickok and Poeppel 2007, and Skeide and Friederici 2016.)

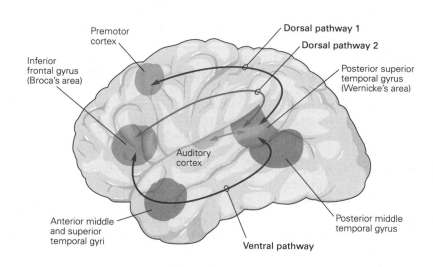

superior temporal gyrus to the premotor cortex, and dorsal pathway 2 connects the posterior superior temporal gyrus to Broca's area. Pathway 2 is involved in higher-order analysis of speech, such as discriminating subtle differences in meaning based on grammar and interpreting language using more complex concepts. The dorsal stream is strongly left hemisphere dominant. The arcuate fasciculus and the superior longitudinal fasciculus are white matter fiber tracts that mediate communication along the dorsal stream.

The ventral stream passes below the Sylvian fissure and is composed of regions of the superior and middle temporal lobes as well as regions of the posterior inferior frontal lobe. This stream conveys information for auditory comprehension, which requires transformation of the auditory signal to representations in a mental lexicon, a "brain-based dictionary" that links individual word forms to their semantic meaning. This stream comprises the inferior fronto-occipital fasciculus, the uncinate fasciculus, and the extreme fiber capsule system and is largely bilaterally represented.

The cortical brain regions included in the dual-stream model also interact with spatially distributed regions throughout both hemispheres of the brain that provide additional information crucial for language processing. These regions include the prefrontal cortex and cingulate cortices, which exert executive control and mediate attentional processes, respectively, as well as regions in the medial temporal, frontal, and parietal areas involved in memory retrieval.

The Neural Architecture for Language Develops Rapidly During Infancy

The study of language development in infancy requires a methodology that documents significant changes in

behavior and links those changes to changes in brain function and morphology over time. Neuroimaging methods for the infant brain have improved substantially over the past decade, allowing for a detailed assessment of the progression of development of the specialized regions and structural connections required by the language network. For example, developmental neuroscientists have created models of the average infant brain and brain atlases for the infant brain at 3 and 6 months of age. These models indicate that brain structures essential to language processing in adulthood, such as the inferior frontal cortex, premotor cortex, and superior temporal gyrus, support speech processing in early infancy. Studies using DTI and tractography indicate that the arcuate fasciculus and the uncinate fasciculus connect language regions by 3 months of age.

The development of the neural substrates for language in 1- to 3-day-old infants has been studied in depth by Daniela Perani using fMRI and DTI. Perani's fMRI work reveals that listening to speech activates the infant superior temporal gyrus bilaterally and that in the left hemisphere this activation extends to the planum temporale, inferior frontal gyrus, and inferior parietal lobe. Perani's DTI studies of the same newborn infants demonstrate weak intrahemispheric connections, but strong connections between the hemispheres. Nevertheless, the ventral fiber tract connecting the ventral portion of the inferior frontal gyrus via the extreme fiber capsule system to the temporal cortex is evident in newborns and in both hemispheres. The dorsal pathway connecting the temporal cortex to the premotor cortex is also present in the newborns, although the dorsal tract that connects the temporal cortex to Broca's area in adults is not detectable in newborns. These early connections between sensory areas and the premotor cortex are important because they may allow the sensory-to-motor mapping essential for the development of early imitation of the sounds and words of the language.

Jens Brauer and colleagues replicated these findings on the development of ventral and dorsal pathways in newborns, revealing the maturational primacy of the ventral connection linking temporal areas to the inferior frontal gyrus. Brauer also verified that the dorsal pathway connects the temporal and premotor cortex at birth and showed that the dorsal pathway to the inferior frontal gyrus develops later. Brauer used the same protocol with children 7 years of age and adults. In 7-year-old, the dorsal pathway fully connects auditory areas and the inferior frontal gyrus, but in adults, it has more extensive and far-reaching connections.

EEG and MEG functional brain imaging studies on young infants as early as 2 months of age show that the inferior frontal and temporal cortices, implicated in both the classical and contemporary models of language processing, are activated bilaterally by speech—syllables, words, and sentences. This finding supports the hypothesis that left hemisphere specialization increases over time, with syllables showing dominant left hemisphere specialization at the end of the first year, words by the age of 2, and sentences in middle childhood.

EEG and MEG studies of young infants in which infants listen passively to native and nonnative syllables have produced results consistent with the behavioral transitions described earlier in this chapter. Several infant laboratories have shown that brain activity in response to speech, measured early in development, provides sensitive markers that predict language skills several years later. These studies hold promise for the eventual identification of brain measures in infants that indicate risk for developmental disabilities involving language, such as autism spectrum disorder, dyslexia, and specific language impairment. Early identification would allow earlier and more effective interventions for these impairments, improving outcomes for these children and their families.

Studies using functional MEG brain imaging of infants show that at 7 months of age, native and nonnative speech syllables activate not only superior temporal regions of the infant brain but also inferior frontal regions and the cerebellum, forging an association between speech patterns they hear and the motor plans they use to babble and imitate. By 12 months of age, language experience alters the patterns of activation in both sensory and motor brain regions.

Auditory activation becomes stronger for *native* sounds, indicating that brain areas have begun to become specialized for native language phonology. In contrast, motor activation in both Broca's area and the cerebellum is increased in response to *nonnative* sounds, because by 12 months infants have sufficient sensorimotor knowledge to imitate native sounds and some words and have linked stored auditory patterns (words like "cup" and "ball") to the motor plans necessary to produce them. But they cannot make the sensorimotor associations for foreign-language sounds and words because the necessary motor plans cannot be generated. Therefore, we see longer and more diffuse activation as infants struggle to create the motor plans for a sound or word they have never experienced. The importance of motor learning in language development is also shown by longitudinal whole-brain voxel-based morphometry studies of 7-month-old infants showing that gray matter concentrations in the cerebellum correlate with the number of words those infants can produce at 1 year of age.

Over the next 5 years, there is likely to be an explosive increase in brain studies focused on development of the language network. In a number of laboratories, these brain measures will be linked to behavioral measures, enabling the creation of models that delineate how language experience alters the infant brain to increase its specialization for the language or languages to which the child is exposed. The finding that the classic brain regions known to be part of the language network in adults—in particular, the left and right temporal cortices and the left inferior frontal cortex—are already activated by speech at birth recalls Chomsky's view of innate language capabilities.

The Left Hemisphere Is Dominant for Language

Current views of language processing agree that while the neural circuitry necessary for transforming speech sounds to meaning may be present in both hemispheres, the left hemisphere is more highly specialized for language processing. This left hemisphere dominance develops with maturation and learning.

Evidence from a variety of sources suggests that left hemisphere specialization for language develops rapidly in infancy. Word learning represents a case in point. Deborah Mills and her colleagues used event-related potentials to track development of the neural signals generated in response to words that children knew. Her studies showed that both age and language proficiency produce changes in the strength of the neural responses to known words, as well as a change in hemisphere dominance between 13 and 20 months of age. At the earliest age studied, known words activate a broad and bilaterally distributed pattern across the brain. As infants approach 20 months and vocabulary grows, the activation pattern shifts to become left hemisphere dominant in the temporal and parietal regions. In late talkers, this shift is delayed to nearly 30 months. In 24-month-old children with autism, the degree to which this left hemisphere dominance is evident predicts children's linguistic, cognitive, and adaptive abilities at age 6.

Several studies show that immersion in a second language in adulthood produces growth in the superior longitudinal fasciculus, a white matter fiber tract that is important for language. Neuroscientist Ping Mamiya, collaborating with geneticist Evan Eichler, demonstrated, using DTI, that white matter integrity of the superior longitudinal fasciculus in the right hemisphere increased in Chinese college students in proportion to the number of days they spent in an English immersion class and decreased after immersion ended. Moreover, analysis of polymorphisms in the catechol-O-methyltransferase (COMT) gene showed an effect on this relationship—students with two of the variants demonstrated these changes, while students with the third variant showed no change in white matter properties with language experience.

There is great interest in brain studies investigating the selectivity of the brain mechanisms underlying language. Studies in the visual system by neuroscientist Nancy Kanwisher led to the suggestion that certain visual areas (the fusiform face area) are highly selective for particular stimuli, such as faces. Similar claims have been advanced for brain areas underlying speech analysis. For example, Kanwisher's group has proposed that Broca's area contains many subregions, each highly selective for particular levels of language. Additional studies on selectivity, particularly during development, will be the focus of future studies.

Helen Neville and Laura-Anne Pettito have shown that the left hemisphere is activated not only by auditory stimuli but also by visual stimuli that have linguistic significance. Deaf individuals process sign language in left hemisphere speech-processing regions. Such studies show that the language network processes linguistic information regardless of modality.

Prosody Engages Both Right and Left Hemispheres Depending on the Information Conveyed

Prosodic cues in language can be linguistic, conveying semantic meaning as tones do in Mandarin Chinese or Thai, as well as paralinguistic, expressing our attitudes and emotions. The pitch of the voice carries both kinds of information, and the brain's processing of each kind of information differs.

Emotional changes in pitch engage the right hemisphere, primarily the right frontal and temporal regions. Emotional information helps convey a speaker's mood and intentions, and this helps interpret sentence meaning. Patients with right hemisphere lesions often produce speech with inappropriate stress, timing, and intonation, and their speech sounds emotionally flat; they also frequently fail to interpret the emotional cues in others' speech.

Semantic changes in pitch involve a different pattern of brain activity, as demonstrated by neuroimaging studies. Jackson Grandour used a novel experimental design using Chinese syllables that carried either their native Chinese tone or the nonnative Thai tone. fMRI results for both Chinese and Thai speakers show higher activation in the left planum temporale for syllables carrying the native tone as opposed to nonnative tone (Figure 55–5). The right hemisphere did not show this double dissociation, supporting the view that language processing occurs in the left hemisphere even for auditory signals typically processed on the right.

Studies of the Aphasias Have Provided Insights into Language Processing

According to recent estimates, there are more than 795,000 strokes per year in the United States. Aphasia occurs in 21% to 38% of acute strokes and increases the probability of mortality and morbidity. In the past decade, the number of individuals with aphasia grew by more than 100,000 per year. Broca's aphasia, Wernicke's aphasia, and conduction aphasia compose the three classical models of clinical aphasia syndromes. Hickok and Poeppel describe each of these subtypes in the context of the dual-stream model. Accordingly, Broca's aphasia and conduction aphasia are due to sensorimotor integration problems related to damage to the dorsal stream of language processing, whereas Wernicke's aphasia, word deafness, and transcortical sensory aphasia are produced by damage to the ventral stream.

Broca's Aphasia Results From a Large Lesion in the Left Frontal Lobe

Broca's aphasia is a disorder of speech production, including impairments in grammatical processing, caused by lesions of the dorsal stream. When we speak, we rely on auditory patterns stored in the brain. Naming a cup when presented with coffee requires a

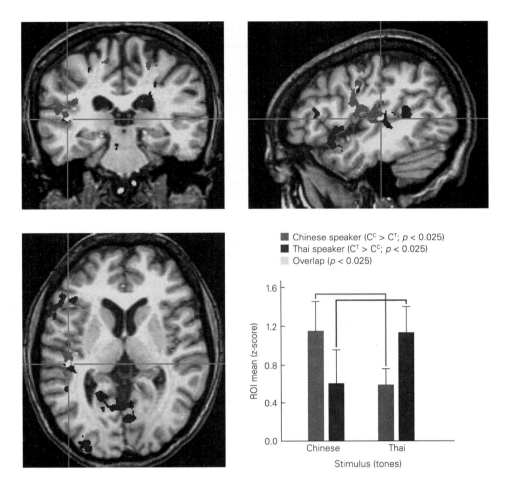

Figure 55–5 Brain activation for Chinese and Thai lexical tones revealed by functional magnetic resonance imaging. Language stimuli were composed of Chinese syllables superimposed with either Thai tones (C^T) or Chinese tones (C^C). Both native Chinese and native Thai speakers demonstrated a left hemisphere (**LH**) dominance when listening to their native tones. In the Chinese speakers, activation of the left hemisphere was stronger for Chinese tones, whereas in the Thai speakers, activation was stronger for Thai tones. Overlap for the two groups occurs in the left planum temporale and the ventral precentral gyrus. In the left planum temporale (**green crosshairs**), a double dissociation was found between tonal processing and language experience (bar charts). The right hemisphere (**RH**) did not show these effects. (*Top left,* coronal section; *top right,* sagittal section; *bottom left,* axial section.) (Abbreviation: **ROI,** region of interest.) (Adapted, with permission, from Xu et al. 2006. Copyright © 2005 Wiley-Liss, Inc.)

patient to connect the stored sensory pattern associated with the word "cup" to the motor plans required to hit that auditory target. With Broca's aphasia, the sensory-motor integration necessary for fluent speech production is damaged. Thus, speech is labored and slow, articulation is impaired, and the melodic intonation of normal speech is lacking (Table 55–2). Yet patients sometimes have considerable success at verbal communication because their selection of certain types of words, especially nouns, is often correct. By contrast, verbs and grammatical words such as prepositions and conjunctions are poorly selected or can be missing altogether. Another major sign of Broca's aphasia is a defect in the ability to repeat complex sentences.

Because most patients with Broca's aphasia give the impression of understanding conversational speech, the condition was initially thought to be a deficit of production only. But Broca's aphasics have difficulty comprehending sentences with meanings that depend mostly on grammar. Broca's aphasics can understand *The apple that the girl ate was green,* but have trouble understanding *The girl that the boy is chasing is tall.* This is because they can understand the first sentence without recourse to grammatical rules—girls eat apples but apples do not eat girls; apples can be green but girls cannot. However, they have difficulty with the second sentence because both girls and boys can be tall, and either can chase the other. To understand

Table 55-2 Examples of Spontaneous Speech Production and Repetition for the Primary Types of Aphasia

Type of aphasia	Spontaneous speech	Repetition
	Stimulus (Western Aphasia Battery picnic picture): What do you see in this picture?	Stimulus: "The pastry cook was elated."
Broca	"O, yea. Det's a boy an' a girl . . . an' . . . a . . . car . . . house . . . light po' (pole). Dog an' a . . . boat. 'N det's a . . . mm . . . a coffee, an' reading. Det's a mm . . . a . . . det's a boy . . . fishin'." (Elapsed time: 1 min 30 s)	"Elated."
Wernicke	"Ah, yes, it's, ah . . . several things. It's a girl . . . uncurl . . . on a boat. A dog . . . 'S is another dog . . . Uh-oh . . . long's . . . on a boat. The lady, it's a young lady. An' a man a They were eatin'. 'S be place there. This . . . a tree! A boat. No, this is a . . . It's a house. Over in here . . . a cake. An' it's, it's a lot of water. Ah, all right. I think I mentioned about that boat. I noticed a boat being there. I did mention that before. . . . Several things down, different things down . . . a bat . . . a cake . . . you have a . . ." (Elapsed time: 1 min 20 s)	"/I/ . . . no . . . In a fog."
Conduction	"Kay. I see a guy readin' a book. See a women /ka . . . he . . . /pourin' drink or something. An' they're sittin' under a tree. An' there's a . . . car behind that an' then there's a house behind th' car. An' on the other side, the guy's flyin' a /fait . . . fait/(kite). See a dog there an' a guy down on the bank. See a flag blowin' in the wind. Bunch of /hi . . . a . . . /trees in behind. An a sailboat on th' river, river . . . lake. 'N guess that's about all. . . . 'Basket there." (Elapsed time: 1 min 5 s)	"The baker was . . . What was that last word?" ("Let me repeat it: The pastry cook was elated.") "The baker-er was /vaskerin/ . . . uh . . ."
Global	(Grunt)	(No response)

the second sentence, it is necessary to analyze its grammatical structure, something that Broca's aphasics have difficulty doing.

Broca's aphasia results from damage to Broca's area (the left inferior frontal gyrus); the surrounding frontal fields; the underlying white matter, insula, and basal ganglia; and a small portion of the anterior superior temporal gyrus (Figure 55–6). A small sector of the insula, an island of cortex buried deep inside the cerebral hemisphere, can also be included among the neural correlates of Broca's aphasia. Broca's aphasics typically have no difficulty perceiving speech sounds or recognizing their own errors and no trouble in coming up with words. When damage is restricted to Broca's area alone or to its subjacent white matter, the result is the condition of Broca's area aphasia, a milder version of true Broca's aphasia, from which many patients are able to recover.

Wernicke's Aphasia Results From Damage to Left Posterior Temporal Lobe Structures

Wernicke's aphasics have difficulty comprehending the sentences uttered by others, and damage occurs in areas of the brain that subserve grammar, attention, and word meaning. Wernicke's aphasia can be caused by damage to different levels of the ventral stream, where auditory information is linked to word knowledge. It is usually caused by damage to the posterior section of the left auditory association cortex, although in severe cases, the middle temporal gyrus and white matter are involved (Figure 55–7).

Patients with Wernicke's aphasia can produce speech at a normal rate that sounds effortless, melodic, and quite unlike that of patients with Broca's aphasia. But speech can be unintelligible as well because Wernicke's aphasics often shift the order of individual sounds and sound clusters. These errors are called *phonemic paraphasias* (a paraphasia is substitution of an erroneous phoneme for the correct one). Even when individual sounds are normally produced, Wernicke's aphasics have great difficulty selecting words that accurately represent their intended meaning (known as a *verbal* or *semantic paraphasia*). For example, a patient might say *headman* when they mean president.

Conduction Aphasia Results From Damage to a Sector of Posterior Language Areas

Conduction aphasia, like Broca's aphasia, is thought to involve the dorsal stream. Speech production and auditory comprehension are less compromised than

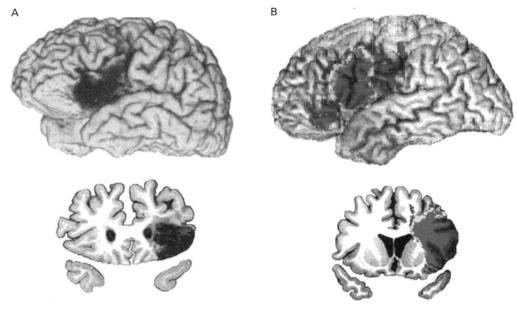

Figure 55–6 Sites of lesions in Broca's aphasia. (Images used with permission of Hanna and Antonio Damasio.)

A. Top: A three-dimensional magnetic resonance imaging (MRI) reconstruction of a lesion (infarction) in the left frontal operculum (**dark gray**) in a patient with Broca's aphasia. **Bottom:** A coronal MRI section of the same brain through the damaged area.

B. Top: A three-dimensional MRI overlap of lesions in 13 patients with Broca's aphasia (**red** indicates that lesions in five or more patients share the same pixels). **Bottom:** A coronal MRI section of the same composite brain image through the damaged area.

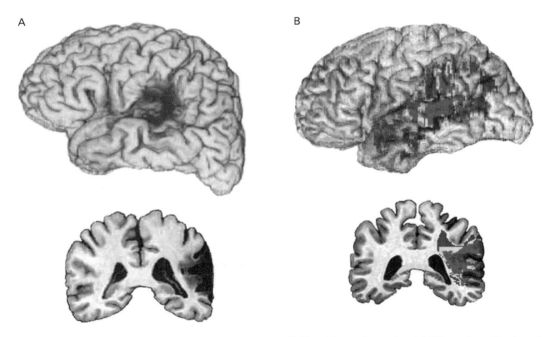

Figure 55–7 Sites of lesions in Wernicke's aphasia. (Images reproduced, with permission, from Hanna and Antonio Damasio.)

A. Top: Three-dimensional magnetic resonance imaging (MRI) reconstruction of a lesion (an infarction) in the left posterior and superior temporal cortex (**dark gray**) in a patient with Wernicke's aphasia. **Bottom:** Coronal MRI section of the same brain through the damaged area.

B. Top: Three-dimensional MRI overlap of lesions in 13 patients with Wernicke's aphasia obtained with the MAP-3 technique (**red** indicates that five or more lesions share the same pixels). **Bottom:** Coronal MRI section of the same composite brain image through the damaged area.

in the two other major aphasias, but patients cannot repeat sentences verbatim, cannot assemble phonemes effectively (and thus produce many phonemic paraphasias), and cannot easily name pictures and objects (Table 55–2).

Conduction aphasia is caused by damage to the left superior temporal gyrus and the inferior parietal lobe. The damage can extend to the left primary auditory cortex, the insula, and the underlying white matter. Large lesions in the Sylvian parietal temporal area, situated in the middle of the network of auditory and motor regions, are consistent with the idea that the damage occurs in the dorsal stream. Damage to left hemisphere auditory regions often produces speech production deficits, supporting the idea that sensory systems participate in speech production. Such lesions interrupt the interfaces linking auditory representations of words and the motor actions used to produce them. The damage compromises white matter (dorsal stream) and affects feedforward and feedback projections that interconnect areas of temporal, parietal, insular, and frontal cortex.

Global Aphasia Results From Widespread Damage to Several Language Centers

Patients with global aphasia are almost completely unable to comprehend language or formulate and repeat sentences, thus combining features of Broca's, Wernicke's, and conduction aphasias. Speech is reduced to a few words at best. The same word might be used repeatedly, appropriately or not, in a vain attempt to communicate an idea. Nondeliberate ("automatic") speech may be preserved, however. This includes stock expletives (which are used appropriately and with normal phonemic, phonetic, and inflectional structures), routines such as counting or reciting the days of the week, and the ability to sing previously learned melodies and their lyrics. Auditory comprehension is limited to a small number of words and idiomatic expressions.

Classic global aphasia involves damage to the inferior frontal and parietal cortices (as seen in Broca's aphasia), the auditory cortex and the insula (as seen in conduction aphasia), and the posterior superior temporal cortex (as seen in Wernicke's aphasia). Subcortical regions, such as the basal ganglia, are often affected as well. Such widespread damage is typically caused by a stroke in the region supplied by the middle cerebral artery. Weakness in the right side of the face and paralysis of the right limbs accompany classic global aphasia.

Transcortical Aphasias Result From Damage to Areas Near Broca's and Wernicke's Areas

Aphasias can be caused by damage not only to speech centers of the cortex but also to pathways that connect those components to the rest of the brain. Transcortical aphasia can be either motor or sensory. Patients with transcortical motor aphasia speak nonfluently, but they can repeat sentences, even very long sentences. Transcortical motor aphasia has been linked to damage to the left dorsolateral frontal area, a patch of association cortex anterior and superior to Broca's area, although there can be substantial damage to Broca's area itself. The left dorsolateral frontal cortex is involved in the allocation of attention and the maintenance of higher executive abilities, including the selection of words.

Transcortical motor aphasia can also be caused by damage to the left supplementary motor area, located high in the frontal lobe, directly in front of the primary motor cortex and buried mesially between the hemispheres. Electrical stimulation of the area in nonaphasic surgery patients causes the patients to make involuntary vocalizations or to be unable to speak, and functional neuroimaging studies have shown it to be activated during speech production. Thus, the supplementary motor area appears to contribute to the initiation of speech, whereas the dorsolateral frontal regions contribute to ongoing control of speech, particularly when the task is difficult.

Transcortical sensory aphasics have fluent speech, impaired comprehension, and great trouble naming things. These patients have deficits in semantic retrieval, without significant disruption of syntactic and phonological abilities.

Transcortical motor and sensory aphasias are caused by damage that spares the arcuate fasciculus and the dorsal stream. Transcortical aphasias are thus the complement of conduction aphasia, behaviorally and anatomically. Transcortical sensory aphasia appears to be caused by damage to the ventral stream, affecting parts of the junction of the temporal, parietal, and occipital lobes, which connect the perisylvian language areas with the parts of the brain responsible for word meaning.

Less Common Aphasias Implicate Additional Brain Areas Important for Language

Several other language-related regions in the cerebral cortex and subcortical structures, for example, the anterior temporal and inferotemporal cortex, have only recently become associated with language. Damage to the left temporal cortex causes severe and pure naming defects—impairments of word retrieval without any accompanying grammatical, phonemic, or phonetic difficulty.

When the damage is confined to the left temporal pole, the patient has difficulty recalling the names of unique places and persons but not the names of common things. When the lesions involve the mid-temporal sector, the patient has difficulty recalling both unique and common names. Finally, damage to the left posterior inferotemporal sector causes a deficit in recalling

words for particular types of items—tools and utensils—but not words for natural or unique things. Recall of words for actions or spatial relationships is not compromised (Figure 55–8).

The left temporal cortex contains neural systems that hold the key to retrieving words denoting various categories of things ("tools," "eating utensils"), but not words denoting actions ("walking," "riding a bicycle"). These findings were obtained not only from studies of patients with brain lesions resulting from

A Defective naming of unique images

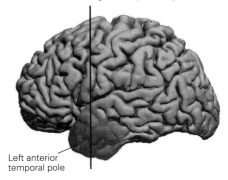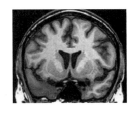

Left anterior temporal pole

Figure 55–8 Regions of the brain other than Broca's and Wernicke's areas involved in language processing. Functional magnetic resonance imaging was used to study patients with selected brain lesions. (Images reproduced, with permission, from Hanna and Antonio Damasio.)

A. The region of maximal overlap of lesions associated with impaired naming of unique images, such as the face of a person, is the left anterior temporal pole.

B. The sites of maximal overlap of lesions associated with impaired naming of nonunique animals are the left anterolateral and posterolateral temporal regions as well as Broca's region.

C. The sites of maximal overlap of lesions associated with deficits in naming of tools are the left sensorimotor cortex and left posterolateral temporal cortex.

B Defective naming of animals

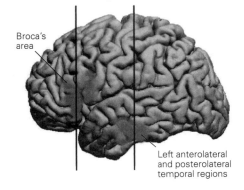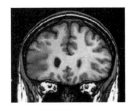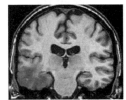

Broca's area

Left anterolateral and posterolateral temporal regions

C Defective naming of tools

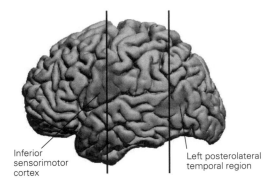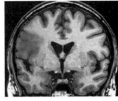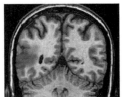

Inferior sensorimotor cortex

Left posterolateral temporal region

stroke, head injury, herpes encephalitis, and degenerative processes such as Alzheimer disease, but also from functional imaging studies of typical individuals and from electrical stimulation of these same areas of temporal cortex during surgery.

Areas of frontal cortex in the mesial surface of the left hemisphere, which include the supplementary motor area and the anterior cingulate region, play an important role in the initiation and continuation of speech. Damage in these areas impairs the initiation of movement (akinesia) and causes mutism, a complete absence of speech. In aphasic patients, the complete absence of speech is a rarity and is only seen during the very early stages of the condition. Patients with akinesia and mutism fail to communicate by words, gestures, or facial expression because the drive to communicate is impaired, not because the neural machinery of expression is damaged as in aphasia.

Damage to the left subcortical gray nuclei impairs grammatical processing in both speech and comprehension. The basal ganglia are closely interconnected with the frontal and parietal cortex and may have a role in assembling morphemes into words and words into sentences, just as they serve to assemble the components of a complex movement into a smooth action.

Highlights

1. Language exists at many levels, each of which has to be mastered during childhood—the elemental phonetic units (vowels and consonants) used to change the meaning of a word, the words themselves, word endings (morphemes) that change tense and pluralization, and the grammatical rules that allow words to be strung together to create sentences with meaning. By the age of 3, young children, regardless of the language(s) they are learning, have mastered all levels and can carry on a conversation with an adult. No artificially intelligent machine can yet duplicate this feat.

2. The learning strategies used by children to master language under 1 year of age are surprising. Language learning proceeds as infants (1) exploit the statistical properties of speech (distributional frequency patterns of sounds to detect relevant phonetic units and transitional probabilities between adjacent syllables to detect likely words), and (2) exploit the social context in which language occurs by following the eye movements of adults as they refer to objects and actions to learn word–object and word–action correspondences.

At early ages, natural language learning requires a social context and social interaction. Infants' strategies are not well described by Skinnerian operant conditioning or by Chomsky's innate representation and selection based on experience. Instead, powerful implicit learning mechanisms that operate in social contexts vault infants forward from the very earliest months of life.

3. Infants' speech production and speech perception skills are "universal" at birth. In speech perception, infants discriminate all sounds used to distinguish words across all languages until the age of 6 months. By 12 months, discrimination for native-language sounds has dramatically increased, whereas discrimination of foreign-language sounds decreases. Production is initially universal as well and becomes language specific by the end of the first year. By the age of 3, infants know 1,000 words. Mastery of grammatical structure in complex sentences continues until the age of 10. Future work will advance the field by linking the detailed behavioral milestones that now exist to functional and structural brain measures to show how the brain's network for language is shaped as a function of language experience.

4. A new "dual-stream" model of language has emerged based on advances in functional neural imaging and structural brain imaging over the past decade. The new model bears similarities to the dual-stream model for the visual system. The dual-stream model for language goes beyond the classic Wernicke-Geschwind model by showing that numerous brain regions and the neural pathways that connect them support sound-to-meaning (ventral) and sound-to-articulation (dorsal) pathways. Refinement in the model will continue as additional studies show relationships between behavioral and brain measures. Future studies will integrate structural and functional brain measures, genetic measures, and behavioral assessments of language processing and of learning, including second language learning in adulthood.

5. Studies on the infant brain reveal a remarkably well-developed set of brain structures and pathways by 3 to 6 months of age. Structural DTI reveals a fully formed ventral pathway at birth and a dorsal pathway that links auditory areas to premotor, but not Broca's, area at birth. EEG and MEG brain imaging studies mirror the transition in phonetic perception between 6 and 12 months of age, a "critical period" for sound learning. MEG

brain scans at this period reveal the co-activation of auditory and motor centers when infants hear speech and show changes in both sensory and motor brain areas as a function of experience. The data indicate that dorsal pathways are sufficiently well formed in the first year to support sensory-to-motor connections and imitation learning during this period.

6. Hemispheric specialization generally increases with age and language experience, with initial representation of the areas and pathways represented bilaterally and dominance emerging with language experience. There are differences in the degree of lateralization, however, for various levels of language. The dorsal stream, which mediates auditory-motor representations of speech, is more left lateralized than the ventral stream, which mediates auditory-conceptual representations of words.

7. The classical aphasias—Broca's, Wernicke's, and conduction aphasia—are well described within the context of the dual-stream model of language. Broca's aphasia, with its emphasis on the inability to produce speech but relatively good speech understanding, is seen as a dorsal stream deficit, whereas Wernicke's aphasia, with its emphasis on speech comprehension deficits, is seen as a ventral stream deficit. Conduction aphasia, like Broca's, is viewed as caused by a dorsal stream deficit, with damage that encompasses auditory and motor regions. Future research on aphasia will benefit from additional studies of functional and structural damage that can be combined with detailed behavioral protocols.

8. Future studies will allow detailed comparisons between human and nonhuman brains to reveal the structures and pathways that are uniquely human and subserve language. Future work will also focus on the degree to which language structures in humans are selectively activated by speech as opposed to other complex auditory sounds and whether adult-level selectivity is present early in development.

9. Human language represents a unique aspect of human cognitive achievement. Understanding the brain systems that allow this cognitive feat in nearly all children, and especially the discovery of biomarkers that identify children who are at risk for developmental disorders of language, will advance brain science and be beneficial for society. Behavioral studies now allow us to connect the dots with regard to how early language experience is linked to advanced language development by the time children enter school. This may lead to language interventions that improve outcomes for all children.

Patricia K. Kuhl

Selected Reading

Brauer J, Anwander A, Perani D, Friederici AD. 2013. Dorsal and ventral pathways in language development. Brain Lang 127:289–295.

Buchsbaum BR, Baldo J, Okada K, et al. 2011. Conduction aphasia, sensory-motor integration, and phonological short-term memory—an aggregate analysis of lesion and fMRI data. Brain Lang 119:119–128.

Chomsky N. 1959. A review of B. F. Skinner's "Verbal Behavior." Language 35:26–58.

Damasio H, Tranel D, Grabowski TJ, Adolphs R, Damasio AR. 2004. Neural systems behind word and concept retrieval. Cognition 92:179–229.

Doupe A, Kuhl PK. 1999. Birdsong and human speech: common themes and mechanisms. Annu Rev Neurosci 22:567–631.

Gopnik A, Meltzoff AN, Kuhl PK. 2001. *The Scientist in the Crib: What Early Learning Tells Us About the Mind.* New York: HarperCollins.

Hickok G, Poeppel D. 2007. The cortical organization of speech processing. Nat Rev Neurosci 8:393–402.

Iverson P, Kuhl PK, Akahane-Yamada R, et al. 2003. A perceptual interference account of acquisition difficulties for non-native phonemes. Cognition 87:B47–B57.

Kuhl PK. 2004. Early language acquisition: cracking the speech code. Nat Rev Neurosci 5:831–843.

Kuhl PK, Rivera-Gaxiola M. 2008. Neural substrates of language acquisition. Annu Rev Neurosci 31:511–534.

Kuhl PK, Tsao F-M, Liu H-M. 2003. Foreign-language experience in infancy: effects of short-term exposure and social interaction on phonetic learning. Proc Natl Acad Sci U S A 100:9096–9101.

Kuhl PK, Williams KA, Lacerda F, Stevens KN, Lindblom B. 1992. Linguistic experience alters phonetic perception in infants by 6 months of age. Science 255:606–608.

Perani D, Saccuman MC, Scifo P, et al. 2011. Neural language networks at birth. Proc Natl Acad Sci U S A 108:16056–16061.

Pinker S. 1994. *The Language Instinct.* New York: William Morrow.

Skeide MA, Friederici AD. 2016. The ontogeny of the cortical language network. Nat Rev Neurosci 17:323–332.

References

Berwick RC, Friederici AD, Chomsky N, Bolhuis JJ. 2013. Evolution, brain, and the nature of language. Trends Cogn Sci 17:89–98.

Broca P. 1861. Remarques sur le siege de la faculte du langage articule, suivies d'une observation d'aphemie (perte de la parole). Bull Societe Anatomique de Paris 6:330–357.

Buchsbaum BR, Baldo J, Okada K, et al. 2011. Conduction aphasia, sensory-motor integration, and phonological short-term memory: an aggregate analysis of lesion and fMRI data. Brain Lang 119:119–128.

Burns TC, Yoshida KA, Hill K, Werker JF. 2007. The development of phonetic representation in bilingual and monolingual infants. App Psycholing 28:455–474.

Damasio AR, Damasio H. 1992. Brain and language. Sci Am 267:88–109.

Damasio AR, Tranel D. 1993. Nouns and verbs are retrieved with differently distributed neural systems. Proc Natl Acad Sci U S A 90:4957–4960.

Dronkers NF, Baldo JV. 2009. Language: aphasia. In: LR Squire (ed). *Encyclopedia of Neuroscience* (Vol. 5), pp. 343–348. Oxford: Academic Press.

Dubois J, Hertz-Pannier L, Dehaene-Lambertz G, Cointepas Y, Le Bihan D. 2006. Assessment of the early organization and maturation of infants' cerebral white matter fiber bundles: a feasibility study using quantitative diffusion tensor imaging and tractography. Neuroimage 30:1121–1132.

Eimas PD, Siqueland ER, Jusczyk P, Vigorito J. 1971. Speech perception in infants. Science 171:303–306.

Fedorenko E, Duncan J, Kanwisher N. 2012. Language-selective and domain-general regions lie side by side within Broca's area. Curr Biol 22:2059–2062.

Ferjan Ramirez N, Ramirez RR, Clarke M, Taulu S, Kuhl PK. 2017. Speech discrimination in 11-month-old bilingual and monolingual infants: a magnetoencephalography study. Dev Sci 20:e12427.

Flege JE. 1995. Second language speech learning: theory, findings, and problems. In: W Strange (ed). *Speech Perception and Linguistic Experience*, pp. 233–277. Timonium, MD: York Press.

Flege JE, Yeni-Komshian GH, Liu S. 1999. Age constraints on second-language acquisition. J Mem Lang 41:78–104.

Friederici AD. 2009. Pathways to language: fiber tracts in the human brain. Trends Cog Sci 13:175–181.

Garcia-Sierra A, Ramirez-Esparza N, Kuhl PK. 2016. Relationships between quantity of language input and brain responses in bilingual and monolingual infants. Int J Psychophysiol 110:1–17.

Geschwind N. 1970. The organization of language and the brain. Science 170:940–944.

Golfinopoulos E, Tourville JA, Guenther FH. 2010. The integration of large-scale neural network modeling and functional brain imaging in speech motor control. Neuroimage 52:862–874.

Hickok G, Okada K, Serences JT. 2009. Area Spt in the human planum temporale supports sensory-motor integration for speech processing. J Neurophysiol 101:2725–2732.

Johnson J, Newport E. 1989. Critical period effects in second language learning: the influence of maturational state on the acquisition of English as a second language. Cognit Psychol 21:60–99.

Knudsen EI. 2004. Sensitive periods in the development of the brain and behavior. J Cogn Neurosci 16:1412–1425.

Kuhl PK. 2000. A new view of language acquisition. Proc Natl Acad Sci U S A 97:11850–11857.

Kuhl PK, Andruski J, Christovich I, et al. 1997. Cross-language analysis of phonetic units in language addressed to infants. Science 277:684–686.

Lenneberg E. 1967. *Biological Foundations of Language.* New York: Wiley.

Lesser RP, Arroyo S, Hart J, Gordon B. 1994. Use of subdural electrodes for the study of language functions. In: A Kertesz (ed). *Localization and Neuro-Imaging in Neuropsychology*, pp. 57–72. San Diego: Academic Press.

Liu H-M, Kuhl PK, Tsao F-M. 2003. An association between mothers' speech clarity and infants' speech discrimination skills. Dev Sci 6:Fl–F10.

Mamiya PC, Richards TL, Coe BP, Eichler EE, Kuhl PK. 2016. Brain white matter structure and COMT gene are linked to second-language learning in adults. Proc Natl Acad Sci U S A 113:7249–7254.

Mills DL, Coffey-Corina SA, Neville HJ. 1993. Language acquisition and cerebral specialization in 20-month-old infants. J Cogn Neurosci 5:317–334.

Miyawaki K, Jenkins JJ, Strange W, Liberman AM, Verbrugge R, Fujimura O. 1975. An effect of linguistic experience: the discrimination of /r/ and /l/ by native speakers of Japanese and English. Percept Psychophys 18:331–340.

Neville HJ, Coffey SA, Lawson D, Fischer A, Emmorey K, Bellugi U. 1997. Neural systems mediating American Sign Language: effects of sensory experience and age of acquisition. Brain Lang 57:285–308.

Newport EL, Aslin RN. 2004. Learning at a distance I. Statistical learning of non-adjacent dependencies. Cogn Psychol 48:127–162.

Peterson SE, Fox PT, Posner MI, Mintun M, Raichle ME. 1988. Positron emission tomographic studies of the cortical anatomy of single-word processing. Nature 331:585–589.

Petitto LA, Holowka S, Sergio LE, Levy B, Ostry DJ. 2004. Baby hands that move to the rhythm of language: hearing babies acquiring sign language babble silently on the hands. Cognition 93:43–73.

Poeppel D. 2014. The neuroanatomic and neurophysiological infrastructure for speech and language. Curr Opin Neurobiol 28:142–149.

Price CJ. 2012. A review and synthesis of the first 20 years of PET and fMRI studies of heard speech, spoken language and reading. Neuroimage 62:816–847.

Pulvermüller F, Fadiga L. 2010. Active perception: sensorimotor circuits as a cortical basis for language. Nat Rev Neurosci 11:351–360.

Raizada RD, Richards TL, Meltzoff A, Kuhl PK. 2008. Socioeconomic status predicts hemispheric specialisation of the left inferior frontal gyrus in young children. Neuroimage 40:1392–1401.

Ramirez-Esparza N, Garcia-Sierra A, Kuhl PK. 2014. Look who's talking: speech style and social context in language input are linked to concurrent and future speech development. Dev Sci 17:880–891.

Rauschecker JP. 2011. An expanded role for the dorsal auditory pathway in sensorimotor control and integration. Hear Res 271:16–25.

Saffran JR, Aslin RN, Newport EL. 1996. Statistical learning by 8-month old infants. Science 274:1926–1928.

Saur D, Kreher BW, Schnell S, et al. 2008. Ventral and dorsal pathways for language. Proc Natl Acad Sci U S A 105:18035–18040.

Silva-Pereyra J, Rivera-Gaxiola M, Kuhl PK. 2005. An event related brain potential study of sentence comprehension in preschoolers: semantic and morphosyntactic processing. Cogn Brain Res 23:247–258.

Skinner BF. 1957. *Verbal Behavior*. Acton, MA: Copley Publishing Group.

Tsao F-M, Liu H-M, Kuhl PK. 2004. Speech perception in infancy predicts language development in the second year of life: a longitudinal study. Child Dev 75:1067–1084.

Weikum WM, Oberlander TF, Hensch TK, Werker JF. 2012. Prenatal exposure to antidepressants and depressed maternal mood alter trajectory of infant speech perception. Proc Natl Acad Sci U S A 109:17221–17227.

Weisleder A, Fernald A. 2013. Talking to children matters: early language experience strengthens processing and builds vocabulary. Psychol Sci 24:2143–2152.

Wernicke C. 1874. *Der Aphasische Symptomenkomplex: Eine Psychologische Studie auf Anatomischer Basis*. Breslau: Cohn und Weigert.

Xu Y, Gandour J, Talavage T, et al. 2006. Activation of the left planum temporale in pitch processing is shaped by language experience. Hum Brain Mapp 27:173–183.

Yeni-Komshian GH, Flege JE, Liu S. 2000. Pronunciation proficiency in the first and second languages of Korean–English bilinguals. Biling Lang Cogn 3:131–149.

Zatorre RJ, Gandour JT. 2008. Neural specializations for speech and pitch: moving beyond the dichotomies. Philos Trans R Soc Lond B Biol Sci 363:1087–1104.

Zhao TC, Kuhl PK. 2016. Musical intervention enhances infants' neural processing of temporal structure in music and speech. Proc Natl Acad Sci U S A 113:5212–5217.

56

Decision-Making and Consciousness

I N THE EARLIER CHAPTERS, WE HAVE SEEN how sensory input is transformed into neural activity that is then processed by the brain to give rise to immediate percepts and how those percepts can be stored as short- and long-term memories (Chapters 52–54). We have also examined in detail how movement is controlled by the spinal cord and brain. Here, we begin to consider one of the most challenging aspects of neuroscience: the transformation of sensory input to motor output through the higher-order cognitive process of decision-making. In doing so, we are afforded a glimpse of the building blocks of higher thought and consciousness.

Outside neuroscience, the term *cognitive* typically connotes some distinction from reflexes and dedicated routines, and yet as we shall see, neuroscience recognizes the rudiments of cognition in simple behaviors that display two types of flexibility—contingency and freedom from immediacy. Contingency means that a stimulus does not command or initiate an action in the way it does for a reflex. A stimulus might motivate a particular behavior, but the action may be delayed, pending additional information, or it may never occur. This freedom from immediacy of action means there are operations that transpire over time scales that are not immediately beholden to changes in the environment or the real-time demands of control of the body.

Both types of flexibility—contingency and time—are on display when we make decisions. Of course, not all decisions invoke cognition. Many behavioral routines—swimming, walking, feeding, and grooming—have branch points that may be called decisions, but they proceed in an orderly manner without much flexibility or control of tempo. They are governed mainly by the time steps of nervous transmission and are dedicated for the most part to particular input–output relationships. The point of drawing these distinctions is not to establish sharp boundaries around decision-making, but to help us focus on aspects of decisions that make them a model for cognition.

For present purposes, we will use the following definition: A decision is a commitment to a proposition, action, or plan based on evidence (sensory input), prior knowledge (memory), and expected outcomes. The commitment is provisional. It does not necessitate behavior, and it can be modified. We can change our mind. The critical component is that some consideration of evidence leads to a change in the state of the organism that we liken to a provisional implementation of an action, strategy, or new mental process.

Such propositions can be represented as a plan of action: I decide to turn to the right, to leave safe shelter, to look for water, to choose a path least likely to encounter a predator, to approach a stranger, or to seek information in a book. The concept of a plan emphasizes freedom from immediacy. Moreover, not all plans come to fruition. Not all thought leads to action, but it is useful to conceive of thought as a type of plan of action. This view invites us to consider knowing as the result of directed—mostly nonconscious—interrogation, rather than an emergent property of neural representations.

Decision-making has been studied in simple organisms, notably worms, flies, bees, and leeches, as well as in mammals from mice to primates. Simpler organisms are appealing because they have smaller nervous systems, but they lack the behavioral repertoire required to study decisions that entail forms of cognition. The hope is that the biological insights from these species will inform our understanding of the processes characterized in mammals, especially primates. This is a laudable goal because, to paraphrase Plato, decision-making offers our best shot at carving cognitive function at its joints—to identify the common principles that support its normal function and to elucidate their mechanisms so they may be repaired in disease.

In this chapter, we focus primarily on perceptual decisions made by primates in contrived settings. The principles extend naturally to reasoning from evidence and to value-based decisions concerning preference. In the last part of the chapter, we derive insights about broader aspects of cognition. Viewed through the lens of decision-making, brain states associated with knowing and being consciously aware may be closer to a neurobiological explanation than is commonly thought.

Perceptual Discriminations Require a Decision Rule

Until recently, decision-making was studied primarily by economists and political scientists. However, psychologists and neuroscientists working in the field of perception have been long concerned with decisions. Indeed, the simplest type of decision involves the detection of a weak stimulus, such as a dim light or a faint sound, odor, or touch. The decision a subject must make is whether or not the stimulus is present—yes or no. In the laboratory, there is no uncertainty about where and when the stimulus is likely to be present. Such experiments were therefore used to infer the fundamental sensitivities of a sensory system from behavior, a subfield of psychology known as psychophysics. Detection experiments played a role in inferring signal-to-noise properties of sensory neurons that transduce light touch, faint sounds, and dim lights. In the last case, such experiments provided evidence that the visual system is capable of detecting the dimmest of light, a single photon, subject to background noise of photoreceptors. In other words, it is as efficient as possible, given the laws of physics.

The psychophysical investigation of perception began with Ernst Weber and Gustav Fechner in the 19th century. They were interested in measuring the smallest detectable difference in intensity between two sensory stimuli. Such measurements can reveal fundamental principles of sensory processing without ever recording from a neuron. It turns out they also lay the foundation for the neuroscience of decision-making, because every yes/no answer is a choice based on the sensory evidence.

In Chapter 17, we learned how psychophysicists conceptualize the detection problem (Box 17–1). On any one trial, the state of the world is either stimulus present or stimulus absent. The decision is based on a sample of noisy evidence. If the stimulus is present, the evidence is a random sample drawn from the probability distribution of signal + noise. If the stimulus is absent, the evidence is a sample from the noise-only distribution (Figure 56–1A). The brain does not directly perceive a stimulus but receives a neural representation of the sample. As a result, some of the noise arises from the neural activity involved in forming this representation. It is the job of the brain to decide from which distribution the sample came, using information encoded in neural firing rates. However, the brain does not have access to the distributions, just the one sample involved in each given decision. It is the separation of these distributions—the degree that they do not overlap—that determines the discriminability of a stimulus from noise. The decision rule is to say "yes" if the evidence exceeds some criterion or threshold.

A Simple Decision Rule Is the Application of a Threshold to a Representation of the Evidence

The criterion instantiates the decision-maker's policy or strategy. If the criterion is lax—that is, the threshold

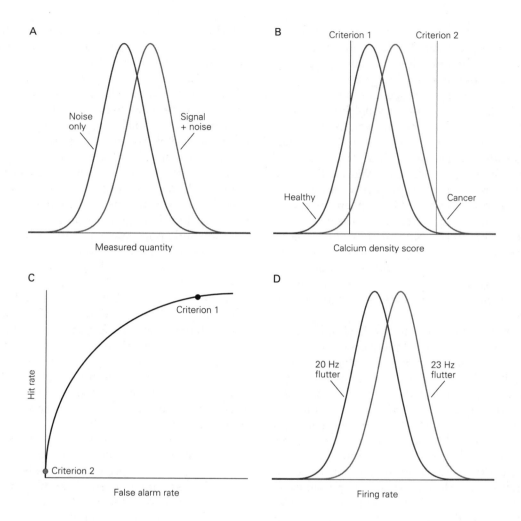

Figure 56–1 The framework of signal detection theory formalizes the relationship between evidence and decisions. In panels **A** through **C**, we consider simple yes–no decisions in which a decision-maker receives just one measurement.

A. The height of the curves represents the probability of observing a measurement on the x-axis (be it number of spikes per second, radioactive counts, or blood pressure) under two conditions: signal present or absent. In both cases, the measurement is variable, giving rise to the spread of possible values associated with the two conditions. If the signal is present, the decision-maker receives a random sample from the **Signal + noise** probability distribution (red). If the signal is absent, the decision-maker receives a sample from the **Noise only** probability distribution (blue). The decision arises by comparing the measure to a criterion, or threshold, and answering *yes* or *no*, signal is present or absent, if the value is greater or less than the criterion.

B. The criterion is an expression of policy, as illustrated in medical decision-making. Suppose the measure is derived from calcifications detected in a screening mammogram—a score combining number, density, and shape. The criterion 1 (**left line**) for interpreting the test as a positive or negative (breast cancer

or not) is liberal. It leads to many false positives (83%), but very few women with cancer receive a negative result. Criterion 2 (**right line**) is conservative. It would miss many cases of cancer, but it would rarely render a positive result to a healthy person. That would make sense if a positive decision were rationale for a dangerous (or painful) procedure.

C. The receiver operating characteristic shows the combination of proportions of "yes" decisions that are correct (hit rate) and incorrect (false alarm rate) for all possible criteria. The liberal and conservative criteria are shown by the **black** and **gray** symbols, respectively.

D. The framework also applies to decisions between two alternatives. Here, the decision is whether a vibration applied to the index finger has a higher frequency than a vibration applied a few seconds before. The same depiction of overlapping distributions might conform to neural responses from some part of the brain that represents a sensory stimulus. For example, a neuron in the somatosensory cortex might respond over many individual trials with a higher average spike rate to vibratory stimulation of the finger at 23 Hz than stimulation at 20 Hz. However, the distributions overlap so that on any given presentation we cannot say with 100% certainty whether the vibration was at 20 Hz or 23 Hz based on the neuron's response.

is low—the decision-maker will rarely fail to detect the stimulus, but they will often respond "yes" on the trials when there was no stimulus because the background noise exceeds the threshold. This type of error is called a *false alarm*. If the criterion is more conservative—that is, the threshold is high—the decision-maker will rarely say "yes" when the stimulus is absent but will often say "no" when the stimulus is present. This type of error is called a *miss*. The appropriate criterion depends on the relative cost of the two types of errors and also on the design of the experiment. For example, if the stimulus is present on 90% of trials, then a lax criterion might be warranted since false alarms will be rare.

The policy ought to be influenced by a value or cost associated with making correct and incorrect decisions. For example, in medical diagnosis, it is often the case that a disease affects only a small fraction of the population, but a diagnostic test does not discriminate perfectly between people with and without the disease. We can illustrate this using the distribution of mammogram calcification scores. The scores are larger in women with breast cancer than in healthy women, but the range of values overlaps to an extent, implying that the test is not perfect (Figure 56–1B).

In this situation, a lax criterion might seem problematic because it would produce a large number of false alarms: patients who are healthy but told they might have a disease based on the test. However, it may well be the case that a miss is life threatening, whereas a false alarm leads to a stressful week as the patient awaits a more decisive test. In this situation, it is actually sensible to apply a lax criterion even if it leads to many false alarms. Alternatively, a false alarm may trigger a painful or risky procedure, in which case a more stringent criterion would be more appropriate. The medical analogy allows us to appreciate the strategic roles of the criterion setting. We praise and criticize decision-makers based on their policy, not on the noisy imperfections of the measurements.

The important point is that the criterion represents a decision rule, which instantiates knowledge about the problem and an attitude about the positive value associated with making correct choices (hits and correct rejections) and the negative value of making errors (misses and false alarms). Note that the application of different criteria does not change the fundamental characteristic of the evidence samples that is responsible for the accuracy of decisions. This is reflected by the overlap between the blue and red distributions, which does not change if a decision-maker adjusts her criterion. The curve in Figure 56–1C, termed the receiver operating characteristic (ROC), shows how changing the criterion affects the accuracy of the decision whether a

stimulus (or cancer) is present or absent for all possible criteria. Each point on the curve is an ordered pair of the probability of a correct "yes" response (hits) versus an erroneous "yes" response (false alarms) associated with a given criterion (threshold). The ROC tells us something about the reliability of the measurement (ie, the separation between the two distributions) regardless of how the decision-maker uses it. The criterion tells us something about the decision-maker's policy. It bears on why two decision-makers receiving the same evidence might reach different decisions. Indeed, it is the policy, not the noise, that the decision-maker controls and for which she may be praised or criticized, that is, held responsible. We will think about this topic again when we discuss the trade-off between speed and accuracy.

The challenge for neuroscience is to relate the terms *signal*, *noise*, and *criterion* to neural representations of sensory information and operations upon those representations that result in a choice. We will develop these connections in subsequent sections. Here, we wish to seed an important insight about the term *noise* as it pertains to the neural representations of evidence. Decision-makers do not make the same decision even when confronting repetitions of identical facts or sensory stimuli. Some variability at some stage must creep into the process. The distinction between signal and noise need not devolve into scholastic arguments about chance and determinism. Any source of variance in the representation of the evidence is effectively noise if it is responsible for errors. If the brain did not distinguish such variability from the signal and thus made a mistake, we would be justified in construing this variability as unaccounted by the decision-maker.

Perceptual Decisions Involving Deliberation Mimic Aspects of Real-Life Decisions Involving Cognitive Faculties

The neural bases for more cognitive decisions have been examined by extending simple perceptual decisions in three ways: first, by moving beyond detection to a choice between two or more competing alternatives; second, by requiring the decision process to take time by involving consideration of many samples of evidence; and third, by considering decisions about matters involving values and preferences.

Vernon Mountcastle was the first to study perceptual decisions as a choice between two alternative interpretations of a sensory stimulus. He trained monkeys to make a categorical decision about the frequency of a fluttering pressure applied lightly to a fingertip (Figure 56–2). Since the *vibratory flutter* has an intensity that is easily detected, the decision is not

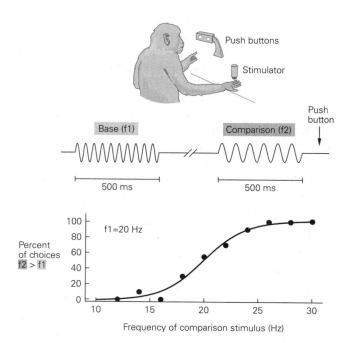

Base (f1)   Comparison (f2)   Push button

500 ms   500 ms

f1 = 20 Hz

Percent of choices f2 > f1

Frequency of comparison stimulus (Hz)

Figure 56–2 The discrimination of flutter-vibration frequency was the first perceptual decision studied in the central nervous system. A 20-Hz vibratory stimulus is applied to the finger on the right hand; following a delay period of several seconds, a second vibratory stimulus is applied. The monkey indicates whether the second vibration (f2) was at a higher or lower frequency than that of the first stimulus (f1) by pushing the left or right button with the other hand. The plots show that the proportion of trials in which the monkey decided that the comparison stimulus was greater than the reference depended on the magnitude and sign of the difference. With larger differences, the monkey almost always chose correctly, but when the difference was small, the choices were often incorrect. (Adapted from Romo and Salinas 2001.)

about whether the stimulus is present or absent but whether the vibration rate is fast or slow. On every trial of the experiment, the monkey experienced a reference frequency, f1, equal to 20 cycles per second (Hz). The pressure cycles are too fast to count; they feel more like a buzzing. The reference was then turned off, and after a few seconds, a second test stimulus, f2, was applied. The frequency of f2 was chosen from a range of values from 10 to 30 Hz. The monkey was rewarded for indicating whether the test frequency was higher or lower than the f1 reference.

We can represent the process conceptually using the same type of signal and noise distributions we drew for the detection problem (Figure 56–1D). Here, the "noise-only" distribution represents a quantity that is sampled in association with the 20-Hz reference, whereas the red distribution represents a quantity that is sampled in association with a test stimulus with a flutter frequency

greater than 20 Hz. Mountcastle favored the idea that the brain obtained two samples of evidence—one accompanying the 20-Hz reference and the second from the test. The decision, higher or lower, could arise by evaluating the inequality—greater than or less than—or, equivalently, by subtracting the two samples and answering based on the sign of the difference. This was a terrific insight, but the neural recordings were out of step with the theory. Mountcastle's neural recordings explained the monkey's ability to detect vibratory stimulation as a function of intensity and frequency (Chapter 17)—a yes/no decision—but they were unable to explain the mechanism for the comparison between the two alternatives, whether f2 is greater or less than f1.

Two key elements were missing. First, to evaluate f2 versus f1, the brain needs a representation of frequency. Mountcastle found neurons in the somatosensory cortex and thalamus with firing rates that were phase-locked to frequencies of the flutter, and they could measure the reliability of this frequency locking, but they did not find neurons that were tuned to particular frequencies less than or greater than 20 Hz. Second, both representations need to be available at the same time in order to compare them. However, the neural responses to f1 lasted only as long as the flutter vibration. Mountcastle failed to observe neural responses that conveyed the representation of the reference frequency through the delay period up to the time that the test stimulus was presented. It was therefore impossible to study the neural operations corresponding to the decision process, which seemed to require some trace of the reference stimulus during analysis of the test.

These obstacles were overcome using a simpler task design and a different sensory modality. Inspired by Mountcastle, William Newsome trained monkeys to decide whether a field of dynamic random dots had a tendency to move in one direction or its opposite (eg, left or right). The random dot motion stimulus is constructed such that at one easy extreme all dots share the same direction of motion, say to the right. At the other easy extreme, all dots move to the left, and in between, the direction can be difficult to discern because many dots contribute only noise (Figure 56–3A).

Unlike the flutter vibration task, where a decision is rendered difficult by making the comparison frequencies more similar, the two directions of motion remain fixed and opposite for all levels of difficulty. The two directions were rendered less distinct by degrading the signal-to-noise ratio of the random dots. Each random dot appears only briefly, and then either reappears at a random location or at a displacement to support a consistent direction and speed. The probability of the latter (displacement) determines the motion

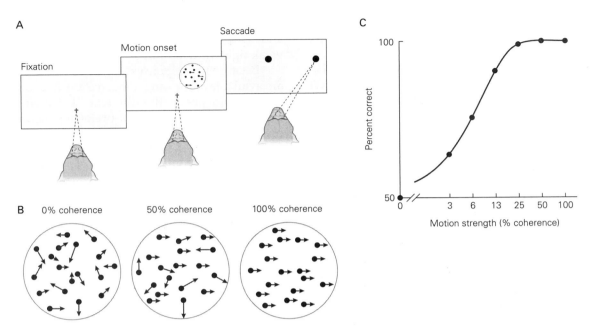

Figure 56–3 In the random dot motion discrimination task, the observer decides if the net motion of dots is in one direction or its opposite (eg, right or left).

A. The monkey maintains its gaze on a cross while viewing the random dot motion display. When the stimulus and fixation cross are extinguished, the monkey indicates its decision by shifting the gaze to the left or right choice targets and receives a reward if the decision is correct.

B. The difficulty of the decision is controlled by the coherence of dot motion. Each dot appears for only a few milliseconds at

a random location and then reappears 40 ms later, either at a new random location or at a displacement consistent with a chosen speed and direction. The probability that a dot present at time t_1 undergoes displacement in the same direction at t_2 establishes the motion strength (% coherence). (Reproduced, with permission, from Britten et al. 1992. Copyright © 1992 Society for Neuroscience.)

C. The decision is more likely to be correct when the motion is stronger.

strength, which is commonly expressed on a 0 to 100 scale, termed the percentage coherence. At the most difficult extreme, 0% coherence, all dots are plotted at random locations in each successive frame, giving the appearance of dancing snowflakes with no dominant direction. At intermediate levels of difficulty, the dancing snowflakes give rise to a weak sense that the wind might be blowing them ever so slightly to the right or left. Any one dot is unlikely to be displaced more than once, so there is no feature to track.

This simple stimulus was originally developed by Anthony Movshon to promote a decision strategy that would benefit from integrating visual information across its spatial extent and as a function of time. Moreover, it satisfied another desideratum: The same neurons should inform the decision at all levels of difficulty. For a left versus right decision, direction-selective neurons in the visual cortex that are, say, sensitive to leftward motion emit signals that are relevant to the decision at all levels of difficulty. That would not be the case if difficulty were controlled by the angular difference between the two directions. Another

advantage of this task over the vibration-flutter task is that there is only one stimulus presentation. There is no need to remember anything between a reference and a test stimulus. Finally, humans and monkeys perform this task at nearly identical levels. They answer perfectly for the strong-motion trials and make more errors when the strength of motion is reduced (Figure 56–3C). This establishes a platform for a quantitative reconciliation of decisions and neural activity. Is there a way to explain the likelihood that a decision will be accurate from measurements of the signal-to-noise ratio in the appropriate sensory neurons?

Neurons in Sensory Areas of the Cortex Supply the Noisy Samples of Evidence to Decision-Making

In higher mammals and primates, neurons that respond differentially to the direction of motion are first encountered in the primary visual cortex (area V1). They are a subset of the orientation-tuned simple and complex cells discovered by Hubel and Wiesel (Chapter 22). These neurons project to a secondary visual cortical area, area MT.[1]

Area MT contains a complete map of the contralateral visual field, and almost all the neurons in area MT are direction selective. Neurons with similar direction preferences cluster together so that MT contains a map of both space and motion direction at each point in the visual field. Their receptive fields are larger than those of V1 neurons, and some manifest properties that are not evident in V1 (eg, pattern motion; Chapter 23), but most respond as if they integrate signals from V1 that share the same direction selectivity over a larger patch of the visual field. In Newsome's experiments, the random dot motion stimulus was contained in a circular aperture that matched the size of an MT neuron's receptive field. It was thus possible to measure the response of a neuron perfectly situated to convey evidence to the decision process on single trials.

It seemed possible that the neurons with receptive fields aligned to the random dot motion stimulus and a firing preference for one or the other direction under consideration might contribute the evidence used to make the decision. Indeed, we can begin to understand the monkey's perception of motion by applying the same signal-to-noise considerations to the MT neural responses. We consider two types of direction-selective neurons (Figure 56–4). One type responds better to rightward motion than to leftward motion, and it yields higher firing rates when the rightward motion is stronger. It also responds above baseline to the 0% coherence stimulus because the random noise contains all motion directions including leftward and rightward, and it yields lower firing rates (compared to 0% coherence) when the leftward motion is stronger (Figure 56–4B). The other type of neuron responds well to leftward motion. It exhibits the same pattern as the right-preferring type, only with the direction preferences reversed. The neural responses are noisy, so the firing rates on any trial or in any epoch may be conceptualized as a random draw from one of the distributions in Figure 56–4C. These distributions can be interpreted in two ways. The two curves might represent the possible firing rates of a rightward-preferring neuron when weak motion is to the right or left, respectively. They

might also represent the possible firing rates of right- and left-preferring neurons, respectively, to the same weak rightward stimulus.

Because the responses of the two classes of neurons are available at the same time, we are able to characterize the evidence as the difference between the firing rates of the left- and right-preferring neurons. (The brain in fact relies on the difference between the averages from many left- and many right-preferring neurons.) We refer to such a quantity as a decision variable because the decision could be made by applying a criterion to this difference. Here, the criterion would be at zero. Thus, if the decision variable is positive, answer right; if it is negative, answer left.

Notice that when the stimulus is purely random (0% coherence), there is no correct answer. The monkey is rewarded randomly by the experimenter on a random half of the trials, and the monkey answers right and left with about equal probability. This is not because the monkey is guessing but because fluctuation in the random dot motion stimulus and the noisy firing rates of the right- and left-preferring neurons lead to variability in the evidence used to make the decision. This makes sense because the right- and left-preferring neurons respond equivalently to this type of stimulus. On some trials, the right-preferring neurons respond more than the left-preferring neurons, and the brain interprets this as evidence for rightward motion. On other trials, the left-preferring neurons respond more and the monkey chooses left.

Neuroscientists have been able to use a network of small populations of neurons to model the relation between the accuracy of an animal's choice versus motion strength, known as the *psychometric function*. The success of such models gives support to the idea that the signal and noise properties of cortical neurons can explain the fidelity of a perceptual decision, just as Mountcastle had hoped. This achievement was possible because of a clever experimental design that allowed the same neuron to participate in decisions across a wide range of difficulty. But are these neurons actually used to make the decision? Do they actually supply the noisy evidence that the monkey uses to make its decision?

We now know that they do. Because of the columnar organization of direction-selective neurons in area MT, it is possible to apply small currents through a microelectrode to excite a cluster of neurons sharing the same receptive field property. Newsome and colleagues placed the electrode in the middle of a cluster of neurons with receptive fields that were exactly aligned to the random dot motion stimulus. He reasoned that at weak stimulating currents the majority of stimulated

---

[1]The letters MT stand for middle temporal, a sulcus in the species of New World monkey in which the area was first discovered. This sulcus does not exist in Old World monkeys and humans, but the homologous area does, and it retains its original name. Area MT is sometimes referred to as area V5 (the fifth visual area) in humans. The name is unimportant, but the area is!

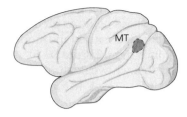

Figure 56–4 Neurons in area MT provide noisy evidence bearing on the direction of motion.

A. Responses from a right-preferring neuron during the discrimination task. The random dot movie is in the receptive field of the neuron. The panels in the left column of the 2 × 2 panel display show the neuron's responses to motion in its preferred direction, and panels in the right column show its responses to the nonpreferred direction. The panels in the top row show the neuron's responses to strongly coherent motion, and the bottom panels show the responses to weakly coherent motion. In each panel, the time of each action potential (spike) is represented by a small vertical tick mark. Each row of spikes in a panel shows the neuron's response to the motion stimulus in a single trial. (Adapted with permission from Mazurek et al. 2003.)

B. The mean firing rate varies as a function of motion strength. The neuron increases its firing rate above baseline even in response to the 0% coherence stimulus because the dynamic random dots contain all directions of motion, including the neuron's preferred direction. The firing rate then increases with stronger rightward motion. It decreases, relative to the response to 0% coherence, with stronger leftward motion. The responses of this right-preferring neuron to leftward motion are mirrored by the responses of a left-preferring neuron to rightward motion.

C. Probability distributions of the firing rates from left-preferring and right-preferring neurons to weak rightward motion. The right-preferring neuron tends to respond more, but the overlap of the distributions shows that it is possible for the left-preferring neuron to respond more than the right-preferring neuron on any given trial. These same considerations apply to the pooled signals from populations of right- and left-preferring neurons. The plot on the right shows the distribution of the difference between firing rates of the left-preferring neuron and the right-preferring neuron measured in response to the same stimulus over many trials. The decision is to choose right if this difference is positive and to choose left if it is negative. This rule would lead to correct rightward choices on 80% of the trials.

A Single trial responses from a right-preferring neuron

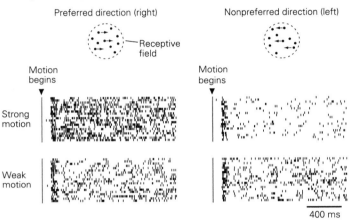

B Firing rate depends on motion strength and direction

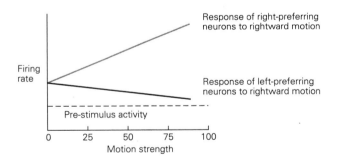

C Noisy evidence for left and right are conceptualized as random samples from probability distributions

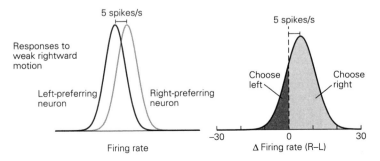

neurons were likely to share the same receptive field and the same direction preference. Newsome had the monkey decide between this direction and its opposite. For example, if these neurons preferred rightward motion, the weak currents caused the monkey to decide more often in favor of right (Figure 56–5).

We now refer to such weak stimulation, designed to affect a cluster of neurons within a 50- to 100-µm radius, as *microstimulation*. Notably, microstimulation did not cause a hallucination of visual motion. It biased the monkey's decisions, which were guided mainly by the random dot motion stimulus. The monkey did not respond when the stimulus was not shown, and microstimulation did not affect the monkey's decisions when the random dots were presented at a location of the visual field outside the receptive field of the stimulated neurons. The microstimulation exerted its largest effect on choices when the motion strength was weakest. The stimulated neurons simply added a small amount of evidence for rightward motion, which is effectively evidence against leftward motion, as discussed below.

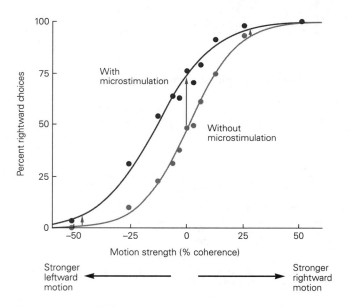

Figure 56–5 Artificial activation of neurons that respond preferentially to rightward motion causes a monkey to decide that motion is rightward. In the experiment, an electrode is placed in the middle of a patch of neurons in area MT that prefer the same direction of motion, say rightward. The random dot motion is shown in the receptive field of these neurons. A weak alternating current is applied on half of the trials during the presentation of the random dots movie. The amount of current activates about 200 to 400 neurons within 50 to 100 µm of the electrode tip. On trials with microstimulation, the monkey is more likely to choose the preferred direction of the simulated neurons. The effect is most pronounced when the decision is more difficult (**middle red arrow**). (Adapted, with permission, from Ditterich, Mazurek, and Shadlen 2003.)

The microstimulation experiment shows that the direction-selective neurons in area MT contribute evidence to the perceptual decision. However, the stimulated neurons do not necessarily need to affect the decision directly; they only have to participate in a neural circuit that lies in a causal chain. In addition, many more neurons in MT were not affected by the electrical stimulation but nonetheless responded to the same random dot patch in the same direction-selective manner. They are in other columns with receptive fields that are not centered on the stimulus but overlap it. If the electrode is moved to stimulate these neurons, they too cause the monkey to choose the preferred direction more often. These findings imply that in any one experiment the microstimulation only affects a small fraction of the neurons that contribute to the decision. Most respond at their usual firing rates to the random dot motion. The microstimulation only changes the total signal that the brain uses to make its decision by a small amount. No wonder the effect is only evident when the decision is difficult.

There is an important principle to be learned here. Had Newsome used only the easier conditions, the electrical stimulation would have yielded a null effect, and thus, the causal relationship between the neural activity and behavior would not have been established. The same pattern of effects has recently been established using techniques to turn neurons off. Silencing induces a bias in choices against the direction of the silenced neurons, but this too is only apparent on trials when the motion is difficult. Without evidence for sufficiency or necessity, a neuroscientist might conclude that the neurons in MT do not cause changes in perceptual decisions. This would be a mistake, notably one that is likely to be made in any experiment in which perturbations are restricted to a subset of the neurons involved in a computation. That is the rule, not the exception, for studies of higher cortical functions. It is only mitigated by studying behavior in conditions when a small difference to the total pool of neural signals might make a difference, as in the difficult (low signal-to-noise) regime employed in Newsome's experiments.

To summarize so far, the perceptual decision arises from a simple *decision rule*: the application of a criterion to the noisy evidence supplied by noisy direction-selective neurons in the visual cortex. We have characterized the noisy evidence as a single number: the difference in the mean firing rates from two opposing pools of direction-selective neurons. This account leaves out two important points: The operations that establish the decision variable must be carried out by neurons that receive information directly or indirectly from area MT, and these operations take time. As we will see, time is the key to understanding decision-making, and it is also the factor that relates decision-making to higher cognitive function.

Accumulation of Evidence to a Threshold Explains the Speed Versus Accuracy Trade-Off

The decision rule considered so far is appropriate if the brain received only a brief snapshot of the motion, say for a tenth of a second. However, decision-making normally takes some time, so that when the viewing duration is longer, decisions tend to be more accurate. In fact, the strength of motion that is required to support 75% accuracy, termed the *sensory threshold*, decreases as a function of viewing duration. With more time, the decision-maker can achieve this level of accuracy with a weaker motion strength. Put another way, the sensitivity to weak motion improves as a function of viewing duration, t. Indeed, the sensitivity improves as a function of the square root of time ($\sqrt{t}$), which is the rate of improvement in the signal-to-noise ratio that one obtains by accumulating or averaging. The suggestion then is that the difference in firing rates of left- and right-preferring direction-selective neurons supplies the momentary evidence to another process that accumulates this noisy evidence as a function of time—in this case, two processes that accumulate evidence for left and right, respectively.

The accumulation of noisy evidence follows a path comprising random steps in both the positive and negative direction on top of a constant bias determined by the coherence and direction of the moving dots. This is termed a *biased random walk* or *drift plus diffusion* process (Figure 56–6). Because evidence for left is evidence against right (and vice versa), the two random walks are anticorrelated, albeit imperfectly so. The accumulations evolve with time and continue to do so until the stimulus is turned off or until one of the accumulations reaches an upper *stopping bound*, which determines the answer, left or right. Even the 0% coherence (pure noise) stimulus will reach a stopping bound eventually, but it is equally likely that the left or right accumulation will do so. When the random dot motion favors one direction, it is more likely that the corresponding accumulation determines the choice, and increasingly so with stronger motion. Such accumulations of noisy evidence are dynamic versions of the decision variable. The decision rule remains similar: Choose right if there is more evidence for right than left, and vice versa. The stopping bounds also explain another important feature of the decision—the time it takes to make it.

This simple idea thus explains the observed trade-off between the speed and accuracy of a decision. It specifies the exact relationship between the probability that each motion strength will lead to a correct choice and the amount of time that is taken, on average, to respond, termed the reaction time (Figure 56–6C). If the stopping bounds are close to the starting point of the accumulation, the decision will be based on very little evidence—fast but error prone. If the stopping bounds are further from the starting point, more accumulated evidence is needed to stop—slower but more likely to be correct. If the flow of information is cut off before either bound has been reached, the decision-maker may feel she has not yet reached an answer, but may nonetheless answer based on the accumulation that is closer to its stopping bound. This mechanism, termed *bounded evidence accumulation*, explains the effect of task difficulty on choice accuracy and the associated reaction times on a variety of perceptual tasks. It explains the degree of confidence that a decision-maker has in a decision and why such confidence depends on both the amount of evidence and deliberation time. It also explains the rate of improvement in accuracy when the experimenter controls viewing duration by $\sqrt{t}$, mentioned above, and it explains why this improvement saturates with longer viewing durations. The brain stops acquiring additional evidence when the accumulated evidence reaches a stopping bound.

Neurons in the Parietal and Prefrontal Association Cortex Represent a Decision Variable

Neurons in several parts of the brain, including the parietal and prefrontal cortices, change their firing rates to represent the accumulation of evidence—in the case of visual motion from area MT—bearing on the direction decision. The neurons that represent the accumulation differ from sensory neurons in two important ways. First, they can continue to respond for several seconds after a sensory stimulus has come and gone. Moreover, they seem to be capable of holding a firing rate at one level and then increasing or decreasing that level when new information arrives. This is exactly the type of feature one would like to see in a neuron that represents the accumulation of evidence. Second, such neurons tend to be associated with circuits that control the behavioral response that the monkey has learned to use to communicate its decision. Such neurons were first identified for their capacity to maintain persistent activity in the absence of a sensory stimulus or ongoing action. They were therefore thought to play a role in working (short-term) memory, planning an action, or maintaining attention at a location in the visual field (Figure 56–7).

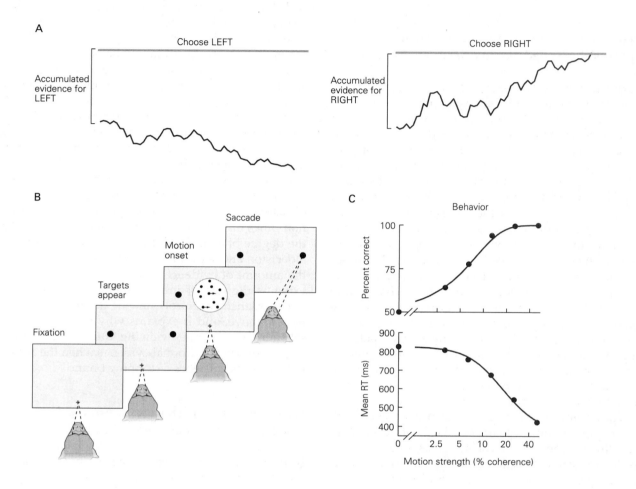

Figure 56–6 The speed and accuracy of a decision are explained by a process of evidence accumulation.

A. A decision and the time it takes to reach it are both explained by the accumulation of evidence, as a function of time, until there is sufficient evidence to terminate the decision in favor of one or the other choice. The cartoon illustrates a decision for rightward motion because the "choose right" accumulation was the first to reach the stopping bound (**thick gray lines**). Because the evidence is noisy, the accumulations resemble biased random walks, also known as drift-diffusion processes. For the decision between left and right motion, there are two accumulations. The one shown on the left accumulates evidence for left and against right. The one shown on the right accumulates evidence for right and against left. For this process, the bias (or drift rate) is the mean of the evidence samples depicted by the distribution of differences (right minus left) in Figure 56–4C. The process is a random walk because even if the motion is rightward, left-preferring neurons in area MT might respond more than right-preferring neurons at any instant. The two processes tend to evolve in an anticorrelated fashion because the random dot motion stimulus supplies the

same noisy samples of evidence to both accumulations via the visual cortex. They are not perfectly anticorrelated because right- and left-preferring neurons introduce additional noise. Were the anticorrelation perfect (eg, if all the noise comes from the motion stimulus), the two processes could be represented by one accumulation that terminates at either an upper or lower stopping bound.

B. In a choice-reaction time task, the decision-maker reports a decision whenever ready with an answer. In this case, the monkey signals its choice by the direction of a saccade.

C. Graphs show a typical data set. In addition to the proportion of correct choices, the reaction time (**RT**), the time from onset of motion to the beginning of the eye movement response, also depends on the strength of motion. The total length of RT is the time to reach a decision, explained by the process in **A**, plus the time required to convey sensory information from the stimulus to the neurons that compute the decision and the time required to convert the decision to a motor response. (Adapted, with permission, from Gold and Shadlen 2007.)

It seemed possible that neurons whose activity represents a plan to act might also represent the formation of that plan during decision making. For example, if a monkey has learned to answer "rightward" by moving its hand to a target on a touch screen, the neurons of interest will tend to be active in association with that movement and they will decrease their activity if the monkey plans to reach to the opposite "leftward" target. Those neurons project to brain areas that command reach movements. If the monkey has learned to answer with an eye movement, the neurons that help to plan eye movements to the choice-target represent the decision variable. Such neurons have been studied extensively in the lateral intraparietal area (LIP). Indeed, these LIP neurons provided neuroscientists with the first view of a decision process as it unfolds in time.

Neurons that represent the evolving decision increase their firing rates gradually as the evidence mounts for one of the choices, and they decrease gradually when the evidence favors the other option (Figure 56–8). Their firing rates, plotted as a function of time, approximate a ramp: a baseline rate plus a constant multiplied by time, where the constant is proportional to the strength of the momentary evidence (eg, the average difference in the firing rates of the right- and left-preferring MT neurons). This captures the average firing rates across many trials, but it leaves out the critical point that the decision variable is an accumulation of both signal and noise. The signal is the mean of the difference. The noise is the variance—that is, the spread around the mean. The accumulated noise is obscured by the averaging in Figure 56–8, but it is apparent in the variability of firing rates across multiple decisions.

The responses start at a common level and evolve as the brain acquires more and more information, until something stops the process. A neural signature of the stopping rule is apparent in the responses aligned to the eye movement itself. The firing rate appears to reach the same level on trials that take as little as a few tenths of a second and trials that take as much as a full second. The level is achieved less than a tenth of a second before the eyes start to move. Of course, it takes less time to achieve this level if the firing rates are increasing at a rapid pace (eg, solid red trace in Figure 56–8). This suggests that the brain terminates the decision when the representation of accumulated evidence reaches a threshold. That is exactly what the bounded accumulation framework predicts. There appears to be no common level of activity in neurons that signal a rightward movement when the monkeys choose the opposite direction. Instead, another population of neurons that accumulate evidence for left (and

against right) reaches their threshold and terminates the decision process when the monkey answers left (Figure 56–6A). The neurons that favor the right choice simply stop accumulating evidence at a time determined by the left choice neurons. This explains why the downward traces in Figure 56–8 do not reach a common level of activity around the time of the eye movement. It is not yet known where in the brain the threshold operation is applied. Computational theorists have proposed that a likely candidate is the striatum, a brain area involved in selecting between competing actions (Chapter 38), but there are many other candidate structures, including movement areas of the cortex and brainstem.

Area LIP is not the only part of the brain that represents the accumulation of evidence toward a decision, and LIP itself is not limited to making decisions about random dot motion. Many neurons in the parietal and prefrontal cortex exhibit persistent firing. In fact, the first brain areas shown to exhibit this type of activity were in the frontal lobe, rostral to the primary motor cortex, and some neurons with this property were found in the motor cortex itself. The persistent activity was thought to represent working memory for a location in space or a rule, category, or plan of action, as discussed in Chapter 52. But these neurons are also capable of representing graded levels of activity, suggesting a capacity to represent more analog quantities, like an evolving decision variable, the expected value of making an action, or working memory of a sensory quality, as we next consider.

Twenty years after Mountcastle published his studies of flutter-vibration discrimination, his student Ranulfo Romo rejuvenated this line of research by focusing on neurons in the prefrontal cortex, which had the kind of persistent activity we have been discussing. Romo modified the task. The monkeys were still presented with two vibrating stimuli, separated by a delay, and were required to decide whether the vibration frequency of the second stimulus (f2) was greater or less than the vibration frequency of the first stimulus (f1). However, instead of using the same 20-Hz reference stimulus on all trials, the flutter frequency was varied across trials. He found that many neurons in the prefrontal cortex respond in a graded and persistent manner to the frequency of the first flutter-vibration stimulus during the delay period while the monkey awaited the second stimulus. Some neurons increased their firing rate as a function of the vibration frequency of f1, while others were more active with lower frequencies. These persistent neural responses were not observed by Mountcastle in his original studies. There is evidence that a decision variable is constructed in the ventral premotor cortex, where neurons respond to the difference, f2 – f1. This is challenging to study because the decision variable does not evolve over a long time scale. There is no need to acquire many samples of evidence. All that is needed is an estimate of f2 and the application of a threshold. The flutter-vibration task complements the motion decision task by demonstrating the diverse functions of persistent activity. In the motion task, the persistence supports the computation of the decision variable—the accumulated evidence bearing on the decision alternatives. In the flutter-vibration task, the persistent activity represents a sensory quality—the frequency of the reference stimulus—through a delay period.

Perceptual Decision-Making Is a Model for Reasoning From Samples of Evidence

Most of the decisions animals and humans make are not about weak or noisy sensory stimuli. They are about activities, purchases, propositions, and menu items. They are informed by knowledge and

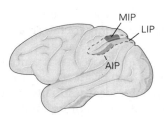

A Lateral intraparietal area

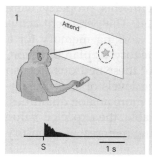 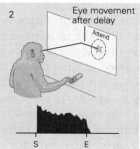 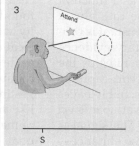 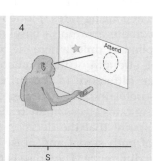

Receptive field characteristics

Attention sensitive, preparation to look

B Medial intraparietal area

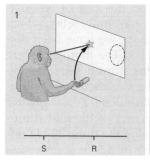 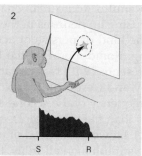 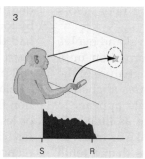 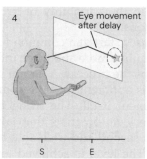

Retina-centered, preparation to reach

C Anterior intraparietal area

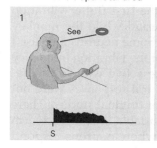 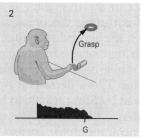 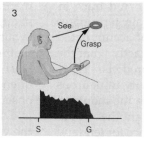 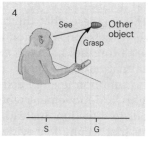

Object-specific viewing, grasping

expectations derived from sources such as personal experience, books, friends, and spreadsheets. Some are based on internal (subjective) valuation or preference. Many involve reasoning from sources of evidence that may differ in reliability and that must be weighed against costs and benefits. To what extent do the neural mechanisms of perceptual decision-making apply to these other types of decisions?

Imagine the following scenario. As you leave your home in the morning, you realize that you will be outdoors from 4 to 5 pm and must decide whether to carry an umbrella. To make this interesting, assume this occurred before the age of the internet and accurate satellite weather prediction. You must decide based on yesterday's forecast of "possible chance of rain", the clear appearance of the sky at 7:00 am, a small drop in the barometric pressure compared to 1 hour before, and the observation that among a dozen pedestrians visible from your window only one seems to be carrying an umbrella. Let us assume further that you have experience with such decisions and have some sense of how reliable these indicators are. Finally, the cumbersomeness of carrying the umbrella is such that your decision boils down to a reasoned assessment of whether rain is more likely than not.

The right way to make this decision is to consider each of the indicators and ask how likely they would be if rain does or does not occur in the afternoon. These likelihoods are learned estimates of conditional probabilities, the probability of observing the indicator when it rains in the afternoon, and the probability of the same observation when it does not rain. For example, suppose through experience you have learned that the forecast, chance of rain, implies a 1 in 4 chance of rain. Then, the conditional probabilities are 1 in 4 and 3 in 4 that it will or will not rain, respectively, given the weather report. The ratio of these two probabilities is termed the likelihood ratio (LR), which is 1 in 3 in this case. If the LR is greater than 1, it favors rain, and if the LR is less than 1, it favors no rain. There is an LR for each of the four indicators. If the product of the four LRs is greater than 1, then you should carry the umbrella.

For reasons that will be clear in a moment, it is useful take logarithms of LRs, termed the log-likelihood ratio (logLR). This provides a more natural scale for belief, and it allows us to replace multiplication with addition [recall that $\log(xy)=\log(x)+\log(y)$]. To appreciate the scale, assume that the one passerby with an umbrella would be equally likely to carry the umbrella whether or not rain is a prospect. Both probabilities are 1 in 2. The LR is therefore 1, and the $\log(1)=0$, which corresponds to the intuition that this observation is uninformative. LRs greater than 1 have positive logarithms, and LRs less than 1 have negative logarithms, consistent with the way they bear on the prediction of rain.

Monkeys can be trained to perform a version of this weather prediction task. In the experiment depicted in

Figure 56–7 (Opposite) **Persistent neural activity maintains working memory, attention, and plans of action.** The monkey is asked to view a scene and respond to a visual stimulus (**S**) by either moving its eyes (**E**) or reaching (**R**) or grasping (**G**) with its hand. Each histogram represents the firing rate of a representative neuron as a function of time following presentation of the visual stimulus. The **dashed circles** show the *response fields*. This term is preferable to receptive and movement field because these neurons are neither purely sensory nor purely motor. The **blue line** shows where on the screen the monkey is asked to initially fixate its gaze.

A. Neurons in the lateral intraparietal area (**LIP**) fire when a monkey is preparing to make an eye movement to an object or when the monkey directs attention to the object's location. Most LIP neurons are not selective for object features such as shape and color. This neuron fires when the object is presented in the neuron's response field, which lies in the circled area to the right of where the monkey is looking (**1**). The neuron's firing is enhanced if the object is presented while the monkey's attention is directed to this location or if the monkey is asked to plan an eye movement to the location (**2**). The firing can persist for several seconds after the stimulus has been removed (**2**), thereby providing a potential mechanism for maintaining a short-term or working memory of its location. The neuron does

not fire if an object is presented outside the neuron's response field (eg, to the left) (**3**) even if the monkey is asked to attend to the location of the neuron's response field (**4**). An object must appear there even if only briefly (**2**).

B. In the medial intraparietal area (**MIP**), neurons fire when the monkey is preparing to reach for a visual target. This neuron starts firing shortly after the appearance of a target in the response field of the neuron, in this case, a fixed angle to the right of where the monkey is looking, whether its gaze is on the left edge (**2**) or the center (**3**) of the screen, and it continues to fire as the monkey waits to reach. The neuron does not fire when the monkey reaches for a target at the center of its gaze (**1**) or when the monkey plans to shift its gaze to a target in the response field, without reaching (**4**). The physical direction of the reach is not a factor in the neuron's firing: It is the same in **1** and **2**, and yet the neuron fires only in **2**.

C. In the anterior intraparietal area (**AIP**), neurons fire when the monkey is looking at or preparing to grasp an object and are selective for objects of particular shapes. This neuron fires when the monkey is viewing a ring (**1**) or making a memory-guided reach to it in the dark (**2**). It fires especially strongly when the monkey is grasping the ring under visual guidance (**3**). It does not fire during viewing or grasping of other objects (**4**).

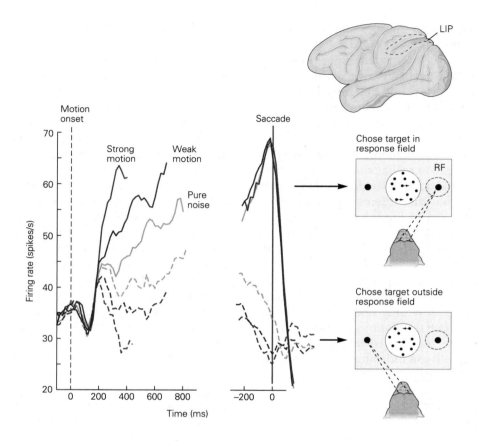

Figure 56–8 Neurons in the lateral intraparietal area (LIP) represent the accumulation of noisy evidence. These neural recordings were obtained while a monkey performed the reaction time version of the motion task. The traces are average firing rates from 55 neurons. The neurons were of the same type shown in Figure 56–7A.

The traces show average responses to three motion strengths: strong (**red**), weak (**purple**), and zero (pure noise, **gray**). The **solid traces** are from trials in which the monkey chose the target in the neuron's response field (**RF**; right choice). The **dashed traces** are from trials in which the monkey chose the target outside the neuron's response field (left choice). For the nonzero strengths, the direction of random dot motion was the direction the monkey chose (ie, only correct choices are shown). The responses in the plot on the left, which are aligned to the start of random dot motion, exhibit a

gradual buildup of activity, leading to rightward choices, and a gradual decline in activity, leading to leftward choices. The rate of this buildup and decline reflects the strength and direction of motion. The responses on the right are to the same dot motion but are now aligned to the moment the monkey makes its eye movement (**saccade**) to indicate its choice and reveal its reaction time. The responses reach a common level just before the monkey makes its choice, consistent with the idea that a threshold applied to the firing rate establishes the termination of these trials. The responses do not reach a common level before leftward choices because these decisions were terminated when a separate population of neurons, with the left choice target in their response fields, reached a threshold firing rate. (Adapted, with permission, from Roitman and Shadlen 2002. Copyright © 2002 Society for Neuroscience.)

Figure 56–9, a monkey had to decide whether to look at a red or a green target, only one of which would lead to a reward. Before committing to red or green, the monkey was shown four shapes. Each served as an indicator about the location of the reward. The monkey had learned to associate predictive value with a total of 10 shapes, half of which favored reward at red, the other half at green. The shapes also differed in the reliability with which they predicted the reward location. The monkey learned to rely on these shapes rationally,

making its decisions by combining evidence from each shape and by giving the more informative shapes more leverage on the choices.

While the monkeys made their decisions, neural activity was recorded from the same parietal area studied in the motion task. As before, the neurons responded in a way that revealed the formation of the decision for or against the choice target in their response field. When the red target was in the response field, the neuron assigned positive values

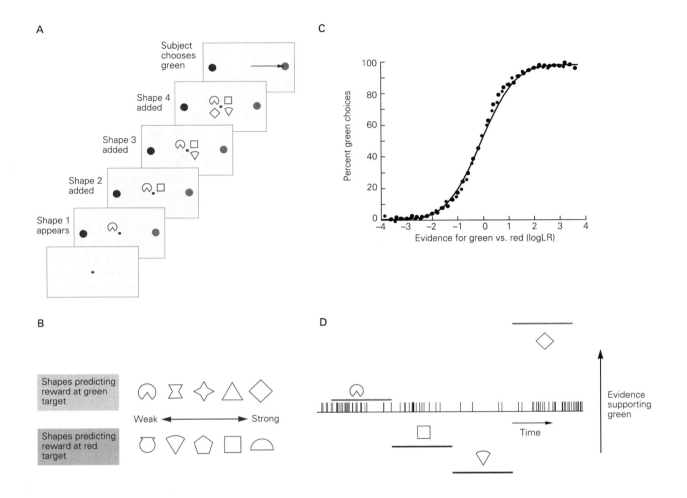

Figure 56–9 Evidence accumulation underlies probabilistic reasoning from evidentiary symbols.

A. A monkey was trained to make decisions based on a sequence of four shapes, drawn randomly with replacement from a set of 10. The shapes were added to the display sequentially every one-half second.

B. Each shape provides a different amount of evidence that a reward is associated with a red or green choice target. Some, like the diamond and semicircle, are highly reliable predictors that a reward will occur if the choice is for green or red, respectively. Others are less reliable predictors. The degree of reliability is quantified by the likelihood ratio or its logarithm. A good decision-maker should base the decision on the product of the likelihood ratios or the sum of their logarithms (**logLR**).

C. The monkey's decisions were guided by the probabilistic evidence from the four shapes. On trials in which the sum of the logLR from the four shapes strongly favored green, the monkey almost always chose green. When the sum was closer to 0, the monkey had to base its decision on weak evidence and chose less consistently. The pattern of choices demonstrates that the monkey assigned greater weight to the shapes that were more reliable (strong versus weak).

D. The same types of parietal neurons studied in the perceptual decision-making task represent the running sum of evidence bearing on the choice target in its response field. The spikes are shown from a single decision when the green target was in the neuron's response field. The **horizontal black line** below the spikes marks the neutral level of evidence for green versus red, such that the two choice targets are equally likely to be rewarded. The vertical position of the green or red lines associated with each successive presentation of an indicated shape show the cumulative evidence conferred by the shapes that the reward was at the green target. The first shape was weak evidence for green. The second and third shapes supplied mounting evidence against green (for red). Note the reduction in firing rate. The final shape provided strong evidence for green, such that the cumulative evidence from all the shapes favored green. Note the increase in firing rate. It is an example of a single neuron in the association cortex using persistent activity to compute quantities useful for decision-making. Based on the firing rates from many trials, it was shown that neurons encode the cumulative sum of the logarithm of the probability ratios—the logLR that a reward is associated with the choice target in the neuron's response field. (Adapted, with permission, from Yang and Shadlen 2007.)

to the red-favoring shapes and negative values to the green-favoring shapes. When the green target was in the response field, the signs were reversed. As shown in the example, the response changed discretely when each of the four shapes was presented, and it did so by an amount commensurate with the degree of reliability. In fact, the increment (or decrement) was proportional to the logLR assigned by the experimenter to the shape! The brain simply adds these logLRs to form a decision. And if the monkey is allowed to view as many shapes as it wants, it will typically stop when the accumulated evidence (in units of logLR) reaches a criterion level. The LIP neurons do the same thing they did in the motion decision. They produce firing rates that represent the cumulative sum of noisy increments and decrements.

By adding in units of logLR, the brain achieves reasoning from probabilistic cues in the way a statistician or actuary combines evidence from multiple sources. The experiment demonstrates that the mechanism used for perceptual decision-making is also at play in more complicated decisions that involve reasoning from more abstract sources of evidence. It speaks to the broader theme of this chapter: The study of decision-making offers insight into how the brain achieves a variety of cognitive functions.

Decisions About Preference Use Evidence About Value

Many, if not most, decisions made by humans and animals are expressions of preference, based on an assignment of value. In some instances, the value is innate. For example, most animals experience sweet as positive and bitter as negative (Chapter 29). In the vast majority of instances, however, value is learned through experience, or it is derived from reasoning based on other preferences. Unlike a decision about the direction of motion, a medical diagnosis, or the weather, a decision about which of a pair of items one prefers is not objectively right or wrong. It can only be said to be consistent or inconsistent with one's expression of value. In fact, our knowledge of a subject's valuation of an object may only be revealed to us by observing her choices.

Yet despite the qualitative difference between subjective and objective evidence, there are parallels between the neural mechanisms that support perceptual and value-based decisions. Decision-makers take more time to choose between items of similar value than items that differ substantially in value, and their choices are less consistent. In a typical experiment, the

participant is asked to indicate the value of each item that they will later make choices about. For example, they might be asked how much they are willing to pay or they are asked to indicate a rating from highly undesirable through neutral to highly desirable. This procedure is typically repeated to provide a subjective value for each item to be used in the experiment.

The participant is then asked to decide between pairs of items. The difference in the subjective values communicated before the experiment provides an index of the difficulty of the decision between the items. It is analogous to motion coherence. A similar approach works with animals. For example, a monkey might demonstrate a preference for grape juice over apple juice, and then be asked to choose between a small volume of grape juice versus a large volume of apple juice. The decision is rendered more difficult by titrating the ratio of volumes to values that lead the monkey to choose either juice with equal tendency.

Two types of neurons associated with this type of value encoding have been identified. The first, typically located in the striatum, encodes the value associated with an action. The second, primarily in the orbitofrontal and cingulate cortex, appears to encode the value associated with specific items. Decisions about preference seem to arise from the same strategy that governs perceptual decisions. Just as a decision between left and right motion is guided by the difference in firing rates of left- and right-preferring sensory neurons, a decision between two items is based on the difference in activity of neurons encoding the values of each item. These neural representations are noisy, and this feature might explain why a decision-maker may make choices that are inconsistent with their values. It might also explain why decisions between items of similar value tend to take more time—a speed–consistency trade-off similar to the speed–accuracy trade-off discussed above.

The analogy to perceptual decision-making is appealing, but it misses the more interesting aspects of value-based decisions. As mentioned above, the value of most items is not given by biology but instead is learned. Further, there is no reason to assume that such value is monovalent. One may value an item differently, based on different qualities and considerations, and one or more of those qualities may dominate under different circumstances. Accordingly, the value of an item could appear to change simply by the occasion of its comparison to another item, which might invite emphasis on a more or less desirable aspect. Novelty, familiarity, and the value of exploration itself might also play a role in modifying a subjective valuation.

These considerations might contribute to the "noisy" representation of value that is thought to explain inconsistencies and long decision times in preference choices. This type of noise belies processes that are far more complex than variability in random dot displays and the noisy spike rates of neurons. Such evaluative processes are likely to involve prospection and memory retrieval, which are only beginning to be understood at the neural level (Chapter 52). In the end, these processes must furnish samples of evidence bearing on the relative value of the items, and this evidence is either accumulated or evaluated individually against a criterion to halt the process with a decision.

Decision-Making Offers a Framework for Understanding Thought Processes, States of Knowing, and States of Awareness

States of knowledge have persistence. Even if they concern information derived from the senses, the knowledge of sensation generally outlasts the sensory activity itself. In this way, the state of knowledge resembles a perceptual decision—a commitment to a proposition about the object, based on sensory evidence. As we have seen, these states are often tied to possible behaviors rather than to the features of the sensory information. This is a position argued by many philosophers and the psychologist James J. Gibson.

This simple point can be made on empirical grounds. Persistent neural activity is not present in sensory areas of the brain unless a stimulus is unchanging and then only if the neurons do not adapt. Naturally, sensory neurons must change their response when the environment changes or the observer moves in the environment, whereas knowledge states persist through sensory changes and without a continuous stream of input. Indeed, persistent activity is apparent in areas of the brain that associate sources of information—from the senses and from memory—with circuits that organize behavior.

In the prefrontal cortex, persistent states represent plans of action, abstract rules, and strategies. In the parietal and temporal lobes, neural representations have the dual character of knowledge and the behavior that knowledge bears upon, such as making an eye movement or reaching, eating, or avoiding. The responses can resemble a spatial representation, as they do in area LIP, if the target of the projection is the eye movement system, but that is only because there is correspondence between space and action. A useful guide is to consider the source and target of the association. If the source is the visual cortex and the targets are premotor areas that control hand posture (eg, grip), as they do in the anterior intraparietal

area (Figure 56–7C), the association area might convey knowledge about curvature, distance, convexity, and texture. One might be inclined to use terms borrowed from geometry to catalogue such knowledge, but it may be simpler to think about the repertory of hand shapes available to the organism. Importantly, the neurons in association cortex do not command an immediate action. They represent the possibility of acting in a certain way—an intention or provisional affordance (Box 56–1).

Let us defer for the moment the aspect of the knowledge state that includes conscious awareness and consider the simpler sense of knowledge as a state of possible utilization. Such preconscious ideation is probably the dominant state in which an animal interacts with the environment. It is arguably also the lion's share of human experience, although because we are not conscious of it, we underestimate its dominance. Two important insights emerge from this perspective. The first is that the correspondence between knowledge and neuronal activity lies at a level of brain organization between sensation and behavior. Although the flow of information from sensory epithelia (eg, the retina) through the primary cortical sensory areas is essential for perception, knowledge resulting from activity in higher brain regions has temporal flexibility and persistence not seen in lower brain regions—what the philosopher Maurice Merleau-Ponty termed the *temporal thickness of the present*.

The second insight is that the computation leading to a knowledge state has the structure of a decision—a provisional commitment to something approximating a possible selection from a submenu of the behavioral repertory. We might say that the parietal association neurons interrogate the sensory areas for evidence bearing on the possibility of a behavior: look there, reach there, posture the hand this way to grasp. Of course, neurons do not ask questions. Nevertheless, we can think of the circuits as if they scan the world looking for evidence bearing on a possible behavior. The type of information they can access is limited by functional and anatomical connectivity. The type of question is framed by the target of the projection, such as regions that control gaze, reaching, and grasping.

Sir Arthur Conan Doyle endowed Sherlock Holmes with the insight that the key to discovery was knowing where to look and what to look for. We acquire knowledge by controlling the brain's interrogation system. Some interrogations are automatic, whereas others are learned. An example of the former is a sudden change of brightness of an object in the visual field; it provides evidence bearing on the possibility of orienting the eyes or body toward it. An example

---

Box 56–1 Affordances, Perception, and Knowledge

James J. Gibson, known for his ecological theory of perception, referred to *affordances* as properties of objects and the environment. The term comes from the verb *afford*. An object affords possible behaviors, such as lifting, grabbing, filling, hiding in, drawing/writing upon (eg, parchment) or with (eg, a brush), or walking upon. The affordance refers to the potential behaviors of the animal. The same object, say a stone, could afford grasping, dropping, breaking (ie, used as a tool), throwing (as a missile weapon), or pinning (as a paperweight).

Gibson was widely criticized for claiming that perceptual processes picked up these affordances directly from the optical array, what he termed "direct perception." The term is commonly misunderstood as antithetical to computational accounts of information processing. By "direct perception," Gibson did not mean that there were no computations on the data received through the senses. He promoted the mathematical understanding of these operations. He meant that we do not perceive the intermediates.

We do perceive the parts of objects that are accidentally occluded by something in our line of sight, and we perceive the back of an opaque object that is occluded by its front. We do not perceive the outlines, the line art, and many other details, but that is not to say that they do not register on the retina and the visual cortex. Gibson held that representation of visual information is not a sufficient condition for perception. From the perspective of the neuroscience of decision-making, one might place emphasis on the representation of potential behavior—something like a provisional commitment to a plan.

Affordance still refers to a category of actions, but it is about the organization of the action (eg, throwing) or strategy, and also—but not necessarily—a quality of the object. The modifier, "provisional," emphasizes that the action may not actually ensue now or ever. This modifier would have been superfluous in Gibson's use of the term *affordance*, because an affordance was a property of the object (in his ecological framework) and therefore had a permanence independent of the perceiver.

---

of the latter draws on learning and foraging; we learn, through play and social interaction (eg, school), how to look for hidden items and how to explore in a goal-directed way.

The beautiful thing about this construction is that an answer to the question confers a kind of meaning. Even for such a mundane question like "Might I look there?," an affirmative answer—a decision to (possibly) look at an as yet undefined object in the periphery of one's visual field—confers a spatial knowledge about the item. Before we have looked directly at it to identify what it is, we know about its *thereness*. From the perspective of decision-making, the location of an object is not perceived because there is a neural activity in a map of the visual field. Rather, the location is perceived because some aspect of the visual field—a fleck of contrast, change in brightness, appearance or disappearance—answered the question above in the affirmative.

This way of thinking helps us understand the disease states known by the term *agnosia*, from the Greek word meaning "absence of knowledge." The classic example is visual hemineglect, which is caused by damage to the parietal lobe (Chapter 59). A patient with a right parietal lesion will ignore the left side of the visual field and also the left side of objects even when the entire object is in the right visual field (Figure 56–10; see

also Figure 59–1). Unlike the left side blindness, called homonymous hemianopsia (or hemiblind), which accompanies damage to the right visual cortex (homonymous because it is the same regardless of which eye is used), the patient with a parietal lesion does not complain of an inability to see. She is unaware of the deficit, so much so that crossing a street is a major hazard.

A hemiblind patient with damage to the right visual cortex still expects to interrogate and receive information from the left visual field. When that patient receives no visual information, he knows to turn to face parallel to the street, thereby placing its contents in the intact right hemifield. In contrast, the patient with hemineglect does not interrogate the left hemifield in the first place. She does not perceive a lack of visual information because the apparatus to conduct the interrogation is not working. Like most deficits, there is enough redundancy in the brain (or the damage partial) that some visual capacities are present. In fact, when confronted with a single spot of light on a dark background, the same patient may report its presence accurately even in the affected hemifield.

There are other versions of hemineglect that involve an absence of knowledge of the body. For example, a patient with a right parietal injury may deny that her left arm is hers. She may recognize it as

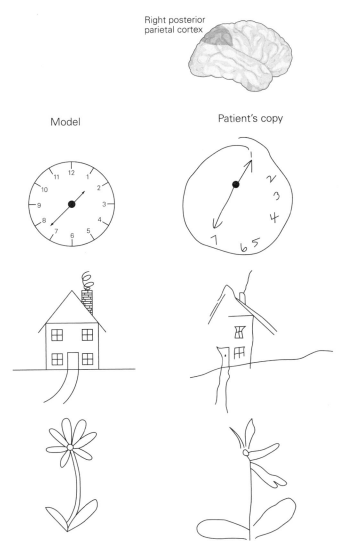

Model   Patient's copy

Figure 56–10 Damage to the parietal or temporal cortices results in agnosia, or deficits in knowing. After damage to the right parietal lobe, many patients are unaware of the left side of space or the left part of objects. The drawings on the right were made by patients with unilateral visual neglect following lesion of the right posterior parietal cortex. Agnosias can also be induced in healthy individuals by diverting attention (see Figure 25–8). (Reproduced, with permission, from Bloom F, Lazerson A. 1988. *Brain, Mind and Behavior*, 2nd ed., p. 300. New York: Freeman.)

consequences of our motor command, and partly through vision. The arm in particular is a common feature of our lower visual field. Indeed, we are accustomed to ignoring it.

These examples are the most common of the agnosias (which are thankfully rare). Other well-known examples involve problems in face recognition (prosopagnosia) and the perception of color (achromatopsia), both associated with lesions of the temporal lobe. The different types of agnosia have loose correspondence to the anatomical specializations we learned about in Chapter 24. In particular, the ventral extension of the central visual pathways into the temporal lobe is referred to as the "what" pathway, which contains circuits that are specialized for processing faces, objects, color, and semantic memory. The dorsal extension, which has been termed the "where" or "how" pathway, seems concerned with representations supporting localization.

An alternative formulation would characterize these streams in terms of associations with behaviorally relevant targets. For the dorsal stream, those are parietal areas with projections to motor systems that reach, look, or grasp. For the ventral stream, those are temporal lobe areas with projections to structures that guide foraging decisions to eat, avoid, hide, approach, mate, and communicate. The last few behaviors are social affordances conferred by color and facial identity/expression. With a small stretch of imagination, the organization of social affordances links to other functions of the temporal lobe (and temporal parietal junction) in humans. For example, these regions are associated with language comprehension and inference about what someone else is thinking. The former is associated with a devastating agnosia, known as Wernicke aphasia (Chapters 1 and 55); the latter is known as theory of mind and will be discussed below.

From the perspective of decision-making, perceiving, believing, and thinking have the character of a provisional commitment to a proposition. Brain states that correspond to a sense of knowing, be it perceiving or believing, share two important aspects with decision-making: an extended temporal profile that withstands changes in the sensory and motor streams (ie, a freedom from immediacy) and a propositional character captured by the term "affordance." Knowing is not solely about the information but is like the outcome of a decision to embrace a proposition: Might I do something, enact something, approach someone, or retain the possibility of trying the option I am not choosing now?

Two caveats deserve mention. This framework does not replace a computational account of

an arm but deny that it is hers. When asked whose arm it is, she may express ignorance bordering on disinterest (personal experience). The syndrome is usually associated with some visual neglect as well and some weakness on the same side of the body suggestive of a more rostral and superior localization. Consider that the body's position is known to us partly through the somatosensory system, partly through the predicted

information processing, nor does it explain the neural mechanisms that support these computations. It mainly tells us about the level of brain organization that carries out these operations. For example, consider the search for the neurons that achieve knowledge about the color red, despite changes in the spectral content of the morning and evening light—a phenomenon known as *color constancy*. Instead of searching in sensory areas for neurons that respond selectively to red in this invariant way, one might look for neurons that guide the choice of ripe fruit. This does not obviate the computations required to recover the surface reflectance properties of the fruit's skin, despite variation in the spectral content of the illuminating light. The raw data for such computations are supplied by sensory neurons that lack color constancy and maintain temporal fidelity with changes in the environment. The knowledge state "red," however, is invariant to the illuminant and likely persistent. In animals that lack language, the knowledge state may not be dissociable from "ripe vegetation."

The second caveat is that we have not distinguished knowledge states that we are consciously aware of from those that we experience unconsciously. For example, as I make my way through the forest trying to find the creek that I hear burbling, my brain might consider locations of objects I pass that are graspable, attached to vegetation, and with color suggesting ripeness. I may be unaware of this consciously. Yet that evening in my search for food, I may return to this part of the forest, guided by these unconscious encounters. I may do this without knowing why, or the memory might pierce consciousness. All that has been said up to now could apply to conscious and nonconscious experience. We are now prepared to elucidate the difference.

Consciousness Can be Understood Through the Lens of Decision Making

Clearly, we are unaware of most of the operations that transpire in our brains, and this is true even for the processes that ultimately pierce consciousness. This is why Freud famously quipped that consciousness is overrated. Every thought that enters our awareness began as neural computation preceding the conscious awareness of that thought. Indeed, the sophistication of nonconscious mental processes, including those leading to "I've got it!" moments and the activities we perform while occupied by a phone call, involves decisions that transpire without conscious awareness.

It is difficult to study nonconscious processing because people deny experience of the process. Indeed, the term nonconscious experience seems like an oxymoron. The experimenter must find a way to prove that information processing has occurred despite the fact that the subject is unaware of it. In recent years, it has become possible to establish conditions whereby information is provided to a human subject that has a high likelihood of going unnoticed but is nonetheless able to influence behavior, thereby permitting scientific characterization of nonconscious mental processing (Chapter 59). This has encouraged neuroscientists to ask what it is about the neural activity that gives rise to the thoughts, perceptions, and movements that do reach conscious awareness. We will not review this vast topic here but instead share a pertinent insight: Viewed through the lens of decision-making, the problem of consciousness may be simpler than imagined.

Broadly speaking, two sets of phenomena fall under the heading consciousness. The first concerns levels of arousal. One is not conscious when one is asleep, under general anesthesia, comatose, or having a generalized seizure. One is fully conscious when awake, and there are levels of consciousness between these extremes. These states are associated with terms such as confusion, dissociation, stupor, and obtundation. Some alterations of consciousness are normal (eg, sleep), whereas others are induced by toxins (eg, alcohol), metabolic disturbances (eg, hypoglycemia), low oxygen, trauma (eg, concussion), or fever (eg, delirium).

The neuroscience underlying these states—and the transitions between them—is immensely important to medicine. We might classify this group of phenomena as neurology-consciousness. However, these topics are not what most people mean when they speak of the mystery of consciousness. This is partly because they are less mysterious but also because their characterization is more objective and the phenomena can be studied in animals. That said, there is much to be learned about the mechanisms responsible for sleep, awakening, anesthesia, and so forth. Much of the neuroscience is unfolding at a rapid pace (Chapter 44).

We will not say more about neurology-consciousness here, except to seed one useful insight. Imagine a mother and father sleeping comfortably in their bedroom as a storm ensues outdoors. There are also traffic sounds and even the occasional thunder. This scene goes on for some time, until the cry of a baby awakens the parents. This common occurrence tells us that the nonconscious brain is capable of processing sounds and deciding to become conscious. It decides, nonconsciously, that some sounds afford an opportunity for more sleep while others sound a call to nurture. This decision is similar to the perceptual decisions considered earlier in this chapter. Both involve nonconscious processing of evidence. However, the commitment to awaken and parent is a decision to engage the environment consciously. This may be a touchstone between neurology-consciousness and the more intriguing consciousness that you are experiencing as you read these words (or so the authors hope).

When neuroscientists, psychologists, and philosophers ponder the mysteries of consciousness, they are referring to loftier themes than wakefulness. This loftier set of phenomena comprises awareness, imagery, volition, and agency. There is a subjective component to all conscious experience. The experience of conscious perception incorporates a sense that it is me that is beholding the content. It parallels the "me" in volition. It is not that my arm moved on its own; I made it move! We used the term deliberation earlier in this chapter to describe the thought process leading to a decision. Our use of the term was metaphorical. It describes a computation and a biological mechanism, but it does not require awareness. Actual deliberation implies conscious intention. We are aware of the steps of reasoning along the way. We could report, were we asked, about the evidence we relied upon—that is, the evidence we were consciously aware of during the decision and possibly some of the evidence we used nonconsciously were it accessible from memory to include in our report. Could the difference between conscious awareness of an item and nonconscious processing of that item be a mere matter of whether the brain has decided on the possibility of reporting? Could it be this simple?

Consider the following scenario. A psychologist concludes that a study participant has seen something nonconsciously because the item affected a subsequent behavior and the participant denies having seen it. Suppose the subsequent behavior involved reaching in the direction of the object. Based on what we know about decision-making, we would conclude that brain circuits like the ones discussed earlier received sufficient evidence to commit to the possibility of looking, reaching, and approaching, but there was insufficient evidence to commit to the possibility of reporting. Just as the brain entertains the possibility of looking, reaching, or grasping, it may also entertain the possibility of reporting. That is, reporting is also a *provisional affordance*.

Events afford the possibility of reporting, and this includes the nonconscious states of knowledge acquired through decision-making. Indeed, the event of having decided may be experienced consciously—the *aha* moment—by virtue of another decision to report. In the study scenario, the participant was not consciously aware of the item because her brain did

not commit to a provisional report. The evidence did not satisfy a decision criterion like the termination bounds in the perceptual decision-making task considered earlier in the chapter.

This account provides a plausible explanation of the failure of the participant to report that she saw the item, but the mere entertaining of the possibility of reporting does not seem to explain the phenomenology of the perceptual experience itself, at least not at first glance. This explanation demands more careful consideration of the character of the report. Just as we attach states of spatial knowledge to configurations of the hand for reaching and grasping, we must consider the knowledge state that accompanies the affordance of reporting. Whether by language or gesture (eg, pointing), the report is a provisional communication with another agent or oneself (eg, in the future). It presumes knowledge about the mind of the receiver.

Cognitive scientists use the term *theory of mind* to refer to this type of knowledge or mental capacity. It can be demonstrated by asking someone to reason about the motivation behind another agent's actions, and it can be studied in animals and preverbal children by examining their reactions to another child or puppet. In one study protocol, two children witness a desired toy placed in a left or right container (Figure 62–2). The test child then witnesses the toy's displacement to the other container while the other child is absent. When that child returns, the experimenter assesses the test child's expectation of which container the returning child will open to find the toy. Children under 3 years old do not exhibit theory of mind by this assay. They think the returning child will open the container that contains the toy, not the one it was in before the transfer. Whether animals other than humans have theory of mind is controversial. We suspect there are inchoate forms of this capacity in the animal kingdom and in children under 3. When adults perform tasks that depend on theory of mind, the right temporal-parietal junction and superior temporal sulcus are active.

Theory of mind—in concert with narrative—has profound consequences for the knowledge state associated with the reporting affordance. Imagine a woman looking at a power drill resting on a table. She experiences the location of the drill, relative to her eyes and hand, as well as its texture and shape. It has a graspable surface that is partly in her line of sight and partly occluded (eg, the back). These are the knowledge states that arise through provisional commitments to look at, reach for, and grasp the drill. They are likely to involve neural activity similar to what is illustrated in Figure 56–7, and they are the outcome of simple decisions. The drill brings to mind other affordances associated with its utility as a tool, its potential to make noise, and the potential danger posed by the sharp bit at one end. This is an elaborate, potentially rich collection of knowledge, but it could all be experienced nonconsciously. For example, if the woman were preoccupied with some other task, such as a phone conversation with her friend, she might nonetheless make use of these knowledge states.

But suppose there is a man on the other side of the table and suppose the woman—her brain, that is—has also reached a provisional commitment to report to the man about the drill between them. Consider the change to her knowledge state. The drill now has a presence not only in her visual field, relative to her gaze, her hand, and her repertory of actions, but also in the man's field of vision and his possible actions. The parts of the drill that are not in plain sight to her are known to be in the line of sight of the man. Indeed, her capacity for "theory of mind" also supplies knowledge that other parts of the drill are seen only by her and that the man could be experiencing those parts just as she experiences the parts that are not in her direct line of sight— that is, both preconsciously as occluded parts of the object and consciously as part of an object that could be seen directly from another vantage point. There is something about the drill that is at once private, public, and in the world—independent of either mind. The drill is there for the next person who enters the room, or an imagined person. The transformation of knowledge of the drill is from a collection of first-person experiences (eg, qualities and affordances) to a thing in the world that possesses an existence unto itself. It is conceivable that this state of knowledge is our conscious awareness of the world, or at least a part of it, for the knowledge state associated with a decision to report is further enriched by content of the report itself.

The report might be simple, like pointing to the location of a tool or a hiding spot, or it might involve narrative. In the case of the hiding spot, additional content might be conveyed to indicate that the enclosure affords safety from a predator or, alternatively, a predator's location. Many simple reports do not require narrative because items such as tools and enclosures persist and theory of mind presumes the affordance of a tool or a hiding place in another's mind, whereas events, which also afford the possibility of reporting, often require narrative because they are transient.

The knowledge state associated with narrative can incorporate history, simulation, prediction, etiology (eg, origin stories), purpose, and consequence. For the drill, narrative might enhance the knowledge state to include memory of the place of purchase, an episode in which it malfunctioned, and the mechanism of its

detachable bit. Narrative allows us to reason in more complex environments than the scenarios considered earlier (eg, the umbrella example and the probabilistic reasoning task; Figure 56–9). We could not reason about science, medical diagnosis, and jurisprudence without origin stories, simulation, hypotheses, prospection, and counterfactuals. The evolutionary advantage of this capacity is obvious (at least for the time being, until it leads us to make the earth uninhabitable).

To summarize, the conscious awareness of an item might arise when the nonconscious brain reaches a decision to report the item to another mind. The intention is provisional in that no overt report—verbal or gesture—need occur, just as no eye movement need ensue for the parietal cortex to engage the possible intention of foveating. Just as the provisional intention to foveate corresponds to preconscious knowledge of the location of an as yet unidentified object in the periphery, the possibility of reporting to another agent (or self), about whom we have theory of mind, corresponds to the knowledge of an item in a way that satisfies most aspects of conscious awareness.

Naturally, our journey from perceptual decision-making through affordances to consciousness is at best incomplete. For example, it does not yet provide a satisfying account of what a conscious experience feels like. But it is a start, as it supplies a coarse explanation of why sensory information acquired through the eyes is experienced differently from auditory or somatosensory experiences, and it provides insight into the private aspects of perceptual awareness as well as our experience of objects as things in the world, independent of what they afford to the perceiver. These last features follow from the consideration of another agent's mind.

The view of consciousness from the perspective of decision-making is, if nothing else, simplifying. There is no reason to search for a special area of the brain that bestows consciousness, or a special neuron type, or a special ingredient in the representation of information (eg, an oscillation or synchronization), or a special mechanism. The mechanism might look like any other kind of provisional commitment—that is, a decision that confers a state of knowing but does not entail conscious awareness. Of course, brain activity itself is not conscious, just as the brain activity supporting a possible hand posture is not the hand posture itself. In this sense, the mechanism of consciousness is only different from other affordances because it involves reporting instead of reaching, looking toward, eating, drinking, hiding from, walking through, and mating. All are likely to involve decision formation and threshold detection.

Thus, by studying the neuroscience of decision-making, we are also studying the neuroscience of consciousness. There is still much to be learned about the mechanisms of the simplest decisions described in the first part of the chapter. For example, we do not know what sets the bounds and how thresholds are implemented in brain circuits. Nevertheless, answers to these and other fundamental questions are in the crosshairs of modern neuroscience, and therefore, so is human consciousness.

Highlights

1. A decision is a commitment to a proposition, action, or plan—among options—based on evidence, prior knowledge, and expected outcomes. The commitment does not necessitate immediate action or any behavior, and it may be modified.

2. Decision-making provides a window on the neuroscience of cognition. It models contingent behavior and mental operations that are free from the immediate demands of sensory processing and control of the body's musculature.

3. A decision is formed by applying a rule to the state of evidence bearing on the alternatives. A simple decision rule for choosing between two alternatives employs a criterion. If the evidence exceeds the criterion, then choose the alternative supported by the evidence; if not, choose the other alternative.

4. For certain perceptual decisions, the source of evidence and its neural representation are known.

5. The accuracy of many decisions is limited by considerations of the signal strength and its associated noise. For neural systems, this noise is attributed to the variable discharge of single neurons, hence the variable firing rate of small populations of neurons that represent the evidence.

6. Many decisions benefit from multiple samples of evidence, which are combined across time. Such decision processes take time and require neural representations that can hold and update the accumulated evidence (ie, the decision variable). Neurons in the prefrontal and parietal cortex, which are capable of holding and updating their firing rates, represent the evolving decision variable. These neurons are also involved in planning, attention, and working memory.

7. The speed–accuracy trade-off is controlled by setting a bound or threshold on the amount of evidence required to terminate a decision. It is an example of a policy that makes one decision-maker different from another.

8. Many decisions are about propositions, items, or goals that differ in value to the organism. Such value-based decisions depend on stored associations between items and valence.

9. The source of evidence for many decisions is memory and active interrogation of the environment—information seeking. These operations come into play when animals forage and explore, and when a jazz musician improvises.

10. Decision-making invites us to consider knowledge not as an emergent property of neural representations but the result of directed, mostly nonconscious interrogation of evidence bearing on propositions, plans, and affordances. The intention is provisional in that no overt action need ensue. Just as the provisional intention to foveate corresponds to preconscious knowledge of the location of an as yet unidentified object in the periphery, the possibility of reporting to another agent (or self), about whom we have theory of mind, corresponds to the knowledge of an item in the ways we are aware of it consciously.

11. Viewed through the lens of decision-making, conscious awareness of an item might arise when the nonconscious brain reaches a decision to report to another mind. The affordance has the quality of narrative, much like silent speech or the idea preceding its expression in language. It also imbues objects with a presence in the environment inhabited by other minds, hence independent of the mind of the perceiver. It confers private and public content to aspects of the object as perceived.

Michael N. Shadlen
Eric R. Kandel

Selected Reading

Clark A. 1997. *Being There: Putting brain, body, and world together again.* Cambridge, MA: MIT Press. 269 pp.

Dehaene S. 2014. *Consciousness and the Brain: Deciphering How the Brain Codes Our Thoughts.* New York: Viking.

Dennett D. 1991. *Consciousness Explained.* Boston: Little, Brown.

Donlea JM, Pimentel D, Talbot CB, et al. 2018. Recurrent circuitry for balancing sleep need and sleep. Neuron 97:378–389.e4.

Gibson JJ. 2015. *The Ecological Approach to Visual Perception.* Classic Edition. New York: Psychology Press.

Graziano MSA, Kastner S. 2011. Human consciousness and its relationship to social neuroscience: a novel hypothesis. Cogn Neurosci 2:98–113.

Green DM, Swets JA. 1966. *Signal Detection Theory and Psychophysics.* New York: John Wiley and Sons, Inc.

Kang YHR, Petzschner FH, Wolpert DM, Shadlen MN. 2017. Piercing of consciousness as a threshold-crossing operation. Curr Biol 27:2285–2295.

Laming DRJ. 1968. *Information Theory of Choice-Reaction Times.* New York: Academic Press.

Link SW. 1992. *The Wave Theory of Difference and Similarity.* Hillsdale, NJ: Lawrence Erlbaum Associates.

Luce RD. 1986. *Response Times: Their Role in Inferring Elementary Mental Organization.* New York: Oxford University Press.

Markkula G. 2015. Answering questions about consciousness by modeling perception as covert behavior. Front Psychol 6:803.

Merleau-Ponty M. 1962. *Phenomenology of Perception.* London: Routledge & Kegan Paul Ltd.

Rangel A, Camerer C, Montague PR. 2008. A framework for studying the neurobiology of value-based decision-making. Nat Rev Neurosci 9:545–556.

Saxe R, Baron-Cohen S. 2006. The neuroscience of theory of mind. Soc Neurosci 1:i–ix.

Shadlen MN, Newsome WT. 1994. Noise, neural codes and cortical organization. Curr Opin Neurobiol 4:569–579.

Vickers D. 1979. *Decision Processes in Visual Perception.* London: Academic Press.

Wimmer H, Perner J. 1983. Beliefs about beliefs: representation and constraining function of wrong beliefs in young children's understanding of deception. Cognition 13:103–128.

References

Albright TD, Desimone R, Gross CG. 1984. Columnar organization of directionally selective cells in visual area MT of macaques. J Neurophysiol 51:16–31.

Andersen RA, Gnadt JW. 1989. Posterior parietal cortex. Rev Oculomot Res 3:315–335.

Born RT, Bradley DC. 2005. Structure and function of visual area MT. Annu Rev Neurosci 28:157–189.

Brincat SL, Siegel M, von Nicolai C, Miller EK. 2018. Gradual progression from sensory to task-related processing in cerebral cortex. Proc Natl Acad Sci U S A 115:E7202-E7211.

Britten KH, Shadlen MN, Newsome WT, Movshon JA. 1992. The analysis of visual motion: a comparison of neuronal and psychophysical performance. J. Neurosci. 12: 4745–65.

Brody CD, Hernandez A, Zainos A, Romo R. 2003. Timing and neural encoding of somatosensory parametric working memory in macaque prefrontal cortex. Cereb Cortex 13:1196–1207.

Constantinidis C, Funahashi S, Lee D, et al. 2018. Persistent Spiking Activity Underlies Working Memory. J Neurosci 38:7020–7028.

Ditterich J, Mazurek M, Shadlen MN. 2003. Microstimulation of visual cortex affects the speed of perceptual decisions. Nat Neurosci 6:891–898.

Fetsch CR, Odean NN, Jeurissen D, El-Shamayleh Y, Horwitz GD, Shadlen MN. 2018. Focal optogenetic suppression in macaque area MT biases direction discrimination and decision confidence, but only transiently. Elife 7:e36523.

Funahashi S, Bruce C, Goldman-Rakic P. 1989. Mnemonic coding of visual space in the monkey's dorsolateral prefrontal cortex. J Neurophysiol 61:331–349.

Gnadt JW, Andersen RA. 1988. Memory related motor planning activity in posterior parietal cortex of monkey. Exp Brain Res 70:216–220.

Gold JI, Shadlen MN. 2007. The neural basis of decision making. Annu Rev Neurosci 30:535–574.

Kiani R, Hanks TD, Shadlen MN. 2008. Bounded integration in parietal cortex underlies decisions even when viewing duration is dictated by the environment. J Neurosci 28:3017–3029.

Kiani R, Shadlen MN. 2009. Representation of confidence associated with a decision by neurons in the parietal cortex. Science 324:759–764.

Mazurek ME, Roitman JD, Ditterich J, Shadlen MN. 2003. A role for neural integrators in perceptual decision making. Cereb Cortex 13:1257–1269.

Mountcastle VB, Steinmetz MA, Romo R. 1990. Frequency discrimination in the sense of flutter: psychophysical measurements correlated with postcentral events in behaving monkeys. J Neurosci 10:3032–3044.

Padoa-Schioppa C. 2011. Neurobiology of economic choice: a good-based model. Ann Rev Neurosci 34:333–359.

Padoa-Schioppa C, Assad JA. 2006. Neurons in the orbitofrontal cortex encode economic value. Nature 441:223–226.

Roitman JD, Shadlen MN. 2002. Response of neurons in the lateral intraparietal area during a combined visual discrimination reaction time task. J Neurosci 22:9475–9489.

Romo R, Salinas E. 2001. Touch and go: decision-making mechanisms in somatosensation. Annu Rev Neurosci 24:107–137.

Salzman CD, Britten KH, Newsome WT. 1990. Cortical microstimulation influences perceptual judgements of motion direction. Nature 346:174–177.

Snyder LH, Batista AP, Andersen RA. 1997. Coding of intention in the posterior parietal cortex. Nature 386:167–170.

Yang T, Shadlen MN. 2007. Probabilistic reasoning by neurons. Nature 447:1075–1080.

Part IX

IX Diseases of the Nervous System

He remembered that during his epileptic fits, or rather immediately preceding them, he had always experienced a moment or two when his whole heart, and mind, and body seemed to wake up to vigour and light; when he became filled with joy and hope, and all his anxieties seemed to be swept away forever; these moments were but presentiments, as it were of the one final second (it was never more than a second) in which the fit came upon him. That second, of course, was inexpressible. When his attack was over, and the prince reflected on his symptoms, he used to say to himself: "These moments, short as they are, when I feel such extreme consciousness of myself, and consequently more of life than at other times, are due only to the disease—to the sudden rupture of normal conditions. Therefore they are not really a higher kind of life, but a lower." This reasoning, however, seemed to end in a paradox, and lead to the further consideration: —"What matter though it be only disease, an abnormal tension of the brain, if when I recall and analyze the moment, it seems to have been one of harmony and beauty in the highest degree—an instant of deepest sensation, overflowing with unbounded joy and rapture, ecstatic devotion, and completest life?" Vague though this sounds, it was perfectly comprehensible to Muishkin, though he knew that it was but a feeble expression of his sensations.*

WHAT, EXACTLY, IS THE NATURE OF THE RELATIONSHIP between the mind and the brain? Dostoevsky's own experience of epilepsy profoundly influenced his writing, and in this passage, he probes some of the most profound questions about human experience. Are our thoughts and moods simply transient combinations of chemicals and electrical signals? Do we have any influence over them? If not, can we be held responsible for our actions? What if some of our peak experiences are just happy chemical accidents? Or, as Prince Muishkin wonders, what if some of our peaks are happy accidents of disease? What, then, would it mean to "get better"? Individuals with bipolar disorder, for example, can have a very difficult time relinquishing the expansive feelings and creative energies that can accompany mania.

Although these profound questions are the purview of philosophers rather than neuroscientists, few circumstances bring the mind–brain relationship into question as sharply as becoming victim to a neurological or psychiatric disorder. The range of these conditions

---

*Dostoevsky F. *The Idiot*. Translated by Eva Martin. Project Gutenberg EBook, last updated May 13, 2017.

is very wide, from motor disturbances to epilepsy, schizophrenia, mood imbalances, cognitive disorders, neurodegeneration, and even aging. The more we learn, the more it becomes apparent that these diseases exert very broad effects that blur the boundaries between their classifications. So-called movement disorders such as Parkinson disease, for example, involve cognitive and affective changes; disorders of cognition such as autism or schizophrenia can have very physical manifestations.

Despite these somewhat fuzzy boundaries, each chapter in this section will examine the principles underlying each major class of disease from the perspective of neuroscience. The emphasis here is on molecular mechanisms, so far as they are currently understood. It is perhaps surprising that so many different disease conditions seem to converge on one physiological point: synaptic function. In autism and several psychiatric disorders, synaptic development goes awry; in epilepsy, abnormal ion channel activity disturbs the balance of synaptic input from excitatory and inhibitory neurons. Aging and neurodegenerative disorders bring about synaptic loss through gradual alterations in protein and RNA homeostasis that tax normal cellular functions.

This observation is offered to help give shape to the material you are about to encounter, but should not be used to oversimplify. Anyone tempted by reductionism would do well to engage with the works of great artists such as Dostoevsky and Van Gogh, who represent the complexities of human experience in all its anguish and glory.

Part Editor: Huda Y. Zoghbi

Part IX

57

Diseases of the Peripheral Nerve and Motor Unit

Disorders of the Peripheral Nerve, Neuromuscular Junction, and Muscle Can Be Distinguished Clinically

A Variety of Diseases Target Motor Neurons and Peripheral Nerves

Motor Neuron Diseases Do Not Affect Sensory Neurons (Amyotrophic Lateral Sclerosis)

Diseases of Peripheral Nerves Affect Conduction of the Action Potential

The Molecular Basis of Some Inherited Peripheral Neuropathies Has Been Defined

Disorders of Synaptic Transmission at the Neuromuscular Junction Have Multiple Causes

Myasthenia Gravis Is the Best-Studied Example of a Neuromuscular Junction Disease

Treatment of Myasthenia Is Based on the Physiological Effects and Autoimmune Pathogenesis of the Disease

There Are Two Distinct Congenital Forms of Myasthenia Gravis

Lambert-Eaton Syndrome and Botulism Also Alter Neuromuscular Transmission

Diseases of Skeletal Muscle Can Be Inherited or Acquired

Dermatomyositis Exemplifies Acquired Myopathy

Muscular Dystrophies Are the Most Common Inherited Myopathies

Some Inherited Diseases of Skeletal Muscle Arise From Genetic Defects in Voltage-Gated Ion Channels

Highlights

... to move things is all that mankind can do, for such the sole executant is muscle, whether in whispering a syllable or in felling a forest.

Charles Sherrington, 1924

A MAJOR TASK OF THE ELABORATE information processing that takes place in the brain is the contraction of skeletal muscles. The challenge of deciding when and how to move is, to a large degree, the driving force behind the evolution of the nervous system (Chapter 30).

In all but the most primitive animals, movement is generated by specialized muscle cells. There are three general types of muscles: Smooth muscle is used primarily for internal actions such as peristalsis and control of blood flow; cardiac muscle is used exclusively for pumping blood; and skeletal muscle is used primarily for moving bones. In this chapter, we examine a variety of neurological disorders in mammals that affect movement by altering either action potential conduction in a motor nerve, synaptic transmission from nerve to muscle, or muscle contraction itself.

In 1925, Charles Sherrington introduced the term *motor unit* to designate the basic unit of motor function—a motor neuron and the group of muscle fibers it innervates (Chapter 31). The number of muscle fibers innervated by a single motor neuron varies widely throughout the body depending on the dexterity of the movements being controlled and the mass of the body part to be moved. Thus, eye movements are finely controlled by motor units with fewer than 100 muscle fibers, whereas in the leg, a single motor unit contains up to 1,000 muscle fibers. In each case, all the muscles innervated by a motor unit are of the same type. Moreover, motor units are recruited in a fixed order for both voluntary and reflex movements. The smallest motor units are the first to be recruited, joined later by larger units as muscle force increases.

The motor unit is a common target of disease. The distinguishing features of diseases of the motor unit vary depending on which functional component is primarily affected: (1) the cell body of the motor or sensory neuron, (2) the corresponding axons, (3) the neuromuscular junction (the synapse between the motor axon and muscle), or (4) the muscle fibers innervated by the motor neuron. Accordingly, disorders of the motor unit have traditionally been grouped into motor neuron diseases, peripheral neuropathies, disorders of the neuromuscular junction, and primary muscle diseases (myopathies) (Figure 57–1).

Patients with peripheral neuropathies experience weakness that arises from abnormal function of motor neurons or their axons, although problems with sensation can also occur since most peripheral neuropathies also involve sensory neurons. By contrast, in motor neuron diseases, the motor neurons and motor tracts in the spinal cord degenerate but sensory nerves are spared. In myopathies, weakness is caused by degeneration of the muscles with little or no change in motor neurons. In neuromuscular junction diseases, alterations in the neuromuscular synapse lead to weakness

that may be intermittent. Clinical and laboratory studies usually distinguish disorders of peripheral nerves from those of the neuromuscular junction or muscle (Table 57–1).

Disorders of the Peripheral Nerve, Neuromuscular Junction, and Muscle Can Be Distinguished Clinically

When a peripheral nerve is cut, the muscles innervated by that nerve immediately become paralyzed and then waste progressively. Because the nerve carries sensory as well as motor fibers, sensation in the area innervated by the nerve is also lost and tendon reflexes are lost immediately. The term *atrophy* (literally, lack of nourishment) refers to the wasting away of a once-normal muscle; because of historical usage the term appears in the names of several diseases that are now regarded as neurogenic.

The main symptoms of the *myopathies* are due to weakness of skeletal muscle and often include difficulty in walking or lifting. Other less common symptoms include inability of the muscle to relax (myotonia), cramps, pain (myalgia), or the appearance in the urine of the heme-containing protein that gives muscle its red color (myoglobinuria). The *muscular dystrophies* are myopathies with special characteristics: The diseases are inherited, all symptoms are caused by weakness, the weakness becomes progressively more

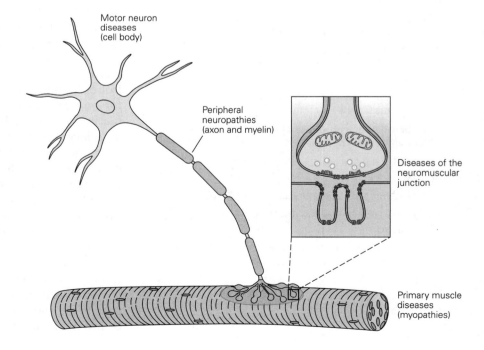

Figure 57–1 The four types of motor unit disorders. Motor unit disorders are categorized according to the part of the motor unit that is affected. Motor neuron diseases affect the cell body of the neuron, while peripheral neuropathies target the axon. Diseases of the neuromuscular junction affect the functioning of the synapse, and myopathies affect muscle fibers.

Motor neuron diseases (cell body)

Peripheral neuropathies (axon and myelin)

Diseases of the neuromuscular junction

Primary muscle diseases (myopathies)

Table 57–1 Differential Diagnosis of Disorders of the Motor Unit

Finding	Nerve	Neuromuscular junction	Muscle
Clinical			
Weakness	++	+	++
Wasting	++	–	+
Fasciculations	+	–	–
Cramps	+	–	+/–
Sensory loss	+/–	–	–
Hyperreflexia, Babinski	+ (ALS)	–	–
Laboratory			
Elevated serum CPK	–	–	++
Elevated cerebrospinal fluid protein	+/–	–	–
Slowed nerve conduction	+	–	–
Response to repetitive stimulation	Normal	Decremental (MG) Incremental (LEMS)	Normal
Electromyography			
Fibrillation, fasciculation	++	–	+/–
Duration of potentials	Increased	Normal	Decreased
Amplitude of potentials	Increased	Normal	Decreased
Muscle Biopsy			
Isolated fiber atrophy	++	Normal	+/–
Grouped fiber atrophy	++	Normal	Normal
Muscle necrosis	Normal	Normal	++

ALS, amyotrophic lateral sclerosis; CPK, creatine phosphokinase; LEMS, Lambert-Eaton myasthenic syndrome; MG, myasthenia gravis.

severe, and signs of degeneration and regeneration are seen histologically.

Distinguishing neurogenic and myopathic diseases may be difficult because both are characterized by weakness of muscle. As a first approximation, weakness of the distal limbs most often indicates a neurogenic disorder, whereas proximal limb weakness signals a myopathy. The main clinical and laboratory features used for the differential diagnosis of diseases of the motor unit are listed in Table 57–1.

One test that is very helpful is needle electromyography (EMG), a clinical procedure in which a small needle is inserted into a muscle to record extracellularly the electrical activity of several neighboring motor units. Three specific measurements are important: spontaneous activity at rest, the number of motor units under voluntary control, and the duration and amplitude of action potentials in each motor unit. (Normal ranges of values have been established for the amplitude and duration of motor unit potentials; the amplitude is determined by the number of muscle fibers within the motor unit.)

In normal muscle, there is usually no activity outside the end-plate in the muscle at rest. During a weak voluntary contraction, a series of motor unit potentials is recorded as different motor units become recruited. In fully active normal muscles, these abundant potentials overlap in an interference pattern so that it is impossible to identify single potentials (Figure 57–2A).

In neurogenic disease, the partially denervated muscle is spontaneously active even at rest. The muscle may still contract in response to voluntary motor commands, but the number of motor units under voluntary control is smaller than normal because some motor axons have been lost. The loss of motor units is evident in the EMG during a maximal contraction, which shows a pattern of discrete motor unit potentials instead of the profuse interference pattern for normal muscles (Figure 57–2B). In recently denervated muscle, the EMG may also show spontaneous low-amplitude electrical potentials that correspond to the firing of a single muscle fiber, known as fibrillation potentials. As the neurogenic disease progresses, the amplitude and duration of individual motor unit potentials may increase because the remaining axons give off small branches that innervate the muscle fibers denervated

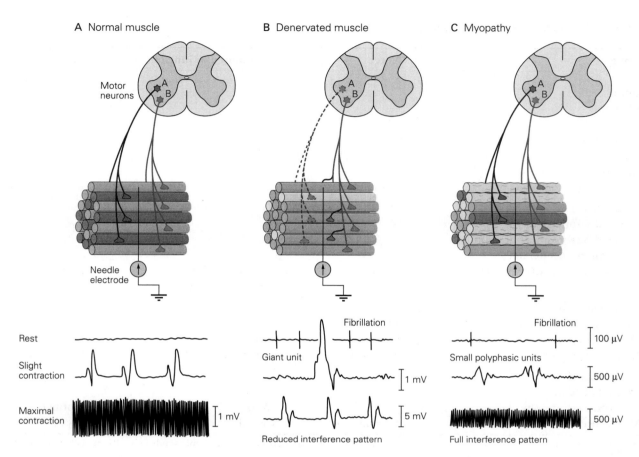

Figure 57–2 Electrical recording from skeletal muscle reveals different profiles in neuropathies and primary muscle diseases.

A. Typical activity in a normal muscle. The muscle fibers innervated by a single motor neuron are usually not adjacent to one another. When a motor unit potential is recorded by a needle electrode inserted into the muscle, the highly effective transmission at the neuromuscular junction ensures that each muscle fiber innervated by the same neuron will generate an action potential and contract in response to an action potential in the motor neuron. In the normal, resting muscle, there is no electrical activity recorded from muscle in the electromyogram (EMG). Slight activation of the muscle by a voluntary movement reveals characteristic extracellular electrical responses in muscle (motor unit potentials (MUPs)). Maximal muscle contraction produces a characteristic complex burst of electrical activity from muscle (the interference pattern).

B. When motor neurons are diseased, the number of motor units under voluntary control is reduced. The muscle fibers supplied by the degenerating motor neuron (cell A) become

denervated and atrophic. However, the surviving neuron (cell B) sprouts axonal branches that reinnervate some of the denervated muscle fibers. Axons of the surviving motor neuron fire spontaneously even at rest, giving rise to fasciculations, another characteristic of motor neuron disease. Single denervated fibers also fire spontaneously, producing fibrillations (top trace). With loss of nerve input from motor neuron A and reinnervation of the denervated fibers by motor neuron B, activation of motor neuron B produces an enlarged MUP (giant motor unit). In this setting, there is simplification of the interference pattern.

C. When muscle is diseased (myopathy), the number of muscle fibers in each motor unit is reduced. Some muscle fibers innervated by the two motor neurons shrink and become nonfunctional. In the electromyogram, the motor unit potentials do not decrease in number but are smaller and of longer duration than normal and are polyphasic. Affected single muscle fibers sometimes contract spontaneously, producing fibrillation. When muscle is mildly activated, the MUPs show reduced amplitudes. After maximal muscle contraction, the interference pattern also shows a reduction in amplitude.

by the loss of other axons. Accordingly, surviving motor units contain more than the normal number of muscle fibers.

In myopathic diseases, there is no activity in the muscle at rest and no change in the number of motor units firing during a contraction. But because there are fewer surviving muscle fibers in each motor unit, the motor unit potentials are of longer duration and more complex, with alternating +/− polarity (polyphasic), and are smaller in amplitude (Figure 57–2C).

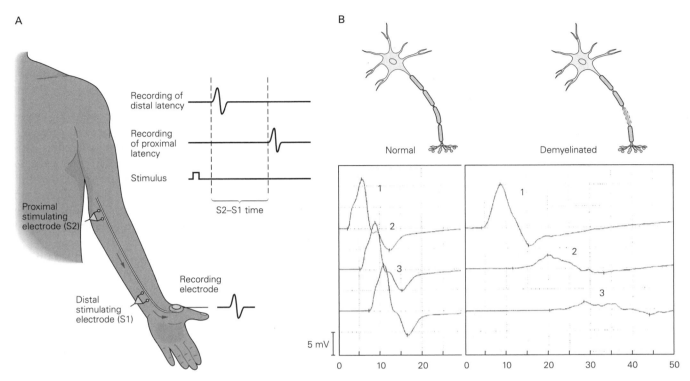

Figure 57–3 Motor nerve conduction velocity can be determined by recording the compound muscle action potential (CMAP) in response to electrical stimulation at different points along the nerve.

A. A shock is applied through a proximal surface stimulating electrode (**S2**) or through a distal stimulating electrode (**S1**), and the extracellular CMAP in the thumb is measured transcutaneously by the recording electrode. The time it takes the action potential to propagate from S2 to the muscle (t_{S2}) is the proximal latency; the time from S1 to the muscle (t_{S1}) is the distal latency. The distance between S1 and S2 divided by ($t_{S2} - t_{S1}$) gives the conduction velocity.

B. The waveforms of the thumb CMAPs elicited by stimulation of the motor nerve at the wrist (**1**), just below the elbow (**2**), and just above the elbow (**3**). In normal subjects (*left*), the waveforms are the same regardless of the site of stimulation. They are distinguished only by the longer time period required for the waveform to develop as the site of the stimulus is moved up the arm (away from the recording site). When the motor nerve is demyelinated between S1 and S2 but above the wrist, the CMAP is normal when stimulation occurs at the wrist (**1**) but delayed and desynchronized when stimulation is proximal to the nerve lesion (**2, 3**). (Adapted, with permission, from Bromberg 2002.)

The conduction velocities of peripheral motor axons can also be measured through electrical stimulation and recording (see Figure 57–3). The conduction velocity of motor axons is slowed in demyelinating neuropathies but is normal in neuropathies without demyelination (axonal neuropathies).

Another test that helps to distinguish myopathic from neurogenic diseases is the measurement of serum enzyme activities. The sarcoplasm of muscle is rich in soluble enzymes that are normally found in low concentrations in the serum. In many muscle diseases, the concentration of these sarcoplasmic enzymes in serum is elevated, presumably because the diseases affect the integrity of surface membranes of the muscle, allowing the enzymes to leak into the bloodstream. The enzyme activity most commonly used for diagnosing myopathy is creatine kinase, an enzyme that phosphorylates creatine and is important in the energy metabolism of muscle.

Muscle histochemical appearance in a biopsy can also provide a useful diagnostic tool. Human muscle fibers are identified by histochemical reactions as type I or type II, which respectively are either aerobic (enriched for oxidative enzymes) or anaerobic (abundant glycolytic enzymes) (Chapter 31). All muscle fibers innervated by a single motor neuron are of the same histochemical type. However, the muscle fibers of one motor unit are normally interspersed among the muscle fibers of other motor units. In a cross section of healthy muscle, enzyme stains show that oxidative or glycolytic fibers are intermixed in a "checkerboard" pattern.

In chronic neurogenic diseases, the muscle innervated by a dying motor neuron becomes atrophic and some muscle fibers disappear. Axons of surviving neurons tend to sprout and reinnervate some of the adjacent remaining muscle fibers. Because the motor neuron determines the biochemical and thus histochemical properties of a muscle fiber, the reinnervated muscle fibers assume the histochemical properties of the innervating neuron. As a result, the fibers of a muscle in neurogenic disease become clustered by type (a pattern called fiber-type grouping).

If the disease is progressive and the neurons in the surviving motor units also become affected, atrophy occurs in groups of adjacent muscle fibers belonging to the same histochemical type, a process called group atrophy. In contrast, in myopathic diseases, the muscle fibers are affected in a more or less random fashion. Sometimes an inflammatory cellular response is evident, and sometimes there is prominent infiltration of the muscle by fat and connective tissue.

Fasciculations—visible twitches of muscle that can be seen as flickers under the skin—are often signs of neurogenic diseases. They result from involuntary but synchronous contractions of all muscle fibers in a motor unit. Fibrillations—spontaneous contractions within single muscle fibers—can also be signs of ongoing denervation of muscle. Fibrillations are not visible but can be recorded with an EMG. The electrical record of a fibrillation is a low-amplitude potential that reflects electrical activity in a single muscle cell. Electrophysiological studies suggest that fasciculations arise in the motor nerve terminal.

In diagnosing motor neuron disorders, clinicians have historically distinguished between so-called lower motor neurons and premotor neurons. Lower motor neurons are motor neurons of the spinal cord and brain stem that directly innervate skeletal muscles. Premotor neurons, also known as "upper" motor neurons, originate in the motor cortex and issue commands for movements to the lower motor neurons through their axons in the corticospinal (pyramidal) tract.

Diseases of upper motor neurons can be distinguished from those affecting lower motor neurons by distinct sets of symptoms. Disorders of lower motor neurons cause atrophy, fasciculations, decreased muscle tone, and loss of tendon reflexes, whereas disorders of upper motor neurons and their axons result in spasticity, overactive tendon reflexes, and abnormal plantar extensor reflex (the Babinski sign).

The primary symptom of disorders of the neuromuscular junction is weakness; in some neuromuscular junction diseases, this weakness is quite variable even over the course of a single day.

A Variety of Diseases Target Motor Neurons and Peripheral Nerves

Motor Neuron Diseases Do Not Affect Sensory Neurons (Amyotrophic Lateral Sclerosis)

The best-known disorder of motor neurons is amyotrophic lateral sclerosis (ALS; Lou Gehrig disease). "Amyotrophy" is another term for neurogenic atrophy of muscle; "lateral sclerosis" refers to the hardness felt when the pathologist examines the spinal cord at autopsy. This hardness results from the proliferation of astrocytes and scarring of the lateral columns of the spinal cord due to degeneration of the corticospinal tracts.

The symptoms of ALS usually start with painless weakness in a single arm or leg. Typically, the patient, often a man in his 40s or 50s, discovers that he has trouble in executing fine movements of the hands—typing, playing the piano, playing baseball, fingering coins, or working with tools. This focal weakness then spreads over 3 or 4 years to involve all four limbs, as well as the muscles of chewing, speaking, swallowing, and breathing.

Most cases of ALS involve both the upper and the lower motor neurons. Some motor neurons are spared, notably those supplying ocular muscles and those involved in voluntary control of bladder sphincters. The typical weakness of the hand is associated with wasting of the small muscles of the hands and feet and fasciculations of the muscles of the forearm and upper arm. These signs of lower motor neuron disease are often associated with hyperreflexia, an over-responsiveness in tendon reflexes characteristic of corticospinal upper motor neuron disease. The cause of most cases (90%) of ALS is not known; the disease is progressive and ultimately affects the muscles of respiration. There is no effective treatment for this fatal condition.

About 10% of cases are inherited in a dominant manner (Table 57–2). In North America, greater than 25% of inherited cases arise from mutations in the gene *C9orf72*. The offending genetic defect is an expansion in an intronic hexanucleotide repeat, from 30 or fewer in normal individuals to hundreds or even thousands in affected individuals. Besides giving rise to conventional ALS, mutations in *C9orf72* can also cause frontotemporal dementia. The toxicity of the mutant C9orf72 protein probably reflects both a reduction in total activity of the mutant protein and toxic effects of the intronic expansion. For example, the expanded intronic segments produce intranuclear deposits of RNA that likely sequester and inactivate important nuclear proteins. In addition, the expanded RNA is translated to

Table 57–2 Selected Amyotrophic Lateral Sclerosis Genes

| Gene | Protein | Protein function | Mutations | Proportion of ALS | |
				Familial	Sporadic
SOD1	Cu-Zn superoxide dismutase	Superoxide dismutase	>150	20%	2%
DCTN1	Dynactin subunit 1	Component of dynein motor complex	10	1%	<1%
ANG	Angiogenin	Ribonuclease	>10	<1%	<1%
TARDBP	TDP-43	RNA-binding protein	>40	5%	<1%
FUS	FUS	RNA-binding protein	>40	5%	<1%
VCP	Transitional endoplasmic reticulum ATPase	Ubiquitin segregase	5	1–2%	<1%
OPTN	Optineurin	Autophagy adaptor	1	4%	<1%
C9orf72	C9orf72	Possible guanine nucleotide exchange factor	Intronic GGGGCC	25%	10%
UBQLN2	Ubiquilin 2	Autophagy adaptor	5	<1%	<1%
SQSTM1	Sequestosome 1	Autophagy adaptor	10	<1%	?
FFN1	Profilin-1	Actin-binding protein	5	<1%	<1%
HNRNPA1	hnRNP A1	RNA-binding protein	3	<1%	<1%
MATR3	Matrin 3	RNA-binding protein	4	<1%	<1%
TUBA4A	Tubulin α-4A chain	Microtubule subunit	7	<1%	<1%
CHCHD10	Coiled-coil-helix-coiled-coil-helix domain-containing protein 10	Mitochondrial protein of unknown function	2	<1%	<1%
TBK1	Serine/threonine-protein kinase TBK1	Regulates autophagy and inflammation	10	1%	<1%

Source: Modified from Taylor, Brown, and Cleveland 2016.

produce peptides composed of repeated couplets of amino acids, such as poly-(glycine-proline) or poly-(proline-arginine); some of these are neurotoxic.

Two other genes commonly mutated in ALS are *SOD1* and *TDP43*. *SOD1* encodes the protein copper/zinc cytosolic superoxide dismutase, whereas *TDP43* encodes a 43-kD, RNA-interacting protein that is normally intranuclear but is mislocalized to the cytosol in most cases of ALS (both inherited and sporadic). Mutations in *SOD1* and several other ALS genes (eg, *ubiquilin-2*) destabilize the conformation of the protein product, promoting misfolding and causing adverse consequences to diverse subcellular processes and compartments. By contrast, mutations in *TDP43* and a few other ALS genes (eg, *FUS*) encoding RNA binding proteins act at the RNA level, impairing RNA homeostasis and perturbing critical processes such as surveillance

of gene splicing. Infrequently, familial ALS is caused by mutations in genes encoding cytoskeletal proteins such as profilin-1, dynactin, or tubulin-A4.

Many studies suggest that mutant ALS-associated proteins tend to aggregate, particularly in membrane-less organelles called stress granules that form in conditions of cellular distress. Several lines of investigation support the view that aggregates migrate and transmit pathology between adjacent cells, accounting for spread of the disease to different brain regions. Strikingly, mice that express high levels of defective SOD1 or profilin-1 proteins develop a lethal, adult-onset form of motor neuron disease, but mice expressing equivalently high levels of normal SOD1 or profilin-1 proteins do not. These findings are consistent with the concept that the defective protein has gained some sort of toxic function.

In the past 10 years, it has also become clear that motor neuron pathophysiology is modulated by the reactions of nonneural cells to degeneration in the motor neuron. Thus, in most cases of ALS, there are varying degrees of proliferation and activation of microglia, astrocytes, and some populations of lymphocytes, which may begin as compensatory responses but can eventually adversely affect the injured motor neurons. Genetic studies have underscored the importance of non–cell-autonomous factors, such as variants that reduce function of the microglial gene TREM-2 and enhance the risk of developing not only ALS but also other neurodegenerative disorders (eg, Alzheimer disease).

Progressive bulbar palsy is a type of motor neuron disease in which damage is restricted to muscles innervated by cranial nerves, causing dysarthria (difficulty speaking) and dysphagia (difficulty swallowing). (The term "bulb" is used interchangeably with "pons," the structure at the base of the brain where motor neurons that innervate the face and swallowing muscles reside, and "palsy" means weakness). If only lower motor neurons are involved, the syndrome is called progressive spinal muscular atrophy.

Progressive spinal muscular atrophy is actually a developmental motor neuron disorder characterized by weakness, wasting, loss of reflexes, and fasciculations. Most cases arise in infancy and are caused by recessively inherited mutations in the gene encoding a protein called survival motor neuron (SMN). Survival in these cases is very short, although there are rare cases that begin in late childhood or even early adulthood and are associated with longer survival of many years. The SMN protein is implicated in trafficking RNA in and out of the nucleus and in the formation of complexes that are important in RNA splicing. The SMN locus on chromosome 5 in humans has two almost identical copies of the SMN gene: SMN1 produces a full-length SMN protein, while alternative splicing of SMN2 causes omission of the seventh exon in the gene, leading to expression of a small amount of full-length SMN and a shortened SMN. The clinical effect of the loss of full-length SMN from mutations at the main locus can be mitigated to some degree by the shortened SMN protein expressed by the SMN2 gene (Figure 57–4A,B).

Two treatment strategies have achieved extraordinary benefits in spinal muscular atrophy. In one, small strings of approximately 20 nucleic acids (antisense oligonucleotides [ASO]s) are administered to alter splicing of the SMN2 gene so that it produces higher levels of the full-length SMN protein (Figure 57-4A). This occurs because the ASO is targeted to bind to the SMN2 RNA and inhibit the action of the RNA binding protein hnRNPA1/A2 that normally leads the splicing machinery to skip exon 7. By blocking the binding of hnRNPA1/A2, the ASO blocks the inhibitory effect of hnRNPA1/A2 on splicing, promoting expression of full-length SMN protein (Figure 57–4B). It seems likely that ASOs will become powerful therapeutic tools with many applications. In this example, ASO is used to promote exon inclusion; as noted below in the discussion on muscle dystrophy, ASO can also be used to promote exon skipping. It can also be used in other paradigms to inhibit or enhance levels of target gene expression.

The second approach to treating spinal muscular atrophy has been to deliver the missing SMN gene to spinal motor neurons and muscle using high doses of intravenously infused adeno-associated virus carrying the SMN1 gene. This, too, dramatically augments survival in infantile spinal muscular atrophy (Figure 57–4B).

ALS and its variants are restricted to motor neurons; they do not affect sensory neurons or autonomic neurons. The acute viral disease poliomyelitis is also confined to motor neurons. These diseases illustrate the individuality of nerve cells and the principle of selective vulnerability. The basis of this selectivity is, in general, not understood.

Diseases of Peripheral Nerves Affect Conduction of the Action Potential

Diseases of peripheral nerves may affect either axons or myelin. Because motor and sensory axons are bundled together in the same peripheral nerves, disorders of peripheral nerves usually affect both motor and sensory functions. Some patients with peripheral neuropathy report abnormal, frequently unpleasant, sensory experiences such as numbness, pins-and-needles prickling, or tingling. When these sensations occur spontaneously without an external sensory stimulus, they are called paresthesias.

Patients with paresthesias usually have impaired perception of cutaneous sensations (pain and temperature), often because the small fibers that carry these sensations are selectively affected. This is not always the case, however. Proprioceptive sensations (position and vibration) can be lost without loss of cutaneous sensation. Lack of pain perception may lead to injuries. The sensory deficits are more prominent distally (called a glove-and-stocking pattern), likely because the distal portions of the nerves are most remote from the cell body and therefore most susceptible to disorders that interfere with axonal transport of essential metabolites and proteins.

Peripheral neuropathy is first manifested by weakness that is usually distal. Tendon reflexes are usually

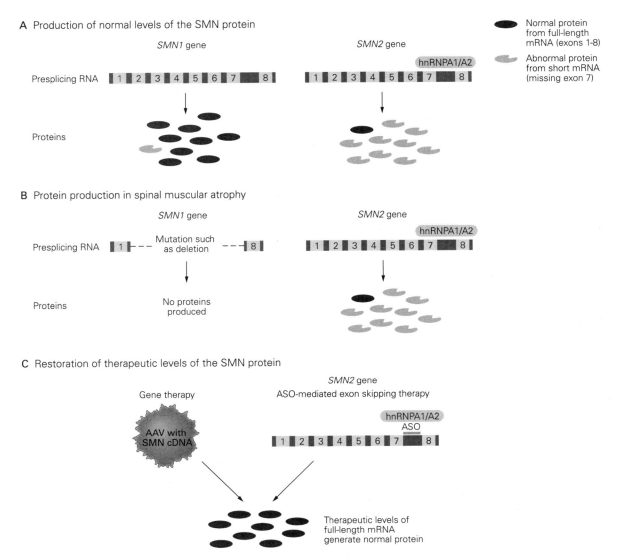

Figure 57–4 Spinal motor atrophy caused by defective survival motor neuron gene (*SMN1*) can be treated by gene replacement therapy or by manipulating splicing of *SMN2*.

A. Normally, most of the survival motor neuron (**SMN**) protein is produced from the *SMN1* gene, whose mRNA is spliced from eight exons. In normal circumstances, about 90% of the mRNA has all eight exons, yielding normal levels of the SMN protein. In the adjacent sister gene, *SMN2*, binding of the protein hnRNPA1/A2 to the *SMN2* transcript excludes exon 7; *SMN2* therefore makes a shortened SMN protein.

B. In spinal muscular atrophy, genetic lesions (commonly deletions) in *SMN1* lead to a marked reduction in levels of total SMN protein.

C. When *SMN1* protein is absent, one therapeutic approach is to replace the missing *SMN1* gene using an adeno-associated viral vector (**AAV**) to deliver the missing gene to the central nervous system and muscle. An alternative approach is to deliver an antisense oligonucleotide (**ASO**) that blocks the effect of hnRNPA1/A2, thereby enhancing production of a full-length mRNA (with all eight exons) from *SMN2*. This restores SMN protein levels.

depressed or lost, fasciculation is seen only rarely, and wasting does not ensue unless the weakness has been present for many weeks.

Neuropathies may be either acute or chronic. The best-known acute neuropathy is Guillain-Barré syndrome. Most cases follow respiratory infection or infectious diarrhea, but the syndrome may occur without apparent preceding illness. The condition may be mild or so severe that mechanical ventilation is required. Cranial nerves may be affected, leading to paralysis of ocular, facial, and oropharyngeal muscles. The disorder is attributed to an autoimmune attack on peripheral nerves by circulating antibodies. It is therefore treated by removing the offending antibodies by infusions of gamma globulin and plasmapheresis (a procedure in which blood is removed from a patient, cells are separated from the antibody-carrying plasma, and the cells alone are returned to the patient).

The chronic neuropathies vary from mild to incapacitating or even fatal conditions. There are many varieties, including genetic diseases (acute intermittent porphyria, Charcot-Marie-Tooth disease), metabolic disorders (diabetes, vitamin B_{12} deficiency), toxicities (lead), nutritional disorders (alcoholism, thiamine deficiency), carcinomas (especially carcinoma of the lung), and immunological disorders (plasma cell diseases, amyloidosis). Some chronic disorders, such as neuropathy due to vitamin B_{12} deficiency in pernicious anemia, are amenable to therapy.

In addition to being acute or chronic, neuropathies may be categorized as demyelinating (in which the myelin sheath breaks down) or axonal (in which the axon is affected). In demyelinating neuropathies, as might be expected from the role of the myelin sheath in saltatory conduction, conduction velocity is slowed. In axonal neuropathies, the myelin sheath is not affected and conduction velocity is normal.

Axonal and demyelinating neuropathies may lead to positive or negative symptoms and signs. The negative signs consist of weakness or paralysis, loss of tendon reflexes, and impaired sensation resulting from loss of motor and sensory nerves. The positive symptoms of peripheral neuropathies consist of paresthesias that arise from abnormal impulse activity in sensory fibers and either spontaneous activity of injured nerve fibers or electrical interaction (cross-talk) between abnormal axons, a process called ephaptic transmission to distinguish it from normal synaptic transmission. It is not known why damaged nerves become hyperexcitable. Even lightly tapping the site of injury can evoke a burst of painful sensations in the region over which the nerve is distributed.

Negative symptoms, which have been studied more thoroughly than positive symptoms, can be attributed to three basic mechanisms: conduction block, slowed conduction, and impaired ability to conduct impulses at higher frequencies. Conduction block was first recognized in 1876 when the German neurologist Wilhelm Erb observed that stimulation of an injured peripheral nerve below the site of injury evoked a muscle response, whereas stimulation above the site of injury produced no response. He deduced that the lesion blocked conduction of impulses of central origin, even when the segment of the nerve distal to the lesion was still functional. Later studies confirmed this conclusion by showing that selective application of diphtheria and other toxins produces conduction block by causing demyelination only at the site of application (Figure 57–5).

Why does demyelination produce nerve block, and how does it lead to slowing of conduction velocity? Conduction velocity is much more rapid in myelinated fibers than in unmyelinated axons for two reasons (Chapter 9). First, there is a direct relationship between conduction velocity and axon diameter, and myelinated axons tend to be larger in diameter. Second, membrane capacitance in the myelinated regions of the axon is lower than at the unmyelinated nodes of Ranvier, greatly speeding up the rate of depolarization and thus conduction. With demyelination, the spatial distribution of ion channels along the denuded axon is not optimal for supporting action potential propagation and may even cause a failure of conduction. When myelin is disrupted by disease, the action potentials in different axons of a nerve begin to conduct at slightly different velocities. As a result, the nerve loses its normal synchrony of conduction in response to a single stimulus. (Figure 57–2 shows how conduction velocities are measured in peripheral nerves.)

This slowing and loss of synchrony are thought to account for some of the early clinical signs of demyelinating neuropathy. For example, functions that normally depend on the arrival of synchronous bursts of neural activity, such as tendon reflexes and vibratory sensation, are lost soon after the onset of a chronic neuropathy. As demyelination becomes more severe, conduction becomes blocked. This block may be intermittent, occurring only at high frequencies of neural firing, or complete (Figure 57–3).

The Molecular Basis of Some Inherited Peripheral Neuropathies Has Been Defined

Myelin proteins are affected in a group of demyelinating hereditary peripheral neuropathies collectively termed Charcot-Marie-Tooth (CMT) disease. CMT is characterized by muscle weakness and wasting, loss of reflexes, and loss of sensation in the distal parts of the limbs. These symptoms appear in childhood or adolescence and are slowly progressive.

One form (type 1) has the features of a demyelinating neuropathy (Figure 57–5). Conduction in peripheral nerves is slow, with histological evidence of demyelination followed by remyelination. Sometimes, the remyelination leads to gross hypertrophy of the nerves. Type 1 disorders are inexorably progressive, without remissions or exacerbations. Another form (type 2) has normal nerve conduction velocity and is considered an axonal neuropathy without demyelination. Both types 1 and 2 are inherited as autosomal dominant diseases.

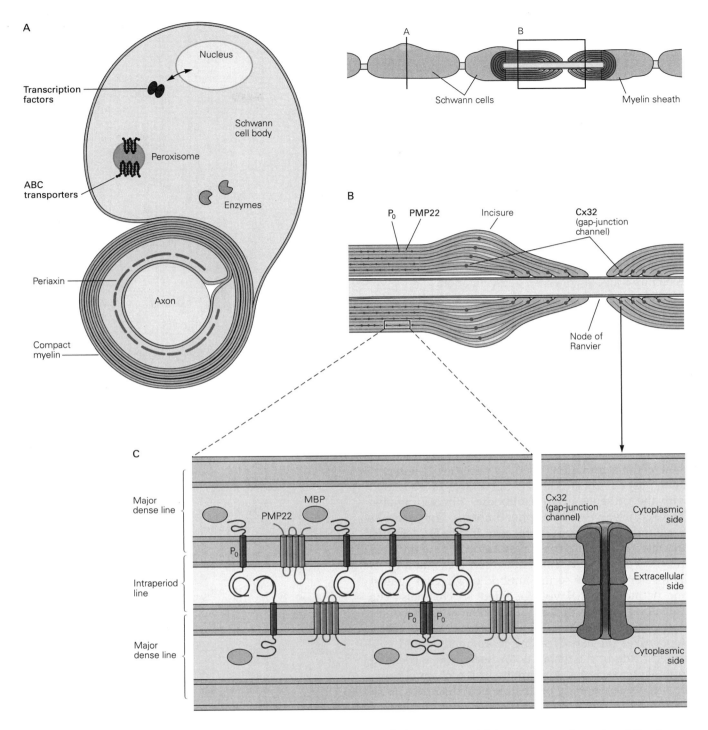

Figure 57–5 Gene defects in components of myelin cause demyelinating neuropathies.

A. Myelin production and function in the Schwann cell can be adversely affected by many genetic defects, including abnormalities in transcription factors, **ABC** (ATP-binding cassette) transporters in peroxisomes, and multiple proteins implicated in organizing myelin. Viewed microscopically at high power, the site of apposition of the intracellular faces of the Schwann cell membrane appears as a dense line, whereas the apposed extracellular faces are described as the "intraperiod line" (see part **C**). (Adapted from Lupski 1998.)

B. Peripheral axons are wrapped in multiple layers of thin sheaths of myelin that are processes of Schwann cells. The myelin is compact and tight except near the nodes of Ranvier and at focal sites described as "incisures" by Schmidt and Lanterman. Three myelin-associated proteins are defective in three different demyelinating neuropathies: P_0 (Dejerine-Sottas infantile neuropathy), peripheral myelin protein 22 (**PMP22**) (Charcot-Marie-Tooth neuropathy type 1), and connexin-32 (**Cx32**) (X-linked Charcot-Marie-Tooth neuropathy). (Adapted from Lupski 1998.)

C. The rim of cytoplasm in which myelin basic protein (**MBP**) is located defines the major dense line, whereas the thin layer of residual extracellular space defines the intraperiod line. Mutations in PMP22 and P_0 genes adversely affect the organization of compact myelin. (Adapted, with permission, from Brown and Amato 2002.)

Type 1 disease is attributed to mutations on two different chromosomes (locus heterogeneity). The more common form (type 1A) is linked to chromosome 17, while the less common form (1B) is localized to chromosome 1. The genes at these loci have been directly implicated in myelin physiology (Figure 57–5). Type 1A involves a defect in peripheral myelin protein 22, and type 1B the myelin protein P_0. Moreover, an X-linked form of demyelinating neuropathy occurs because of mutations in the gene expressing connexin-32, a subunit of the gap-junction channels that interconnect myelin folds near the nodes of Ranvier (Figure 57–5B,C). Still other genes have been implicated in inherited demyelination.

Some of the genes and proteins implicated in axonal neuropathies are shown in Figure 57–6 and Table 57–3. Genes encoding the neurofilament light subunit and an axonal motor protein related to kinesin, which is important for transport along microtubules, are mutated in two types of axonal neuropathies. Defects in these genes are associated with peripheral neuropathies with prominent weakness. The mechanisms by which genes alter axonal function in other axonal neuropathies are less evident.

As noted above, a wide range of problems other than genetic mutations lead to peripheral neuropathies. Particularly striking are nerve defects associated with the presence of autoantibodies directed against ion channels in distal peripheral nerves. For example, some individuals with motor unit instability (cramps and fasciculations), as well as sustained or exaggerated muscle contractions caused by hyperexcitability of motor nerves, have serum antibodies directed against one or more axonal voltage-gated K^+ channels. The prevailing view is that binding of the autoantibodies to the channels reduces K^+ conductance and thereby depolarizes the axon, leading to augmented and sustained firing of the distal motor nerve and associated muscle contractions. Alterations in ion channel function underlie a variety of neurological disorders, as in acquired disorders of channels in the neuromuscular junction and inherited defects in voltage-gated channels in muscle (discussed below).

Disorders of Synaptic Transmission at the Neuromuscular Junction Have Multiple Causes

Many diseases involve disruption of chemical transmission between neurons and their target cells. By analyzing such abnormalities, researchers have learned a great deal about the mechanisms underlying normal synaptic transmission as well as disorders caused by dysfunction at the synapse.

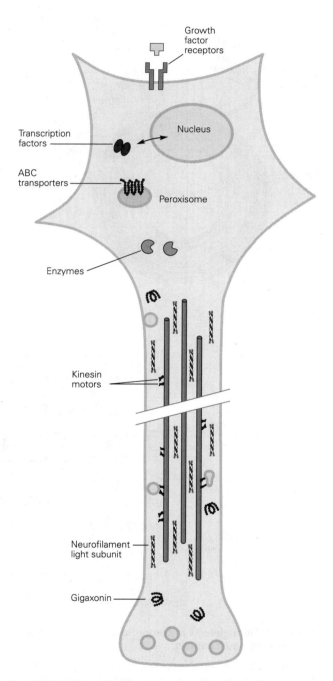

Figure 57–6 Gene defects that cause axonal neuropathies. These include defects in receptors for growth factors, **ABC** (ATP-binding cassette) transporters in peroxisomes, cytosolic enzymes, microtubule motor proteins like the kinesins, neurofilament proteins, and other structural proteins such as gigaxonin. (Adapted, with permission, from Brown and Amato 2002.)

Table 57–3 Representative Peripheral Neuropathy Genes

Site of primary defect	Protein	Disease
Myelin	Proteolipid myelin protein 22	Charcot-Marie-Tooth disease (CMT)
	Proteolipid protein P_0	Infantile CMT (Dejerine-Sottas neuropathy)
	Connexin-32	X-linked CMT
Axon	Kinesin KIF1Bβ motor protein	Motor predominant neuropathy
	Heat shock protein 27	Motor predominant neuropathy
	Neurofilament light subunit	Motor predominant neuropathy
	Tyrosine kinase A receptor	Congenital sensory neuropathy
	ABC1 transporter	Tangier disease
	Transthyretin	Amyloid neuropathy

Diseases that disrupt transmission at the neuromuscular junction fall into two broad categories: those that affect the presynaptic terminal and those that primarily involve the postsynaptic membrane. In both categories, the most intensively studied cases are autoimmune and inherited defects in critical synaptic proteins.

Myasthenia Gravis Is the Best-Studied Example of a Neuromuscular Junction Disease

The most common and extensively studied disease affecting synaptic transmission is myasthenia gravis, a disorder at the neuromuscular junction in skeletal muscle. Myasthenia gravis (the term means severe weakness of muscle) has two major forms. The most prevalent is the autoimmune form. The second is congenital and heritable; it is not an autoimmune disorder and is heterogeneous. Fewer than 500 of these congenital cases have been identified, but they have provided information about the organization and function of the human neuromuscular junction. This form is discussed later in the chapter.

In autoimmune myasthenia gravis, antibodies are produced against components of the postsynaptic end-plate in muscle, such as the nicotinic acetylcholine (ACh) receptor and muscle-specific tyrosine kinase

(MuSK). Anti–ACh receptor antibodies interfere with synaptic transmission by reducing the number of functional receptors or by impeding the interaction of ACh with its receptors. As a result, communication between the motor neuron and the skeletal muscle becomes weakened. This weakness always affects cranial muscles—eyelids, eye muscles, and oropharyngeal muscles—as well as limb muscles. Its severity of symptoms varies over the course of a single day, from day to day, or over longer periods (giving rise to periods of remission or exacerbation), making myasthenia gravis unlike most other diseases of muscle or nerve. The weakness is reversed by drugs that inhibit acetylcholinesterase, the enzyme that degrades ACh. As one example, when patients are asked to look upward in a sustained gaze, the eyelids tire after several seconds and droop downward (ptosis). Like decremental responses on EMG, this fatiguability and drooping reverse after treatment with inhibitors of acetylcholinesterase (Figure 57–7).

When a motor nerve is stimulated at rates of two to five stimuli per second, the amplitude of the compound

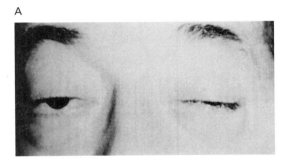

Figure 57–7 Myasthenia gravis often selectively affects the cranial muscles. (Reproduced, with permission, from Rowland, Hoefer, and Aranow 1960.)

A. Severe drooping of the eyelids, or ptosis, is characteristic of myasthenia gravis. This patient also could not move his eyes to look to either side.

B. One minute after an intravenous injection of 10 mg of edrophonium, an inhibitor of acetylcholinesterase, both eyes are open and can be moved freely.

action potential evoked in normal human muscle remains constant. In myasthenia gravis, the amplitude of the evoked compound action potential decreases rapidly. This pattern of decremental response of the compound muscle action potential to repetitive stimulation of the motor nerve mirrors the clinical symptom of fatiguability in myasthenia. Moreover, this abnormality resembles the pattern induced in normal muscle by d-tubocurarine (the active compound in curare), which blocks nicotinic ACh receptors and inhibits the action of ACh at the neuromuscular junction. Neostigmine (Prostigmin), which inhibits acetylcholinesterase and thus increases the duration of action of ACh at the neuromuscular junction, reverses the decrease in amplitude of evoked compound action potentials in myasthenic patients (Figure 57–8).

About 15% of adult patients with myasthenia have benign tumors of the thymus (thymomas). As the symptoms in myasthenic patients are often improved by removal of these tumors, some element of the thymoma may stimulate autoimmune pathology. Indeed, myasthenia gravis often affects people who have other autoimmune diseases, such as rheumatoid arthritis,

systemic lupus erythematosus, or Graves disease (hyperthyroidism).

Normally, an action potential in a motor axon releases enough ACh from synaptic vesicles to induce a large excitatory end-plate potential with an amplitude of about 70 to 80 mV relative to the resting potential of −90 mV (Chapter 12). Thus, the normal end-plate potential is greater than the threshold needed to initiate an action potential, about −45 mV. In normal muscle, the difference between the threshold and the actual end-plate potential amplitude—the safety factor—is therefore quite large (Figure 57–8). In fact, in many muscles, the amount of ACh released during synaptic transmission can be reduced to as little as 25% of normal before it fails to initiate an action potential.

The density of ACh receptors is reduced over time in myasthenia. This reduces the probability that a molecule of ACh will find a receptor before it is hydrolyzed by the acetylcholinesterase. In addition, the geometry of the end-plate is also disturbed in myasthenia (Figure 57–9). The normal infolding at the junctional folds is reduced and the synaptic cleft is enlarged. These morphological changes increase the diffusion of

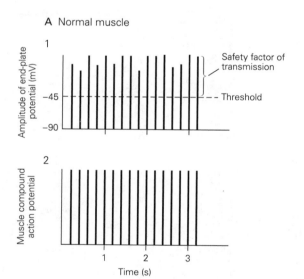
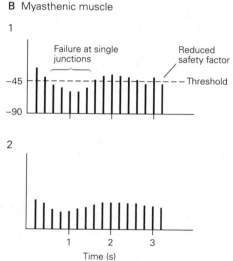

Figure 57–8 Synaptic transmission at the neuromuscular junction fails in myasthenia gravis. (Reproduced, with permission, from Lisak and Barchi 1982.)

A. In the normal neuromuscular junction, the amplitude of the end-plate potential is so large that all fluctuations in the efficiency of transmitter release occur well above the threshold for a muscle action potential. This results in a large safety factor for synaptic transmission (1). Therefore, during repetitive stimulation of the motor nerve, the amplitude of the compound action potentials, representing the contributions from all muscle fibers in which synaptic transmission is

successful in triggering an action potential, is constant and invariant (2).

B. In the myasthenic neuromuscular junction, postsynaptic changes reduce the amplitude of the end-plate potential so that under optimal circumstances the end-plate potential may be just sufficient to produce a muscle action potential. Fluctuations in transmitter release that normally accompany repeated stimulation now cause the end-plate potential to drop below this threshold, leading to conduction failure at that junction (1). The amplitude of the compound action potentials in the muscle declines progressively and shows only a small and variable recovery (2).

Figure 57–9 Morphological abnormalities of the neuromuscular junction are characteristic of myasthenia gravis. At the neuromuscular junction, acetylcholine (ACh) is released by exocytosis of synaptic vesicles at active zones in the nerve terminal. Acetylcholine flows across the synaptic cleft to reach ACh receptors that are concentrated at the peaks of junctional folds. Acetylcholinesterase in the cleft rapidly terminates transmission by hydrolyzing ACh. The myasthenic junction has reduced numbers of ACh receptors, simplified synaptic folds, a widened synaptic space, but a normal nerve terminal.

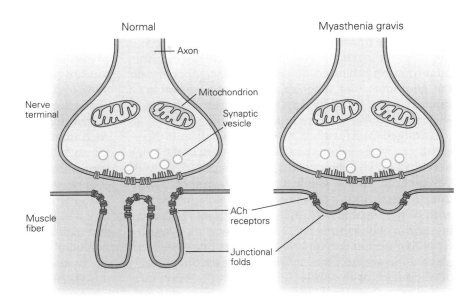

ACh away from the synaptic cleft and further reduce the probability of ACh interacting with the few remaining functional receptors. As a result, the amplitude of the end-plate potential is reduced to the point where it is barely above threshold (Figure 57–8).

Thus, in myasthenia, synaptic transmission is readily blocked even though the vesicles in the presynaptic terminals contain normal amounts of ACh and the process of transmitter release is intact. Both the physiological abnormality (the decremental response) and the clinical symptoms (muscle weakness) are partially reversed by drugs that inhibit acetylcholinesterase. This is because the released ACh molecules remain unhydrolyzed for a longer time, and this increases the probability that they will interact with receptors.

How do antibodies cause the symptoms of myasthenia? The antibodies do not simply occupy the ACh binding site. Rather, they appear to react with epitopes elsewhere on the receptor molecule. This increases the turnover of nicotinic ACh receptors, probably because myasthenic antibodies bind and cross-link the receptors, triggering their degradation (Figure 57–9). In addition, some myasthenic antibodies bind proteins of the complement cascade of the immune system, causing lysis of the postsynaptic membrane.

Despite the evidence documenting the primary role of antibodies against the nicotinic ACh receptor in myasthenia, about one-fifth of patients with myasthenia do not have these antibodies—including some who respond to anti-immune therapy like plasmapheresis. Instead, the majority of these patients have antibodies to other postsynaptic proteins, such as MuSK (*muscle-specific trk-related receptor with a Kringle domain*)

and lipoprotein-related protein 4 (LPR4), which is an activator of MuSK. MuSK is a muscle-specific receptor tyrosine kinase that interacts with another postsynaptic protein, agrin, to organize the nicotinic ACh receptors into clusters at the neuromuscular junction (Chapter 48); it appears to be functionally important both during development and in the adult. The anti-MuSK antibodies block some of the normal clustering of the nicotinic ACh receptors following the interaction of agrin with MuSK. Anti-LPR4 antibodies also block ACh receptor clustering.

Treatment of Myasthenia Is Based on the Physiological Effects and Autoimmune Pathogenesis of the Disease

Anticholinesterases, especially pyridostigmine, provide some symptomatic relief but do not alter the basic disease. Immunosuppressive therapies such as corticosteroids and azathioprine or related drugs suppress antibody synthesis. Intravenous infusions of pooled immunoglobulins reduce levels of the pathogenic autoantibodies and ameliorate symptoms, often within a few days. An analogous benefit is achieved by plasmapheresis, which involves filtering the plasma. Although the benefit of these interventions is short-lived, it may be sufficient to prepare a patient for thymectomy or to support the patient through more severe episodes.

There Are Two Distinct Congenital Forms of Myasthenia Gravis

In two distinct types of myasthenia, symptoms may be present from birth or shortly thereafter. In neonatal myasthenia, the mother herself has autoimmune myasthenia that is transmitted passively to the newborn via the immune system. In congenital myasthenia, the infant has an inherited defect in some component of the neuromuscular junction, rather than an autoimmune disease, and thus does not have serum antibodies to the nicotinic ACh receptor or MuSK.

Congenital myasthenic syndromes fall into three broad groups based on the site of the defect in the neuromuscular synapse: presynaptic, synaptic cleft, and postsynaptic forms. Clinical features common to all three types include a positive family history, weakness with easy fatigability (present since infancy), drooping of the eyelids (ptosis), a decremental response to repetitive stimulation on EMG, and negative screening for anti-nicotinic ACh receptor antibodies. Subnormal development of the skeletal muscles reflects the fact that normal function at the neuromuscular synapse is required to maintain normal muscle bulk.

In one presynaptic form of congenital myasthenia, the enzyme choline acetyltransferase is absent or reduced in the distal motor terminal. This enzyme is essential for the synthesis of ACh from choline and acetyl-CoA (Chapter 16). In its absence, the synthesis of ACh is impaired. The result is weakness that usually begins in infancy or early childhood. In another presynaptic form of congenital myasthenia, the number of quanta of ACh released after an action potential is less than normal; the molecular basis for this defect is not known.

Congenital myasthenia may also result from the absence of acetylcholinesterase in the synaptic cleft. In this circumstance, end-plate potentials and miniature end-plate potentials are not small, as in autoimmune myasthenia, but are markedly prolonged, which may explain the repetitive response of the evoked muscle potential in those patients. Cytochemical studies indicate that ACh-esterase is absent from the basement membranes. At the same time, nicotinic ACh receptors are preserved.

The physiological consequence of ACh-esterase deficiency is sustained action of ACh on the end-plate and ultimately the development of an end-plate myopathy. This myopathy indicates that skeletal muscle can react adversely to excessive stimulation at the neuromuscular junction. In treating this disorder, it is critical to avoid using agents that inhibit ACh-esterase, which can increase the electrical firing at the end-plate and thereby exacerbate the muscle weakness.

The majority of congenital myasthenia cases are caused by primary mutations in the genes encoding different subunits of the ACh receptor. The *slow channel syndrome* is characterized by prominent limb weakness but little weakness of cranial muscles (the reverse of the pattern usually seen in autoimmune myasthenia, where muscles of the eyes and oropharynx are almost always affected). End-plate currents are slow to decay, and there is abnormal prolongation of channel opening. The mutations probably act both by increasing the affinity of the nicotinic ACh receptor for ACh, thereby prolonging the effects of this transmitter, and by directly slowing the channel closing rate. In some instances, quinidine is effective therapy for slow channel syndrome because it blocks the open receptor-channel. As with ACh-esterase mutations, end-plate function degenerates due to excessive postsynaptic stimulation, so anticholinesterase medications are potentially dangerous.

In the fast channel syndrome, a different set of mutations in one or more nicotinic ACh receptor subunits leads to an accelerated rate of channel closing and end-plate current decay. The fast channel syndrome may respond to either acetylcholinesterase inhibitors or 3,4-diaminopyridine. The latter blocks a presynaptic potassium conductance and thereby increases the probability of quantal release of ACh, probably by prolonging the action potential.

Lambert-Eaton Syndrome and Botulism Also Alter Neuromuscular Transmission

Some patients with cancer, especially small-cell cancer of the lung, have a syndrome of proximal limb weakness and a neuromuscular disorder with characteristics that are the opposite of those seen in myasthenia gravis. Instead of a decline in synaptic response to repetitive nerve stimulation, the amplitude of the evoked potential increases; that is, neuromuscular transmission is facilitated. Here, the first postsynaptic potential is abnormally small, but subsequent responses increase in amplitude so that the final summated potential is two to four times the amplitude of the first potential.

This disorder, *Lambert-Eaton syndrome*, is attributed to the action of antibodies against voltage-gated Ca^{2+} channels in the presynaptic terminals. It is thought that these antibodies react with the channels, degrading the channels as the antibody–antigen complex is internalized. Calcium channels similar to those of presynaptic terminals are found in cultured cells from the small-cell carcinoma of the lung; development of antibodies against these antigens in the tumor might be followed by pathogenic action against nerve terminals, another kind of molecular mimicry.

A facilitating neuromuscular block is also found in human botulism, as the botulinum toxin also impairs release of ACh from nerve terminals. Both botulism and Lambert-Eaton syndrome are ameliorated by administration of calcium gluconate or guanidine, agents that promote the release of ACh. These drugs are less effective than immunosuppressive treatments for long-term control of Lambert-Eaton syndrome, which is chronic. Botulism, on the other hand, is transient, and if the patient is kept alive during the acute phase by treating symptoms, the disorder disappears in weeks as the infection is controlled and botulinum is inactivated.

Diseases of Skeletal Muscle Can Be Inherited or Acquired

The weakness seen in any myopathy is usually attributed to degeneration of muscle fibers. At first, the missing fibers are replaced by regeneration of new fibers. Ultimately, however, renewal cannot keep pace and fibers are progressively lost. This leads to the appearance of compound motor unit potentials of brief duration and reduced amplitude. The decreased number of functioning muscle fibers then accounts for the diminished strength, whether the skeletal muscle disease is inherited or acquired.

Dermatomyositis Exemplifies Acquired Myopathy

The prototype of an acquired myopathy is dermatomyositis, defined by two clinical features: rash and myopathy. The rash has a predilection for the face, chest, and extensor surfaces of joints, including the fingers. The myopathic weakness primarily affects proximal limb muscles. The rash and weakness usually appear simultaneously and become worse in a matter of weeks. The weakness may be mild or life-threatening.

This disorder affects children or adults. About 10% of adult patients have malignant tumors. Although the pathogenesis is not known, dermatomyositis is thought to be an autoimmune disorder of small intramuscular blood vessels.

Muscular Dystrophies Are the Most Common Inherited Myopathies

The best-known inherited muscle diseases are the muscular dystrophies; several major types are distinguished by clinical and genetic patterns (Table 57–4). Some types are characterized by weakness alone (Duchenne, facioscapulohumeral, and limb-girdle dystrophies); others (eg, the myotonic muscular dystrophies) have additional clinical features. Most are recessively inherited and begin in early childhood (Duchenne, Becker, and limb-girdle dystrophy); less

frequently, the dystrophies are dominantly inherited (facioscapulohumeral or myotonic dystrophy). The cardinal trait of limb-girdle dystrophies is slowly progressive proximal weakness; in the myotonic muscular dystrophies, progressive weakness is accompanied by severe muscle stiffness.

Duchenne muscular dystrophy affects only males because it is transmitted as an X-linked recessive trait. It starts in early childhood and progresses relatively rapidly, so that patients are in wheelchairs by age 12 and usually die in their third decade. This dystrophy is caused by mutations that severely reduce levels of dystrophin, a skeletal muscle protein that apparently confers tensile strength to the muscle cell. In a related inherited muscle disorder, Becker muscular dystrophy, dystrophin is present but is either abnormal in size or reduced in quantity. Becker dystrophy is thus typically much milder, although there is considerable clinical variability according to how much dystrophin is retained; individuals with Becker dystrophy typically are able to walk well into adulthood, albeit with weakness of the proximal leg and arm muscles.

Dystrophin is encoded by the *DMD* gene, the second largest human gene, spanning about 2.5 million base pairs, or 1% of the X chromosome and 0.1% of the total human genome (Figure 57–10A). It contains at least 79 exons that encode a 14-kb mRNA. The inferred amino acid sequence of the dystrophin protein suggests a rod-like structure and a molecular weight of 427,000, with domains similar to those of two cytoskeletal proteins, alpha-actinin and spectrin. Dystrophin is localized to the inner surface of the plasma membrane. The amino terminus of dystrophin is linked to cytoskeletal actin, whereas the carboxy terminus is linked to the extracellular matrix by transmembrane proteins (Figure 57–11).

The majority of boys with Duchenne muscular dystrophy have a deletion in the *DMD* gene; about a third have point mutations. In either case, these mutations introduce premature stop codons in the mutant RNA transcripts that prevent synthesis of full-length dystrophin. Becker dystrophy is also caused by deletions and missense mutations, but the mutations do not introduce stop codons. The resulting dystrophin protein is nearly normal in length and can at least partially substitute for normal dystrophin (Figure 57–10B). Some boys with Duchenne dystrophy benefit from treatment with ASOs that cause skipping of specific mutant exons, generating a shortened but partially functional dystrophin protein (Figure 57–10C). Another promising approach is to deliver a form of the *DMD* gene to the muscle using adeno-associated virus. While the full-length *DMD* gene is too large to fit

Table 57–4 Representative Muscular Dystrophy Genes

Site of primary defect	Protein	Disease
Extracellular matrix	Collagen VI α1, α2, and α3	Bethlem myopathy
	Merosin laminin α2-subunit	Congenital myopathy
Transmembrane	α-Sarcoglycan	LGMD-2D
	β-Sarcoglycan	LGMD-2E
	χ-Sarcoglycan	LGMD-2C
	σ-Sarcoglycan	LGMD-2F
	Dysferlin	LGMD-2B, Miyoshi myopathy
	Caveolin-3	LGMD-1C, rippling muscle disease
	α7-Integrin	Congenital myopathy
	XK protein	McLeod syndrome
Submembrane	Dystrophin	Duchenne, Becker dystrophies
Sarcomere/myofibrils	Tropomyosin B	Nemaline rod myopathy
	Calpain	LGMD-2A
	Titin	Distal (Udd) dystrophy
	Nebulin	Nemaline rod myopathy
	Telethonin	LGMD-2G
	Skeletal muscle actin	Nemaline rod myopathy
	Troponin	Nemaline rod myopathy
Cytoplasm	Desmin	Desmin storage myopathy
	αβ-Crystallin	Distal myofibrillar myopathy
	Selenoprotein	Rigid spine syndrome
	Plectin	Epidermolysis bullosa simplex
Sarcoplasmic reticulum	Ryanodine receptor	Central core disease, malignant hyperthermia
	SERCA1	Brody myopathy
Nucleus	Emerin	Emery-Dreifuss dystrophy
	Lamin A/C	Emery-Dreifuss dystrophy
	Poly A binding protein, repeat	Oculopharyngeal dystrophy
Enzymes/miscellaneous	Myotonin kinase, CTG repeat	Myotonic dystrophy
	Zinc finger 9, CCTG repeat	Proximal myotonic dystrophy
	Epimerase	Inclusion body myositis
	Myotubularin	Myotubular myopathy
	Chorein	Chorea-acanthocytosis
Golgi apparatus	Fukutin	Fukuyama congenital dystrophy
	Fukutin-related peptide	Limb-girdle dystrophy
	POMT1	Congenital muscular dystrophy
	POMGnT1	Congenital muscular dystrophy

LGMD, limb-girdle muscular dystrophy.

within that virus, there is evidence that some truncated versions of dystrophin retain partial function; indeed, severely shortened dystrophins have been discovered in patients with very mild forms of Becker dystrophy. Packaging of genes encoding mini-dystrophins into the adeno-associated virus is feasible, permitting delivery to skeletal muscle and improvement of the dystrophic process (Figure 57–10C).

The discovery of the affected gene product in Duchenne muscular dystrophy by Louis Kunkel in the mid-1980s stimulated rapid discovery of numerous other novel muscle proteins, some with an intimate relationship to dystrophin. As a result, the primary genetic and protein defects underlying most major muscular dystrophies have now been identified (Figure 57–11). From these, several themes have emerged in our understanding of the biology of the muscular dystrophies.

First, and perhaps most important, is the concept that normal muscle requires a functional unit linking the contractile proteins through dystrophin to a complex of dystrophin-associated transmembrane proteins (sarcoglycans, β-dystroglycan) that, in turn, are linked to proteins at the membrane surface (eg, α-dystroglycan) and the extracellular matrix (eg, laminin). Disruption of this linked network due to a mutation in one of the proteins leads to reductions in levels of many of the proteins (Table 57–4).

Second, some of these proteins have attached sugar groups that are critical for binding the extracellular matrix proteins. Genetic defects in several of the intracellular Golgi proteins (fukutin, fukutin-related peptide, POMT1, POMTGn1) impair the deposition of the sugars (glycosylation) of the transmembrane proteins, often leading to aberrant muscle development and pronounced clinical pathology, not only in muscle but sometimes in the brain.

Third, the integrity of the extracellular matrix is essential for normal muscle function: Defects in extracellular matrix proteins (laminin α2- or α7-integrin) also cause muscular dystrophies.

Fourth, other proteins (eg, dysferlin), distinct from those complexed with dystrophin, mediate membrane repair after injury. Whereas dystrophin is important in maintaining the tensile strength and integrity of the muscle membrane, dysferlin and its binding partner caveolin-3 are central to generating rafts of vesicles that coalesce and heal breaches that occur in the muscle membrane.

It is of clinical interest that disorders due to defects in many of these proteins are less aggressive and more slowly disabling than those in Duchenne dystrophy. Defects in this diverse group of skeletal muscle proteins lead to the limb-girdle phenotype, characterized by slowly progressive proximal weakness of the arms and legs. Most are recessively inherited; mutations in both copies of a particular gene prevent expression of the normal protein product and lead to loss of function of that protein. Some limb-girdle genes are dominantly transmitted; mutations in only one copy of the gene in a pair can cause pathology. As in most primary muscle diseases, in the limb-girdle phenotype, weakness is prominent in the torso and in proximal muscles of the arms and legs. Why this pattern is so common is not known, especially since the affected proteins are expressed in both distal and proximal muscles. The pattern of degeneration most likely reflects muscle use. The proximal muscles are, on average, more subject to low-level but chronic contractile activity because they serve as antigravity muscles.

Myotonic dystrophy has several distinctive features including an autosomal inheritance pattern, weakness that is predominantly distal, involvement of nonmuscle tissues, and striking muscle stiffness (*myotonia*). The stiffness is induced by excessive electrical discharges of the muscle membrane associated with voluntary muscle contractions or percussion or electrical stimulation of the muscle. It is most intense within the first few movements after a period of rest and improves with continued muscular activity ("warm-up" phenomenon). Patients typically have difficulty relaxing the grip of a handshake for several seconds, opening the eyelids after forceful squinting, or moving their legs with the first few steps after rising from a chair. EMG demonstrates that the muscle cell membrane is electrically hyperexcitable in myotonic dystrophy; after a contraction, bursts of repetitive action potentials wax and wane in amplitude and frequency (20–100 Hz) over several seconds and thereby delay relaxation (Figure 57–12A). This sustained contraction is truly myogenic and independent of nerve supply because it persists after blockade of either the incoming motor nerve or neuromuscular transmission with agents such as curare.

The manifestations of myotonic dystrophy are not confined to muscles, however. Almost all patients have cataracts; affected men commonly have testicular atrophy and baldness and often develop cardiac conduction system defects that lead to irregularities in the heartbeat. The primary genetic defect is a dominantly transmitted expansion of a triplet of base pairs (CTG) in a noncoding region of a gene (myotonin kinase) on chromosome 19. RNA transcripts of the expanded CTG segments accumulate in the nucleus and alter splicing of several critical genes, including the ClC-1 Cl⁻ channel. Loss of function of this channel leads to

A The *DMD* gene

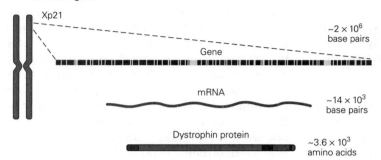

Xp21

Gene   ~2 × 10⁶ base pairs

mRNA   ~14 × 10³ base pairs

Dystrophin protein   ~3.6 × 10³ amino acids

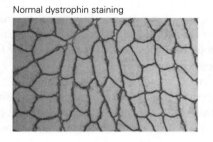

Normal dystrophin staining

B Effects of deletion

1 Deletion of single exon results in severe (Duchenne) dystrophy

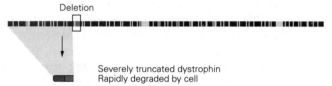

Deletion

Severely truncated dystrophin
Rapidly degraded by cell

Dystrophin staining in Duchenne dystrophy

2 Deletion of four exons results in milder (Becker) dystrophy

Deletion

Internally deleted,
semifunctional dystrophin
Allowed to persist by cell

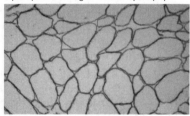

Dystrophin staining in Becker dystrophy

C Exon skipping and mini gene replacement therapies in Duchenne dystrophy

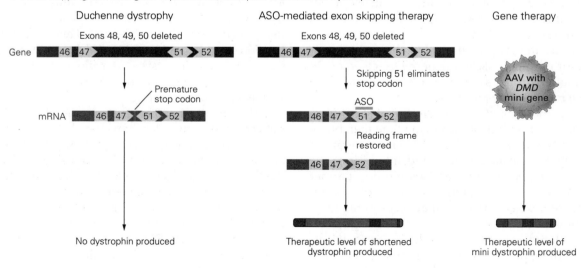

Duchenne dystrophy

Exons 48, 49, 50 deleted

Gene   46 47   51 52

Premature stop codon

mRNA   46 47 51 52

No dystrophin produced

ASO-mediated exon skipping therapy

Exons 48, 49, 50 deleted

46 47   51 52

Skipping 51 eliminates stop codon

ASO

46 47 51 52

Reading frame restored

46 47 52

Therapeutic level of shortened dystrophin produced

Gene therapy

AAV with *DMD* mini gene

Therapeutic level of mini dystrophin produced

excessive electrical activity in skeletal muscle and, as a consequence, myotonia. As discussed below, direct mutations in the same Cl⁻ channel gene can lead to a similar abnormal pattern of muscle activity.

Some Inherited Diseases of Skeletal Muscle Arise From Genetic Defects in Voltage-Gated Ion Channels

The electrical excitability of skeletal muscle is essential to the rapid and nearly synchronous contraction of an entire muscle fiber. The depolarizing end-plate potential at the neuromuscular junction triggers an action potential that propagates longitudinally along the surface of the muscle fiber and radially inward along the transverse tubules, invaginations of the fiber membrane in apposition with the sarcoplasmic reticulum (Chapter 31).

Depolarization of the transverse tubules induces a conformational change in L-type voltage-gated Ca^{2+} channels that is directly transmitted to Ca^{2+} release channels (the ryanodine receptors) in the sarcoplasmic reticulum, causing the channels to open. The release of Ca^{2+} from the sarcoplasmic reticulum raises myoplasmic Ca^{2+} and thus activates adenosine triphosphate (ATP)-dependent movement of actin-myosin filaments.

Normally, one action potential is generated in a muscle fiber for each end-plate potential. Repolarization of the muscle action potential depends on inactivation of Na^+ channels and the opening of delayed-rectifier voltage-gated K^+ channels similar to those in axons. This repolarization is also augmented by Cl^- influx through the ClC-1 Cl⁻ channels. Inherited muscle diseases arise from mutations in any one of these channels.

The electrical coupling of the end-plate potential to depolarization of the transverse tubules is disrupted in several inherited diseases of muscle. These disorders reflect a variety of defects in excitability, ranging from complete failure of action potential generation to prolonged bursts of repetitive discharges in response to a single stimulus (Figure 57–12). The derangements of muscle fiber excitability are transient and result in periodic paralysis from reduced excitability or myotonia from hyperexcitability. Between episodes, muscle function is normal. These are rare diseases of skeletal muscle, with a prevalence of 1 per 100,000 or less. Inheritance is autosomal dominant, except for one form of myotonia.

Weakness may be so severe during an attack of periodic paralysis that a patient is bedridden for hours, unable to raise an arm or leg off the bed. Fortunately, during such attacks, the muscles of respiration and swallowing are spared, so life-threatening respiratory arrest does not occur; consciousness and sensation are also spared. Attack frequency varies from almost daily to only a few in a lifetime.

During an attack, the resting potential of affected muscles is depolarized from a normal value of –90 mV

Figure 57–10 (Opposite) Two forms of muscular dystrophy are caused by deletion mutations in the dystrophin gene. (After Hoffman and Kunkel 1989.)

A. The relative position of the *DMD* gene within the Xp21 region of the X chromosome. An enlargement of this locus shows the 79 exons (**light blue lines**) and introns (**dark blue lines**) defining the gene with about 2.0×10^6 base pairs. Transcription of the gene gives rise to mRNA (about 14×10^3 base pairs), and translation of this mRNA gives rise to the protein dystrophin (molecular weight 427,000).

B. A deletion that disrupts the reading frame results in the clinically severe Duchenne muscular dystrophy, whereas a deletion that preserves the reading frame usually results in the clinically milder Becker muscular dystrophy. In both cases, the gene is transcribed into mRNA and the exons flanking the deletion are spliced together. **1.** If the borders of neighboring exons do not maintain the translational reading frame, then incorrect amino acids are inserted into the growing polypeptide chain until an abnormal stop codon is reached, causing premature termination of the protein. The truncated protein may be unstable, may fail to be localized in the membrane, or may fail to bind to glycoproteins.

Functional dystrophin is then almost totally absent. **2.** If the deletion preserves the reading frame, a dystrophin molecule is produced with an internal deletion but intact ends. Although the protein is smaller than normal and may be present in less than normal amounts, it can often suffice to preserve some muscle function.

C. One approach to correcting a deletion of the *DMD* gene is to induce formation of an mRNA transcript that skips one or more exons to restore the reading frame. For example, when there is a deletion of exons 48, 49, and 50, the splicing of exon 47 to exon 51 yields a transcript that is out of frame, in which a stop codon is introduced, preventing production of dystrophin. However, addition of an antisense oligonucleotide (**ASO**) that binds exon 51 and prevents its splicing will promote the in-frame splicing of exon 47 to exon 52. Although this transcript is slightly shorter than normal, as is the resulting dystrophin protein, the protein will nonetheless function well enough to ameliorate the muscle degeneration. Another therapeutic approach is to deliver a short form of the dystrophin gene (mini- or micro-dystrophin, ~30% of the full-length protein) to the muscle using adeno-associated virus (**AAV**); full-length dystrophin is too large to be delivered within the AAV.

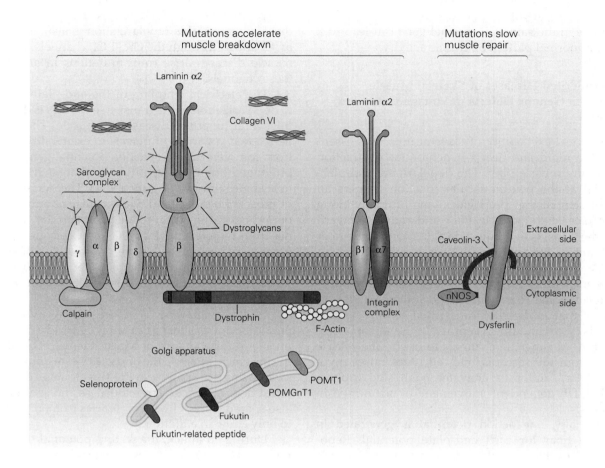

Figure 57–11 In muscular dystrophy, mutated proteins either weaken the muscle cell membrane or slow its repair after injury. For example, a deficiency of dystrophin, a submembrane protein, causes Duchenne muscular dystrophy. Dystrophin interacts with complexes of other membrane proteins that are mutated in other dystrophies, including the dystroglycans and the sarcoglycans, which are closely associated with extracellular proteins such as laminin α2 and collagen. Several other proteins mutated in different forms of muscular dystrophy are normally present in the Golgi apparatus, where they are essential for adding sugar groups to membrane proteins. These include **POMT1** (protein-O-mannosyl transferase 1), **POMGnT1** (protein-O-mannosyl α-,2-N-acetylglucosaminyl transferase), fukutin, fukutin-related peptide, and a selenoprotein. Dysferlin, which is mutated in still other dystrophies, is involved in the repair of skeletal muscle membrane after injury. (Adapted, with permission, from Brown and Mendell 2005.)

to about –60 mV. At this potential, most Na⁺ channels are inactivated, rendering the muscle fiber chronically refractory and thus unable to generate action potentials. Recovery of strength occurs spontaneously and is associated with repolarization to a resting potential within a few millivolts of normal and recovery of excitability.

Two variants of periodic paralysis have been delineated. Hyperkalemic periodic paralysis attacks occur during periods of high venous K⁺ (≥6.0 mM versus normal levels of 3.5–4.5 mM). Ingesting foods with high K⁺ content such as bananas or fruit juice may trigger an attack. Conversely, hypokalemic periodic paralysis presents as episodic weakness in association with low blood K⁺ (≤2.5 mM). Affected muscle is paradoxically depolarized in the setting of reduced extracellular K⁺, which shifts the reversal potential for K⁺ to more negative values. Both forms are inherited as autosomal dominant traits.

Hyperkalemic periodic paralysis is caused by missense mutations in a gene that encodes the pore-forming subunit of a voltage-gated Na⁺ channel

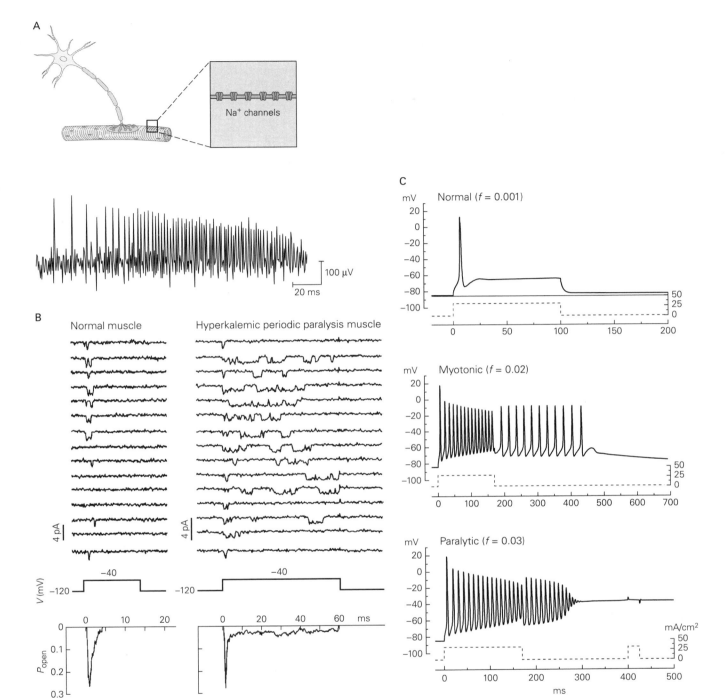

Figure 57–12 Myotonia or paralysis can result from genetically altered function in ion channels in skeletal muscle.

A. The electrical signature of myotonia (muscle stiffness) is a rapid burst of action potentials in response to a single stimulus. The action potentials, here shown in extracellular recordings, vary in amplitude and wax and wane in frequency. Such a burst may follow a voluntary muscle contraction or a mechanical stimulus, such as percussion of the muscle.

B. Cell-attached patch recordings from cultured human muscle cells. In normal muscle, the Na$^+$ channels open early and briefly in response to a 60 msec voltage-clamp depolarization from –120 mV to –40 mV. In muscle from patients with hyperkalemic periodic paralysis (defective M1592V Na$^+$ channel), the prolonged openings and reopenings indicate impaired inactivation. The probability of channel opening (obtained by averaging individual records) persists in the hyperkalemic muscle following inactivation. (Reproduced, with permission, from Cannon 1996.)

C. Even modest disruption of Na$^+$ channel inactivation is sufficient to produce bursts of myotonic discharges or depolarization-induced loss of excitability. These computer simulation records show muscle voltage in response to depolarizing current injection (**dashed line**). Among the total pool of mutant channels, a small fraction (*f*) fails to inactivate normally. In these simulations, *f* was varied from normal to values appropriate for myotonic or paralytic muscle. (Reproduced, with permission, from Cannon 1996.)

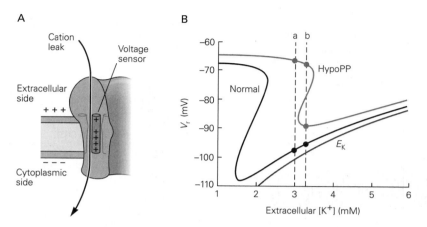

Figure 57–13 Hypokalemic periodic paralysis (HypoPP) is caused by leaky ion channels.

A. In HypoPP, missense mutations in the voltage-sensor domains create leaky Ca^{2+} or Na^+ channels that allow cation influx via an anomalous pathway separate from the channel pore.

B. Although this leak is small (~0.5% of the total resting membrane conductance), model simulations show that it causes an increased susceptibility to depolarization of resting potential (V_r), resulting in inexcitability and weakness as the external $[K^+]$ is lowered. This paradoxical depolarization of V_r diverges from the Nernst potential for K^+ (E_K) because of loss of the contribution from the inward rectifier K^+ channel in low $[K^+]$. Normally, this depolarization occurs only at extremely low $[K^+]$ (<2 mM) and is not seen in healthy people, but for patients with HypoPP, the cation leak shifts the depolarization point into the physiological range of $[K^+]$. For this simulation, in 3.3 mM $[K^+]$ (**line b**), excitability is preserved for normals ($V_r = -95.6$ mV), whereas HypoPP fibers may be excitable ($V_r = -89$ mV) or refractory and inexcitable ($V_r = -67.7$ mV). Reduction of $[K^+]$ to 3.0 mM (**line a**) results in complete loss of excitability for all HypoPP fibers (–66.3 mV) and retained excitability for normal fibers ($V_r = -97.8$ mV). (Adapted, with permission, from Cannon 2017.)

expressed in skeletal muscle. The resulting mutant Na^+ channels have inactivation defects. Subtle inactivation defects produce myotonia, whereas more pronounced defects result in chronic depolarization and loss of excitability with paralysis (Figure 57–12A–C). Hypokalemic paralysis is caused by missense mutations in the voltage-sensor domains of either Ca^{2+} channels or Na^+ channels in skeletal muscle. Disruption of the voltage-sensor domain allows an influx of ion current through an anomalous pathway, separate from the channel pore (Figure 57–13). This current "leak" in resting fibers produces a susceptibility to depolarization and loss of

excitability in low extracellular K^+. A rare form of periodic paralysis that is characterized by weakness, developmental defects, and cardiac irritability is caused by primary mutations in an inwardly rectifying K^+ channel important for the resting potential (Figure 57–13).

In myotonia congenita, muscle stiffness is present from birth and is nonprogressive. Unlike myotonic dystrophy, there is no muscle wasting, permanent muscle weakness, or other organ involvement. Congenital myotonia is a consequence of mutations in the gene coding for the ClC-1 Cl^- channel in skeletal muscle membrane (Figure 57–14). The resultant decrease

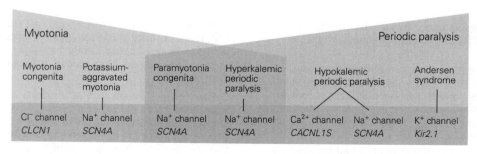

Figure 57–14 The myotonias and periodic paralyses are caused by mutations in genes that code for diverse voltage-gated ion channels in the skeletal muscle membrane. Some of these channel disorders are characterized only by myotonia, some by periodic paralysis without myotonia, and some by both myotonia and paralysis. Some clinical disorders (eg, hypokalemic periodic paralysis) can arise from defects in different channels in different individuals.

in Cl⁻ influx leads to membrane depolarization and repetitive firing. The disease is inherited as a dominant, semi-dominant, or recessive trait.

Highlights

1. Distinct disorders arise from pathology in different components of the motor unit. Pure motor diseases such as amyotrophic lateral sclerosis or spinal muscular atrophy are caused by loss of motor neurons, whereas combined motor and sensory features are present in most peripheral nerve disorders. These disorders usually spare eye movements and the eyelids.

2. Pure motor weakness, sometimes highly variable in severity over time, is also caused by disorders of the neuromuscular junction, which may begin early in life (congenital or neonatal myasthenia) or in childhood or adulthood (commonly autoimmune myasthenia gravis). The latter often involves eyelids and facial muscles.

3. Many forms of weakness are caused by mutations in genes that are important in skeletal muscle. These disorders usually become evident in infancy or childhood, involve the proximal more than the distal muscles, and progress relentlessly. Some (eg, Duchenne muscular dystrophy) also entail degeneration of cardiac muscle.

4. Inherited diseases of skeletal muscle with transient episodes of weakness (periodic paralysis) or involuntary after-contractions lasting seconds (myotonia) are caused by missense mutations in voltage-gated ion channels. During an episode of weakness, muscle fibers are depolarized and refractory from conducting action potentials. This intermittent failure to maintain the resting potential may arise from gain-of-function mutations in Na⁺ channels, loss-of-function mutations in K⁺ channels, or anomalous leakage currents in Na⁺ or Ca²⁺ channels. Myotonia is a hyperexcitable state of skeletal muscle caused by Cl⁻ channel loss-of-function or Na⁺ channel gain-of-function mutations.

5. Studies of the diseases of the peripheral nervous system show the powerful synergy between clinical and basic neuroscience. For most of the disorders inherited as Mendelian traits, molecular genetic analyses have led to the description of causative defects in muscle and nerve proteins, beginning only with the clinical data in affected families and DNA from family members.

6. Small animal models of many of these disorders, with precisely defined genetic defects, are proving invaluable for the analysis of mechanisms of disease pathogenesis and studies of new treatments. Combined with innovation in new biological therapies (gene therapy, gene silencing), these models have led to transformative successes in human trials (eg, spinal muscular atrophy).

7. In several of these disorders, a new generation of molecular therapies (eg, antisense oligonucleotides or viral-mediated gene delivery) that augment function of the mutant genes is substantially improving clinical outcomes.

Robert H. Brown
Stephen C. Cannon
Lewis P. Rowland

Selected Reading

Brown RH, Al-Chalabi A. 2017. Amyotrophic lateral sclerosis. N Engl J Med 377:162–172.

Cannon SC. 2015. Channelopathies of skeletal muscle excitability. Compr Physiol 5:761–790.

Engel AG, Shen X-M, Selcen D, et al. 2015. Congenital myasthenic syndromes: pathogenesis, diagnosis, and treatment. Lancet Neurol 14:420–434.

Fridman V, Reilly MM. 2015. Inherited neuropathies. Semin Neurol 35:407–423.

Gilhus NE, Verschuuren JJ. 2015 Myasthenia gravis: subgroup classification and therapeutic strategies. Lancet Neurol 14:1023–1036.

Ranum LP, Day JW. 2004. Pathogenic RNA repeats: an expanding role in genetic disease. Trends Genet 20:506–512.

References

Bromberg MB. 2002. Acute and chronic dysimmune polyneuropathies. In: WF Brown, CF Bolton, MJ Aminoff (eds). *Neuromuscular Function and Disease*, p. 1048, Fig. 58–2. New York: Elsevier Science.

Bromberg MB, Smith A, Gordon MD. 2002. Toward an efficient method to evaluate peripheral neuropathies. J Clin Neuromuscular Dis 3:172–182.

Brown RH Jr, Amato AA. 2002. Inherited peripheral neuropathies: classification, clinical features and review of molecular pathophysiology. In: WF Brown, CF Bolton, MJ Aminoff (eds). *Neuromuscular Function and Disease*, p. 624, Fig. 35–2. New York: Elsevier Science.

Brown RH, Mendell J. 2005. Muscular dystrophy. In: *Harrison's Principles of Internal Medicine*. New York: McGraw-Hill.

Cannon SC. 2010. Voltage-sensor mutations in channelopathies of skeletal muscle. J Physiol (Lond) 588:1887–1895.

Cannon SC. 1996. Ion channel defects and aberrant excitability in myotonia and periodic paralysis. Trends Neurosci 19:3–10.

Cannon SC. 2017. Sodium channelopathies of skeletal muscle. Handb Exp Pharm 246:309–330.

Cannon SC, Brown RH Jr, Corey DP. 1991. A sodium channel defect in hyperkalemic periodic paralysis: potassium-induced failure of inactivation. Neuron 64:619–626.

Cannon SC, Brown RH Jr, Corey DP. 1993. Theoretical reconstruction of myotonia and paralysis caused by incomplete inactivation of sodium channels. Biophys J 66:270–288.

Chamberlain JR, Chamberlain JS. 2017. Progress toward molecular therapy for Duchenne muscular dystrophy. Mol Ther 25:1125–1131.

Cull-Candy SG, Miledi R, Trautmann A. 1979. End-plate currents and acetylcholine noise at normal and myasthenic human endplates. J Physiol (Lond) 86:353–380.

Drachman DB. 1983. Myasthenia gravis: immunology of a receptor disorder. Trends Neurosci 6:446–451.

Finkel RS, Mercuri E, Darras BT, et al. 2017. Nusinersen versus sham control in infantile-onset spinal muscular atrophy. N Engl J Med 377:1723–1732.

Gilhus NE, Verschuuren JJ. 2015. Myasthenia gravis: subgroup classification and therapeutic strategies. Lancet Neurol 14:1023–1036.

Hoffman EP, Brown RH, Kunkel LM. 1987. Dystrophin: the protein product of the Duchenne muscular dystrophy locus. Cell 51:919–928.

Hoffman EP, Kunkel LM. 1989. Dystrophin in Duchenne/Becker muscular dystrophy. Neuron 2:1019–1029.

Lisak RP, Barchi RL. 1982. *Myasthenia Gravis*. Philadelphia: Saunders.

Lupski JR. 1998. Molecular genetics of peripheral neuropathies. In: JB Martin (ed). *Molecular Neurology*, pp. 239–256. New York: Scientific American.

Mendell JR, Al-Zaidy S, Shell R, et al. 2017. Single-dose gene-replacement therapy for spinal muscular atrophy. N Engl J Med 377:1713–1722.

Mendell JR, Goemans N, Lowes LP, et al. 2016. Longitudinal effect of etiplersen versus historical control on ambulation in Duchenne muscular dystrophy. Ann Neurol 79:257–271.

Milone M. 2017. Diagnosis and management of immune-mediated myopathies. Mayo Clin Proc 92:826–837.

Newsom-Davis J, Buckley C, Clover L, et al. 2003. Autoimmune disorders of neuronal potassium channels. Ann NY Acad Sci 998:202–210.

Patrick J, Lindstrom J. 1973. Autoimmune response to acetylcholine receptor. Science 180:871–872.

Rahimov F, Kunkel LM. 2013. Cellular and molecular mechanisms underlying muscular dystrophy. J Cell Biol 201:499–510.

Rosen DR, Siddique T, Patterson D, et al. 1993. Mutations in Cu/Zn superoxide dismutase gene are associated with familial amyotrophic lateral sclerosis. Nature 362:59–62.

Rowland LP, Hoefer PFA, Aranow H Jr. 1960. Myasthenic syndromes. Res Publ Assoc Res Nerv Ment Dis 38:548–600.

Taylor JP, Brown RH, Cleveland DW. 2016. Decoding ALS: from genes to mechanisms. Nature 539:197–206.

58

Seizures and Epilepsy

UNTIL QUITE RECENTLY, THE FUNCTION and organization of the human cerebral cortex—the region of the brain concerned with perceptual, motor, and cognitive functions—has eluded both clinicians and neuroscientists. In the past, the analysis of brain function relied largely on observations of loss of brain functions resulting from brain damage and cell loss caused by strokes or trauma. These natural experiments provided much of the early evidence that distinct brain regions serve specific functions, or as the famous American neurologist C. Miller Fisher said, "We learn about the brain 'stroke by stroke.'" Observation of patients with seizures and epilepsy has been equally important in the study of brain function because the behavioral consequences of these disorders of neural *hyperactivity* inform clinicians how activation affects the brain regions from which they originate.

Temporary disruptions of brain function resulting from abnormal, excessive neuronal activity are called seizures, whereas the chronic condition of repeated seizures is called epilepsy. For centuries, understanding the neurological origins of seizures was confounded by the dramatic, and sometimes bizarre, behaviors associated with seizures. The chronic condition of epilepsy was widely associated with possession by evil spirits, yet seizures also were thought to be a sign of oracular, prescient, or special creative powers.

The Greeks in the time of Hippocrates (circa 400 BC) were aware that head injuries to one side of the brain could cause seizure activity on the opposite side of the body. In those earlier times, the diagnosis of epilepsy was probably much broader than the contemporary definition. Other causes of episodic unconsciousness, such as syncope as well as mass hysteria and psychogenic seizures, were almost certainly attributed to epilepsy. Moreover, historical writings typically describe generalized convulsive seizures involving both cerebral hemispheres; thus, it is likely that seizures involving a very limited area of the brain were misdiagnosed or never diagnosed at all. Even today, it can be difficult for physicians to distinguish between episodic loss of consciousness and the various types of seizures. Nevertheless, as our ability to treat and even cure epilepsy continues to improve, these diagnostic distinctions take on increasing significance.

The early neurobiological analysis of epilepsy began with John Hughlings Jackson's work in London in the 1860s. Jackson realized that seizures need not involve loss of consciousness but could be associated with localized symptoms such as the jerking of an arm. His observation was the first formal recognition of what we now call partial (or focal) seizures. Jackson also observed patients whose seizures began with focal neurological symptoms, then progressed to convulsions with loss of consciousness by steadily involving adjacent regions in an orderly fashion (the so-called Jacksonian march). His observations gave rise to the concept of the motor homunculus (the anatomical map representing the body organization or "wiring diagram" over the cortical surface) long before functional organization was established using electrophysiological techniques (Chapter 4).

Another pioneering development that presaged modern therapy was the first surgical treatment for epilepsy in 1886 by the British neurosurgeon Victor Horsley. Horsley resected cerebral cortex adjacent to a depressed skull fracture and cured a patient with focal motor seizures. Related medical innovations include the first use of phenobarbital as an anticonvulsant in 1912 by Alfred Hauptmann, the development of electroencephalography by Hans Berger in 1929, and the discovery of the anticonvulsant properties of phenytoin (Dilantin) by Houston Merritt and Tracey Putnam in 1937. The birth of routine surgical treatment for epilepsy dates to the early 1950s, when Wilder Penfield and Herbert Jasper in Montreal stimulated the cortex and pinpointed the motor and sensory maps before removing the epileptic focus. As in any chronic disease, the physiological features of seizures are not the only consideration in the care and management of patients with epilepsy. Psychosocial factors are also extremely important. The diagnosis of epilepsy has consequences that can affect all aspects of everyday life, including educational opportunities, driving, and employment. Although many societal limitations imposed on epileptics are appropriate—most would agree that patients with epilepsy should not be commercial pilots—a diagnosis of epilepsy can result in inappropriately negative effects on educational opportunities and employment. To improve this situation, physicians have a duty to educate themselves and the public on the underlying science of epilepsy and its major comorbidities, including cognitive problems and depression.

Classification of Seizures and the Epilepsies Is Important for Pathogenesis and Treatment

Not all seizures are the same. Thus, the pathogenesis and classification of seizures must take into account their clinical characteristics as well as acquired and genetic factors in each patient. Seizures and the chronic condition of repetitive seizures (epilepsy) are common. Based on epidemiological studies in the United States, 1% to 3% of all individuals living to the age of 80 will be diagnosed with epilepsy. The highest incidence occurs in young children and the elderly.

In many respects, seizures represent a prototypic neurological disease in that the symptoms include both "positive" and "negative" sensory or motor manifestations. Examples of positive signs that can occur during a seizure include the perception of flashing lights or the jerking of an arm. Negative signs reflect impairments of normal brain function such as an impairment of consciousness and cognitive awareness or even transient blindness, speech arrest, or paralysis. These examples underscore a general feature of seizures: The signs and symptoms depend on the location and extent of brain regions that are affected. Finally, the manifestations of seizures result in part from synchronous activity triggered in surrounding tissue with normal cellular and network properties. The latter activity is particularly important in the spread of a seizure beyond its original boundaries—seizures quite literally hijack the normal functions of the brain.

Seizures Are Temporary Disruptions of Brain Function

Seizures have been classified clinically into two categories, focal or generalized, based on their onset (Table 58–1). This classification is conceptually simple, but because several terms have been used over the years to refer to the same condition, the binary nature may have been obscured. Nonetheless, this classification of seizures has proven extremely useful to clinicians, and anticonvulsant medications are targeted to one or the other type.

Table 58–1 International Classification of Seizures

Seizures

Focal onset
  Aware versus impaired awareness
  Motor versus nonmotor onset
  Focal to bilateral tonic-clonic

Generalized onset
  Motor
    Tonic-clonic (formerly grand mal)
    Other motor
  Nonmotor (absence)

Unknown onset
  Motor
    Tonic-clonic
    Other motor
  Nonmotor

Unclassified

Source: Commission on Classification and Terminology of the International League Against Epilepsy, 2017.

Focal onset (also called partial) seizures originate in a small group of neurons (the seizure focus), and thus the symptoms depend on the location of the focus within the brain. Focal onset seizures can occur either without alteration of consciousness (often called simple partial) or with alteration of consciousness (often called complex partial). A typical focal onset seizure might begin with jerking in the right hand and progress to clonic movements (ie, jerks) of the entire right arm. If a focal onset seizure progresses further, the patient may lose consciousness, fall to the ground, rigidly extend all extremities (tonic phase), then have convulsive jerking in all extremities (clonic phase).

A focal onset seizure can be preceded by telltale symptoms called *auras*. Common auras include unprovoked and often vivid sensations such as a sense of fear, a rising feeling in the abdomen, or even a specific odor. The novelist Fyodor Dostoyevsky described his auras as a "feeling ... so strong and sweet that for a few seconds of such bliss I would give ten or more years of my life, even my whole life perhaps." The aura is a product of electrical activity in the seizure focus and thus represents the earliest seizure manifestation. The time after a seizure but before the patient returns to his or her normal level of neurological function is called the post-ictal period.

Generalized onset seizures constitute the second main category. They begin without an aura or focal onset and involve both hemispheres from the onset. Thus, they are sometimes called primary generalized seizures to avoid confusion with seizures that secondarily generalize following a focal onset. Generalized onset seizures can be further divided into motor (convulsive) or nonmotor types depending on whether the seizure is associated with tonic-clonic movements.

The prototypic nonmotor generalized onset seizure is the *typical absence seizure* in children (formerly called petit mal). These seizures begin abruptly, usually last less than 10 seconds, are associated with staring and sudden cessation of all motor activity, and result in loss of awareness but not loss of posture. Patients appear as if in a trance, but the episodes are so brief that their occurrence can be missed by a casual observer. Unlike a focal onset seizure, there is no aura before the seizure or confusion after the seizure (the post-ictal period). Patients may exhibit mild motor manifestations such as eye blinking, but do not fall or have tonic-clonic movements. Typical absence seizures have very distinctive electrical characteristics on the electroencephalogram (EEG) known as a spike-and-wave pattern.

Some generalized onset seizures involve only abnormal (myoclonic, clonic, or tonic) movements or a sudden loss of motor tone (atonia). The most common motor type of generalized onset seizure is the tonic-clonic (formerly called *grand mal*) seizure. Such seizures begin abruptly, often with a grunt or cry, as tonic contraction of the diaphragm and thorax forces expiration. During the tonic phase, the patient may fall to the ground in a rigid posture with clenched jaw, lose bladder or bowel control, and become blue (cyanotic). The tonic phase typically lasts 30 seconds before evolving into clonic jerking of the extremities lasting 1 to 2 minutes. This active phase is followed by a post-ictal phase during which the patient is sleepy, disoriented, and may complain of headache and muscle soreness.

A generalized onset tonic-clonic seizure can be difficult to distinguish on purely clinical grounds from a focal seizure with a brief aura, which then rapidly progresses to a generalized tonic-clonic seizure. This distinction is not academic, as it can be vital to pinpointing the underlying cause and choosing the proper treatment. However, some seizures are simply difficult to classify because of undetermined onset.

Epilepsy Is a Chronic Condition of Recurrent Seizures

Recurrent seizures constitute the minimal criterion for the diagnosis of epilepsy. The oft-quoted clinical rule, "A single seizure does not epilepsy make," emphasizes this point, and even repeated seizures in response to a provocation such as alcohol withdrawal are not considered epilepsy. Various factors that contribute to a clinical pattern of recurrent seizures—the underlying etiology of the seizures, the age of onset, or family history—are ignored in the classification scheme for seizures in Table 58–1. The classification of the epilepsies evolved primarily based on clinical observation rather than a precise cellular, molecular, or genetic understanding of the disorder. The factors influencing seizure type and severity can often be recognized as patterns of signs and symptoms, referred to as *epilepsy syndromes.* Such factors include the age of seizure onset, whether the seizures are inherited, and certain patterns on the EEG. The recognition of these syndromes has played a role in the recent discovery of single gene mutations as a cause of seizure disorders.

The primary variables in the classification of the epilepsies are whether or not a focal brain abnormality can be identified (localization-related versus generalized epilepsies) and whether there is an identifiable cause (symptomatic) or not (unknown, often called idiopathic). The great majority of adult-onset epilepsies are classified as symptomatic localization-related epilepsies. This category includes such causes as trauma, stroke, tumors, and infections. A large number of individuals have adult-onset epilepsies without a clearly defined cause.

Unfortunately, despite the usefulness of this classification scheme, many epilepsy syndromes do not fit neatly. One expects (and hopes) that this classification will be greatly refined as criteria include the underlying etiologies rather than just clinical phenotype.

The Electroencephalogram Represents the Collective Activity of Cortical Neurons

Because neurons are excitable cells, it should not be surprising that seizures result directly or indirectly from a change in the excitability of single neurons or groups of neurons. This view dominated early experimental studies of seizures. To study such effects, electrical recordings of brain activity can be made with intracellular or extracellular electrodes. Extracellular electrodes sense action potentials in nearby neurons and can detect the synchronized activity of ensembles of cells called *field potentials.*

At the slow time resolution of extracellular recording (hundreds of milliseconds to seconds), field potentials can appear as single transient changes called spikes. These spikes reflect action potentials in many neurons and should not be confused with spikes in recordings of single neurons, which are individual action potentials that last only 1 or 2 ms. The EEG thus represents a set of field potentials as recorded by multiple electrodes on the surface of the scalp (Figure 58–1).

Because the electrical activity originates in neurons in the underlying brain tissue, the waveform recorded by the surface electrode depends on the orientation and distance of the electrical source with respect to the electrode. The EEG signal is inevitably distorted by the filtering and attenuation caused by intervening layers of tissue and bone that act in the same way as resistors and capacitors in an electric circuit. Thus, the amplitude of EEG signals (measured in microvolts) is much smaller than the voltage changes in a single neuron (millivolts). High-frequency activity in single cells, such as action potentials, is filtered out by the EEG signal, which primarily reflects slower voltage changes across the cell membrane, such as synaptic potentials.

Although the EEG signal is a measure of the extracellular current caused by the summated electrical activity of many neurons, not all cells contribute equally to the EEG. The surface EEG reflects predominately the activity of cortical neurons in close proximity to each of the set of EEG electrodes on the scalp. Thus, deep structures such as the base of a cortical gyrus, mesial walls of the major lobes, hippocampus, thalamus, or brain stem do not contribute directly to the surface EEG. The contributions of individual nerve cells to the EEG are discussed in Box 58–1.

The surface EEG shows patterns of activity—characterized by the frequency and amplitude of the electrical activity—that correlate with various stages of sleep and wakefulness (Chapter 44) and with some pathophysiological processes such as seizures. The normal human EEG shows activity over the range of 1 to 30 Hz with amplitudes in the range of 20 to 100 μV. The observed frequencies have been divided into several groups: alpha (8–13 Hz), beta (13–30 Hz), delta (0.5–4 Hz), and theta (4–7 Hz).

Alpha waves of moderate amplitude are typical of relaxed wakefulness and are most prominent over parietal and occipital sites. During intense mental activity, beta waves of lower amplitude are more prominent in frontal areas and over other regions. Alerting relaxed subjects by asking them to open their eyes results in so-called desynchronization of the EEG with a reduction in alpha activity and an increase in beta activity (Figure 58–1B). Theta and delta waves are normal during drowsiness and early slow-wave sleep; if they are present during wakefulness, it is a sign of brain dysfunction.

As neuronal ensembles become synchronized, as when a subject relaxes or becomes drowsy, the summated currents become larger and can be seen as abrupt changes from the baseline activity. Such "paroxysmal" activity can be normal, eg, the episodes of high-amplitude activity (1–2 seconds, 7–15 Hz) that occur during sleep (sleep spindles). However, a sharp

A Standard electrode placement

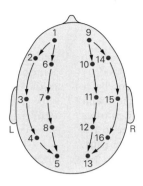

B EEG of awake human

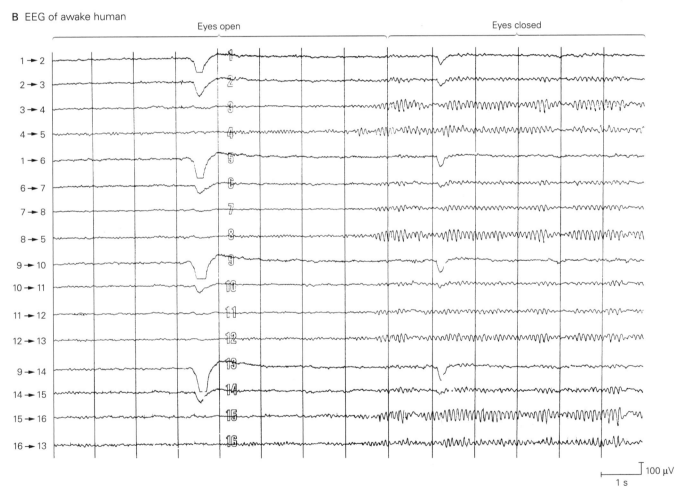

Figure 58–1 The normal electroencephalogram (EEG) in an awake human subject.

A. A standard set of placements (or montage) of electrodes on the surface of the scalp. The electrical response at each site reflects the activity between two of the electrodes.

B. At the beginning of the recording, the EEG shows low-voltage activity (~20 μV) over the surface of the scalp. The **vertical lines** are placed at 1-second intervals. During the first 8 seconds, the subject was resting quietly with eyes open, and then the subject was asked to close his eyes. With the eyes closed, larger-amplitude activity (8–10 Hz) develops over the occipital region (sites 3, 4, 8, 12, 15, and 16). This is the normal alpha rhythm characteristic of the relaxed, wakeful state. Slow large-amplitude artifacts occur at 3.5 seconds when the eyes blink and at 9 seconds when the eyes close.

Box 58–1 The Contribution of Individual Neurons to the Electroencephalogram

The contribution of the activity of single neurons to the electroencephalogram (EEG) can be understood by examining a simplified cortical circuit and some basic electrical principles. Pyramidal neurons are the major projection neurons in the cortex. The apical dendrites of these cells, which are oriented perpendicular to the cell surface, receive a variety of synaptic inputs. Thus, synaptic activity in the pyramidal cells is the principal source of EEG activity.

To understand the contribution of a single neuron to the EEG, consider the flow of charge produced by an excitatory postsynaptic potential (EPSP) on the apical dendrite of a cortical pyramidal neuron (Figure 58–2). Ionic current enters the dendrite at the site of generation of the EPSP, creating what is commonly called a current sink. It then must complete a loop by flowing down the dendrite and back out across the membrane at other sites, creating a current source.

The voltage signal created by a synaptic current is approximately predicted by Ohm's law ($V = IR$, where V is voltage, I is current, and R is resistance). Because the membrane resistance (R_m) is much larger than that of the salt solution that constitutes the extracellular medium (R_e), the voltage recorded across the membrane with an

Figure 58–2 The pattern of electrical current flow for an excitatory postsynaptic potential (EPSP) initiated at the apical dendrite of a pyramidal neuron in the cerebral cortex. Activity is detected by three electrodes: an intracellular electrode inserted in the apical dendrite (**1**), an extracellular electrode positioned near the site of the EPSP in layer II of the cortex (**2**), and an extracellular electrode near the cell body in layer V (**3**). At the site of the EPSP (current sink), positive charge flows across the cell membrane (I_{EPSP}) into the cytoplasm, down the dendritic cytoplasm, and then completes the loop by exiting through the membrane near the cell body (current source). The potentials recorded by the extracellular electrodes at the sink and at the source have opposite polarity; the potentials recorded by the intracellular electrode have the same polarity regardless of the site. R_m, R_a, and R_e are the resistances of the membrane, cytoplasm, and extracellular space, respectively.

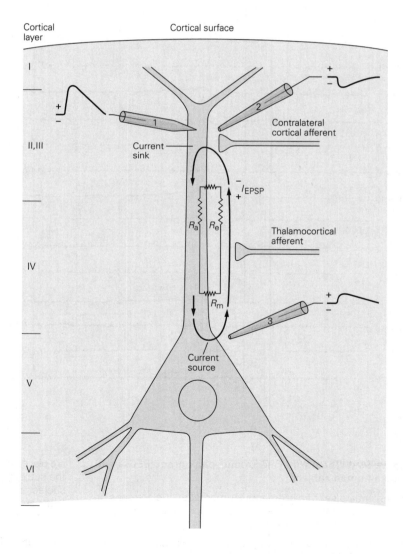

intracellular electrode (V_m) is also larger than the voltage at an extracellular electrode positioned near the current sink (V_e).

At the site of generation of an EPSP, the extracellular electrode detects the voltage change due to charge flowing away from the electrode into the cytoplasm as a negative voltage deflection. However, an extracellular electrode near the current source records a signal of opposite polarity (compare electrodes 1 and 3 in Figure 58–2). The situation is reversed if the site of the EPSP generation is on the basal segment of the apical dendrites.

In the cerebral cortex, excitatory axons from the contralateral hemisphere terminate primarily on dendrites in layers II and III, whereas thalamocortical axons terminate in layer IV (Figure 58–2). As a result, the activity measured by a surface EEG electrode will have opposite polarities for these two inputs even though the electrical event (membrane depolarization) is the same.

Similarly, the origin or polarity of cortical synaptic events cannot be unambiguously determined from surface EEG recordings alone. EPSPs in superficial layers and inhibitory postsynaptic potentials (IPSPs) in deeper layers both appear as upward (negative) potentials, whereas EPSPs in deeper layers and IPSPs in superficial layers have downward (positive) potentials (Figure 58–3).

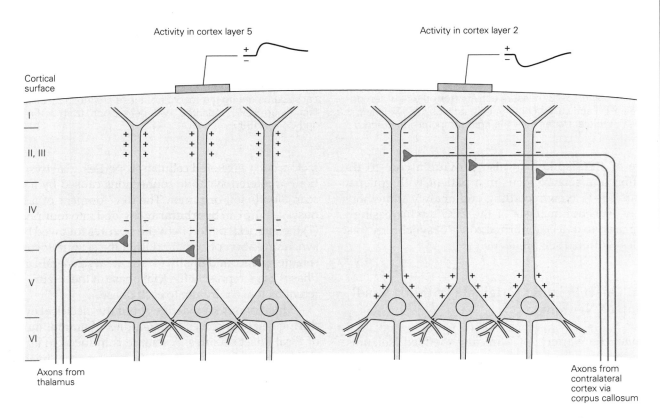

Figure 58–3 Surface electroencephalogram (EEG) recordings do not unambiguously indicate the polarity of synaptic events. The polarity of the surface EEG depends on the location of the synaptic activity within the cortex. A thalamocortical excitatory signal in layer V causes an upward voltage deflection at the surface EEG electrode because the electrode is nearer the current source. In contrast, an excitatory signal from the contralateral hemisphere in layer II causes a downward deflection because the electrode is nearer the sink.

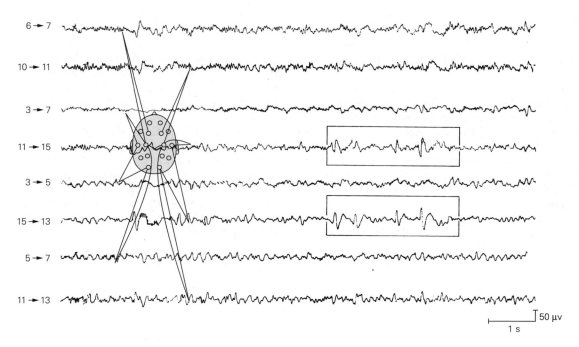

Figure 58–4 The electroencephalogram (EEG) can provide clues to the location of a seizure focus. Each trace represents the electrical activity between pairs of scalp electrodes as indicated in the electrode map. For example, electrode pairs 11–15 and 15–13 measure activity from the right temporal area. EEG activity in a patient with epilepsy shows sharp waves in the electrodes over the right temporal area (record enclosed in boxes). Such paroxysmal activity arises suddenly and disrupts the normal background EEG pattern. The focal abnormality may indicate that the seizure focus in this patient is in the right temporal lobe. Because the patient had no clinical seizures during the recording, these are interictal spikes (see Figure 58-7). (Adapted, with permission, from Lothman and Collins 1990.)

wave or EEG spike can also provide a clue to the location of a seizure focus in a patient with epilepsy (Figure 58–4). New recording and analytical methods such as spectral analysis of the EEG are increasingly being used to detect abnormal zones of synchrony (fast ripples) within a seizure focus.

Focal Onset Seizures Originate Within a Small Group of Neurons

Despite the variety of clinically defined seizures, important insights into the generation of seizure activity can largely be understood by comparing the electrographic patterns of focal onset seizures with those of generalized onset seizures.

The defining feature of focal onset seizures is that the abnormal electrical activity originates from a *seizure focus*. The seizure focus is considered to be nothing more than a small group of neurons, perhaps 1,000 or so, that have enhanced excitability and the ability to occasionally spread that activity to neighboring regions and thereby cause a seizure. The enhanced excitability (epileptiform activity) may result from many different

factors such as altered cellular properties, glial dysfunction, or altered synaptic connections caused by a local scar, blood clot, or tumor. The development of a focal onset seizure can be arbitrarily divided into four phases: (1) the interictal period between seizures followed by (2) synchronization of activity within the seizure focus, (3) seizure spread, and finally, (4) secondary generalization. Phases 2 to 4 represent the ictal phase of the seizure. Different factors contribute to each phase.

Much of our knowledge about the electrical events during seizures comes from studies of animal models of focal onset seizures. A seizure is induced in an animal by focal electrical stimulation or by acute injection of a convulsant agent. This approach along with in vitro studies of tissue from these animal models has provided a good understanding of electrical events within the focus during a seizure as well as during the onset of the interictal period.

Neurons in a Seizure Focus Have Abnormal Bursting Activity

How does electrical activity in a single neuron or group of neurons lead to a focal onset seizure? Each neuron within a seizure focus has a stereotypic and synchronized electrical response, the paroxysmal depolarizing shift, a depolarization that is sudden, large (20–40 mV), and long-lasting (50–200 ms), and that triggers a train of action potentials at its peak. The paroxysmal depolarizing shift is followed by an afterhyperpolarization (Figure 58–5A).

The paroxysmal depolarizing shift and afterhyperpolarization are shaped by the intrinsic membrane properties of the neuron (eg, voltage-gated Na^+, K^+, and Ca^{2+} channels) and by synaptic inputs from excitatory and inhibitory neurons (primarily glutamatergic and GABAergic, respectively). The depolarizing phase results primarily from activation of α-amino-3-hydroxy-5-methyl-4-isoxazolepropionic acid (AMPA)- and N-methyl-D-aspartate (NMDA)-type glutamate receptor-channels (Figure 58–5A), as well as voltage-gated Na^+ and Ca^{2+} channels. NMDA-type receptor-channels are particularly effective in enhancing excitability because depolarization relieves Mg^{2+} blockage of the channel. Removal of the blockage increases current through the channel, thus enhancing the depolarization and allowing additional Ca^{2+} to enter the neuron (Chapter 13).

The normal response of a cortical pyramidal neuron to excitatory input consists of an excitatory postsynaptic potential (EPSP) followed by an inhibitory postsynaptic potential (IPSP) (Figure 58–5B). Thus,

the paroxysmal depolarizing shift can be viewed as a massive enhancement of these depolarizing and hyperpolarizing synaptic components. The afterhyperpolarization is generated by voltage-dependent and Ca^{2+}-dependent K^+ channels as well as by a γ-aminobutyric acid (GABA)-mediated Cl^- conductance (ionotropic $GABA_A$ receptors) and K^+ conductance (metabotropic $GABA_B$ receptors) (Figure 58–5A). The Ca^{2+} influx through voltage-dependent Ca^{2+} channels and NMDA-type receptor-channels triggers the opening of calcium-activated channels, particularly K^+ channels. The afterhyperpolarization limits the duration of the paroxysmal depolarizing shift, and its gradual disappearance is the most important factor in the onset of a focal onset seizure, as discussed later.

Thus, it is not surprising that many convulsants act by enhancing excitation or blocking inhibition. Conversely, anticonvulsants can act by blocking excitation or enhancing inhibition. For example, the benzodiazepines diazepam (Valium) and lorazepam (Ativan) enhance $GABA_A$-mediated inhibition and are used in the emergency treatment of prolonged repetitive seizures. The anticonvulsants phenytoin (Dilantin) and carbamazepine (Tegretol) and several others reduce the opening of voltage-gated Na^+ channels that underlie the action potential. Molecular models of the Na^+ channel indicate that these drugs are more effective when the channel is in the open or activated state.

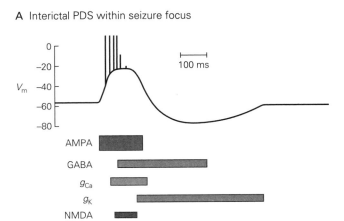

A   Interictal PDS within seizure focus

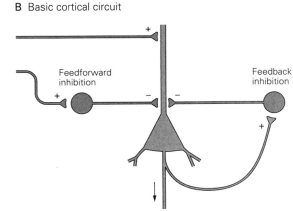

B   Basic cortical circuit

Figure 58–5 **The conductances that underlie the paroxysmal depolarizing shift of a neuron in a seizure focus.**

A. The paroxysmal depolarizing shift (**PDS**) is largely dependent on α-amino-3-hydroxy-5-methyl-4-isoxazolepropionic acid (**AMPA**)- and N-methyl-D-aspartate (**NMDA**)-type receptor-channels whose effectiveness is enhanced by the opening of voltage-gated Ca^{2+} channels (g_{Ca}). Following the depolarization, the cell is hyperpolarized by activation of γ-aminobutyric acid (**GABA**) receptors (both ionotropic $GABA_A$

and metabotropic $GABA_B$) as well as by voltage-gated and calcium-activated K^+ channels (g_K). (Adapted, with permission, from Lothman 1993a.)

B. Recurrent axon branches activate inhibitory neurons and cause feedback inhibition of the pyramidal neuron. Extrinsic excitatory inputs can also activate feedforward inhibition. The PDS represents exaggerated excitation in a seizure focus, whereas the inhibitory circuitry forms the basis of surround inhibition, important in restricting interictal activity to the seizure focus.

Thus, fittingly, the ability of these drugs to block Na⁺ channels is enhanced by repetitive activity associated with seizures; that is, the greatest effect is in those neurons that need to be silenced the most.

The Breakdown of Surround Inhibition Leads to Synchronization

As long as the abnormal electrical activity is restricted to a small group of neurons, there are no clinical manifestations. The synchronization of neurons in a seizure focus is dependent not only on the intrinsic properties of each individual cell but also on the number and strength of connections between neurons. During the interictal period, the abnormal activity is confined to the seizure focus by inhibition of the surrounding tissue.

This "inhibitory surround," initially described by David Prince, is particularly dependent on feedforward and feedback inhibition by GABAergic inhibitory interneurons (Figure 58–6A). Although inhibitory circuits in the cerebral cortex are often represented by simple diagrams (Figure 58–6B), the morphology and connectivity of cortical inhibitory neurons are actually quite complex and a topic of continuing investigation with many new methods such as cell type–specific viral labeling and optogenetic stimulation.

During the development of a focal seizure, the excitation in the circuit overcomes the inhibitory

Figure 58–6 The spatial and temporal organization of a seizure focus depends on the interplay between excitation and inhibition of neurons in the focus.

A. The pyramidal cell *a* shows the typical electrical properties of neurons in a seizure focus (see part **B**). Excitation in cell *a* activates another pyramidal cell (*b*), and when many such cells fire synchronously, a spike is recorded on the electroencephalogram. However, cell *a* also activates γ-aminobutyric acid (GABA)-ergic inhibitory interneurons (**gray**). These interneurons can reduce the activity of cells *a* and *b* through feedback inhibition, thus limiting the seizure focus temporally, as well as prevent the firing of cells outside the focus, represented here by cell *c*. This latter phenomenon creates an inhibitory surround that acts to contain the hyperexcitability to the seizure focus during interictal periods. When extrinsic or intrinsic factors alter this balance of excitation and inhibition, the inhibitory surround begins to break down and the seizure activity spreads, leading to seizure generation. (Adapted, with permission, from Lothman and Collins 1990.)

B. The synaptic connections and activity patterns for cells *a*, *b*, and *c* shown in part **A**. Cells *a* and *b* (within the seizure focus) undergo a paroxysmal depolarizing shift, whereas cell *c* (in the inhibitory surround) is hyperpolarized due to input from GABAergic inhibitory interneurons.

A

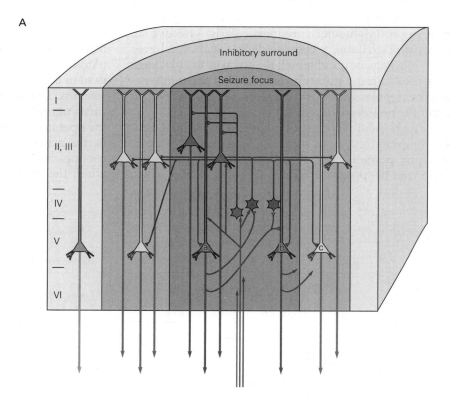

B

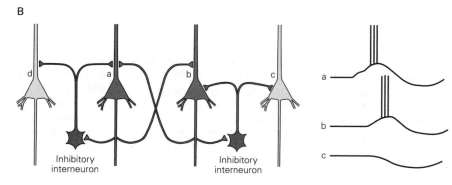

Figure 58–7 A focal onset seizure begins with the loss of the afterhyperpolarization and surround inhibition. (Adapted, with permission, from Lothman 1993a.)

A. With the onset of a seizure (**arrow**), neurons in the seizure focus depolarize as in the first phase of a paroxysmal depolarizing shift. However, unlike the interictal period, the depolarization persists for seconds or minutes. The γ-aminobutyric acid (**GABA**)-mediated inhibition fails, whereas excitatory activity in the α-amino-3-hydroxy-5-methyl-4-isoxazolepropionic acid (**AMPA**)- and *N*-methyl-D-aspartate (**NMDA**)-type glutamate receptors is functionally enhanced. This activity corresponds to the tonic phase of a secondarily generalized tonic-clonic seizure. As the GABA-mediated inhibition gradually returns, the neurons in the seizure focus enter a period of oscillation corresponding to the clonic phase.

B. As the surround inhibition breaks down, neurons in the seizure focus become synchronously excited and send trains of action potentials to distant neurons, thus spreading the abnormal activity from the focus. Compare this pattern of activity in cells *a* to *c* with that during the interictal period (Figure 58–6B).

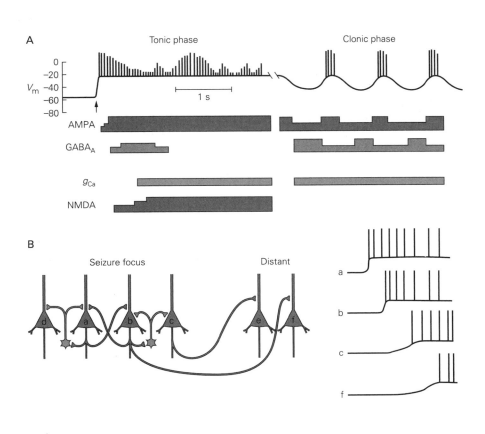

surround, and the afterhyperpolarization in the neurons of the original focus gradually disappears. As a result, a nearly continuous high-frequency train of action potentials is generated, and the seizure begins to spread beyond the original focus (Figure 58–7).

An important factor in the spread of focal onset seizures appears to be that the intense firing of the pyramidal neurons results in a relative decrease in synaptic transmission from the inhibitory GABAergic interneurons, although the interneurons remain viable. Whether this decrease results from a presynaptic change in the release of GABA or a postsynaptic change in GABA receptors is still not understood and may not be the same in all cases. Other factors that may contribute to the loss of the inhibitory surround over time include changes in dendritic morphology, the density of receptors or channels, or a depolarizing shift in E_K caused by extracellular K^+ ion accumulation. Prolonged firing also transmits action potentials to distant sites in the brain, which in turn may trigger trains of action potentials in neurons that project back to neurons in the seizure focus (backpropagation). Reciprocal connections

between the neocortex and thalamus may be particularly important in this regard.

Despite our understanding of such mechanisms, we still do not know what causes a seizure to occur at any particular moment. The inability to predict when a seizure will occur is perhaps the most debilitating aspect of epilepsy. New approaches to this dilemma are discussed in Box 58–2. Some patients learn to recognize the triggers most critical for them, such as sleep deprivation or stress, and thus adjust their lifestyle to avoid these circumstances. But in many individuals, seizures do not follow a predictable pattern.

In a few patients, sensory stimuli such as flashing lights can trigger seizures, suggesting that repeated excitation of some circuits causes a change in excitability. For example, NMDA-type glutamate receptor activity and GABAergic inhibition can undergo changes dependent on the frequency of firing of the presynaptic neuron. This provides one possible molecular mechanism for such changes in network excitability. On a longer time scale, circadian rhythms and hormonal patterns may also influence the likelihood

Box 58–2 New Approaches to Real-Time Seizure Detection and Prevention

Perhaps the most disabling aspect of seizures and epilepsy is the uncertainty of it all—when will the next seizure occur? As you can imagine, this impacts employment, driving, and recreation and often prevents the development of an individual's full potential. Patients with epilepsy sometimes have a brief warning or aura, but rarely do they have enough time to institute a therapeutic intervention such as a pill or an injection in order to abort the seizure.

Clinicians and epilepsy researchers have long recognized the importance of real-time seizure *detection* and real-time seizure *prevention* as a goal of therapy. Of course, acute detection must precede an acute treatment. However, in general, this approach has only been possible in patients undergoing EEG monitoring either with surface EEG electrodes or implanted electrodes. Several technologies are now emerging that allow new hope for detection, and thus enable efforts to abort or prevent an imminent seizure. Most are still in the experimental phase in animal models, but a few have reached clinical trials and even, in the case of vagal nerve stimulators, clinical practice. Seizure prevention can be imagined in two general ways: either altering the excitability of large regions of brain or somehow interrupting activity within a seizure focus. These two approaches can also be considered in engineering terms as open-loop or closed-loop strategies, respectively.

The first approach led to the development in 1997 of the vagal nerve stimulator, implanted in the neck and powered by a pacemaker-style battery (Figure 58–8). The

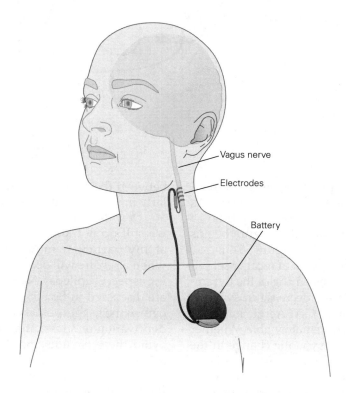

Vagus nerve

Electrodes

Battery

Figure 58–8 Vagal nerve stimulation. Schematic of placement of electrodes on the left vagus nerve powered by a battery implanted subcutaneously in the chest wall. The stimulation can be programmed at regular intervals (eg, every 30 seconds) and can also be activated on demand by placing a magnet over the chest. (Adapted from Stacey and Litt 2008.)

resulting chronic, intermittent stimulation of the vagus nerve has been effective in reducing seizure frequency in some patients. The patient can also activate the stimulator with a hand-held magnet during an aura to see if acute stimulation can prevent a seizure. The exact mechanism of seizure reduction by vagal nerve stimulation is still unclear, but presumably involves activation of the autonomic nervous system, and thus, this form of stimulation has limited specificity to particular brain regions.

Because many patients with intractable epilepsy have seizures that originate from one or more discrete foci in the brain, it would obviously be ideal to be able to detect abnormal activity within a seizure focus, and thus through some sort of feedback mechanism deliver a stimulus that would abort the spread of epileptiform activity from that focus. This goal has been an active area of investigation over the past decade, leading to a clinical trial the results of which were recently published. The device tested was a chronically implanted neurostimulator (RNS System, Neuropace) that directly stimulates the seizure focus when epileptiform activity is detected (Figure 58–9).

In this multicenter double-blinded trial, the device was implanted in patients with intractable focal onset seizures with one or two seizure foci. The patients were monitored for an average of 5 years. The device can be programmed by the clinician to match characteristics for each patient. The patients were randomized into two groups, responsive stimulation or sham stimulation groups, for the first 5 months, and then followed for up to 2 years. There was a 44% reduction in seizure frequency after 1 year and a 53% reduction after 2 years, suggestive of a progressive effect. The device was generally well tolerated. Thus, this approach has therapeutic potential for some patients and provides proof-of-concept evidence for closed-loop seizure detection and stimulation.

The RNS System uses electrical stimulation, but other strategies being studied in animals promise to refine methods of seizure prevention. These include neuronal stimulation or silencing using viral-mediated delivery of opto- or chemogenetic probes. In general, a replication-defective virus can be targeted to a specific cell type within a brain region. In the optogenetic approach, the virus is engineered to express ion channels or pumps that reduce neuron excitability when

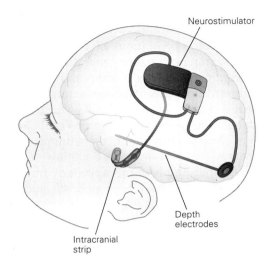

Figure 58–9 Closed-loop seizure detection and prevention. This schematic diagram of the closed-loop RNS System shows the intracranial strip and depth electrodes that detect seizure activity and subsequently deliver programmed stimulation to the seizure focus. (Adapted, with permission, from Heck et al. 2014.)

exposed to light. In the chemogenetic approach, a chemical is delivered systemically. This strategy has now been successfully employed in animal models of epilepsy.

The optogenetic strategy is similar to the neurostimulator except that stimulation is delivered through a fiber optic light guide implanted near the seizure focus. The advantage of this approach is that the virus is engineered to deliver stimulation to a specific population of neurons. The chemogenetic approach has the advantage of noninvasive delivery of the chemical, but lacks the speed that can be achieved with optical or electrical stimulation. Even when optimized and tested in clinical trials, these invasive approaches are likely to be useful only in a subset of focal onset epilepsies that have stable and well-defined seizure foci. Thus, continuing and complementary efforts to understand the genetic mechanisms of epileptogenesis, as well as new technologies such as stem cell therapies, remain essential.

of seizures, as demonstrated by patients who have seizures only while sleeping (nocturnal epilepsy) or during their menstrual period (catamenial epilepsy). If we could develop continuous monitoring methods to predict the timing of seizure generation (Box 58–2), acute intervention to deliver a drug or change neural activity patterns to prevent seizures might become a therapeutic option. However, EEG studies reveal great variability between patients in pre-ictal patterns. Continuous chronic stimulation of neural circuits is another method of modifying the excitability of epileptic circuits. As an example of this approach, implanted vagal nerve stimulators have been modestly successful in treating pharmaco-resistant epilepsy that does not respond to other treatments.

The Spread of Seizure Activity Involves Normal Cortical Circuitry

If activity in the seizure focus is sufficiently intense, the electrical activity begins to spread to other brain regions. Spread of seizure activity from a focus generally follows the same axonal pathways as does normal cortical activity. Thus, thalamocortical, subcortical, and transcallosal pathways can all become involved in seizure spread. Seizure activity can propagate from a seizure focus to other areas of the same hemisphere or across the corpus callosum to involve the contralateral hemisphere (Figure 58–10). Once both hemispheres become involved, a focal onset seizure has become secondarily generalized. At this point, the patient generally experiences loss of consciousness. The spread of a partial seizure usually occurs rapidly over a few seconds, but can also evolve over many minutes. Rapid generalization is more likely if a focal onset seizure begins in the neocortex than if it begins in the limbic system (in particular, the hippocampus and amygdala).

An interesting unanswered question is what terminates a seizure. Remarkably, few mechanisms for the self-limiting return to the interictal state have been defined with certainty. One definite conclusion at this point is that termination is not due to cellular metabolic exhaustion, because under severe conditions clinical seizures may continue for hours (see below). During the initial 30 seconds or so of a focal onset seizure that secondarily generalizes, neurons in the involved areas undergo prolonged depolarization and fire continuously (due to loss of the afterhyperpolarization that normally follows a paroxysmal depolarizing shift). As the seizure evolves, the neurons begin to repolarize and the afterhyperpolarization reappears. The cycles of depolarization and repolarization correspond to the clonic phase of the seizure (Figure 58–7A).

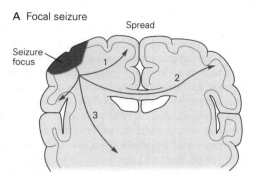

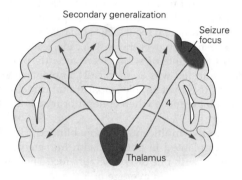

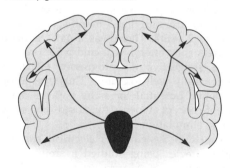

Figure 58–10 Focal and generalized onset seizures propagate via several pathways. (Adapted, with permission, from Lothman 1993b.)

A. Focal onset seizures can spread locally from a focus via intrahemispheric fibers (**1**) and more remotely to homotopic contralateral cortex (**2**) and subcortical centers (**3**). The secondary generalization of a focal onset seizure spreads to subcortical centers via projections to the thalamus (**4**). Widespread thalamocortical interconnections then contribute to rapid activation of both hemispheres.

B. In a generalized onset seizure, such as a typical absence seizure, interconnections between the thalamus and cortex are a major route of seizure propagation.

The seizure is often followed by a period of decreased electrical activity, the postictal period, which may be accompanied by symptoms of confusion, drowsiness, or even focal neurological deficits such as a hemiparesis (Todd paralysis). A neurological exam in the postictal period can lead to insights about the locus of the seizure focus when there is prolonged depression of one brain region or function, once other brain regions have regained normal function.

Generalized Onset Seizures Are Driven by Thalamocortical Circuits

Unlike the typical focal onset seizure, a generalized onset seizure abruptly disrupts normal brain activity in both cerebral hemispheres simultaneously. Generalized onset seizures and their associated epilepsies vary both in their manifestations and etiologies. Although the cellular mechanisms of generalized onset seizures differ in a number of interesting respects from those of focal onset or secondarily generalized seizures, a generalized onset seizure can be difficult to distinguish clinically or by EEG from a focal onset seizure that rapidly generalizes.

The most studied type of generalized onset seizure is the typical absence seizure (petit mal), whose characteristic EEG pattern (the 3-Hz spike-and-wave pattern in Figure 58–11A) was first recognized by Hans Berger in 1933. F. A. Gibbs recognized the relationship of this EEG pattern to typical absence seizures (he aptly described the pattern as "dart and dome") and attributed the mechanism to generalized cortical disturbance. The distinctive clinical features of typical absence seizures have a clear correlation with the EEG activity.

The typical absence seizure begins suddenly, lasts 10 to 30 seconds, and produces impaired awareness with only minor motor manifestations such as blinking or lip smacking. Unlike a focal onset seizure that secondarily generalizes, generalized onset seizures are not preceded by an aura or followed by postictal symptoms. The spike-wave EEG pattern can be seen in all cerebral areas abruptly and simultaneously and is immediately preceded and followed by normal background activity. Very brief (1–5 seconds) runs of 3-Hz EEG activity without apparent clinical symptoms are common in patients with absence seizures, but if frequent, they can affect their ability to carry out normal activities such as school performance.

In contrast to Gibbs's hypothesis of diffuse cortical hyperexcitability, Penfield and Jasper noted that the EEG in typical absence seizures is similar to rhythmic EEG activity in sleep, so-called sleep spindles (Chapter 44). They proposed a "centrencephalic" hypothesis in which

generalization was attributed to rhythmic activity (pacing) by neuronal aggregates in the upper brain stem or thalamus that project diffusely to the cortex.

Research on animal models of generalized onset seizures and studies of the genetics of generalized epilepsy suggest that elements of both hypotheses are correct. In cats, parenteral injections of penicillin, a weak $GABA_A$ antagonist, produce behavioral unresponsiveness associated with an EEG pattern of bilateral synchronous slow waves (generalized penicillin epilepsy). During such a seizure, thalamic and cortical cells become synchronized through the same reciprocal thalamocortical connections that contribute to normal sleep spindles during slow-wave sleep.

Such seizures could in theory represent a form of diffuse hyperexcitability in the cortex. Recordings from individual cortical neurons show an increase in the rate of firing during a depolarizing burst that in turn produces a powerful GABAergic inhibitory feedback that hyperpolarizes the cell for approximately 200 ms after each burst (Figure 58–11C). This depolarization followed by inhibition differs fundamentally from the paroxysmal depolarizing shift in focal onset seizures in that GABAergic inhibition is preserved. In the typical absence seizure, the summated activity of the bursts produces the spike while the summated inhibition produces the wave of the spike-wave EEG pattern.

What are the properties of cells and networks that facilitate this generalized and synchronous activity? An early clue came from studies of the intrinsic bursting of thalamic relay neurons. Henrik Jahnsen and Rodolfo Llinas found that these neurons robustly express the T-type voltage-gated Ca^{2+} channel that is inactivated at the resting membrane potential but becomes available for activation when the cell is hyperpolarized (Chapter 10). A subsequent depolarization then transiently opens the Ca^{2+} channel (thus its name, T-type), and the Ca^{2+} influx generates low-threshold Ca^{2+} spikes. Consistent with the hypothesis that T-type channels contribute to absence seizures, certain anticonvulsant agents that block absence seizures, such as ethosuximide (Zarontin) and valproic acid (Depakote), also block T-type channels. T-type channels are encoded by three related genes (*Cav3.1–Cav3.3*), with *Cav3.1* the predominant type in the thalamus.

The circuitry of the thalamus seems ideally suited to the generation of generalized onset seizures. The pattern of thalamic neuron activity during sleep spindles suggests a reciprocal interaction between thalamic relay neurons and GABAergic interneurons in the thalamic reticular nucleus and perigeniculate nucleus (Figure 58–11B). Studies of thalamic brain slices by David McCormick and his colleagues indicate that

A Spike and wave activity in typical absence seizure

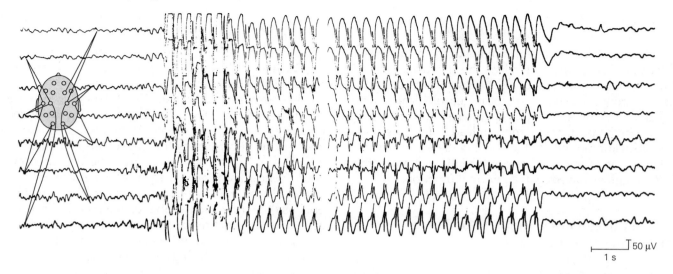

$$\begin{array}{c} \underset{1\ s}{\vdash\!\!\!\dashv} \quad \overline{\top}\,50\ \mu V \end{array}$$

B Thalamocortical projections

C Synchrony of neuronal activity in primary generalized (spike-wave) seizure

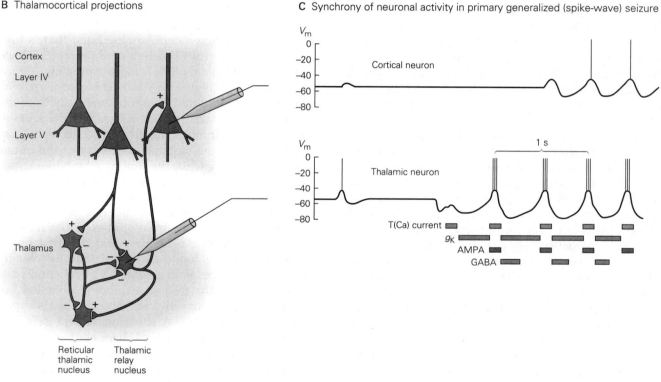

Figure 58–11 Generalized onset seizures have distinctive electroencephalogram (EEG) and single-neuron patterns.

A. This EEG from a 12-year-old patient with typical absence (petit mal) seizures shows the sudden onset of synchronous spikes at a frequency of 3 per second and wave activity lasting approximately 14 seconds. The seizure clinically manifested as a staring spell with occasional eye blinks. Unlike a focal onset seizure, there is no buildup of activity preceding the seizure and the electrical activity returns abruptly to the normal background level following the seizure. The discontinuity in the trace is due to removal of a 3-second period of recording. (Reproduced, with permission, from Lothman and Collins 1990.)

B. Thalamocortical connections that participate in the generation of sleep spindles (Chapter 44) are thought to be essential for the generation of generalized onset seizures. Pyramidal cells in the cortex are reciprocally connected by excitatory synapses with thalamic relay neurons. GABAergic inhibitory interneurons in the reticular thalamic nucleus are excited by pyramidal cells in the cortex and by thalamic relay neurons and inhibit the thalamic relay cells. The interneurons are also reciprocally connected.

C. Neuronal activity of cortical and thalamic neurons becomes synchronized during a generalized onset seizure. The depolarization is dependent on conductances in α-amino-3-hydroxy-5-methyl-4-isoxazolepropionic acid (**AMPA**)-type glutamate receptor-channels and T-type voltage-gated Ca^{2+} channels. The repolarization is due to γ-aminobutyric acid (**GABA**)-mediated inhibition as well as voltage- and calcium-dependent K^+ conductances (g_K). (Adapted, with permission, from Lothman 1993a.)

the interneurons hyperpolarize the relay neurons, thus removing the inactivation of T-type Ca^{2+} channels. This action leads to an oscillatory response: A rebound burst of action potentials following each IPSP to which the T-type Ca^{2+} channels contribute stimulates the GABAergic interneurons, resulting in another round of relay neuron rebound firing. The relay neurons also excite cortical neurons, manifested in the EEG by a "spindle." Both the T-type Ca^{2+} channel and the $GABA_B$ receptor-channel play an important role in the generation of this activity, which resembles human absence seizures (Chapter 44).

Mutations in voltage-gated Ca^{2+} channels have produced several mouse models of generalized epilepsy, including the so-called *totterer* mouse, which bears a mutation in the P/Q-type calcium channels involved in neurotransmitter release. Studies of these mutants by Jeffrey Noebels and his colleagues have revealed that the animals develop generalized onset seizures when they reach adolescence. EEGs in these animals show a paroxysmal spike-wave discharge and seizures that are characterized by an arrest of behavior and blockade by ethosuximide, similar to typical absence seizures in children. Thalamic neurons in these mice have elevated T-type Ca^{2+} channels that favor rebound bursting. Mutations of over 20 different genes for this phenotype have now been described in mice. Remarkably, many encode ion channel subunits or proteins involved in presynaptic transmitter release.

Locating the Seizure Focus Is Critical to the Surgical Treatment of Epilepsy

The pioneering studies of Wilder Penfield in Montreal in the early 1950s led to the recognition that removal of the temporal lobe in certain patients with focal onset seizures of hippocampal origin could reduce the number of seizures or even cure epilepsy. As surgical treatment for such patients became more common, it became clear that the surgical outcome is directly related to the adequacy of the resection. Thus, precise localization of the seizure focus in cases of focal onset seizures is essential. Electrical mapping of seizure foci originally relied on the surface EEG, which we have seen is biased toward particular sets of neurons in the cortex immediately adjacent to the skull. However, seizures intractable to conventional medical management often begin in deep structures that show little or no abnormality on the surface EEG at the onset of the seizure. Thus, the surface EEG is somewhat limited in identifying the location of the seizure focus.

The development of magnetic resonance imaging (MRI) markedly improved the noninvasive anatomical mapping of seizure foci. This technique is now routine in the evaluation of epilepsies involving the temporal lobe, but also shows increasing promise for identifying seizure foci in other locations. The scientific basis of anatomical mapping of seizure foci by MRI was the observation that a majority of patients with intractable focal onset seizures with impaired awareness have atrophy and cell loss in the mesial portions of the hippocampal formation. There is a dramatic loss of neurons within the hippocampus (mesial temporal sclerosis), changes in dendritic morphology of surviving cells, and collateral sprouting of some axons. The anatomical resolution of modern MRI machines has allowed a noninvasive, quantitative assessment of the size of the hippocampus in epilepsy patients. Loss of volume of the hippocampus on one or another side of the brain generally correlates well with the localization of seizure foci in the hippocampus as determined by functional criteria using implanted depth electrodes.

The typical patient with mesial temporal epilepsy has unilateral disease, which leads to shrinkage of the hippocampus on one side that can be associated with apparent dilatation of the temporal horn of the lateral ventricle. Such a case is illustrated in Box 58–3. However, in many patients, abnormalities cannot be detected using anatomical MRI; thus, nonanatomical (functional) imaging techniques (fMRI) are used as well (Chapter 6).

Functional neuroimaging takes advantage of the changes in cerebral metabolism and blood flow that occur in the seizure focus during the ictal and interictal periods. The electrical activity associated with a seizure places a large metabolic demand on brain tissue. During a focal onset seizure, there is an approximately three-fold increase in glucose and oxygen utilization. Between seizures, the seizure focus often shows decreased metabolism. Despite the increased metabolic demands, the brain is able to maintain normal adenosine triphosphate (ATP) levels during a focal onset seizure. On the other hand, the transient interruption of breathing during a generalized motor seizure causes a decrease in oxygen levels in the blood. This results in a drop in ATP concentration and an increase in anaerobic metabolism as indicated by rising lactate levels. This oxygen debt is quickly replenished in the postictal period, and no permanent damage to brain tissue results from a single generalized seizure.

Positron emission tomography (PET) scans of patients with focal onset seizures originating in the mesial temporal lobe frequently show interictal hypometabolism, with metabolic changes extending to the lateral temporal lobe, ipsilateral thalamus, basal ganglia, and frontal cortex. PET scans using nonhydrolyzable glucose analogs have been particularly helpful

Box 58–3 Surgical Treatment of Temporal Lobe Epilepsy

A 27-year-old woman had episodes of decreased responsiveness beginning at age 19. At first, she would stare off and appear confused during the episodes. Later, she developed an aura consisting of a feeling of fear. This fear was followed by altered consciousness, a wide-eyed stare, tightening of the left arm, and a scream that lasted for 14 to 20 seconds (Figure 58–12).

These spells were diagnosed as complex partial seizures. The seizures occurred several times a week despite treatment with several antiepileptic drugs. She was unable to work or drive due to frequent seizures. She had a history of meningitis at age 6 months, and throughout childhood she had experienced brief episodes of altered perception described as "like someone threw a switch."

Based on an evaluation summarized in Figures 58–13 and 58–14, a right amygdalohippocampectomy was performed. The patient was seizure-free following the operation and returned to full-time employment.

A   B   C

Figure 58–12 The patient is shown reading quietly in the period preceding the seizure (A), during the period when she reported a feeling of fear (B), and during the period when there was alteration of consciousness and an audible scream (C). (Reproduced, with permission, from Dr. Martin Salinsky.)

---

in identifying seizure foci in patients with normal MRI scans and in some early childhood epilepsies. Unfortunately, for unclear reasons, PET has been less reliable in localizing seizure foci in extratemporal areas such as the frontal lobe. An additional limitation is the expense of the PET scan and the short half-life of the isotopes (a nearby cyclotron is required). PET scanning can also be used to look for functional changes in neurotransmitter receptor binding and transport related to seizure activity.

A related technique that measures cerebral blood flow, single-photon emission computed tomography (SPECT), has been used more frequently than PET. SPECT does not have the resolution of PET but can be performed in the nuclear medicine department of many large hospitals. Injection of radioisotopes and SPECT imaging at the time of a seizure (ictal SPECT) reveal a pattern of hypermetabolism followed by hypometabolism in the seizure focus and surrounding tissue. Magnetoencephalography and functional MRI also offer further advantages in the mapping of seizure foci.

With rigorous selection of patients for epilepsy surgery, the cure rate for epilepsy with a well-defined

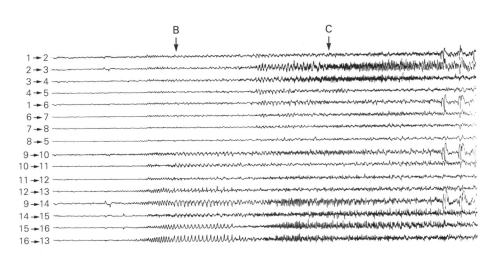

Figure 58–13 The electroencephalogram (EEG) at the time of the photographs in Figure 58–12. Low-amplitude background rhythms occur in the beginning (*left*). At the point when the patient reported fear (**B**), there is a buildup of EEG activity at the onset of a focal onset seizure with impaired awareness, but this activity is confined to the EEG electrodes over the right hemisphere (electrodes 9–16). At the point awareness is altered (**C**), the seizure activity has spread to the left hemisphere (electrodes 1–8). EEG spike-waves are particularly prominent in lead 9 over the right anterior temporal region. (Reproduced, with permission, from Dr. Martin Salinsky.)

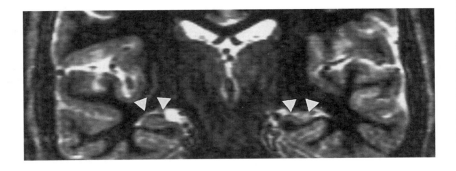

Figure 58–14 Enhanced magnetic resonance imaging reveals atrophy of the right hippocampus (arrows on the right) and a normal left hippocampus (arrows on the left). (Reproduced, with permission, from Dr. Martin Salinsky.)

seizure focus in the temporal lobe can approach 80%. Patients with complicating factors (eg, multiple foci) have lower success rates. However, even among these patients, the number and severity of seizures are usually reduced. Patients who have been "cured" of seizures may still experience cognitive problems such as memory loss and social problems such as adjustments to more independent living and limited employment opportunities. These factors emphasize the need for treatment as early in life as feasible.

Prolonged Seizures Can Cause Brain Damage

Repeated Convulsive Seizures Are a Medical Emergency

As noted above, brain tissue can compensate for the metabolic stress of a focal onset seizure or the transient decrease in oxygen delivery during a single generalized tonic-clonic seizure. In a generalized seizure, stimulation of the hypothalamus leads to massive activation of the "stress" response of the sympathetic nervous system. The increased systemic blood pressure and serum glucose initially compensate for increased metabolic demand, but these homeostatic mechanisms fail during prolonged seizures. The resulting systemic metabolic derangements, including hypoxia, hypotension, hypoglycemia, and acidemia, lead to a reduction in high-energy phosphates (ATP and phosphocreatine) in the brain and thus can be devastating to brain tissue.

Systemic complications such as cardiac arrhythmias, pulmonary edema, hyperthermia, and muscle breakdown can also occur. The occurrence of repeated generalized seizures without return to full consciousness between seizures, called *status epilepticus*, is a true medical emergency. This condition requires aggressive seizure management and general medical support because 30 or more minutes of continuous convulsive seizures leads to brain injury or even death. Status epilepticus can involve nonconvulsive seizures for which the metabolic consequences are much less severe.

In addition to the dangers of status epilepticus, patients with poorly controlled seizures are also at risk for sudden death (sudden unexpected death in epilepsy [SUDEP]), the leading cause of death in patients with uncontrolled seizures. The underlying mechanisms for SUDEP are not completely understood, but recent studies by Richard Bagnall and colleagues as well as others suggest that cases of SUDEP have clinically relevant mutations in the genes implicated in cardiac arrhythmia and epilepsy. Such data support an association between SUDEP and cardiac arrhythmias or interruption of brain stem circuits involved in respiratory control. This topic is appropriately the focus of intense current investigation.

Excitotoxicity Underlies Seizure-Related Brain Damage

Repeated seizures can damage the brain independently of cardiopulmonary or systemic metabolic changes, suggesting that local factors in the brain can result in neuronal death. The immature brain appears particularly vulnerable to such damage, perhaps because of greater electrotonic coupling between neurons in the developing brain, less effective potassium buffering by immature glia, and decreased glucose transport across the blood–brain barrier.

In 1880, Wilhelm Sommer first noted the vulnerability of the hippocampus to such insults, with preferential loss of the pyramidal neurons in the CA1 and CA3 regions. This pattern has been duplicated in experimental animals by electrical stimulation of afferents to the hippocampus or by injection of excitatory amino acid analogs such as kainic acid. Interestingly, kainic acid causes local damage at the site of injection and also at the site of termination of afferents originating at the injection site.

These observations suggest that release of the excitatory transmitter glutamate during excessive stimulation such as a seizure can itself cause neuronal damage, a condition termed *excitotoxicity*. Because it has been difficult to detect increases in extracellular glutamate during status epilepticus, it appears that excitotoxicity results more from excessive stimulation of glutamate receptors than from tonic increases in extracellular glutamate. The histological appearance of acute excitotoxicity includes massive swelling of cell bodies and dendrites, the predominant locations of glutamate receptors and excitatory synapses.

Although the cellular and molecular mechanisms of excitotoxicity are still not fully understood, several features are clear. Overactivation of glutamate receptors leads to an excessive increase in intracellular Ca^{2+} that can activate a self-destructive cellular cascade involving calcium-dependent enzymes, such as phosphatases, proteases, and lipases. Lipid peroxidation can also cause production of free radicals that damage vital cellular proteins and lead to cell death. The role of mitochondria in Ca^{2+} homeostasis and in control of free radicals may also be important. The pattern of cell death was first thought to reflect necrosis due to the autolysis of critical cellular proteins. However, the activation of "death genes," characteristic of programmed cell death (apoptosis), may also be involved.

Seizure-related brain damage or excitotoxicity can be specific to certain types of cells in particular brain regions, perhaps due to protective factors, such as calcium-binding proteins in some cells and sensitizing factors, such as the expression of calcium-permeable glutamate receptors in other cells. For example, excitotoxicity induced in vitro by excessive activation of AMPA-type glutamate receptors preferentially affects interneurons that express AMPA-type receptors that have high Ca^{2+} permeability, providing a possible mechanism for their selective vulnerability.

Several outbreaks of "amnestic" shellfish poisoning provide a vivid example of the consequences of overactivation of glutamate receptors. Domoic acid, a glutamate analog not present in the brain, is a natural product of certain species of marine algae that flourish during appropriate ocean conditions. Domoic acid can be concentrated by filter feeders such as shellfish. Ingestion of domoic-contaminated shellfish sporadically causes outbreaks of neurological damage, including severe seizures and memory loss (amnesia).

The area most sensitive to damage is the hippocampus, providing further support for the excitotoxicity hypothesis and the critical role of the hippocampus in learning and memory.

The Factors Leading to Development of Epilepsy Are Poorly Understood

A single seizure does not warrant a diagnosis of epilepsy. Normal people can have a seizure under extenuating circumstances such as after drug ingestion or extreme sleep deprivation. Clinicians look for possible causes of seizures in such patients but usually do not begin treatment with anticonvulsants following a single seizure. Unfortunately, our understanding of what factors contribute to susceptibility to epilepsy is still rudimentary. However, progress on this front is increasing rapidly with the advent of experimental mutagenesis in animal models and clinical neurogenetics in patients including whole-exome sequencing.

Some forms of epilepsy have long been considered to result in part from a genetic predisposition. For example, infants with febrile seizures often have a family history of similar seizures. The role of genetics in epilepsy is supported by the existence of familial epileptic syndromes in humans as well as seizure-prone animal models with such exotic names as *Papio papio* (a baboon with photosensitive seizures), audiogenic mice (in which loud sounds induce seizures), and spontaneous single-locus mutations such as *reeler* and *totterer* mice (names alluding to the clinical manifestations of cerebellar mutations in these animals). Even with a genetic predisposition or a structural lesion, the evolution of the epileptic phenotype often involves maladaptive changes in brain structure and function.

Mutations in Ion Channels Are Among the Genetic Causes of Epilepsy

Recent studies have provided a wealth of new information concerning the molecular genetics of epilepsy. At present, more than 120 genes have been linked to an epileptic phenotype; approximately half of these were discovered in humans and the others in animals, mostly mice. The affected proteins include ion channel subunits, proteins involved in synaptic transmission such as transporters, vesicle proteins, synaptic receptors, and molecules involved in Ca^{2+} signaling. For example, seizures in the *totterer* mutant mouse are due to a spontaneous mutation in the gene that encodes the

$Ca_V2.1$ or α_{1A}-subunit of the P/Q-type voltage-gated Ca^{2+} channel. That a mutation in these classes of proteins can cause epilepsy is perhaps not unexpected given the dependence of seizures on synaptic transmission and neuronal excitability.

Some of the other genes linked to epilepsy in mice have been more surprising, such as the genes for centromere BP-B, a DNA binding protein, and the sodium/hydrogen exchanger, which is affected in the slow-wave epilepsy mouse. A wide variety of human genes cause neurological disorders, of which epilepsy is only one manifestation. For example, Rett syndrome, a disease associated with intellectual disability, autism, and seizures, is caused by mutations in *MECP2* (methyl-CpG-binding protein-2), a regulator of gene transcription. Although the exact links are not known, it is clear that mutations in many different genes may result in epilepsy.

In most cases, genetic epilepsy syndromes in humans have complex rather than simple (Mendelian) inheritance patterns, suggesting the involvement of many, rather than single, genes. Nevertheless, a number of monogenic epilepsies have been identified in studies of families with epilepsy. Ortrud Steinlein and colleagues reported in 1995 that a mutation in the α4-subunit of the nicotinic acetylcholine receptor-channel is responsible for autosomal dominant nocturnal frontal lobe epilepsy (ADNFLE), the first example of an autosomal gene defect in human epilepsy. Subsequently, other voltage- and ligand-gated channel proteins have been identified as critical genes for epilepsy. Mutations in ion channel genes (channelopathies) constitute a major cause of known monogenic epilepsies (Figure 58–15). Many more genes are being discovered by clinical exome analysis for de novo mutations. The large number of genes for K^+ channels and the critical role of these channels in balancing excitation and inhibition are important reasons for the expanding epilepsy genome.

In voltage-gated channels, mutations largely involve the main pore-forming subunit(s), but there are also examples of epilepsy-causing mutations in regulatory subunits. When examined in vitro, the mutant channel proteins are most commonly associated with either reductions in the expression of the channel on the surface of the plasma membrane (due to reduced targeting to the membrane or premature degradation) or altered kinetics of the channels. It is straightforward to consider how changes in ion channel gating might affect the excitability of neurons and their synchronization during seizure generation. However, ion channel mutations may also affect neuronal development and thus exert their epileptogenic effects through a

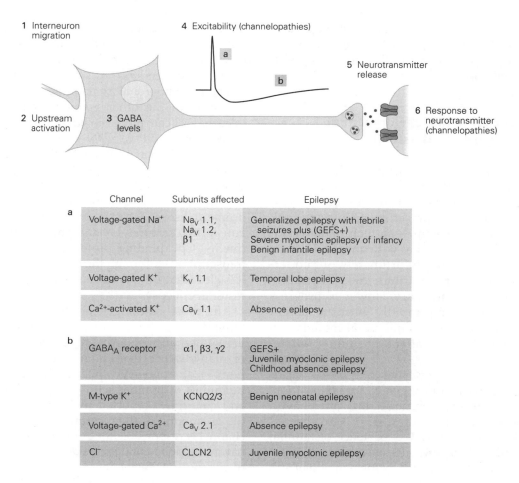

Figure 58–15 Channelopathies are a major, but not the only, cause of monogenic human epilepsies. The human epilepsy genes discovered so far can affect multiple phases of synaptic transmission including the migration of interneurons (1), upstream activation of interneurons (2), γ-aminobutyric acid (GABA) levels within interneurons (3), the excitability of excitatory and inhibitory neurons (4), the release of neurotransmitters (5), and the postsynaptic response to neurotransmitters (6). The inset shows that the impact of mutations in these genes on neuronal excitability can affect the shape of the action potential as well as the afterpotentials and synaptic events that follow. Mutations indicated near the spike (**a**) affect the repolarization of the action potential. Other mutations shown in (**b**) affect the afterhyperpolarization, synaptic conductances, or interspike interval.

secondary action on cell migration, network formation, or patterns of gene expression.

In the early days of research on epilepsy genes, it was widely expected that the genes would mostly underlie generalized epilepsies, based on the idea that a gene mutation (eg, in an ion channel) would be expected to affect most neurons. However, the very first autosomal dominant epilepsy gene discovered by Steinlein and colleagues caused a focal onset (frontal lobe) epilepsy, and another gives rise to seizures originating in the temporal lobe with an auditory aura. In retrospect, this should not be so surprising because channel subunits are rarely expressed uniformly in the brain, and some brain regions are more likely to generate seizures than other regions.

Timing of gene expression is also important. For example, *totterer* mice with mutations in the pore-forming $Ca_V2.1$ subunit of P/Q-type Ca^{2+} channels show spike-wave–type seizures that begin in the third postnatal week, presumably because N-type Ca^{2+} channels are the predominant functional isoform earlier in development, whereas P/Q-type Ca^{2+} channels predominate later. The neurological phenotype begins once the mutant channel is functionally required during development.

Moreover, one mutation can give rise to different epilepsy phenotypes, or different mutant genes can cause the same epilepsy phenotype. As an example of the latter, the ADNFLE syndrome, first discovered as a mutation in the α4-subunit of the nicotinic

ACh receptor, can also be caused by a mutation in the α2-subunit. But not all family members who carry this autosomal dominant mutation have epilepsy, indicating that even in this form of monogenic epilepsy other genes as well as nongenetic factors can influence the phenotype. The GEFS+ syndrome (generalized epilepsy with febrile seizures plus) is a good example of this heterogeneity. It is a childhood syndrome and can involve different seizure types in different family members. GEFS+ is seen in families with mutations in the genes for one of three different Na⁺ channel subunits or one of two GABA$_A$ receptors. Family studies of generalized onset epilepsy suggest that seizure types may be heritable within families. These findings indicate that even monogenic epilepsies are likely modified by other genes, environmental influences, and even experience-dependent changes in synapses.

Altered cortical development may be a common cause of epilepsy. The increased resolution of MRI scans has revealed an unexpectedly large number of cortical malformations and localized areas of abnormal cortical folding in patients with epilepsy. Thus, mutations that disturb the normal formation of the cortex or network wiring are candidate genes for epilepsy. This idea is supported by the mapping of two X-linked cortical malformations with epileptic phenotypes: familial periventricular heterotopia and familial subcortical band heterotopia. The genes responsible for these two disorders that encode filamin A and doublecortin, respectively, are presumably important in neuronal migration. Small focal cortical dysplasias can function as seizure foci that give rise to partial and secondarily generalized seizures, whereas more extensive cortical malformations can cause a variety of seizure types and usually are associated with other neurological problems.

Another X-linked gene, *aristaless related homeobox* (*ARX*), is an example of a cell type–specific transcription factor altering migration, because it is expressed only in interneuron precursors. A particularly instructive example is the association of epilepsy with tuberous sclerosis complex (TSC), an autosomal dominant genetic disorder that results from the lack of the functional Tsc1-Tsc2 complex, leading to hyperactivity of the mammalian target of rapamycin (mTOR) complex 1 (mTORC1) signaling pathway. Early clinical trials of mTOR inhibitors as treatment for refractory epilepsy in these patients have been promising. Such examples provide hope for linking the underlying biology of epilepsy syndromes to clinically relevant treatments.

The epilepsy genome is rapidly expanding, driven by clinical exome sequencing and an appreciation of the biological pathways leading to neural network instability. Unfortunately, the vast majority of cases of epilepsy cannot yet be explained by even the recent surge in the identification of epilepsy genes. The identification of large numbers of patients through online registries may provide the population samples needed to evaluate susceptibility genes that underlie complex inheritance patterns.

The Genesis of Acquired Epilepsies Is a Maladaptive Response to Injury

Epilepsy often develops following a discrete cortical injury such as a penetrating head wound. This injury serves as the nidus for a seizure focus, leading at some later point to seizures. This has led to the idea that the early insult triggers a set of progressive physiological or anatomical changes that lead to chronic seizures. That is, the characteristic "silent" interval (usually months or years) between the insult and the onset of recurrent seizures may reflect progressive maladaptive molecular and cellular changes that might be amenable to therapeutic manipulation. Although an attractive hypothesis, a unified picture of this process has yet to emerge. The most promising evidence has come from studies of tissue removed from patients undergoing temporal lobectomy and rodent models of limbic seizures.

In one experimental model, hyperexcitability is induced by repeated stimulation of limbic structures, such as the amygdala or hippocampus. The initial stimulus is followed by an electrical response (the afterdischarge) that becomes more extensive and prolonged with repeated stimuli until a generalized seizure occurs. This process, called *kindling*, can be induced by electrical or chemical stimuli. Many investigators believe that kindling may contribute to the development of epilepsy in humans.

Kindling is thought to involve synaptic changes in the hippocampal formation that resemble those important in learning and memory (Chapters 53 and 54). These include short-term changes in excitability and persistent morphological changes, including generation of adult-born neurons, axonal sprouting, and synaptic reorganization. Rearrangements of synaptic connections have been observed in the dentate gyrus of patients with long-standing temporal lobe seizures as well as following kindling in experimental animals. In addition to axonal sprouting (Figure 58–16), changes include alterations in dendritic structure, control of transmitter release, and novel expression and alterations in subunit stoichiometry of ion channels and pumps.

The long-term changes that lead to epilepsy also are likely to involve specific patterns of gene

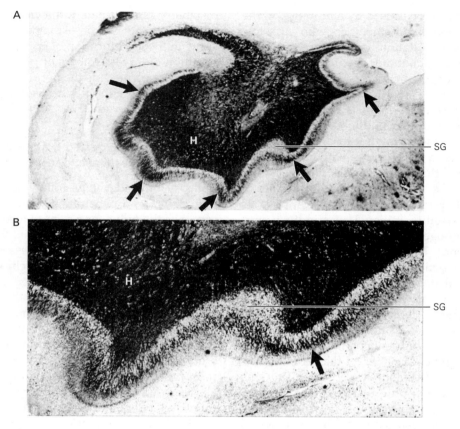

Figure 58–16 Mossy fiber synaptic reorganization (sprouting) in the human temporal lobe may cause hyperexcitability. (Reproduced, with permission, from Sutula et al. 1989. Copyright © 1989 American Neurological Association.)

A. Timm stain of a transverse section of hippocampus removed from a patient with epilepsy at the time of temporal lobectomy for control of epilepsy. The stain appears black in the axons of the dentate granule cells (mossy fibers) due to the presence of zinc in these axons. The mossy fibers normally pass through the dentate hilus (**H**) on their way to synapse on CA3 pyramidal cells. In the epileptic tissue shown here, stained fibers appear in the supragranular layer of the dentate gyrus (**SG, arrowheads**), which now contains not only the granule cell dendrites but also newly sprouted mossy fibers. These aberrant sprouts of mossy fibers form new recurrent excitatory synapses on dentate granule cells.

B. This high magnification of a segment of the supragranular layer shows the Timm-stained mossy fibers in greater detail.

expression. For example, the proto-oncogene *c-fos* and other immediate early genes as well as growth factors can be activated by seizures. Because many immediate early genes encode transcription factors that control other genes, the gene products that result from epileptiform activity could initiate changes that contribute to or suppress the development of epilepsy by altering such mechanisms as cell fate, axon targeting, dendritic outgrowth, and synapse formation.

Highlights

1. Seizures are one of the most dramatic examples of the collective electrical behavior of the mammalian brain. The distinctive clinical pattern of partial seizures and generalized seizures can be attributed to the distinctly different patterns of activity of cortical neurons.

2. Studies of focal onset seizures in animals reveal a series of events—from the activity of neurons in the seizure focus to synchronization and subsequent spread of epileptiform activity throughout the cortex. The gradual loss of GABAergic surround inhibition is critical to the early steps in this progression. In contrast, generalized onset seizures are thought to arise from activity in thalamocortical circuits, perhaps combined with a general abnormality in the membrane excitability of all cortical neurons.

3. The electroencephalogram (EEG) has long provided a window on the electrical activity of the cortex, both in normal phases of arousal and during abnormal activities such as seizures. The EEG

can be used to identify certain electrical activity patterns associated with seizures, but it provides limited insight into the pathophysiology of seizures. Several much more powerful and noninvasive approaches are now available to locate the focus of a partial seizure. This has led to the widespread and successful use of epilepsy surgery for selected patients, particularly those with complex partial seizures of hippocampal onset. The promise of invasive approaches to seizure detection and seizure prevention provides additional hope for improved control of seizures.

4. The increasing power of genetic, molecular, and modern cell-physiological approaches applied to the study of seizures and epilepsy also gives new hope that an understanding of these disruptions of normal brain activity will provide new therapeutic options for patients afflicted with epilepsy, as well as new insights into the function of the mammalian brain.

5. Further neurobiological studies of the progression from an acute seizure to the development of epilepsy should provide alternative strategies for treatment beyond the standard options of anticonvulsants or epilepsy surgery.

Gary Westbrook

Selected Reading

Cascino GD. 2004. Surgical treatment for epilepsy. Epilepsy Res 60:179–186.

Engel J. 1989. *Seizures and Epilepsy*. Philadelphia: Davis.

Kleen JK, Lowenstein DH. 2017. Progress in epilepsy: latest waves of discovery. JAMA Neurol 74:139–140.

Krook-Magnuson E, Soltesz I. 2015. Beyond the hammer and the scalpel: selective circuit control for the epilepsies. Nat Neurosci 18:331–338.

Krueger DA, Wilfong AA, Holland-Bouley K, et al. 2013. Everolimus treatment of refractory epilepsy in tuberous sclerosis complex. Ann Neurol 74:679–687.

Kullmann DM, Schorge S, Walker MC, Wykes RC. 2014. Gene therapy in epilepsy—is it time for clinical trials? Nat Rev Neurol 10:300–304.

Lennox WG, Lennox MA. 1960. *Epilepsy and Related Disorders*. Boston: Little, Brown.

Lennox WG, Mattson RH. 2003. Overview: idiopathic generalized epilepsies. *Epilepsia* 44(Suppl 2):2–6.

Lerche H, Shah M, Beck H, Noebels J, Johnston D, Vincent A. 2013. Ion channels in genetic and acquired forms of epilepsy. J Physiol 591:753–764.

Lowenstein DH. 2015. Decade in review-epilepsy: edging toward breakthroughs in epilepsy diagnostics and care. Nat Rev Neurol 11:616–617.

Maheshwari A, Noebels JL. 2014. Monogenic models of absence epilepsy: windows into the complex balance between inhibition and excitation in thalamocortical microcircuits. Prog Brain Res 213:223–252.

Noebels J. 2015. Pathway-driven discovery of epilepsy genes. Nat Neurosci 18:344–350.

Paz JT, Huguenard JR. 2015. Optogenetics and epilepsy: past, present and future. Epilepsy Curr 15:34–38.

Penfield W, Jasper H. 1954. *Epilepsy and the Functional Anatomy of the Human Brain*. Boston: Little, Brown.

Snowball A, Schorge S. 2015. Changing channels in pain and epilepsy: exploring ion channel gene therapy for disorders of neuronal hyperexcitability. FEBS Letters 589:1620–1624.

Stables JP, Bertram EH, White HS, et al. 2002. Models for epilepsy and epileptogenesis: report from the NIH workshop. Epilepsia 43:1410–1420.

Stafstrom CE, Carmant L. 2015. Seizures and epilepsy: an overview for neuroscientists. Cold Spring Harb Perspect Med 5:a022426.

References

Bagnall RD, Crompton DE, Petrovski S, et al. 2016. Exome-based analysis of cardiac arrhythmia, respiratory control and epilepsy genes in sudden unexpected death in epilepsy. Ann Neurol 79:522–534.

Berenyi A, Belluscio M, Mao D, Buzsaki G. 2012. Closed-loop control of epilepsy by transcranial electrical stimulation. Science 337:735–737.

Bergey GK, Morrell MJ, Mizrahi EM, et al. 2015. Long-term treatment with responsive brain stimulation in adults with refractory partial seizures. Neurology 84:810–817.

Biervert C, Schroeder BC, Kubisch C, et al. 1998. A potassium channel mutation in neonatal human epilepsy. Science 279:403–406.

Fisher RS, Cross JH, D'Souza C, et al. 2017. Instruction manual for the ILAE 2017 operational classification of seizure types. Epilepsia 58:531–542.

Gadhoumi K, Lina J-M, Mormann F, Gotman J. 2016. Seizure prediction for therapeutic devices: a review. J Neurosci Methods 260:270–282.

Haug K, Warnstedt M, Alekov AK, et al. 2003. Mutations in CLCN2 encoding a voltage-gated chloride channel are associated with idiopathic generalized epilepsies. Nat Genet 33:527–532.

Heck CN, King-Stephens D, Massey AD, et al. 2014. Two-year reduction in adults with medically intractable partial onset epilepsy treated with responsive neurostimulation: final results of the RNS system pivotal trial. Epilepsia 55:432–441.

Kätzel D, Nicholson E, Schorge S, Walker MC, Kullmann DM. 2013. Chemical-genetic attenuation of focal neocortical seizures. Nat Commun 5:3847.

Kramer MA, Eden UT, Kolaczyk E, et al. 2010. Coalescence and fragmentation of cortical networks during focal seizures. J Neurosci 30:10076–10085.

Lothman EW. 1993a. The neurobiology of epileptiform discharges. Am J EEG Technol 33:93–112.

Lothman EW. 1993b. Pathophysiology of seizures and epilepsy in the mature and immature brain: cells, synapses and circuits. In: WE Dodson, JM Pellock (eds). *Pediatric Epilepsy: Diagnosis and Therapy*, pp. 1–15. New York: Demos Publications.

Lothman EW, Collins RC. 1990. Seizures and epilepsy. In: AL Pearlman, RC Collins (eds). *Neurobiology of Disease*, pp. 276–298. New York: Oxford University Press.

Mulley JC, Scheffer IE, Harkin LA, Berkovic SF, Dibbens LM. 2005. Susceptibility genes for complex epilepsy. Hum Mol Genet 14:R243–R249.

Santhakumar V, Aradi S, Soltesz I. 2005. Role of mossy fiber sprouting and mossy cell loss in hyperexcitability: a network model of the dentate gyrus incorporating cell types and axonal topography. J Neurophysiol 93:437–463.

Spencer WA, Kandel ER. 1968. Cellular and integrative properties of the hippocampal pyramidal cell and the comparative electrophysiology of cortical neurons. Int J Neurol 6:266–296.

Stacey WC, Litt B. 2008. Technology insight: neuroengineering and epilepsy-designing devices for epilepsy control. Nat Clin Pract Neurol 4:190–201.

Steinlein OK, Mulley JC, Propping P, et al. 1995. A missense mutation in the neuronal nicotinic acetylcholine receptor alpha 4 subunit is associated with autosomal dominant nocturnal frontal lobe epilepsy. Nat Genet 11:201–203.

Sutula T, Cascino G, Cavazos J, Parada I, Ramirez L. 1989. Mossy fiber synaptic reorganization in the epileptic human temporal lobe. Ann Neurol 26:321–330.

Teitelbaum J, Zatorre RJ, Carpenter S, et al. 1990. Neurologic sequelae of domoic acid intoxication due to ingestion of contaminated mussels. N Engl J Med 322:1781–1787.

Tung JK, Berglund K, Gross RE. 2016. Optogenetic approaches for controlling seizure activity. Brain Stimul 9:801–810.

von Krosigk M, Bal T, McCormick DA. 1993. Cellular mechanisms of a synchronized oscillation in the thalamus. Science 261:361–364.

Walsh CA. 1999. Genetic malformations of the human cerebral cortex. Neuron 23:19–29.

Wiebe S, Blume WT, Girvin JP, Eliasziw M. 2001. Effectiveness and efficiency of surgery for temporal lobe epilepsy study. A randomized, controlled trial of surgery for temporal lobe epilepsy. N Engl J Med 14:211–216.

Winawer MR, Marini C, Grinton BE, et al. 2005. Familial clustering of seizure types within the idiopathic generalized epilepsies. Neurology 65:523–528.

Zhao M, Alleva R, Ma H, Daniel AGS, Schwartz TH. 2015. Optogenetic tools for modulating and probing the epileptic network. Epilepsy Res 116:15–26.

59

Disorders of Conscious and Unconscious Mental Processes

ALTHOUGH COGNITIVE NEUROSCIENCE emerged at the end of the 20th century as a major new discipline, a precise meaning of the term *cognition* remains elusive. The term is used in different ways in different contexts. At one extreme, the term *cognitive* in cognitive neuroscience connotes what was meant by the older term *information processing*. In this sense, cognition is simply what the brain does. When cognitive neuroscientists say that visual features or motor acts are *represented* by neural activity, they are using concepts from information processing. From this point of view, the language of cognition provides a bridge between descriptions of neural activity and behavior because the same terms can be applied in both domains.

At the other extreme, the term *cognition* refers to those higher-level processes fundamental to the formation of conscious experience. This is what is meant by the term *cognitive therapy*, an approach to treatment pioneered by Aaron Beck and Albert Ellis and developed from behavior therapy. Rather than trying to change a patient's behavior directly, cognitive therapy has the aim of changing the patient's attitudes and beliefs (Box 59–1).

In common parlance, the term *cognition* means thinking and reasoning, a usage closer to its Latin root *cognoscere* (getting to know or perceiving). Thus, the *Oxford English Dictionary* defines it as "the action or faculty of knowing." Indeed, we know the world by applying thinking and reasoning to the raw data of our senses.

This idea is implicit in our characterization of many kinds of disorders of cognition. After brain damage, some patients can no longer process the input supplied by the senses. This type of disorder was first delineated by Sigmund Freud, who called it agnosia, or loss of knowledge (Chapter 17). Agnosias can take many forms. A patient with visual agnosia can see perfectly well but is no longer able to recognize or make sense of what he sees. A patient with prosopagnosia has a specific problem recognizing faces. A patient with auditory agnosia might hear perfectly well but is unable to recognize spoken words.

Cognition is sometimes impaired from birth so that a person has difficulty in acquiring knowledge. This might lead to general mental retardation or, if the problem is more localized, to specific learning difficulties such as dyslexia (difficulty learning about written

---

Box 59–1 Cognitive Therapy

Dissatisfaction with psychological treatments based on Freud's theories of unconscious motivation intensified in the middle of the 20th century. Not only did these theories have no relevance to experimental psychology, but there was no empirical evidence that psychodynamic treatments actually worked.

The first form of alternative psychological therapy to emerge from laboratory studies is known as *behavior therapy*. The fundamental assumption of this approach is that maladaptive behavior is learned and can therefore be eliminated by applying the Pavlovian and Skinnerian principles of stimulus-response learning. So, for example, a child who has been attacked by a dog can become fearful of all dogs, but this fearful response can be extinguished if the child learns that the conditioned stimulus (the sight of a dog) is not followed by the unconditioned stimulus (being bitten).

Behavior therapy was shown to be quick and effective for phobias, but many mental disorders are better characterized in terms of maladaptive thinking rather than maladaptive behavior. In the 1960s, Aaron Beck and Albert Ellis initiated a new kind of therapy in which the principles of learning are used to change thoughts rather than behavior. This is known as *cognitive therapy* or *cognitive behavior therapy*.

This form of therapy has been particularly successful in the treatment of depression. Depression is typically associated with negative thoughts (eg, a person remembering only the bad things that have happened to him/her) and negative attitudes (eg, a person believing that he/she will never achieve his/her goals). Cognitive therapists teach their clients methods for reducing the frequency of negative thoughts and changing their negative attitudes into positive ones.

---

language) or autism (difficulty in learning about other minds). Finally, cognition can become dysfunctional so that the knowledge acquired about the world is false. These disorders of thinking lead to the sort of false perceptions (hallucinations) and false beliefs (delusions) associated with major mental illnesses such as schizophrenia.

Conscious and Unconscious Cognitive Processes Have Distinct Neural Correlates

Cognition—deriving knowledge through thinking and reasoning—is one of three components of consciousness (see Chapter 42 for discussion of the conscious aspects of emotions, often called feelings). The other two are emotion and will. It used to be taken for granted that thinking and reasoning were under conscious voluntary control and that cognition was not possible without consciousness. By the end of the 19th century, however, Freud developed a theory of unconscious mental processes and suggested that much human behavior is guided by internal processes of which we are not aware.

Of more direct importance for neuroscience was the idea of *unconscious inference*, originally proposed by Helmholtz. Helmholtz was the first to carry out quantitative psychophysical experiments and to measure the speed with which afferent signals in peripheral nerves are conducted. Prior to these experiments,

sensory signals were assumed to arrive in the brain immediately (with the speed of light), but Helmholtz showed that nerve conduction was actually quite slow. He also noted that reaction times were even slower. These observations implied that a great deal of brain work intervened between sensory stimuli and conscious perception of an object. Helmholtz concluded that much of what goes on in the brain is not conscious and that what does enter consciousness (ie, what is perceived) depends on unconscious inferences. In other words, the brain uses evidence from the senses to decide on the most likely identity of the object that is causing activity in the sensory organs but does this without our awareness.

This view was extremely unpopular with Helmholtz's contemporaries and, indeed, still is today. Most people believe that consciousness is necessary for making inferences and that moral responsibility can be assigned only to decisions that are based on conscious inference. If inferences could be made without consciousness, there could be no ethical basis for praise or blame. Helmholtz's ideas about unconscious inferences were largely ignored.

Nevertheless, by the middle of the 20th century, evidence began to accumulate in favor of the idea that most cognitive processing never enters consciousness. After the development of electronic computers and the emergence of the study of artificial intelligence, researchers began to study how, and to what extent, machines could perceive the world beyond themselves.

It rapidly became clear that many perceptual processes that at first seem simple are actually very complex when defined as a set of computations.

Visual perception is the prime example. In the 1960s, almost no one realized how difficult it would be to build machines that could recognize the shape and appearance of objects, because it seems so easy for us. I look out of the window and I see buildings, trees, flowers, and people. I am not aware of any mental processes behind this perception; my awareness of all these objects seems instantaneous and direct. It turns out that teaching a machine how to work out which edges go with which object in a typical cluttered visual scene containing many overlapping objects is exceptionally difficult. The computational approach to vision revealed the underlying neural processes on which our seemingly effortless perception of the world depends. Similar processes underlie all sensory perception and especially the perception of sounds as speech. Most neuroscientists now believe that we are not conscious of cognitive processes, only our perceptions.

The evidence for unconscious cognitive processes comes not only from artificial intelligence studies but also studies of cognition in people with brain damage. The effects of unconscious processes on behavior can be demonstrated most strikingly in certain patients with "blind sight," a disorder first delineated in the 1970s by Lawrence Weiskrantz. These patients have lesions in the primary visual cortex and claim to see nothing in the part of the visual field served by the damaged area. Nevertheless, when asked to guess, they are able to detect simple visual properties such as movement or color far better than is expected by chance. Despite having no sensory-based perception of objects in the blind parts of the visual field, these patients do possess unconscious information about the objects, and this information is available to guide their behavior.

Another example is unilateral neglect caused by lesions in the right parietal lobe (Chapter 17). Patients with this disorder have normal vision, but they seem unware of objects on the left side of the space in front of them. Some patients even ignore the left side of individual objects. In one experiment by John Marshall and Peter Halligan, patients were shown two drawings of a house. The left side of one house was on fire (Figure 59–1). When asked if there were any differences between the houses, patients replied "no." But when asked which house they would prefer to live in, they chose the house that was not burning. This choice was thus made based on information that was not represented in consciousness. Blind sight and unilateral neglect are just two examples of the abundant

Figure 59–1 Unconscious processing in cases of spatial neglect. After damage to the right parietal lobe, many patients seem to be unaware of the left side of space (unilateral neglect syndrome). When such patients are shown the two drawings reproduced here, they say that the two houses look the same. However, they also say that they would prefer to live in the lower house, indicating that they have unconsciously processed the image of the fire in the other house. (Adapted from Marshall and Halligan 1988.)

empirical evidence for the existence of unconscious cognitive processes, evidence not available to us through introspection.

Currently, one of the most exciting areas of investigation in neuroscience concerns the search for the *neural correlates of consciousness* initiated by Francis Crick and Christopher Koch. The aim is to demonstrate qualitative differences between the neural activity associated with conscious and unconscious cognitive processes. This research is important not only because it may give us answers to the difficult question of the function of consciousness but also because it is relevant to our understanding of many neurological and psychiatric disorders. The weird experiences and delusional beliefs of patients with certain cognitive disorders were once dismissed as beyond understanding. Cognitive neuroscience provides us with a framework for understanding how these experiences and beliefs can arise from specific alterations in normal cognitive mechanisms.

Differences Between Conscious and Unconscious Processes in Perception Can Be Seen in Exaggerated Form After Brain Damage

The relationship between sensory stimulation and perception is far from direct. Perception can change without any change in sensory stimulation, as illustrated by ambiguous figures such as the Rubin figure and the Necker cube (Figure 59–2). Conversely, a big change in sensory stimulation can occur without the observer being aware of this change—the perception remains constant. A compelling example of this is change blindness.

To demonstrate change blindness, two versions of a complex scene are constructed. In one well-known example developed by Ron Rensink, the picture consists of a military transport plane standing on an airport runway. In one of the two versions, an engine is missing. If these two pictures are shown in alternation on a computer screen, but critically interspersed with a blank screen, it can take minutes to notice the difference even though it is immediately obvious when pointed out. (See Figure 25–8 for another example.)

In light of these phenomena, we can explore the neural activity associated with changes in perception when there is no change in sensory stimulation. Likewise, we can discover whether changes in sensory input are registered in the brain even if not represented in consciousness. We can ask whether there is some qualitative difference between the neural activity associated with conscious as opposed to unconscious processes.

Two important results have emerged from studies of the neural activity associated with specific types of conscious percepts. First, certain kinds of percepts are related to neural activity in specific areas of the brain. Those brain areas that are specialized for recognition of certain kinds of objects (eg, faces, words, landscapes) or for certain visual features (eg, color, motion) are more active when the object or the feature is consciously perceived (Figure 59–3). For example, when we perceive the faces in the Rubin figure, there is more activity in the area of the fusiform gyrus, which is specialized for the processing of faces.

This observation also applies to deviant perception (hallucinations). After degeneration of the peripheral visual system leading to blindness, some patients experience intermittent visual hallucinations (Charles Bonnet syndrome). These hallucinations vary from one patient to another: Some patients see colored patches, others see grid-like patterns, and some even see faces. Dominic ffytche found that these hallucinations are associated with increased activity in the secondary visual cortex, and the content of the hallucination is related to the specific locus of activity (Figure 59–4). Schizophrenic patients frequently experience complex auditory hallucinations, which usually have the form of voices talking to or about the patient. These

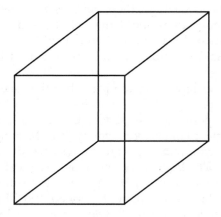

Figure 59–2 Ambiguous figures. If you stare at the figure on the left (the Rubin figure), you sometimes see a vase and sometimes two faces looking at each other. If you stare at the figure on the right (the Necker cube), you see a three-dimensional cube, but the front face of the cube is sometimes seen at the bottom left and sometimes at the top right. In each figure, the brain finds two equally good, but mutually exclusive, interpretations of what is there. Our conscious perception spontaneously alternates between these two interpretations.

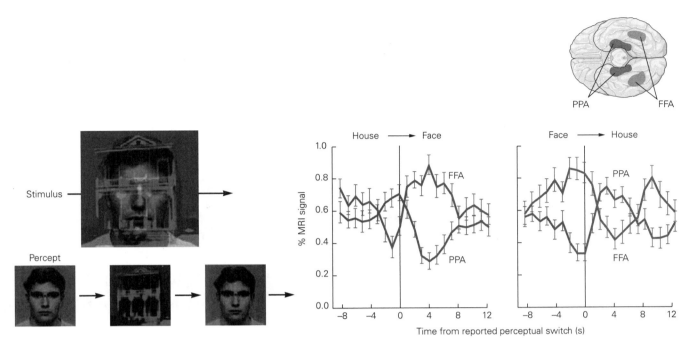

Figure 59–3 Neural activity associated with ambiguous visual information. An ambiguous stimulus was created by simultaneously presenting a face to one eye and a house to the other eye. Brain activity was measured while subjects observed these images. Subjects were instructed to press a button whenever a spontaneous switch in perception occurred (because of binocular rivalry). When the face is perceived (*left*), activity increases in the fusiform face area (**FFA**); when the house is perceived (*right*), activity increases in the parahippocampal place area (**PPA**). (Abbreviation: **MRI**, magnetic resonance imaging.) (Reproduced, with permission, from Tong et al. 1998. Copyright © 1998 by Cell Press.)

hallucinations are associated with activity in the auditory cortex.

These observations suggest that conscious experience may result from activity in certain cortical regions. This idea is difficult to test experimentally, but in the 1950s, the neurosurgeon Wilder Penfield found that electrical stimulation of the cortex in patients undergoing neurosurgery can generate a conscious experience. More recently, it has been found that transcranial magnetic stimulation of the cortex in the region of V5/MT can lead to seeing moving light flashes.

The second important conclusion drawn from studies that seek to correlate neural activity and specific percepts is that activity in a specialized area is necessary but not sufficient to yield conscious experience. For example, in the change blindness paradigm, subjects are often unaware of large changes in the picture they are viewing. If the change involves a face, activity is elicited in the fusiform gyrus whether or not the subject is aware of the change. But when the sensory change is also perceived consciously, there is, in addition, activity in the parietal and frontal cortices (Figure 59–5).

These observations are relevant to our understanding of unilateral neglect. Since objects on the left side

still elicit neural activity in the visual cortex, it may be that the damage in the right parietal cortex simply prevents the formation of *conscious* representations of objects on the left side of space. Nevertheless, this sensory activity can support an unconscious inference in patients that they would not want to live in the house that is burning on the left side.

Stimuli that do not enter awareness can also elicit overt responses. A face with a fearful expression elicits a fear response in the autonomic nervous system, measured as an increase in skin conductance (galvanic response) because of sweating. This response occurs even if the face is immediately followed by another visual stimulus, such that the face is not consciously perceived. There may be an advantage to having a rapid but low-resolution system for recognizing dangerous things. We jump first; only later, on the basis of a slow, high-resolution system, are we able to identity the object that made us jump (Chapter 48). Damage in one or the other of these two recognition systems can explain certain otherwise puzzling neurological and psychiatric disorders.

Prosopagnosia is a perceptual disorder in which faces are no longer recognizable. The patient knows

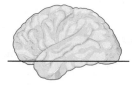

Subject 1

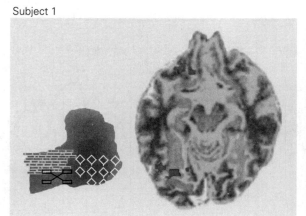

Subject 2

Subject 3

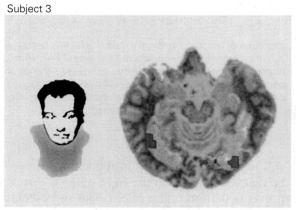

Subject 4

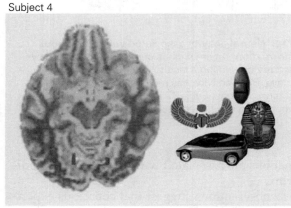

**Figure 59–4 Neural activity associated with visual halluci-
nations.** Some patients with damage to the retina experience
visual hallucinations. The location of the neural activity and
the content of the hallucination are related. The experience of
colors, patterns, objects, or faces is associated with heightened
activity (**red**) in specific regions of inferior temporal cortex. The
blue area is the fusiform gyrus. (Reproduced, with permission,
from ffytche et al. 1998. Copyright © 1998 Springer Nature.)

he is looking at a face but cannot recognize the face,
even a beloved face known for years. The problem
is specific to faces, since the patient may still be able
to recognize the person from their clothes, gait, and
voice. However, patients with prosopagnosia are able
to identify faces unconsciously. They show autonomic
responses to familiar faces and do better than chance
when asked to guess whether or not a face shown to
them belongs to a person who is familiar. In fact, their
awareness of the autonomic (emotional) responses
elicited by a face may enable them to judge familiarity.

Capgras syndrome, a delusion that is occasionally observed in schizophrenic patients and in some
patients suffering from brain injury or dementia,
produces a more unsettling experience. These patients
firmly believe that someone close to them, usually a
husband or wife, has been replaced by an impostor.
They claim that the person, although similar if not
identical in appearance, is in fact someone else. Often,
this delusion is acted on with the demand that the
impostor leave the house.

Hadyn Ellis and Andy Young have suggested that
this bizarre delusion is the mirror phenomenon of
prosopagnosia. According to this view, the circuitry
for face recognition is intact, but the circuitry that
mediates the emotional response to the face is not.
As a result, patients recognize the person in front of
them but, because the emotional response is lacking,

A Unconscious detection

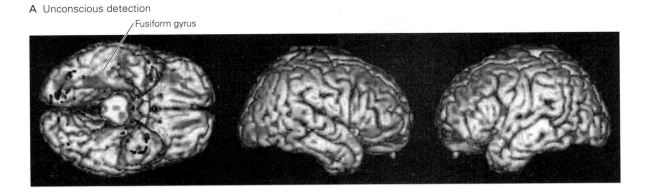

B Conscious report

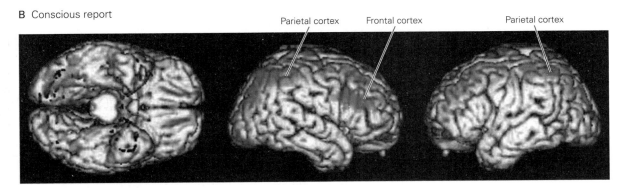

Figure 59–5 Brain activity with and without aware-ness. Activity in the fusiform face area increases when the face viewed by subjects changes, whether subjects are unaware of the change or conscious of it. When subjects are aware of the change, activity in parietal and frontal cortex also increases. (Reproduced, with permission, from Beck et al. 2001.)

feel that there is something fundamentally wrong. This account has been partially confirmed by the observation that these patients do not have normal autonomic responses to familiar faces.

This explanation implies that Capgras delusions are not the consequence of disordered thinking but of disordered experience. A patient sees the face of his wife without having the normal emotional response. The conclusion that this is not his wife but an impostor is a cognitive response to this abnormal experience, the mind's attempt to explain the experience.

The Control of Action Is Largely Unconscious

The sense that we are in control of our own actions is a major component of consciousness. But are we aware of all aspects of our own actions? David Milner and Mel Goodale studied a patient known as D.F. who demonstrates a striking lack of awareness of certain aspects of her own actions. As a result of damage to her inferior temporal lobe caused by carbon monoxide poisoning,

D.F. suffers from *form agnosia*—she is unable to identify the shapes of things. She cannot distinguish a square from an oblong card and cannot describe the orientation of a slot. Yet when she picks up the oblong card to place it through the slot, she orients her hand and forms her grasp appropriately because of the unconscious operation of visuomotor circuits (Figure 59–6).

This sort of unconscious guidance is not unique to patients with brain damage. It is simply revealed more starkly in the case of D.F. because the system that normally brings visual information about shape into consciousness is impaired. Indeed, we can all make rapid and accurate grasping movements without being aware of the perceptual and motor information that is being used to control these movements. Sometimes, we are not even aware of having made the movement. This largely unconscious system for visually guided reaching and grasping is analogous to, and probably overlaps with, the rapid but poor-resolution system associated with fear responses.

Although we may not be aware of the perceptual and motor details of actions like reaching and grasping,

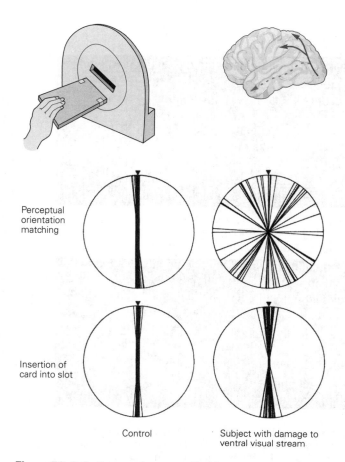

Perceptual
orientation
matching

Insertion of
card into slot

Control

Subject with damage to
ventral visual stream

Figure 59–6 Action can be controlled by unconscious stimuli. A patient, D.F., with damage to the inferior temporal cortex, is unable to recognize objects based on their shape (form agnosia). She cannot align the tablet with the orientation of the slot (perceptual matching) because she is not consciously aware of the orientation of either the tablet or the slot. However, when she is asked to put the tablet through the slot in a quick movement, she orients her hand rapidly and accurately. Presumably, the movement is driven by visuomotor computations of which the subject is unaware. (Adapted, with permission, from Milner and Goodale 1995.)

we are vividly aware of being in control of some of our actions—we are aware of a difference between actions that we cause and those that happen involuntarily. Benjamin Libet studied the phenomenon of voluntary action in controlled experiments. He asked his subjects to lift a finger "whenever they felt the urge to do so" and to report the time at which they had this urge. His subjects had no difficulty in reliably reporting the time of this subjective experience. At the same time, Libet used electroencephalography to measure the "readiness potential," a change in brain activity that occurs up to 1 second before a subject makes any voluntary movement. The time at which subjects reported feeling the urge to lift a finger occurred hundreds

of milliseconds *after* the beginning of this readiness potential. This result has generated much discussion among philosophers as well as neuroscientists concerning the existence of free will. If brain activity can *predict* an action before a person is aware of having the urge to perform that action, does this mean that our experience of freely willing actions is an illusion?

Although Libet's result has been widely replicated, the relevance of his experimental protocol for our understanding of free will remains controversial. Lifting one finger is not an action that we often perform. Actions usually have goals. For example, we might press a button in order to ring a bell. When our actions are followed by the goal we expect, we feel that we are in control of our actions. It is this subjective experience that gives us a sense of agency, of being the cause of events. Applying Libet's paradigm to such actions, Patrick Haggard discovered the phenomenon of "intentional binding." When a deliberate movement (pressing a button) is followed by its intended goal (hearing a tone), these events are experienced subjectively as bound together in time (Figure 59–7).

This temporal binding of our actions to their goals provides an empirical marker of our sense of agency, since a stronger sense of agency is associated with a greater degree of binding. If a movement occurs passively, caused for example by magnetic stimulation to the brain, then intentional binding is decreased; we actually perceive the time between movement and outcome as longer than the actual physical time.

Our sense of agency is closely linked to our belief in free will and to the idea that people can be held

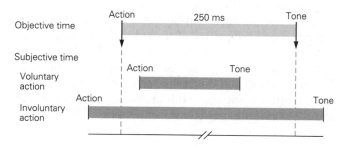

Figure 59–7 We experience our actions and their effects as bound together in time. When subjects are asked to press a button that triggers a sound 250 ms later, they experience their action and the sound as occurring closer together (subjective time) than they actually are (objective time). In contrast, when their finger moves involuntarily through trans cranial magnetic stimulation (TMS) of motor cortex, the movements and the sound are experienced as further apart compared to objective time. Temporal binding occurs only when the movement is intended and deliberate and thus is a marker of the experience of agency. (Based on Haggard, Clark, and Kalogeras 2002.)

responsible for their actions when these are performed deliberately. Intentional binding is increased when associated with outcomes that have moral consequences. It is reduced for actions that have been commanded by others, rather than performed freely. These results do not address the question of whether or not free will exists, but they suggest that our conscious experience of acting freely has a major role in creating social norms of responsibility. Such norms are critical for maintaining social cohesion.

Unconscious inference occurs in the motor domain as well as the sensory domain. Our experience of agency is created from two components: our prior expectations and the sensory consequences of the outcome of the action. We are surprised if the actual sensations do not match what we expect, as when we pick up an object that is much lighter than anticipated (Chapter 30). If the outcome confirms our expectations, however, we pay little attention to the actual sensory evidence—we experience what we expected to happen rather than what actually happened.

Pierre Fourneret and Marc Jeannerod asked subjects to draw a vertical line using a computer's mouse. The subjects could not see their hand and so could not see that the computer created a distortion in the line displayed on the screen. The striking result was that subjects were not aware that they had moved their hand at an angle of 10° to the left to produce the vertical

line on the screen (Figure 59–8). This lack of awareness occurred for deviations of up to 15°. When subjects were instructed not to look at the screen but simply repeat the movement they had just made, they did not reproduce the deviant movement they had made but instead drew the straight-ahead movement that they believed they had made. It would seem that as long as the goal is realized (drawing a straightforward line), we experience the expected sensory feedback, not the actual sensory feedback.

This phenomenon helps us understand some otherwise bizarre experiences. For example, after the amputation of a limb, some patients may experience a phantom limb. They still experience the urge to move the missing limb, and they can select specific movements they want the missing limb to make. Their sensorimotor systems predict the proprioceptive sensations they would feel if they were to move an intact limb, and it is these predicted sensations that underlie the sensation of a moving phantom limb.

After a limb has been paralyzed due to stroke, some patients believe that they are still able to move the limb (anosognosia for hemiplegia). Here, again, such patients can select the movements they want to make and are aware of their expectations about the movement. Despite the lack of sensory evidence that follows their attempt to initiate the movement, they believe that the movement did occur.

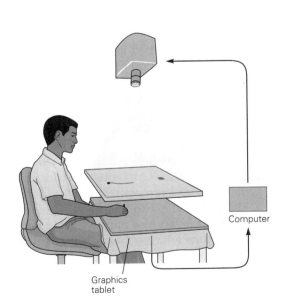

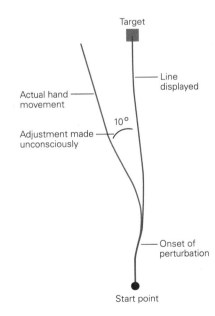

Figure 59–8 Actions can be modified unconsciously. Subjects are asked to draw a straight line with a computer mouse. They can see the line on the screen but not their hand movement. The computer is programmed to systematically distort the line displayed on the screen. In the result shown here, the subject had to move his hand 10° to the left to produce a vertical line on the screen. Subjects are not aware of making such adjustments. (Adapted, with permission, from Fourneret and Jeannerod 1998. Copyright © 1998 Elsevier Science Ltd.)

The Conscious Recall of Memories Is a Creative Process

For most of us, memory is the conscious imaginative reliving of a past experience. If we take no account of subjective experience (the behaviorist stance), however, memory is a process by which our past experience alters future behavior. Our behavior is often affected by past experience, but without conscious recall of the memory or awareness of the influence it is having on us. Once again, this type of experience is seen most strikingly in patients with damage to specific areas of the brain.

Some patients become densely amnesic after damage to the medial regions of the temporal lobe. They show no decline in intellect as measured by IQ tests but cannot remember anything for more than a few minutes. Although devastating, this memory impairment is actually rather circumscribed. The problem is largely manifested in *declarative memory*, and most severely in a type of declarative memory called *episodic memory*, the ability to recollect events in one's life (Chapter 54). *Procedural memory*, in which consciousness has a minor role (Chapter 53), remains intact. Thus, patients can still remember motor skills such as riding a bicycle and can often learn new motor skills at a normal rate. This selective effect of brain damage can lead to dramatic dissociations. A patient who has been learning some new skill every day for a week will deny ever having performed the task before. He is then surprised to find how skillful he has become.

A widely used protocol tests subjects' ability to recall lists of words they have memorized, a task that taps a form of declarative memory. In the recall phase, a subject is presented with a list of the words that were on the study list plus new words. An amnesic patient has great difficulty with this type of task and may misclassify most of the previously seen words as new since she cannot recall seeing them before. Nevertheless, the brain activity elicited by reading old words is different from that elicited by the new words: There is unconscious recognition of a difference, equivalent to that shown by patients with unilateral neglect or prosopagnosia. Normal subjects usually find this task easy, but they too will occasionally misclassify old words as new; as with amnesiacs, evoked brain responses in normal subjects register the distinction lost to conscious recall (Figure 59–9).

Occasionally, a subject misclassifies a new word as an old one. This misclassification amounts to a false memory. Such misclassifications are most likely to occur when the new word is semantically related to one or more of the old words. If the list of old words contained *big*, *great*, *huge*, then the new word *large* is

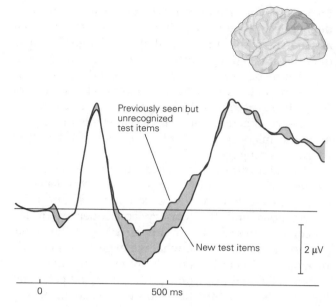

Figure 59–9 Brain activity shows the imprint of forgotten memories. Subjects were presented with a list of words, including some that had been presented earlier and some that were new. When asked to identify the words presented earlier, subjects correctly identified some of the old words but forgot others. Immediately after the visual presentation of a word, there is a brief fluctuation in the evoked potential in the brain. Evoked responses in the parietal region of the brain reflect whether or not the words had been seen before, even when subjects did not consciously recognize the words. The pattern produced by old words, whether recognized or not, is different from that produced by the new words. (Reproduced, with permission, from Rugg et al. 1998. Copyright © 1998 Springer Nature.)

likely to be identified as old. One explanation for this is that the perception of the new word *large* has been unconsciously primed by the previous presentation of the old words. Thus, the new word *large* is processed easily and quickly, and because the subject is aware of this, he concludes the word must be familiar and classifies it as old.

This observation emphasizes that memory is a creative process. Our conscious memories are constructed from both conscious recall and unconscious knowledge. To guard against false memories, as with false percepts, we use our knowledge about the world to determine which memories are plausible.

In some patients, the process by which memories are screened can become dramatically disturbed. If asked what happened yesterday, most patients with amnesia will say that they cannot remember, but a few will give elaborate accounts that do not correspond to reality. Such false memories are called confabulations and can sometimes be extremely implausible. For example, one patient said that he had met Harold Wilson (a former British Prime Minister) and discussed a building job they were both working on.

The creative mechanisms needed to reconstruct memories of past episodes are also involved when imagining events that might happen in the future. In amnesic patients with damage to the hippocampus, the ability to imagine new events is markedly impaired.

Behavioral Observation Needs to Be Supplemented With Subjective Reports

By the middle of the 20th century, it had become clear that the classic behaviorist approach was inadequate for the exploration of many psychological processes. Language acquisition, selective attention, and working memory cannot be understood in terms of relations between stimuli and responses, however complex the relationships postulated.

The demonstration that some cognitive processes are unconscious requires that we move even further from behaviorism. If we want to explore the whole range of conscious and unconscious cognitive processes, we will not be able to do so by focusing on overt behavior alone. We cannot assume that a subject making purposeful, goal-directed actions is necessarily aware of the stimuli eliciting the action or even of the action itself. We must supplement behavioral observations with subjective reports. We have to ask the subject, "Did you see the stimulus? Did you move your hand?"

One hundred years ago, introspection was the major method for obtaining data in psychology. How else could one study consciousness? But different schools of psychology obtained different results and, as John B. Watson emphasized, there seemed to be no objective way of deciding who was right. How can you independently confirm subjective experience? Thus, the method fell into disrepute. During the decades in which psychology was dominated by behaviorism, subjective reports were not considered an appropriate source of data. As a result, methods for recording subjective reports lag far behind methods for recording overt behavior. Regrettably, many studies of cognitive processes still do not require reports of subjective experience from subjects because of the long tradition of excluding such reports.

The one domain of psychology in which subjective reports continued to be used was psychophysics, the study of the relationship between sensation (physical energy) and perception (psychological experience) introduced by Fechner in 1860. Such studies give robust and reliable results and have created some of the few laws in psychology, such as Weber's law (the just-noticeable difference between two stimuli is proportional to the magnitude of the stimuli). In these studies, subjects are typically asked "Did you see the stimulus?" or "How confident are you that you saw the stimulus?"

Signal detection theory, developed in the 1950s, provides a robust methodology for measuring the ability to detect a stimulus (discriminability, d') independently of any reporting biases (Chapter 17). If your discriminability is high, then you will successfully detect small changes in the stimuli. More recently, there has been increasing interest in the second question, "How confident are you that you saw the stimulus?" Reporting one's confidence requires *metacognition*, the ability to reflect on our cognitive processes. This ability has an important role in the control of behavior. For example, if we realize that we are not performing some task very well, we might slow down and pay more attention to what we are doing.

The ability to reflect on our perception can be measured objectively. Likewise, the ability to reflect on the quality of our cognitive processes can also be assessed quantitatively. If your metacognitive accuracy is high, then you will successfully discriminate between your right and wrong answers. In other words, a correct detection will usually be associated with a high degree of confidence, whereas an incorrect detection will be associated with a low degree of confidence. However, your metacognitive accuracy need not be related to your signal detection ability. You could be good at detecting signals while at the same time poor at knowing whether your answers are likely to be right or wrong. In fact, patients with damage to anterior prefrontal cortex retain the ability to detect visual signals but show a marked deficit in metacognitive accuracy.

Verbal reports cannot, of course, be used in signal detection experiments with laboratory animals or preverbal infants. One alternative is to identify aspects of behaviors that reflect confidence. For example, if we are confident that we left our keys somewhere in the living room, we will spend more time looking there before we switch to the hall. Louise Goupil and Sid Kouider applied this insight to the study of metacognition in preverbal infants. The infants had to remember which of two boxes had contained a toy that was later removed without their knowledge. They spent more time searching inside the correct box. The infants were also more likely to ask an adult for help to open the correct box. These effects did not occur after long intervals. This behavior suggests that the infants had some insight into their current state of knowledge. They knew when they could no longer remember which was the correct box. Similar experiments suggest that rats and monkeys also have some metacognitive abilities.

Verification of Subjective Reports Is Challenging

Reports of subjective experience, such as confidence, serve like a meter. Just as an electrical meter converts electrical resistance into the position of a pointer on a dial (reading 100 ohms), so a subject converts a light stimulus into the report of a color ("I see red"). But there is a critical way in which the meter is not like a person. The meter does not experience red and cannot communicate meaning. And, although the meter might be faulty, it can never pretend to see red when it is really seeing blue. Most of the time, we presume that subjective reports are true, that is, the subject is trying as far as possible to give an accurate description of his experience. But how can we be sure that we can rely on these subjective reports?

The problem of verifying subjective reports can partially be addressed with the use of brain imaging. Brain imaging studies have shown that neural activity occurs in localized areas of the brain during mental activity that is not associated with any overt behavior. The content of such mental activity, such as imagining or daydreaming, can be known only from the subject's reports.

If we scan a subject while he says he is imagining moving his hand, activity will be detected in many parts of the motor system. In most motor regions, this activity is less intense than the activity associated with an actual movement, but it is well above resting levels. Similarly, if a subject reports that she is imagining a face she has recently seen, activity can be detected in the fusiform gyrus, the "face recognition area" (Figure 59–10). In these examples, the location of the observed neural activity detected by the scanner

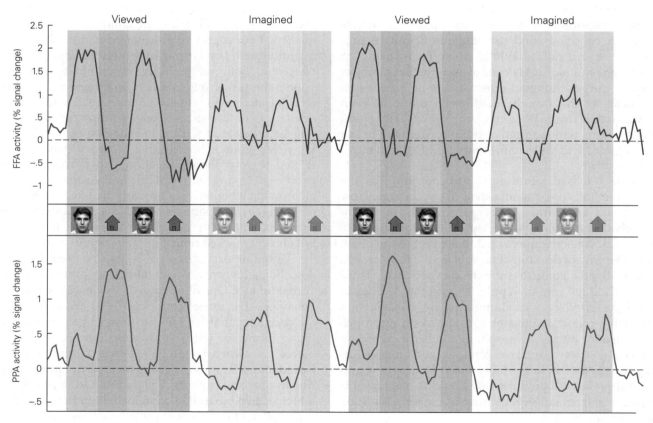

Figure 59–10 Imagining a face or a place correlates with activity in specific areas of the brain. Subjects were scanned while they viewed or imagined faces and houses. In the first block of trials, subjects alternately viewed a face or a house. When viewing a face, brain activity increases in the fusiform face area of the inferior temporal lobe (**FFA**). When viewing a house, brain activity increases in the parahippocampal place area of the inferior temporal cortex (**PPA**). In the next block of trials, subjects alternately *imagined* a face and a house. The same brain regions are active during both the imagining and direct viewing of faces and houses, although the activity is less pronounced during the imagined viewing. (Reproduced, with permission, from O'Craven and Kanwisher 2000. Copyright © 2000 MIT.)

provides independent confirmation of the content of the experience reported by the subject. The content of consciousness can, in certain limited cases, be inferred from patterns of neural activity.

Malingering and Hysteria Can Lead to Unreliable Subjective Reports

What if a subject reports seeing "blue" even though what they experienced was red? How could this arise, and what is the status of the subjective report in such cases?

Consider a patient who has become amnesic as a result of extensive damage to the medial temporal cortex. Shown a photograph of someone whom he sees every day on the ward, the patient denies ever having seen this person, even while physiological measurements (electroencephalogram or skin conductance) show a response to this photo (but not to photos of people he has not seen before). We conclude that conscious memory processes have been damaged while unconscious processes remain intact. This patient's subjective report is an accurate account of what he knows *consciously*, but it excludes those things he "knows" that have not entered consciousness.

Another patient, found wandering on the street, shows no evidence of brain damage but reports he cannot remember anything about himself or his history. When shown photographs of people from his past, he denies any knowledge of them, but at the same time, he shows physiological responses to the photos. In this case, because of the lack of detectable brain damage (and other features of the memory loss), we begin to wonder about the truthfulness of his statements. Perhaps the physiological responses indicate that he does consciously recognize people. Subsequently, the patient is identified by the police, and we discover that he is wanted for a serious crime committed in the neighboring county. Our doubts about the reliability of his reports increase. Finally, our suspicions are confirmed when he foolishly tells a fellow patient, "It's so easy to fool those clinical psychologists."

In this case, we have direct evidence that the patient was deliberately misleading others about himself. To deceive others, we must be conscious not only of our own mental state but also that of others. Is there some way we can test for deceit? One approach is to use a memory test of the kind discussed earlier. The patient studies a list of words. He is then shown a new list consisting of the words he has just studied and new words, and he must decide whether each word is old or new. A genuine amnesic would not recognize any of the words; he would have to guess, but through

unconscious priming effects, he would perform better than chance. The malingering patient can recognize the old words but will have a strong tendency to deny that he has seen them before. Unless he is very sophisticated, he may perform worse than chance. It seems we should be able to distinguish between the genuine amnesic and the malingerer.

A third kind of patient also simulates amnesia (or some other disorder) but does so unconsciously and thus is not a malingerer. Such a case would be called hysterical or psychogenic amnesia. Like the malingerer, his performance on the recognition test is worse than chance. Nonetheless, he is not aware of his simulation. The same mechanism occurs in normal people who have been hypnotized and then told that they will have no memory for what has just happened. This phenomenon is sometimes referred to as a dissociated state: That part of the mind that records experiences and makes verbal reports has become dissociated from the part that is creating the simulation. Hysterical simulations can also create sensory loss, such as hysterical blindness, and motor disorders, such as hysterical paralysis or hysterical dystonia.

We are still a long way from understanding the cognitive processes or underlying physiology of these disorders. A key problem is how to distinguish hysteria from malingering. From the standpoint of conscious experience, the two disorders are quite different: The malingerer is aware that he is simulating, whereas the hysterical patient is not. Yet the patients' subjective reports and overt behavior in the two cases are very similar. Is there no measure that can distinguish between these different disorders? Perhaps the only way to demonstrate the critical distinction between these different states of consciousness is through neuroimaging studies.

Highlights

1. The study of mental disorders forces us to confront the conceptual gap between the mental and the physical. It is no longer possible to maintain that mental disorders have mental causes, whereas physical disorders have physical causes.
2. Cognitive neuroscience has had a major impact on our attempts to bridge this gap because its descriptive language, the language of information processing, can be applied simultaneously to psychological and neural processes. Information theory and the development of the computer hint at how science can address the question of how subjective experience can emerge from activity in a physical brain.

3. It is now clear that perception, action, and memory are the result of many parallel processes and that, although some of these processes support conscious experience, the majority occur below the level of awareness.

4. Striking abnormalities occur when some of these processes are damaged while others remain intact. One patient, D.F., with damage to the inferior temporal cortex, is no longer consciously aware of the shape of an object and hence cannot describe it or recognize what it is. She can nevertheless form her hand into the appropriate shape to pick up the object.

5. We have very little awareness of the details of our actions, but we are vividly aware of being in control (the sense of agency). In extreme cases, this sense of agency can become detached from the control of action. After limb amputation, many people experience having a phantom limb that they can move, and after a limb has been paralyzed due to a stroke, some patients believe that they can still move the limb.

6. Recollection of the past is not like replaying a video. Memory is a creative process based on imperfect recall filled out with general knowledge. Through loss of this creativity, patients with amnesia have difficulty with imagining the future as well as remembering the past.

7. Subjective experience is an important part of human life. When we make a decision, our choice is indicated by our behavior, but our confidence in that choice is a subjective experience. We can study such experiences through verbal report. Confidence in our choices is an example of *metacognition* (ie, the ability to reflect on our cognitive processes). Damage to the frontal cortex can impair metacognition, while leaving decision-making intact.

8. Verbal reports are not always reliable. People can fake memory loss in order to escape justice. Malingering of this kind is very difficult to detect, since it closely resembles disorders such as hysterical amnesia, in which the patient is not aware that he is simulating the disorder. The challenge for cognitive neuroscience is to distinguish these cases.

Christopher D. Frith

Selected Reading

Dehaene S. 2014. *Consciousness and the Brain: Deciphering How the Brain Codes Our Thoughts*. New York: Viking.

Frith CD. 2007. *Making Up the Mind: How the Brain Creates Our Mental World*. Oxford: Blackwell.

Gazzaniga MS (ed). 1995. *Cognitive Neuroscience: A Reader*. Oxford: Blackwell.

Marr D. 1982. *Vision: A Computational Investigation into the Human Representation and Processing of Visual Information*. San Francisco: Freeman.

McCarthy R, Warrington EK. 1990. *Cognitive Neuropsychology: A Clinical Introduction*. London, San Diego: Academic Press.

Sacks O. 1970. *The Man Who Mistook His Wife for a Hat and Other Clinical Tales*. New York: Touchstone.

References

Bauer RM. 1994. Autonomic recognition of names and faces in prosopagnosia: a neuropsychological application of the Guilty Knowledge test. Neuropsychology 22:457–469.

Beck DM, Rees G, Frith CD, Lavie N. 2001. Neural correlates of change and change blindness. Nat Neurosci 4:645–650.

Beck JS. 1995. *Cognitive Therapy: Basics and Beyond*. New York: Guilford Press.

Burgess PW, Baxter D, Rose M, Alderman N. 1986. Delusional paramnesic syndrome. In: PW Halligan, JC Marshall (eds). *Method in Madness: Case Studies in Cognitive Neuropsychiatry*, pp. 51–78. Hove, UK: Psychology Press.

Caspar EA, Christensen JF, Cleeremans A, Haggard P. 2016. Coercion changes the sense of agency in the human brain. Curr Biol 26:585–592.

Dierks T, Linden DE, Jandl M, et al. 1999. Activation of Heschl's gyrus during auditory hallucinations. Neuron 22:615–621.

Ellis HD, Young AW. 1990. Accounting for delusional misidentification. Br J Psychiatry 157:239–248.

ffytche DH, Howard RJ, Brammer MJ, David A, Woodruff P, Williams S. 1998. The anatomy of conscious vision: an fMRI study of visual hallucinations. Nat Neurosci 1:738–742.

Fleming SM, Ryu J, Golfinos JG, Blackmon KE. 2014. Domain-specific impairment in metacognitive accuracy following anterior prefrontal lesions. Brain 137:2811–2822.

Fotopoulou A, Tsakiris M, Haggard P, Vagopoulou A, Rudd A, Kopelman, M. 2008. The role of motor intention in motor awareness: an experimental study on anosognosia for hemiplegia. Brain 131:3432–3442.

Fourneret P, Jeannerod M. 1998. Limited conscious monitoring of motor performance in normal subjects. Neuropsychology 36:1133–1140.

Frith CD. 2011. Explaining delusions of control: the comparator model 20 years on. Conscious Cogn 21:52–54.

Frith CD. 2013. Action, agency and responsibility. Neuropsychologia 55:137–142.

Glinsky EL, Schacter DL. 1988. Long-term retention of computer learning in patients with memory disorders. Neuropsychology 26:173–178.

Goupil L, Romand-Monnier M, Kouider S. 2016. Infants ask for help when they know they don't know. Proc Natl Acad Sci U S A 113:3492–3496.

Haggard P, Clark S, Kalogeras J. 2002. Voluntary action and conscious awareness. Nat Neurosci 5:382–385.

Hassabis D, Kumaran D, Vann SD, Maguire EA. 2007. Patients with hippocampal amnesia cannot imagine new experiences. Proc Natl Acad Sci U S A 104:1726–1731.

Jacoby LL, Whitehouse K. 1989. An illusion of memory: false recognition influenced by unconscious perception. J Exp Psychol Gen 118:126–135.

Kopelman MD. 1995. The assessment of psychogenic amnesia. In: AD Baddeley, BA Wilson, FN Watts (eds). *Handbook of Memory Disorders*. New York: Wiley.

Libet B, Gleason CA, Wright EW, Pearl DK. 1983. Time of conscious intention to act in relation to onset of cerebral activity (readiness potential). The unconscious initiation of a freely voluntary act. Brain 106:623–642.

Maniscalco B, Lau H. 2011. A signal detection theoretic approach for estimating metacognitive sensitivity from confidence ratings. Conscious Cogn 21:422–430.

Marshall JC, Halligan PW. 1988. Blindsight and insight in visuo-spatial neglect. Nature 336:766–767.

Milner AD, Goodale MA. 1995. *The Visual Brain in Action*. Oxford: Oxford Univ. Press.

Moore J, Haggard P. 2008. Awareness of action: inference and prediction. Conscious Cogn 17:136–144.

Moretto G, Walsh E, Haggard P. 201 Experience of agency and sense of responsibility. Conscious Cogn 20:1847–1854.

O'Craven KM, Kanwisher N. 2000. Mental imagery of faces and places activates corresponding stimulus-specific brain regions. J Cogn Neurosci 12:1013–1023.

Öhman A, Soares JJ. 1994. "Unconscious anxiety": phobic responses to masked stimuli. J Abnorm Psychol 103:231–240.

Penfield W, Perot P. 1963. The brain's record of auditory and visual experience: a final summary and discussion. Brain 86:595–696.

Rensink RA, O'Regan JK, Clark JJ. 1997. To see or not to see: the need for attention to perceive changes in scenes. Psychol Sci 8:368–373.

Rugg MD, Mark RE, Walla P, Schloerscheidt AM, Birch CS, Allan K. 1998. Dissociation of the neural correlates of implicit and explicit memory. Nature 392:595–598.

Schurger A, Mylopoulos M, Rosenthal D. 2016. Neural antecedents of spontaneous voluntary movement: a new perspective. Trends Cogn Sci 20:77–79.

Shepherd J. 2012 Free will and consciousness: experimental studies. Conscious Cogn 21:915–927.

Smith CH, Oakley DA, Morton J. 2013. Increased response time of primed associates following an "episodic" hypnotic amnesia suggestion: a case of unconscious volition. Conscious Cogn 22:1305–1317.

Stewart L, Battelli L, Walsh V, Cowey A. 1999. Motion perception and perceptual learning studied by magnetic stimulation. Electroencephalogr Clin Neurophysiol Suppl 51:334.

Swets JA, Tanner WPJ, Birdsall TG. 1959. Decision processes in perception. Psychol Rev 68:301–340.

Tong F, Nakayama K, Vaughn JT, Kanwisher N. 1998. Binocular rivalry and visual awareness in human extrastriate cortex. Neuron 21:753–759.

Watson JB. 1930. *Behaviorism*. Chicago: Univ. of Chicago Press.

Weiskrantz L. 1986. *Blindsight: A Case Study and Its Implications*. Oxford: Oxford Univ. Press.

60

Disorders of Thought and Volition in Schizophrenia

I**N THIS CHAPTER AND THE NEXT**, we examine disorders that affect perception, thought, mood, emotion, and motivation: schizophrenia, depression, bipolar disorder, and anxiety disorders. These have been challenging to understand, but recent progress in genetic analysis has begun to yield significant clues to their pathogenesis.

Mental illness has damaging effects on individuals, families, and society. The World Health Organization reports that mental illnesses, in the aggregate, constitute the leading cause of disability worldwide and are the leading risk factors for the 800,000 annual suicides reported by the World Health Organization. In addition, depression and anxiety disorders frequently co-occur with and worsen the outcomes of diabetes mellitus, coronary artery disease, stroke, and several other illnesses.

Medications such as antipsychotic drugs, lithium, and antidepressant drugs discovered during the mid-20th century made it possible to close large and often substandard mental hospitals; however, halfway houses and other less restrictive treatment settings did not materialize in sufficient numbers. As a result, many people with schizophrenia and severe bipolar disorder become homeless at some time in their lives, and in many countries, individuals with severe mental disorders compose a large fraction of prison populations.

In addition, although antipsychotic drugs, lithium, and antidepressant drugs have played important roles in controlling symptoms of mental disorders, significant limitations in treatment efficacy remain. For example, there are no effective treatments for the highly disabling cognitive impairments and deficit symptoms of schizophrenia. Even for symptoms that benefit from existing medications, such as hallucinations and delusions, residual symptoms remain and relapses are the rule. Because of significant scientific challenges posed by the human brain and limitations in animal models of mental disorders, there has been little advance in the efficacy of psychiatric drugs for more than 50 years. However, recent progress in human genetics and neural science has created significant opportunities to improve upon this unfortunate state of affairs.

Schizophrenia Is Characterized by Cognitive Impairments, Deficit Symptoms, and Psychotic Symptoms

In medicine, the understanding of a disease, and therefore its diagnosis, is ultimately based on identification of two features: (1) etiological factors (eg, microbes, toxins, or genetic risks) and (2) mechanism of pathogenesis (the processes by which etiologic agents produce disease). While human genetics and neural science are beginning to provide insights into the etiology and pathogenesis of disorders such as schizophrenia, bipolar disorder, and autism spectrum disorders, this research has not yet yielded objective diagnostic tests or biomarkers. As a result, psychiatric diagnoses still rely on a description of the patient's symptoms, the examiner's observations, and the course of the illness over time.

Schizophrenia is a very severe illness. Its symptoms can be divided into three clusters: (1) cognitive symptoms; (2) deficit, or negative, symptoms; and (3) psychotic symptoms. These symptom clusters exhibit different temporal patterns of onset—with cognitive impairments and deficit symptoms typically the earliest. The different timing of onset and the precise symptoms of each cluster are thought to result from the effects of developmental pathogenic mechanisms on different neural circuits and brain regions. As a result,

existing treatments such as antipsychotic drugs, which act on one "downstream" aspect of the disease process, exert no beneficial effects on cognitive impairments or deficit symptoms.

At the beginning of the 20th century, Emil Kraepelin in Germany recognized that cognitive decline was a distinguishing feature of schizophrenia, because psychotic symptoms occur in a variety of psychiatric conditions. Indeed, Kraepelin's term for what later came to be called schizophrenia was *dementia praecox*, a term that highlighted the early onset of cognitive loss. Cognitive impairments in schizophrenia target working memory and executive function, declarative memory, verbal fluency, the ability to identify the emotions conveyed by facial expressions, and other aspects of social cognition. These impairments do not significantly improve with existing medications, but ongoing research shows promising, albeit still modest, benefits from psychological therapies aimed at cognitive remediation.

Deficit symptoms include blunted emotional responses, withdrawal from social interaction, impoverished content of thought and speech, and loss of motivation. Psychotic symptoms include hallucinations, delusions, and disordered thought such as loosening of association (Box 60–1). Psychotic symptoms of schizophrenia are responsive to antipsychotic drugs. These drugs also reduce psychotic symptoms that

---

Box 60–1 Thought Disorder

The structure of a psychotic person's speech may range from wandering to incoherence, a symptom commonly referred to as loosening of association. Other examples of schizophrenic speech include neologisms (idiosyncratically invented words), blocking (sudden spontaneous interruptions), or clanging (associations based on the sounds rather than the meanings of words, such as, "If you can make sense out of nonsense, well, have fun. I'm trying to make cents out of sense. I'm not making cents anymore. I have to make dollars.")

Examples of loosening of associations are:

"I'm supposed to be making a film but I don't know what is going to be the end of it. Jesus Christ is writing a book about me."

"I don't think they care for me because two million camels . . . 10 million taxis . . . Father Christmas on the rebound."

Question: "How does your head feel?" Answer: "My head, well that's the hardest part of the job. My memory is just as good as the next working man's. I tell you what my trouble is, I can't read. You can't learn anything if you can't read or write properly. You can't pick up a nice book, I don't just mean a sex book, a book about literature or about history or something like that. You can't pick up and read it and find things out for yourself."

Several types of loosening of association have been described (eg, derailment, incoherence, tangentiality, or loss of goal). However, it remains unclear whether these reflect disturbances in fundamentally different mechanisms or different manifestations of a common underlying disturbance, such as the inability to represent a "speech plan" to guide coherent speech. A disturbance of such a mechanism would be consistent with, and may parallel, impairment of control of other cognitive functions in schizophrenia, such as deficits in working memory.

occur in other neuropsychiatric disorders, including bipolar disorder, severe depression, and neurodegenerative disorders such as Parkinson disease, Huntington disease, and Alzheimer disease.

Schizophrenia Has a Characteristic Course of Illness With Onset During the Second and Third Decades of Life

Schizophrenia affects 0.25% to 0.75% of the population worldwide, with only modest regional differences. Males are more commonly affected than females, with the sex ratio estimated to be 3:2, and onset is often earlier in males. Schizophrenia typically begins during the late teen years or the early to mid-twenties. Enduring cognitive and deficit symptoms generally begin months and sometimes years prior to the onset of psychotic symptoms. This period is referred to as the ultra-high-risk state by some researchers and as the schizophrenia prodrome by others.

Individuals in this risk state generally have measurable declines in cognitive functioning accompanied by such symptoms as social isolation, suspiciousness, and decreased motivation to engage in school work or other tasks. Attenuated psychotic symptoms often follow, including transient and mild hallucinations. Not every teen with such symptoms progresses to develop the full spectrum of symptoms warranting a diagnosis of schizophrenia. A small fraction recovers; others develop serious psychiatric conditions other than schizophrenia. Antipsychotic medications do not appear to benefit individuals in the risk state, nor do they delay the onset of schizophrenia. However, talk therapies and therapies delivered via computer-based approaches aimed at cognitive remediation show promise in delaying the onset of psychosis.

The Psychotic Symptoms of Schizophrenia Tend to Be Episodic

Psychotic symptoms, including hallucinations and delusions, are the most dramatic manifestations of schizophrenia. Hallucinations are percepts that occur in the absence of appropriate sensory stimuli, and they may occur in any sensory modality. In schizophrenia, the most common hallucinations are auditory. Typically, an affected person hears voices, but noises and music are also common. Sometimes, the voices will carry on a dialog and frequently are experienced as derogatory or bullying. Occasionally, voices will issue commands to the affected individual that can create a high risk of harm to self or others.

Delusions are firm beliefs that have no realistic basis and are not explained by the patient's culture, nor are they amenable to change by argument or evidence. Delusions may be quite varied in form. For some affected individuals, reality is significantly distorted: The world is full of hidden signs meant only for the affected person (ideas of reference), or the person believes that he is being closely watched, followed, or persecuted (paranoid delusions). Others may experience bizarre delusions; for example, they may believe that someone is inserting thoughts into or extracting thoughts from their minds or that their close relatives have been replaced by aliens from another planet. In addition to the person's enduring cognitive impairments, psychotic episodes are frequently accompanied by disordered thought and odd patterns of speech (Box 60–1).

Psychotic symptoms may also occur in other neuropsychiatric disorders, such as bipolar disorder, major (unipolar) depression, various neurodegenerative disorders, and drug-induced states. However, these other conditions can usually be distinguished from schizophrenia by associated symptoms and age of onset. Once schizophrenia has become fully manifest, psychotic symptoms tend to be episodic. Periods of florid psychosis accompanied by markedly disordered thinking, emotion, and behavior are interspersed with periods in which psychotic symptoms are milder or even absent. Psychotic episodes typically require hospitalization; the severity and duration of such episodes are markedly shortened by antipsychotic drugs. First and second episodes of psychosis often respond fully to antipsychotic drugs, but cognitive impairments and deficit symptoms typically persist. After the first few psychotic relapses, people with schizophrenia typically suffer residual psychotic symptoms even between their acute relapses and suffer these symptoms despite treatment with antipsychotic drugs. Cognitive and social functioning typically continue to deteriorate over several years until they reach a plateau well below the person's premorbid level of functioning.

The Risk of Schizophrenia Is Highly Influenced by Genes

As early as 1930, Franz Kalman in Germany studied familial patterns of schizophrenia and concluded that genes contribute significantly. To separate genetic from environmental influences more clearly, Seymour Kety, David Rosenthal, and Paul Wender examined children who were adopted at or shortly after birth in Denmark. They found that the rate of schizophrenia in the biological family of the adoptee was much more strongly predictive of schizophrenia than the rate of schizophrenia in the adoptive family.

Kety and his colleagues also observed that some of the biological relatives of adoptees with schizophrenia exhibited milder symptoms related to schizophrenia, such as social isolation, suspiciousness, eccentric beliefs, and magical thinking, but not frank hallucinations or delusions. Since Kety's time, it has been observed that such relatives may also exhibit cognitive impairments that are intermediate between unaffected individuals and those with schizophrenia. They also may exhibit thinning of the cerebral cortex observed by magnetic resonance imaging (MRI) that is also intermediate between healthy individuals and those with schizophrenia. (Cortical thinning in schizophrenia is discussed below.) Such individuals are now diagnosed with schizotypal disorder, which appears to be the milder end of the schizophrenia spectrum of psychotic disorders. The severity and nature of symptoms appear to be influenced by the individual's overall burden of risk-associated genetic variants as well as exposure to environmental risk factors.

Irving Gottesman's studies of extended pedigrees of Danish patients with schizophrenia supported the importance of genes. Gottesman noted the correlations between the risk of schizophrenia in relatives and the degree to which they shared DNA sequences with an affected person. He found a greater lifetime risk of schizophrenia among first-degree relatives (including parents, siblings, and children, who share 50% of

DNA sequences with the patient) than among second-degree relatives (including aunts, uncles, nieces, nephews, and grandchildren, who share 25% of their DNA sequences). Even third-degree relatives (who share only 12.5% of the patient's DNA sequences) were at higher risk for schizophrenia than the approximately 1% of the general population at risk for this disease (Figure 60–1).

Based on the differences in levels of risk Gottesman measured in these pedigrees, he recognized that schizophrenia risk was not transmitted within families as Mendelian dominant or recessive traits (ie, it was not caused by a single genetic locus). He predicted correctly that schizophrenia is a polygenic trait, involving a large number of loci throughout the human genome. This genetic architecture underlies many human phenotypes, including disease phenotypes, and may involve many hundreds of loci within the genome. In polygenic traits, variants at each disease-associated locus contribute small, additive effects to the phenotype. Genetic risk variants act together with environmental factors to produce the schizophrenia phenotype.

In 2014, a large global consortium reported on a genome-wide association study of more than 35,000 individuals with schizophrenia. The study identified 108 genome-wide significant loci associated with schizophrenia that were distributed across the genome. The research continues, and the number of known loci is already greater than 250. Each of these loci represents a segment of DNA identified by a single

Figure 60–1 The lifetime risk of schizophrenia increases as a function of genetic relatedness to a person with schizophrenia. The risk of schizophrenia rises with genetic relatedness to an affected individual and, therefore, with increased sharing of DNA sequences. However, the pattern of segregation in families does not follow simple Mendelian ratios; rather, inheritance reflects genetic complexity. In addition, risk varies within categories of relatedness (first- and second-degree relatives), suggesting a role for unshared developmental or environmental effects. (Reproduced, with permission, from Gottesman 1991.)

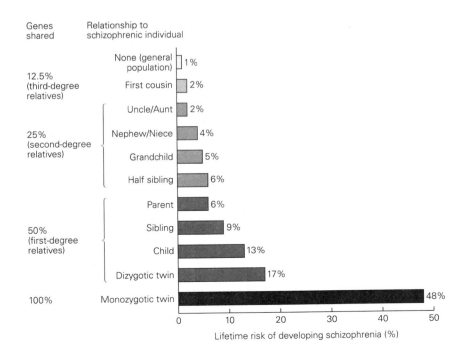

nucleotide polymorphism that confers a small increment in risk (typically 5%–10%) for schizophrenia. The value of such allelic variants is as a tool to identify genes that play a role in the molecular mechanism of disease. In turn, the implicated genes help identify molecular pathways that can potentially be exploited in the development of therapeutic drugs.

In addition to the utility of genetics for discovering biological processes involved in disease, it can also contribute to the stratification of study populations in epidemiological and clinical studies. A person's risk of schizophrenia or other disorders can be estimated by calculating his or her total burden of common risk alleles for the condition. The result is a polygenic risk score, a measure that is increasingly being used to stratify populations by genetic susceptibility to schizophrenia in both clinical studies and in epidemiologic studies of environmental risk factors.

Environmental risk factors for schizophrenia that have been replicated across studies include nutrient deprivation in utero (notably in studies following famines), season of birth (winter and early spring birth), urban birth, and migration. The analysis of causal factors within such broad categories of exposure is likely to benefit from knowing who is susceptible. Moreover, clues to environmentally induced causal pathways may be found in the risk genotypes of those with schizophrenia who have had a particular exposure.

Given the lack of objective diagnostic tests, current diagnostic criteria, such as those within the fifth edition of the *Diagnostic and Statistical Manual of Mental Disorders*, are based on clinical observation and course of illness. As a result, individuals currently diagnosed with schizophrenia are highly heterogeneous. Polygenic risk scores can explain only a portion of the variance in schizophrenia cohorts, and the scores provide only probabilistic information. However, they represent the first objective tool that permits stratification of subjects diagnosed with schizophrenia. As such, the application of such scores may begin to diminish heterogeneity in clinical studies ranging from neuroimaging to neurophysiological studies to treatment trials.

Although almost all cases of schizophrenia reflect polygenic risks, as predicted by Gottesman, a small percentage of cases are highly influenced by the presence of a penetrant mutation that typically exerts pleiotropic effects, including intellectual disability, resulting in what is often called syndromic schizophrenia. Most of these penetrant mutations are copy number variants: deletions, duplications, or sometimes triplications of a particular segment of a chromosome.

The most common and best studied cause of syndromic schizophrenia is the 22q11.2 microdeletion, which accounts for approximately 1% of patients diagnosed with schizophrenia. The microdeletion typically occurs de novo and results in loss of one of two copies of 38 to 44 genes. As is typical for such copy number variations, those affected suffer from a complex of symptoms. The syndrome accompanying the 22q11.2 microdeletion, sometimes called velocardiofacial or DiGeorge syndrome, includes cognitive disability, cardiovascular defects, and facial dysmorphology. The penetrance of each of these symptoms and signs is independent of the others; thus, affected individuals have different combinations of phenotypes. Individuals with the 22q.11.2 microdeletion have a 25% to 40% risk of schizophrenia and a 20% risk of autism. Other syndromic forms of psychosis are similarly variable.

Syndromic forms of schizophrenia can provide important windows into the biology of psychosis, even if their similarities to common polygenic types of schizophrenia are still a matter of study. One powerful advantage of penetrant mutations is the ability to generate cellular and animal models in order to characterize their effects on brain structure and function. A second advantage is the ability to prospectively study individuals carrying these mutations. Studying syndromic schizophrenia, therefore, has the potential to reveal much about basic pathophysiological mechanisms. One important area of investigation is how copy number variations and other high-penetrance mutations that lead to psychosis manifest based on a person's genetic background, specifically the many common DNA variants that influence risk. To this point, recent findings suggest that the propensity in individuals carrying a copy number variation to develop psychotic symptoms may result from a strong interaction of the copy number variation with the person's polygenic background risk for schizophrenia, suggesting significant shared mechanisms between schizophrenia associated with single genetic mutations and that associated only with polygenic variants.

Schizophrenia Is Characterized by Abnormalities in Brain Structure and Function

Abnormalities in the structure and function of the brain have been identified in schizophrenia both by postmortem examination and by a variety of noninvasive technologies in living patients. The best replicated finding, both by postmortem study and by structural MRI, is loss of gray matter in prefrontal, temporal, and parietal regions of cerebral cortex (Figure 60–2) with counterbalancing increases in the size of the cerebral ventricles (Figure 60–3). Thinning of the cerebral cortex

Figure 60–2 Gray matter loss in schizophrenia. Gray matter loss is well documented in schizophrenia. First-degree relatives who do not have a diagnosis of schizophrenia still often exhibit cortical gray matter loss intermediate between healthy individuals and those diagnosed with schizophrenia. Consistent with this, a study that examined losses of cortical gray matter in monozygotic and dizygotic twin pairs discordant for schizophrenia compared to healthy matched control twins found significant losses in those at genetic risk for schizophrenia but without the disease. Those members of twin pairs diagnosed with schizophrenia demonstrated additional, disease-specific cortical thinning in dorsolateral prefrontal, superior temporal, and superior parietal association areas. These additional defects appear to reflect the influence of nongenetic factors involved in pathogenesis (eg, developmental or environmental factors). The disease-specific gray matter loss correlates with the degree of cognitive impairment rather than with duration of illness or drug treatment. The images here show regional deficits in gray matter in monozygotic twins with schizophrenia relative to their healthy co-twins (n = 10 pairs) viewed from the right, left, and right oblique perspectives. Differences in twins are illustrated by the pseudocolor scale superimposed on cortical surface maps, with **pink** and **red** indicating the greatest statistical significance. (Reproduced, with permission, from Cannon et al. 2002.)

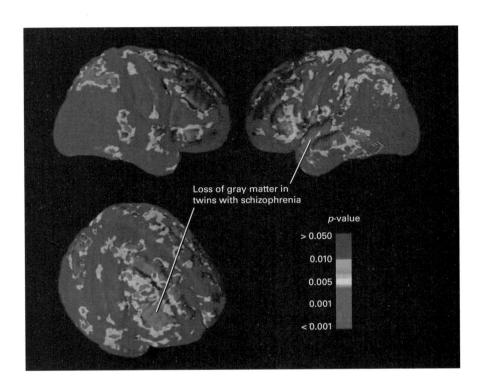

Figure 60–3 Enlargement of lateral ventricles in schizophrenia. Magnetic resonance imaging comparison of monozygotic twins discordant for schizophrenia. The affected member of the twin pair has the enlarged ventricles characteristic of schizophrenia. Because there is a wide range of normal ventricular volumes in the population, an unaffected monozygotic twin serves as a particularly appropriate control subject. Because monozygotic twins have identical genomes, this comparison also illustrates the role of nongenetic factors in schizophrenia.

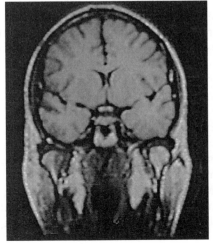
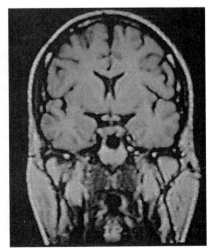

Unaffected twin   Schizophrenic twin

is most pronounced in the dorsolateral prefrontal cortex, a brain region critical for working memory and thus cognitive control of thought, emotion, and behavior.

Loss of gray matter in the superior temporal gyrus, temporal pole, amygdala, and hippocampus in schizophrenia has also been correlated with impairments in cognition, recognition of emotions in others, and regulation of emotion in the affected person. Functional neuroimaging using positron emission tomography and functional MRI (fMRI) has demonstrated that patients' deficits in performing working memory tasks while being imaged are associated with decrements in the activation of dorsolateral prefrontal cortex, a brain region known to play a critical role in working memory (Figure 60–4).

There is also growing recognition that schizophrenia is characterized by disruptions in connectivity between brain regions (Figure 60–5). Anatomical connectivity can be measured by diffusion tensor imaging, which identifies major axon tracts as they course between brain regions. Functional connectivity between brain regions can be estimated physiologically by measuring the degree to which activity patterns in different brain regions correlate with each other, using such approaches as resting state fMRI and electrophysiology. Both imaging and physiological methods reveal that individuals with schizophrenia have deficits in the connections between brain regions. Weaker connections would likely impair cognition and complex behaviors.

Loss of Gray Matter in the Cerebral Cortex Appears to Result From Loss of Synaptic Contacts Rather Than Loss of Cells

Postmortem studies have examined the cellular abnormalities that underlie the gross anatomical findings and functional deficits in schizophrenia. These studies have revealed that gray matter loss in the prefrontal and temporal cortical regions is not the result of cell death but rather a reduction in dendritic processes. As a consequence, the packing density of cells in the cerebral cortex increases. More cells per unit volume and less total gray matter contribute to enlargement of the ventricular spaces.

A reduction in dendrites and dendritic spines on pyramidal neurons, the most common type of excitatory neuron in the neocortex (Figure 60–6), would likely signify a loss of synaptic contacts in affected brain regions in individuals with schizophrenia. The loss of synaptic connections could underlie abnormalities in long-range functional connectivity and

failures to recruit prefrontal cortical regions during tasks that require working memory (Figures 60–4 and 60–5).

Abnormalities in Brain Development During Adolescence May Be Responsible for Schizophrenia

Schizophrenia exhibits a stereotypic onset between late adolescence and early adulthood, with cognitive decline and negative symptoms occurring months or years before the onset of psychosis. This timing suggests the pathogenesis of schizophrenia might involve abnormalities in the late stages of brain development during adolescence, when cognitive function, emotion regulation, and executive function normally mature.

Throughout development, neurons elaborate an excessively large number of synaptic connections. Generally, synapses are strengthened and preserved when they are utilized, while weak or inefficient synapses are eliminated through a process called pruning. The process of synaptic refinement, which involves both synaptogenesis and pruning, results in neural computations that are efficient and adapted to the environment. Experience-dependent synaptic refinement was first described in the visual cortex, where pruning of weak connections is necessary for the emergence of binocular vision (see Chapter 49). Synaptogenesis and pruning continue throughout life, making possible new learning and updating of older memories. However, superimposed on such local events are significant waves of synaptic pruning that are spatially specific and developmentally timed. The last such wave in human brain maturation occurs during adolescence and early adulthood, with pruning in the temporal and prefrontal association cortex. This late wave of pruning is followed by myelination of many axons in these areas of cortex.

In the early 1980s, Irwin Feinberg hypothesized that schizophrenia might result from abnormal and excessive synaptic pruning during adolescence. Postmortem examination of the brains of persons with schizophrenia subsequently demonstrated a reduction of dendritic spines, and of synapses, in prefrontal and temporal cortices. Studies in nonhuman primates, taken together with human postmortem and neuroimaging studies, suggest that loss of dendritic arbors does not result from antipsychotic medications taken by many individuals with schizophrenia. The onset of cognitive impairment and negative symptoms during this period is consistent with the idea that synaptic pruning somehow goes haywire, damaging the ability of the cerebral cortex to process information. When the overpruning hypothesis was first enunciated in the

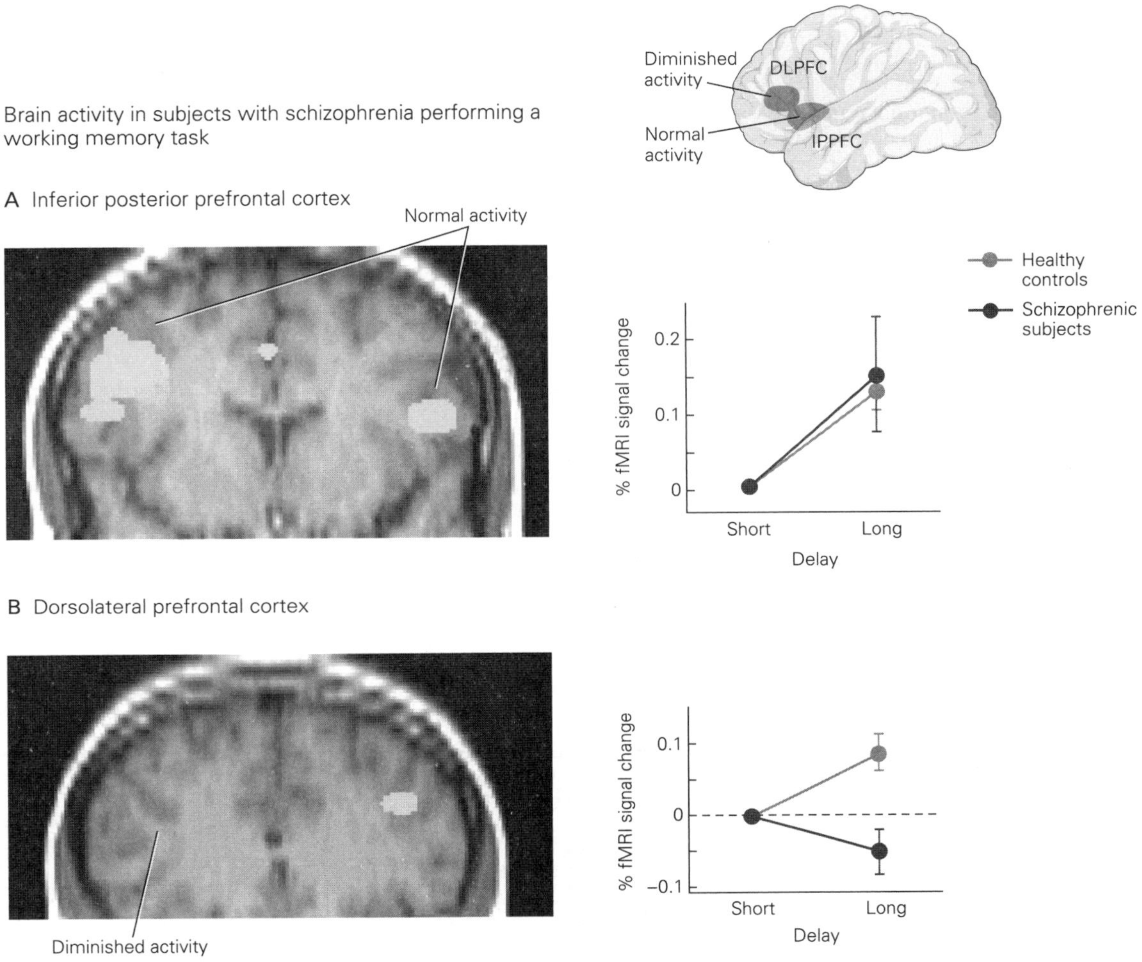

Figure 60–4 Deficits in the function of prefrontal cortex in schizophrenia. Functional magnetic resonance imaging (fMRI) was used to test the hypothesis that in patients with schizophrenia working memory engages circuits in the prefrontal cortex differently than in controls. Activity in the prefrontal cortex of two groups—patients with schizophrenia (first-episode patients who had never been given antipsychotic drugs) and healthy controls—was examined while subjects performed a working memory task. Subjects were presented with a sequence of letters and instructed to respond to a particular letter (the "probe" letter) only if it immediately followed another specified letter (the "contextual cue" letter). Demands on working memory were increased by increasing the delay between the cue and the probe letters. The greater demand on working memory requires greater activation of prefrontal cortical circuits. (Adapted, with permission, from Barch et al. 2001.)

A. In both patients with schizophrenia and controls, normal increases in activation within inferior posterior regions of prefrontal cortex (**IPPFC**; Brodmann's area 44/46) as a function of demand on working memory suggest that the function of these regions remains intact in schizophrenia. The plot shows the fMRI signal change that occurs in the right side of the prefrontal cortex in the long-delay and short-delay conditions in healthy controls and in patients with schizophrenia. Similar effects were observed for the left side.

B. There is less activity in Brodmann's area 46/49, a region of dorsolateral prefrontal cortex (**DLPFC**), in patients with schizophrenia relative to healthy controls. Unlike Brodmann's area 44/49 (shown in part **A**), Brodmann's area 46/49 is not activated normally in subjects with schizophrenia, consistent with the deficit in working memory seen in patients with schizophrenia. Selective impairment of one region of prefrontal cortex alongside other regions that appear to have normal function suggests that the impairment is due to a regionally specific process rather than a diffuse and nonspecific pathological process.

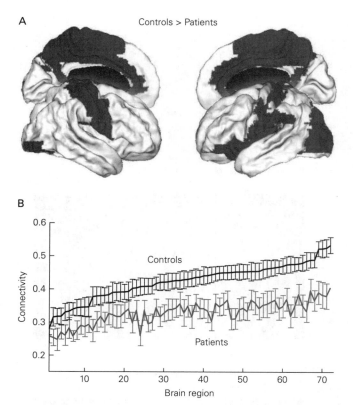

Figure 60–5 Decreased functional connectivity in schizophrenia. Correlations in neural activity between 72 defined brain regions were measured in patients with schizophrenia and in control subjects by resting state functional magnetic resonance imaging. (Reproduced, with permission, from Lynall et al. 2010.)

A. Brain regions that showed statistically significant reductions in functional connectivity in patients compared to controls are highlighted in **red**.

B. Mean (+/– standard error of the mean) functional connectivity between each brain region and the rest of the brain for patients and healthy controls.

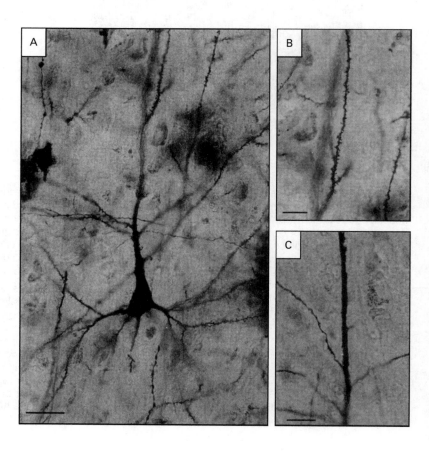

Figure 60–6 Photomicrographs of pyramidal neurons from the cerebral cortex from human brains stained by the Golgi method.

A. A layer III pyramidal neuron from a control brain, showing its morphology and its dendrites which are studded with spines.

B. A higher power view showing spines on a dendrite of a pyramidal neuron from a control brain.

C. A segment of a dendrite devoid of spines from the cerebral cortex of a person who had schizophrenia. (Scale: **A:** 30 µm; **B:** 20 µm; **C:** 15 µm.) Spine numbers are a rough proxy for the number of synaptic contacts onto the dendrite from other neurons; thus the paucity of spines in schizophrenia is consistent with fewer synaptic contacts than are found in the cerebral cortex of healthy brains.

(Reproduced, with permission, Garey et al. 1998. With permission from BMJ Publishing Group Ltd.)

1980s, it lacked a plausible molecular or cellular mechanism that might explain how synaptic pruning might go awry in schizophrenia. Recent genetic analysis may have provided a solution.

Unbiased, large-scale genetic studies have found that the strongest association with risk for schizophrenia lies within the major histocompatibility (MHC) locus on chromosome 6. The MHC locus encodes many proteins involved in immune function. Fine mapping of the locus pinpointed the largest genetic association signal to the genes encoding complement factor C4, a component of the classic complement cascade that, outside the brain, is involved in tagging microbes and damaged cells for engulfment and destruction by phagocytic cells. Subsequent analysis showed that the risk for schizophrenia is elevated as a function of increased expression in the brain of C4A (one of two isoforms). This finding adds support to the overpruning hypothesis because one function of the complement system in brain is to tag weak or inefficient synapses for removal by microglia (Figure 60–7).

Elevated expression of the complement factor C4A involved in synaptic pruning is certainly not the only mechanism leading to schizophrenia. As with any polygenic disorder, no one gene is necessary or sufficient for the disease phenotype. Thus, not everyone with schizophrenia has a high-risk C4A genotype, and not everyone with a high-risk C4A genotype develops schizophrenia. Many other genes are implicated in the risk for schizophrenia. Several such risk factors other than C4 are involved in regulation of the complement cascade, but the vast majority are not. Many of the genes associated with schizophrenia that have been identified to date are involved in various aspects of the structure and function of synapses; several encode ion channels. Thus, it seems likely that the genetic risk for schizophrenia involves, at least in part, synaptic function, synaptic plasticity, and synaptic pruning, and overpruning of synapses during adolescence is one plausible mechanism that should be explored further in studies of youth at high risk for schizophrenia. Nevertheless, other pathways, as yet less well characterized, may also turn out to be important. We have a long way to go in understanding the pathogenesis of schizophrenia.

Antipsychotic Drugs Act on Dopaminergic Systems in the Brain

All current antipsychotic drugs produce their therapeutic effects by blocking D_2 dopamine receptors in the forebrain. These drugs have many other effects at

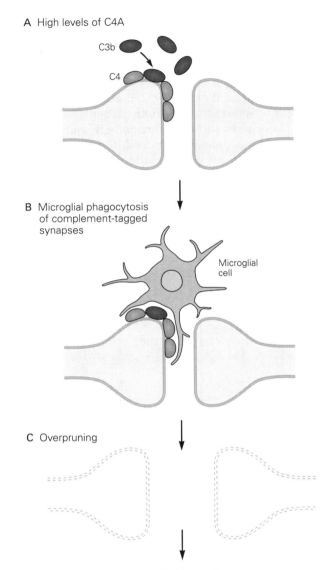

A High levels of C4A

C3b

C4

B Microglial phagocytosis of complement-tagged synapses

Microglial cell

C Overpruning

Development of schizophrenia symptoms

Figure 60–7 Complement factors and microglia have a role in synapse elimination. Maturation and plasticity of the nervous system involve both synaptogenesis and elimination of weak synapses. Complement factor 3b (**C3b**) is thought to serve as a "punishment signal" that identifies weak synapses for phagocytosis by microglia. Complement factor 4 (**C4**), a component of the complement cascade, is synthesized by neurons and astrocytes and recruits C3b to weak synapses. In humans, a complex genomic locus on chromosome 6 contains varying numbers of copies of the genes that encode the complement factor C4 proteins, C4A and C4B. Variants within this locus that give rise to high levels of C4A expression in brain increase schizophrenia risk. (Reproduced, with permission, from Christina Usher and Beth Stevens.)

various neurotransmitter receptors and intracellular signaling pathways, but these other actions primarily influence their side effects, not their main therapeutic mechanisms (Figure 60–8).

The first effective antipsychotic drug, chlorpromazine, was developed for its antihistaminic and sedating effects and was first investigated as a surgical preanesthetic by Henry Laborit in 1952. Based on its sedating effects, it was tested in psychotic patients soon thereafter. These tests showed, surprisingly, reduced hallucinations and delusions; indeed, the sedative effect of chlorpromazine is now considered a side effect. The success of chlorpromazine led to attempts to discover other antipsychotic drugs. Although many chemically diverse antipsychotic drugs are now in use, all share the same initial action of chlorpromazine in the brain, the ability to block the D_2 dopamine receptor. As a class, these drugs ameliorate psychotic symptoms not only in schizophrenia, but also in bipolar disorder, severe depression, and various neurodegenerative disorders. None of the antipsychotic drugs provide effective treatment for the cognitive impairments or deficit symptoms of schizophrenia.

Among their side effects, chlorpromazine and related drugs caused Parkinson-like motor symptoms. Because Parkinson disease is caused by the loss of dopaminergic neurons in the midbrain, the occurrence of Parkinson-like side effects suggested to Arvid Carlsson that these drugs acted by decreasing dopaminergic transmission. Following up on this idea, Carlsson established that the antipsychotic drugs block dopamine receptors. Two families of dopamine receptors are known. The D_1 family, which in humans includes D_1 and D_5, are coupled to stimulatory G proteins that activate adenylyl cyclase. The D_2 family, which includes D_2, D_3, and D_4, are coupled to the inhibitory G protein (G_i) that inhibits the cyclase and activates a hyperpolarizing K^+ channel. A second signaling pathway for D_2 receptors is mediated by β-arrestin. The D_1 receptor is expressed in the striatum and is the major class of dopamine receptor in the cerebral cortex and hippocampus. The D_2 receptor is expressed most densely in the striatum, cerebral cortex, amygdala, and hippocampus. Correlations between receptor binding studies and clinical efficacy on psychotic symptoms indicated that the D_2 family is the molecular target for the therapeutic actions of antipsychotic drugs.

Clozapine, an antipsychotic drug discovered in 1959, had a low liability for causing Parkinson-like motor side effects. However, because it had some severe side effects, including a small chance of causing a potentially lethal loss of blood granulocytes, its use was discontinued until a clinical trial in the late 1980s clearly showed that it had greater efficacy than other antipsychotic drugs. Clozapine caused improvement in some individuals who had not responded to other antipsychotic drugs. It was reintroduced in conjunction with weekly monitoring of white cell counts; attempts to equal the efficacy of clozapine also motivated the development of second-generation antipsychotic drugs that mimicked some of its receptor binding properties, notably the ability to block serotonin $5\text{-}HT_{2A}$ receptors, an action that appears to diminish motor side effects. Large-scale clinical trials of the second-generation antipsychotic drugs have shown that their efficacy is no greater than the first-generation drugs, with none

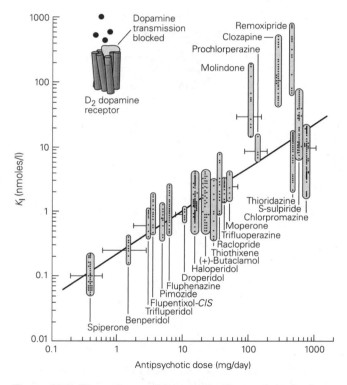

Figure 60–8 The potency of first-generation antipsychotic drugs in treating psychotic symptoms correlates strongly with their affinity for D_2 dopamine receptors. On the horizontal axis is the average daily dose required to achieve similar levels of clinical efficacy. On the vertical axis is K_i, the concentration of drug required to bind 50% of D_2 receptors in vitro. The higher the drug concentration required, the lower is the affinity of the drug for the receptor. One caveat is that the measurements on the two axes were not entirely independent of each other, as the ability of a drug to block D_2 receptors in vitro was often used to help determine doses used in clinical trials. Clozapine, which does not fall on the line, has significantly greater efficacy than the others. The mechanism of its greater efficacy is not understood. (Adapted, with permission, from Seeman et al. 1976.)

having efficacy equal to clozapine. The liability of second-generation drugs for causing Parkinson-like motor side effects is lower than that of the first generation, but they typically cause more severe weight gain and other metabolic problems.

Because drugs that reduce psychotic symptoms do so by blocking D_2 receptors, investigators have asked: What is the role of dopamine in the symptoms of schizophrenia? Although some drugs that block D_2 receptors reduce psychotic symptoms, other drugs that increase dopamine at synapses, such as amphetamine and cocaine, can produce psychotic symptoms when taken chronically at high doses. Thus, Carlsson suggested that dopaminergic systems are hyperactive in schizophrenia. Evidence for this hypothesis has been difficult to obtain. The most direct evidence for this idea comes from studies begun in the mid-1990s that found that amphetamine-induced increases in dopamine release were greater in patients with schizophrenia than in healthy subjects. These studies suggest that abnormalities in amphetamine-sensitive processes—such as dopamine storage, vesicular transport, dopamine release, or dopamine reuptake by presynaptic neurons—may lead to hyperactivity in the subcortical dopaminergic systems and could contribute to the psychotic symptoms of schizophrenia, the symptoms that respond to antipsychotic drugs.

Although dopamine activity may increase in subcortical regions of the brain in schizophrenia, it may decrease in cortical regions; such a decrease might contribute to the cognitive impairments seen in schizophrenia. In particular, there may be fewer D_1 receptors in the prefrontal cortex in schizophrenia, which would be consistent with the observation that D_1 receptors in the prefrontal cortex normally play a role in working memory and in executive functions.

Highlights

1. Schizophrenia is a chronic, profoundly disabling disorder characterized by dramatic psychotic symptoms as well as deficits in emotion, motivation, and cognition.
2. Risk for schizophrenia is an inherited, polygenic trait.
3. Antipsychotic drugs are effective in reducing hallucinations, delusions, and thought disorder but do not benefit the cognitive and deficit symptoms of schizophrenia.
4. Cognitive impairments reduce the ability of people with schizophrenia to regulate their behavior in accordance with their goals. As a result, people

with schizophrenia are frequently unable to succeed in school or to hold down jobs, even at times when antipsychotic drugs effectively control their hallucinations and delusions.
5. Postmortem and neuroimaging studies document loss of gray matter in the prefrontal and temporal cerebral cortex in a pattern that is consistent with cognitive impairments, such as deficits in working memory.
6. The gray matter loss results from decreased dendritic arborization and decreased dendritic spines, which implies that synaptic connections are also reduced. One hypothesis consistent with these anatomic findings and with the typical age of onset in adolescence is that schizophrenia is triggered by excessive and inappropriate synaptic pruning in the prefrontal and temporal cerebral cortices during adolescence and young adulthood.
7. Progress in the genetic analysis of schizophrenia combined with the use of new tools to study systems-level neuroscience promises to help attain the much needed advances in understanding disease mechanisms and in discovering new therapeutics.

Steven E. Hyman
Joshua Gordon

Selected Reading

Barch DM. 2005. The cognitive neuroscience of schizophrenia. Annu Rev Clin Psychol 1:321–353.

Nestler EJ, Hyman SE, Holtzman D, Malenka RJ. 2015. *Molecular Neuropharmacology: Foundation for Clinical Neuroscience*, 3rd ed. New York: McGraw-Hill.

Owen MJ, Sawa A, Mortensen PB. 2016. Schizophrenia. Lancet 388:86–97.

Stephan AH, Barres BA, Stevens B. 2012. The complement systems: an unexpected role in synaptic pruning during development and disease. Annu Rev Neurosci 35:369–389.

References

Addington J, Heinssen R. 2012. Prediction and prevention of psychosis in youth at clinical high risk. Annu Rev Clin Psychol 8:269–289.

Barch DM, Carter CS, Braver TS, et al. 2001. Selective deficits in prefrontal cortex function in medication-naïve patients with schizophrenia. Arch Gen Psychiatry 58:280–288.

Brans RG, van Haren NE, van Baal GC, et al. 2008. Heritability of changes in brain volume over time in twin pairs discordant for schizophrenia. Arch Gen Psychiatry 65:1259–1268.

Cannon TD, Thompson PM, van Erp TG, Toga AW. 2002. Cortex mapping reveals regionally specific patterns of genetic and disease-specific gray-matter deficits in twins discordant for schizophrenia. Proc Natl Acad Sci U S A 99:3228–3233.

Feinberg I. 1983. Schizophrenia: caused by a fault in programmed synaptic elimination during adolescence? J Psychiatr Res 17:319–324.

Fisher M, Loewy R, Hardy K, Schlosser D, Vinogradov S. 2013. Cognitive interventions targeting brain plasticity in the prodromal and early phases of schizophrenia. Annu Rev Clin Psychol 9:435–463.

Fusar-Poli P, Borgwardt S, Bechdolf A, et al. 2013 The psychosis high-risk state: a comprehensive state-of-the-art review. JAMA Psychiatry 70:107–120.

Garey LJ, Ong WY, Patel TS, et al. 1998. Reduced dendritic spine density on cerebral cortical pyramidal neurons in schizophrenia. J Neurol Neurosurg Psychiatry 65:446–453.

Glantz LA, Lewis DA. 2000. Decreased dendritic spine density on prefrontal cortical pyramidal neurons in schizophrenia. Arch Gen Psychiatry 57:65–73.

Gottesman II. 1991. *Schizophrenia Genesis: The Origins of Madness.* New York: Freeman.

Gur RE, Cowell PE, Latshaw A, et al. 2000. Reduced dorsal and orbital prefrontal gray matter volumes in schizophrenia. Arch Gen Psychiatry 57:761–768.

Kambeitz J, Abi-Dargham A, Kapur S, Howes OD. 2014. Alterations in cortical and extrastriatal subcortical dopamine function in schizophrenia: systematic review and meta-analysis of imaging studies. Br J Psychiatry 204:420–429.

Kane J, Honigfeld G, Singer J, Meltzer H. 1988. Clozapine for the treatment-resistant schizophrenic. A double-blind comparison with chlorpromazine. Arch Gen Psychiatry 45:789–796.

Kety SS, Rosenthal D, Wender PH, Schulsinger F. 1968. The types and prevalence of mental illness in the biological and adoptive families of adopted schizophrenics. J Psych Res 6:345–362.

Lesh TA, Niendam TA, Minzenberg MJ, Carter CS. 2011. Cognitive control deficits in schizophrenia. Mechanisms and meaning. Neuropsychopharmcology 36:316–338.

Lieberman JA, Stroup TS, McEvoy JP, et al. 2005. Effectiveness of antipsychotic drugs in patients with chronic schizophrenia. N Engl J Med 353:1209–1223.

Lynall M-E, Bassett DS, Kerwin R, et al. 2010. Functional connectivity and brain networks in schizophrenia. J Neurosci 30:9477–9487.

McGrath J, Saha S, Welham J, El Saadi O, MacCauley C, Chant D. 2004. A systematic review of the incidence of schizophrenia: the distribution of rates and the influence of sex, urbanicity, migrant status, and methodology. BMC Med 2:13.

Mortensen PB, Pedersen CB, Westergaard T, et al. 1999. Effects of family history and place and season of birth on the risk of schizophrenia. N Engl J Med 340:603–608.

Rapoport JL, Giedd JN, Blumenthal J, et al. 1999. Progressive cortical change during adolescence in childhood-onset schizophrenia. A longitudinal magnetic resonance imaging study. Arch Gen Psychiatry 56:649–654.

Schizophrenia Working Group of the Psychiatric Genomics Consortium. 2014. Biological insights from 108 schizophrenia-associated genetic loci. Nature 511:421–427.

Seeman P, Lee T, Chau-Wong M, Wong K. 1976. Antipsychotic drug doses and neuroleptic/dopamine receptors Nature 261:717–9.

Sekar A, Bialas AR, de Rivera H, et al. 2016. Schizophrenia risk from complex variation of complement component 4. Nature 530:177–183.

Suddath RL, Christison GW, Torrey EF, Casanova MF, Weinberger DR. 1990. Anatomical abnormalities in the brains of monozygotic twins discordant for schizophrenia. N Engl J Med 322:789–794.

Thompson PM, Vidal C, Giedd JN, et al. 2001. Mapping adolescent brain change reveals dynamic wave of accelerated gray matter loss in very early-onset schizophrenia. Proc Natl Acad Sci USA 98:11650–11655.

Vidal CN, Rapoport JL, Hayashi KM, et al. 2006. Dynamically spreading frontal and cingulate deficits mapped in adolescents with schizophrenia. Arch Gen Psychiatry 63:25–34.

61

Disorders of Mood and Anxiety

EPRESSION, BIPOLAR DISORDER, AND ANXIETY DISORDERS have been well documented in medical writings since ancient times. In the fifth century BC, Hippocrates taught that moods depended on the balance of four humors—blood, phlegm, yellow bile, and black bile. An excess of black bile (*melancholia* is the ancient Greek term for black bile) was believed to cause a state dominated by fear and despondency. Robert Burton's *Anatomy of Melancholy* (1621) was not only an important medical text but also viewed literature and the arts through the lens of melancholia. Such texts describe symptoms that remain familiar today; they also recognized that symptoms of depression and of anxiety often occur together.

In this chapter, we discuss mood and anxiety disorders together, not only because they frequently co-occur but also because of overlapping genetic and environmental risk factors and some shared neural structures, including regions of the amygdala, hippocampus, prefrontal cortex, and insular cortex.

Mood Disorders Can Be Divided Into Two General Classes: Unipolar Depression and Bipolar Disorder

There are no objective medical tests for mood and anxiety disorders. Thus, diagnosis depends on observation of symptoms, behavior, cognition, functional impairments, and natural history (including age of onset, course, and outcome). Patterns of familial transmission and response to treatment can also inform diagnostic classification. Based on such factors, it is possible to distinguish between two major groupings of mood disorders: unipolar depression and bipolar disorder. Unipolar depression, when severe and pervasive, is classified as major depression or major depressive disorder. Major depression is diagnosed when people suffer from depressive episodes alone. Bipolar disorder is diagnosed when episodes of mania also occur.

The lifetime risk of major depressive disorder in the United States is approximately 19%. Within any 1-year period, 8.3% of the population suffers major depression. The prevalence of depression differs in different countries and cultures; however, in the absence of objective medical tests, such epidemiologic data are subject to diagnostic and reporting biases, and thus, it is difficult to draw comparative conclusions. The World Health Organization reports that depression is a leading cause of disability worldwide, and other studies find it to be a leading cause of economic loss from noncommunicable disease. These dire social and economic consequences occur because depression is common, often begins early in life, and interferes with cognition, energy, and motivation, which are all necessary to learn in school and to work effectively.

Bipolar disorder is less common than unipolar depression, with a prevalence of approximately 1% worldwide. Its symptoms are relatively constant across countries and cultures. The incidence of bipolar disorder is equivalent in males and females.

Major Depressive Disorder Differs Significantly From Normal Sadness

Several factors distinguish major depression from transient periods of sadness that may occur in everyday life and from the grief that often follows a personal loss. These include the life context in which symptoms occur, their duration and pervasiveness, and their association with physiological, behavioral, and cognitive symptoms (Table 61–1). In healthy people, mood alternates between low and high, with timing and intensity phased appropriately with interpersonal interactions and life events. Mood states that are contextually inappropriate, extreme in amplitude, rigid, or prolonged are suggestive of either depression or mania, depending on their valence.

Depressive episodes, whether associated with unipolar or bipolar illness, are characterized by negative mood states such as sadness, anxiety, loss of interests, or irritability lasting for most of the day, day in and day out, and unrelieved by events that were previously enjoyable. This loss of interest is well expressed

Table 61–1 Symptoms of Mood Disorders

Five or more of the following symptoms have been present during the same 2-week period and represent a change from previous functioning. At least one of the symptoms is either (1) depressed mood or (2) loss of interest or pleasure.

1. Depressed mood most of the day, nearly every day, as indicated by either subjective report (eg, feels sad, empty, hopeless) or observations made by others (eg, appears tearful).
2. Markedly diminished interest or pleasure in all, or almost all, activities most of the day, nearly every day (as indicated by either subjective account or observation made by others).
3. Significant weight loss when not dieting, or weight gain (eg, a change of >5% of body weight in a month), or decrease or increase in appetite nearly every day.
4. Insomnia or hypersomnia nearly every day.
5. Psychomotor agitation or retardation nearly every day (observable by others, not merely subjective feelings of restlessness or being slowed down).
6. Fatigue or loss of energy nearly every day.
7. Feelings of worthlessness or excessive or inappropriate guilt (which may be delusional) nearly every day (not merely self-reproach or guilt about being sick).
8. Diminished ability to think or concentrate, or indecisiveness, nearly every day (either by subjective account or as observed by others).
9. Recurrent thoughts of death (not just fear of dying), recurrent suicidal ideation without a specific plan, or a suicide attempt or a specific plan for committing suicide.

Source: Adapted from the American Psychiatric Association. 2013. *Diagnostic and Statistical Manual of Mental Disorders*, 5th ed. Washington, DC: American Psychiatric Association.

by Hamlet's complaint, "How weary, stale, flat, and unprofitable seem to me all the uses of this world!" When depression is severe, individuals may suffer intense mental anguish and a pervasive inability to experience pleasure, a condition known as anhedonia.

Physiologic symptoms of depression include sleep disturbance, most often insomnia with early morning awakening, but occasionally excessive sleeping; loss of appetite and weight loss but occasionally excessive eating; decreased interest in sexual activity; and decreased energy. Some severely affected individuals exhibit slowed motor movements, described as psychomotor retardation, whereas others may be agitated, exhibiting such symptoms as pacing. Cognitive symptoms are evident in both the content of thoughts (hopelessness, thoughts of worthlessness and guilt, suicidal thoughts and urges) and in cognitive processes (difficulty concentrating, slow thinking, and poor memory).

In the most severe cases of depression, psychotic symptoms may occur, including delusions (unshakable false beliefs that cannot be explained by a person's culture) and hallucinations. When psychotic symptoms occur in depression, they typically reflect the person's thoughts of being undeserving, worthless, or bad. A severely depressed person might, for example, believe that he is emitting a potent odor because he is rotting from the inside.

The most severe outcome of depression is suicide, which represents a significant cause of death worldwide; the World Health Organization estimates that there are 800,000 deaths by suicide annually. More than 90% of suicides are associated with mental illness, with depression being the leading risk factor, especially when accompanied by substance use disorders.

Major Depressive Disorder Often Begins Early in Life

Major depressive disorder often begins early in life, but first episodes do occur across the life span. Those who have had a first episode in childhood or adolescence often have a family history of the disorder and have a high likelihood of recurrence. Once a second episode has occurred, a pattern of repeated relapse and remission often sets in. Some people do not recover completely from acute episodes and have chronic, albeit milder, depression, which can be punctuated by acute exacerbations. Chronic depression, even when symptoms are less severe than those of an acute episode, can prove extremely disabling because of long-term erosion of a person's ability to function in life roles. Major depressive disorder in childhood occurs equally in males and females. After puberty, however,

it occurs more commonly in females; the ratio of females to males is approximately 2:1 across countries and cultures.

A Diagnosis of Bipolar Disorder Requires an Episode of Mania

Bipolar disorder is named for its chief symptom, swings of mood between mania and depression; indeed, the influential 19th-century psychiatrist Emil Kraepelin called this condition the manic-depressive insanity. By convention, a diagnosis of bipolar disorder requires at least one episode of mania. Mania is typically associated with recurrent episodes of depression, whereas mania without depression is distinctly uncommon.

Manic episodes are typically characterized by elevated mood, although some individuals are predominantly irritable. During manic episodes, individuals have markedly increased energy, a decreased need for sleep, and occasionally a decreased desire for food (Table 61–2). People with mania are typically impulsive and engage excessively in reward-directed

---

Table 61–2 Symptoms of a Manic Episode

A. A distinct period of abnormally and persistently elevated, expansive, or irritable mood, and abnormally and persistently increased goal-directed activity or energy, lasting at least 1 week (or any duration if hospitalization is necessary).

B. During the period of mood disturbance and increased energy or activity, three (or more) of the following symptoms (four if the mood is only irritable) have persisted and have been present to a significant degree:

1. Inflated self-esteem or grandiosity.
2. Decreased need for sleep (eg, feels rested after only 3 hours of sleep).
3. More talkative than usual or pressure to keep talking.
4. Flight of ideas or subjective experience that thoughts are racing.
5. Distractibility (ie, attention too easily drawn to unimportant or irrelevant external stimuli).
6. Increase in goal-directed activity (either socially, at work or school, or sexually) or psychomotor agitation (ie, purposeless non–goal-directed activity).
7. Excessive involvement in pleasurable activities that have a high potential for painful consequences (eg, engaging in unrestrained buying sprees, sexual indiscretions, or foolish business investments).

Source: Adapted from the American Psychiatric Association. 2013. *Diagnostic and Statistical Manual of Mental Disorders*, 5th ed. Washington, DC: American Psychiatric Association.

behaviors, often with poor judgment characterized by extreme optimism. For example, a person may go on spending sprees well beyond his or her means or on uncharacteristic binges of drug and alcohol use or sexual behavior. Self-esteem is typically inflated, often to delusional levels. For example, an individual might falsely believe himself to have extensive influence on events or to be a significant religious figure. In antiquity, mania was described as "a state of raving madness with exalted mood." However, such elevated mood may be brittle, with sudden intrusions of anger, irritability, and aggression.

Mania, like depression, affects cognitive processes, often impairing attention and verbal memory. During a manic episode, a person's speech is often rapid, profuse, and difficult to interrupt. The person may jump quickly from idea to idea, making comprehension of speech difficult. Psychotic symptoms commonly occur during manic episodes and are generally consistent with the person's mood. For example, people with mania may have delusions of possessing special powers or of being objects of adulation.

The depressive episodes that occur in bipolar disorder are symptomatically indistinguishable from those in unipolar depression, but are often more difficult to treat. For example, they are often less responsive to antidepressant medications. Longitudinal studies have found that the most common affective state of bipolar patients between severe acute episodes of mania or depression is not healthy mood (euthymia), as was often taught in older textbooks, but a state of chronic depression.

Historically, the concept of bipolar disorder described patients who experienced full manic episodes, which often included psychotic symptoms and necessitated hospitalization (Table 61–2). In recent decades, diagnostic classifications have added type 2 bipolar disorder in which mild manias (also called hypomanias) alternate with depressive episodes. The manic episodes of type 2 bipolar disorder are, by definition, not accompanied by psychosis or severe enough to require hospitalization. Whether this represents a variant of classic (type 1) bipolar disorder or some other pathophysiology is not yet known, although genetic dissection of mood disorders may offer some clarification in the near future.

Bipolar disorder generally begins in young adulthood, but the onset may occur earlier or as late as the fifth decade of life. Many manic episodes often lack an obvious precipitant; however, sleep deprivation can initiate a manic episode in some individuals with bipolar disorder. For such individuals, travel across time zones or shift work represents a risk. The rate of cycling among mania, depression, and periods of normal mood varies widely among bipolar patients. Individuals with short, rapid cycles tend to be less responsive to mood-stabilizing drugs.

Anxiety Disorders Represent Significant Dysregulation of Fear Circuitry

Anxiety disorders are the most common psychiatric disorders worldwide. In the United States, 28.5% of the population suffers from one or more anxiety disorders over the course of their lifetimes. Some anxiety disorders are mild, such as the simple phobias that involve rarely encountered stimuli; others, such as panic disorder or posttraumatic stress disorder, are often highly debilitating based on the severity of symptoms, interference with functioning, and chronicity.

Anxiety and fear are related emotional states; both are critical to surviving dangers that might be encountered throughout life. The major distinction is that fear is a response to threats that are present and clearly signify danger, whereas anxiety is a state of readiness for threats that are less specific either in proximity or timing. The neural circuits of fear and anxiety strongly overlap, as do their physiological, behavioral, cognitive, and affective aspects.

Fear is normally a transient adaptive response to danger that, like pain, serves as a survival mechanism. Like pain, fear is alerting and aversive and motivates more or less immediate behavioral responses. Thus, fear interrupts ongoing behaviors, supplanting them with such responses as avoidance or defensive aggression. To prepare the body to cope physiologically, fear circuitry activates the sympathetic nervous system and causes release of stress hormones. This "fight or flight" response facilitates blood flow to skeletal muscle, increases metabolic activity, and elevates pain thresholds. Like reward and other survival-relevant emotional responses, fear strongly facilitates the encoding and consolidation of both implicit and explicit memories that prepare an organism to respond rapidly and effectively to future predictive cues. (Fear circuitry is described in Chapter 42.)

Many cognitive and physiological components of anxiety are similar to fear, but typically exhibit lower intensity and a more protracted time course. Anxiety is adaptive when proportionate to the probability and likely severity of a threat, leading to appropriate levels of arousal, vigilance, and physiological preparedness. Given the dangerous, indeed potentially lethal, consequences of ignoring even ambiguous threat cues, failure to mount appropriate anxiety responses can prove highly maladaptive. However, excessive contextually inappropriate and prolonged vigilance, tension, and physiological activation can be the basis of distressing and disabling anxiety disorders or anxiety symptoms that may accompany depression. Risk factors for anxiety disorders include a person's genetic background, developmental experience, and lessons learned not only from direct experience but also taught by families, peers, schools, and other institutions.

Cues that elicit anxiety may be environmental or interoceptive (ie, arising from within the body, such as abdominal discomfort or heart palpitations). Social cues and social situations can be a major source of anxiety. In humans, anxiety states can also be initiated by trains of thought that elicit memories or imagination of danger. Anxiety can also arise from stimuli that are processed unconsciously because of their brevity or ambiguity, and the resulting emotion might then be experienced as arising spontaneously. In contrast to fear, which is initiated and terminated by the presence or termination of clear stimuli denoting threat, anxiety has a more variable time course. Anxiety states may be prolonged if the potential for danger or harm is long-lasting or if there is no clear safety signal.

Anxiety disorders and the anxiety that may accompany major depression are associated with diverse symptoms. Affected individuals may develop excessive preoccupation with possible threats and attentional biases toward cues interpreted as threatening. Such cognitive states are often associated with persistent worry, tension, and vigilance. Common physiological symptoms include hyperarousal, as evidenced by a low threshold for being startled, difficulty sleeping, and sympathetic nervous system activation, including a rapid, pounding heartbeat. Individuals with anxiety may become exquisitely aware of their heartbeat or breathing, which can become a source of preoccupation and worry in their own right. Sympathetic nervous system activation may reach extreme levels of intensity during a panic attack, one of the most severe manifestations of anxiety.

In anxiety disorders, cognitive, physiological, and behavioral responses that would be adaptive in the face of a serious threat may be maladaptively activated by innocuous stimuli, may be inappropriately intense for the situation, and may have a protracted time course in which safety signals fail to terminate the symptoms. Affected individuals may avoid places, people, or experiences that, although objectively safe, have become associated with perceptions of threats or the experience of anxiety. When severe, such avoidance can impair the ability of affected individuals to function in different capacities or roles.

Because there are no biomarkers or objective medical tests for particular constellations of anxiety symptoms, current psychiatric classifications such as the fifth edition of the *Diagnostic and Statistical Manual of Mental Disorders* (DSM-5) classify anxiety disorders based on clinical histories, such as the nature, intensity, and time course of symptoms, the role of external cues in triggering episodes, and associated symptoms. The DSM-5 divides pathological anxiety syndromes into several distinct disorders: panic disorder, posttraumatic stress disorder, generalized anxiety disorder, social anxiety disorder (previously called social phobia), and simple phobias. For heuristic purposes, these disorders are discussed below, but current evidence from long-term clinical observation and from family, twin, and epidemiological studies does not support dividing anxiety symptoms into discrete nonoverlapping categories. Rather, the evidence suggests that pathological anxiety symptoms and symptoms of depression might be better conceptualized as a continuum or spectrum in which individuals experience varying symptoms that cross current DSM boundaries.

Consistent with the concept of a symptom spectrum, anxiety disorders and depression do not often occur together across generations in families as distinct DSM-5 categories; instead, diverse patterns of anxiety and depressive symptoms are typically observed among affected family members. Twin studies that compare concordance for traits in monozygotic and dizygotic twin pairs find significant shared genetic risk across multiple anxiety disorders and major depression. In addition, epidemiological studies find that individuals diagnosed with one categorical DSM-5 anxiety disorder, during, for example, teen years, have a high probability of developing new anxiety or depressive symptoms over the next decade that could result in the person being diagnosed with multiple disorders based on DSM-5 classifications. The high frequency at which putatively distinct DSM-5 anxiety disorders and depression co-occur and the results of family and twin studies suggest significant sharing of etiologic factors and pathogenic mechanisms among anxiety disorders and major depression. Nevertheless, individual disorders that are listed in DSM-5 are briefly described below.

Panic attacks are a severe manifestation of anxiety. They are characterized by discrete periods (that can last for many minutes) of intense foreboding, a sense of doom, fear of losing control over oneself, or fear of death. They are associated with prominent bodily symptoms such as heart palpitations, inability to catch one's breath, sweating, paresthesias, and dizziness (Table 61–3).

Table 61–3 Symptoms of a Panic Attack

A discrete period of intense fear or discomfort in which four (or more) of the following symptoms develop abruptly and reach a peak within 10 minutes.

1. Palpitations, pounding heart, or accelerated heart rate
2. Sweating
3. Trembling or shaking
4. Sensations of shortness of breath or smothering
5. Feeling of choking
6. Chest pain or discomfort
7. Nausea or abdominal distress
8. Feeling dizzy, unsteady, lightheaded, or faint
9. Chills or heat sensations
10. Paresthesias (numbness or tingling sensations)
11. Derealization (feelings or unreality) or depersonalization (being detached from oneself)
12. Fear of losing control or "going crazy"
13. Fear of dying

Source: Adapted from the American Psychiatric Association. 2013. *Diagnostic and Statistical Manual of Mental Disorders*, 5th ed. Washington, DC: American Psychiatric Association.

Panic attacks often give rise to anxiety about future episodes such that the contexts in which attacks have occurred can become phobic stimuli that trigger subsequent attacks (fear conditioning). As a result, some severely affected individuals restrict their activities to avoid situations or places in which panic attacks have occurred or from which they fear they might not be able to escape should they experience an attack. The most severely affected may develop generalized phobic avoidance, leading them to become housebound, a state described as agoraphobia. Current diagnostic classification systems such as the DSM-5 define panic disorder based on the number and frequency of attacks and whether or not a phobic trigger can be identified. Such detailed criteria lack a strong empirical basis, but it is certainly the case that individuals who have recurrent panic attacks along with other anxiety symptoms are not only highly distressed but may also be significantly disabled.

Posttraumatic stress disorder (PTSD) follows an experience of severe danger or injury. Under different names and descriptions, including shell shock, a term coined during World War I, PTSD has long been recognized as a result of combat. More recently, civilian traumas such as assault, rape, or automobile crashes have been recognized as potential causes of PTSD. The current approach to PTSD was formalized by the American Psychiatric Association based on the experience of Vietnam War veterans.

PTSD is initiated by a traumatic experience. Its cardinal symptoms include intrusive reexperiencing of the traumatic episode, typically initiated by cues such as sounds, images, or other reminders of the trauma. For example, a person who has been assaulted might respond potently to an unexpected touch from behind. Such episodes are often characterized by activation of the sympathetic nervous system and, when severe, may be characterized by "fight or flight" responses. The reexperiencing of a traumatic event may also occur in the form of nightmares. Other symptoms of PTSD include emotional numbness that may interfere with relationships and social interactions, insomnia, chronic hyperarousal including excessive vigilance, sympathetic nervous system activation, and an exaggerated startle response to an innocuous stimulus such as a touch or sound.

Generalized anxiety disorder (GAD) is diagnosed when a person suffers chronic worry and vigilance not warranted by circumstances. The worry is accompanied by physiological symptoms such as heightened sympathetic nervous system activation and motor tension. GAD commonly co-occurs with major depressive disorder.

Social anxiety disorder is characterized by a persistent fear of social situations, especially situations in which one is exposed to the scrutiny of others. The affected person has an intense fear of acting in a way that will prove humiliating. Stage fright is a form of social anxiety that is limited to circumstances of performance, such as public speaking. Social anxiety disorder can lead to avoidance of verbal classroom participation or communicating with others at work and can therefore prove disabling as well as distressing.

Simple phobias consist of intense and inappropriately excessive fear of specific stimuli, such as elevators, flying, heights, or spiders.

Both Genetic and Environmental Risk Factors Contribute to Mood and Anxiety Disorders

Bipolar disorder, major depression, and anxiety disorders all run in families. Twin studies that compare the rate of concordance of monozygotic and dizygotic twin pairs demonstrate significant heritabilities among these disorders, where heritability represents the percentage of the variation in a phenotype explained by genetic variation. Among mood and anxiety disorders, bipolar disorder has the highest heritability (70%–80%); major depression and anxiety disorders exhibit lower but still significant heritabilities (approximately 35%), with greater roles for developmental and environmental risk factors. Although there is an important role for genes in the pathogenesis of mood and anxiety disorders, all of them exhibit non-Mendelian patterns of transmission across generations, including frequent co-occurrence of major depression and anxiety disorders. Such patterns reflect the complexity of genetic and nongenetic risk factors.

Molecular genetic studies aimed at discovering the precise DNA sequence variants (alleles) that predispose to mood and anxiety disorders have been initiated. Such studies are challenging because the risk architecture of these, and indeed all, common psychiatric disorders is highly polygenic, meaning that population risk appears to involve many thousands of common and rare alleles linked to or contained within many hundreds of genes. Unlike some neurologic disorders such as Huntington disease, there is no "depression gene" or "anxiety gene." Disease-associated alleles confer small additive effects on the risk of an illness. The risk for any given individual results from genetic loading (comprised of diverse combinations of disease-associated alleles) acting in concert with developmental and environmental factors. This polygenic architecture explains non-Mendelian patterns of transmission and the diverse combinations of depressive and anxiety symptoms observed within families and across populations.

The lack of objective diagnostic tests for mood and anxiety disorders means that any study cohort is likely to have some proportion of diagnostic misclassification. As a result, the search for common disease-associated variants by genome-wide association studies (GWAS) and rare disease-associated variants by DNA sequencing requires significant statistical power conferred by very large cohorts and by meta-analyses conducted across multiple cohorts. Early results of GWAS have been reported for major depression and bipolar disorder; in both cases, several significant genome-wide loci have been found to date, but not yet enough to identify molecular pathways of pathogenesis with any certainty. Whole-exome sequencing (ie, DNA sequencing of all genomic regions that encode proteins) and whole-genome sequencing are being conducted for bipolar disorder.

The highly polygenic risk architecture of mood and anxiety disorders means that there is no diagnostic value in testing for one or a few risk gene variants that might be associated with these disorders. Rather, polygenic risk scores (PRS), based on the sum of all genetic risk variants for a trait, are emerging as useful tools to stratify individuals in epidemiological and clinical studies by severity of genetic risk. A discrepant PRS within a clinical cohort, eg, showing low depression

risk in a study of people with major depression, would suggest misclassification. It is important to emphasize that the polygenic nature of risk for mood and anxiety disorders and the significant contribution of environmental risk factors mean that, like any genetic test, the PRS provides only a probability.

As more is learned, the PRS can be combined with other measures to yield a more predictive risk score, just as modern cardiac risk models increasingly include genetic measures, smoking history, lipid levels, and blood pressure. For mood and anxiety disorders, one type of measure that shows early promise is identification of intrinsic patterns of neural connectivity derived from resting-state functional magnetic resonance imaging (fMRI; imaging conducted when subjects are not engaged in task performance). Differing patterns of connectivity could potentially distinguish among different forms of disorder.

Epidemiological evidence has identified significant developmental risk factors for major depression and anxiety disorders. The best documented is a history of physical or sexual abuse early in life, serious child neglect, or other early, severe stressors. Investigations of such early stressors have focused on possible roles for altered reactivity of the hypothalamic-pituitary-adrenal (HPA) axis. Studies of early stress in animal models suggest that epigenetic regulation of gene expression may have a role in altering developmental trajectories. Such results cannot be readily followed up in humans because of lack of access to human brain tissue and thus remain hypothetical.

Other risk factors for depression and anxiety disorders include alcohol and other substance use disorders and the presence of other psychiatric disorders, such as attention deficit hyperactivity disorder, learning disorders, and obsessive-compulsive disorder. There is also evidence that alcoholism and other substance abuse disorders may be initiated by misguided attempts at self-medication of depression or anxiety, in turn worsening the underlying condition.

Environmental factors that may trigger new episodes of depression or anxiety include life transitions such as marriage, a new job, or retirement. Serious illness, whether acute or chronic, is also associated with the onset of major depression and anxiety. Some neurological disorders are associated with an elevated risk of depression, including Parkinson disease, Alzheimer disease, multiple sclerosis, and stroke. Some prescribed medications, such as interferons, also frequently trigger depression. When major depression accompanies a chronic illness such as type 2 diabetes or cardiovascular disease, the overall medical outcomes are worse, as a result of both the physiological effects of depression, such as increased release of stress hormones (see below) and decreased motivation to engage in rehabilitative regimens.

Depression and Stress Share Overlapping Neural Mechanisms

Depression and responses to stress exhibit complex but significant interactions. As already noted, severe childhood adversity is a developmental risk factor for depression; moreover, depressive episodes may be initiated by a stressful experience. Conversely, the experience of depression is itself stressful because of the suffering it causes and its negative effects on functioning. Symptomatically, depression shares several physiological features with chronic stress, including changes in appetite, sleep, and energy. Both major depression and chronic stress are associated with persistent activation of the HPA axis (Figure 61–1).

Many but not all individuals with major depression and many in the depressed phase of bipolar disorder exhibit excess synthesis and secretion of the glucocorticoid stress hormone cortisol and the factors that regulate it, corticotropin-releasing hormone (CRH) and adrenocorticotropic hormone (ACTH). In a healthy state, a *transient* increase in cortisol secretion, as occurs in response to acute stress, shifts the body to a catabolic state (making glucose available to confront the stressor or threat), increases subjective energy levels, sharpens cognition, and may increase confidence. However, a *chronic* increase in glucocorticoids may contribute to depression-like symptoms. For example, many people with Cushing disease (in which pituitary tumors secrete excess ACTH, leading to excess cortisol) experience symptoms of depression.

Feedback mechanisms within the HPA axis normally permit cortisol (or exogenously administered glucocorticoids) to inhibit CRH and ACTH secretion and therefore to suppress additional cortisol synthesis and secretion. In approximately half of people with major depression, this feedback system is impaired; their HPA axis becomes resistant to suppression even by potent synthetic glucocorticoids such as dexamethasone. Although readily measurable disturbances of the HPA axis have not proven sensitive or specific enough to be used as a diagnostic test for depression, the observed abnormalities suggest strongly that a pathologically activated stress response is often an important component of depression.

The relationship of stress with depression has led to the development of several chronic stress paradigms in rodent models of depression. The reliance on

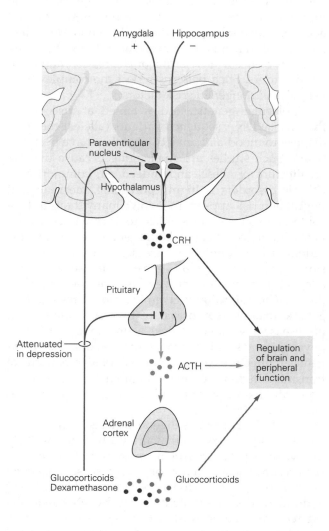

Figure 61–1 The hypothalamic-pituitary-adrenal axis. Neurons in the paraventricular nucleus of the hypothalamus synthesize and release corticotropin-releasing hormone (**CRH**), the key regulatory peptide in the hormonal cascade activated by stress. The CRH neurons have a circadian pattern of secretion, and the stimulatory effects of stress on CRH synthesis and secretion are superimposed on this basal circadian pattern. Excitatory fibers from the amygdala convey information about stressful stimuli that activates CRH neurons; inhibitory fibers descend from the hippocampus onto the paraventricular nucleus. CRH enters the hypophyseal portal system and stimulates the corticotropic cells in the anterior pituitary that synthesize and release adrenocorticotropic hormone (**ACTH**). The released ACTH enters the systemic circulation and stimulates the adrenal cortex to release glucocorticoids. In humans, the major glucocorticoid is cortisol; in rodents, it is corticosterone. Both cortisol and synthetic glucocorticoids such as dexamethasone act at the level of the pituitary and hypothalamus to inhibit further release of ACTH and CRH, respectively. The feedback inhibition by glucocorticoids is attenuated in major depression and the depressed phase of bipolar disorder. (Adapted, with permission, from Nestler et al. 2015.)

stress-induced syndromes in these animal models has been strengthened by the observation that many antidepressant drugs reverse stress-induced changes in physiology or behavior in these animals. However, the degree to which animals subjected to diverse chronic stressors actually model the disease mechanisms underlying depression in human beings remains unknown. Concern about overreliance on stress-based and other rodent models is indicated by the failure to identify new antidepressant mechanisms despite more than 50 years of trying. Drug screens using such models have only identified molecules with actions similar to prototype antidepressant drugs that were first identified by their unexpected psychotropic effects on humans.

Dysfunctions of Human Brain Structures and Circuits Involved in Mood and Anxiety Disorders Can Be Identified by Neuroimaging

Investigation of human brain regions and the neural circuitry involved in mood and anxiety disorders has relied on noninvasive structural and functional neuroimaging, neurophysiologic testing, and postmortem analyses. More recently, information is being gleaned from neuroimaging of patients being treated with deep brain stimulation.

Identification of Abnormally Functioning Neural Circuits Helps Explain Symptoms and May Suggest Treatments

Functional neuroimaging and electrophysiological studies are being pursued in order to elucidate abnormalities in circuit activity and in patterns of intrinsic connectivity in mood and anxiety disorders. Given the heterogeneity of major depression, bipolar disorder, and anxiety disorders defined by current diagnostic methods, it has been challenging to identify robust and replicable abnormalities. In addition, the use of diverse cognitive and emotional tasks to experimentally probe mood and anxiety disorders has limited researchers' ability to replicate and confirm findings. Overcoming the resulting uncertainties will require larger numbers of subjects, application of data standards that permit meta-analyses, and increasingly, methods such as use of the PRSs to stratify subjects.

Despite current limitations, fMRI and electrophysiological studies of mood and anxiety disorders have begun to provide initial empirical leads about circuit abnormalities in mood and anxiety disorders. Resting-state fMRI studies comparing subjects with major depression and healthy control subjects suggest differences in patterns of intrinsic connectivity, specifically within neural circuits that regulate "top down" control of cognition and emotion—the "cognitive control network"—and in circuits that process significant emotional and motivational stimuli—"the salience network" (Figure 61–2). Despite the need for replication, these findings are noteworthy because they are consistent with results from task-based imaging studies of humans (eg, studies of fear conditioning) and animal studies that investigate responses to aversive stimuli.

In healthy human subjects, regions of the amygdala are activated by threatening stimuli and during fear conditioning, such as pairing a previously neutral tone with a mild shock. Beginning with the work of Charles Darwin, human faces expressing fear have been recognized to elicit anxiety responses across diverse human cultures, presumably as a mechanism to communicate the presence of danger among members of a group.

The effects of fearful and other emotion-expressing faces on measurements of autonomic activity and brain activity measured by fMRI or by electroencephalography have been studied in subjects with anxiety disorders or with major depression. In one such paradigm, fearful faces are shown very briefly (33 ms) while the subject is in an MRI scanner. This presentation is followed by a neutral face (referred to as backward masking). Under such circumstances, subjects report that they have no awareness of having seen the fearful face. Yet they exhibit an altered galvanic skin response, a measure of sympathetic activation, as well as activation of the basal amygdala, the amygdala region that processes sensory inputs and that responds selectively to threat. Several functional neuroimaging studies of individuals with PTSD, other anxiety disorders, and major depression have demonstrated heightened activity in the amygdala, activation even to innocuous stimuli, and persistence of amygdala activity in contrast to normal patterns of adaptation (Figure 61–3).

Functional neuroimaging studies of anxiety disorders and major depression have also found decreased activity in prefrontal cortical regions that are interconnected with the basal amygdala. Studies of animals with prefrontal cortical lesions demonstrate that projections from the prefrontal cortex to the basal amygdala are necessary for cognitive control over aversive information. In individuals suffering from anxiety disorders or major depression, reduced activation of the prefrontal cortex by aversive stimuli is consistent with cognitive testing that demonstrates decreased cognitive control and might contribute to excessive and persistent anxiety and other negative emotions.

Electrophysiological and functional neuroimaging studies of both major depression and bipolar disorder

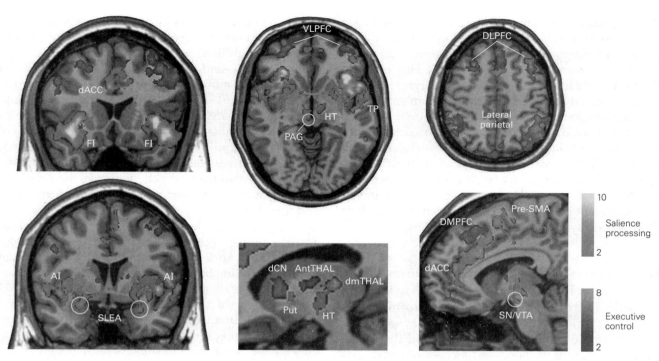

Figure 61–2 Mood disorders involve independent neural networks associated with processing of emotional salience and cognitive control. Statistical analysis (independent component analysis) applied to resting-state functional magnetic resonance imaging data identifies separable networks that compute emotional salience (**red-orange**) and regulate cognitive control/executive function (**blue**). The emotional salience network links dorsal anterior cingulate cortex (**dACC**) and frontoinsular cortex (**FI**) with subcortical structures involved in emotion. The cognitive control network links the dorsolateral prefrontal (**DLPFC**) and parietal cortices and several subcortical structures. The

brain regions shown to be networked in this study have been implicated in major depression by multiple independent studies. (Abbreviations: **AI**, anterior insula; **antTHAL**, anterior thalamus; **dCN**, dorsal caudate nucleus; **DMPFC**, dorsomedial prefrontal cortex; **dmTHAL**, dorsomedial thalamus; **HT**, hypothalamus; **PAG**, periaqueductal gray matter; **Pre-SMA**, pre–supplementary motor area; **Put**, putamen; **SLEA**, sublenticular extended amygdala; **SN/VTA**, substantia nigra and ventral tegmental area of the midbrain; **TP**, temporal pole; **VLPFC**, ventrolateral prefrontal cortex.) (Reproduced, with permission, from Seeley et al. 2007. Copyright © 2007 Society for Neuroscience.)

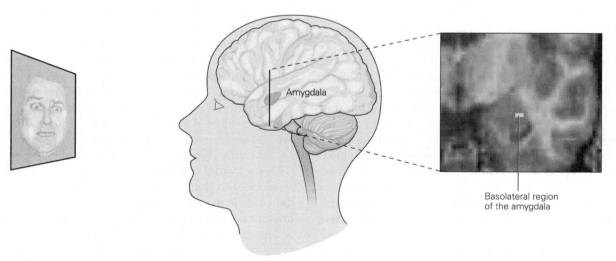

Figure 61–3 Amygdala activation in response to a masked presentation of a fearful stimulus. A human subject observes projected images while being scanned by magnetic resonance imaging. When a fearful face is presented for a very brief time followed by presentation of a neutral face, a protocol called

backward masking, the subject is not consciously aware of the fearful face. Under these conditions, the basolateral region of the amygdala is more strongly activated in individuals with anxiety disorder than in normal individuals. (Reproduced, with permission, from Etkin et al. 2004.)

have shown abnormal functioning of the rostral and ventral subdivisions of the anterior cingulate cortex (ACC), a region of prefrontal cortex that participates in the emotional salience network. The rostral and ventral ACC have extensive connections with the hippocampus, amygdala, orbital prefrontal cortex, anterior insula, and nucleus accumbens and are involved in the integration of emotion, cognition, and autonomic nervous system function. The caudal subdivision of the ACC is involved in cognitive processes involved in control of behavior; it has connections with dorsal regions of the prefrontal cortex, secondary motor cortex, and posterior cingulate cortex.

Although abnormal function in both subdivisions of the ACC has been observed in depressive episodes, the most consistent abnormality observed in major depression and in the depressed phase of bipolar disorder is increased activity in the rostral and ventral subdivisions, especially in the subgenual region ventral to the genu (or "knee") of the corpus callosum. In a study using positron emission tomography, effective treatment of major depression with selective serotonin reuptake inhibitor antidepressants was correlated with decreased activity in the rostral ACC, whereas self-induced sadness in healthy subjects increased activity (Figure 61–4). Based on such studies, the rostral anterior (subgenual) cingulate cortex has been used as a target for electrode placement in deep brain stimulation for treatment-resistant major depression, which is operationally defined as depressive illness that has

been unresponsive to antidepressant medication and psychotherapy.

Functional abnormalities of brain reward circuitry may also play a role in the symptoms of mood disorders. The reward circuitry comprises the dopaminergic projections from the ventral tegmental area of the midbrain to forebrain targets, including the nucleus accumbens, habenula, prefrontal cortex, hippocampus, and amygdala (Chapter 43). Under normal conditions, these pathways are involved in the valuation of rewards (eg, palatable food, sexual activity, and social interactions) and in motivating the necessary behavior to obtain them. Reward processing appears to be abnormal in depression, based on such symptoms as decreased interest in previously pleasurable activities, decreased motivation, and, when depression is severe, the inability to experience pleasure (anhedonia). Although less well studied, reward processing is also likely abnormal in mania, which is characterized by excessive engagement in goal-directed behaviors, even when they are maladaptive, such as uncontrolled spending, dangerous drug use, and promiscuous sexual activity.

In a recent analysis of resting-state fMRI, data showed that patients with major depression could be stratified based on connectivity patterns that correlated with their degree of anhedonia and anxiety. However, although modulation of the reward circuitry has been considered as a possible treatment for major depression, it has proven difficult in practice. For example, drugs known to activate this circuitry by

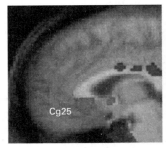

Induced sadness in healthy subjects

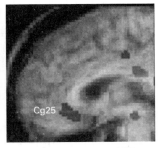

Depression recovery with SSRI

Figure 61–4 Activity in the rostral anterior (subgenual) cingulate cortex is increased by sadness and decreased by successful treatment of major depression with an antidepressant. (Reproduced, with permission, from Mayberg et al. 1997.)

Left. Healthy volunteers provided a script of their saddest memory that was later used to generate transient sadness while undergoing positron emission tomography (PET). The rostral anterior cingulate cortex was activated (**red** pseudo-color in the sagittal section of the human brain) was activated when the sad story was read. **Cg25** is an alternative nomenclature

for the cingulate gyrus, Brodmann area 25. The PET ligand was oxygen-15–labeled water, used to measure cerebral blood flow as a proxy for brain activity.

Right. Elevated metabolism in the rostral anterior cingulate cortex was confirmed in subjects with major depression. Following successful treatment with a selective serotonin reuptake inhibitor (**SSRI**) antidepressant, brain activity in Cg25 decreased (**blue** pseudo-color in the sagittal section of the human brain). The PET ligand was 2-deoxyglucose, used to measure cerebral metabolism as a proxy for brain activity.

increasing synaptic dopamine, such as amphetamine and cocaine, pose a high risk of overuse and addiction. More recently, tests of drugs that release reward circuits from inhibitory control, such as kappa opiate receptor antagonists, have been initiated in patients with major depression.

A Decrease in Hippocampal Volume Is Associated With Mood Disorders

The best-established structural abnormality in mood disorders is decreased hippocampal volume in individuals with major depression compared with healthy subjects. Recent studies of patients with major depression and bipolar disorder have found hippocampal volume loss in unmedicated subjects in regions of the cerebral cortex associated with the control of emotion. Such studies, which still need replication, show both overlapping and nonoverlapping patterns of volume loss in patients with major depression compared with bipolar disorder. Volume reductions observed in patients with major depression correlate with the duration of depressive episodes when controlling for duration of medication use. These findings suggest that in major depression the volume losses result from persistent illness and do not represent an antecedent risk factor. Some researchers have hypothesized that elevated cortisol levels in patients with major depression might be associated with reduced hippocampal volumes.

Reduced hippocampal volume has also been reported in cases of PTSD. In contrast with major depression, studies of monozygotic twins discordant for PTSD suggest that small hippocampi precede onset of the disorder and may thus represent a risk factor instead of a result of the disorder.

The acquired loss of hippocampal volume in major depression could result from loss of dendrites and dendritic spines, from decreased cell numbers (neurons or glia), or both. Given the relationship of stress and depression, excessive cortisol secretion could play a causal role in either type of loss. A decrease in hippocampal cell number could be explained by the fact that stress and elevated glucocorticoid levels suppress adult hippocampal neurogenesis, as shown in studies of several animal species.

In several mammals, including humans, new granule cells within the dentate gyrus of the hippocampus are produced during adult life. Studies of rodents have shown that these new neurons can be incorporated into functional neural circuits where they initially exhibit heightened structural and synaptic plasticity. A role for cell death as a balance to adult neurogenesis is less well studied.

In rodents, stressful or aversive treatments or administration of glucocorticoids inhibits the proliferation of granule cell precursors and thus suppresses normal rates of neurogenesis in the hippocampus. Antidepressants, including the selective serotonin reuptake inhibitors, exert an opposite effect, increasing the rate of neurogenesis. Thus, excess secretion of glucocorticoid stress hormones, as occurs in depression, could cause hippocampal volume loss by inhibiting neurogenesis over time. Because glucocorticoid receptors in the hippocampus are required for inhibitory feedback to hypothalamic neurons that synthesize and release CRH, impairments of hippocampal function could further impair feedback regulation of the HPA axis, creating a vicious cycle.

The hippocampus permits the brain to resolve differences among closely related stimuli (pattern separation) and provides contextual information that facilitates interpretation of the survival significance of a stimulus. Such information is needed by the organism to accurately identify threats that are signaled within a stream of complex sensory inputs. In animal studies, hippocampal lesions increase anxiety responses; it is thought that the resulting impairment of pattern separation and processing of contextual information permits threat-related memories to generalize inappropriately and thus to become associated with innocuous stimuli. Physiological and behavioral evidence suggests that newborn neurons within the dentate gyrus of the hippocampus play a particularly important role in pattern separation. Thus, inhibition of neurogenesis might contribute to anxiety symptoms that often accompany major depression, and abnormally low hippocampal volumes might increase the risk of PTSD.

Major Depression and Anxiety Disorders Can Be Treated Effectively

Major depressive disorder can be treated effectively with antidepressant drugs, cognitive psychotherapy, and electroconvulsive therapy. Major depressive disorder refractory to other interventions is being treated experimentally with deep brain stimulation targeted to the subgenual prefrontal cortex and other targets, including the nucleus accumbens.

Current Antidepressant Drugs Affect Monoaminergic Neural Systems

Named for their first clinical indication, the antidepressant drugs have broader utility than suggested by their name. Indeed, antidepressants are also the first-line drugs for the treatment of anxiety disorders. Along with frequent co-occurrence and sharing of risk factors and some neural circuits, the overlap in effective treatment modalities is further evidence that mood and anxiety disorders are related.

All widely used antidepressant drugs increase activity in monoaminergic systems in the brain, most significantly serotonin and norepinephrine, although some antidepressants exert modest effects on dopamine as well. The relevant monoamine neurotransmitters—serotonin, norepinephrine, and dopamine—are synthesized by cells that reside within brain stem nuclei (Chapter 40). Serotonergic and noradrenergic neurons in the pons and medulla project widely to highly diverse terminal fields in brain regions that include the hypothalamus, hippocampus, amygdala, basal ganglia, and cerebral cortex (Figures 61–5 and 61–6). Dopaminergic

neurons in the ventral tegmental area and substantia nigra pars compacta of the midbrain project to somewhat less widespread areas. Ventral tegmental neurons project to the hippocampus, amygdala, nucleus accumbens, and prefrontal cortex; substantia nigra neurons innervate the caudate and putamen. The widely divergent projections of these monoaminergic neurons permit them to influence functions such as arousal, attention, vigilance, motivation, and other cognitive and emotional states that require integration of multiple brain regions.

Serotonin, norepinephrine, and dopamine are synthesized from amino acid precursors and packaged into synaptic vesicles for release. Monoamines in the cytoplasm that are outside of vesicles are metabolized by the enzyme monoamine oxidase (MAO), which is associated with the outer leaflet of mitochondrial

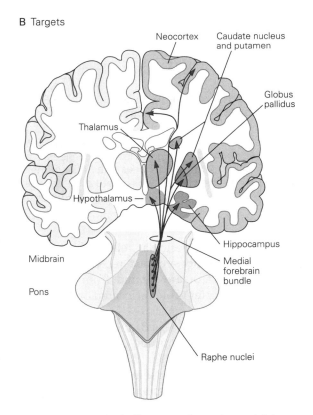

Figure 61–5 The major serotonergic systems in the brain arise in the raphe nuclei of the brain stem. Serotonin is synthesized in a group of brain stem nuclei called the raphe nuclei. These neurons project throughout the neuraxis, ranging from the forebrain to the spinal cord. The serotonergic projections are the most massive and diffuse of the monoaminergic systems, with single serotonergic neurons innervating hundreds of target neurons. (Adapted, with permission, from Heimer 1995.)

A. A sagittal view of the brain illustrates the raphe nuclei. In the brain, these nuclei form a fairly continuous collection of cell groups close to the midline of the brain stem and extending along its length. In the drawing here, they are shown in more distinct rostral and caudal groups. The rostral raphe nuclei project to a large number of forebrain structures.

B. This coronal view of the brain illustrates some of the major structures innervated by serotonergic raphe nuclei neurons.

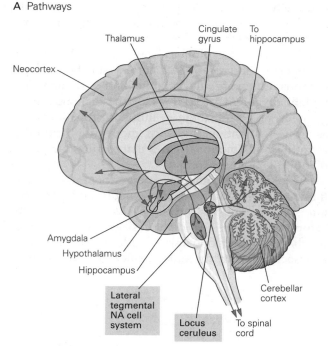

A Pathways

B Targets

membranes. After vesicular release, monoamine neurotransmitters bind synaptic receptors to exert their biological effect or are cleared from the synapse by specific transporter proteins located on the presynaptic cell membrane.

The most widely used antidepressant drugs fall into several major groupings, which affect monoaminergic neurons and their targets (Figure 61–7). The *MAO inhibitors* discovered in the 1950s, such as phenelzine and tranylcypromine, are effective against both depression and anxiety disorders but are rarely used today because of their side effects. MAO inhibitors block the capacity of MAO to break down norepinephrine, serotonin, or dopamine in presynaptic terminals, thus making extra neurotransmitter available for packaging into vesicles and for release.

Two forms of MAO, types A and B, are found in the brain. Type A is also found in the gut and liver, where it

catabolizes bioactive amines that are present in foods. Inhibition of MAO-A permits bioactive amines such as tyramine to enter the bloodstream from foods that contain it in high concentrations, such as aged meats and cheeses. Transporters shuttle these amines into the terminals of sympathetic neurons, where they can displace endogenous vesicular norepinephrine and epinephrine into the cytoplasm, leading to nonvesicular release that causes significant elevations of blood pressure.

The *tricyclic antidepressants*, also first identified in the mid-1950s, include imipramine, amitriptyline, and desipramine; these block the norepinephrine transporter (NET), the serotonin reuptake transporter (SERT), or both. These drugs are effective in treating both depression and anxiety disorders. However, in addition to their therapeutic targets, the older tricyclic drugs also block many neurotransmitter receptors, including the muscarinic acetylcholine, histamine H_1, and α_1 noradrenergic receptors, producing a panoply of side effects.

The *selective serotonin reuptake inhibitors* (SSRIs) such as fluoxetine, sertraline, and paroxetine, first approved in the 1980s, have no greater efficacy than the older tricyclic antidepressants and MAO inhibitors but are widely used because they have milder side effects and are far safer if taken in overdose. As their name implies, they selectively inhibit SERT. They are effective for major depressive disorder and many anxiety disorders. In high doses, selective serotonin reuptake inhibitors are also effective for symptoms of obsessive-compulsive disorder. Selective norepinephrine and serotonin-norepinephrine reuptake inhibitors have also been developed; these drugs have side effect profiles similar to those of selective serotonin reuptake inhibitors but are useful for some patients who do not benefit from inhibition of SERT alone.

Despite knowledge of the initial molecular targets that mediate the effects of antidepressant drugs, MAO or monoamine transporters, the ultimate molecular mechanisms by which they relieve depression remain unknown. One major challenge to understanding the therapeutic action of these drugs is the delay in their therapeutic effects. Although antidepressant drugs bind to and inhibit MAO, NET, or SERT with their first dose, several weeks of treatment are typically required to observe a lifting of depressive symptoms.

Several hypotheses have been put forward to explain this delay. One is that a slow buildup of newly synthesized proteins alters the responsiveness of neurons in a manner that treats the depression. Another is that increases in the levels of synaptic transmission of serotonin or norepinephrine rapidly increase plasticity in different emotion-processing circuits and that the latency to therapeutic benefit reflects the time it takes for new experiences to alter synaptic weights. A third hypothesis is that antidepressant efficacy is mediated in part by enhancement of hippocampal neurogenesis. Narrowing down the possible therapeutic mechanisms is challenging because of the lack of good animal models of depression. Without an animal model, it is not possible to know which of the many observable molecular, cellular, and synaptic changes cause depression or underlie the therapeutic actions of effective antidepressants.

Ketamine Shows Promise as a Rapidly Acting Drug to Treat Major Depressive Disorder

Ketamine, which blocks the *N*-methyl-ᴅ-aspartate (NMDA) glutamate receptor, is currently used in pediatric anesthesia for its ability to produce dissociative experiences as well as analgesia. It has been studied in randomized clinical trials with subjects suffering from major depression. In the trials, ketamine was administered by intravenous infusion; it produced an antidepressant effect within 2 hours, a significant advantage over existing antidepressant drugs that typically take weeks to show benefit. The therapeutic effects of ketamine last for approximately 7 days, after which second and third doses may continue to be effective. If such results become widely replicated, ketamine would represent the first antidepressant drug that does not exert its primary action on monoamine neurotransmission. Studies to identify mechanisms by which ketamine relieves depression, like those for older antidepressants, are challenging in part because of the lack of good animal models of depression.

At higher doses, ketamine is misused as a recreational drug to produce euphoria, dissociation, depersonalization, and hallucinations. Ketamine has also been used in laboratory settings to induce cognitive symptoms reminiscent of schizophrenia in human subjects. Although the advantages of a rapidly acting antidepressant would be significant, for example in treating acutely suicidal individuals, the unwanted psychotropic effects of ketamine make its use problematic. Attempts to develop alternative NMDA receptor blockers in which the antidepressant effects might be separated from psychotropic side effects are under way.

Psychotherapy Is Effective in the Treatment of Major Depressive Disorder and Anxiety Disorders

Short-term symptom-focused psychotherapies have been developed for depression and anxiety and tested in clinical trials. The best-studied psychotherapies are the cognitive behavioral therapies. Cognitive therapies that might be used to treat major depression focus on identifying and correcting excessively negative interpretations of events and of interactions with other people. For example, many depressed people exhibit a strong attentional bias toward negative information, automatically interpret neutral events as negative, and read evidence of disapproval into the behavior of others. Such automatic negative thinking, which can initiate or perpetuate depressed mood, can be much improved through cognitive psychotherapies.

Therapies with a more behavioral component have proven useful in the treatment of anxiety disorders such as phobias or PTSD. In exposure therapy, the affected individual is directed to vividly recollect phobic stimuli that trigger anxiety or avoidance. The therapist provides a safe context for such experiences

A Serotonergic neurons

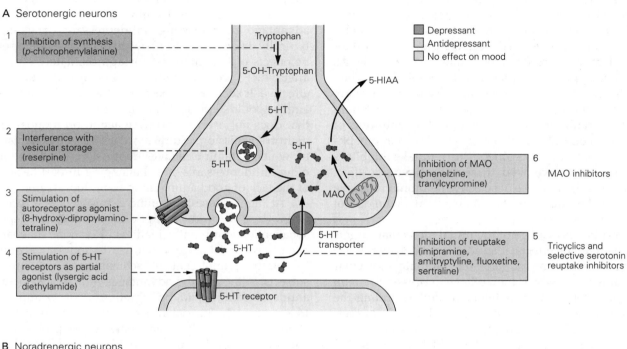

B Noradrenergic neurons

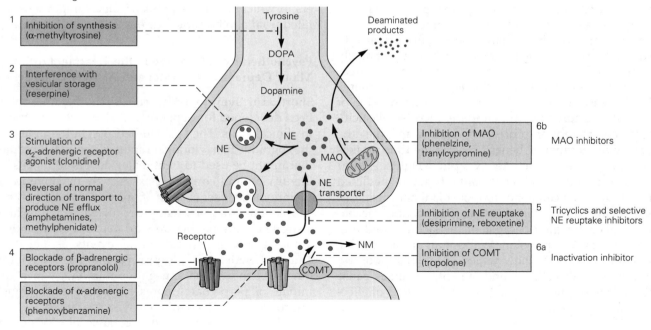

Figure 61–7 (Opposite) **Actions of antidepressant drugs at serotonergic and noradrenergic synapses.** The figure shows the pre- and postsynaptic sides of serotonergic and noradrenergic synapses. Serotonin and norepinephrine are synthesized from amino acid precursors by enzymatic cascades. The neurotransmitters are packaged in synaptic vesicles; free neurotransmitter within the cytoplasm is metabolized by monoamine oxidase (**MAO**), an enzyme associated with the abundant mitochondria found in presynaptic terminals. Upon release, serotonin and norepinephrine interact with several types of pre- and postsynaptic receptors. Each neurotransmitter is cleared from the synapse by a specific transporter. The serotonin and norepinephrine transporters and MAO are targets of antidepressant drugs.

A. Important sites of drug action at serotonergic synapses. Not all actions described are shown in the figure.

1. *Enzymatic synthesis.* Inhibition of synthesis of the rate-limiting enzyme tryptophan hydroxylase by *p*-chlorophenylalanine initiates the cascade that converts tryptophan to 5-OH-tryptophan, the precursor of 5-hydroxytryptophan (**5-HT**, serotonin).

2. *Storage.* Reserpine and tetrabenazine interfere with the transport of serotonin and catecholamines into synaptic vesicles by blocking the vesicular monoamine transporter VMAT$_2$. As a result, cytoplasmic serotonin is degraded (see step 6 below), and thus, the neuron is depleted of neurotransmitter. Reserpine was used as an antihypertensive drug but commonly caused depression as a side effect.

3. *Presynaptic receptors.* Agonists at presynaptic receptors produce negative feedback on neurotransmitter synthesis or release. The agonist 8-hydroxy-diprolamino-tetraline (8-OH-DPAT) acts on 5-HT$_{1A}$ receptors on the presynaptic neuron. The antimigraine triptan drugs (eg, sumatriptan) are agonists at 5-HT$_{1D}$ receptors.

4. *Postsynaptic receptors.* The hallucinogen lysergic acid diethylamide (LSD) is a partial agonist at 5-HT$_{2A}$ receptors on postsynaptic serotoninergic neurons. Second-generation antipsychotic drugs, such as risperidone and olanzapine, are antagonists at 5-HT$_{2A}$ receptors in addition to their ability to block D$_2$ dopamine receptors. The antiemetic compound ondansetron is an antagonist at 5-HT$_3$ receptors, the only ligand-gated channel among the monoamine receptors. Its key site of action is in the medulla.

5. *Uptake.* The selective serotonin reuptake inhibitors, such as fluoxetine and sertraline, are selective blockers of the serotonin transporter. The tricyclic drugs have mixed actions; some, such as clomipramine, are relatively selective for the serotonin transporter. Uptake blockers increase synaptic concentrations of serotonin. Amphetamines enter monoaminergic neurons via the uptake transporter and bind to the vesicular transporter

found on the membranes of synaptic vesicles, causing reverse transport of the monoamine neurotransmitter into the cytoplasm. The neurotransmitter is then reverse-transported out of the neuron into the synapse via the uptake transporter.

6. *Degradation.* Phenelzine and tranylcypromine, both of which are effective for depression and panic disorder, block MAO-A and MAO-B. Moclobemide, effective against depression, is selective for MAO-A; selegiline, which has been used to treat Parkinson disease, is selective for MAO-B in low doses. (Abbreviation: **5-HIAA**, 5-hydroxyindoleacetic acid.)

B. Important sites of drug action at noradrenergic synapses.

1. *Enzymatic synthesis.* The competitive inhibitor α-methyltyrosine blocks the reaction catalyzed by tyrosine hydroxylase that converts tyrosine to DOPA. A dithiocarbamate derivative, FLA-63 (not shown), blocks the reaction that converts DOPA to dopamine.

2. *Storage.* Reserpine and tetrabenazine interfere with the transport of norepinephrine (**NE**), dopamine, and serotonin into synaptic vesicles by blocking the vesicular monoamine transporter VMAT$_2$. As a result, the cytoplasmic neurotransmitter is degraded (see below), and thus the neuron is depleted of neurotransmitter.

3. *Presynaptic receptors.* Agonists at presynaptic receptors produce negative feedback on neurotransmitter synthesis or release. Clonidine is an agonist at α$_2$-adrenergic receptors, inhibiting NE release. It has anxiolytic and sedative effects and is also used to treat attention deficit hyperactivity disorder. Yohimbine is an antagonist at α$_2$-adrenergic receptors; it induces anxiety.

4. *Postsynaptic receptors.* Propranolol is an antagonist at β-adrenergic receptors that blocks many effects of the sympathetic nervous system. It is used to treat some forms of cardiovascular disease but is commonly used to block anxiety during performance situations. Phenoxybenzamine is an agonist at α-adrenergic receptors.

5. *Uptake.* Certain tricyclic antidepressants, such as desipramine, and newer NE selective reuptake inhibitors, such as reboxetine, selectively block the NE transporter, thus increasing synaptic NE. Amphetamines enter monoaminergic neurons via the uptake transporter and interact with the vesicular transporter (the transporter on synaptic vesicles) to release neurotransmitter into the cytoplasm. The neurotransmitter is then pumped out of the neuron into the synapse via the uptake transporter acting in reverse.

6. *Degradation.* At the postsynaptic neuron, tropolone inhibits the enzyme catechol *O*-methyltransferase (**COMT**), which inactivates NE (step 6a). Normetanephrine (**NM**) is formed by the action of COMT on NE. At the presynaptic neuron, degradation by MAO is blocked by the MAO inhibitors phenelzine and tranylcypromine.

and also suggests new interpretations of such stimuli that help the patient cope with the experience. Where possible and when tolerable to patients, gradual transition to real-world exposures to phobic stimuli can be employed.

Exposure therapy produces extinction learning in analogy with studies of animal behavior. The memory of the phobic stimulus is not erased, but the fearful response is suppressed by new information that the stimulus and the context in which it is experienced are not dangerous. Animal physiology and lesioning studies and human imaging studies demonstrate that the prefrontal cortex is required for extinction learning and that the hippocampus is required for learning new contexts for familiar events or stimuli (eg, that a helicopter flying overhead does not portend an attack).

Electroconvulsive Therapy Is Highly Effective Against Depression

Although it still conjures up negative images in the popular imagination, electroconvulsive therapy (ECT) administered with modern anesthesia is medically safe and a tolerable patient experience, and it remains a highly effective intervention for the acute treatment of serious major depressive disorder. It is most often used when depressive symptoms are severe and medications and psychotherapies have proven ineffective. It is also effective in both the depressed and manic phases of bipolar disorder. It is not effective for anxiety disorders in the absence of a mood disorder and is not used to treat them clinically.

Generally, six to eight treatments are given, most commonly on an outpatient basis. Patients are anesthetized, and electrical stimulation is administered just above the threshold to produce electroencephalographic evidence of a generalized seizure. The major side effect is a variable degree of anterograde and retrograde amnesia. Amnesia can be minimized, but not eliminated, by placing the electrodes unilaterally and using the lowest level of electrical stimulation needed. Rodents given ECT exhibit massive release of neurotransmitters, which causes significant activation of gene expression, presumably leading to large-scale neural plasticity. However, the precise molecules, cells, and circuits involved in the therapeutic response remain unknown.

Newer Forms of Neuromodulation Are Being Developed to Treat Depression

Other forms of therapeutic electrical stimulation of the brain are being explored, motivated by the desire to improve upon the therapeutic effects of ECT while diminishing its side effects. These approaches are often described as "neuromodulation."

Transcranial magnetic stimulation (TMS) employs a device on the scalp to deliver brief pulses of rapidly alternating magnetic stimulation. This induces currents to flow within axons in regions of cerebral cortex beneath the device. Daily administration of TMS over the left prefrontal cortex is safe and was effective enough to have received regulatory approval by the US Food and Drug Administration. Nonetheless, in subsequent trials, its efficacy appears to be only modest. Additional clinical experiments are under way aimed at improving efficacy.

Alternative therapies under development include magnetic seizure therapy, an alternative to ECT in which a magnetic field is the used to produce a seizure. The hope for this experimental therapy is to reproduce the efficacy of ECT with less anterograde and retrograde amnesia.

Deep brain stimulation (DBS), mentioned above, is an invasive neuromodulatory treatment in wide use for treatment of the motor symptoms of Parkinson disease and of essential tremor. For treatment of Parkinson disease, an electrode is typically placed within the subthalamic nucleus, a component of basal ganglia circuitry involved in motor control that is well understood compared with circuits that regulate mood. A DBS electrode is connected by a wire that exits the skull and travels under the scalp and skin of the neck to a controller and battery pack that resides in the chest, much like a cardiac pacemaker battery. The rate at which the electrode stimulates its target can be controlled externally and is typically adjusted by the treatment team to optimize the therapeutic response. During the past decade, clinical trials of DBS have been extended from Parkinson disease and other movement disorders to psychiatric disorders. In addition to its use in treatment-refractory depression, DBS is being studied for the treatment of obsessive-compulsive disorder.

Several locations in the brain have been targets for DBS to treat depression. As described in Figure 61–4, the rostral anterior (subgenual) cingulate cortex is activated by sadness. Accordingly, it has been used as a DBS target for treatment-resistant depression (Figure 61–8). In some clinical series, 60% of treatment-resistant patients achieved stable improvement with stimulation of the subgenual cingulate cortex. However, similar levels of efficacy using this target could not be replicated in a large multisite clinical trial. Differences in patient selection, interindividual differences in brain anatomy, or small differences in electrode placement may account for the disparate results seen to date. To

Figure 61–8 Electrode placement for deep brain stimulation (DBS) in the rostral anterior cingulate cortex and measurement of response by [18F] fluoro-2-deoxyglucose positron emission tomography (PET). (Reproduced, with permission, from Helen Mayberg.)

A. *Left:* The rostral anterior (subgenual) cingulate cortex, Brodmann area 25 (**Cg25**), is an anatomic target for DBS for patients with treatment-resistant depression. (Sagittal section; electrode site in **red**; corpus callosum is just superior and shown in **white**; **dotted line**, position of the electrode relative to the AC-genu line.) (Abbreviations: **AC**, anterior commissure; **Mid-SCC**, mid-subcallosal cingulate.) *Right:* A PET scan shows placement of the electrodes in the brain of a patient undergoing stimulation of the rostral anterior cingulate cortex. (Sagittal section.)

B. PET scans show the changes in activity in patients with treatment-resistant depression who have improved with stimulation of the rostral anterior cingulate cortex. The **top** panels are sagittal sections; the **bottom** panels are coronal sections. *Left:* Pretreatment metabolic activity in patients with treatment-resistant depression. **Red** pseudo-color denotes elevated metabolic activity compared with healthy control subjects (note elevated activity in Cg25 before DBS); **blue** denotes lower metabolic activity. *Right:* Averages of patients who have improved at 3 or 6 months after initiation of DBS. Activity in Cg25 is decreased (**blue**) in patients who have had a positive response to stimulation. (Abbreviations: **ACC**, anterior cingulate cortex; **BS**, brain stem; **F9**, dorsolateral prefrontal cortex; **F46**, prefrontal cortex; **F47**, ventrolateral prefrontal cortex; **HT**, hypothalamus; **Ins**, insula; **mF10**, medial frontal cortex; **MCC**, middle cingulate cortex; **OF11**, orbital frontal cortex; **SN**, substantia nigra; **vCD**, ventral caudate.)

A Surgical procedure

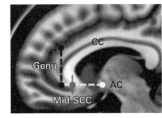 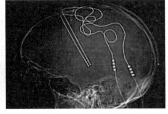

Anatomical target for electrode   Bilateral DBS electrode placement

B Change in PET activity in DBS responders

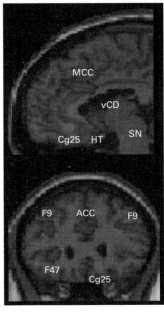 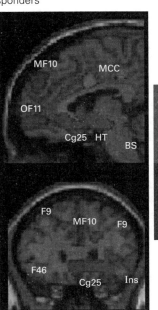

PET baseline patients vs. healthy control subjects   PET in improved patients after 3 or 6 months

put it simply, depression is highly heterogeneous, and it should not be surprising that a single DBS target is not useful for all treatment-resistant patients.

Lacking good animal models of mood disorders, human DBS treatment trials may provide a particularly important source of information about the brain circuitry responsible for the symptoms of mental disorders. Although careful attention must be paid to obtaining informed consent and to safety, especially when the judgment of patients is influenced by severe depression, DBS may provide an opportunity to learn about mood regulation. In particular, newly developed electrodes not only stimulate a DBS target but can also record extracellular neuronal activity. Such "read–write" electrodes, currently being used in research settings only, may not only improve clinical results but also advance our knowledge of circuit dysfunction and therapeutic modulation in psychiatric disorders.

Bipolar Disorder Can Be Treated With Lithium and Several Anticonvulsant Drugs

In 1949, John Cade discovered the calming effects of lithium in guinea pigs and, soon thereafter, in a small clinical trial in bipolar patients. Cade's observations initiated the modern era of psychopharmacology in which drugs, ultimately subjected to randomized, blinded clinical trials, were used to treat specific symptoms of mental disorders. Lithium eventually proved to be effective in treating acute episodes of mania and in stabilizing mood by reducing the frequency of cycling into mania and depression.

Several drugs initially developed to treat epilepsy, such as valproic acid and lamotrigine, have also been shown to be effective in treating acute mania and for mood stabilization and can serve as substitutes for lithium. In addition, antipsychotic drugs effectively ameliorate symptoms of acute mania and, at low doses, can also help stabilize mood. None of these drugs exerts therapeutic effects rapidly; improvements in mental state and behavior may take several weeks.

The mechanisms by which lithium and anticonvulsant drugs exert beneficial effects on mania and on mood cycling are not known. Unlike the antidepressant and antipsychotic drugs, however, there remain open questions about the initial molecular target of lithium in the nervous system relevant to the initiation of its therapeutic effects. This lack of certainty reflects the many actions of lithium at therapeutic concentrations in the brain. The most likely molecular target is inhibition of glycogen synthase kinase type 3β (GSK3β), a component of the Wnt signaling pathway that has many functions in the nervous system. As in the case of other drugs to treat psychiatric disorders, investigation of the therapeutic mechanism of lithium and of the mood-stabilizing properties of anticonvulsants is impeded by the lack of an animal model of bipolar disorder.

Whatever the molecular mechanisms of lithium or the anticonvulsants, mood stabilizers appear to dampen the dynamics of mood regulatory systems. Mood is regulated by the external environment as well as several internal inputs, including the internal hormonal milieu, immune modulators, and circadian controls (eg, both the serotonergic and noradrenergic systems show diurnal variations closely coupled with the sleep–wake cycle). The integration of these systems is complex, involving dynamic interactions that are still poorly understood.

Second-Generation Antipsychotic Drugs Are Useful Treatments for Bipolar Disorder

All antipsychotic drugs act by blocking D_2 dopamine receptors, but these drugs have long been recognized to have therapeutic effects not only in the treatment of the psychotic symptoms of schizophrenia, severe mood disorders, and many other conditions, but also in the treatment of acute manic episodes. The side effects of first-generation antipsychotic drugs are severe, most prominently Parkinson-like motor side effects that result from D_2 dopamine receptor antagonism.

Most second-generation drugs have somewhat lower affinity for D_2 dopamine receptors than first-generation drugs and, in addition, have other receptor effects, such as blocking serotonin 5-HT$_{2A}$ receptors, resulting in a lower liability for severe motor side effects. These drugs are by no means free of serious side effects; most cause weight gain and associated metabolic conditions. However, their relative tolerability and their effects on serotonin receptors have made

them an important treatment for the depressed phase of bipolar disorder as well as the treatment of acute mania. They have gained an important role in therapeutics because bipolar depression is less likely to respond to antidepressant drugs than unipolar depression.

Highlights

1. Mood disorders are divided into unipolar and bipolar disorder based on whether depression occurs alone (unipolar) or whether a person also suffers from episodes of mania. Unipolar and bipolar disorders have different familial patterns of transmission.

2. Clinically significant unipolar depression, often denoted as major depressive disorder (major depression), differs from normal sadness by its persistence, pervasiveness, and association with physiological, cognitive, and behavioral symptoms.

3. Major depression is common (15%–20% lifetime prevalence) and disabling, making it a leading cause of disability worldwide. Bipolar disorder is less common (1% lifetime prevalence worldwide) but tends to produce severe symptoms that often require hospitalization.

4. Anxiety disorders are the most common psychiatric disorders. They range in severity from highly disabling cases of panic disorder and posttraumatic stress disorder (PTSD) to simple phobias. They often co-occur with major depression.

5. Mood and anxiety disorders have both genetic and nongenetic components of risk. Bipolar disorder is more heritable than major depression or anxiety disorders. Childhood adversity and later environmental stressors play a significant role in susceptibility to major depression and anxiety disorders. Genetic analyses of bipolar disorder, major depression, and PTSD are beginning to yield molecular clues to pathogenesis.

6. The neural circuitry of fear and anxiety disorders involves the amygdala and its interconnections with the prefrontal cortex. The neural circuitry of major depression and bipolar disorder is less well understood. However, neuroimaging in humans with major depression implicated circuits involved in the processing of emotional salience and in cognitive control.

7. Bipolar disorder can be treated with lithium, certain anticonvulsant drugs such as valproic acid, and second-generation antipsychotic drugs, although many patients have residual symptoms, most commonly depression.

8. Major depression and anxiety disorders can be treated with diverse antidepressant drugs and by cognitive and behavioral therapies. Electroconvulsive therapy is effective for major depression that is unresponsive to medications.

9. Experimental treatments such as deep brain stimulation are being investigated for treatment of major depression and other psychiatric disorders. The development of electrodes that can record as well as stimulate promise greater insight in human neural circuit function in disease and its treatment.

Steven E. Hyman
Carol Tamminga

Selected Reading

Nestler EJ, Hyman SE, Holtzman D, Malenka RJ. 2015. *Molecular Neuropharmacology: Foundation for Clinical Neuroscience*, 3rd ed. New York: McGraw-Hill.

Otte C, Gold SM, Penninx BW, et al. AF. 2016. Major depressive disorder. Nat Rev Dis Primers 2:16065.

Sullivan PF, Daly MJ, O'Donovan M. 2012. Genetic architecture of psychiatric disorders: the emerging picture and its implications. Nat Rev Genet 13:537–551.

Yehuda R, Hoge CW, McFarlane AC, et al. 2015. Post-traumatic stress disorder. Nat Rev Dis Primers 1:15057.

References

Adhikari A, Lerner TN, Finkelstein J, et al. 2015. Basomedial amygdala mediates top-down control of anxiety and fear. Nature 527:179–185.

American Psychiatric Association. 2013. *Diagnostic and Statistical Manual of Mental Disorders*, 5th ed. Washington, DC: American Psychiatric Association.

Anacker C, Hen R. 2017. Adult hippocampal neurogenesis and cognitive flexibility: linking memory and mood. Nat Rev Neurosci 18:335–346.

Bagot RC, Cates HM, Purushothama I, et al. 2016. Circuit-wide transcriptional profiling reveals brain region-specific gene networks regulating depression susceptibility. Neuron 90:969–983.

Besnard A, Sahay A. 2016. Adult hippocampal neurogenesis, fear generalization, and stress. Neuropsychopharm 41:24–44.

Cade JFJ. 1949. Lithium salts in the treatment of psychotic excitement. Med Australia 2:349–352.

Clementz BA, Sweeney JA, Hamm, JP, et al. 2015. Identification of distinct psychosis biotypes using brain-based biomarkers. Am J Psychiatry 173:373–384.

Cross-Disorder Group of the Psychiatric Genomics Consortium, Lee SH, Ripke S, et al. 2013. Genetic relationship between five psychiatric disorders estimated from genome-wide SNPs. Nat Genet 45:984–994.

Davidson RJ, Pizzagalli D, Nitschke JB, Putnam K. 2002. Depression: perspectives from affective neuroscience. Annu Rev Psychol 53:545–574.

Dayan P, Huys QJ. 2009. Serotonin in affective control. Annu Rev Neurosci 32:95–126.

Drysdale AT, Grosenick L, Downar J, et al. 2017. Resting-state connectivity biomarkers define neurophysiological subtypes of depression. Nat Med 23:28–38.

Etkin A, Klemenhagen KC, Dudman JT, et al. 2004. Individual differences in trait anxiety predict the response of the basolateral amygdala to unconsciously processed fearful faces. Neuron 44:1043–1055.

Fettes P, Schulze L, Downar J. 2017. Cortico-striato-thalamic loop circuits of the orbitofrontal cortex: promising therapeutic targets in psychiatric illness. Front Syst Neurocsci 11:25.

Fornaro M, Stubbs B, De BD, et al. 2016. Atypical antipsychotics in the treatment of acute bipolar depression with mixed features: a systematic review and exploratory meta-analysis of placebo-controlled clinical trials. Int J Mol Sci 17:241.

Heimer L. 1995. *The Human Brain and Spinal Cord*, 2nd ed. New York: Springer-Verlag.

Holtzheimer PE, Mayberg HS. 2011. Deep brain stimulation for psychiatric disorders. Annu Rev Neurosci 34:289–307.

Hui PS, Sim K, Baldessarini RJ. 2015. Pharmacological approaches for treatment-resistant bipolar disorder. Curr Neuropharmacol 13:592–604.

Hyde CL, Nagle MW, Tian C, et al. 2016. Identification of 15 genetic loci associated with risk of major depression in individuals of European descent. Nat Genet 48:1031–1036.

Ivleva EI, Morris DW, Moates AF, et al. 2010. Genetics and intermediate phenotypes of the schizophrenia: bipolar disorder boundary. Neurosci Biobehav Rev 34:897–921.

Johansen JP, Cain CK, Ostroff LE, LeDoux JE. 2011. Molecular mechanisms of fear learning and memory. Cell 47:509–524.

Kendler KS, Prescott CA, Myers J, Neale MC. 2003. The structure of genetic and environmental risk factors for common psychiatric and substance use disorders in men and women. Arch Gen Psychiatry 60:929–937.

Kessler RC, Bromet EJ. 2013. The epidemiology of depression across cultures. Annu Rev Public Health 34:119–138.

Kreuger RF, Markon KE. 2006. Reinterpreting comorbidity: a model-based approach to understanding and classifying psychopathology. Annu Rev Clin Psychol 2:111–133.

Mayberg HS, Brannan SK, Mahurin RK, et al. 1997. Cingulate function in depression: a potential predictor of treatment response. NeuroReport 8:1057–1061.

Mayberg HS, Liotti M, Brannan SK, et al. 1999. Reciprocal limbic-cortical function and negative mood: converging PET findings in depression and normal sadness. Am J Psychiatry 156:675–682.

Mayberg HS, Lozano AM, Voon V, et al. 2005. Deep brain stimulation for treatment-resistant depression. Neuron 45:651–660.

McClintock SM, Reti IM, Carpenter LL, et al. 2018. Consensus recommendations for the clinical application of repetitive transcranial magnetic stimulation (rTMS) in the treatment of depression. J Clin Psychiatry 79:1. doi:10.4088/JCP.16cs10905.

Miller BR, Hen R. 2015. The current state of the neurogenic theory of depression and anxiety. Curr Opin Neurobiol 30:51–58.

Moussavi S, Chatterji S, Verdes E, et al. 2007. Depression, chronic diseases, and decrements in health: results from the World Health Surveys. Lancet 370:851–858.

Muller VI, Cieslik EC, Serbanescu I, et al. 2017. Altered brain activity in unipolar depression revisited. Meta-analyses of neuroimaging studies. JAMA Psychiatry 74:47–55.

Neal, BM, Sklar P. 2015. Genetic analysis of schizophrenia and bipolar disorder reveals polygenicity but also suggests new directions for molecular interrogation. Curr Opin Neurobiol 30:131–138.

Nock MK, Borges G, Bromet EJ, et al. 2008. Cross-national prevalence and risk factors for suicidal ideation, plans and attempts. Br J Psychiatry 192:98–105.

Pizzagalli D, Pascual-Marqui RD, Nitschke JB, et al. 2001. Anterior cingulate activity as a predictor of degree of treatment response in major depression: evidence from brain electrical tomography analysis. Am J Psychiatry 158:405–415.

Ripke S, Wray NR, Lewis CM, et al. 2013. A mega-analysis of genome-wide association studies for major depressive disorder. Mol Psychiatry 18:497–511.

Seeley WW, Menon V, Schaztzberg AF, et al. 2007. Dissociable intrinsic connectivity networks for salience processing and executive control. J Neurosci 27:2349–2356.

Sheline YI, Sanghavi M, Mintun MA, Gado MH. 1999. Depression duration but not age predicts hippocampal volume loss in medically healthy women with recurrent major depression. J Neurosci 19:5034–5043.

Stoddard J, Gotts SJ, Brotman MA, et al. 2016. Aberrant intrinsic functional connectivity within and between corticostriatal and temporal-parietal networks in adults and youth with bipolar disorder. Psychol Med 46:1509–1522.

Trivedi MH, Rush AJ, Wisniewski SR, et al. 2006. Evaluation of outcomes with citalopram for depression using measurement-based care in STAR*D: implications for clinical practice. Am J Psychiatry 163:28–40.

Tye KM, Prakash R, Kim SY, et al. 2011. Amygdala circuitry mediating reversible and bidirectional control of anxiety. Nature 471:358–362.

Whiteford HA, Degenhardt L, Rehm J, et al. 2013. Global burden of disease attributable to mental and substance use disorders: findings from the global burden of disease study 2010. Lancet 382:1575–1586.

Zarate CA Jr, Singh JB, Carlson PJ, et al. 2006. A randomized trial of an N-methyl-D-aspartate antagonist in treatment-resistant major depression. Arch Gen Psychiatry 63:856–864.

62

Disorders Affecting Social Cognition: Autism Spectrum Disorder

MENTAL RETARDATION, now referred to widely as *intellectual disability*, is currently defined as having an IQ below 70 accompanied by marked deficits in adaptive functioning. Both terms have been broadly used to label a variety of cognitive impairments linked to prenatal or early postnatal brain abnormalities. For decades, subsets of individuals with rare intellectual disability syndromes, such as Rett syndrome or fragile X syndrome, have been characterized by their genetic etiologies. We are now beginning to elucidate the complex genetics of more prevalent neurodevelopmental disorders without distinct physical features that distinguish them, including so-called *idiopathic* or *nonsyndromic* forms of autism spectrum disorder (ASD). The combination of insights resulting from the intensive study of rare genetic syndromes coupled with successes in unraveling the genetics underlying idiopathic ASD has transformed our understanding of normal and pathological development of the human brain.

Common to all of these disorders are mental impairments that persist throughout life, hampering development and learning. Generally speaking, even if all mental functions seem to be affected, conditions with distinct etiologies and natural histories can be differentiated because some cognitive domains tend to be more impaired than others. And indeed, these differences are reified in diagnostic schemes that draw distinctions between developmental abnormalities that affect primarily general cognition, social cognition, or perception. These differential cognitive and behavioral vulnerabilities may provide useful clues about the origin and developmental time course of specific mental functions in normal development.

In this chapter, we focus principally on neurodevelopmental disorders that include abnormalities in social functioning, including ASD, fragile X syndrome, Williams syndrome, Rett syndrome, and Angelman and Prader-Willi syndromes. These conditions all impair highly sophisticated brain functions including social awareness and communication. ASD is a prime focus for several reasons: the high prevalence in the population; the overlap in genetic risks with other common neuropsychiatric conditions, including schizophrenia; and the absence of a defining neuropathology. They are also exemplars of the etiological and phenotypic heterogeneity common to many psychiatric syndromes. In this respect, ASD is a paradigmatic neuropsychiatric syndrome.

Autism Spectrum Disorder Phenotypes Share Characteristic Behavioral Features

Profound social disability has probably always been with us, but the characterization of autism as a medical syndrome was first described in the literature in 1943 by Leo Kanner and in 1944 by Hans Asperger. Today, clinicians and researchers think of autism as a spectrum of disorders with two defining but highly variable diagnostic features: impaired social communication and stereotyped behaviors with highly restricted interests.

Until recently, the term "Asperger syndrome" was used to describe individuals who met these two diagnostic criteria, but in whom language acquisition was not delayed and IQ was in the normal range. In the most recent edition of the standard psychiatric diagnostic manual, *Diagnostic and Statistical Manual of Mental Disorders, Fifth Edition* (DSM-5), Asperger syndrome along with a distinct disorder known as pervasive developmental disorder not otherwise specified—designed to capture individuals with deficits in social communication who did not meet full criteria in other areas—were eliminated in favor of including variations within a single spectrum construct.

Autism spectrum disorder is present in at least 1.5% of the population. Rigorous epidemiological studies estimate prevalence as high as 2.6% for the full

spectrum of social disability, far higher than estimated only decades ago. The reasons for the increase in the prevalence over a relatively short time frame are of considerable interest and active debate, particularly among the lay public. Within the scientific community, a consensus has emerged that this increase reflects a combination of changing diagnostic criteria, increased awareness among families and health care professionals, "diagnostic substitution" (in which individuals who formerly would have been diagnosed with intellectual disability are now more likely to be identified as socially disabled), and some true increase in incidence. These issues will be discussed below with regard to genetic risks.

Autism spectrum disorders occur predominantly in males, although the typically cited 4:1 male-to-female ratio has recently been called into question based on concerns about male bias in the approaches used to ascertain the diagnosis, including the diagnostic instruments. Even accounting for these challenges, however, the cumulative evidence suggests a ratio bias of at least 2:1 to 3:1 male excess. Individuals across the IQ spectrum are affected, and based on current diagnostic practices, about half of all individuals with ASD also have intellectual disability. By definition, ASD must be detectable before 3 years of age, but recent studies have shown that it is possible to identify affected children in high-risk families well within the first year of life. ASD occurs in all countries and cultures and in every socioeconomic group.

Although ASD clearly affects the brain, no definitive biological markers have yet been identified; thus, diagnosis is based on behavioral criteria. This does not mean that there are not strong biological correlates, including specific gene mutations and neuroimaging findings, but none of these are sufficiently specific or predictive to be useful as an alternative to the gold standard of clinical assessment. Moreover, because behavior is variable during development and depends on a number of factors—age, environment, social context, and availability and duration of remedial help—no single behavior is likely ever to be conclusively diagnostic.

Like other neurodevelopmental syndromes, ASD typically endures throughout life. However, in recent longitudinal studies, approximately 10% of clearly affected children showed improvement, with little or no evidence of social disability later in life. Autism is not progressive. On the contrary, special educational programs and professional support often lead to improvements in behavior and adaptive functioning with age.

Autism Spectrum Disorder Phenotypes Also Share Distinctive Cognitive Abnormalities

Social Communication Is Impaired in Autism Spectrum Disorder: The Mind Blindness Hypothesis

One cognitive theory of social communication postulates that humans have a particularly well-developed ability to understand the mental states of others in an intuitive and fully automatic fashion. Watching a young person surreptitiously trying to open a car door without a key, you instantly understand that she believes she can break in while being unobserved,

and you expect her to run away as soon as she realizes someone is watching. Thus, you explain and predict her behavior by inferring her mental states (desires, intentions, beliefs, knowledge) from her overt behavior. This so-called mentalizing ability, termed a *theory of mind*, is thought to depend on specific brain mechanisms and circuits underlying social cognition (Figure 62–1). Further, it is postulated that mentalizing is impaired in ASD, with profound effects on social development.

It is now generally agreed that insight into the mental state of others depends on the capacity to mentalize spontaneously. Spontaneous mentalizing allows us to appreciate that different people have different

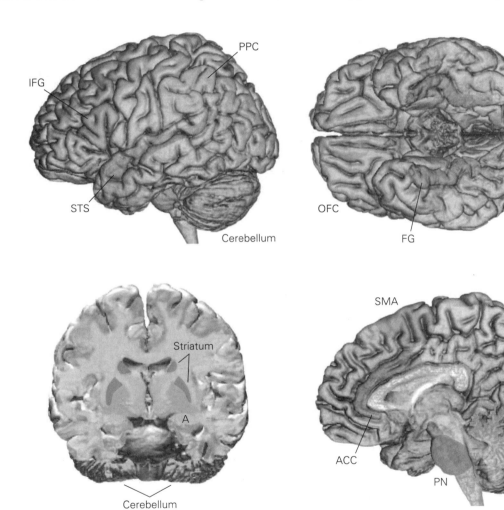

Figure 62–1 Brain areas implicated in the three core deficits characteristic of autism: impaired social interaction, impaired language and communication, and severely restricted interests with repetitive and stereotyped behaviors. Areas implicated in social deficits include the orbitofrontal cortex (**OFC**), the anterior cingulate cortex (**ACC**), and the amygdala (**A**). Cortex bordering the superior temporal sulcus (**STS**) has been implicated in mediating the perception that a

living thing is moving and gaze perception. Face processing involves a region of the inferior temporal cortex within the fusiform gyrus (**FG**). Comprehension and expression of language involve a number of regions including the inferior frontal region, the striatum, and subcortical areas such as the pontine nuclei (**PN**). The striatum has also been implicated in the mediation of repetitive behaviors. (Abbreviations: **IFG**, inferior frontal gyrus; **PPC**, posterior parietal cortex; **SMA**, supplementary motor area.)

thoughts and that thoughts are internal and different from external reality.

The inability to mentalize, or "mind blindness," was first tested in children with autism using a simple puppet game, the Sally-Anne test. Young children with ASD, unlike those with Down syndrome or typically developing 4-year-olds, cannot predict where a puppet will first look for an object that was moved while the puppet was out of the room. They are not able to imagine that the puppet will "think" that the object will be where the puppet had left it (Figure 62–2). Many children with ASD eventually do learn to pass this task, but on average with a 5-year delay. Mentalizing acquired so slowly remains effortful and error-prone even in adulthood.

At the same time, young children with ASD show excellent appreciation of physical causes and events. For instance, a child who is incapable of falsely telling another that a box is locked is quite capable of locking the same box to prevent its contents from being stolen.

Variations of the Sally-Anne test and other mentalizing tasks have been used with children and adults with ASD since the mid-1980s (Figure 62–3).

Figure 62–2 The Sally-Anne test. This first test of the "theory of mind" begins with a scripted performance using two dolls. Sally has a basket; Anne has a box. Sally puts a ball into her basket. She goes for a walk and leaves the room. While Sally is outside, naughty Anne takes the ball out of the basket and puts it into her box. Now Sally comes back from her walk and wants to play with her ball. Where will she look for the ball, the basket or the box? The answer, the basket, is obvious to most typically developing 4-year-olds but not to autistic children of the same or even higher mental age. (Adapted from original artwork by Axel Scheffler.)

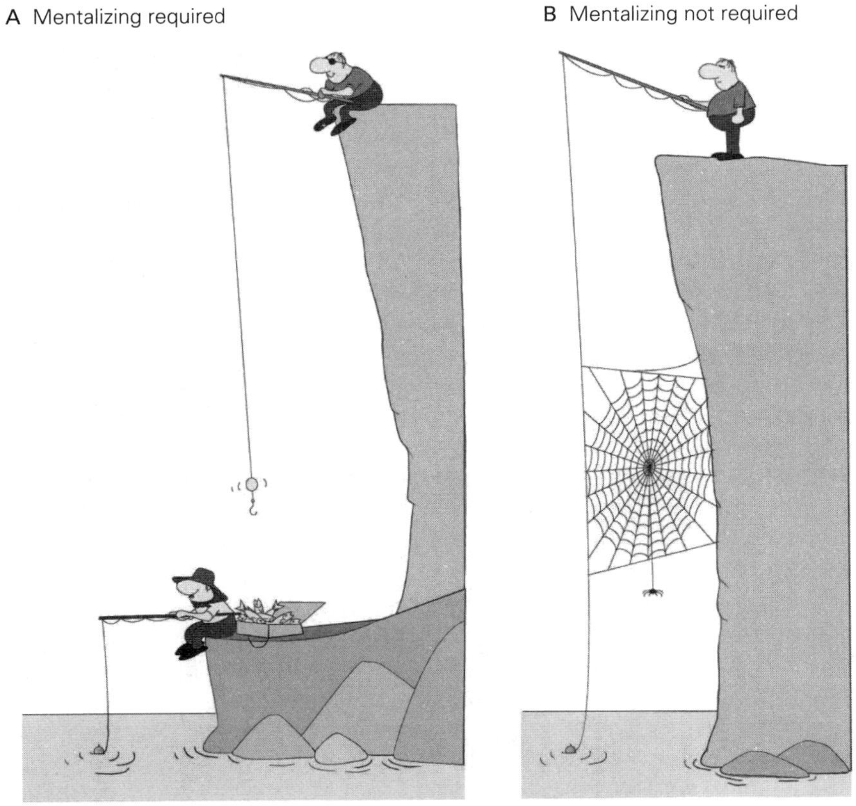

A Mentalizing required

B Mentalizing not required

Figure 62–3 Examples of cartoons used in imaging studies of "mentalizing." Participants were asked to consider the meaning of each picture (silently) and then to explain them. In a functional magnetic resonance imaging study, normal adults passively viewed cartoons that require mentalizing versus those that do not. A characteristic network of brain regions is activated in each subject (see Figure 62–4). (Adapted from Gallagher et al. 2000.)

Functional neuroimaging has been used to examine activity in the brain of healthy subjects while they are engaged in tasks that necessitate thinking about mental states. A wide range of tasks using visual and verbal stimuli has been used in these studies. In an early positron emission tomography study, adults in a control cohort viewed silent animations of geometric shapes. In some of the animations, the triangles move in scripted scenarios designed to evoke mentalizing (eg, triangles tricking each other). In other animations, the triangles move randomly in a manner that does not evoke mentalizing. Comparison of the scans made while subjects viewed each type of animation reveals a specific network of four brain centers involved in mentalizing (Figure 62–4). Functional magnetic resonance imaging (MRI) studies using the same animations have shown that activity in this network is reduced in subjects with ASD.

This network has four components. The first, in the medial prefrontal cortex, is a region thought to be involved in monitoring one's own thoughts. A second component, in the temporoparietal region of the superior temporal lobe, is known to be activated by eye gaze and biological motion. Patients with lesions in this area in the left hemisphere are unable to pass

the Sally-Anne test. The third region is the amygdala, which is involved in the evaluation of social and nonsocial information for indications of danger in the environment. The fourth region is an inferior temporal region involved in the perception of faces.

Recent studies have used stimuli intended to capture more nuanced and naturalistic social content, for example, using movies of actual social encounters as opposed to static pictures of facial expressions. These studies have identified, among other things, the role of the orbital frontal cortex in social cognition.

Other Social Mechanisms Contribute to Autism Spectrum Disorder

From birth, normal infants prefer to attend to people rather than other stimuli. An absence of this preference could lead to an inability to understand and interact with others. Indeed, the absence of preferential attention to social stimuli and mutual attention are widely acknowledged as early signs of ASD. These deficits may not involve problems with mentalizing, given that mutual attention normally appears toward the end of the first year when signs of mentalizing are still sparse.

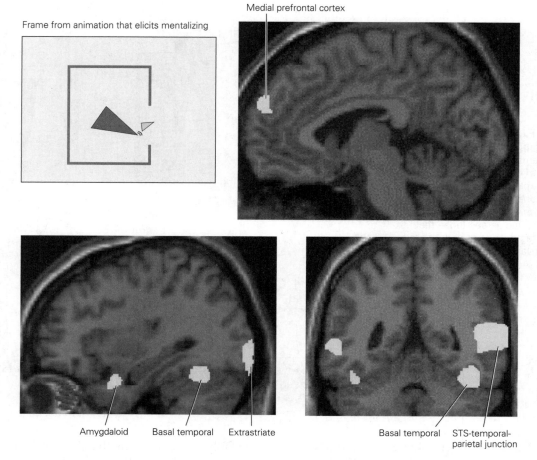

Figure 62–4 The mentalizing system of the brain. Healthy volunteers were presented with animated triangles that moved in such a way that viewers would attribute mental states to them. In the sample frame shown, the larger triangle was seen as encouraging the smaller triangle to leave the enclosure. They were also presented with animated triangles that moved in a more or less random fashion and thus would not elicit mentalizing. The highlighted areas show differences in the positron emission tomography scans of brain activation when these two viewing conditions were compared. (Abbreviation: **STS**, superior temporal sulcus.) (Reproduced, with permission, from Castelli et al. 2002. Copyright © 2002, Oxford University Press.)

Researchers have long considered the possibility that a specific neural mechanism underlies attention to social stimuli, such as faces, voices, and biological motion. In favor of this hypothesis, researchers found that the gaze of individuals with ASD is abnormal when watching social scenes. For example, multiple studies have found that individuals with ASD fixate on people's mouths instead of showing the normal preference for eyes (Figure 62–5).

People With Autism Show a Lack of Behavioral Flexibility

Repetitive and inflexible behavior in ASD may reflect abnormalities in frontal lobe executive functions, a wide array of higher cognitive processes that include the ability to disengage from a given task, inhibit inappropriate responses, stay on task (plan and manage sequences of deliberate actions), keep multiple task demands in working memory, monitor performance, and shift attention from one task to another.

Even ASD individuals with IQs in the normal range have problems in planning, organizing, and flexibly switching between behaviors. Irrespective of IQ, affected individuals have difficulties suggesting various different uses of a single object such as a handkerchief (used to block a sneeze, to wrap loose objects, etc.). Flexible thinking is also poor in patients with acquired damage to the frontal lobe.

Some Individuals With Autism Have Special Talents

A particularly fascinating feature of ASD in some individuals is "savant syndrome," defined by the presence

Figure 62–5 Individuals with autistic disorder often do not look into the eyes of others. Patterns of eye movements in individuals with autism were studied while the subjects watched clips from the film *Who's Afraid of Virginia Wolf?* When looking at human faces, the subjects tended to look at the mouth rather than the eyes, and in scenes of intense interaction between people, they tended to look at irrelevant places rather than at the faces of the actors. (Reproduced, with permission, from Klin et al. 2002. Copyright © 2002 American Psychiatric Association.)

Typically developing viewer
Viewer with autism

of one or more exceptional skills that are in marked contrast to the individual's overall disability but also rare in the population at large. The most widely cited estimate is that 10% of individuals with ASD demonstrate such exceptional abilities compared to about 1 in 1,000 individuals with other forms of intellectual disability.

In the largest ASD cohort surveyed by self-reporting to date (about 5,000 families), 531 individuals were reported to have exceptional abilities in the following 10 areas (listed in descending frequency): music, memory, art, hyperlexia, mathematics, mechanical, coordination, directions, calendar calculating, and extrasensory perception. Subsequent small-scale studies have placed the prevalence of savant skills in ASD at between 13% and 28%.

A recently established savant syndrome registry now includes more than 400 people from 33 countries. Among a group of 319 individuals who met some criteria that earned them a savant diagnosis based on family or caregiver reports or self-reporting, 75% who showed savant skills in childhood were diagnosed with ASD. Approximately half reported a single exceptional skill and half reported multiple skills. Music was the most commonly reported exceptional skill, followed by art, memory, and mathematics. Calendar calculating, while present in many savants along with another skill, was the sole skill in only about 5% of the sample. Among this self- or family-selected group, the overall sex distribution mirrored that reported for ASD in general, with a male-to-female ratio of approximately 4:1.

One explanation for savant syndrome is that information processing is preferentially geared to tiny details at the cost of seeing the bigger picture. (For example, the drawing by the gifted artist with high-functioning autism in Figure 62–6 shows remarkably detailed cityscapes, as well as detailed numerical patterns and dates.) A similar hypothesis is that brain regions involved in perception are overfunctioning; another possibility is that there is a preference for manipulating the bits of information that fit within a strict framework such as calendar knowledge or a bus timetable. Neuropsychological data support both explanations, but decisive experiments to distinguish between them remain to be done.

Genetic Factors Increase Risk for Autism Spectrum Disorder

The earliest evidence that genes contribute to ASD arose from studies of twin pairs as well as familial aggregation. The former show from 60% to 90% concordance among monozygotic twin pairs; this wide range is due in part to previously used diagnostic criteria and classifications. For example, the highest estimates of monozygotic concordance are derived from observations of twins with any of three diagnoses that made up the social disability spectrum prior to the reformulations in the DSM-5. Only approximately 60% of monozygotic twins were found to be concordant for the "full diagnosis" of autism, which was defined at

Figure 62–6 Strikingly beautiful art work by George Widener. George is a highly accomplished and much-admired outsider artist. In the attention to detail, this drawing resembles the drawings of other autistic savant artists. The intricate topographical detail of a symmetrically arranged city, with rivers, bridges, and tall buildings, is combined with minutely executed and seemingly abstruse calendar sequences. Mastery of the calendar and the ability to name the day of the week for any given date has often been described for autistic savants. The viewer of this drawing can partake in an otherwise very private world of space and time, numbers, and patterns. (Reproduced, with permission, from the Henry Boxer Gallery, London.)

the time as comprising fundamental impairments in each of three categories: social communication, language development, and restricted interests or repetitive behaviors. In contrast, dizygotic twins show 10% to 30% concordance—again with the lower number estimating concordance for the diagnosis of isolated autism, while the larger number encompasses any of three diagnoses on the autism spectrum.

This difference between the rates at which monozygotic and dizygotic twins share an ASD phenotype is attributed to differences in the amount of shared genetic material between the two types of twin pairs. Monozygotic siblings share all their DNA, whereas dizygotic twins share as much DNA as any sibling pair. In addition to these types of data, it has long been observed that ASD runs in families: Current estimates are that if parents have one child with ASD, the risk that a second child will be affected increases approximately 5- to 10-fold over the population base rate.

The most generous estimates of genetic contribution do not explain all risk for ASD in the population. Some contribution from the environment is a certainty. However, given the well-known public debate on the issue of whether immunization is a factor in ASD, it is important to note that there is no credible evidence that the increase in ASD prevalence is due to immunizations. The initial study that raised the issue of the contribution of the trivalent measles-mumps-rubella (MMR) vaccine has been retracted and thoroughly repudiated by the editors of the journal in which the article appeared, as well as by 10 of 12 of the original authors. A wide range of subsequent investigations, both of the MMR vaccine and of vaccines with the mercury-containing preservative thimerosal, has found no evidence for association with ASD risk.

The counter argument that certain rare individuals may be predisposed to a vulnerability to vaccines leading to ASD is nonfalsifiable. However, three lines of evidence suggest that such a contribution, if present, is likely to be quite small. First, it is important to recall that the basis for the MMR hypothesis has been thoroughly debunked, and consequently, the prior probability that vaccines are major etiological factors is extremely low. Second, even in very large research cohorts, it has so far not been possible to detect a risk signal. Third, although there is a subset of children with ASD who show developmental regression in the second year of life, there is often evidence on careful examination of preexisting delay. In the final analysis, although the current level of understanding of pathophysiological mechanisms makes it impossible to definitively exclude any etiological contributor in a single individual, what is incontrovertible is that the risks to children of not receiving vaccinations are clear, measurable, and far greater overall than the role vaccines might play in ASD risk.

Although the evidence for a predominantly genetic contribution has been consistent, until recently, the search for risk genes contributing to nonsyndromic forms of ASD proved to be extremely challenging. Now, as will be discussed below, technological advances and changes in research culture have transformed the field. Moreover, critically important initial insights into both the genetics and neurobiology of ASD have emerged from the investigation of well-characterized genetic neurodevelopmental disorders, sometimes referred to as Mendelian syndromes (those with a single causative gene or genomic locus to the condition). These disorders typically manifest with intellectual disability, often with evidence of social impairment. Several of these syndromes, including fragile X, Rett, Williams, and Prader-Willi/Angelman syndromes, have been particularly important in beginning to elaborate the biology of ASD.

Rare Genetic Syndromes Have Provided Initial Insights Into the Biology of Autism Spectrum Disorders

Fragile X Syndrome

Fragile X syndrome is a common form of chromosome X–linked intellectual disability. Patients display a range of behavioral abnormalities including poor eye contact, social anxiety, and repetitive behaviors. In addition, approximately 30% of boys with fragile X meet the all diagnostic criteria for ASD. Moreover, in research with multiple cohorts, up to 1% of participants with apparently idiopathic ASD also carried fragile X mutations. The overall prevalence is approximately 1 in 4,000 boys and 1 in 8,000 girls.

The fragile X mutation is quite remarkable. The *FMR1* gene on the X chromosome includes the nucleotide triplet CGG. In normal individuals, this triplet is repeated in approximately 30 copies. In fragile X syndrome patients, the number of repeats is more than 200, with approximately 800 repeats being most common. This expansion of trinucleotide repeats has since been observed in other genes leading to neurological diseases, such as Huntington disease (Chapters 2 and 63). When the number of CGG repeats exceeds 200, the *FMR1* gene regulatory region becomes heavily methylated, and gene expression is shut off. Consequently, in these children, the fragile X mental retardation protein (FMRP) is lacking.

Lack of functional FMRP is considered responsible for fragile X syndrome. FMRP is a selective RNA-binding protein that blocks translation of messenger RNA until protein synthesis is required. It is found with ribosomes at the base of dendritic spines, where it regulates local dendritic protein synthesis that is needed for synaptogenesis and certain forms of long-lasting synaptic changes associated with learning and memory (Chapters 52 and 53). Interestingly, long-term depression of excitatory synaptic transmission, a form of long-lasting synaptic change that requires local protein synthesis, is enhanced in a mouse model of fragile X syndrome in which the gene encoding FMRP has been deleted. Loss of FMRP may enhance long-term depression by allowing excess translation of the messenger RNAs important for synaptic plasticity.

An exciting implication of these data is that antagonists of the type 5 metabotropic glutamate receptor (mGluR5), the activation of which is required for the enhanced protein synthesis underlying long-term depression, may lessen the excess protein translation. In fact, compounds with this activity have been found to rescue the mutant phenotype in mouse and fruit fly models. Thus far, clinical trials of mGluR5 antagonists for individuals with fragile X with ASD have not shown efficacy against the defined clinical end points. However, it is still too early to tell whether these initial forays into rational drug design for neurodevelopmental disorders may or may not be promising in the long run. A range of challenges have confronted these pioneering efforts, including measuring change in individuals with ASD, identifying ideal clinical end points, and determining the best age for evaluating interventions.

Rett Syndrome

Another single-gene disorder showing overlap with ASD is Rett syndrome, a devastating disorder that primarily affects girls. Affected females have normal development from birth until 6 to 18 months of age, when they regress, losing speech and hand skills that they had acquired. Rett syndrome is progressive, and initial symptoms are followed by repetitive hand movements, loss of motor control, and intellectual disability. Often young girls will display symptoms indistinguishable from ASD early in the course of the syndrome, although social communication frequently improves later in childhood. Its prevalence is approximately 1 in 10,000 live female births.

Rett syndrome is an X-linked inherited disease caused by loss-of-function mutations in the *MECP2* gene, which encodes a transcriptional regulator that binds to methylated cytosine bases in DNA, regulating gene expression and chromatin remodeling. The gene product was initially thought to act predominantly as a transcriptional repressor, but studies of both the mouse model and human induced pluripotent stem cells have shown that overall gene expression is reduced when the gene is knocked out. Among the genes that have reduced expression in neurons is *BDNF*, encoding brain-derived neurotrophic factor. Studies in mouse models of Rett have found that overexpression of *BDNF* improves the knock-out phenotype. Other growth factors that increase gene expression but have more favorable neuropharmacological profiles, including insulin-like growth factor-1 (IGF-1), have also improved aspects of the mouse phenotype, leading to optimism about clinical trials of related compounds. Phase II human trials with both molecules are currently underway.

One might think that such a global abnormality in gene expression would lead to a very severe phenotype, but because females are mosaic, with approximately half of their brain cells expressing one normal copy of *MECP2* (due to random X-inactivation), they are viable but manifest the devastating Rett phenotype. Boys, who have a single X chromosome and thus a single copy of *MECP2*, typically die soon after birth or in infancy if they carry a loss-of-function mutation in *MECP2*.

The role of X-inactivation in the survival of female mutation carriers and the observation that favorable skewing (a shift toward preferential silencing of the mutant X) leads to a less severe clinical course have generated considerable interest in therapeutic strategies aimed at reactivating the normal but silenced X chromosomes in females with Rett syndrome. Although one can imagine considerable challenges resulting from the reactivation of many genes on a normally silenced chromosome, a recent study has reported a mouse mutation that leads to both alleles expressing MeCP2 without wholesale activation of genes on the X chromosome.

Interestingly, in 2005, duplications spanning *MECP2* were identified in males with severe intellectual disability. This condition, called *MECP2* duplication syndrome (MDS), includes autistic features, hypotonia, epilepsy, gait abnormalities, and recurrent infections. Like Rett syndrome, it has also been productively modeled in rodents. However, unlike Rett, the majority of identified cases are familial and not sporadic in nature. In these cases, female carriers are often healthy enough (due to favorable X-inactivation) to reproduce and transmit the duplication to boys with only a single X chromosome.

Williams Syndrome

Williams syndrome is caused by a segmental deletion of about 27 genes on the long arm of chromosome 7 and is characterized by mild to moderate intellectual disability, connective tissue abnormalities, cardiovascular defects, distinctive facies, and a behavioral phenotype characterized by increased sociability, preserved language abilities, affinity for music, and impaired visuospatial capabilities. The disorder occurs in 1 in 10,000 live births. The connective tissue and key cardiovascular symptoms have been attributed to the loss of the gene *ELN* (*elastin*), although no specific genes within the deleted interval have yet been definitively shown to result in the behavioral phenotype. Nonetheless, the social cognitive features of Williams syndrome are particularly intriguing: The degree of interest in social interaction is striking, leading to a nearly universal loss of reticence with strangers in children with the syndrome. In contrast to the almost complete absence of social anxiety, individuals with Williams syndrome have a high degree of general anxiety and isolated phobias. Finally, the affinity for and interest in music among a very large percentage of 7q11.23 deletion carriers, although less well characterized, are striking.

Conversely, duplication of the identical region of chromosome 7, including the same 26 to 28 genes, is a significant risk factor for ASD and other neurodevelopmental syndromes apart from Williams syndrome. The observation of contrasting social phenotypes depending on whether there is loss or gain of a small region of the genome is fascinating. Whether social functioning in William syndrome is truly the opposite of that seen in ASD, as is sometimes argued, seems less interesting than the conclusion that this region of the genome must contain one or more genes that modulate social affiliation. Consequently, the molecular characterization of these deletion and duplication syndromes and intensive investigation of their impact on the development of molecular, cellular, and circuit properties in the central nervous system are particularly important.

Angelman Syndrome and Prader-Willi Syndrome

Angelman and Prader-Willi syndromes are paradigmatic examples of genetic syndromes that result from mutations in genes subject to parental imprinting. To understand these conditions, one must not only know the associated DNA lesion but also its parental origin.

For example, both syndromes most often result from the loss of the identical region of chromosome 15 (15q11-q13) but have readily distinguishable phenotypes. Angelman syndrome is characterized by severe intellectual disability, epilepsy, absence of speech, hyperactivity, and inappropriate laughter. In contrast, Prader-Willi is characterized by infantile hypotonia, mild to moderate intellectual disability, obesity, highly perseverative behavior, social disability, and diminished or absent satiety.

How these contrasting phenotypes result from the loss of the identical set of genes confounded medical geneticists until about the year 2000. The mystery was solved by the discovery that the chromosomal interval is imprinted. Specifically, within this region, multiple genes are expressed only on the paternally inherited chromosome (*maternal* imprinting), whereas at least two genes, *UBE3A* and *ATP10C*, are expressed only on the maternally inherited chromosome (*paternal* imprinting) (Figure 62–7).

This discovery, along with a series of studies that allowed for fine mapping of the interval, provided a parsimonious explanation for the clinical observations. If the deletion of proximal chromosome 15 involved the maternal chromosome, the patient would suffer the loss of the protein product of *UBE3A*, a ubiquitin-protein ligase that stimulates the degradation and turnover of other proteins, leading to Angelman syndrome. Alternatively, if the paternal chromosome carried the deletion, *UBE3A* would be expressed normally, but a series of other genes, including several strongly implicated in Prader-Willi syndrome, would be lost.

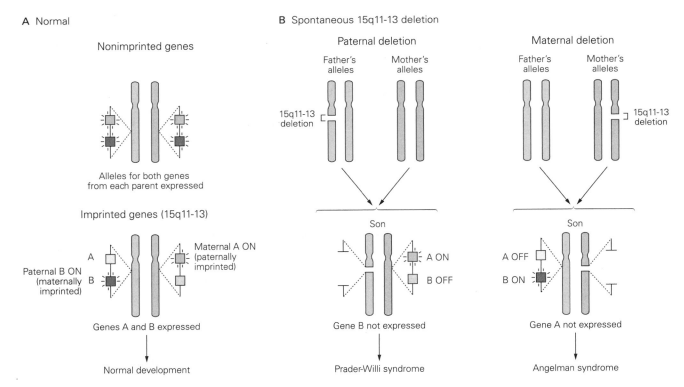

Figure 62–7 Imprinting in Prader-Willi and Angelman syndromes. Approximately 70% of Prader-Willi and Angelman syndrome patients inherit chromosome 15 from one parent with spontaneous (noninherited) deletions of the q11-13 interval. This interval contains imprinted genes with alleles that are either expressed or not depending on whether the chromosome was inherited from the father or mother. If the chromosome with the deletion is from the father, Prader-Willi syndrome occurs because maternally imprinted genes on the corresponding interval of the intact maternal chromosome (gene B, for example) are not expressed. If the chromosome with the deletion is from the mother, the gene for ubiquitin ligase (*UBE3A*) will not be expressed in offspring because of its normal inactivation on the paternal chromosome caused by imprinting; loss of expression of this gene leads to Angelman syndrome.

The solution to the phenotypic complexity seen in 15q11-13 deletion also led to a series of observations that revealed other previously unappreciated genetic mechanisms of behavioral pathology. For example, deletions on the maternal chromosome not directly involving the *UBE3A* gene were also observed in rare patients with Angelman syndrome, contributing to the identification of an Angelman syndrome *imprinting control* region mapping some distance from UBE3A but within the deletion interval. Similarly, the discovery of both Prader-Willi and Angelman syndromes in patients without deletions of any kind led to the recognition that in a small percentage of both conditions two copies of a chromosome from the same parent were present (with no representation from the other parent), a phenomenon called *uniparental disomy*.

Both syndromes have complex behavioral phenotypes. Social disability is characteristic of Prader-Willi; with Angelman syndrome, the overlap with ASD has been more difficult to establish because of the marked intellectual disability associated with the syndrome. Differentiating intellectual disability from ASD in individuals with very low IQ can be quite challenging. Nonetheless, there are multiple clear molecular and behavioral links with ASD. For example, duplications of the 15q11-13 region are a well-established risk factor for nonsyndromic ASD (see below), and functional de novo missense mutations in the gene *UBE3A* have been found in individuals with ASD without all of the features of Angelman syndrome.

Neurodevelopmental Syndromes Provide Insight Into the Mechanisms of Social Cognition

Although the fragile X, Rett, Williams, Angelman, and Prader-Willi syndromes collectively account for a small fraction of the burden of social disability in the population, studies of these disorders have contributed to major advances in the understanding of normal brain development, neurodevelopmental syndromes in general, and the mechanisms underlying social disability in particular. A number of biological processes identified in the study of these disorders—including the contribution of epigenetic mechanisms and chromatin dynamics, synaptic dysfunction, and the role of aberrant local protein synthesis—have all turned out to be important initial clues to the biological and developmental mechanisms underlying nonsyndromic forms of ASD. Moreover, characterization of the genetics underlying certain neurodevelopmental syndromes provided some of the earliest examples of a phenomenon that is now well accepted in ASD—either losses or gains of identical risk genes or regions may lead to

neurodevelopmental disorders, sometimes with overlapping and sometimes contrasting phenotypes.

Importantly, in addition to the first clues regarding molecular mechanisms, recent studies of a number of Mendelian syndromes have challenged conventional wisdom by highlighting, in model systems, the potential reversibility of developmental phenotypes, even into adulthood. These observations, particularly with regard to Rett, Angelman, MDS, and fragile X syndromes, defied the long and generally held belief that the deficits associated with these types of severe syndromes are unchangeable. Moreover, the relevant studies have underscored the fact that a range of manipulations—from genetic, to pharmacological, to the more recent use of antisense oligonucleotides (in the case of *MEC2* duplication and Angelman syndromes)—have all been successful in reversing phenotype.

These findings provide not only an avenue forward for the development of rational therapies in humans but also a critical antidote to the penchant for nihilistic views of therapeutics development in neurodevelopmental disorders. In short, these findings have collectively, and now repeatedly, reinforced the notion that rationally designed therapies may reverse key symptoms long after initial pathology has begun to unfold in brain development. The question of how much of the core symptomatology seen in nonsyndromic ASD is a consequence of ongoing functional derangements, versus what would more traditionally be considered developmental pathology, remains to be clarified. One should note, however, that even with the limited treatments available, the observation that some children improve years after the onset of symptoms suggests that aspects of ASD pathology are not entirely static and may ultimately yield to the development of novel biologically driven treatment approaches.

The Complex Genetics of Common Forms of Autism Spectrum Disorder Are Being Clarified

The recent discovery of genes causing idiopathic ASD—once a scientific quagmire—has been among the most dramatic success stories in the field of human genetics. The combination of high-throughput genomic technologies—including the ability to assay common and rare variations in both the sequence and structure of DNA—the consolidation of large patient cohorts, and considerable investment in ASD research has transformed the field.

Initial breakthroughs can be traced to studies of the genes encoding the family of neuroligins—cell adhesion molecules found at postsynaptic densities of glutaminergic synapses (Chapter 48). At the beginning of this century, the group led by Thomas Bourgeron, a geneticist at the Pasteur Institute, first identified putatively deleterious coding mutations in the genes coding for neuroligin 4X (loss-of-function) and neuroligin 3X (missense). About 6 months after the initial report on the loss-of-function mutation in *NLGN4X*, a nearly identical loss-of-function mutation in the same gene was found linked to both intellectual disability and ASD in a large pedigree. The relevance of the neuroligin 3X mutation to ASD has taken longer to clarify. Contemporary studies provide statistical evidence that *NLGN3X* is a probable, but not yet definitive, ASD risk gene. Additional studies of large cohorts will clarify this question.

In retrospect, these findings were prescient. The two papers on neuroligins pointed to the importance of loss-of-function heterozygous mutations leading not only to ASD but to a wide range of neurodevelopmental phenotypes and highlighted a role for synaptic proteins at the excitatory synapse. Moreover, in addition to being a harbinger of the contributions of both rare and de novo mutations (Chapter 2), the reported findings from Bourgeron's group also hinted, in retrospect, at a female protective effect as well as a paternal origin of de novo point mutations. In the initial report, the unaffected mother carried a de novo loss-of-function mutation on her paternally inherited X chromosome, which she passed to two affected sons.

Several years later, two key findings further ushered in the modern age of reliable and reproducible genetic studies in ASD. First, papers in 2006 and 2007 reported on the observation of rare de novo heterozygous copy number variations (Chapter 2) in children with ASD and intellectual disability. These studies focused specifically on idiopathic, nonsyndromic ASD and on families with only a single affected individual (simplex families). Both papers reported high rates of relatively large copy number variations among individuals with both intellectual and social disability. Second, it was not clear if individuals with ASD simply had more chromosomal abnormalities than those without. However, this question was soon answered by studies from multiple laboratories. De novo copy number variations did not appear to be distributed randomly throughout the genome but tended to cluster in distinct regions of the genome, suggesting that the increased rate in such cases was a consequence of an accumulation of specific risk events. Moreover, as higher-resolution genomic assays began to be applied, similar results emerged: Only certain subsets of mutations (eg, point mutations that disrupt gene function) proved to be elevated in individuals with autism, pointing to the aggregation of causal mutations in affected individuals, not hypermutability, as an explanation for the excess rate(s) of de novo events in affected individuals.

A considerable investment in studying copy number variations in simplex families has resulted in a steadily expanding list of copy number variations that clearly and dramatically increase the risk for ASD. At present, about a dozen genomic intervals reach genome-wide significance based on genome-wide screening of cases for de novo mutations (Figure 62–8). As a result, the American College of Medical Genetics now considers screening for copy number variations the standard of care for an individual presenting with ASD of unknown etiology.

Studies of de novo mutations have advanced throughout the second decade of this millennium, leading to the discovery that, similar to de novo copy number variations, de novo changes in the sequence of DNA—both single nucleotide variants and insertions or deletions (indels)—also contribute to ASD risk and can similarly be used to identify specific risk genes. Recent reports have now leveraged this approach to include more than 100 genes carrying large-effect single nucleotide variants and indel mutations that disrupt the function of the encoded protein (ie, likely gene-disruptive [LGD] mutations) (Chapter 2).

Several associated findings deserve mention here. First, although the contribution of de novo mutations to the risk for ASD in the total population is quite small (in the neighborhood of 3%), the proportion of individuals with large-effect de novo mutations who are seen in clinical settings and recruited for genetic studies is quite significant, as high as 40% of girls. The reason for this apparent contradiction is that most of the risk to the population writ large is carried in small-effect common variations that in most individuals are not sufficient to result in them crossing a diagnostic threshold for ASD. In short, most individuals carrying some degree of risk in the population never show overt social impairment and do not come to clinical attention. Conversely, individuals with large-effect de novo copy number variations, single nucleotide variants, and indels are much more likely to have significant clinical manifestations and seek medical attention.

Second, studies of de novo single nucleotide variants and indels in ASD using exome sequencing have found that the rate of de novo mutations increases with the father's age. Consistent with this observation is the finding that the vast majority of deleterious de novo sequence mutations in ASD cases are present on the paternally inherited chromosome. Although the

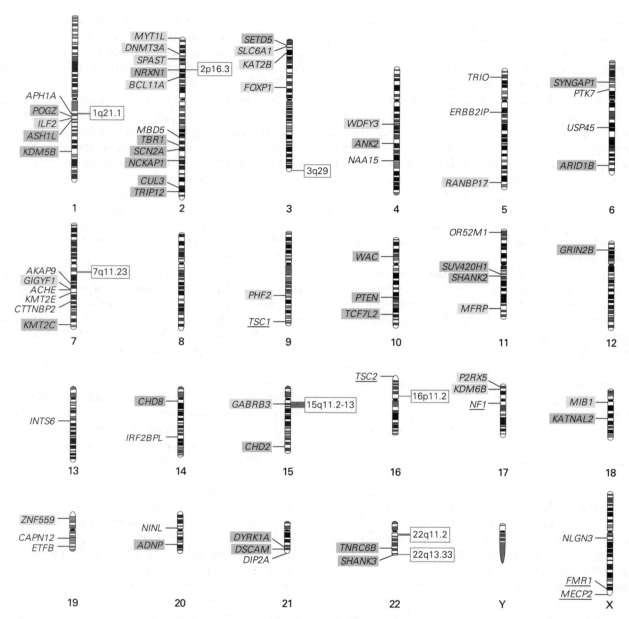

Figure 62–8 Multiple genes and copy number variations that have been strongly associated with idiopathic risk for autism spectrum disorders (ASDs). The figure identifies 71 genes and copy number variations (CNVs) associated with risk for ASDs based predominantly on the recurrence of de novo mutations. Abbreviations in **blue shading** denote genes with a false discovery rate (**FDR**) less than 0.01; abbreviations in **yellow shading** denote an FDR between 0.01 and 0.05; and abbreviations with no shading denote an FDR greater than 0.05 and less than 0.1. **Green bars** identify CNVs with an FDR less than 0.05. Data from Sanders et al. 2015. Statistical analysis was performed using the methods described in Sanders et al. 2015. Five additional genes with names underlined cause the syndromic forms of ASD discussed in the chapter text. Gene identification in ASD is continuing at a rapid pace. Up-to-date lists of associated genes and genomic regions can be found at https://gene.sfari.org.

absolute increase in risk with age is small, this observation nonetheless provides a conceptual framework for understanding secular increases in ASD prevalence. It also sets the stage for further studies of the impact of environmental factors in increasing de novo mutations and thereby potentially increasing the true incidence of clinical ASD cases.

The relationship of de novo large-effect mutations to intellectual disability has been the subject of considerable discussion, with some contending that de novo large-effect mutations are typically seen in ASD with intellectual disability. Although it is the case that de novo mutations that are damaging (either copy number variations or single nucleotide variants) are more prevalent in ASD patients with lower IQ, it is also the case that mutations that confer ASD risk are found across the entire IQ spectrum, reinforcing the idea that domains of cognitive and social functioning are to some degree separable.

One notable difference between the genetics of ASD and other common disorders, including schizophrenia, has been the lack of progress using genome-wide association studies (Chapter 2). To date, only a handful of common genetic variants have been found that are significantly and reproducibly associated with ASD risk. Moreover, earlier conventional wisdom about the contribution of candidate genes such as *5-HTT*, *MTHFR*, or *OXT* polymorphisms is highly uncertain, based on the lack of findings from genome-wide association studies and the fact that these associations derived from an approach that has been shown empirically to be unreliable for gene discovery in complex common disorders. On the other hand, as noted, there is very strong evidence that common variation plays a substantial role in the population risk for ASD. Indeed, genome-wide association studies that have inferred the degree of contribution from this type of variation agree that the lion's share of vulnerability resides in common variation. One can reconcile these observations by noting that for those disorders, such as ASD, that markedly impair reproductive fitness, only common genetic variations carrying small effects remain in the human population over many generations. Those with larger effects would either be driven to low frequencies or removed entirely by natural selection. In addition, sample sizes for case-control genome-wide association studies of ASD to date have been more modest than those that have led to the marked success in identifying common polymorphisms associated with schizophrenia. In short, limited power is almost certainly a key limitation in identifying the common small-effect alleles contributing to ASD.

The relative success in identifying de novo mutations does not suggest that these are the only important mechanisms of ASD risk. Recent progress in this area is the result of a fortuitous combination of the large-effect size of the mutations, their location in the most easily interpreted portion of the genome (the coding region), and their low base rate in typically developing individuals. With regard to population risk, however, common noncoding, small-effect alleles are likely to collectively account for a greater overall proportion of the liability for ASD compared to rare high-effect

variants. Moreover, there is evidence for recessive forms of ASD as well. These have initially been identified predominantly in consanguineous populations via the identification of homozygous loss-of-function mutations—that is, the identical damaging allele on both the paternally and the maternally inherited chromosomes—including in the genes *CNTNAP2*, *BCKDK*, and *NHE9*. Moreover, several recent studies have highlighted the contribution of compound heterozygote mutations to ASD risk—that is, different mutations mapping to the same gene on the maternally and paternally inherited chromosome—in populations with low rates of consanguinity.

A key point is that the pursuit and discovery of different types of mutations may help advance the science in different ways. For example, rare de novo high-effect mutations can be quickly studied in model systems. Also, common variants provide an opportunity to assess overall polygenic risk in research cohorts, an approach that may be highly useful for multimodal studies, such as those that integrate neuroimaging with genetic data, or other investigations linking human behaviors to genotypes. Finally, very rare homozygous/recessive variants mitigate some of the challenges of modeling haploinsufficiency.

Even though heritability—the proportion of the phenotypic variance due to genetic factors—is very high for ASD, environmental factors also play a role, although few specific environmental factors have been conclusively identified. Infections by viruses (eg, rubella, measles, influenza, herpes simplex, and cytomegalovirus) in utero may contribute to the etiology of ASD. There is substantial evidence that mediators of immune functions also play a role in brain development including synaptogenesis. Given the complexity of ASD and its various forms, it is likely that a variety of etiologies will ultimately be discovered—some purely genetic, others that depend on combinations of genetic risk factors and environmental factors, and some purely environmental causes.

Genetics and Neuropathology Are Illuminating the Neural Mechanisms of Autism Spectrum Disorder

Genetic Findings Can Be Interpreted Using Systems Biological Approaches

The recent advances in gene discovery are a particularly exciting development, offering many opportunities for biological analyses using an increasing armamentarium of in vitro and in vivo methods. In addition, contemporary genomic approaches, examining large sections of the genome simultaneously, allow for unbiased approaches to be used in the study of groups of risk genes in an effort to identify points of convergence among disparate ASD genes.

To date, biological approaches that examine multiple systems have been divided roughly into two types of efforts: those that attempt to identify the types of biological processes reflected in the growing list of ASD genes and those that attempt to identify biological points of convergence at either the molecular or cellular level. The latter approach is based on the notion that multiple types of genetically driven perturbations in differing pathways may lead to a common phenotype due to their convergence on specific cell types, regions, or circuits at specific time points during the development of the human brain.

Biological processes or pathways in which ASD risk genes are present in a greater proportion than expected under the null hypothesis include chromatin modification, synaptic function, the WNT signaling pathway, and targets of FMRP. This list is clearly not exhaustive. For example, what we learned from genes involved in genetic syndromes as well as genes that cause nonsyndromic or idiopathic ASD implicates synaptic local protein synthesis as well as neurogenesis as points of potential biological convergence of disparate risk mutations.

Some variability in these findings is almost certainly attributable to different selection criteria for ASD risk genes as well as differences in the data used to annotate their function. This latter issue is important to keep in mind as numerous confounding factors are inherent in current efforts to annotate the biological processes assigned to a given gene or protein. These include the sources of data. For example, the assigned function of a gene can be markedly influenced by publication bias, whether in vitro or in vivo assays were employed, and what types of tissues and model systems were used to generate the data. Moreover, most functional annotations provide limited information on the time course of function for genes that may be developmentally regulated and biologically pleiotropic. Nonetheless, it is the increasing consistency in findings that is most striking. Despite the varying approaches, the biological processes noted above have been repeatedly identified among diverse rigorous studies.

As noted, an alternative approach to determining where multiple autism risk genes overlap involves examining not just their function but also their developmental expression pattern. Such studies are predicated on the notion that multiple risk genes may have different overt functions but share the ability to disrupt the same circuit, cellular, or developmental process. For example, a mutation in a gene encoding a protein known to mediate synaptic adhesion and a separate mutation in a gene encoding a chromatin modifier may both lead to identical abnormalities in the development of early cortical striatal connections. In such cases, the timing and location of the perturbation may be as relevant as a specific molecular pathway or the assigned molecular function of the individual gene(s). These studies have also tended to rely on assaying developmental expression trajectories genome-wide to minimize some of the confounding factors associated with other available annotation systems. For instance, it is now possible to assay essentially every gene in the genome simultaneously—eliminating the need to rely on prior research to assign a specific function to a gene. Moreover, such studies increasingly examine gene expression in human and/or nonhuman primate brain, mitigating some of the challenges of relying on in vitro data. Of course, such studies must still contend with the limits of resolution of expression analyses as well as a less than complete (and potentially biased) representation of different brain regions. Nonetheless, to date, the degree of agreement among varying studies is reassuring.

Despite differences in analytical and statistical approaches used in these types of studies, there has been general agreement to date that ASD risk genes point to vulnerability in human mid-fetal cortical development. There is also emerging evidence that these genes point to the involvement of both deep and upper layer projection neurons in cerebral cortex and of striatum and cerebellum (although the data on developmental expression in these regions remain limited in publicly accessible databases compared to cortical regions).

Autism Spectrum Disorder Genes Have Been Studied in a Variety of Model Systems

As a result of the tremendous progress of late, even a cursory description of the literature on the study of ASD in animal models is beyond the scope of this chapter. In part, this is a consequence of the sheer number of studies; in part, it is a product of the marked differences in the type of perturbation studied (eg, well-validated genetic models, "candidate gene" models, pharmacological models such as valproate exposure, or maternal immune activation). Moreover, differences in brain regions, cell types, developmental periods, and the biological processes assayed make summary generalizations problematic. In short, no consensus has yet been reached regarding the range of pathophysiological mechanisms relevant to ASD.

Nonetheless, given recent progress in human genetics, it is increasingly important to distinguish between models based on reproducible genetic findings, including those leading to syndromic ASD, and models based on unreliable candidate gene loci or solely on behaviors (ie, those that appear to reproduce human symptoms). Given the multiple options now available to study genetic variations that demonstrably increase the risk in humans for ASD in the phenotypes of interest, the study of models with more tenuous links to human pathophysiology is increasingly difficult to justify.

Many publications reporting rodent models of ASD, regardless of their origins, focus on phenotypes that resemble symptoms seen in the human syndrome, including changes in social interactions, vocalizations, and behaviors reminiscent of human anxiety or aggression. Even with a bias toward publishing positive findings, results vary dramatically. Of note, there has been a long-standing debate over the relevance of animal models for ASD, given the important differences in brain development, organization, and function between humans and the most commonly used experimental animals. Nevertheless, unbiased assessments of a wide range of animal behaviors—not necessarily prioritizing those that "look" like core ASD symptoms—may well provide a valuable window into pathophysiological mechanisms. For example, some of the most commonly observed phenotypes reported to date across various ASD genetic models (and by various laboratories) involve motor behavior. In this case, it seems far less important that the observed behaviors are reminiscent of core diagnostic features in humans than that the observations suggest an important point of biological convergence, providing clues to cell types, circuits, and processes involved in ASD.

Although rodent models continue to dominate the ASD literature, a wide range of other models have already provided important insights into biology. These include the fruit fly, worm, zebrafish, frogs, voles, nonhuman primates, induced pluripotent stem cells and brain organoids, and human postmortem samples. Given the complexity of the problems at hand, the differing strengths and limitations of various models, and the important differences in brain structure and development across species, continued progress will likely require integrating data across a wide range of existing models, from flies and worms to humans.

Postmortem and Brain Tissue Studies Provide Insight Into Autism Spectrum Disorder Pathology

The neuropathology of autism at a microscopic level is also not yet clear, but several studies provide evidence for the potential for multiple anatomical correlates. The multiple correlates may in part be due to the small number of brains available for pathological analysis. Moreover, only a small fraction of these have undergone quantitative analysis. Another problem is the frequent occurrence of epilepsy. Approximately 30% of individuals with autism also have seizure disorders, and seizures may damage the amygdala and many other brain regions implicated in ASD.

One of the earliest and most consistent anatomical findings in ASD has been the lower number of Purkinje cells in the cerebellum in some individuals. When neural stains are used to mark cell bodies, gaps in the orderly arrays of Purkinje cells are noticeable. Whether this reduction in cell number is because of ASD, epilepsy, or the co-occurrence of both disorders is not clear. It is also not clear whether the reduced number of Purkinje cells is characteristic of ASD in particular or neurodevelopmental disorders more generally. A wide variety of cerebellar changes were identified in cases of idiopathic intellectual disability, in Williams syndrome, and in other neurodevelopmental disorders. A few cases of alterations of brain stem nuclei that are connected to the cerebellum, such as the olivary complex, have also been reported. Finally, contemporary analyses have found considerable heterogeneity in cell number, with only a subset of samples showing a decrease in the number of Purkinje cells.

Microscopic abnormalities have also been observed in the autistic cerebral cortex, including defects in the migration of cells into the cortex, such as ectopias (nests of cells in white matter that failed to enter the cortex). It has also been proposed that the columnar organization of the autistic cortex is abnormal. These findings still await confirmation in larger studies using quantitative strategies. Finally, one study found fewer neurons in the mature amygdala of people with ASD without epilepsy.

In one of the few reported descriptions of live pathological tissue samples from patients with ASD (removed from three patients during surgery for intractable epilepsy), multiple cytoarchitectural abnormalities were identified in the temporal cortices. These individuals all carried rare recessive loss-of-function mutations in the gene *contactin associated protein-like 2*. Multiple histological abnormalities were observed in these patients, including areas of cortical thickening and blurring of the boundary between gray matter and white matter. Moreover, the authors described neurons in multiple cortical regions that were abnormally organized into tightly packed columns or clusters. In both the hippocampus and temporal cortex, the number of neurons was increased, and many of the neurons had abnormal shapes instead of their pyramidal morphology. Given the presence of gross temporal lobe abnormalities visible on MRI in two of the three patients, the rare recessive genetic contribution, and the particularly severe seizure disorder, the generalizability of these findings to idiopathic ASD remains in question.

The notion overall that there are neuroanatomical changes in some ASD patients is supported by several other lines of evidence. A number of well-supported and well-characterized ASD risk genes (eg, *PTEN* mutations) are associated with increases in brain size ranging from modest (eg, *CHD8* loss-of-function mutations) to frank macrocephaly. In addition, ASD is often associated with microcephaly. Girls with Rett syndrome have acquired microcephaly, suggesting, not surprisingly, that multiple anatomical derangements may occur in social disability phenotypes.

Advances in Basic and Translational Science Provide a Path to Elucidate the Pathophysiology of Autism Spectrum Disorder

A full understanding of the neurobiological basis of the many neurodevelopmental disorders that lead to social and intellectual disability will require the convergence of neuroscience, other medical disciplines, computational biology, and genomics. A bottom-up approach—progressing from the identification of genes responsible for cognitive and behavioral disorders to an understanding of their effects on brain development—is already providing some key insights. At the same time, a top-down approach may also be highly productive by identifying and defining critical neural circuits involved in social function and dysfunction.

Fortunately, the tools available to pursue both approaches are increasingly accessible, from high-throughput whole-genome sequencing to rapidly advancing informatics pipelines, genome editing, optogenetics and other methods to study circuits in vivo, single cell technologies, improved neuroimaging methods and technologies, and the development of tractable human and nonhuman primate neural models, including brain organoids.

Although there has been great progress in elaborating the genetics and biology of ASD and other neurodevelopmental disorders, the findings from genomic studies have also pointed to some key challenges: At the most basic level, the translation of these discoveries to an understanding of pathophysiology is limited by the current state of knowledge regarding brain organization and development. It seems likely

that without a detailed cellular understanding of the brains of humans, nonhuman primates, and other model systems, it will be challenging to interpret the wide variety of genetic perturbations and move from an understanding of the biology to any understanding of the pathogenesis. It is also reasonable to presume that to be most useful for the disorders of the type discussed in this chapter, this type of map will have to capture developmental dimensions. It is exciting and heartening, then, that the recent BRAIN Initiative, other large-scale governmental efforts, and the efforts of private foundations have all highlighted foundational knowledge as a key to success.

There is little doubt that the distance between our knowledge of clinical phenomenology, genetics, imaging, and neuropathology on the one hand and, on the other, the development of novel treatments that will profoundly improve the lives of severely affected individuals can seem daunting. At the same time, it is heartening to see the progress with Mendelian neurodevelopmental disorders, where some clinical trials of rational therapies have been completed and others are currently underway. Although some of the early results have been disappointing, the mere fact that the understanding of these syndromes has advanced to this point is cause for continued optimism. Along these lines, it is useful to consider the required extent of revision of this chapter from the prior volume to the current one. The ability to confidently assign large-effect genetic risk at nearly 100 genomic loci and genes, the emerging consensus regarding what types of molecular processes and pathways are involved, the first glimpses of the developmental characteristics, and the initiation of biologically driven therapeutic trials have all emerged over a relatively short period of time. It is exciting to speculate where the field could be by the publication of the next revision of this book.

Highlights

1. Neurodevelopmental syndromes can involve varying degrees of impairment in different cognitive domains. Syndromes that involve dysfunction in the social realm, with or without involvement of general cognition or perception, are the focus of this chapter.

2. Autism is the paradigmatic social disability syndrome, first described in the literature in 1943 by Leo Kanner. Today, autism is considered a spectrum of disorders with two defining diagnostic features: fundamentally impaired social communication and stereotyped behaviors and/

or highly restricted interests. The prevalence of autism spectrum disorder (ASD) is estimated to be at least 1.5% in developed countries and is much more frequent in males than females.

3. Both environmental factors and myriad genes contribute to ASD risk. This genetic complexity resulted for several decades in scant progress in efforts to map specific ASD risk genes and genomic regions (loci).

4. The earliest clues to both the genetics and neurobiology of ASD emerged from early investigations of neurodevelopmental syndromes that manifest both with intellectual disability and social impairment. These include, among others, fragile X syndrome, Rett syndrome, Williams syndrome, and Prader-Willi and Angelman syndromes.

5. High-throughput genomic technologies, the consolidation of large patient cohorts, and considerable investment in ASD research have transformed the field of gene discovery in idiopathic ASD. At present, dozens of specific genes and genomic regions have been reliably and reproducibly associated with risk for ASD.

6. Recent progress in the genetics of common forms of ASD has emerged from a focus on rare and sporadic (de novo) mutations in the coding portion of the genome. On average, these mutations carry much larger biological effects than have been identified in studies of other psychiatric disorders, such as schizophrenia, where many common genetic risk variants have been identified, each with a small effect.

7. Studies of both genetic syndromes and idiopathic ASD have begun to reveal processes, pathways, and developmental epochs involved in pathophysiology. These include epigenetic mechanisms and chromatin dynamics, synaptic dysfunction, and the role of aberrant local protein synthesis. Recent studies of genetically determined ASDs have also shown that human mid-fetal cortical development and glutamatergic neurons are particularly vulnerable.

8. The current availability of a significant number of confirmed ASD loci, both for syndromic as well as idiopathic forms of the disorder, provides a solid foundation for neurobiological studies. These advances provide a strong link to human pathophysiology, including potential traction on the question of cause versus effect, given that germline genetic changes are present prior to the earliest stages of brain development.

9. In addition to providing some of the first clues regarding molecular mechanisms of idiopathic ASD, studies of Mendelian syndromes have challenged conventional wisdom by highlighting the potential reversibility of developmental phenotypes. These observations, particularly with regard to Rett syndrome and fragile X syndrome, have generated renewed optimism about the opportunities for rational development of therapeutic treatments.

10. Multiple methods are now converging to elaborate the pathology underlying ASD, including gene discovery and systems biology, model systems approaches, neuroimaging studies, and neuropathological studies. The key challenge going forward will be to move from a general understanding of biology to an actionable understanding of pathophysiology.

Matthew W. State

Selected Reading

de la Torre-Ubieta L, Won H, Stein JL, Geschwind DH. 2016. Advancing the understanding of autism disease mechanisms through genetics. Nat Med 22:345–361.

Frith U. 2008. *Autism: A Very Short Introduction*. Oxford: Oxford Univ. Press.

Happé F, Frith U (eds). 2010. *Autism and Talent*. Oxford: Oxford Univ. Press. (First published as a special issue of *Philosophical Transactions of the Royal Society, Series B*, Vol. 364, 2009.)

Klin A, Jones W, Schultz R, Volkmar F, Cohen D. 2002. Defining and quantifying the social phenotype in autism. Am J Psychiatry 159:895–908.

Sesan N, State MW. 2018. Lost in translation: traversing the complex path from genomics to therapeutics in autism spectrum disorders. Neuron 100:406–423.

Zoghbi HY, Bear MF. 2012. Synaptic dysfunction in neurodevelopmental disorders associated with autism and intellectual disabilities. Cold Spring Harb Perspect Biol 4:a009886.

References

Amaral DG, Schumann CM, Nordahl CW. 2008. Neuroanatomy of autism. Trends Neurosci 31:137–145.

Anderson DK, Liang JW, Lord C. 2014. Predicting young adult outcome among more and less cognitively able individuals with autism spectrum disorders. J Child Psychol Psychiatry 55:485–494.

Baron-Cohen S, Cox A, Baird G, et al. 1996. Psychological markers in the detection of autism in infancy in a large population. Br J Psychiatry 168:158–163.

Baron-Cohen S, Leslie AM, Frith U. 1985. Does the autistic child have a "theory of mind"? Cognition 21:37–46.

Bear MF, Huber KM, Warren ST. 2004. The mGluR theory of fragile X syndrome. Trends Neurosci 27:370–377.

Cassidy SB, Morris CA. 2002. Behavioral phenotypes in genetic syndromes: genetic clues to human behavior. Adv Pediatr 49:59–86.

Castelli F, Happé F, Frith CD, Frith U. 2002. Autism, Asperger syndrome and brain mechanisms for the attribution of mental states to animated shapes. Brain 125:1839–1849.

De Rubeis S, He X, Goldberg AP, et al. 2014. Synaptic, transcriptional and chromatin genes disrupted in autism. Nature 515:209–215.

Deuse L, Rademacher LM, Winkler L, et al. 2016. Neural correlates of naturalistic social cognition: brain-behavior relationships in healthy adults. Soc Cogn Affect Neurosci 11:1741–1751.

Dolen G, Bear MF. 2009. Fragile x syndrome and autism: from disease model to therapeutic targets. J Neurodev Disord 1:133–140.

Ecker C, Bookheimer SY, Murphy DG. 2015. Neuroimaging in autism spectrum disorder: brain structure and function across the lifespan. Lancet Neurol 14:1121–1134.

Gallagher HL, Happé F, Brunswick N, et al. 2000. Reading the mind in cartoons and stories: an fMRI study of "theory of mind" in verbal and nonverbal tasks. Neuropsychologia 38:11–21.

Gaugler T, Klei L, Sanders SJ, et al. 2014. Most genetic risk for autism resides with common variation. Nat Genet 46:881–885.

Grove J, Ripke S, Als TD, et al. 2019. Identification of common genetic risk variants for autism spectrum disorder. Nat Gen 51:431–444.

Halladay AK, Bishop S, Constantino JN, et al. 2015. Sex and gender differences in autism spectrum disorder: summarizing evidence gaps and identifying emerging areas of priority. Mol Autism 6:36.

Happe F, Ehlers S, Fletcher P, et al. 1996. "Theory of mind" in the brain. Evidence from a PET scan study of Asperger syndrome. Neuroreport 8:197–201.

Hill E. 2004. Executive dysfunction in autism. Trends Cogn Sci 8:26–32.

Iossifov I, O'Roak BJ, Sanders SJ, et al. 2014. The contribution of de novo coding mutations to autism spectrum disorder. Nature 515:216–221.

Jacquemont ML, Sanlaville D, Redon R, et al. 2006. Array-based comparative genomic hybridisation identifies high frequency of cryptic chromosomal rearrangements in patients with syndromic autism spectrum disorders. J Med Genet 43:843–849.

Jamain S, Quach H, Betancur C, et al. 2003. Mutations of the X-linked genes encoding neuroligins NLGN3 and NLGN4 are associated with autism. Nat Genet 34:27–29.

Jin P, Alisch RS, Warren ST. 2004. RNA and microRNA in fragile X syndrome. Nat Cell Biol 6:1048–1053.

Kana RK, Keller TA, Cherkassky VL, Minshew NJ, Just MA. 2009. Atypical frontal-posterior synchronization of theory of mind regions in autism during mental state attribution. Soc Neurosci 4:135–152.

Kim YS, Leventhal BL. 2015. Genetic epidemiology and insights into interactive genetic and environmental effects in autism spectrum disorders. Biol Psychiatry 77: 66–74.

Klei L, Sanders SJ, Murtha MT, et al. 2012. Common genetic variants, acting additively, are a major source of risk for autism. Mol Autism 3:9.

Koldewyn K, Yendiki A, Weigelt S, et al. 2014. Differences in the right inferior longitudinal fasciculus but no general disruption of white matter tracts in children with autism spectrum disorder. Proc Natl Acad Sci U S A 111:1981–1986.

Kovács ÁM, Téglás E, Endress AD. 2010. The social sense: susceptibility to others' beliefs in human infants and adults. Science 330:1830–1834.

Kumar RA, Marshall CR, Badner JA, et al. 2009. Association and mutation analyses of 16p11.2 autism candidate genes. PLoS One 4:e4582.

Laumonnier F, Bonnet-Brilhault F, Gomot M, et al. 2004. X-linked mental retardation and autism are associated with a mutation in the NLGN4 gene, a member of the neuroligin family. Am J Hum Genet 74:552–557.

Lombardi LM, Baker SA, Zoghbi HY. 2015. MECP2 disorders: from the clinic to mice and back. J Clin Invest 125:2914–2923.

Marshall CR, Noor A, Vincent JB, et al. 2008. Structural variation of chromosomes in autism spectrum disorder. Am J Hum Genet 82:477–488.

Morrow EM, Yoo SY, Flavell SW, et al. 2008. Identifying autism loci and genes by tracing recent shared ancestry. Science 321:218–223.

Nakamoto M, Nalavadi V, Epstein MP, et al. 2007. Fragile X mental retardation protein deficiency leads to excessive mGluR5-dependent internalization of AMPA receptors. Proc Natl Acad Sci U S A 104:15537–15542.

Neale BM, Kou Y, Liu L, et al. 2012. Patterns and rates of exonic de novo mutations in autism spectrum disorders. Nature 485:242–245.

Novarino G, El-Fishawy P, Kayserili H, et al. 2012. Mutations in BCKD-kinase lead to a potentially treatable form of autism with epilepsy. Science 338:394–397.

Ozonoff S, Iosif AM, Baguio F, et al. 2010. A prospective study of the emergence of early behavioral signs of autism. J Am Acad Child Adolesc Psychiat 49:256–266.

Ozonoff S, Macari S, Young GS, Goldring S, Thompson M, Rogers SJ. 2008. Atypical object exploration at 12 months of age is associated with autism in a prospective sample. Autism 12:457–472.

Parikshak NN, Luo R, Zhang A, et al. 2013. Integrative functional genomic analyses implicate specific molecular pathways and circuits in autism. Cell 155:1008–1021.

Pinto D, Delaby E, Merico D, et al. 2014. Convergence of genes and cellular pathways dysregulated in autism spectrum disorders. Am J Hum Genet 94:677–694.

Raznahan A, Wallace GL, Antezana L, et al. 2013. Compared to what? Early brain overgrowth in autism and the perils of population norms. Biol Psychiatry 74:563–575.

Samson D, Apperly IA, Chiavarino C, Humphreys GW. 2004. Left temporoparietal junction is necessary for representing someone else's belief. Nat Neurosci 7:499–500.

Sanders SJ, Ercan-Sencicek AG, Hus V, et al. 2011. Multiple recurrent de novo CNVs, including duplications of the 7q11.23 Williams syndrome region, are strongly associated with autism. Neuron 70:863–885.

Sanders SJ, He X, Willsey AJ, et al. 2015. Insights into autism spectrum disorder genomic architecture and biology from 71 risk loci. Neuron 87:1215–1233.

Sanders SJ, Murtha MT, Gupta AR, et al. 2012. De novo mutations revealed by whole exome sequencing are strongly associated with autism. Nature 485:237–241.

Satterstrom FK, Kosmicki JA, Wang J, Breen MS, et al. 2020. Large-scale exome sequencing study implicates both developmental and functional changes in the neurobiology of autism. Cell 180:568–584.

Schultz RT, Grelotti DJ, Klin A, et al. 2003. The role of the fusiform face area in social cognition: implications for the pathobiology of autism. Philos Trans R Soc Lond B Biol Sci 358:415–427.

Sebat J, Lakshmi B, Malhotra D, et al. 2007. Strong association of de novo copy number variation with autism. Science 316:445–449.

Senju A, Southgate V, White S, Frith U. 2009. Mindblind eyes: an absence of spontaneous theory of mind in Asperger syndrome. Science 325:883–885.

State MW, Sestan N. 2012. Neuroscience. The emerging biology of autism spectrum disorders. Science 337:1301–1303.

Strauss KA, Puffenberger EG, Huentelman MJ, et al. 2006. Recessive symptomatic focal epilepsy and mutant contactin-associated protein-like 2. N Engl J Med 354:1370–1377.

Sztainberg Y, Chen HM, Swann JW, et al. 2015. Reversal of phenotypes in MECP2 duplication mice using genetic rescue or antisense oligonucleotides. Nature 528:123–126.

Sztainberg Y, Zoghbi HY. 2016. Lessons learned from studying syndromic autism spectrum disorders. Nat Neurosci 19:1408–1417.

Weiss LA, Shen Y, Korn JM, et al. 2008. Association between microdeletion and microduplication at 16p11.2 and autism. N Engl J Med 358:667–675.

Willsey AJ, Sanders SJ, Li M, et al. 2013. Coexpression networks implicate human midfetal deep cortical projection neurons in the pathogenesis of autism. Cell 155:997–1007.

Yang DY, Beam D, Pelphrey KA, Abdullahi S, Jou RJ. 2016. Cortical morphological markers in children with autism: a structural magnetic resonance imaging study of thickness, area, volume, and gyrification. Mol Autism 7:11.

63

Genetic Mechanisms in Neurodegenerative Diseases of the Nervous System

THE MAJOR DEGENERATIVE DISEASES of the nervous system—Alzheimer, Parkinson, and the triplet repeat diseases (Huntington disease and the spinocerebellar ataxias)—afflict more than six million people in the United States and more than 25 million throughout the world. Although this is a relatively small percentage of the population, these diseases bring a disproportionate amount of suffering and economic hardship, not only to their victims but also to the families and friends of the afflicted.

Most of these disorders strike in mid-life or later. Aging itself may contribute to susceptibility. The first symptoms to appear often involve loss of fine motor control. Huntington disease can first manifest itself in cognitive deficits, and this is certainly the case for Alzheimer disease. Nevertheless, the end result is the same: A period of slow deterioration, usually 10 to 20 years, robs afflicted patients of their abilities and eventually their lives.

The late-onset neurodegenerative diseases can be divided into two categories: inherited and sporadic (ie, of unknown etiology). Alzheimer and Parkinson diseases are predominantly sporadic; nevertheless, inherited forms, which afflict only a small number of patients, have provided some insight into the pathophysiology of these diseases. Huntington disease, the spinocerebellar ataxias, dentatorubropallidoluysian atrophy, and spinobulbar muscular atrophy are inherited, the result of polyglutamine or CAG triplet repeat diseases.

The triplet repeat diseases are notable for being caused by a "dynamic" mutation: The disease proteins contain a CAG repeat tract that codes for glutamine and can undergo expansion during DNA replication. Unfortunately, the longer the CAG tract, the more likely it is to further expand, which accounts for the striking phenomenon of *anticipation*: Younger generations within a family have longer repeats and develop more severe symptoms at an earlier age than their parents. Identification of the molecular basis of these disorders has facilitated diagnosis and provides hope for eventual treatment.

Huntington Disease Involves Degeneration of the Striatum

Huntington disease usually strikes in early or middle adulthood and affects 5 to 10 people per 100,000. Symptoms include loss of motor control, cognitive impairment, and affective disturbance. Motor problems most commonly manifest first as chorea (involuntary, jerky movement that involves the small joints at first but then gradually affects the legs and trunk, making walking difficult). Fast, fluid movements are replaced by rigidity and bradykinesia (unusually slow movements).

Cognitive impairment—especially difficulty in planning and executing complex functions—may be detected by formal neuropsychological testing even prior to motor dysfunction. Affected individuals may also have disordered sleep and affective disturbances such as depression, irritability, and social withdrawal. About 10% of patients experience hypomania (increased energy), and a smaller percentage experience frank psychosis.

In adult patients, the disease progresses inexorably to death some 17 to 20 years after onset. Juvenile-onset patients suffer a more rapid course and typically develop bradykinesia, dystonia (spasms of the neck, shoulders, and trunk), rigidity (resistance to the passive motion of a limb), seizures, and severe dementia within only a few years.

The pathological hallmark of Huntington disease is degeneration of the striatum, which can show up in neuroimaging as much as a decade prior to the onset of symptoms. The caudate nucleus is more affected than the putamen. Loss of the medium spiny neurons, a class of inhibitory interneurons in the striatum, reduces inhibition of neurons in the external pallidum (Chapter 38). The resulting excessive activity of the pallidal neurons inhibits the subthalamic nucleus, which could account for the choreiform movements. As the disease progresses and striatal neurons projecting to the internal pallidum degenerate, rigidity replaces chorea. Disruption of the corticostriatal projections

leads to thinning of the cortex. In addition to this central nervous system pathology, patients can suffer from immune system and metabolic disturbances, testicular atrophy, cardiac failure, osteoporosis, and skeletal muscle wasting. Cases of juvenile Huntington disease are more severe, and the pathology progresses more rapidly and broadly; for example, degeneration of cerebellar Purkinje cells can occur.

Huntington disease is an autosomal dominant disorder and one of the first human diseases whose gene was mapped using polymorphic DNA markers. It is caused by expansion of a translated CAG repeat that encodes a glutamine tract in the huntingtin protein. Normal or wild-type alleles have 6 to 34 repeats, whereas disease-causing alleles typically have 36 or more repeats that are quite unstable when transmitted from one generation to the next, especially through paternal germ cells. Disease severity, age of onset, and speed of progression correlate with repeat length; individuals with 36 to 39 repeats have later onset and milder disease, while those with more than 40 repeats will have earlier onset and a more severe course. Those carrying more than 75 repeats will develop the disease as juveniles.

The expanded glutamine tract causes a gain of function in huntingtin, a 348-kDa protein that is well conserved in nature from invertebrates to mammals. It is expressed throughout the brain as a soluble cytoplasmic protein, with a minor fraction present in cell nuclei. It is particularly abundant in somatodendritic regions and axons and has been found to associate with microtubules. Although its precise functions are not fully understood, huntingtin is essential for normal embryonic development. Based on a wide array of protein interactors that function in metabolism, protein turnover, cargo trafficking, and gene expression, it has been postulated that huntingtin functions as a molecular scaffold. Its large size, stability, and ability to switch between multiple conformations suggest it brings together multiple proteins into macromolecular complexes.

Huntingtin has multiple protein domains, the best studied of which is the N-terminal region, which contains the polyglutamine expansion and a nuclear localization signal. The N-terminal region consists of an amphipathic α-helix, which creates a structure critical for the protein's retention in the endoplasmic reticulum. The N-terminus undergoes extensive posttranslational modification by acetylation, ubiquitination, phosphorylation, and sumoylation, all of which affect huntingtin clearance and subcellular localization. Interestingly, the polyglutamine repeats in exon 1 are followed by a proline-rich domain, which, unlike the other exons, has been poorly conserved during evolution.

The remaining 66 exons outside the N-terminus, which account for about 98% of the protein, are far less well characterized. Several HEAT repeats are important for protein–protein interactions. These interactions allow the huntingtin protein to adopt a large number of three-dimensional conformations (up to 100 in vitro). Furthermore, the *HTT* gene produces two different mRNA transcripts, a short and long form. The long form contains an additional 3' untranslated region and is enriched in the brain. Rare alternative splicing produces isoforms that skip exons 10, 12, 29, and 46 or include exon 41b or a fragment of intron 28, but their significance has not been determined. The diversity of these isoforms might be important during development and could expand the variety of protein interactions available to huntingtin.

Spinobulbar Muscular Atrophy Is Caused by Androgen Receptor Dysfunction

Spinobulbar muscular atrophy (SBMA, also known as Kennedy disease) is the only X-linked disorder among the neurodegenerative diseases discussed in this chapter. It is caused by expansion of a translated CAG repeat in the androgen receptor protein, a member of the steroid hormone receptor family. Only males manifest symptoms: The mutant androgen receptor is toxic only when localized to the nucleus, and this localization is dependent on the hormone androgen.

Proximal muscle weakness is usually the presenting symptom; eventually, the distal and facial muscles weaken as well. Muscle wasting is prominent, secondary to degeneration of motor neurons. Loss of androgen function typically leads to gynecomastia (growth of breast tissue in men), late hypogonadism, and sterility. Because individuals who lose androgen receptor function from other causes do not develop motor neuron degeneration, it seems that the glutamine expansion in SBMA causes both a partial loss of function that accounts for the secondary sexual characteristics and a partial gain of function that damages neurons and produces the neurological dysfunction.

Hereditary Spinocerebellar Ataxias Share Similar Symptoms but Have Distinct Etiologies

The spinocerebellar ataxias (SCAs) and dentatorubro-pallidoluysian atrophy (DRPLA) are characterized by dysfunction of the cerebellum, spinal tracts, and various brain stem nuclei. The basal ganglia, cerebral cortex, and peripheral nervous system can also be affected (Table 63–1).

Two clinical features common to all the SCAs, ataxia and dysarthria, are signs of cerebellar dysfunction. These typically appear in mid-adulthood and gradually worsen, eventually making walking impossible and speech incomprehensible. Brain stem dysfunction in advanced disease causes difficulties in keeping the airway clear; patients often die of aspiration pneumonia. Some SCAs are associated with additional symptoms such as chorea, retinopathy, or dementia, but these are too variable to support a differential diagnosis. Even individuals within the same family can present quite different clinical pictures. Thus, although the SCAs are single-gene Mendelian disorders, individual genetic makeup and environmental influences affect the clinical-pathological picture.

For example, Machado-Joseph disease and SCA type 3 (SCA3) had been regarded clinically as distinct diseases before it was discovered that they are caused by mutations in the same gene. The clinical confusion arose by historical accident. The most prominent features of the families of Azorean descent who were first studied were bulging eyes, faciolingual fasciculations, Parkinsonism, and dystonia; this syndrome was named Machado-Joseph disease. Subsequently, a group of European geneticists studied patients who had symptoms more reminiscent of SCA1—hypermetric saccades and brisk reflexes in addition to the characteristic ataxia and dysarthria. This constellation of symptoms was therefore called SCA3. It took several years before it became clear that the genetic locus of the two diseases was the same, but still both names (Machado-Joseph disease and SCA3) are used. We now know that the differences observed in the original two groups of patients are at least partially attributable to differences in length of the CAG repeats. Nonetheless, differences in the activity of other proteins caused by genetic variations are probably also at play.

The age of onset within each type of ataxia depends on the number of CAG repeats in the gene (Figure 63–1), although the toxicity of different repeat lengths depends on the protein context. For example, the CAG expansion in SCA6 is the shortest of all the SCAs: Normal alleles have fewer than 18 repeats, and pathological repeats have only 21 to 33 repeats. Yet, tracts of the very same length are completely nonpathogenic in other SCAs. In fact, the gene responsible for SCA7 normally tolerates a few dozen CAG repeats and, in the disease state, can undergo expansion to hundreds of CAGs, the largest expansions seen in any SCA.

Table 63–1 Pattern of Inheritance and Main Clinical Features of Neurodegenerative Diseases Caused by Unstable CAG Trinucleotide Repeats

Disease	Inheritance	Typical presenting features	Principal regions affected
SBMA	X-linked recessive	Muscle cramps, weakness, gynecomastia	Lower motor neurons and anterior horn cells
Huntington	AD	Cognitive impairment, chorea, depression, irritability	Striatum, cortex
Huntington-like 2	AD	Cognitive impairment, chorea, depression, irritability	Striatum, cortex
SCA1	AD	Hypermetric saccades, ataxia, dysarthria, balance, nystagmus	Purkinje cells, brain stem
SCA2	AD	Ataxia, hyporeflexia, slow saccades	Purkinje cells, granule cells, inferior olive
SCA3	AD	Ataxia, gaze-evoked nystagmus, bulging eyes, dystonia, spasticity	Pontine neurons, substantia nigra, anterior horn cells
SCA6	AD	Ataxia, late onset (>50 years of age)	Purkinje cells, granule cells
SCA7	AD	Ataxia, visual loss due to retinal degeneration, hearing loss	Purkinje cells, retina (cone-rod degeneration)
SCA8	AD	Scanning dysarthria, ataxia	Purkinje cells
SCA10	AD	Ataxia and seizures	Purkinje cells
SCA12	AD	Early arm tremor, hyperreflexia, ataxia	Purkinje cells, cortical and cerebellar atrophy
SCA17	AD	Dysphagia, intellectual deterioration, ataxia, absence seizures	Purkinje cells, granule layer, upper motor neurons
DRPLA	AD	Dementia, ataxia, choreoathetosis	Dentate nucleus, red nucleus, globus pallidus, subthalamic nucleus, cerebellar cortex, cortex

AD, autosomal dominant; DRPLA, dentatorubropallidoluysian atrophy; SBMA, spinobulbar muscular atrophy; SCA, spinocerebellar ataxia.

Besides tolerating different CAG repeat lengths, the gene products of mutated genes in polyglutamine diseases vary widely in function:

- The gene product in SCA1, ataxin-1 (ATXN1), is predominantly a nuclear protein that forms a complex with the transcriptional repressor Capicua (CIC). The expanded glutamine tract alters ATXN1's interaction with CIC in the cerebellum, which helps explain this region's vulnerability to SCA1 pathophysiology.
- SCA2 is caused by a CAG trinucleotide expansion in *ATXN2*. Genetic ablation of *Atxn2* increases global transcript abundance, indicating that it may work as an RNA-binding protein. More recent studies revealed that it interacts with TDP43, a protein involved in amyotrophic lateral sclerosis (ALS10), and mutations in *ATXN2* may contribute to amyotrophic lateral sclerosis.

- Impaired protein clearance is a theme among the SCAs, insofar as elevated levels of the disease-causing protein seem to drive pathogenesis. In the case of SCA3, the relationship is more direct in that ataxin-3 (ATXN3) is a deubiquitinating enzyme, and the expanded version cannot remove ubiquitin from proteins slated for clearance. More recently, ATXN3 has been linked to DNA damage repair.
- The affected gene product in SCA6, CACNA1A, is the α_{1A}-subunit of the voltage-gated Ca^{2+} channel; interestingly, loss-of-function mutations in the gene (not gain of function caused by CAG repeats) have been reported in patients with episodic ataxia and familial hemiplegic migraine.

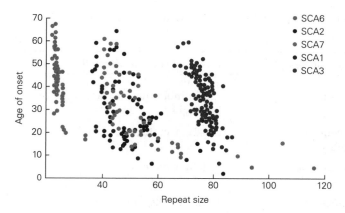

Figure 63–1 The length of the CAG repeat and age of onset in spinocerebellar ataxia (SCA) are inversely correlated. The longer the CAG tract, the earlier is the onset for a given disease. Specific repeat lengths, however, have different results depending on the host protein. For example, a 52-repeat of CAG causes juvenile onset of symptoms in spinocerebellar ataxia type 2 (**SCA2**), adult onset in spinocerebellar ataxia type 1 (**SCA1**), and no disease in spinocerebellar ataxia type 3 (**SCA3**).

- In SCA17, the affected gene product is the TATA box-binding protein, an essential transcription factor.
- Atrophin-1, the disease-causing protein in DRPLA, is thought to be a corepressor based on functional studies of its probable ortholog in *Drosophila*.

Despite these differences, some pathogenetic mechanisms may be common to the polyglutamine diseases, as discussed later in this chapter.

CAG repeats in coding regions are not the only dynamic mutations occurring in the SCAs (Table 63–2). SCA8 involves expansion of both a CAG tract and its complementary CTG repeat on the opposite strand in the 3' untranslated region of a transcribed RNA with no open reading frame. The mutation responsible for SCA12 is a CAG repeat, but it occurs in a noncoding region 5' upstream of a brain-specific regulatory subunit of the protein phosphatase 2A. SCA10 is caused by massive expansion of a pentanucleotide (ATTCT) repeat in the intron of a novel gene.

So far, a total of 33 SCAs have been identified. For the SCAs whose underlying pathogenesis is better understood, the most promising therapeutic approach seems to be to reduce the levels of the disease-driving protein. In the SCA7 mouse model, reducing the amount of both mutant and wild-type ATXN7 by RNA interference greatly improves the behavioral and pathological signs of disease. Likewise, in both *Drosophila* and mouse models of SCA1, genetic or pharmacological downregulation of several components of the

RAS-MAPK-MSK1 pathway decreases ATXN1 levels and suppresses neurodegeneration.

Parkinson Disease Is a Common Degenerative Disorder of the Elderly

Parkinson disease, one of the more common neuro-degenerative disorders, affects approximately 3% of the population older than age 65 years. Patients with Parkinson disease suffer from a resting tremor, bradykinesia, rigidity, and impairment in their ability to initiate and sustain movements. Affected individuals walk with a distinctive shuffling gait, and their balance is often precarious. Spontaneous facial movements are greatly diminished, creating a mask-like, expressionless appearance. The pathological hallmarks of Parkinson disease are the progressive loss of dopaminergic neurons, mainly in the substantia nigra pars compacta (Chapter 38), and the accumulation of proteinaceous aggregates termed Lewy bodies and Lewy neurites throughout the brain.

Although most cases of Parkinson disease are sporadic, studies of rare familial cases, which can be either autosomal dominant or recessive, have provided insight into the pathophysiology of this disorder and revealed novel risk factors for disease. To date, several genetic loci have been mapped (designated *PARK1–PARK22*), and the genes for all but four of these loci (*PARK3, PARK10, PARK12,* and *PARK16*) have been identified (Table 63–3). Of these mapped loci, the most studied and characterized are *PARK1/4, PARK2, PARK6,* and *PARK7*. Here, we focus on how the genetic basis of some forms of Parkinson disease provides insight into sporadic Parkinson disease.

Parkinson disease type 1/4 (4q2-22) is the locus for the dominantly inherited Parkinson disease caused by mutations in the gene *SNCA* encoding for α-synuclein. (As with Machado-Joseph disease and SCA3, Park1 and Park4 were initially thought to be two distinct variants.) Variants in the *SNCA* locus have been associated with increased risk of sporadic Parkinson disease, and several mutations in *SNCA* alter the conformation of the membrane-bound portion of the α-synuclein protein and cause it to aggregate. Duplications and triplications of *SNCA* have also been identified as causes of autosomal dominant Parkinson disease, indicating that elevated levels of even wild-type α-synuclein can cause disease. Patients with *SNCA* duplication have a disease course that resembles sporadic cases, but patients with triplication manifest an earlier-onset, more rapidly progressing disease with atypical features such as dementia and hallucinations.

Table 63–2 Hereditary Ataxias Caused by Expansion of Unstable CAG Trinucleotide Repeats

					Repeat lengths	
Disease	Gene	Locus	Protein	Mutation	Normal	Disease
SCA1	SCA1	6p23	Ataxin-1	CAG repeat in coding region	6–44[1]	39–121
SCA2	SCA2	12q24.1	Ataxin-2	CAG repeat in coding region	15–31	36–63
SCA3 (Machado-Joseph disease)	SCA3, MJD1	14q32.1	Ataxin-3	CAG repeat in coding region	12–40	55–84
SCA6	SCA6	19p13	α_{1A} subunit of voltage-dependent calcium channel	CAG repeat in coding region	4–18	21–33
SCA7	SCA7	3p12-13	Ataxin-7	CAG repeat in coding region	4–35	37–306
SCA8	SCA8	13q21	None	CTG repeat in the 3′ terminal exon (antisense)	16–37	110–250
SCA10	SCA10	22q13ter	Ataxin-10	Pentanucleotide (ATTCT) repeat in the intron	10–20	500–4,500
SCA12	SCA12	5q31-33	Protein phosphatase 2A	CAG repeat in 5′ UTR	7–28	66–78
SCA17	TBP	6qter	TATA-binding protein	CAG repeat in coding region	29–42	47–55
DRPLA	DRPLA	12q	Atrophin-1	CAG repeat in coding region	6–35	49–88
FXTAS	FMR1	Xq27.3	FMRP	CGG repeat in 5′ UTR	6–60	60–200

[1]Alleles with 21 or more repeats are interrupted by one to three CAT units; disease alleles contain pure CAG tracts.
DRPLA, dentatorubropallidoluysian atrophy; FXTA, fragile X–associated tremor ataxia; SCA, spinocerebellar ataxia.

Patients with *SNCA* mutations differ from those with sporadic Parkinson disease in that the age of onset is earlier (a mean of 45 years), and they exhibit fewer tremors and more rigidity, cognitive decline, myoclonus, central hypoventilation, orthostatic hypotension, and urinary incontinence.

Autosomal recessive juvenile parkinsonism is characterized by early-onset dystonia, brisk deep tendon reflexes, and cerebellar signs in addition to the classic signs of Parkinson disease, all as early as 3 years of age. Mutations in *PARK2*, *PARK6*, and *PARK7*—which encode parkin, PTEN-induced putative kinase 1 (PINK1), and protein deglycase DJ-1, respectively—have been confirmed as causes of this disease. Mutations in *PARK2* are much more frequent than mutations in *PARK6* and *PARK7*, and more than 60 different inactivating mutations have been identified; autosomal recessive juvenile Parkinsonism is thus caused by loss of function of the gene product rather than a gain of function. The pathology is also characterized by loss of dopaminergic neurons, but Lewy bodies are not as common as in sporadic or *PARK1/4* cases. Parkin is

an E3 ubiquitin ligase of the RING finger family that transfers activated ubiquitin to lysine residues in proteins destined for degradation by proteasomes. Studies in the fruit fly *Drosophila melanogaster* revealed that parkin and PINK1 work together to promote healthy mitochondria. Interestingly, DJ-1, the third cause of autosomal recessive juvenile parkinsonism, is also involved in mitochondrial function, acting as an oxidative stress sensor.

Not all genetic causes of Parkinson disease exhibit complete penetrance. Such is the case with mutations in the gene encoding the leucine-rich repeat kinase 2 (*LRRK2*, *PARK8*). Interestingly, *LRRK2* mutations are a risk factor for sporadic Parkinson disease. Another genetic risk factor for Parkinson disease is the gene coding for glucocerebrosidase-1 (*GBA1*): Heterozygous carriers of *GBA1* mutations are at increased risk of developing Parkinson disease later in life, whereas homozygous carriers develop a recessive disorder known as Gaucher disease. There are undoubtedly additional genetic risk factors for Parkinson disease, and efforts to identify them are ongoing.

Table 63–3 Genetics and Main Clinical Features of Inherited Parkinson Disease

Disease	Locus map	Inheritance pattern	Gene	Main features
PARK1/4	4q21	AD	SNCA	Early onset, rigidity, and cognitive impairment
PARK2	6q26	AR	PARKIN	Juvenile onset and dystonia
PARK3	2p13	AD	Unknown	Adult onset, dementia
PARK5	4p13	AD	UCHL1	Adult onset
PARK6	1p36.12	AR	PINK1	Early onset, dystonia
PARK7	1p36.21	AR	DJ1	Early onset, behavioral disturbance, dystonia
PARK8	12q12	AD	LRRK2	Classic PD
PARK9	1p36.13	AR	ATP13A2	Juvenile or early onset, cognitive impairment
PARK10	1p32	AD	Unknown	Classic PD
PARK11	2q37.1	AD	GIGYF2	Adult onset, cognitive impairment
PARK12	Xq21-25	X-linked	Unknown	Unknown
PARK13	2p13.1	AD	Omi/HtrA2	Classic PD
PARK14	22q13.1	AR	PLA2G6	Early onset, cognitive impairment, dystonia.
PARK15	22q12.3	AR	FBXO7	Juvenile onset or early onset
PARK16	1q32	Unknown	Unknown	Unknown
PARK17	16q11.2	Unknown	VPS35	Adult onset, cognitive impairment, dystonia
PARK18	6p21.3	Unknown	EIF4G1	Classic PD
PARK19a/b	1p31.3	AR	DNAJC6	Juvenile or early onset, cognitive impairment
PARK20	21q22.11	AR	SYNJ1	Early onset, seizures
PARK21	3q22	AD	DNAJC13	Classic PD
PARK22	7p11.2	AD	CHCHD2	Classic PD

AD, autosomal dominant; AR, autosomal recessive; PARK, PD, Parkinson disease.

Selective Neuronal Loss Occurs After Damage to Ubiquitously Expressed Genes

One perplexing aspect of these neurodegenerative diseases is that the altered gene products are widely and abundantly expressed not only in the nervous system but also in other tissues, yet the phenotypes are predominantly neurological. Moreover, the phenotypes usually reflect dysfunction in only specific groups of neurons (Figure 63–2), a phenomenon referred to as neuronal selectivity.

Why are striatal neurons the most vulnerable in Huntington disease, whereas the Purkinje cells are targeted in the SCAs? Why are the dopaminergic neurons in the substantia nigra pars compacta primarily affected in Parkinson disease even though α-synuclein, parkin, DJ-1, PINK1, and LRRK2 are abundant in many other neuronal (and even nonneuronal) groups?

Although definitive answers are not yet available, some hypotheses have been advanced. One possibility was suggested by the finding that the dopaminergic neurons that are vulnerable in Parkinson disease exhibit an unusual physiological characteristic: They depend on Ca^{2+} channels to fire in a rhythmic pattern. This dependence on Ca^{2+} influx in the neuron is thought to cause baseline mitochondrial stress, which could explain why these neurons are so vulnerable to direct insults to mitochondrial recycling, such as caused by parkin, DJ-1, and PINK1 dysfunction, as well as additional stress caused by LRRK2 dysfunction and α-synuclein accumulation.

In the polyglutamine diseases, the selectivity of the cellular pathology diminishes as the length of the glutamine tract increases: The more severe the mutation, the greater is the number of neuronal groups affected. This is especially evident in the early-onset forms

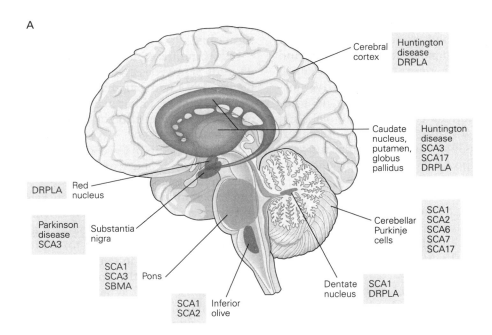

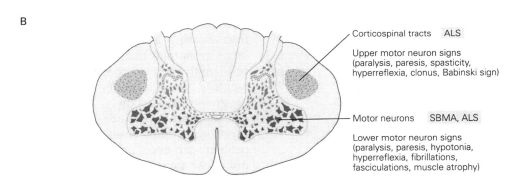

Figure 63–2 Neuronal selectivity illustrated by the primary sites of neuronal degeneration in the trinucleotide repeat diseases and Parkinson disease.

A. Brain regions most typically affected by adult-onset disease (see Table 63–1). (Abbreviations: **DRPLA**, dentatorubropallidoluysian atrophy; **SCA**, spinocerebellar ataxia.)

B. Comparison of neuropathology of amyotrophic lateral sclerosis (**ALS**) and spinobulbar muscular atrophy (**SBMA**).

characterized by extremely long repeats. Juvenile SCA1 can involve oculomotor abnormalities, for example, or cause dystonia, rigidity, and cognitive impairment, features that overlap with Huntington disease and DRPLA; death usually occurs within 4 to 8 years of symptom onset. Juvenile SCA7 patients can suffer seizures, delusions, and auditory hallucinations, and infantile disease also produces somatic features such as short stature and congestive heart failure. Infantile SCA7 causes progressive blindness by destroying both rods and cones; interestingly, infants with SCA2 can also suffer retinal degeneration. Such observations suggest that different cell types have different thresholds of vulnerability to toxic proteins with expanded glutamine tracts. Retinal cells, for example, seem more resistant to polyglutamine toxicity than cerebellar neurons, but more vulnerable than cardiac myocytes. Once the number of glutamines in the tract expands beyond a certain length—which varies from one protein to the next—no cell is safe.

Studies using mouse models suggest that protein misfolding is responsible for polyglutamine disorders. The longer the glutamine tract, the more severe the misfolding, and the more resistance there is to clearance; thus, the slow accumulation of higher-than-normal protein levels is a feature common to neurodegenerative diseases. As the tracts become very long, even cells with lower concentrations of disordered gene product become vulnerable. Indeed, studies of animal models show that even a doubling in concentration can be the difference between phenotypic manifestation and apparent normality. It is therefore conceivable that the neurons affected in each disease have more dysfunctional protein than do the less vulnerable neurons. Although not detectable by current immunolabeling techniques, this incremental increase would nevertheless be sufficient to interfere with cellular function if the neuron was exposed to the toxic protein over decades.

Other major contributors to selective vulnerability might be variations in the levels of proteins that interact with or help dispose of the mutant proteins. Variations in the genes encoding such proteins could contribute to the clinical variability that is so prominent among ataxia families.

Why are neurons affected before other cells? As the organism ages, slight insults that have small detrimental effects could be exacerbated by the extra challenge the toxic protein presents to the protein-folding machinery. Because neurons are postmitotic, they might be especially sensitive to perturbations in the balance of intracellular factors. If the organism could survive the neurological assault long enough, other tissues might also eventually show signs of distress.

Animal Models Are Productive Tools for Studying Neurodegenerative Diseases

Animal models have proven extremely valuable for probing the pathogenesis of various neurodegenerative diseases and investigating therapies. The mouse has been the favored animal for modeling neurological disorders, but the *Drosophila* fly and the worm *Caenorhabditis elegans* have also proven useful in delineating genetic pathways.

Mouse Models Reproduce Many Features of Neurodegenerative Diseases

With the exception of the autosomal recessive juvenile forms of Parkinson disease, the neurodegenerative diseases discussed here primarily reflect gain-of-function mutations. Thus, most of the genetically engineered mice that model these diseases are created using one of two techniques. In the transgenic approach, an allele harboring the mutant gene is overexpressed, whereas in the knock-in approach, a human mutation, such as an expanded CAG tract, is inserted into an endogenous mouse locus to promote expression of the gene

product at the correct time in development and in the right cells.

In some transgenic models, such as those generated for SCA types 1, 2, 3, and 7 and DRPLA, a full-length cDNA with either wild type or expanded alleles is overexpressed either in a particular class of neurons or in a larger population of cells (Figure 63–3). In other transgenic models of SCA3 and in models of SBMA, truncated versions of the coding regions are expressed. Both full-length and truncated huntingtin have been used in transgenic models.

Knock-in mice have been generated for Huntington disease, SBMA, and SCA types 1 and 7. These models confirm that sequences other than the expanded glutamine tract can produce toxic protein. Moreover, the same expansion in two different host proteins can affect cells differently. For example, in humans, 33 repeats cause SCA2, whereas 44 to 52 CAG repeats may or may not develop SCA3 (Figure 63–1 and Table 63–2). In mouse models, however, the relationship of the length of the tract to the rest of the protein is a good predictor of toxicity: Severe, widespread, non-selective neuronal dysfunction occurs in transgenic mice bearing a truncated protein with a relatively large glutamine tract. In contrast, mice that express full-length proteins containing the same CAG repeat length develop a milder neurological syndrome that progresses more slowly. Weakly expressing promoters also tend to produce more selective neuronal dysfunction. In some cases, expression of the full-length protein with even a moderately large expansion does not cause neurological dysfunction, but a truncated version bearing a similar repeat size does produce the disease phenotype. In sum, a glutamine tract of a certain length is more toxic when expressed in isolation or flanked by short peptide sequences, that is, when it occupies a larger proportion of the protein.

In SCA1 knock-in mice with 78 glutamine repeats, neurological dysfunction is barely detectable; only when the repeat length is expanded to approximately 154 glutamines does a neurological phenotype become apparent. Longer repeats are necessary to see a phenotype during the short life span of a mouse because polyglutamine toxicity takes time to exert its effects. In transgenic mice, however, massive overproduction of the mutant protein compensates for a moderate repeat length and brevity of exposure. Indeed, in mice, even overproduction of wild type ataxin-1 results in mild neurological dysfunction, and overexpression of wild type human α-synuclein is enough to cause parkinsonian symptoms.

Analysis of brain tissue from humans and various experimental mice reveals that misfolded proteins

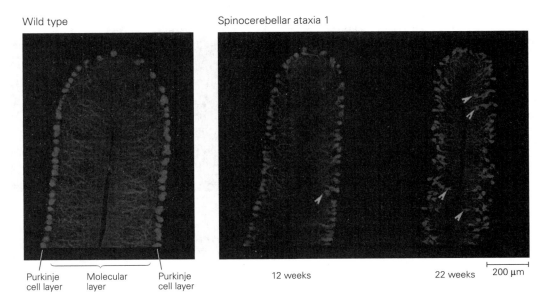

Wild type

Spinocerebellar ataxia 1

Purkinje cell layer   Molecular layer   Purkinje cell layer

12 weeks   22 weeks   200 μm

Figure 63–3 Progressive Purkinje cell pathology in spinocerebellar ataxia type 1 transgenic mice. Cerebellar sections from a wild type mouse and mice expressing a spinocerebellar ataxia type 1 (SCA1) transgene with 82 glutamines in Purkinje cells at 12 and 22 weeks of age. Calbindin immunofluorescence staining marks the Purkinje cells and their extensive dendritic arbors. In SCA1, there is progressive loss of dendrites, thinning of the molecular layer, and Purkinje cell displacement (**arrowheads**). (Images reproduced, with permission, from H.T. Orr.)

tend to accumulate in various neurons, often forming visible aggregates (Figure 63–4). Lewy bodies and abnormal accumulation of α-synuclein develop in mouse models of Parkinson disease, just as in humans. Although protein accumulation is common to all these neurodegenerative disorders, the location of the accumulated protein in the cell varies, and location within the cell is a factor in the protein's pathogenicity. For example, mutant ataxin-1 that accumulates in the cytoplasm instead of the nucleus (because its nuclear localization signal is disabled) exerts no detectable toxic effects.

The fact that mutant proteins accumulate both in mouse models that do not overproduce the proteins and in human patients who carry a single mutant allele suggests that neurons have difficulty clearing the proteins. This hypothesis is supported by the finding that ubiquitin and proteasome components, the machinery of protein degradation, are found with protein aggregates in both human and mouse tissues.

Invertebrate Models Manifest Progressive Neurodegeneration

Several invertebrate models have been used to study polyglutamine proteins, α-synuclein, parkin, and PINK1. The similarities in the pathogenic effects of these proteins across species are remarkable.

Flies with high levels of human α-synuclein develop progressive degeneration of dopaminergic neurons and have α-synuclein–immunoreactive cytoplasmic aggregates reminiscent of Lewy bodies. As in the mouse model, high levels of α-synuclein in flies carrying either the wild-type or mutant allele induce this phenotype. Additionally, flies with PINK1 or parkin mutations have dopaminergic neuron defects and motor abnormalities. Overexpression of wild type or mutant ataxin-1 in flies induces progressive neuronal degeneration that correlates with protein levels, but of course is more severe for flies with the mutant protein.

Polyglutamine toxicity has also been evaluated in the nematode *C. elegans* by expressing an amino terminal fragment of huntingtin containing glutamine tracts of different lengths. Neuronal dysfunction and cell death occur in worms expressing expanded tracts embedded within a truncated protein.

The Pathogenesis of Neurodegenerative Diseases Follows Several Pathways

Protein Misfolding and Degradation Contribute to Parkinson Disease

The gradual accumulation of neurodegenerative disease proteins along with chaperones and components

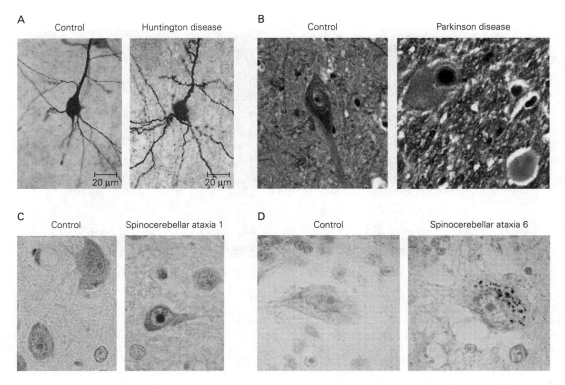

Figure 63–4 Neuropathological features of selected neuro-degenerative disorders.

A. Comparison of a normal spiny neuron from the caudate nucleus and a spiny neuron affected by Huntington disease. Note the marked recurving of terminal dendritic branches in the diseased neuron. (Image of Huntington neuron reproduced, with permission, from Marian Di Figlia and J.-P. Vonsattel.)

B. A pigmented dopaminergic neuron in the substantia nigra with a classic cytoplasmic inclusion (Lewy body). The circular cytoplasmic inclusion is surrounded by a clear halo. Recent electron-microscopic and biochemical evidence indicates that the primary components of Lewy bodies

are synuclein, ubiquitin, and abnormally phosphorylated neurofilaments that form a nonmembrane-bounded compacted skein in the cell body. Extracellular Lewy bodies occur following neuronal cell death and disintegration.

C. A neuron with a typical nuclear inclusion, almost as large as the nucleolus, and another Purkinje cell with a sizable vacuole and axonal swelling known as a torpedo.

D. Because spinocerebellar ataxia type 6 results from a repeat expansion in *CACNA1A*, which encodes a calcium channel, *CACNA1A* labeling occurs diffusely throughout the cytoplasm rather than in the nucleus.

of the ubiquitin-proteasome degradation pathway suggests that expansion of the glutamine tract alters the folding state of the native protein, which in turn recruits the activity of the protein-folding and degradation machinery. When that machinery cannot process the protein molecules, they accumulate, eventually forming aggregates. Evidence in support of this idea first came from observations in cell culture that over-production of chaperones reduces protein aggregation and mitigates the toxicity of expanded glutamine tracts in proteins. In contrast, blocking the proteasome inhibits protein degradation and thus enhances aggregation and toxicity. Genetic studies in flies and mice provide even more compelling evidence. Overproduction of at least one chaperone, such as Hsp70, Hsp40, or tetratricopeptide protein 2, suppresses polyglutamine toxicity

in *Drosophila* and reduces degeneration in mouse models of Parkinson disease and several types of ataxia. Conversely, loss of chaperone function worsens the neurodegenerative phenotypes (Figure 63–5).

The importance of the ubiquitin-proteasome pathway and protein degradation in the SCAs is further supported by genetic modification in animal models. In a *Drosophila* model of SCA1, haploinsufficiency for ubiquitin, ubiquitin carrier enzymes, or a ubiquitin carboxyl-terminal hydrolase worsens neurodegeneration. It appears that inclusions are part of the cell's attempt to sequester the mutant protein and thereby limit its toxic effects. Cells that are unable to form aggregates suffer the worst damage from polyglutamine toxicity. Indeed, the knock-in mouse models of SCA types 1 and 7 show conclusively that cells that

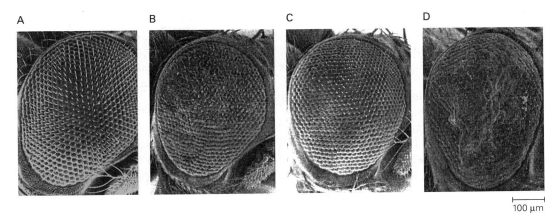

100 μm

Figure 63–5 Polyglutamine-induced degeneration in the *Drosophila* eye and the effect of modifiers. (Images reproduced, with permission, from J. Botas.)

A. A scanning electron micrograph depicts the eye of a fly with normal ommatidia.

B. Ommatidia of a transgenic fly bearing a protein with expanded glutamine repeats.

C. Owing to the mitigating effect of over-producing a heat shock protein on the polyglutamine-induced phenotype, the ommatidia appear almost normal.

D. Absence of another heat shock protein aggravates the polyglutamine-induced phenotype.

form aggregates survive longer; cerebellar Purkinje cells, the prime targets in this disease, are the last to form nuclear aggregates.

In Huntington disease, the expanded huntingtin protein is readily cleaved by proteases, but the fragments are toxic to the cell, interfering with transcription and dysregulating the activity of dynamin-1. Expanded huntingtin activates the autophagy pathway via repression of mammalian target of rapamycin (mTOR) kinase, but the resulting autophagosomes are defective and cannot help the neurons degrade aggregated proteins. Finally, the expanded huntingtin protein forms aggregates. Some studies suggest these aggregates are harmful to the cell, while other studies have shown them to be protective because they reduce the circulating level of the soluble toxic protein. Their role in pathology might depend on the disease stage and which interacting proteins co-accumulate in the aggregates.

Studies of Parkinson disease further underscore the importance of the ubiquitin-proteasome pathway and reveal additional parallels with the polyglutamine diseases. First, studies focused on α-synuclein have shown that ubiquitinated forms of the protein accumulate in Lewy bodies and that ubiquitination regulates α-synuclein stability. Second, recent studies have identified NEDD4 as an E3 ubiquitin ligase that targets α-synuclein, and USP9X as a de-ubiquitinating enzyme that removes the modification. Other studies have identified parkin as an E3 ubiquitin ligase that

targets many different mitochondrial proteins and is important for mitochondrial quality control.

How do misfolded α-synuclein or expanded glutamine tracts disrupt neuronal function? A protein that resists degradation might linger too long in the cell, performing its normal function longer than it should; the altered conformation might also cause it to favor certain protein interactions over others. This is what happens with glutamine-expanded ataxin-1: Part of the gain of toxic function involves prolonged binding with Capicua and subsequent alterations in its transcriptional activity.

Protein Misfolding Triggers Pathological Alterations in Gene Expression

One of the key consequences of misfolding as a result of expanded glutamine tracts is alteration in gene expression. This was first suspected when it was realized that most of the mutant proteins accumulate in the cell nucleus and that they interact with or affect the function of key transcriptional regulators. For example, huntingtin interacts with the transcription factors CREB-binding protein, NeuroD, specificity protein-1, nuclear factor-κB, and tumor suppressor protein 53 (p53), among others. Disruption of these interactions secondary to the polyglutamine expansion leads to myriad transcriptional changes seen in the disease state.

Alterations in gene expression are among the earliest events in pathogenesis, occurring within days of expression of the mutant transgene in mouse models of SCA1 and Huntington disease. Many of the genes whose expression is altered are involved in Ca^{2+} homeostasis, apoptosis, cell-cycle control, DNA repair, synaptic transmission, and transduction of sensory events into neural signals. In fly models of SCA1, several modifiers of the neurodegenerative phenotype are transcriptional cofactors. Overproduction of polyglutamine proteins also can reduce levels of histone acetylation in cells, an effect that can be reversed by overproduction of CREB-binding protein. Finally, ataxin-1 is in a native complex with the transcriptional repressor Capicua; thus, some of the gain-of-function effects involve a gain of enhanced Capicua-mediated repression.

Mitochondrial Dysfunction Exacerbates Neurodegenerative Disease

Both morphological and functional studies provide evidence of mitochondrial dysfunction in polyglutamine disorders and Parkinson disease. Lymphoblast mitochondria from patients with Huntington disease as well as brain mitochondria from a transgenic mouse model for Huntington disease have a lower membrane potential and depolarize at lower Ca^{2+} loads than do control mitochondria.

Several proteins implicated in Parkinson disease affect mitochondrial function and integrity. For example, studies in *Drosophila* showed that the loss of PINK1 leads to mitochondrial dysfunction, dopaminergic neuron impairment, and motor abnormalities that can be rescued by parkin. These studies led to the finding that PINK1 and parkin regulate mitochondrial turnover in the cell in a process termed mitophagy. Thus, given the functions and interactions of these proteins, mitochondrial dysfunction is likely to be a key contributor to the Parkinson disease phenotype.

Apoptosis and Caspases Modify the Severity of Neurodegeneration

Although studies of most neurodegenerative diseases demonstrate that symptoms appear long before detectable cell death, loss of neurons is a hallmark of the end stage of all these disorders. Two major factors are implicated in the death of neurons: altered Ca^{2+} homeostasis and decreased induction of neuronal survival factors, such as brain-derived neurotrophic factor in Huntington disease. There is, however, specific evidence that the caspase activity critical for apoptosis is a contributing factor in neurodegenerative diseases. Some of the polyglutamine proteins, such as huntingtin, androgen receptor, ataxin-3, and atrophin-1, are substrates for caspases

in vitro. This raises the possibility that caspase liberates the fragments of these proteins with expanded glutamine tracts. As noted above, such fragments are even more damaging than the full-length protein.

Intranuclear huntingtin increases production of caspase-1 in cells; this could lead to apoptosis and caspase-3 activation. Hip-1, a protein that interacts with huntingtin, forms a complex that activates caspase-8. This process might be enhanced by the glutamine expansion in huntingtin because Hip-1 binds less avidly to mutant huntingtin than to the wild type protein. In *Drosophila*, production of the antiapoptotic protein p35 results in partial rescue of the pigment loss induced by mutant ataxin-3.

In summary, expansions of polyglutamine tracts as well as several missense mutations in proteins implicated in neurodegenerative diseases alter the host protein, leading to its accumulation or abnormal interactions. The neuronal dysfunction results from the downstream effects of such abnormal interactions (Figure 63–6).

Understanding the Molecular Dynamics of Neurodegenerative Diseases Suggests Approaches to Therapeutic Intervention

The discovery of the genetic bases and pathogenic mechanisms of various neurodegenerative diseases offers us hope that therapies for these diseases will soon emerge. Dopamine replacement therapy has so far been the only pharmacological option for Parkinson disease, but it is not ideal. Patients tend to develop tolerance and require higher and higher doses of the drugs, which in turn cause a side effect known as levodopa-induced dyskinesias. The uncontrollable movements of dyskinesia soon become as disruptive as the motor symptoms originally being treated. Advancements in deep brain stimulation are promising, but the procedure is invasive and therefore reserved for medication-refractory Parkinson disease.

Patients with Huntington disease and SCA are worse off. No treatments that slow the progressive loss of motor coordination are currently available. However, several exciting therapeutic approaches that show great promise are under investigation. The most exciting therapeutic advances are those related to gene silencing of pathogenic products, including editing the genome, turning down transcription, or reducing expression of the protein. The most promising of these approaches in Huntington disease is the use of antisense oligonucleotides (ASOs). ASOs are small single-stranded molecules designed to bind to complementary sequences found within the mRNA product

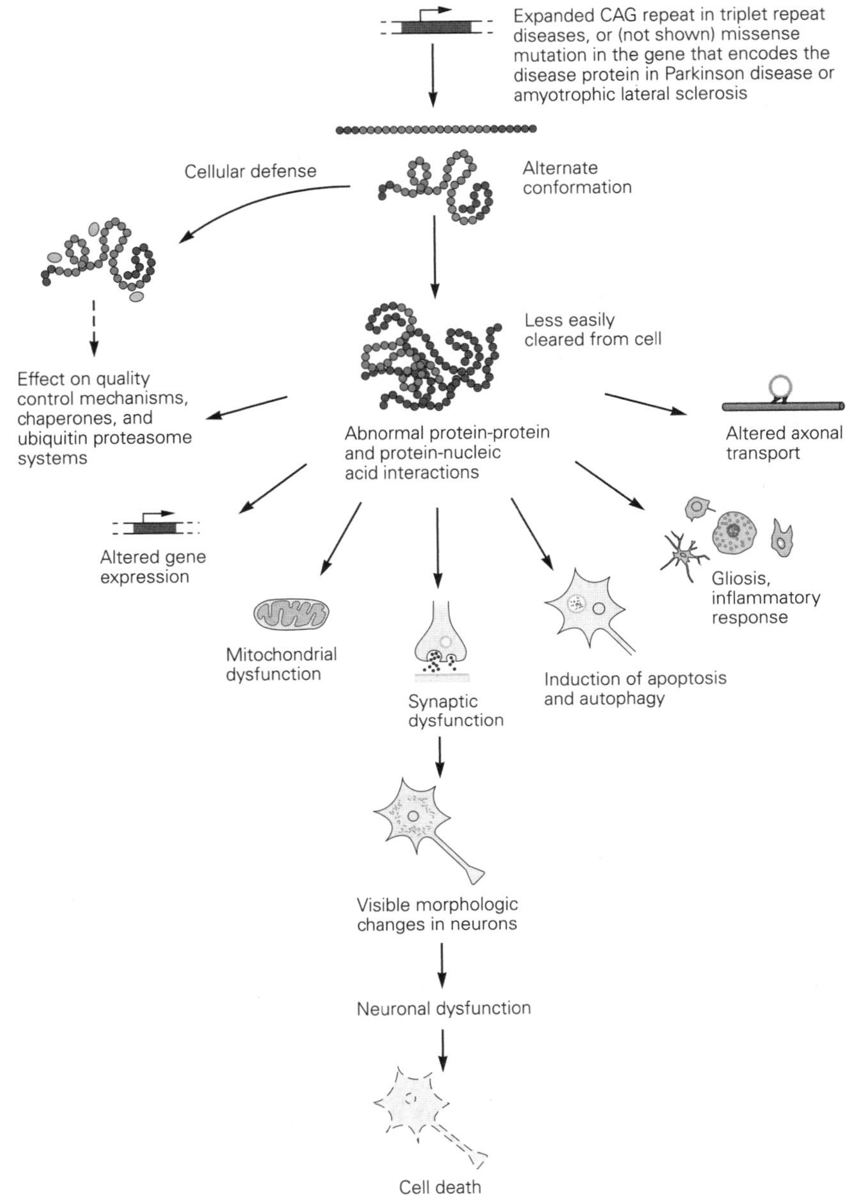

Figure 63–6 Current model for pathogenesis of the proteinopathies. The disease-causing protein adopts an alternative conformation that changes its interactions with other proteins, DNA, or RNA, altering gene expression and perhaps generating an inflammatory response. These early events in pathogenesis occur years before symptoms appear. Since this alternative conformation is more difficult for the cell to refold or degrade, steady-state levels of the mutant protein rise slowly over a period of decades. As levels of the mutant protein rise, the neuron attempts to sequester the mutant protein and forms aggregates. As disease progresses, these proteinaceous deposits themselves may affect protein interactions or compromise the protein quality-control system.

one wants to downregulate. When an ASO binds its mRNA target, it triggers degradation of the mRNA through RNAse H activity while at the same time sparing the ASO itself, thereby allowing it to bind to another mRNA molecule. In Huntington disease, several ASOs have been successfully used to reduce huntingtin protein levels. In fact, this approach is currently being utilized in clinical trials and holds promise for other diseases such as Parkinson disease.

Ideally, therapies should be targeted at some of the earliest pathogenic stages, when intervention could in theory halt the disease or even allow recovery of function. Indeed, studies of mouse models of Huntington disease and SCA1, in which expression of the mutant gene can be turned off, have shown that the neuronal dysfunction is reversible. When expression of the transgene is turned off, the neurons have a chance to clear the mutant polyglutamine protein and regain normal activity.

Because most neurodegenerative diseases progress over a period of decades, pharmacological interventions that even slightly modulate one or more of the pathways described above could delay disease progression or improve function, which would greatly enhance the quality of life for patients suffering from these devastating disorders.

Highlights

1. Late-onset neurodegenerative diseases collectively afflict more than 25 million people throughout the world, and it is anticipated that the prevalence of Alzheimer and Parkinson diseases will rise, given the increasing trend in life expectancy.

2. The identification of genes causing several forms of Parkinson disease and the various polyglutamine neurodegenerative diseases has allowed the accurate diagnosis and classification of these clinically heterogeneous disorders.

3. Although the gene product driving disease is widely expressed in the brain, there is selective neuronal vulnerability in all adult-onset neurodegenerative disorders. Perhaps a slight increase in abundance of the disease-driving protein and/or its interactors might explain such selective vulnerability.

4. Mitochondrial dysfunction is common in Parkinson disease; some of the genes mutated in Parkinson disease regulate mitochondrial turnover.

5. Studies in cell culture and model organisms have revealed a pathogenic mechanism common to adult-onset neurodegenerative diseases: protein misfolding. Mutations that cause the respective proteins to adopt an altered conformation gradually induce neuronal dysfunction either because of abnormal protein interactions or because of intracellular protein accumulation and altered activity.

6. The accumulation of polyglutamine-expanded proteins causes a variety of molecular changes in the cells, including alterations in gene expression, alterations in Ca^{2+} homeostasis, mitochondrial dysfunction, and activation of caspases.

7. The discovery that many adult neurodegenerative disorders are reversible in mouse models gives hope that some of the neuronal dysfunction can be rescued if a treatment is implemented early enough in the disease course before cell death occurs.

8. The identification of pathways that mediate some of the pathogenic effects is likely to lead to the discovery of drugs that can first be tested in animals and then applied in humans.

9. Lowering the levels of disease-driving proteins can ameliorate their toxic effects. This opens the way for therapeutic strategies that either employ antisense oligonucleotides that target the toxic RNA or that use small molecules to target regulators of the toxic protein.

Huda Y. Zoghbi

Selected Reading

Gatchel JR, Zoghbi HY. 2005. Diseases of unstable repeat expansion: mechanisms and common principles. Nat Rev Genet 6:743–755.

Gusella JF, MacDonald ME. 2000. Molecular genetics: unmasking polyglutamine triggers in neurodegenerative disease. Nat Rev Neurosci 1:109–115.

Haelterman NA, Yoon WH, Sandoval H, et al. 2014. A mitocentric view of Parkinson's disease. Annu Rev Neurosci 37:137–159.

Laforet GA, Sapp E, Chase K, et al. 2001. Changes in cortical and striatal neurons predict behavioral and electrophysiological abnormalities in a transgenic murine model of Huntington's disease. J Neurosci 21:9112–9123.

Moore DJ, West AB, Dawson VL, Dawson TM. 2005. Molecular pathophysiology of Parkinson's disease. Annu Rev Neurosci 28:57–87.

Pickrell AM, Youle RJ. 2015. The roles of PINK1, parkin, and mitochondrial fidelity in Parkinson disease. Neuron 85:257–273.

Sherman MY, Goldberg AL. 2001. Cellular defenses against unfolded proteins: a cell biologist thinks about neurodegenerative diseases. Neuron 1:15–32.

Steffan JS, Bodai L, Pallos J, et al. 2001. Histone deacetylase inhibitors arrest polyglutamine-dependent neurodegeneration in *Drosophila*. Nature 413:739–743.

Wong Y, Krainc D. 2017. α-Synuclein toxicity in neurodegeneration: mechanism and therapeutic strategies. Nat Med 23:1–13.

Zoghbi HY, Orr HT. 2009. Pathogenic mechanisms of a polyglutamine-mediated neurodegenerative disease, spinocerebellar ataxia type 1. J Biol Chem. 284:7425–7429.

References

Alexopoulou, Z, Lang J, Perrett RM, et al. 2016. Deubiquitinase Usp8 regulates α-synuclein clearance and modifies its toxicity in Lewy body disease. Proc Natl Acad of Sci U S A 113:4688–4697.

Alves-Cruzeiro JM, Mendonça L, Pereira de Almeida L, Nóbrega C. 2016. Motor dysfunctions and neuropathology in mouse models of spinocerebellar ataxia type 2: a comprehensive review. Front Neurosci 10:572.

Auluck PK, Chan HY, Trojanowski JQ, Lee VM, Bonini NM. 2002. Chaperone suppression of α-synuclein toxicity in a *Drosophila* model for Parkinson's disease. Science 295:865–888.

Bonifati V. 2012. Autosomal recessive parkinsonism. Parkinsonism Relat Disord 18:S4–S6.

Bonini NM, Gitler AD. 2011. Model organisms reveal insight into human neurodegenerative disease: ataxin-2 intermediate-length polyglutamine expansions are a risk factor for ALS. J Mol Neurosci 45:676–683.

Burré, J. 2015. The synaptic function of α-synuclein. J Parkinson Dis 5:699–713.

Chai Y, Koppenhafer SL, Bonini NM, Paulson HL. 1999. Analysis of the role of heat shock protein (Hsp) molecular chaperones in polyglutamine disease. J Neurosci 19:10338–10347.

Chesselet M.-F, Richter F, Zhu C, et al. 2012. A progressive mouse model of Parkinson's disease: the thy1-aSyn ('Line 61') Mice. Neurotherapeutics 9:297–314.

Cummings CJ, Mancini MA, Antalffy B, et al. 1998. Chaperone suppression of ataxin-1 aggregation and altered subcellular proteasome localization imply protein misfolding in SCA1. Nat Genet 19:148–154.

Cummings CJ, Sun Y, Opal P, et al. 2001. Over-expression of inducible HSP70 chaperone suppresses neuropathology and improves motor function in SCA1 mice. Hum Mol Genet 10:1511–1518.

Davies SW, Turmaine M, Cozens BA, et al. 1997. Formation of neuronal intranuclear inclusions underlies the neurological dysfunction in mice transgenic for the HD mutation. Cell 90:537–548.

Feany MB, Bender WW. 2000. A Drosophila model of Parkinson's disease. Nature 404:394–398.

Fernandez-Funez P, Nino-Rosales ML, de Gouyon B, et al. 2000. Identification of genes that modify ataxin-1-induced neurodegeneration. Nature 408:101–106.

Fryer JD1, Yu P, Kang H, et al. 2011. Exercise and genetic rescue of SCA1 via the transcriptional repressor Capicua. Science. 334:690–3.

Fujioka S, Wszolek ZK. 2012. Update on genetics of parkinsonism. Neurodegener Dis 10:257–260.

Gennarino VA, Singh RK, White JJ, et al. 2015. Pumilio1 haploinsufficiency leads to SCA1-like neurodegeneration by increasing wild-type Ataxin1 levels. Cell 160:1087–1098.

Hagerman RJ, Hagerman PJ. 2002. The fragile X premutation: into the phenotypic fold. Curr Opin Genet Dev 12:278–283.

Holmes SE, O'Hearn EE, McInnis MG, et al. 1999. Expansion of a novel CAG trinucleotide repeat in the 5' region of PPP2R2B is associated with SCA12. Nat Genet 23:391–392.

Huynh DP, Del Bigio MR, Ho DH, Pulst SM. 1999. Expression of ataxin-2 in brains from normal individuals and patients with Alzheimer's disease and spinocerebellar ataxia 2. Ann Neurol 45:232–241.

Huynh DP, Figueroa K, Hoang N, Pulst SM. 2000. Nuclear localization or inclusion body formation of ataxin-2 are not necessary for SCA2 pathogenesis in mouse or human. Nat Genet 26:44–50.

Kegel KB, Kim M, Sapp E, McIntyre C, Castano JG, Aronin N, DiFiglia M. 2000. Huntingtin expression stimulates endosomal-lysosomal activity, endosome tubulation, and autophagy. J Neurosci 20:7268-7278.

Koob MD, Moseley ML, Schut LJ, et al. 1999. An untranslated CTG expansion causes a novel form of spinocerebellar ataxia (SCA8). Nat Genet 21:379–384.

Kruger R, Kuhn W, Muller T, et al. 1998. Ala30Pro mutation in the gene encoding α-synuclein in Parkinson's disease. Nat Genet 18:106–108.

La Spada AR, Fu YH, Sopher BL, et al. 2001. Polyglutamine-expanded ataxin-7 antagonizes CRX function and induces cone-rod dystrophy in a mouse model of SCA7. Neuron 31:913–927.

Leroy E, Boyer R, Auburger G, Leube B, et al. 1998. The ubiquitin pathway in Parkinson's disease. Nature 395: 451–452.

Lucking CB, Durr A, Bonifati V, et al. 2000. Association between early-onset Parkinson's disease and mutations in the parkin gene. N Engl J Med 342:1560–1567.

Luthi-Carter R, Strand A, Peters NL, et al. 2000. Decreased expression of striatal signaling genes in a mouse model of Huntington's disease. Hum Mol Genet 9:1259–1271.

Masliah E, Rockenstein E, Veinbergs I, et al. 2000. Dopaminergic loss and inclusion body formation in α-synuclein mice: implications for neurodegenerative disorders. Science 287:1265–1269.

Matsuura T, Yamagata T, Burgess DL, et al. 2000. Large expansion of the ATTCT pentanucleotide repeat in spinocerebellar ataxia type 10. Nat Genet 26:191–194.

McCampbell A, Taye AA, Whitty L, Penney E, Steffan JS, Fischbeck KH. 2001. Histone deacetylase inhibitors reduce polyglutamine toxicity. Proc Natl Acad Sci U S A 98:15179–15184.

Miller J, Arrasate M, Shaby BA, Mitra S, Masliah E, Finkbeiner S. 2010. Quantitative relationships between huntingtin levels, polyglutamine length, inclusion body formation, and neuronal death provide novel insight into Huntington's disease molecular pathogenesis. J. Neurosci. 30:10541–10550.

Nakamura K, Jeong SY, Uchihara T, et al. 2001. SCA17, a novel autosomal dominant cerebellar ataxia caused by an expanded polyglutamine in TATA-binding protein. Hum Mol Genet 10:1441–1448.

Nalls MA, Pankratz N, Lill CM, et al. 2014. Large-scale meta-analysis of genome-wide association data identifies six new risk loci for Parkinson's disease. Nat Genet 46:989–993.

Nucifora FC, Sasaki M, Peters MF, et al. 2001. Interference by huntingtin and atrophin-1 with cbp-mediated transcription leading to cellular toxicity. Science 291:2423–2428.

Orr HT, Zoghbi HY. 2007. Trinucleotide repeat disorders. Annu Rev Neurosci 30:575–621.

Panov AV, Gutekunst CA, Leavitt BR, et al. 2002. Early mitochondrial calcium defects in Huntington's disease are a direct effect of polyglutamines. Nat Neurosci 5:731–736.

Park J, Al-Ramahi I, Tan Q, et al. 2013. RAS-MAPK-MSK1 pathway modulates ataxin 1 protein levels and toxicity in SCA1. Nature 498:325–331.

Piedras-Renteria ES, Watase K, Harata N, et al. 2001. Increased expression of alpha 1A Ca^{2+} channel currents arising from expanded trinucleotide repeats in spinocerebellar ataxia type 6. J Neurosci 21:9185–9193.

Polymeropoulos MH, Lavedan C, Leroy E, et al. 1997. Mutation in the α-synuclein gene identified in families with Parkinson's disease. Science 276:2045–2047.

Ramachandran PS, Boudreau RL, Schaefer KA, La Spada AR, Davidson BL. 2014. Nonallele specific silencing of ataxin-7 improves disease phenotypes in a mouse model of SCA7. Mol Ther 22:1635–1642.

Ravikumar B, Vacher C, Berger Z, et al. 2004. Inhibition of mTOR induces autophagy and reduces toxicity of polyglutamine expansions in fly and mouse models of Huntington disease. Nat Genet 36:585–595.

Rott R, Szargel R, Haskin J. 2011. α-Synuclein fate is determined by USP9X-regulated monoubiquitination. Proc Natl Acad Sci U S A 108:18666–18671.

Saudou F, Humbert S. 2016. The biology of huntingtin. Neuron 89:910–926.

Sidransky E, Lopez G. 2012. The link between the GBA gene and parkinsonism. Lancet Neurol 11:986–998.

Singleton AB, Farrer M, Johnson J, et al. 2003. α-*Synuclein* locus triplication causes Parkinson's disease. Science 302:841.

Smith WW, Pei Z, Jiang H, et al. 2005. Leucine-rich repeat kinase 2 (LRRK2) interacts with parkin, and mutant LRRK2 induces neuronal degeneration. Proc Natl Acad Sci U S A 102:18676–18681.

Surmeier JD, Guzman JN, Sanchez-Padilla J, Schumacker PT. 2011. The role of calcium and mitochondrial oxidant stress in the loss of substantia nigra pars compacta dopaminergic neurons in Parkinson's disease. Neuroscience 198:221–231.

Valente EM, Abou-Sleiman PM, Caputo V, et al. 2004. Hereditary early-onset Parkinson's disease caused by mutations in PINK1. Science 304:1158–1160.

Vonsattel JP, DiFiglia M. 1998. Huntington's disease. J Neuropathol Exp Neurol 57:369–384.

Warrick JM, Chan HY, Gray-Board GL, Chai Y, Paulson HL, Bonini NM. 1999. Suppression of polyglutamine-mediated neurodegeneration in *Drosophila* by the molecular chaperone HSP70. Nat Genet 23:425–428.

Wyant KJ, Riddler AJ, Dayalu P. 2017. Huntington's disease—update on treatments. Curr Neurol Neurosci Rep 17:1–11.

Zhang S, Xu L, Lee J, Xu T. 2002. *Drosophila* atrophin homolog functions as a transcriptional corepressor in multiple developmental processes. Cell 108:45–56.

Zu T, Duvick LA, Kaytor MD, et al. 2004. Recovery from polyglutamine-induced neurodegeneration in conditional SCA1 transgenic mice. J Neurosci 24:8853–8861.

64

The Aging Brain

THE AVERAGE LIFE SPAN IN THE UNITED STATES in 1900 was about 50 years. By 2015, it was approximately 77 years for men and 82 for women (Figure 64–1). The average is even higher in 30 other countries. These increases result largely from a reduction in infant mortality, the development of vaccines and antibiotics, better nutrition, improved public health measures, and advances in the treatment and prevention of heart disease and stroke. Because of increased life expectancy, along with the large cohort of "baby boomers" born soon after World War II, the elderly are the most rapidly growing segment of the US population.

Increased longevity is a double-edged sword since age-related cognitive alterations are increasingly prevalent. The magnitude of the change varies widely among individuals. For many, the alterations are mild and have relatively little impact on the quality of life—the momentary lapses we jokingly call "senior moments." Other cognitive impairments, although not debilitating, are troubling enough to hinder our ability to manage life independently. The dementias, however, erode memory and reasoning and alter personality. Of these, Alzheimer disease is the most prevalent.

As the population ages, neuroscientists, neurologists, and psychologists have begun to devote more energy to understanding age-related changes in the brain. The primary motivation has been to find treatments for Alzheimer disease and other dementias, but it is also important to understand the normal process of cognitive decline with age. After all, age is the greatest risk factor for a wide variety of neurodegenerative disorders. Understanding what happens to our brains as we age may not only improve the quality of life for the general population but may also provide clues that will eventually help us vanquish seemingly unrelated pathological changes.

With this in mind, we begin this chapter with a consideration of the normal aging of the brain. We then turn to the broad range of pathological changes in cognition, and finally focus on Alzheimer disease.

The Structure and Function of the Brain Change With Age

As we grow old, our bodies change—our hair thins, our skin wrinkles, and our joints creak. It is no surprise then that our brain also changes. Indeed, the widespread behavioral alterations that occur with age are

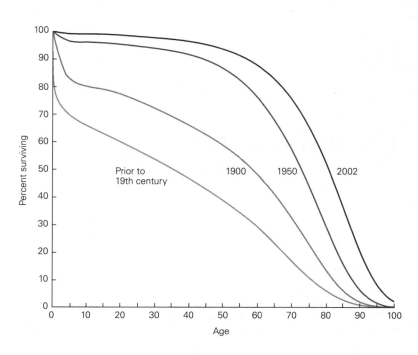

Figure 64–1 The human life span is increasing. The average life span in the United States has increased rapidly over the past 100 years. (Adapted from Strehler 1975; Arias 2004.)

signs of underlying alterations in the nervous system. For example, as motor skills decline, posture becomes less erect, gait is slower, stride length is shorter, and postural reflexes often become sluggish. Although muscles weaken and bones become more brittle, these motor abnormalities result in large part from subtle processes that involve the peripheral and central nervous systems. Sleep patterns also change with age; older people sleep less and wake more frequently. Mental functions ascribed to the forebrain, such as memory and problem-solving abilities, also decline.

Age-related declines in mental abilities are highly variable, in both rate and severity (Figure 64–2A). Although most people experience a gradual decline in mental agility, for some, the decline is rapid, whereas others retain their cognitive powers throughout life—Giuseppe Verdi, Eleanor Roosevelt, and Pablo Picasso are well-known examples of the latter category. Titian continued to paint masterpieces in his late 80s, and Sophocles is said to have written *Oedipus at Colonus* in his 92nd year. The fact that elderly people with completely preserved mental function are rare suggests that there may be special properties in the life experiences or genes of these people. Accordingly, there has been great interest in studying individuals who retain nearly intact cognition into their tenth or even eleventh decade. These centenarians may provide insight into environmental or genetic factors that protect against normal age-related cognitive decline or the more devastating pathological descent into dementia. One

protective gene variant, discussed below, is the epsilon 2 allele of the apolipoprotein E gene.

An interesting finding that has emerged from studies of many individuals is that some cognitive capacities decline significantly with age while others are largely spared (Figure 64–2B). For example, working and long-term memories, visuospatial abilities (measured by arranging blocks into a design or drawing a three-dimensional figure), and verbal fluency (measured by rapid naming of objects or naming as many words as possible that start with a specific letter) usually decline with old age. On the other hand, measures of vocabulary, information, and comprehension often show minimal decline in normal individuals well into the 80s.

Age-related changes in memory, motor activity, mood, sleep pattern, appetite, and neuroendocrine function result from alterations in the structure and function of the brain. Even the healthiest 80-year-old brain does not look like it did at the age of 20. Elderly people exhibit mild shrinkage in the volume of the brain and a loss in brain weight, as well as enlargement of the cerebral ventricles (Figure 64–3A). The decreases in brain weight average 0.2% per year from college age onward, and about 0.5% per year in the 70s.

These changes could result from death of neurons. Indeed, some neurons are lost with age. For example, 25% or more of the motor neurons that innervate skeletal muscles die in generally healthy elderly individuals. As we will see, neurodegenerative diseases such as Alzheimer disease markedly accelerate the

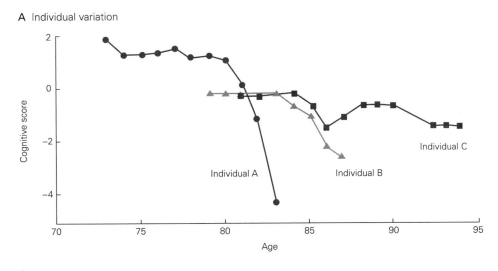

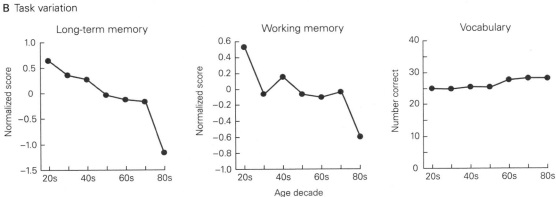

Figure 64–2 There is variation in age-related cognitive decline.

A. Scores of three people who were given a battery of cognitive tests annually for decades. Person A declined rapidly. Persons B and C showed similar cognitive performances into their 80s but then diverged. (Adapted from Rubin et al. 1998.)

B. Average scores on several cognitive tests administered to a large number of people. Long-term declarative memory and working memory decline throughout life and more so in advanced age. In contrast, knowledge of vocabulary is maintained. (Adapted from Park et al. 1996.)

death of neurons (Figure 64–3B). In most parts of the healthy brain, however, there is minimal to no neuronal loss simply because of age, so brain shrinkage must arise from other factors.

In fact, analysis of the brains of humans and experimental animals reveals structural alterations in both neurons and glia. Myelin is fragmented and lost, compromising the integrity of white matter. At the same time, the density of the dendritic arbors of cortical and other neurons decreases, resulting in shrinkage of neuropil. Levels of enzymes that synthesize some neurotransmitters, such as dopamine, norepinephrine, and acetylcholine, decrease with age, and this decline presumably results in functional defects in synapses

that use these transmitters. Synapse structure is also altered, at least at the neuromuscular junction (Figure 64–4), raising the possibility that structural changes also lead to functional deficits at central synapses. Finally, the number of synapses in the neocortex and many other regions of the brain declines (Figure 64–5).

These cellular changes interfere with the integrity of the neural circuits that mediate our mental activities. Age-related loss of synapses along with impairment in function of remaining synapses are thought to be important contributors to cognitive decline. Changes in white matter are widespread but are especially notable in the prefrontal and temporal cortex. They may underlie alterations in executive functions

A Age-related changes

Normal 22-year old

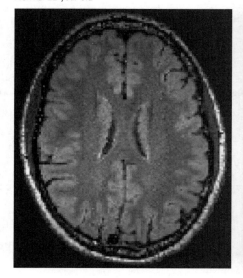

Normal 89-year old

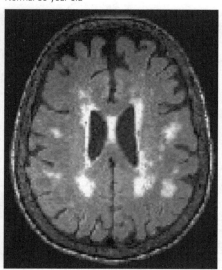

Figure 64–3 Changes in brain structure with age and at the onset of Alzheimer disease. (Also see Figure 64–8.)

A. Images of normal 22- and 89-year-old brains reveal changes in the structure of the living brain. (Reproduced, with permission, from R. Buckner.)

B. Images of the same individual over a 4-year period illustrate the progressive shrinking of cortical structures and the beginning of ventricular enlargement (**red**). These structural changes are evident prior to the onset of behavioral symptoms. (Reproduced, with permission, from N. Fox.)

B Changes with Alzheimer disease

Asymptomatic 45-year old

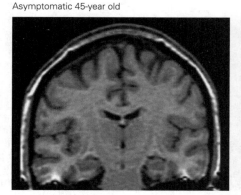

Onset of behavioral symptoms 4 years later

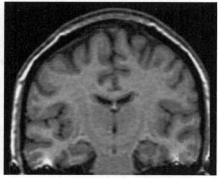

and the ability to focus attention and encode and store memory, functions that are localized in frontal-striatal systems and the temporal lobes. The loss of white matter may also help explain the recent finding that the elderly brain is less able to support synchronization of activity in widely separated areas that normally work together to carry out complex mental activities. Disruption of these large-scale networks could be an important cause of cognitive decline.

It was long thought that aging resulted from progressive deterioration of cells and tissues due to accumulated genetic damage or toxic waste products. In support of this idea was the finding that mitotic cells removed from animals and placed in a tissue culture dish divide only for a limited number of times before they age and die. This view of "preordained" aging has

changed radically over the past 10 to 20 years, primarily as a result of the discovery in model organisms of mutations that significantly extend life span (Figure 64–6).

Such dramatic discoveries established that the aging process is under active genetic control. One such regulatory pathway that has been characterized includes insulin and insulin-like growth factors, their receptors, and the signaling programs they activate. Disruption of these genes actually increases the resistance of cells to lethal oxidative damage. It is thought that the normal forms of these genes have been selected through evolution because they benefit the organism during the reproductive period. Their deleterious effects on longevity, once the animals are past reproductive age, may be an unfortunate side effect about which evolution cares little.

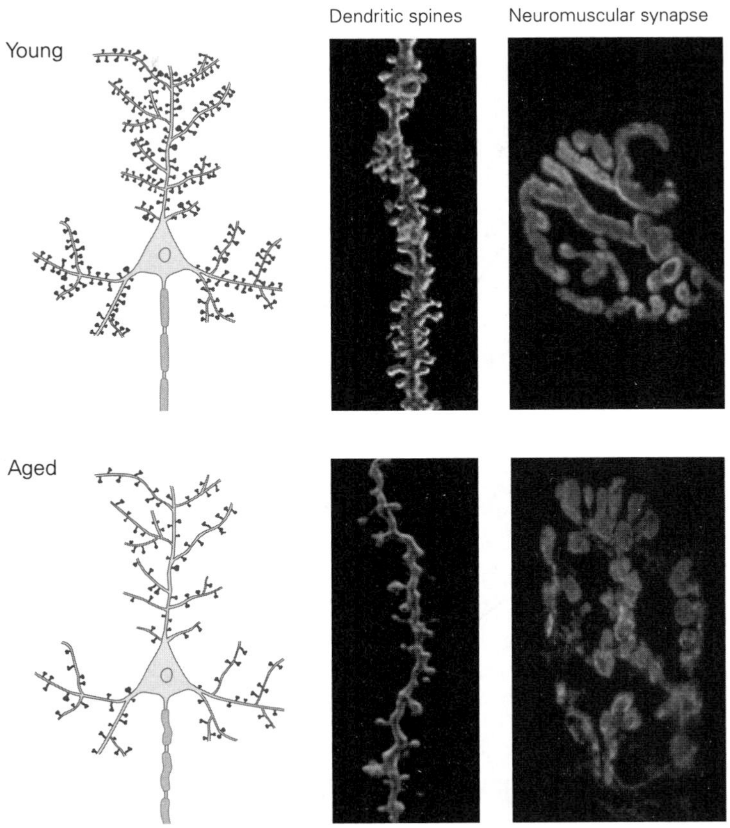

Young

Dendritic spines   Neuromuscular synapse

Aged

Figure 64–4 Age-related changes in dendritic and synaptic structure. Cortical pyramidal neurons in rodents lose dendritic spines with age. Neuromuscular synapses in rodents also exhibit age-related changes in structure. (Spine images reproduced, with permission, from J. Luebke; synapse images reproduced, with permission, from G. Valdez.)

These findings have two major implications for understanding how aging affects the nervous system. First, the biochemical mechanisms that lead to, or protect us from, the ravages of age are likely contributors to the changes in neurons that lead to age-related cognitive decline. Research to explore this link between cellular change and cognitive functioning is now underway in model organisms. Second, and perhaps more exciting, research on the pathways uncovered by genetic studies can identify pharmacological

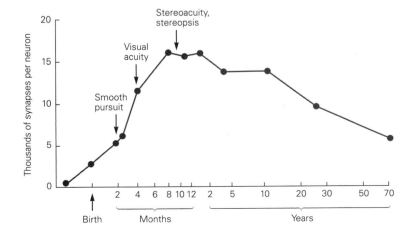

Figure 64–5 Age-related changes in synaptic density. Early cognitive development is accompanied by a marked increase in synapse density in different regions of the human cerebral cortex. Developmental landmarks through age 10 months are indicated. The density of cortical synapses declines with age. (Adapted from Huttenlocher 2002.)

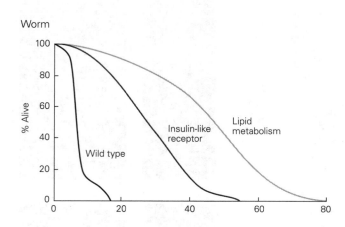

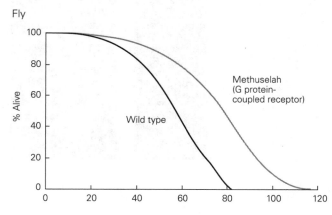

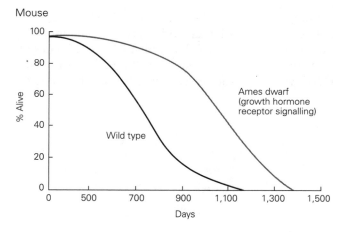

Figure 64–6 Life span can be increased through genetic mutation. Genetic mutations in specific receptors and signaling proteins markedly enhance life span in mutant strains of the worm, fly, and mouse, indicating that genetic regulatory mechanisms affect aging and life span. (Top adapted from Hekimi and Guarente 2003; middle reproduced, with permission, from Yi, Seroude, and Benzer 1998. Copyright © 1998 AAAS; bottom adapted from Brown-Borg et al. 1996.)

or environmental strategies for extending life span or health span (the period during which one remains generally healthy).

To date, the best-validated environmental strategy for extending life span (in organisms ranging from yeast to worms to primates) is caloric restriction. It appears that caloric restriction acts through genes in the insulin pathway mentioned above and may involve a set of enzymes called *sirtuins*. The sirtuins are activated by the compound *resveratrol*, originally isolated from red wine. Resveratrol, in turn, retards some aspects of aging, including cognitive decline, when administered to mice. While it is unlikely that resveratrol will serve as a fountain of youth in humans, it nevertheless exemplifies the new chemistries that are currently under consideration. These chemical strategies use model organisms to explore not only the positive factors that lead to aging but also the constraints that prevent model organisms, and presumably humans, from remaining generally healthy throughout their life span.

Cognitive Decline Is Significant and Debilitating in a Substantial Fraction of the Elderly

In most people, age-related cognitive changes do not seriously compromise the quality of life. In some elderly people, however, cognitive decline reaches a level that can be viewed as pathological. At the lower end of the abnormal range is a constellation of changes known as mild cognitive impairment (MCI). This syndrome is characterized by memory loss with or without other cognitive impairments that go beyond what is seen in normal aging. Individuals with MCI may be able to carry out most activities of daily living, although the impairments are noticeable to others and often influence the ability of the affected person to carry out certain activities that are important or pleasurable to them, such as managing finances or playing word games.

Importantly, MCI is a syndrome, not a diagnosis. Many underlying problems such as depression, overmedication, strokes, and neurodegenerative diseases can contribute to MCI. Approximately half of individuals with MCI have underlying Alzheimer disease, and more than 90% of this group will progress to full-blown dementia within 5 years from the time of diagnosis of MCI (Figure 64–7). As discussed below, there are now biomarkers that can suggest the presence of underlying Alzheimer disease pathology. As yet, however, there are no good biomarkers for predicting progression to dementia in people with MCI resulting from diseases other than Alzheimer.

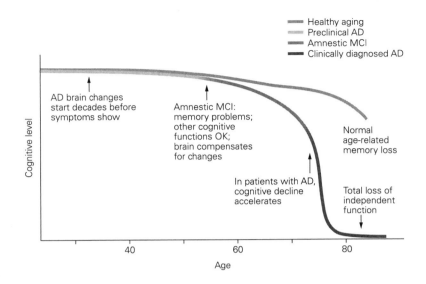

Figure 64–7 Cognitive performance can vary widely with age. The chart shows current thinking about the etiology of Alzheimer disease (**AD**). This gradual process, which results from a combination of biological, genetic, environmental, and lifestyle factors, eventually sets some people on a course to mild cognitive impairment (**MCI**) and then dementia. Other people, with a different genetic makeup or a different combination of factors over a lifetime, continue on a course of healthy cognitive aging. (From the National Institute on Aging: http://www.nia.nih.gov/alzheimers/publication/part-2-what-happens-brain-ad/changing-brain-ad.)

Like MCI, dementia is also a syndrome that involves progressive impairment of memory as well as other cognitive abilities such as language, problem solving, judgment, calculation, or attention. It is associated with a variety of diseases. The most common is Alzheimer disease, as discussed below. The second most common cause in the elderly is cerebrovascular disease, particularly strokes that lead to focal ischemia and consequent infarction in the brain.

Large lesions in the cortex are often associated with language disturbances (aphasia), hemiparesis, or neglect syndromes, depending on which portions of the brain are compromised. Small infarctions in white matter or deeper structures of the brain, termed *lacunes*, also occur as a consequence of hypertension and diabetes. In small numbers, these infarctions may be asymptomatic, or they may contribute to what appears to be normal age-associated cognitive decline or certain cases of MCI. As vascular lesions increase in number and size, however, their effects accumulate, and eventually, they can lead to dementia.

Numerous other conditions can lead to dementia, including Parkinson disease, Lewy body dementia, frontotemporal dementia, alcoholism, drug intoxications, infections such as HIV and syphilis, brain tumors, subdural hematomas, repeated brain trauma, vitamin deficiencies (notably lack of vitamin B$_{12}$), thyroid disease, and a variety of other metabolic disorders. Repeated brain trauma can result in what is termed chronic traumatic encephalopathy (CTE). Numerous cases of CTE in American professional athletes have recently been reported. In some patients, schizophrenia or depression may mimic a dementia syndrome. (Emil Kraepelin chose the term "dementia praecox" to

describe the cognitive disease that we now call schizophrenia.) Because some dementias can be treated, it is important for the physician to probe differential diagnoses of dementia based on clinical history, physical examinations, and laboratory studies.

Alzheimer Disease Is the Most Common Cause of Dementia

In 1901, Alois Alzheimer examined a middle-aged woman who had developed a progressive loss of cognitive abilities. Her memory became increasingly impaired. She could no longer orient herself, even in her own home, and she hid objects in her apartment. At times, she believed that people intended to murder her.

She was institutionalized in a psychiatric hospital and died approximately 5 years after she was first seen by Dr. Alzheimer. After death, Alzheimer performed an autopsy that revealed specific alterations in the cerebral cortex, described below. The constellation of behavioral symptoms and physical alterations was subsequently given the name Alzheimer disease (AD).

This case caught Alzheimer's attention because it occurred in middle age; the initial clinical manifestations of AD (usually memory loss and decreased executive function) most commonly appear after age 65. The prevalence of AD at age 70 is about 2%, whereas after age 80, it is greater than 20%. Early-onset cases before age 65 are often familial (autosomal dominant AD), and gene mutations have been discovered in many of these patients, as we shall discuss below. In fact, new genetic tests on preserved brain samples from Alzheimer's first case recently showed that her disease resulted from a mutation of a gene called presenilin-1, the most common cause of familial or dominantly inherited AD. Late-onset AD (onset at age 65 or greater) is more often sporadic, implying that there is no single causative gene as occurs in dominantly inherited AD. Nonetheless, it is clear that genetics contribute greatly to risk for even late-onset AD more likely through variants that affect susceptibility, along with environmental and other contributing factors that are just now being uncovered.

Both early-onset and late-onset varieties of AD usually present with a selective defect in episodic memory and executive function. At first, language, strength, reflexes, and sensory abilities and motor skills are nearly normal. Gradually, however, memory and attention are lost, along with cognitive abilities such as problem solving, language, calculation, and visuospatial perception. Unsurprisingly, these cognitive losses lead to behavioral alterations, and some patients develop psychotic symptoms such as hallucinations and delusions. All patients suffer progressive impairment of mental functions and activities of daily living; in the late stages, they become mute, incontinent, and bedridden.

Alzheimer disease affects approximately one-eighth of people older than 65 years. More than 5 million people in the United States now suffer from dementia due to AD. Because the elderly population is increasing rapidly, the population at risk for AD is growing rapidly. During the next 25 years, the number of people with AD in the United States is expected to triple, as will the cost of caring for patients no longer able to care for themselves. Thus, AD is one of society's major public health problems.

The Brain in Alzheimer Disease Is Altered by Atrophy, Amyloid Plaques, and Neurofibrillary Tangles

Three categories of brain abnormalities are found in AD. First, because of neuronal and synaptic loss, the brain is atrophied, with narrowed gyri, widened sulci, reduced brain weight, and enlarged ventricles (Figure 64–8). These changes are also seen in milder forms in cognitively intact elderly people who die from other causes. Thus, AD is a neurodegenerative disease.

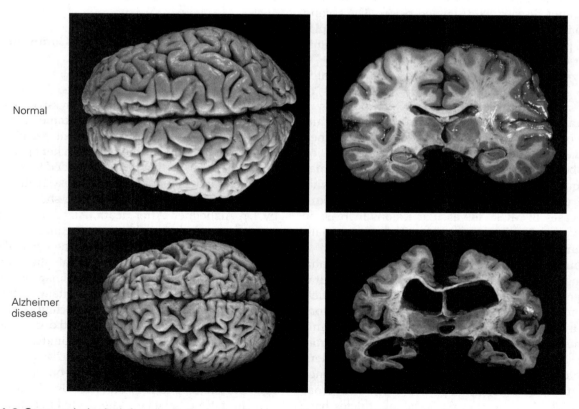

Normal

Alzheimer disease

Figure 64–8 Overt pathological changes in the brain of individuals with Alzheimer disease. When compared to age-matched normal brains, the brain of an Alzheimer patient displays marked shrinkage and ventricular enlargement. (See also Figure 64–3.) (Whole brain photos reproduced, with permission, from University of Alabama at Birmingham Department of Pathology © PEIR Digital Library [http://peir.net]; brain slice photos reproduced, with permission, from A.C. McKee.)

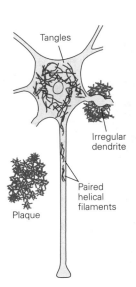
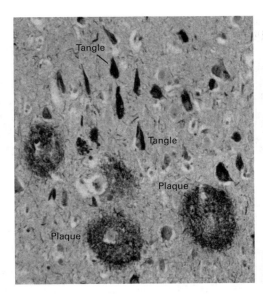
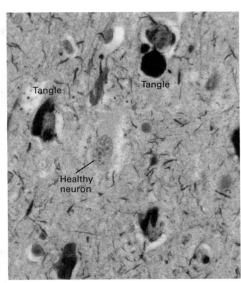

Figure 64–9 Plaques and tangles in the Alzheimer brain. A section of cerebral cortex from the brain of an individual with severe Alzheimer disease shows characteristic plaques and neurofibrillary tangles. (Images reproduced, with permission, from James Goldman.)

Left: The diagram shows a neuron containing neurofibrillary tangles in the cell body and axon. Amyloid plaques are shown in the neuropil; one of them surrounds a dendrite, which displays an altered, swollen shape. Tangles, composed of bundles of paired helical filaments, are comprised of abnormal polymers of hyperphosphorylated tau protein, and amyloid plaques are extracellular deposits of polymers of the amyloid-β (A4) peptide.

Middle: A section of neocortex from a patient with Alzheimer disease treated with a silver stain shows neuronal cell bodies containing neurofibrillary tangles and neuropil containing amyloid plaques.

Right: A higher magnification of the cortex shows neurofibrillary tangles in neuronal cell bodies and a healthy neuron without a tangle. Many thin silver-positive cell processes are seen in the neuropil.

Second, the brains of AD patients contain extracellular plaques composed predominantly of an aggregated form of a peptide called amyloid-β, or Aβ, which is cleaved from a normally produced protein. Aggregates of Aβ are called amyloid plaques. Much of the Aβ in plaques is fibrillar; aggregates of Aβ appear in a β-pleated sheet conformation along with other proteins that co-aggregate with Aβ (Figure 64–9). Amyloid can be detected when stained with dyes such as Congo red, and is refractive when viewed in polarized light or when stained with thioflavin S and viewed with fluorescence optics. The extracellular deposits of amyloid are surrounded by swollen axons and dendrites (neuritic dystrophy). These neuronal processes in turn are surrounded by the cell processes of activated astrocytes and microglia (inflammatory cells). Aβ can also form amyloid deposits in the walls of arterioles in the brain, producing what is known as cerebral amyloid angiopathy. This occurs to varying extents in up to 90% of patients who develop AD, but it can also occur independently of AD. Cerebral amyloid angiopathy can lead to ischemic stroke, and it is a common cause of hemorrhagic stroke in the elderly.

Third, many neurons that are affected by Alzheimer pathology but still alive have cytoskeletal abnormalities, the most dramatic of which is the accumulation of neurofibrillary tangles and neuropil threads (Figure 64–9). The tangles are filamentous inclusions in the cell bodies and dendrites that contain paired helical filaments and 15-nm straight filaments. These filaments are made up of an aggregated form of the normal microtubule-associated protein tau.

In AD, tangles do not occur uniformly throughout the brain, but rather affect specific regions. The entorhinal cortex, the hippocampus, parts of the neocortex, and the nucleus basalis are especially vulnerable (Figure 64–10). Alterations in the entorhinal cortex and hippocampus likely underlie the problems with episodic memory that are among the first symptoms of AD. Abnormalities in the basal forebrain cholinergic systems may contribute to cognitive difficulties and attention deficits. These cholinergic abnormalities contrast with those in frontostriatal circuits that correlate with age-related cognitive decline in normal subjects. The combination of anatomical differences, pathological changes, widespread neuronal death, and genetic

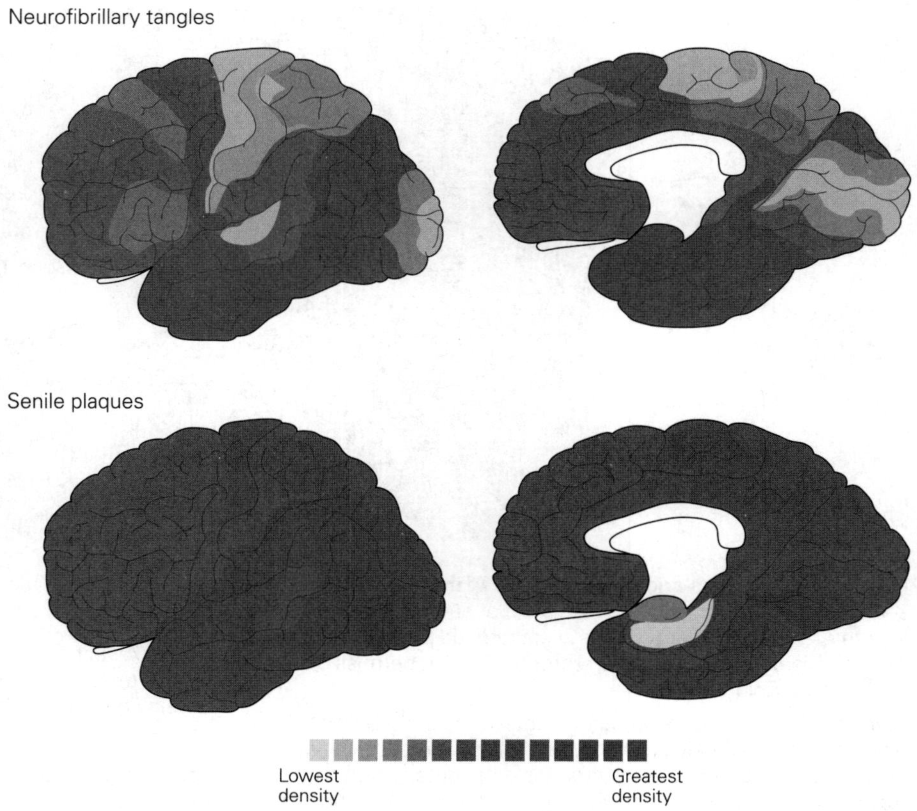

Neurofibrillary tangles

Figure 64–10 Neurofibrillary tangles and senile plaques are concentrated in different regions of the Alzheimer brain. (Adapted, with permission, from Arnold et al. 1991. Copyright © 1991, Oxford University Press.)

Senile plaques

Lowest density   Greatest density

mutations (see below) argue against the idea, once prevalent, that AD is an aberrant form of normal aging processes.

Amyloid Plaques Contain Toxic Peptides That Contribute to Alzheimer Pathology

The main constituent of amyloid plaques, aggregates of Aβ peptides, were first isolated in the early 1980s by centrifugation, based on their low solubility. The predominant peptides were 40 and 42 amino acids in length (the 40 residues plus two additional amino acids at the carboxy terminal end). Biochemical studies showed that the Aβ42 peptide nucleates more rapidly than Aβ40 into amyloid fibrils.

Considerable experimental evidence indicates that Aβ42 drives the initial aggregation, although Aβ40 also accumulates to a significant extent, especially in cerebral amyloid angiopathy. For neurons in culture, the forms of the Aβ42 peptide that are larger than a monomer are generally more toxic than aggregated forms of Aβ40. These results implicate Aβ42 as a key driver of amyloid formation as well as Aβ toxicity.

Once it was discovered that Aβ peptides 38 to 43 amino acids in length are formed by cleavage of

a precursor protein, researchers set out to isolate the precursor. The precursor was found in the mid-1980s, molecularly cloned, and named the *amyloid precursor protein* (APP). It is a large transmembrane glycoprotein that is present in all types of cells but is expressed at its highest levels in neurons. The normal functions of APP in the brain are not understood.

How is APP processed to form Aβ peptides? The answer has turned out to be complex. Three enzymes, α-, β-, and γ-secretase, cut APP into pieces. The β- and γ-secretases cleave APP to generate soluble extracellular fragments that are released into the interstitial fluid. These are the Aβ peptides, which include part of the transmembrane segment of APP (Figure 64–11). The cleavage by γ-secretase is unusual in that it occurs in a membrane-spanning portion of APP, a region long thought to be immune from hydrolysis because it is surrounded by lipids rather than water. Cleavage by α-secretase in the middle of the Aβ sequence prevents the formation of Aβ peptides.

The enzymes that account for α-, β-, and γ-secretases have been isolated and characterized. The enzyme α-secretase is a member of a large family of extracellular proteases called ADAM (a disintegrin and metalloproteinase) that are responsible

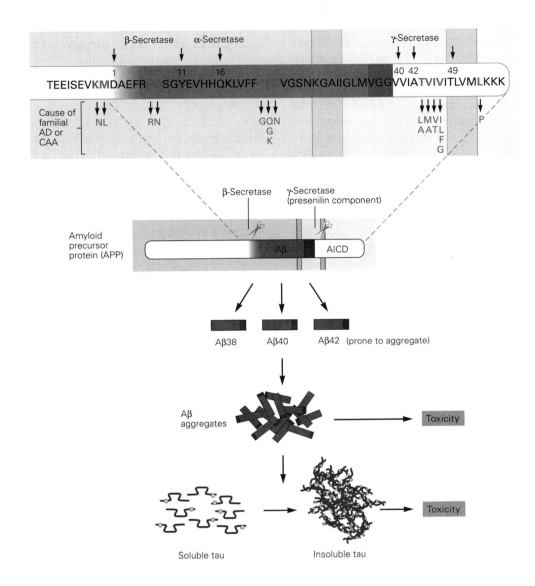

Figure 64–11 Processing of the amyloid precursor protein, generation of the Aβ peptide, and downstream effects on tau aggregation. The Aβ peptide is produced from the amyloid precursor protein (**APP**), a transmembrane protein, via cleavage by two enzymes, β-secretase and γ-secretase. (Cleavage by α-secretase prevents Aβ production.) Presenilin is the active enzymatic component of the γ-secretase complex and cleaves APP at several sites within the membrane to produce Aβ peptides of different lengths such as Aβ38, Aβ40, and Aβ42. Several mutations in APP that are just outside of the Aβ region or within the coding sequence of Aβ cause forms of autosomal dominant Alzheimer disease (**AD**). The amino acids (**blue**) in the APP/Aβ amino acid sequence represent the normal amino acids in APP; amino acids in **green** (below the normal sequence) are those that cause familial AD or cerebral amyloid angiopathy (**CAA**). Aβ is predominantly produced from APP within endosomes. A variety of molecules and synaptic activity regulate Aβ levels. There is evidence that Aβ aggregation is influenced by the Aβ-binding molecules ApoE and clusterin, which likely interact in the extracellular space of the brain. A variety of molecules and processes affect Aβ clearance from the interstitial fluid (ISF) that is present in the extracellular space of the brain, including neprilysin and insulin-degrading enzyme (IDE), as well as cerebral spinal fluid and interstitial fluid bulk flow. LRP1 and RAGE (receptor for advanced glycation end products) appear to influence Aβ transport across the blood–brain barrier. The concentration and type of Aβ influence aggregation (Aβ42 is more fibrillogenic). Once it aggregates into oligomers and fibrils, it can be directly toxic to cells, induce inflammation, and exacerbate the conversion of soluble tau to aggregated tau through mechanisms that remain unclear. In addition to Aβ, a variety of factors influence tau aggregation and toxicity, including tau levels, sequence, and phosphorylation state. (Abbreviation: **AICD**, APP intracellular domain.)

for degrading many components of the extracellular matrix. β-Secretase, called BACE1 (β-site APP cleaving enzyme 1), is a transmembrane protein in central neurons that is concentrated in synapses. Brain cells derived from mutant mice lacking BACE1 do not produce Aβ peptides, proving that BACE1 is indeed the neuronal β-secretase. γ-Secretase, the most complicated of the three, is actually a multiprotein complex that cleaves several different transmembrane proteins. As would be expected, given its peculiar ability to act within the membrane, γ-secretase itself includes several transmembrane proteins. Two of these are called presenilin-1 and presenilin-2, reflecting their association with AD. Other components of the complex include the transmembrane proteins nicastrin, Aph-1, and Pen-2.

Although the biochemical properties of Aβ and APP are interesting, the critical question is whether they have a part in the debilitating symptoms of AD. The disease might be caused by Aβ accumulation, but Aβ might itself be a result of another pathological process or even be an innocuous correlate. Genetic evidence in humans and experimental animals has been critical in demonstrating that APP and, specifically, Aβ play a central role in AD.

The first clue came from the observation that the APP gene lies on chromosome 21, which is present in three copies rather than the normal two in people with Down syndrome (also known as trisomy 21). All people with Down syndrome who live to middle age develop AD pathology and dementia, with onset around 50 years. This association is consistent with the idea that APP predisposes to AD by overproducing APP and Aβ by 50% throughout life. Nevertheless, copies of many genes are present in three copies in individuals with trisomy 21, and initially, it was not clear that triplication of APP in Down syndrome was responsible for AD in this population. Subsequently, rare families were found in which both AD and cerebral amyloid angiopathy developed in the absence of Down syndrome due to duplication of just the APP locus on human chromosome 21. This is strong evidence that overexpression of APP alone is enough to lead to AD and cerebral amyloid angiopathy.

More direct genetic evidence came from analysis of the rare patients with dominantly inherited AD, in whom the onset of symptoms is usually between 30 and 50 years of age. In the late 1980s, several research groups began using methods of molecular cloning to identify the genes mutated in dominantly inherited AD. Remarkably, the first three genes identified were those encoding the proteins APP, presenilin-1, and presenilin-2 (Figure 64–12). Many different mutations in

these three genes have been found, and the majority influence cleavage of APP, increasing the production of Aβ peptides or specifically the proportion of the more aggregation-prone Aβ42 species. Interestingly, some APP mutations occur within the Aβ sequence itself and do not affect Aβ production but do affect Aβ aggregation and clearance from the brain.

Some APP mutations are amino acid substitutions flanking the Aβ region. Cells that express a double mutation at the β-secretase cleavage site (the so-called Swedish mutation), which is required for Aβ formation, secrete several-fold more Aβ peptide than cells expressing wild type APP. Interestingly, another mutation in APP adjacent to the β-secretase site was recently discovered. This mutation appears to protect against AD by decreasing Aβ production. Yet another APP mutation causes γ-secretase to generate a greater proportion of longer Aβ species, such as Aβ42, in relation to shorter species such as Aβ40. Likewise, in most presenilin mutants the mutant γ-secretase has higher than normal activity or generates peptides with an increased ratio of Aβ42 to Aβ40.

These human genetic studies offer compelling evidence that (1) cleavage of APP to generate Aβ and the propensity of Aβ to aggregate play key instigating roles in some cases of dominantly inherited early-onset AD and (2) less Aβ production decreases the risk for late-onset AD. Genetic studies in mice have also strengthened the case that APP cleavage and specifically Aβ aggregation contribute to AD. Transgenic expression or knock-in of mutant APP forms identical to those found in autosomal dominant AD leads to the appearance of amyloid plaques in the hippocampus and cortex, dystrophic neurites in proximity to Aβ deposits, decreased density of synaptic terminals around amyloid plaques, and impairments in synaptic transmission. Several mouse models develop functional abnormalities such as deficits in spatial and episodic-like memory. Alterations are more severe in transgenic mice that express altered forms of both APP and presenilin-1. It is important to note that although these mice do not develop tau aggregation or neurofibrillary tangles, lesions believed to be important in the cognitive decline seen in AD, they remain invaluable models for addressing the mechanistic role of Aβ and related pathology in the pathogenesis of AD, especially the role of Aβ, and for testing potential therapies.

Given the strong evidence that APP cleavage is involved in the pathogenesis of AD, the next question is: How does the accumulation of cleavage products contribute to symptoms and ultimately dementia? There are three sets of cleavage products: the secreted extracellular region (ectodomain), the Aβ peptide, and

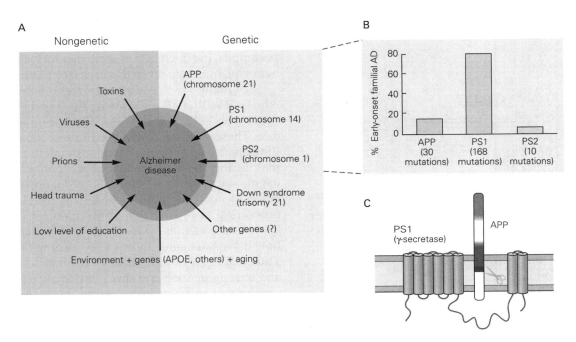

Figure 64–12 Environmental and genetic factors play a role in Alzheimer disease.

A. Environmental and genetic factors. (Abbreviations: **APOE**, apolipoprotein E; **APP**, amyloid precursor protein; **PS1**, presenilin-1; **PS2**, presenilin-2.)

B. Specific genes involved in early-onset Alzheimer disease (AD).

C. Presenilin-1 (a component of the gamma secretase enzyme complex) is associated with the APP protein within the plasma membrane.

cytoplasmic fragment. Although all three fragments can have deleterious effects on neurons in experimental animals, the Aβ peptides have received the most attention, and evidence for their involvement is strongest. There is evidence that different aggregated forms of Aβ such as oligomers, protofibrils, and fibrils can lead to synaptic and neuronal damage that might contribute to AD.

Neurofibrillary Tangles Contain Microtubule-Associated Proteins

Until around 2005, most research on the molecular and cellular basis of AD focused on Aβ peptides and amyloid plaques, but tau aggregation in neurofibrillary tangles appears to play a key role in the progression of AD (Figure 64–9). Molecular analysis revealed that these abnormal inclusions in cell bodies and proximal dendrites contain aggregates of hyperphosphorylated isoforms of tau, a microtubule-binding protein that is normally soluble (Figure 64–13). The tau protein plays a key role in intracellular transport, particularly in axons, by binding to and stabilizing microtubules. Impairments in axonal transport compromise synaptic stability and trophic support. While the mechanism by which aggregation and hyperphosphorylation of tau lead to toxicity is still not understood, tau accumulation is clearly associated with neuronal degeneration.

Although tangles are a defining feature of AD, it was initially unclear what role tangles and hyperphosphorylated forms of tau play in the pathogenesis of the disease. Whereas mutations of APP and presenilin genes can lead to AD, no mutations of the tau gene have been found in familial AD. Nevertheless, there is now a great deal of evidence indicating that tau aggregation is a key factor in the neurodegeneration that occurs in AD.

First, filamentous deposits of hyperphosphorylated tau are seen in a variety of neurodegenerative disorders, including AD, forms of frontotemporal dementia, progressive supranuclear palsy, corticobasal degeneration, and CTE. Second, mutations in the tau gene have been found to underlie another form of autosomal dominant neurodegenerative disease: frontotemporal dementia with Parkinson disease type 17 (FTPD17). These patients develop tau aggregation together with brain atrophy in specific brain regions in the absence of Aβ deposition. Third, progressive symptoms of AD correlate much better with the number and distribution of tangles than with the amyloid plaques

A Healthy neuron

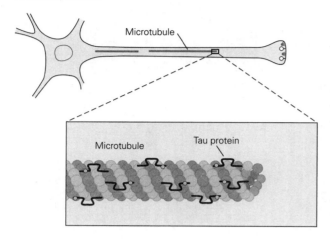

B Alzheimer neuron

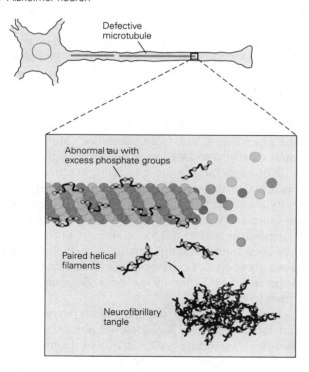

Figure 64–13 Formulation of neurofibrillary tangles.

A. In healthy neurons, tau protein associates with normal microtubules but not as paired helical filaments, and contributes to the structural integrity of the neuron.

B. In a diseased neuron, the tau protein becomes hyperphosphorylated and loses its association with normal microtubules, which begin to disassemble. It then forms paired helical filaments, which become sequestered in neurofibrillary tangles.

seen in autopsy. For example, tangles are usually first evident in neurons of the entorhinal cortex and hippocampus, the likely site of early memory disturbance, before plaques appear in this area (see Figure 64–16).

For many years, controversy raged between those who believe that Aβ is the main causal agent of AD and those who believe that tau-rich tangles play a major role. These partisans have been called "Baptists" and "Tauists," respectively. Baptists pointed to the fact that during the development of AD pathology, which begins about 15 years before symptom onset, accumulation of neocortical Aβ precedes the development of neocortical tau pathology. More recent evidence suggests, however, that Aβ accumulation appears to somehow drive tau aggregation and spreading in the brain. Thus, Aβ aggregation probably instigates the disease and tau aggregation and spreading likely contribute in a major way to neurodegeneration. For example, transgenic mice that express both mutant APP and mutant tau develop much worse tau pathology.

There appears to be interplay between plaques and tangles. Injection of Aβ42 into specific brain regions of transgenic mice that express a mutant tau protein increases the number of tangles in nearby neurons. Further, a manipulation that reduces the number and size of plaques leads to a decrease in levels of hyperphosphorylated tau. Importantly, recent experiments suggest that Aβ deposition in some way promotes spreading of tau aggregates from one brain region to another, possibly transsynaptically in a prion-like fashion. The details of this process remain to be worked out and are likely to be extremely important.

There is now abundant evidence from cell culture and studies in animal models that several proteins that aggregate in neurodegenerative diseases, including tau and synuclein, can spread from cell to cell in a prion-like manner. This is particularly important as a potential disease mechanism. For example, if the cell-to-cell spreading of misfolded proteins occurs in the extracellular space, this process might be interrupted with antibodies directed against the appropriate disease-associated protein. In fact, this now serves as the basis of several clinical trials in humans targeting tau and synuclein.

Risk Factors for Alzheimer Disease Have Been Identified

Very few individuals develop AD because they bear autosomal dominant mutant alleles of the APP or presenilin genes, and these are generally of the early-onset variety. Hardly any cases of late-onset AD are due to mutations in APP or presenilin genes. Can we, then, predict AD in such individuals?

The major risk factor is age. The disease is present in a vanishingly small fraction of people younger than age 60 (many of those being autosomal dominant cases), 1% to 3% of those between ages 60 and 70, 3% to 12% of those between ages 70 and 80, and 25% to 40% of those older than age 85. Knowing that elderly people are prime candidates for AD is of little therapeutic use, however, because modern medicine can do nothing to slow the passage of time. Therefore, there has been intense interest in other factors that affect the incidence of AD.

To date, the most significant genetic risk factors discovered for late-onset AD are the alleles of the gene *APOE*. The ApoE protein is an apolipoprotein. In the blood, it plays an important role in plasma cholesterol metabolism. It is also expressed at high levels in the brain, most prominently by astrocytes and to some extent by microglia. In the brain, where its normal function has not been clarified, it is secreted as a component of high density-like lipoproteins. In humans, there are three alleles of the *APOE* gene, *APOE2*, *APOE3*, and *APOE4*, which differ from each other by at most two amino acids. People with the *APOE4* allele are at risk for AD, whereas those with the *APOE2* allele are protected against AD relative to people who have the most

common *APOE3/APOE3* genotype. The *APOE4* allele is present in about 25% of the general population but present in as many as 60% of those with AD. One copy of the *APOE4* allele increases the risk of AD by about 3.7-fold, and two copies by about 12-fold, relative to someone who is *ApoE3/E3* (Figure 64–14). One copy of the *APOE2* allele decreases the risk for AD by about 40% relative to being *APOE3/APOE3*.

The mechanism by which *APOE4* predisposes to AD and *APOE2* protects against AD is uncertain, but ApoE4 clearly promotes Aβ aggregation by diminishing Aβ clearance and promoting fibrillization (ApoE4 > ApoE3 > ApoE2). It may also act through additional mechanisms such as influencing tau, the innate immune system, cholesterol metabolism, or synaptic plasticity, although these pathways remain to be worked out.

A number of other genes and genetic loci influence risk for late-onset AD. Some are common variants that alter risk only slightly, whereas other rarer variants increase risk to a greater extent (Figure 64–14). For example, relatively rare mutations in the gene *TREM2* double or triple the risk for AD, similar to having one copy of the *APOE4* allele. This is interesting because *TREM2* as well as another gene associated with risk

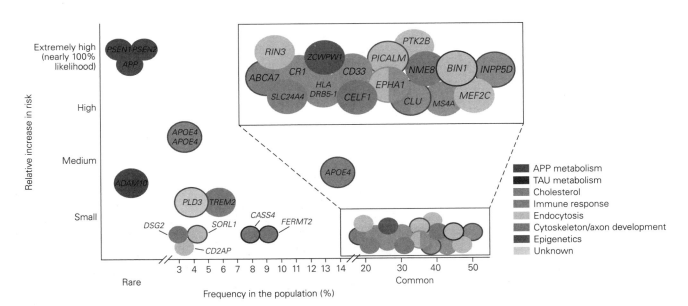

Figure 64–14 The risk for Alzheimer disease due to rare and common genetic variants. Data are from genome-wide association studies (GWAS). Mutations in the three genes that cause early-onset familial Alzheimer disease (*PSEN1*, *PSEN2*, and *APP*) are rare but result in Alzheimer disease in virtually 100% of people with these mutations if they live to middle age. There have been a number of common genetic changes found located in regions around or in genes that are relatively

frequent in the population (eg, *ABCA7, CLU, BIN1*) that affect risk for Alzheimer disease but to a very small degree. The one common and strong genetic risk factor for Alzheimer disease that is present in about 20% to 25% of the population (allele frequency ~15%) is *APOE4*. One copy of *APOE4* increases risk approximately 3.7-fold and two copies increase risk approximately 12-fold relative to people who are homozygous for *APOE3*. (Adapted, with permission, from Karch and Goate 2015.)

for AD, *CD33*, are expressed only in microglia. Along with other emerging cellular and animal model data, this finding suggests that the innate immune system is involved in AD pathogenesis. A number of other rare variants that increase risk to varying degrees are under investigation. It seems likely that these developments will ultimately result in a more personalized clinical approach to determining risk for AD, especially as treatments for the disease emerge.

Alzheimer Disease Can Now Be Diagnosed Well but Available Treatments Are Unsatisfactory

Diagnosing AD at its earliest stages in the absence of biomarkers can be challenging, as its initial symptoms can be similar to those of normal age-related cognitive decline or of other related diseases. Nevertheless, diagnosis of mild to moderate dementia due to AD is usually fairly accurate. In fact, during the past few decades, the ability to accurately diagnose the disease has improved, largely because of three factors.

First, protocols for physical, neurological, and neuropsychological examination have become more sophisticated and standardized. Second, increased knowledge of the structural changes revealed by magnetic resonance imaging (MRI) have helped in diagnosing AD at early stages. For example, it is now possible to predict, with approximately 80% accuracy, which patients with MCI will develop AD based on the cortical thinning and ventricular enlargement visible by MRI. These imaging and diagnostic methods also assist in distinguishing dementia syndromes from

each other and relating structural to functional defects. For example, patients with the disease known as *behavioral variant of frontotemporal dementia* experience personality changes early on, and MRI at that stage reveals atrophy of the frontal and/or temporal lobes. Likewise, initial difficulties in AD usually center on memory and attention, and MRI reveals initial alterations in the medial temporal cortex and hippocampus.

Third, and perhaps most promising, amyloid plaques and neurofibrillary tangles can be visualized by positron emission tomography (PET) using compounds that avidly bind fibrillar forms of Aβ or aggregated forms of tau. The first of these, Pittsburgh compound B (PIB), binds with high affinity to fibrillar Aβ; its radioactive form, labeled with short-lived isotopes of carbon or fluorine, is readily detected by PET (Figure 64–15). The US Food and Drug Administration (FDA) has approved three amyloid imaging agents: florbetapir (Amyvid), flutemetamol (Vizamyl), and florbetaben (Neuraceq).

The availability of safe molecular markers of AD allows early stages of the disease to be identified before clinical symptoms are present. Of equal importance, it allows for improved selection of patients for clinical trials and keener selection of subjects for detailed analyses of normal aging. It is important to note that these changes can also be detected in the cerebrospinal fluid, where the level of Aβ42 drops when amyloid deposition is present and total tau and phosphorylated forms of tau increase with neurodegeneration and tau aggregation.

Of course, improved diagnosis of AD is most useful if treatments are available that can halt or slow its progression at an early stage. While we still do not

Normal   Alzheimer disease

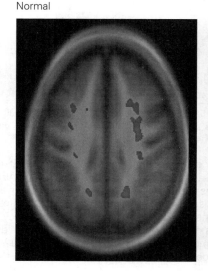

Figure 64–15 Positron emission tomography scans can visualize amyloid plaques in the living brain. The density of Aβ plaques is indicated by the **red** regions in these images made after administration of Pittsburgh compound B (PIB), a fluorescent analog of thioflavin T. (Images reproduced, with permission, from R. Buckner.)

have a treatment that delays the onset or slows the progression of AD, there is hope that we are not too far off from being able to mitigate symptoms. Although there is no definitive proof, there is good evidence that a variety of lifestyle factors decrease risk for AD. These include high levels of education, cognitive stimulation, staying socially engaged, regular exercise, not being overweight, and getting appropriate amounts of sleep. Present-day therapies focus on treating associated symptoms such as depression, agitation, sleep disorders, hallucinations, and delusions.

One of the principal therapeutic targets to date has been the cholinergic system in the basal forebrain, a region of the brain that is damaged in AD and that contributes to attention. Acetylcholinesterase inhibitors increase levels of acetylcholine by inhibiting its breakdown and represent one of the few drug classes approved by the FDA for treatment of AD. Another drug, the N-methyl-D-aspartate (NMDA) receptor antagonist memantine, also improves symptoms in individuals with mild to moderate dementia due to AD. It is believed that memantine's action modulates glutamate-mediated neurotransmission. Nevertheless, these drugs exert only a modest effect on cognitive functions and the activities of daily living.

Recent advances in our understanding of the cell-biological basis of AD have produced several promising new therapeutic targets, all of which are being explored intensively. One approach is to develop drugs that reduce or modulate the activity of the β- and γ-secretases that cleave APP to generate Aβ peptides and the associated soluble extracellular and intracellular fragments. In fact, decreasing either β- or γ-secretase levels in transgenic mice that overexpress mutant APP decreases Aβ deposition and, in some cases, functional abnormalities.

Accordingly, pharmaceutical companies have developed drugs that decrease or modulate levels of β- and γ-secretases in humans. An obstacle to this approach is that the secretases also act on substrates other than APP, so decreasing their levels can have deleterious side effects. This is especially true for γ-secretase, whose inhibition has led to toxicity in human trials for AD. There are now several β-secretase inhibitors in clinical trials for AD, and it is likely such drugs will also move into trials for what is called preclinical AD, when AD pathology is accumulating but there is no sign yet of cognitive decline (Figure 64–16). The goal of this therapy would be to delay or prevent the onset of cognitive decline and dementia.

Another approach is to decrease levels of Aβ through immunological means. Both immunization with Aβ, which leads to generation of antibodies to Aβ, and passive transfer of Aβ antibodies have been

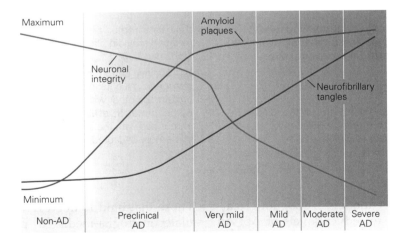

Figure 64–16 Relationship of biomarker changes to cognitive and clinical changes in Alzheimer disease (AD). In cognitively normal people who are going to develop AD dementia, one of the first physical signs is the initiation of Aβ aggregation in the brain in the form of amyloid plaques. While people are still cognitively normal, amyloid plaques continue to accumulate. At some point, about 5 years before any clear-cut cognitive decline, tau accumulation begins to increase in the neocortex, inflammation and oxidative stress increase, and brain network connections and metabolism begin to decline. Neuronal and synaptic loss and brain atrophy also begin. This period—when the patient remains cognitively normal but AD-type pathology is building up—is termed preclinical AD. Once there is enough neuronal and synaptic dysfunction as well as cell loss, very mild dementia and mild cognitive impairment become detectable. At that time, amyloid deposition has almost reached its peak. As dementia worsens to mild, moderate, and severe stages, neurofibrillary tangles form, and neuronal and synaptic dysfunction, inflammation, cell death, and brain atrophy worsen. (Adapted, with permission, from Perrin, Fagan, and Holtzman 2009.)

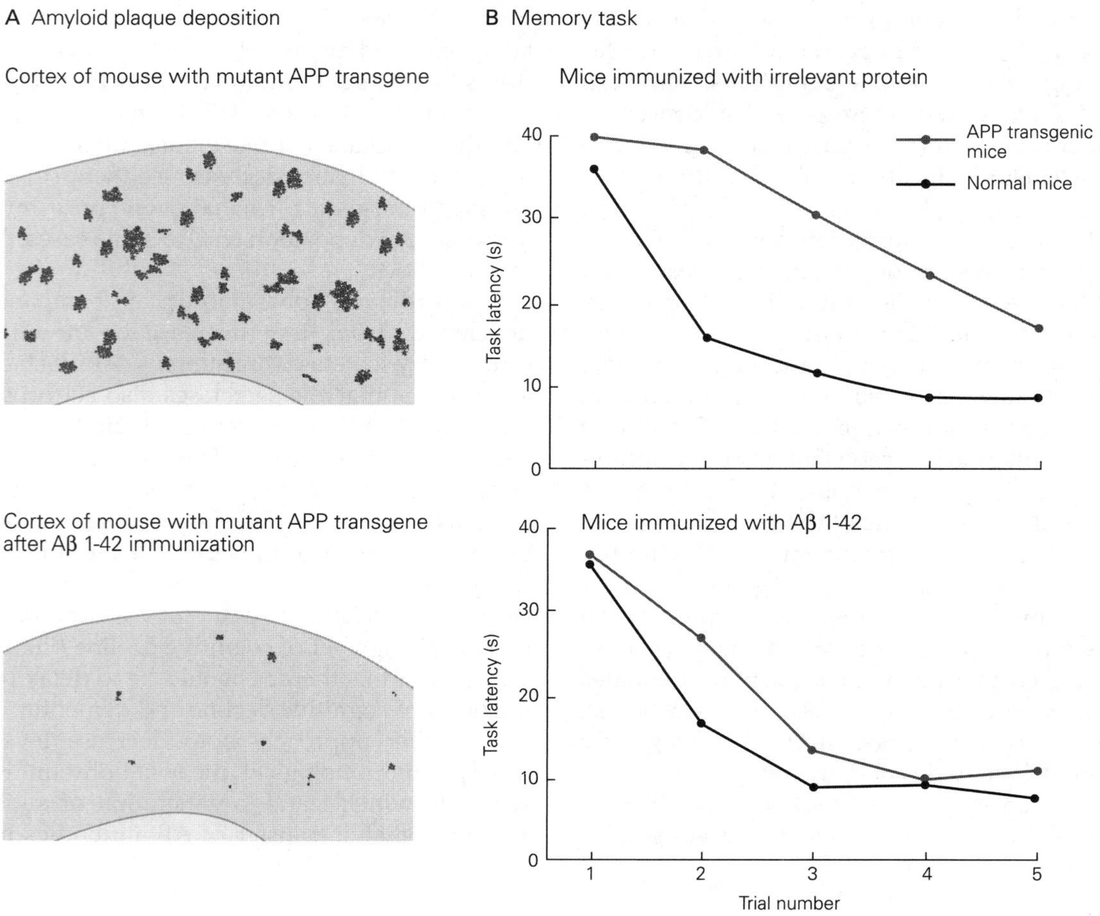

A Amyloid plaque deposition

Cortex of mouse with mutant APP transgene

Cortex of mouse with mutant APP transgene
after Aβ 1-42 immunization

B Memory task

Mice immunized with irrelevant protein

Mice immunized with Aβ 1-42

Figure 64–17 Immunization with antibodies to the Aβ peptide clears amyloid plaques and preserves cognitive performance in mice expressing the peptide. Mice that develop Aβ deposition in the form of amyloid plaques were immunized with the Aβ peptide. This led to production of antibodies against Aβ.

A. Comparison of amyloid plaque deposition in the cerebral cortex of mice overexpressing a mutant APP transgene (APP transgenic mice) that develop amyloid plaques. The mice

that were immunized with the Aβ peptide have substantially reduced amyloid plaque deposition. (Adapted from Brody and Holtzman 2008.)

B. Cognitive performance (a memory test) in two groups of APP transgenic mice. One group was immunized with an irrelevant protein, the other with the Aβ peptide. The mice vaccinated with Aβ performed at levels close to normal animals, whereas mice immunized with the irrelevant protein showed severe impairment in memory. (Adapted, with permission, from Janus et al. 2000.)

tested in transgenic mouse models of AD. Both treatments have been shown to reduce levels of Aβ, Aβ toxicity, and plaques (Figure 64–17). The mechanisms of enhanced Aβ clearance are not completely clear. Serum antibodies likely serve as a "sink," resulting in Aβ peptides with low molecular weight being cleared more extensively from the brain into the circulation, thus changing the equilibrium of Aβ in different compartments and promoting removal of Aβ from the brain.

It is also clear that in the brain several anti-Aβ antibodies bind either soluble or fibrillar Aβ, or both.

Those that bind to aggregated forms of Aβ can stimulate microglia-mediated phagocytosis to remove Aβ, although there is also plaque removal that is not dependent on microglial-mediated phagocytosis. Antibodies to soluble Aβ that enter the brain may decrease soluble Aβ toxicity. These findings suggest that immunotherapeutic strategies may be successful in AD patients, especially if they are given early enough in the disease course, prior to significant neuronal damage and loss. There are multiple human trials underway using active and passive immunotherapies against Aβ both in preclinical and mild AD.

In addition to targeting Aβ, clinical trials have also begun targeting tau. This is being done with active and passive immunization against tau as well as with small molecules that, in cell culture and animal models, can decrease tau aggregation. A number of studies in animal models have shown that certain anti-tau antibodies can decrease the amount of aggregated, hyperphosphorylated tau in the central nervous system and in some cases improve function. Although tau is predominantly a cytoplasmic protein, one of the reasons that anti-tau antibodies may be having an effect is that, as discussed above, tau aggregates may spread from cell to cell in the extracellular space in a prion-like fashion. It is in this space that an antibody may be able to interact with tau and block this process.

Highlights

1. It is only in the past 50 years that a large percentage of the population has lived into the eighth to tenth decades of life. With this increase, neuroscientists have been able to study changes in the brain that occur with normal aging as well as in individuals who develop age-related brain disorders.

2. Subtle changes in a variety of brain functions occur with age, including declines in speed of processing and memory storage and changes in sleep. The underlying basis for these changes is likely brain atrophy and loss of white matter integrity. In general, however, there is not a significant decrease in neuronal number that accounts for changes in brain function that occur with normal aging.

3. The changes in cognition that occur in normal aging are not disabling. When memory and often other areas of cognitive function decline more than is expected with age such that it is noticeable to others and mildly affects one's day-to-day life, this syndrome is called mild cognitive impairment (MCI).

4. MCI is not a disease, it is a syndrome. About 50% of individuals with MCI have Alzheimer disease (AD) as the underlying cause of the MCI. Other conditions that can cause MCI include depression, cerebrovascular disease, Lewy body disease, metabolic disorders, and drugs, prescribed for other diseases, that cause central nervous system side effects.

5. AD is the most common cause of dementia and manifests as loss of memory and other cognitive abilities sufficient to impair social and occupational functions. AD accounts for about 70% of cases of dementia in the United States, with the remainder caused primarily by cerebrovascular disease, Parkinson and Lewy body dementia, and frontotemporal dementia.

6. The pathology of AD is characterized by the accumulation of aggregated forms of two proteins in the brain, the Aβ peptide and tau. Aβ accumulates in a fibrillar form in extracellular structures called amyloid plaques in the brain parenchyma as well as in the walls of arterioles (where it is called cerebral amyloid angiopathy). Tau accumulates in neurofibrillary tangles in cell bodies and dendrites.

7. In addition to the accumulation of protein aggregates in the Alzheimer brain, marked brain atrophy as well as synaptic and neuronal loss occurs as the disease progresses. There is also a strong neuroinflammatory response, especially around amyloid plaques, which involves microglia and astrocytes.

8. The pathology of AD begins about 15 years prior to the onset of cognitive decline or the MCI phase of the disease. Aβ accumulation in the neocortex appears to initiate the disease with markedly abnormal levels, followed by the spread of tau aggregates from the medial temporal lobe to other regions of the neocortex. This phase of Alzheimer pathology prior to symptom onset is known as preclinical AD.

9. Significant data suggest that certain aggregated forms of the Aβ peptide lead to synaptic and neuronal damage in the Alzheimer brain, but a much better correlate of the cognitive decline is the presence and accumulation of aggregated forms of the tau protein.

10. There are two major forms of AD. The first is a dominantly inherited AD, which accounts for less than 1% of Alzheimer patients and is caused by mutations in one of three genes encoding the proteins APP, PS1, and PS2; this form leads to clinical disease onset between the ages of 30 and 50. Genetic, biochemical, and other studies have shown that the genes that cause autosomal dominant AD do so through early accumulation of the Aβ peptide in the brain. The second form, late-onset AD, with an age of onset of 65 years or later, accounts for more than 99% of cases. Although age is the greatest risk factor for late-onset AD, genetics also contribute. The *APOE* gene is by far the biggest genetic contributor to AD, with the *APOE4* variant increasing risk and the *APOE2* variant decreasing risk. There are a number of

other common genetic variants in other genes that influence risk. There are also rare variants in other genes such as *TREM2* that increase risk to a level similar to that associated with one copy of *APOE4*. Nonetheless, there is general agreement that major features of pathogenesis are similar in sporadic and familial AD.

11. Besides clinical symptoms and signs of AD, amyloid and tau imaging and cerebrospinal fluid markers can determine that Alzheimer pathology is present in a living person with or without cognitive decline.

12. There are currently only symptomatic therapies for AD that have modest benefit at best. A number of potential disease-modifying therapies that influence the production, clearance, and aggregation of either Aβ or tau are being tested in humans. Although none of these therapies is yet approved, there is hope that over the next several years one or more of these therapies will begin to show a clear benefit.

Joshua R. Sanes
David M. Holtzman

Selected Reading

Brody DL, Holtzman DM. 2008. Active and passive immunotherapy for neurodegenerative disorders. Annu Rev Neurosci 31:175–193.

Buckner RL. 2004. Memory and executive function in aging and AD: multiple factors that cause decline and reserve factors that compensate. Neuron 44:195–208.

Goedert M, Eisenberg DS, Crowther RA. 2017. Propagation of tau aggregates and neurodegeneration. Annu Rev Neurosci 40:189–210.

Haass C, Selkoe DJ. 2007. Soluble protein oligomers in neurodegeneration: lessons from the Alzheimer's amyloid beta-peptide. Nat Rev Mol Cell Biol 8:101–112.

Holtzman DM, Herz J, Bu G. 2012. Apolipoprotein E and apolipoprotein E receptors: normal biology and roles in Alzheimer disease. Cold Spring Harb Perspect Med 2:a006312.

Holtzman DM, Morris JC, Goate AM. 2011. Alzheimer's disease: the challenge of the second century. Sci Transl Med 3:77sr1.

Kenyon C. 2005. The plasticity of aging: insights from long-lived mutants. Cell 120:449–460.

Musiek ES, Holtzman DM. 2015. Three dimensions of the amyloid hypothesis: time, space and "wingmen." Nat Neurosci 18:800–806.

Sanders DW, Kaufman SK, Holmes BB, Diamond MI. 2016. Prions and protein assemblies that convey biological information in health and disease. Neuron 89:433–448.

References

Andrews-Hanna JR, Snyder AZ, Vincent JL, et al. 2007. Disruption of large-scale brain systems in advanced aging. Neuron 56:924–935.

Arias E. 2004. United States Life Tables, 2001. *National Vital Statistics Reports,* Vol. 52, No. 14. Hyattsville, MD: National Center for Health Statistics.

Arnold SE, Hyman BT, Flory J, Damasio AR, Van Hoesen GW. 1991. The topographical and neuroanatomical distribution of neurofibrillary tangles and neuritic plaques in the cerebral cortex of patients with Alzheimer's disease. Cereb Cortex 1:103–116.

Bard F, Cannon C, Barbour R, et al. 2000. Peripherally administered antibodies against amyloid beta-peptide enter the central nervous system and reduce pathology in a mouse model of Alzheimer disease. Nat Med 6:916–919.

Bateman RJ, Xiong C, Benzinger TL, et al. 2012. Clinical and biomarker changes in dominantly inherited Alzheimer's disease. N Engl J Med 367:795–804.

Bishop NA, Lu T, Yankner BA. 2010. Neural mechanisms of ageing and cognitive decline. Nature 464:529–535.

Brown-Borg H, Borg K, Meliska C, Bartke A. 1996. Dwarf mice and the ageing process. Nature 384:33.

Cai H, Wang Y, McCarthy D, et al. 2001. BACE1 is the major beta-secretase for generation of A–beta peptides by neurons. Nat Neurosci 4:233–234.

Choi SH, Kim YH, Hebisch M, et al. 2014. A three-dimensional human neural cell culture model of Alzheimer's disease. Nature 515:274–278.

Cleary JP, Walsh DM, Hofmeister JJ, et al. 2005. Natural oligomers of the amyloid-beta protein specifically disrupt cognitive function. Nat Neurosci 8:79–84.

Cohen E, Dillin A. 2008. The insulin paradox: aging, proteotoxicity and neurodegeneration. Nat Rev Neurosci 9:759–767.

Corder EH, Saunders AM, Strittmatter WJ, et al. 1993. Gene dose of apolipoprotein E type 4 allele and the risk of Alzheimer disease in late onset families. Science 261:921–923.

De Strooper B, Saftig P, Craessaerts K, et al. 1998. Deficiency of presenilin-1 inhibits the normal cleavage of amyloid precursor protein. Nature 391:387–390.

Dickstein DL, Kabaso D, Rocher AB, Luebke JI, Wearne SL, Hof PR. 2007. Changes in the structural complexity of the aged brain. Aging Cell 6:275–284.

Fitzpatrick AWP, Falcon B, He S, et al. 2017. Cryo-EM structures of tau filaments from Alzheimer's disease. Nature 547:185–190.

Glenner GG, Wong CW. 1984. Alzheimer's disease: initial report of the purification and characterization of a novel cerebrovascular amyloid protein. Biochem Biophys Res Commun 120:885–890.

Goate A, Chartier-Harlin MC, Mullan M, et al. 1991. Segregation of a missense mutation in the amyloid precursor protein gene with familial Alzheimer's disease. Nature 349:704–706.

Guerreiro R, Wojtas A, Bras J, et al. 2013. TREM2 variants in Alzheimer's disease. N Engl J Med 368:117–127.

Hansson O, Zetterberg H, Buchhave P, Londos E, Blennow K, Minthon L. 2006. Association between CSF biomarkers and incipient Alzheimer's disease in patients with mild cognitive impairment: a follow-up study. Lancet Neurol 5:228–234.

Hebert LE, Scherr PA, Bienias JL, Bennett DA, Evans DA. 2003. Alzheimer disease in the US population: prevalence estimates using the 2000 census. Arch Neurobiol 60: 1119–1122.

Hekimi S, Guarente L. 2003. Genetics and the specificity of the aging process. Science 299:1351–1354.

Hsiao K, Chapman P, Nilsen S, et al. 1996. Correlative memory deficits, Aβ elevation, and amyloid plaques in transgenic mice. Science 274:99–102.

Huttenlocher PR. 2002. *Neural Plasticity: The Effects of Environment on the Development of the Cerebral Cortex.* Cambridge, MA: Harvard Univ. Press.

Janus C, Pearson J, McLaurin J, et al. 2000. Abeta peptide immunization reduces behavioural impairment and plaques in a model of Alzheimer's disease. Nature 408:979–982.

Johnson KA, Schultz A, Betensky RA, et al. 2016. Tau positron emission tomographic imaging in aging and early Alzheimer disease. Ann Neurol 79:110–119.

Jonsson T, Atwal JK, Steinberg S, et al. 2007. A mutation in APP protects against Alzheimer's disease and age-related cognitive decline. Nature 488:96–99.

Kang J, Lemaire HG, Unterbeck A, et al. 1987. The precursor of Alzheimer's disease amyloid A4 protein resembles a cell-surface receptor. Nature 325:733–736.

Kang JE, Lim MM, Bateman RJ, et al. 2009. Amyloid-beta dynamics are regulated by orexin and the sleep-wake cycle. Science 326:1005–1007.

Karch CM, Goate AM. 2015. Alzheimer's disease risk genes and mechanisms of disease pathogenesis. Biol Psychiatry 77:43–51.

Klunk WE, Engler H, Nordberg A, et al. 2004. Imaging brain amyloid in Alzheimer's disease with Pittsburgh Compound-B. Ann Neurol 55:306–319.

Lesne S, Koh MT, Kotilinek L, et al. 2006. A specific amyloid-beta protein assembly in the brain impairs memory. Nature 440:352–357.

Levy-Lahad E, Wasco W, Poorkaj P, et al. 1995. Candidate gene for the chromosome 1 familial Alzheimer's disease locus. Science 269:973–977.

Morgan D, Diamond DM, Gottschall PE, et al. 2000. A beta peptide vaccination prevents memory loss in an animal model of Alzheimer's disease. Nature 408:982–985.

Morris JC, McKeel DW Jr, Storandt M, et al. 1991. Very mild Alzheimer's disease: informant-based clinical, psychometric, and pathologic distinction from normal aging. Neurology 41:469–478.

Oddo S, Caccamo A, Shepherd JD, et al. 1996. Mediators of long-term memory performance across the life span. Psychol Aging 11:621–637.

Park DC, Smith AD, Lautenschlager G, et al. 1996. Mediators of long-term memory performance across the life span. Psychol Aging. 11:621–37.

Perrin RJ, Fagan AM, Holtzman DM. 2009. Multimodal techniques for diagnosis and prognosis of Alzheimer's disease. Nature 461:916–922.

Price JL, Davis PB, Morris JC, White DL. 1991. The distribution of tangles, plaques and related immunohistochemical markers in healthy aging and Alzheimer's disease. Neurobiol Aging 12:295–312.

Rubin EH, Storandt M, Miller JP, et al. 1998. A prospective study of cognitive function and onset of dementia in cognitively healthy elders. Arch Neurol 55:395–401.

Sanders DW, Kaufman SK, DeVos SL, et al. 2014. Distinct tau prion strains propagate in cells and mice and define different tauopathies. Neuron 82:1271–1278.

Schenk D, Barbour R, Dunn W, et al. 1999. Immunization with amyloid attenuates Alzheimer-disease-like pathology in the PDAPP mouse. Nature 400:173–177.

Sevigny J, Chiao P, Bussière T, et al. 2016 The antibody aducanumab reduces Aβ plaques in Alzheimer's disease. Nature 537:50–56.

Shi Y, Yamada K, Liddelow SA, et al. 2017. ApoE4 markedly exacerbates tau-mediated neurodegeneration in a mouse model of tauopathy. Nature 549:523–527.

Sperling RA, Aisen PS, Beckett LA, et al. 2011. Toward defining the preclinical stages of Alzheimer's disease: recommendations from the National Institute on Aging-Alzheimer's Association workgroups on diagnostic guidelines for Alzheimer's disease. Alzheimers Dement 7:280–292.

Strehler BL. 1975. Implications of aging research for society. Fed Proc 34:5–8.

Valdez G, Tapia JC, Kang H, et al. 2010. Attenuation of age-related changes in mouse neuromuscular synapses by caloric restriction and exercise. Proc Natl Acad Sci U S A 107:14863–14868.

Van Broeckhoven C, Haan J, Bakker E, et al. 1990. Amyloid beta protein precursor gene and hereditary cerebral hemorrhage with amyloidosis (Dutch). Science 248:1120–1122.

Yanamandra K, Kfoury N, Jiang H, et al. 2013. Anti-tau antibodies that block tau aggregate seeding in vitro markedly decrease pathology and improve cognition in vivo. Neuron 80:402–414.

Yi L, Seroude L, Benzer S. 1998. Extended life-span and stress resistance in the *Drosophila* mutant Methuselah. Science 282:943–946.

Index

The letters b, f, and t following a page number indicate box, figure, and table.

W